Third Edition
Electromagnetic Compatibility

Methods, Analysis, Circuits, and Measurement

Third Edition
Electromagnetic Compatibility

Methods, Analysis, Circuits, and Measurement

David A. Weston

EMC, Consulting Inc.,
Merrickville, Ontario, Canada

CRC Press is an imprint of the
Taylor & Francis Group, an **informa** business

CRC Press
Taylor & Francis Group
6000 Broken Sound Parkway NW, Suite 300
Boca Raton, FL 33487-2742

First issued in hardback 2019

ISBN-13: 978-1-4822-9950-2 (hbk)

Printed and bound by CPI Group (UK) Ltd, Croydon, CR0 4YY

Library of Congress Cataloging-in-Publication Data

Names: Weston, David A., 1942- author.
Title: Electromagnetic compatibility : methods, analysis, circuits, and
measurement / David A. Weston.
Description: Third edition. | Boca Raton : Taylor & Francis, CRC Press,
[2016] | Includes bibliographical references and index.
Identifiers: LCCN 2016003888 | ISBN 9781482299502 (alk. paper)
Subjects: LCSH: Electromagnetic compatibility. | Electronic apparatus and
appliances--Design and construction.
Classification: LCC TK7867.2 .W46 2016 | DDC 621.382/24--dc23
LC record available at http://lccn.loc.gov/2016003888

Visit the Taylor & Francis Web site at
http://www.taylorandfrancis.com

and the CRC Press Web site at
http://www.crcpress.com

Contents

Preface

This third, revised and updated, edition of the book contains approximately 65% more information than the first and second editions. This includes a review of computer modeling programs, additional PCB-level crosstalk calculations, many new signal and power filter designs in Chapter 5, and antenna to antenna coupling in Chapter 10. Chapter 9 contains a complete update to commercial and military electromagnetic interference (EMI) test methods and requirements. New data on enclosure aperture coupling and USB and Ethernet connector/cable shielding are also included. All electronic and electrical equipment is a potential source of EMI. Similarly, such equipment will not function as designed at some level of electromagnetic ambient. The problems associated with EMI can range from simple annoyance (e.g., static on telecommunications equipment or increased bit error rates on digital equipment) to catastrophe (e.g., inadvertent detonation of explosive devices).

Electromagnetic compatibility (EMC) can be achieved by evaluating the electromagnetic environment (often characterized by standards or requirements) to which equipment/systems are exposed and then designing and building equipment/systems to function correctly in the operational environment without itself creating EMI.

This book is written for the design/systems engineer, technologist, technician, or engineering manager who designs, maintains, or specifies equipment either to meet an EMC requirement specification or to function safely in a given electromagnetic environment.

Many engineers do not have, or need, radio frequency (RF) experience. However, in operation, digital control or switching power equipment functions as an RF system. Therefore, an understanding of the high-frequency characteristics of components, simple radiators, and wave theory is imperative in achieving an understanding of EMC.

One aim of the book is to teach EMI prediction and enable the reader to build EMC into equipment and systems without overdesign. By achieving EMC, the designer averts the program delay and additional cost of fixing EMI after the equipment is built. With the recognition that EMI problems exist, we present EMI diagnostic techniques and cost-effective solutions with practical implementation and options.

The book discusses typical sources of EMI and characteristics of the radiated and conducted emissions that might be expected in a given electromagnetic environment and reviews ways of decreasing electromagnetic emissions as well as the susceptibility of equipment to EMI. Some books on EMI/EMC contain equations that are theoretically sound but may not be useful in practical EMI/EMC problems. All equations in this book have been found to be invaluable in EMI prediction and EMC design and in most cases have been validated by measurements. Free computer programs containing aperture coupling, antenna to antenna coupling predictions, etc., are available from emcconsultinginc.com. In most instances, theory is substantiated by measured data, and where anomalies exist, most probable reasons are offered. Where the reader may wish to pursue a given subject area further, numerous references are provided. Worked examples of the equations are given in predictions and case studies throughout the book. Physical geometry and frequency limitations exist in the application of all wave or circuit theory, including the effect of parasitic components, and these limits are discussed.

The apparent anomalies that have given EMC a reputation for "black magic" are explained. For example, the case where the addition of shielding, a filter, or grounding increases either the level of EMI emissions or the susceptibility during EMC tests is examined. The major reason these results are apparently inexplicable is that the underlying theory is not well understood. The approach used in the book is to provide an understanding of the theory with an emphasis on its applicability in the practical realization of EMC design and EMI solutions, including implementation and maintenance.

The intent is that information contained herein have a practical application or be required for an understanding of the principles of EMC. For example, calculated or published data on attenuation or shielding effectiveness are of little use unless its application is explained. Therefore, it must be used in conjunction with the worst-case levels of radiated or conducted noise that may be expected in a given environment. Any practical limitation on the achievable attenuation or shielding must then be accounted for, after which the noise levels applied to the system or circuit and its immunity may be predicted. The aim has been to avoid the overly simplistic cookbook approach with its inherent errors and yet to limit the mathematics to that used by the practicing engineer or technician.

Simple measurement techniques that are possible with standard electronic measurement equipment are described. These are useful for EMI diagnostic measurements as well as a "quick look" at equipment that must meet EMC requirements such as the commercial FCC, DO-160, and EN and the military/aerospace MIL-STD-461. Also, the correct measurement techniques and possible errors encountered using more sophisticated equipment required for certification and qualification EMC testing are introduced.

The book is based on the experience gained in EMC consulting and on the course notes of one- to four-day EMC seminars presented over a 12-year period. Many questions posed by attendees of the seminars and clients have been answered in this book.

Many thanks to David Viljoen who worked extensively on the first edition of the book and without whom the book would not have been possible in its present form.

For the third edition of the book, I thank Jennifer Gale and Kerri McDougal for performing excellent R&D and report writing, Lorne Kelly for writing all of the measurement and the analysis software described in the book, and also Robert Gray for excellent drawings and reviews. Special thanks go to Calvin Gale for his dedication and high-quality work in performing all of the recent R&D and in the book layout, generation of plots, use of EM software, and editing. Finally, I thank the organizations specifically acknowledged in the figures and in the text, especially the Canadian Space Agency.

Author

David A. Weston is an electromagnetic compatibility (EMC) consultant and certified National Association of Radio and Telecommunications Engineers (iNARTE) EMC engineer at EMC Consulting Inc., Merrickville, Ontario, Canada. A life member of the Institute of Electrical and Electronics Engineers, Weston has worked in electronic design for 44 years, specializing in the control, prediction, measurement, problem solving, and design aspects of EMC for the last 34 years. Weston studied at Croydon Technical College from 1960 to 1965 and in 1965 received the radio and television certificate from the City and Guilds of the London Institute, both in England.

1 Electromagnetic Compatibility

1.1 INTRODUCTION TO ELECTROMAGNETIC INTERFERENCE

This chapter introduces a few of the important coupling modes involved in electromagnetic interference (EMI) as well as some EMI regulations. Both of these topics will be addressed in more detail in subsequent chapters. In addition, this chapter presents data on average and worst-case electromagnetic emissions found in the environment—the electromagnetic ambient.

The concern is the susceptibility of components, devices, equipment, subsystems, and systems to nearby sources of electromagnetic energy both radiated and conducted.

These include intentional radiators such as AM, FM, and television (TV) transmitters, cell phone base stations, remote control toys, small FM transmitters, heating and ventilation, and appliance wireless controls, as well as portable sources such as cellular phones; walkie-talkies; personal communications services (PCS); wireless LANs, for example, Wi-Fi, Bluetooth, and Zigbee; and cordless phones.

These also include unintentional sources of emissions such as information technology equipment and industrial, scientific, and medical (ISM) as well as susceptibility to magnetic fields from typical AC power transformers and high-voltage transmission lines and wiring.

1.1.1 EFFECTS OF ELECTROMAGNETIC INTERFERENCE

The effects of EMI are extremely variable in character and magnitude, ranging from simple annoyance to catastrophe. Some examples of the potential effects of EMI are

- Interference to TV and radio reception
- Loss of data in digital systems or in transmission of data
- Delays in production of equipment exhibiting intraunit, subsystem, or system-level EMI
- Malfunction of medical electronic equipment (e.g., neonatal monitor, heart pacemaker)
- Malfunction of automotive microprocessor control systems (e.g., braking or truck antijack-knife systems)
- Malfunction of navigation equipment
- Inadvertent detonation of explosive devices
- Malfunction of critical process control functions (e.g., oil or chemical industry)

To correct EMI problems that occur after equipment is designed and in production is usually expensive and results in program delays, which may adversely affect the acceptance of a new product. It is preferable to follow good electromagnetic compatibility (EMC) engineering practice during the equipment design and development phases. Our goal should be to produce equipment capable of functioning in the predicted or specified electromagnetic environment and that does not interfere with other equipment or unduly pollute the environment—that is, to achieve EMC.

The techniques of EMC prediction described in subsequent chapters will aid in meeting the goal of EMC when applied at the design stage. These same techniques of analysis and modeling are applicable to EMI control and problem solving or in the location of out-of-specification emissions. It is in the area of emission reduction where analysis is most likely to be supplemented by measurement and diagnostic intervention. However, the value of simple EMI measurements made as early as feasible in the design, breadboard, and prototype phases cannot be emphasized enough.

1.1.2 ELECTROMAGNETIC INTERFERENCE COUPLING MODES

For EMI to exist, there must be a source of emission, a path for the coupling of the emission, and a circuit, unit, or system sensitive to the received noise. Figures 1.1 through 1.3 illustrate that two modes of coupling, radiated and conducted, can exist. In the near field, the radiated coupling may be either a predominantly magnetic (H) field coupling or an electric (E) field coupling, whereas the coupling in the far field will be via electromagnetic waves exhibiting a fixed ratio of the E to H field strengths. A more rigorous definition of near and far field is presented in Section 2.2.1. Suffice it to

FIGURE 1.1   Possible sources of ambient noise and how they may be coupled into a receiver.

FIGURE 1.2   Some of the possible interference coupling modes within a system.

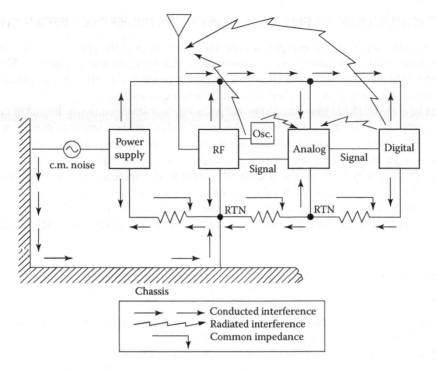

FIGURE 1.3 A few of the intraequipment (unit) coupling modes.

say here that the near field is in close proximity to a source and the far field is beyond some determined distance from the source.

For circuits and conductors in close proximity, we consider the coupling or crosstalk to be via mutual inductance and intercircuit capacitance, although one of these modes usually predominates. However, even in close proximity, the radiated coupling may predominate. The sources of noise may be power lines, signal lines, logic (especially clocks and data lines), or current-carrying ground connections. The conducted path may be resistive or contain inductance or capacitance, intentional or otherwise, and it is often a combination of these. The reactive components often result in resonances, with their concomitant increase or decrease in current at the resonant frequencies.

1.1.3 GENERAL

The following summary of measured spectrum occupancy and radiated emission levels from unintentional and intentional sources is intended to provide guidelines to manufacturers whose equipment must achieve EMC in these harsh environments. One important piece of information presented with the data is the measuring distance from the source. In measurements, it has been observed that the low-frequency component falls more rapidly with an increasing distance than those at VHF and UHF. The mechanism for E-field reduction with distance and frequency is discussed in Chapter 2. When predicting the electromagnetic ambient level at a given location, an important consideration is not only the distance from potential sources of EMI but also the presence of conductive structures close to the emission source, such as walls, buildings, and ground topology. When structures are behind or to the side of the source, they act as reflectors; they act as shields when the structures are intervening. Chapter 6 discusses the shielding effectiveness of structures and soil.

In Chapter 6, we see that the field behind an isolated conductive plate or PCB shield may be higher than the incident field.

1.2 INTRODUCTION TO ELECTROMAGNETIC INTERFERENCE REGULATIONS

The level of immunity built into equipment depends on how critical the correct functioning of the equipment is and on the electromagnetic environment in which it is designed to operate. Many EMI requirements take the criticality and environment into account by classifying equipment and by imposing different susceptibility test levels on the different classes.

EMI can be considered a form of environmental pollution; in order to reduce the impact of this pollution, some control on the environmental level of conducted and radiated emissions of noise is necessary.

Many countries impose commercial regulations on the emissions from data processing equipment, ISM equipment, vehicles, appliances, etc. In some instances, standards are developed by a nongovernmental agency, such as the Society of Automotive Engineers, and are not necessarily mandatory. The majority of military regulations and standards and some commercial specifications also require that equipment be demonstrated immune to susceptibility test levels.

Chapter 9 describes the typical EMI regulations and requirements and EMI measurement techniques.

The Federal Communications Commission (FCC) in 47 CFR Part 15 places limits on licensed and unlicensed transmissions as well as intentional and unintentional radiators.

Part 15

Subpart B places limits on unintentional radiators, those that produce electromagnetic waves by the nature of their operation, such as computers.

Subpart C deals with intentional radiators that place limits on devices that are specifically designed to produce electromagnetic waves, such as transmitters, and these can be licensed or unlicensed.

Subpart D covers unlicensed PCS. PCS is a wireless phone service similar to digital cellular that was developed for extensive mobility.

Subpart E is about unlicensed National Information Infrastructure Devices.

Subpart F deals with ultra-wideband operation used typically for ground-penetrating radar, imaging systems, surveillance, and vehicular radar systems.

Subpart G is about access broadband over power line.

Subpart H covers TV band devices (TVBDs) that are unlicensed intentional radiators operating on available channels in the broadcast TV frequency bands.

1.2.1 MILITARY REGULATIONS

The options are limited for manufacturers of equipment that must meet specified requirements. The military requirements are intended to be tailored to the specific electromagnetic environment by the procuring agency; however, this is seldom implemented. Should equipment fail specified military requirements and, after analysis or measurement, the environment be found more benign than the specified levels indicate, then the possibility exists for the procuring agency to grant a waiver on the specification limit. A more satisfactory approach is to specify realistic limits in the first place. The difficulty here is that the requirements are location dependent. That is, the proximity of equipment to transmitting antennas or other equipment or the number of units connected to the same power supply varies from case to case. Where equipment is intended for operation in a known location, the limits may be readily tailored to the environment.

1.2.2 COMMERCIAL REGULATIONS

The manufacturers of equipment that must meet the commercial requirements are seldom if ever awarded a waiver, and the limits are inflexible. To date, only the countries of the European Union (EU) require immunity testing. Some manufacturers who want to market in non-EU countries may consider this an advantage until the equipment is found to be susceptible in a typical environment.

1.2.3 UNREGULATED EQUIPMENT

For the manufacturers of equipment to which no regulations apply but who want to achieve EMC either for the sake of customer satisfaction or safety or to minimize the risk of a lawsuit, the choice is either to design for a realistic worst-case environment or to define the environment with an existing EMI standard. We define a realistic worst case as either a measured maximum environment in a large sample of similar environments or a predicted maximum where all the mitigating factors have been considered.

In an ideal world, specified limits would be close to the realistic worst-case environment, whereas, as we shall see in Section 1.3, this is not always true.

1.3 ELECTROMAGNETIC ENVIRONMENT

The information in this section is intended to provide a comparison of the various worst-case environments and to provide guidelines to equipment designers and those writing procurement specifications.

The sources of EMI can be divided into natural and man made, with, in most cases, the natural sources of radiation present at a much lower level than the man made. The majority of unintentional emissions occupy a wide range of frequencies, which we may call broadband in a nonstrict sense of the term. Intentional emissions, such as radio and TV transmissions, are termed narrowband (NB) and in the strictest sense of the term are emissions that occur at a single frequency or are accompanied by a few frequencies at the sidebands. The strict definitions of NB and broadband, as used in EMI measurements and addressed in Chapter 9, are dependent on both the receiver bandwidth and the pulse repetition rate of the source. The emissions from intentional radiators are typically much higher than from nonintentional such as informational technology when measured at the same distance.

Electric field strength is measured in V/m, as described in Section 2.1. Another unit of measurement is the dBμV/m. The unit of broadband field strength used in earlier versions of military requirements was the dBμV/m/MHz. Here, the reference bandwidth of 1 MHz is included in the unit. Another unit is the dBμV/m/kHz, where the reference bandwidth is 1 kHz.

For the sake of comparison to other sources of broadband noise, we will arbitrarily use the military MIL-STD 461F R1E02 limit for emission from equipment measured at a distance of 1 m from the source. Figure 1.4 is a reproduction of the RE102 limit for spacecraft. The RE102 limit is more stringent than commercial limits on broadband noise. For example, the EU limits on ISM, information technology, and generic for Class B devices (the most stringent limit) are 40 dBμV/m from 30 to 88 MHz at 3 m. The FCC limit for Class B is also 40 dBμV/m from 30 to 80 MHz.

When scaled to a 1 m measuring distance, the limit is 49.54 dBμV/m.

The MIL-STD-461F limit for fixed-wing external aircraft and spacecraft and for ground-based Navy Mobile and Army is 24 dBμV/m. converted to 3 m = 14.5 dBμV/m. These commercial limits are 25.5 dB above the MIL-STD limit.

These limits are presented in greater detail in Section 9.4.

The remainder of this chapter deals with radiated and conducted components of the electromagnetic environment (ambient).

1.3.1 NATURAL SOURCES OF ELECTROMAGNETIC NOISE

Natural sources of electromagnetic noise are

- Atmospheric noise produced by electrical discharges occurring during thunderstorms
- Cosmic noise from the Sun, the Moon, stars, planets, and galaxy

Atmospheric noise is produced predominantly by local thunderstorms in the summer and by tropical-region thunderstorms in the winter. The electromagnetic emissions from thunderstorms are

FIGURE 1.4   MIL-STD-461F RE102 limit used as comparison.

propagated over distances of several thousand kilometers by an ionospheric skywave, and thus, potential EMI effects are not localized. In the time domain, atmospheric noise is complex, but it may be characterized by large spikes on a background of short random pulses or smaller pulses on a higher continuous background.

The upper and lower limits for atmospheric radio noise displayed in the frequency domain are shown in Figure 1.5d, ranging in level from a maximum of 108 dBµV/m/MHz (0.25 V) to a minimum of −6 dBµV/m/MHz (0.5 µV) at 100 kHz. Figure 1.5e also plots the atmospheric noise measured in the Arctic, which describes a yearly variation of approximately 7 dB as well as a systematic daily and seasonal variation. From measurements in Canada, the daily and seasonal variation is from 91 to 106 dBµV/m/MHz. Compared to our arbitrarily chosen RE02 broadband reference limit, contained in earlier versions of MIL-STD-461, the upper-limit atmospheric noise may be close to the RE02 limit at 100 kHz and 20 dB below the limit at 10 MHz.

Cosmic noise is a composite of noise sources comprising sky background radio noise, which is caused by ionization and synchrotron radiation (which undergoes daily variation), and solar radio noise, which increases dramatically with an increase in solar activity and the generation of solar flares. The secondary cosmic noise sources are the Moon, Jupiter, and Cassiopeia A. At 30 MHz, the average cosmic noise is 34 dB below the RE02 limit.

A comparison of the relative intensities of atmospheric and cosmic sources and the frequency range of emissions are shown in Figure 1.6. The average daily upper and lower limits of the normal cosmic noise are shown in Figure 1.5a. Additional sources of emissions exist at lower levels, including the thermal background. The theoretical thermal background from the Earth's surface is shown in Figure 1.5c with the internal thermal noise of a well-designed receiver for comparison purposes.

The EMI effect on radio communications is often described as "static" due to the impulsive nature of atmospheric noise. The second source of transient EMI, which may be incorrectly attributed to atmospheric noise, is precipitation static discharge in the proximity of the receiving antenna. A static discharge on the ground is caused by a buildup of charge on the surface, resulting in a corona discharge. Correct grounding and bonding of conductive elements, use of high-breakdown-voltage dielectrics, static discharge coating, and transient suppression devices are some of the methods used to avoid static charge buildup and protect the receiver input circuit.

Because of the wide frequency span over which natural emissions occur, they may cause EMI in HF/VHF/UHF/SHF transmissions.

FIGURE 1.5 Atmospheric, cosmic, and thermal noise levels. (From Cottony, H.V., *Science*, 111(2872), 41, January 1950.) (a) Average daily upper and lower limits of normal cosmic radio noise field intensities. (b) Noise field intensities corresponding to the internal noise of well-designed receiver. (c) Noise field intensities (in one plane of polarization) produced by "black-body" radiation at 300 K. (d) Upper and lower limits of atmosphere noise intensities (National Bureau of Standards Circular No. 462). Also radio propagation unit report RPU-5. (e) Atmospheric radio noise intensities measured in the Arctic.

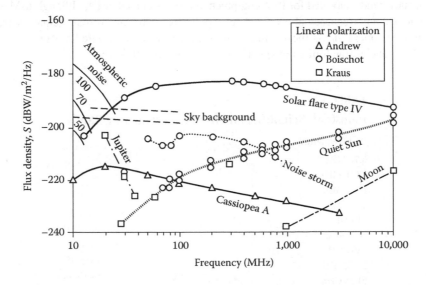

FIGURE 1.6 Comparative levels of cosmic noise sources. (From Skomal, E.N., The dimension of radio noise, *1969 IEEE Symposium on Electromagnetic Compatibility*, Vol. 69, C3-EMC, June 17–19, 1969. Copyright 1969 IEEE.)

1.3.2 MAN-MADE ELECTROMAGNETIC NOISE

Some of the major sources of manmade electromagnetic noise are

- Arc welders
- RF heaters
- ISM equipment
- AC high-voltage transmission line
- Automotive ignition
- Fluorescent lamps
- Microwave ovens
- Hospital equipment
- Diathermy equipment
- Communication transmitter intentional and spurious radiation
- Electric motors

Each of these sources will be discussed with reference to the RE102 specification limit.

1.4 INDUSTRIAL, SCIENTIFIC, AND MEDICAL EQUIPMENT

1.4.1 FCC PART 18

The ISM devices fall in the unlicensed spectrum category that includes 2.4 GHz (802.11b and 802.11g) and 5 GHz (802.11a). This includes cordless phones 900 MHz, 2.4 GHz, and 5 GHz and microwave ovens (2.4 GHz). These devices include nonlicensed toys, wireless security systems, wireless telemetry, and automatic meter reading, all operating in the 902–928 MHz ISM band.

The FCC also allows unlicensed wireless equipment to operate in some of the ISM bands. In the 5.8 GHz band, the maximum EIRP is 53 dBm (30 dBm output power plus 23 dB antenna gain). This generates a 25.8 V/m field at 3 m and 2.58 V/m at 30 m. In the 2.4 GHz band, the antenna gain can be increased to get an EIRP above 36 dBm but for every 3 dBi increase in antenna gain, the transmit power must be reduced by 1 dBm. Thus, with an antenna gain of 30 dB, the transmit power is limited to 22 dBm. This generates an E field of 23 V/m at 3 m and 2.3 V/m at 30 m.

The frequency bands allocated for ISM equipment are shown in Table 1.1, although ISM equipment may be operated at any frequency above 9 kHz except for the search and rescue frequency bands.

ISM equipment operating on these bands is permitted unlimited radiated power.

At other frequencies, the limits are specified at 30, 300, and 1600 m.

TABLE 1.1
Industrial, Scientific, and Medical Frequency Bands

ISM Frequency

6.78 MHz
13.56 MHz
27.12 MHz
40.68 MHz
915 MHz
2450 MHz
5800 MHz
24.125 GHz
61.25 GHz
122.5 GHz
245 GHz

1.4.2 MEASURED FIELDS FROM INDUSTRIAL EQUIPMENT

Heliarc welders use an RF arc at a typical fundamental frequency of 2.6 MHz. The spectrum occupancy of the heliarc welder emission covers the frequency range from 3 kHz to 120 MHz and thus contains frequencies lower than 2.6 MHz. The typical EMI effect on radio is a "frying" noise. The following are the representative levels of radiation from a substantial population of RF-stabilized arc welders measured at a distance of 305 m [1]:

Frequency	Radiation Level
0.7 MHz	75 dBμMHz
25 MHz	82 dBμV/m/MHz and
	80 dBμV/m/MHz (10 mV/m/MHz)
30 MHz	70 dBμV/m/MHz

The levels are in the broadband specification. Assuming that the noise does cover a frequency range of 1 MHz and is at a uniform magnitude over this bandwidth, the 25 MHz 82 dBμV/m/MHz reduces to 42 dBμV/m at 25 MHz when measured using the MIL-STD-461F bandwidth of 10 kHz. This is 18 dB above the MIL-STD-461F RE102 limit for ground applications at 25 MHz and is naturally much higher at distances closer than 305 m. For example, at a distance of 2 m from an arc welder, the E-field level at 30 MHz is approximately 124 dBμV/m/MHz (1.5 V/m). In extrapolating for distance at these frequencies, either a l/d or $l/d^{1.5}$ law is used (d = distance), depending on the frequency and other criteria. From measurements at 14 manufacturing plants, the peak level of emission from arc welders was found to be 0.1 V/m at a measuring distance of 1–3 m [5]. Unfortunately, the bandwidth of the measuring instrument was not given, so it is not possible to know if this was an NB or broadband measurement.

In measurements described in Reference [5], a heliarc welder exhibited significant emissions from 14 kHz to 240 MHz. However, when measured at a distance of a few miles, only the fundamental frequency of 2.6–3 MHz remained at a significant level. In measurements at different locations on 152 welders, comprising 54 different models, the highest emission measured at 305 m was 64 dBμV/m. Of the 152 welders, 31 produced emissions above 40 dBμV/m and only 6 above 54 dBμV/m. Some welders generated levels as low as 0 dBμV/m (1 μV). The low-level emitters were characterized by one or more of the following: short welding leads; low-impedance ground; shielded wiring, including supply; and enclosure in a shielded building. High-level emitters, in contrast, were characterized by one or more of the following: poor grounding, unshielded wires, and proximity to power lines, which picked up the emissions and reradiated them.

The FCC in Document 47 CFR Part 18 places a limit of 10 μV/m at 1600 m on industrial heaters and RF-stabilized arc welders below 5.725 MHz. Converting this to the 305 m distance used in the preceding survey and applying a $1/d$ law result in a limit of 34.4 dBμV/m. From the survey, it is clear that many existing arc welder installations do not comply with the FCC limit.

The fields generated by induction and dielectric-type RF heaters are principally NB, with peaks extending up to about the ninth harmonic. Induction heaters are used for forging, case hardening, soldering, annealing, float zone refining, etc., while dielectric-type heaters are typically used to seal plastic packages. The fundamental operating frequency for induction heaters is from 1 kHz to 1 MHz and for dielectric heaters is from 13 MHz to 5.8 GHz. In measurements on 36 induction heaters, the minimum and maximum emissions at a distance of 30 m varied from 30 dBμV/m (31.6 μV) to 114 dBμV/m (0.75 V) [4]. Since these emissions are NB, the 1 MHz reference bandwidth is omitted in the measurement unit used.

We have so far used the relatively low-level military broadband radiated emission limit (RE102). The emissions from four different manufacturers and 10 different models of dielectric heater when measured at a distance of 30 m ranged from a maximum of 98.8 dBμV/m (87 mV) and a minimum of 75 dBμV/m (5.6 mV) at the fundamental (27 MHz) and reduced to a maximum of 84 dBμV (15.8 mV) and a minimum of 38 dBμV (79 μV) at the sixth harmonic (162 MHz).

No measurements of the conducted noise placed on the power line by these devices are available. However, the radiated emissions at 30 m were found to be above the typical ambient levels of 0.15–0.9 mV/m measured in offices, electronic laboratories, and computer facilities, but they do not pose a severe EMI threat. This is illustrated by a cursory survey of the 35,434 complaints of interference with radio communication lodged with the UK regulatory authority more than a 12-month period. This survey reveals that 143 complaints were attributed to ISM sources, 11 to medical apparatus, and 66 to RF devices not tuned to designated frequencies, but not one complaint was attributed to induction or dielectric heating equipment.

The levels of noise measured at 14 manufacturing sites from a variety of ISM equipment are shown in Figure 1.7. The 14 manufacturing plants included discrete and continuous production plants, an automotive tool and die shop, a chemical plant, a heavy equipment manufacturer, an aerospace manufacturing plant, newspaper printers, paper and pulp plants, and metal smelting plants. The levels are given in volts per meter, with no information as to the bandwidth of the measuring instrument. Assuming the bandwidth used was narrow enough to capture only one of the spectral lines of emission, then the electromagnetic field levels contained in Figure 1.7 will have the same magnitude as an NB emitter of the same field strength, at the specified frequency of maximum emission.

When a broadband field is expressed in broadband units, the magnitude is invariably higher than the same field expressed in an NB unit. Here, we use the term *broadband field* loosely to indicate a field comprising a number of frequencies. A broadband unit of measurement uses a reference bandwidth, typically 1 MHz (e.g., dBμV/m/MHz), whereas an NB unit does not specify a bandwidth (e.g., dBμV/m). However, the bandwidth used in an NB measurement is always less than 1 MHz or equal to 1 MHz above 1 GHz. In a broadband measurement of broadband noise, many of the random or harmonically related spectral lines are captured in the receiver bandwidth. In an NB measurement, only one of the spectral lines is captured in the receiver bandwidth.

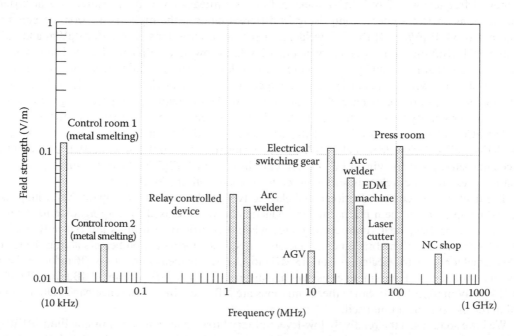

FIGURE 1.7   Peak field measurements from industrial, scientific, and medical sources. (From Nielson, R., *Control Eng.*, 35(10), 1988.)

For example, assume a broadband signal source generating harmonically related spectral lines at 1 kHz intervals apart with a constant amplitude over a 1 MHz span. Assume the field strength at some distance from the source, for the sake of our example, is 50 dBμV/m/MHz. If we were to make an NB measurement with a 1 kHz measuring bandwidth, then only one spectral line is measured and the field amplitude is reduced to −10 dBμV/m, expressed in NB terms. In this way, coherent broadband noise decreases as a function of 20 dB/decade of bandwidth. It is therefore incorrect to directly compare the magnitude of NB sources, for example, those "intentional emitters" contained in Section 1.3.3, to broadband sources. The subject of broadband and NB measurements is dealt with in Chapter 9 and sources of broadband noise in Chapter 3.

The susceptibility of equipment and cables to an impinging broadband or NB field is dependent, among other factors, on cable and enclosure resonance effects, the bandwidth of the equipment (including signal interfaces), and the transient response of the cable and structure. These factors are considered in subsequent chapters.

1.4.3 INTERFERENCE FROM HIGH-VOLTAGE TRANSMISSION LINES AND *E* AND *H* FIELD CLOSE TO THE LINE

It is common to experience EMI to AM reception in cars that are driven in close proximity to or under high-voltage transmission lines. Figure 1.8 illustrates the spectrum occupancy of transmission line noise with, as expected, a maximum at the power line frequency of either 50 or 60 Hz. The curves in Figure 1.8 are from several sources, and the numbers in brackets indicate the distance (meters) from the line at which measurements were made.

1.4.3.1 Magnetic Field

The magnitude of the magnetic field close to a power line, as seen in Figure 1.9, depends on the distance, the current flow, and the conductor configuration.

In Reference [2], the average for 138, 230, and 500 lines is provided at 0, 10, 20, 30, and 40 m from the line. These values are for average power loads and average tower heights. For the 500 kV line,

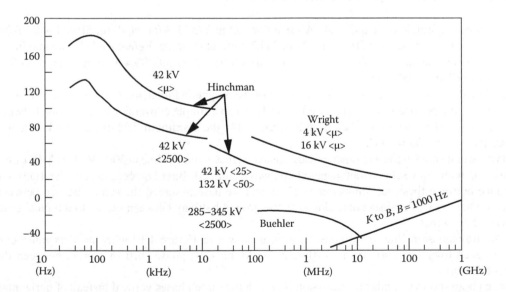

FIGURE 1.8 Transmission line noise (numbers in brackets are measuring distances in meters). Observation point location relative to the sources in feet <25>, <50>, etc.; <μ> beneath conductor. (From Skomal, E.N., *The dimension of radio noise, 1969 IEEE Symposium on Electromagnetic Compatibility*, Vol. 69, C3-EMC, June 17–19, 1969. Copyright 1969 IEEE.)

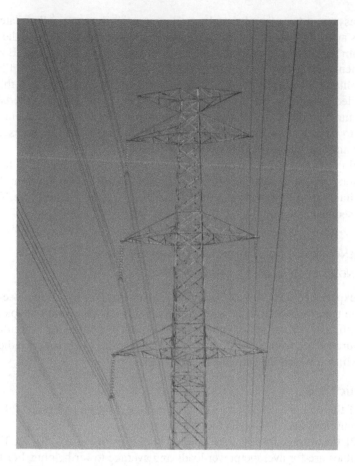

FIGURE 1.9   Typical high-power transmission line and tower.

the average magnitude is 81 mG (6.43 A/m) at 0 m, 72 mG (5.71 A/m) at 10 m, 51 mG (4.04 A/m) at 20 m, 33 mG (2.6 A/m) at 30 m, and 21 mG (1.7 A/m) at 40 m. In Reference [14], the maximum magnetic field under a 275 or 400 kV line is quoted as 100 µT (997 mG, 79.4 A/m) with a typical field of 10 µT (99.7 mG, 7.94 A/m).

Although the gauss is strictly a measurement of the magnetic flux density when the field in air is considered, the permeability of air is a constant 1, and so a simple conversion can be made between gauss, tesla, and amps/meter. Then from Reference [14], the maximum field directly under the line is 100 µT = 79.3 A/m = 10 G.

For a worst case (highest magnetic field), some distance from the line, a 500 kV 3000 MW three-phase line with three subconductors per phase, was examined. Based on corona considerations, the distance between these subconductors is 17 m, and we have assumed the same distance between phases. Due to sag, the lines come closest to the ground halfway between spans. In a typical case, this height is 9 m.

The transmission line under consideration is shown in Figure 1.10 with the maximum current in each line. The currents are different due to the 120° phase shift in current between the subconductors.

The photo shows a similar transmission line with the three phases vertical instead of horizontal, but the composite fields from the lines will be very similar.

The maximum current in any one phase is the maximum power of the transmission divided by the three lines divided by the three phases divided by the peak voltage and multiplied by the square root of 3.

FIGURE 1.10 Configuration of the three phases in the line and the current distribution.

That is, (3000e6/3/3)/500e3 ∗ 01.732 = 1154 A, and for a worst case, we assume this is the current in the subconductors located closest to the building. In the next set of subconductors 17 m away from the first set, the current is −770 A (due to the 120° phase shift), and in the third set of subconductors, the current is −384 A at the same moment in time. Figure 1.10 shows the distance between the conductors in each phase and the current.

The magnetic field from an infinitely long conductor is provided by $H = I/2\pi d$, where I is the current in amps and d is the distance from the conductor in meters.

Using this equation, the composite H field (addition of fields from all the conductors) at different distances from the source was calculated. As the conductors increase in height above ground, the distance to the measuring point increases and the magnetic field decreases. From the equation for the H field, we see that reductions are inversely proportional to distance ($1/d$), although this is not strictly true for the composite field. Table 1.2 shows this worst case (highest level of magnetic field) at 5, 10, 20, and 30 m along the ground from all of the conductors. The field directly under a power line has been reported in Reference [14] to be at 79.4 A/m, but it is unlikely that a building is directly under the line.

Table 1.2 assumes a constant distance from the conductor down the length of the line, whereas the lines increase in height on either side of the span center (lowest point). However, as the span between lines is typically 400 m and the longest building considered is 61 m, any error in the calculation is considered negligible.

TABLE 1.2
Worst-Case Magnetic Fields from a 3000 MW Power Line

Distance from Closest Conductors	Magnetic Field (A/m)	Magnetic Field (Magnetic Flux Density) (T)
5	11.5	1.445e−5
10	8.5	1.07e−5
20	3.85	4.85e−6
30	1.55	1.95e−6

The magnetic and electric fields from the power line result in a current flow in a conductive building. The skin depth of a conductor is defined as the depth at which 63.2% of the current flows and is dependent on the permeability and conductivity of the material as well as the frequency. In the case of aluminum, the skin depth at 60 Hz is 11 mm, while the thickness of #18 gauge aluminum is 1.02 mm. The thickness of #18 gauge steel is 1.2 mm and the skin depth is 6.3 mm. Therefore, the current on the inside of the aluminum or the steel building at 50/60 Hz is almost the same as the outside.

Even though the current flow on the inside of the building is virtually the same as the outside, some attenuation of the internal magnetic field occurs above some frequency due to field cancelation. The mechanism for this is described in Section 6.5.1.

However, at 50 or 60 Hz, the attenuation of the practical large building made of overlapping steel or aluminum panels is effectively zero. Even for the smaller building, the attenuation is negligible. Thus, the magnetic field within the building is practically the same as the magnetic field outside of the building. However, either building very effectively shields against the electric field component of the incident wave electromagnetic wave as described in Section 6.1.4.

1.4.3.2 Electric Field

In addition to the horizontally polarized electric field described in Section 1.4.3.3, the electric field under or close to a power line comprises a vertical field due to the potential between the lines and ground or between the lines and a conductive building in close proximity to the lines. From Reference [3], the maximum E field for a 400 kV power line 10 m from the line is 11 kV/m similar to the 6.2 kV calculated for our example power lines. The vertical field results in displacement current flow due to the capacitance between the lines and the ground/building. Based on the 11 kV/m field, the displacement current flow in the 61 m long side of the large metal building has been estimated and is very low, approximately 0.15 A. Also, the vertical field reduces very quickly with distance from the line.

Although the magnetic field is virtually unattenuated by a metal building, the electric field is very effectively attenuated, especially at the low 50/60 Hz frequency. The level of attenuation of the E field is described in Figure 6.5, Sections 6.2 for an infinite sized plate and 6.3.1.2 for an enclosure, and indicates that even aluminum foil or aluminized Mylar is an effective shield against low-frequency E fields.

1.4.3.3 Currents Induced into Metal Buildings Close to the Power Line

Any current varying in time generates not only a magnetic field but also an electric field. Thus, as current flows down the line, it generates a horizontally polarized electric field as well as a magnetic field that circles the line.

The concern is that as the incident magnetic field induces eddy current flow in a conductive building and that as a conductive door either opens or closes, sparking can result. Reference [15] provides a simple equation to calculate this current flow in conductive plates. The input data are the magnitude and the rate of change in the magnetic field, the thickness of the metal, the conductivity of the metal, the aspect ratio of the plate, and the surface area of the plate.

The equation for the total current in a rectangular plate is

$$I = \frac{-1}{4\sqrt{\pi}} \times \frac{\sqrt{\gamma_e}}{\left(1.5\left(\gamma_e + 1\right)\right) - \sqrt{\gamma_e}} \times \frac{(\gamma+1)\sqrt{\gamma}}{\left(\gamma^2 + 1\right)} \sigma h S B(t)$$

where
  π is the PI ≈ 3.142
  γ is the length/width, which is the aspect ratio
  σ is the conductivity in Siemens/m
  h is the thickness of the material in m
  S is the area of the plate in m^2
  $B(t)$ is the uniform time varying magnetic field in T/s

The equation was checked by entering into the Excel spreadsheet the same input data as provided in Reference [15], and the same results were obtained.
  The data for the end of the small building shown in Figure 1.11 (most likely location for door) were entered as follows:

Height 2.44
Width 2.62 m
Conductivity of steel = 1e7 S/m

FIGURE 1.11   Smaller building.

Aspect ratio = length/width = 2.62/2.44 = 1.07
Incident field at 10 m = 6e−4 T/s
Area = 6.4 m²

Assuming the door seam has the same conductivity as the wall, then the current flow across the door with the building 10 m from the high-voltage lines is 16 A. At 30 m, this reduces to 2.9 A.

For the large building, the input data are

Height 18.2 m
Width 60.9 m
Conductivity of steel 1e7 S/m
Aspect ratio 3.35
Incident field at 10 m = 6e−4 T/s
Area = 1108 m²

The predicted total current in the side of the building is 366 A/m. However, the door height is likely to be less than 18.2 and 3 m has been assumed.

The predicted current at 10 m from the high-voltage lines is therefore 366 A × 3 m/18.2 m = 60 A. This reduces to 10.9 A at 30 m.

By assuming the conductivity of the seam is the same as the solid steel, the currents are very much a worst case. In reality, the seam is likely to have a resistance of at least 1 mΩ/m = 1000 S/m.

The current flow with 1 mΩ seams is negligible. However, on a new door with new hinges and new latch, a resistance lower than 1 mΩ is possible, and so it is wise to use the worst-case numbers.

Measurements were made of sparking due to 60 Hz current in steel-to-steel and aluminum-to-aluminum metal plates as they contacted, and the results are shown in Table 1.3.

With the trailer 10 m from the power line and assuming a very much worst case, the prediction is that light sparking can occur and, for safety, the building should be at least 30 m from the power line. As the power line under consideration is a high-voltage and a high-power installation, this distance may be less for lower-power and lower-voltage lines. For the large building, and worst case again, the distance should be greater than 50 m. If the resistance of the seam, hinges, and latch can be shown to be 1 mΩ or greater, then either building can be 10 m from the power line. The recommended distance for the safety of personnel is at least 22 m from the 3000 MW line.

In addition to the current induced by the magnetic field when a conductive building is close to the power line, it can become charged up, and when a person touches the building, he or she can experience a nuisance shock. However, this is no greater than when a person is charged by friction (electrostatic discharge), typically when the air is dry.

TABLE 1.3

Level of Sparking as Current Flows across a Moving Surface

Length of Spark (cm) or Level of Sparking	Current Flow (A)
Steel-to-steel interface	
2.5	15
5	22.5
10–13	30
Steel to aluminum	
Light sparking	10.5
2.5	23
10–13	28.2

1.5 NOISE FROM TRAFFIC, FLUORESCENT TUBES, MICROWAVE OVENS, AND MAGNETIC FIELDS IN THE HOME AND OFFICE

The ignition noise level is dependent on traffic density and proximity. In the time domain, ignition noise is characterized by bursts of short-duration (ns) pulses with a burst duration from a microsecond to milliseconds. The repetition rate of the bursts is dependent on the RPM, the number of cylinders of the motor, and the number of cars. When measured in close proximity to the road, these bursts vary in amplitude and direction with traffic flow. In cosmopolitan areas, ignition noise is a major contributor to the electromagnetic environment. The average of measurements taken in three cities at a roadside location for a traffic flow of 30 autos per minute is 60 dBµV/m/MHz at 100 MHz decreasing to 50 dBµV/m/MHz at 1 GHz. Ignition noise is also present from 10 kHz upward, and levels as high as 80 dBµV/m/MHz have been measured at 10 kHz.

Fluorescent and gaseous discharge tubes produce impulsive radio noise similar to power line noise in its waveform characteristics. The level of noise from 10 kHz to 1 MHz was measured in a shielded room with the lights on and the ambient noise with the lights off. The highest fluorescent light emission was at 300 kHz, measured at 89 dBµV/m/MHz, and the RE02 BB limit at 300 kHz is 98 dBµV/m/MHz. Thus, although the ambient is just below the specification limit, all subsequent measurements were made with the fluorescent lights off. For MIL-STD 461 measurements, the ambient should be at least 6 dB below the specification limit. The remaining incandescent lamps in the shielded room did not add to the electromagnetic ambient. The average level of radiation from fluorescent lamp fittings measured at a distance of 1 m is shown in Figure 1.12.

Field-strength measurements of microwave ovens operating in the ISM band at 915 MHz were measured in the laboratory at distances of 3.05 and 305 m. Measurements were also made outside a large condominium containing 385 ovens. The measurements were made on one make of oven and similar models. The maximum field strength measured in the laboratory was 1.5 V/m at 3.05 m and 11 mV/m at 305 m. The field strength measured outside the condominium building would have been altered by some shielding due to the building structure and reflections from the ground and nearby buildings. The maximum field strength measured was 8.9 mV/m at 920 MHz at a measuring distance of 152 m from the two buildings comprising the condominium [6].

In addition to RF electromagnetic fields, magnetic fields at power line frequencies are ever present. Due to the concern for the potential health hazard of magnetic fields, a number of measurements were made in a home and in an office building. The magnetic field strength in one room on the first

FIGURE 1.12 Average level of radiation from fluorescent lamps.

TABLE 1.4
Magnitude of Magnetic Fields Measured in a Home

Location of Magnetic Field Measurements in a Home	Magnetic Field (A/m)
20 cm from top of stove	4.86
20 cm from oven door	0.4
20 cm from electric food mixer	4
20 cm from electrical panel	0.82
20 cm from 120 to 250 V autotransformer	46.2
2 m from Hi-Fi setup	0.1
Center of kitchen; all appliances off except for refrigerator	0.1
20 cm from toaster	0.81

---

floor in the office building was sufficient to result in distortion of computer monitor displays. The displays were susceptible to magnetic fields with magnitudes of 1.3–3 A/m. The magnetic fields in this first floor varied from a low 0.016 A/m to a maximum of 3 A/m at 60 Hz. On the fifth floor of the same building, the magnetic field varied from 0.00485 to 0.0428 A/m. The highest magnetic field measured in the basement of the same building was 1.63 A/m, close to a power transformer. A survey was also conducted at several locations in a home; the result of these measurements is shown in Table 1.4.

One technique for reducing 60 Hz magnetic fields from wiring is to keep line and neutral in any two-phase circuit or the A, B, C, and neutral conductors in close proximity or preferably twisted together. Section 2.1.3 discusses the reduction effect seen with twisted-pair cable in detail. An alternative is to run the cables in a high-permeability conduit. Measurements were made on a galvanized cold-rolled-steel seamless conduit with a wall thickness of approximately 1.2 mm (15/32 in.), and an attenuation of approximately 16 dB at 60 Hz was seen. Section 6.2 describes low-frequency magnetic field shielding in detail.

1.6 HOSPITAL ENVIRONMENT

The electromagnetic environment in hospitals has been of growing concern as the number of EMI problems experienced in hospitals has increased. Measurements have been made in 10 American hospitals in a number of locations, including operating rooms, intensive care units, chemistry labs, special procedure rooms, and physical therapy facilities. In the majority of locations, E fields as high as 3 V/m/MHz were measured. It should be emphasized that the measurement results are displayed in broadband units. If the measurements had been made with an NB bandwidth and displayed in NB units, the majority of emissions would have reduced in amplitude but with peaks at 70 kHz and 1–5 MHz remaining at approximately 3 V/m.

Figure 1.13 is a composite of the worst-case field levels measured in all locations at the 10 hospitals surveyed [7].

When these unintentional man-made high-level sources of radiation are compared to our standard R1E02 level, the sources may have intensities of 60 dB above (i.e., 1000 times) the RE102 limit.

The FCC limit for medical diathermy is a maximum of 25 μV/m at 300 m at an ISM frequency.

Diathermy is dielectric heating, whereas electrosurgery is used in surgical procedures such as cutting and coagulation and operates by using alternating current to directly heat the tissue itself. The commonly used frequencies are 333 kHz, 350 kHz, 2.2 MHz, and 4 MHz. The electrosurgical device can typically generate 80–400 W as a sine wave or pulse with a duration of tens of microseconds.

During electrosurgery, the high voltage with high frequency and the high current through the body generates an electromagnetic field in the room.

FIGURE 1.13 Composite worst-case field levels in hospitals. (From Cottony, H.V., Radio noise of iono-spheric origin, *Science*, 111(2872), 41, January 1950. Copyright 1975 IEEE.)

As a worst case, the surgeon's hands may be exposed to a magnetic field of 0.75 A/m and an *E* field of 400 V/m. At the patient grounding plate, the maximum *E* field is 450 V/m and the *H* field 5.8 A/m. These data are from Reference [11]. Measurements were made with source to detector distances of 5 and 50 cm. The *E* field over this increase in distance was measured for a number of modes of operation and two devices. For one device, the reduction was 0.133 and for the other device, it was 0.2. The *E* field at 3 and 30 m was not measured. Assuming that at some distance, the field would reduce as 1/*r*, where *r* is the distance; very approximately, the *E* field at 3 m would be 75 V/m and at 30 m 7.5 V/m. At 30 m, the field may be reduced due to intervening walls.

It is surprising that sensitive electronics, such as cardiac monitors, are immune to these high fields.

E and *H* fields were measured in the corridors of a hospital as well as outside nurseries, CT and MRI scanners, emergency unit, and the operating theater. The fields were measured using an NB measurement. The *E* and *H* fields are shown in Table 1.5.

TABLE 1.5
Typical Hospital Ambient

Frequency (MHz)	*H* Field (dBµA/m)	*E* Field (dBµV/m)
0.09		50
0.11	52	51
0.25		33.6
0.4		55
0.5	30	55
4		80
217		69
900		122
1800		122
2400		118
5300		106

1.7 INTENTIONAL EMITTERS

The electromagnetic environment is also crowded with the intentional emissions from radio, TV, and radar transmitters, all of which can interfere with equipment that is not intended for any form of reception as well as receivers tuned to a different frequency. One of the most frequent causes of EMI is electromagnetic field produced by radio transmitters. The EMI effect may be confined to annoyance, for example, due to the spurious operation of a garage door caused by the use of a CB radio. There also exists documentation of far more serious effects, such as the destruction of a warship attributed indirectly to EMI (due to interference with communication) or the crash of an aircraft as the direct result of EMI (flying in close proximity to a high-power transmitter).

Radio transmissions are NB, with limited emissions at the harmonics of the fundamental continuous wave (CW). If these waves are modulated, then the resulting sidebands are also transmitted. In addition to frequencies related to the fundamental, a transmitter may radiate the local oscillator frequency and broadband noise generated within the stages of the transmitter. The composite of spurious, broadband, and harmonically related noise from a transmitter is typically at least 70 dB down from the fundamental. The radiated fields from transmitting antennas are dependent on the proximity, transmitter output power, directivity of the antenna, relative height between the antenna and measuring point, and proximity of reflecting or intervening absorbing material or structures, etc. Measurements have been made in two major urban centers in Canada (Montreal and Toronto, [7]). From a large number of measurements made close to ground level, a typical field-strength value developed at usual transmitter–receiver distances was computed. The results are shown in Table 1.6. The maximum typical E field at a fixed distance of 100 m is shown in Figure 1.14.

1.8 LOW-POWER INTENTIONAL RADIATORS

These include cordless phones, baby monitors, garage door openers, wireless home security systems, and keyless automobile entry systems.

In FCC CFR47 Subpart B, nonlicensed transmitters are prohibited from causing interference with licensed transmitters and must accept any interference that they receive.

Generally speaking, Subpart B limits are those for intentional radiators. The limits are specified at 300 and 3 m, and in order to normalize levels, these have been converted where necessary to 3 and 30 m and are shown in Figure 1.15 and Table 1.7.

The measurement detector is either average or quasi peak, and the difference in these is described in Section 9.2.2.

TABLE 1.6
Typical Ambient Field Strengths

Frequency (MHz)	Field Strength[a] (V/m)
0.5–1.6	0.6
26.9–27.4	<0.1
54–88	0.07
88–108	0.15
108–174	0.05
174–216	0.07

Source: Adams, J., Electric field strength measured near personal transceivers, *IEEE Electromagnetic Compatibility Symposium Record*, National Institute of Standards and Technology, 1993.

[a] Typical field strength in major urban centers from broadcast transmitters.

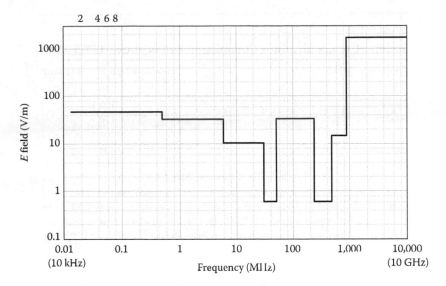

FIGURE 1.14 Average and peak fields from radio transmitters at a distance of 100 m.

FIGURE 1.15 Federal Communications Commission limits for intentional radiators below 1.705 MHz at 3 m.

However, the limits in Table 1.7 may be higher for cable-locating equipment; telephone company electronic detectors, when the 6 dB bandwidth is equal to or greater than 10% of the carrier; intermittent control signals (1250 μV/m at 3 m); cordless telephones; auditory assistance devices (80,000 μV/m at 3 m); biomedical telemetry (50,000 μV/m 2.465–2.4835 MHz at 3 m); and automatic vehicle identification systems.

The FCC has published a document entitled "Understanding the FCC Regulations For Low Power, Non-Licensed Transmitters, OET Bulletin No 63," which contains the limits for these devices.

The devices that use frequency hopping and digitally modulated radiators such as 802.11 WLANs (2.4 and 5 GHz) and 802.15 Bluetooth (2400–2483.5 MHz and Zigbee) have limits contained in FCC Section 15.247. The Bluetooth power limits are low and typically 100 mW from 2.4 to 2.835 GHz.

TABLE 1.7

Federal Communications Commission Limits above 1.705 MHz

Frequency (MHz)	Field Strength at 3 m (μV/m)
1.705–30	30
30–88[a]	100
88–216[a]	150
216–960[a]	200
Above 960	500

The conversion from dBμV/m to V/m is shown in Table 1.8.

[a] Except as provided for perimeter protection systems that may operate in the 54–72 and 76 MHz bands, fundamental emissions from intentional radiators shall not be located in the frequency bands 54–72, 76–88, 174–216, or 470–806 MHz. However, operation within these frequency bands is permitted under sections 15.231 Periodic Operation and 15.241 Biomedical Telemetry Devices, etc.

TABLE 1.8

Typical Field Strength at 20 cm from Cellular Phones

Phone	E Field (dBμV/m)	E Field (V/m)
Motorola i730	146	20
LG TG800	141–145.6	11–19
Nokia	136.7	7

With certain exceptions, the limits from 2400 to 2483.5 MHz, 902 to 928 MHz, and 5725 to 5850 MHz are the maximum peak conducted output power of 1 W. The maximum antenna gain is 6 dBi. Assuming no loss, the maximum E field at 3 m is 3.64 V/m and at 30 m 0.364 V/m.

PCS is a wireless phone service very similar to digital cellular. Limits on unlicensed PCS devices are contained in 15.301–15.323. The highest allowable emissions are the general emission limits shown in Table 1.7. Cordless telephones using DECT 6.0 standards use the PCS band. The DECT operates from 1.88 to 1.9 GHz at typically 250 mW.

The maximum output power for the Global System for Mobile Communications (GSM) system, which is a 2G digital cellular network, at 900 MHz is 8 W and at 1800 MHz 4 W. Assuming antenna gain of 0 dBi, the E field at 3 m at 8 W is 5.17 V/m and at 30 m is 0.524 V/m. At 4 W and 3 m, it is 3.65 V/m and at 30 m 0.36 V/m.

For the Wideband Code Division Multiple Access, which is a 3G mobile communications network system, in the universal Mobile Telecommunications (2.0 GHz) band, the maximum power is 2 W. Assuming 0 dBi antenna gain, the E field at 3 m is 2.58 V/m and at 30 m 0.26 V/m.

The limits for 800 and 1900 MHz PCS band Code Division Multiple Access mobile transmitters are 1 W. Assuming 0 dBi antenna gain, the maximum E fields are 1.8 V/m at 3 m and 0.18 V/m at 30 m.

The TETRA/TETRAPOL operates from 380 to 676 MHz at typically 10 W(rms).

WLAN operates from 2.4 to 2.835 GHz at typically 100 mW and from 5.15 to 5.725 GHz at typically 1 W.

The FCC controls the radiation from ultra-wideband operation transmitters, such as ground-penetrating radars, wall imaging systems, through-wall imaging systems, and medical imaging systems in Subpart F of FCC Part 15.

The highest EIRP for these transmitters is −41.3 dBm from 3,100 to 10,600 MHz. This converts to 7.4×10^{-8} W. The E field at 3 m is therefore only 4.97 V/m (0.497 mV/m).

FIGURE 1.16 **(See color insert.)** Field strength from three cellular phones measured at 20 cm from antenna.

The concern with cellular phones is the power induced into the human head and the FCC limits are in specific absorption rate, described in Section 2.7.3.

The power generated by a cellular phone at a distance of 2.5 cm from the human head is absorbed by the head at the same rate as an incident field of 41 V/m [10].

To calculate the field incident on electronics in close proximity to phones, the E field at 20 cm was measured using a calibrated bow tie antenna. The transmit link was made poor to obtain the maximum power from the phone, and the results are shown in Figure 1.16 from three manufacturers of phones.

A measurement was made of the change in E field with change in distance over 10–50 cm, and this was used to predict the E field at 5 and 2.5 cm.

Extrapolating for distance at 5 cm, the E field will be approximately 148 dBμV/m = 25 V/m and at 2.5 cm = 148.3 dBμV/m = 26 V/m.

Transceivers (walkie-talkies) can generate fields as high as 55 V/m at 823 MHz at a distance of 12 cm from a 5 W transceiver [8].

The maximum measured E field from two different walkie-talkies tested on different channels was 9.5 V/m at a distance of 1 m from the device. At 20 cm, the predicted E field from 161 to 161.5 MHz is 47.5 V/m, and at 5 cm, it is 62 V/m.

The General Mobile Radio Service uses repeater output frequencies from 462.55 to 462.725 MHz. The power limit is 50 W with the antenna at a height of 60 ft (18.3 m) above ground. Assuming a 15 dBi gain antenna, the E field directly under the antenna is approximately 12 V/m.

The Family Radio Service operates from 467.562 to 467.7125 MHz. The maximum ERP is 5 W, and so using the equation in Section 2.1.1, the E field at 3 m is 3.83 V/m and, at 30 m, it is 0.38 V/m.

The FCC controls TVBDs in Subpart H. The maximum limits are the same as for general requirements shown in Figure 1.15 and Table 1.7.

1.9 HIGH-POWER INTENTIONAL RADIATORS

These include FM, AM, TV, and cellular phone and PCS transmitters.

An electromagnetic RF ambient survey is often made before erecting a receiving antenna or before the installation of potentially susceptible equipment. Examples of peak emitters encountered in such measurements are a radar signal measured at a site in Goose Bay, Labrador, which exhibited a frequency of 1280 MHz and a level of −21 dBW/m^2 (i.e., an E field of 1.73 V/m); an AM radio transmitter in Alice Springs, Australia, generated E fields of 4 V/m at 4.83 MHz incident on nearby receiving equipment; and, at a proposed site in Hong Kong, E fields of 3.75 V/m at 12 MHz were measured at the proposed location of receiving equipment. Many apartment and office buildings have antennas mounted on the roof. In one apartment building, Bell cellular antennas were mounted on the roof and the E fields, at 880 MHz, measured in two apartments were 7.86 and 6.48 V/m, respectively. On the balconies of these same apartments, the maximum E field was at 12 and 12.9 V/m, respectively.

In surveys of the fields generated by rooftop-mounted cellular phone antennas on the top floor balcony of one apartment block, a field of 45 V/m was measured at 900 MHz. The Health Canada limit for the general public is 47.5 V/m, and so although just below the safety limit, occupants of the apartment and any portable electronic device are exposed to this high level when on the balcony.

The ambient inside a household that contains an amateur radio transmitter with the antenna mounted on the roof was measured at 30 V/m inside the home. In Reference [10], ambient measurements were made on apartment buildings in 15 different cities in the United States. The median exposure for the inhabitants of all cities was 0.137 V/m, and 99.9% of the inhabitants were exposed to fields of less than 1.94 V/m. Thus, we can conclude that ambients with magnitudes of volts per meter are uncommon but are increasingly the cause of EMI in apartment and office buildings. Electric fields having volts-per-meter magnitudes are potential sources of interference to equipment even when the passband of the equipment is far from the interfering frequency. Case studies for this type of EMI are contained in Chapter 12.

Additional tests in 2007 of both the E field and H field close to transmitter sites in the United Kingdom show similar levels for the E field. The concern was the health effects of these fields on workers in the broadcast industry. The tests are described in Reference [12]. The transmitters measured were

Analog TV
Digital TV
National FM radio
Local FM radio
Digital audio broadcasting

A summary of the fields and locations is shown in Table 1.9.

TABLE 1.9
Worst-Case (Highest) Fields Close to Transmitters

Transmitter	Location	E Field (V/m)	H Field (A/m)
UHF TV	Between two antennas	60–100	
VHF radio	1 m outside of antenna structure	95	0.4
Telecommunication antennas	Near pager antenna	350	0.57
Telecommunication antennas	Near VHF/UHF antenna	270	0.62
Telecommunication antennas	In front of GSM antenna	72 V	
AM transmitter	250 kW sender close to window	300	1.0
MF transmitter	Transmitter 5 cm from window	160	
MF transmitter	Near protective fence	430–700	
VFL/LF transmitter (60 kHz)	Outside main building	200–400	0.3–1.4
Radar site	Outdoor on site thoroughfare	240–560 peak	
Nondirectional beacon	Outside mast enclosure, 0.5 m from fence	70–120	

Source: Cooper, T.G. et al., *Occupational Exposure to Electromagnetic Fields at Radio Transmitter Sites,* Health Protection Agency, Didcot, U.K., 2007.

These fields were measured close to the transmitters in areas we assume are out of bounds to the general public.

However, the general public may be exposed to high fields from the roof-mounted cellular base antennas as discussed earlier.

1.10   CONDUCTED NOISE ON POWER LINES

The electromagnetic environment is confined not only to electromagnetic fields but also to signals/noise existing in a transmission medium. Thus, the noise conducted on equipment power lines must be accounted for when characterizing an electromagnetic ambient. For example, a computer on a machine shop floor, located at some distance from numerical control machines, may be immune to the radiated field levels but susceptible to conducted noise. Therefore, in addition to immunity to electromagnetic fields, achieving EMC entails immunity to conducted noise, present predominantly not only on power lines but also on signal interfaces and grounds. To predict susceptibility for equipment on which no EMC requirements are placed, data on expected maximum amplitude, frequency content, and waveform type (e.g., CW or spike) that may exist on the power supply in a given electromagnetic environment are required.

The conducted noise currents flowing out of medical equipment into a known load connected to the power line were measured in the survey made in 10 hospitals. The impedance of the load is not given in Reference [11]; however, a typical value is 50 Ω. A 400 Hz high-pass filter was included in the measurement setup to reduce the magnitude of load current measured at 60 Hz and its harmonics by the current probe. Figure 1.17 is a composite of the maximum conducted noise currents measured in all locations at the 10 hospitals [7]. In other measurements in a hospital environment, voltage spikes as high as 3000 V have been measured on the AC power line.

Unfortunately, the amount of data available from measurements of conducted noise current or voltage is limited. An alternative approach is to use the specified susceptibility test levels described in Chapter 9 for the relevant type of equipment (i.e., military, ISM, or vehicle-mounted equipment). Where susceptibility test levels have not been developed for the equipment type, then the regulations concerning conducted emission limits for equipment such as digital, household appliance,

FIGURE 1.17   Composite of conducted AC power line noise currents measured at 10 hospitals. (From Cottony, H.V., Radio noise of ionospheric origin, *Science*, 111(2872), 41, January 1950. Copyright 1975 IEEE.)

and portable tools may be used to predict the conducted power line noise level. When using emission limits in a prediction, some correction must be made for the specific or worst-case power line impedance and the number of devices sharing the same power line. For example, in a laboratory environment, several computers, hand tools, RF sources, and a refrigerator may share a common power line. A worst-case AC, 120 V, 60 Hz power line source impedance plotted against frequency is shown in Figure 5.28. Taking a composite of the noise currents from all sources, a composite noise voltage developed across the worst-case source impedance may be calculated to arrive at susceptibility test/design levels.

REFERENCES

1. E.N. Skomal. The dimension of radio noise. *1969 IEEE Symposium on Electromagnetic Compatibility*, Vol. 69, C3-EMC, June 17–19, 1969.
2. H.V. Cottony. Radio noise of ionospheric origin. *Science*, January 1950, 111(2872), 41.
3. H.V. Cottony and J.R. Johler. Cosmic radio noise intensities in the VHF band. *Proceeding of the IRE*, September 1952, 40(9), 1050–1060.
4. R. Nielson. Ethernet performance in harsh industrial environments. *Control Engineering*, 1988, 35(10).
5. A.S. McLachlan. Radio frequency heating apparatus as a valuable tool of industry and a potential source of radio interference. *Seventh International Zurich Symposium and Technical Exhibition on Electromagnetic Compatibility*, Zurich, Switzerland, 1987.
6. A. Tell. Field strength measurements of microwave oven leakage at 915 MHz. *IEEE Transactions on Electromagnetic Compatibility*, May 1978, EMC-20(2), 341–346.
7. Government of Canada Department of Communications. EMCAB 1, Issue 2.
8. J. Adams. Electric field strength measured near personal transceivers. *IEEE Symposium on Electromagnetic Compatibility Record*, National Institute of Standards and Technology, 1993.
9. P. Bernardi, M. Cavagnaro, and S. Pisa. Evaluation of the SAR distribution in the human head for cellular phones used in a partially closed environment. *IEEE Transactions on Electromagnetic Compatibility*, August 1996, 38(3), 357–366.
10. D.D. Holihan. A technical analysis of the United States environmental protection agency's proposed alternatives for controlling public exposure to radiofrequency radiation. *IEEE Electromagnetic Compatibility Symposium Record*, 1987.
11. R.J. Hoff. EMC measurements in hospitals. *1975 IEEE International Symposium on Electromagnetic Compatibility*, 1975.
12. T.G. Cooper, S.M. Mann, R.P. Blackwell, and S.G. Allen. *Occupational Exposure to Electromagnetic Fields at Radio Transmitter Sites*. Health Protection Agency, Didcot, U.K., 2007.
13. Understanding Electric and Magnetic Fields. BC Transmission Corporation at https://www.bchydro.com/BChydro/customer-portal/documents/corporate/community/understanding-electric-magnetic-fields2013.pdf. Accessed August 9, 2016.
14. Electric and Magnetic Fields, National Grid EMF overhead power lines. https://www.nationalgridus.com/non_html/shared_env_emfs.pdf. Accessed August 9, 2016.
15. Siakavellas N.J. Two simple methods for analytical calculations of eddy currents in thin conductive plates. *IEEE Transactions on Magnetics*, 33(3), 1997.

2 Introduction to *E* and *H*, Near and Far Fields, Radiators, Receptors, and Antennas

Judging by the response from attendees of seminars, this chapter may be the least popular with design and system engineers.

The concept of radiation from, and coupling to, interface cables, printed circuit board (PCB) tracks, wiring, etc., is generally foreign to engineers, despite involvement with equipment containing digital, analog, radio frequency (RF), and control circuits. The reason may be that it is difficult to envisage interconnections as antennas or circuit elements or to see the potential for crosstalk between conductors. This is particularly true when equipment exhibits EMI or fails an EMC requirement, and the engineer is under the pressure of schedule to find the quick fix usually demanded by the management. In order to make simple EMC predictions or solve an EMI problem efficiently, an understanding of the principles of radiation and coupling, including frequency dependency and resonance effects, is essential.

Even the choice of an effective diagnostic test or the evaluation of a problem for either a radiated or a conducted source is difficult without this understanding. The title of this chapter may appear to imply that radiators and receptors are somehow different from antennas. Of course, this is not true. One antenna engineer, when asked to define an *antenna*, replied, "What is not an antenna?" It used to be said by TV repair technicians, not totally in jest, of a high field strength location that a piece of wet string would serve as an antenna.

One investigator has examined the resonant frequency and receiving properties of leaves and fir cones, modeled as log periodic antennas, and questioned the effects of RF currents flowing, as a result of electromagnetic fields, on their surfaces. The differentiation made in this chapter between radiators, receptors, and antennas is merely to distinguish between structures intentionally designed to radiate and receive and those not. It will be seen that much of the antenna theory is applicable to nonintentional antennas, for example, in investigating the magnitude of current flow caused by an incident field on a structure and the radiation from a current-carrying conductor. Many good books exist on the subject of wave theory and antennas. Then why is there a need for this chapter? The magnitude of an electromagnetic field at some distance from an AC source varies with time. In the measurement and prediction of field magnitude for EMI analysis or in meeting a limit, it is the peak magnitude that is required. Therefore, it has been possible to simplify the field equations presented in this book by eliminating the time-dependent terms. In EMC, we are interested both in highly efficient resonant antennas and in the nonintentional radiation and coupling, to nonresonant structures, with impedances not matched to the termination impedance. Most antenna books are confined to the analysis of antennas that are highly efficient and relatively narrowband. However, many of the equations developed for antennas are quite applicable to any conductive structures.

The formulas contained in this chapter are used in the predictions contained in subsequent chapters and are presented as *magnitudes* (i.e., they may be used directly). Derivations for the equations are omitted and may be found in the reference material. In order to make life as simple as possible, the equations for coupling to and radiation from wires and cables as well as crosstalk between cables have been incorporated into computer programs written in BASIC and provided at the end of this and other chapters. These programs will also be available in C+ or C++ to download from www.emcconsultinginc.com. However, an understanding of the correct model to use and the limits on its applicability are still required.

In addition, this chapter will enable the reader to make, calibrate, and correctly use simple magnetic field probes and E-field antennas, and it introduces the meaning of terms, such as *gain* and *antenna factor*, used with EMI measuring antennas.

2.1 STATIC AND QUASI-STATIC FIELDS

In order to understand electromagnetic waves in proximity to current-carrying conductors as well as in free space, it is helpful to examine static electric and magnetic fields. In addition, at low frequency and at close distances to the source, it is often the reactive near-field or quasi-static field that couples strongly, and these fields can be computed using the following DC analyses.

2.1.1 DC Electric Field

An electric field exists between the plates of a capacitor to which a potential V_0 is applied (Figure 2.1). The electric field intensity is

$$E = \frac{V_0}{h} \ \left[\text{V/m}\right] \tag{2.1}$$

where h is the distance between the plates, measured in meters. Thus, E is a magnitude per unit length. For example, if 10 V were applied to plates 10 cm apart, the field strength between the plates would be 100 V/m. The electric field has electric lines of force associated with it that are tangent to the electric field and are proportional in strength to the electric field strength. The electric field intensity can be measured by inserting a small dipole into the field (Figure 2.1). The component of the electric force tangential to the two conductive arms of the probe moves electrons in the conductors, and a voltage appears across the arms.

The voltage induced into a thin probe is substantially independent of the radius and proportional to the length of the probe. In practice, the conductive wires connected to the probe would disturb the electric field and Equation 2.1 would not be strictly correct. The field between two infinite parallel plates is uniform, and in the field between the plates, we find

$$V_1 = E h_d \tag{2.2}$$

where
 E is the electric field strength [V/m]
 h_d is the effective dipole length

In a nonuniform field, the electric probe measures the average intensity of the field occupied by the probe; hence, h_d should be as small as possible. The electric field probe described is the simplest example of a receiving antenna. Electric field lines may start on a positive charge and end on a negative charge, as shown in Figure 2.1, or they may start on a positive charge and end at infinity or start at infinity and end on a negative charge or with time-varying fields forming closed loops that neither start nor end on a charge.

FIGURE 2.1   A cross section between parallel plates.

2.1.2   DC Magnetic Field

An electric current is surrounded by a magnetic field of force. The magnetic field around a very long wire carrying a constant current is given by

$$H = \frac{I}{2\pi r} \quad \left[\text{A/m}\right] \tag{2.3}$$

and illustrated in Figure 2.2.

Magnetic field lines always form closed loops around current-carrying conductors because magnetic charges do not exist. The magnetic field is, like an electric field, a magnitude per unit length. For example, the magnetic field 10 cm from a wire carrying 10 A is 16 A/m and reduces to 1.6 A/m at a distance of 1 m. Expressed in another way, if the circumference of the field line at 10 cm distance is 0.628 m, the magnetic field strength is then 10 A/0.628 m = 16 A/m. Equation 2.3 may be found by dividing up the conductor into infinitesimal lengths (current elements) through which the current flows. The total field is then the sum of the contributions from all of the current elements stretching out to infinity on both sides of the measuring point. When the length of the wire is not infinite but much longer than the measuring distance *r*, Equation 2.3 may still be used with a normally acceptable small error in the computed field strength. If the wire changes direction, then the equation for computing the field from a current loop may be applicable.

In practice, it is uncommon to find short lengths of isolated current-carrying conductors. One exception may be where a conductor enters a space through a shield and then exits the space after a short distance via a second shield. The approximate magnetic field from this specific example is then

$$H = \frac{I}{4\pi r^2} \quad \left[\text{A/m}\right] \tag{2.4}$$

It is very common to find a second return conductor in close proximity to the supply conductor, with the return conductor carrying exactly equal and opposite current to the supply.

This configuration is shown in Figure 2.3a. Because the field magnitude for the two conductors is not the same in the plane of the loop, the *x*-axis, as it is tangential to the plane of the loop, the *z*-axis, two field magnitudes H_z and H_x must be computed.

The derivation of the following equations for H_z and H_x may be found in Reference [1] as well as in other textbooks on electromagnetic theory:

$$H_z = \frac{I}{2\pi} \left(\frac{x + S/2}{r_1^2} - \frac{x - S/2}{r_2^2} \right) \tag{2.5}$$

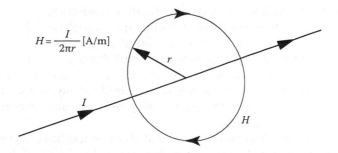

FIGURE 2.2   Magnetic field around a long conductor.

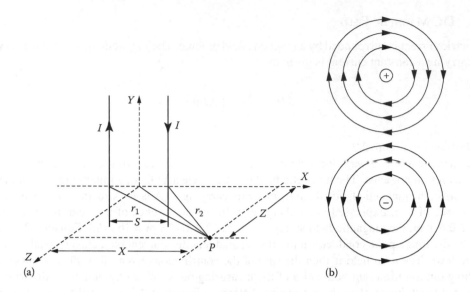

FIGURE 2.3   (a) Configuration of a two-wire line. (b) Field distribution around a two-wire line.

$$H_x = \frac{I}{2\pi}\left(\frac{z}{r_1^2} - \frac{z}{r_2^2}\right) \tag{2.6}$$

All distances, such as x, z, r_1, and r_2, are in meters.

From Equations 2.5 and 2.6, we see that the field on the x-axis at the center of the two conductors is zero, and on the z-axis, it is twice the magnitude computed for the z-component on either side of the center. This may be seen graphically in Figure 2.3b, where the fields from the two conductors are in antiphase on the x-axis and in phase on the z-axis at the center of the conductors.

When probing around the two conductors with a magnetic field measuring probe, the change in measured field may be seen with the change in probe location and orientation in the x- and z-axes around the conductors.

Twin conductor cables carrying 50–20 kHz AC power are potential sources of high magnetic fields at close proximity to the conductor. The closer together the conductors are, the lower is the resultant H field. Where the conductors are separated and, for example, routed around the inside of an enclosure, the magnetic field is at a maximum, and the correct prediction method for magnetic field from a current loop is Equation 2.7.

2.1.3 TWISTED-PAIR WIRES

In order to minimize the generation of magnetic fields from conductors, twisted-pair wire is used. The magnetic fields from each loop of the twisted wires are in antiphase and therefore tend to cancel. At the exact center of the loop and equidistant from all loops, assuming all loops are of equal size, the total magnetic field is zero. The realistic situation is that not all loops are equidistant from the measuring point and that the sizes of the loops are not exactly equal. Figure 2.4 illustrates the resultant fields and their directions, which may be computed for any location by computing the individual fields from each loop. Then, the total field may be graphically estimated from the vector product or vector analysis may be used.

The magnetic field from a twisted pair contains a radial component, x; a component down the axis of the pair, z; and a circumferential component, φ, due to the helical nature of the twist. The field at some distance from the wire, p, is dependent not only on p but on the pitch distance, h (distance for

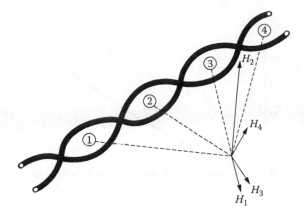

FIGURE 2.4 Resultant fields and their directions for an individual loop of a twisted pair.

one cycle of twist), as well as the distance between the center of the twisted pair and the center of one conductor of the pair.

At distances very close to the twisted pair, the field may be higher than for the two-wire line. However, the field very quickly reduces in magnitude with increasing distance p. From measurements in Reference [2], it was found that the magnetic field strength from a twisted pair reduces more rapidly, as a function of $1/r^3$, than the field from an untwisted pair or a coaxial cable in which the center conductor is not concentric or on which an unbalanced current flows, which reduce as a function of $1/r^2$.

We use the term $1/r^n$ throughout this chapter. In order to illustrate the meaning, imagine that the magnetic field has been measured at a distance of 1 cm from a twisted wire pair and that the field at 10 cm is required. The change in distance between the measuring point and the predicted point is 10 cm/1 cm = 10; the reduction in field strength is therefore $1/10^3 = 1/1000$.

From measurements and predictions on twisted-pair cable with a large pitch distance of 3 in. (i.e., only 0.33 twists/in.) from Reference [3], the reduction in field for the twisted-pair cable is approximately $1/r^{10}$, or 60 dB, for a doubling in distance p. From the same reference, the reduction in field for an untwisted cable is 12 dB (i.e., $1/r^2$) for a doubling of distance. A comparison between the reduction in field versus distance for both a twisted pair and an untwisted pair is shown in Figure 2.5. In Reference [4], generalized curves using the ratios of a/h and p/h have been constructed for B_r, B_ϕ, and B_z. These generalized curves are reproduced in Figure 2.6. The y-axis of Figure 2.6 is magnetic field strength in decibels/gauss for 1 A of current flow. To calculate the field components from the figure, perform the following:

1. Find the number of decibels for the ratio a/h.
2. Add the decibels for the ratio of p/h.
3. For the B and B_z fields, add $20\log(1/2.54xh)$ [cm] or l/h [in.] to the number of decibels obtained in step 2. For the field, add $20\log(1/2.54p)$ [cm] or l/p [in.] to the number of decibels obtained in step 2.
4. To correct for the actual current flow I, add $20\log I/l$ to the number obtained in step 3.

The field magnitude at some point or points in close proximity to a twisted pair is of interest when the voltage induced into either a loop or a wire, which crosses the pair at some angle, is required. Where the concern is not the electromagnetic field incident on a point but on a cable or wire running the length of the twisted wire pair, then a current is induced into the receptor wire. Again, these currents tend to cancel, but not exactly due to varying distance and orientation between the twisted wire source and the receptor wire, unequal loop areas, and an odd number of loops.

FIGURE 2.5 Comparison of the reduction in field versus distance curves for both a twisted pair and an untwisted pair. (From Moser, J., Predicting the magnetic fields from a twisted pair cable, *IEEE Trans. Electromagn. Compat.*, 10(3), September 1968. Copyright 1968 IEEE.)

FIGURE 2.6 Field reductions as a function of a/h and p/h.

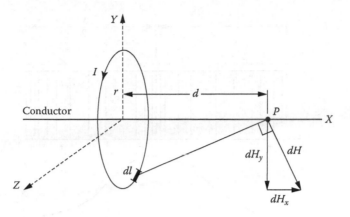

FIGURE 2.7 Configuration for the magnetic field from a loop.

End effects where large loops may exist due to the termination of the twisted pair must also be considered. For the evaluation of crosstalk, the worst-case magnetic field generated down the length of the twisted pair may be estimated for a given distance from the twisted pair and used with the cable crosstalk equations and computer program found in Chapter 4. This program allows the prediction of coupling to either a single wire above a ground plane, a two-wire line, or a shielded cable.

The susceptibility and radiation from twisted pairs from 20 kHz to GHz (i.e., above the frequency where the quasi-static equations are valid for long distances from the source) are considered in Section 2.2.6.

2.1.4 DC AND QUASI-STATIC FIELDS FROM A LOOP

The radial magnetic field from a loop (i.e., in the *x*-axis) at a point coaxial to the loop at DC and low frequency in close proximity is given by

$$H_x = \frac{Ir^2}{2(r^2 + d^2)^{1.5}} \quad \left[\text{A/m}\right] \tag{2.7}$$

where
d is the distance from the loop [m]
r is the radius of the loop [m]
I is the current carried by the loop [A]

The loop and measuring point are shown in Figure 2.7. The components of the magnetic field in the *y*-axis at point *p* cancel, and thus, H_y is zero.

2.2 ELECTRIC WAVES ON WIRES AND IN FREE SPACE

Connecting a sine wave generator to parallel conductors of length greater than the wavelength, λ, results in the generation of a wave along the length of the conductors. Parallel conductors can be considered chains of small inductors in series and small capacitors in parallel. We thus expect a delay in the transmission of voltage down the length of the conductors. The wave profile is shown in Figure 2.8.

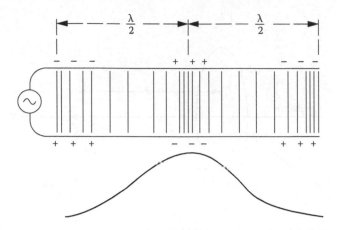

FIGURE 2.8 Distribution of charge and waves on parallel plates.

The distance between the crests of the sine wave is called the *wavelength*, λ. If the wave moves down a conductor so that the time between successive crests is t, then the velocity of the wave is

$$v = \frac{\lambda}{t}$$

The frequency is given by

$$f = \frac{1}{t}$$

Hence,

$$v = \lambda f$$

The velocity of a wave in free space or air is a constant 3.0×10^8 m/s, and the relationship between frequency and wavelength is

$$f\,[\text{MHz}] = \frac{300}{\lambda\,[\text{m}]}$$

$$\lambda\,[\text{m}] = \frac{300}{\lambda\,[\text{MHz}]}$$

The velocity for a medium is given by

$$\frac{1}{\sqrt{\mu\mu_r \varepsilon\varepsilon_r}}$$

where
   μ is the permeability of free space = $4\pi \times 10^{-7}$ [H/m] = 1.25 [μH/m]
   μ_r is the relative permeability
   ε is the dielectric constant of free space = $1/36\pi \times 10^9$ = 8.84 [pF/m]
   ε_r is the relative permittivity

FIGURE 2.9 Lines of force around divergent conductors.

The wave velocity of a transmission line is given by

$$\frac{1}{\sqrt{LC}}$$

where
　L is the inductance of the line per unit length
　C is the capacitance per unit length

Figure 2.8 shows the distribution of charge in the form of electric lines of force. Lines of force are considered to be tangential to the electric field *E*. Clusters of dense lines of force exist where the charge density is high, and in the next cluster, the lines of force are oppositely directed. Loops of current are also present, formed partly on the conductors and partly in the space between the plates.

Thus, a magnetic field is present around the parallel plates. The current flow in the space between the conductors is named the *displacement current*, and, as we shall see in Section 7.13.3.1, it is important when considering the common-mode current on a line. The open circuit transmission line of Figure 2.8 results in standing waves along the length of the line, due to reflection of the wave arriving at the end of the line. If the conductors are not parallel but rather divergent, then the lines of force are as shown in Figure 2.9.

2.2.1 RADIATION

The dipole antenna depicted in Figure 2.10 is said to be short, because charge reaches the ends of the antenna in much less than a period; expressed differently, the length of the antenna is much less than a wavelength. The current flow in conductors of the two-wire transmission line driving the antenna is of the same magnitude but 180° out of phase, and standing waves are created on the transmission line. When the length of the antenna is less than the wavelength, the standing wave current along the length of each arm of the antenna is in phase, and the fields radiated from both arms will reinforce. As the discharge begins, the lines of force diminish and after a half period are at zero. During this period, the lines of force reaching out to *P* are canceled; however, a fraction of the lines of force spread out to *Q*, and this cluster of lines becomes detached at the end of the first

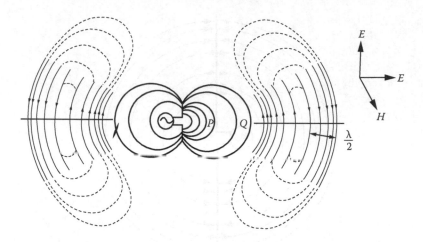

FIGURE 2.10 Detached lines of force around a short dipole antenna.

half period. The cluster of lines of force moves on, and new lines of force take their place. The width of the configuration of lines remains $\lambda/2$, with the area over which the lines of force spread increasingly with increasing distance r from the antenna. The area can be proven to equal $\pi\lambda r$.

In accordance with our conception of lines of force, the density of the lines of force is proportional to the electric intensity. The number of lines of force issuing from the antenna is proportional to the charge and, therefore, to the current in the generator. The fraction of the lines of force detached from the antenna is proportional to the length $2l$ of the antenna. Magnetic and electric fields exist in the equatorial plane of the antenna (Figure 2.10). In the equatorial plane, the magnetic field, in amps per meter, is proportional (where I is amperes) to

$$\frac{2Il}{\pi\lambda r}$$

The radial electric field is the field that has the same direction as the static or quasi-static field, and it is given, in volts per meter, by

$$\frac{2Il}{4\pi r^2}$$

and therefore diminishes as a function of $1/r^2$ with distance r. The electric field in the meridian (i.e., in the plane of the antenna arms) is given by

$$\frac{2IlZ_w}{\pi\lambda r}\cos\theta$$

where θ is the angle between the direction of the antenna arm and the measuring point.

The ratio of E/H has the dimension of an impedance and is often referred to as the wave impedance, Z_w, used in the preceding expression. The electric intensity E and the magnetic intensity H in free space take the place of the voltage and current at the terminals of a circuit. Therefore, $E = Z_wH$ and $H = E/Z_w$.

The physical dimensions of E and H are those of voltage and current per unit length. The ratio of E/H is dependent on the proximity of the wave to the source of emission. Very close to the source, where the field contains radial components, it is called the *induction field*. Further away, where some field components decrease as $1/r^3$, it is called the *Fresnel region*. Still further from the source, where the fields fall off as $1/r$, it is called the *far-field* or *Fraunhofer region*. The fields in the Fresnel and Fraunhofer regions both radiate. At low frequency and in close proximity to the dipole antenna, the reactive or quasi-static field exists and does not radiate.

The distances at which far-field conditions occur are dependent, among other factors, on the size of the antenna. Where the size D of an antenna is less than $\lambda/2$, then the near-field/far-field interface distance is defined as

$$r = \frac{\lambda}{2\pi} \qquad (2.8)$$

where $D > \lambda/2\pi$ the interface distance is $r = D/2\pi$.

This definition is useful in EMC because it describes the distance at which the magnetic and electric fields from an electric current element or current loop begin to reduce as a function of $1/r$. When D is much less than $\lambda/2\pi$, the antenna approximates a point source, and the phase is a function of the distance r from the source. When the dimension D approaches $\lambda/2\pi$, a correction for phase error may be made, as described in Section 2.2.2. The phase error expressed, in degrees, can be defined as the difference in the following path lengths: between the closest point of the antenna to the measuring point and the furthest point of the antenna to the measuring point.

For a high-gain antenna, the interface distance is often considered to be

$$r > \frac{4D^2}{\lambda}$$

at which distance the phase error is less than $22.5°$. For the equation to be useful, the aperture dimension of the antenna, D, should be larger compared to the wavelength. In the far-field region, Z_w is equal to the intrinsic impedance of free space, which is commonly expressed as either $R_c Z_c$, n, or p.

The intrinsic impedance of free space is equal to

$$\sqrt{\frac{\mu_0}{\varepsilon_0}}$$

which is $376.7 \ \Omega$, or very nearly 120π or $377 \ \Omega$. As we have seen, μ_0 is the permeability of free space ($44\pi \times 10^{-7}$ [H/m]) and ε_0 is the dielectric constant of free space ($1/36\pi \times 10^9$ [F/m]).

At large distances from a short antenna, phase differences can be ignored and the resultant field can be termed a *plane wave* with an electric field intensity proportional to

$$\frac{2(377)\mathrm{II}}{\pi\lambda r}$$

For a plane wave, the electric and magnetic field intensities in free space are always in phase and perpendicular to each other and to the direction of propagation. Often, the electric wave from an antenna is more complex with regard to phase and doubly so in the near or induction field of an antenna.

In problems of EMC, unintentional sources of electromagnetic emissions are primarily from current loops or current elements, whether from PCB tracks, intraunit, interunit, or system wiring and cables. Thus, we shall concentrate on fields from current loops and current elements and in the current induced into loops and short lengths of conductor by electromagnetic fields.

2.2.2   CURRENT ELEMENTS AS RADIATORS

A *current element* is defined, for the purpose of EMC, as an electrically short length (i.e., $l < \lambda$) of current-carrying conductor connected to the current source at one end and disconnected, or at least connected via a high impedance, from ground at the other end. Strictly, the current source should be disconnected from ground at both ends. The practical use of an isolated wire to which no generator or return current path is connected as a model in EMC predictions may appear extremely limited. However, a practical example is in the modeling of a shielded cable on the center conductor of which a current flows and returns on the inside of the shield. In our example, the shielded cable is connected to a well-shielded enclosure at each end. If the shield of the cable is not perfect, then some small fraction of the current flowing on the inside of the shield will diffuse through the shield to the outside. If the length of the cable is less than the wavelength, λ, then the current flow on the outside of the shield will reduce to zero at the termination with the enclosures at both ends and remain relatively constant over the length of the cable.

For certain cable and measuring distance configurations, discussed later, the electric current element model may be used with either one or both ends connected to a return path. In addition, the electric current element model is strictly valid when the distance from a plane conductor or ground plane is much greater than the wavelength. When the proximity is closer, the current element equations may be used and corrected for the effect of reflections in the plane conductor, as discussed in Section 9.3.2. When the proximity of the ground plane is much less than the wavelength, λ, and the length of the current element is equal to or greater than λ, then the equation for radiation from a transmission line contained in Section 7.13 is applicable.

The current on a current element is assumed to be constant over its length. When applying the formulas for the current element to an electrically short length of wire or cable (i.e., $l < 0.1\lambda$) disconnected from ground at both ends, then the current distribution is as shown in Figure 2.11a, and

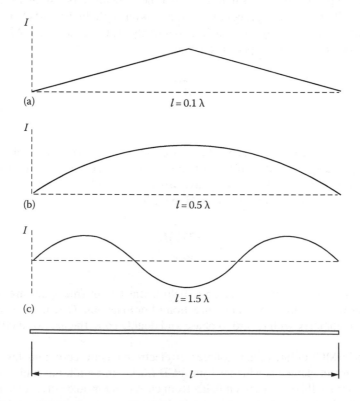

FIGURE 2.11   Current distribution on a wire of length l.

the electromagnetic radiation from the short wire is approximately half of the value of the current element. The current distribution for a length of wire that is resonant (i.e., $l = 0.5\ \lambda$) is sinusoidal, as shown in Figure 2.11b, and the radiation is approximately 0.64 that of the current element. A more accurate approach that is particularly useful when the length of the wire is greater than the wavelength, as shown in Figure 2.11c, is to break up the wire into a number of current elements and compute the composite field from all sources. If the distance from the wire, at which the magnitude of the field is required, is much closer than the wavelength, then the current element equations may be used with no correction, because the contribution to the total field from the current elements at some distance from the point of calculation is negligibly small. Even where the wire is terminated at one or both ends, the current element equation may still be used at close proximity, as long as the return current path is at a large electrical distance from the wire. If, as often happens, the return path is in close proximity, then the formulas for fields from either a transmission line or a current loop, whichever is applicable, should be used. In calculating the field strength, the magnitude of the current flow on the wire must be known. The current may be calculated; however, a simpler approach is to measure by means of a current probe.

Thus, we see from the foregoing discussion that the current element, as a model, has more practical applications than may at first be realized.

2.2.3 CURRENT LOOPS

The current loop is a valuable model for use in predicting radiation and coupling. The cable connecting two units in a rack or two units mounted one upon the other often forms a loop. Likewise, where a cable is routed along the ground or in a cable tray and then is connected to equipment, a loop is often formed. The wiring inside an enclosure (or the tracks on a PCB) often takes the shape of a loop. Another use of the current loop model is in the design of a simple antenna, which may be constructed and used as either a source of, or to measure, magnetic fields.

2.2.4 SPHERICAL WAVES

To understand spherical waves (i.e., those present close to an antenna or an electric current element or loop), we shall use the spherical coordinate system. Figure 2.12 shows the field vector around an

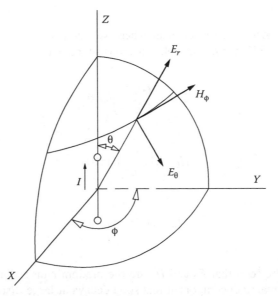

FIGURE 2.12   Field vector around electric current element.

electric current element, sometimes referred to as an infinitesimal dipole. The simplified field equations for a nondissipative medium such as air are

$$E_\theta = \left(jRc\,\frac{I_s}{2\lambda r} \right)\left(1 - \frac{\lambda^2}{4\pi^2 r^2} - j\frac{\lambda}{2\pi r} \right)\sin\theta \qquad (2.9)$$

$$H_\phi = \left(j\frac{I_s}{2\lambda r} \right)\left(1 - j\frac{\lambda}{2\pi r} \right)\sin\theta \qquad (2.10)$$

$$E_r = \left(\frac{Z_w I_s}{2\pi r^2} \right)\left(1 - j\frac{\lambda}{2\pi r} \right)\cos\theta \qquad (2.11)$$

Expressed as magnitudes, these become

$$|E_\theta| = 377\,\frac{I_s}{2\lambda r}\sqrt{\left[1 - \left(\frac{\lambda^2}{4\pi^2 r^2} \right)^2 \right]^2 \left(\frac{\lambda}{2\lambda r} \right)^2}\,\sin\theta \qquad (2.12)$$

$$|H_\phi| = \frac{I_s}{2\lambda r}\sqrt{\left[1 - \left(\frac{\lambda}{2\pi r} \right)^2 \right]^2}\,\sin\theta \qquad (2.13)$$

$$|E_r| = 377\,\frac{I_s}{\left(2\pi r\right)^2}\sqrt{\left[1 - \left(\frac{\lambda}{2\pi r} \right)^2 \right]^2}\,\cos\theta \qquad (2.14)$$

In the far field of the electric current element, where we assume $r/\lambda \gg \pi/2$, and using the intrinsic impedance of free space (377) for Z_w, the field equations simplify to

$$E_\theta = j\left(\frac{60\pi I_s}{\lambda r} \right)\sin\theta \qquad (2.15)$$

$$H_\theta = j\left(\frac{I_s}{2\lambda r} \right)\sin\theta \qquad (2.16)$$

$$E_r = 60\,\frac{I_s}{r^2}\cos\theta \qquad (2.17)$$

In the far field, it can be seen that E_θ and H_ϕ are the predominant radiation components, with E_s as the induction (or reactive) component that soon decays at large distances from the current element.

The open circuit voltage induced in the loop is

$$V_{ab} = -j2\pi f\mu_0 H_\phi S \tag{2.18}$$

where

μ_0 is the permeability or inductivity of air or vacuum, $4\pi \times 10^{-7}$ [H/m]
S is the area of the loop
$\mu_0 H_\phi$ is B, which is the flux density

The value of H_ϕ can be found from the current I flowing in the current element at position P_0 in Figure 2.13 and from Equations 2.10 and 2.18 as follows:

$$V_{ab} = -j\omega\mu\left(\frac{j\beta I_s}{4\pi r}\right)\left(1 + \frac{1}{j\beta r}\right)S \tag{2.19}$$

where $\beta = 2\pi/\lambda$. As a magnitude, this is

$$|V_{ab}| = 2\pi f\mu\left(\frac{\beta I_s}{4\pi r}\right)\sqrt{1 + \left(\frac{1}{\beta r}\right)^2}\,S \tag{2.20}$$

2.2.5 RECEIVING PROPERTIES OF A LOOP

From Equation 2.18, the receiving properties of the loop can be expressed as

$$V = 2\pi f\mu_0 H_\theta S \cos\theta = \frac{2\pi Z_w H_\theta S \cos\theta}{\lambda} = \frac{2\pi E_\phi S \cos\theta}{\lambda} \tag{2.21}$$

where θ is the angle between the normal or perpendicular to the plane of the loop and the magnetic intensity H_θ as shown in Figure 2.13.

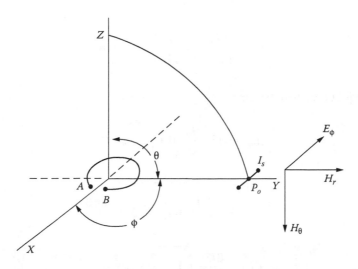

FIGURE 2.13 Field vector around a current loop depicts a small loop with a current element at position P_0.

Equation 2.21 gives the open circuit–receiving properties of a loop. The impedance of the loop and the impedance across which the voltage is developed must be accounted for when predicting either the voltage developed across the load or the current flowing in the loop. In Chapter 7, on cable coupling, the EMI voltage induced in a loop formed by a wire or cable by a field is examined further.

The inductance of a loop of wire or a wire above a ground plane, which forms a loop with its electromagnetic image in the ground plane, is given by

$$L = \frac{\mu_0}{\pi}\left[l \ln\frac{2hl}{a(l+d)} + h\ln\frac{2hl}{a(h+d)} + 2d - \frac{7}{4}(l+h)\right] \tag{2.22}$$

where

    a is the radius of the wire, in meters
    l is the length of the wire, in meters
    h is either the distance between the wires or twice the distance between the wire above a ground plane to account for the electromagnetic image of the wire in the ground plane and

$$d = \sqrt{(h^2 + l^2)}$$

The impedance of the loop and load as a magnitude is

$$Z = \sqrt{\left(Z_L^2 + 2\pi f L^2\right)}$$

With current applied to the loop and assuming even current distribution around the loop (i.e., $S \ll \lambda$), the field components at any point around the loop can be found from

$$E_\phi = \frac{R_c \beta^2 IS}{4\pi r}\left(1 + \frac{1}{j\beta r}\right)e^{-j\beta r}\sin\theta \tag{2.23}$$

$$H_\theta = \frac{\beta^2 IS}{4\pi r}\left(1 + \frac{1}{j\beta r} - \frac{1}{\beta^2 r^2}\right)e^{-j\beta r}\sin\theta \tag{2.24}$$

$$H_r = \frac{j\beta IS}{2\pi r^2}\left(1 + \frac{1}{j\beta r}\right)e^{-j\beta r}\cos\theta \tag{2.25}$$

Expressed as magnitudes, these become

$$|E_\phi| = 377\frac{\beta^2 IS}{4\pi r}\sqrt{1 + \frac{1}{\beta^2 r^2}}\sin\theta \tag{2.26}$$

$$|H_\theta| = \frac{\beta^2 IS}{4\pi r}\sqrt{\left(1 + \frac{1}{\beta^2 r^2}\right) + \frac{1}{\beta^2 r^2}}\sin\theta \tag{2.27}$$

$$|H_r| = \frac{\beta IS}{2\pi r^2}\sqrt{\left(1 + \frac{1}{\beta^2 r^2}\right)}\cos\theta \tag{2.28}$$

These equations are useful in computing the fields generated by a current loop in the induction field, in the near field, and in the far field. For example, Equation 2.28 may be used to obtain the approximate value of the radial induction/near field when the radius of the loop is less than the measuring distance. Equation 2.28 should always be used when the measuring distance is greater than six times the radius of the loop. At low frequencies and for distances closer than the radius of the loop, Equation 2.7 or 11.1, which provides the DC or quasi-static radial field, should be used.

In order to examine the ratio between *E* and *H* fields generated by a loop, as shown in Figure 2.13, the values of *E* and *H* fields from a 0.01 m circular loop with a generator current of 1 mA at 10 MHz are plotted against distance from the loop in Figure 2.14. In Figure 2.16, the ratio *E*/*H* (i.e., the wave impedance) has been plotted. It can be seen that the near-field/far-field interface for the small current loop is at approximately 4 m, which corresponds well with the definition of $\lambda/2\pi = 30$ m/6.28 = 4.7 m. The wave impedance for the small current loop is given by

$$Z_w = \frac{Z_c 2\pi r}{\lambda} \leq 377 \ \Omega \tag{2.29}$$

(e.g., at 0.1 m, $Z_w = 7.9$, and this is called a *magnetic* or *low-impedance field*).

Also, from Figures 2.14 and 2.15, it can be seen that in the near field, H_θ reduces as a function of $1/r^3$ as does H_r, whereas E_ϕ reduces as a function of $1/r^2$ (where *r* is the distance). Had we been examining a small dipole or current element (i.e., $1 \ll \lambda$), then E_θ would have followed a $1/r^3$ law with distance and H_ϕ, a $1/r^2$. The wave impedance would then equal

$$Z_w = \frac{Z_c 2\pi r}{\lambda} \geq 377 \ \Omega \tag{2.30}$$

and the field is then called an *electric* or *high-impedance field*. It appears from Figure 2.16 that the wave impedance increases above 377 close to the $\lambda/2\pi$ transition zone, which is not true. Instead, a transition zone exists, and the $(1/r^2)$-to-$(1/r)$ and the $(1/r^3)$-to-$(1/r)$ transitions are not abrupt. Numerous measurements on a single current-carrying cable and on multiple cables located on a nonconductive table 1 m above the floor in a shielded room have confirmed that the current element

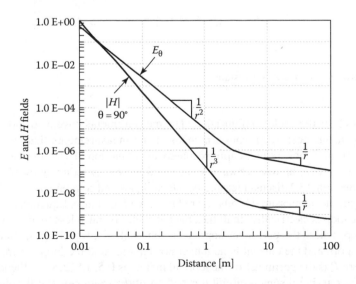

FIGURE 2.14   *E* and *H* fields versus distance.

FIGURE 2.15 H_θ and H_r versus distance.

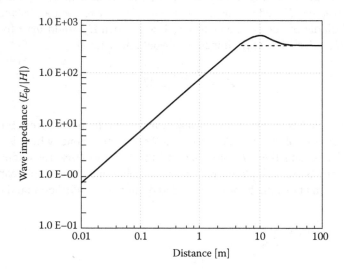

FIGURE 2.16 Wave impedance versus distance.

is the correct model for the cable configuration. The measurements have also shown that the transition zone exists. It has been seen from a number of the measurements that the circumferential magnetic field H_θ from a cable reduces approximately as a function of $1/r^2$ up to a certain distance, thereafter reducing by $1/r^{1.5}$ up to a certain distance, and thereafter reducing by $1/r$.

Chapter 9 and Section 9.3.2 discuss measurement techniques and compare predictions, based on the cable current magnitude, to measured levels and reduction with distance. Equations 2.26 through 2.28 are valid when the loop is electrically small (i.e., the perimeter is much less than the wavelength).

For a large loop, where the perimeter approaches one-half the wavelength, standing waves are generated on the loop, and the current is no longer constant around the loop. At frequencies where the ratio P/λ, where P is the perimeter of the loop, in meters is 0.5, 1.5, 2.5, ..., the loop exhibits the characteristics of a parallel resonant circuit (termed an *antiresonance*), and the input current and radiation from the loop are greatly reduced.

FIGURE 2.17   (a) Radiation pattern for a small loop, E_ϕ, $P < 0.1\lambda$. (b) Radiation pattern for a large loop, E_ϕ, $P = \lambda$. (c) Radiation pattern for a large loop, E_θ, $P = \lambda$.

At frequencies where the ratio P/λ is 1, 2, 3, …, the loop behaves as a series resonant circuit, and the current is limited only by the DC and radiation resistance.

The input impedance of the loop is a minimum at resonance and a maximum at an antiresonance. For example, when the circumference of the loop P is equal to λ, the input impedance is approximately 100 Ω, and when $P = 2\lambda$, the input impedance is approximately 180 Ω. At the first antiresonance, when $P = 0.5\lambda$, the input impedance is approximately 10 kΩ, and when $P = 1.5\lambda$, the input impedance is from 500 to 4000 Ω, depending on the ratio of the radius of the loop to the radius of the conductor.

In addition, the radiation pattern changes, with reduced radiation from the plane of the large loop and maximum radiation from the plane of the small loop; these are the field of Figure 2.13. Figure 2.17a and b illustrates the E_ϕ radiation pattern for the small loop and the large loop, where $P = 0.1\lambda$ and λ, respectively. With the large loop in addition to the E_ϕ field, which is small in the vertical plane, a second component E_ϕ, which has a maximum value in the vertical plane and is zero in the horizontal plane, is generated. Figure 2.17c shows the radiation pattern for the E_θ field. The directivity in the vertical plane (at 90° to the plane of the loop, where $\theta = 0$) with sinusoidal current distribution around it is provided in Table 2.1. The power gain of the loop, when the ratio of the radius of the loop to the radius of the conductor is 30, is 1.14 when $P = 0.6\lambda$ and 3.53 when $P = 1.2\lambda$.

The radiation resistance of the large loop, compared to 31,200 $(\pi a^2/\lambda^2)^2$ for a small circular or rectangular loop, is shown in Figure 2.18 from Reference [5].

2.2.6 FAR-FIELD RADIATION FROM A TWISTED WIRE PAIR

The EMC literature contains a model for the prediction of radiation from a twisted pair based on the model of a helix described in Reference [5]. For a helix whose dimensions are much less than the wavelength, the radiation is primarily from the sides of the helix, and the axial-mode radiation is at a minimum.

TABLE 2.1
Directivity of the E_θ Field
for a Large Loop

P/λ	Directivity
0.2	0.7
0.4	0.9
0.6	1.12
0.8	1.3
1.0	1.5
1.2	1.58
1.4	1.67
1.6	1.58
1.8	1.25
2.0	0.75

FIGURE 2.18   Radiation resistance of large loop.

The helix may be modeled as a series of small current loops and current elements, as shown in Figure 2.19a, and this model may be of use in an EMC prediction where a single wire takes the form of a helix. In this case, the far-field radiation is found from the sum of the loop and element sources. The physical layout of a twisted pair is shown in Figure 2.19b, from which it is seen that the twisted pair forms a bifilar helix in which both helices are identical but displaced in position axially down the length of the pair. The axial displacement is determined by the distance between the two wires. An important parameter in controlling the radiation from the pair is the distance between the wires and the pitch of the twist. Decreasing both the distance and the pitch will decrease the radiation. For commercially available twisted-pair wire made up as a cable, the distance between the two wires is approximately 0.3 cm, and the pitch is from 3.8 to 4.4 cm. The current elements in the equivalent circuit are therefore longer than the diameter of the loops in this practical example.

Because the two helices are axially in the same plane and the current flow is in opposite directions, the equivalent circuit is as shown in Figure 2.19c. Assuming an even number of twists with the total length of the pair less than 0.5 λ, we may see from the equivalent circuit that the fields from the current elements and current loops tend to cancel at a considerable distance from the source.

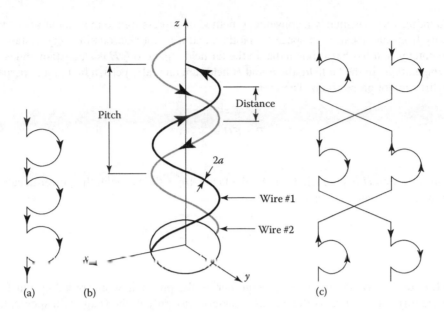

FIGURE 2.19   (a) Equivalent circuit of a helix comprising current elements and current loops. (b) Physical layout of a twisted wire pair illustrating the bifilar helix form of the cable. (c) Equivalent circuit of a twisted pair.

This is not surprising, based on the physical layout of the pair. Measurements from a twisted pair have confirmed that the far-field radiation is lower than for an untwisted pair.

Where the total length of the twisted-pair line is greater than 0.5 λ, the current flow in the two wires changes direction at some point down the length of the pair, and the cancelation of fields from the loops and elements is not as complete. In the limiting case where the pitch is equal to the wavelength, the axial radiation from two adjacent loops and current elements tends to add, and the level of radiation is higher than for an untwisted-pair cable. This occurs for the majority of twisted pairs in the gigahertz frequency range, which is typically, but not exclusively, above the frequency of emissions from unintentional sources, such as logic and converters.

The major source of radiation from the twisted pair is not from loops and elements but from the common-mode current, which inevitably flows on both conductors of the cable. One source of common-mode current is displacement current, which flows in the space between the two conductors; this source, along with others, is discussed in Section 7.13.

Likewise, when a field cuts a twisted-pair cable, the induced differential-mode currents are at a relatively low magnitude, whereas the common-mode current is identical to that of an untwisted pair. Calculations for differential-mode voltage and common-mode current flow resulting from an incident field and the effect of unbalanced input impedance on the noise voltage developed are discussed in subsequent chapters.

2.3   RADIATED POWER

In addition to *E* and *H* fields, we may consider antenna transmitted and received power. *Radiated power* is defined as the average power flow per unit area and is expressed in watts per square meter. In a perfect, lossless, isotropic antenna, the power is radiated equally in all directions and must pass through every sphere of space. The relationship between the power and the *E* and *H* fields is given by the Poynting vector cross product

$$\vec{P} = \vec{E} \times \vec{H} \qquad (2.31)$$

The Poynting vector is merely a convenient method of expressing the movement of electromagnetic energy from one location in space to another, and it is not rigorous in every instance when applied to antennas. It has been shown that in the far field, the ratio E/H is a constant and is termed the *wave impedance*. In the far field, the E and H fields are mutually perpendicular and perpendicular to the direction of propagation. Therefore,

$$P = EH = \frac{E^2}{Z_w} \tag{2.32}$$

Because the E and H fields each decay by the factor $1/r$ in the far field, the power decays by the factor $1/r^2$ and the radiated power is then

$$W = \frac{P_{in}}{4\pi r^2} \quad \left[W/m^2 \right] \tag{2.33}$$

The $1/4\pi r^2$ term may be visualized by considering the power flow as spreading out from the source in the form of a sphere. If the radiation pattern was truly in the shape of a sphere, then the source would be an isotropic radiator. No practical antenna is an isotropic radiator; the least directive radiator is a current element and a current loop, both of which have the same radiation pattern.

The antenna gain is a measure of the "directivity" of the antenna. In antenna terminology, the term *directivity* is reserved to describe the ratio of power at beam peak, assuming no losses, to the power level realized with the same input power radiated isotropically. Where ohmic losses in the antenna are low and cable/antenna impedances are matched, the directivity D and the gain G are virtually the same. The gain is defined as the ratio

$$\frac{4\pi W_{max}}{P_{in}}$$

where P_{in} is the antenna input power. The gain of the perfect isotropic antenna is thus 1. Neither the equation for radiated power nor that for antenna gain is valid in the near field of the antenna. The received power of a receiving antenna is likewise not constant in the near field nor equal to the gain in the far field, because near-field contributions radiating from the antenna, which would be in phase in the far field, can cancel in the near field. The radiated power at a distance r from an antenna with gain G is given by

$$P_d = G - LP_{in}/4\pi r^2 \quad \left[W/m^2 \right] \tag{2.34}$$

$$\text{in } dBW/m^2 \quad 10\log(G-L) + 10\log(P_{in}) - 10\log(4\pi r^2)$$

where
P_d is the power density
G is the antenna gain
L is the cable and mismatch loss
P_{in} is the antenna input power
$G = P_d 4\pi r^2 / P_{in}$
$P_d = EH = E^2/377$

The EIRP is the equivalent radiated power that a theoretical isotropic antenna (which radiates power equally in all directions) would radiate to achieve a measured level, which is the peak power in the direction of maximum antenna gain:

$$EIRP = P_t - L + G$$

where
  P_t is the transmitted power
  L is the cable and antenna cable mismatch losses
  G is the antenna gain

Thus,

$$P_d = EIRP/4\pi r^2$$

$$W = \frac{GP_{in}}{4\pi r^2} \tag{2.35}$$

The input voltage to the antenna for a given input power is

$$V_{in} = \sqrt{P_{in}R_{in}}$$

where R_{in} is the antenna input or radiation resistance.

The *E* field generated by the antenna, using V_{in} and assuming the antenna input impedance is 50 Ω and equals the radiation resistance, is given by

$$E = \frac{V_{in}}{r\sqrt{628/Z_w G}} \tag{2.36}$$

where r is the distance from the antenna, in meters.

The *radiation resistance* is defined as that part of the antenna impedance that contributes to the power radiated by an antenna. In the case of a resonant antenna, such as a dipole, the antenna input impedance equals the radiation resistance. For the tuned dipole, the input impedance equals 70–73 Ω, and for a tuned monopole, it equals 30–36 Ω, depending on the radius of the conductors used and assuming the conductor radius is small compared to the antenna length. We use the term *tuned* to denote that the antenna length has been adjusted so that the antenna is resonant at the frequency of interest.

For an electrically small dipole or monopole antenna, where l, which is the length of one arm, is much less than $\lambda/4$, the input impedance is predominantly capacitive and is higher than the resonant value. The radiation resistance is then only a small fraction of the input impedance.

The *effective area* of an antenna (A_e) is defined as the ratio of the power received at the antenna load resistance (P_r) to the power per unit area of incident wave (*W*). If the antenna impedance is matched to the load impedance, then $P_r = WA_e$.

The effective capture area, or *aperture*, of a receiving antenna (A_e) is dependent on the actual area and efficiency of the antenna. The physical area of a rectangular-aperture antenna, such as the horn antenna, is the length times the width. For a circular antenna, such as a dish, the area is πr^2. The effective aperture is the physical aperture times the aperture efficiency, η_a.

For example, η_a is approximately 0.5 for a dish antenna. The effective area can be expressed in terms of gain; thus,

$$A_e = \frac{G\lambda^2}{4\pi} \tag{2.37}$$

So

$$G = \frac{4\pi A_e}{\lambda^2} \quad \text{and} \quad P_r = \frac{WG\lambda^2}{4\pi} \quad P_d = (P_r 4\pi)/(G\lambda^2)$$

The $\lambda^2/4\pi$ term does not imply that with increasing frequency, waves decrease in magnitude. It means that at higher frequency, the area over which a given power flow occurs is smaller.

The equation for gain must also include a correction for losses. In the specified aperture efficiency for an antenna, losses due to reflections caused by a mismatch between the transmission line driving the input terminals of the antenna and the antenna input impedance should be included. Where aperture efficiency or realized gain is not specified and Equation 2.37 is used to determine the gain of an antenna, any loss due to transmission-line-to-antenna impedance mismatch must be calculated to arrive at the realized gain (G_{re}). G_{re} equals $G \times L$, where L is the loss. The loss may be found from the transmission line/antenna reflection coefficient, K, using $L = 1 - K^2$:

$$K = \frac{1 - Z_{ant}/Z_L}{1 + Z_{ant}/Z_L} \tag{2.38}$$

where
    Z_{ant} is the antenna input impedance
    Z_L is the impedance of the transmission line connected to the antenna

For two antennas in free space in the far-field region, both polarization matched and ignoring losses in cable baluns and due to impedance mismatch, the received power is given by combining Equation 2.34 for transmitted power, using the antenna input power, with the equation for received power (Equation 2.37), which gives

$$P_r = \left(\frac{G_t P_{in}}{4\pi r^2}\right)\left(\frac{G_r \lambda^2}{4\pi}\right) \tag{2.39}$$

which may be simplified to

$$P_r = P_{in} G_t G_r \left(\frac{\lambda^2}{4\pi}\right)^2 \tag{2.40}$$

where
    P_r is the received power at the input of the receiving device
    P_{in} is the input power to transmitting antenna
    G_t is the gain of transmitting antenna
    G_r is the gain of receiving antenna

If we express the received power in dBW and the antenna gains in decibels, then Equation 2.40 is

$$P_r = P_t + G_t + G_r - 20 \log \left(\frac{4\pi r}{\lambda} \right) \tag{2.41}$$

The received power is readily converted to voltage at the input of the receiver when the receiver input impedance is known (assuming the cable impedance matches the terminating impedance) by $V = \sqrt{P_r Z_{in}}$. The power equation is useful in calculating free space propagation loss in the basic antenna-to-antenna EMC prediction. For this prediction, additional factors due to the unintentional nature of the coupling have to be accounted for. Some of these additional factors, which are discussed in Chapter 10, are

1. Non-line-of-sight coupling
2. Antenna polarization and alignment losses
3. Intervening atmospheric effects
4. Frequency misalignment losses

2.4 UNITS OF MEASUREMENT

The unit dBW was used to describe power; the following are some definitions of commonly used units.

dB: The decibel (dB) is a dimensionless number that expresses the ratio of two power levels. It is defined as

$$dB = 10 \log \frac{P_2}{P_1}$$

The two power levels are relative to each other. If power level P_2 is higher than P_1, then dB is positive; otherwise, dB is negative. Since

$$P = \frac{V_2}{R}$$

when voltages are measured across the same or equal resistors, the number of decibels is given by

$$dB = 20 \log \frac{V_2}{V_1}$$

A rigid voltage definition of dB has no meaning unless the two voltages under consideration appear across equal impedances. Thus, above some frequency where the impedance of waveguides varies with frequency, the decibel calibration is limited to power levels only.

dBW: The decibel above 1 W (dBW) is a measure for expressing power level with respect to a reference power level P_1 of 1 W. Similarly, if the power level P_2 is lower than 1 W, the dBW is negative.

dBm: This is the decibel above 1 mW in 50 Ω. dBm = 1 mW = 225 mV (i.e., 225 mV²/50 = 1 mW). Since the power level in receivers is usually low, dBm is a useful measure of low power.

dB μV: The decibel above 1 μV is a dimensionless voltage ratio in decibels referred to *a* reference voltage of 1 μV and is a commonly used measure of EMI voltage.

μV/m: Microvolts per meter are units used in expressing the electric field intensity.

dB μV/m: The decibel above 1 μV/m (dB μV/m) is also used for field intensity measurement.

μV/m/MHz: The microvolt per meter per megahertz is a broadband field intensity measurement.

dB μV/m/MHz: This is the decibel above 1 μV/m/MHz.

μV/MHz: Microvolts per megahertz are units of broadband voltage distribution in the frequency domain. The use of this unit is based on the assumption that the voltage is evenly distributed over the bandwidth of interest.

The following log relationships, which have been used in this chapter to convert magnitude to decibels, are useful to remember:

$$\log(AB) = \log A + \log B$$

$$\log\left(\frac{A}{B}\right) = \log A - \log B$$

$$\log(A^n) = n \log A$$

Appendices B and C show the conversion between electric fields, magnetic fields, and power densities.

2.5 RECEIVING PROPERTIES OF AN ANTENNA

In accordance with the reciprocity theorem, certain characteristics of a receiving antenna, such as pattern shape and return loss, remain the same when the antenna is used for transmitting. Likewise, the power transfer between two antennas, not necessarily identical, is the same regardless of which is transmitting and which is receiving. One important consideration when applying the reciprocity theorem is that it applies to the terminals to which the voltage is applied in the case of the source and to the terminals to which the voltage is measured in the case of the receptor. The pattern of power reradiated by a receiving antenna is different from its radiation when used as a transmitting antenna. In problems of EMI, it is often the current flow on a structure, cable, or wire that is required, and it should be remembered that the current distribution on a radiating structure is often different from that on the receptor structure, even when both structures are identical, and the power level most certainly is. An explanation for the coupling of an *E* field to a wire receiving antenna follows.

A field incident on the receiving antenna is assumed to have an angle of incidence that results in a voltage developed across the terminals of the antenna. This open circuit voltage *V* is proportional to the effective height, sometimes referred to as the effective length, of the antenna, which is seldom equal to the physical height of the antenna; thus, $V = h_{eff}E$. For aperture antennas, the ratio between the power developed in the antenna load and the incident power is known as the *effective aperture* of the antenna.

When the antenna input impedance equals the load impedance, a division of 2 occurs in the antenna-received voltage. When the antenna impedance is greater than the load impedance, the voltage division is greater than 2. The loss due to antenna/load impedance mismatch is accounted for in the equation for the gain of the antenna, which is a term in the equation for h_{eff}.

A second potential source of loss is due to a mismatch between the wave impedance and the radiation resistance of the antenna, and this is accounted for in determining the effective height of the antenna. The physical length of the wire antenna and the wavelength of the incident field both determine the radiation resistance and, therefore, the voltage developed.

2.5.1 Conversion of Power Density to Electric Field Intensity

For EMI measurements, the value of *E*- and *H*-field intensity is normally required, and it is possible to find these intensities from the power density at some distance from the source. To convert from power density P_d in the far-field to electric field intensity at the measuring point,

$$P_d = \frac{E^2}{Z_w} \quad [\text{W/m}^2]$$

(2.42)

and

$$E = \sqrt{Z_w P_d} = \sqrt{377 P_d}$$

(2.43)

2.5.2 Conversion of Power Density to Electric Field Intensity in terms of Antenna Gain

A receiving antenna is often used in EMI as a measuring device to find either the power density or the incident *E* field. The power density may be found from the received power developed in the antenna load resistance using Equation 2.23:

$$P_d = \frac{4\pi P_r}{\lambda^2 G_r}$$

(2.44)

where
   P_d is the power density at the receiving antenna
   P_r is the power into the receiver
   G_r is the gain of the receiving antenna

Therefore,

$$P_r = \lambda^2 G_r P_d / 4\pi$$

The value of the *E* field may be found from the power density using $P_d = E^2/Z_w$; the *E*-field-to-power relationship is then $E = \sqrt{P_d Z_w}$. Using Equation 2.44 for P_d and assuming $Z_w = 377\ \Omega$, the electric field intensity *E* in the far field is given by Equation 2.24:

$$E = \frac{68.77}{\lambda} \sqrt{\frac{P_r}{G_r}} \quad [\text{V/m}]$$

(2.45)

Assuming the field intensity measuring instrument has an input impedance of 50 Ω and V is the voltage measured by the instrument, P_r can be found from

$$P_r = \frac{V^2}{50}$$

Then, E can be expressed as

$$E = \frac{9.7V}{\lambda} \sqrt{\frac{1}{G_a}} \quad [\text{V/m}] \tag{2.46}$$

The Friis transmission equation provided the ratio between received powers and is transmitted between two antennas. It is a very useful sanity check on electromagnetic analysis programs used for antenna-to-antenna coupling analysis.

The equation assumes free space conditions. If both antennas are over a conductive surface, then a correction must be made for the field reflected from the conductive surface that may add or subtract from the direct radiated field. Section 10.2.9.3 provides an equation to compute the combined direct and reflected field. The Friis transmission equation assuming the receiving antenna is in the far field and ignoring cable and antenna mismatch losses is

$$P_r/P_t = G_t G_r \left(\lambda/4\pi r\right)^2$$

$$P_r = G_t G_r \left(\lambda/4\pi r\right)^2 P_t$$

In dB,

$$10\log P_t + 10\log G_t + 10\log G_r + 20\log\left(\lambda/4\pi r\right).$$

2.5.3 Antenna Factor

Antenna factor AF or K is an important calibration term; it is defined as the ratio of the electric field to the voltage developed across the load impedance of the measuring antenna:

$$\text{AF} = \frac{E}{V} \tag{2.47}$$

where
  AF is the antenna factor numeric
  E is the field strength in volts/meter
  V is the voltage developed across the load

The voltage developed is given by

$$V = \frac{Eh_{eff}}{(Z_{ant} + Z_L)/Z_L} \tag{2.48}$$

Thus, for an antenna where $Z_{ant} = 50\ \Omega$, $V = Eh_{eff}/2$. For a resonant dipole where $Z_{ant} = 72\ \Omega$, $V = Eh_{eff}/2.46$; for a resonant monopole $Z_{ant} = 42\ \Omega$, $V = Eh_{eff}/1.84$. The equation for h_{eff} is

$$h_{eff} = \frac{Z_{ant} + Z_L}{Z_L}\sqrt{\frac{A_{max}R_r}{Z_w}} \qquad (2.49)$$

where

$$A_{max}\ \text{is the maximum effective aperture} = \frac{G\lambda^2}{4\pi}$$

$$= 0.135\lambda^2\ \text{for a resonant dipole}$$
$$= 0.1191\lambda^2\ \text{for a small dipole} \qquad (2.50)$$

Z_w is the wave impedance
R_r is the radiation resistance

At resonance, the radiation resistance equals the antenna impedance. For a short dipole, from Reference [5],

$$R_r = 197\left(\frac{2H}{\lambda}\right)^2$$

And for a monopole,

$$R_r = \frac{197\left(2H/\lambda\right)^2}{2}$$

H is the physical height of one arm of the dipole or of the rod of the monopole. The gain in Equation 2.50 must include the loss correction for any Z_{ant}-to-Z_L impedance mismatch. This is true when Z_{ant} is higher than Z_L. However, when the load impedance is higher than the antenna impedance, the voltage developed across the load is either the same magnitude or higher than when Z_{ant} and Z_L are matched, even though reflections occur. In determining the AF, it is the ratio E/V that is of interest and not reflected power. Therefore, when the load impedance is higher than the antenna impedance, K (given by Equation 2.38) is set to zero, the loss is zero, and the gain is unchanged.

The $Z_{ant} + Z_L$ term is included in both Equation 2.48 for V and Equation 2.49 for h_{eff}, and thus, Equation 2.49 may be expressed as

$$V = E\sqrt{\frac{A_{max}R_r}{Z_w}} \qquad (2.51)$$

Assuming the antenna is matched to the terminating impedance,

$$V = \frac{h_{eff}E}{2} \qquad (2.52)$$

In some textbooks, either the antenna-to-load mismatch is ignored or an unterminated antenna is examined, in which case $V = h_{eff}E$. Assuming $R_r = Z_{ant} = Z_{term}$ (i.e., impedance of the antenna termination), for a dipole,

$$h_{eff} = 2\sqrt{\frac{A_{max}Z_{term}}{Z_W}}$$ (2.53)

Thus, for $Z_{term} = 50\ \Omega$ and $Z_w = 377\ \Omega$,

$$h_{eff} = \frac{\lambda\sqrt{G}}{4.87}$$ (2.54)

For a monopole, the effective height is half the value given by Equations 2.53 and 2.54. Using Equations 2.46 through 2.54 for antenna factor and antenna height gives us

$$AF = \frac{9.73}{\lambda\sqrt{G}}$$ (2.55)

Expressed in decibels, this is

$$AF_{dB} = 20\log\left(\frac{9.73}{\lambda}\right) - G\ \left[dB\right]$$

and, from Equation 2.55,

$$G = \left(\frac{9.73}{AF\lambda}\right)^2$$

or

$$G = 20\log\left(\frac{9.73}{\lambda}\right) - AF_{dB}\ \left[dB\right]$$

The relationship between AF and frequency is plotted in Figure 2.20a for a number of constant-gain antennas. One example of a constant-gain antenna is the half-wave tuned resonant dipole in which the physical length of the arms is adjusted to ensure that the total length of both arms is nearly $\lambda/2$.

For example, the effective height of a half-wave dipole is given by

$$h_{eff} = \frac{\lambda}{\pi}$$

For a monopole, it is

$$h_{eff} = \frac{\lambda}{2\pi}$$

where h is the physical height of both arms of the antenna = $\lambda/2$ in free space (assuming sinusoidal current distribution and an unterminated antenna).

FIGURE 2.20   (a) AF versus *f* for constant-gain antennas. (b) AF versus *f* for a λ/2 dipole and a 50 Ω receiver.

When the physical height of an antenna is much smaller than λ/2, the current distribution is not sinusoidal but rather of a triangular shape, with maximum current at the center and zero current at either end, and the effective height is one-half of the physical length of both arms of the dipole. For a monopole, h_{eff} is one-half of the physical length of the monopole rod.

For a true electric current element or infinitesimal dipole, $h_{eff} = h$. An equation that may be used when the physical length of the antenna is not λ/2 but is approaching this length is

$$h_{eff} = \frac{\lambda}{\pi} \tan\left(\frac{H}{2}\right)$$

where

$H = \beta h = 2\pi h/\lambda$

h is the physical length of the dipole antenna or the length of the rod of the monopole antenna

H_{eff} for the monopole is one-half the value given by Equation 2.56. The foregoing values of h_{eff} for unterminated antennas do not include losses due to impedance mismatch, so Equation 2.49, using gain and maximum effective area, has a more general application, especially in EMC prediction.

Consider the following example: At 100 MHz, the effective height of a half-wave dipole is 3 m/3.142 = 0.955. From Equations 2.47 and 2.48, and assuming far-field conditions and $Z_{term} = 50\ \Omega$ and $R_r = Z_{ant} = 70\ \Omega$,

$$\text{AF} = \frac{2.4}{h_{eff}} = 2.5$$

which, expressed in decibels, is 8 dB. Figure 2.20b shows AF versus frequency for a $\lambda/2$ dipole; here, the loss from the mismatch between the antenna impedance (70 Ω) and the terminating impedance (50 Ω) has been included. Manufacturers of antennas designed for EMI measurements usually publish gain and AF figures, whereas antennas designed for communications are typically supplied with gain calibration figures only. The antenna factor graphs published by manufacturers of broadband antennas are measured either on an open-field site or in a semianechoic chamber. In either case, the results have been corrected for the effects caused by the reflection from the ground or the floor of the chamber. However, should the antenna then be used in a shielded room, as recommended in MIL-STD-462 EMC test methods, where the antenna is positioned 1 m away from the equipment under test (EUT) and the ground plane, then the actual antenna factor can be greatly different from the published figures. This error is caused primarily by reflections from ceiling, walls, and floor, standing waves in the shielded room, and capacitive loading on the antenna due to ground plane proximity. Section 9.5.1 includes a description of techniques available to reduce these errors.

Where large standing waves exist, the ratio of E to H fields may vary from nearly zero to very high values (theoretically from zero to infinity), depending on the region (enclosure or cavity). In practice, the value of magnetic field measured in a shielded room is more constant than the E field.

One method of measuring the antenna factor is to use two identical antennas in the test setup shown in Figure 2.21. Two paths for the received voltage are obtained, one directly via the matching network into 50 Ω (V_{50dir}) and the other one the radiated path (V_{50rad}). For two identical antennas, where $G_t = G_r$, and using Equation 2.39, the product of the two gains is given by

$$G_t G_r = \frac{P_{rad}}{P_{dir}} \left(\frac{4\pi r}{\lambda} \right)^2$$

where
  P_{rad} is the radiated power measured by the receiver
  P_{dir} is the power measured by direct connection

FIGURE 2.21   Test setup for the two-antenna method of determining AF.

Because $G_t = G_r$, the antenna realized gain, G_{re}, is

$$G_{re} = \sqrt{\frac{P_{rad}(50)}{P_{dir}(50)}\left(\frac{4\pi r}{\lambda}\right)^2}$$

Now,

$$\frac{P_{rad}}{P_{dir}} = \frac{V_{rad}(50)^2}{V_{dir}(50)^2}$$

where $P(50)$ is the power into 50 Ω and $V(50)$ is the voltage developed across 50 Ω. Therefore,

$$G_{re} = \frac{V_{rad}(50)}{V_{dir}(50)}\frac{4\pi r}{\lambda}$$

and from Equation 2.29,

$$\text{AF} = \frac{9.73}{\lambda\sqrt{G_{re}}}$$

If this calibration is conducted in a shielded room where subsequent EMI measurements are to be made and the receiving antenna is located at a fixed and marked location, which is then used for subsequent measurements, then the measured AF will have compensated, to a large extent, for the errors caused by reflections inherent in the test location.

Because the realized gain has been determined using this method, no further compensation due to wave-impedance-to-load-impedance mismatch is required and the near-field AF for the antenna may be obtained. It is therefore important that the distance between the two antennas, used in obtaining the AF, be duplicated when using the calibrated antenna for EMI measurements, for at different distances the AF calibration is not valid. Any loss due to the mismatch between the antenna impedance and load is also included in the measured G_{re}. It is important to ensure that the generator is able to produce the same voltage across the antenna impedance as across the 50 Ω load. When this is not the case and the reason is a high-input-impedance antenna, then a 50 Ω load at the generator end of the cable feeding the transmitting antenna may be included during the radiated path measurement and removed during the direct measurement. This ensures that the generator load is approximately the same for the radiated and direct path measurements. When the antenna impedance is high, the receiving and transmitting antenna connecting cables should be kept short and of equal length. For maximum accuracy, the same cable should be used in measurements with the antenna. Any attenuation due to the cables is not included in the two-antenna measurement technique, and the cables should be calibrated separately.

The antenna factor for a tuned dipole using the two-antenna calibration method in a shielded room, with the manufacturer's AF curve, is plotted in Figure 2.22. By plotting the AF curve over a narrower frequency range, Figure 2.23, the effect of resonances, antiresonances, and reflections may be clearly seen, and the AF curve may be used to calculate the field magnitude with higher accuracy than when the manufacturer's curve is used. The accuracy of the AF curve is sensitive to exact antenna location and to the presence of equipment or personnel in the shielded room. Techniques that are helpful in achieving good repeatability include the installation of a plumb bob directly above the receiving antenna location used in the two-antenna test setup. This location should be chosen for its usefulness. For example, MIL-STD-462 test requirements specify a distance for the

Frequency [MHz]

Dipole antenna factor [dB]

Antenna height above ground = 1.22 m

Antenna separation from signal source = 1 m

FIGURE 2.22   AF, in decibels, of a tuned dipole measured in a shielded room.

measuring antenna of 1 m from the EUT, which is placed 0.05 m from the edge of a table covered with a ground plane that is bonded to the shielded room wall. Thus, where MIL-STD measurements are made, one or more calibration points down the length of the table at a distance of 0.95 m from the edge of the ground plane would be useful.

The antenna calibration and subsequent measurements using the antenna should be made in a room clear of equipment and personnel, or at least the same room configuration should be maintained for both.

If the two-identical-antenna test method is used for calibrating antennas to be used on an open area test site (OATS) or in an anechoic chamber, then the calibration should be conducted on an OATS. The site should preferably be without a ground plane and within the area between the two antennas covered in a combination of ferrite tile and foam absorber material to absorb the reflected ground wave. Horizontal polarization of the antenna should be used to reduce the impact of the antenna feed cables, which are oriented vertically. The 6–12 dB 50 Ω attenuators should be included, as shown in Figure 2.21. Both antennas should be placed at least 4 m above the ground to reduce antenna-to-ground coupling. The distance between the antennas should be 10 m to eliminate antenna-to-antenna coupling.

Alternative antenna calibration techniques are the reference OATS and the reference dipole calibration contained in American National Standards Institute (ANSI) C63.5 and described in Section 9.4.1. If correctly constructed, the four reference dipoles, designed to cover the 25–1000 MHz frequency range, will have antenna factors within 0.3 dB of what is theoretical. An alternative test method uses the GTEM cell, as described in Section 9.5.2.2. The antenna factor can also be calculated using either the "induced emf method," numerical electromagnetics codes (NEC), MININEC, GEMACS, or the site attenuation. In the induced emf method, mutual coupling between antennas is calculated by formulas given by S.A. Schelkunoff and H.E. King. These formulas usually replace the presence of the ground plane by images of the transmit and receive antenna. NEC, MININEC, and GEMACS use the moment method (MOM) to calculate the coupling between antennas, and with GEMACS using the MOM and GTD hybrid, the presence of a ground plane can be included. Reference [16] compares the accuracy of these analysis techniques to OATS, anechoic chamber, and

FIGURE 2.23   AF, numeric, for the same tuned dipole.

GTEM measurements for symmetrical dipoles and horn and waveguide antennas. With symmetrical dipoles in horizontal polarization, the maximum deviation between the site attenuation model and measurements was 3 dB. The maximum deviation between measurements and the induced emf method, MOM, MININEC, and other calculations was 2.7 dB.

2.5.4   Receiving Properties of an Isolated Conductor/Cable

An *isolated cable* is defined as one disconnected from ground at both ends and at some distance from a ground plane. Although this configuration used as a model has limited practical use, one practical, albeit unusual, example of this cable configuration was, in a case of EMI, caused by coupling to a shielded cable. The cable was connected to a helicopter at one end and to equipment contained in an enclosure towed beneath the helicopter at the other end. With the helicopter air-borne, the cable was isolated from ground at both ends. The more common cable configurations are discussed in detail in Chapter 7, on cable coupling; however, for the sake of completeness, here, we examine coupling to an isolated cable. When a cable is disconnected from ground at both ends, the *E*-field component of a wave that cuts the cable at an angle of 90° to its axis does not induce a current flow. In the case where a wave is incident on the cable such that the magnetic field component cuts it at an angle of 90° and the *E* field is in the plane of the cable, a current is induced. The magnitude of the current is determined by the length and impedance of the cable and the wavelength of the field. If the length of the cable is less than 0.1λ (i.e., nonresonant), then the average flow per second in the conductor of the cable is given approximately by

$$\frac{4\pi fBl}{Z_c}\left(\frac{\pi}{\lambda}\right)^2 \frac{\left(l/2\right)^2 - \frac{1}{2}l^2}{2}$$

where
  f is the frequency, in hertz
  B is the magnetic flux density
  l is the physical length of the conductor

$$Z_c = \sqrt{R^2 + 2\pi fL^2}$$

where
  R is the total resistance of the conductor
  L is the total inductance of the conductor

Values of resistance and inductance for single conductors and for the shields of shielded cables are provided in Tables 5.1 and 4.2, respectively.

When the length of the cable equals λ/2, it is resonant and the characteristics of a resonant short circuit dipole, sometimes referred to as *a parasitic element*, may be used. In a resonant dipole terminated in its radiation resistance and ignoring losses, half of the received power is delivered to the load, and half the power is reradiated from the antenna. In a shorted resonant dipole, four times the power is reradiated, compared to the antenna-matched dipole. The radiation resistance of the resonant dipole is approximately 70 Ω, and this is true regardless of the load. The reradiated power from the dipole is therefore $I^2 70$. Because the reradiated power for the short circuit dipole is four times of that for the matched dipole, the current flow is twice the value of that for a matched dipole.

The receiving characteristics of the dipole, described in the preceding section, may be used to find the current *I* in the load of a dipole. The average current in the same length of isolated cable (i.e., the short circuit dipole) at the same frequency then has the value $2I \times 0.64$.

2.5.5 Monopole Antenna as a Measuring Device and in Prediction of Electromagnetic Compatibility

Few practical cable, wire, or structure configurations look like a dipole antenna, and thus, it is not particularly useful as a model in EMC predictions. The monopole antenna is in principle a vertical wire above a ground plane, with the ground plane sometimes referred to as a *counterpoise*. The monopole receiving properties and to a lesser extent its transmitting properties are useful in EMC prediction. A cable connected to a metal enclosure, with the cable routed over a nonconductive table, may be represented by a monopole antenna as long as the far end of the cable does not connect to ground. That the cable is horizontal to the floor of the room does not change the validity of the antenna model. The far end of the cable may terminate in a small metal enclosure, in which case the model is similar to a top-loaded monopole antenna. If the far end of the cable loops down to a conductive floor, such as is found in shielded rooms, or terminates in equipment connected to ground, then the monopole model is not appropriate. Instead, the model described in Section 7.7 for coupling to loops should be used. The cable, which we model as the rod of the monopole antenna, may be a shielded cable, with the shield terminating on the enclosure or a twin or multiconductor cable that terminates inside the enclosure. Whichever it is, we require the termination impedance in order to apply the monopole model. One additional good reason to examine the receiving properties of a monopole antenna is its usefulness as an EMC measurement antenna and its ease of construction. A 41 in. (1 m) monopole or rod antenna over a square-meter counterpoise is a common measurement antenna. Its useful frequency range is 14–30 MHz when it is not used as a resonant antenna. When the rod is adjustable from 1 m to approximately 0.2 m, the antenna may be used as a resonant monopole over the range 75–375 MHz. To cover the 300–1500 MHz, it is more convenient to manufacture a monopole with a smaller counterpoise, typically 30 cm × 30 cm, and a rod adjustable from 30 to 5 cm. For use in EMC predictions and in EMC measurements, the resonant and nonresonant characteristics of the monopole are required. When the length of the cable or rod is equal to λ/4, where λ is the wavelength of the incident field, then the antenna is resonant, the input impedance is 35–42 Ω, and the gain is 1.68 numeric. When the length of the rod is much less than λ, the gain is the same as the current element, which is 1.5, and the antenna input impedance is high.

For an electrically short monopole, the input impedance is approximately

$$Z_o = 10H^2 - j\frac{30}{H}\left(\frac{\omega - 2}{1 + \dfrac{2\ln 2}{\omega - 2}}\right)$$

or, as a magnitude,

$$|Z_o| = \sqrt{\left(10H^2\right)^2 + \left[\frac{30}{H}\left(\frac{\omega - 2}{1 + \dfrac{2\ln 2}{\omega - 2}}\right)\right]^2}$$

where

$$\omega = 2\ln\left(\frac{2h}{a}\right)$$

and *h* is the physical height of the rod and *a* is its radius, β/*h* is defined as 2π*h*/λ.

As β*h* approaches 1, the antenna impedance tends to become resonant and at values of β/*h* = 1.5, 2.5, 3, 3.5, 4, 4.5, …, the monopole is resonant. Curves for the reactance and resistance of the dipole, for β/*h* = 0.5–7, are provided in Figures 2.24 and 2.25 from Reference [6]. The reactance and resistance of a monopole are one-half of the values for a dipole. The magnitude of the impedance of the monopole is

$$Z_o = \sqrt{Z_{oim}^2 + Z_{ore}^2}$$

where
Z_{oim} is the reactance of the antenna
Z_{ore} is the resistance of the antenna

FIGURE 2.24 Input reactance of a dipole (monopole is half the dipole value). (From King, R.W.P. et al., *Transmission Lines, Antennas, and Wave Guides*, Dover, New York, 1965.)

The maximum open circuit gain of the resonant monopole is 1.68; for the short monopole, it is 1.5. By using Equation 2.38 to obtain the reflection coefficient due to the antenna-to-load mismatch, the realized gain G_{re} may be obtained. From Equation 2.50 and using G_{re}, the value of A_{max} may be determined. Using A_{max} with the wave and radiation resistance in Equation 2.51 gives us the voltage developed across the load, from which the load current may be calculated. The average current flow in the rod of the monopole is then equal to one-half the load current for the short monopole and 0.64 times the load current for the resonant monopole.

Consider an example of the calculations of antenna factor for a small tunable rod antenna at a frequency of 550 MHz. The AF of the antenna is

$$\frac{E}{V} = \frac{1}{h_{eff}} = \frac{1}{\lambda/2\pi}$$

Therefore, the AF at 550 MHz is $1/(0.545/12.56) = 23 = 27$ dB.

The antenna factor of a tunable monopole over the 300–800 MHz frequency range is plotted in Figure 2.26. The calibration was made in a shielded room using the two-antenna test method. At 550 MHz, the measured AF = 26.2 dB, whereas the predicted AF is 27 dB. At other frequencies, a difference of ±3 dB between predicted and measured AF is apparent. The measured antenna factor and gain for a 1 m rod antenna, calibrated at 1 m distance from a second identical antenna, over a 10 kHz to 30 MHz frequency range are shown in Figures 2.27 and 2.28. Consider the antenna

FIGURE 2.25 Input resistance of a dipole (monopole is half the dipole value). (From King, R.W.P. et al., *Transmission Lines, Antennas, and Wave Guides*, Dover, New York, 1965.)

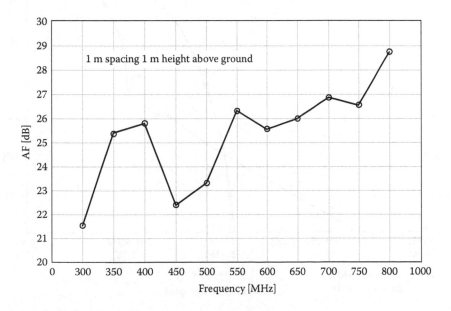

FIGURE 2.26 Measured AF, in decibels, of a resonant monopole.

FIGURE 2.27   Measured AF, numeric, for a 1 m monopole.

FIGURE 2.28   Measured gain of a 1 m monopole.

factor at 20 MHz, where the antenna impedance is approximately 500 Ω and the wave impedance at a distance of 1 m is approximately 900 Ω. The reflection coefficient, from Equation 2.38, is

$$K = \frac{1 - 500/50}{1 + 500/50} = \frac{-9}{11} = -0.818$$

The loss is $1 - (-0.818)^2 = 0.33$, so the gain is $1.5 \times 0.33 = 0.496$.

The first term in Equation 2.49 appears in the equation for the AF and is, in our example, omitted from both equations. Therefore, from Equations 2.47 and 2.51, the AF (*E/V*) is $1/0.065 = 15.4$. From the plot of the measured antenna factor for a monopole, Figure 2.27, the AF at 20 MHz is 6.2 and the difference between measured and predicted AF is 8 dB. The measurements were made in a shielded room, and thus, some difference between measured and predicted AF is to be expected due to the proximity of the rod of the antenna to the conductive ceiling and due to reflections. An additional potential source of error is that the measurements were made in the near field, where curvature of the radiated *E* field will reduce the coupling to the receiving antenna.

The monopole in our example is neither an effective *E*-field measurement nor a communication antenna, due to the large mismatch between the antenna input impedance and the 50 Ω load impedance. The inclusion of a high input impedance, for example, FET, buffer, or amplifier positioned under the counterpoise and connected directly at the base of the antenna, will greatly increase the sensitivity of the antenna at low frequency. The two monopoles, located 1 m apart during measurements, are in the near field, and thus, the quasi-static radial field terminates on the counterpoises of both antennas as well as on the rod of the receiving antenna.

The quasi-static coupling is via the very low ($\cong 0.5$ pF) mutual capacitance of the antennas. The mutual reactance of the antennas is therefore very high at the frequencies of interest, and minimal voltage is developed across the 50 Ω receiving antenna load impedance, due to the radial *E* field. When the antenna termination impedance is high, the quasi-static field will develop an appreciable voltage across the load and change the AF calibration of the antenna.

When the monopole model is used to predict the current flow on a shielded cable connected to a shielded enclosure, the load impedance is equal to the termination impedance of the shielded cable, typically the sum of the shield-to-backshell, male-to-female connector and connector-to-bulkhead interface impedances. This termination impedance may then be used in the equations for the short nonresonant antenna to find the load voltage, from which the load current may be found. The current flowing on the shield of the cable is then one-half of the load current.

Where the shielded cable length is resonant, the characteristics of a resonant short circuit monopole may be used to find the shield current. First, the receiving properties of the resonant antenna into a matched load are used to find the load current. The average matched antenna current is then 0.64 times the load current, and the current flow on the short circuit resonant antenna (cable shield) is twice that for the matched antenna.

A simple circuit comprising a MOSFET input stage followed by a single transistor gain stage, with an input impedance of 100 kΩ, was built and powered by a 9 V battery. The increase in gain at 10–100 kHz was measured at 83–78 dB, respectively. The noise floor increased by 55 dB at 10 kHz and 42 dB at 100 kHz, with a resultant increase in signal-to-noise ratio of 28–36 dB.

2.6   SIMPLE, EASILY CONSTRUCTED *E*- AND *H*-FIELD ANTENNAS

This section presents the design of a number of easily constructed antennas for the measurement of either *E* field or *H* field. When considering time-varying waves, *E* and *H* fields are always present and either may be measured. One important aspect of the design of a measurement antenna is that the antenna differentiates between *E* and *H* fields. Thus, an antenna designed to measure *H* fields should reject the influence of the *E*-field component of the electromagnetic wave.

The advantage of physically small antennas is that the localized E and H fields in close proximity to sources such as enclosures, apertures, and cables may be readily measured. The larger dipole and commercial broadband antennas also have a place in measuring the composite field from all sources. Measurement techniques using both types of antenna are discussed in Chapter 9.

2.6.1 Shielded Loop Antenna

The simple loop antenna connected to a shielded cable is unbalanced with respect to the shield of the cable and therefore responds to both E and H fields. One technique used to reduce the influence of the E field is to shield the loop.

The schematic of a shielded loop antenna is shown in Figure 2.29a, with a photograph of an 11–6 cm antenna given in Figure 2.30. The antenna is constructed from semirigid cable that is

FIGURE 2.29 Shielded loop antennas. (a) Single shielded loop antenna and (b) multiturn shielded loop antenna.

FIGURE 2.30 Photo of shielded loop antennas.

FIGURE 2.31   Calibration curves for a 6 cm shielded loop antenna, upper limit 15 MHz.

available in a number of different diameters. The shielded loop antenna becomes resonant above a frequency determined by the inductance of the loop and the capacitance between the center conductor and the outer sheath. The lower the capacitance of the semirigid cable used in the manufacture of the loop, the higher the useful frequency response. Calibration curves for the 6 cm loop, which has a useful upper frequency of greater than 15 MHz, and the 11 cm loop, which has a useful upper frequency of 10 MHz, are shown in Figures 2.31 and 2.32. The receiving properties of the 6–11 cm loops are close to the predicted properties when terminated in a 50 Ω load, and so calibration is necessary only to determine the upper frequency limit or where maximum accuracy is required. A word of warning: the measurement technique, unless it is carefully controlled, may result in less accuracy than the use of the predicted characteristics alone. Where the sensitivity of the single loop at low frequency is not sufficient, a multiturn shielded loop may be constructed. The multiturn antenna may be shielded by wrapping the *turns* with a conductive tape. The tape is connected to the outer surface of the coaxial connector as shown in Figure 2.29b. The shield must contain a gap around the circumference of the loop in order to avoid the shielding of the loop against magnetic fields.

The multiturn loop exhibits a lower resonance frequency than a single turn, due to its higher inductance and intrawinding capacitance.

2.6.2 BALANCED LOOP ANTENNA

A second extremely useful *H*-field antenna is a 6 cm balanced loop antenna, the schematic of which is shown in Figure 2.33. The balanced loop antenna contains a balun constructed from a two-hole ferrite bead. Baluns are used in a number of balanced antennas, both *E*- and *H*-field types, to match the antenna to the unbalanced coaxial cable. The cable and associated equipment are unbalanced because the shield of the cable and the outer shell of the coaxial cable are connected to ground, via the enclosure that is connected to safety ground on the majority of equipment. In the balanced loop antenna, the *E* field induces equal and opposite voltages into the primary (loop side) of the balun, and thus, ideally, zero *E*-field-induced voltage appears across the secondary of the balun.

In practice, some capacitive imbalance exists both in the intrawinding capacitance of the balun and between the loop and the metal enclosure housing the balun. Any imbalance results in an incomplete cancellation of the *E*-field-induced voltage. In addition, the capacitance, with the inductance

FIGURE 2.32   Calibration curve for an 11 cm shielded loop antenna, upper limit 10 MHz.

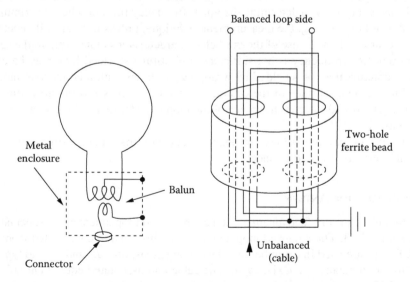

FIGURE 2.33   Schematic of a balanced loop antenna and construction details of the balun.

of the loop, determines the resonant frequencies of the loop. In a carefully designed antenna, the connections from the loop to the primary of the balun should have a characteristic impedance of 200 Ω. The secondary connection to the coaxial connector mounted on the enclosure should have an impedance of 50 Ω. The center tap of the primary of the balun is connected to ground via the enclosure. Ideally, this connection should be made via a low impedance, such as a short length

FIGURE 2.34 Calibration curve of a balanced loop antenna.

of wide PCB material. The calibration curve of a balanced loop antenna is shown in Figure 2.34. This antenna was not constructed with maximum care in the layout, and some of the kinks in the calibration curve may be the result. Nevertheless, the measured characteristics are within 4 dB of what is predicted. The sensitivity of this simple antenna is approximately 46 dB above that of the Hewlett-Packard HP 11940A near-field magnetic field probe, which costs almost 100 times as much. It should be added that the HP probe, due to its very narrow tip, is invaluable for locating emissions from PCB tracks and integrated circuits (ICs). In addition, the HP probe covers the wider, 30–1000 MHz frequency range, compared to the 20–200 MHz for the simple loop. One potential source of measurement error in the small loop antennas is produced by *E*-field-induced current flow on the shield of the coaxial interconnection cable. The shield current induces a voltage in the center conductor of the cable, which adds to the signal from the antenna.

In other chapters, we will see that *H*-field measurements are typically more reliable in locations where reflections cause measurement errors. We shall see that measurement errors may be reduced by computing the magnitude of the *E* field, at some distance from the measuring point, from the magnitude of the *H* field at the point. The calibration of the example *H*-field antennas is in mV/mA/m. However, an alternative is to use the *H*-field antenna factor, defined as $H(A/m)/V$.

2.6.3 *E*-Field Bow Tie Antenna

The balun incorporated into the balanced loop antenna may also be used to match a broadband *E*-field "bow tie" antenna to the coaxial cable. The schematic diagrams of a bow tie antenna for the 20–250 MHz and 200–600 MHz frequency ranges are shown in Figure 2.35. The impedances of the 20–250 MHz and 200–600 MHz antennas with 2/1 voltage ratio and 4/1 impedance ratio baluns are shown in Figure 2.36 and are derived from Reference [7].

The bow tie is useful in measuring both the radiation from an EUT in close proximity and the level of the *E* field in a radiated susceptibility (immunity) test. The small bow tie, with 10 cm long elements, although designed for 200–600 MHz, has been calibrated from 25 to 1000 MHz. At low frequency, the antenna factor is high, in common with all physically small antennas, and great care has to be exercised, in both calibration and use, to ensure that the coupling to the feed cable is not

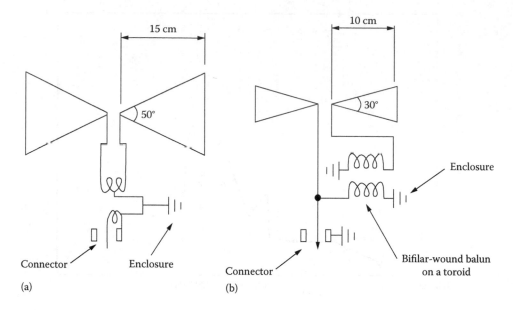

FIGURE 2.35   Bow tie antennas. (a) 20–250 MHz bow tie antenna and (b) 200–600 MHz bow tie antenna.

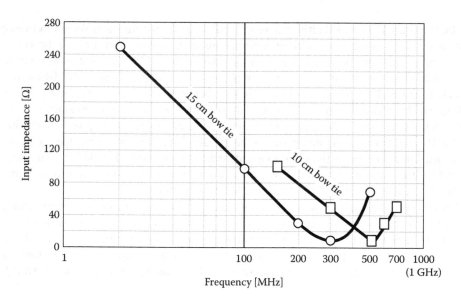

FIGURE 2.36   Input impedance calibration curves for the 15 and 10 cm bow tie antennas. (From Brown, G.H. and Woodward, O.M., Jr., *RCA Rev.*, 13(4), 425, December 1952.)

higher than that to the antenna. To reduce the coupling, use a double-braid shielded cable with many ferrite baluns mounted on the cable; when monitoring susceptibility test levels, orient the cable vertically for horizontally polarized fields and antenna, and orient it horizontally for vertically polarized fields and antenna. Figure 2.37 shows the 10 cm bow tie calibration at a distance of 0.2 m; Figure 2.38 shows it at a distance of 1 m, with the antenna horizontally polarized and 1 m above a ground plane. The antenna was constructed on a rectangular-shaped PCB with the two bow tie elements etched out of the PCB material, with a BNC connector mounted in the center of the antenna and with the balun located close to the connector as shown in Figure 2.38.

FIGURE 2.37   Bow tie antenna factor at 0.2 m.

FIGURE 2.38   Bow tie antenna factor at 1 and 1 m above a ground plane.

2.6.4 MONOPOLE ANTENNAS

The 1 m and the tunable resonant monopole antennas described in Sections 2.5.5 and 2.5.6 cover the 14 kHz–30 MHz and 300–800 MHz frequency ranges, respectively, and the simple construction of these antennas is illustrated in Figure 2.39. The monopoles with ground plane do not require a balun to match the antenna to the cable, which is one reason for their widespread use.

2.6.5 TUNED RESONANT DIPOLE ANTENNAS

The schematic of the dipole with a typical balun used over the 40–300 MHz frequency range is shown in Figure 2.40. The calibration curve for this antenna in a shielded room and the theoretical open-field test site calibration are shown in Figure 2.22. The construction details of a similar balun, showing how the coaxial cable is wound on formers and attached to the connector, are presented in Figure 2.41. The construction details are reproduced by the courtesy of the Electro-Metrics Corporation.

2.6.6 HELICAL SPIRAL ANTENNAS

Helical spiral antennas, which cover the 800 MHz–18 GHz frequency range, as shown in Figure 2.42, are relatively simple to construct. The larger antenna is 26.5 cm long and 12 cm in diameter at the base, tapering to 10 cm at the top. The larger antenna is wound as a helical log periodic, with the angle between the base of the antenna and the spiral at approximately 20°. The distances between the windings of the spiral are not critical, and in the larger antenna, shown in Figure 2.42, the ratio of the distance between the windings of the lower spiral and the distance between the windings of

FIGURE 2.39   1 m and resonant monopole antennas.

Metal enclosure

9 turns on 1/2 in. dia.,
1-3/4 in. long former

Connector

FIGURE 2.40   30–500 MHz balun.

FIGURE 2.41   Construction details of a similar balun. (Courtesy of Electro-Metrics Corporation, New York.)

the next higher spiral is approximately 1.25. The important criterion is that the two antennas be made as similar as possible when the two-antenna test method is used for calibration.

The calibration curve for the 26.5 cm log periodic helix is shown in Figure 2.43. The design frequency range for the antenna is 1–4.77 GHz, based on the circumferences of the lowest and highest spiral. From the calibration curve, it is seen that the antenna is usable down to 800 MHz; a recent recalibration was made up to 4 GHz.

In order to simplify the construction of the two smaller antennas, covering the 3–18 GHz frequency range, only the diameter of the helix is changed from top to bottom. The intermediate-sized spiral has a maximum circumference of 8.96 cm and a minimum of 3.49 cm; thus, the design frequency range is 3.34–8.6 GHz. The distance between the spirals on the intermediate-sized antenna is a constant 0.476 cm (3/16 in.). The small antenna has a design frequency range of 8–18 GHz. The material on which the helix is wound should exhibit a low permittivity and low loss at high frequency; here, expanded polystyrene is a good choice. However, the antennas illustrated were wound on cardboard formers impregnated with epoxy, and acceptable results were achieved in the calibration of the antenna.

FIGURE 2.42   Helical antennas covering the 800 MHz–18 GHz frequency range.

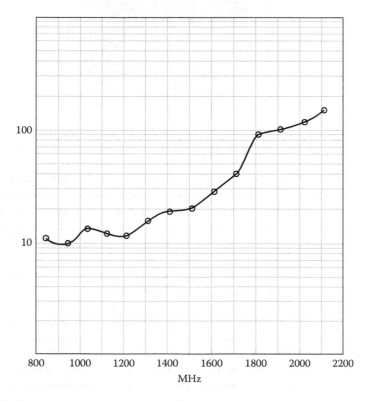

FIGURE 2.43   Calibration curve for the 26.5 cm helical log spiral antenna.

2.6.7 SMALL TOROID ANTENNA

An inductor wound on a toroid core does not contain the field inside the core particularly well, and, as described in Section 2.5, any inductor wound on a core will have a current induced in it when an electromagnetic field impinges on the inductor. Thus, in certain filter applications, it is imperative to shield the inductor to minimize this radiated coupling effect. This coupling effect is used in the toroid antenna, which is constructed from a small iron–dust toroid type T12-3 from Micrometals. The T12-3 core has a diameter of only 3.18 mm and an A_L value of 6 nH/N²; to construct the antenna, 10 turns of magnet wire are wound on the core. The toroid is mounted at the end of a brass tube, with one side of the inductor connected to the tube and the second side connected to a center conductor in the tube. At the other end of the tube, an SMA connector is soldered, as shown in Figure 2.44. The toroid antenna shows no resonances over the 20–1000 MHz frequency range. It is sensitive to very localized fields and is invaluable for measuring PCB emissions down to a single trace or to differentiate between emissions from the PCB and the ICs. It is used at very close proximity to sources, where the field reduces rapidly with distance, for comparative measurements, and so it makes no sense to calibrate the antenna. If quantitative measurements are required, then the balanced loop and bow tie antenna at distances of approximately 0.2 m are the correct antennas to use.

2.6.8 0.1–1000 MHz *H*-FIELD PROBE

An *H*-field probe, which is very useful for probing seams or apertures in an enclosure but can also be used by placing the tip flat on cables, is shown in Figure 2.50. It has been calibrated from 0.1 to 1000 MHz using the test jig shown in Figure 2.55h.

The calibration is shown in Figure 2.45 to show the frequency response as the probe would normally be used for relative measurements, that is, before and after modifications, and the absolute level of *H* field is not important.

Figure 2.46a and b shows the construction and schematic.

2.6.9 CALIBRATION

When care is taken in the construction of two antennas to make them as similar as possible, then the two-antenna test method described in Section 2.5.3 is recommended. Alternative antenna calibration techniques are the reference OATS and the reference dipole calibration, both of which are contained in ANSI C63.5 and described in Section 9.4.1. If correctly constructed, the four reference

FIGURE 2.44 Small toroid antenna.

FIGURE 2.45   Near-field probe calibration.

FIGURE 2.46   (a) *H*-field probe 0.1–1000 MHz construction.   (*Continued*)

(b)

FIGURE 2.46 (*Continued*)   (b) *H*-field probe 0.1–1000 MHz schematic.

dipoles, described in ANSI C63.5 and designed to cover the 25–1000 MHz frequency range, will have antenna factors within 0.3 dB of theoretical. An alternative test method uses the GTEM cell, as described in Section 9.5.2.2. The antenna factor can also be calculated using either the "induced emf method," NEC, MININEC, or GEMACS or from site attenuation. In the induced emf method, mutual coupling between antennas is calculated via formulas given by S.A. Schelkunoff and H.E. King. These formulas usually replace the presence of the ground plane by images of the transmit and receive antennas. NEC, MININEC, and GEMACS use the MOM to calculate the coupling between antennas; with GEMACS, using the MOM and GTD hybrid, the presence of a ground plane can be included.

One alternative test method is to use a calibrated gain transmitting antenna. A measured and controlled input power is applied, and therefore, a known *E* field is generated at the distance from the transmitting antenna at which the receiving antenna is placed. The measured output voltage from the antenna under calibration can then be used, with the value of the *E* field, to obtain the antenna factor, from which the gain of the antenna under calibration may be calculated. The second test method is to use the antenna under calibration as the transmitting antenna and the antenna factor of the calibrated receiving antenna to calculate the gain of the antenna under test, from which the antenna factor may be calculated. The mutual coupling between antennas can be minimized by using a 10 m distance between the antennas, and, as described in Section 2.5.3, the effect of the ground reflected wave has to be either minimized or calculated.

One technique for the calibration of *H*-field antennas is shown in Section 9.6.7.8. Although the calibration is for an RS01 transmitting antenna, it is identical to the calibration of a receiving loop antenna, if the reference transmitting loop is constructed.

When measuring an *E* field using a small antenna, an attached conductive cable distorts the measured field as well as forms part of the antenna. This is true even with a small-diameter cable covered in absorber or ferrite baluns.

The uses of a physically small antenna include the measurement of the shielding effectiveness of an enclosure or close to a source of emissions.

With the antenna connected to a small circuit that drives data on a fiber-optic cable, the effect of an attached conductive cable can be eliminated. The antenna was connected to a wide dynamic range RF/IF detector with a high input impedance of 1.73 kΩ/1.45 pF, followed by an operational amplifier, and thus, the circuit close to the antenna can be very small. The DC supply to the circuit and the DC output from the amplifier are connected to a small enclosure using a thin #30 AWG wire. The box contains a fiber-optic transmitter and A/D converter and a 9 V battery and regulators.

The antenna is a bow tie with tip-to-tip dimensions of 10 cm that was used for measuring aperture coupling in a cylinder and box and *E*-field shielding of conductively coated plaques, both of which are described in Chapter 6.

The 10 cm antenna was an ideal size for these tests, but a larger antenna can be used when appropriate. At the far end of the fiber-optic cable was a fiber-optic receiver connected to a PIC that was programmed to output to the USB input of a computer. Thus, a digital level indicated the level measured by the antenna. The frequency response of the detector is 10 MHz–1 GHz, which was adequate for the tests mentioned earlier. The schematic of the antenna, detector, and amplifier is shown in Figure 2.47.

Figures 2.48 and 2.49 show the lowest and highest *E*-field levels that can be measured by the detector circuit, although if higher fields need to be measured an input attenuator can be used with care to ensure that the detector itself is not susceptible to the field.

To increase the sensitivity of the circuit, the size of the bow tie antenna can be increased.

A similar detector was designed and built with operational amplifiers and detector diodes, for the 10 kHz–10 MHz frequency range, using the same A/D, digitizer, fiber driver, and PIC.

FIGURE 2.47 Schematic of antenna, detector, and amplifier.

FIGURE 2.48 The lowest level of *E* field detected by the circuit in Figure 2.47.

FIGURE 2.49 The highest level of *E* field detected before saturation by the circuit in Figure 2.47.

The tip of the probe is cast in epoxy to provide protection and this is shown in Figure 2.50.

The probe measures surface current as well as *H* field. If a well-shielded enclosure has cables connected to it on which RF current flows, then the fields generated by these cable currents will induce current on the enclosure. This is illustrated in Figure 9.23. The *H*-field probe will then measure these currents, and the investigator of the source of emissions may incorrectly assume that it is the enclosure that is radiating. Despite that, if the emissions measured around connectors, at apertures and seams, are higher than the body of the enclosure, then it is safe to assume that these locations are leaking.

FIGURE 2.50   *H*-field probe enclosure.

Generating very high levels of *E* field (200–500 V/m) in a MIL-STD-RS103, DEF STAN DRS02, or DO-160 radiated susceptibility test requires high-power amplifiers and large antennas that can withstand the power when the complete EUT is tested.

Also, the *E* and *H* fields from an indirect lightning strike require high-power equipment when the complete EUT must be exposed. Sometimes, a test of components that are exposed, with little or no shielding, and that are attached to the EUT such as microphone, speaker, LED array, key pad, memory chip, and PCB can be tested alone. These can be attached to either the EUT or a simulator, which can monitor performance during the susceptibility test with both of these isolated from the *E* or *H* field. The advantage to this type of measurement is that very small injection probes are required when the component is small and the required input power is low.

One test was of the immunity of a memory chip used in a deployable black box. The chip was exposed to the fields radiated by an adjacent lightning strike. The test was performed with the chip unpowered and powered by a small battery. A lightning generator was used to generate a field of 200,000 V/m in a small plate antenna and 20,000 A/m in the center of a coil with the field waveform close to that typical for a lightning strike.

For passports and RF ID devices, these antennas may be used to generate an excitation field for a duration and then, after cessation of the field, measure the response from the device.

Antennas that simulate the *E* and *H* field from high-power transmission lines at 50 and 60 Hz have been designed and tested. For the generation of the *E* field, a step-up transformer is used connected to a plate antenna and for the *H* field a step-down transformer connected to a coil. These are used to test a substance, which contains small metal particles. The concern is that high *H* field–induced eddy currents could result in heating of the substance or that high electric field gradient could result in breakdown between the particles. Due to the high voltages connected to the plate antenna, a proximity switch is used to switch off power to the generator when the protective plastic cover is removed.

Figure 2.51 shows the *H*-field schematic and the *E* field.

Figure 2.57b shows the *E*-field test jig and Figure 2.57c the *H* field.

The *E*-field antenna can generate up to 45,000 V/m and the *H*-field antenna above 800 A/m.

When very high currents at a specific frequency are required to generate an *H* field, then the injection coil can form part of a resonant circuit. The trimmer capacitor is the capacitance between two parallel insulated wires that can be cut to length, and the fixed capacitors are high voltage types. A step-down transformer was used to match the circuit more closely to the power amplifier. The length of the coil is 6.8 cm, the width 11.5 cm, and the height 1.2 cm and the gap in the middle for the card or passport sheet is 6 mm.

Figure 2.52 shows the schematic and Figure 2.57a the probe.

When a high level of *E* field is required, the use of a step-up transformer reduces the output voltage needed from the power amplifier for a given *E* field.

When high currents are required in a magnetic field antenna, a step-down transformer may be used and may more closely match the load to the power amplifier.

FIGURE 2.51 *H*-field and *E*-field schematic.

FIGURE 2.52 Schematic of the injection coil in a resonant circuit.

A suitable transformer was built using a ferrite pot core. A single turn was used on one side and three turns on the other. This can be used as a step-up or step-down determined by which side is connected to the source. Figure 2.53 shows the circuit and the primary and secondary inductances and Figure 2.57i the transformer.

The transformer was tested with the results shown in Figures 2.54 and 2.55.

When used as a step-up, the load was 450 Ω and when used as a step-down 5.6 Ω.

The measured results are shown in Figures 2.52 and 2.53.

The probes shown in Figure 2.57a through g may be used for *E*-field or *H*-field injection or as an *E*- or *H*-field measurement antenna.

Figure 2.57a is the series resonant circuit, Figure 2.57b is the 50/60 Hz *E*-field probe, and Figure 2.57c is the 50/60 Hz *H*-field probe.

The *H* field probe in Figure 2.57a may also be used as a broadband injection probe when the capacitor is shorted and the calibration is shown in Figure 2.56. The probe is capable of higher *H* fields at all frequencies. For example, the maximum with a 75 W power amplifier at 30 MHz is typically 1.5 A/m.

FIGURE 2.53   Schematic of step-up/step-down transformer.

FIGURE 2.54   Frequency response when used as a step-up transformer.

FIGURE 2.55   Frequency response when used as a step-down transformer.

FIGURE 2.56   Calibration of probe in Figure 2.57a

The 0.1 m stripline in Figure 2.57e is capable of *E* fields as high as 614 V/m when used with a step-up transformer, and the calibration is shown in Table 2.2. The same calibration was made with the 8 mm stripline shown in Figure 2.57d without a transformer, and here, the maximum *E* field is limited by the maximum breakdown voltage between the upper plate and the ground plane. The breakdown could occur theoretically at 1500 V rms depending on the shape of the edge of the upper strip, the humidity, and the dust particles.

FIGURE 2.57   (a) *H*-field resonant probe. (b) 50/60 Hz *E*-field parallel plate antenna. (c) 50/60 Hz solenoid. (d) 8 mm stripline antenna.   (*Continued*)

FIGURE 2.57 (*Continued*)   (e) 10 cm stripline antenna. (f) High-frequency *H*-field antenna. (g) 16 cm stripline antenna. (h) Calibration jig for 0.1–1000 MHz *H*-field probe. (i) Step-up/step-down transformer.

TABLE 2.2

0.1 m Stripline and 8 mm Stripline Calibration

Frequency [MHz]	*E*-Field Strength Rms [V/m]
0.3–30	614
3.6	510
5	370
10	184
15	123
30–300	62.4

The 0.1 m diameter high-frequency *H*-field injection loop, shown in Figure 2.57f, has been calibrated to generate from 0.49 to 0.1163 A/m from 10 to 300 MHz, although higher levels are possible above 30 MHz, typically 2 A/m.

2.7 NONIONIZING ELECTROMAGNETIC FIELD EXPOSURE SAFETY LIMITS

The term *nonionizing radiation* is used to denote radiation from such sources as microwave ovens, antennas, and wires carrying high-RF currents. Infrared and visible light are also nonionizing radiation. Ionizing radiation, which includes x-rays, gamma rays, and electromagnetic particles, is so named because it can cause ionization of gas molecules, whereas nonionizing radiation cannot.

The maximum allowable or recommended levels of exposure to nonionizing radiation are not the same for every country or for different organizations within the same country. One reason for the inconsistency may be that, unlike ionizing radiation, the amount of data on the biological effects on human beings and other warm-blooded creatures is limited; also, different criteria are used by the various regulatory organizations to arrive at safe levels. Generally speaking, the effects of nonionizing radiation on the living tissue depend upon the wavelength (frequency), which, with the conductivity of the tissue, determines the depth of tissue penetration. As a general guideline for those working with high-power microwave radiation, special precautions must be taken at frequencies where an ungrounded person is half a wavelength or less from a radiating source.

Caution: Although every effort has been made to verify the accuracy and the currency of the safety levels published in this section, they are intended for information only. If any of these limits are to be applied, the responsible organization must be contacted to obtain the current frequency band and levels. Also, separate limits apply to controlled environments and to RF workers.

2.7.1 CLINICAL STUDIES ON HUMAN BEINGS

Since the first edition of this book, many articles and books have been published on human exposure to electromagnetic fields, especially the fields generated by 50–60 Hz power lines. Many of the data are contradictory, and so a review of the data is well beyond the scope of this edition. Information on this topic may be obtained from the Internet and from the database used in the IEEE C95.1—1991 document.

2.7.2 CANADIAN LIMITS

In the early limits, both Health Canada and the IEEE American National Standard confined themselves to the thermal effect in the interaction of RF and microwave fields with biological systems.

The basal metabolic rate (BMR), or the dissipation of an average man resting, is 100 W, and this figure can double or even triple during activity. It was assumed that the human body could easily cope with a 100% increase in the BMR (i.e., an additional 100 W). The earliest widely adopted limit in Western Europe, the United States, and Canada was 10 mW/cm^2 based on the exposure of one-half the cross-sectional area of an average man to an incident plane wave with a field intensity of 10 mW/cm^2. This was chosen because it is the level sufficient to increase the dissipation by an additional 100 W.

By calculation and experimentation, using saline-filled human phantoms, it was subsequently found that the average-sized man, grounded, resonates at 31–34 MHz (the resonant frequency for a 6 ft-high shorted monopole). The resonance frequency is, of course, higher for children. Taking into account the resonance and other frequency effects as well as the possibility of reflection [8], the Canadian exposure limit for the general public was revised down to 1 mW/cm^2 continuous in the frequency range 3 MHz–100 GHz. In the latest version of this document [10], the power limit is further reduced to 2 mW/m^2 (0.2 mW/cm^2) from 30 to 300 MHz and is f (MHz)/150 (mW/m^2) from 300 to 1,500 MHz and 10 mW/m^2 from 1,500 to 300,000 MHz.

TABLE 2.3

Canada Safety Code 6, 1999: Exposure Limits for Persons Other than RF Workers (the General Public)

1 Frequency [MHz]	2 Electric Field Strength Rms [V/m]	3 Magnetic Field Strength Rms [A/m]	4 Power Density [W/m²]	5 Averaging Time [min]
0.003–1	280	2.19		6
1–10	280/f	2.19/f		6
10–30	28	2.19/f		6
30–300	28	0.073	2[a]	6
300–1,500	1.585$f^{0.5}$	0.0042$f^{0.5}$	f/150	6
1,500–15,000	61.4	0.163	10	6
15,000–150,000	61.4	0.163	10	616.000/$f^{1.2}$
150,000–300,000	0.158$f^{0.5}$	4.21 × 10⁻⁴$f^{0.5}$	6.67 × 10⁻⁵f	616.000/$f^{1.2}$

Notes:

1. Frequency, f, is in MHz.
2. A power density of 10 W/m² is equivalent to 1 mW/cm².
3. A magnetic field strength of 1 A/m corresponds to 1.257 μT or 12.57 mG.

[a] Power density limit is applicable at frequencies greater than 100 MHz.

The detailed derivation of the limits in Reference [10] that are reproduced in Table 2.3 for the general public has not been provided. Reference [10] contains different exposure limits for RF workers and persons other than RF workers. The power limit shall not be exceeded when averaged over any 0.1 h period and when spatially averaged.

Where the electromagnetic radiation consists of frequencies from more than one frequency band in the first column of Table 2.3, the fraction of the actual radiation (power density or square of the field strength) in a frequency band relative to the value given in column 2, 3, or 4 shall be determined and the sum of all fractions of all frequency bands shall not exceed unity when time and spatially averaged.

Time-averaged values need only be calculated from multiple measurements if the field changes significantly (more than 20%) within a period of 0.1 h; otherwise, a single measurement suffices. Reference [10] provides formulas for time averaging. Spatial averaging refers to the uniformity of the exposure field. For portable transmitters and other devices that produce highly spatially nonuniform fields, the levels in Table 2.3 may be exceeded, but the following values shall not be exceeded:

Conditions [W/kg]	SAR Limit
The SAR[a] averaged over whole-body mass	0.08
The local SAR[a] for the head, neck, and trunk, averaged over any 1 gram [g] of tissue	1.6
The SAR[a] in the limbs, as averaged over 10 g of tissue	4

[a] SAR, specific absorption rate = power absorbed per weight of tissue, in watts per kilogram.

Although not a requirement of the code, it is suggested that whenever possible, the organ-averaged SAR for the eye shall not exceed 0.2 W/kg.

In addition, a limit is placed on contact current. For free standing individuals (no contact with metallic objects), the current induced into the human body by electromagnetic radiation shall not exceed the value in column 2 of Table 2.4. No object with which a person may come in

TABLE 2.4

Canada Safety Code 6: Induced and Contact Current Limits for Persons Other than RF Workers (the General Public)

1	2 Rms Induced Current [mA] through		3	4
Frequency [MHz]	Both Feet	Each Foot	Rms Contact Current [mA] Hand Grip and through Each Foot	Averaging Time
0.003–0.1	900*f*	450*f*	450*f*	1 s
0.1–110	90	45	45	0.1 h (6 min)

Notes:

1. Frequency, *f*, is in MHz.
2. The limits mentioned earlier may not adequately protect against startle reactions and burns caused by transient spark discharges for intermittent contact with energized objects.

contact will be energized by electromagnetic radiation in the frequency band listed in column 1 of Table 2.4 to such an extent that the current flowing through a circuit having the impedance of the human body exceeds the value given in column 3 of Table 2.4 as measured with a contact current meter, where the electromagnetic radiation consists of frequencies from more than one frequency band of column 2 of Table 2.4, the ratio of the square of the measured current at each frequency to the square of the limit at that given frequency shown in column 2 or 3 (whichever is applicable).

2.7.3 AMERICAN STANDARDS

One of the present standards is contained in Reference [9], IEEE C95.1—1991. The present standards include revisions in the limits from the 1982 to the 1998 version based on additional data. From 1988 to 1991, the revisions resulted in an expanded frequency range, limits on induced body current to prevent RF shock or burn, a relaxation of limits on exposure to magnetic fields at low frequencies, and exposure limits and averaging time at high frequencies that are compatible at 300 GHz with existing infrared maximum permissible exposure (MPE) limits.

Different MPEs apply to controlled environments, where exposure is incurred by persons who are aware of the potential for exposure, such as RF workers, and to uncontrolled environments, where exposure to individuals who have no knowledge or control of their exposure occurs, such as the general public. The following limits are for the uncontrolled environment.

The MPEs refer to exposure values obtained by spatially averaging over an area equivalent to the vertical cross section of the human body (projected area). In the case of partial-body exposure, Reference [9] describes some relaxation. In nonuniform fields, spatial peak values of field strengths may exceed the MPEs if the spatially averaged value remains within the specified limits. The MPEs may also be relaxed by reference to SAR limits in Reference [9]. For near-field exposure at frequencies less than 300 MHz, the applicable MPE is in terms of rms electric and magnetic field strength. The MPE refers to values averaged over any 6–30 min period for frequencies up to 3000 MHz and over shorter periods for higher frequencies down to 10 s at 300 GHz, as indicated in Table 2.5.

For mixed or broadband fields at a number of frequencies for which there are different values of the MPE, the fraction of the MPE (in terms of E^2, H^2, or power density [S]) incurred within each frequency interval should be determined, and the sum of all such fractions should not exceed unity. In a similar manner for mixed or broadband induced currents, the fraction of the induced current limits (in terms of I^2) incurred within each frequency interval should be determined, and

TABLE 2.5

Action Level (Maximum Permissible Exposure for the General Public When a Radio Frequency Safety Program Is Unavailable)

1 Frequency Range [MHz]	2 Electric Field Strength (E)[a] [V/m]	3 Magnetic Field Strength (H)[a] [A/m]	4 Power Density (5) E Field, H Field [mW/cm²]	5 Averaging Time[b] [min] |E²|, S, or |H²|	
0.1–1.34	614	16.3/fM	(1000, 100,000/fM^2)[c]	6	6
1.34–3	823.8/fM	16.3/fM	(1800/fM^2, 100,000/fM^2)	fM^2/0.3	6
3–30	823.8/fM	16.3/fM	(1800/fM^2, 100,000/fM^2)	30	6
30–100	27.5	158.3/$fM^{1.668}$	(2, 9,400,000/$fM^{3.336}$)	30	0.0636$f^{1.337}$
100–400	27.5	0.0729	2	30	30
400–2,000	—	—	fM/200	30	
2,000–5,000	—	—	10	30	
5,000–30,000	—	—	10	150/fG	
30,000–100,000	—	—	10	25.24/$fG^{0.476}$	
100,000–300,000	—	—	(90fG–7000)/200	5048/[(9fG–700)$fG^{0.476}$	

Source: IEEE C95.1, IEEE standard for safety levels with respect to human exposure to radio frequency electromagnetic fields, 3 kHz to 300 GHz, 1991.

Notes:

1. fM is frequency in MHz.
2. fG is frequency in GHz.

[a] For exposures that are uniform over the dimensions of the body, such as certain far-field plane wave exposures, the exposure field strengths and power densities are compared with the MPEs in the table. For nonuniform exposures, the mean values of exposure fields, as obtained by spatially averaging the squares of the field strengths or averaging the power densities over an area equivalent to the vertical cross section of the human body (projected area) or a smaller area depending on the frequency (see Notes to Tables 8 and 9 in IEEE Std C95.1), are compared with the MPEs in the table.

[b] The left column is the averaging time for |E²|; the right column is the averaging time for |H²|. For the frequencies greater than 400 MHz, the averaging time is for power density S.

[c] These plane wave equivalent power density values are commonly used as a convenient comparison with MPEs at higher frequencies and are displayed on some instruments in use.

the sum of all such fractions should not exceed unity. For exposures to pulsed RF fields in the range 0.1–300,000 MHz, the peak (temporal) value of the MPE in terms of the E field is 100 kV/m.

For exposures to pulsed RF fields of pulse durations less than 100 ms and for a single pulse, the MPE is given by the MPE from Table 2.5 (E field equivalent power density) multiplied by the averaging time, in seconds, and divided by five times the pulse width, in seconds. That is,

$$\text{Peak MPE} = \frac{\text{MPE} \times \text{Avg time (s)}}{5 \times \text{Pulsewidth (s)}}$$

A maximum of five such pulses, with a pulse repetition rate of at least 100 ms, is permitted during any period equal to the averaging time. If there are more than five pulses during any period equal to the averaging time, or if the pulse durations are greater than 100 ms, normal averaging time calculations apply, except that during any 100 ms period, the energy density is limited by the preceding formula, namely,

$$\text{Peak MPE} \times \text{Pulsewidth (s)} = \frac{\text{MPE} \times \text{Avg time (s)}}{5}$$

At frequencies between 100 kHz and 6 GHz, the MPE in uncontrolled environments for electromagnetic field strengths may be exceeded if the exposure conditions can be shown by appropriate techniques to produce SARs below 0.08 W/kg, as averaged over the whole body, and spatial peak SAR values not exceeding 1.6 W/kg, as averaged over any 1 g of tissue (defined as a tissue volume in the shape of a cube), except for the hands, wrists, feet, and ankles, where the spatial peak SAR will not exceed 4 W/kg, as averaged over any 10 g of tissue, and the induced currents in the body conform to Table 2.4.

A second source of limits is the FCC RF exposure guidelines contained in 47 C.F.R. 1.1310, Reference [17]. The criteria listed in Table 2.6 will be used to evaluate the environmental impact of human exposure to RF radiation as specified in 1.1307(b), except in the case of portable devices, which shall be evaluated according to the provisions of 2.1093 of the 47 C.F.R. document. Further information on evaluating compliance with these limits can be found in the FCC's OET/OET Bulletin Number 65, "Evaluating Compliance with FCC-Specified Guidelines for Human Exposure to Radiofrequency Radiation." The FCC limits are generally based on recommended exposure guidelines published by the National Council on Radiation Protection and Measurements (NRCP) in "Biological Effects and Exposure Criteria for Radiofrequency Electromagnetic Fields" (NCRP Report No. 86, Sections 17.4.1, 17.4.1.1, 17.4.2, and 17.3. Copyright NRCP, 1986, Bethesda, MD, 20814). In the frequency range from 100 to 1500 MHz, exposure limits are also generally based

TABLE 2.6

International Commission on Nonionizing Radiation Protection Reference Levels for General Public Exposure to Time-Varying Electric and Magnetic Fields[a]

Frequency Range	E-Field Strength [V/m]	H-Field Strength [A/m]	B Field [μT]	Equivalent Plane Wave Power Density S_{eq} [W/m]
1–8 Hz	5000	$3.2 \times 10^4/f^2$	$0.04/f^2$	—
8–25 Hz	5000	$4{,}000/f$	$0.005/f$	—
25–50 Hz	5000	160	2×10^{-4}	—
50–400 Hz	$2.5 \times 10^5 f$	160	2×10^{-4}	—
0.4–3 kHz	$2.5 \times 10^5 f$	$64{,}000/f$	0.08	—
0.003–10 MHz	83	21	2.7×10^{-5}	—
10–400 MHz	28	0.073	0.092	2
400–2000 MHz	$1.375 f^{1/2}$	$0.0037 f^{1/2}$	$0.0046 f^{1/2}$	$f/200$
2–300 GHz	61	0.16	0.20	10

[a] *Note:*

1. *f* is as indicated in the frequency range column.
2. Provided that the basic restrictions are met and adverse indirect effects can be excluded, field strength values can be exceeded.
3. For frequencies between 100 kHz and 10 GHz, S_{eq}, E^2, H^2, and B^2 are to be averaged over any 6 min period.
4. For peak values at frequencies up to 100 kHz (see Table 4, note 3 in the ICNIRP Guidelines for limiting exposure to time-varying electric, magnetic, and electromagnetic fields).
5. For peak values at frequencies exceeding 100 kHz (see Figures 1 and 2 in the ICNIRP Guidelines for limiting exposure to time-varying electric, magnetic, and electromagnetic fields) between 100 kHz and 10 MHz, peak values for the field strengths are obtained by interpolation from the 1.5-fold peak at 100 kHz to the 32-fold peak at 10 MHz. For frequencies exceeding 10 MHz, it is suggested that the peak equivalent plane wave power density, as averaged over the pulse width, does not exceed 1000 times the S_{eq} restrictions, or that the field strength does not exceed 32 times the field strength exposure levels given in the table.
6. For frequencies exceeding 10 GHz, S_{eq}, E^2, H^2, and B^2 are to be averaged over any $68/f^{1.05}$ min period (*f* in GHz).
7. No *E*-field value is provided for frequencies <1 Hz, which are effectively static electric field; perception of surface electric charges will not occur at field strengths less than 25 kV/m. Spark discharges causing stress or annoyance should be avoided.

on guidelines recommended by the ANSI in Section 4.1 of IEEE Standard for Safety Levels with Respect to Human Exposure to Radio Frequency Electromagnetic Fields, 3 kHz–300 GHz, ANSI/IEEE C95.1—1992.

2.7.4 EAST EUROPEAN, EUROPEAN, AND OTHER STANDARDS

The philosophy of the standard setting in East European countries, at least in 1977, appears to be that any effect due to microwave radiation, whether subjective, objective, or reversible, should be avoided, and thus, the Russian and Czechoslovakian standards were as low as 1 $\mu W/cm^2$ (i.e., 1/2000th of the Canadian level).

In May 1997, the Hungarian Standards Institution released the European Pre-Standard (ENV 50166-1/2) as Hungarian Pre-Standard MSZ ENV 50166-1/2. It is the general policy of the Hungarian Parliament that confirmed EU standards shall be considered as Hungarian National Standards.

The Russian limits have remained the same for more than 40 years and are based on intensity and time of exposure: 10 $\mu W/cm^2$ for 8 h, 100 $\mu W/cm^2$ up to 2 h, and 1000 $\mu W/cm^2$ for up to 20 min per working day.

New general public limits were issued in Poland in August 1998 and, although increased, are still lower than the proposals of the European Committee on Electrotechnical Standardization (CENELEC), International Radiation Protection Association (IRPA), or ANSI for the same frequency range (0.1 Hz–300 GHz). It would almost certainly be difficult to increase the public values too drastically, at least initially.

The European Commission has proposed basic restrictions on SAR designed to prevent whole-body heat stress and excessive localized heating of tissues. The basic restrictions are set to account for uncertainties related to individual sensitivities and environmental conditions and for the fact that the age and health status of members of the public varies. The SAR limits over the 10 MHz–10 GHz frequency range are 0.08 W/kg for the whole body; the localized SAR for the head and trunk is 0.08 W/kg and for the limbs 4 W/kg. The CENELEC accepted the ENV 50166 at the end of November 1994 for provisional application for three years, which has been extended to the year 2000.

In 1997, the 26th EMF Ordinances came into effect in Germany and are based on the IRPA, WHO, and International Commission on Non-Ionizing Radiation Protection (ICNIRP) guidelines. The EMF Ordinances is limited to the frequency range 10–300,000 MHz at the levels shown in Table 2.6.

In the United Kingdom, the National Radiological Protection Board revised its advice in 1993 and published the new advice 8.

In Japan, the Ministry of Posts and Telecommunications (MPT) issued revised rules to establish Radio Radiation Protection for Human Exposure to Electromagnetic Fields from 10 kHz to 300 GHz. These guidelines contain levels that are a mix of the Canada safety code 6 and the IEEE C95.1 but are not the same as either. In 1997, the MPT slightly revised the guidelines in changing the local SAR limit from 1.6 W/kg for 1 g of tissue to 2 W/kg for 10 g of tissue for the general environment.

The Australasian Standards Committee created a joint Australian and New Zealand standard and published a draft of NZS/AS2772 Part 1 in 1998. At present, the UHF reference level remains capped at 200 $\mu W/cm^2$, which is more conservative than the ICNIRP that allows 450 $\mu W/cm^2$ at 900 MHz.

The Italian limits for the general public are 60 V/m and 0.2 A/m from 100 kHz to 3 MHz, 20 V/m and 0.05 A/m from 3 MHz to 3 GHz, and 40 V/m and 0.1 A/m in the frequency range 3–300 MHz. These limits are to be averaged over 6 min. "Precautionary levels" are 6 V/m and 0.016 A/m for any frequency.

2.7.5 ICNIRP, CENELEC, IRPA, AND CEU LIMITS

In 1998, the ICNIRP published guidelines for limiting exposure to time-varying electric, magnetic, and electromagnetic fields. They concluded that any established biological and health effects in the frequency range from 10 MHz to a few GHz are consistent with responses to a body temperature rise of more than 1°C. This level of temperature increase results from exposure of individuals

TABLE 2.7

The European Committee on Electrotechnical Standardization ENV 50166, International Radiation Protection Association, and Commission of the European Communities Guidelines for Workers (Not the General Public) over a Full Working Day

(a) Power Frequency 50/60 Hz	CENELEC	IRPA	CEU
Induced current density, head and trunk [mA/m^2]	10	10	10
Electric field strength [kV/m]	10	10	19.6;12.3;6.1
Magnetic flux density [mT]	1.6	0.5	0.64;0.4;0.2
Contact current [mA]	3.5		1.5
(b) High Frequencies	CENELEC	IRPA	CEU
Frequency range [Hz]	1.0×10^4	1.0×10^7	1.0×10^5
	3.0×10^{11}	3.0×10^{11}	3.0×10^{11}
Specific absorption rate SAR (6 min) [W/kg]			
Whole body	0.4	0.4	0.4
Extremities (averaged over 10 g tissue)	20	20	20
Head and trunk (average over 10 g)	10	10	10
Peak specific absorption rate SA [mJ/kg]	10		10
Contact current (0.1–3 (100) MHz) [mA]	35	(50)	50

under moderate environmental conditions to a whole-body SAR of approximately 4 W/kg for about 30 minutes. A whole-body average of 0.4 W/kg has been chosen as the restriction that provides adequate protection for occupational exposure. An additional safety factor of 5 has been introduced for exposure of the public, giving an average whole-body SAR limit of 0.08 W/kg. The lower basic restriction for exposure of the general public takes into account the fact that their age and health status may differ from those of workers. The reference levels (*E*, *H*, and power density) are obtained from the basic SAR restrictions by mathematical modeling and by extrapolation from the results of laboratory investigations at specific frequencies. These reference levels for the general public are provided in Table 2.6. The ICNIRP developed basic restriction guidelines with the IRPA, and these are shown in Table 2.7 for the exposure of workers.

The guideline limit values for the CENELEC contained in the European Pre-Standard ENV 50166, the IRPA, and the Commission of the European Communities (CEU) are provided in Table 2.7. These values of SAR are applicable to the exposure of workers only, over a full working day, and include limits on the exposure to fields at 50/60 Hz.

However, in common with all other organizations, the committee did not take into account the effects of

1. Drugs
2. Chemical or physical agents in the environment
3. Modulated microwave effects
4. Possible nonthermal very-long-term effects
5. High-level low-repetition-rate sources

2.7.6 MEASUREMENT OF ELECTROMAGNETIC FIELD LEVELS

A broadband isotropic probe is an instrument designed for the measurement of EM fields. However, such instruments are calibrated in the far field (i.e., illuminated with a near plane wave).

When measuring in the near field, it is important to use the correct probe, and a magnetic (*H*) field or electric (*E*) field probe is available for most instruments. The near-field/far-field transition for radiation from a point source is defined as $\lambda/2\pi$; however, for an antenna, the transition is dependent on antenna aperture. For example, the transition is at 25 ft for a specific helical antenna [11] that has an operating frequency of 250–400 MHz.

As discussed in preceding sections and of use as a guideline only, the type of field generated by a dipole is predominantly an E field, as is the field from a current-carrying cable, and the field from a loop is predominantly an H field. In the case of one type of helical antenna, neither the electric nor the magnetic fields predominate in the proximity of the antenna; therefore, either an E-field or H-field probe may be used for measurements [11]. To illustrate the importance of using the correct probe: If a magnetic field probe is used to measure the near field of a dipole and it registers the maximum recommended level of magnetic field (i.e., 0.073 A/m) and we assume at the measurement location that the wave impedance Z_W is 6000 Ω, then the electric field would be

$$Z_W \times A/m = V/m$$

$$6000 \ \Omega \times 0.073 \ A/m = 438 \ V/m \quad \text{(i.e., 16 times the recommended level)}$$

Potentially hazardous fields can be generated from unlikely, or at least unexpected, sources. In the course of EMI investigations, a broadband E field of 200 V/m from a shielded box (without apertures) and its power cable have been measured. Similarly, narrowband magnetic fields as high as 7 A/m at 30 MHz have been measured in the vicinity of an instrument containing a plasma source.

2.7.7 DC AND POWER FREQUENCY FIELDS

Those living in an industrial society are constantly exposed to electric and magnetic fields from power wiring, appliances, and high-voltage transmission lines at frequencies in the 50–60 Hz range. For the sake of efficiency, overhead power lines have increased in voltage over the years, with a recent increase from 345 to 765 kV. As a result of public concern, a number of studies on possible hazards have been undertaken.

It is clear from earlier studies that 60 Hz electric and magnetic fields can affect biological systems; however, due to the conflicting results of the recent studies on the effects on humans, the author considers that no conclusions can be drawn at the time of publication of this book. Health and Welfare Canada does not specify a maximum exposure to magnetic fields below 10 kHz. However, a contact there has provided a reference to an IRPA report entitled "Interim Guidelines for Exposure to 60 Hz Magnetic Fields," published in the *Health Physics Journal* in 1990. The recommended maximum exposure over a period of 24 h is 1000 mG = 79.36 A/m. (Flux density limits in gauss or tesla can be converted to magnetic field strength when the relative permeability is 1. The relationship is $1 \ T = 7.936 \times 10^5 \ A/m$ and $1 \ T = 10^4 \ G$. Therefore, $1 \ G = 79.36 \ A/m$.)

The ICNIRP has provided "Guidelines on Limits of Exposure to Static Magnetic Fields." These guidelines have resulted in the limits for exposure to static magnetic fields shown in Table 2.8. The guidelines state that those with cardiac pacemakers should be discouraged from inadvertently entering areas with fields large enough in dimension to include most of a person's trunk at magnetic flux densities greater than 0.5 mL (396 A/m).

Reports [12–14] have raised the issue of the extremely high LF or DC electric fields (up to 10,000 V/M) in the proximity of video display terminals (VDTs), whereas one report [13] indicates that ionizing radiation measured from a number of VDTs is at an insignificantly low level. Several manufacturers have produced conductive antistatic plastic material that may be placed under the bezel surrounding a cathode ray tube screen. One disadvantage of this material is that when the screen is curved, wrinkles form in the material. Another disadvantage with the use of this material

TABLE 2.8

International Commission on Nonionizing Radiation Protection Limits for Exposure[a] to Static Magnetic Fields

Exposure Characteristics	Magnetic Flux Density
Occupational[b]	
Whole working day (time-weighted average)	200 mT (1.58×10^5 A/m)
Head and trunk	2 T
Limbs[c]	8 T
General Public[d]	
Continuous exposure	40 mT (3.17×10^4 A/m)
Any part of the body	

[a] ICNIRP recommends that these limits should be viewed operationally as spatial peak exposure limits.
[b] For specific work applications, exposures up to 8 T can be justified, if the environment is controlled and appropriate work practices are implemented to control movement-induced effects.
[c] Not enough information is available to base exposure limits beyond 8 T.
[d] Because of potential indirect adverse effects, ICNIRP recognizes that practical policies need to be implemented to prevent inadvertent harmful exposure of persons with implanted electronic medical devices and implants containing ferromagnetic material, and dangers from flying objects, which can lead to much lower restriction levels such as 0.5 mT.

is reflections, which can be alleviated if the material is antiglare. The antistatic material is ineffective unless it is grounded to some metal structure within the monitor that is in turn connected to AC safety ground. As typically only one side of the material is conductive, it is important to choose the correct side when making the ground connection.

Far better are antiglare/antistatic screens that are specifically designed for different makes of monitor. The screens are available from computer accessory suppliers and, according to the advertising material from one manufacturer, reduce the field strength by 80%–90%.

2.8 COMPUTER PROGRAMS

The following computer programs were written using Microsoft's QuickBASIC Version 3.0. Unlike other common versions of compiled BASIC, the QuickBASIC compiler does not require line numbers and allows the use of alphanumeric labels.

The programs incorporate some of the equations contained in this chapter and Chapter 7 and are useful in EMC prediction and problem solving as well as antenna calibration using a PC. The programs are available as freeware from EMC Consulting Inc.

2.8.1 COMPUTER PROGRAMS FOR RADIATION FROM WIRES

These programs may be used for the calculation of the electric field or magnetic field or power radiated from a wire or cable, depending on the equations used in the program. The equations cover the radiation from a current element or electric dipole, a current loop, and a wire above a ground plane. These equations may be used to model the following cable configurations:

1. An electrically short length of wire/cable either that is some distance from a ground plane or for which the fields in close proximity are required. The wire/cable must be disconnected at one or both ends.

2. A wire or cable forming a loop with either a return conductor or its electromagnetic image in a ground plane. The length of the wire and the distance between the wire and the return path must be much less than a wavelength. When modeling the wire or cable above a ground plane, enter twice the height of the wire or cable above a ground plane for the width of the frame antenna.

3. A transmission line comprising a wire/cable and a return path. The return path may be either a conductor or formed by the wire/cable electromagnetic image in a ground plane. The cable/wire must be of a resonant length, that is, 1/4, 3/4, 1-1/4, ... wavelength when only one end is connected to ground and 1/2, 1, 1-1/2, ... wavelength when both are connected to ground. The distance between the wires/cables must be much less than a wavelength. When the return path is the electromagnetic image in a ground plane, use twice the height of the wire above the ground plane when entering the distance between conductors.

2.8.1.1   Computer Program for the Electric Field, Magnetic Field, and Wave Impedance for a Current Element (Electric Dipole)

```
'Initialize constants
C=3E+08
PI=3.14159
'Initialize input variables
F=2000: CUR=1: R = 1: THETA=90
GOTO ComputeCurrentElement:
CurrentElementMenu:
CLS
PRINT
PRINT "CURRENT ELEMENT
PRINT "Schelkunoff & Friis, Antennas, Theory and Practice
PRINT "Wiley, 1952, p120
PRINT
PRINT "INPUTS:"
PRINT "[F] FREQUENCY (F):",F;" Hz"
PRINT "[C] LOOP CURRENT (1): ",CUR;" amps"
PRINT "[D] DISTANCE FROM ELEMENT:",R;" meters"
PRINT "[A] ANGLE:",THETA;" degrees"
PRINT "[X] End program"
PRINT "Which variable do you wish to change (F,C,etc)?"
PRINT
PRINT "OUTPUTS:"
PRINT "MAGNETIC FIELD     :",HH,"A/m"
PRINT "ELECTRIC FIELD     :",EE,"V/m"
PRINT "WAVE IMPEDANCE     :",ZW,"Ohms"
CurrentElementLoop:
SEL$=1NKEY$
IF LEN(SEL$)>0 THEN LOCATE 12,1: PRINT SPACE$(70): LOCATE 12,4
IF (SEL$="F") OR (SEL$="f") THEN INPUT" Enter frequency ",F
IF (SEL$="C") OR (SEL$="c") THEN INPUT" Enter loop current ",CUR
IF (SEL$="D") OR (SEL$="d") THEN INPUT" Enter distance from element ",R
IF (SEL$="A") OR (SEL$="a") THEN INPUT" Enter angle ",THETA
IF (SEL$="X") OR (SEL$="x") THEN END
IF LEN(SEL$)>0 THEN GOTO ComputeCurrentElement:
GOTO CurrentElementLoop:
ComputeCurrentElement:
```

```
LAMBDA=C/F : BETA=2*PI/LAMBDA
ERE=1-LAMBDA^2/(4*PI^2*R^2): EIM=LAMBDA/(2*PI*R)
EMAG = SQR(ERE^2 + EIM^2)
ET =377*CUR/(2*LAMBDA *R)*EMAG*SIN(THETA *PI/180)
XMAG=SQR(1 +(LAMBDA/(2*PI*R)Y2)
HH = CUR/(2*LAMBDA*R)*XMAG*SIN(THETA *PI/180)
ER =377*CUR/(2*PI*R"2)*XMAG*COS(THETA *PI/180)
IF THETA=90 THEN ZW=ET/HH ELSE ZW=O
EE = SQR(ET^2 + ER^2)
GOTO CurrentElementMenu:
```

2.8.1.2　Computer Program for the Electric Field, Magnetic Field, and Wave Impedance for a Current Loop (Frame Antenna)

```
'Initialize constants
C=3E +08: RC=377: PI =3.14159
'Initialize variables
F=20000: 1=1: L=1: W=1: R=1: THETAD=90
GOTO ComputeCurrentLoop:
CurrentLoopMenu:
CLS
PRINT
PRINT "ELECTRIC FIELD FROM FRAME ANTENNA"
PRINT "Schelkunoff, Antennas, Theory and Practice, pg. 320"
PRINT
PRINT "INPUTS:"
PRINT "[F] FREQUENCY:",F;" Hz"
PRINT "[C] LOOP CURRENT:",!;" amps"
PRINT "[L] LENGTH OF FRAME ANTENNA:",';" m"
PRINT "[W] WIDTH OF FRAME ANTENNA:",W;" m"
PRINT "[D] DISTANCE FROM FRAME TO MEAS. PT.:",R;" m"
PRINT "[A] ENTER FRAME ANGLE:",THETAD;" degrees"
PRINT "[X] End program
PRINT "Which variable do you wish to change (F,C,etc)?"
PRINT
PRINT "OUTPUTS:"
PRINT "ELECTRIC FIELD=",EPHI;" V/m"
PRINT "MAGNETIC FIELD=",HMAG;" A/m"
PRINT "WAVE IMPEDANCE=",ZW;" Ohms"
'
CurrentLoopLoop:
SEL$=1NKEY$
IF LEN(SEL$)>0 THEN LOCATE 13,1
IF LEN(SEL$)>0 THEN PRINT SPACE$(70)
LOCATE 13,4
IF (SEL$="C") OR (SEL$="c") THEN INPUT" Enter loop current", I
IF (SEL$="F") OR (SEL$="f") THEN INPUT" Enter frequency ",F
IF (SEL$="L") OR (SEL$="1") THEN INPUT" Enter length of frame antenna ",L
IF (SEL$="W") OR (SEL$="w") THEN INPUT" Enter width of frame antenna ",W
IF (SEL$="D") OR (SEL$="d") THEN INPUT" Enter distance ",R
IF (SEL$="A") OR (SEL$="a") THEN INPUT" Enter frame angle ",THETAD
IF (SEL$="X") OR (SEL$="x") THEN END
IF LEN(SEL$)>0 THEN GOTO ComputeCurrentLoop:
GOTO CurrentLoopLoop:
ComputeCurrentLoop:
```

```
S=L*W
BETA=2*PI*F/C
THETAR=THETAD*PI/180
EPHI = RC*BETA^2*1*S*SQR(1 + (1/(BETA*R)^2))*SIN(THETAR)/(4*PI*R)
HTHETA      BETN2*1*S*SQR((1-
(1/(BETA*R)^2))^2+(1/(BETA*R)^2))*SIN(THETAR)/(4*PI*R)
HR=BETA*I*S*SQR(1+1/(BETA*R)^2)*COS(THETAR)/(2*PI*R^2)
HMAG=SQR(HTHETA^2+HR^2)
IF THETAD=90 THEN ZW=EPHI/HMAG ELSE ZW=O
GOTO CurrentLoopMenu:
```

2.8.1.3 Computer Program for the Radiation from a Resonant Transmission Line

```
WireAboveGroundPlane:
'CLS:LOCATE 4,1
'PRINT "RADIATION FROM A TRANSMISSION LINE"
'PRINT
'PRINT" P=30*BETA^2*BB^2*1^2"
'PRINT " H=SQR(P*K/(4*PI*RA2*ZW}"
'PRINT" BETA=2*PI*F/C
'PRINT" PI=3.14159
'PRINT" K =DAVID'S CONSTANT= 1.5
'PRINT" SPEED OF LIGHT IN VACUUM, (C = 3E08 m/s)"
'PRINT
'GOSUB Spacebar:
GOTO ComputeWireAboveGroundPlane:
WireAboveGroundPlaneMenu::
CLS: PRINT
PRINT "WIRE ABOVE A GROUND PLANE
PRINT
PRINT "INPUTS:"
PRINT
PRINT "[F] FREQUENCY (F):",F;" Hz"
PRINT "[R] DIAMETER OF WIRE (D):",D;" metres"
PRINT "(C) LINE CURRENT (I):",I;" amps"
PRINT "[D] DISTANCE FROM WIRE TO MEAS. PT.:",R;" metres"
PRINT "[H] DISTANCE BETWEEN CONDUCTORS"
PRINT" OR HEIGHT ABOVE GROUND PLANE:",BB;" metres"
PRINT "[X] Exit to Main Menu"
PRINT "Which parameter would you like to change (A,B, .. )?"
PRINT
PRINT "OUTPUTS:"
PRINT
PRINT "POWER= ",P;"W/m^2"
PRINT "MAGNETIC FIELD= ",H;" A/m"
PRINT "ZW= ",ZW;"Ohms"
PRINT
PRINT
WireAboveGroundPianeLoop::
SEL$=1NKEY$
IF LEN(SEL$)>0 THEN LOCATE 12,1
IF LEN(SEL$)>0 THEN PRINT SPACE$(70)
LOCATE 12,4
IF (SEL$="F") OR (SEL$="f") THEN INPUT" Enter frequency ",F
IF (SEL$="C") OR (SEL$="c") THEN INPUT" Enter loop current ",1
IF (SEL$="R") OR (SEL$="r") THEN INPUT " Enter diameter of wire ",D
```

```
IF (SEL$="D") OR (SEL$="d") THEN INPUT" Enter distance ",R
IF (SEL$="H") OR (SEL$="h") THEN INPUT" Enter distance or height ",BB
IF (SEL$="X") OR (SEL$="x") GOTO MainMenu:
IF LEN(SEL$)>0 GOTO ComputeWireAboveGroundPlane:
GOTO WireAboveGroundPlaneLoop::
ComputeWireAboveGroundPlane:
GOSUB CheckVariables:
K=1.5
LAMBDA=C/F
IF R>LAMBDA/(2*PI} THEN
    ZW=377
ELSE
    ZO= 138*(LOG((4*BB)/D)/LOG(1 0))
    ZW=(R/(LAMBDA/(2*PI}})*(ZO-377)+377
END IF
P=30*BETA^2*BB^2*I^2
H=SQR(P*K/(4*PI*R^2*ZW))
GOTO WireAboveGroundPlaneMenu:
```

2.8.2 COMPUTER PROGRAMS FOR COUPLING TO WIRES/CABLES

These programs calculate the current flow in the victim wire/cable based on an incident *E* or *H* field. The equations used in the programs cover the following cable configurations:

1. A wire or cable forming a loop with either a return conductor or the electromagnetic image of the cable or wire in the ground plane. Where a wire above a ground plane is modeled, enter the height of the cable above the ground plane. For a wire or cable, enter one-half the distance between the two conductors for the height. Where the height (*H*) or length (*L*) is much less than the wavelength, use the inductance of a rectangular loop and enter (*I*) in the program. (*I*) may also be used to calculate the receiving properties of a loop antenna when the load impedance is set to zero. Where the length is much greater than the height and the height is less than half the wavelength, use transmission line theory and enter (*T*) in the program.
2. Where a cable is less than one-fourth of a wavelength in length, is disconnected from ground at one end, and terminates on a metal enclosure at the other end, enter (*M*) in the program. (*M*) may be used to calculate the receiving properties of a nonresonant monopole antenna.

A computer program for the coupling between wires or cables is provided at the end of Chapter 4 on cable coupling and crosstalk.

2.8.2.1 Computer Program for the Receiving Properties of a Loop (I)

```
'Initialize constants
C=3E+08: RC=377:PI=3.14159
'Initialize variables
Z0=50: Z1 =50: H = 1: LL=1: A=.006: F=20000: IO=1
GOTO INDcompute:
InductanceInputs:
CLS
PRINT "LINE CURRENTS USING INDUCTANCE"
PRINT "Equation 8 from Taylor & Castillo, IEEE, Vol. EMC-20, No. 4, Nov. 1978"
PRINT "INPUTS:
PRINT "[F] FREQUENCY",,F;"Hz
PRINT "[S] SOURCE IMPEDANCE",,Z0;"Ohms
PRINT "[T] LOAD IMPEDANCE",,Z1 ;"Ohms
PRINT "[Z] HEIGHT",,"H;metres
```

```
PRINT "[L] LENGTH",„LL;"metres
PRINT "[D] DIAMETER OF WIRE"„A;"metres
IF XX$="?" THEN
    PRINT "[C] CURRENT IN TERMINATION (Amps or?):","?= Refer to OUTPUT"
    PRINT "[H] MAGNETIC FIELD",MF;"Amps/m"
ELSE
    PRINT "[C] CURRENT IN TERMINATION (Amps or ?):", 10;" Amps
END IF
PRINT "[X] End program"
PRINT "Which parameter would you like to change (A,B, .. )?"
PRINT
PRINT
PRINT "OUTPUT:"
PRINT "INDUCTANCE OF RECTANGULAR LOOP= ",L;"Henrys"
PRINT "FLUX DENSITY = ",B;"Webers/mA2"
MF=B/UO
IF XX$="?" THEN PRINT "CURRENT IN TERMINATION (Amps) = ",IO;"A"
IF XX$<>"?" THEN PRINT "MAGNETIC FIELD = ",MF;"A/m"
INDLoop:
SEL$=1NKEY$
IF LEN(SEL$)>0 THEN LOCATE 13,1: PRINT SPACE$(70):LOCATE 13,1
IF (SEL$="S") OR (SEL$="s") THEN INPUT "Enter source impedance ",Z0
IF (SEL$="T") OR (SEL$="t") THEN INPUT "Enter source impedance ",Z1
IF (SEL$="8") OR (SEL$="b") THEN GOTO CNGEXX::
IF (SEL$="Z") OR (SEL$="z") THEN INPUT "Enter height ",H
IF (SEL$="L") OR (SEL$="1") THEN INPUT "Enter length ",LL
IF (SEL$="D") OR (SEL$="d") THEN INPUT "Enter diameter of wire ",A
IF (SEL$="F") OR (SEL$="f") THEN INPUT "Enter Frequency ",F
IF (SEL$="C") OR (SEL$="c") THEN GOTO CNGEXX:
IF (SEL$="H") OR (SEL$="h") THEN INPUT "Enter magnetic field ",MF
IF (SEL$="X") OR (SEL$="x") THEN END
IF LEN(SEL$)>0 THEN GOTO INDcompute:
GOTO INDLoop:
CNGEXX:
INPUT "Enter CURRENT IN TERMINATION (Amps or ?): ",XX$
IF XX$="?" THEN
    LOCATE 13,1
    PRINT "Current value for MAGNETIC FIELD is:";MF;"Amps/m"
    INPUT "Enter new value:";MF
    GOTO INDcompute:
END IF
IO=VAL(XX$)
GOTO INDcompute:
CNGEMF:
    LOCATE 13,1
    PRINT "Current value for MAGNETIC FIELD is:";MF;"Amps/m"
    INPUT "Enter new value:";MF
GOTO INDcompute:
INDcompute:
    UO=4*PI*.0000001
    D=SQR((2*H)^2+LL^2)
    AA=LL*LOG(4*H*LL/(A*(LL+D)))
    BB=2*H*LOG(4*H*LLI(A*(2*H+D)))
    CC=2*D-7/4*(LL+2*H)
    L=UO/PI*(AA+BB+CC)
    ZZ= SQR((2*(ZO+ Z1) Y2 + (2*PI*F*LY2)
```

```
    IF XX$="?" THEN
        B=MF*UO
        IO= B*2*PI*F*4 *H*LL/ZZ
ELSE
        B= IO*ZZ/(2*PI*F*4*H*LL)
    END IF
GOTO InductanceInputs:
```

2.8.2.2 Computer Program for Calculating the Receiving Properties of a Transmission Line (T)

```
'Initialize constants
C=3E+08: RC=377:PI=3.14159
'Initialize variables
IO=1: EO=1: ZI=SO: H=1: F=20000: A=.003: IsItCurrent=1
GOTO ComputeTransmissionLine:
TransmissionLineInputs:
CLS
PRINT "LINE CURRENTS USING TRANSMISSION LINE THEORY"
PRINT "Taylor & Castillo: Eqn. 6,10 from IEEE, Vol. EMC-20, No. 4, Nov,
        1978"
PRINT
IF lsltCurrent= 1 THEN
    PRINT "[A] WIRE CURRENT (Amps or ?) :";10
ELSE
    PRINT "[A] ELECTRIC FIELD (V /m or ?) :";EO
END IF
PRINT "[R] RESISTANCE OF LINE:"; ZI;"Ohms"
PRINT "[H] HEIGHT :";H;"metres"
PRINT "[F] FREQUENCY :";F;"Hz"
PRINT "[D] DIAMETER OF WIRE :";A;"metres"
PRINT "[X] End program
PRINT "Which parameter would you like to change (A,B, .. )?"
PRINT
PRINT
PRINT "OUTPUT:"
IF lsltCurrent=1 THEN PRINT "ELECTRIC FIELD (V/m) = ";EO;" V/m"
IF lsltCurrent=O THEN PRINT "WIRE CURRENT = ";10;" Amps"
PRINT "CHARACTERISTIC IMPEDANCE = ";ZC;"Ohms"
TLoop:
SEL$=1NKEY$
IF LEN(SEL$)>0 THEN LOCATE 10,1: PRINT SPACE$(70): LOCATE 10,1
IF (SEL$="A") OR (SEL$="a") THEN GOTO ALTERXXX:
IF (SEL$="R") OR (SEL$="r") THEN INPUT "Enter impedance of line ",ZI
IF (SEL$="H") OR (SEL$="h") THEN INPUT "Enter height ",H
IF (SEL$="F") OR (SEL$="f") THEN INPUT "Enter frequency ",F
IF (SEL$="D") OR (SEL$="d") THEN INPUT "Enter diameter ",A
IF (SEL$="X") OR (SEL$="x") THEN END
IF LEN(SEL$)>0 THEN GOTO ComputeTransmissionline: ·
GOTO TLoop:
ALTERXXX:
LOCATE 10,1
IF lsltCurrent= 1 THEN GOTO ALTERXXX2:
INPUT "ELECTRIC FIELD (V/m or?): ",XXX$
IF XXX$="?" THEN
    PRINT "Current value for WIRE CURRENT is:";IO;"Amps"
```

```
    INPUT "Enter new value:",IO
    IsItCurrent= 1
ELSE
    EO=VAL(XXX$)
END IF
GOTO ComputeTransmissionline:
ALTERXXX2:
LOCATE 10,1
INPUT "WIRE CURRENT is (Amps or?): ",XXX$
IF XXX$="?" THEN
    PRINT "Current value for ELECTRIC FIELD is:";EO;"V /m"
    INPUT "Enter new value:",EO
    IsItCurrent=0
ELSE
    IO= VAL(XXX$)
END IF
GOTO ComputeTransmissionline:
ComputeTransmissionline:
UO=4*PI*.0000001
LE= (UO/PI)*LOG(H/ A)
K=2*PI*F/C
ZC1 = (ZI/1000 +2*PI*F*LE)*2*PI*F*LE/(K^2)
ZC=SQR(ZC1)
IF IsItCurrent= 1 THEN EO=IO*ZC/4/H
IF IsItCurrent=0 THEN IO=4*EO*H/ZC
GOTO Transmissionlinelnputs:
```

2.8.2.3   Computer Program for Calculating the Receiving Properties of a Nonresonant Monopole Antenna

```
CLS
PRINT "Open circuit voltage or current into a load caused by an E field
PRINT "incident on a short length of wire/cable (length << lambda) located
PRINT "at least 1 m from a ground plane and terminated to ground at one end
PRINT "only.
GOSUB Spacebar:
'Initialize constants
C=3E+08: Pl=3.14159
'Initialize variables
E0=1: F=2000: H=1: A=.001: RL=50: ZW=377
GOTO Compute:
Inputs:
CLS
PRINT " INPUTS:
PRINT" [A] Electric field",EO;"V/m"
PRINT " [B] Frequency",F;"Hz"
PRINT " [C] Length of wire",H;"metres"
PRINT " [D] Diameter of wire",A;"metres"
PRINT" [E] Load Resistance",RL;"Ohms"
PRINT " [F] Wave lmpedance",ZW;"Ohms"
PRINT " [X] End program
PRINT" Please select variable you wish to change (A,B,etc)
PRINT"
PRINT " OUTPUTS:"
IF 2*PI*H/LAMBDA<1 THEN GOSUB Outputs1:
IF (2*PI*H/LAMBDA>1) AND (2*PI/LAMBDA<2.75) THEN GOSUB Outputs2:
```

```
IF 2*PI*H/LAMBDA>2.75 THEN GOSUB Outputs3:
LOCATE 9,3
SelectLoop:
SEL$=1NKEY$
IF LEN(SEL$)>0 THEN PRINT SPACE$(80)
LOCATE 9,3
IF (SEL$="A") OR (SEL$="a") THEN INPUT "Enter electric field",EO
IF (SEL$="8") OR (SEL$="b") THEN INPUT "Enter Frequency",F
IF (SEL$="C") OR (SEL$="c") THEN INPUT "Enter Length of wire",H
IF (SEL$="D") OR (SEL$="d") THEN INPUT "Enter Diameter of wire",A
IF (SEL$="E") OR (SEL$="e") THEN INPUT "Enter Load Resistance",RL
IF (SEL$="F") OR (SEL$="f") THEN INPUT "Enter Wave lmpedance",ZW
IF (SEL$="X") OR (SEL$="x") THEN END
IF LEN(SEL$)>0 THEN GOTO Compute:
GOTO SelectLoop:
Compute:
SEL$=""
LAMBDA=C/F
BETA=2*PI/LAMBDA
HH=BETA*H
OMEGA=2*(LOG(2*H/ A))/LOG( 1 0)
RR = 197 .5*( (2*H)/LAMBDAY2
ZORE=10*(HW2)
TEMP1 =(OMEGA-2t2
TEMP2=2*(LOG(2)/LOG(10))
ZOIM=(30/HH)*(TEMP1/(1 + TEMP2))
ZO=SQR(ZORE^2 + ZOIM^2)
K=(1-(ZO/RL))/(1 +(ZO/RL))
ALPHA=1-K^2
G=1.5*ALPHA
Amax=G*(LAMBDA^2/(4*PI))
HE1 =(Amax*RR)/(2*ZW)
HE2=(EO*LAMBDA)/(2*PI)*TAN(PI*H/LAMBDA)
IF RL>1000000! THEN VO=EO*HE2 ELSE VO=EO*HE1
IF RL>1000000! THEN HE=HE2 ELSE HE=HE1
VL=EO*HE1
IL=VL/RL
IS=VO*TAN(2*PI*H/LAMBDA)/60
GOTO Inputs:
Outputsl:
PRINT " Open circuit voltage",ABS(VO);"Volts"
PRINT " Short circuit current",ABS(IS);"Amps"
PRINT " Effective height (He)",ABS(HE);"metres"
PRINT " Self impedance (Zo)",ABS(ZO);"Ohms"
PRINT " Current in load (IL)",ABS(IL);"Amps"
PRINT " Voltage across load (VL)",ABS(VL);"Volts"
RETURN
Outputs2:
PRINT " **RESONANT**
PRINT " See Transmission Lines, Antennas and Waveguides
PRINT" (pp. 98,166) PRINT "
PRINT " Omega=";OMEGA
PRINT " Beta=";HH
PRINT " Open Circuit Voltage-";VO
RETURN
Outputs3:
```

```
PRINT "
PRINT " The frequency and length of wire are such that:"
PRINT " 2 *PI*H/LAMBDA>2.75
PRINT
PRINT " This program is not applicable for such conditions.
RETURN
Spacebar:
LOCATE 15,1
PRINT "Press the spacebar to continue or X to end
SpacebarLoop:
Spcbr$=INKEY$
IF Spcbr$=CHR$(32) THEN GOTO Returner:
IF (Spcbr$="x") OR (Spcbr$="X") THEN END
GOTO SpacebarLoop:
Returner:
```

Return

These and many of the equations for antenna gain and AF are available in a computer program from emcconsultinginc.com.

REFERENCES

1. M. Zaret. *Outline of Electromagnetic Theory*. Regents, New York, 1965.
2. J.E. Bridges. Study of low-frequency fields for coaxial and twisted pair cables. *Proceedings of the 10th Tri-Service Conference on Electromagnetic Compatibility*, Chicago, IL, November 1964.
3. J. Moser. Predicting the magnetic fields from a twisted pair cable. *IEEE Transactions on Electromagnetic Compatibility*, September 1968, 10(3).
4. S. Shenfield. Magnetic fields of twisted wire pairs. *IEEE Transactions on Electromagnetic Compatibility*, November 1969, 11(4).
5. A. Richtscheid. Calculation of the radiation resistance of loop antennas with sinusoidal current distribution. *IEEE Transactions on Antennas and Propagation*, November 1976.
6. R.W.P. King, H.R. Mimno, and A.H. Wing. *Transmission Lines, Antennas, and Wave Guides*. Dover, New York, 1965.
7. G.H. Brown and O.M. Woodward Jr. Experimentally determined radiation characteristics of conical and triangular antennas. *RCA Review*, December 1952, 13(4), 425–452.
8. Health aspects of radio frequency and microwave exposure. Part 2. Health and Welfare Canada, 78-EHD-22, Ottawa, Ontario, Canada, 1978.
9. IEEE C95.1. IEEE standard for safety levels with respect to human exposure to radio frequency electromagnetic fields, 3 kHz to 300 GHz, 1991.
10. Safety Code 6. Canada Health and Welfare, 99-EHD-237, 1999.
11. N. Sultan. Private Communication, formerly with Canadian Astronautics.
12. B. Spinner, J. Purdham, and K. Marha. The case for concern about very low frequency fields from visual display terminals: The need for further research and shielding of VDT's. From the Canadian Centre for Occupational Health and Safety, Hamilton, Ontario, Canada.
13. K. Marha. Emissions from VDTs: Possible biological effects and guidelines. Canadian Centre for Occupational Health and Safety, Hamilton, Ontario, Canada.
14. K. Marha. VLF—Very low frequency fields near VDTs and an example of their removal. Canadian Centre for Occupational Health and Safety, Hamilton, Ontario, Canada.
15. E.B. Larsen. NBS, calibration and meaning of antenna actors and gain for EMI antennas. ITEM, 1986.
16. H. Garn. A comparison of electric field-strength standards for the frequency range of 30–1000 MHz. *IEEE Transactions on Electromagnetic Compatibility*, November 1997, 39(4), 397–403.
17. Code of Federal Regulations 47. Chapter 1, Part 1, 1.1310. October 1, 1997; S.A. Schelkunoff and H.T. Friis. *Antennas Theory and Practice*. Wiley, New York, 1952.

3 Typical Sources and Characteristics of Radiated and Conducted Emissions

3.1 INTRODUCTION TO NOISE SOURCES

Some typical sources of electromagnetic emissions are transmitters, pulse generators, oscillators, digital logic circuits, switching power supplies and converters, relays, motors, and line drivers. The majority of unintentional emissions are composed of a great number of frequencies typically caused by switching and transient noise. Design engineers usually measure digital and impulsive signals in the time domain with an oscilloscope or a logic analyzer, whereas EMI specifications and measurements are made in the frequency domain. Either Fourier analysis or Laplace transform may be used to compute the amplitude of the frequencies contained within the impulsive noise (the spectrum occupancy), or a spectrum analyzer or receiver may be used to measure the frequency components. Making a measurement is easier and more accurate, especially when the waveform is complex in shape.

During breadboarding and when the first prototype of equipment is available, an assessment of the level and frequencies generated by the circuit is important, both in achieving intraequipment EMC and in meeting EMC specifications. The same information is also useful when locating the source in an EMI investigation. Assuming a spectrum analyzer is not available, some simple technique for assessing the frequency occupancy based on the time-domain signal is desirable. This chapter illustrates the frequency occupancy of common waveforms with simple formulas for calculating the amplitude versus frequency characteristics, introduces a "quick transform technique," and presents a case study of measurements made on a voltage converter (which is a common source of noise).

3.1.1 HARMONICALLY RELATED NOISE FROM SINGLE AND PERIODIC PULSES

For a periodic pulse, the low-frequency spectrum is consisted of the pulse repetition rate (PRR) and its harmonics. Figure 3.1 shows a symmetrical trapezoidal periodic pulse train where $t_0 + t_r = T/2$. Figure 3.2 shows the envelope approximation of the spectral lines shown at 1/7, 3/T, and 5/T (the spectral lines at 2/T and 4/T are at zero amplitude).

Figure 3.3 shows the spectral lines for a pulse where $t_0 + t_r = T/5$. The negative amplitudes of the sin x/x envelope shown in Figures 3.2 through 3.4 may be inverted, because we are only interested in the amplitude and frequency and an envelope approximation made of the decrease in the amplitude of the spectral line with increasing frequency, as shown in all the subsequent figures.

The spectrum analyzer inverts the negative amplitude and, depending on the resolution bandwidth and sweeptime settings, may display the spectral lines. When the pulse train is unmodulated, the scan starts at 0 Hz.

The pulse width can be found from the frequency at the sidelobe corresponding to $1/(t_r + t_0)$ and the PRR from the spacing between the spectral lines corresponding to $1/T$, as shown in Figure 3.3. Assuming $T = 1$ μs for the pulse illustrated in Figure 3.1, $t_r + t_0 = 0.5$ μs (i.e., $T/2$) and the sidelobe frequency will be 2 MHz (i.e., 1/0.5 μs).

By Fourier analysis, the amplitude at the frequency of the nth harmonic is given by

$$C_n = 2A \cdot \frac{(t_0 + t_r)}{T} \cdot \frac{\sin\left[\pi n\,(t_0 + t_r)/T\right]}{\pi n\,(t_0 + t_r)/T} \cdot \frac{\sin\left(\pi n t_r/T\right)}{\pi n t_r/T} \quad \text{the argument of sine is in radians}$$

FIGURE 3.1 Symmetrical trapezoidal periodic pulse. (From Kann, E., Design guide for electromagnetic interference (EMI) reduction in power supplies, MIL-HDBK-241B, Power Electronics Branch, Naval Electronic Systems Command, Department of Defense, Washington, DC, 1983.)

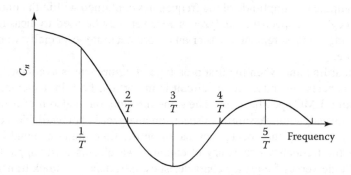

FIGURE 3.2 Low-frequency harmonic spectrum for $t_r + t_0 = T/2$ trapezoidal pulse. (From Kann, E., Design guide for electromagnetic interference (EMI) reduction in power supplies, MIL-HDBK-241B, Power Electronics Branch, Naval Electronic Systems Command, Department of Defense, Washington, DC, 1983.)

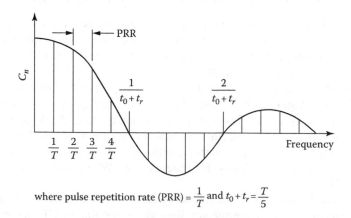

where pulse repetition rate $(\text{PRR}) = \dfrac{1}{T}$ and $t_0 + t_r = \dfrac{T}{5}$

FIGURE 3.3 Low-frequency harmonic spectrum for trapezoidal periodic pulse (narrowband). (From Kann, E., Design guide for electromagnetic interference (EMI) reduction in power supplies, MIL-HDBK-241B, Power Electronics Branch, Naval Electronic Systems Command, Department of Defense, Washington, DC, 1983.)

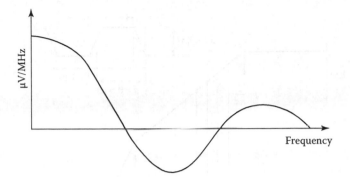

To convert to coherent broadband:

$\quad$ Let $t_0 + t_r = d$

$$f = \frac{n}{T}$$

$$\frac{\text{Narrowband } C_n}{\text{PRR}} = 2Ad \ \frac{\sin \pi fd}{\pi fd} \text{ broadband}$$

or in dB and MHz:

$\quad\quad$ dBμV (narrowband) − 020 log (PRR in MHz) = dBμV/MHz (broadband)

FIGURE 3.4 Low-frequency harmonic envelope for single-pulse broadband. (From Kann, E., Design guide for electromagnetic interference (EMI) reduction in power supplies, MIL-HDBK-241B, Power Electronics Branch, Naval Electronic Systems Command, Department of Defense, Washington, DC, 1983.)

One common error is often made in finding the PRR from a spectrum analyzer measurement for a waveform with a mark-to-space ratio of unity, that is, the symmetrical pulse of Figure 3.1. The sine x/x function envelope is shown for convenience only in Figure 3.2 and is not displayed in a spectrum analyzer measurement; therefore, the location of the sidelobe is not clear from a spectrum analyzer display. The spectral lines at the sidelobes are at zero amplitude, and the negative line at $3/T$ is inverted by the spectrum analyzer. Thus, the PRR is commonly taken as the spacing between the two positive spectral lines, which is $2/T$, that is, a PRR twice as high as the signal under measurement. Section 9.2.2 deals with the use of a spectrum analyzer in some detail. The best method of familiarization with the spectrum analyzer and its representation of a periodic pulse train, and one I know the customer support people at Hewlett-Packard have used, is to apply a known signal from a pulse or square wave generator and examine the resultant spectral lines.

The periodic pulse train illustrated in Figure 3.1 generates narrowband frequencies harmonically related to the PRR. The equation that may be used to find the amplitude at the frequency of the nth harmonic is shown under the waveform in Figure 3.1. An example of the use of this equation is given in calculating the harmonic components of the 1 MHz transistor–transistor logic (TTL) clock pulse illustrated in Figure 3.5.

A single pulse and pulses with low repetition rates, relative to the pulse width, generate coherent broadband noise. This noise is coherent because the frequencies generated are related to the rise and fall times and width of the pulse. A definition of *coherent* is a signal or emission in which the neighboring frequency elements are related or well defined in both amplitude and phase.

The definition of *broadband* as a single pulse or low-repetition-rate pulse is sometimes used in commercial EMC specifications with a specified maximum repetition rate, typically 10 kHz, below which a source is considered broadband.

A commonly accepted definition of *broadband* relates to a measurement bandwidth: noise that has a spectrum broad in width compared to the nominal bandwidth of a receiver or spectrum analyzer. Chapter 9 discusses the definition and measurement of narrowband and broadband noise in some detail.

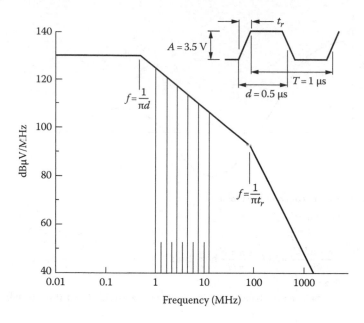

FIGURE 3.5 High-frequency spectrum of a transistor–transistor logic level 1 MHz clock pulse.

The equation shown in Figure 3.4 allows the conversion of the narrowband harmonic amplitude from the periodic pulse into the broadband coherent amplitude, measured in μV/MHZ, generated by a single pulse.

Figure 3.5 shows the high-frequency spectrum of a trapezoidal periodic pulse train. The first breakpoint comes at $1/\pi d$ and the second at $1/\pi t_r$. The amplitude below $1/\pi d$ is equal to $(2A) \times d/T$.

The high-frequency spectrum of a 3.5 V 1 MHz clock with 3.5 ns rise and fall times, which is a typical TTL level, is illustrated in Figure 3.5. We will use this 1 MHz clock as an example in calculating the amplitude at the harmonics and converting them into a coherent broadband-equivalent level. It should be noted that due to the relatively high PRR, the clock is almost certainly classified as a narrowband source, under the definitions given in Chapter 9. From the following equation, the amplitude of the narrowband harmonics may be calculated:

$$C_n = (2A)\frac{(t_0 + t_r)}{T} \cdot \frac{\sin[\pi n(t_0 + t_r)/T]}{\pi n(t_0 + t_r)/T} \cdot \frac{\sin(\pi n t_r/T)}{\pi n t_r/T} \quad \left(\text{arguments of sine in radians}\right)$$

The narrowband amplitude (C_n) at the first harmonic (the fundamental) using this equation is 2.226 V. The equations in Figure 3.4 are used to find the broadband-equivalent amplitude as follows: from narrowband C_n/PRR, we obtain 2.226 V/1 MHz = 2.226 μV/Hz. To express the broadband level in the common broadband unit of dBμV/MHz, we convert 2.226 μV to dBμV (i.e., 6.9 dBμV) and add a 120 dB correction for the conversion from the 1 Hz to the 1 MHz reference bandwidth, which results in 127 dBμV/MHz.

A second equation using dBμV (narrowband) and PRR in megahertz is as follows: dBμV (narrowband) − 20log (PRR in MHz). The amplitude at 1 MHz (the first harmonic) equals 2.226 V, which, expressed in dBμV, is 127 dBμV. The broadband level is therefore 127 dBμV − 20log 1 = 127 dBμV/MHz. The broadband amplitudes at other harmonics are 53 dBμV/ MHz at the second harmonic, 117 dBμV/MHz at the third harmonic, 110 dBμV/MHz at the seventh harmonic, and 53 dBμV/MHz at the eighth harmonic, as plotted along with the envelope in Figure 3.6.

FIGURE 3.6 Interference level for a 1 V, 1 μs rectangular pulse. (From Kann, E., Design guide for electromagnetic interference (EMI) reduction in power supplies, MIL-HDBK-241B, Power Electronics Branch, Naval Electronic Systems Command, Department of Defense, Washington, DC, 1983.)

The equations for the calculation of the spectrum occupancy for square, rectangular, and many other single-pulse or low-repetition-rate event pulse shapes are shown in Figures 3.7 through 3.13. The curves in these figures result from connecting the maxima of the spectral lines envelope, as shown in Figure 3.6. These figures also contain information useful in achieving a design that limits the amplitude of the low- and high-frequency emissions from pulses. Figure 3.8, for example, shows

FIGURE 3.7 Interference level for a 1 V, 1 μs trapezoidal pulse. (From Kann, E., Design guide for electromagnetic interference (EMI) reduction in power supplies, MIL-HDBK-241B, Power Electronics Branch, Naval Electronic Systems Command, Department of Defense, Washington, DC, 1983.)

FIGURE 3.8 Trapezoidal pulse interference. (From Kann, E., Design guide for electromagnetic interference (EMI) reduction in power supplies, MIL-HDBK-241B, Power Electronics Branch, Naval Electronic Systems Command, Department of Defense, Washington, DC, 1983.)

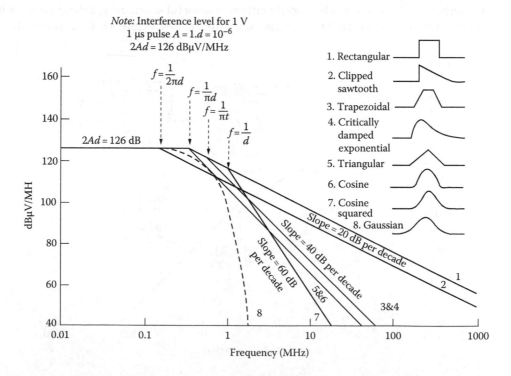

FIGURE 3.9 Interference levels for eight common pulse shapes (dBμV/MHz vs. f). (From Kann, E., Design guide for electromagnetic interference (EMI) reduction in power supplies, MIL-HDBK-241B, Power Electronics Branch, Naval Electronic Systems Command, Department of Defense, Washington, DC, 1983.)

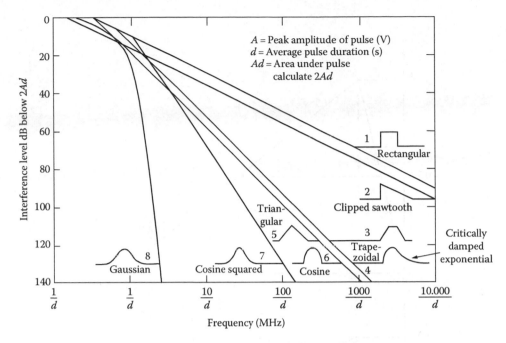

FIGURE 3.10 Interference levels for various pulse shapes (dB below 2Ad vs. f). (From Kann, E., Design guide for electromagnetic interference (EMI) reduction in power supplies, MIL-HDBK-241B, Power Electronics Branch, Naval Electronic Systems Command, Department of Defense, Washington, DC, 1983.)

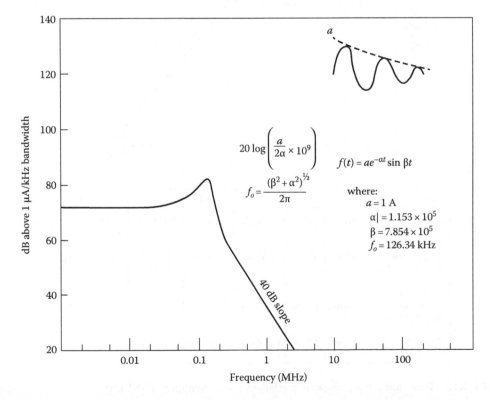

FIGURE 3.11 Normalized frequency spectrum of damped sine wave pulse. (From Kann, E., Design guide for electromagnetic interference (EMI) reduction in power supplies, MIL-HDBK-241B, Power Electronics Branch, Naval Electronic Systems Command, Department of Defense, Washington, DC, 1983.)

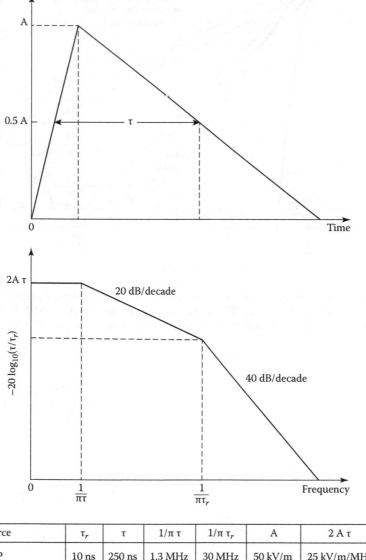

Source	τ_r	τ	$1/\pi\tau$	$1/\pi\tau_r$	A	$2A\tau$
EMP	10 ns	250 ns	1.3 MHz	30 MHz	50 kV/m	25 kV/m/MHz (25 V/m/kHz)
Space plasma arc discharge (MIL-STD-1541)	15 ns	40 ns	8 MHz	20 MHz	1200 V (at 30 cm)	96 V/m/MHz
Human body ESD model (MIL-STD-883c)	15 ns	100 ns	3.2 MHz	20 MHz	6 A	1.2 A/MHz
Lightning (MIL-STD-5087B)	2 µs	25 µs	12.7 MHz	159 kHz	200 kA	10 kA/kHz

FIGURE 3.12 Time- and frequency-domain relations due to lightning, EMP, ESD, and space plasma arc discharge.

FIGURE 3.13   The relative spectral content of a 1 MHz transistor–transistor logic clock source, the resultant interface cable common-mode current flow, and the *E* field generated by the cable.

that the low-frequency component of a trapezoidal pulse may be reduced by decreasing the duration of the pulse. However, when the pulse width is decreased, the maximum value of the rise and fall times may be limited by the pulse width. Increasing the rise and fall times decreases the frequency of the breakpoint and reduces the amplitude of the high-frequency components.

To summarize: for any pulse shape, the interference level at low frequencies depends only on the area under the pulse, while at higher frequencies the level depends on the number (PRR) and steepness (rise and fall times) of the slopes.

Figure 3.9 illustrates the rate of reduction in amplitude for frequencies above the breakpoint for eight common pulse shapes. The Gaussian pulse has the most rounded contours and exhibits the steepest slope in amplitude reduction with increasing frequency beyond the breakpoint. Thus, it is not the rise time alone that controls the amplitude of the high-frequency components.

A logic or similar pulse may often exhibit rounding at the start and end of the rising/falling edge, with a subsequent reduction in high-frequency emissions, compared to a pure trapezoidal pulse. Conversely, a pulse with overshoot and/or undershoot or with a section of the transition that exhibits a faster rate of *dV/dt* than the overall edge will result in high-frequency emissions at greater amplitude. Where possible, the edges of pulses should be rounded and the rise and fall times increased. In a printed circuit board (PCB) loaded with logic, this is a near impossible task, and other reduction techniques described in subsequent chapters may be applicable. However, for a clock line or data bus with a high fan-out (i.e., driving a large number of logic inputs), it may be practical to slow down and round edges by the inclusion of a series resistor or a series resistor–parallel capacitor combination. Adding the resistor in TTL-type logic will decrease the *high*-level voltage and increase the *low*-level voltage and must be limited in value to typically 100 Ω or less to ensure the correct operation of the logic and maintain some level of DC noise immunity. An alternative to the resistor is a ferrite bead, which introduces some inductance. Assuming the edges are not slowed down to the extent that the maximum *high* and minimum *low* levels are not reached at the repetition rate of the signal, the inclusion of the ferrite bead will not degrade the *high* and *low* logic magnitudes. One disadvantage with the bead is the possibility of resonance and peaking in the logic levels. The addition of a 1–10 Ω resistor in series with the inductor or a resistor across the inductor will reduce the inductor Q and minimize peaking. The characteristic of a ferrite bead, resonance, and peaking and the choice of resistor value to minimize these effects are discussed in Chapter 5.

Figure 3.10 is the same curve as in Figure 3.9 except that the y-axis is in "dB below $2Ad$" and the x-axis is plotted in frequency as a function of d. Thus, Figure 3.9 may be conveniently used for any pulse amplitude and pulse duration. The curves of Figure 3.10 may be used for other waveshapes by considering the waveshape as made up of a number of different waveshapes, such as rectangular and triangular, and finding the total area to obtain the low-frequency content and the number and steepness of the slopes to find the high-frequency content. From the envelopes of typical waveshapes, it is clear that the high-frequency components are at a lower magnitude than the fundamental but can exist at amplitudes that may be problematic to other sections of a system or equipment. It is therefore not surprising that designers are often amazed when equipment fails radiated or conducted emission requirements at frequencies orders of magnitude above the fundamental frequencies used in the equipment. When we look at the level of emissions from a 1 MHz clock with a typical TTL waveshape at frequencies of 100–300 MHz, they are indeed 40–46 dB below the fundamental.

Then why are measured emissions often at a maximum in this 100–300 MHz frequency range? Typically, a combination of factors is involved. One is that common-mode noise developed across a PCB tends to increase with increasing frequency as PCB track/ground plane impedance increases. In addition, the coupling of common-mode noise may be via the capacitance between the board and the enclosure, and this coupling path decreases in impedance with increasing frequency. Where radiated emissions are caused by common-mode current flow on a cable, the radiation will typically increase with increasing frequency until cable resonances occur, and at the same time, the common-mode voltage driving the current increases, due to increasing circuit impedance. Where a shielded interface cable is used, the shielding effectiveness typically decreases with increasing frequency, and the radiation, caused by current flow on the outside of the shield, increases. These coupling mechanisms, including the effect of PCB and cable resonance, are discussed in subsequent chapters.

Consider a simple model of an unshielded cable exiting a piece of equipment and terminating in a second piece of equipment, as illustrated in Figure 7.92. Let us assume that the common-mode voltages in each piece of equipment, which cause the current flow, and its coupling to the enclosure result in a 26 dB increase in current per decade increase in frequency. The envelope of the frequency components of the cable current flow, plotted for the sake of comparison against the source voltage of Figure 3.5, are as illustrated in Figure 3.13. The radiated emissions from the cable increase by approximately 20–30 dB/decade increase in frequency, dependent on the frequency, until the resonant frequency of the length of cable is reached, after which emissions either level out or reduce, as discussed in Section 7.13. This may be confirmed by the use of the loop equations given in Chapter 2. The envelope of the frequency components of the resultant E field at a fixed distance from the cable is plotted for the sake of comparison in Figure 3.13. It should be emphasized that the current and E-field plots are dimensionless.

Figure 3.13 illustrates that the maximum E field in the 50–500 MHz frequency range may be at an amplitude 36 dB above the fundamental, something commonly seen in practice. Often, circuit and cable resonances occur and we see one or more frequencies at which the measured field is higher than surrounding frequencies.

Sometimes, the repetition rate of the impulsive noise coincides with a resonance frequency, and a sine wave or damped sine wave is generated. When a continuous sine wave is generated in this manner, it gives rise to an emission that is considered narrowband in nature. It is not uncommon to find narrowband and broadband emissions together resulting from a single source of noise.

We have assumed thus far that the source of noise is the logic voltage transition. Although this voltage plays an important role in determining the radiated emission and in the crosstalk and conducted noise voltage generated by the logic, it is often the current pulse that is the predominant source of emissions.

High-current spikes flow in the logic signal and return interconnections, as the output device changes state and charges/discharges the input capacitance of the load device and removes the

TABLE 3.1

Input Current during Switching Transient, Output Rise and Fall Times, and Typical Input Capacitance of Different Logic Types

Characteristic	TTL	CMOS 5 V[a]	CMOS 10 V[a]	CMOS 15 V[a]	TTL 74 L	TTL 74 S	TTL 74 H	ECL 10 K	ECL 100 K
Propagation delay (ns)	10	150	65	50	33	3	6	2	0.75
Approximate rise and fall times (ns)	3.5	100	40	30	7	1.7	2	1.8	0.75
Input current pulse (mA) (fan-out = 1)	6.5	0.16	0.9	1.6	1	—	9.2	16	14
Max. input capacitance (pF)	6	4	4	4	6	4	6	0.1[b]	0.1[b]

[a] B-type CMOS.
[b] Dependent on PCB layout.

charge from either a base/emitter junction or a diode. During the transition, the load on the output device is therefore predominantly capacitive, and, depending on the geometry of the interconnections and the logic type, the source of radiated emissions is typically a low impedance (i.e., a predominantly magnetic field source). Even though the current pulse is the predominant source of emissions, those created by the voltage transition should not be ignored in an EMC prediction, especially when the load has high impedance with low capacitance.

Thus, in predicting the radiated emission from a PCB track or the voltage drop in either signal power or return connections, the typical magnitude of the current spikes must be known. The switching currents from various logic types and the rise and fall times with single input load are shown in Table 3.1.

Using the current pulse for a fan-out of 1, as shown in Table 3.1, the current spike due to a maximum fan-out of 15 on a 74 series TTL output device is 97 mA. Large-scale integrated circuits typically incorporate buffered clock and data inputs; however, input current spikes as high as 500 mA at the logic transition have been measured. When the noise source is the current spikes at the transition and the current flow during the steady-state logic level is relatively low, the peak current pulse duration is short (typically 7 ns for TTL), and the first breakpoint in Figure 3.5, at $f = 1/\pi d$, moves from 636 kHz to 45 MHz. At the same time, the low-frequency amplitude reduces by 37 dB. This represents, therefore, another reason for the relatively high magnitude of emissions at high frequencies (above the fundamental).

3.1.2 FREQUENCY SPECTRUM OCCUPANCY OF A STEP FUNCTION

From the Laplace transform, the amplitude generated by a step function (i.e., a single event such as caused by a control line or a switch changing state) is given, for the case where the frequency of interest, f, is higher than $1/t_r$ approximately by

$$V = \frac{0.05V_1 - V_2}{t_r f^2} \quad \text{for } f > \frac{1}{t_r}$$

where
   V_1 is the initial voltage
   V_2 is the voltage after step
   t_r is the rise time [s]
   f is the frequency [Hz]

The envelope of the resultant curve (V vs. f) decreases at a rate proportional to the frequency squared (40 dB/decade); the magnitude is in V/Hz, in order to express this in the common broadband unit of dBµV/MHz; 120 dB must be added to the magnitude expressed in dBµV. For example, if Equation 3.1 yields a value of 0.1 µV/Hz, then this can be converted to dBµV/Hz,

$$0.1 \ \mu V/Hz = 20 \log\left(\frac{0.1}{1}\right) = -20 \ dB\mu V/Hz$$

and then into dBµV/MHz,

$$-20 \ dB\mu V/Hz + 120 \ dB = 100 \ dB\mu V/MHz$$

3.2 FOURIER TRANSFORM METHODS AND COMPUTER PROGRAMS

"Practicing engineers tend to shun the application of complex mathematics. In EMC work, this tendency surfaces in a reluctance to perform Fourier Transforms when investigating complex waveforms" [2]. This book has been written for "practicing engineers" and by one for whom this reluctance is well understood. The figures and equations contained in the previous sections of this chapter are useful in obtaining the frequency occupancy of fairly simple waveforms. What is needed, however, is a simple method of obtaining the same information for the complex waveforms so often encountered in practice. One such method is the "quick transform procedure" by Toia [2]. This is a graphical method of obtaining the harmonics and their amplitudes from a complex time-domain waveform. The following is merely a brief review of this method. In order to use the method, the reader must, almost certainly, refer to the paper by Toia [2].

In this method, the time-domain waveform is sketched over time in cycles, the x-axis, with the amplitude of the waveform plotted on the y-axis. The time-derivative spikes are then sketched and, from a number of simple properties of delta chains, a harmonic table is constructed. In summary, the steps involved in the "quick transform" procedure are the following:

1. Sketch the time-variant waveform.
2. If it contains no spikes, sketch its time derivative.
3. Continue sketching derivatives until spikes appear. Record the highest order of derivatives sketched as m.
4. In tabular form, record the relative amplitude and phase of each harmonic using the known spectral property of delta-function chains (i.e., for the positive chain that all frequencies are present and of the same magnitude and for the negative chain the same but at a different position in time).
5. Add spectral components of the table vectorially.
6. Divide the nth harmonic's amplitude by $n^m(2\pi/T)$ for each harmonic.
7. Retard the nth harmonics phase by ($n \times 90°$) for each harmonic.
8. Remove the spikes from the sketch. If a residue remains, continue this procedure.

An example for a waveform more complex than a square wave is given in Reference [2] along with an example for a smooth waveform that contains no spikes. Although the quick transform possesses shortcomings, it is a useful procedure that reduces the math to a matter of addition.

A number of programs exist for use on both mainframe and personal computers that perform fast Fourier transforms (FFTs). A typical program enables the FFT of a vector of real data representing measurements at regular intervals in the time domain. The result is a vector of complex coefficients in the frequency domain. The phase information is retained, and thus, the inverse Fourier transform can be used to convert from the frequency domain back into the time domain. The real data contain magnitude and phase or time information. The program can also accept complex data containing real and imaginary magnitudes, and the phase information is then implicit in the complex data. An inverse Fourier transform of the resultant vector of data representing values in the frequency domain returns a vector representing values in the time domain.

Common sources of noise at the unit level are AC-to-DC converters, DC-to-DC converters, and switching power supplies as described in Section 3.3.

3.3 CASE STUDY 3.1: NOISE LEVELS GENERATED BY DC-TO-DC CONVERTERS

This section supplies data on the magnitude and type of waveform, including frequency-domain components, of the noise generated by DC-to-DC converters. The emissions from two typical high-reliability converters used in military and space applications are then compared to the military standard MIL-STD-461F limits on conducted and radiated emissions. In the investigation, the input noise voltage and output noise voltage were measured from the two converters and in the worst case, that is, the highest level, data are supplied. One converter was rated at 72 W and the second at 50 W. The characteristics and magnitude of noise from the different manufacturers' converters were relatively similar. Differences were confined to the fundamental converter switching frequency and the maximum frequency at which emissions were still significant. As expected, both the converters produced coherent harmonically related emissions covering frequencies from the fundamental up to 599 MHz.

The switching repetition rate of converter #1 was measured at close to 500 kHz.

3.3.1 GENERAL TEST SETUP AND METHOD

Two converters, one with 28 V input and 24 V output and the second with 24 V input and 5 V output, were tested for the following:

Conducted emission from the input terminals
Conducted emission from the output terminals
Radiated emission from a 2 m long cable connected to the output terminals

Conducted emissions were tested in accordance with the methods and to the limits specified by MIL-STD-46F CE102. For radiated emissions, MIL-STD-461 RE102 was used. The CE102 represents limits on the noise voltage developed across the calibrated impedance of the line impedance stabilization network (LISN) produced by the device under test at its input connections.

Each converter was mounted in a shielded enclosure with the load resistors also contained in the enclosure. Both converters isolated the input and output power lines. The input power was not connected to the enclosure in the initial measurements as this floating grounding scheme is common in military and space applications. Less common is where input return is grounded to the enclosure; nervertheless, this configuration was measured for CE102 and RE102. The output power was also isolated with the exception of when the common-mode noise voltage was measured with an oscilloscope.

The manufacturer of the first 24–24 V converter recommends that the converter be connected to a PCB plane connected to chassis. The converter has a five-sided metal enclosure. By mounting this enclosure directly to a PCB ground plane making a six-sided enclosure from measurements with a small probe, the emissions from the converter enclosure are reduced. Also, the connection of a 10 µF capacitor across the input power line is required and is especially important when the power is supplied via the impedance of the LISN.

In earlier test to MIL-STD-461A, it was found during testing that the quality of the electrical connection of the converter case to the PCB ground plane was an important factor in the level of attenuation achieved in the high-frequency components of the output noise. Filters were included on the output power but not on the input power. With the case clamped to the PCB ground plane with a moderate

FIGURE 3.14 CE102 test setup.

clamping pressure, the output noise voltage with the least effective output filter in the circuit was 40 mV. At a higher clamping pressure, the noise voltage is reduced to 5 mV. For the most effective filter, the GK2AA-SO8, the high-frequency component is reduced to 0.25 mV, and this residual level may be attributed to radiated coupling between the input and the output wiring.

The MIL-STD-461F CE0102 test method was used to measure the conducted input noise on the two present converters without input filters. The voltage developed across the LISN impedance was measured.

The CE102 measurements were made in the frequency domain by the use of a spectrum analyzer.

Measurement was also made in the time domain by the use of an oscilloscope. The CE0102 test setup and test equipment are shown in Figure 3.14 and the RE102 in Figure 3.15.

As the levels were very high, no preamplifier was required for the CE102 measurement.

As a test of the type of noise emitted by the converter, the resolution bandwidth of the spectrum analyzer was increased. As expected, the measured level also increased, which indicates that the noise may be classified as broadband.

3.3.2 SUMMARY OF +24 TO 24 V CONVERTER CONDUCTED EMISSIONS (CE102) AND RADIATED EMISSIONS (RE102) FROM THE INPUT POWER TEST RESULTS

The test results for CE102 and RE102 for a power converter that must comply with military standard 461-F are shown in Tables 3.2 through 3.6. The converter metal case is connected to chasis ground.

Although most military and space requirements mandate that the input power be isolated from chassis ground, in some cases, the input return is connected to chassis. To determine if this connection makes a difference to conducted and radiated emissions, the measurements were repeated with the input power return connected to chassis. The results are shown in Tables 3.7 through 3.9, and the emissions are shown visually in Figures 3.16 through 3.21.

Users who place the inverters on a PCB where a chassis ground connection is difficult to make often do not connect the case to chassis ground, and measurements were made with this configuration as shown in Tables 3.10 through 3.13 with the emissions shown in Figures 3.22 through 3.24.

FIGURE 3.15   RE102 test setup.

TABLE 3.2
Worst-Case Emissions Compared to the Limit

Frequency Units	Type	Frequency	Corrected (dBµV)	Limit (dBµV)	Margin (dB)
MHz	CE	1.580	107.0	60.0	−47.0
MHz	CE	0.792	106.0	60.0	−46.0
MHz	CE	1.176	106.0	60.0	−46.0
MHz	CE	1.990	103.0	60.0	−43.0
MHz	RE V	77.000	57.2	24.0	−33.2
MHz	RE H	63.507	56.7	24.0	−32.7
MHz	RE	15.130	52.5	24.0	−28.5
MHz	RE	17.100	52.4	24.0	−28.4
MHz	RE	17.800	50.4	24.0	−26.4
MHz	RE	2.337	43.8	24.0	−19.8
MHz	RE	156.974	38.4	27.9	−10.5
MHz	CE	0.392	71.0	62.1	−8.9
MHz	RE	65.200	31.4	24.0	−7.4
MHz	CE	0.150	75.7	70.5	−5.2
MHz	CE	9.580	64.0	60.0	−4.0
MHz	RE	300.000	33.7	33.5	−0.2
MHz	RE	411.000	31.0	36.2	5.2
kHz	CE	122.400	47.0	72.2	25.2
kHz	CE	39.800	45.0	82.0	37.0
MHz	CE	2.390	0.0	60.0	60.0

CE = conducted emission, RE = radiated emission, RE V = radiated emission vertical, and RE H = radiated emission horizontal.

TABLE 3.3
Test Result Sheet

Sheet 3-3   Date 15-1-2015

Customer R&D _____   Equipment converter _____   SN _____

Specification MIL-STD-461F _____   Test CE102–28 V limit _____

Equipment MIL-STD/FCC LISN _____

BW 1 kHz _____   VBW _____   Reference _____   Span _____

Notes: Converter case connected to chassis ground plane

Plot	Frequency (kHz)	Measured (dBµV)	Corrected (dBµV)	Limit (dBµV)	Margin (dB)
	39.800	45.0	45.0	82.0	37.0
	122.400	47.0	47.0	72.2	25.2

Test result sheet 3–1, 10–150 kHz.

TABLE 3.4
Test Result Sheet

Sheet 3-4   Date 15-1-2015

Customer R&D _____   Equipment converter _____   SN _____

Specification MIL-STD-461F _____   Test CE102–28 V limit _____

Equipment MIL-STD/FCC LISN _____

BW 9 kHz _____   VBW _____   Reference _____   Span _____

Notes: Converter case connected to chassis ground plane

Plot	Frequency (kHz)	Measured (dBµV)	Corrected (dBµV)	Limit (dBµV)	Margin (dB)
	0.792	106.0	106.0	60.0	−46.0
3–16	1.176	106.0	106.0	60.0	−46.0
	1.580	107.0	107.0	60.0	−47.0
	1.990	103.0	103.0	60.0	−43.0
	2.390		0.0	60.0	60.0

Test result sheet 3–2, 150 kHz–10 MHz.

Table 3.14 shows the comparison of the case and return chassis connection and the isolated emissions to the configuration where only the case of the converter is connected to the chassis.

The manufacturer of converter #1 provides a schematic for a power line filter that is designed to ensure meeting the MIL-STD-461C conducted and radiated emission limits although the class of equipment is not specified.

When more stringent requirements must be met, a custom EMI filter may be designed as described in Section 5.2.1.

3.3.3 DIFFERENTIAL-MODE AND COMMON-MODE CONDUCTED NOISE AT THE OUTPUT OF A 24–5 V CONVERTER

We will use the terms *common-mode* and *differential-mode voltage* or *current* throughout the book. The concepts of *differential mode* and *common mode* are often difficult to comprehend,

TABLE 3.5
Test Result Sheet

Sheet 3-3 _____   Date 18-1-2015 _____

Customer R&D _____   Equipment converter _____ SN _____

Specification MIL-STD-461F _____   Test RE102–ground navy mobile and army _____

Equipment 1 m mono, EMC 0.25–1 GHz preamp _____

BW 9 kHz _____ VBW _____ Reference _____ Span _____

Notes: Converter case connected to chassis ground plane

Plot	Frequency (MHz)	Measured (dBμV)	Antenna Factor (dB)	Preamp Gain (dB)	Corrected (dBμV/m)	RE102 Limit (dBμV/m)	Margin (dB)
3–17	2.337	71.9	4.4	−32.5	43.8	24.0	−19.8
	17.800	74.0	8.6	−32.2	50.4	24.0	−26.4

Test result sheet 3–3, 2–30 MHz.

TABLE 3.6
Test Result Sheet

Sheet 3-6 _____   Date 15-1-2015 _____

Customer R&D _____ Equipment converter _____ SN _____

Specification MIL-STD-461E _____   Test RE102–ground navy mobile and army _____

Equipment LPB, EMC 0.25–1 GHz preamp _____

BW 120 kHz _____ VBW _____ Reference _____ Span _____

Notes: Converter case connected to chassis ground plane

Plot	Frequency (MHz)	Measured (dBμV)	Antenna Factor (dB)	Preamp Gain (dB)	Corrected (dBμV/m)	RE102 Limit (dBμV/m)	Margin (dB)
	65.200	52.0	12.2	−32.8	31.4	24.0	−7.4
3–18	156.974	59.4	12.0	−33.0	38.4	27.9	−10.5
	300.000	54.0	12.9	−33.2	33.7	33.5	−0.2
	411.000	48.0	15.8	−32.8	31.0	36.2	5.2

Test result sheet 3–4, 30 MHz–1 GHz.

but an understanding of the difference between the two modes is of great importance when attempting to analyze a source or in applying reduction techniques, such as those described in Chapters 5 and 11.

We will use an isolating DC-to-DC converter to illustrate the different modes of noise voltages, since such converters are notorious generators of both modes of noise. The converter shown in Figure 3.25 has two output connections, both of which are isolated from the enclosure of the converter and from the chassis ground. Had one or the other of the output connections been made to the chassis ground, and assuming an infinitesimally low-impedance ground connection, then no common-mode voltage would exist. A common-mode voltage exists between each output connection and the chassis. In Figure 3.25, this common-mode voltage is shown between the lower connection and the chassis only. In addition, a differential-mode noise voltage exists between the two output connections. Thus, the total noise voltage appearing on the output connections of Figure 3.25

TABLE 3.7
Test Result Sheet

Sheet 3-7   Date 15-1-2015

Customer R&D _____ Equipment converter _____ SN _____

Specification MIL-STD-461F _____ Test CE102–28 V limit _____

Equipment MIL-STD/FCC LISN _____

BW 9 kHz _____ VBW _____ Reference _____ Span _____

Notes: Converter and input power return connected to chassis ground plane

Plot	Frequency (MHz)	Measured (dBµV)	Corrected (dBµV)	Limit (dBµV)	Margin [dB]
3–19	0.150	75.7	75.7	70.5	−5.2
	0.392	71.0	71.0	62.1	−8.9
	9.580	64.0	64.0	60.0	−4.0

Test result sheet 3–5, 150 kHz–10 MHz.

TABLE 3.8
Test Result Sheet

Sheet 3-8   Date 18-1-2015

Customer R&D _____ Equipment converter _____ SN _____

Specification MIL-STD-461F _____ Test RE102–ground navy mobile and army _____

Equipment 1 m mono, EMC 0.25–1 GHz preamp _____

BW 9 kHz _____ VBW _____ Reference _____ Span _____

Notes: Converter and input power return connected to chassis ground plane

Plot	Frequency (MHz)	Measured (dBµV)	Antenna Factor (dB)	Preamp Gain (dB)	Corrected (dBµV/m)	RE102 Limit (dBµV/m)	Margin [dB]
3–20	15.130	79.1	5.5	−32.1	52.5	24.0	−28.5
	17.100	77.0	7.5	−32.1	52.4	24.0	−28.4

Test result sheet 3–6, 2–30 MHz.

---

is the sum of the common-mode voltage appearing between the connection and the chassis and the differential noise appearing between the two connections.

An oscilloscope was used to measure the common-mode and differential-mode noise at the input of the 24–5 V converter as well as the common-mode and differential-mode noise at the output of the converter, and these waveforms are shown in Figures 3.26 and 3.29. The input impedance was that of the LISN.

3.3.4 Radiated Emissions from the Output Power of the Converter

The level of radiated emissions from the secondary of the converters is dependent on the geometry of the wiring between the output of the converter and the load.

The MIL-STD-461F (RE102) radiated emissions test setup (10 kHz–18 GHz) requires that interface cabling be included in the test setup and that any cables be located 5 cm from the front of the ground plane at a height of 5 cm above the ground plane and run for a length of at least 2 m.

TABLE 3.9
Test Result Sheet

Sheet 3-9 _____   Date 15-1-2015 _____

Customer R&D _____ Equipment converter _____ SN _____

Specification MIL-STD-461E _____ Test RE102–ground navy mobile and army _____

Equipment LPB, EMC 0.25–1 GHz preamp _____

BW 120 kHz _____ VBW _____ Reference _____ Span _____

Notes: Converter case connected to chassis ground plane

Plot	Type	Frequency (MHz)	Measured (dBµV)	AF (dB)	Preamp Gain (dB)	Corrected (dBµV/m)	RE102 Limit (dBµV/m)	Margin (dB)
	H	63.507	76.6	12.9	−32.8	56.7	24.0	−32.7
3–21	V	77.000	76.6	13.4	−32.8	57.2	24.0	−33.2

Test result sheet 3–7, 30 MHz–1 GHz.

FIGURE 3.16 CE102 emission case connected to chassis. (Courtesy of EMC Consulting Inc., Merrickville, ON.)

The radiated emissions from the output cable from converter number 2 with a 2 m long cable in the RE012 test setup are shown in Tables 3.15 and 3.16 and Figures 3.30 and 3.31.

In earlier measurements on converters, the radiated emissions test setup complied with the basic RE102 configuration.

The major contributor to radiated emissions is the common-mode currents flowing on the two-/three-wire line, with negligible contribution from the differential-mode current flow. Thus, by

FIGURE 3.17 RE102 2–30 MHz case connected to chassis. (Courtesy of EMC Consulting Inc., Merrickville, ON.)

FIGURE 3.18 RE102 30–470 MHz case connected to chassis. (Courtesy of EMC Consulting Inc., Merrickville, ON.)

FIGURE 3.19   CE102 case and input power return connected to chassis. (Courtesy of EMC Consulting Inc., Merrickville, ON.)

FIGURE 3.20   RE102 2–30 MHz case and input power return connected to chassis. (Courtesy of EMC Consulting Inc., Merrickville, ON.)

FIGURE 3.21   RE102 30–470 MHz case and input power return connected to chassis. (Courtesy of EMC Consulting Inc., Merrickville, ON.)

TABLE 3.10
Worst-Case Emissions Compared to the Limit

Frequency (MHz)	Corrected (dBμA)	Limit (dBμA)	Margin (dB)
0.742	113.5	60.0	−53.5
0.747	113.0	60.0	−53.0
1.540	108.0	60.0	−48.0
0.364	110.0	62.8	−47.2
1.120	107.0	60.0	−47.0
2.310	97.0	60.0	−37.0
2.393	58.2	24.0	−34.2
2.000	56.0	24.0	−32.0
3.970	88.0	60.0	−28.0
25.500	50.2	24.0	−26.2
5.570	84.0	60.0	−24.0
59.980	46.6	24.0	−22.6
0.010	108.0	94.0	−14.0
0.044	87.0	81.1	−5.9
0.062	80.0	78.1	−1.9

Worst-case emissions compared to the limit.

TABLE 3.11
Test Result Sheet

Sheet 3-11 _____   Date 9-1-2015 _____

Customer R&D _____ Equipment converter _____ SN _____

Specification MIL-STD-461F _____ Test CE102–28 V limit _____

Equipment MIL-STD/FCC LISN _____

BW 1 kHz and 9 kHz _____ VBW _____ Reference _____ Span _____

Notes: Case isolated

Plot	Frequency (MHz)	Measured (dBµV)	Corrected (dBµV)	Limit (dBµV)	Margin (dB)
	0.010	108.0	108.0	94.0	−14.0
	0.044	87.0	87.0	81.1	−5.9
	0.062	80.0	80.0	78.1	−1.9
	0.364	110.0	110.0	62.8	−47.2
3.22	0.747	113.0	113.0	60.0	−53.0
	1.120	107.0	107.0	60.0	−47.0
	1.540	108.0	108.0	60.0	−48.0
	2.310	97.0	97.0	60.0	−37.0
	3.970	88.0	88.0	60.0	−28.0
	5.570	84.0	84.0	60.0	−24.0

Test result sheet 2–2, 10 kHz–10 MHz.

TABLE 3.12
Test Result Sheet

Sheet 3-12 _____   Date 9-1-2015 _____

Customer R&D _____ Equipment converter _____ SN _____

Specification MIL-STD-461F _____ Test RE102–ground navy mobile and army _____

Equipment 1 m mono, EMC 0.25–1 GHz preamp, 1 m H&S, 2 m H&S _____

BW 9 kHz _____ VBW _____ Reference _____ Span _____

Notes: Case isolated

Plot	Frequency (MHz)	Measured (dBµV)	Antenna Factor (dB)	Preamp Gain (dB)	Cable Atten (dB)	Cable Atten (dB)	Corrected (dBµV/m)	RE102 Limit (dBµV/m)	Margin (dB)
	2.000	79.0	9.4	−32.5	0.0	0.1	56.0	24.0	−32.0
2–23	2.393	81.0	9.6	−32.5	0.0	0.1	58.2	24.0	−34.2
	25.500	65.0	17.5	−32.4	0.0	0.1	50.2	24.0	−26.2

Test result sheet 2–3, 150 kHz–30 MHz.

measuring the common-mode current and using the equation for radiation from a resonant transmission line, the E field at a distance of 1 m from the line may be predicted. Alternatively, the 4Nec2d MOM program can be used with the cable isolated from the ground at the far end (the use of this program and the cable configuration are discussed in Section 7.8).

A second measurement was made using an 11 cm balanced loop antenna, which measures H field. Once the H field at 1 m distance from the line is known, the E field may be calculated.

TABLE 3.13
Test Result Sheet

Sheet 3-13   Date 9-1-2015

Customer R&D   Equipment converter   SN

Specification MIL-STD-461F   Test RE102–ground navy mobile and army

Equipment Equipment LPB, EMC 0.25–1 GHz preamp, 1 m H&S, 2 m H&S

BW 120 kHz   VBW   Reference   Span

Notes: Case isolated

Plot	Frequency (MHz)	Measured (dBμV)	AF (dB)	Preamp Gain (dB)	Cable Atten (dB)	Cable Atten (dB)	Corrected (dBμV/m)	RE102 Limit (dBμV/m)	Margin (dB)
3.24	59.980	65.9	13.4	−32.8	0.0	0.1	46.6	24.0	−22.6

Test result sheet 2–4, 30 MHz–1 GHz.

FIGURE 3.22 CE102 converter case isolated from chassis. (Courtesy of EMC Consulting Inc., Merrickville, ON.)

The methods using the resonant line prediction and *H*-field measurement result in a relatively good correlation, as described in Chapter 9.

In summary, the level of radiated emissions from the line was found by

1. The measurement using an 11 cm balanced loop antenna
2. The measurement of common-mode current flow with a current probe around the line and the use of a program for the calculation of radiation from a resonant transmission line.

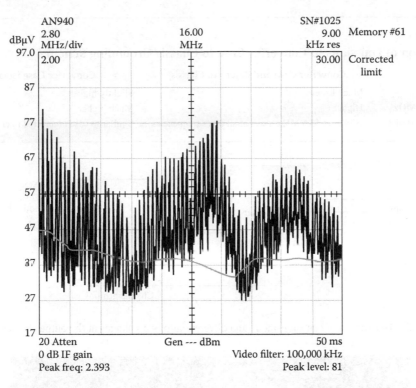

FIGURE 3.23 RE102 2–30 MHz converter case isolated from chassis. (Courtesy of EMC Consulting Inc., Merrickville, ON.)

FIGURE 3.24 RE102 30–470 MHz converter case isolated from chassis. (Courtesy of EMC Consulting Inc., Merrickville, ON.)

TABLE 3.14

Comparison of Emissions to Converter Case to Chassis Grounding Scheme

Test	Converter Case and Return to Chassis	Converter Case Isolated
CE102	Much lower	Slightly higher
RE102 2–30 MHz	Higher	Much higher
RE102 30–470 MHz	Higher at 63.5 MHz lower at all other frequencies	Similar to case connected to chassis alone

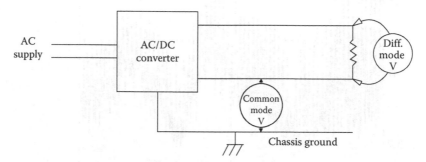

FIGURE 3.25   Illustration of common-mode and differential-mode voltages at the output of a converter.

FIGURE 3.26   Differential-mode transient at the input of the converter.

Shielding the 2 m long line reduced the radiated emission level to below either the ambient level of the shielded room or the measuring instrument's sensitivity. An obvious question is, why not connect the return of either the +5 or ±15 V supply to the ground plane and thereby reduce the common-mode noise and radiation from the line? Indeed, this may be practical and result in a significant reduction in radiation. However, where many converters are used and the returns are connected together, multiple connections of the returns to the chassis are made. This may violate the grounding scheme or result in unacceptably high levels of RF current induced via the loops formed between the commoned returns and chassis.

FIGURE 3.27 Common-mode transient at the input of the converter.

FIGURE 3.28 Common-mode transient at the output of the converter.

In the case study, a requirement existed for a single-point ground that was made at some considerable distance from the equipment. An additional requirement was for a reference ground linking each piece of equipment and was isolated from chassis at the equipment. The reference ground was used to connect the signal ground to the single-point ground. The +5 and ±15 V return was connected to signal ground and thus must be isolated from chassis to comply with the ground isolation requirement. This type of grounding scheme is illustrated in Chapter 8, on grounding.

FIGURE 3.29 Differential-mode noise at the output of the converter.

TABLE 3.15
Test Result Sheet

Sheet 3-15 _____   Date 2-2-2015 ___

Customer R&D _____ Equipment 2nd converter _____ SN _____

Specification MIL-STD-461F _____ Test RE102–ground navy mobile and army _____

Equipment 1 m mono, EMC 0.25–1 GHz preamp, 1 m H&S, 2 m H&S _____

BW 9 kHz _____ VBW _____ Reference _____ Span _____

Notes: Converter #2   output power cable 2 m long in RE102 test setup

Plot	Frequency (MHz)	Measured (dBµV)	Antenna Factor (dB)	Preamp Gain (dB)	Cable Atten (dB)	Cable Atten (dB)	Corrected (dBµV/m)	RE102 Limit (dBµV/m)	Margin (dB)
	8.000	63.0	15.6	−31.9	0.0	0.1	46.8	24.0	−22.8
2–1	28.541	76.3	18.2	−32.5	0.0	0.1	62.1	24.0	−38.1

Test result sheet 2–1, 2–30 MHz.

A total of 17 similar converters were used in an instrument, with five located in close proximity in a compartment in one of the pieces of equipment making up the instrument. Based on the investigation, it was decided, for the sake of intraequipment EMC and to limit the levels of conducted noise placed on the common +28 V supply, that the converter inputs and outputs should be filtered. The choice of filter and attenuation or gain achieved when six different filters were tested is described in a continuation of this case study in Section 5.2.5, on power-line filters.

Digital logic and converters are common and major sources of noise, and so these have been the focus in the preceding sections. However, the equations and figures presented are applicable to other sources of transient intraequipment noise, such as stepper motors, CRT deflection drivers, relays, and switches.

TABLE 3.16
Test Result Sheet

Sheet 3-16   Date 2-2-2015

Customer R&D _____ Equipment 2nd converter   SN _____

Specification MIL-STD-461F _____ Test RE102–ground navy mobile and army _____

Equipment LPB, EMC 0.25–1 GHz preamp, 1 m H&S, 2 m H&S _____

BW 120 kHz _____ VBW _____ Reference _____ Span _____

Notes: Converter #2   output power cable 2 m long in RE102 test setup

Plot	Frequency (MHz)	Measured (dBμV)	AF (dB)	Preamp Gain (dB)	Cable Atten (dB)	Cable Atten (dB)	Corrected (dBμV/m)	RE102 Limit (dBμV/m)	Margin (dB)
	62.400	57.0	12.9	−32.8	0.0	0.1	37.2	24.0	−13.2
2–2	96.573	65.3	9.9	−32.8	0.0	0.1	42.5	24.0	−18.5
	163.000	51.0	10.7	−33.0	0.0	0.1	28.8	28.2	−0.6
	223.000	47.5	11.9	−33.1	0.0	0.1	26.4	30.9	4.5
	292.000	47.0	12.8	−33.2	0.0	0.1	26.7	33.3	6.6

Test result sheet 2–2, 30 MHz–1 GHz.

FIGURE 3.30 RE102 2–30 MHz from the output power cable of converter 2. (Courtesy of EMC Consulting Inc., Merrickville, ON.)

FIGURE 3.31   RE102 30–330 MHz from the output power cable from converter 2. (Courtesy of EMC Consulting Inc., Merrickville, ON.)

DC motors with brushes are notorious sources of low-frequency radiated and conducted emissions, due to the arcing at the brushes. Nevertheless, in some cases, DC brushless motors have an even higher level of noise. The DC brushless motor has a PCB inside the motor that switches power sequentially to coils that attract the magnet/s. This pulse is the source of the noise. For MIL-STD 461 and DEF STAN requirements, it is sometimes necessary to shield the motor using mesh and to shield the wires up to the motor. For an EMC prediction external to any equipment (e.g., at the subsystem or system level), the electromagnetic environment may be found from the typical or maximum levels from ISM sources, examined in Chapter 1, as well as the typical levels of E field at the carrier frequency of transmitters.

3.4 TRANSMITTER-GENERATED NOISE

A transmitter is designed typically to generate a specific frequency or a number of adjacent frequencies. When the transmitter is used to convey information, one or more modulation techniques are used that result in sidebands (i.e., frequency bands on either side of the carrier). For example, when amplitude modulation is used, the upper sideband is the carrier frequency plus the modulating frequencies, and the lower sideband is the carrier frequency minus the modulating frequencies. In addition to the frequencies used in the desired transmission, a transmitter generates a number of unwanted, spurious emissions. These include harmonics of the carrier and sidebands and the master oscillator. In addition, the transmitter generates nonharmonically related frequencies and broadband noncoherent noise.

Broadband noncoherent noise includes *thermal noise*, sometimes called *Gaussian noise* or *white noise*. The assumption made is that the thermal noise exhibits uniform power spectral density. An additional source of noise is known as *flicker* or $1/f$ noise. Thermal noise, $1/f$ noise, and popcorn noise are of particular concern in analog and video circuits; Section 5.5.1 discusses typical levels and evaluation techniques. The major potential source of EMI to equipment and cabling in proximity to an antenna is at the carrier frequency, especially when the E field at the equipment location is in the volts-per-meter range. For receiving equipment tuned to a different frequency and incorporating a filter after the antenna, immunity at the carrier and sideband frequencies generated by a transmitter may be achieved. Case Study 10.2 describes an EMI problem due to equipment, including a receiver that is located in close proximity to a transmitter. When the transmitter is licensed, the frequencies required by the transmitter are clearly denied to a receiver in the area. In addition, channels may be denied due to the spurious emissions from a transmitter. Typical spurious transmitter output spectrums are contained in Reference [3].

Transmitters in the past almost exclusively used tuned amplifiers in the driver and final output stage, and thus, an effective output filter existed. With the advent of broadband solid-state amplifiers, the useful frequency range of the amplifier was extended to decades. For example, amplifiers are available with frequency ranges of 1–100 MHz, 10 kHz–1 GHz, and 1–10 GHz.

The use of wideband amplifiers has typically resulted in increased levels of noncoherent broadband noise. The only filtering effect may be due to the reduction in out-of-band transmitting antenna gain, which is discussed in Chapter 10, on system-level EMC. Modern resonant antennas, which exhibit a high gain at the resonant frequency and a rapid reduction in gain on either side of resonance, may mitigate the EMI potential due to broadband noise. Other antennas are designed to be broadband and so the gain is relatively constant over a wide range of frequencies. The effect of broadband noise and other spurious emissions is to increase the level of ambient noise and possibly deny a number of additional channels in the area. When the noise level at the output of the amplifier, the gain of the transmitting antenna, the distance and relative location, and the gain of the receiving antenna, reflections, or shielding caused by buildings or topography are all known, the potential for EMI may be predicted. An example of an antenna-to-antenna EMI prediction and subsequent measurement at the site is included in Case Study 10.2. The level of thermal noise power, P_n, generated by a transmitter, ignoring input-phase noise and internally generated $1/f$ noise, may be found from the following equation:

$$P_{n(out)} = F(KT)BG \text{ W}$$

where
  F is the transmitter noise figure (numeric)
  KT is the thermal noise power density (at 290 K, KT = 4×10^{-21} W/Hz = −144 dBW/Hz)
  B is the bandwidth under consideration, in hertz
  G is the gain of the amplifier = P_{out}/P_{in} (P_{out} and P_{in} in watts)

Figure 3.32 illustrates measured transmitter broadband noise from a tuned amplifier transmitter and a transmitter with a wideband amplifier for which noise levels almost 10 dB (power) above the levels for the tuned amplifier transmitter are apparent.

Some of the measures that may alleviate EMI caused by transmitter broadband noise are the inclusion of an antenna output filter, the choice of a low-noise power amplifier, or a low-gain amplifier with a corresponding higher-level, low-noise input.

REFERENCES

1. E. Kann. Design guide for electromagnetic interference (EMI) reduction in power supplies, MIL-HDBK-241B. Power Electronics Branch, Naval Electronic Systems Command, Department of Defense, Washington, DC, 1983.
2. M.J. Toia. Sketching the Fourier transform: A graphical procedure. *1980 IEEE International Symposium on Electromagnetic Compatibility*, New York, 1980.
3. P.N.A.P. Rao. The impact of power amplifiers on spectrum occupancy. Theoretical models and denied channel calculations. *IEEE International Conference on Electromagnetic Compatibility*, London, U.K., 1984.

4 Crosstalk and Electromagnetic Coupling between Printed Circuit Board Tracks, Wires, and Cables

4.1 INTRODUCTION TO CROSSTALK AND ELECTROMAGNETIC COUPLING

Crosstalk is defined as the interference to a signal path from other localized signal paths. The term *crosstalk* is often limited in its use to circuits/conductors in close proximity, where the coupling path may be characterized by the mutual capacitance and mutual inductance of the circuits. This restricted definition will be used throughout this chapter. Crosstalk prediction applies irrespective of whether the source is an intentional signal or noise; thus, *crosstalk* is used in this chapter in a broader sense to describe the coupling from either a signal or noise source into a signal path. The terms *electromagnetic coupling* and *crosstalk* are used interchangeably in this chapter. However, in general, *electromagnetic coupling* includes the coupling of an electric, magnetic, or electromagnetic field from one conductor into another, which is not necessarily in close proximity.

As with all electromagnetic compatibility (EMC) predictions or electromagnetic interference (EMI) investigations, the first step is to identify the source of emissions and to determine if the path is conducted, radiated, or crosstalk. Crosstalk should be a prime suspect in an EMI investigation, or a candidate for examination in a prediction, if high transient current or fast rise-time voltages are present on conductors in close proximity to signal-carrying conductors. An example is a cable that contains conductors that carry control or drive levels to a relay motor, lamps, etc., or digital transmission and also contains conductors used for analog, RF, or video signals. The sharing of a cable with digital signals or supply as well as analog signals is also likely to result in crosstalk. Likewise, when a printed circuit board (PCB) track carrying control or logic levels is in close proximity to a second track, which is used for low-level signals, over a distance of 10 cm or more, crosstalk may be expected. Case Study 11.1 (Section 11.12 on PCB layout case studies) provides an example of crosstalk between a logic supply track on a PCB and an analog signal track that are in close proximity over a distance of only 6 cm.

Long cables that carry several serial channels or parallel bits of high-speed data or telemetry are also prime candidates for a crosstalk prediction or candidates in an EMI investigation.

Crosstalk between conductors and cables in close proximity can be caused by an electric field coupling via mutual capacitance or a magnetic field coupling via mutual inductance, as illustrated in Figure 4.1. A third form of coupling is via common impedance, which is covered in Section 8.3.3. It is important to discriminate between EMI caused by common impedance coupling and that resulting from electromagnetic coupling or crosstalk. An example in which either common impedance or crosstalk may be the culprit is as follows: Assume high current flows on a conductor driving filament lamps and that an adjacent signal conductor shares the same return connection. When a common connection of the two returns is made at the lamp end of the cable, EMI may result either from crosstalk or due to the high voltage drop caused by the high lamp current flowing in the DC resistance of the returns.

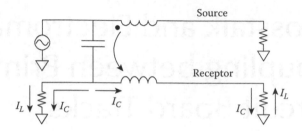

FIGURE 4.1   Electric field (capacitive) and magnetic field (inductive) crosstalk.

Before attributing EMI to crosstalk, it is important to ensure that the data on the receptor line are noise-free at the source end and that the input circuit, at the far end of the line from the source, does not generate noise.

Another potential cause of incorrect reception of data on an interface is reflections due to a line-to-load impedance mismatch, which may be mistaken for crosstalk.

One practical method of determining the presence of crosstalk–induced EMI is to separate the source and receptor conductors. Where this is not feasible, the temporary slowing down of the edges of the source waveform or a reduction in the frequency may be feasible. Where intermittent EMI exists and the suspected cause is crosstalk, the source signal may be increased in amplitude, frequency, or speed, or an additional noise source may be injected via, for example, a transformer or a capacitor. When the incidence of EMI is thereby increased, the likelihood exists that the source of crosstalk has been found.

The crosstalk values in the following sections may be computed, or they may be found in tables or plots, where they are given for specified frequencies or rise times. Generally, when the frequency is given, the rise time for which the same crosstalk occurs is given by $1/\mu f$. Likewise, when the rise time is specified, the frequency may be found from $1/\mu t$.

The crosstalk predictions in this chapter assume lossless transmission lines. One source of losses in transmission lines is *skin effect* (i.e., the current flows predominantly in the surface of the conductor). The skin effect in a current-carrying conductor is characterized by the AC resistance of the conductor. The confinement of the current flow to a small fraction of the cross-sectional area of a conductor is increased in the case where two conductors carrying currents in opposite directions are in close proximity. The current flow then tends to crowd away from the area close to the adjacent conductor, which increases the AC resistance by a factor of 0.6–0.7. A second loss mechanism is in the dielectric material separating the conductors. In practice, the loss due to skin effect tends to predominate in cable transmission lines, such as shielded twisted pairs.

The error due to the exclusion of transmission line loss in the prediction of crosstalk is significant only with long cables and fast rise times and is generally much less than the prediction error. For example, the error due to ignoring loss in the crosstalk prediction for a 60 m long cable with a 100 ns rise-time source is only 7%, whereas the prediction techniques presented in this chapter do not result in errors of much less than 15%.

When source and receptor circuits are not in close proximity, the equations presented in Sections 2.2.4 and 2.2.5 for the generation and reception of electromagnetic fields may conveniently be used to predict the coupling, as described in Section 4.4.6.

When considering crosstalk between PCB tracks, conductors in a cable, or wires and cables in close proximity, it is important to determine if it is predominantly electric field (capacitive) or magnetic field (inductive) coupling. If neither mode predominates, then either both must be examined separately or, more conveniently, the characteristic impedance between the source and receptor circuit and the receptor circuit and ground may be used in a crosstalk prediction, as described in Section 4.4.1. Which of the two modes of coupling predominates depends on circuit impedances, frequency, and other factors. A rough guideline for circuit impedance (in ohms) is as follows:

When source and receptor-circuit impedance products are less than 300^2 Ω, the coupling is primarily magnetic field.
When the products are above 1000^2 Ω, the coupling is predominantly electric field.
When the products lie between 300^2 and 1000^2 Ω, the magnetic or electric field coupling may predominate, depending on geometry and frequency.

However, these guidelines are not applicable in all situations (e.g., crosstalk between PCB tracks located over a ground plane). When the characteristic impedance of the tracks above a ground plane is relatively low (e.g., 100 Ω) and the load and source impedances on the receptor track are higher than the characteristic impedance, the crosstalk is predominantly capacitive.

4.2 CAPACITIVE CROSSTALK AND ELECTRIC FIELD COUPLING BETWEEN WIRES AND CABLES

Figure 4.2 shows a four-wire arrangement of parallel signal lines. If the circuits are unbalanced (i.e., when the return conductor in each circuit is at the same ground potential), then the four-wire arrangement may be replaced by two wires above a ground plane for prediction purposes. The capacitance between the wire connected to one circuit and the ground plane we will call C_1, and the wire connected to the second circuit and the ground plane we will call C_2, as shown in Figure 4.2. When the distances between each wire and its ground plane are equal, $C_1 = C_2$ and $L2 = L1$. The capacitance can be obtained from the inductance of a wire above a ground plane, upcoming Equation 4.8, as shown in Reference [1]:

$$C = \frac{3.38}{L1}$$

$$\therefore C = \frac{3.38}{0.14 \log\left(\dfrac{4h}{d}\right)} \text{ pF/m} \tag{4.1}$$

where
  h is the height above the ground plane
  d is the diameter of the wire

Any unit may be used for the height and the diameter as long as they are both expressed in the same units.

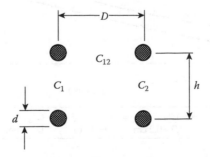

FIGURE 4.2 Four-wire capacitive arrangement.

The mutual capacitance between the two-wire circuits, C_{12}, is computed from the mutual inductance (M), upcoming Equation 4.9, and the equation for the inductance of a wire above a ground plane, Equation 4.8, in accordance with Reference [1]:

$$Cm = \left[\frac{0.07 \cdot \log\left[1+\left(\dfrac{2h}{D}\right)^2\right]}{0.14 \cdot \log\left(\dfrac{4h}{d}\right)\, 0.14 \cdot \log\left(\dfrac{4h}{d}\right)} \right] 3.28 \ \text{pF/m} \qquad (4.2)$$

where

    D is the distance between the two circuits
    h is the height above the ground plane
    d is the diameter of the wires

Although any unit of measurement may be used, D, h, and d must be in the same units for all of these computations. Figure 4.3 shows how the geometry of the circuits affects the self-capacitance and mutual capacitance for 22 AWG wires, which have a diameter of 0.8 mm (0.032 in.). For small h/D ratios, C_{12} grows slowly with an increase in the height of the wire above the ground plane, h. For large values of h/D, C_{12} remains more or less unchanged and then begins to decrease with the wire height. It can be seen that the self-capacitance of the receptor circuit, C_2, plays a role in the coupling. This can be shown by considering the coupling coefficient, K.

$$K = \frac{C_{12}}{C_2} + C_{12} \qquad (4.3)$$

From Figure 4.3, it can be seen that C_2 increases with a decrease in height, and from Equation 4.3, it can be seen that K decreases with an increase in C_2. Thus, the level of crosstalk can be decreased by

    Reducing the height of the wires above the ground plane or decreasing the distance between wires in the same circuit
    Increasing the distance between the two circuits

FIGURE 4.3   Variation of C_1, C_2, and C_{12} with h and D. (From Mohr, R.J., Coupling between open wires over a ground plane, *IEEE Electromagnetic Compatibility Symposium Record*, Seattle, WA, July 23–25, 1968.)

FIGURE 4.4   Unbalanced circuit exhibiting crosstalk.

Figure 4.4a shows two unbalanced circuits exhibiting capacitive crosstalk, with the equivalent circuit in Figure 4.4b. The self-capacitance of the source circuit does not play a role in the crosstalk and is therefore ignored. The self-capacitance of the receptor circuit, C_2, is in parallel with the source impedance, R_S, and load impedance, R_L, of the receptor circuit.

When the line length is longer than the wavelength (i.e., $l \gg \lambda$), the voltage down the length of the line is not constant and a more complex analysis is required. However, for the simple case, where the wavelength is such that the voltage is assumed to be constant for the entire length of the line, the crosstalk voltage V_c is given by

$$V_c = V_1 \frac{Z_2}{Z_1 + Z_2} \tag{4.4}$$

where
$Z_2 = XC_2$ in parallel with R_L and R_S

$$Z_1 = XC_{12} = \frac{1}{2\pi f C_{12}}$$

The total resistance R_t equals R_L in parallel with R_S and is given by

$$R_t = \frac{1}{1/R_L + 1/R_S} \tag{4.5}$$

$$Z_2 = \frac{1}{\sqrt{\left(1/R_t\right)^2 + 2\pi f C_2^2}} \tag{4.6}$$

If the receptor wire of Figure 4.4a is replaced by a twisted pair in which one conductor is connected to the ground plane at each end, as shown in Figure 4.4c, then the capacitive crosstalk is the same as for an untwisted pair of receptor wires.* Assuming an untwisted pair of receptor wires in close proximity, the untwisted pair will exhibit reduced crosstalk, compared to a single wire above a ground plane, simply because of the increase in capacitance C_2 due to the close proximity of the wires. If the twisted pair is connected to a balanced circuit, illustrated in upcoming Figure 4.8, then the capacitive crosstalk is theoretically zero. In practice, the level of crosstalk is very sensitive to irregularity in the twist and variation in the distance between the source and receptor wires down the length of the wires. A twisted pair connected to a balanced circuit typically exhibits a lower level of crosstalk than an untwisted pair. The level of crosstalk voltage induced in an untwisted pair connected to a balanced circuit is dependent on the relative distances between the source wire and the two receptor wires, as discussed later in the evaluation of crosstalk in a four-conductor cable. If the distances between the source and receptor wires are known or if a worst case can be assumed, then the mutual and self-capacitances of the wires may be calculated, an equivalent circuit constructed, and the capacitive crosstalk predicted. When crosstalk occurs between conductors inside a shielded cable, the capacitance between the center conductors and the shielding may become a significant factor in assessing the level of crosstalk. For example, consider a four-conductor shielded cable connected to two unbalanced circuits in which the return conductors are connected at some point to the shield of the cable (i.e., the two return conductors and the shield are at the same potential). The cable configuration and capacitances are shown in Figure 4.5a, and the equivalent circuit

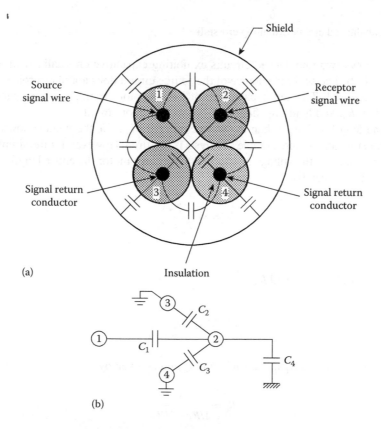

FIGURE 4.5 Four-conductor shielded-cable configuration.

---

* This is because one wire is connected to ground and the induced voltage is close to zero whereas the second wire is connected to a source and load impedance and a crosstalk voltage is induced.

of the unbalanced arrangement is shown in Figure 4.5b. In this configuration, conductor 1 is the source-signal wire, conductor 2 is the receptor-signal wire, and conductors 3 and 4 are the signal return conductors. The crosstalk coefficient K of this arrangement is given by

$$K = \frac{C_1}{C_1 + C_2 + C_3 + C_4} \qquad (4.7)$$

where
   C_1 is the mutual capacitance between signal conductors of the coupled pair
   C_2 and C_3 are the capacitances between the receptor-signal wire and the signal returns
   C_4 is the shield capacitance

Equations 4.4 through 4.6 may be used for the shielded cable by replacing C_2 by the sum of capacitances C_2, C_3, and C_4. A practical example of crosstalk in a four-conductor shielded cable is given in the following.

The cable used in this example contains 20-gauge stranded (27 × 34 AWG) conductors with a diameter of 0.035 in. (0.89 mm), an insulation thickness of 0.0156 in. (0.4 mm), and an inter-axial spacing diagonally between the insulated conductors of approximately 0.008 in. The relative permittivity of the polyethylene insulation is 2.3. The mutual capacitance between conductors 1 and 2, 2 and 4, 4 and 3, and 3 and 1 is 11 pF; diagonally, between 1 and 4 and 2 and 3, it is 5 pF. The conductor to shield capacitance is 20 pF. Assuming that the cable is used for two unbalanced circuits, as shown in Figure 4.4a, that the receptor source and load impedances are each 20 kΩ, and that 1 V AC is applied to the source circuit, the crosstalk plotted against frequency is as shown in Figure 4.6. In our example circuit, we see that the crosstalk voltage increases linearly with frequency until approximately 160 kHz, after which the slope decreases until the crosstalk voltage levels out at 1 MHz. With the load and source resistors of the receptor circuit reduced to 2 kΩ each, the high-frequency breakpoint occurs at 10 MHz. The amplitude above the breakpoint is given by the crosstalk coefficient and is independent of frequency. The crosstalk coefficient in our example circuit from Equation 4.7 is

$$K = \frac{11 \text{ pF}}{11 \text{ pF} + 11 \text{ pF} + 5 \text{ pF} + 20 \text{ pF}} = 0.234$$

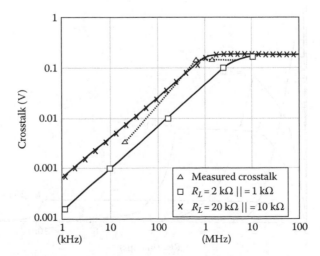

FIGURE 4.6 Measured and predicted crosstalk on the four-conductor shielded cable in an unbalanced circuit.

The maximum crosstalk voltage is therefore 1 V × 0.234 = 0.234 V or 23% of the voltage applied to the source circuit. Measurements on the four-conductor cable are plotted against the theoretical values in Figure 4.6 for the R_L = 20 k ∥ 20 k = 10 kΩ case.

When the source voltage is a step function and the rise or fall time is equal to or less than 0.2 times the time constant of the receptor circuit, the crosstalk peak voltage may be found from the crosstalk coefficient. The time constant is the sum of the cable capacitances times the receptor-circuit load and source resistors in parallel. Thus, in our example cable, the time constant is

$$\tau = \frac{20 \text{ k}\Omega \times 20 \text{ k}\Omega}{20 \text{ k}\Omega + 20 \text{ k}\Omega}\left(11 \text{ pF} + 11 \text{ pF} + 5 \text{ pF} + 20 \text{ pF}\right) = 0.47 \text{ }\mu s$$

When the rise or fall time of the source step function is greater than the time constant, the peak crosstalk voltage is given approximately by $K\tau/t$. The rise or fall time of the crosstalk voltage is defined in this case as the time between the 10% and 90% voltage levels and is given approximately by 2τ. The duration of the crosstalk voltage is equal to the source rise time, t_r. Figure 4.7 illustrates the measured crosstalk voltage for τ = 0.47 μs, R_L = R_S = 20 kΩ, and t = 0.1, 2, 5, and 10 μs.

Using the equivalent circuit approach for multiconductor cables becomes unwieldy. A simpler method is to use published data, which include both the inductive and capacitive crosstalk contributions, as described in Section 4.4.2. A worst-case example of capacitive crosstalk in a multiconductor shielded cable is when two conductors are located in the center of the cable (i.e., furthest from the shield) and are surrounded by other signal conductors. Assuming two centrally located 24 AWG signal conductors connected to unbalanced circuits and surrounded by other 24 AWG signal conductors in a 50-conductor cable with a mutual capacitance between conductors of 14 pF, the crosstalk coefficient is approximately 25%. These very high levels of crosstalk may be considerably reduced by the correct choice of cable, as described in Section 4.4.2. The choices include twisted pair, cross-stranded twisted pair, and shielded twisted pair. The use of balanced circuits, as shown in Figure 4.8, will theoretically reduce the crosstalk to zero. However, when using our example four-conductor cable in balanced circuits, we see from the equivalent circuit of Figure 4.9a that this will not be the case. The measured coupling coefficient for this configuration is 0.1 or 10%, which is only 13% better than for the unbalanced circuits.

FIGURE 4.7   Crosstalk–induced voltages for rise times of 0.1, 2, 5, and 10 μs.

FIGURE 4.8 Balanced circuit configuration.

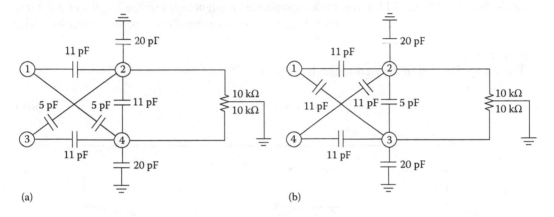

(a)   (b)

FIGURE 4.9 (a) Equivalent circuit of the four-conductor cable in a balanced circuit. (b) Equivalent circuit of the cable when the diagonal conductors are used for the source and receptor circuits.

By using conductors 1 and 4 in the source circuit and 2 and 3 in the receptor circuit, the theoretical crosstalk is zero and the measured crosstalk coefficient is 6.8 mV or 0.68%. Figure 4.9b illustrates the arrangement and why the voltages at conductors 2 and 3 should be at zero. In practice, crosstalk voltages do occur on the receptor conductors due to capacitive imbalances between the four conductors. The crosstalk may be considerably reduced by the use of the balanced circuit with a cable that contains twisted pairs, as discussed in Section 4.4.2.

4.3 INDUCTIVE CROSSTALK AND MAGNETIC FIELD COUPLING BETWEEN WIRES AND CABLES

At high frequencies, it is normally the capacitive coupling that predominates. However, if either the source or receptor lines, or both, are shielded, with the shield connected to ground at both ends, as is often the case, then the coupling will be magnetic. Also at low frequency and where low circuit impedances are encountered, the crosstalk is likely to be inductive. The inductance of a wire over a ground plane is given by

$$L = 0.14\log\left(\frac{4h}{d}\right)\ [\mu H/ft] \qquad (4.8)$$

where
  h is the height above a ground plane
  d is the diameter of the wire

The mutual inductance between two wires over a ground plane is given by

$$M = 0.07 \log\left[1 + \left(\frac{2h}{D}\right)^2\right] \ [\mu H/ft] \tag{4.9}$$

where D is the distance between the two wires.

Figures 4.10 and 4.11 show mutual and self-inductance as a function of h/D and h/d. When the height is much greater than the diameter but the diameter of the cable is large, the inductance of the wire may be limited by the value for a wire far removed from ground, as given in Section 5.1.2.

A practical circuit, which we will use to demonstrate magnetic coupling between unshielded wires, is shown in Figure 4.12. Here, cable capacitances are ignored, and the lengths of the wires are less than a wavelength. The generator voltage e_1 causes a current flow i_1, in the emitter loop. The current i_1 is determined by the voltage e_1, the source impedance R_s, the reactance of L_1, and the load impedance R_b.

The voltage induced in the victim loop is given by

$$e_2 = j\omega M i_1 \tag{4.10}$$

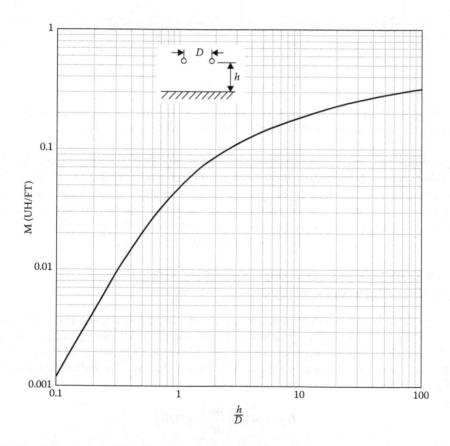

FIGURE 4.10 Inductance between the wires. (From Mohr, R.J., Coupling between open and shielded wire lines over a ground plane, *IEEE Trans. Electromagn. Compat.*, 9(2), 34–45, September 1967. Copyright 1967 IEEE.)

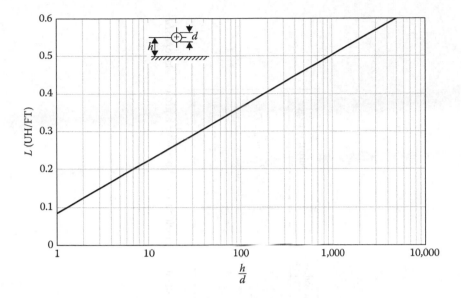

FIGURE 4.11 Inductance of a wire above a ground plane. (From Mohr, R.J., Coupling between open and shielded wire lines over a ground plane, *IEEE Trans. Electromagn. Compat.*, 9(2), 34–45, September 1967. Copyright 1967 IEEE.)

and the voltage across the receiver load resistance R_d is given by

$$V_d = 2\pi f M i_1 \frac{R_d}{R_c + R_d} \frac{1}{\left(1 + \dfrac{j\omega L_2}{R_c + R_d}\right)} \tag{4.11}$$

The last term is replaced in the following discussion by the factor aL_2, giving

$$V_d = 2\pi f M i_1 \frac{R_d}{R_c + R_d} aL_2$$

Equation 4.11 expressed as a magnitude is

$$|V_d| = 2\pi f M i_1 \frac{R_d}{R_c + R_d} \frac{1}{\sqrt{1 + \dfrac{\omega^2 L_2^2}{\left(R_c + R_d\right)^2}}} \tag{4.12}$$

From Equation 4.12, we see that the load voltage is divided not merely by the source and load resistance but by the inductive reactance of the victim wire L_2. The interfering source voltage e_1 may be a sine wave, in which case the open-circuit voltage e_2 induced in the receptor loop is

$$e_2 = 2\pi f M i_1 \tag{4.13}$$

FIGURE 4.12 Magnetic coupling, schematic, and equivalent circuit. (From Mohr, R.J., Coupling between open and shielded wire lines over a ground plane, *IEEE Trans. Electromagn. Compat.*, 9(2), 34–45, September 1967. Copyright 1967 IEEE.)

or a transient interference when

$$e_2(t) = M \frac{I_1}{\tau} A e^{-t/\tau} \tag{4.14}$$

where
  M is the mutual inductance between circuits
  i_1 is the source AC current
  I_1 is the peak value of source transient voltage
  τ is the time constant of source current
  a and A are the factors that account for the division of induced voltage

Where shielded cable is used, factors a and A must be included in Equations 4.15 and 4.16 to account for the additional attenuation of the shields. The peak voltage across the receiver load resistance V_d for the transient case is

$$V_{d(\max)} = M \frac{I_1}{\tau} \frac{R_d}{R_c + R_d} \frac{1}{1 - \dfrac{L_2}{(R_c + R_d)\tau}} \left[e^{-t/\tau} - e^{-(R_c + R_d)t/L_2} \right] \tag{4.15}$$

The equation can be simplified by the use of aL_2 and AL_2 factors so that

$$V_{d(\max)} = M \frac{I_1}{\tau} \frac{R_d}{R_c + R_d} AL_2$$

The factors aL_2 and $AL_2(t)$ are plotted in Figures 4.13 and 4.14, from Reference [2], along with the factors for the shielded case. In the transient case, the induced voltage has a peak value equivalent to

$$M \frac{I_1}{\tau} \quad \text{occurring at } t = 0$$

If the time constant of the receptor circuit is very short compared to the time constant of the source circuit, that is, if

$$\frac{L_2}{R_c + R_d} \ll \tau$$

then the induced voltage will effectively divide between only R_c and R_d. When it is not much smaller than τ, the voltage drop across L_2 must be considered.

Where the coupling is from an unshielded current-carrying wire into a shielded wire, the return current in the victim circuit via the ground plane is attenuated by the factor

$$\frac{1}{1 + j \left(\dfrac{\omega L_S}{R_S} \right)} \tag{4.16}$$

where
  L_S is the shield inductance
  R_S is the resistance

Thus, at high frequency, the return path for a large proportion of the current induced in the shielded victim cable is via the shield. In Table 4.1, the values of the shield inductance L_S for several types of shielded cable located 2 in. above a ground plane have been computed based on cable diameter,

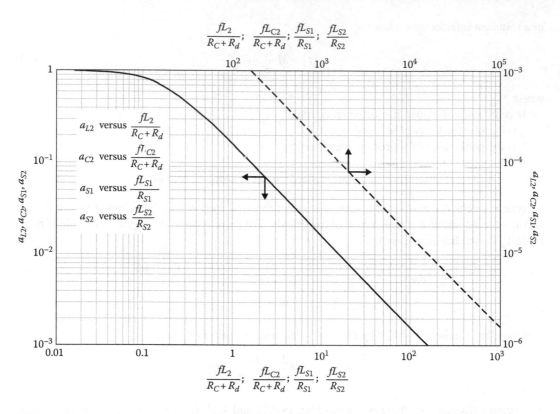

FIGURE 4.13   AC, inductive, and shield attenuation factors. (Copyright 1967, IEEE.)

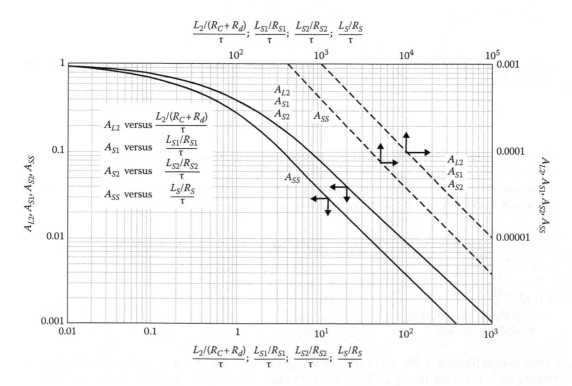

FIGURE 4.14   Transient, inductive, and shield attenuation factors (Copyright 1967, IEEE.)

TABLE 4.1
Coaxial Cable and Shielded Wire Parameters

Type RG()/U	Z_0 (Ω)	ds (in.)	di (in.)	Ls[a] (μH ft)	Li[a] (μH ft)	L_c (μH ft)	R_s (mΩ ft)	Ls/Rs[a] (μs)
Single braid								
8A	50.0	0.285	0.086	0.20	0.27	0.07	1.35	150
58C	50.0	0.116	0.035	0.26	0.33	0.07	4.70	55
141	50.0	0.116	0.036	0.26	0.33	0.07	4.70	55
122	50.0	0.096	0.029	0.27	0.34	0.07	5.70	47
188	50.0	0.060	0.019	0.30	0.36	0.07	7.60	40
29	53.5	0.116	0.032	0.26	0.33	0.07	4.75	55
59A	75.0	0.146	0.022	0.24	0.36	0.12	3.70	65
180	95.0	0.103	0.011	0.26	0.40	0.14	6.00	43
A"	—	0.057	0.023	0.30	0.36	0.06	16.00	19
B"	—	0.056	0.030	0.30	0.34	0.04	6.20	48
Double braid								
9B	50.0	0.280	0.085	0.20	0.28	0.08	0.80	250
5B	50.0	0.181	0.053	0.23	0.31	0.08	1.20	190
55A	50.0	0.116	0.035	0.26	0.33	0.07	2.55	100
142	50.0	0.116	0.039	0.26	0.33	0.07	2.20	118
223	50.0	0.116	0.035	0.26	0.33	0.07	2.55	100
13A	75.0	0.280	0.043	0.20	0.32	0.10	0.95	210
6A	75.0	0.185	0.028	0.23	0.34	0.11	1.15	200

[a] Valid for $h = 2$ in. For $h \neq 2$ in., use Figure 4.17.
A" and B" are shielded hook-up wires.

from Reference [2]. Table 4.1 also shows the values of core inductance L and the difference, which is the cable inductance $L_C (L_C = L_i - L_s)$, as well as the resistance of the shield for lengths of 1 ft. The equivalent circuit of the unshielded-to-shielded wire coupling is shown in Table 4.2. For convenience, the current flow i_1 is shown attenuated by the attenuation factor of the shielded cable:

$$\frac{i_1}{1 + j\left(\dfrac{\omega L_{S2}}{R_{S2}}\right)} \tag{4.17}$$

whereas in reality, the attenuation takes place in the receptor loop, not at the source, and the mechanism is, as discussed in Section 7.2.1, on shielded cables.

Due to the mutual inductance between the shield and the center conductor, equal noise voltages are induced in both, and therefore, the noise current through R_d is reduced as compared to the unshielded case. The voltage induced in the load resistor R_d is now divided not by the core inductance but by the difference between the shield inductance L_s and the core inductance L_i (i.e., by L_c). The voltage V_d is given by

$$V_d = 2\pi f M i_1 \frac{R_d}{R_c + R_d} a_{S2} a_{C2} \tag{4.18}$$

where
   a_{S2} is the attenuation due to the shielded cable
   a_{C2} is the factor due to the division of voltage in the victim loop

Both of these factors are plotted in Figure 4.13.

TABLE 4.2

Summary of Open-Wire and Shielded-Wire Induced Interference

Interference Case	Schematic	AC Equivalent Circuit	AC Equivalent Circuit Of Victim Loop	Output Voltage
(1) Open wire to open wire			$e_2 = jwMi_1$	*AC RMS* $$V_d = 2\pi f m i_1 \frac{R_d}{R_c + R_d} a_{L2}$$ *Peak transient* $$V_{d\max} = \frac{MI_1}{\tau} \frac{R_d}{R_c + R_d} A_{L2}$$
(2) Shielded wire to open wire			$$e_2 = jwMi_1 \frac{1}{1 + j\frac{wL_{S1}}{R_{S1}}}$$	*AC RMS* $$V_d = 2\pi f m i_1 \frac{R_d}{R_c + R_d} a_{S1} a_{L2}$$ *Peak transient* $$V_{d\max} = \frac{MI_1}{\tau} \frac{R_d}{R_c + R_d} A_{S1}$$
(3) Open wire to shielded wire			$$e_2 = jwMi_1 \frac{1}{1 + j\frac{wL_{S2}}{R_{S2}}}$$	*AC RMS* $$V_d = 2\pi f m i_1 \frac{R_d}{R_c + R_d} a_{S2} a_{C2}$$ *Peak transient* $$V_{d\max} = \frac{MI_1}{\tau} \frac{R_d}{R_c + R_d} A_{S2}$$
(4) Shielded wire to shielded wire			$$e_2 = jwMi_1 \frac{1}{1 + j\frac{wL_{S1}}{R_{S1}}} \frac{1}{1 + j\frac{wL_{S2}}{R_{S2}}}$$	*AC RMS* $$V_d = 2\pi f m i_1 \frac{R_d}{R_c + R_d} a_{S1} a_{S2} a_{C2}$$ *Peak transient* $$V_{d\max} = \frac{MI_1}{\tau} \frac{R_d}{R_c + R_d} A_{SS}$$

Source: Mohr, R.J., Coupling between open and shielded wire lines over a ground plane, *IEEE Trans. Electromagn. Compat.*, 9(2), 34–45, September 1967. Copyright 1967 IEEE.

For the transient case, V_d is given by

$$V_{d(\text{max})} = M \frac{I_1}{r} \frac{R_d}{R_c + R_d} A_{S2} \tag{4.19}$$

A_{S2} is plotted in Figure 4.14.

Table 4.2 includes the additional equivalent circuits for shielded-to-unshielded and shielded-to-shielded coupling modes and the equations for both the ac_{RMS} and peak transient voltages. The voltage induced in the victim loop versus frequency is shown in Figure 4.15 for both calculated and experimental values. The voltage for an open-wire to open-wire coupling mode increases linearly with the frequency due to the $2\pi f M$ expression in the equation. The induced voltages for open wire to two types of shielded cable are shown. Here, above a certain frequency, a large proportion of the induced current flows on the shield; the inductive reactance of the shield is high, and so the $2\pi f M$ term is balanced by the term

$$\frac{1}{1 + j\left(\dfrac{\omega L_s}{R_s}\right)}$$

and the induced voltage levels out.

The difference between the two types of shielded cable is that the RG58C/U is a single-braid cable exhibiting a resistance of 4.70 mΩ/ft and the RG223/U is a double braid with a resistance

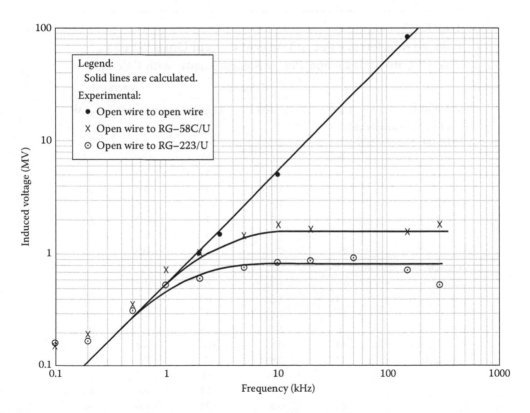

FIGURE 4.15 Coupling from open-wire to open-wire, or RG58C/U to RG223/U, shielded cables. (Copyright 1967, IEEE.)

of 2.55 mΩ/ft. The experimental and calculated results for the shielded cables begin to diverge above 200 kHz because the transfer impedance of the cables no longer equals the DC resistance, due to skin effect and braid leakage factors. The preceding equations may be used for shielded cables above 200 kHz by modeling them as unshielded and finding the current flow on the shield. The load and source resistances are then the shield-termination impedances. The current is found from the voltages developed across the low impedances. Once the current flow is known, the transferred voltage, V_i, may be found from the transfer impedance of the cable as described in Section 7.2.2.

In the inductive crosstalk calculations, no account has been made for the effect of current flow in the receptor circuit on the source circuit. Little error will result in ignoring the interaction between the source magnetic field and the receptor magnetic fields when the circuits are not closely coupled. However, when close coupling exists, the additional load represented by the receptor circuit may be reflected back into the source. Using the equations presented here for close coupling to a receptor circuit, with a low load and source impedance or a shielded cable, may result in a predicted current that is above the source current.

A very approximate limit to the magnitude of the receptor-circuit current is given by measurements using the MIL-STD-462 RS02 test setup, in which the source cable is wrapped around the receptor cable for a distance of 2 m. In performing the RS02 test, currents of approximately 50% of the source current have been measured in the receptor circuit.

Table 4.3 shows the comparison of the measured and calculated transient-induced voltage for the two types of shielded cable previously considered, and Figure 4.16 shows the induced waveforms for transient currents of various rise times. It should be noted that the inductances of the two cables are identical; only the DC resistances are different (refer to Table 4.1).

TABLE 4.3
Induced Interference due to Transient Current Open Wire to Coaxial Cable and Comparison of Measurements with Calculations

Cable Type	τ (μs)	Calculated[a]		Measured[a]	
		e_m (mV)	t_m (μs)	e_m (mV)	t_m (μs)
Open wire to coaxial cable					
RG58C/U	1	2.7	4	2.7	3
	18	1.8	31	1.9	28
	34	1.5	44	1.6	40
	50	1.3	55	1.4	50
RG223/U	1	1.6	4.6	1.6	5.5
	18	1.2	39	1.2	35
	34	1	58	1	50
	50	0.9	72	0.9	65
Coaxial to coaxial cable					
RG58C/U	0.2	1.16	50	1.2	50
	8	1.08	60	1.2	50
	17	1.06	77	1.1	60
	26	1	89	1	70
	54	0.86	108	0.8	80
RG223/U	0.2	0.65	80	0.7	70
	8	0.64	96	0.7	80
	17	0.61	114	0.6	80
	26	0.6	133	0.6	90
	54	0.54	167	0.56	130

[a] $l = 24$ in., $h = 2$ in., $D = 1$ in., $I = 1$ A; e_m is the induced voltage and t_m is the pulse width.

(a)   (b)

(c)

FIGURE 4.16   Scope traces of induced voltage, open wire to coaxial cable. (Copyright 1967, IEEE.) 100 μs/cm on *x*-axis. Waveforms for 0.8, 13, and 50 μs rise times. (a) Open-wire input voltage across 50 Ω, 20 V/cm on *y*-axis. (b) Open-circuit induced voltage in RG58C/U, 1 mV/cm on *y*-axis. (c) Open-circuit induced voltage in RG233/U, 1 mV/cm on *y*-axis.

Due to the number of calculations, the numerous attenuation factors, and the amount of cable data involved in calculating the induced voltage in the victim load resistance, this type of prediction lends itself to the use of a computer program such as the one at the end of this chapter (Section 4.4.6). The program is also available in C+ or C++ from http://www.emcconsultinginc.com. The limitations on the use of these equations are as follows:

When the cable length is long compared to 0.4 times a wavelength.
When the frequency is above 200 kHz for shielded cables, after which the cable may be treated as unshielded and the shield current may be computed.

4.4   COMBINED INDUCTIVE AND CAPACITIVE CROSSTALK

Referring to Figure 4.1, we see that the current flow due to capacitive coupling in the resistor of the receptor circuit close to the source-voltage end of the lines (near end) adds to the inductively induced current, whereas, in the resistor at the far end of the receptor circuit, the capacitive and inductively induced currents tend to cancel. The reason for this difference can be found by considering the capacitive and inductive coupling modes separately. The voltage induced in the receptor circuit by capacitive coupling results in a current that flows in both directions on the line, due to the line-to-ground capacitance distributed down the length of the line, as described in Section 4.4.3. When examining inductive coupling, we may consider the source conductor as the primary of a transformer and the receptor conductor as the secondary. The transformer-coupled voltage appears down the length of the receptor conductor and develops a current flow in the load, composed of the near-end and far-end resistances. The inductively induced current thus flows in one direction only in the receptor circuit.

When the coupling impedance, denoted as Z_{12}, which equals $\sqrt{L_{12}/C_{12}}$, equals $\sqrt{RL_s RL_N}$, the inductively and capacitively induced currents through the far-end receptor load resistor exactly cancel.

Combining the equations for capacitive and inductive coupling, described in Sections 4.2 and 4.3, gives the following expressions for the total interference in the circuit at either end of the receptor circuit and for AC and transient coupling.

For the AC coupling,

$$V_c = 2\pi f M 1 I_1 \left(\frac{R_c}{R_c + R_d}\right) \times \left[\left(3.38 \times 10^{-6}\right)\left(\frac{R_b R_d}{L_1 L_2}\right) - 1\right]$$

$$V_d = 2\pi f M 1 I_1 \left(\frac{R_d}{R_c + R_d}\right) \times \left[\left(3.38 \times 10^{-6}\right)\left(\frac{R_b R_c}{L_1 L_2}\right) + 1\right]$$

For the transient case,

$$V_{c(\text{pk})} = \left(\frac{M 1 I_2}{\gamma}\right)\left(\frac{R_c}{R_c + R_d}\right) \times \left[\left(3.38 \times 10^{-6}\right)\left(\frac{R_b R_d}{L_1 L_2}\right) - 1\right]$$

$$V_{d(\text{pk})} = \left(\frac{M 1 I_2}{\gamma}\right)\left(\frac{R_d}{R_c + R_d}\right) \times \left[\left(3.38 \times 10^{-6}\right)\left(\frac{R_b R_c}{L_1 L_2}\right) + 1\right]$$

where
   V_c and V_d are the AC voltages across R_c and R_d, respectively
   $V_{c(\text{pk})}$ and $V_{d(\text{pk})}$ are the peak transient voltages across R_c and R_d, respectively
   f is the source frequency, in MHz
   M is the mutual inductance, in µH/ft
   L_1 and L_2 are the self-inductance of source and receptor lines, in µH/ft
   I_1 is the AC current in source circuit, in A
   I_2 is the peak value of current in source circuit, in A
   γ is the time constant of source current, in µs

4.4.1 Use of the Characteristic Impedance of PCB Tracks and Wires over a Ground Plane for Predicting Crosstalk

One conductor of a pair of conductors located over a ground plane exhibits a characteristic impedance to ground when both conductors are driven at the same potential, termed the even-mode impedance, Z_{oe}. When driven with different potentials, the impedance of one conductor to ground is termed the odd-mode impedance, Z_{oo}, which has a different value than Z_{oe}. The characteristic impedance between an isolated conductor and the ground Z_o for a number of different conductor configurations can be found in Appendix A. When the conductors are terminated with an impedance that approximates their characteristic impedance, Z_o, the magnitudes Z_{oo} and Z_{oe} may be used to calculate crosstalk for AC voltages using the following relationship [3]:

$$C = \frac{\dfrac{Z_{oe}}{Z_{oo}} - 1}{\dfrac{Z_{oe}}{Z_{oo}} + 1} \tag{4.20}$$

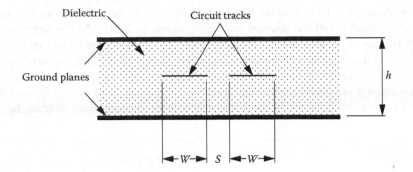

FIGURE 4.17 Coupled stripline configuration.

Consider the configuration shown in Figure 4.17, which is termed an open coupled stripline by the microwave community. The practical application of the coupled stripline for the nonmicrowave engineer is in crosstalk predictions in ribbon cables or in multiconductor PCBs, where the signal conductors are immersed in a homogeneous dielectric and are sandwiched between ground planes. Admittedly, this configuration is not too practical in a PCB, due to the difficulty of tracing signals on the board. However, sandwiching circuit tracks in a closed stripline is a very effective method of reducing radiation from a board. Section 11.4.7 compares the radiated emissions from a practical stripline to a microstrip. We may extend the usefulness of the model by including the case where signal tracks are sandwiched between a plane connected to a supply rail and a ground plane, when a low impedance exists between the supply and the ground, typically via decoupling capacitors. The crosstalk found from Equation 4.20 applies only to near-end crosstalk in the coupled stripline, since the capacitively and inductively coupled voltages exactly cancel at the far end.

The characteristic impedance of the coupled stripline may be found from the relationship

$$Z_o = \sqrt{Z_{oo}Z_{oe}}$$

The relative permittivity of epoxy glass PCBs is approximately 4, and the relative permittivity of the insulation of a ribbon cable ranges from 2 to 5. Typical PCB dimensions are 0.02 in. (0.5 mm) for the width of the tracks, W, 0.03 in. (0.75 mm) for the distance between the tracks, S, 0.034 in. (0.86 mm) for the height of a single-layer board, and 0.005 in. (0.12 mm) for the height of the layers in a multilayer board. The near-end crosstalk, C, for a number of ratios of S/h and W/S and $\varepsilon_r = 4$ and 2 may be obtained from Table 4.4.

TABLE 4.4
Z_{oo}, Z_{oe}, and Crosstalk, C, for the Stripline Configuration

		$\varepsilon_r = 4$				$\varepsilon_r = 2$	
W/h	S/h	Z_{oe}	Z_{oo}	C	Z_{oe}	Z_{oo}	C
0.3	0.01	360	104	0.55	253	73	0.55
0.3	0.10	340	172	0.38	239	121	0.38
0.3	0.50	284	236	0.09	200	166	0.09
0.5	0.01	266	92	0.48	187	65	0.48
0.5	0.10	244	140	0.27	172	99	0.27
0.5	0.50	216	186	0.08	152	130	0.08
1.0	0.01	154	74	0.35	108	52	0.35
1.0	0.10	150	100	0.20	106	70	0.20
1.0	0.50	138	124	0.05	97	87	0.05

A more common PCB configuration, in which the tracks are on the surface of the PCB material, may be modeled by the coupled microstrip shown in Table 4.5. The impedances Z_{oo} and Z_{oe} for permittivities of 2 and 4 and the crosstalk as a coefficient, C, are shown in Table 4.6. From Equation 4.20, as demonstrated in Tables 4.4 and 4.6, it is seen that the crosstalk coefficient is independent of the values of ε_r and may be obtained simply from the ratios W/h and S/h.

Tables 4.4 and 4.6 provide the near-end crosstalk. Unlike the stripline, the far-end crosstalk in the microstrip is not zero, due to the inhomogeneous dielectric composed of both the air and the

TABLE 4.5
Coupled Microstrip Parameters with Predicted and Measured Crosstalk for 50 ns Rise Time and a 2.5 V Transition

	Predicted Crosstalk (mV)	Measured Crosstalk* (mV)	Predicted Crosstalk (μWave SPICE) (mV)
50 Ω termination			
VNEXT(max)	4.98	4.5	5.2
VFEXT(max)	−2.35	−2	−2.25
1000 Ω termination			
VNEXT(max)	24.3	25	29
VFEXT(max)	24.1	25	22

Per-unit length parameters

$C_{11} = 137.7$ pF/m $C_{21} = 5.15$ pF/m
$C_{12} = 5.15$ pF/m $C_{22} = 137.7$ pF/m
$L_{11} = 0.314$ μH/m $L_{21} = 0.036$ μH/m
$L_{12} = 0.036$ μH/m $L_{22} = 0.314$ μH/m

TABLE 4.6
Z_{oo}, Z_{oe}, and Crosstalk, C, for the Coupled Microstrip

W/h	S/h	$\varepsilon_r = 4$			$\varepsilon_r = 2$		
		Z_{oe}	Z_{oo}	C	Z_{oe}	Z_{oo}	C
0.2	0.01	360	104	0.55	253	73	0.55
0.2	0.05	202	56	0.56	310	86	0.56
0.2	0.20	190	77	0.42	292	119	0.42
0.2	0.50	171	98	0.27	263	150	0.27
0.2	1.00	157	116	0.15	240	177	0.15
0.5	0.05	141	45	0.52	216	69	0.52
0.5	0.20	136	60	0.38	208	93	0.38
0.5	0.50	125	73	0.26	191	113	0.26
0.5	1.00	116	85	0.15	177	130	0.15
1.0	0.05	97	37	0.45	149	56	0.45
1.0	0.20	95	47	0.33	145	73	0.33
1.0	0.50	89	56	0.23	137	86	0.23
1.0	1.00	84	63	0.14	129	96	0.14
2.0	0.20	61	28	0.37	94	43	0.37
2.0	0.50	61	35	0.26	92	53	0.26
2.0	0.50	58	40	0.18	88	61	0.18
2.0	1.00	55	43	0.12	85	66	0.12

PCB material. Far-end crosstalk is typically, but not always, lower than near-end crosstalk and may be zero, depending on the track layout, which influences the magnitudes of Z_{oo} and Z_{oe} and the receptor and source load resistor values. The typical values of far-end crosstalk are a factor of 1.5–0.37 of the near-end values for microstrip lines [4].

A comparison of Tables 4.4 and 4.6 illustrates that, for the same dimensions of h, S, and W, the crosstalk is lower for the stripline configuration, due to the increased capacitance between the center conductors and the ground planes. For either configuration, increasing the distance between the tracks (the width of the tracks) or decreasing the height between the tracks and the ground will decrease the crosstalk. Increasing the spacing between tracks and inserting a shield track, which is connected to ground, will decrease the level of crosstalk. However, measurements have shown that the presence of the shield conductor reduces the crosstalk by only 6–9 dB. The wider the shield conductor, the greater the reduction. The use of Equation 4.20 is not limited to PCBs but may be used in a wire-to-wire crosstalk prediction. The odd- and even-mode impedances for pairs of wires above a ground plane may be obtained from the following equations:

$$Z_{oo} = 276\log\left(\frac{2D}{d}\right)\sqrt{\frac{1}{1+\left(\frac{D}{2h}\right)^2}}$$

$$Z_{oe} = 69\log\left(\frac{4h}{d}\right)\sqrt{\frac{1}{1+\left(\frac{2h}{D}\right)^2}}$$

where
   h is the height above the ground plane
   D is the distance between the wires
   d is the diameter of the wires

The same units must be used for h, D, and d. The crosstalk may be found by the use of Equation 4.20.

4.4.2 CROSSTALK IN TWISTED-PAIR, CROSS-STRANDED TWISTED-PAIR, SHIELDED TWISTED-PAIR, AND RIBBON CABLES

The very high levels of crosstalk measured in cables manufactured from untwisted conductors and used in either balanced or unbalanced circuits may be greatly reduced by replacement with twisted-pair or shielded twisted-pair cables. To reduce problems due to reflections, such as pulse degradation, long cables used for communications or cables used for transmission of data with short rise and fall times are usually terminated in the characteristic impedance of the lines. These terminated cables are the focus of attention here. In a properly designed, appropriately terminated communication signal interface, the characteristic impedance of the cable will be extended to the back plane inside the equipment and even to the tracks of the PCB on which the interface circuits are mounted. The characteristic impedance of a number of different cables and PCB track configurations may be computed from the equations contained in Appendix A. A statistical model has been developed for predicting the near-end and far-end crosstalk in cross-stranded multiple twisted-pair cable versus measured crosstalk [5]. A cross-stranded multiple twisted-pair cable comprises a number of twisted pairs randomly positioned within the cable during manufacture. The advantage of the cross-stranded over the more typical construction, in which the pairs rotate around a fixed point, is that the same pairs are not always in close proximity down the length of the cross-stranded cable. Assuming that the group twist frequency is the same (i.e., the twist period is the same for all twisted pairs in the cable), the near-end crosstalk isolation varies between 60 and 70 dB and the far-end crosstalk isolation between 45 and 57 dB for a 10-pair cross-stranded cable.

The crosstalk, in decibels, is defined as either the ratio of the crosstalk–induced voltage to the source voltage, in which case the crosstalk is negative, or the ratio of the source voltage to the induced voltage, in which case the crosstalk is positive and is more accurately termed the *crosstalk isolation*. The latter of these two definitions will be used here.

By twisting the pairs within the cable at a different twist frequency, the crosstalk may be further reduced; however, this type of construction is rarely available in commercial cables. Measured data are available on crosstalk in unshielded twisted-pair cable and shielded twisted-pair cable [6]. The CR-CS test method, contained in the U.S. government document CR-CS-0099–000, was used in the test. The CR-CS method uses the cable in a balanced circuit configuration, so the test results are valid only for balanced circuits.

The mean values of near-end and far-end crosstalk for 19-pair shielded and unshielded cables and 27-pair shielded and unshielded cables are reproduced in Figure 4.18. The pairs in the shielded type B cable have a characteristic impedance of 50 Ω and a polypropylene insulation with a thickness of 0.01 in. In type C cable, the pairs have a characteristic impedance of 100 Ω and a foam polyethylene insulation with a thickness of 0.023 in.

Ribbon or flat cables exhibit a stable characteristic impedance, a controlled distance between conductors, and, depending on type, a low crosstalk. The following crosstalk figures for various types of flat or ribbon cable are published by, and reproduced here with the kind permission of, the Belden Cable Company.

The characteristic impedance of the cables is specified either with one conductor connected to ground and the adjacent conductor used for the signal (GS configuration) or with the signal conductor sandwiched between two ground conductors (GSG configuration). The typical unbalanced crosstalk is expressed in percentages, with a rise time of between 3 and 7 ns. The frequency at which the unbalanced frequency–domain crosstalk equals the pulse crosstalk may be very approximately obtained by the relationship $f = 0.5 \, \pi t$.

The Belden 9L280XX Series is a flat 0.05-pitch unshielded cable with a GS characteristic impedance of 150 Ω and a GSG impedance of 105 Ω. The cable is available with 10–64 conductors. The cable dimensions and adjacent conductor unbalanced pulse crosstalk results are reproduced in Figure 4.19. The balanced crosstalk in the frequency domain is provided in Figure 4.20.

The 9L283XX Series is a flat 0.05-pitch shielded cable with a GS impedance of 45 Ω and a GSG impedance of 50 Ω. The cable is available with 9–64 conductors. The cable dimensions and adjacent conductor pulse crosstalk are reproduced in Figure 4.21.

The 9GP10XX Series is a 0.05-pitch flat cable that incorporates a copper ground plane and is available with 20–60 conductors, of which one conductor is ground. The cable has a GSG impedance of 60 Ω. The dimensions and unbalanced pulse crosstalk are given in Figure 4.22.

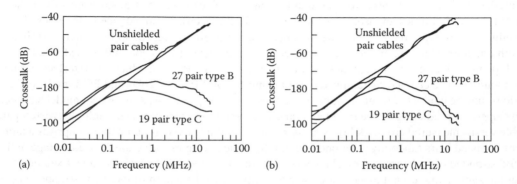

FIGURE 4.18   Mean values of (a) far-end and (b) near-end crosstalk for shielded and unshielded multipair cables.

FIGURE 4.19 (a) Belden 9L280XX flat cable dimensions and (b) unbalanced pulse crosstalk.

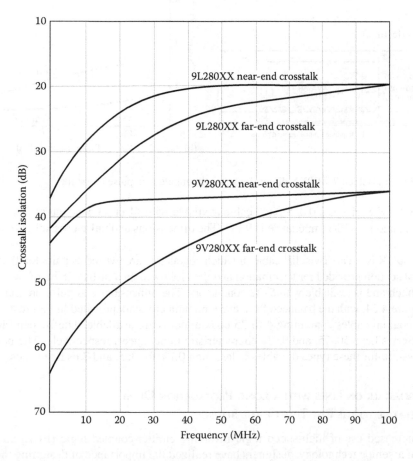

FIGURE 4.20 Belden 9L280XX and 9V280XX balanced frequency–domain crosstalk.

FIGURE 4.21   (a) Belden 9L283XX cable dimensions and (b) unbalanced pulse crosstalk.

FIGURE 4.22   (a) Belden 9GP10XX dimensions and (b) unbalanced pulse crosstalk.

The 9L320XX Series is a 0.025-pitch cable available with 41 or 53 conductors. The GS impedance is 135 Ω and the GSG impedance is 93 Ω. The dimensions and pulse crosstalk are reproduced in Figure 4.23.

The 9V280XX is a Vari-Twist flat cable in which conductors are twisted as pairs for 18 in. followed by a 2 in. flat section provided for termination and then followed by a further 18 in. of twist. The cable has a 0.05 pitch and is available with 5–32 conductors. The dimensions and pulse crosstalk are illustrated in Figure 4.24, with the balanced frequency–domain crosstalk provided in Figure 4.20.

Ribbon coaxial cables containing 4 to 25 coaxial cores are available. The Belden 9K50, 9K75, and 9K93 Series have 50, 75, and 93 Ω characteristic impedances, respectively. The near-end and far-end crosstalk for these types of cables is less than 0.1% for 3, 5, and 7 ns rise times.

4.4.3   CROSSTALK ON LINES WITH A LONG PROPAGATION DELAY
RELATIVE TO THE RISE TIME OF THE SOURCE

With the increased use of high-speed logic involving emitter-coupled logic (ECL) and Schottky and gallium arsenide technology, designers have realized the importance of designing the interconnections as transmission lines terminated in their characteristic impedance. When the rise time

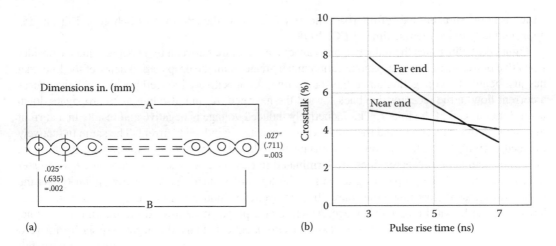

FIGURE 4.23 (a and b) Belden 9L320XX unbalanced pulse crosstalk.

FIGURE 4.24 (a) Belden 9V280XX dimensions and (b) unbalanced pulse crosstalk.

of the logic is equal to or shorter than the propagation delay of the line, the crosstalk prediction is somewhat different than described thus far in this chapter. These high-speed terminated source transmission lines are the focus here.

The propagation delay of two PCB tracks side by side is given by

$$t_{pd} = 1.017\sqrt{0.475\varepsilon_r + 0.67} \ \text{(ns/ft)}$$

where ε_r is the relative dielectric constant of the PCB. Thus, for a typical glass epoxy PCB, for which the dielectric constant (£) ranges from 4 to 5, the propagation delay (t_d) is 1.62–1.77 ns/ft, which is approximately half of the rise time of TTL, equals the rise time of Schottky TTL, and is approximately twice the rise time of ECL100K.

Figure 4.25 illustrates the inductive and capacitive crosstalk between two coupled lines. Consider a positive pulse, with rise time t_r and duration much greater than the propagation delay of the line, that has just begun its propagation down the source line. A capacitively induced positive voltage causes a current flow in the receptor line back toward the near end, N, and also begins to propagate down toward the far end of the line, F. The inductively induced voltage is negative and results in a current that flows in the N direction and thus adds to the capacitively induced current. The same inductively induced current, I_L, subtracts from the capacitively induced current, I_c, flowing toward the far end. When the source and receptor lines are terminated in their characteristic impedances, I_L is greater than I_c and the voltage propagating toward F, the forward voltage, has a negative polarity. As the pulse propagates further down the source line, the voltages induced in the receptor line add cumulatively to the forward crosstalk voltage, which is also propagating toward the far end of the line, assuming the propagation delays of the two lines are identical. Thus, the negative-polarity forward crosstalk voltage, V_L, increases in direct proportion to the length of the line and has a pulsewidth approximately equal to the rise time of the source voltage. The backward voltage propagated toward the N end of the line results in a continuous, constant current flow, so the voltage developed across the near-end termination impedance is constant, regardless of the length of the line and of a duration longer than the rise time of the source voltage. As the source voltage reaches the end of the source line, the forward crosstalk voltage reaches the F end of the receptor line and appears across the F termination impedance. However, the backward crosstalk voltage must still propagate toward the N end of the line in a time equal to the propagation delay of the line. Therefore, the voltage appearing across the terminating resistor at the N end has a duration equal to twice the propagation delay of the line, since it was developed from the moment the source voltage began its propagation down the line until the last backward voltage returned from the far end of the line.

Logic does not typically respond to an input voltage pulse with a duration equal to the rise time of the logic, even when the voltage level is above the logic threshold. The input pulse duration must be several times the rise time of the logic before an output response occurs. It is for this reason that

FIGURE 4.25 Near-end and far-end crosstalk in a line where the propagation delay is longer than t_r.

near-end crosstalk is more likely to result in EMI than is far-end crosstalk. A method of computing crosstalk in high-speed digital interfaces when the mutual inductance, mutual capacitance, and propagation delay of the line are known is provided in Reference [7]. An alternative to this is the use of one of the computer programs described in the next section.

4.4.4 CROSSTALK COMPUTER PROGRAMS

Many of the latest computer-aided design PCB layout programs do allow crosstalk, and these include the "incases" and "applied simulation technology" programs described in Section 11.10.

In addition to full PCB layout programs, other programs allow the characteristics of high-speeds transmission on PCB tracks, such as impedance, crosstalk, ringing, and time delay. One such program is GREENFIELD 2 TM, from Quantic Laboratories. Also, µWave SPICE, from EESOF, allows the computation of crosstalk, delay, and ringing. This program was used to compile the tables, presented in this section, of near-end and far-end crosstalk voltage for a number of PCB layouts and pulse rise times. Clayton R. Paul has developed a simple model that may be used with SPICE or PSPICE to predict crosstalk in coupled transmission lines. SPICE allows the transient analysis of transmission lines with a specified characteristic impedance and propagation delay but not the analysis of coupled transmission lines. The model developed by C.R. Paul extends the SPICE capability to coupled lines and allows the modeling of a line immersed in an inhomogeneous medium, such as tracks on a PCB [11]. A comparison of the crosstalk prediction with measured crosstalk data from PCBs shows excellent prediction accuracy.

In order to examine the effect on crosstalk voltage, a number of PCB configurations as well as source and receptor-circuit impedances were modeled, using µWave SPICE. The initial configuration was that described in Reference [9] and shown in Figure 4.26, in which the PCB track length was 7860 mil (20 cm); the track width, W, and distance between tracks, S, were 100 mil (2.5 mm);

FIGURE 4.26 Tracks on a PCB forming a coupled microstrip configuration.

the PCB material had a relative permittivity of 5; and the height of the tracks above the ground plane was 62 mil (1.6 mm). The predicted near-end and far-end crosstalk voltage and circuit coupling parameter [9], the measured voltage [10], and the predicted crosstalk using μWave SPICE are shown in Table 4.5. The rise time of the source signal is 50 ns, with a peak-to-peak voltage of 2.5 V. The source-circuit termination impedance is 50 Ω, and the receptor-circuit termination impedances are 50 Ω in case 3 and 1000 Ω in case 4. A very good correlation exists between the predicted crosstalk [9] and the measured crosstalk, whereas an acceptable correlation exists between the μWave SPICE predicted values and the measured. In the following μWave SPICE plots, $v[1]$ is the near-end source waveform, $v[2]$ is the far-end source waveform, $v[3]$ is the near-end receptor waveform, and $v[4]$ is the far-end receptor waveform. In the plots these are labeled $v[1]$, $v[2]$, $v[3]$, and $v[4]$.

The waveforms for correctly terminated (50 Ω) source and receptor circuits with a source rise time of 2 ns and a peak voltage of 0.8 V are shown in Figure 4.27. The positive near-end crosstalk exhibits an amplitude approximately twice the value of the negative far-end crosstalk, with a slightly greater pulsewidth, as expected.

Decreasing the distance S between the two tracks to 13 mil (0.33 mm) not only increases the crosstalk voltage but also increases the near-end and far-end crosstalk voltage pulsewidths, as shown in Figure 4.28. The pulsewidth increase is due to a modification in the microstrip characteristic impedance caused by the proximity of the two tracks.

Retaining the 50 Ω terminations but removing both source and receptor tracks from close proximity to the ground plane also changes the characteristic impedance of the line, resulting in an increase in the amplitude and pulsewidths of the near-end and far-end crosstalk, as shown in Table 4.7. Figure 4.29 illustrates the waveforms. The rise time of the voltage at the far end of the source line is degraded by reflections due to the mismatch between the line and the load impedance.

The case where the characteristic impedance of the source line matches the source load impedance (50 Ω) but the receptor terminations are 1000 Ω and are mismatched to the line is shown in

FIGURE 4.27 Crosstalk with $V_s = 0.8$ V, $t_r = 2$ ns, $W = 100$ mil, $S = 100$ mil, $l = 7860$ mil, $h = 62$ mil, $t = 0.001$ mil, and $R_L = R_s = 50$ Ω.

FIGURE 4.28 Crosstalk as a result of decreasing track spacing to 13 mil (0.33 mm).

TABLE 4.7
Crosstalk Values for the Coupled Microstrip

W (mil)	S (mil)	L (mil)	L (cm)	H (mil)	R_{L1} (Ω)	R_{L2} (Ω)	t_r (ns)	N_E (V)	F_E (V)
Source voltage = 0.8 V into R_o = 50 Ω, ε_1 = 5									
100	100	7,860	20	62	50	50	2	+0.03	−0.019
100	13	7,860	20	62	50	50	2	+0.13	−0.11
100	13	7,860	20	62	1000	1000	2	+0.23	+0.2
100	100	15,720	40	62	50	50	2	+0.03	−0.027
100	100	7,860	20	62	1000	1000	2	+0.6	−0.3
									+0.5
100	500	7,860	20	62	50	50	2	+0.0024	−0.0025
100	100	7,860	20	62	50	50	4	+0.02	−0.01
100	100	7,860	20	62	50	50	8	+0.01	+0.005
100	100	7,860	20	500	50	50	2	+0.12	−0.08
100	100	7,860	20	5,000	1000	1000	2	+0.4	+0.06
Source voltage = 3.6 V into R_o = 50 Ω, ε_r = 5									
100	100	7,860	20	62	50	50	2	+0.14	−0.08

Figure 4.30. Here, we see that although the far-end crosstalk is initially negative, it changes sign and equals the near-end crosstalk in amplitude. Due to the multiple reflections, which are clearly visible, the pulsewidths at 30 ns are much greater than in the matched case.

The potential for EMI to analog-signal lines may not be high with a 30 ns crosstalk pulsewidth. However, when the pulse repetition rate is in megahertz, a very real potential for EMI exists.

FIGURE 4.29   Crosstalk with $S = 100$ mil, $W = 100$ mil, and $H = 1000$ mil.

FIGURE 4.30   Crosstalk with $W = 100$ mil, $S = 100$ mil, $H = 62$ mil, and $Z = 50\ \Omega$ but with the receptor load resistors changed to $1000\ \Omega$.

In examining the original matched configuration and extending the length of the line from 20 to 40 cm, we see that the far-end crosstalk amplitude and the near-end crosstalk pulsewidth are both twice the value of the 20 cm long line. Table 4.7 contains the predicted crosstalk for a number of risetimes, receptor-circuit termination impedances, distances between tracks, and heights above the ground plane. As expected, the crosstalk is greater where the tracks are closer together and/or higher above the ground plane.

The microstrip model terminated in passive loads, used thus far, is useful in evaluating factors that mitigate crosstalk, such as using ground planes under coupled tracks, increasing the distance between tracks, and matching the termination impedance to the line impedance. In addition, the crosstalk with 1000 Ω receptor-circuit termination impedance may apply to the prediction of cross-talk from a logic-to-analog circuit, because analog-circuit input and source impedances are often close to 1000 Ω. The crosstalk voltage may be scaled when the source voltage is other than 0.8 V. For example, with a logic voltage of 3.6 V at a 2 ns risetime, a 1000 Ω termination impedance at both ends of the receptor track, $l = 20$ cm, $W = 2.5$ mm, and $S = 2.5$ mm, the far-end crosstalk voltage is $(3.6/0.8) \times 0.6$ V $= 2.7$ V.

Further information on modeling crosstalk on PCBs is provided in Reference [10]. Reference [13] describes a finite element method that may be used for modeling crosstalk on multiple PCB tracks in either a microstrip or a stripline configuration.

In predicting the crosstalk to a line connected to IC gates, the line source and load impedances are not the same and are complex (i.e., with real and imaginary components). Also, they change with input or output voltage level. Consider the receptor circuit shown in Figure 4.31, in which the input of a TTL gate is connected at the near end of the line and the output of a TTL gate is connected at the far end. For the positive excursion of the source-voltage step, the configuration shown in Figure 4.31 may be considered the worst case for crosstalk. The input of a TTL gate sources current that flows out of the gate and is sunk by the driver gate at the far end of the line. Thus, a TTL input tends to rise to a "1" level when disconnected from a low impedance. The positive near-end crosstalk is therefore developed across an impedance that is effectively connected to the logic supply

FIGURE 4.31 Worst-case configuration for crosstalk to TTL gates.

and not to ground. The far-end impedance is low referenced to ground, for it is formed by the "on" impedance of a transistor.

μWave SPICE has the advantage that discrete, logic, and analog ICs and reactive loads may be modeled directly, whereas GREENFIELD offers only a SPICE interface. The worst-case TTL crosstalk configuration shown in Figure 4.31 was modeled using μWave SPICE. To ensure that this is indeed a worst-case configuration, the crosstalk was compared to that predicted for a negative source-voltage step and when the driver and input gate locations were interchanged. The configuration shown in Figure 4.31 was found to be indeed the worst case. This configuration was used for cases where the track widths were fixed at 12 mil (0.3 mm) and the distance between the tracks was varied, as were the lengths for two values of board thickness (i.e., 34 mil [0.86 mm] and 71 mil [1.8 mm]).

The near-end and far-end waveforms for two 96 cm tracks located 0.86 mm above the ground plane with a spacing of 0.6 mm are shown in Figure 4.32. The amplitude of the near-end crosstalk is higher than that of the far end, and the far end exhibits both a negative and positive voltage excursion, which is common with an unmatched load impedance. The far-end positive excursion is clamped to between 0.4 and 0.6 V by the "on" transistor in the driver output circuit. In the μWave SPICE plot, $v[7]$ is the output of the near-end TTL gate. A summary of the near-end and far-end crosstalk for line lengths up to 192 cm and tracks as close as 0.3 mm are shown in Table 4.8. Very often, the receptor track is sandwiched between two or more source tracks, especially true for a data bus, and the impact of this configuration is shown in Table 4.8, where the far-end crosstalk voltage is seen to be approximately 1.62 times and the near-end 1.76 times that for a single source track. When the receptor track is on the outside of the two source tracks, the far-end crosstalk is 1.62 times and the near end 1.17 times that for a single source track.

Electromagnetic interference does not occur until the length of the track is 384 cm for a distance between tracks of 0.3 mm, as shown in Figure 4.33. The input voltage is 1.7 V, with a pulsewidth of 45 ns. The measured input current spike is 11 mA, so the input noise energy at which EMI occurs, given by tIV (where t is the pulsewidth, I is the current, and V is the voltage), is 0.87 nJ. At a line

FIGURE 4.32   Crosstalk to TTL gates with $S = 0.6$ mm, $l = 96$ cm, $h = 0.86$ mm, $V_s = 5$ V, and $t_r = 2$ ns.

TABLE 4.8
Crosstalk on a TTL Signal Track in a Microstrip Configuration

H (mm)	L (cm)	S (mm)	N_E (V)	N_E (t_{P_W} ns)	F_E (V)	F_E (t_{P_W} ns)
0.86	96	0.3	+1.7	13	−0.37	4
					+0.38	13
0.86	96	0.6	+1.0	13	−0.46	4
					+0.2	13
0.86	96	7.2	+0.65	13	−0.35	4
					—	—
0.86	96	2.5	+0.42 to +0.70	13	−0.25	4
					+0.05	13
0.86	96	12.5	+0.008	13	−0.0035	4
					+0.0015	13
0.86	192	0.3	+1.7 (2.1 pk)	26	−0.8	4
					+0.38	26
0.86	384	0.3	+1.7	47	−0.82	4
					+0.4	47
1.8	96	0.3	+2.3	13	−0.3	4
					+0.35	13
1.8	96	1.2	+1.1	13	−0.3	4
					+0.15	13
1.8	96	2.5	+0.4 to +0.6	—	−0.25	4
					+0.05	13
Two source tracks, receptor on outside						
0.86	96	0.3	+2		−0.6	
					+0.4	
Two source tracks, receptor on inside						
0.86	96	0.3	+3		−0.6	
					+0.6	

length of 192 cm, the input level is well above the TTL DC "1" level voltage of 1 V but with a pulse width of only 26 ns, the TTL gate does not have time to respond. Had we modeled a faster type of logic, the permissible line length before EMI occurs would almost certainly be less than 384 cm. The near-end crosstalk voltage is independent of line length, and the 1.7 V level at a pulse width of 13 ns, developed with a 96 cm line length, may upset some modern fast logic types.

When the ground plane is removed from under the source and receptor tracks, source line-to-load impedance mismatch is created. This results in an increase in the rise time of the source voltage developed across the far-end termination impedance and in an increase in the crosstalk pulsewidth, as illustrated in Figure 4.34 for 96 cm tracks with 2.5 mm spacing and a 5 V source voltage with a 2 ns risetime.

4.4.5 Crosstalk between PCB Traces and Coupling around Components on a PCB

The analysis of crosstalk between short lengths of PCB traces (lands) connected, for example, to components in power line or signal line filters and between the input and output connections of an operational amplifier or RF circuit is often required.

Reference [14] provides equations for the mutual capacitance between parallel PCB traces over a ground plane, that is, between microstrip traces. It also provides the coupling between traces

FIGURE 4.33   Crosstalk–induced EMI to a TTL signal track. $S = 0.3$ mm, $h = 0.86$ mm, $l = 384$ cm, $V = 5$ V, $t_r = 2$ ns.

FIGURE 4.34   Crosstalk to a TTL signal track when the ground plane is removed from under both source and receptor tracks.

without a ground plane but omits the signal return path. Often, the signal return is in close proximity to the signal trace and then the equation in Reference [14] is not valid.

In a microstrip with ground plane, the electric flux lines are not confined between the trace and the ground plane, instead the electric flux lines spread and in Reference [14], this is termed the fringing factor. For parallel traces, there is both an inductive and capacitive fringing factor. Reference [14] shows this in a plot against the ratio of gap distance/height (d/w) of a trace above a ground plane and a substrate relative permittivity of 4.5, reproduced in Table 4.9. As the distance between traces increases for a given width, these fringing factors increase.

The mutual capacitance between the two parallel microstrips is given, from Reference [14], by

$$0.0276 * er * K_{L1} * K_{C1} * (w/d)^2 \text{ pF/cm} \tag{4.21}$$

$$0.07 * er * K_{L1} * K_{C1} * (w/d)^2 \text{ pF/in.} \tag{4.22}$$

where
  er is the relative permittivity
  K_{L1} is the inductive fringing factor
  K_{C1} is in capacitive fringing factor
  W is the width of the trace
  D is the distance between the two traces

Equations 4.21 and 4.22 are only accurate for gaps where $2h/d$ is less than 0.3, h is the height and d is the distance between the two traces.

TABLE 4.9
Two Parallel Traces Fringing Factors

d/w	K_{C1}	K_{L1}
1.0	1.6	2.0
1.5	1.8	2.2
2.0	2.1	2.8
3.0	2.7	3.6
4.0	3.1	4.4
5.0	3.6	5.2
6.0	4.1	6.0
7.0	4.5	6.6
8.0	4.9	7.3
9.0	5.3	7.9
10	5.7	8.5
11	6.1	9.1
12	6.5	9.7
13	7.0	10.3
14	7.4	10.9

Source: Reproduced from Walker, C.S., *Capacitance, Inductance, and Crosstalk Analysis*, Artech House Inc., Norwood, MA, 1990. With permission. Copyright by Artech House, Inc.

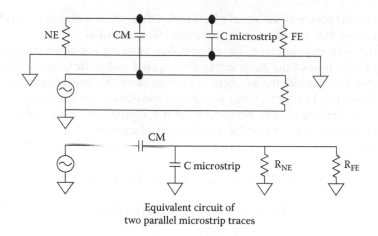

Equivalent circuit of
two parallel microstrip traces

FIGURE 4.35 The equivalent circuit of two parallel microstrip traces. The receptor trace loaded with resistors at the near and far ends.

In measurements, the receptor microstrip was loaded with a resistive load at the near end and a resistive load at the far end. This, in parallel with the capacitance between the receptor microstrip trace and ground plane, forms a divider with the mutual capacitance between the source and receptor microstrips. The equivalent circuit is shown in Figure 4.35. The load impedance is provided in Equations 4.5 and 4.6.

The equation for the microstrip capacitance is given by

$$0.0884 * er * K_{C1}(w/h) \text{ pF/cm} \tag{4.23}$$

$$0.225 * er * K_{C1} (w/h) \text{ pF/in} \tag{4.24}$$

where
er is the effective permittivity
K_{C1} is the capacitive fringing factor from Table 4.10 [14].

Microstrips are shown in Figure 4.36 (the mutual capacitance) with their associated parameters. $K_{L1} * K_{C1}$ coupling factors and the microstrip capacitance is shown in Table 4.11.

In addition to the capacitive coupling, a mutual inductance exists between the two microstrip traces whether inductive or capacitive crosstalk predominately depends on frequency and load impedances.

Two PCBs with the geometry shown in Figure 4.36 were made with $d = 1$ mm and $d = 1$ cm.

The PCBs were tested with a signal generator set to 0 dBm level and with a 50 Ω load on the source microstrip. The loads on the receptor microstrips are shown in Figure 4.35 with the values in Table 4.12 where R_{NE} is the near-end load and R_{FE} is the far-end load.

The 3 Ω R_{NE} represents a low impedance driver output impedance and at low frequency, the crosstalk level is higher than with the load at 50 Ω predominates. Thus, keeping receptor impedances low does not necessarily reduce the crosstalk.

The near-end crosstalk levels with these loads for the microstrip where $d = 1$ mm is shown in Figure 4.37 and for the 1 mm microstrip far-end load in Figure 4.38. The near-end crosstalk for the microstrip where $d = 1$ cm is shown in Figure 4.39 and the far end in Figure 4.40.

The measured level was in –dBm, and this is shown as attenuation in the figures.

TABLE 4.10
Microstrip Fringing Factor

$2h/w$	K_{C1}
1	1.5
2	2.1
3	2.5
4	3.1
5	3.6
6	4.0
7	4.5
8	4.9
9	5.3
10	5.7
11	6.0
12	6.4
13	6.9
14	7.3

Source: Reproduced from Walker, C.S., *Capacitance, Inductance, and Crosstalk Analysis*, Artech House Inc., Norwood, MA, 1990. With permission. Copyright by Artech House, Inc.

$w = 2.8$ mm
$d = 1$ mm or 1 cm
$h = 1.6$ mm

Ground plane

FIGURE 4.36 Coupled microstrip geometry.

TABLE 4.11
Mutual Capacitance, Microstrip Capacitance, and the Coupling Factors for the Geometry Shown in Figure 4.36

d (mm)	Mutual Capacitance (pF)	K_{L1} Mutual	K_{C1} Mutual	Microstrip Capacitance (pF)	K_{C1} Microstrip
10	1.143	3	3	10.9	1.6
15	0.9	3.8	5.6	10.9	1.6
20	0.71	4.4	6.8	10.9	1.6
25	0.646	5.3	8	10.9	1.6
30	0.57	6.1	8.8	10.9	1.6

TABLE 4.12

Loads Used on the Receptor Microstrip in the Tests

R_{NE} (Ω)	R_{FE} (Ω)
3	50
50	50
1000	50
50	3
50	50
50	10,000

FIGURE 4.37   Near-end crosstalk between microstrip traces where d = 1 mm.

The measured far-end and near-end attenuation of the coupled microstrips with 1 mm and 10 mm distance between traces are substantially identical. Likewise the near-end and far-end crosstalk with d = 1 cm are very similar. At 400 MHz, it is the capacitive crosstalk that predominates in these two configurations. Above 400 MHz, for the microstrip with a 1 cm distance between traces, the far-end attenuation increases slightly, that is, the induced voltage reduces and the near-end attenuation reduces slightly, that is, the induced voltage increases. This indicates both capacitive and inductive crosstalk. Above 700 MHz, the crosstalk appears to be primarily capacitive.

For the microstrip with 1 mm distance between traces, the far-end and near-end attenuations are substantially the same. However, at 60 MHz, the near-end attenuation, with a 3 Ω load at the far end, increases. This is the opposite of expected if the crosstalk was both inductive and capacitive.

The mutual capacitance between the two traces forms a voltage divider with the receptor microstrip capacitance in parallel with the R_{NE} and R_{FE}. The impedance is provided in Equations 4.5 and 4.6. The attenuation in Figures 4.37 through 4.40 also contains the attenuation of the 0 dBm signal in both source and receptor traces.

FIGURE 4.38 Far-end crosstalk between microstrip traces where $d = 1$ mm.

FIGURE 4.39 Near-end crosstalk between microstrip traces where $d = 1$ cm.

For the traces 10 cm apart, the total attenuation due to the voltage divider is 24.8 dB. The reduction in signal at 800 MHz is 11.2 dB. Table 4.13 plots the measured values at $d = 1$ cm and the theoretical attenuation at 1, 1.5, 2, 2.5, and 3 cm.

Adding a shield trace between the source and receptor trace and keeping the distance between the two, the same only reduces the crosstalk from between 6 and 9 dB.

Comparing the measured attenuation of the microstrip with $d = 1$ mm, w/h = 1.75, s/h = 0.625 with a similar geometry with w/h = 2 and s/h = 0.5 shown in Table 4.6, the correlation above 800 MHz is very good.

FIGURE 4.40   Far-end crosstalk between microstrip traces where $d = 1$ cm.

TABLE 4.13
Theoretical and Measured Attenuation for the Two Microstrip Traces

Distance between Traces (cm)	Attenuation (dB)	NE Measured (dB)	FE Measured (dB)
1	36	37.8	35
1.5	37.9		
2	39.8		
2.5	40.6		
3.0	41.6		

There are two parallel traces for the source and two for the receptor without ground plane.

One other geometry is where two traces are used for the source circuit and two traces used as the receptor circuit with no ground plane.

Figure 4.41 illustrates this geometry. Measurements were made of the crosstalk level with $d = 1$ mm and $d = 10$ mm.

Some of the crosstalk between traces without a ground plane may be due to radiated coupling as well as capacitive and inductive. The traces in Figure 4.41 were modeled using the method of moments (MOM) program 4NEC2 using the geometry of the upcoming Figure 4.60 for the traces. However, the input impedance of the source trace was 700 Ω and the coupled level was approximately 20 dB lower than shown in Figures 4.42 through 4.45. This error has been seen earlier when using NEC for closed circuits and even with a monopole attached to a metal enclosure, as described in Section 12.4.

Very often, the coupling between traces that are in, or close to, a straight line such as a gap in a PCB trace is required. The coupling between a gap in a 10.5 cm long microstrip, where the height of the trace above the ground plane is 1.6 mm, the width of the trace is either 1, 2.8, or 4.8 mm, and the substrate relative permittivity is 4.2, has been measured for gaps from 1 to 30 mm.

FIGURE 4.41 Parallel two source traces and two receptor traces, no ground plane.

FIGURE 4.42 Far-end attenuation traces 1 mm apart (parallel traces).

The mutual capacitance increases as a function of $(w/d)^2$ and the microstrip capacitance after the gap by (w/h). The width of the traces reduces, so the fringing factor K_{C1} shown in Tables 4.14 through 4.16 increases. The 2.8 and 4.5 mm traces are typical for power-line filter traces but wide compared to modern digital PCB layout, and so, a 1 mm wide microstrip with gap was made.

The input level during the test was 0 dBm and the output was negative; this is shown as an attenuation in Figures 4.46 through 4.48.

The microstrip was loaded with a 50 Ω load, and this is in parallel with the capacitance between the trace and ground plane after the gap and forms a divider with the mutual capacitance across

FIGURE 4.43 Near-end attenuation traces 1 mm apart (parallel traces).

FIGURE 4.44 Far-end attenuation traces 1 cm apart (parallel traces).

the gap, and the equivalent circuit is shown in Figure 4.49. Figure 4.50 shows the microstrip tested. Equations 4.5 and 4.6 provide the impedance of the load.

As the frequency is increased, the capacitive reactance of the mutual capacitance reduces to below 50 Ω and at some frequency, the attenuation levels out.

Equation 4.23 has been modified to calculate the approximate mutual capacitance of the gap in the microstrip.

FIGURE 4.45 Near-end attenuation traces 1 cm apart.

TABLE 4.14
Data from Equations 4.21 and 4.23 for 1 mm Wide Trace Microstrip

Gap (mm)	K_{C1}	Mutual Capacitance (pF)	Microstrip Capacitance (pF)
10	5.7	0.0036	1.91
15	7.7	0.0032	1.75
20	9.3	0.0022	1.75
25	11.5	0.0018	1.75

TABLE 4.15
Data from Equations 4.21 and 4.23 for 2.8 mm Wide Microstrip

Gap (mm)	K_{C1}	Mutual Capacitance (pF)	Microstrip Capacitance (pF)
10	3	0.0148	5.35
15	3.7	0.012	4.9
20	4.5	0.008	4.9
25	5.3	0.0065	4.34
30	5.9	0.0058	4.34

The fringing factor K_{C1} shown in Figures 4.51 through 4.53 applies to this configuration.

The capacitance between square lands with the dimensions $w \times w$ is calculated for the distance d. To this is added the capacitance of lands $d + 2 \times w$, $d + 4 \times w$, $d + 6 \times w$ until the total length of the traces has been calculated. This is shown in Equation 4.25.

With the dimensions shown in Figure 4.50, the equation is only valid when d is 1 cm or greater.

TABLE 4.16

Data from Equations 4.21 and 4.23 for 4.5 mm Wide Microstrip

Gap (mm)	K_{C1}	Mutual Capacitance (pF)	Microstrip Capacitance (pF)
10	2.2	0.028	8.32
15	2.8	0.024	8.32
20	3.2	0.015	8.32
25	3.8	0.012	7.61
30	4.2	0.011	7.61

FIGURE 4.46   Attenuation due to a gap in a 1 mm wide microstrip.

The equation for the mutual capacitance of a gap in a ground planes is approximately

$$Cm = 0.07 * er * K_{C1} * (w/d)^2 * w/25.4 + 0.07 * er * K_{C1} * (w/d + 2 * w)^2 * w/25.4$$

$$0.07 * er * K_{C1} * (w/d + 4 * w)^2 * w/25.4 + 0.07 * er * K_{C1} * (w/d + 6 * w)^2$$

$$0.07 * er * K_{C1} * (w/d + 8 * w)^2$$

until $d + n * w$ is the total length of trace with the gap (mm). (4.25)

where
  d is the gap in the trace (mm)
  w is the width of the trace (mm)
  er is the relative permittivity of the substrate
  $K_{C1} = 10$ mm = 3, 15 mm = 3.7, 20 mm = 4.5, 25 mm = 5.3, 30 mm = 5.9

At high frequencies, the mutual capacitance and microstrip capacitance provide a divider, which provides a limit to the attenuation of the input test level. If no resistive load exists, then this limit

FIGURE 4.47   Attenuation due to a gap in a 2.8 mm wide microstrip.

FIGURE 4.48   Attenuation due to a gap in a 4.8 mm wide microstrip.

appears also at low frequency. Equation 4.25 provides the capacitance of a microstrip and in this case, it is the length of microstrip after the gap.

The fringing factor K_{C1}, mutual capacitance, and microstrip capacitance for the 1, 2.8, and 4.5 mm wide traces are shown in Figures 4.51 through 4.53.

As the trace width decreases, we see that mutual capacitance decreases as a function of $(w/d)^2$ and the microstrip capacitance after the gap by (w/h). However, the fringing factor increases and

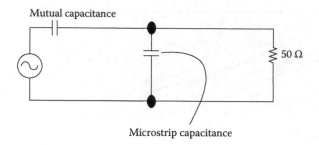

FIGURE 4.49　Equivalent circuit of the test microstrip.

FIGURE 4.50　Test microstrip with a gap.

FIGURE 4.51　Difference between the measured and the theoretical coupling level for the 1 mm wide microstrip.

FIGURE 4.52   Difference between the measured and the theoretical gap for the 2.8 mm wide trace.

FIGURE 4.53   Difference between the measured and the theoretical gap for the 4.5 mm wide trace.

so the attenuation at the frequency limit for the three trace widths is similar and this is shown in Figures 4.46, 4.47, and 4.48.

Figure 4.51 shows the comparison between the measured and calculated values for the 1 mm wide trace, Figure 4.52 for the 2.8 mm, and Figure 4.53 for the 4.5 mm.

Figure 4.54 shows a comparison between the calculated level from Equation 4.25 and the measured level at 700 MHz for the 2.8 mm wide trace.

FIGURE 4.54   Difference between the measured and the calculated level from Equation 4.25 for the 4.8 mm wide trace.

Where the gap is in a microstrip with a zigzag section as shown in Figure 4.55, the microstrip mutual capacitance after the gap is increased as there is also capacitive coupling between the parallel sections of the trace. This results in a lower attenuation, that is, increased coupling, and the leveling due to the capacitor divider effect occurs at a lower frequency. Figure 4.56 illustrates this.

Another geometry is where there is no ground plane and the signal return is on a parallel trace to the signal trace with a gap in the signal trace as shown in Figure 4.57.

FIGURE 4.55   Zigzag microstrip with gap.

FIGURE 4.56 Attenuation of zigzag microstrip with gap.

FIGURE 4.57 Geometry of two parallel traces with a gap.

The circuits were measured with 0 dBm applied to the trace and with the trace loaded with 50 Ω. The attenuation with gap length and frequency is shown in Figure 4.58 for the 2.8 mm wide $d = 2$ mm PCB and Figure 4.59 for the 5 mm wide $d = 4$ mm PCB.

The attenuation does not level out due to the capacitor divider effect and the attenuation is much lower than predicted using the mutual capacitance and microstrip capacitance. This means that the coupling is radiated and not inductive or capacitive crosstalk.

One method for calculating the coupled level is to use an electromagnetic modeling program. The MOM program in 4NEC2 was used to model the two parallel traces with a gap and no ground plane. The microstrip was modeled using thin wires due to the limit of wire length to radius inherent in the MOM formulation. Two wires were used to model the 2.8 mm thick traces. The wires at the source and load could not be as short as in the actual PCB, and so one of the wires making up the trace was tapered at the load and source end as shown in Figure 4.60.

FIGURE 4.58   **(See color insert.)** Attenuation for gap in 2.8 mm wide trace *d* = 2 mm.

FIGURE 4.59   **(See color insert.)** Attenuation for gap in 5 mm wide trace *d* = 4 mm.

FIGURE 4.60 NEC model of the traces.

The NEC input file for 2 cm is described in the following:

CM	Adjacent PCB traces-2 cm gap							
CE								
GW	1	10	−0.053	0.0	0.0	−0.01	0.0	0.0	0.0002
GW	2	1	−0.053	0.0	0.0	−0.048	0.0024	0.0	0.0002
GW	3	10	−0.048	0.0024	0.0	−0.01	0.0024	0.0	0.0002
GW	4	10	0.01	0.0	0.0	0.053	0.0	0.0	0.0002
GW	5	10	0.01	0.0024	0.0	0.048	0.0024	0.0	0.0002
GW	6	1	0.053	0.0	0.0	0.048	0.0024	0.0	0.0002
GW	7	20	−0.048	0.0044	0.0	0.048	0.0044	0.0	0.0002
GW	8	1	−0.048	0.0044	0.0	−0.053	0.0068	0.0	0.0002
GW	9	1	0.053	0.0068	0.0	0.048	0.0044	0.0	0.0002
GW	10	20	−0.053	0.0068	0.0	0.053	0.0068	0.0	0.0002
GW	11	1	−0.053	0.0	0.0	−0.053	0.0068	0.0	0.0002
GW	12	1	0.053	0.0	0.0	0.053	0.0068	0.0	0.0002
GW	13	1	−0.01	0.0	0.0	−0.01	0.0024	0.0	0.0002
GW	14	1	0.01	0.0	0.0	0.01	0.0024	0.0	0.0002
GE								
LD	0	12	0	0	50.	0.0	0.0	
FR	0.	0.	0.	0.	700	0.		
EX	0.	11.	1.	0	0.23	0.0		
XQ								
PQ								
EN								

FIGURE 4.61 Comparison of measured NEC predicted and using mutual capacitance.

FIGURE 4.62 Delta between measured NEC predicted and mutual capacitance calculation.

A comparison between the measured induced level and NEC predicted is shown in Figure 4.61 with the delta in Figure 4.62.

4.4.5.1 Stripline Coupling

The stripline PCB geometry is where a trace is imbedded between two ground planes is shown in Figure 11.5. When two adjacent traces are imbedded, it is named a balanced stripline.

In the balanced stripline, a mutual capacitance appears between the two traces and a stripline capacitance between a trace and the two ground planes.*

---

* Reproduced by permission from Walker, C.S., *Capacitance, Inductance, and Crosstalk Analysis*, Artech House Inc., Norwood, MA, 1990. Copyright by Artech House, Inc.

The equation for the stripline capacitance is given approximately by

$$0.45 \times er \times K_{C2}\ (w/0.5h)\ \text{pF/in.}$$

$$0.177 \times er \times K_{C2}\ (w/0.5h)\ \text{pF/cm}$$

The equation for the mutual capacitance is as follows:

$$0.07 \times er \times (K_{C2})^2\ (w/d)^2\ \text{pF/in.}$$

$$0.028 \times er \times (K_{C2})^2\ (w/d)^2\ \text{pF/cm}$$

The values of K_{C2}, mutual and capacitance for different ratios of $w/0.5h$, and distance between traces are shown in Table 4.17 and the stripline capacitance in Table 4.18.

The mutual capacitive reactance forms a divider with the stripline capacitive reactance that is in parallel with the near-end and far-end load resistance. The equations for calculating the load impedance are 4.5 and 4.6.

The capacitive crosstalk level between the source and the receptor trace can be calculated for different frequencies. The reciprocal of this is the attenuation and this is plotted in Figure 4.63 for a number of gap distances with a stripline length of 10 cm, trace width of 1.25 mm, and near-end and far-end resistances of 50 Ω.

In the stripline described in Reference [14] and shown in Figure 4.64, the capacitive crosstalk is much lower than the inductive crosstalk, and therefore, a calculation of the inductive crosstalk is advisable.

TABLE 4.17
Mutual Capacitance between Two Traces in a Stripline

W (mm)	h (mm)	K_{c2}	er	Distance (mm)	Mutual Capacitance
0.8	1.6	2.1	4.2	1	0.332
0.8	1.6	2.1	4.2	2	0.083
0.8	1.6	2.1	4.2	5	0.013
0.8	1.6	2.1	4.2	10	0.0033

TABLE 4.18
Data for Stripline Capacitance

W/h	h (mm)	w (mm)	K_{c2}	er	Capacitance (pF/cm)
3.35	1.6	5.36	1.2	4.2	5.98
2	1.6	3.2	1.3	4.2	3.86
1	1.6	1.6	1.6	4.2	2.37
0.5	1.6	0.8	2.1	4.2	1.56
0.25	1.6	0.4	3.2	4.2	1.19
0.166	1.6	0.266	4.2	4.2	1.04
0.125	1.6	0.2	5.1	4.2	0.95
0.1	1.6	0.16	6.1	4.2	0.91
0.083	1.6	0.1328	7	4.2	0.86

FIGURE 4.63   Reciprocal of capacitive crosstalk level between the source and the receptor trace with stripline length of 10 cm for multiple gap lengths.

FIGURE 4.64   Dimensions of the stripline in Reference [14].

Reference [14] provides equations for the mutual inductance and trace inductance, and Equations 4.10 and 4.12 were used to predict the induced voltage based on the geometry shown in Reference [14] and reproduced in Figure 4.64.

The predicted voltage was significantly lower than the modeled and measure voltage in Reference [14]. The predicted crosstalk of 28.6 dB was closer to the 22.6 dB predicted using the data in Table 4.4 for a similar but not identical geometry.

Reference [14] describes a 2 m long stripline PCB with the dimensions shown in Figure 4.64. In a simulation, the applied voltage is 2 V rms, scaling this down to 1 V rms, the induced voltage between the source trace and the adjacent trace is 0.033 V at 15 MHz, that is, −29 dBV. Theoretically, as the substrate is homogeneous, unlike the microstrip, the far-end crosstalk should be zero. Reference [14] shows that for more than two traces, this is not true. The far-end crosstalk from trace 1 to 2 is lower than the near end but from trace 1 to 3, it is the same.

When the center trace (2) is loaded at both ends with 50 Ω, the crosstalk to trace 3 is 27 dB lower. When the center trace is grounded at both ends, it is approximately 25 dB. Only with the center trace grounded at one end and open at the other does the crosstalk increase.

Vias are another potential source of crosstalk. The equations for mutual inductance and mutual capacitance for wires in Sections 4.2 and 4.3 may be applicable to via crosstalk. The use of a via connected to ground between two signal vias will reduce crosstalk and is recommended when the vias penetrate power planes or other planes both to reduce radiation and to maintain signal integrity.

4.4.5.2 Radiated Coupling

The equations for the radiated E field from transmission line geometries and loops, either open circuit or terminated in an impedance, can be found in Sections 7.13.2 and 7.13.3.

Once the E field at a distance is calculated, the voltage induced into a transmission line can be found from Section 7.7, a loop from Section 2.2.5, and an isolated conductor from Section 2.5.4.

4.4.6 ELECTROMAGNETIC COUPLING

When the distance between conductors or the distance between the signal conductors and a return conductor or ground plane is large, the coupling prediction may be more accurately accomplished using the transmitting and receiving properties of electric current elements and loops, as described in Sections 2.2.2 through 2.2.6. The signal tracks on a PCB that are terminated in a high impedance and are far removed from a ground plane may be conveniently modeled by the electric current element using either the electric field receiving or the electric and magnetic field radiation characteristics of the element. When a track forms a loop, either circular or rectangular, then the receiving or transmitting properties of the loop may be used in a coupling prediction. The coupling may be loop to loop, loop to element, or element to element. When two loops are formed on the same PCB surface, the loop coupling is coplanar; when the coupling is from a loop on one PCB to a loop on a second, adjacent board, the coupling is coaxial. The impedance of the loop and the load impedance must be included when predicting the loop current flow and the voltages developed around the loop. Chapter 7 addresses electromagnetic field-to-cable coupling and provides an equation for the inductance of a loop that may be used to calculate the loop impedance. Often, parallel impedances formed by IC loads or decoupling capacitors exist around the loop, in which case the loop may be divided into a number of loops sharing a load impedance. One rather idealized example of loop-to-loop coupling, with the loop dimensions, is illustrated in Figure 4.65. The receiving loop layout of Figure 4.65 is optimized for minimum error in the predicted received voltage by making the width of the loop small and the radius of the loop follow the radius of equipotential electric and magnetic fields from the source loop.

Equations 2.26 through 2.28 are used to compute the radiation from a current loop, and Equation 2.21, which gives the receiving properties of a loop, is used to predict the induced voltage. Figure 4.66 shows the measured and calculated values for receptor loops short-circuited at one end and open at the other [11]. The resonance at approximately 70 MHz would typically be caused by cable-to-source loop impedance mismatch. The more realistic situation, where driver output impedance and receiver input impedance form the loads on either end of the loop, is found in Equation 2.22.

The electric and magnetic quasi-static near fields, as we have seen in Chapter 2, are extremely nonuniform, reducing as a function of $1/r^2$ or $1/r^3$ with distance from the source. When the receptor loop is in the form of a circle or square with an area larger than the source loop, the nonuniformity of the field across the receptor loop and the change in the angle of the incident field as it cuts the receptor loop result in an error in the predicted voltage received by the loop.

The error for both coaxially and coplanar-oriented loops has been calculated [12]. Figure 4.67 presents the error in the open-circuit voltage developed at the terminals of the receiving loop as a

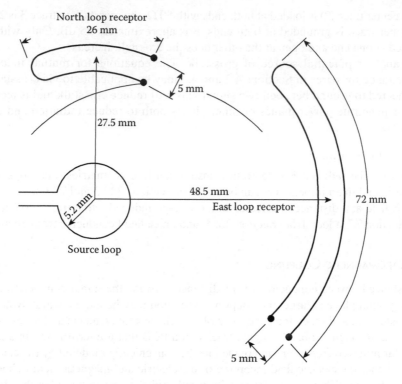

FIGURE 4.65 Dimensions of source and receptor current loops.

FIGURE 4.66 Measured and calculated coupling voltages. (Copyright 1985, IEEE.)

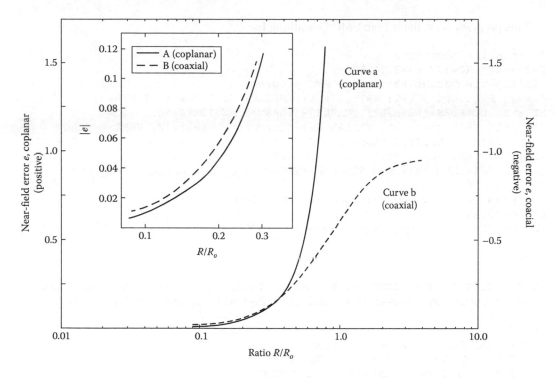

FIGURE 4.67 Near-field coupling error for coaxial and coplanar loops as a function of R/R_n. (Copyright 1989, IEEE.)

function of the ratio R/R_0, where R_0 is the distance between the center of the loops and R is the radius of the receiving loop, assuming the source loop is much smaller. The error is defined as follows:

$$e = \frac{V_{near} - V_{uniform}}{V_{uniform}}$$

where

V_{near} is the voltage induced in a nonuniform near field

$V_{uniform}$ is the voltage induced into the loop in a uniform field

The measured voltage must be corrected as follows:

$$V = \frac{V_{near}}{1+e}$$

When the error is negative, which is true of the coaxially oriented loop, V is greater than V_{near}; when the error is positive, which is true for the coplanar-oriented loops, V is less than $V_{near}+$.

4.4.7 COMPUTER PROGRAM FOR EVALUATING SHIELDED AND UNSHIELDED CABLE/WIRE-TO-CABLE/WIRE COUPLING

The following computer program was written using Microsoft's QuickBASIC Version 3.0. As with the programs in Chapter 2, there are no line numbers and alphanumeric labels are used to identify subroutines, etc.

This program is available from EMC Consulting Inc.

```
'Initialize constants
C=3E-f08: RC=377:PI=3.14159
'Initialize variables used in EMC program
1 = 1: F=100000!: L=1: W=1: R = 1: THETAD=90: CUR = 1: THETA=90
BB = 1: CCABLE$="RG5B/U": VCABLE$="10AWG": ACOTEXT$="AC": ICAC = 1
VARIABLE1$="Frequency": VARIABLE2$="Current": VAR1 = F: VAR2 = ICAC
UNITS1$="Hz": UNITS2$="Amps": RCC = 200: D = 1: LW=1: ZO = 200: ZI = 50:
LL=1: A=.006: IO=1: EO=1: RL=50: RD=50: H = 1: ZW=377
CZ1=50: DS1=.181: DI1=.053: RS1=1.2: CZ2 = 1: DS2 =.1019: DI2 =.1019: RS2 = 1.018
ICP = 1: TR =.001
H = H*39.37   'Convert from meters to inches
D = D*39.37
LW=LW*3.28    'Convert from meters to feet
GOTO ComputeCoupling:
WireToWireMenu:
CLS
PRINT "WIRE TO WIRE COUPLING-RJ. Mohr, IEEE, Vol.EMC-9 No.2, Sept, 1967:
PRINT "'Coupling Between Open and Shielded Wire Lines Over a Ground
Plane'"
PRINT
PRINT" [A] CULPRIT wire: ",, CCABLE$
PRINT "   [B] VICTIM wire:" „ VCABLE$
PRINT "   [T] Type of analysis:" „ACOTEXT$
PRINT "   [1] ";VARIABLE1$;" = ":LOCATE 7,43:PRINT VAR1;UNITS1$
PRINT "   [2] ";VARIABLE2$;" = ":LOCATE 8,43:PRINT VAR2;UNITS2$
PRINT " [S] Source Impedance:"„RCC;" ohms"
PRINT " [L] Load Impedance:"„RD; " ohms"
PRINT " [H] Height above ground plane:",H;" inches"
PRINT " [D] Distance between wires:",D;" inches"
PRINT " [C] Coupled Length:"„LW;" feet"
PRINT " [X] End program"
PRINT " Which parameter would you like to change (A,B,C...)?"
PRINT "- - - - - - - - - - - - - - - - - - - - - "
PRINT " ";OUTtext$
PRINT " ";IDTEXT$;" = ";OUT1;"V"
PRINT "- - - - - - - - - - - - - - - - - - - - - "
IF PEAKI>ICAC THEN
PRINT "TIGHTLY COUPLED, VICTIM SHIELD CURRENT APPROACHES CULPRIT
CURRENT"
'IF UTU$="Y" THEN LOCATE 21,1: PRINT "VICTIM SHIELD CURRENT =",PEAKI END
IF
'Test for f>200kHz
FTEST=F/1000/100
IF FTEST>2 THEN
PRINT "* WARNING, METHOD NOT VALID ABOVE 200kHz FOR SHIELDED CABLES *"
'PRINT "Victim shield current calculated using unshielded-unshielded
coupling
PRINT "Load and source impedance are the victim shield termination
impedance"
END IF
IF (CCABLE$=VCABLE$) AND (CCNUM<16) THEN
PRINT " *PROGRAM ASSUMES SAME CABLE FOR SHIELDED TO SHIELDED
CASE*
```

```
END IF
CouplingInputLoop:
SEL$=INKEY$
IF LEN(SEL$)>0 THEN LOCATE 15,1
IF LEN(SEL$)>0 THEN PRINT SPACE$(70)
LOCATE 15,4
IF (SEL$="A") OR (SEL$="a") THEN GOSUB CulpritCable:
IF (SEL$="B") OR (SEL$="b") THEN GOSUB VictimCable:
IF (SEL$="T") OR (SEL$="t") THEN
INPUT "Enter A for AC or T for Transient ",ACOT$
IF (ACOT$="a") OR (ACOT$="A") THEN ACOT$="A" ELSE ACOT$="T"
END IF
IF (SEL$="S") OR (SEL$="s") THEN INPUT "Enter source impedance ", RCC
IF (SEL$="L") OR (SEL$=l) THEN INPUT "Enter load impedance", RD
IF (SEL$="H") OR (SEL$="h") THEN INPUT "Enter height above ground plane", H
IF (SEL$="D") OR (SEL$="d") THEN INPUT "Enter distance between wires", D
IF (SEL$="C") OR (SEL$="c") THEN INPUT "Enter coupled length", LW
IF (ACOT$="A") AND (SEL$="1") THEN INPUT "Enter frequency", F
IF (ACOT$="A") AND (SEL$="2") THEN INPUT "Enter current", ICAC
IF (ACOT$="T") AND (SEL$="1") THEN INPUT "Enter peak current", ICP
IF (ACOT$="T") AND (SEL$="2") THEN INPUT "Enter risetime", TR
IF (SEL$="X") OR (SEL$="x") THEN GOTO WireToWireExit:
IF LEN(SEL$)>0 THEN GOTO ComputeCoupling:
GOTO CouplingInputLoop:
CulpritCable:
GOSUB Database:
PRINT "Choose the CULPRIT wire from the above list [Number]:";CCNUM
INPUT X$:IF LEN(X$)>0 THEN CCNUM = VAL(X$)
RESTORE
FOR N = 1 TO 27
READ CCABLE$,CZ1,DS1,DI1,RS1
IF CCNUM=N THEN GOTO ComputeCoupling:
NEXT N
PRINT "Must be a number from 1 to 27"
GOTO CulpritCable:
VictimCable:
GOSUB Database:
PRINT "Choose the VICTIM wire from the above list [NUMBER]:";VCNUM
INPUT X$:IF LEN(X$)>0 THEN VCNUM = VAL(X$)
RESTORE
FOR N = 1 TO 27
READ VCABLE$,CZ2,DS2,DI2,RS2
IF VCNUM = N THEN GOTO ComputeCoupling:
NEXT N
PRINT "Must be a number from 1 to 27"
GOTO VictimCable:
RETURN
ComputeCoupling:
IF ACOT$="A" THEN
ACOTEXT$="AC"
VARIABLE1$="Frequency"
VAR1=F
UNITS1$="Hz"
VARIABLE2$="Current"
VAR2=ICAC
```

```
UNITS2$="Amps"
OUTtext$="Peak AC Voltage"
ELSE
ACOTEXT$="Transient"
VARIABLE1$="Peak Current"
VAR1=ICP
UNITS1$="Amps"
VARIABLE2$="Interference signal rise time"
VAR2=TR
UNITS2$ = "sec"
OUTtext$="Peak Transient Voltage"
END IF
THAU=TR/2.2
RS1C=RS1*LW*.001
RS2C = RS2*LW*.001
M=7E-08*LOG(1+(2*H/D) 2)/LOG(10)*LW
IF CCNUM<16 THEN
IF VCNUM<16 THEN
VCABLE$=CCABLE$
CZ2 = CZ1: DS2 = DS1: DI2 = DI1: RS2=RS1
GOTO ShieldedToShielded:
ELSEIF VCNUM > 15 THEN
GOTO ShieldedToOpen:
END IF
ELSEIF CCNUM>15 THEN
IF VCNUM<16 THEN
GOTO OpenToShielded:
ELSEIF VCNUM>15 THEN
GOTO OpenToOpen:
END IF
END IF
OpenToOpen:
IF ACOT$="T" THEN GOTO OpenToOpenTransient:
IDTEXT$="AC UNSHIELDED-UNSHIELDED Coupling"
LS2=1.4E-07*LOG(4*H/DI2)/LOG(10)
AL2=1/(SQR(1+(2*PI*F*LS2/(RCC + RD))^2))
VD=2*PI*F*M*ICAC*(RD/(RCC + RD))*AL2
PEAKI=VD/RD
OUT1=VD
GOTO WireToWireMenu:
OpenToOpenTransient:
IDTEXT$="Transient UNSHIELDED-UNSHIELDED Coupling"
LS2= 1.4E-07*LOG(4*H/DI2)/LOG(10)
TM=THAU*LOG((RCC + RD)*THAU/LS2)/((RCC + RD)*THAU/LS2-1)
AAL2 = 1/(1-LS2/((RCC + RD)*THAU))*(EXP(-TM/THAU)-EXP(-(RCC+RD)*TM/LS2))
VDMAX=M*ICP/THAU*RD/(RCC + RD)*AAL2
OUT1=VDMAX
GOTO WireToWireMenu:
OpenToShielded:
IF ACOT$="T" THEN GOTO OpenToShieldedTransient:
IDTEXT$="AC UNSHIELDED-SHIELDED Coupling"
LC2=1.4E-07*LOG(4*H/DI2)/LOG(10)*LW
LS2 = 1.4E-07*LOG(4*H/DI2)/LOG(10)
AS2=1/(SQR(1 + (2*PI*F*LS2/(RS2C))^2))
AC2 = 1/(SQR(1+(2*PI*F*LC2/(RCC + RD))^2))
VD=2*PI*F*M*ICAC*(RD/(RCC + RD))*AS2*AC2
```

```
OUT1=VD
GOTO WireToWireMenu:
OpenToShieldedTransient:
IDTEXT$="Transient UNSHIELDED-SHIELDED Coupling"
LS2 = 1.4E-07*LOG(4*H/DS2)/LOG(10)*LW
TM=THAU*LOG(THAU/(LS2/RS2C))/(THAU/(LS2/RS2C)-1)
AAS2 = 1/(LS2/RS2C/THAU-1)*(EXP(-RS2C*TM/LS2)-EXP(-TM/THAU))
VDMAX-M*ICP/THAU*RD/(RCC + RD)*AAS2
OUT1=VDMAX
GOTO WireToWireMenu:
ShieldedToOpen:
IF ACOT$="T" THEN GOTO ShieldedToOpenTransient:
IDTEXT$="AC SHIELDED-UNSHIELDED Coupling"
LS1 = 1.4E-07*LOG(4*H/DI1)/LOG(10)
AL2=1/(SQR(1+(2*PI*F*LS2/(RCC+RD))^2))
AS1 = 1/(SQR(1 + (2*PI*F*FS1/RS1 C) 2))
VD = 2*PI*F*M*ICAC*(RD/(RCC + RD))*AL2*AS1
OUT1=VD
GOTO WireToWireMenu:
ShieldedToOpenTransient:
IDTEXT$="Transient SHIELDED-UNSHIELDED Coupling"
LS1 = 1.4E-07*LOG(4*H/DI1)/LOG(10)
TM=THAU*LOG(THAU/(LS1/RS1C))/(THAU/(LS1/RS1 C)-1)
AAS1 = 1/(LS1/RS1C/THAU-1)*(EXP(-RS1C*TM/LS1)-EXP(-TM/THAU))
VDMAX=M*ICP/THAU*RD/(RCC + RD)*AAS1
OUT1=VDMAX
GOTO WireToWireMenu:
ShieldedToShielded:
IF ACOT$="T" THEN GOTO ShieldedToShieldedTransient:
IDTEXT$="AC SHIELDED-SHIELDED Coupling"
LS2 = 1.4E-07*(LOG(4*H/DS2)/LOG(10))*LW
LS1 = 1.4E-07*(LOG(4*H/DS1)/LOG(10))*LW
AS1 = 1/(SQR(1 + (2*PI*F*LS1/(RS1C*RS1 C))))
AS2 = 1/((1 + (2*PI*F*LS2/RS2C^2))^5)
AC2 = 1/((1+(2*PI*F*LS2/((RCC + RD))^2))^.5)
VD=2*PI*F*M*ICAC*(RD/(RCC + RD))*AS1*AS2*AC2
OUT1=VD
GOTO WireToWireMenu:
ShieldedToShieldedTransient:
IDTEXT$ = "Transient SHIELDED-SHIELDED Coupling"
LS2=1.4E-07*LOG(4*H/DS2)/LOG(10)*LW
E2#=0
TM = LS2/RS2C
TransientConvergence:
DD#=1-THAU*RS2C/LS2
E#=EXP(-RS2C/LS2*TM)
E2NEW#=M*ICP*RS2C/LS2*((THAU*RS2C/LS2/DD#^2)*(EXP(-TM/THAU)-E#)+TM*(RS2C/
LS2)/DD#*E#)
IF E2NEW#-E2#>0 THEN TM=TM +.05*LS2/RS2C ELSE GOTO
OutOfConvergenceLoop:
E2#=E2NEW#
GOTO TransientConvergence:
OutOfConvergenceLoop:
VDMAX=RD/(RCC + RD)*E2#
OUT1=VDMAX
GOTO WireToWireMenu:
```

```
Database:
CLS:PRINT " DATA BASE": PRINT
PRINT "Number RG#/U Type Number ##AWG Type
PRINT "[1] RG5B/U [16] 10AWG"
PRINT "[2] RG6A/U [17] 12AWG"
PRINT "[3] RG8A/U [18] 14AWG"
PRINT "[4] RG9B/U [19] 16AWG"
PRINT "[5] RG13A/U [20] 18AWG"
PRINT "[6] RG29/U [21] 22AWG"
PRINT "[7] RG55A/U [22] 24AWG"
PRINT "[8] RG58C/U [23] 26AWG"
PRINT "[9] RG59A/U [24] 28AWG"
PRINT í£[10] RG122/U [25] 30AWG"
PRINT "[11] RG141/U [26] 32AWG"
PRINT "[12] RG142/U [27] 34AWG"
PRINT "[13] RG180/U"
PRINT "[14] RG188/U"
PRINT "[15] RG223/U":PRINT ""
RETURN
'CABLE DATA FOR PROGRAM
'RG(#)/U,Char. Imp..Shield Diameterjnner Conducter Diameter.Resistance
'
   DATA RG5B/U,50,.181,.053,1.2
   DATA RG6A/U,75,.185,.028,1.15
   DATA RG8A/U,50,.285,.086,1.35
   DATA RG9B/U,50,.280,.085,.8
   DATA RG13A/U,75,.280,.043,.95
   DATA RG29/U,53.5,.116,.032,4.75
   DATA RG55A/U,50,.116,.035,2.55
   DATA RG58C/U,50,.116,.035,4.7
   DATA RG59A/U,75,.146,.022,3.7
   DATA RG122/U,50,.096,.029,5.7
   DATA RG141/U,50,.116,.036,4.7
   DATA RG142/U,50,.116,.039,2.2
   DATA RG180/U,95,.103,.011,6.0
   DATA RG188/U,50,.06,.019,7.6
   DATA RG223/U,50,.116,.035,2.55
   DATA 10AWG,1,.1019,.1019,1.018
   DATA 12AWG,1,.0808,.0808,1.619
   DATA 14AWG,1,.0641,.0641,2.575
   DATA 16AWG,1,.0508,.0508,4.099
   DATA 18AWG,1,.0403,.0403,6.510
   DATA 20AWG,1,.0320,.0320,10.35
   DATA 22AWG,1,.0253,.0253,16.46
   DATA 24AWG,1,.0201,.0201,26.17
   DATA 26AWG,1,.0159,.0159,41.62
   DATA 28AWG,1,.0126,.0126,66.17
   DATA 30AWG,1,.0100,.0100,105.2
   DATA 32AWG,1,1,.203,
   DATA 34AWG,1,1,.160,
   DATA ***
WireToWireExit:
H = H/39.37 'Convert from meters to inches
D = D/39.37
LW=LW/3.28 'Convert from meters to feet
GOTO ENDER:
```

```
Spacebar:
PRINT "Press the spacebar to continue"
SBLOOP:
SPCBR$=INKEY$
IF SPCBR$=" " THEN GOTO Returner:
GOTO SBLOOP:
Returner:
CLS
RETURN
ENDER:
END
```

REFERENCES

1. R.J. Mohr. Coupling between open wires over a ground plane. *IEEE Electromagnetic Compatibility Symposium Record*, Seattle, WA, July 23–25, 1968.

2. R.J. Mohr. Coupling between open and shielded wire lines over a ground plane. *IEEE Transactions on Electromagnetic Compatibility*, September 1967, 9(2), 34–45.

3. G.L. Matthaei, L. Young, and E.M.T. Jones. *Microwave Filters, Impedance-Matching Networks and Coupling Structures*. Artech House, Dedham, MA, 1980.

4. J.A. Defalco. Predicting crosstalk in digital systems. *Computer Design*, June 1973, 12(6), 69–75.

5. N. Holte. A crosstalk model for cross-stranded cables. *Proceedings of the International Wire and Cable Symposium*, Cherry Hill, NJ, 1982.

6. J.A. Krabec. Crosstalk and shield performance specifications for aluminum foil shielded twisted pair cable (MIL-STD-49285). *Proceedings of the International Wire and Cable Symposium*, Arlington, VA, 1987.

7. A. Feller, H.R. Kaupp, and J.J. Digiacomo. Crosstalk and reflections in high speed digital systems. *Proceedings of the Fall Joint Computer Conference*, 1965.

8. C.R. Paul. A simple SPICE model for coupled transmission lines. *Proceedings of the IEEE International Symposium on Electromagnetic Compatibility*, Atlanta, GA, 1988.

9. R.L. Khan and G.I. Costache. Considerations on modeling crosstalk on printed circuit boards. *Proceedings of the IEEE International Symposium on Electromagnetic Compatibility*, Zurich, Switzerland, 1987.

10. C.R. Paul and W.W. Everett. Modeling crosstalk on printed circuit boards. RAdc-TR-85-107. Rome Air Development Center, Griffiss AFB, New York, Phase Report, July 1985.

11. W.J. Adams, J.G. Burbano, and H.B. O'Donnell. SGEMP induced magnetic field coupling to buried circuits. *IEEE Electromagnetic Compatibility Symposium Record*, Wakefield, MA, 1985.

12. S. Iskra. H field sensor measurement error in the near field of a magnetic dipole source. *IEEE Transactions on Electromagnetic Compatibility*, August 1989, 31(3), 306–311.

13. R.L. Khan and G.I. Costache. Finite element method applied to modeling crosstalk problems on printed circuit boards. *IEEE Transactions on Electromagnetic. Compatibility*, February 1989, 31(1), 5–15.

14. C.S. Walker. *Capacitance, Inductance, and Crosstalk Analysis*. Artech House Inc., Norwood, MA, 1990.

5 Components, Emission Reduction Techniques, and Noise Immunity

5.1 COMPONENTS

5.1.1 INTRODUCTION TO THE USE OF COMPONENTS IN ELECTROMAGNETIC COMPATIBILITY

Components are used either to reduce noise current by increasing the impedance of the current path or to reduce noise voltage by decreasing the impedance of a shunt path. A combination of these functions is often used in the design of a filter.

All components and interconnections contain unintentional (parasitic) circuit elements, such as inductance, capacitance, and resistance, and often a combination of these. An understanding of the magnitude of these parasitic elements and the characteristics of components over a range of frequencies will ensure the correct choice and application of a component. Some components are lossy and have the property of converting radio frequency (RF) noise energy into heat. Loss may be preferable to shunting the noise energy into an alternative path. All components and interconnections resonate, and an understanding of the effect of resonance is often vital in achieving EMC or solving an EMI problem.

Other uses of components in EMC are to slow down the edges of a waveform (thereby to reduce emissions) and to clamp voltages to a specific level and thus suppress transients.

5.1.2 IMPEDANCES OF WIRES, PRINTED CIRCUIT BOARD TRACKS, AND GROUND PLANES

A common mistake in an EMC/EMI evaluation is not including the impedance of a conductor, especially when the conductor is used to make a ground connection or a cable shield termination. Conductors exhibit an intrinsic or internal impedance composed of an internal inductance, due to the internal magnetic flux; an AC resistance, due to the concentration of current near the outer periphery of the conductor, known as the *skin effect*; and a DC resistance. In addition, conductors exhibit an external inductance, giving rise to an external magnetic flux. This external inductance is often referred to as *self-* or *partial inductance* and, when it is unintentional, as *parasitic inductance*. Above some frequency, the AC resistance is higher than the DC resistance of the conductor and increases as the square of the frequency. At the same frequency, the internal inductance begins to reduce by the square of the frequency. The external inductance is frequency independent.

By definition, inductance must form a loop in which current flows, for the inductance of a current-carrying loop is the ratio of the total magnetic flux penetrating the surface of the loop and the current in the loop that produced the flux. A wire or component has no inductance until it is part of a loop. This loop may be part of a circuit comprising resistance capacitance, and the predominant inductance may be that of the wire and a ground plane. Although inductance is defined only for complete loops, the assigning of partial inductance values to a section of a current loop, such as the ground plane, and signal trace inductances of PCB layouts is described in Reference [8].

In the case of a number of wires making the loop, each side of the loop has a partial inductance and a partial mutual inductance between the sides. The total inductance is the sum of sides 1, 2, 3, and 4 minus the mutual inductance between the two sides.

Thus,

$$L_t = L_1 + L_2 + L_3 + I_4 - M_{1-2} - M_{3-4}$$

The inductance of wires is contained in Equations 5.1 through 5.4 with the inductance of a wire above a ground plane in Equation 4.8 and the mutual inductance of two wires above a ground plane in Equation 4.9.

Wires with a circular cross section are commonly used as interconnections. In circular wires, the frequency above the AC resistance begins to increase, and the internal inductance decreases when the skin depth approaches the radius of the wire. The AC resistance of a copper wire is given by the following equation:

$$R_{AC} = \left(0.244 d \sqrt{f} + 0.26\right) R_{DC}$$

where
  d is the diameter of the wire [cm]
  f is the frequency [Hz]
  R_{DC} is the DC resistance [Ω]

Wires invariably exhibit higher inductive reactance, due to self-inductance, than resistance, even at low frequencies. The self (external)-inductance of a wire is usually higher than the internal inductance. The inductance of a wire at high frequencies with a circular cross section at least 15 cm from a ground structure is given by

$$L = 0.002l \left(2.303 \log\left(\frac{4l}{d}\right) - 0.75\right) [\mu H] \tag{5.1}$$

where
  l is the length of wire [cm]
  d is the diameter [cm]

Equation 5.1 assumes that the relative permeability of the conductor is 1. When μ_γ is not 1, the self-inductance is given by

$$L = 0.002l \left(2.303 \log\left(\frac{4l}{d}\right) - 1 + \frac{\mu_r}{4}\right) [\mu H] \tag{5.2}$$

In comparing the inductances of a square conductor and a circular conductor, we find that when the diameter of a circular wire equals the dimension of the sides of a square conductor, the square conductor exhibits a lower self-inductance and AC resistance. When the cross-sectional area of the square and circular conductors are the same, the inductance is the same. The inductance of a square conductor is approximately

$$L = 0.002l\left(2.303\log\left(\frac{2l}{W+t}\right) + 0.498\right) [\mu H] \tag{5.3}$$

where
  W is the width of the conductor [cm]
  t is the thickness of the conductor [cm]

Equation 5.3 applies approximately to a rectangular conductor. For a PCB or similar ground plane where $t < 0.025\,W$, the inductance approximately equals

$$L = 0.002l\left(2.303\log\left(\frac{2l}{W}\right) + 0.5\right) [\mu H] \tag{5.4}$$

N.B (All logs are to the base 10)
  Mutual inductance is

$$0.002[1 \times 2.303 \times \log(1 + \text{sqrt}((l^2 + d^2)/d)) - \text{sqrt}((l^2 + d^2) + d)] \,(\mu H) \tag{5.5}$$

where
  d is the distance between wires (cm)
  l is the length of wires (cm)

We see that replacing a circular wire with a strap that has a thickness approximating the diameter of the wire but that has a much greater width will result in reduced inductance. Table 5.1 illustrates the impedance of some straight copper wires with circular cross section. At 10 Hz, the impedance equals the DC resistance of the wire, but even at frequencies as low as 10 kHz, the impedance is significantly higher than the DC resistance due to the self-inductance. The impedance values in Table 5.1 are for wires at some great distance from the ground and below the frequency at which the length of the wire equals a quarter wavelength ($\lambda/4$). In practice, conductors such as wires, PCB tracks, and ground straps are typically close to grounded conductive structures or other wires. One effect on the inductance of the conductor due to proximity effects is to reduce the inductance, as shown in Figure 4.11.

Published data, even in the most recent publications, on the impedance of a square ground plane based exclusively on the intrinsic or metal impedance of the 1 oz copper foil conductor give the impedance as follows (the unit Ω/square is explained in Section 6.1.3):

10 Hz	0.812 mΩ/square
10 MHz	1.53 mΩ/square
100 MHz	3.72 mΩ/square

To disregard the self-inductance of the plane, which presents an impedance to an AC current flowing from one end of the plane to the other, is to minimize the impedance. Table 5.2 shows the impedance for a number of ground straps and a square 1 oz (0.03 mm thick) copper ground plane based on Equation 5.4 and the measured values. The voltage across these structures, due to a controlled current flow at frequencies of 2, 10, and 20 MHz, when measured differentially, results in good correlation between predicted and measured values. The measurement setup must ensure that the conductors

TABLE 5.1

Impedance of Straight Circular Copper Wires

	Impedance of Wire (Ω)							
	AWG #2, d = 0.65 cm				AWG #10, d = 0.27 cm			
Freq.	1 cm	10 cm	1 m	10 m	1 cm	10 cm	1 m	10 m
10 Hz	5.13 μm	51.4 μm	517 μm	5.22 m	32.7 μm	327 μ	3.28 m	32.8 m
1 kHz	18.1 μm	429 μm	7.14 m	100 m	42.2 μ	632 μ	8.91 m	116 m
100 kHz	1.74 m	42.6 m	712 m	10	2.66 m	54 m	828 m	11.1
1 MHz	17.4 m	426 m	7.12	100	26.6 m	540 m	8.28	111
5 MHz	87.1 m	2.13	35.5	500	133 m	2.7	41.3	555
10 MHz	174 m	4.26	71.2	—	266 m	5.4	82.8	—
50 MHz	870 m	21.3	356	—	1.33	27	414	—
100 MHz	1.74	42.6	—	—	2.66	54	—	—
150 MHz	2.61	63.9	—	—	4.0	81	—	—

	Impedance of Wire (Ω)							
	AWG #22, d = 0.065 cm				AWG #26, d = 0.04 cm			
Freq.	1 cm	10 cm	1 m	10 m	1 cm	10 cm	1 m	10 m
10 Hz	529 μ	5.29 m	52.9 m	529 m	1.33 m	13.3 m	133 m	1.33
1 kHz	531 μ	5.34 m	53.9 m	545 m	1.38 m	14 m	144 m	1.46
10 kHz	681 μ	8.89 m	113 m	1.39	1.81 m	21 m	239 m	2.68
100 kHz	4.31 m	71.6 m	1.0	12.9	6.1 m	90.3 m	1.07	14.8
1 MHz	42.8 m	714 m	10	129	49 m	783 m	10.6	136
5 MHz	214 m	3.57	50	645	241 m	3.86	53	676
10 MHz	428 m	7.14	100	—	481 m	7.7	106	—
50 MHz	2.14	35.7	500	—	2.4	38.5	530	—
100 MHz	4.28	71.4	—	—	4.8	77	—	—
150 MHz	6.42	107	—	—	7.2	115	—	—

carrying current to and from the conductor under test are shielded and are far removed from the conductor under test. In a practical PCB layout, current may flow in a conductor that is in close proximity to the ground plane and return in the ground plane. In this configuration, the return current tends to be confined to the ground plane immediately below the supply conductor. This current concentration may be expected to increase the impedance of the ground plane, whereas above approximately 2 MHz, the impedance is lower than when the supply conductor is far removed from the ground plane.

In Reference [9], the inductance of the ground plane with a signal trace located above it is determined using Maxwell's equations. The value of ground plane inductance when it is used as the signal return is given by

$$L_{ground} = (0.2 \times l) \times ln\left(\frac{h \times \pi}{w} + 1\right) (\mu H) \tag{5.6}$$

where
　　l is the length of the ground plane under the trace (i.e., the current return path) in meters
　　h is the height in meters
　　w is the width of the ground plane in meters

The inductance of the trace above the ground plane can be found by replacing the width of the ground plane (w) with the width of the trace, assuming the thickness of the trace is much smaller

TABLE 5.2

Impedance of Ground Straps and a 1 oz Copper Ground Plane

Frequency (MHz)	Measured Impedance (Ω)[a]	Calculated Impedance (Ω)	Measured 1 cm above GP[b]	Calculated 1 cm above GP	Calculated Trace 1 mm above GP
		1 oz copper ground plane 12.6 cm long by 10 cm wide			
2	0.312	0.45			
10	1.67	2.25	0.64	0.43	0.045
20	2.46	4.5	1.14	0.86	0.09
50			1.76	2.15	0.225
		10 oz copper ground plane 3.2 cm long by 3.2 cm wide			
10	0.36	0.47			
20	0.56	0.94			
		Impedance of a 12.6 cm long by 2 cm wide 3 mm thick braid			
10	2.84	4.77			
20	7.3	9.5			
		Impedance of a 12.6 cm long by 0.8 cm wide, 1.5 mm thick tinned copper wire			
10	5.5	6.2			
20	14.3	12.4			

[a] Measured impedance with the supply conductor far removed from the ground plane.
[b] Measured inductance with the supply conductor 1 cm above the ground plane.

than the width. Equation 5.6 is used to calculate the impedance of the ground plane with the trace located at 1 cm and 1 mm above the ground plane. The calculated impedance at 1 cm is close to the measured impedance. A measurement made at 2 mm does not agree well with the calculation; the reason is almost certainly due to radiated coupling to the measurement probe in the test setup, resulting in a higher measured voltage across the ground plane.

The reason for the reduction in the ground plane impedance is the mutual inductance between the supply conductor and the ground plane, which facilitates the return current flow in the plane. Table 5.2 illustrates the reduction in the impedance of a 12.6 cm long by 10 cm wide ground plane with the supply conductor 1 cm and 1 mm above the plane compared to the return conductor far removed from the ground plane. The reduction in impedance due to the proximity of a wire, used for either a supply or signal, to a return wire is discussed in further detail in Chapter 7, on cable shielding.

The importance of including conductor impedance in any EMC prediction or EMI investigative work is illustrated in Case Study 12.1. One common practice is the temporary grounding of circuits or structures by the use of an approximately 3 ft long wire in order to determine if a ground connection is required. If no detectable change occurs as a result of grounding, the reason may be that the impedance of the conductor, which may be up to 300 Ω, or open circuit at resonance and is just too high to function as an effective ground connection.

The influence of conductor impedance on the performance of a component can also be beneficial. For example, high-frequency (HF) attenuation in a filter may be higher than predicted, due to the self-inductance of a wire. It can also be detrimental, as in the case of increased radiation from an enclosure when a small-diameter high-impedance ground wire is replaced with a lower-impedance ground connection. Thus, for the purpose of EMC, all conductors should be considered as components exhibiting impedance.

In equipment designed for space applications, the need may arise for a high thermal resistance between structures and, at the same time, a low-impedance bond. One technique that may be applicable when the structures are close together is the use of a shorted transmission line and increased parasitic capacitance between the two structures. A practical example is shown in Figure 5.1. Decreasing the distance between the tracks of the line and increasing the length of the adjacent filled-out sections of the line reduce the line impedance and increase the parasitic capacitance, respectively. Unfortunately, the thermal resistance decreases as well. An equation for the impedance of a short-circuited transmission line is [1]

$$Z_{sc} = Z_o \tan\left(\pi l f \sqrt{\varepsilon_r}/150\right) \tag{5.7}$$

where
Z_o is the characteristic impedance of the line
l is the length of the line
f is the frequency [MHz]
ε_r is the relative permittivity of the material between the conductors of the line
$\sqrt{\varepsilon_r}$ may be replaced by $\left(3\times10^8\right)\sqrt{LC}$

The measured impedance of the capacitor and shorted transmission line with a characteristic impedance of 215 Ω and the geometry as shown in Figure 5.1 is compared to the same length of straight track in the following table:

	Impedance (Ω)	
Frequency (MHz)	**Shorted Transmission Line**	**Straight Track**
15	—	19
54	20	—
63	—	50
108	40	—
162	—	447
287	—	350
364	200	—
1000	—	125
1100	51	—
1200	10	—

FIGURE 5.1   Shorted transmission-line ground connection.

A conductor used as an interconnection is usually only part of a larger circuit that contains a return path. Intentional or parasitic capacitance, typically across a load or between the conductor and a ground structure, may form a series resonant circuit with the inductance of the conductor. The input impedance of a series resonant circuit is low, and the output voltage may be higher than the input voltage, depending on the Q of the circuit. The Q will be high when the load resistance is high. An equation for the calculation of the Q of a series resonant circuit is provided in Figure 5.54. Due to the high currents and high voltages present, a series resonance is very often the cause of high levels of radiated or conducted emissions.

Conductors in close proximity to ground may also form a quasi-parallel resonant circuit that exhibits a higher impedance of the wire, and these effects may be measured on a network analyzer. When the distance between a conductor and a return path is not much less than the length of the conductor, the circuit may be modeled as a loop. In practice, it is the parasitic capacitance to a grounded structure or across a load included in series with the loop that dictates the resonant frequency. If we assume a loop far removed from a grounded structure and terminated in a low value of resistance, the resonance may coincide with the distance around the perimeter of the loop. When the perimeter of a loop reaches a resonant length, the input impedance varies from a very high to a very low value, as described in Section 2.2.5. The resonant frequency and impedance of LC circuits with and without resistance are provided in Appendix D.

When the return path, or a grounded structure, is closer to a conductor than the length of the conductor, a transmission line is formed. For most practical applications, the transmission line can be considered lossless, above approximately 100 kHz, and the characteristic impedance is

$$Z_o = \sqrt{\frac{L}{C}}$$

The characteristic impedance of a number of conductor and PCB track configurations is provided in Appendix A.

The velocity of a wave traveling down a transmission line is

$$v = \frac{1}{\sqrt{LC}} \quad \text{or} \quad \frac{3 \times 10^8}{\sqrt{\varepsilon_r}} \left[\text{m/s} \right]$$

where ε_r is the relative permittivity of the material between the conductors of the line.

The propagation delay of a line is

$$\tau = l\sqrt{LC}$$

where l is the line length in meters. And the propagation constant per unit length of the line is

$$\gamma = \sqrt{LC}$$

When a transmission line is terminated in its characteristic impedance, Z_o, the input impedance equals Z_o.

When a transmission line is formed unintentionally, the terminating impedance is unlikely to equal the characteristic impedance of the line. The input impedance of a short-circuited line is given by Equation 5.7. When the short-circuited line length equals $\lambda/4$, $3\lambda/4$, 1.25λ, …, the input impedance is theoretically infinite, and when the line length is $\lambda/2$, λ, 1.5λ, …, it is zero. The measured impedance of the shorted transmission line is lower at frequencies that correspond to a line length of $\lambda/4$ or multiples thereof because of the intentionally increased capacitance at the input of the line.

The input impedance of an open-circuited line is

$$Z_{oc} = \frac{Z_o}{\tan\left(\dfrac{\pi l f \sqrt{\varepsilon_r}}{150}\right)}$$

The input impedance is theoretically zero when the line length equals $\lambda/4$ and infinite when the line length is $\lambda/2$. When the termination impedance of a transmission line is higher than the characteristic impedance, the output voltage will be higher than the input voltage when the line length equals $\lambda/4$, $3\lambda/4$, etc.

For a transmission line terminated in an impedance Z_t that does not equal Z_o, the input impedance of the line, Z_i, equals

$$Z_i = Z_o \frac{Z_t + Z_o \tan(\pi l f \sqrt{\varepsilon_r}/150)}{Z_o + Z_t \tan(\pi l f \sqrt{\varepsilon_r}/150)}$$

Low-impedance power distribution and returns on PCBs are important for both digital and analog circuits. For example, a lower-impedance connection results in a smaller voltage developed by noise currents flowing on the line. A useful method of achieving a low-impedance power distribution for digital logic, plus some built-in decoupling capacitance, is by the use of a "Q Pac" bus bar containing $+V_{cc}$ and 0 V tracks soldered either perpendicular to the PCB or horizontally under the integrated circuits (ICs). The use of such bus bars does not fully eliminate the need for decoupling capacitors located close to ICs and designed to reduce the area of the "current loop"; however, the number of decoupling capacitors required may be reduced.

5.1.3 GENERAL WIRING GUIDELINES

These wiring guidelines are useful for design engineers and for incorporation into EMC control plans but are no substitute for a detailed prediction, which should be carried out for potentially critical designs and layouts. Many exceptions to the following guidelines exist!

5.1.4 CIRCUIT CLASSIFICATION

Circuits that interface between units or subsystems should be allocated a classification. For example, the following classification scheme might be used:

Class	Type of Signal
I	Digital, low current control, and filtered/regulated power
II	Analog or video
III	High-current control, for example, relay switching
IV	AC and DC unfiltered line voltage

5.1.5 WIRE SEPARATION

Wire bundles of differing EMC classification should be physically separated from each other. Where no metallic barrier exists between them, the following minimum separation distance should be maintained:

A 10 cm minimum between Class III wire bundles and Class II wire bundles
A 5 cm minimum between all other categories of wire bundles

Maximum separation must be maintained between "clean" wires and "noisy" wires. This is especially important in wiring of filters, where the input and output wires of the filter must be isolated as much as possible.

5.1.6   INTERNAL UNIT/EQUIPMENT WIRING

The area of current loops on PCBs, board interconnections, and unit interconnections should be the practical minimum. The design aim should be that the maximum loop area be less than 4 cm². Larger loops should be subdivided into smaller loops arranged so that their generated fields cancel. Loops and wires should cross at right angles to reduce coupling. Analog wiring, digital wiring, and RF wiring should not be bundled together and should be separated as far as possible. Power lines should be twisted two-wire circuits or in some cases twisted shielded pair. Care must be taken to ensure that the distance between the twisted pairs is at least 1.5 times the twist length for proper isolation. *Twist length* is defined as the reciprocal of the number of full twists per inch. Compliance with this guideline usually requires that separate harnesses be used for each of the twisted pairs. If the distance cannot be increased, this criterion can be met by increasing the number of twists per inch, thereby reducing the twist length.

5.1.7   EXTERNAL UNIT/EQUIPMENT WIRING

Circuits that interconnect equipment should provide minimum loop coupling and maximum field cancelation by twisting the return with the high side wires. Both signal and power circuits should be twisted with their respective returns and assigned adjacent pins in connectors when safety permits. Multiple circuits using a common return should be twisted as a group. Twists, where possible, should be as high as 16 per foot. Wires between units should follow the most direct route, since interference coupling is a direct function of length. Twisted-pair wiring should be used for balanced or quasi-balanced circuits to reduce magnetic field coupling, as well as in energized circuits requiring a high DC current, in order to reduce the DC magnetic field. Coupling caused by common-mode (C/M) current flow is not reduced by twisting; instead, source reduction or shielding must be used.

5.1.8   WIRE SHIELDING

When unit/subsystem designers use wire shielding to meet emission requirements or to protect susceptible circuits, these shields should cover the twisted pair or twisted group rather than individual wires. The shield should be connected concentrically around the backshell to minimize RF potentials at the cable termination. If one outer shield is used (e.g., twisted shielded pair), the shield should be connected to chassis at both ends. High E-field susceptibility circuits may include an additional inner electrostatic shield connected to signal ground at the analog source end only, but isolated from the chassis. No shield should intentionally carry current except for coax cables used for RF.

High-voltage and HF high-power circuits are likely to be E-field generators. Circuits that carry large currents at low frequencies are likely to be H-field generators. These circuits should be shielded to control emissions. Sensitive circuits having low-current, high-impedance inputs are likely to be susceptible to E fields and should be shielded. Shielding effectiveness is a positive function of a shield's thickness, conductivity, and percent coverage, as described in Chapter 7.

When circuits are transformer coupled, C/M noise can be reduced by using transformers with shielded primary and secondary windings. The reduction in C/M noise due to the use of a single shield connected to case ground is approximately 20 dB from 100 Hz to 1 MHz. With three shields connected in a single transformer—one connected to the input ground, the second to case ground, and the third to output ground—the reduction in C/M noise may be as high as 80 dB at 20 kHz, reducing to 35 dB at 1 MHz. The shield must be connected to a quiet ground, such as a chassis, for if a single shield is used that is connected to a noisy signal ground, the C/M coupling may increase with the shield in place.

Cable shield continuity should be maintained over the entire length of the cable. When a shielded wire goes through a junction box, the wire should be shielded inside the junction box. Unshielded segments of shielded wires should be kept to an absolute minimum. Peripheral shield termination hardware at the connector or conductive epoxy potting of connector backshells can be used to obtain this objective.

5.1.9 Radio Frequency Shielding

Where high-RF currents have to be carried and it has been determined that an EMI emission problem may arise, the following shielding recommendation should be followed. A second shield over the shield carrying RF current return should be used, as is the case in a triax cable, or the RF return may be via a second center conductor, as in a twinax cable. Multiple connections down the length of triax shield introduce ground current loops or C/M currents and therefore should not be used to shield sensitive input circuits, although reduced emissions may result in high-power RF when the connections to the ground plane are made at distances less than $\lambda/4$. If multiple grounding of triax is impractical, apparent continuous RF grounding may be possible. Apparent continuous RF grounding of a triax cable is achieved by routing the triax cable, with its insulating outer sheath, close to case and structure, thus causing significant capacitance between the triax shield and structure.

Cables connecting sensitive RF input circuits should be shielded by the use of semirigid coax where possible. If this is not possible, then a triax cable should be used with the outer shield connected to chassis at both ends. The inner shield should be used for the signal current return. When such cables are routed external to a unit and are located unavoidably close to a high-power RF emitter (e.g., an antenna), the cable should be routed as close to a metal ground plane as feasible.

5.1.10 Components Used in Emission Control and to Increase Noise Immunity

5.1.10.1 Capacitors

A capacitor is commonly formed between two parallel plates. As voltage is applied across the plates, a displacement current flows and an electric field exists between the plates. With the voltage applied, negative charge is removed from one plate of the capacitor and placed on the other side leaving one side with a negative charge and the other with a positive charge although the net charge is zero. With the voltage removed, the charge distribution in the capacitor remains.

If an atom has more electrons than protons, it is negatively charged, and if an atom has more protons than electrons, it is positively charged. In an object comprising many electrons, the charge is the arithmetic sum of all the positively and negatively charged atoms. The unit of charge Q is the coulomb.

The relationship between capacitance voltage and charge is

$$Q = CV$$

When an electrostatic discharge (ESD) is applied through a capacitor (such as in an ESD gun) and charges up a capacitor, the higher the capacitance, the lower the voltage across the capacitor. This capacitance divider effect is useful in ESD protection.

Capacitors in series add as $1/Ct = !/C1 + 1/C2 + 1/Cn$ and in parallel by $C1 + C2 + Cn$.

After the plates have established the field, the current reduces to zero. The parallel plate capacitor has the lowest inductance of any capacitor, and for this reason, a practical application is in embedded capacitance, used on PCBs as part of the decoupling capacitor. The embedded capacitor uses the intrinsic capacitance formed between a power plane and a ground plane. For maximum capacitance, these should be as close together as possible. The capacitance of a parallel plate capacitor is given by

$$C = e \times w \times (l/d)/100 \text{ F} \tag{5.8}$$

where
    l is the length of one side [cm]
    w is the width of other side [cm]
    d is the distance between plates [cm]
    e is the dielectric constant of the slab $= er \times 8.84e - 12$
    er is the relative permittivity
    F is the capacitance Farads

Appendix F contains the relative permittivity of common materials.

Capacitors are used for filtering, for control of rise and fall times, and for providing a low-impedance path at high frequencies. The impedance of an ideal capacitor decreases linearly with increasing frequency. The AC equivalent circuit of a practical capacitor, shown in Figure 5.2, contains an equivalent series resistance (ESR) and an equivalent series inductance (ESL). At some frequency, a capacitor is a series resonance circuit with an impedance equal to the ESR. Above this resonance, the impedance of the capacitor increases with frequency due to the ESL. The length of the leads is an important factor in assessing the self-inductance, and thus the resonant frequency, of a capacitor. The following are examples of the typical self-inductance for a number of low-inductance-type capacitors:

    Metallized polyester lead length = 6 mm each, $L = 12$ nH
    MKT Radial polyester (Mylar) capacitor stacked film

Lead spacing (mm)	7.5	10	15
Self-inductance (nH)	5	6	7

Polypropylene capacitors—radial

Capacitor's body height (h)	9	12	15
Self-inductance (nH)	15	17	20

High rel. polyester (Mylar)
Lead length = 3 mm each, $L = 20$ nH

FIGURE 5.2 AC equivalent circuit of a capacitor.

The self-resonant frequencies (SRF) of a number of values of ceramic CCR series capacitors, with lead length of 1.6 mm each, were measured, with the following results:

Capacitance	SRF (MHz)	L (nH)
1 µF	1.7	9
0.1 µF	4	16
0.01 µF	12.6	16
3300 pF	19.3	16
1800 pF	25.5	16
1100 pF	33	21
820 pF	38.5	21
680 pF	42.5	21
560 pF	45	22
470 pF	49	22
390 pF	54	22
330 pF	60	21

The ESR of a capacitor is given by the Q of the capacitor, which is typically quoted for a frequency of 1 MHz. The Q of a capacitor is given by X_c/ESR, where X_c equals $1/2\pi fC$, with f in Hz and C in F or f in MHz and C in µF. Thus for a typical 30 pF capacitor with a Q of 931 at 1 MHz, the ESR is 5.7 Ω, for a typical 0.01 = µF capacitor with a Q of 168 at 1 MHz, the ESR is 95 mΩ.

When the noise voltage to be attenuated covers a wide frequency range, the impedance of a single capacitor may not be sufficiently low, especially above and below the resonant frequency of the capacitor. The use of two, three, or even four different capacitances in parallel will ensure a low impedance over a wide range of frequencies. For example, a 10 µF tantalum, a 1 µF ceramic, a 1000 pF ceramic, and a 200 pF ceramic all in parallel will ensure an impedance lower than 50 Ω from 350 Hz to 100 MHz. The resonant frequencies of capacitors are valid for the extremely short lead lengths quoted. As seen in Section 5.1.2, the impedance of a conductor due to self-inductance is appreciable even when the length is short. Thus, the inductance of conductors used to connect capacitors will reduce the resonant frequency of the capacitor in circuit and therefore increase the HF impedance. When an EMI/EMC problem occurs at a specific frequency, the value of the capacitor should be chosen so that the resonant frequency of the capacitor and its connections match the problem frequency, thereby achieving the lowest possible impedance. This technique is used in cellular phones where the decoupling capacitors are resonant at the transmit frequencies.

For cables that fail radiated emission limits or conducted emission limits at a specific frequency or narrowband of frequency, the addition of a C/M capacitor to chassis is often recommended. With the capacitor chosen to have a resonant frequency matching the problem frequency, the results are often a surprisingly high reduction in the emissions. However, location is everything and the capacitors should be located as close to the point where the cables exit the enclosure as possible. If the inductance can be determined from manufacturer's data or Table 5.3 or 5.4, then the capacitance for a given resonant frequency can be calculated from

$$C = 1/(39.4 \times f^2 \times L)$$

In one case of radiated emissions from a short length of unshielded cable, the emission at 160 MHz exceeded the limit by 14.3 dB. The source was a 5 V supply connected to one of the wires even though the 5 V supply was well filtered at the power supply board. The wires were connected to pins on a small PCB that was attached to chassis by screws. One of the wires and pin was connected to chassis via a trace captured under one screw. It was possible to connect a 0805 capacitor directly between the 5 V pin and the chassis pin resulting in negligible inductance. The inductance of a

TABLE 5.3
Capacitor Characteristics with Ground Braid Connection

Resonant Frequency (MHz)	Capacitance (pF)	Attenuation (dB)	Q	Inductance (nH)	Type of Capacitor
250	82	52	6.2	4.95	0603 MLCC
225	110	52.5	5.42	4.57	0603 MLCC
230	110	52	5	4.37	0402 MLCC A
192	150	53	5	4.6	0603 MLCC
158	220	52	3.64	4.6	0603 MLCC A
154	220	54	4.7	4.9	0603 MLCC
105	470	54	3.23	4.9	0603
109	470	54	3.11	4.55	0402 MLCC A
108	470	53	2.8	4.62	0402 MLCC B
92.3	1000	38	2.71	3	0603
47.9	2200	58	2.4	5.03	0603 MLCC A
48.7	2200	57	2.1	4.86	0603 MLCC A repeat[a]
54	2200	47	5.97	3.96	0603 MLCC B
53.6	2200	48	6.76	4.02	0603 MLCC B repeat[a]
60.7	2200	42	2.98	3.13	0603
54.8	2200	47	5.89	3.84	0402 MLCC B
45	3300	42	2.68	3.8	0805 MLCC
43.9	3300	48	5.5	3.99	0603 MLCC A
40.4	3300	59	21.3	4.72	0603 MLCC B
39.6	3300	60	24.3	4.9	0603 MLCC B repeat[a]
33	4700	60	20.5	4.96	0.063 MLCC B
33	4700	60	20.5	4.96	0603 MLCC C
39	4700	46	3.45	3.55	0805
36.6	4700	51	6.55	4.03	0402 MLCC B
31.2	6800	52	5.96	3.83	0603
27.8	10000	49.5	3.41	3.28	0603
82.5	750	55	29.8	4.97	1812 MICA
70.7	1000	58	35.8	5.08	2220 MICA

[a] Repeat = different samples of the same value from the same manufacturer.

220 pF 805 from Table 5.4 is 1.27 nH and of the 470 pF capacitor is 1.26 nH. The required capacitance at 160 MHz is thus

$$780 \text{ pF} = 1/(39.4 \times (160e^6)^2 \times 1.27e^{-9})$$

Using a 720 pF capacitor, the emission reduced to 8.7 dB below the specification limit. A reduction of 23 dB.

The resonance of a capacitor can also be used with a series inductor as described later in the use of the Laird (Steward) 4 line C/M choke or ferrite toroid in an LC C/M filter.

To illustrate the effect of additional inductance on the resonant frequency of a capacitor, to measure the Q, and to confirm that chip size does affect the self-inductance of the device measurement were made.

In the measurements, the inductance of the capacitor is derived from the value of the capacitor and the resonant frequency.

TABLE 5.4

Capacitors Characteristics with Negligible Additional Inductance in the Test Setup

Resonant Frequency (MHz)	Capacitance (pF)	Attenuation (dB)	Q	Inductance (nH)	Type of Capacitor
301	220	65	86	1.27	0805 MLCC
207	470	62	41	1.26	0805 MLCC
94.4	2200	48	3.8	1.29	0805 MLCC
86	3300	49	3.15	1.04	0805 MLCC
68	4700	49	2.8	1.17	0805 MLCC
190	1000	44	2.64	0.7	0603 MLCC
97	2200	65	27	1.23	0603 MLCC
109	2200	54	6.64	0.97	0603 MLCC
87	3300	50	3.51	1.03	0603 MLCC A
93	3300	49.5	3.09	0.89	0603 MLCC
67	4700	66	20	2.01	0603 MLCC B
62.2	4700	66	21.7	1.4	0603 MLCC C
81.4	4700	52	3.31	0.820	0603 MLCC
195	470	60	35	1.42	0402 MLCC
201	470	63	47.6	1.34	0402 MLCC
104	4700	50	2.1	0.5	0402 MLCC
80	3300	50	3.8	1.2	0402 MLCC

The Q is obtained from the DC resistance at resonance derived from the attenuation. When the Q is high, as in the two preceding examples, the resonant frequency can be determined reasonably accurately, whereas when the Q is low, this is not so easy.

The measurement was of a capacitor connected to a microstrip and to a 2 cm long, 5 mm wide, and 0.4 mm thick braid to ground. The PCB at the signal generator input end contained a series 50 Ω resistor, and the load in parallel with the capacitor to ground was also 50 Ω.

Based on Equation 5.3, the braid ground connection was added approximately with 2.5 nH to the self-inductance of the capacitor.

A number of 0603 sized capacitors from different manufacturers were tested, and these are labeled A, B, or C.

The total inductance from Table 5.3 varied from 3.13 to 4.97 nH.

We see from Table 5.4 that the Q was either in the range of 20–86 or 2–3.8. In case this was an anomaly, the tests were repeated with new samples with the same results. In the case of the 0603 4700 pF from manufacturers B and C, the Q was 20.1 or 21.7 and the inductance was 1.19 or 1.4 nH.

The capacitors with the lowest Q were the non-MLCC type, whereas the expectation is that these capacitors would have the highest inductance.

To eliminate the inductance of the ground braid, the top of the stripline was covered by a copper foil shield connected to the lower ground plane. A small aperture was cut in the shield, and the capacitor was contained in the aperture and soldered between the signal trace and the shield. Thus, the measurement was made without the addition of a via.

Table 5.4 shows the resonant frequency, the Q, and the inductance. As expected for a given capacitor size, the resonant frequency was higher and the inductance lower, as shown in Table 5.3.

The inductance varied from 0.5 to 1.42 nH. For the high Q examples where the measurement was more accurate, the inductance varied from 1.19 to 1.42 nH.

In none of the measurements of inductance were the expected differences between the three package sizes seen.

From the measurements, it seems advisable to measure a number of capacitors for Q and resonant frequency to select the most appropriate.

AVX manufactures a "CU" series C0G (NPO) RF/Microwave chip capacitor with ultralow ESR designed for the communications market and UHF applications (300 MHz–3 GHz).

In size 01005, the capacitance range is 0.5–6.2 pF and in 0201 size from 0.5 to 22 pF.

The typical Q is 290 at 1 GHz for a 4 pF capacitor, and the resonant frequency of the 01005 6.2 pF capacitor is 4.5 GHz, the 0.201 4.7 pF is 4.8 GHz, and the 0201 22 pF is 2.2 GHz.

The series resonant frequency of decoupling capacitors on a PCB with no overall ground and power plane can be used to determine the impedance between power and ground. However, on PCBs where significant power-line decoupling is achieved due to capacitance between power and ground, parallel resonances can occur resulting in higher impedances than a bare PCB with no decoupling capacitors. This is due to the PCB capacitance in parallel with the combined decoupling capacitor, trace, and via inductances.

In some instances, increasing the inductance of the connection to the capacitor and thereby decreasing the value of the capacitor for a given resonant frequency is beneficial.

For example, an electronic writing board was located close to an AM wireless transmitter operating at 1 MHz. When drawing on the board, a wavy line was displayed on the board due to EMI. The equipment was in production, and so minimum modifications were possible. The circuit included a capacitor at the equipment analog input from the writing board.

On increasing this value, the wavy line disappeared but the line drawn on the board lagged the pen, and so a gap existed between the pen and the line. The capacitor was a leaded device, and by raising the capacitor on standoffs, the inductance of the leads was increased. This meant that the capacitor value could be decreased to achieve resonance at 1 MHz. With a lower value capacitor, the lag in writing the line was solved.

An interesting corollary is that during production, the apparently redundant standoffs were removed and the problem reappeared until the standoffs were reinstated.

The traces of a PCB should go to the capacitor connection and then continue from there as shown in the correct connection of Figure 5.3. In the incorrect connection of Figure 5.3, the PCB trace inductance is in series with the capacitor self-inductance, which lowers the resonant frequency. Better still is the connection in which the capacitor is directly between the traces, as shown for surface-mount (SM) capacitors in Figure 5.4, without the use of vias. If vias must be used, they should have a large diameter or use multiple vias. SM chip capacitors may have extremely low inductance when the traces connecting them are properly laid out, as shown in Figure 5.4.

FIGURE 5.3 Correct and incorrect PCB trace connections for a capacitor.

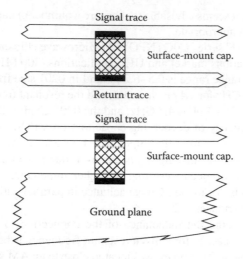

FIGURE 5.4   Ideal PCB connection, shown for a surface-mount capacitor.

Theoretically, the total inductance of a chip capacitor is determined by its length and width ratio and by the mutual inductance coupling between its electrodes. Thus, a 1210 chip size has a lower inductance than a 1206 chip. An 0805 chip size, NPO-type ceramic chip capacitor with values from 0.5 pF to 0.033 μF manufactured by AVX has a resonant frequency of 150 MHz for a 1000 pF value; thus, the self-inductance is 1.2 nH. A 100 pF 0805 capacitor has a resonant frequency of 380 MHz; thus, the self-inductance is 1.76 nH. Both the resonant frequency and the impedance above resonance vary with chip size. Thus, the 1210 has a lower impedance above resonance than does the 1206 size, although this is not obvious in the measurements in Table 5.5. The ceramic formulation also affects the impedance at resonance, so the NSPO 1000 pF 0805-sized chip has an impedance of 0.1 Ω at the resonant frequency of 150 MHz, whereas an 1000 pF with X7R ceramic formulation has the same resonant frequency, but the impedance is 0.4 Ω.

TABLE 5.5

Connection Inductance for Typical Capacitor Configurations with 10 Mil from Capacitor to Via Pad

Distance from Board to Plane	0805 Typical 148 Mil between Via Barrels (nH)	0603 Typical 128 Mil between Via Barrels (nH)	0402 Typical 106 Mil between Via Barrels (nH)
10	1.2	1.1	0.9
20	1.8	1.6	1.3
30	2.2	1.9	1.6
40	2.5	2.2	1.9
50	2.8	2.5	2.1
60	3.1	2.7	2.3
70	3.4	3.0	2.6
80	3.6	3.2	2.8
90	3.9	3.5	3.0
100	4.2	3.7	3.2

Source: Archambeault, B., Diepenbrock, J., IBM, RTP, NC PCB power decoupling myths debunked, Available on the internet.

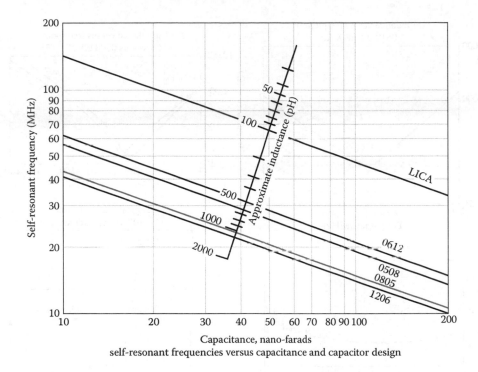

self-resonant frequencies versus capacitance and capacitor design

FIGURE 5.5 Self-resonant frequency and inductance for the low-inductance chip array versus other chip capacitors. (Courtesy of AVX.)

AVX has manufactured a low-inductance capacitor, the low-inductance chip array (LICA); the SRF and the approximate inductance of the LICA versus other capacitor designs are shown in Figure 5.5.

Multilayer chip (MLC) capacitors are constructed of a monolithic block of ceramic containing two sets of offset, interleaved planar electrodes that extend to two opposite surfaces of the ceramic dielectric. The great advantage of the MLC capacitor is its low ESR and low ESL, making it ideal for use in the input and output filters of switching power supplies. High-value MLC capacitors have an inductance of approximately 3 nH. A 24 µF MLC capacitor manufactured by AVX has an ESR of 0.001 Ω at the resonant frequency of 50 kHz compared to an ESR of 0.1 Ω for a low-ESR tantalum. Low-inductance MLC capacitors have a resonant frequency of 25 MHz for a 0.1 µF with an ESR of 0.06 Ω at resonance. The typical inductance of a 0.1 µF with a 0.508 chip size is 0.6 nH, and for the 0612 chip size, it is 0.5 nH. Thus, these 0.1 µF low-inductance MLC capacitors are an ideal choice as decoupling capacitors.

Microwave capacitors are designed with extremely high resonant frequencies and therefore extremely low self-inductance. Although the series resonant frequency is invariably lower, a parallel resonant frequency also exists. Figure 5.6 shows the variation of impedance with capacitor value, chip size, and ceramic formulation.

For an AVX AQ11-14 capacitor with a 100 pF capacitance, the series inductance is 0.4 nH, and the ESR is 0.034 Ω at 100 MHz and 0.062 Ω at 500 MHz. Figure 5.7 shows the series resonant frequency of a single-layer capacitor (SLC), which has a series inductance of an incredibly low 0.035 nH. ATC manufactures a low-ESR microwave capacitor with a series inductance of 0.129 nH and, for a 20 pF capacitor, an ESR of 0.04 Ω at 500 MHz and 0.055 Ω at 1000 MHz.

Another make of ultrahigh Q, low ESR capacitor is the CBR series from KEMET. The SRF in the 0201 package and for a 30 pF capacitor is 1.7 GHz. For the 0603, it is 1 GHz for 30 pF,

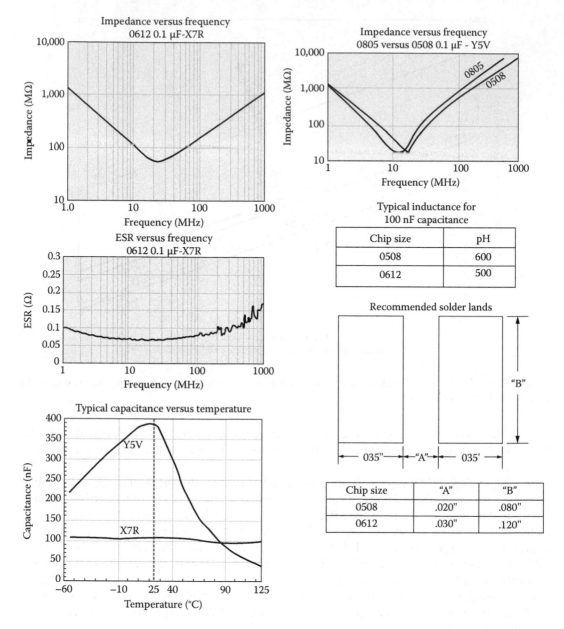

FIGURE 5.6 Variation of impedance with capacitor value, chip size, and ceramic formulation from AVX. (Courtesy of AVX.)

and for the 0805, it is 1 GHz for 30 pF and 600 MHz for the 100 pF. For the 0603 47 pF capacitor, the Q is 600 at 150 MHz and 35 at 1.2 GHz and the ESR is 0.03 Ω at 150 MHz and 0.08 Ω at 1.2 GHz.

AVX manufactures a 0.01 μF 10 V at 85°C broadband capacitor in a 0201 footprint. It has a resonance-free insertion loss of less than 0.5 dB from 0.4 to 40 GHz. A typical value for a through hole via is 2.5 nH to 10 NH, results depending on the trace length, which adds to the parasitic inductance of the capacitor.

The inductance of these connections for 0805, 0603, and 0402 size capacitors is shown for different distances into the board to a plane for a distance of 10 mil from the capacitor pad to a via pad

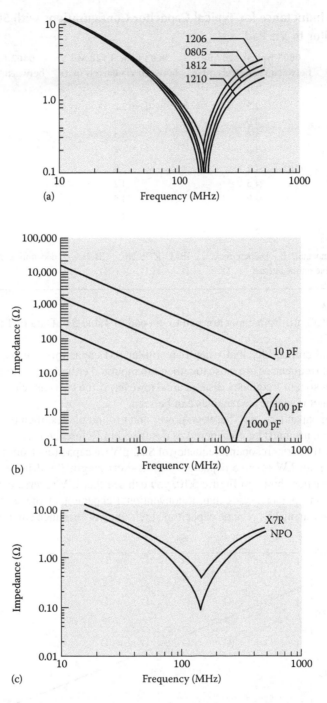

FIGURE 5.7 Series resonant frequency versus capacitance for an AVX single-layer capacitor. (a) Variation of impedance with chip size (1000 pF-NPO), (b) variation of impedance with cap value (0805-NPO) 10 pF versus 100 pF versus 1000 pF, and (c) variation of impedance with ceramic foundation (1000 pF-NPO vs. X7R 0805). (Courtesy of AVX.)

TABLE 5.6

Connection Inductance for Typical Capacitor Configurations with 50 Mil from Capacitor to Via Pad

Distance from Board to Plane	0805 Typical 148 Mil between Via Barrels (nH)	0603 Typical 128 Mil between Via Barrels (nH)	0402 Typical 106 Mil between Via Barrels (nH)
10	1.7	1.6	1.4
20	2.5	2.3	2.0
30	3.0	2.8	2.5
40	3.5	3.2	2.8
50	3.9	3.5	3.1
60	4.2	3.9	3.5
70	4.5	4.2	3.7
80	4.9	4.5	4.0
90	5.2	4.7	4.3
100	5.5	5.0	4.6

Source: Archambeault, B., Diepenbrock, J., IBM, RTP, NC PCB power decoupling myths debunked, Available on the internet.

in Table 5.5 and for 50 mil from capacitor pad to via pad in Table 5.6. Data in Tables 5.5 and 5.6 are from Reference [12].

It has been found empirically that when two similar SM capacitors, for which the capacitance values and resonant frequencies are close together, are mounted either one on top of the other or side by side, the upper resonant frequency disappears. However, if the two capacitors are moved further apart, the two distinct resonant frequencies can be seen.

With decoupling capacitors used between power and ground planes, then the distributed capacitance between the planes must be considered as described in Reference [12].

Figure 5.8 shows the series resonant frequency of SLC μWave capacitors from 1 pF to over 5000 pF.

Figure 5.9 shows an LW reverse capacitor with a short length *l* and large width *w*. From the frequency characteristics shown in Figure 5.10, you can see that LW reverse capacitors have lower impedance and better characteristics than a conventional capacitor of the same capacity above a certain frequency. By using LW reverse capacitors, the same performance can be achieved as that of

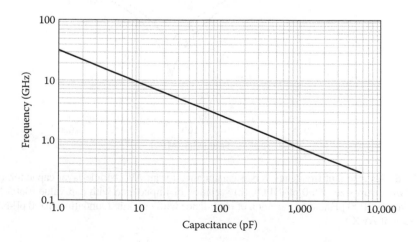

FIGURE 5.8 Series resonant frequencies of single-layer capacitors. (Courtesy of AVX.)

FIGURE 5.9   External appearance of an LW reverse capacitor.

FIGURE 5.10 |Z|/equivalent series resistance of an LW reverse capacitor and a general-purpose capacitor.

conventional capacitors with a fewer number of units. The reduction of unit number enables reduced costs and a reduction of mounting space.

The feedthrough type of capacitor exhibits characteristics closest to those of an ideal capacitor. The construction and insertion loss for tubular and discoidal feedthroughs are shown in Figures 5.11 through 5.14. The insertion loss is valid only when used with a 50 Ω series resistance. When the

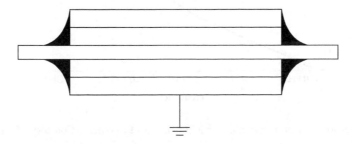

FIGURE 5.11   Tubular ceramic feedthrough capacitor construction.

FIGURE 5.12   Insertion loss for typical ceramic feedthrough capacitor. (Courtesy of Erie Technological Products.)

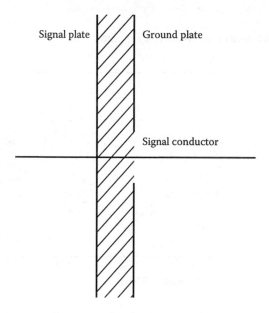

FIGURE 5.13   Typical construction of a discoidal capacitor.

FIGURE 5.14   Insertion loss of a discoidal capacitor in a 50 Ω system. (Courtesy of Erie Technological Products.)

noise voltage source impedance is low, the attenuation will be much reduced and is limited at resonance by the ESR of the capacitor. In these situations, a series resistance and/or inductance may be required to achieve sufficient attenuation. When both source and load impedances are low, the capacitor may not be the correct component to use to achieve a significant attenuation. When using a capacitor to attenuate LF noise or noise from a low-impedance source, values of one to hundreds of microfarads may be required.

The effectiveness of adding an inductor but still using the capacitor resonant frequency to obtain maximum attenuation is illustrated in Section 5.2.3.

The feedthrough capacitor, due to its construction, has a very low self-inductance. The correct method of mounting the feedthrough is to bolt or solder the case directly to the chassis, bulk-head, etc. When a length of wire is used to connect the case to ground, the performance is severely compromised, as shown in Figure 5.15.

The discoidal capacitor exhibits a lower impedance than the feedthrough type and has found a use in filtered connectors.

For modern LSI or VLSI IC packages, the use of SM capacitors between VCC planes and ground planes, as discussed in Chapter 11, is the correct decoupling method.

Figure 5.6 shows that the use of a 0.1 μF capacitor with an X7R formulation has a low impedance and very low ESR over the 20–100 MHz frequency range and would be ideal if used as decoupling for digital circuits with clock and data frequencies operating over this range. For other materials and capacitors with higher inductance, a lower value of capacitance should be used over the same frequency range, but this will result in an increase in ESR.

Sufficient decoupling capacitors, at least one per three ICs when they are in close proximity and one per IC when they are spaced apart, should be the goal.

For SM capacitors connected directly, or through vias, to the VCC and ground plane, an added trace between the capacitor and the VCC pin of the IC can reduce radiation from the IC and reduce ground bounce if the trace is long enough. The additional trace does not reduce the resonant frequency of the capacitor. Although this practice is controversial, in many practical cases, it has been slightly beneficial. The addition of a resistor between the capacitor and the VCC pin of the device will increase the bounce in the VCC at the IC but will reduce ground bounce and reduce emissions from the IC. Figure 5.16a illustrates the recommended decoupling capacitor connection and Figure 5.16b the resistor.

FIGURE 5.15 Effect of adding a lead to a feedthrough capacitor. (Courtesy of Erie Technological Products. Copyright MuRata Erie.)

FIGURE 5.16 (a) Adding a trace between the VCC pin and the decoupling capacitor, (b) Adding a resistor between the VCC pin and the decoupling capacitor.

The choice of the correct value of decoupling capacitor is important both in reducing supply and return-line noise voltage and in reducing radiation from the supply and return connections. Consider a case where the combined inductance of the supply and return tracks is 100 nH and the combined series resistance is 10 mΩ. Two TTL gates are connected at the end of the supply and return connections. The following table shows the influence of the decoupling capacitor on the magnitude of a 5 ns current pulse, $i_{S/R}$, and of a 5 ns negative voltage spike, $v_{S/R}$, on the supply and return connections that are generated by the TTL gates changing states at the same time:

C	ESL (nH)	$i_{S/R}$ (mA)	$v_{S/R}$ (V)
—	—	16	0.15
300 pF	22	16	0.13
0.1 μF	16	2	0.03

The 0.1 μF decoupling capacitor supplies the remaining 14 mA during the gate transition, and the reduced current flow in the supply connections results in less radiation from the circuit.

The importance of locating a decoupling or reservoir capacitor close to the IC, which requires or supplies the transient current, is illustrated in the following example. A MOSFET analog switch connects a 3 V supply to a device that has a capacitance of 5800 pF. A large transient on the 3 V supply is observed as the analog switch connects the 3 V to the load device. The circuit and transient voltages are shown in Figure 5.17a. The reason for the transient was that the reservoir capacitor was located approximately 10 cm from the analog switch. By moving the capacitor to the pins of the analog switch, the transient voltage was reduced, as shown in Figure 5.17b, and the load device functioned correctly.

The feedthrough capacitor is designed for use between compartments of equipment or as the power-line connection through the case of a filter, as shown in Figure 5.18a. Nevertheless, the low impedance of the feedthrough capacitor may be applied to an RF ground connection, as shown in Figure 5.18b–d. The ground plane may be a copper PCB or the metal wall of the enclosure, and the electrical connection is made to one terminal of the feedthrough only. When the use of a feedthrough is impractical, an axial or radial lead capacitor, as shown in Figure 5.19a, may be used.

FIGURE 5.17 (a) Reservoir capacitor far removed from an analog switch. (b) Reservoir capacitor located close to an analog switch.

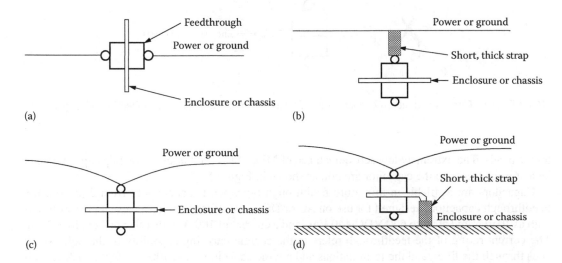

FIGURE 5.18 Proper installation configurations for a feedthrough capacitor.

Figure 5.19b and c illustrates incorrect connection of the capacitor, since the impedance of the wire between the power and signal ground reduces the effectiveness of the capacitor. In Figure 5.19a, the impedance of the wire is in series with the noise currents and increases the effectiveness of the capacitor. In reality, the inductances of the wires form a T-type filter with the capacitor. Figures 5.3 and 5.4 show the correct and incorrect PCB trace layout for a capacitor. Erie has manufactured a three-wire capacitor in which the inductance of the wires is further increased by the inclusion of

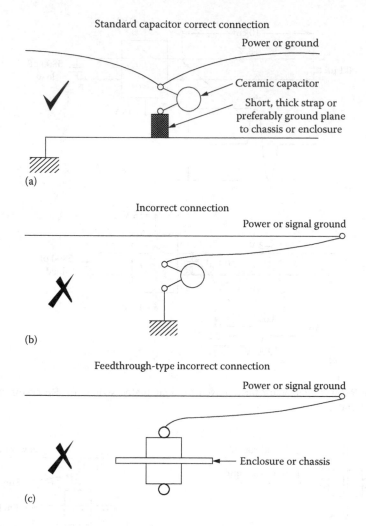

FIGURE 5.19   Correct and incorrect connection of (a) axial, (b) radial lead, and (c) feedthrough capacitors.

ferrite beads. The extremely low self-inductance of MLC, SLC, and microwave chip capacitors can be lost if the traces to the capacitor are not, as shown in Figure 5.4.

Capacitors are available in miniature feedthrough types with capacitances from 2 pF to 2 μF. Feedthrough capacitors designed for use on AC or DC power lines may have capacitance values as high as 10 μF rated at up to 600 VDC/440 VAC and a current of 100 A with a size of 7 cm by 8.2 cm. The current rating of the feedthrough refers to the current-carrying capability of the bolt, which goes through the filter, and the terminations and not the capacitance. A 100 V, 1.5 μF, 50 A capacitor feedthrough, or a filter feedthrough, may conveniently be used as the terminations of a power supply. When a feedthrough cannot be accommodated in an existing design, a PCB may be used to connect capacitors between the terminals and the metal case of the supply. Figure 5.20 illustrates this method, which achieves the lowest possible impedance.

A capacitor does not dissipate noise energy; rather, it shunts it to another path, typically ground. When the ground impedance is high, an additional noise voltage appears across the ground. When the connection is made to an imperfectly shielded enclosure, the radiation from the enclosure may increase. Thus, the inclusion of a capacitor between power or signal ground and chassis, to reduce C/M voltages, may not be without a secondary effect on EMC.

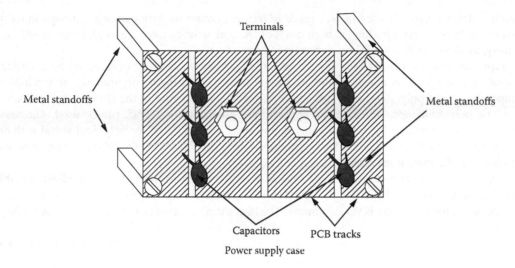

FIGURE 5.20 Low-impedance connection of capacitors between terminals and chassis.

5.1.10.2 Inductors

Here, we consider inductors as an isolated component, whereas in reality inductance requires a loop in which current flows and the inductor is only part of the loop. When inductance is measured on an inductance meter, the two test probes are shorted together and the internal meter inductance is canceled out.

The impedance of an ideal inductor increases linearly with increasing frequency. The AC equivalent circuit of a practical inductor, shown in Figure 5.21, contains series resistance and parallel capacitance. At some frequency, an inductor becomes a parallel resonant circuit with a very high impedance. Above this resonance, the impedance decreases due to the parallel capacitance.

The inductance (L), resistance (R), DC current-carrying capacity (I_{DC}), and the resonant frequency (f_o) of a number of heavy-duty hash chokes wound on molded powdered iron forms are as follows:

L (μH)	R (Ω)	f_0 (MHz)	I_{DC} (A)
3.35	0.010	45	20
8.8	0.021	28	10
68	0.054	5.7	5.0
125	0.080	2.6	3.5
500	0.260	1.17	2.0

FIGURE 5.21 AC equivalent circuit of an inductor.

Iron powder is invariably used in preference to ferrite, because its permeability, although initially lower than ferrite's, remains high at high levels of DC magnetizing force and high levels of AC flux density, as shown in Section 5.2 on power-line filters.

Typically, magnet wire is used for the windings, which is coated or covered by an electrical insulating material. This material must be removed when soldering a polyurethane or at a higher temperature using polyester–imide insulation. When the temperature in the choke is high, then an aromatic polyimide tape, which has an upper temperature limit of 220°C, can be used. The most common is a polyimide film with an upper temperature limit of 220°C or polyvinyl acetal with an upper temperature limit of 175°C. Polyester/amide–imide is good to 200°C–220°C and has superior toughness and chemical resistance.

When very-large-diameter/low AWG magnet wire is needed for high-current applications, the wire becomes difficult to bend. Then a flexible bonding braid covered with a glass or glass/polyester tape or heat-shrink sleeving is very common. The heat shrink is rated at 135°C or 150°C depending on the type.

Litz wire is made up of numerous thin isolated wires that are connected together at their ends to reduce the skin effect and the AC resistance of the wire. It is commonly used at frequencies up to 1 MHz.

The Q of an inductor is usually more important in signal filter circuits. However, if a power-line filter utilizes a series or parallel resonant circuit by design, then the Q may be required.

For air core inductors, the Q may be very high. It is determined by the ratio of

$$2\pi f L / R \tag{5.9}$$

For a core with permeability, it is

$$2\pi f L / R_{DC} + R_{ac} + R_d \tag{5.10}$$

where
R_{DC} is the DC winding resistance
R_{ac} is the resistance due to core loss
R_d is the resistance due to winding dielectric loss

and so Q is proportional to frequency for an inductor and is specified at a frequency, such as 1 kHz. At other frequencies, the Q will be different due to changes in core loss.

For inductors wound on a high-permeability core, the losses in the core will reduce the Q significantly. The iron powder cores with high permeability (up to 100) have low loss at low frequency (LF) but high loss at HF. The carbonyl iron powder with a permeability of 4–35 has higher Qs.

An example of the Q number of Micrometals is provided in Figure 5.22.

The inductance of a wound core is

$$L = \frac{0.4\pi\mu N^2 A \times 10^2}{l}$$

where
L is the inductance [μH]
μ is the core permeability
N is the number of turns
A is the effective cross-sectional area [cm²] (same as the cross-sectional area)
l is the mean magnetic path length [cm]

FIGURE 5.22 Q of different Micrometals cores with frequency. (Courtesy of Micrometals.)

Inductance at N turns

$$L_N = A_L \times N^2 \times 10^{-3}$$

where
   A_L is the nominal inductance [nH/N^2]
   L_N is the inductance [μH]

The mean magnetic path length is

$$L = \pi(OD - ID)/\ln\,(OD/ID)\ \text{[cm]}$$

where
   OD is the outside diameter of core [cm]
   ID is the inside diameter of core [cm]

The mean magnetic path length is illustrated in Figure 5.23.

Mean magnetic path length

FIGURE 5.23   Mean magnetic path length.

FIGURE 5.24   Common-mode choke.

Although the inductor is wound as a C/M choke, as shown in Figure 5.24, the permeability of the material, the number of turns, and therefore the inductance of the choke may be very high without the danger of saturating the core material due to very high DC currents. Both supply and return conductors should be threaded through the inductor in a bifilar fashion. Therefore, the DC flux generated by the supply and return tend to cancel in the core material. The disadvantage with the C/M inductor is that no appreciable attenuation of differential-mode (D/M) noise currents is achieved when bifilar wound; that is, both conductors wound together through the core. If the conductor is an enamel-coated magnet wire, then the insulation of the enamel may not be considered adequate when the inductor is used on AC or DC power lines. In reality, this is true only if the enamel has become abraded during the manufacturing process. To increase the insulation, thread the magnet wire through a Teflon tube or use Teflon-insulated wire. If the C/M windings are separated around the circumference of the core, then the increased insulation has been achieved and some D/M (leakage) inductance. Be aware that with very high inductance, C/M coils wound on high-permeability cores and used on high-DC-current supplies that the leakage inductance may be enough to reduce the permeability of the core. This is because the DC flux is not totally canceled unless the supply and return conductors are kept very close together using a bifilar winding. Some leakage inductance can be introduced when the windings are spaced apart around the toroid. This will add some D/M inductance but make the core inductance more sensitive to DC current and can degrade HF signals.

The use of a coil wound in sections, each exhibiting a different inductance and, therefore, a different capacitance, has the effect of increasing the useful frequency range of the inductor. The alternative is to use a number of individual inductors in series, each with a different inductance value and different resonance frequencies. When the problem frequency or frequencies are known, the resonant frequency of the inductor/s may be chosen or tailored by additional capacitance in order to match the problem frequency and achieve maximum impedance.

Molded RF coils are available in a form similar to a resistor. These use either phenolic (air), iron, or ferrite as the core material and are available from 0.022 µH (resonance frequency 50 MHz, DC resistance 0.01 Ω, and current capacity of 3.8 A) to 10 mH (0.25 MHz, 72 Ω, and 48 mA).

The permeability of a core material is often quoted in inductance per 100 turns or inductance per turn. To obtain the number of turns for a given inductance when the inductance is specified per 100 turns, use the following formula:

$$T = 100\sqrt{\dfrac{\text{Required inductance}\ \left[\mu H\right]}{\text{Inductance per 100 turns}\ \left[\mu H\right]}}$$

When the inductance is specified in mH/1000 turns, use

$$T = 1000 \sqrt{\frac{\text{Required inductance } [\text{mH}]}{\text{Inductance per } 1000 \text{ turns } [\text{mH}]}}$$

When the inductance is specified per turn or inductance per N^2, use

$$T = \sqrt{\frac{\text{Required inductance } [\mu\text{H}]}{\text{Inductance per turns } [\mu\text{H}]}}$$

When the source and load impedances are high, the attenuation of the inductor may not be adequate and a high-value resistor or a combination of series resistor and shunt capacitor or series inductor and shunt capacitor may be more effective. This type of filter is described in detail in Section 5.2.

The inductor will reduce noise current flow and may be used in reducing ground or supply currents. Having said that, a low-loss high-Q inductor does not dissipate noise energy but rather develops a noise voltage across it. Thus, the inclusion of an inductor may increase the noise voltage at the source of the noise. When the source is digital logic or load switching and the inductor is in the supply line, the potential increase in noise voltage across the logic or load may result in EMI. When the inductor forms a filter with load or onboard decoupling capacitance, the LC combination may exhibit insertion gain, as described in Section 5.2 on filters.

5.1.10.3 Ferrite Beads

Ferrite material exhibits a frequency-sensitive permeability and quality factor ($Q = X_L/R_S$). In this section, beads or baluns are assumed to have either a wire passing through the material or a limited number of turns of wire through the material. As frequency is increased, the reactance of the bead increases, as does its residual core loss represented by a resistance. Above a given frequency, the inductive reactance (permeability) X decreases and the resistance R continues to increase, and so does the impedance Z. When a wire is taken through the center of a bead, we refer to it as a *half-turn*. The ferrite bead is one of the few components in which the noise energy is converted into heat within the component. Figure 5.25a–f shows the impedance versus frequency characteristics for a number of ferrite materials made by the Laird (Steward) Technologies and Fair-Rite Products.

As the number of turns in the core is increased, there is an increase in the interwinding capacitance and the impedance at higher frequencies is reduced. Figure 5.25c shows impedance versus frequency for 2-1/2 turns of wire as compared to a half-turn and 1-1/2 turns. The permeability of ferrite, in addition to its frequency dependence, is very sensitive to line DC current, shown in Figure 5.26.

Due to the relatively low impedance (30–800 Ω) at most frequencies above 1 MHz, ferrite beads are most effective in low-impedance circuits or in conjunction with shunt capacitance. Also at LF, the losses are small and the Q of the bead may result in the resonance effects described in Section 5.2.

The inductance of a single wire through a bead is given by

$$L = \frac{\mu_r \mu_o 0.01 s}{2\pi} \ln\left(\frac{r_2}{r_1}\right)$$

where
  s is the length of the bead [cm]
  r_1 is the inner diameter [cm]
  r_2 is the outer diameter [cm]
  $\mu_o = 4\pi \times 10^{-7}$ [H/m]
  μ_r is the relative permeability

FIGURE 5.25 (a) Impedance versus frequency characteristics for different ferrite materials and configurations. (b) Low frequency 75 material. *(Continued)*

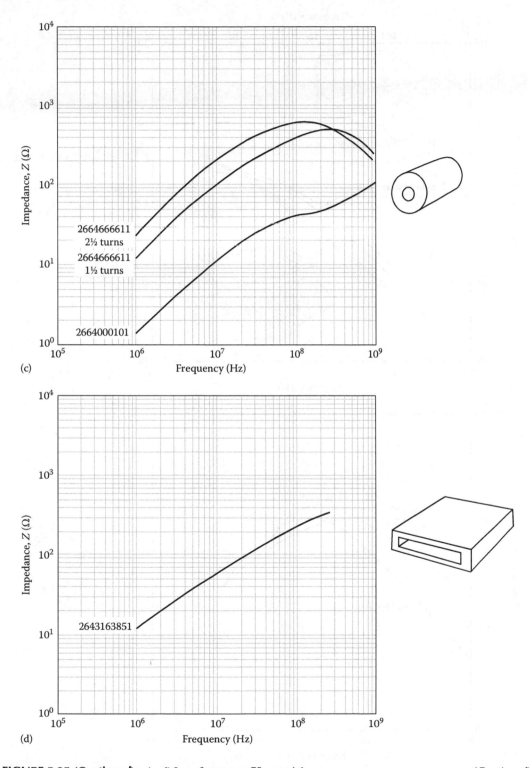

(c)

(d)

FIGURE 5.25 (*Continued*) (c–d) Low frequency 75 material. (*Continued*)

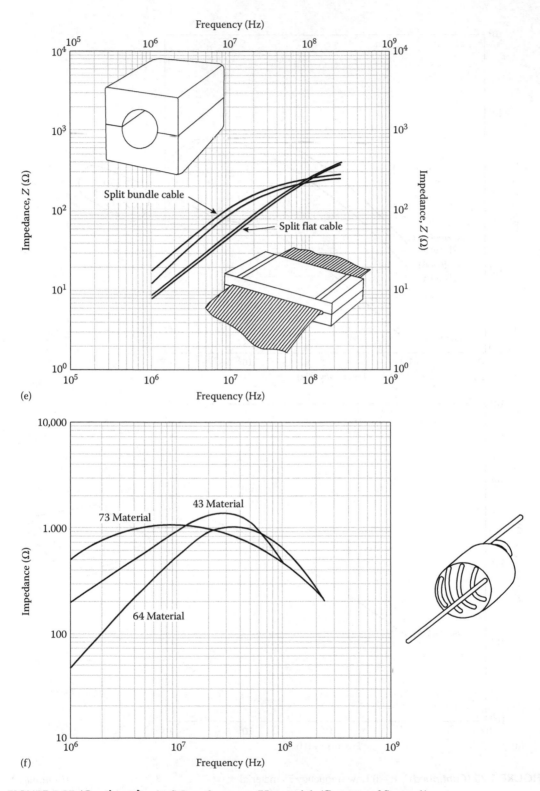

FIGURE 5.25 (*Continued*)   (e–f) Low frequency 75 material. (Courtesy of Steward.)

FIGURE 5.26   Permeability versus DC current. (Reproduced from Cowdell, R.B., *Electron. Design*, 12, June 7, 1969.)

FIGURE 5.27   Relevant dimensions of a cylindrical bead. (Reproduced with kind permission from Cowdell, R.B., *Electron. Designs*, 12, June 7, 1969.)

The physical parameters included in this equation are shown in Figure 5.27. The magnetic force within the bead is given by

$$H = \frac{I}{p}$$

where
    I is the magnetizing current [mA or A]
    $P = r_2 - r_1$ [cm or m]

To obtain the magnetizing force in oersteds, divide the force in amps per meter by 79.6.

The number of ferrite beads on a line also affects the HF performance, as can be seen from Figure 5.28. Although the LF impedance of the line with multiple beads is higher than that for the single bead, the impedance at higher frequency is much reduced.

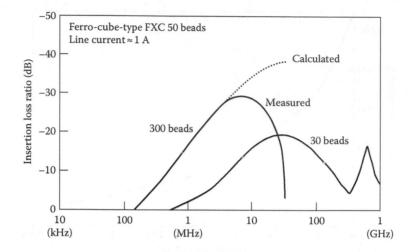

FIGURE 5.28 Effect of 30 and 300 ferrite beads on a wire. (Reproduced with kind permission from Cowdell, R.B., *Electron. Designs*, 12, June 7, 1969.)

One material that should have a very useful application at microwave frequencies is "lossy sleeving," manufactured by Capcon International Inc. This material is flexible and can be used over a temperature range from −55°C to +250°C. It comes with an internal diameter of from 0.04 to 1.25 in. and in a shielded or unshielded version. The specification shows an attenuation of 45 dB at 1 GHz and 100 dB at 9 GHz for a 1 in. length of the unshielded type and 100 dB at 1 GHz for a 1 ft length and remaining at 100 dB up to 100 GHz. Although this material should have many applications at gigahertz frequencies, it has not proven so in practice. In one case, a ¼ in. length was placed over a pair of conductors, used for a differential signal, to function as a balun to reduce C/M currents at 2.5–10 GHz. This was not as effective as a ¼ in. bifilar wound and twisted magnet wire balun, with the magnet wire wound with one turn through a two-hole core made of a material rated to be effective up to only 3 GHz.

Fair Rite has introduced a new 75 LF material. An example of a round cable core is the 2675540002, which is 14 mm OD and 6.35 mm ID and has a length of 28.6 mm. The 26750230002 is shown in Figure 5.25a and Table 5.7.

A new product is ferrite sheet either thin solid or flexible. The theory is that at lower frequencies, it shields and at higher frequency it absorbs. The most common application is glued to logic ICs or areas of a PCB with high-level emissions. Although when effective they do reduce emissions, in many cases, the attenuation is low or negligible. A flexible sheet is available from Fair Rite in six material grades and four standards of thickness. The sheets are manufactured with 0.01 mm thick PET film on one side and 0.02 mm thick adhesive tape on the opposite side. The permeability of these sheets ranges from 20 to 220 μm. The 20 μm material maintains the permeability

TABLE 5.7

Impedance of Fair-Rite 2675540002 Core

Frequency	Impedance (Ω)
100 kHz	18
1 MHz	139
1.1 MHz	160
1.5 MHz	95
13 MHz	59

out to 300 MHz and then it drops to 1 at 1 GHz. The 220 μm material has a permeability of 100 at 30 MHz, 30 at 100 MHz, 10 at 200 MHz, and 1 at 1 GHz.

Kitagawa Industries (Intermark) supply a number of ferrite baluns including thin rectangular baluns, either in a single piece or in two pieces, designed for flex print, flat cable or ribbon cables, and some of these are shown in Figure 5.29 along with split cores and a grounding braid.

The impedance of the small flat print 1 piece GFPC type and 2 piece type GFPH and GFPO is 20–50 Ω at 100 MHz with a typical impedance curve as shown in Figure 5.30. Intermark also offers the MRFC range of split ferrite clamps for LF with a typical impedance as shown in Figure 5.31.

FIGURE 5.29   Kitagawa Industries (Intermark) ferrite products.

FIGURE 5.30   Typical impedance of the GFPC flat cable-type ferrite. (Courtesy of Katagawa Industries, San Jose, CA (Intermark).)

FIGURE 5.31 Typical impedance of the low-impedance MRFC split cores. (Courtesy of Katagawa Industries, San Jose, CA (Intermark).)

A SD sintered solid ferrite tile is also available from Kitagawa Industries (Intermark).

When modeling a circuit in SPICE, a SPICE model of a ferrite bead is often required.

Some manufacturers provide a SPICE model of their products. However, they use a simple model and are often highly inaccurate especially at HF. One simple equivalent circuit for a ferrite bead or SM bead is shown in Figure 5.32.

An alternative simple model is shown in Figure 5.33.

Reference [13] describes a method of modeling a ferrite bead for SPICE simulation using a frequency-dependent component, which is a voltage-controlled current source with transconductance, but this again results in significant errors at high frequencies. The second approach taken in Reference [14] is to include measured data in the frequency-dependent component.

Spectrum Software (www.spectrum-soft.com/news/winter2011/ferritebead.shtm) offers an integrated schematic editor and mixed analog/digital simulator entitled Micro-Cap. This uses in its software the more sophisticated model of the ferrite shown in Figure 5.34. The program is used to

FIGURE 5.32 Simple SPICE model of ferrite bead.

FIGURE 5.33 Alternative simple SPICE model of a ferrite bead.

FIGURE 5.34 More accurate SPICE model of ferrite bead used by Spectrum Software.

create a basic template of the ferrite bead model, and then a schematic is made that simulates the impedance versus frequency using Micro-Cap. The impedance of the ferrite bead is then optimized using the "Model" program.

One problem with the use of baluns at gigahertz frequencies is coupling around the balun. Ideally, the traces up to PCB-mounted baluns used at HF should be in stripline, and the connections should be routed through a PCB-mounted local shield, typically used to shield one section of a circuit from a different section (either higher or lower power or a different frequency).

5.1.10.4 Baluns

The *balun* is a wideband transformer used to transfer energy from a balanced to an unbalanced line. A common application for a balun is in matching a balanced antenna to an unbalanced cable. In this section, we use the term *balun* to describe a special use of a ferrite or metal oxide bead or toroid, and in this application, it is commonly referred to as a choke, although in many people's minds, a choke is a multiturn inductor. The simple case we will discuss is a balun core with two wires passing through its center. When the two wires carry differential currents, the impedance to these currents at all frequencies is effectively zero. Thus, where the differential currents are caused by a signal, neither the signal amplitude nor the rise time is affected by the presence of the core, although C/M current due to either current induced by an incident electromagnetic field or noise voltage present between the grounds of two systems is reduced by the series impedance of the balun. A differential-input amplifier with C/M noise rejection is often compromised in its rejection by the conversion of C/M currents to D/M voltage. For example, when twisted shielded pair cable is used to carry the differential signal, an imbalance in the pair to shield impedance may result in the conversion of C/M noise to D/M noise. When the circuit impedances are sufficiently low, the use of a balun will reduce the C/M current and thereby any D/M-induced noise. It is imperative that the two wires be threaded through the same hole in the bead when using a multihole bead. When a single hole is used, the differential signal remains unaffected by the presence of the bead up to 1 GHz and above. Having said that, when the two wires are threaded through different holes, signals with repetition rates as low as 1 MHz are degraded by the inclusion of the balun. The HF C/M noise rejection of an amplifier is invariably compromised by capacitive imbalance to ground on input wiring and PCB tracks and by the poor C/M noise rejection of operational amplifiers (op amps) at HF. Here, the addition of a balun over input wiring may cure the problem by a decrease in the C/M current flow. When the input impedance of the differential circuit is too high, the inclusion of two capacitors, matched as closely as possible in value, located after the balun and between each of the inputs and ground may be the solution.

Figure 5.35 illustrates the use of the balun where the input is unbalanced and the C/M current flowing in the ground results in a low voltage, whereas the current flow in the high input impedance of the amplifier results in a much higher voltage (i.e., a D/M voltage is developed). Decreasing the magnitude of C/M current at the source is a viable option in the configuration shown in Figure 5.35a. Also, increasing the number of turns to between 1 and 5 may improve the performance of the balun. The effectiveness of the balun on the shielded cable in Figure 5.35b will be higher than that of the Figure 5.35a configuration due to the low impedance of the shield to the equipment case connection below resonant frequencies.

5.1.10.5 Common-Mode Chokes

5.1.10.5.1 Commercial Common-Mode Chokes

Wurth manufactured a range of C/M chokes. The published impedances for these signal line/power-line C/M chokes may be used in circuits to calculate the C/M attenuation.

The DC current rating, inductance, leakage inductance, resistance, and typical impedances of these chokes are shown in Table 5.8.

Wurth also published a design guide entitled "Trilogy of Inductors" that covers basic principles, components, filter circuits, and applications.

FIGURE 5.35   Uses of a balun.

TABLE 5.8
Published Data on Inductors

Wurth Part Number	Rated Current (A)	DC Resistance (Ω)	Inductance (µH)	Leakage Inductance (nH)	Typical Impedance
WE-SL2.744 224	1.2	0.13	2 × 250	60	1,800
WE-SL2.744 226	1.6	0.08	2 × 10	850	920
WE-SL2.744 227	1	0.16	2 × 51	85	5,500
WE-SL2.744 222	0.8	0.32	2 × 1000	90	6,000
WE-SL2.744 220	0.5	0.75	2 × 4700	180	20,000

Source: Courtesy of Wurth Industries, Pune, India.

The impedance change with frequency is shown in Figure 5.36a–e.

The size of the choke is shown in Figure 5.37.

When measured with a 50 Ω load, chokes that exhibit an impedance of up to 3000 Ω will exhibit up to 35 dB of attenuation with no DC current flow in the choke. Very often, these chokes are used to reduce C/M currents on wires/cables. Measurements were made of the C/M RF current generated

FIGURE 5.36 (a) WE-SL2. 744 224 Wurth common-mode (C/M) choke impedance. (b) WE-SL2. 744 226 Wurth C/M choke impedance. (*Continued*)

FIGURE 5.36 (*Continued*)   (c) WE-SL2. 744 227 Wurth C/M choke impedance. (d) WE-SL2. 744 222 Wurth C/M choke impedance.   (*Continued*)

F1 typical impedance characteristics

Impedance (Ω) vs *Frequency (MHz)*

Legend: —— Z (comm) - - - Z (diff)

(e)

FIGURE 5.36 (*Continued*)   (e) WE-SL2. 744 220 Wurth C/M choke impedance. (a–d: Courtesy of Wurth Industries, Pune, India.)

by switching power supplies on a small PCB, conducted out on short, 0.23, 0.49, and 0.62 m long wires. The DC current flow on the three different power-line outputs was 0.5 A each. The measurement is relative, that is, a C/M current without the WE-SL2 744 226 choke and with the choke and the difference is shown in Tables 5.9 through 5.11:

If the PCB is contained in a conductive enclosure and the RF currents flowing on the inside of the enclosure are low, adding a capacitor between one line on the cable side of the choke to the enclosure and a second capacitor on the second line to the enclosure may improve the level of attenuation, and this is shown in Figure 5.38. However, if high RF currents flow in the enclosure, then these may be injected onto the output cables and if so the effectiveness of the choke is compromised.

The inductor may also be used as a differential inductor as shown in Figure 5.39. In either C/M or D/M, the impedance will change with DC current through the inductor.

This was tested in the D/M at 12 MHz with both inductors in parallel in the test setup shown in Figure 5.40. The impedance drops from 393 Ω at 0 A DC to 50 Ω at 0.8 A. When the inductor is connected in series as shown in Figure 5.41, the impedance drops from 2867 Ω at 0 A DC to 55 Ω at 0.6 A.

When connected as a C/M choke, the impedance is constant up to 0.26 A but reduces to 83 Ω at 0.55 A even though rated at 1 A.

These test results are shown in Table 5.12.

The impedance change with DC current is shown in Table 5.12.

Another manufacturer of C/M chokes is Murata. The DL W21SN_HQ2 series have similar characteristics to the Wurth chokes. The characteristics of these chokes are shown in Table 5.13.

Murata also provides the attenuation for differential signals as shown in Figure 5.42.

FIGURE 5.37 Dimensions of 220, 222, 224, 226, and 227 chokes. (Courtesy of Wurth Industries, Pune, India.)

TABLE 5.9
Output 1

Frequency (MHz)	No Choke	With Choke	Delta (dB)
31.6	50	48	—
134.6–166	70	47	23
177	60	47	13
202	57	47	10
277–281	51	47	4

TABLE 5.10
Output 2

Frequency (MHz)	No Choke	With Choke	Delta (dB)
26.9–31.6	47	47	—
134.6	57	45	12
269	57	Ambient	>30

TABLE 5.11
Output 3

Frequency (MHz)	No Choke	With Choke	Delta (dB)
28–31.64	57	47	10
88.8	67	—	>30
134.6	54	41	13
202–270	57	51	6

FIGURE 5.38 Adding common-mode capacitors to a choke.

FIGURE 5.39 Differential-mode DC, common-mode AC test setup.

FIGURE 5.40 Common-mode inductors in parallel test setup.

FIGURE 5.41   Common-mode inductors in series test setup.

TABLE 5.12
Effect of Common-Mode and Differential-Mode
DC Current on the WE-SLT 744.227 Choke

Test Configuration

DC Current	DC Current (A)	Impedance (Ω)
C/M in parallel	0	393
C/M in parallel	0.1	389
C/M in parallel	0.2	389
C/M in parallel	0.3	302
C/M in parallel	0.4	238
C/M in parallel	0.5	150
C/M in parallel	0.6	117
C/M in parallel	0.7	73
C/M in parallel	0.8	50
C/M in series	0	2867
C/M in series	0.1	2523
C/M in series	0.2	1164
C/M in series	0.3	543
C/M in series	0.4	302
C/M in series	0.5	83
C/M in series	0.6	55
D/M	0	320
D/M	0.26	320
D/M	0.4	283
D/M	0.55	83

Another manufacturer of C/M chokes is Coilcraft. These have current rating as high as 3.16 A and inductances of up to 900 μH, but the dimensions are larger. Information on current rating, inductance, dimensions, and typical D/M, C/M attenuation, and impedance is shown in Table 5.14 and Figures 5.43 and 5.44.

EPCOS and Panasonic manufacture C/M chokes in the footprint shown in #9 in Figure 5.45.

The Panasonic range is from 82 mH, 0.8 A, 0.892 Ω to 0.15 mH 10 A, 0.0085 Ω, and the EPCOS range is from 3.3 mH, 1.8 A, 0.14 Ω to 100 mH, 0.35 A, 4.5 Ω.

The EPCOS choke is described as the current compensated *D* core double choke.

TABLE 5.13
Characteristics of Murata Chokes

Part Number	Common-Mode Impedance (at 100 MHz/20°C)	Rated Current (mA)	Rated Voltage (Vdc)	Insulation Resistance (min) (MΩ)	Withstand Voltage (Vdc)	DC Resistance (Ωmax)	Operating Temperature Range (°C)
DLW21SN501SK2#	500 Ω ± 25%	250	50	10	125	0.5	−40 to +85
DLW21SN670SQ2#	67 Ω ± 25%	400	50	10	125	0.25	−40 to +85
DLW21SN900SQ2#	90 Ω ± 25%	330	50	10	125	0.35	−40 to +85
DLW21SN121SQ2#	120 Ω ± 25%	370	50	10	125	0.30	−40 to +85
DLW21SN181SQ2#	180 Ω ±25%	330	50	10	125	0.35	−40 to +85
DLW21SN261SQ2#	260 Ω ± 25%	300	50	10	125	0.40	−40 to +85
DLW21SN371SQ2#	370 Ω ± 25%	280	50	10	125	0.45	−40 to +85
DLW21SN670HQ2#	67 Ω + 25%	320	20	10	50	0.31	−40 to +85
DLW21SN900HQ2#	90 Ω ± 25%	280	20	10	50	0.41	−40 to +85
DLW21SN121HQ2#	120 Ω ± 25%	280	20	10	50	0.41	−40 to +85
DLW21SR670HQ2#	67 Ω ± 25%	400	20	10	50	0.25	−40 to +85

Number of circuit: 1

Source: Courtesy of Murata.

FIGURE 5.42   Differential-mode characteristics. (Courtesy of Murata.)

5.1.10.5.2 Custom-Designed Common-Mode Chokes

As the supply current on in the C/M choke flows in one direction on one conductor and returns on the other and assuming 100% of the current does this, then theoretically the flux in the core will be zero and the permeability of the core is unchanged. In practice, the leakage inductance of the choke means that some D/M current flows, and in high permeability, cores can result in saturation. The solution is to bifilar wind the choke and keep the two wires in very close proximity.

The most common core material for C/M chokes is high-permeability ferrite.

TABLE 5.14

Coilcraft Common-Mode Inductor Data

Part Number[1]	Inductance[2] Min (µH)	DCR Max[3] (Ω)	Isolation (*Vrms*)	Imfis[4] (A)
LPD3015-391MR_	0.31	0.071	100	1.45
LPD3015-103MR_	8.0	1.04	100	0.38
LPD3015-104MR_	80.0	8.54	100	0.13
LPD3015-334MR_	264	27.7	100	0.07
LPD4012-331NR_	0.23	0.042	100	1.87
LPD4012-103MR	8.0	0.638	100	0.48
LPD4012-104MR_	80.0	4.76	100	0.18
LPD5010-681ME_	0.54	0.070	100	1.95
aLPD5010-103ME_	8.0	0.78	100	0.50
LPD5010-104ME_	80.0	8.0	100	0.15
LPD5030-102NE_	0.70	0.042	100	2.60
LPD5030-103ME_	8.0	0.21	100	1.05
LPD5030-104ME_	80.0	1.79	100	0.35
LPD5030-105ME_	800	16.5	100	0.11
MSD7342-252ML_	2.00	0.033	200	2.17
MSD7342-103ML_	8.00	0.10	200	1.24
MSD7342-104ML_	80.0	0.77	200	0.45
MSD7342-824ML_	656	660	200	0.15
MSD1260-472ML_	3.76	0036	500	3.16
MSD1260-103ML_	8.0	0.060	500	2.45
MSD1260-104ML_	80.0	0.321	500	1.06
MSD1260-105KL_	900	3.06	500	0.34
MSD1278-472ML_	3.76	0.040	500	3.16
MSD1278-103ML_	8.0	0.058	500	2.56
MSD1278-104ML_	80.0	0.30	500	1.13
MSD1278-105KL_	900	2.87	500	0.37

Source: Courtesy of Coilcraft.

One disadvantage ferrite cores have over metal cores are their brittleness but ferrite is very hard and so hard wearing.

Magnetics is a manufacturer of ferrite cores. A summary of the characteristics of the different materials most suitable for use in EMI filters is shown in Table 5.15.

The Curie temperature is the critical temperature where magnetism is lost.

The maximum useable frequency is where there is a 50% roll-off in permeability although the material is still somewhat useful at higher frequencies. For example, although the W material has a maximum usable frequency of ≤ 0.5 MHz, a coil wound with this material still has significant reactance at 1 MHz. The reason is that as the permeability reduces with increasing frequency, so the reactance increases with increasing frequency and the two tend to cancel.

The C/M choke may be wound on a toroid or on a pot core. The advantage of the pot core is that the choke is shielded.

As with iron dust cores, the ferrite core permeability increases with increasing AC flux level until after a critical flux it reduces rapidly to 0% increase. For example, a core may exhibit a 25% increase at 1000 G (0.1 T) and 40% at 2000 G (0.2 T) and 0% at 3000 G (0.3 T).

The permeability and flux density change with temperature, and the permeability changes with flux density as shown in Figures 5.46 through 5.48, reproduced from the Magnetics ferrite core catalog.

LPD 3015/LPD4012/LPD5010/LPD5030   MSD7342/MSD1260/MSD1278

Part series	A max	B max	C max	D	E	F	G	H	I	J	K		Weight
LPD3015	0.121	0.121	0.059	0.039	0.130	0.094	0.040	0.022	–	–	–	inches	45–52 mg
	3,076	3,076	1,50	0,99	3,30	2,30	1,02	0,54	–	–	–	mm	
LPD4012	0.158	0.158	0.047	0.060	0.174	0.103	0.049	0.031	–	–	–	inches	54–64 mg
	4,016	4,016	1,20	1,52	4,40	2,606	1,24	0,78	–	–	–	mm	
LPD5010	0.192	0.192	0.039	0.060	0.197	0.134	0.060	0.030	0.090	0.138	–	inches	60–70 mg
	4,876	4,876	1,0	1,52	5,0	3,4	1,5	0,75	2,3	3,5	–	mm	
LPD5030	0.192	0.192	0.118	0.060	0.197	0.134	0.060	0.030	0.090	0.138	–	inches	0.21–0.23 g
	4,876	4,876	3,0	1,52	5,0	3,4	1,5	0,75	2,3	3,5	–	mm	
MSD7342	0.295	0.295	0.181	0.028	0.091	0.063	0.150	0.083	0.110	0.130	0.043	inches	0.76–0.87 g
	7,5	7,5	4,6	0,70	2,3	1,6	3,8	2,1	2,8	3,3	1,10	mm	
MSD1260	0.484	0.484	0.25	0.059	0.197	0.138	0.197	0.157	0.217	0.177	0.079	inches	2.8–3.2 g
	12,3	12,3	6,2	1,5	5,0	3,5	5,0	4,0	5,5	4,5	2,0	mm	
MSD1278	0.484	0.484	0.317	0.059	0.197	0.138	0.197	0.157	0.217	0.177	0.079	inches	3.7–4.4 g
	12,3	12,3	8,05	1,5	5,0	3,5	5,0	4,0	5,5	4,5	2,0	mm	

FIGURE 5.43 Coicraft common-mode inductor dimensions. (Courtesy of Coilcraft.)

Typically, the current flow at the input of equipment with switch-mode power supplies (SMPS) is D/M at the switching frequency of the supplies and C/M at high frequencies; however, there are exceptions. In one example of a high-current design, the problem frequency was at 0.2 MHz and the current was C/M. A 4.8 mH C/M inductor was required in a filter rated at 20 A DC. The W material from Magnetics with a relative permeability of 10,000 is still very effective at this frequency.

It was found that 7 bifilar turns of braid on three Magnetics W44932-TC cores glued together provided the 4.8 mH. There was a limit on the maximum value of C/M capacitance that dictated the 4.8 mH to achieve a resonance of 15 kHz.

Due to the bifilar winding, no change was seen in the core performance at 20 A DC.

By gluing the three cores face to face three times, the inductance is achieved compared to a single core. The braid used was the Belden 8668, which is rated at 36 A. The predicted temperature rise in the core at 20 A is 66°C. The upper temperature limit imposed by the core Curie temperature and the heat-shrink sleeving were set conservatively to 105°C, which means the maximum ambient must be 39°C, and this was acceptable. The maximum before damage is 65°C.

As with all inductors, unshielded C/M chokes function as receiving antennas. If placed close to SMPS, the magnetic field from the magnetics may couple to the C/M choke and induce C/M currents in the lines. This mechanism resulted in exceeding MIL-STD-461F CE102 limits.

FIGURE 5.44 Coicraft common-mode (C/M) choke typical characteristics. (a) Typical differential-mode attenuation (ref: 50 Ω), (b) typical C/M attenuation (ref: 50 Ω), and (c) typical impedance versus frequency. (Courtesy of Coilcraft.)

The use of shielded inductors is always recommended when inductors are close to high-level noise sources.

Typically, the current flow at the input of equipment with SMPS is D/M at the switching frequency of the supplies and C/M at high frequencies; however, there are exceptions. In one example of a high-current design, the problem frequency was at 0.2 MHz and the current was C/M.

5.1.10.6 Commercial Power Inductors

These are available in a great number of sizes, ratings, and performances.

Figure 5.45 shows samples of these. And Table 5.16 shows the characteristics.

FIGURE 5.45 Commercial inductors.

The ranges and specifications for some of the inductors shown in Figure 5.45 are as follows:

(1) Delevan 4922 series SM inductance range 0.22 µH, 7 A, 0.008 Ω to 22,000 µH 0.05 A, 160 Ω.
(2) Delevan 5500/R series inductance range 3.9 µH, 11.9 A to 100,000 µH 0.11 A 76 Ω.
(13) Coiltronics FP3 series inductance range 0.1 µH, 19 A, Isat 10% = 27 A, Isat 15% = 34.7 A, 1.21 mΩ to 14.9 µH 2.22 A Isat 10% = 2 A, Isat 155 = 2.5 A.
(4, 5, and 6) Hammond Heavy duty Hash chokes inductance range 3.35 µH, 20 A 0.01 Ω SRF 45 MHz, 500 µH 2 A 0.26 Ω SRF 1.17 MHz, 1000 µH,1.25 A, 0.83 Ω.
(3) Vishay IHSM-5832 SM inductance range 1 µH, 9 A, 0.01 Ω to 4700 µH, 0.14 A, 23.1 Ω.
(9) EPCOS current compensated *D* core double chokes inductance range 100 mH, 0.35 A, 4.5 Ω to 3.3 mH, 1.8 A, 0.14 Ω.
(10,11 and 12) Coilcraft SM inductance range 10,000 µH at 0.09 A to 1.2 µH at 5.4 A, 1 µH at 7.5 A to 0.18 µH at 14.3 A.
(7) Pulse inductance range 40 µH at 0 A DC, 17 µH at 17 A DC, 0.065 Ω, 688 µH at 0 A DC, 400 µH at 3.6 A 0.13 Ω.
(14) ACT high-current and power-line 0603, 0805 impedance range Z = 12 Ω1.5 A 0.05 Ω to Z = 600 Ω 1.5 A 0.1 Ω. See Table 5.17.

It would appear that the ACT 0605 and 0805 inductors have a very-high-rated current for such a small device, but this does not mean that they achieve their rated impedance at this current.

The change in impedance of two of these inductors was measured at 11–12 MHz with DC current, and from Table 5.17, it can be seen that a significant reduction occurs at currents as low as 2.8 mA.

To obtain the SRF of the Coiltronics and Vishay inductors, the impedance was measured, and these are shown in Figures 5.49 and 5.50.

TABLE 5.15

Magnetics Material Characteristics

Material			Inductors and Power Transformers and Power EMI Line Filters					EMI/RFI Filters and Broadband Transformers		Linear Filters and Sensors		
			L	R	P	F	T	J	W	C	E	V
Initial permeability	μ		900%	2300%	2500%	3000%	3000%	5000%	10,000%*	900%	2000%	2300%
			±25%	±25%	±25%	±20%	±25%	±20%	±25%	±25%	±25%	±25%
Maximum useable frequency (50% roll off)	f	(MHz)	3	≤1.8	≤1.8	≤1.5	≤1.5	≤0.7	≤0.5	<8	<3	<1.5
Relative loss factor ×10⁶, 25°C								≤15 100 kHz	<7 10 kHz	10 at 300 kHz	3 at 100 kHz	≤5 100 kHz
Curie temperature	T_C	°C	>300	>210	>210	>210	>220	>145	>135	>200	>160	>170
Flux density at 1194 A/m (0.15 Oe), 25°C	B_m	G	4200	470	4700	4700	5300	4300	3,900	3800	3600	4400
		mT	420	4700	470	470	530	430	390	380	360	440

FIGURE 5.46 Permeability versus temperature. (Courtesy of Magnetics.)

FIGURE 5.47 Permeability versus flux density. (Courtesy of Magnetics.)

5.1.10.7 Resistors

Resistors are also frequency dependent, exhibiting capacitance and resistance as shown in Figure 5.51.

The inductance and capacitance of several types of resistors, reproduced from Reference [10], are shown in Table 5.18. The inductance of a normal metallic film (MR25) resistor is reported to be 20 nH of the 1206 chip resistor 2 nH and 0,4 nH of the 0603 chip resistor. For low-resistance values, the series inductance has the predominant effect of modifying the impedance at HF, and for high-resistance values, the parallel capacitance has the largest effect. For example, a 50 Ω MR25 resistor will have an impedance of approximately 170 Ω at 1 GHz, and a 50 Ω 1206 chip resistor has an impedance of approximately 62 Ω at 1 GHz due to inductance. Due to the parasitic capacitance,

FIGURE 5.48   Flux density versus temperature. (Courtesy of Magnetics.)

TABLE 5.16

Characteristics of the Inductors in Figure 5.45

Number in Photo	Make	L	I	R (Ω)	SRF	Q
1	Pulse	120 µH 0 A 60 µH at 16 A	16 A	0.012		
2	Unknown	250 µH		0.2		
3	Vishay	2,700 µH	0.2 A	11.29	1.0 MHz	
4	Hammond	1 mH	1 A	0.83		7.4 at 1 kHz
5	Hammond	750 µH	2 A	0.46		10 at 1 kHz
6	Hammond	500 µH	3.2 A	0.26		11.6 at 1 kHz
7	Delevan	1,800 µH	0.157	16		
8	Delevan	22,000 µH	0.19 A	18		
9	EPCOS	10 mH +10 mH C/M	1.1 A	0.4		
10	Coilcraft	1,000 µH, 1.1 A 1 µH, 28.8 A	See L	0.062		
11	Coilcraft	1,000 µH, 0.8 A 10 µH, 8A	See L	0.141		
12	Coilcraft	6,800 µH, 0.15 A 1500 µH, 0.3 A	See L	0.184		
13	Coiltronics	4.7 µH	3.25 A 10% sat 3.5 A 15% sat 4.2 A	0.04	50 MHz	
14	ACT 0605 package (see Table 5.17)	600 Ω	1 A (see Table 5.17)	0.2		

a 2.2 kΩ metal oxide resistor will have an impedance of approximately 320 Ω at 1 GHz, whereas a typical chip resistor will maintain an impedance just below 2.2 kΩ.

When a short on a component is required, a 0 Ω SM resistor may be used. In measurements on two manufacturers 0603 and 0805 size 0 Ω resistors, no impedance was measured up to 1 GHz indicating zero or low inductance.

TABLE 5.17
ACT 0605 and 0805 Surface-Mount
Impedance Change with DC Current

Current (mA)	Impedance (Ω)
ACT CBC0603-600-102	
0	89
6	12.5
150	8.6
300	0
ACT CBC0805-600-102	
0	99
2.8	63
6	33
150	19
300	6.5
400	2.8

FIGURE 5.49 Coiltronics inductor impedance.

5.2 POWER-LINE FILTERS

Power-line filters are available in a number of different configurations, including feedthrough types used for a single AC or DC line and chassis-mounted types for either single- or double-DC line or two- or three-phase AC applications. The supply connection may be via an integral AC line socket or connector or via terminals. Figure 5.52 illustrates the packaging of typical power entry filters. The current ratings of standard filters range from 0.2 to 200 A, with voltage ratings from 50 VDC to 440 Vac.

Some manufacturers of SMPS offer an input filter specifically for that supply. These filters may be active and in one model are designed to meet the transient immunity and 0.15–30 MHz conducted emission requirements of Bellcore, FCC, ETSI, and European Norms. In another model, the C/M and D/M emissions are attenuated to comply with MIL-STD-461F.

FIGURE 5.50   Vishay inductor impedance.

FIGURE 5.51   AC equivalent circuit of a resistor.

TABLE 5.18

Parasitic Capacitance and Inductance of Different Types of Resistor

Type of Resistor	Inductance (nH)	Capacitance (pF)
Wire wound, axial leads	100–1000	0.5–1
Composite, axial lead	2–10	0.05–0.3
Carbon film, axial lead	5–200	0.3–1
Metal film, axial lead	5–200	0.3–1
Metal oxide, chip	0.5	0.05–0.1

Source: Johansson, L.O., EMC fundamentals passive and their EMC characteristics, Compliance Engineering, Knoxfield, Victoria, Australia, July/August 1998.

These filters are not designed to meet the radiated emissions requirements and so when used with unshielded input power leads will typically require an additional HF filter. The manufacturer typically designs a filter for each SMPS, and so, if a number of supplies are connected to the same power line and the additional HF filter is required, a single custom input filter should be considered. As mentioned later, the location of these filters in the equipment is crucial to their performance.

If the load on the SMPS contains high-current transients, then these can be reflected from the load to the input power of the SMPS and the filter will have to attenuate these. With a custom filter, the design can be based on the predicted transient characteristics.

FIGURE 5.52   Typical power entry modes. (Courtesy of Corcom.)

The attenuation of a line filter is usually measured in accordance with MIL-STD-220A, which specifies 50 Ω source and load impedance. MIL-STD-220A allows either a C/M or D/M attenuation measurement and does not require that the attenuation measurement be made with the supply current flowing through the filter. It is typically easier to achieve a high level of C/M attenuation by the use of a high-permeability C/M inductor of small dimension, and the manufacturer's data sheet may not indicate if the specified attenuation is for D/M or C/M noise.

When a manufacturer of a filter of small dimensions and a current capacity of 5 A or greater claims a greater than 20 dB level of attenuation over the LF range (10–50 kHz), the attenuation may apply to C/M noise only. When a manufacturer does not specify attenuation above 30 MHz, one of two assumptions may be made. The first is that the filter is specifically designed to meet a commercial conducted-emission requirement, such as contained in FCC Part 15, in which the maximum specified test frequency is 30 MHz. The second possibility is that the filter exhibits a parallel resonance close to or just below 30 MHz, and so, attenuation above 30 MHz decreases linearly.

H.M. Schlicke quite rightly exhorts us to throw away our conventional filter books, because they illustrate filter performance in highly idealized and simplified conditions that are not valid for many EMC situations [3]. Figure 5.53a through c illustrates how far load and source impedances may deviate

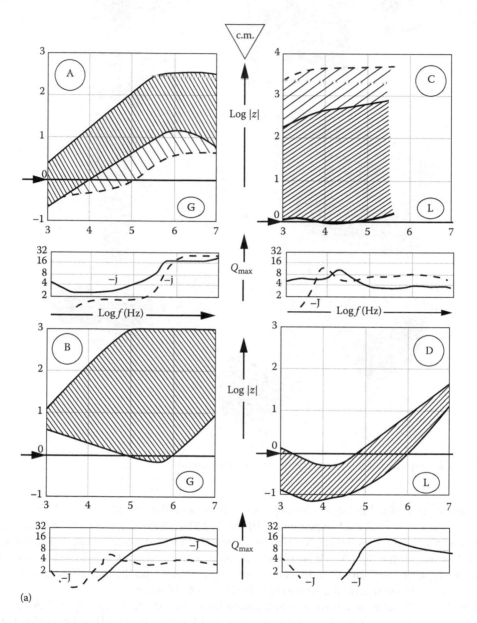

FIGURE 5.53 (a) Differential-mode interface impedances (United States) 120 V, 60 Hz. (A) Unregulated outlets (generators), (B) regulated outlets (generators), (C) 60 Hz loads (with the top less densely shaded area representing high-voltage, low-current power supplies; the middle range pertaining to motors, TV sets, and so on; and the low limit dominated by high-current devices), (D) regulated power supplies.   (*Continued*)

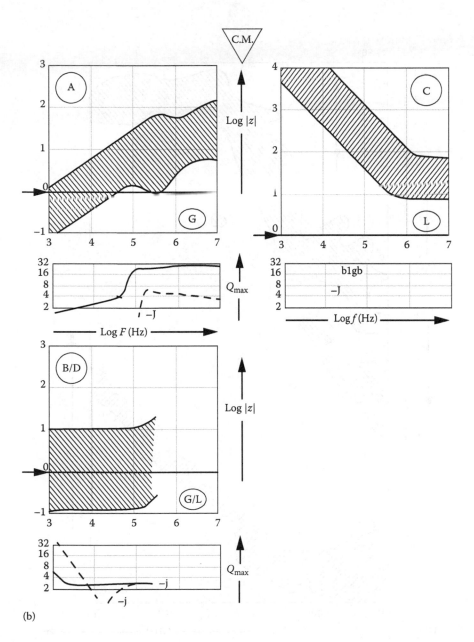

FIGURE 5.53 (*Continued*)   (b) Common-mode interface impedances (United States) 120 V, 60 Hz. (A) 60 Hz outlets (generators). (B/D) Outlets and loads in heavily filtered installations. (C) Loads.   (*Continued*)

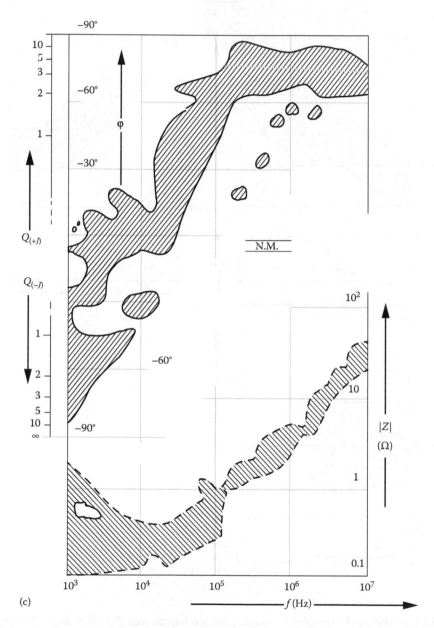

(c)

FIGURE 5.53 (*Continued*)   (c) Impedances and Q of DC power supplies loaded and unloaded.

from 50 Ω. The graphs in Figure 5.53 are based on statistical data obtained from many hundreds of outlets and devices used in households, factories, and laboratories and aboard naval vessels. We see that impedances vary from 0.1 Ω to above 1000 Ω; with the exception of C/M loads, low Q's are dominant.

Other measurements on the source impedance of AC power lines above 10 MHz have resulted in a maximum of 1000 Ω, average of 100 Ω, and minimum of 25 Ω at 15 MHz decreasing to a maximum of 100 Ω, an average of 40 Ω, and a minimum of 10 Ω at 400 MHz. A more realistic filter source and load impedance for testing and analysis purposes are 1 and 100 Ω, respectively.

During MIL-STD-461 CE01- and CE03-conducted emission measurements, the equipment power-supply source impedance below 50 kHz is typically the output impedance of the supply connected to the input of the equipment under test (EUT). Above 50 kHz, the impedance is limited by the impedance of the 10 μF capacitors specified between the line and the ground plane on which the EUT is mounted and between the return and the ground plane. As the frequency of the emissions increases, the source impedance decreases. The filters procured or designed for equipment that must meet MIL-STD-461 requirements must therefore function adequately with the normal variation of supply impedances and the very low impedances of the 10 μF capacitors at RF. Adding lower than 10 μF C/M or D/M capacitors on filters that will be tested using the 10 μF capacitor has no value, for the RF high-quality 10 μF capacitor in the test setup will carry the bulk of the current, and it is this noise current that is the subject of the test. Adding low-value C/M capacitors (approximately 0.1 μF) does have a value, because the power lines between the filter and the 10 μF test capacitor exhibit some impedance at high frequencies, and the filter capacitors will shunt the RF current at the filter. This is a case of designing a power-line filter around a test setup and incorporating the 10 μF into the design of the filter. This is surely valid, because the specified test setup should be representative of the real world. An alternative to the 10 μF capacitor, which is proposed for the latest version of MIL-STD-461 CE01 and CE03 measurements, is a line impedance stabilization network (LISN) with an impedance of 50 Ω above some specified frequency.

This section is intended to aid in either the design or the selection of a filter that is adequate for the job but not overdesigned. The required level of attenuation for impulsive noise generated by digital logic and converters, for example, may only be achieved by careful selection of a filter design tailored to the specific type and level of emissions and susceptibility test levels. When designing or procuring a filter for equipment, it is advisable to wait for the breadboard or engineering model and then measure the level and type of emissions as well as the response to the susceptibility levels. The filter may then be an optimum EMC design and of minimum size and weight. In general, series inductors are used for low source or load impedance. Thus, the following types of filter are most suitable for the various combinations of source and load impedance:

Impedance		
Source	Load	Filter Type
Low	Low	T
Low	High	L
High	Low	π

In designing filters, it must be understood that components do not maintain their identities over wide frequency ranges. Components have natural resonant frequencies beyond which they no longer follow their LF impedance characteristics. Capacitors become inductive and inductors become capacitive beyond their natural resonant frequencies.

It is usually impossible to achieve a power-line filter design that achieves high attenuation over a frequency range of 10 kHz to 1 GHz with a single filter section. A combination of sections or filters is usually required. Feedthrough capacitors and low-value bypass capacitors and inductors must be used at high frequencies in order to reduce the self-capacitance of inductors and the self-inductance of capacitors. The Q of the filter must be low under all load and source impedances to minimize insertion gain. When a transient noise source is applied to a filter with a high Q and the transient is at the resonance frequency or a harmonic thereof, the filter output waveform may be an undamped sine wave!

The equivalent circuit and equations for Q for a single-section filter are shown in Figure 5.54 [2]. In a multisection filter, where the series inductances have different values and the shunt capacitances have different values, the equations will provide an approximate value for Q when the highest value of inductance (L) and capacitance (C) are used.

The input impedance of an SMPS is active and so a filter designed with a resistive load may experience a different Q when used with an SMPS. For example a filter modeled in SPICE with a resistive load predicted a 15 dB insertion again whereas when tested with the power supply the insertion gain was 6 dB. If possible it advisable to perform a CS101 test on the prototype filter and power supply before deciding that additional damping components are required.

The use of an AC and transient circuit analysis program like SPICE is invaluable in the design or selection of filters. The program allows the modeling of the filter and noise sources and is especially useful in the analysis of complex load and source impedances. As an example, consider a single-section filter with a 500 µH inductance, a 2 µF capacitance, and a 72 Ω load. The filter response to an AC waveform, supplied from a low-impedance source, peaks (resonates) at 5 kHz. At this frequency, it exhibits 13 dB of insertion gain. Changing the filter to a three-section type with equal-value inductances and capacitances does not change the insertion gain but changes the frequencies at which insertion gain is a maximum (i.e., 2–6.3 kHz). When a second section is added, with a 100 µH inductance and a 1 µF capacitance, the original resonance occurs at 4 kHz, with a second resonance at 20 kHz. In filter selection, the magnitude of the load capacitance must be considered. For example, a low-Q filter may exhibit a high Q when used at the output of a supply that is loaded by a number of onboard decoupling capacitors.

The Q of a filter such as the one in Figure 5.54 may be reduced by increasing the value of R_e or R_c via the addition of a resistor or via placing a resistor across the inductor, L. Increasing R_c is the best option, for it does not degrade the attenuation performance of the filter. In high-current, low-voltage

$$Q = \frac{R_o}{R_{te}} \text{ degree of damping at } \omega_o$$

$$\text{where } R_o = \sqrt{\frac{L}{C}}$$

$$\omega_o = \frac{1}{2\pi\sqrt{LC}}$$

$$R_{te} = R_e + R_c + \frac{R_o^2}{R}$$

$$= \text{total effective damping resistance}$$

FIGURE 5.54 Q of an LC filter. (From Kami, E., Design guide for electromagnetic interference (EMI) reduction in power supplies, MIL-HDBK-241B, Power Electronics Branch, Naval Electronic Systems Command, Department of Defense, Washington, DC, 1983.)

(a)

(b)

FIGURE 5.55   (a) Reducing the Q of a filter with an additional capacitor and damping resistor. (b) Resistor in parallel with an additional inductor for damping.

supply applications, the voltage drop across the additional resistor, R_e, is probably unacceptable. Despite that, in applications with low current requirements and voltages above 12 V, a loss of 0.5–1 V may be acceptable. Another practical option is shown in Figure 5.55a.

Here, a second, larger-value capacitor with an appreciable value of added series resistance is placed in parallel with the existing capacitor. Very often, the combined load capacitance formed by decoupling capacitors on PCBs is higher than the highest value of filter capacitor that will fit into the filter available footprint. In this case, an additional series inductor with the damping resistor in parallel with the inductor, as shown in Figure 5.56b, is very effective, although the filtering effectiveness of the inductor is lost. The sole purpose of the inductor is to reduce the DC voltage drop and to present a higher impedance than the resistor at the resonant frequency of the filter. If damping cannot be incorporated into the filter design, ensure that the resonant frequency is no higher than 1/10 of the frequency of switching power supplies (typically 100 kHz) and at least 10 times the AC power-line frequency. Also, if the frequencies of high-level noise sources sharing the same power line are known, try to shift the filter resonance away from these frequencies.

The reason all filters perform well in an MIL-STD-220 test is that the input attenuator effectively increases the value of R_e and decreases the Q of the filter. Although a ferrite bead has a low Q at HF, when used to filter a supply on a board that contains decoupling capacitors, the Q of the resultant LC circuit at LF may result in a sine wave impressed on the board supply. Here, the addition of a series resistor or replacement of the bead by a resistor may be the solution.

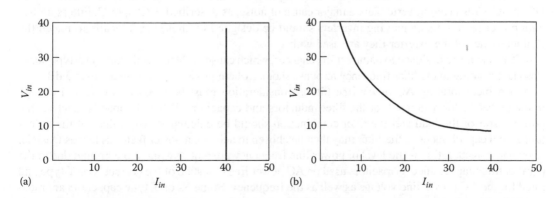

(a)   (b)

FIGURE 5.56   Input characteristics of (a) linear-mode and (b) switch-mode regulators.

Manufacturers of commercial filters are invariably willing to provide the component values and schematics of their filters. From this information and a knowledge of the minimum and maximum load and source impedances, a filter with a low Q and, therefore, either minimum or no insertion gain may be chosen using either the equations of Figure 5.54 or a circuit analysis program. The design of a filter exhibiting C/M attenuation and a low Q when used with a supply or load isolated from enclosure ground is difficult. In this application, the C/M LC combination is not damped by a load resistance (due to the isolation requirement). Very high levels of undamped sine wave may then be generated due to the high Q of the filter. If the input of the power supply is isolated for the output and from chasis ground then the voltage appears as common mode on both power and return and so the supply may be immune. One solution is to move the resonance to a frequency at which the load is least susceptible or at which the conducted emission limit is at a maximum.

Lossy line or dissipative filters perform well at HF (500 MHz to 10 GHz) and have the great advantage of removing the noise energy in the form of heat. In lumped π or L-type filters, the noise energy flows in the capacitors and into the structure or power or signal ground, and this in turn can cause EMI/EMC problems. Where T-type filters are used, the magnitude of capacitor ground current is reduced. A potential disadvantage is that the noise energy generated—by logic, for example— typically results in a higher, and sometimes unacceptably high, level of noise voltage across the load (noise source) due to the input or output impedance of the T-type filter. Some examples of lossy components and filters are

Ferrite beads
Ferrite rods
EMI suppressant tubing (permeable)
"Lossy line" EMI absorptive filters

The disadvantage of lossy filters is their typically poor performance at LF.

The remainder of this section provides a set of general filter guidelines. Supply-line filters should have their output lines shielded from the supply line, preferably by mounting the filter/s outside the unit enclosure or within a shielded compartment inside the unit enclosure.

The first choice in filter selection is the type that has built-in connectors for input power and terminals for connection to the load. With this type, the filter case may be mounted on the exterior of the enclosure or to the rear wall, with the connectors accessible through an aperture in the wall. When the filter must be located inside the enclosure, shielded cable should be used, with the shield connected to the filter case at one end and to the wall of the enclosure, as close to the rear of the connector as feasible, at the other end. In designing a filter, the output terminals should be feedthrough capacitors or a feedthrough filter with a low value of inductance.

Converters are SMPS which convert from AC to DC or DC to DC and inverters convert from DC to AC. Converters/inverters are a major cause of noise, as described in Chapter 2. Filters for use with power regulators employing inverters should be designed or chosen specifically to match the characteristics of the inverter they are used with.

Filters should be designed to damp self-resonances, which cause problems with impulsive-type noise. Resonant peaking in the filter frequency response should, where possible, be no greater than 3 dB.

When incorporating AC power-line filters, consideration must be given to a potential change in the power factor as a result of the filter inductors and capacitors. When the load is reactive, the power factor of the load and the filter combination should be calculated. The values of the inductances and capacitances in the filter may then be chosen to achieve a power factor as low as feasible. A second consideration is the level of power-line frequency current flowing in the ground due to the parallel filter capacitance. Capacitors used on AC power lines must be of the correct X or Y type and rated for the AC power-line voltage as well as the frequency. Some X- or Y-type capacitors are rated at 250 V AC, 60 Hz, and are not suitable for use at 250 V and 400 Hz. Some EMC requirements limit the value of filter capacitance used on AC power lines. When the supply to a DC power-line filter is switched via a relay or contactor and the filter input component is an inductor, it is good practice to incorporate a diode or rectifier between the supply line and the case ground to clamp the voltage spike as the input power is interrupted.

Where the filter contains capacitors connected to the metal enclosure of the filters, designed to reduce C/M noise voltages, the enclosure of the filter must be connected back to the ground at the noise source via a low-impedance connection.

Filters used with switching power supplies or converters may compromise the stability of the supply unless certain precautions are taken. The input power of a linear regulator power supply increases with increasing input voltage, unlike a converter, where the input power remains relatively constant. Figure 5.56 shows the comparison of the input characteristics and illustrates that the converter input is represented by a negative resistance.

When proposing the addition of a filter to the input of a converter, or where the addition of a filter results in a change in the converter line or load regulation, the stability criteria may be checked by referring to Appendix B in Reference [2].

The usefulness of power-line filters is not confined to reducing conducted noise levels. When MIL-STD 461 RE102, DO-160, or similar radiated emission levels are exceeded due to radiation from shielded cables, either a filter is required in order to reduce the radiation from the cable or source reduction must be used. We will see in Chapter 7 that the use of double-braid, braid-and-foil, or triple-braid shielded cables brings only a limited reduction in radiation compared to a single-braid cable. The commercially available and relatively inexpensive power-entry filters are designed to reduce conducted emissions over the range 150 kHz to 30 MHz, and the attenuation is typically specified over this range. However, these power-line filters do provide significant attenuation at higher frequencies, and if this is not adequate and the noise has been determined as C/M, then an additional ferrite balun on the wires inside the equipment to the filter may be all that is required. Alternatively C/M capacitors with 100 pF between power and chassis and between power return and chassis may be successful. Also in shielded cables, the transfer impedance of the connector and its interface to the enclosure contributes to the radiation from the cable. One case is where noise is generated on a shared power line by equipment contained in one enclosure or rack, with the power line connected to equipment in a second rack. The resultant noise currents flowing on the inside of the braid of the shielded cable diffuse through to the outside or couple through the apertures, resulting in an above-specification level of radiated emissions. The solution is one or more power-line filters. Only one filter may be required when the noise source is confined to equipment in one enclosure. When both enclosures contain equipment that places noise on the shared power line, the use of one filter may still suffice. The filter should be of the L or T type and contained, where possible, in the enclosure/rack that houses the source of the highest noise. The inductors in the filter should be the load presented to the equipment in enclosure 2, without filter.

The C/M noise currents generated in enclosure 2, which flow out on the power-line conductors and return on the inside of the shielded cable, are reduced in magnitude due to the high impedance presented by the filter inductors.

Filters may be designed that are extremely effective at reducing switching power supply output noise. This enables a switching power supply to be used in sensitive electronic equipment, where in the past only a linear supply provided sufficiently low noise. Typical applications for this high level of filtering have been found in satellite communication, lasers, sonar buoys, and microwave simulators.

Also by the use of these filters, a single switching power supply can be used for both digital boards and low-signal-level analog and RF circuits. In this case, the supply output is used directly for the digital boards, and the filter is placed between the supply and the sensitive. These filters will typically attenuate noise spikes (as high as 40 V) and ripple (as high as 1.5 V) to an output ripple of 30 μV at the switching frequency of the converter. No measurable HF component is seen when measured with a number of different manufacturers' switching power supplies. The full attenuation effectiveness of the chassis-mounted filter can be achieved only if it is mounted directly to a dividing bulkhead or shielded cables are used for both input and

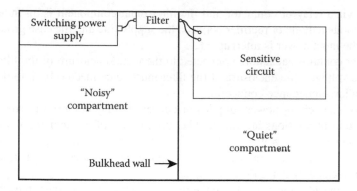

FIGURE 5.57 Bulkhead mounting of a filter to maintain the compartmentalization of equipment.

output connections. Figure 5.57 shows the use of a bulkhead-mounted filter to maintain isolation between a "noisy compartment" and a "quiet compartment" in a piece of equipment. The PCB-mount version must have the input and output connections via striplines at different levels in the PCB to achieve maximum effectiveness. The effectiveness of these filters is measured by mounting them to the wall of a shielded room, with the signal source outside the room and the measuring device inside the room. The room is used to isolate the input test signal from the filter output level.

If C/M noise appears on the power return, then this may require it be filtered. Figure 5.58 shows the topography of typical filters. If no impedance is possible between the unfiltered supply-power return and the filtered-power return, then only D/M filtering is required, as shown in the single-line and two-line D/M filters of Figure 5.58. Also, if the power return is connected to chassis at the power supply as well as on the load PCB, or more likely on a signal interface connection, then adding return-line filtering would be redundant. Although if the supply return is isolated from chassis at the switching power, then power return filtering not only is possible but should be included, as shown in the two-line D/M and C/M filter and the three-line D/M and C/M filter. Very often, the value of the C/M capacitor is limited, due to the required voltage rating and the limit on leakage current and because nonpolarized capacitors may not be allowed, even on DC power lines, since the polarity of the C/M DC voltage is typically unknown. This excludes the use of high-value tantalum

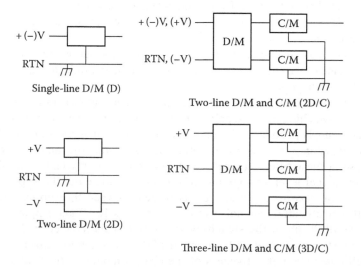

FIGURE 5.58 Number of secondary power lines filtered and attenuation mode.

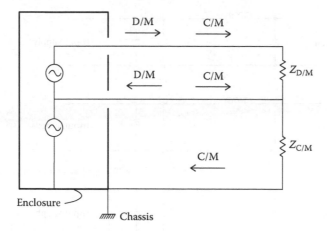

FIGURE 5.59 Differential-mode and common mode voltages and currents flowing in load impedances and example is a line impedance stabilization network.

or aluminum electrolytic capacitors unless they are unpolarized or placed back to back. Luckily, the permeability of the C/M inductor core is not affected by the power supply current, especially when the inductor is bifilar wound. Typically, a high-permeability ferrite core is used for the C/M inductor/s, resulting in a high (500 μH to 10 mH) inductance, and the C/M components' resonant frequency can be optimized.

As well as containing either differential mode (D/M) or common modse (C/M) or both components, it may contain more than one stage.

The concept of differential and C/M may be unfamiliar to circuit designer, and so, these voltages and currents are illustrated in Figure 5.59.

Examples of single-stage filters are shown in Figure 5.60a through d. These are designed to both reduce radiated emissions from the equipment and supply D/M attenuation of input noise.

A SM pi filter that may be used as a power-line or signal-line filter is manufactured by Tusonix. This 4700 series filter is available with capacitors from 100 to 8200 pF and with a typical inductance of 100 nH. The performance of all filters is compromised by coupling between the input and output connections. The Tusonix filter is designed to be shielded by, typically, a PCB-mount shielded compartment, as illustrated in Figure 5.61. The insertion loss of the shielded filter is shown in Figure 5.62 and for the unshielded filter in Figure 5.63. A potential insertion loss of 70 dB above 1 GHz for the 8200, 5000, and 2000 pF versions when shielded is degraded to 30 dB when the filter is used unshielded.

5.2.1 CUSTOM-DESIGNED FILTERS

In designing a filter, a number of considerations apply:

1. Is the RF current D/M or C/M or both and does this vary with frequency?
2. How much current must the windings in the coils carry and is it AC or DC current?
3. What is the maximum temperature rise allowed in the inductors?
4. What is the maximum voltage that the capacitors must withstand?
5. What is the physical foot print that the filter must occupy?
6. What is the input source impedance and load impedance?
7. What type of load current, that is, transients from SMPS, motor drivers, RF power circuits, is predicted?
8. Is inrush current limiting required?
9. Is there a limit on the value of the C/M capacitance?

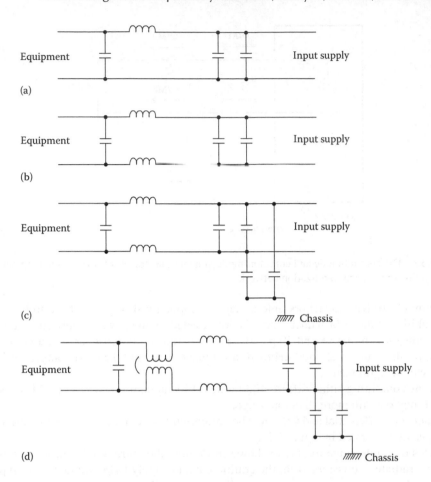

FIGURE 5.60   Single-stage power-line filter configurations. (a) Differential-mode (D/M) unbalanced. (b) D/M balanced. (c) D/M + common mode (C/M). (d) D/M and enhanced C/M.

FIGURE 5.61   Connectorized, terminal, and PCB-mount filter configurations. (Courtesy of EMC Consulting Inc.)

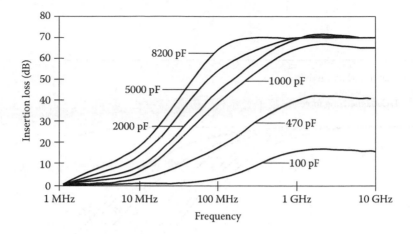

FIGURE 5.62 Typical shielded insertion loss versus frequency for the Tusonix 4700 series surface-mount filter.

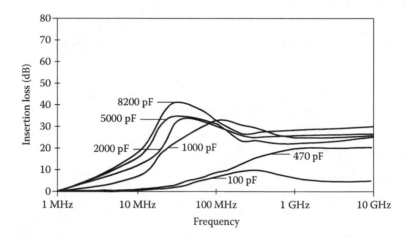

FIGURE 5.63 Typical unshielded insertion loss versus frequency for the Tusonix 4700 series surface-mount filter.

Ideally, the filter should be custom designed around the test setup it will be used in, that is, for MIL-STD-461 a 2 m long power cable 5 cm above a ground plane and 10 cm from the edge and connected to an MIL-STD-461 LISN. For commercial EMI tests, no tabletop ground plane is used instead a nonconductive table with the cables looping down to an FCC-type LISN mounted on the floor ground plane. The cables to be 0.40 cm from a vertical conducting surface bonded to the floor ground plane.

If used with SMPS, the emissions from the supply should be tested in the MIL-STD or commercial test setup. A conducted emission may be characterized as D/M by clamping a current probe around first the power and then the return and then around both cables. If the level measured with the probe around a single cable is higher than around both, then the source is predominantly differential, and if the level is approximately 6 dB higher with the current probe around both, it is C/M. The reason for the 6 dB difference is that one-half of the C/M current and all of the D/M current are measured with the current probe on one conductor and none of the D/M and all of the C/M with the current probe around both conductors.

When DC current flows through an inductor wound on a core with permeability greater than 1, the permeability will reduce as shown in Figure 5.64a for the Micrometals cores. Although referred

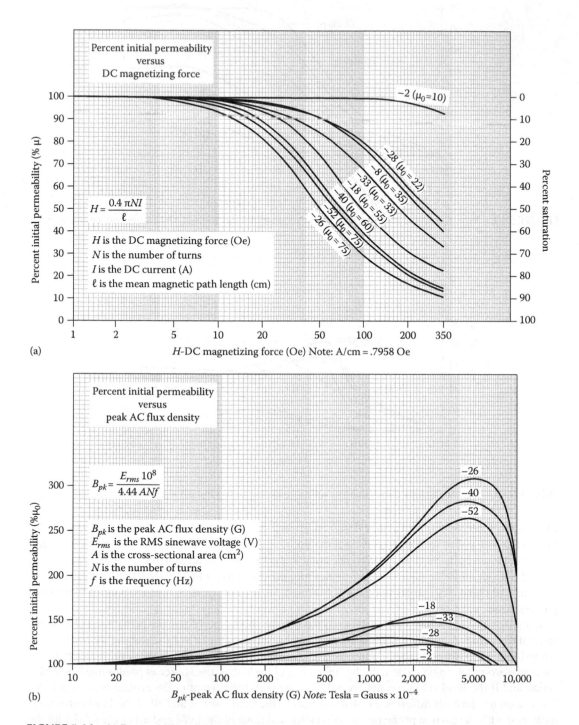

FIGURE 5.64 (a) Permeability of the metal oxide toroidal-core versus DC magnetizing force. (b) Variation in permeability with AC flux density for Micrometals cores. *(Continued)*

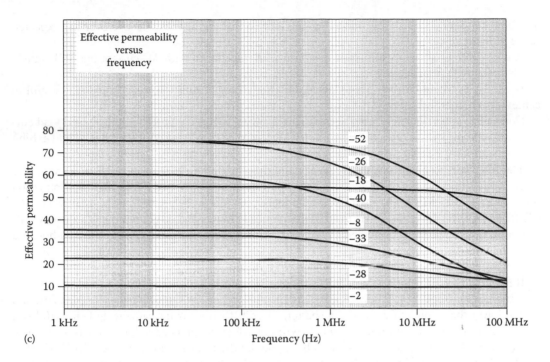

FIGURE 5.64 (Continued) (c) Variation in permeability with frequency for Micrometals cores. (a–c: Courtesy of Micrometals.)

to as saturation using a core with a DC, current that results in some reduction in permeability is very common. When modeling the filter, the change in attenuation and the change in filter resonant frequency with DC current should be part of the analysis.

In many cases, an inductor that at the maximum DC current exhibits a known change in permeability is optimum. For example, with 30 turns of #22AWG magnet wire on a Micrometals T68-52D core, the inductance is

$$L = 30^2 \cdot 80 \text{ nH} = 72 \text{ } \mu\text{H}$$

The maximum current in the coil is 3 A, the resistance is 0.055 Ω, and the temperature rise is 13°C.

The magnetic path length of the core is 4.25 cm, and the magnetizing force from Figure 5.64a is

$$0.4 \times 3.142 \times 30 \times 3/4.25 = 22.6 \text{ Oe}$$

The change in permeability is 16%, and so, the inductance at 3 A is 84% of 72 μH = 60 μH.

For the core permeability to change to only 88%, the turns will be 20 and the inductance with no DC current will be 32 μH. At 3 A, the inductance will be 88% of 32 μH = 28 μH.

If the inductance is only allowed to change to 95%, then the maximum turns are 11 and the inductance with no current is only 9.7 μH.

Clearly, if the 16% change in inductance from no load to full load is acceptable, then the higher inductance at lower current is a benefit.

When AC flows through the same core, the initial permeability rises, and at some high level of flux density, it will rapidly decrease as shown in Figure 5.64b. This increase in permeability should also be taken into account when modeling the filter so that the filter is effective over the expected range of input current.

A core with an air gap will reduce this change in impedance with AC current but will significantly reduce the permeability.

If a fairly constant current is maintained, this rise in permeability with AC current is useful in achieving a higher inductance for a given core size and number of turns.

Conventional wisdom says that the emissions from an SMPS increases with increased load current, and although this is often the case, it is not always true. This means that the filter should exhibit sufficient attenuation with light or heavy loads.

The filter will resonate at some frequency and this should be chosen if possible at a frequency where EMI limits do not apply. For example, the European (EN)- and FCC-conducted emission limits start at 0.15 MHz, and the resonant frequency should be below 0.15 MHz. MIL-STD-461 CE101 and CE102 cover the 30 Hz to 10 MHz frequency range, and so the resonant frequency will be within this range. If conducted emission measurements have been made on the EUT without the filter, then the resonant frequency should be chosen to lie at a frequency where emission is very low or nonexistent, at least 1/10th of the frequency of high emissions.

If MIL-STD-461F applies, then the CE102 emission limit is higher at 10 kHz and then reduces with increasing frequency. Likewise, the CE101 limit is typically higher at lower frequencies, and so, the lower the resonant frequency, the better. The CS101 limit is the same from below 10 kHz, and so, choosing a resonance below 10 kHz will not affect the induced voltage due to resonance. The Q of the filter will, and so it is advisable to model the filter in SPICE in the CE101, CE102, and CS101 setup. In reality, a number of factors mean that the frequency must be higher, and these are a limit of the inductor and capacitor values that can be used. For example, MIL-STD-461F places a limit on C/M capacitor values in Navy equipment of 0.1 μF for 60 Hz and 0.02 μF for 400 Hz and for submarine DC-powered equipment and Navy aircraft DC-powered equipment of each line to ground of 0.075 μF/kW. For DC loads less than 0.5 kW, the filter capacitance shall not exceed 0.03 μF.

If the filter is designed for use on AC power, then the current flowing in the filter capacitance at the power-line frequency must be considered. If the capacitance is high, the current will add to the power-line current and the capacitor will heat. For C/M capacitors, the AC current in the chassis or safety ground may adversely affect circuits.

In the simple single-stage filter of Figure 5.65a, the load resistance, inductance, and inductor resistance were used in the equations of Figure 5.54 to calculate the Q of the filter. The source was at the input of the filter, and the load was the input voltage divided by the current = 3 Ω.

In Figure 5.65b, the source is now on the load side and the impedance is approximately that of the $LISN = 6$ Ω. By changing LC values, we can see the change in Q, which is only from 7.5 to 14.3 dB,

(a)   (b)

FIGURE 5.65   (a) Source at input of filter with 8 Ω load. (b) Source at load side of filter with 6 Ω line impedance stabilization network impedance.

FIGURE 5.66   Plot of attenuation using SPICE. Minus attenuation = insertion gain.

TABLE 5.19
Q of Simple Filter

F (Hz)	RL	RC	Ro	L (µH)	C (µH)	R	Rt	Q	Q (dB)
839	0.055	0.001	0.038	72	500	8	0.074	5.13	14
1500	0.055	0.001	0.53	45	200	8	0.15	5.8	15.3
2130	0.055	0.001	0.75	56	100	6	0.15	5	14
1880	0.055	0.001	0.85	72	100	6	0.18	4.8	13.7
2770	0.055	0.001	0.57	33	100	6	0.11	5.2	14
3390	0.055	0.001	0.47	22	100	6	0.09	5	14
5040	0.055	0.001	0.32	10	100	6	0.07	4.35	12.8
3560	0.055	0.001	0.22	10	200	6	0.064	3.5	11
2250	0.055	0.001	0.14	10	500	6	0.059	2.4	7.5
2250	0.055	0.001	0.14	10	500	8	0.058	2.42	7.7

and so, it is difficult to significantly reduce the Q with component value change. Instead, the damping resistors described in Figure 5.55a and b may be used.

The plot of the attenuation for these combinations of L and C using SPICE and including parasitic components is shown in Figure 5.66. The insertion gain at resonance has exactly the same value as the Q in Table 5.19. When the attenuation with the 10 µH inductor is adequate, then this can be wound on the smaller T50–52B core with 15 turns. The magnetizing force is then 17.7 Oe, and the inductance is 90%, that is, 9 µH.

When designing a C/M coil for a filter, the AC and DC current will flow through the core in one direction and then returns in the other. This means that theoretically the flux density is zero and no change in core permeability and no saturation occur. In reality due to leakage inductance, often some flux is generated and the core permeability may change resulting in a change in impedance as shown in Table 5.12 for the WE-SLT 744.227 choke.

The desired amount of attenuation achieved by the filter at the problem frequencies should be calculated. Although it would seem the more, the better, experience has shown that managers and customers complain of over design if the safety margin is too high.

In the production of one filter, two inductors were arbitrarily removed on the advice of an engineering manager. Luckily, the filter then failed to meet the EMI requirements proving the inductors were required! The lesson is, listen to the engineers!

Once you have the perfect filter, it is important to consider where it is to be installed. As with real estate, it is a question of location, location, location.

In one case, a filter that met all of the EMI requirements when the prototype was located in the power lines just outside the EUT enclosure failed miserably the conducted emissions and radiated emissions requirements when the filter was mounted on a power supply board. The problems were that the wiring from the power connector to the board was unshielded, the power traces were then run alongside signal traces in the back plane, and on the board the filter components were exposed to fields from SMPS.

Although shielded cable was tried between the power connector and the backplane, the out-of-specification emissions remained. It was only when the filter in a metal enclosure mounted behind the input power connector that all the requirements were met.

When the traces carrying the supply power and return to the filter components are long, then crosstalk/radiated coupling can occur around the filter at very high frequency (VHF). To demonstrate this, the PCB layout and test circuit as shown in Figure 5.67 were tested. The source was a

FIGURE 5.67   Filter with long traces test setup.

FIGURE 5.68 Attenuation of filter with long versus short traces.

signal generator and the load the 50 input impedance of a spectrum analyzer. In the first measurement, the trace before the inductor was 5 cm long and the trace after the capacitors was 3.3 cm long. The traces were then made as short as possible and the attenuation measured. A plot of the two measurements is shown in Figure 5.68, and it can be seen that with the longer trace, the attenuation is 10–14 dB lower at some frequencies.

C/M chokes with a high inductance typically have many turns and tend to resonate at an LF, after which the impedance at higher frequencies is low.

5.2.2 Common-Mode Filter with Resonant Capacitor

C/M chokes with a high inductance typically have many turns and tend to resonate at an LF, after which the impedance at higher frequencies is low.

If a small C/M choke is required with a high impedance at higher frequencies and the windings can be the power-line conductors, one solution is to bifilar wind the power and return, each having three turns, on a Laird Technologies (Steward) 28B0500-100 Broadband solid Ferrite Toroid Core. The dimensions of the core are 13 mm OD, 8 mm ID, and 6.3 mm in height. The inductance of the three turns is 8.9 µH. This core is shown in the photo in Figure 5.69.

The impedance with frequency is shown in Figure 5.70.

By adding a 1000 pF capacitor on each line, the attenuation is much increased from 10 to 100 MHz as shown in Figure 5.71. The resonance at 96 MHz is not due to the self-capacitance of the inductor but the 3.8 nH inductance of the 1000 pF capacitor and some small effect that the inductor has. The inductor does change the resonant frequency of the capacitor slightly, which is 92.7 MHz, when measured alone. Thus, the resonant frequency (the frequency of maximum attenuation) of the circuit may be changed merely by changing the value of the capacitor.

When an SM or through hole mount C/M inductance is required, a Laird Technologies (Steward) four-line EMI filter array 29F0430-2SR-10 shown in Figure 5.69, which can be connected as shown in Figure 5.72.

The total inductance of one line is 9 µH, and the impedance is shown in Figure 5.73.

FIGURE 5.69   Photos of the two common-mode inductors.

FIGURE 5.70   Impedance of the 28B0500-100.

FIGURE 5.71   Attenuation of 28B500-100 with 1000 pF capacitors.

FIGURE 5.72   Connection of the 29F0430-2SR-10 four-line inductor as a common-mode choke.

FIGURE 5.73   Impedance of four-line 29F430-2SR-10.

FIGURE 5.74   Attenuation of four-line of 29F0430-2SR-10 with 1000 pF capacitors.

Adding the 1000 pF capacitor on each line increases the LF attenuation as shown in Figure 5.74. Again the resonance at 82 MHz is the inductance of the capacitance and some small effect from the inductor.

5.2.3   POWER-LINE FILTER TO MEET SPACE REQUIREMENTS

A typical filter was designed to meet Space EMI requirement with 4 SMPS modules as the load. The input voltage is 28 V rising to 42 V with a current of 9 A.

FIGURE 5.75 Filter that met SPACE EMI requirements with inrush current limiting.

The filter met the conducted emission requirements with two makes of SMPS and a safety margin of 12 dB. The CS101 test was met with an insertion gain of 9.8 dB when the CS101 input was at 0.9 V pk–pk and 4 dB when the CS101 input level was 3.8 V pk–pk.

The filter is enclosed in a metal enclosure with the power connector an integral part of the enclosure and with the output terminals unfiltered. Inrush current limiting was a requirement, and this was achieved with a radiation hard MOSFET, as shown in the schematic.

The filter schematic is shown in Figure 5.75. For the inrush current limiting to work, the input of the load, typically SMPS, must be isolating, and this is often a requirement in equipment designed for SPACE.

5.2.4 Separating the Low- and High-Frequency Filter Components

When the filter cannot be incorporated into the connector or located close to the connector but must be located on the power supply PCB, then the filter should be split into an LF section located on the PCB and HF section located at the connector.

The input RF current was D/M at the switching frequency of the SMPS, as is often but not always the case, and C/M at HF.

The schematic is shown in Figure 5.76.

FIGURE 5.76 Filter with D/M common-mode (C/M) low-frequency components on PCB and high-frequency C/M at connector.

The D/M inductance of 2.5 μH was the inductance at the maximum current of 18 A.

The resonant frequency of the D/M components is 2.25 kHz and that of the C/M is 7.12 kHz, both of which are well below the switching frequency of the SMPS.

The 10 mH C/M inductor, 2.5 μH inductors and 1000 μF components are on the PCB and the ferrite balun and two 1000 pF capacitors are at the connector and these are separated by power-line wiring. All the same backplane is not used for the power lines, and the wiring and PCB components are kept away from high-level sources of emission to mitigate against radiated coupling.

5.2.5 Case Study 5.1: Filter Design

We will examine the design of a +28 V power-line filter designed to attenuate the MIL-STD-461 conducted susceptibility test levels CS01, CS02, and CS06 and to provide adequate reduction of transient noise from 17 converters and 12 boards of logic in order to meet CE01 and CE03 conducted emission levels. The filter schematic is shown in Figure 5.77. The design was made with the aid of the SPICE computer program using the equivalent circuit of the filter shown in Figure 5.78. A large ferrite bead with 1 in. length and ½ in. diameter was used on the +28 V input line and another on the return line. The first iron dust toroidal-core inductor was wound with 70 turns of No. 18 AWG magnet wire with a method designed to reduce interwinding capacitance. The capacitance of this core is 30 pF, and the permeability at a DC current of 6 A

FIGURE 5.77 +28 V DC line filter schematic and construction.

FIGURE 5.78 Equivalent circuit of one line of the +28 V power-line filter.

(magnetizing force = 81 Oe) results in a minimum inductance of 230 µH. The SRF of this inductor is 1.2 MHz.

Figure 5.64a illustrates how the permeability of the core decreases with an increase in DC current flow (the permeability of a core may increase with high levels of AC current flow). If the permeability of the core reduces to below 50% at the maximum current, the use of a core with a lower initial permeability that is less sensitive to DC current may be advisable Although iron dust cores exhibit a high permeability at HF with some materials, the permeability does reduce. Figure 5.64c shows the permeability at HF for the micrometals cores.

The second inductor exhibited 30–42 µH of inductance and an SRF of 17 MHz and was wound in a manner to achieve the low self-capacitance of 3 pF. The feedthrough filter is either a Capcon LMP-50 500 or a US µTeK GK3AA-P11, both of which exhibit a high level of attenuation at high frequencies (up to 1 GHz).

The filter was analyzed with the configurations corresponding to the MIL-STD CS01, 2, 6, CE01, 3, and MIL-STD-220 test setups. These configurations, as well as the attenuation characteristics or transient response as computed using SPICE, are shown in Figures 5.79 through 5.86 and in the following tables. In the conducted emission CE01 configuration, the transient response to a

FIGURE 5.79 Configuration for the analysis of the +28 V power-line filter response to CS01 and CS02 input levels using "SPICE."

FIGURE 5.80 CS01 and CS02 frequency response of the power-line filter.

FIGURE 5.81 Configuration for the analysis of the +28 V power-line filter response to converter-generated O/P noise in accordance with the CE01, 3 test setups using "SPICE."

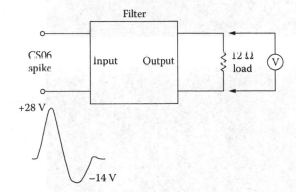

FIGURE 5.82 Configuration for the analysis of the +28 V power-line filter response to the CS06 + 28 V spike using "SPICE."

FIGURE 5.83 Result of the "SPICE" transient analysis of the +28 V power-line filter to the CS06 spike.

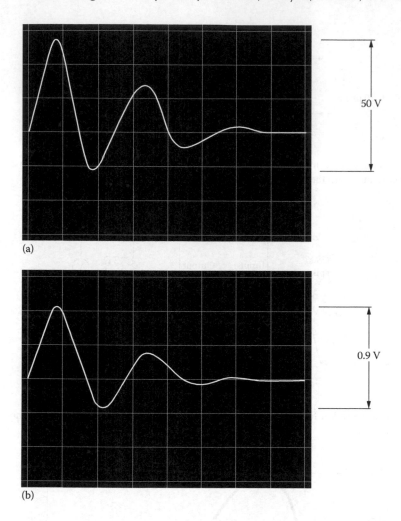

FIGURE 5.84 (a) Measured CS06 input to the filter. (b) Measured output response from the power-line filter.

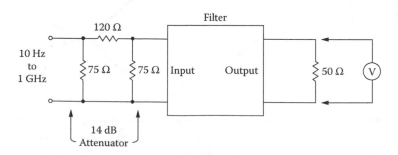

FIGURE 5.85 Configuration for the analysis of +28 V power-line filter response to the MIL-STD-220A test using "SPICE."

FIGURE 5.86 Result of the "SPICE" AC analysis of the + 28 V power-line filter response to the MIL-STD-220-A test.

typical load noise waveform was analyzed. The test setup is shown in Figure 5.81. A summary of the filter performance is given in the following tables:

MIL-STD-462-CS01 Conducted Susceptibility Test Levels (5 Vrms Input, 30 Hz–1.5 kHz)		
Frequency	Output Referenced to Input (dB)	Output (V)
30 Hz	0.0	5.0
960 Hz	2.4	6.6
1.3 kHz	0.0	5.0
1.5 kHz	−2.0	4.0
50 kHz	−40	0.05

MIL-STD-462-CS02 Conducted Susceptibility Test Levels (1 Vrms Input)		
Frequency	Output Referenced to Input (dB)	Output
50 kHz	−50	10 mV
400 MHz	−99	11 µV

MIL-STD-220A Insertion Loss (Attenuation) Test	
Frequency	Insertion Loss (dB)
100 Hz	4
500 Hz	15
1 kHz	21
10 kHz	28
100 kHz	40
1 MHz	80
100 MHz	90
1 GHz	94

The attenuated conducted emission level, measured in accordance with CE01 and CE02 test methods, from the simulated input current from one 15 W converter power supply of 0.2 mA into

10 µF gives 46 dBµA. The input current from 17 converters must be attenuated. Assuming a worst-case addition of input currents gives 59 dBµA. The specification limit at the frequency of maximum emission is 90 dBµA. Thus, the filter provides adequate attenuation for this source of emission.

We see that the attenuation of the filter used in a realistic test setup exhibits an insertion gain of 2.4 dB, whereas in the MIL-STD-220 test an attenuation of 21 dB is predicted at the same frequency.

5.2.6 CONTINUATION OF CASE STUDY 3.1

The first part of Case Study 3.1, in Chapter 3, presented the measured waveforms and noise currents generated by +5 V and ±15 V DC-to-DC converters. In order to choose a suitable commercially available filter, six different miniature feedthrough filters were tested. During the test, the filters were connected through the sides of a small enclosure on which the converter was mounted, as shown in Figure 5.87.

The input noise current with and without the filters in the circuit was measured in accordance with MIL-STD-462, CE01, and CE03. In addition, the D/M and C/M noise at the output of the supply with filters was measured in both the frequency domain and the time domain. The test setup is described in Section 3.3.1. The published minimum insertion loss of the filters measured in accordance with MIL-STD-220A is shown in Table 5.20.

In the CE01 and CE03 input test on filters 1, 2, 3, and 6, only certain frequencies of input noise current were reduced at or close to the major frequency of emission, which was 250 kHz (twice the switching frequency of the converter). Only filters 4 and 5 reduced the emissions close to the

FIGURE 5.87 Mounting method for the filters described in Case Study 3.1. (Courtesy of Canadian Astronautics Ltd., Ottawa Ontario, Canada.)

TABLE 5.20
Published Minimum Insertion Loss of Filters Measured in Accordance with MIL-STD-220A

		Frequency						
Filter	Current (A)	30 kHz	150 kHz	300 kHz	1 MHz	10 MHz	1 GHz	10 GHz
1	3	9	25	45	70	70	70	—
2	15	15	25	34	44	60	70	—
3	15	20	32	36	46	64	80	80
4	1	18	50	66	80	80	80	80
5	1	15	31	42	63	70	70	—

TABLE 5.21
Measured Insertion Loss When Used at Input and Output Terminals of a 15 W + 5 V Converter

	Input			Output	
Filter	CE01 210 kHz	CE03 36 MHz	D/M 206 263 MHz	C/M at 250 kHz	Wideband
1				−11 dB	30 dB
2				−12 dB	
3				−5 dB	
4	38 dB	30 dB	6.5 dB to −3.8 dB	8.7 dB	39 dB
5	18 dB	34 dB	0 dB	27 dB	
6				−2.6 dB	42 dB

switching frequency and at or close to 36 MHz. The magnitude and frequency components of the noise currents without filters are shown in Plots 1–4 of Section 3.3.3.

Table 5.21 provides a summary of the performance for the six filters. Only filters 4 and 5 achieved any C/M insertion loss at 250 kHz; the remaining filters exhibited insertion gains as high as 12 dB. The output of the majority of the filters measured in the time domain was either an undamped or a damped 250 kHz sine wave. These measurements on available filters highlight the difficulty of choosing a suitable filter based on the manufacturer's published data, whereas using the equations of Figure 5.54 with a 12.5 or 25 Ω load resistor, as used in the measurement setup, results in an accurate prediction of those filters that exhibited insertion gain.

5.2.7 AC Power-Line Filter

When used with AC power, the permeability of an iron dust core increases at high AC flux density, and this increase is applicable at all frequencies, not just the power-line frequency. This means the inductance may be higher than predicted and any filter resonance lower. Figure 5.64b shows the variation in permeability with AC flux density for the Micrometals cores.

In the design of a 20 A 230 V AC power-line filter, the use of two 100 μH inductors one in the line and the second in the neutral followed by two D/M 1 μF safety capacitors in parallel will provide high D/M attenuation, especially at 100 kHz. Based on measurements for this filter design, the high frequencies were C/M, and so, a 3.2 mH C/M choke followed by 0.01 μF C/M capacitors was used.

The 100 µH inductor was constructed using 2 × T157-52 Micrometals cores wound with 22 turns of 10 AWG magnet wire or the equivalent insulated braid. The inductance at 20 A was approximately 140 µH and at 8.2 A 250 µH due to the increase in permeability. This means that the resonant frequency can reduce by up to 36%.

5.2.8 Output Power-Line Filters

SMPS are replacing linear supplies in almost all applications due to their small size, efficiency, and low-temperature rise. When the output load on an SMPS is digital logic, then the additional ripple placed on the IC power supply, compared to the noise generated by the logic, is typically negligible and no EMI results. When the load circuit is very low-level analog, control, or RF, then the output noise of an SMPS may be unacceptable. One alternative is a linear power supply but even these generate broadband noise as the rectifiers commutate. Filters with a very high level of attenuation are sometimes required at the output power lines of an SMPS that are extremely effective at reducing switching power supply output noise. Typical applications for this high level of filtering have been found in satellite communication, lasers, sonar buoys, and microwave simulators.

As mentioned previously these filters can attenuate noise spikes for 40 V and ripple as high as 1.5 V to an output ripple of only 30 µV. Also by using these filters, a single switching power supply can be used for both digital boards and low-signal-level analog and RF circuits. In this case, the supply output is used directly for the digital boards, and the filter is placed between the supply and the sensitive circuits. These filters will typically attenuate noise spikes (as high as 40 V) and ripple (as high as 1.5 V) to an output ripple of 30 µV at the switching frequency of the converter. No measurable HF component is seen when measured with a number of different manufacturers' switching-power supplies. The full attenuation effectiveness of the chassis-mounted filter can be achieved only if it is mounted directly to a dividing bulkhead or shielded cables are used for both input and output connections. Figure 5.57 shows the use of a bulkhead-mounted filter to maintain isolation between a "noisy compartment" and a "quiet compartment" in a piece of equipment. The PCB-mount version must have the input and output connections via striplines at different layers in the PCB to achieve maximum effectiveness. The effectiveness of these filters is measured by mounting them to the wall of a shielded room, with the signal source outside the room and the measuring device inside the room. The room is used to isolate the input test signal from the filter output level.

The typical performances of these filters in an MIL-STD-220A test are shown in Table 5.22.

A typical schematic for this filter is shown in Figure 5.88.

The filter is contained in a metal enclosure as shown in Figure 5.61.

The filters are rated at 28 V 4 A. The 100 µH is made from 34 turns of #20 AWG magnet wire on a T68-52D core, and the 19 µH is made from 20 turns of #20 AWG on a T50-52B core. The voltage drop across one line of the filter is 0.5 V at 3.6 A.

The feedthrough capacitor is a Tusonix 0.12 µF 70 V P/N 4151-000.

TABLE 5.22
Typical Filter Performance in an MIL-STD-220A Test

Mode	Minimum Insertion Loss (dB) per MIL-STD-220A				
	20 kHz	100 kHz	1 MHz	10 MHz	100 MHz
D/M	80	90	100	100	80
C/M	25	50	100	100	80

FIGURE 5.88   Schematic of differential-mode and common-mode output filter.

5.3   SIGNAL FILTERS

Commercial filters are available covering frequencies from audio up to tens of gigahertz. These filters are designed and specified for a specific load impedance, typically 50, 75, 125, or 600 Ω. The filters may be active or passive and of the following types: *LC* high pass, low pass, band pass, bandsplitter, diplexer, band reject or notch, tunable notch, tunable preselectors, narrowband helical resonator filters, comb-line, interdigital cavity resonator, and digital electronic. The amplitude and delay response of filters may be chosen and include delay equalized low pass, Bessel, Butterworth, elliptic, Chebyshev, transitional Gaussian low pass, transitional Gaussian linear-phase band pass, Gaussian linear-phase band pass, and delay equalized band pass. The relative merits and weaknesses of the different filter types are described in books on filter design. Simple filter facts are contained in a catalog of precision *LC* filters published by Allen Avionics, Inc. In addition to books, computer programs and nomographs are readily available and ensure a suitable design for all but complex and variable source and load impedances. Very often, a single-section or double-section *LC* circuit is unnecessary and a simple *RC* filter will suffice.

At frequencies higher than the maximum operating frequency of the circuit, the noise immunity of control, analog, video, or digital interface circuits may be increased by the inclusion of an *RC* filter without the resonance problems encountered with the *LC* filter. For differential or balanced input circuits, the values of resistance and capacitance in each of the input lines should be matched, to minimize the conversion of C/M noise to differential at the input.

As with all filters, the radiated coupling between the input wiring/PCB tracks and the output wiring/PCB tracks should be minimized to the extent feasible. With SM filters, the coupling between input and output traces can be reduced by imbedding the traces in a stripline PCB configuration. Regardless of filter type, in applications where the filter is designed to exhibit high insertion loss, the components should be contained in a metal enclosure with either a central compartment and feedthrough capacitor or the use of a feedthrough capacitor mounted in the enclosure wall to bring the "clean" signal out of the enclosure. The "noisy" input signals should be connected via a shielded cable, or, where the environment is noisy, the "clean" signal should be shielded.

The commercial filters that are available may not fit the footprint available. For example, many μWave filters although very effective use tuned cavities and are physically large such as the filter described in Section 10.2.6.1. Also the exact frequency and attenuation may not be available.

5.3.1 ACTIVE FILTERS

It is easy to design active high-pass, low-pass, band-pass and band-stop filters using online software design tools.

Two of these are the Texas Instruments "FilterPro" and the Analog Devices "Filter Wizard."

Active filters using op amps are limited in the upper frequency possible usually limited by the gain bandwidth product (GBW) of the op amp.

Texas Instruments provides a very useful FilterPro™ User's Guide. This guide explains the terms passband, stopband, cutoff frequency, stopband bandwidth, and bandwidth and describes the difference between the Butterworth, Chebyshev, Bessel, Gaussian, and linear phase filters.

The Sallen–Key topology is a noninverting circuit and is useful when adding different sources together including the input signal to the filter as shown in Figure 5.89.

One of the parameters that may be important is the ripple. This is the variation in gain after the cutoff frequency. The Texas Instruments tool allows selection of Gaussian 6 dB ripple, Gaussian 12 dB ripple, Linear phase 0.5°, Linear phase 0.05°, Butterworth, Bessel, Chebyshev 1 dB, and Chebyshev 0.5 dB.

The component values specified are often nonstandard values, and so, a number of resistors may be required in series and capacitors in parallel as shown in Figure 5.90.

It is often useful to model the filter in a circuit analysis tool such as SPICE as a sanity check on the frequency response.

Figure 5.90 shows the schematic of a typical 30 dB amplifier with a passband frequency of 100 kHz at −1 dB, that is, 29 dB. The frequency at 0 dB gain is approximately 9 kHz. The frequency response is shown in Figure 5.91.

If the passband frequency using Filter Wizard is increased to 10 MHz with the gain at 30 dB, then the message is posted that no op amps are available, whereas using the FilterPro software, the GBW of the op amp is shown as 1.5 GHz.

If a gain of 30 dB at 10 MHz is needed, then three circuits in series each having a gain of 10 dB and using an AD8039 op amp can be used.

FIGURE 5.89 Schematic of composite high-pass filter.

FIGURE 5.90 Schematic of a practical 30 dB 100 kHz high-pass filter.

FIGURE 5.91 Response of 30 dB 100 kHz high-pass filter.

Although stable circuits have been tested using the Texas Instruments FilterPro software, one Texas design that oscillated and another that resulted in a distorted signal were OK when using an Analog Devices design. These filters should not be used in a final circuit without making, and testing a prototype for the old adage still holds good "an oscillator won't and an amplifier will oscillate."

If in a high-pass filter the lower frequencies, below the frequency where the filter gain is 0 dB, need to be added to the filter output, then the circuit in Figure 5.89 can be used. As an example assume a high-pass filter where the gain in the passband needs to change.

Here, the gain from DC to 100 kHz is 6 dB, at 1 MHz 18 dB, at 2 MHz 24 dB, and at 3 MHz 21 dB. The circuit is shown in Figure 5.89 and the frequency response in Figure 5.92.

5.3.2 PASSIVE FILTERS

Many of the filters include a parallel resonant circuit, a rejector, or a series resonant circuit, an acceptor. The parallel resonant circuit has a high impedance at resonance and the series a low impedance. The impedance of a capacitor with its parasitic or self-inductance forms a series resonant circuit, and its impedance is shown in Figure 5.6.

The use of low-pass, band-pass, and band-stop filters at the input of receivers can be effective at reducing compression, generation of spurious response, and desensitization of the receiver.

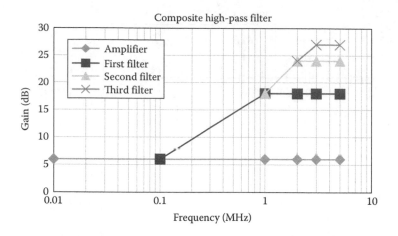

FIGURE 5.92   Frequency response of composite high-pass filter. Gain = 6 dB DC–100 kHz, 18 dB at 1 MHz, 24 dB at 2 MHz, and 27 dB at 3 MHz.

5.3.3 MICROWAVE FILTERS

Microwave filters are commercially available but are usually large as they are constructed using resonant cavities. The performance of one such filter is described in Section 10.2.6.1.

The filters described here are shown in small metal enclosures and have either BNC or TNC connectors. The connectors are selected for the intentional frequency of use. For the BNC, typically up to 700 MHz and the TNC 1.03 GHz and not for the microwave frequency.

One common application for a microwave filter is to attenuate a specific transmitter frequency from affecting LF, HF, VHF, and UHF communication, and this requires a band-stop filter.

For filters such as those shown in Figures 5.93 and 5.94 that are designed to be connected at the antenna, additional attenuation due to the connector and cable is typical. In some cases, the attenuation due to the cable connecting the antenna to the receiver exhibits so much attenuation at μWave frequencies that no filter may be required. The cable attenuation is described in Section 10.2.6.6.

The filter shown in Figure 5.93 is a band-stop filter tuned to 8.3 GHz, with its attenuation shown in Figure 5.95 and that in Figure 5.94 is at 9.375 GHz, with its attenuation shown in Figure 5.96.

FIGURE 5.93   8.3 GHz band stop.

FIGURE 5.94   9.375 GHz band stop.

FIGURE 5.95   8.3 GHz band-stop filter attenuation.

FIGURE 5.96   9.375 GHz band-stop filter attenuation.

These include a parallel resonant circuit (rejector) that exhibits a high impedance at resonance. The capacitance is formed between the two disks, and the inductance is the inductance of the wire connecting the two disks. The inductance of the wire is provided in Table 5.1.

The formula (5.10) for capacitance between two disks where d is the distance between the disks and A is the area is reasonably accurate when d is much smaller than A but ignores fringing fields and nonuniformity of the field. If the distance d becomes larger, as is true for the filter, then the equation is less accurate. Many methods exist for accurately calculating the capacitance although Equation 5.1 is adequate if only a starting point for the disk diameter is required The design allows tuning of the filter by moving the disks further apart or closer. By moving the plates closer together, the capacitance is increased and the inductance is decreased, which makes the adjustment less sensitive but at the same time provides an adequate adjustment, possibly because the capacitance is no longer directly proportional to d.

The capacitance is:

$$C = 8.84 \times 10^{-12} \, A/d \tag{5.11}$$

where
  A is the area = (πr^2) where r is the radius of the disk in m
  d is the distance between disks in m

One possibility for tuning is to use a threaded rod and to solder a brass nut to one of the disks as shown in Figure 5.97. This can then be moved and once tuned soldered in place.

The 8.3 GHz filter has one resonant circuit, and the 9.375 GHz filter has two resonant circuits. Unlike the SM component filters, these can be used on 100 W transceivers at LF, HF, VHF, and UHF. The lid of the 9.375 GHz filter is loaded with Cuming Corporation C-RAM-GDSS0.06 absorber material and the 8.3 GHz filter with C-RAM FFS-125 absorber.

5.3.4 PCB Microwave Filters

Filters at GHz frequencies may also be built into a microstrip PCB using traces. See Section 11.3.6.1 for a description of a microstrip and Appendix A for the characteristic impedance.

These microstrip PCB traces used in a filter should be kept far away from conductive structures as the parasitic capacitance and the field coupling to the structure change the filter characteristics. One filter can be constructed using a gap in the microstrip as shown in Figure 5.98 with its equivalent circuit. The microstrip radiates, and as we show later in this chapter, the radiated coupling between

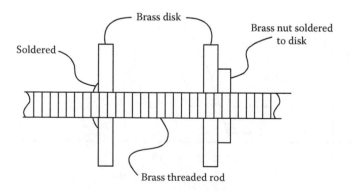

FIGURE 5.97 Proposed method of tuning disk and rod filter.

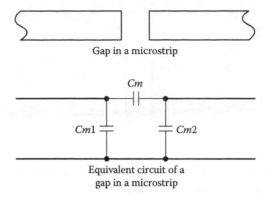

FIGURE 5.98 Gap in a microstrip.

the two traces occurs and can be dominant when the gap in the microstrip is large. Due to the cou pling capacitance C_g, the coupling in the microstrip with gap increases with frequency, and it acts as a high-pass filter. The gap can be replaced with a capacitor and usually is at lower frequencies, but even a microwave capacitor exhibits inductance and becomes a series resonant circuit at the resonant frequency. This resonance due to a parasitic component can be useful as shown in Section 5.2.2.

For large gaps, C_g becomes very low and the discontinuity becomes equivalent to an open circuit. In this case, the radiated coupling predominates.

A step in the PCB trace width as shown in Figure 5.99 is also similar to an open circuit with an equivalent circuit also shown in Figure 5.99. Due to the two inductances, this becomes a low-pass filter.

A cross-junction is shown in Figure 5.100 and acts as a notch filter.

The cross-junction can be considered two T junctions in parallel. A T junction known as a stub is useful in filters when the length of the stub is a quarter wavelength as shown in Figure 5.101.

FIGURE 5.99 Step discontinuity.

FIGURE 5.100 Cross-junction.

$\lambda/4$

FIGURE 5.101   Quarter wave stub.

For microwave frequencies, the quarter wave stub is usually left open circuit. Under this condition, the open circuit $1/4\lambda$ stub acts as a series LC resonator.

The impedance of the stub is theoretically zero, and the stub attenuator acts as a notch filter. It attenuates not only when the stub is $1/4\lambda$ but also when the stub is 0.75λ and 1.25λ and at other odd harmonics.

A ring resonator is a shorted transmission line where standing wave patterns form around the ring. The ring resonator may be end coupled as shown in Figure 5.102 or edge coupled as shown in Figure 5.103. The gap between the edge-coupled resonator is important. When the gap is small, the losses will be lower but the fields in the resonator will be affected. With a larger gap, the fields are less affected but the losses will be greater.

Increasing the ring circumference, which is in proximity to the exciting trace, as shown in Figure 5.104, should improve coupling but does add a discontinuity, and to our knowledge, this has not been tried.

FIGURE 5.102   End-coupled ring resonator.

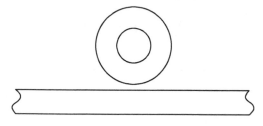

FIGURE 5.103   Edge-coupled ring resonator.

FIGURE 5.104   Edge coupled with increased coupling.

The end-coupled resonator acts like a band-pass filter, and the edge-coupled resonator acts as a band stop when the median circumference is 1, 2, 3,… wavelengths.

The mean radius is thus $\lambda/2\pi$.

For low-impedance and low-permittivity PCB substrate, the PCB trace is wide, and it becomes difficult to control the mean radius when the resonant frequency is high. Ring resonators at the higher microwave frequencies are difficult to realize as the resonant frequency when tested can be very far away from the design frequency.

Figure 5.105 shows a PCB trace with a 1/4λ stub followed by an edge-coupled ring and again a 1/4λ stub. The resonant frequency of this configuration is designed to be 11.6 GHz.

A filter was built using Rogers RT/duroid 5880 substrate that has a dielectric constant of 2.2 and is specifically for HF exacting stripline and microstrip PCBs.

Using the "Line Calc" utility, the width of the microstrip trace for a 50 Ω impedance was calculated at 2.4 mm.

The performance of a similar 14 GHz filter is shown in Figure 5.106. The stopband is narrow, and the dimensions are then critical.

By resonating the stub and the ring at different frequencies, a wider bandwidth can be achieved as shown in Figure 5.107.

FIGURE 5.105 11.6 GHz PCB notch filter.

FIGURE 5.106 Measured performance of a 14 GHz notch filter.

FIGURE 5.107   Notch filter with different stub and ring resonant frequencies.

5.3.5 CONNECTORIZED FILTERS

A small PCB with a wire inductor and a tuning capacitor can be used at lower frequencies, and the example is a notch filter at 1.03 GHz as shown in Figure 5.108.

The metallization is on one side to which the loop and one terminal of the trimmer capacitor are soldered. On the other side of the PCB, two traces join the other end of the loop and the other end of the trimmer together and to the connector pin. The tube in the photo is filled with absorber.

The construction is shown in Figure 5.109.

It is important to use a trimmer capacitor designed for high frequencies with either an air dielectric or a ceramic dielectric designed for VHF to μWave. A trimmer with a plastic dielectric was tested, and the Q was so low that no resonant peak was seen.

The performance of the filter is shown in Figure 5.110.

Other nonresonant PCB traces less than 0.1λ in length may be used as inductors and capacitors. The inductance of a straight length of PCB trace over a ground plane is given in Section 5.1.2, Equation 5.4. The inductance may be increased by making a circular or square spiral. The capacitance between two parallel traces over a ground plane is given in Section 5.1.10.1, and a typical PCB band-stop filter made from a spiral for the inductance and added traces for the capacitance is shown in Figure 5.111.

FIGURE 5.108   1.03 GHz filter.

FIGURE 5.109   1.03 GHz notch filter.

FIGURE 5.110   1.03 GHz filter attenuation.

The capacitance between the two traces can be increased by the addition of an SM capacitor although even a microwave capacitor has parasitic inductance (see Section 5.1.10.1), which must be taken into account when designing the filter.

5.3.6   Low-Frequency Passive High-Pass, Low-Pass, Notch, and Band-Pass Filters

Connectorized filters were designed, built, and tested to cover the 2–512 MHz frequency range. These filters are made from combinations of inductor and capacitor, and the components used in these filters are critical. Its Q is determined by not only the DC resistance of the wire used in an

FIGURE 5.111   PCB band-stop filter.

inductor but also the loss in the core at high frequencies. Ferrite is inherently lossy, and sometimes, inductors designed for power applications have a much higher Q.

Some of the inductors used in the following filter designs are

    B82496C356 J 56 nH
    MLG1005SR24 0.24 μH multilayer
    MLF1608DR33M 330 nH multilayer
    MLF 1608DR33M 0.33 μF multilayer
    LMQ21NNR47K10D 0.47 μF
    L-14C12NJV4T 12 nH
    744765119A 19 nH
    MLG1005SR12J 120 nH
    ATFC-0402-7N2-BT 7.2 nH thin film
    SD6020-4R1-R 4.1 μH

The SRF for some of the inductors are shown in Table 5.23 with their use in the filters.

TABLE 5.23
Self-Resonant Frequency for Inductors 7 Used in the Filters

Filter	Parasitic Capacitance	Inductance (nH)	SRF (GHz)
200	0.206	12	3.2
200	0.0864	47	2.5
2	0.878	1000	0.17
2	1.72	4100	0.06
30	0.423	240	0.5
30	0.0725	56	2.5
406	0.12	5.9	6
406	0.135	24	2.8
512	0.235	4.7	4.8
512	0.144	19	3

FIGURE 5.112 Attenuation of the 406 MHz low-pass filter.

For the higher-frequency filters, the capacitors also need to exhibit low inductance and high Q. Capacitors specifically designed for μWaves should be chosen. One capacitor type was the GQM series monolithic ceramic capacitors 0.1–100 pF. Also the GRM for values to 1800 pF.

For higher values, the C0G capacitor can withstand high temperature with a capacitance value of up to 0.1 μF.

One excellent passive filter design software is available in www-usres.cs.york.ac.uk/~fisher/cgi-bin/lcfilter, and this was used for all of the designs.

For all of the filters, the in-band insertion loss was less than 1 dB.

The measured attenuation above 406 MHz is not as high as predicted as shown in Figure 5.112.

This is not because the software is faulty but because the parasitic inductance of the capacitors, the parasitic capacitance of the inductors and the Q of the components is not included.

The filters that were designed and tested were

2 MHz low pass
30 MHz low pass
108–110 MHz band pass
136–174 MHz band pass
156–163 MHz band pass
200 MHz low pass
406 MHz low pass
700 MHz high pass

The measured attenuations of a 2 and a 30 MHz low-pass filter, designed with the software, in which the attenuation also decreases at higher frequencies due to the parasitics are shown in Figures 5.113 and 5.114, and a photo of the 30 MHz is shown in Figure 5.115. The attenuation of a 136–174 MHz band-pass filter created is shown in Figure 5.116. The plots show reduction in level from 0 dBm. Thus, −90 dBm equals 90 dB of attenuation. The PCB is universal for all the low-pass filters, and so only the values change.

5.3.7 Example of Filter Design

A helmet containing a microphone and a microphone preamplifier was tested for RS103 from 14 kHz to 18 GHz at 50 V/m. The RS103 signal is amplitude modulated with a frequency of 1 kHz, and this could be heard. At 30 MHz and other frequencies, the 1 kHz tone could be heard. The filter

FIGURE 5.113   2 MHz low-pass filter reduction in level.

FIGURE 5.114   30 MHz low-pass filter reduction in level.

FIGURE 5.115   30 MHz low-pass filter with TNC connectors.

FIGURE 5.116 136–174 MHz band-pass filter attenuation.

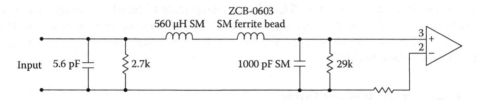

FIGURE 5.117 Microphone preamplifier input filter.

FIGURE 5.118 Microphone filter attenuation.

in Figure 5.117 was implemented, and the preamplifier was immune from 0.014 to 350 MHz. The response of this filter is shown in Figure 5.118. However, from 350 MHz to 18 GHz, the tone was heard between 1.77 and 3.16 GHz.

A 10 pF capacitor was connected between the input of the filter and ground. This reduced the tone but did eliminate it. A 5.6 pF 1206 size capacitor should resonate with its parasitic inductance at 2 GHz, and when this replaced the 10 pF, a significant improvement was achieved. The schematic is shown in Figure 5.117.

FIGURE 5.119   Ferrite at microphone.

The remaining problem was traced to the microphone itself. The microphone had a metal case, and this was connected to a shielded enclosure as shown in Figure 5.119.

Most ferrites materials are ineffective above approximately 1 GHz, and when several beads made of 25 and 28 material were tried on the short unshielded section of the cable, no improvement was seen. Despite that when a SFC-3 ferrite clamp manufactured by Kitagawa Industries (Intermark) was tried close to the microphone, the 1.77 to 3.16 GHz problem was solved. Digital microphones are available which are typically more immune, although the problem may simply be switched from RS103 to RE102!

5.3.8   COMMERCIALLY AVAILABLE FILTERS

A great number of companies manufacture signal filters in a range of footprints and specifications. Some of these who complied with the requirements in Table 5.24 are

    Mercury microwave
    Spectrum microwave

Allen Avionics Inc. quoted on a low-pass filter rated at 200 W with BNC connectors suitable for transceivers.

5.3.9   CASE STUDY 5.2

Case Study 5.2 illustrates the effectiveness of a simple *RC* filter. In this case, we have a 50 kW transmitter, at a frequency of either 4.8 or 2.2 MHz, located approximately 40 m from a 10–14 kHz

TABLE 5.24
Filter Specifications

Passband	DC–30 MHz	30–600 MHz	118–137
Passband insertion loss	1 dB max	1 dB max	1 dB max
Rejection	40 dB min at 100 MHz 60–80 dB at 1030 MHz	20 dB min at 1030 MHz	40 dB min at 1030 MHz
Power handling	150 W CW	150 W	20 W
Operating temperature	0°C–60°C	0°C–60°C	0°C–60°C
Package	Connectorized-type BNC or TNC	Connectorized-type BNC or TNC	Connectorized-type BNC or TNC

FIGURE 5.120 Signal-line filter and construction.

Omega time code receiver antenna. A 4 V/m E-field incident on the time code rod antenna was measured. The antenna is connected at the base to a high-input-impedance preamplifier. The preamplifier gain reduces sharply above 50 kHz, and the preamplifier normally functions as an effective filter. Yet, the voltage generated at the input of the preamplifier by the interfering transmitter is approximately 26 V peak to peak, which is above the preamplifier supply voltage. The interfering voltage resulted in compression in the first stage of the preamplifier and amplitude modulation of the time code signal, thereby destroying its integrity.

The antenna impedance is 1.18 MΩ at 10 kHz, and the input impedance of the preamplifier is 3 MΩ; thus, a standard 50 Ω low-pass filter cannot be used in this application. The RC filter shown in Figure 5.120 achieved an attenuation of 22 dB at 2.2 MHz and 28 dB at 4.83 MHz. The filter is contained in a small metal enclosure connected between the base of the antenna and the preamplifier. The inside of the filter was compartmentalized to avoid radiated coupling between the input and the output, which was almost inevitable due to the very high level of the interfering signal as well as the high input impedance and low capacitance of the preamplifier. The feedthrough is of a standard type, with no intentional capacitance. However, the measured parasitic capacitance was 1.5 pF, and this was accounted for in the filter design along with the 1.5 pF of preamplifier input capacitance. An additional 4.7 pF capacitor was included in the filter, resulting in the approximately 7.7 pF capacitance shown in the schematic. The insertion loss at the time code signal frequency of 10 kHz was an acceptable 0.3 dB, and the EMI problem was solved by the inclusion of the filter.

5.3.10 Filter Connectors

Filter connectors are connectors with intrinsic filter components such as capacitors or capacitors and inductors in T, L, or π configurations or simply bulk ferrite forming a balun. Filter connectors may be used to filter either signal or power up to 15 A and 200 VDC or 120 VAC. The connector types available include D type and the majority of military types, such as MIL-C-389999, 5015, 83723, 26482, TKJL, TKJ, and TKJA, manufactured by Amphenol and ITT Cannon, and BNC and TNC types manufactured by Erie. Filter connectors are available with moderate levels of attenuation at LF, suitable for power or control lines, or with minimum attenuation at 1, 10, or 100 MHz, resulting in negligible pulse degradation in signals.

API Technologies (Spectrum Control) manufactures Micro D, Series 400, 500, and 700 filter-cons or filter adapters depending on the type of result in good signal pulse performance at 0.13, 0.16, 0.26, 0.64, 1.3, 2.8, 3.4, 6.4, and 130 MHz.

Typical performance curves for the 700 series are shown in Figure 5.121.

FIGURE 5.121   API Technologies (Spectrum Control) 700 Series performance.

The performance of the Micro D series is shown in Table 5.25.

One disadvantage of the filter connector when used in balanced, differential circuits is the 20% or so tolerance on the capacitor value, which effectively unbalances the input and converts C/M noise to D/M.

There are filter adapters available that are inserted between the existing unfiltered plug and socket and are useful in a diagnostic test or to cure an EMI problem. In the MIL-STD-462 test setup for the radiated emission (RE02) test, the EUT in the shielded room may be controlled by a test set, computer, or ground-based equipment located outside of the room. The RE02 limits are very often exceeded when the interface cables entering the shielded room are connected to the test set outside the room but disconnected from the EUT. The reason is typically C/M noise on signal or supplies sourced by the test set, resulting in radiation from the interface cables, which are usually shielded. The solution is to use D-type connectors on the test set and include filter adapters where required.

TABLE 5.25
API Micro D Filtered Connector Performance

Filter Designation	Type	Capacitance Value (pF)	Tolerance (%)	Dielectric Withstanding Voltage (V)	Working Voltage DC −55°C to +125°C (V)	Minimum Insertion Loss—Decibels (dB) 50 Ω System per MIL-STD-220 (No Load)							
						5 MHz	10 MHz	20 MHz	50 MHz	100 MHz	200 MHz	500 MHz	1 GHz
High performance													
01	FT	100	±20	300	100	—	—	—	—	1	6	14	20
02	FT	470	±20	300	100	—	—	2	8	14	20	28	34
03	FT	820	±20	300	100	—	2	6	13	19	25	33	39
04	FT	1500	±20	300	100	—	5	10	18	24	30	38	44
05	FT	4700	±20	300	100	8	14	20	28	34	40	48	54
Standard performance													
10	COB	100	±20	300	100	—	—	—	—	1	6	14	20
11	COB	470	±20	300	100	—	—	2	8	14	20	28	32
12	COB	820	±20	300	100	—	2	6	13	19	25	32	32
13	COB	1500	±20	300	100	—	5	10	13	24	30	32	32
14	COB	4700	±20	300	100	8	14	20	23	32	32	32	32

Note: The EESeal filter seal made by Quell is a thin rubber insert easily slipped over connector pins in D type, ARINC and circular connectors and contain 1 pF to 1 μF capacitors.

FIGURE 5.122   Filter box enclosing the rear of a D-type connector.

When filter connectors are too expensive or unavailable "space qualified," or the capacitance must be matched closely for differential circuits, the filter box arrangement illustrated in Figure 5.122 may be used. Flexible wafers containing filter SM components, available from TRW and Metatech, fit over the pins of existing connectors.

5.4   EMISSION REDUCTION TECHNIQUES

5.4.1   Signal and Power Generation Characteristics

Pulsewidths and rise times of signals as well as power control and conversion should be selected so as to use no more of the frequency spectrum than required. This means that pulsewidths should be as narrow as possible and that the rise and decay times should be as long as possible within the limits of the required signal characteristics. High-current pulses due to switching charge in diodes and other semiconductor devices as well as the rapid charge and discharge of capacitors should be minimized, typically by the inclusion of series impedance.

The rise and fall times of switching transistor and rectifier diodes, usually tailored for the fastest possible switching times to improve efficiency, will produce an incredibly rich frequency spectrum of emissions. Selection of "soft" fast-recovery rectifiers and slowing down the transistor rise and fall times (at the expense of efficiency) will significantly suppress HF (greater than 1 MHz) components. Where possible, the use of thyristors or triacs with their associated high dV/dt or dI/dt characteristics should be replaced with transistors or power MOSFETS, where the transition from "off" to "on" may be slowed down.

The heater switching circuit shown in Figure 5.123, without the additional 0.01 μF miller capacitor between Q16 base and collector, switched "on" in 1.5 μs and "off" in 1.4 ms. The capacitor slowed the switching speed to 4 ms "on" and 4 ms "off". The reduction in the level of broadband conducted noise on the line is shown in Figure 5.124.

Broadband noise is often generated by linear power supplies at the AC power-line frequency and appears on both the input and output terminals. The reason for the noise voltage/current is the

FIGURE 5.123 Heater switching circuit.

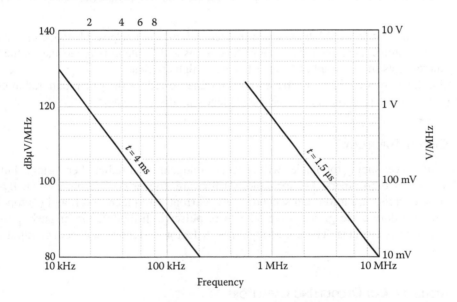

FIGURE 5.124 Reduction in the level of broadband emissions due to the inclusion of C2 (0.01 μF).

commutation of the rectifiers from a conducting to a nonconducting state. This commutation may be speeded up by the use of fast-recovery rectifiers, although some time is still required to extract the charge from the rectifier junction during turnoff. During the charge switching time, a short circuit appears across the AC input terminals. The use of series inductance, as shown in Figure 5.125a, will very effectively reduce the input-current spike causing broadband noise.

The inclusion of a series inductor between the bridge and the smoothing capacitor may be used to isolate the supply from current spikes generated by the load switching or to reduce the input ripple current, as shown in Figure 5.125b. The trade-off is that the inclusion of the inductor reduces the maximum voltage developed across the smoothing capacitor. The ripple current at the input power frequency may be further reduced by adding a capacitor across the inductor and thereby creating a parallel tuned circuit at the power-line frequency.

FIGURE 5.125   Inductor between bridge and capacitor used to reduce ripple and transient current spikes.

It is understood that at very high switching frequency, it may not be possible to increase rise and fall times. For example, the trend toward converter designs with switching frequencies in the megahertz region requires short rise and fall times to ensure high efficiency.

However, there are techniques available to the designer to reduce the generation, radiation, and conduction of unwanted noise. Four of these techniques are discussed next.

5.4.2   CIRCUIT TOPOLOGY

All switching power supply topologies will produce conducted and radiated emissions, but some are better choices than others. For example, (1) forward converters are better than flybacks, because peak currents are lower and secondary AC currents are much lower, and (2) topologies that operate at a duty cycle greater than 50% (e.g., pulsewidth modulation or push–pull) also result in lower peak currents. The use of input power factor correction will reduce input spikes on AC power.

5.4.3   RESERVOIR AND DECOUPLING CAPACITORS

It is better to provide transient current from a capacitor, with a low ESL and low ESR, mounted as close to the switching transistors or MOSFET as feasible, than from the supply. Series resistance or inductance may further isolate the supply to the converter from transient current requirements. The conductive paths for transient current flow should cover as small an area as possible.

5.4.4   HEAT SINKS

The major cause of high C/M noise in converters is the switching waveform on the collector/source of the switching transistor/MOSFET, which is usually connected to the case of the device. The case of the switching device is then typically heatsunk to the enclosure. Thus, the noise source is referenced to the enclosure via the high-quality capacitance formed between the case of the switching transistor or MOSFET and the enclosure, with the heat sink insulator acting as the dielectric. The capacitance between the device and the enclosure may be reduced by using a beryllium oxide heatsink insulator (capacitance 18 to 41 pF assembled) instead of a mica insulator (capacitance 150,

156 pF assembled). A better solution is the use of an insulator that contains an imbedded shield. With the shield connected to the return of the power supply of the switching devices, the capacitively coupled transient currents flow back into the supply and not into the enclosure, and the C/M noise voltages are considerably reduced.

5.4.5   CIRCUIT LAYOUT

Confining high transient or RF current-carrying conductors and devices into a small loop area with the use of short low-impedance connections will reduce both conducted and radiated emissions. The circuit should be as far removed from the enclosure as feasible in order to reduce capacitive coupling. The use of a small secondary enclosure, shielding the source of noise, or compartmentalization with the use of feedthrough capacitors or filters may be required for sources of high radiation. When switching inductive loads by either a relay or a semiconductor device, high voltages may be generated during switch off. The magnitude of the voltage is determined by the rate of change of current as follows: $V = L(dI/dt)$. The high voltage developed when dI/dt is high may couple capacitively to circuit conductors, to the inside of an enclosure, or across the capacitance of the semiconductor device. When the inductive load is switched by a relay, damage to the relay contacts may result due to arcing. The snubber circuits shown in Figure 5.126 may be used to reduce the voltage generated. Omission of the resistor that is in series with the diode in Figure 5.126b and d decreases the voltage further but will result in a high transient current flowing in the diode, which in turn may result in an EMI problem. When switching a load with high in-rush current, such as a converter or filament lamps, a series inductor may be used to reduce the magnitude of the transient.

It is not uncommon in an EMI/EMC investigation to find that in switching a load, the resultant voltage or current transients are the cause of EMI or emissions above a specification limit. Despite that, the following case concerns a load that would not be expected to cause a problem. A piece of commercial equipment made up of a mix of TTL and CMOS logic on a single PCB contained in a nonshielded enclosure FCC-radiated emission requirements, but only when a beeper sounded. The cause of the problem was the piezoelectric beeper. When the transistor switching the beeper at an audio frequency switched off, the beeper resonated at a frequency above 30 MHz, and a 100 Vpk–pk voltage was developed across the beeper. Adding a resistor in parallel with the beeper dampened the resonance but reduced the audio output to a lower, albeit acceptable, level. Chapter 11 describes PCB layout for reduced emissions and further circuit-level reduction techniques.

FIGURE 5.126   Snubber circuits.

5.5 NOISE IMMUNITY

5.5.1 INTERFACE-CIRCUIT NOISE IMMUNITY

It is important to assess the EM environment before deciding on the level of D/M and C/M immunity required of an interface. It is not unusual for C/M noise between the grounds of systems separated by appreciable distances to reach levels of several volts. This type of noise can be considerably reduced by the correct grounding scheme, carefully realized.

Digital equipment communicates with low error using standard interfaces in an office environment. However, when a computer is located in such an electromagnetically benign environment and interfaces to equipment located in a hostile environment, such as the shop floor in a process and control facility, the noise immunity of standard interfaces may be insufficient. When the computer is located in the hostile environment, EMI is likely caused by power-line transients.

As described in Chapter 7, the level of achievable shielding of an incident field with flexible shielded cables is limited. It is therefore not uncommon to experience failures during a radiated susceptibility test due to interface circuits with insufficient noise immunity. When EMI occurs in the passband of the interface, filtering as a solution is precluded.

After the EM environment has been estimated, using the examples of Chapter 2, the specified susceptibility test limits, or the techniques described in Chapter 11, an interface can be chosen that, in addition to meeting speed and power consumption constraints, will meet the EMI immunity requirements.

Manufacturers of commercial equipment exported to the European Community must make the EMC Declaration of Conformity, and this will typically be made after emission and immunity testing. One of the signal cable tests is the C/M capacitive injection of an electrical fast transient (EFT) pulse of 500 V for domestic equipment and 1 kV for nondomestic equipment. The EFT pulse when measured into a 50 Ω load has a rise time of 5 ns and a pulse width of 50 ns at the 50% amplitude value. The pulsewidth induced in a cable may be significantly longer than 50 ns, dependent not only on the C/M load seen by the EFT generator but also on the EFT generator design.

Another test is bulk C/M injection into a cable of a 1 kHz 80% am signal from 150 kHz to 80 MHz.

Standard EIA interfaces using unshielded cables, or shielded cables with a poor shield termination technique, commonly fail the EFT test, and less commonly the C/M signal injection. It is important to ensure that sufficient noise immunity is built into the interfaces. LF unbalanced inputs may require HF filtering and HF. Balanced input lines should have equal path lengths and present equal impedances to ground to very fast pulses.

The noise immunity of standard TTL, CMOS, and ECL logic is shown in Table 5.26. The noise immunity of standard logic can be increased by the use of low-impedance drivers and terminating networks as well as Schmitt trigger receivers, as shown in Figure 5.127.

Standard Line Drivers and Receivers, RS232, RS423A, RS422A, RS485, RS449, RS530, IEEE (488–1978) (GPIB), 10BASE TX, 100BASE TX, 155Mb/s ATM, USB, SSTL, and MIL-STD-1553 Interfaces

Line driver and receiver ICs are available that have some or all of the following features:

Built-in termination resistors
Clamp diodes at input
Clamp diodes at output
High-current outputs
Individual frequency response control
Input hysteresis
High DC noise margin
High C/M noise rejection ratio

TABLE 5.26

Noise Immunity of Standard TTL, CMOS, and ECL Logic

	TTL 64	HTL	CMOS 5 V	CMOS 10 V	CMOS 15 V	TTL 74 L	TTL 74 H	EC[a] 10 k	ECL 100 k
Typical propagation delay (ns)									
tPHL	12	85	150	65	50	30	8	2	0.5
tPLL	17	130	150	65	50	35	9	2	0.5
DC-noise margin (minimum)									
VNL (V)	0.4	5	1	2	2.5	0.4	0.4	0.155	0.155
VNH (V)	0.4	4	1	2	2.5	0.4	0.4	0.125	0.125
Typical output impedance (Ω)									
Low	30	140	400	400	400	30	30	8	8
High	140	1.6k	400	400	400	510	55	6	6
Typical noise energy immunity									
ENL (nJ)	1,7	60	1	3.1	7.2	—	—	—	—
Pulsewidth (ns)	20	125	155	70	50	—	—	—	—
ENH (nJ)	1	5	0.9	3.1	8.5	—	—	—	—
Pulsewidth (ns)	25	145	280	90	75	—	—	—	—

[a] $R_L = 50\ \Omega$.

FIGURE 5.127 Standard TTL, CMOS, and ECL line drivers/receivers.

EIA standard
GPIB standard
Single-ended transmission
Differential data transmission
Input threshold reference input

The noise immunity of standard digital logic may be increased when the signal path is electrically short, that is, propagation delay is short and when crosstalk is capacitive by low-impedance drivers and terminating networks. When the signal path is electrically long or crosstalk is inductive using low impedance, drivers may be either non- or counterproductive.

5.5.2 RECEIVERS AND DRIVERS

Typical differential line driver and receiver configurations are shown in Figure 5.128. When more than one transceiver shares the line shown in Figure 5.128, this configuration is termed a *party line* or *bus* system. The transmission line is typically a twisted pair individually shielded or with an overall shield over a number of twisted pairs. The differential configuration has the advantage of high C/M noise immunity below 5 MHz, increased immunity to crosstalk from pair to pair, and the possibility of added "in-band" C/M noise filtering. The input impedance to ground in the lower circuit in Figure 5.128 is typically in the kiloohm range due to the relatively low input impedance of most commercially available line receiver ICs. Thus, these devices are not suitable for the isolation of equipment–equipment or system–system grounds. Conversely, the input impedance to ground of the device in the lower circuit of Figure 5.128 is too high to achieve effective C/M filtering by

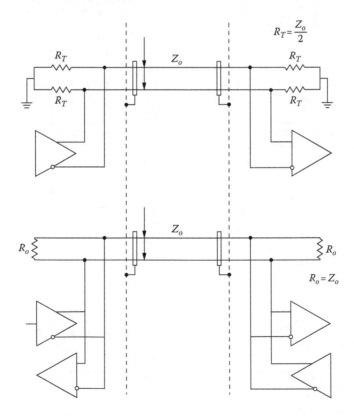

FIGURE 5.128 Typical differential line drivers and receivers.

the use of a balun. The impedance of a ferrite balun is approximately 50–1000 Ω, and with a C/M input impedance of 1–10 kΩ, the C/M attenuation is likely to be less than 6 dB. Baluns are often placed on line driver and receiver connections to the interface cable to reduce radiated emissions, and these will have very little effect on increasing C/M noise immunity when used in the Figure 5.128 circuit. Although the terminating resistor is split as shown in the upper circuit of Figure 5.127, significant C/M attenuation is achieved. For example, assume Z_o is 100 Ω and the impedance of the balun shown in Figure 5.129a is 500 Ω over a wide frequency range. The impedance line to ground R_T is 50 Ω, and the C/M noise voltage attenuation is 21 dB. The C/M noise current divides between the two lines and flows in the two 50 Ω impedances, so the total impedance seen by the C/M noise source is 25 Ω. The C/M power noise immunity is therefore high.

Noise voltages are typically generated C/M due to a ground noise potential between the driver and the receiver ground and due to C/M injection onto the signal cable.

Differential receiver inputs that are well balanced to ground and exhibit high C/M noise rejection over the in-band frequencies should be chosen. Where out-of-band signals are a problem, the addition of a high-pass filter for LF noise or a low-pass filter for HF noise can be added to the circuit, as shown in Figure 5.129b.

The noise immunity to in-band noise of a single-ended system is much lower than the differential input, because C/M-induced noise is converted to D/M noise at the receiver input by the unbalanced impedance of the input. Additional filtering can be added to a single-ended receiver input for out-of-band noise rejection.

A receiver's differential input impedance is typically dependent on the input frequency and the input signal. Adding a termination resistor to the receiver input will reduce the reflections and improve the signal quality and often the noise immunity of the circuit. This increased noise immunity is achieved because noise spikes with voltages lower than the signal level and impressed on an undistorted signal are less likely to exceed the receiver input switching threshold than spikes impressed on a distorted signal. The noise input power required to generate voltages above the threshold is also much higher for a 45–460 Ω input impedance than a 1–10 kΩ input impedance. The increased noise immunity achieved with a termination resistor is offset to some extent by the

FIGURE 5.129 (a) Increasing differential receiver common-mode input noise immunity, (b) Low-speed single-ended receiver input filter.

reduction in signal level due to the loading. To maintain noise immunity, choose a driver with a low output impedance. In a party line system, the driver output is often open collector with a low-value pull-up resistor. The pull-up resistor should be no higher than 20 Ω when the receiver input resistance is 50 Ω. If the interface is not party line, then an active pull-up should be used, because as a "high" output is pulled down, more current flows out of the active driver.

The DC noise margin and the power noise margin shown in Table 5.26 assume no AC or DC loss between the driver and the receiver, which may be true for very short line lengths. Where longer line lengths are used at HF, the cable attenuation plays an important role in reducing the signal level and decreasing the signal-to-noise ratio. The 50 Ω impedance RG188A/U and the unshielded twisted pair have a higher attenuation than the 50 Ω RG58/U, and the 75 Ω RG59/U is even lower.

The DS26LS32 receiver, manufactured by National Semiconductor, has a D/M or C/M input range of ±7 V and an input sensitivity of ±0.2 V and is therefore ideal for the EIA RS-422, RS-423, RS-499, RS-530, and federal standards 1020 and 1030 for balanced and unbalanced data transmission. The DS26LS32 has a ±15 V D/M and C/M voltage range and a ±0.5 V differential sensitivity, and this device, although it does not conform to the EIA standards, will exhibit a much higher noise immunity. Table 5.27 shows the characteristics of drivers and Table 5.28 those of receivers, reproduced from the AN781A application note from Motorola Inc.

Figure 5.129a illustrates the C/M filtering circuit. For LF single-ended signals, such as the RS232C, the HF noise immunity can be increased by adding a filter or filter connector at the receiver input. Filter connectors containing a π-circuit can be selected with a 3 dB frequency appropriate for the data rate of the signal. The only problem with filter connectors is the capacitive imbalance between the pins. Very high levels of C/M noise applied to the connector are converted to differential noise by this imbalance.

Figure 5.129b illustrates a filter suitable for a low-speed single-ended receiver. The input impedance is maintained at a low 100 Ω to increase the amount of input noise power required at in-band frequencies before an upset occurs. The values of L and C are chosen to produce a low-pass filter with a frequency response that does not degrade signal quality but may increase the signal rise and

---

TABLE 5.27

EIA-Compatible Line Drivers from the Motorola Application Note AN781A

RS Std.	Device #	Drivers per Package	Power Supplies	Prop, Delay (ns)	Rise/Fall Time	Hi-Z Output Comments
232-C	MC1488	4	±9 to ±13.2 V	175/350	Adjustable	No inverting
232-C	MC3488	2	±10.8 to ±13.2 V	20	Adjustable	No inverting
422	MC3487	4	+5 V	20	20 ns	Yes
423	MC3488	2	+10.8 to 13.2 V	—	Adjustable	No inverting
485	SN75172	4	+5 V	25/65	75 ns	Yes
485	SN75174	4	+5 V	25/65	75 ns	Yes

TABLE 5.28

EIA-Compatible Line Receivers from Motorola AN781A

RS Std.	Device #	Drivers per Package	Input Hysteresis	Input Threshold	Prop, Delay (ns)	Comments
232-C	MC1489	4	0.25–1.15 V	3 V	85	
422/423	MC3486	4	30 mV typical	200 mV	35	±7 V C/M input range
485	SN7513	4	50 mV typical	200 mV	35	

fall times within acceptable limits. The input impedance of the receiver is high, and so, the 2 Ω resistor is required to damp the filter resonance. The resonance frequency of the filter should be far removed from the maximum clock rate of the signal.

Where in-band D/M noise at a differential receiver input, or C/M noise, which is converted to D/M at a single-ended receiver input, is causing a problem, the solution is to use a shielded cable for the interface. If the cable is already shielded, then it is important to ensure the shield termination is made via a low impedance to the equipment enclosure or to use a double-braid shielded cable or at least a foil/braid shielded cable.

Line drivers are often capable of driving a large number of receivers and have sink and source capabilities of typically 40 mA and up to 200 mA. Load impedances of 78–250 Ω are common, and the interface may be used with a twisted pair that exhibits an impedance of 80–130 Ω or a twisted shielded pair of 70–120 Ω impedance. The DS26S31 line driver meets all the requirements of EIA RS-422, RS-530, and RS-449 and federal standard 1020. The DS26S31 can supply a minimum high output voltage of 2.5 V at an output current of –20 mA and a maximum low voltage of 0.5 V at 20 mA. The output short-circuit current is limited to between –30 mA and a maximum 150 mA. Table 5.26 shows the characteristics of the EIA standard interfaces. The RS449 standard was developed to provide a new definition for RS232-C interconnection lines. RS449 uses 30 lines in a 37-pin connector, whereas RS232 uses 20 lines in a 25-pin connector. RS449 does not define electrical characteristics, because those defined for RS422 and RS423 are to be used. RS530 is just like RS422 and uses differential signaling on a DB25–RS232 format.

To decrease crosstalk in a cable containing a number of interface circuits, a maximum dV/dt is specified, as is the transition time.

The maximum cable length for any interface is dependent on the attenuation characteristics of the cable: higher resistance shortens the maximum permissible length and higher capacitance shortens the maximum length at the higher baud rates. The required noise immunity and maximum bit error rate also limit either the cable length or the baud rate. The maximum cable lengths recommended in RS232, RS422, and RS423 are recommendations only; longer lengths are possible with high-quality cable and/or using signal waveshaping, signal distortion compensation, or interface conditioning techniques, which may also be used to improve noise immunity. The min/max cable lengths for the RS422 in Table 5.29 are based on the use of twisted-pair, 24 AWG cable and a balanced interface.

Table 5.30 shows the characteristics of a typical bidirectional instrumentation bus (IEEE-488, GPIB) transceiver. The input termination resistors of 2.4 kΩ to VCC and 5 kΩ to ground do not represent a low-impedance input, and thus, the coupling, either capacitively line to line or from an incident electric field, is higher than expected for a low (50–130 Ω)-input-impedance receiver.

Although the GPIB is a single-ended transmission line, the immunity of the cable and receiver to magnetic fields is surprisingly high. Using a 6.7 m single shielded GPIB cable with plastic backshells, the D/M noise detected at the receiver input with the cable exposed over a 1 m length to 0.7 A/m at 30 MHz was approximately 600 mV (i.e., 200 mV above the receiver noise threshold). When the GPIB cable was changed to a double-braid, single-foil type with metal backshell (the so-called EMI hardened), the D/M-induced voltage was reduced to 150 mV (i.e., –12 dB), and the receiver was immune to the reduced level of combined common and differential noise. The local area network (LAN) "Ethernet" has been tested and shown immunity to high levels of E field (20 V/m from 10 kHz to 1 GHz) (Chapter 4, Reference [4]).

The MIL STD 1553B data bus exhibits a C/M noise immunity of 10 Vpk–pk, 1 Hz to 2 MHz, and a D/M noise immunity of 140 mVrms (Gaussian 1 kHz to 4 MHz) for a transformer-coupled stub and 200 mVrms for a direct-coupled stub.

The 10BASE-TX interface transmits ≈2.5 Vpk Manchester encoded data at 10 Mb/s. For 100BASE-TX, a 1 Vpk MLT-3 encoded data stream at 100 Mb/s is transmitted. A suitable receiver circuit for either 10BASE-TX or 100BASE-TX is the National Semiconductor DP83223 TWISTER High Speed Networking Transceiver Device. The DP83223 allows links of up to 100 m over shielded

TABLE 5.29

ETA Standard Interfaces

	Specification	RS232-C 1969	RS423-A 1978	RS422-A 1978	RS485 1983
General	Type	Single-ended	Single-ended	Differential	Differential
	Line length	50 ft (15 m)	1200 m (4000 ft)	1200 m (4000 ft)	Application dependent (4000 ft)
	Maximum frequency	20 kbaud	10 Mh at 10 m (33 ft); 80 kbaud at 1200 m (3937 ft)	100 kbaud	10 baud
	Transition time	<4% of γ or < 1 ms (in undefined area between "0" and "1")	Lesser of 0.3γ and 300 μs (time for 10–90% of final value)	Greater than 20 ns or <0.1 γ (time for 10%–90% of final value)	<0.3γ (54 Ω, 50 pF load (time for 10%–90% of final value)
Drivers	No. of drivers	1	1	1	32
	Open-circuit output	$3.0\,V < |V_0| < 15\,V$	$4.0\,V < |V_0| < 6\,V$	$|V_0| < 6\,V\ |V_{0a}|,\ |V_{0b}| < 6\,V$	$1.5\,V < |V_0| < 6\,V$
	V_1, (loaded output)	$5\,V < |V_0| < 15\,V$	$|V_1| > 0.9|V_0|450\,Ω$ load	$>2\,V$ or $\frac{1}{2}V_0,$ $< |V_1| < 6\,V$ 100 Ω balanced	$1.5\,V < V_1 < 5\,V$
	Short-circuit I	<500 mA	<150 mA	<150 mA	<250 mA
	Output Z	—	<50 Ω	<100 Ω balanced	—
	Driver load	3–7 kΩ	460 Ω	100 Ω	54 Ω
Receivers	No. of receivers	1	10	10	32
	Input range	+15 V	+12 V	−7 to +7 V	−7 to +12 V
	Minimum receiver response	$>\pm3.0\,V$	200 mV differential	200 mV differential	200 mV differential
	Input resistance	3–7 kΩ, 2500 pF	4.0 kΩ min	4.0 kΩ min	15 kΩ, dynamic 12 kΩ (+12 V) 8.75 kΩ at −7 V
	C/M voltage for balanced receiver	—	—	7.0 V, $V_{cm} < 7\,V$	7.0 V, $V_{cm} < 7\,V$
	DC noise margin	2 V	3.4 V	1.8 V	1.3 V
	Power noise margin	1.3 mW	25 mW	32 mW	31 mW

twisted pair (Type-1A S TP) and category-5 datagrade unshielded twisted pair (Cat-5 UTP) or equivalent. The receiver is specified with a maximum 700 mV signal detect assert threshold. The signal detect threshold is a measure of the pk–pk differential signal amplitude at the RXI ± inputs at the RJ45 connector of Figure 5.130a required to cause the signal detect differential output to assert. The signal detect de-assert threshold is minimum 200 mV and the signal de-assert is defined as a measure of the pk–pk differential signal amplitude at the RX1 ± inputs at the RJ45 connector of Figure 5.130a required to cause the signal detect differential output to de-assert. The typical hysteresis is therefore 500 mV. The RXI differential input resistance is 7–9 kΩ, and the C/M impedance is not defined. A typical receiver circuit recommended by National is the DP83223 for a 100BASE-TX physical medium–dependent circuit, shown in Figure 5.130a. The data stream is AC coupled from the RJ45-8 media connector to the DP3223 twisted-pair receiver by an isolation transformer. The DP83223 then equalizes the receiver signal to compensate for signal degradation caused by the nonideal

TABLE 5.30

Characteristics of Typical Bidirectional Instrumentation Bus (IEEE-488, GPIB) Transceiver

GPIB Input (Receiver)	GPIB Output (Driver)	Logic State	Voltage or Current	
			Minimum	Maximum
V low		"1"	−0.6 V	+0.8 V
V high		"0"	+2.0 V	+5.5 V
	V low	"1"	0.0 V	+0.4 V
	V high	"0"	+2.4 V	+5.0 V
I low		"1"		−1.6 mA
I high		"0"		+50 μA
	I low	"1"	+48 mA	
	I high	"0"		−5.2 mA

Note: Input impedance terminated = 1.6 kΩ. Input hysteresis = 600 mV. Typical noise margin: high to low, 1.5 V; low to high, 1.2 V.

transmission-line properties of the twisted-pair cable. The resistor values shown in the schematic are chosen to match an insertion loss of approximately 0.4 dB, which with the 12.1 and 37.9 Ω dividers will result in a final amplitude of approximately 1.45 V pk–pk differential as seen across the RXI inputs for zero meters of cable and with a standard 2 V input. The noise immunity at the RXI input is 0.75 V and at the RJ-45 input is 1.03 V for zero meters of cable. An 0.01 μF capacitor is included as a C/M noise filter. This, along with the 37.9 Ω resistors and the isolating transformer primary to secondary capacitance, will determine the level of C/M voltage developed at the receiver input. If available, an isolating transformer with an electrostatic shield between primary and secondary and connected to ground will reduce the capacitive coupling significantly. The alternative receiver circuit described by National Semiconductor and required to allow full support of autonegotiation is shown in Figure 5.130b. In this circuit, the C/M termination is immediately after the RJ45 connector. The impedance to C/M currents presented by the transformer winding on the RJ45 connector side is low, for the C/M currents flow out of phase through the transformer. The total receive C/M impedance is therefore a 49.9 Ω resistor and a 0.01 μF capacitor. The C/M noise power required to generate a specific voltage is almost certainly much higher than Figure 5.130a, where the C/M input impedance through the primary–secondary capacitance should be high. The problem with the Figure 5.130b circuit is that the receiver-to-ground impedance is unknown, and probably high, which means that neither the isolating transformer nor the common inductor is as effective at reducing C/M noise as in the Figure 5.130a circuit. At first glance, the transformer and C/M inductor in Figure 5.130b should effectively attenuate C/M noise currents flowing out on the interface, and this is true when the 0.01 μF 2 kV capacitor is connected to the enclosure close to the RJ45 connector. On no account should the 0.01 μF capacitor be connected to digital ground, which will have C/M noise impressed upon it and through the 0.01 μF and 49.9 Ω resistor out on the twisted-pair cable.

The DP83223 driver transmit current is from 38.2 to 41.8 mA, and this is a measure of the total differential current present at the TXO ± outputs into a standard 50 Ω differential load. In Figure 5.130a, the R_{ref} resistor is adjusted to achieve the standard specific 2.0 V differential TXO ± output amplitude as measured across the RJ45 transmit pins.

DP83223A Twister can also be used for binary or three-level signaling at data rates up to 155 Mb/s. The ATM Forum Physical Layer Subworking Group has drafted a working document that specifies the protocol for 155 Mb/s STS-3c signaling over 100 Ω Cat-5 UTP (unshielded twisted pair) or 150 Ω STP (shielded twisted pair) cabling, and the DP83223A TWISTER transceiver is

FIGURE 5.130 (a) 100Base-TX driver and receiver connections to a Cat-5 UTP 100 Ω cable. (b) 100BaseTX driver and receiver circuit for full support of autonegotiation. (c) 155-Mb/s ATM recommended transmitter and receiver using the National Semiconductor DP83223. (Courtesy of National Semiconductor.)

compliant with this document. Figure 5.130c provides a typical circuit used for either 100 Ω UTP or 150 Ω STP cable. The step-up transformer in the 150 Ω version of the receiver matches the 100 Ω receiver input resistance and the 100 Ω transmitter output impedance to the 150 Ω STP cable. The receiver must exhibit a C/M noise immunity of 1 V pk–pk from 0 to 155 MHz when injected via a 75 Ω resistor into the center tap of the receiver side of the isolation transformer. With the 1 V C/M test level, the bit error rate will be less than $1 - 10^{-10}$. The transmitter differential output with the 100 Ω UTP load will be from 960 to 1060 mV and for the 150 Ω STP load will be from 1150 to 1300 mV. A channel reference model is used to characterize the link between the transmitter and the receiver, and the receiver will function and meet link attenuation and a maximum 20 mV near-end crosstalk requirements with this mode. For the category-5 100 Ω UTP system, the channel reference model is defined as 90 m of category-5 UTP cable, 10 m of category-5 flexible cords, and four category-5 connectors. The channel reference model for the 150 Ω STP is defined as 90 m of STPA cable, 10 m of STP-A patch cord, and four STP-A connectors. The unshielded UTP system can meet FCC and CISPR Class B limits when the transmitted signal imbalance (C/M component) is minimized, when the category-5 cable is used, and when each line is referenced to a low noise ground. As with all interfaces, C/M noise current flowing out on the cable can be reduced by adding the balun at the transmitter, as shown in Figure 5.130a. The C/M immunity of the receiver can be improved by adding the balun at the receiver, as shown in Figure 5.130c. Additional receiver C/M noise immunity and reduced C/M current on the interface can be achieved by connecting all unused pairs in the cable to a low-noise chassis ground via a capacitor. This configuration is also effective at reducing radiation due to C/M currents on the cable.

Where a very high-EM ambient is predicted or where there is a requirement for ground-to-ground isolation in the megaohms in circuits with multiple interfaces, either interface circuit with very high input impedance and high noise immunity or fiber-optic links must be used.

There are commercially available fiber-optic links that provide a direct conversion of a single serial TTL data channel into a fiber-optic link and conversion back to a TTL level at the receiver. Other fiber-optic links are available with additional associated electronics that are transparent to the GP-IB or RS232 data buses. No current will flow on a fiber-optic link due to radiated or conducted noise. However, if the EM environment is very severe, it may be important to reduce the radiated coupling through the aperture in the enclosure provided for the fiber-optic cable. This reduction may be achieved by making the aperture into a waveguide below cutoff, as described in Chapter 6 on shielding.

The following examples are of non-fiber-optic analog and digital interface circuits. The general requirements for these interfaces are

Low power consumption
Very high C/M noise immunity (for use in a high-noise environment or over long distances)
Good input isolation and in some instances well-balanced differential input and output circuits

All of the circuits are short-circuit proof to 0 V, with some circuits, designated *protected*, short-circuit protected to +10 V at 200 mA. The circuits operate at different speeds, ranging from DC-50 kHz, DC-200 kHz, or DC-2 MHz to 150 kHz–17 MHz.

5.5.2.1 VMOSFET Driver and Optocoupler Receiver PLD2 and PLR2 DC-50 kHz Interface

The circuit in Figure 5.131 provides the maximum degree of isolation between grounds with 1.5 GΩ at 1500 V, input–output capacitance of 1 pF, and a C/M transient noise immunity of 1000 V/μs. Although designed to use 125 Ω characteristic impedance twisted shielded pair cable, the interface circuit receiver load impedance is not an exact match to 125 Ω. Even so, because of the slow and controlled driver rise and fall times, no reflections were seen when using the interface circuit with 100 m of deliberately mismatched 75 Ω impedance cable. The relatively slow driver rise and fall

FIGURE 5.131 PLR2 and PLD2 circuit. (Courtesy of Canadian Astronautics Ltd., Ottawa Ontario, Canada and the Canadian Space Agency.)

times of 2 µs have the secondary advantage of reducing radiated emission from the interface cable. The PLR2/PLD2 interfaces are ideally suited for critical control lines or slow data buses used in a very-high-noise environment or over long distances. The interface has been tested with 100 m of interconnection cable and remains within the following specifications over the temperature range 0°C to + 70°C.

PLD2 and PLR2 Typical Performance

Note: All parameters tested from 0°C to + 70°C.

C/M noise immunity: A Gaussian noise generator was connected between the driver and receiver grounds in an attempt to induce errors in the output signal from the LM 139 comparator. The output waveform in particular was examined to ensure that "double edging" did not occur at the output waveform transition points. With the following peak-to-peak noise voltages covering the corresponding bandwidths, neither double edging nor jitter was observed.

Noise bandwidth pk–pk voltage measured between *Tx* and *Rx* grounds is as follows:

20 kHz	25 Vpk–pk
500 kHz	14 Vpk–pk
5 MHz	9 Vpk–pk

Maximum speed: 100 kHz with $C_1 = 0.01$ µF and 100 m of cable

Driver output: Single ended into 300 Ω impedance

PLD2 S/C protected to +10 V and ground

Receiver input: Opto-coupler PLR2 S/C protected to +10 V.

5.5.2.2 Transformer Coupled 10 MHz Interface HAD-1 and HAR-1

The HAD-1 and HAR-1 interface, shown in Figure 5.132, is transformer coupled, and thus, the data/clock must achieve a specific mark–space ratio to avoid saturation of the input transformer. The CLC103 AM device is no longer available and can be replaced by a video amplifier IC followed by one of the many high-speed video buffer ICs that are specifically designed to drive coaxial cables. This interface is suitable for the transmission of pulse width modulated or Manchester coded data over long distances.

5.5.2.3 VMOSFET Driver, Differential Input Receiver, 2 MHz Interface: HD-1, HR-1, and PHR-1

This circuit, shown in Figure 5.133, provides for high-speed transmissions over long distances with high noise immunity. A well-balanced input providing 1 MΩ isolation in parallel with approximately 10 pF and 1 kΩ in series to ground and C/M voltage capability of +10 V characterizes this circuit. It is important to use 120–130 Ω twisted shielded pair cable to ensure minimal line reflections as both the source and load impedances of this circuit are 125 Ω. Maximum cable length tested was 100 m. Because of the inherent high-current capacity of the 2N6661 VMOSFET, it is necessary only to increase the wattage rating of the two 62 Ω resistors to 2 W to achieve protection against a S/C to +10 V as well as to ground. The maximum speed of operation with 100 m of cable is 2 MHz with an adequate safety margin. The measured rise and fall times

FIGURE 5.132 HAD-1 and HAR-1 circuit. (Courtesy of Canadian Astronautics Ltd., Ottawa Ontario, Canada and the Canadian Space Agency.)

FIGURE 5.133 HD-1 and HR-1 circuit. (Courtesy of Canadian Astronautics Ltd., Ottawa Ontario, Canada.)

of the driver output were both 20 ns when loaded with the receiver input via 3 m of 120 Ω twisted shielded pair cable. The receiver contributes 150 ns to the overall transmitter input–receiver output delay, which was measured at 180 ns.

D/M voltage noise immunity as follows:

High = 2.5 V
Low = 2.5 V

Input power noise immunity is 50 mW.

C/M voltage noise immunity: The C/M noise level at which double edging occurred is recorded in the following table (the cause of double edging was a combination of the C/M and approximately 100 mV of different mode noise present).

The broadband noise voltage level is as follows:

Noise Bandwidth	pk–pk Voltage
20 kHz	20 V
500 kHz	10 V
5 MHz	8 V

Driver output, differential above ground into 125 Ω load impedance protected against +10 V and short-circuited to ground with 62 Ω resistors rated at 2 W. The 2 MHz receiver may be used as an analog receiver with the LM119 replaced with a differential amplifier. A differential video amplifier version of the interface has been built, with a measured CM noise rejection of 40 dB at 30 MHz.

5.5.3   TYPICAL INTEGRATED CIRCUIT RESPONSE TO NOISE AND IMMUNITY TEST LEVELS

The available data on the immunity level of ICs describe the input voltage or power at which upset occurs, that is, the device malfunctions or some parameter such as DC offset, gain, frequency changes, or a spurious response occurs.

With the conducted C/M RF and radiated RF immunity tests specified in EN 50082-1 and EN 50082-2, the equipment is exposed to an incident RF field modulated at 1 kHz and 80% and to a C/M RF (bulk current) cable injected test level, also modulated at 1 kHz and 80%. Either of these injection methods will typically result in a C/M RF current flow across a PCB. Despite the C/M nature of the current, IC junctions can be affected by the radiation setup by the current flow, which in turn results in current flow in the semiconductor junction. Also, C/M current is easily converted to D/M voltage when the circuit is unbalanced, and the C/M current can induce a noise voltage into the device.

The spurious response is typically demodulation of the audio frequency used to modulate the incident field or the bulk RF current injected into cables. In RF circuits, the spurious response appears as either FM of the received or transmitted carrier, or "close in" or "far out" (from the carrier frequency) responses or an increase in broadband noise, or receiver desensitization. Often, a combination of these problems is seen.

Any semiconductor device can rectify an RF signal, and devices can be characterized by their rectification efficiency. For example, bipolar devices, rectification efficiency is proportional to $1/\sqrt{2\pi f}$, whereas MOS devices exhibit an efficiency of $(1/2\pi f^2)$. Therefore, due to the higher rectification efficiency of the bipolar device for the same emitter perimeter size, it tends to be more susceptible than MOS devices.

The rectification effect is also a function of the f_T of the device (f_T is the frequency at which the small-signal gain of the device equals unity). Above f_T, EMI through the rectification effect of the device manifests itself as audio rectification, in which the output of the device follows the envelope of the input RF.

Also, frequencies well above the f_T of the device may merely be transmitted through the device and can cause EMI to a device further down the chain.

If a PCB is not contained inside a shielded enclosure, the EN 50082 radiated immunity test field is incident on the PCB. We assume an 18 cm long PCB layout in the form of a transmission line (without ground plane) and an 18 cm long microstrip PCB (signal trace above a ground plane). If the incident E field is 3 V/m amplitude modulated at 1 kHz and 80%, what is the effect on typical digital and analog devices? The first step is to predict the power level and voltage level induced in the transmission line and microstrip PCB layout. Table 5.31 shows the induced power calculated from the measured gain of the 18 cm microstrip and the 18 cm transmission line. Table 5.32 shows the induced voltage based on the induced power, the structure, and the load impedance. The load impedance affects the voltage induced in the circuit. The load impedance for the microstrip is calculated as 50 Ω at 30 MHz, 47 Ω at 160 MHz, and 40 Ω at 1 GHz, which is approximately correct for RF and video circuits but is almost certainly too low for analog circuits. Table 5.32 shows the induced voltage for both the ≈50 Ω load and for a 10 kΩ load in parallel with a 0.4 pF capacitor. The 10 kΩ in parallel with the 0.4 pF load has an impedance of 10 kΩ at 30 MHz, 2 kΩ at 160 MHz, and 390 Ω at 1 GHz.

TABLE 5.31

Power Induced in Microstrip and Transmission-Line PCB Layouts with an Incident 3-V/m Field

	Power (W)	
f (MHz)	T.X. Line	Microstrip
30	9.6×10^{-8}	9.6×10^{-10}
160	84×10^{-6}	8.4×10^{-8}
1000	89×10^{-6}	1.1×10^{-6}

TABLE 5.32

Voltage Induced in Microstrip and Transmission-Line PCB Layouts

f (MHz)	T.X. Line	Microstrip
Voltage induced in 50 Ω load		
30	2.3 mV	0.26 mV
160	20 mV	1.4 mV
1000	6.3 mV	9 mV
Voltage induced in 10 kΩ ‖ 0.4 pF		
30	31 mV	0.26 mV
160	320 mV (520 mV)[a]	13 mv
1000	80 mV	19 mV

[a] This is a corroborating value obtained by the use of a high-level electromagnetic analysis program.

5.5.4 DIGITAL LOGIC NOISE IMMUNITY

Logic noise immunity is usually expressed as a DC voltage, which is a valuable measure of noise immunity for a nonimpulsive LF-induced noise. The voltage noise immunity of a device is often specified as the voltage at which the device becomes susceptible, and therefore, immunity is achieved at levels below this voltage.

Where the induced noise consists of voltage spikes with a duration equal to or just a little longer than the rise and fall times of the logic (which is typical of the crosstalk between logic tracks on a PCB), the input noise energy is a useful immunity criterion. In addition, the immunity of logic to an RF signal is often required. The input noise energy immunity (defined as the voltage multiplied by the current for a specific pulse duration, i.e., that required to change the logic output state) is an important factor in comparing the noise immunity of different logic types. Typically, voltages much higher than the DC noise immunity level are required for short-duration pulses before an output response is evoked. Table 5.26 shows DC noise margin, the noise energy immunity, and signal-line impedance or the input/output impedance in parallel.

The energy noise immunity expressed in Table 5.26 is valid only for the specified pulsewidth. For example, from the predictions of crosstalk to a TTL gate in Chapter 4, the gate responded to a 2.5 V, 80 ns pulse, which produced a 1.6 mA current at the TTL input. Thus, the input noise energy immunity of the gate with an 80 ns input pulsewidth is 0.24 nJ, whereas, from Table 5.26, the input noise energy immunity at a 20 ns pulsewidth is 1.7 nJ.

From measurements, the immunity of a CMOS gate, type CD4013B, and an advanced low-power Schottky, type 54ALS74A, and to an RF sine wave signal from 1.2 to 200 MHz has been determined [6]. The RF was injected into the Vcc pin, the data input pin, and the clock pin of the IC. In the case of the data and clock pins, the RF was combined with data and clock waveforms, so the RF was riding on these waveforms. The ICs were included in a test setup that ran a functional test on the IC. At the same time, the level of RF voltage injected into the IC was increased until the functional test program failed. Table 5.33 shows the RF voltage at which an upset occurred (i.e., a change in the normal operating mode of the IC). Two units of each device type were tested for RF upset susceptibility.

Figure 5.134 shows both the calculated and the measured input power required to upset digital devices with different values of f_T.

The induced power calculated for the transmission line and microstrip PCB layouts shown in Table 5.28 is also plotted in Figure 5.134. The predicted induced power is far below that required to cause an upset to TTL or ECL 10K, although, for VHSIC TTL or ECL, the power induced into the transmission line or into the microstrip PCB at 3 V/m could cause an upset. The major difference between the standard TTL and the phase I VHSIC is that the TTL has an emitter perimeter of 12.5 µm and the VHSIC I devices have emitter perimeters of 1.25 µm. The phase II VHSICs have linewidths of 0.5 µm. To reduce the probability of an upset to VHSICs at 3 V/m, reduce the length of the microstrip or use a stripline PCB layout.

Open pins on logic ICs degrade noise immunity. For example, with a 2 V 2 ns wide pulse applied to the input of a 7420 device, the output pulse is 2.2 V with a 10 ns pulsewidth. Clamping the unused inputs reduces the output pulse to 1 V with an 8 ns pulsewidth. For standard saturated logic types, the power induced into the 18 cm long PCB s when illuminated at 3 V/m will not result in an upset.

Some of the latest logic types include ABT, LVT, LCX, LVX, LVO, AC, ACT, ACQ, ACTQ, VHC, HC, HCT, GTLP, and FAST. No RF or noise energy immunity data are readily available for these types of logic; however, we have seen the importance of the emitter perimeter size in determining noise immunity. If a number of devices are suitable for an application and noise immunity is a potential problem, a comparative test of the noise energy immunity can be performed. The test involves varying the high- and low-input voltage level at different input pulsewidths and recording the energy at which an upset occurs.

The noise immunity of standard digital logic may be increased when the signal path is electrically short, that is, propagation delay is short and when crosstalk is capacitive by low-impedance

TABLE 5.33

Radio Frequency Upset Voltage of a CMOS and ALS Devices

	ALS 74		CD4013	
f (MHz)	S/N1 (V)	S/N4 (V)	S/N1 (V)	S/N2 (V)
Vcc pin upset voltage				
1.2	0.44	0.34	0.006	0.006
5	0.4	0.4	0.4	0.28
10	0.44	0.4	0.44	0.4
50	0.44	0.4	1.1	1.1
100	0.6	0.6	2.8	3.4
200	0.44	a	a	a
Data pin upset voltage	(mV)	(mV)	(mV)	(mV)
1.2 MHz	0.0011	0.0010	0.0004	0.0005
5 MHz	0.0022	0.0022	0.0009	0.0012
10 MHz	0.0017	0.0031	0.0024	0.0028
50 MHz	0.0044	0.0040	0.0120	0.0140
100 MHz	0.0070	0.0100	0.0220	0.0400
200 MHz	0.0070	a	a	a
Clock pin upset voltage				
1.2 MHz	0.0008	0.0008	0.00044	0.00035
5 MHz	0.0008	0.0008	0.0015	0.0012
10 MHz	0.00085	0.00085	0.0028	0.0040
50 MHz	0.0022	0.0020	0.0170	0.0200
100 MHz	0.0044	0.0040	0.0200	0.0440
200 MHz	0.0070	0.0055	a	a

a Denotes that no upset could be induced.

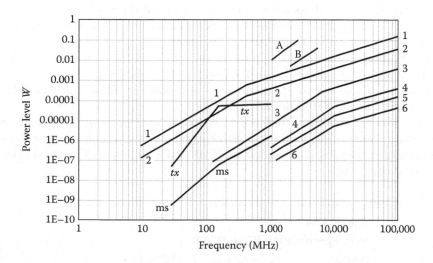

FIGURE 5.134   A, measured TTL level; B, measured ECL 10 K level. 1, Calculated TTL f_T = 350 MHz; 2, calculated ECL f_T = 350 MHz; 3, calculated VHSIC I TTL f_T = 350 MHz; 4, calculated VHSIC I TTL f_T = 5 GHz; 5, calculated VHSIC I ECL f_T = 5 GHz; 6, calculated VHSIC II ECL f_T = 5 GHz. tx, Induced power into 18 cm transmission PCB, E = 3 V/m; ms, induced power into 18 cm microstrip PCB, E = 3 V/m.

drivers and terminating networks. The signal path is electrically long or crosstalk is inductive using low-impedance drivers may be either non- or counterproductive.

The SSTL differential input logic interface standard is intended for operating frequencies of up to 200 MHz. A typical output buffer will have a 25 Ω series resistor and a 25 Ω resistor termination resistor.

A 2.5 V supply SSTL is defined in JEDEC JC-16-97-58.

Table 5.34 shows the convention for 3.3 and 2.5 V.

BiCMOS integrates bipolar junction transistor and the CMOS transistor in the device. The BiCMOS family includes ABT, ALB, ALVT, BCT, and LVT logic.

The BiCMOS Heterojunction Bipolar Transistor (HBT) allows operation at much higher frequencies, depending on gate density from 43 to >300 GHz. The HD74B series Bi-CMOS.

Low-voltage differential signaling is used for high-speed digital data transfer. It is terminated at the receiver by a resistor. The resistors used at input receivers are typically from 25 to 100 Ω and increase the receivers noise energy immunity = $V^2/R*t$ where t is the pulse duration required for the IC to change state.

The propagation delay of the device does not tell us directly the maximum pulse width before the device changes state but is an indication.

Typically, a current source at the driver develops a voltage across the terminating resistor. As the current source develops a positive and a negative current, a differential voltage is developed across the resistor. For example, with a ±3.5 mA current and a 100 Ω resistor, the voltage is ±350 mV.

Other types of device are the low-voltage TTL (LVTTL, the low-voltage CMOS (LVCMOS), the low-voltage positive emitter coupled logic, the high-speed transceiver logic (HSTL), and the HMC logic.

The Schottky logic family includes advanced ALS, AS, and low-power LS.

The DC noise margin, output impedance, and typical propagation delay are shown in Table 5.35. The dynamic characteristics of the devices are shown in Section 11.6.

The USB operates with differential signals and has three speeds. The full-speed 1.1 is 12 Mbps, the high-speed 2.0 is 480 Mbps, and the superspeed 3.0 is 5 Gbps.

The characteristics of the full-speed and high-speed specification USB driver are shown in Table 5.36 and the receiver in Table 5.37.

As described in Section 7.16, the shielding effectiveness of the USB cable and connector is limited and varies from 0 to 37 dB over the 6–1100 MHz frequency range.

The USB 3.0 superspeed specification at $Vcc = 1.8$ V is shown in Table 5.38.

The Ethernet signal network specified in DIX Ethernet and IEEE 802.3 is designed for LANs and larger networks. The 10base5 uses RG8 cable at a speed of 10 Mbps and can communicate over 500 m/1640 ft per segment with a maximum of 5 segments.

The Ethernet 10base2 uses thin coax over a distance of 185 m/606 ft and the 10baseT UTP uses twisted (T) unshielded 100–150 Ω cable over a distance of 100 m/328 ft.

The categories of cables used and bandwidths are shown in Table 5.39 where UTP is the unshielded twisted pair and STP is the shielded twisted pair.

The Gigabit Ethernet 1000base-T may use unshielded or shielded twisted pair at 1 Gbps.

As described in Section 7.16, the shielded Ethernet cable and RJ45 connector combination only achieve a maximum shielding effectiveness of 35 dB and a minimum of 0 dB over the 30 MHz to 1.2 GHz frequency range.

TABLE 5.34
SSTL Specification JC-16-97-58

Parameter	$VCC = 3.3$ V	$VCC = 2.5$ V
VREF	1.5	1.25
VNH	VREF + 0.2 = 1.7 V	VREF + 0.18 = 1.43 V
VNL	VREF − 0.2 = 1.3 V	VREF − 0.18 V = 1.07 V
DC noise margin	0.4 V	0.36 V

TABLE 5.35
Noise Immunity of Modern Digital ICs

	Typical Propagation Delay	DC Noise Margin (Minimum)	Typical Output Impedance (Ω)		Notes
			Low	High	
LVT TTL/LVCMOS 3.3 V	1.5–4.4 ns	1.2 V	23	100	
HSTL 3–3.6 V	0.85 ns	0.2–1.3 V		130	Differential output Into 50 Ω
LVP ECL 3.3 V	0.1–0.6 ns	0.55 V	51.5	136	Common-mode range min 0.6 V Differential output into 50 Ω
74LVT (BiCMOS) D flip flop 3.3 V	2.9–3.9 ns	1.2 V	20.8	0.6	
74LVT buffer 3.3 V	3.2 ns	1.2 V	66.7	83	With 30 Ω output resistors
HD 74 BC buffer (BiCMOS) 5 V	6 ns	1.2 V	8.6	866	
HMC 3.3 V	98 ps	0.1–2 V differential			Differential output or single ended output
AVC	0.2 ns	1.8 V	Vcc 1.1 267	100	Can be loaded with 50 Ω Inputs terminated in 50 Ω
			1.4 175	66.7	
			1.65 150	56.3	
			2.3 194	61.1	
			3.0 192	58.3	
GTLP	1.5	1.2		125	
AC					
$Vcc = 3.3$ V	5.6				
$Vcc = 5$ V	4.3				
CBT	0.13 ns $Vcc = 2.3$–2.7 V 0.2 ns $Vcc = 3$–3.6 V	1 V	Low = 4.2–40 $Vcc = 2.3$–2.7 V 4–15 $Vcc = 3$–3.6 V		Bus switch
ABT	1–4.5 ns	1.2	Low 67 High 94d		

(Continued)

TABLE 5.35 (*Continued*)
Noise Immunity of Modern Digital ICs

	Typical Propagation Delay	DC Noise Margin (Minimum)	Typical Output Impedance (Ω)		Notes
			Low	High	
ALVT	0.5–3 ns	1 V at Vcc 2.5 ± 0.2 V 1.2V at Vcc 3.3 ± 0.3 V	21	88 Vcc = 2.5 ± 0.2 V 63 Vcc = 3 ± 0.3 V	Primarily bus interface
ALVC	1.5–5.3 ns Vcc = 2.3 V 1.5–4 ns Vcc = 3.3 V	1V Vcc = 2.3–2.7 V 1.2V Vcc = 2.7–3.6 V	23*–83	41*–100	Primarily bus interface *Same drivers
ABTE	1.5–6.4 ns	1.2 V	8.6	8.6	
GTL	2–6.6 ns	H = 1.5 V L = 0.24 0.1 V	A port 23 B port 5	28.6 28.6	Back plane driver Bus transceiver GTL/GTL+ to LVTTL translater
BCT	0.5–10.8 ns	1.2 V	Vcc = 4.5 V 67	167	Line driver
AC	1.5–8 ns	Vcc = 3 V, H = 1.4V, L = 1.5 V Vcc = 4.5, H = 2.11 V, L = 2.25 V	Vcc = 3 V 8.3 Vcc = 5 V 4.2	8.3 4.2	Buffer line driver
ALS	1–10.5	1.2 V	10.5–33.3	150–162	
LS	8–40 ns	1.2 V	21	216	
AHCT	3.8–10 ns	1.2 V	55	70	

TABLE 5.36
USB Driver Characteristics

Driver	VoH	VoL	Crossover Voltage (V)	Output Impedance	Rise Time	Fall Time
Low-speed driver	2.8–3.6 V	0–0.3 V	1.3–2.0	40.5–49.5	75–300 ns	75–300 ns
Full-speed driver	2.8–3.6 V	0–0.3 V	1.3–2.0	40.5–49.5	4–20 ns	4–20 ns
High-speed driver	380–440 mV	−10 to 10 mV			500 ps	500 ps

TABLE 5.37
USB Receiver Characteristics

Receiver	Voltage Low to High (V)	Voltage High to Low (V)	HS Squelch Detector Threshold (mV)	HS Disconnect Detection Threshold (mV)	Data Line Pull-Up Resistance (kΩ)	Data Line Pull-Down Resistance (Ω)
Full speed/low speed	0.8–2.0	0.8–2			1.425	14.25–24.8
High-speed receiver			100–150	−10 to 10	1.425	14.25–24.8

TABLE 5.38
USB 3.0 Superspeed Characteristics

Vin H Min	Vin L Max	VoH Min	VoL Max	Differential Impedance
1.17 V	0.63 V	1.35 V	0.45 V	72–120

TABLE 5.39
Cable Determined Speed and Bandwidth

Category	Shielded or Unshielded	Speed (Mbps)	Bandwidth (MHz)
3	UTP	10	16
4	UTP	20	
5	UTP	1,000	100
6	UTP to STP	1,000	250
6a	STP	10,000	500
7	SSTP	10,000	600

TABLE 5.40
Ethernet Characteristics

Ethernet	Description	Volts	Rise and Fall Times
10Base-T	Input differential accept voltage	585–3100 mV	
	Input differential reject voltage	0–585 mV	
	Vout peak differential	2200–2800 mV	
100Base-T	Input differential accept voltage	200–1000 mV	
	Input differential reject voltage	0–200 mV	
	Output peak differential voltage	950–1050 mV	
1000Base-T	Transmitter differential input volts	500–2400 mV	
	Receiver differential output volts	5500–2000 mV	
	Rise and fall times		100–250 ps
	Input and output differential load	100 Ω	

TABLE 5.41
Firewire Characteristics

Characteristic	Voltage (mV)	Output Impedance (Ω)	Differential Input Impedance	Notes
Differential output voltage	172–265			Output load = 56 Ω
Common-mode signaling S200		36–102		
Common-mode signaling S400		14–32		
Receiver			4–7 kΩ	
SP200 speed signal threshold	49–131			
SP400 speed signal threshold	314–396			

The Ethernet characteristics are shown in Table 5.40.

As described in Section 7.16, the shielding effectiveness of the Ethernet cable and connector is limited and varies from 0 to 50 dB over the 6–1100 MHz frequency range depending on the manufacturer.

The characteristics of the IEEE1394 Firewire interface are shown in Table 5.41.

5.5.5 Analog Video and Radio Frequency Circuit Noise and Immunity

5.5.5.1 Thermal Noise

Analog, video, and RF circuits generate noncoherent noise that is caused by thermal noise in resistors, semiconductors, and ICs as well as $1/f$ noise in semiconductors. Analog circuit noise may be modeled as an input noise current and an input noise voltage and is typically specified by manufacturers at 1 kHz in units of $nV/\sqrt{MHz}$ and $pA/\sqrt{MHz}$. Thermal rms noise voltage in resistors is given by the following equation:

$$E_{nt} = \sqrt{4kTRBW}$$

where
$k = 1.38 \times 10^{-23}$
T is the temperature in degrees Kelvin
BW is the bandwidth of the signal chain in which the resistor is used
R is the value of the resistor [Ω]

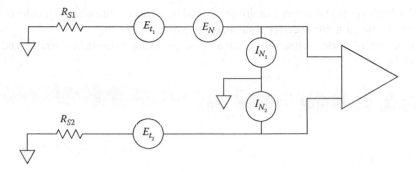

FIGURE 5.135 Noise sources referenced to the input of an amplifier.

The noise at the output of an amplifier may be found by computing the total resistance of the resistors included in each input and calculating the thermal noise voltage developed by the resistors, the noise voltage developed across the resistors by the specified maximum amplifier input noise current, and the specified input noise voltage. Figure 5.135 illustrates the noise sources referenced to the input of an amplifier. The noise sources are considered noncoherent, and the rms sum of the noise sources is found by taking the square root of the sum of the squares of the individual noise sources. The noise is referenced to the input of the amplifier, so the output noise is the input noise times the gain of the amplifier.

In cascaded (i.e., a chain of) amplifiers, the maximum output noise may be computed by finding the noise per stage of amplification and multiplying by the remaining gain of the chain. The sum of the noise sources at the output, N_t, is again the square root of the sum of the squares:

$$N_t = \sqrt{\left(N_1 G_1 G_2\right)^2 + \left(N_2 G_2\right)^2}$$

where

N_1 is the total noise of the first stage referenced to the input
G_1 is the gain of the first stage
G_2 is the gain of the second stage
N_2 is the noise of the second stage referenced to the input

The $1/f$ noise in op amps is significant only at very low frequency, because it reduces to the thermal noise level at typically 6 Hz and, worst case, 30 Hz.

The rms noise voltage is less commonly used by RF engineers to characterize the intrinsic, component, and induced noise in RF circuits; more often, noise power is used. In calculating the total noise power developed by an RF circuit or receiver, it is only necessary to add noise powers algebraically, multiplied by the power gain of amplifier stages and divided by losses, where applicable. One measure of noise is the noise factor, defined as

$$F = \frac{N_o/N_i}{P_o/P_i}$$

where

P_o/P_i is the available power gain of the receiver
N_o/N_i is the ratio of available noise output power and noise input power

The *available* power is that power density developed by a source into a conjugate load. When the ratio N_o/N_i equals the gain, the receiver contributes no noise and $F = 1$.

In measurements of noise factor, very often a noise generator is used as a source and the noise factor is given by

$$F = \frac{P_{i(1/1)}}{kT_oBW}$$

where

$P_{i(1/1)}$ is the available power required from the noise generator to produce a power carrier-to-noise ratio (power in/noise in) of 1

BW is the bandwidth [Hz]

T_o is the approximately 290 K[Å 17°C]

k is Boltzmann's constant = 1.38×10^{-23}

The noise figure of a circuit is either the noise factor expressed in decibels (thus, F (db) = 10 log F) or synonymous with the noise factor, as defined by the IEEE.

An alternative measurement of RF circuit and receiver noise is noise temperature. The noise temperature of a receiver, in kelvin, is given by

$$T_r = FT_o - T_o$$

Therefore, the noise factor is

$$F = 1 + \frac{T_r}{T_o}$$

The limiting sensitivity of a receiving system is determined by the ratio of received power to noise power that allows acceptable reception of information. In satellite communications, a major source of noise is external atmospheric, cosmic, and ground radiation. The antenna is then considered a noise source, due largely to the noise flux incident on the antenna.

The total noise temperature of a receiving system is made up of the antenna noise temperature and the noise temperature of the receiving system up to the demodulator.

Passive components such as connectors, cables, and filters introduce a loss into the receiving system and increase the effective noise temperature. The effective noise temperature due to loss L is

$$T_e = \left[\left(1 - \frac{1}{L}\right)L\right]T_o$$

and the output noise is

$$N_o = \left(1 - \frac{1}{L}\right)T_o$$

A figure of merit used to describe satellite communication systems is gain over temperature (G/T), which is defined as

$$\frac{G}{T} = \frac{(P_o - N_o)/(P_{in} - N_{in})}{FT_o - T_o}$$

Once the antenna noise temperature is known, the receiver noise factor and input losses may be used to obtain G/T. In a cascaded receiver, G/T is constant regardless of the position in the cascade at which G/T is calculated.

As an example, we will calculate G/T of a system after the antenna connecting cable and input filter and at the input of the receiver. Assume an antenna noise temperature T_a of 50 K, a cable loss of 0.15 dB (1.035), a filter loss of 0.3 dB (1.072), and a receiver noise factor of 0.5 dB (1.122). The total loss is 0.45 dB (1.109), and the noise temperature at the receiver input due to antenna and losses is

$$T_i = \frac{T_a}{L} + \left(1 - \frac{1}{L}\right)T_o$$

$$= \frac{50}{1.109} + (1 - 0.902)290 = 45\ \text{K} + 28.42\ \text{K} = 73.42\ \text{K}$$

The noise temperature due to the receiver is

$$T_r = (1.122 \times 290) - 290 = 35.38\ \text{K}$$

Thus, the total noise temperature is $73.42 + 35.38 = 108.8$ K

The gain of the antenna is 30 dBi, from which the filter and cable losses are subtracted, resulting in a gain of 30 dB − 0.45 dB = 29.55 dB (901). G/T is therefore 901/108.8 = 8.28/K = 18.3 dB/K.

Very often, the design engineer makes a careful worst-case calculation of the maximum intrinsic noise of an amplifier at the highest operating temperature but ignores induced noise from the electromagnetic ambient in which the circuit must function. For example, when the circuit is located in an enclosure that contains digital circuits or is powered by a switching power supply, the induced noise may be many orders of magnitude higher than the intrinsic noise.

5.5.5.2   Coupling Modes to Analog/Video and Radio Frequency Circuits

When analog/video/RF circuitry shares a board with digital circuits, the probability of EMI is high. Chapter 11, Case Study 11.1, examines coupling modes between digital logic and analog circuits contained on the same board. When the noise of an analog circuit must be restricted to the intrinsic noise level, careful design and layout are imperative. Ideally, power supplies to analog circuits should be dedicated to the analog circuit and not shared with digital circuits. Linear supplies should be used and not switching power supplies. Where the main power comes from a switching supply used for a number of different circuits, the most susceptible analog circuits should be supplied by a dedicated linear supply. For example, when the switching supply provides ±15 V, this may be used as the input of a ±10–12 V linear supply. Radiated coupling between the input and the output of the linear supply must be controlled in order to reduce HF noise on the output of the supply. It is, nevertheless, possible to use a switching power supply for the most sensitive circuits with the correct secondary power-line filter.

As an alternative to a large and inefficient linear supply, bulkhead-mounted and PCB-mount filters have been designed to allow the use of switching power supplies for critical applications. These filters will reduce switching power supply C/M and D/M noise to typically 30 μV of LF ripple. With this type of filter, noise levels lower than the average linear supply can be achieved. The potential for crosstalk between wires, cables, and PCB tracks, as discussed in Chapter 4, should be examined.

In a very harsh EM environment, compartmentalization of the analog and digital circuits with the use of feedthrough filters may be required. The calculation of noise effects on analog circuits might be considered simply a function of circuit gain and bandwidth, and this is true of noise at in-band frequencies.

The potential for EMI at higher frequencies (i.e., at which the open-loop gain of the amplifier is zero) should not be ignored. Nonlinearities due to input diodes, transistor junctions, or saturation of an input stage due to very high levels of RF can result in demodulation of AM or FM RF. Where the source of RF is unmodulated, DC offsets or gain may be affected.

Analog circuits are not immune to RF noise that is far out of band. For example, the output level of a 1.5 GHz high-power amplifier (HPA) was controlled by the output of a detector. The detector was contained in a shielded enclosure, but an op amp with an f_T of 100 kHz was located outside of the enclosure. The unmodulated field generated by the HPA was rectified by the op amp's input circuit, which changed the output power level. The solution to the problem was to include a ground plane under the op amp and to connect SM microwave capacitors with a low impedance at 1.5 GHz between each of the op amp's input pins and the ground plane. This solution has been effective in a number of similar EMI cases.

As an example of analog-circuit susceptibility to RF, a 741 op amp configured with a gain of 20 and with RF injected directly into either the inverting or noninverting input first exhibits a change in the output level at an input power of −20 dBm (22.5 mV) and is saturated (i.e., close to the rail voltage) at 0 dBm (225 mV).

A common practical example of demodulation EMI is when a CB radio is operated in close proximity to wireless devices or audio preamplifiers. Often, the amplitude-modulated RF of the CB communication is demodulated by the input stage of the phonograph preamplifier, and the communication may be heard by the listener. The Nonlinear Circuit Analysis Program has been used to predict the demodulation EMI effects in bipolar, JFET-bipolar, and MOSFET-bipolar ICs [7].

The addition of capacitors on analog or video circuit pins has the effect of reducing the impedance of the load at the problem frequencies. In conjunction with series resistance or inductance, an L-type filter can be formed and the attenuation is higher, especially when the noise source impedance is low. Very often, the addition of a capacitor does not reduce the EMI response of the circuit. There are a number of possible reasons for this. The most common are as follows: a different component or part of the circuit is susceptible; the capacitor is not the right value or type for the problem; the noise source impedance is too low; the capacitor is too far away from the rectifying semiconductor junction (it must be connected directly to the pin/s of the device); and the grounding around the device/s is not a sufficiently low impedance. The capacitor is most effective when its resonance frequency is at, or close to, the problem frequency.

Different analog devices exhibit different levels of immunity, and a comparison can be made based on measured data. The analog devices tested were the 741, OP27, LM10, LF355, and CA081 op amps. These devices were connected as an inverting amplifier with a gain of 10. The value of the input resistor used was 10 kΩ and that of the feedback resistor was 100 kΩ. Above some frequency, the input capacitance of the device will reduce the level of RF, due to the low impedance of the capacitance, or when a series resistance is placed between the RF source and the input pin of the device, the resistance/input capacitance will act as a filter.

Figure 5.136 (from Reference [11]) shows the comparison of the response of the inverting input to the noninverting input of a 741 device. Figure 5.136 plots $1/x$, where x is the circuit response. Therefore, the higher the value in decibels, the lower the circuit response. Figures 5.136, 5.139, and 5.140 are provided for comparison purposes. Adding additional capacitors around the op amp, as shown in Figure 5.137, will decrease the response of the circuit to RF. An alternative filter for both C/M and D/M noise is shown in Figure 5.138.

FIGURE 5.136 Comparison of 20 log $1/x$ for inverting and noninverting inputs of the 741. x is the operational amplifier response and is dimensionless. (Copyright 1991, IEEE.)

FIGURE 5.137 Additional capacitors around an operational amplifier circuit.

Figure 5.139 shows the comparison of the demodulated output of the device with different values of capacitor at locations C_1 and C_2.

The lower the value of the response shown in the figures, the more effective the filtering. Figure 5.140 shows the comparison of the input RF voltage for a fixed-amplitude demodulated response for different devices, and here the higher the level of RF voltage for a standard response, the more immune is the op amp type.

FIGURE 5.138 A common-mode and differential-mode filter at the inverting input of an operational amplifier.

Most wireless designs have a performance specification that includes phase noise, FM, and spurious response. For example, a commercial aeronautical system requirement specifies that all fixed spurs (those that do not change with the carrier frequency) and all spurs remaining within 42 kHz of the carrier frequency must be at least 70 dB below the carrier (−70 dBc). Any spurs outside of the 42 kHz band that move with the carrier frequency are limited to −55 dBc. Other communication requirements (such as those for EW simulators) place the spurious limit at approximately −60 dBc. The EIA/TIA-250C specification on the ratio of video signal to periodic noise from 300 Hz to 4.2 MHz is 67 dB for short-haul links, 63 dB for satellite links, and 58 dB for long-haul links.

The use of digital receivers and digital signal processing (DSP) and switching power supplies for the RF circuits makes the requirements increasingly difficult to meet. Generally speaking, if spurs close to the carrier must be 30–40 dB below the carrier (−30 to −40 dBc) and the RF circuit is in close proximity to sources of emission such as switching power supplies and PCBs containing logic, then care must be taken in the design of the wireless for compatibility. In the case where the specification requires spurs to be at −50 to −70 dBc, extreme care must be taken, especially if signal levels are low. The following guidelines are based on achieving compatibility for RF circuits that can be located as close as 2 cm to sources of high-level emissions and where spurs must be at levels between −30 and −40 dBc.

High-level in-band spurs, typically generated by clock and data frequencies, will often make communications on the chosen channel unreliable to the point of being unusable (a denied channel). If other channels are available, then they can be selected. The alternative is to move the clock frequency out of band or to wiggle, sweep, or hop the frequency, as described later. Another problem caused by broadband noise sourced by switching power supplies or data with a HF component at a low repetition rate is the desensitization of the receiver. In one example, fire trucks and ambulances fitted with a new data terminal were found that voice communication by radio to the base station was impossible when the vehicle was in the mountains. Yet, with the data terminal switched off, normal voice communication resumed. The problem is that in the mountains the radio's received level was low, and the broadband emissions from the data terminal, picked up by the vehicle-mounted antenna, caused the radio to desense, resulting in the loss of communication.

With PCMCIA wireless cards, the inclusion of the card in a PC can result in receiver desensitization due to broadband noise generated by the PC. Motorola has developed a tool that, unlike spectrum analysis measurements, provides an accurate measurement for wireless devices that account for the effects of the air protocol, the PC card antenna, and the interaction between the card and the host, typically a laptop computer. The system measures how much RF level is required to produce successful massaging. It emulates wireless wide-area network operation using packet data to determine the receiver sensitivity of the wireless card while inserted into the host PC card slot. As with system coverage, if the host computer is noisy, more RF signal is required to receive a message.

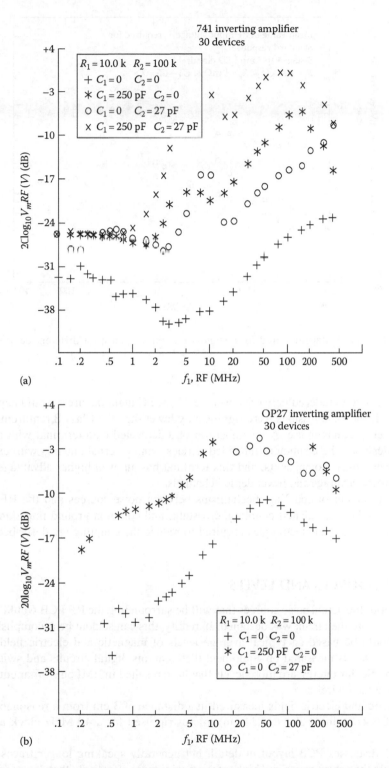

FIGURE 5.139 (a) Comparison of the demodulated output of a 741 with and without additional capacitors. (b) Demodulated response at the output of an op 27 with and without additional capacitors. (Copyright 1991, IEEE.)

FIGURE 5.140   Input voltage required for a standard output response for different devices. (Copyright 1991, IEEE.)

This would translate to reduced network coverage. Thus, PC manufacturers would have a competitive advantage if the PC emissions were significantly lower than the Class B requirements over the wireless frequency bands in use. A manufacturer of a dedicated data terminal with an imbedded wireless has designed the computer, imbedded wireless, and external antenna with extremely low coupling between these components, and this terminal has an even higher advantage in terms of increased network coverage and fewer denied channels.

There are a number of coupling mechanisms between noise sources and the RF circuit, and these include conducted, radiated coupling, crosstalk, and common ground impedance coupling. These potential sources and techniques required to reduce the coupling are described throughout this book.

5.6   NOISE SOURCES AND LEVELS

In an ideal world, the actual noise sources that will be surrounding the RF PCB would be measured for emissions before the circuit was designed. In reality, this can seldom be accomplished; instead, the design should be based on typical average levels of magnetic and electric field and typical conducted noise levels measured on equipment that contains digital circuits and switching power supplies. Of particular interest are those levels that have resulted in EMI on similar equipment containing an imbedded wireless.

The magnetic and electric fields measured in a distance of 3 cm from a reasonably well-laid-out digital PCB containing short (2.54 cm) microstrip used for a 40 MHz clock are shown in Table 5.42.

Chapter 11 discusses PCB layout in detail, but generally speaking longer traces will radiate more effectively than short traces and higher frequencies more effectively than lower frequency (see Chapter 11 for the limits of frequency and trace length). For example, a 5 cm long microstrip with a 10 MHz clock radiated at a lower level than the 2.54 cm track with the 40 MHz clock. Compared to the microstrip, the use of a stripline will reduce the level of the emissions from 15 to 25 dB over

---

TABLE 5.42
Levels of *E* and *H* Fields Close to a "Typical" Digital PCB

f (MHz)	E_0 (mV/m)	H_ϕ (mA/m)
Emissions 3 cm from a PCB microstrip layout, 40 MHz clock, 2.54 cm length		
100	25	0.23
360	83	1.0
950	8.5	—
5 cm of ribbon cable with 10 MHz clock		
360	301	3.6

---

the 20–600 MHz frequency range. In practice, where a stripline PCB layout is used with the upper and lower ground plane stitched together around the board with vias 3 mm apart, the radiation from the PCB is typically dominated by emissions from the ICs and PCB connectors. If the PCB located close to the wireless PCB has yet to be designed and is under the control of the wireless manufacturer, the "good" PCB layout described in Chapter 11 should be followed. One additional tip is to add a low-value decoupling capacitor in parallel with all of the existing 0.01–1 μF capacitors in the digital section of a PCB or adjacent digital PCB. The additional decoupling capacitors should be chosen to resonate at the center frequency of the wireless. Digital sections on the wireless PCB, such as DSP, should be laid out using the following considerations:

1. Keep oscillators, clock and data bus traces, and high-speed LSI chips as far away from the RF section as possible. Locate ICs so that the interconnects are as short as possible. Locate oscillators close to ICs that use the clock. Keep LSI devices, which have fast data buses and clocks, and switching power supplies as far away as possible from signal *and* power connectors and ribbon cable.
2. Use stripline over the digital section. If that is not possible, use localized stripline to imbed any digital and power interface traces routed to the wireless section.
3. Follow the grounding rules in Section 11.8 and avoid slots in the ground plane located directly under signal and clock traces, as shown in Figure 5.141.

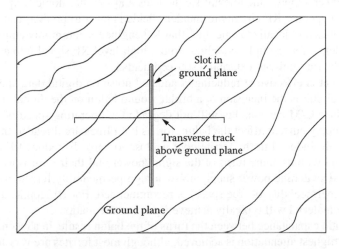

FIGURE 5.141 Ground plane with slot, formed by a row of vias, directly under data bus or clock.

4. Keep traces between ICs in the DSP section as short as possible. Minimize path lengths between sources and loads and between loads sharing a common clock or bus.

5. Keep "hot" signals as close to the signal return plane as feasible and do not "weave" these traces through the ground plane or power plane from layer to layer except for short and necessary sections at ICs. When penetrating a power plane add ground vias around the signal via.

6. Reduce the drive current on tracks by minimizing loading, typically by the use of buffers at the load end or series resistance at the source end.

7. Reduce the number of signal tracks that change state simultaneously.

8. Design logic and software such that data are flowing on data buses intermittently, with as low a duty cycle as possible, and not continuously. If data are transmitted and received in packets with considerable gaps, then, if possible, cease all unnecessary computer operations during transmissions.

9. If digital clocks are designed to wiggle in frequency, sweep or hop then the occurrence of in-band spurs can be reduced. Alternatively, if a clock is derived from a DCO, VCO, or phase-locked loop and the tuned frequency of the wireless is known, choose the clock frequency to generate spurs out of the wireless band or tuned channel.

10. Use an overall shield over the RF section, with a separate "enclosure" ground plane as described in Section 5.5.2 and Chapter 11.

5.6.1 RADIO FREQUENCY AND WIRELESS

The levels described in Table 5.42 can create a problem with an RF circuit; this was illustrated by a receiver that exhibited spurs at 70 kHz and at 28 MHz (with a repetition rate of 70 kHz). The source of these spurs was a switching power supply that generated a magnetic field incident on the RF section of the board of 13 mA/m at 70 kHz and 0.2 mA/m at 28 MHz. The problem was a 2 cm by 2 cm loop of PCB trace on the RF PCB that connected the oscillator signal to a mixer. The current flow in this 2 cm loop was calculated at approximately 30 μA at 70 kHz and 17 μA at 28 MHz. Shielding of the mixer and oscillator signal traces was only partially successful, and not at all at 70 kHz. The potential solution to this and other susceptibility problems are to keep PCB signal traces close together, to minimize loop area, or, better still, to imbed the oscillator traces in stripline. The direct coupling into the loop at 70 kHz can be reduced by capacitively coupling the oscillator so that the capacitor presents a low impedance to RF and a high impedance to audio frequencies and to the switching power supply fundamental frequencies and low harmonic frequencies. Although the susceptibility of specific RF circuits to incident fields is hard to predict, using all the "good" PCB layout techniques and localized shielding that RF engineers commonly employ to avoid self-compatibility problems, for example, coupling from a high-level RF signal into a low-level signal, will also increase the immunity to externally sourced fields.

One technique that is effective at reducing conducted noise on digital data and clock lines that interface to an RF circuit is the inclusion of a bifilar wound balun on the digital lines and return/s. The balun will reduce C/M currents but will not degrade the operating speed of the digital lines. Nor does the DC supply current affect the balun, for, as both lines are threaded through the balun, the DC magnetic field induced in the ferrite core is close to zero. To reduce LF currents, the ferrite should be wound with as many turns of the signal/power and their returns as possible. For LF NRZ data noise and switching power supply noise, a high-permeability ferrite of typically 10,000 μ_o (where μ_o is the permeability of free space) is recommended. For HF applications (>5 MHz), a lower-permeability material will typically achieve a greater attenuation.

When the parasitic capacitance between the turns of the balun results in a resonance at the problem frequency, the highest attenuation is achieved, although most ferrites are very low Q (lossy) and this is not a very significant effect.

If digital control and data lines are connected via a D subconnector to RF components or subassemblies contained in a conductive enclosure, the use of a filtered connector or filter adapter can reduce both C/M and D/M noise.

AC power at 50–400 Hz and their harmonics can also result in spurs, typically diagnosed as FM in a carrier. No effective shielding exists at these frequencies, short of high-permeability steel or mu-metal, and the best approach is to move the RF circuit as far away from transformers, AC fans, and AC power wiring as possible. All power wiring should be twisted to cancel the magnetic fields generated. As an extreme measure, the AC field may be canceled or bucked by the use of a coil through which the AC power current flows. The coil is then orientated to cancel the AC field incident on a susceptible component or circuit. The use of coupling capacitors that have a low impedance at RF and a high impedance at the power-line frequency should also be used. In one case, the AC power current picked up on a loop formed by a semirigid cable, and an RF circuit mounted on a metal enclosure was shunted to chassis by a clamp placed over the semirigid cable. This measure reduced the AC power current flow into the SMA connector mounted on the enclosure.

5.6.2 SHIELDING

Past experience has shown that when RF circuits are in very close proximity to the fields generated by logic and switching power supplies, EMI can and often does exist. Shielding can, when implemented in conjunction with signal and power frequency filtering, be very effective. However, when signal and power connections are routed from the "noisy" external ambient into the shielded RF section of the circuit without filtering, the results may be no increase in immunity at all! Thus, it is not enough simply to shield a circuit; it invariably requires additional filtering at the shield interface, as described in the filtering section.

All digital logic that switches nonreturn to zero data will radiate fields with a low-audio-frequency component, and thin, small, shielded enclosures will provide 0 dB attenuation at these frequencies. The only effective technique is "good" RF PCB layout and the use of low-value capacitive coupling between stages of the RF. Chapter 11 describes shielding a PCB in detail. We see that for a small enclosure made of thin material, the enclosures may be adequate at switching power supply frequencies of around 100 kHz. The weak link in any shield is at the seam between the separate components of the shield. In the PCB-mounted shield, this is the electrical connection between the shielded enclosure and the ground plane on the opposite side of the board. When the enclosure is constructed using a shield fence into which lids are clipped and the connection to the ground plane relies on the tabs in the fence material, the shielding effectiveness is compromised. The best solution to this problem is to include an upper "enclosure" ground plane in the form of a loop approximately 3–5 mm wide on the fence material side of the board. This upper loop is not connected to any other (digital/RF/analog) ground plane but is connected around the circumference of the loop to the lower PCB ground plane, which completes the enclosure. Vias are used to tack the upper loop down to the lower ground plane, and these should be located at intervals of approximately 3 mm. Then if the tabs on the shield fence do not provide a low enough "contact impedance" with the lower ground plane, which forms the full enclosure, then the fence can be soldered to the upper loop. This configuration is shown in Figure 5.142.

Some of the types of material commonly used for the shield are 0.1–0.25 mm thick beryllium copper or 0.2–0.5 mm thick tin-plated steel. The magnetic field shielding effectiveness of these types of shield with a low contact impedance seam (described earlier) is shown in Table 5.43. It is important to state that Table 5.43 applies only to an enclosure that is constructed completely of the materials shown. In the case of many PCB-mount enclosures, the loop that forms the enclosure is completed by an enclosure PCB ground plane. It is not, however, generally necessary to go to the expense of "buried vias" for this lower "enclosure" ground plane, for up to 1 GHz of the ground plane can contain many small (2 mm) apertures without unduly reducing shielding effectiveness. At all costs, avoid long slots (2 cm) in this enclosure ground plane, for this will begin to further reduce the shielding effectiveness of the enclosure. In one instance small apertures, added for thermal equalization, around the solder

(Upper ground trace connected to lower ground trace with vias at 1–3 mm intervals.
Do not rely on tabs in shield to make the connection!)

FIGURE 5.142 Location of filter capacitor to the enclosure, and connection of radio frequency signals through the shield into the RF section.

TABLE 5.43

Magnetic Field Shielding Effectiveness of 0.127 m × 0.101 m × 0.01 m Enclosure

f	0.5 mm Thick Tin-Plated Steel	0.2 mm Thick Tin-Plated Steel	0.2 mm Thick Beryllium Copper	1 oz Copper PCB Material
4 kHz	6	4	0	0
10 kHz	14	14	1	0
100 kHz	33	33	25	10
1 MHz	47	47	46	34
10 MHz	54	54	54	52
100 MHz	56	56	56	56

tabs of a coaxial connector attached to the printed circuit resulted in out-of-band spurs. In this case, a −50 dBm RF signal at 1.9 GHz was brought into the RF section via an external coaxial connector. When the small apertures around the connector tabs were filled with copper shim soldered in place, the spurs reduced to within specification. Some of the manufacturers of this type of shield is Boldt Metal Industries A.K. Stamping with laminates made by Flexlan, Orion, and Insul-Fab.

When RF current flows at gigahertz frequencies on, typically, semirigid cable, data cables, or well-shielded enclosures, where the current flow is a problem and additional shielding is not possible, the use of absorber material over the enclosures or cables can be very effective. Gigahertz frequency absorber is available in rubber sheets that can be wrapped or glued or held in place by Velero tape. As an alternative, potable epoxies may be used to coat assemblies.

5.6.3　Radio Frequency Grounding

Chapter 11 discusses PCB grounding in detail. But, in general, the main goal in grounding an RF circuit is to avoid power and signal RF currents generated in the digital section or C/M currents on signal interfaces from flowing on the RF section power/signal ground plane. Two options exist: One is to ensure that no currents can flow through the RF ground plane, either by choosing the locations of the different grounding points or by covering the RF circuit and ground on both sides with a shield that is terminated either around the periphery to the RF ground or at one or two points only. A second grounding scheme is to isolate the digital and RF grounds. Either solution will work when implemented with care, although isolating the grounds is typically more difficult.

The goal in the isolated grounding scheme is to isolate all of the signal and power connections referenced to the digital ground from the RF ground. RF signals carried on coaxial cables must remain isolated from digital return as they exit the board or the equipment. This grounding scheme will provide DC isolation between the digital ground and the RF ground and allow the RF ground to be connected to chassis. That RF ground is connected to chassis somewhere is usually unavoidable. This is often due to the connection of shields of coaxial cables to chassis or building ground when they interface to the outside world. For example, the shield of an antenna cable located outside of a building and brought into an equipment rack should be grounded to the rack at the entry point. This provides a connection for secondary lightning currents and ensures that RF currents picked up on the antenna cable shield do not enter the equipment enclosure.

The isolated grounding scheme used in a typical wireless PCB is illustrated in Figure 5.143. The power supply must be of the isolating type. If it is a switching power supply, the ripple and spikes at the output will be attenuated by the filters, as shown in Figure 5.143. RF C/M currents will also flow in the interwinding capacitance of the TX I and Q transformers. The solution is to add capacitors, as illustrated in Figure 5.143.

If these capacitors cannot be added, one technique to reduce coupling is to add a capacitor between the digital ground at the location of the entry of the I and Q signals into the enclosure, between the digital ground and the enclosure, which is in turn connected to chassis. This capacitor may be a ceramic having a 3–6 kV voltage rating if a hi-pot test is required. If a hi-pot test is applicable, then the isolating switching power supply must also have an adequate voltage rating between primary to secondary, and this includes primary power return to secondary power return, and so must the I and Q transformers. Ensure that opto-isolators are located at the shield and not inside the shielded enclosure as the traces to the LEDs can radiate.

The second grounding scheme must be used if one of the signal or control interfaces that reference digital ground or a power-supply ground cannot be isolated from the RF ground. In this case,

FIGURE 5.143 Grounding scheme with isolation between digital and radio frequency ground.

the RF ground is connected to the digital ground, and both will be connected to the chassis of the equipment and thereby to safety ground.

In this scheme, the physical location of this common ground connection is important. In the equipment under consideration, the digital ground is connected to the equipment chassis ground and to RF ground at the location where the digital and I and Q signals interface to the RF section. The result is that RF currents set up by C/M noise voltages and by radiated emissions tend to flow back to the source via the chassis and not through the RF ground plane. In the instances where a signal return cannot be connected directly to chassis, isolation or filtering is required. If a DC connection cannot be made to the equipment chassis ground, then one or more high-quality capacitors can be used to provide an RF ground.

In the fully isolated grounding scheme, the PCB plane, which is used as part of the enclosure, and the RF ground plane should if possible be separate and should be isolated as far as possible from each other. In practice, a connection must be made from the enclosure to the shield of shielded cables that enter the RF section—in the example, the TX IF O/P and the RX IF. It would be counterproductive to bring shields into the RF section without a connection to the enclosure, because any RF currents flowing on the shield, due to electromagnetic coupling from the digital section and the switching power supply, in the equipment and from external sources such as AM and FM transmitters will reradiate inside the "quiet" shielded RF section. The reason to isolate the RF ground plane from the enclosure is that RF currents are set up in the enclosure by the electromagnetic ambient. In a perfect enclosure, these currents would be confined to the outside surface, although in a practical enclosure a voltage is set up over the contact impedance of the seams and this results in an internal current across the ground plane. Even in the scheme shown in Figure 5.144, in which the digital and RF grounds are connected through a balun, the RF ground should be connected to the enclosure at one point only.

FIGURE 5.144 Grounding scheme with radio frequency and digital ground tied together through a balun.

In the grounding scheme where RF and digital ground are tied together at one point only, it is important that the RF ground plane section not be in the path of power return currents from the digital section back to the power supply via a connector. If the RF ground plane is in the path between the digital section and a connector used for the power and return, the RF ground plane should be isolated from this path. An alternative is to ground the RF ground around the periphery to the all-encompassing shield enclosure, in which case the RF noise currents flow through the shield back to the source. As mentioned in the PCB layout section in Chapter 11, the locations of the various sections of the board and the power and signal connectors play a vital role in achieving EMC.

Some of the latest RF circuits are differential. One of these devices is the HELA-10 manufactured by Mini-Circuits®. This device has a 50–1000 MHz bandwidth, a gain of 10–13 dB, an NF of 3.5–4.7 dB, and a 26–30 dBm output. To use this device with an unbalanced input and to provide an unbalanced output, balun transformers with either 50 or 75 Ω impedances are used at the device input and output. These baluns are also available from Mini-Circuits and can be used to provide C/M noise rejection and RF isolation, at least at certain frequencies.

If there is a golden rule to grounding, it is to build some flexibility into the grounding scheme, especially in the breadboard, prototype, or first production models. For example, if a chassis plane is included in a motherboard or back plane. The PCB should be laid out to allow a direct connection or disconnection of the chassis plane from any analog/RF/digital ground and permit connection of the chassis plane through a low-impedance direct connection or capacitors to any or all of the remaining grounds at several locations around the board.

5.6.4 FILTERING

Using either the isolated grounding scheme or the common grounding scheme, every connection between RF circuits inside the shielded enclosure and external circuits must have some level of filtering or C/M isolation. In the case where HF signals enter the RF section, a number of options exist. For balanced transformer-coupled signals, in our example, the TX, I, and Q, the parasitic coupling is almost certainly C/M via the interwinding capacitance of the transformer. If a transformer with an internal electrostatic shield is available, this should be used. Failing that, provision should be made for a low-value capacitor to be placed between the upper winding of the secondary and the enclosure and a second capacitor between the lower winding and the enclosure. These can be removed if found unnecessary or if they interfere with the transformer coupling or balance.

Digital signals should preferably be connected via optoisolators when the digital and RF grounds are isolated, and these should be located outside of, but tight up against, the shield. For return to zero data, a transformer or a differential-input or quasi-differential-input amplifier may be used.

If digital signals cannot be connected through optoisolators, transformers, or differential amplifiers, then they can, in common with other single-ended signals, be routed through a ferrite balun. The signal return must always be routed through the balun along with the signal/s, and this almost certainly requires that the digital and RF ground plane be split as shown in Figure 5.145. One technique is to have slots in the PCB either side of the traces. A split balun can then be placed on either side of the board if this is found to be a requirement. The balun is shown in Figure 5.146.

FIGURE 5.145 Connection of ground through the balun.

FIGURE 5.146   Split ferrite balun PCB mounting technique.

In addition to presenting an impedance to C/M RF currents, the balun acts as a receiving and a transmitting antenna, and the best location is outside the shield, unless some source of high-level emission exists in close proximity to the balun, in which case it should be located just inside the shield, but still maintaining the split between digital and RF ground, as shown in Figure 5.145.

Although a signal is designated as an output, C/M currents can flow on the output connection and reradiate inside the shield or result in RF currents on the ground plane. Therefore, filtering, isolation, or a balun should also be placed on outputs. In our example, when digital ground and RF ground are tied together, the RX1 and RX2 connections to the primary of the transformers should be routed through baluns, with the ground planes connected through the balun as shown in Figure 5.145. A better solution is to connect the shield of the RX1 and RX2 to the enclosure, in which case the balun is not required. If the RX1 and RX2 grounds are isolated from RF ground, then the transformer will reduce the flow of C/M RF current, with the addition of an RF (capacitive) chassis ground connection to digital ground, if required.

As described in Section 5.2, PCB-mounted switching power supply regulators and filters can, when properly designed, reduce switching power supply noise, which typically exhibits 400 mV short-duration spikes and 100 mV ripple, down to approximately 5 mV ripple with negligible HF component. Where extremely low ripple is required, typically satellite communication, radar simulators, and laser power supplies, small shielded PCB-mount filters have been designed, built, and tested that reduced ripple down to between 30 and 300 µV. Large-value D/M SM inductors should be located just outside of the shielded enclosure, followed by capacitors connected directly to the shielded enclosure, as shown in Figure 5.142. A C/M ferrite should also be located outside of the shielded enclosure over all of the power and return lines. However, if the C/M ferrite is located close to a high-level noise source, then RF currents will be induced into the ferrite. In this case, shield the ferrite or move it just inside the shield. The components inside the enclosure should be the smaller D/M capacitors and the linear regulators. With this configuration, the shield is used to minimize radiated coupling between the input components of the filter and the output components.

The RX IF input coax should have the coaxial shield tied to chassis as it enters the equipment and then tied to the shielded enclosure, as shown in Figure 5.142. A connection must also be made to the RF ground plane at that location. The RX IF input filter should attenuate a broad range of noise, including 60 Hz and its harmonics. This is typically achieved at power-line and switching power supply frequencies by simply by using a coupling capacitor with a low series impedance from 400 to 900 MHz and a high impedance at LF. For this to work, the input filter and amplifier after the coupling capacitor must also maintain a low, ≈50 Ω, input impedance at LF. The TX IF O/P connection must also contain a similar filter.

If the preceding guidelines are followed, then equipment has a very high probability of achieving the goal of EMC. Some of the additional components may be unnecessary, such as the baluns; having said that, if provision is made for them, they can easily be removed. Alternative techniques for grounding, shielding, isolation, and several combinations of power filtering can be designed to achieve the same goal, and these can be discussed. The most common problems usually occur in the practical realization of the design; for this reason, the schematic and PCB should be reviewed prior to commitment to a PCB layout.

5.7 RADIATED EMISSION REDUCTION

Printed circuit board emission is very dependent on the layout and use of ground planes or transmission-line principles. The reduction techniques using PCB layout and circuit design are discussed in Chapter 11. The reduction in radiated emissions from wire and cables is discussed in Chapter 7.

5.8 TRANSIENT PROTECTION

Typical transients are due to lightning, ESD, power-line surges and high-voltage breakdown in adjacent equipment or cables.

These are characterized in different test requirements:

Human body model (HBM) ESD in IEC61000-4-2.
EFTs are specified in IEC 61000-4-4.
Power-line surges arc specified in IEC 61000-4-5.

These documents are under the European Union harmonized standards used under the EMC Directive and the Medical Directive.

Other transient specifications are

The Bellcore TR-NWT-001089 surges 10/1000, 10/360, 2/10, 10/560, and 10/160 μs.
FCC part 68 10/560 μs, 10/160 μs.
MIL-STD-461C 0.15, 5, and 10 μs
MIL-STD-461F CS106 5 μs.
DO-160 10 μs spikes as well as lightning induced 0.1, 6.4, 120, and 500 μs spikes and damped sinusoids.

Transient protection is most commonly required to protect the equipment from damage when powered, unpowered, or both.

The other, more difficult, requirement is that the equipment continues to function with either no error or acceptable errors after cessation of the transient. It is seldom the requirement that equipment should continue to function during the transient; it is more common that it should fail or is safe or recover correct operation after cessation of the transient.

One approach to recover after a transient is to make the equipment robust from a system point of view. That is, make reset circuits and power-line monitoring, which is a common cause of failure, only immune to transients. An alternative is to ensure that the software resumes the correct function after the event without loss or change in data and without human intervention, such as a manual reset.

Components may still be required to protect against damage. One very simple component that has a number of applications is the capacitor.

Protection circuits and techniques exist for protection against high-level transients caused by a direct ESD, the EFT, lightning strike, or the electromagnetic pulse (EMP) ambient. Protection may also be required against the lower-level secondary effects of lightning, EMP, and power-line-conducted transients. AC power may be protected as it enters the building or at outlets. Power strips and power-line filters may be purchased with built-in transient protection.

Most transient protection devices turn into a low impedance above some turn on voltage. The resultant clamping voltage may be too high to avoid damage; especially as in many devices, the clamping voltage is twice or more the turn on voltage. Also, the return path to ground must have a low impedance to avoid a higher voltage. In the case of high-current transients such as lightning, very high currents can flow and PCB traces and wiring must be rated accordingly, otherwise they may fuse.

If the transient protection is between signal ground and chassis or safety ground, depending on the resistance and impedance of the path a high C/M voltage is developed. C/M capacitors may reduce this voltage and must be capable of withstanding it.

When a device is connected to ground, the current develops a voltage across any inductance the voltage given by $v = L(di/dt)$ that can appear across the load.

All devices should be connected with a low-inductance connection to chassis/earth ground.

5.8.1 Transient Protection Devices

The gas discharge tube can take very high currents, up to 50,000 A, and they have very low capacitance and thus are ideal for RF circuits. Often, when lightning current is higher than the device rating they fail safe, that is, breakdown, and continue conducting after the lightning strike. This means that the equipment is protected during the lightning strike although the device needs replacing. The striking voltage is between 200 V and 10,000 V, depending on the specified breakdown and the rate of rise of current. Thus, they are often used with filters or secondary protection devices such as varistors, transient voltage suppressors (TVSs), or diodes. They are available alone or with filter components in a coaxial tube with RF connectors on both ends. They can therefore be directly connected in a coaxial cable with the metal case bonded to chassis. One disadvantage of the gas discharge tube is that it is a crowbar-type device, and if used on a DC supply that is higher than its arc voltage (typically 20 V), it will not extinguish until the supply is turned off or a fuse blows or breaker opens.

On AC power the tube extinguishes as the current through it reverses.

A metal oxide varistor (MOV) is made of grains of zinc oxide or other metal oxides. Each grain boundary is the equivalent of a junction with a breakdown of 2–3 V. This results in hundreds of junctions in series and parallel resulting in a high-current clamping capability, up to 70,000 A and a high capacitance. It is slower than the TVS and has a clamping voltage of between 36 and 1800 V. However, as with all protection devices, this is highly dependent on the peak current. The MOV is too slow for clamping ESD or EFT events but useful for lightning and surge protection. The MOV conducts surges for a limited number of events after which it may degrade. Clamping voltage to breakdown voltage is about a factor of four.

The thyristor surge protection device (TSPD) is a "crowbar"-type device. This means that after triggering it goes into a low on state, which provides a much higher surge current capability for a relative chip size. For a 1500 W bidirectional clamp, that can handle 5 A for a 10/1000 µs, the equivalent size TSPD can conduct 200 A for the same surge duration.

The TSPD can be used on signal and communication lines but when used on a DC supply they present a short after the transient has passed. A resettable fuse may be used to interrupt the current below the holding current or a resonant circuit on the line that reverses the current through the TSPD. The holding current is typically 50–250 mA at 25° and 30–150 mA at 100°. Power-line filter with active tracking, designed for spike and transient suppression, with the trademark Islatrol is manufactured by Control Concepts. Protection against transients with voltages less than 1 kV, a rise time of 10 µs or longer, and pulse duration of 1000 µS may be provided on data lines by devices manufactured by MCG electronics. The data lines that may be protected include modems, RS232, RS422, RS423, 20 mA loop, telephone, video, and high-speed lines.

Unipolar and bipolar silicon TVSs are available in DIP, SM, radial- or axial-lead, and modular forms. SM devices are available rated at 1500 W of peak power and clamping voltages from 9.2 to 189 V from Protek devices. Leaded devices, from General Instrument, can take 5000 W of peak power at 1 ms, can switch as fast as 1.0 ps, with clamping voltages of from 6.4 to 122 V. Modular TVSs from Protek can take up to 15,000 W of peak pulse power dissipation at the 1 ms pulsewidth and clamp in subnanosecond times. California Micro Devices manufactures 17 and 18 channel ESD protection arrays that divert the ESD current pulse to either the positive or negative supply via diodes.

The Sidactor is a transient surge suppressor in a modified TO-220, TO-92, and SM DO-214AA package, which, according to the manufacturer, has the following features: bidirectional protection, breakdown voltages from 27 to 720 V, clamping speed of nanoseconds, and surge current capability up to 500 A. The AVX Transguard® TVS is available in SM (0603–1210) or pill configurations, with a peak current of 150 A for the 1206 chip. The MOV is capable of high peak currents (up to 70,000 A in a disk type, formerly from Siemens, for a very short duration), but it is slower than a TVS, at less than 15 ns. The lowest clamping voltage is higher than that of the TVS and for the Siemens disk type covers a range of 36–1815 V, and so, it has the possible advantage of a much higher maximum clamping voltage than the TVS. The peak clamping current for a 8/20 µs pulse is 2500 A for the 1100 V MOV, but the continuous current must be limited to achieve the 0.4 W average power dissipation. The average power must be limited to between 0.01 W and 0.6 W, depending on the size of the varistor. The Sidactor and TVS and many diodes and zeners are fast enough to protect against the ESD and EFT test levels, although the varistor will not be fast enough unless a filter is used between the source of the fast pulse and the circuit to be protected by the varistor. Thyristors and triacs have also been used in circuits designed to clamp transients to a very low level, but, when compared to the TVS, these devices are typically relatively slow. ABB manutactures a solid-state transient surge suppressor based on the thyristor diode structure. It is designed for lightning protection on telephone transmission lines. The smallest devices are available in a DIP or SOIC-16 package and are capable of 150 A pulse, a breakdown voltage of 60–80 V, and a nanosecond response time. ABB also manufactures a silicon unidirectional transient suppressor consisting of a PNPN thyristor diode integrated with a fast zener that has a short-circuit failure mode when overloaded. The peak current for an 8/20 µS pulse is 500 A, the static breakover voltage is 75 V, and the dynamic is 95 V at $dV/dt = 1.5$ kV/µs. The on-state voltage is 3 V at 2 A, and the turn-on time is 20 ns, which is OK for lightning protection but too slow for EFT and ESD protection. A higher-current device from ABB can take 1000 A for the 8/20 µs pulse, has an on-state voltage of 6 V at 20 A, and has a breakover voltage of 280 V and a dynamic breakover voltage of 375 V with $dV/dt = 1.5$ kV/µs.

TVS with capacitances as low as 0.35 pF and clamping voltages down to 8 V at 16 A made by ON Semiconductor are suitable for ESD protection on circuits operating at data rates of greater than 10 Gb/s such as USB 3.0, USB 3.1, and HDMI.

SM devices are available for lightning protection of Ethernet, video, etc., and these include the Littelfuse SR70 device that contains four diodes and is suitable for two channels. It is specified for EFT 5/50 ns 80 A. ESD protection ±30 kV contact, ±30 kV air. Lightning 8/20 µs 40 A.

Littelfuse offers ZA varistors with operating voltages of AC 4–460 V and DC 5.5–615 V and a peak current from 250 A (8 × 20 µs) at 5.5 V to 2000 A at 22 V.

The CH Varistor series is SM and has operating voltages of 14 to 275 V AC and 18–369 V DC. The peak current (8 × 20 µs) is from 250 to 500A.

Microsemi also offers varistors with AC 4–460 V, DC 5.5–615 V, and lightning 8 × 20 µs of 50–2000 A.

Microsemi offers a monolithic flip chip TSPD array, standoff 5–8.5 V, breakdown = 8–12 V, clamping 3 V at 1 A, 6 V at 8 A, and peak current 30 A with a ton of 5.0 ns. The Microsemi TVS RT65KP48A has a peak power capability of 65,000 W and a maximum standoff/working voltage of 48 V and a clamping voltage of 17.7 V. It is a unidirectional device designed to meet RTCA/DO-60E section 22 and is compatible with Section 16.6.2.4 category A for 46.3 V, category B for 60 V, and category Z for 80 V. Microsemi offers a 500 W SM TVS with standoff voltages of 5–170 V and breakdown voltages of 6.4–189 V, maximum clamping voltage 9.6–275 V, and peak pulse current of 52–1.8 A. The ESD0P8RFL ESD protection device contains two diodes in a 1.2 × 0.8 × 0.39 mm SM package. It has a low line capacitance of 0.8 pF at 1 GHz, ultralow 0.4 nH series inductance, and a clamping voltage of 3 V at 2 A and 11.5 V at 10 A. ESD contact discharge of 20 kV, EFT 5/50 ns 40 A, and lightning 8/20 µs current of 10 A. It is ideal for ESD protection of RF antenna and/or ultrahigh-speed lines such as HDMI, S-ATA, and Gbit Ethernet. The EESeal filter seal for connector pins in D type, ARINC, and circular connectors. These are normally available with 1 pF to 1 μF capacitors, but they are also available with a wide range of AVX Transguard® Voltage suppressors. API Industries (Spectrum control) manufactures the series USB-E ESD-protected USB connectors that contain diodes between line and ground. The working voltage is 5 V DC, maximum current 1 A, and clamping voltage at 1 A 8/20 μs of 9.8 V.

Table 5.44 lists the characteristics of a number of devices.

5.9 LIGHTNING PROTECTION

A direct lightning strike can be modeled as a constant-current source with typical peak current of 20,000 A. However, approximately 10% of strikes have peak currents of 60,000 A or greater. If a 200,000 A lightning strike occurs on a metal aircraft frame, any electronic components mounted external to the frame can be exposed to magnetic fields as high as 318,000 A/m close to the surface and E fields as high as 250,000 V/m, assuming the strike does not occur directly to the externally mounted component. The preferred approach in protection is to avert a direct strike to cables, antennas, equipment, or equipment huts. Lightning rods or wires strung between poles create the so-called zone of protection beneath the rod or wire that extends in an arc from the top of the conductor to the ground. This form of protection is described in Section 8.6.1.

When antenna, control, AC or DC power, telephone wires, and other cables cannot be protected by lightning rods, then primary lightning strike arrestor devices are available, which are designed to protect the input of equipment. Varistors and gas discharge arrestors are capable of handling currents up to 60,000 A. At these high currents, the protection is usually one shot, and the arrestor fails safe (i.e., with a permanent short). This means the arrestor device must then be replaced. Even so, the cost of lightning arrestors is relatively low when compared to the cost of repairing receiving/transmitting equipment. The surge current handling capability of common protection devices versus the voltage developed across the device during a strike is shown in Figure 5.147. The striking voltage of a gas discharge arrestor is dependent on the rise time of the transient. Figure 5.148 illustrates the breakdown voltage versus the transient gradient for very fast gas gaps with DC breakdown voltages from 75 to 7500 V. These fast arrestors are used by Fischer Custom Communications Inc. in a number of different products, both with leads and mounted with coaxial connectors, as described later.

One use for a simple capacitor is as a coupling capacitor at the input of an RF circuit. If the input frequency is high enough and the input impedance is 50 or 100 Ω, then the network shown in Figure 5.149 in combination with a gas discharge tube acts as a high-pass filter, and the relatively slow lightning pulse is attenuated. Even if the lightning excites a long cable the resultant damped sine wave is typically LF. Assuming a 1000 pF coupling capacitor a 1000 V spike with a rise time of 1 μs will develop a voltage of 50 V across the 50 Ω load resistor shown in Figure 5.149. If the rise time is slower, then this voltage will be lower. If 50 V is too high, then a faster acting gas discharge tube with a lower striking voltage must be used. For example, Fischer Custom Communications Inc. provides a gas discharge tube with a striking voltage of 75 V. At a rise time of 1000 v/μs, the peak voltage will be 200 V and the voltage across the 50 Ω load will be 10 V. The 1000 pF will add some attenuation in the RF signal, but it may be acceptable for frequencies above 150 MHz. The capacitor voltage rating must be high enough to withstand the striking voltage of the gas discharge tube.

Figure 5.149 shows the gas discharge circuit with series coupling capacitance.

The advantage of the capacitor solution compared to a parallel protection device is that no high currents are shunted to ground, although cables and PCB tracks must be capable of sustaining voltages of 1000 V or greater. With additional diodes incorporated after the capacitor, higher currents flow, but the magnitude is limited by the time constant of the capacitor-and-diode combination.

For signals of a single frequency, a band-pass filter may be incorporated. The schematic in Figure 5.150 is designed for a 10 MHz IF signal that is sourced by a radio outside of the building mounted on a mast. The circuit passes the 10 MHz with negligible attenuation, whereas the 1000 V transient is reduced to 5.5 V.

TABLE 5.44

Characteristics of Protection Devices

Device Type	Manufacturer	Application	Breakdown Voltage	Clamping Voltage at I Peak	Peak Current	Rise Time and Capacitance	Peak Power
Transient voltage suppressor	Protek	ESD and lightning	6 V bidirectional	24 V at 45 A 8/20 μs	45 A at 5 V	$T_{on} = 5$ ns with 10 kV HBM ESD	8/20 μs
LCD05C/ SMADA05LCC Surface mount				34.8 V with 10 kV HBM ESD		$C = 15$ pF	800 W
Transient voltage suppressor Surface mount	Microsemi	ESD and lightning	6 V bi- or unidirectional	9.6 V at 52 A	10/1000 μs = 10 A 52 A at 5 V 1.8 A at 170 V	Depends on circuit configuration	8/20 μS 800 W
Transient voltage suppressor Surface mount	Microsemi	ESD and lightning 10 base T, Rs485, USB, bidirectional	18.9–311 V	For 17 V device standoff voltage = 32.3 V at 464 A	464 A at 17 V standoff to 33 A at 28C V		15,000 W 10/1,000 μs
Transient voltage suppressor	Protek	ESD and lightning Unidirectional and bidirectional	18.9–311 V 6 V	20 V at 5 A for 6 V varistor 50 V at 20 A for 23 V varistor	250 A 8/20 μs 6 V 2000 A 8/20 μs for 23 V	6 V = 1400 pF 675 V = 30 pF	0.8 J for 6 V 120 J for 23 V

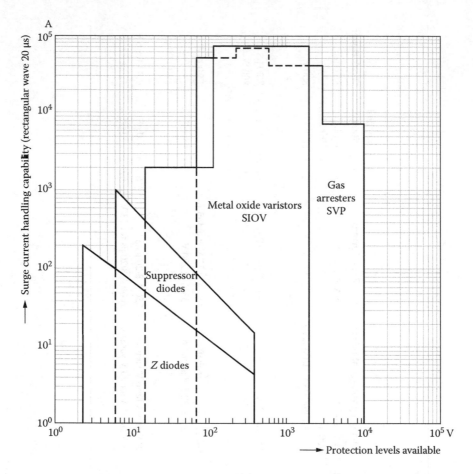

FIGURE 5.147   Surge current handling capabilities versus the voltage protection levels available. (Courtesy of Siemens Ltd.)

FIGURE 5.148   Striking voltage versus rise time for gas discharge arrestors with DC breakdowns from 75 to 7500 V. (Courtesy of Fisher Custom Communications Inc.)

FIGURE 5.149   Antenna lightning protection using a gas discharge tube with series coupling capacitor.

FIGURE 5.150   Signal lightning protection.

A very low level of current flows in the gas discharge arrestor until the striking voltage is reached, after which the current increases to the milliamp region and the voltage drops to the glow voltage. As the current continues to increase, an arc occurs and the voltage drops to the arc voltage (approximately 20 V). On cessation of the strike, the current through the arrestor decreases until the arrestor extinguishes, assuming the supply current is less than the arc current and the supply voltage is less than the glow voltage. When used on an AC supply, the gas discharge device may not extinguish until the AC voltage waveform changes phase. On a DC supply, the arrestor may not extinguish until the supply is disconnected or the short causes a contactor, cutout, or fuse to open. One solution to the problem of not extinguishing on a 20–40 V supply is to place two arrestors in series across the supply. The disadvantage here is that the voltage spike generated across the arrestors, until the striking voltage is reached, and the arc voltage will be twice as high as for a single arrestor.

A new range of gas discharge arrestors designed for the much faster rise time of an EMP pulse may be used for lightning protection, with a resultant shorter striking time and strike voltage. Many lightning protection devices are manufactured specifically for use with coaxial or twisted-pair cables and in a bulkhead panel for mounting at entry points into buildings. The coaxial type will typically have a specified characteristic impedance, such as 50 or 75 Ω over the frequency range of interest. Two basic types exist. Fischer Custom Communications Inc. manufactures a range of "Spikeguard" nanosecond transient protection designed for lightning and EMP. They exhibit fast response through the UHF region and are constructed from transmission-line, gaseous discharge components and silicon components.

The Spikeguard Lightning arrestors are available in metal enclosures designed for insertion in antenna, control, or telephone cables. The enclosures have attached grounding lugs and are designed for mounting on an antenna mast or ground rod. Typical VSWR characteristics for the Spikeguard and type N, UHF, and C coaxial connectors are 100 MHz 1.2:1, 200 MHz 1.4: 1, 300 MHz, 1.6:1, and 400 MHz 1.8:1.

The FCC-350 series is rated at 20 kA peak for a 10 µs duration and with breakdown voltages from 90 V to 20 kV. For DC breakdown voltage of 1 kV or higher, the overshoot voltages of transients having a rise time of 1 kV/ns is minimal, since the breakdown voltage is achieved in approximately 2 ns. The insertion loss to RF is approximately 0.2 dB. The FCC 450 series Spikeguards have been designed to provide transient protection for receivers and transmitters up to 100 W of output power and are capable of clamping in a few nanoseconds and dissipating 15 kW of power for transients having a pulse width of 10 µs and 1.5 kW for 1 ms transients. Other versions of the Spikeguard are available to protect HF transmitters to 1 and 2 kW. The FCC-550-10-BNC hybrid suppressors are designed for the transient protection of signal and control circuits that are digital or analog with operating voltages of 6 V or less. Transients will be clamped to between 10 and 20 V for the duration of transients with rise times of the order of 100 kV/µs and 1 MV/µs. Series impedance is 18 Ω, with a shunt impedance equal to or greater than 5 MΩ in parallel with no more than 10 pF of capacitance. These units come with BNC connectors.

Other versions of the FCC-550 suppressor wire connections are designed for 1 V control, analog, and digital signals and will limit the transient to 8.5 V for the duration of transients that are several thousand volts. On these suppressors, the series impedance is 80 Ω. The shunt impedance is several kiloohms in parallel with no more than 500 pF, except during transient suppression, when the impedance approaches zero ohms. The 550 series may be specified with clamping voltages between 8.5 and 200 V.

Fischer also has Spikeguards designed for 60–400 Hz AC power-line protection.

Arrestors and transient filters made by the Polyphaser Corporation, and referred to as "protectors," are available for use in receiving or transmitting antenna cables and antenna control cables as well as in the protection of twisted-pair, combiner, and AC power. The protectors are selected for use over a specific frequency range, for example, for baseband (DC: 50 MHz), HF/VHF/UHF 50–550 MHz, 450–900 MHz, and other much narrower frequency band protectors, such as 800–980 MHz for cellular/paging combiner protection. The 800-MHz and up, DC-blocked, non-gas-tube type are referred to as *microwave filters*. Polyphaser also manufactures telephone semiconductor protectors, motor control, GPS, and PCS tower-top pre-amp protectors for 1.2–2 GHz, DC injection/DC path protectors, IBM twinax series protectors, and a range of GPS/PCS/cellular/pager/microwave protectors. The protectors either block DC or allow DC through, and come in weather-resistant, for outdoor mounting, and nonweather-resistant, for indoor mounting, versions and are for either transmit and/or receive-only applications. One Polyphaser high-power coaxial protector has a DC to 220 MHz frequency range and a 25 kW transmit power rating.

A Polyphaser protector designed for wireless local loop protection is the LSX series shown in Figure 5.151. The LSX series provides surge protection for a 20 kA waveform with an 8 µs rise time and a 20 µs pulsewidth (8/20 µs) at the 50% amplitude level. The throughput energy with a 3 kA at 8/20 µs waveform is less than or equal to 0.5 µJ and the let-through voltage is ±3 V for a 3 kA at 8/20 µs waveform. The frequency range is 4.2–6 GHz with an insertion loss of 0.1 dB typical, and the transmit power rating is 10 W continuous. The protector includes DC blocking and has a temperature range of −40°C to +85°C.

Another one of the many available Polyphaser protectors of this type is the PSX-D series, which provides surge protection against a 30 kA waveform with an 8 µs rise time and a 20 µs pulsewidth (8/20 µs) at the 50% amplitude level. The throughput energy with a 3 kA at 8/20 µs waveform is less than or equal to 0.5 µJ, and the let-through voltage is ±3 V for a 3 kA at 8/20 µs waveform. The frequency range is 1.7–2.3 GHz, with an insertion loss of 0.1 dB typical, and the transmit power rating is 500 W continuous. The protector includes DC blocking and has a temperature range of −40°C to +85°C. Polyphaser also manufactures a range of grounding and cable grounding systems and cable entrance panels, as well as provides courses and publications covering lighting protection and grounding solutions for communications sites.

FIGURE 5.151   Arrestor manufactured by the Polyphaser Corporation.

EPCOS (Siemens) also makes a range of gas discharge tubes in addition to MOVs for lighting protection. Coto Technology manufactures mini- to small gas discharge surge arrestors, and data and application notes can be found at www.cpclare.com.

The advantage of gas discharge arrestors is the low self-capacitance, which allows their use at such high frequencies. Due to the shunting effect of lightning arrestors, it is imperative that a low-impedance ground connection be made to the case of the arrestor. For example, a 16 mm diameter rod will develop, due to its self-inductance of 1.22 µH/m, 12 kV/m length during an average strike with a 2 µs rising edge, and a peak current of 20,000 A. Despite a low-impedance ground, it is inevitable that some current will flow on the shields of cables connected to a lightning arrestor or on cables connected to an antenna protected by a lightning rod, due to potential gradients along the ground.

Case Study 8.1 describes a grounding scheme that reduces cable-induced currents during a lightning strike. A large metal oxide core through which one to four turns of the interconnection cable/s are wrapped will increase the impedance of the cables and thereby reduce the magnitude of the cable current flow. Where possible, the core should be located close to the equipment to be protected, but, certainly after the lightning arrestor, where used.

The voltage spike generated before the arrestor breaks down and clamps the voltage may damage transmitters, receivers, and telephone equipment. Semiconductor devices or varistors may be used to clamp the spike from the gas discharge arrestor or the voltage induced as a secondary effect of a strike to safe levels.

Some combinations of primary arrestor and secondary transient protection devices are shown in Figure 5.152. When secondary protection devices, such as zener diodes, diodes, and transient absorber zener diodes, are mounted on PCB, the ground path on the PCB back to the enclosure should be a low impedance and must be capable of carrying the high transient current. The location of the protection device in the circuit is very important. When the device is placed after capacitors and inductors, the voltage developed across the device may be higher than the impressed transient and in the form of a damped sinusoid. The voltage developed across transient protection zeners, when tested using a generator capable of producing the MIL-STD-461 CS06-specified spike, can exhibit forward voltage drops two or three times higher than specified for the device, even when the current through the device is below the maximum rated.

SM devices are available for lightning protection of Ethernet, video, etc.

Microsemi offers a monolithic flip Chip TSPD array, standoff 5–8.5 V, breakdown = 8–12 V, clamping 3 V at 1 A, 6 V at 8 A, and peak current 30 A with a ton of 5–00 ns.

The Microsemi TVS RT65KP48A has a peak power capability of 65,000 W and a maximum standoff/working voltage of 48 V and a clamping voltage of 17.7 V. It is a unidirectional device designed to meet RTCA/DO-60E section 22 and is compatible with Section 16.6.2.4 category A for 46.3 V, category B for 60 V, and category Z for 80 V.

Microsemi offers a 500 W SM TVS with standoff voltages of 5–170 V and breakdown voltages of 6.4–189 V maximum clamping voltage 9.6–275 V and peak pulse current of 52–1.8 A.

The ESD0P8RFL ESD protection device contains two diodes in a 1.2 × 0.8 × 0.39 mm SM package. It has a low line capacitance of 0.8 pF at 1 GHz, ultralow 0.4 nH series inductance, and clamping voltage of 3 V at 2 A and 11.5 V at 10 A. ESD contact discharge of 20 kV, EFT 5/50 ns 40 A and lightning 8/20 µ current of 10 A. It is ideal for ESD protection of RF antenna and/or ultrahigh-speed lines such as HDMI, S-ATA and Gbit Ethernet.

The EESeal filter seal for connector pins in D type, ARINC, and circular connectors. These are normally available with 1 pF to 1 µF capacitors, but they are also available with a wide range of AVX Transuard® Voltage suppressors.

API Industries (Spectrum control) manufacture the series USB-E ESD-protected USB connectors that contain diodes between line and ground. The working voltage is 5 V DC, maximum current 1 A, and clamping voltage at 1 A 8/20 µs of 9.8 V.

The disadvantage in the use of both semiconductors and varistors at the input of HF receivers or interface circuits is their high intrinsic capacitance. This capacitance is in parallel with the input

FIGURE 5.152 Combination of primary and secondary protection devices.

and may degrade the signal. The RTCA/DO-160D requirements for airborne equipment contains, in section 23, tests for direct lightning effects and in section 22 lightning-induced transient susceptibility. The tests in section 22 are designed to simulate the current and voltages induced in cables contained within aircraft or helicopters. Different test levels apply to cables that are protected by airframes composed of metal and composite skin panels and equipment in carbon fiber composite airframes whose major surface areas have been protected with metal meshes or foil. For this type of aircraft, the apertures are the main source of induced transients. In other aircraft (i.e., carbon fiber composite), the structural resistance of the airframe is also a significant source of induced transients. DO-160D describes both direct pin injection tests and induced cable bundle tests. Five different test levels are applicable to the pin injection tests and five different levels for the cable bundle tests. Both pulse and damped sinusoid test waveforms are specified. All pulse waveforms have a double exponential shape. Current waveform 1 is a waveform with a rise time to peak current of 6.4 μs and a fall time, to 50% amplitude, in 69 μs. Voltage waveform 2 has a rise time of maximum 100 ns and a fall time, to 50% amplitude, of 6.4 μs. Voltage/current waveform 3 is a damped sinusoid.

Voltage waveform 4 has a rise time of 6.4 µs and a fall time, to 50% amplitude, in 69 µs. Current/ voltage waveform 5a has a rise time of 40 µs and 5b 50 µs, and waveform 5a has a fall time, to 50% amplitude, of 120 µs and 5b 500 µs. The voltage waveforms are specified as open circuit and the current into a short circuit. The voltage and current at other loads can be calculated.

Table 5.45 shows the level 1 and level 5 voltage/current test levels for pin injection and Table 5.46 for the cable bundle injection.

In numerous measurements and analysis, the currents and voltages induced in shielded cables based on the cable-induced test levels can be relatively high. For example, with waveform 1 (6.4/70 µs pulse) at test level 4, the test level is 750 V open circuit and 1500 A short circuit. In tests on a 9 m long RG108A/U, twisted shielded pair cable, the internal (center conductor) open-circuit voltage at the load end of the cable was 264 V and the short-circuit current 825 A. The current is C/M, and if a number of circuits are connected to the conductors in the cable bundle, the current will divide between the circuits, with a magnitude dependent on the impedance of each circuit to chassis or safety ground. Traditional miniature secondary transient suppressors will either not carry the test current, or the voltage will rise across them to levels approaching the 264 V level. Some series impedance is necessary on each line before the transient suppressor. The minimum value for this series impedance depends on the test level, how the current is shared between the circuits, and the peak current-handling capabilities of the suppressor. The most common series impedance would be the resistor, although high-value C/M inductors are also useful, and for pulse and audio circuits, an isolating transformer will present a high C/M impedance. If a transformer is used, standard silicon or Schottky diodes may provide sufficient protection due to the low transient current coupled, primarily through the interwinding capacitance, with the transformer.

The level of voltage protection required depends on the maximum rating of the circuit to be protected. In some instances, Schottky diodes are used to limit the voltage to safe levels. The protection shown in Figures 5.153 through 5.155 is primarily designed to protect the circuit, and some disruption of data is to be expected during the application of the lighting test level. Transformer coupling

---

TABLE 5.45

Lowest and Highest Pin Injection Test Levels

	Waveforms		
	Waveform 3: Damped Sinusoid	Waveform 4: 6.4/69 µs	Waveform 5A: 40/120 µs
Level	V_{oc}/I_{sc}	V_{oc}/I_{sc}	V_{oc}/I_{sc}
1	100/4	50/10	50/50
5	3200/128	1600/320	1600/1600

TABLE 5.46

Lowest and Highest Cable Bundle Injection Test Levels

	Waveform				
Level	Waveform 1: 6.4/69 µs V_L/IT_T	Waveform 2: 100 ns/6.4 µs V_T/I_L	Waveform 3: Damped Sinusoid	Waveform 4: 6.4/69 µs V_T/I_L	Waveform 5A: 40/120 µs V_L/I_T
1	50/100	50/100	100/20	50/100	50/150
5	1600/3200	1600/3200	3200/640	1600/3200	1600/5000

FIGURE 5.153 Secondary lightning protection of a single-ended (RS232 type) signal interface.

is the best technique if data integrity is important, although this can only be used on "return to zero" type of data communications.

5.10 ELECTROSTATIC PROTECTION

It is a myth to assume that once an IC is mounted on a PCB; it is immune to an ESD even when the IC input connections are not brought out to a connector. There is no better approach to protection than to avoid an ESD event by correct procedures, such as the use of a ground strap, antistatic bags, grounded mats, and deionizers. ICs mounted on PCBs where the input pins are brought out to an edge connector are just as sensitive to ESD as non-board-mounted ICs. Additional protection for CMOS logic and FET input op amps is relatively easy to achieve by the inclusion of series resistance between the input and the device and either resistance or diodes/zener diodes between the input and the ground.

The simple capacitor to ground may limit the voltage to ground as it provides a capacitor divide between the ESD gun internal capacitance and the input to be protected. Neither commercially nor military ESD requirements normally specify direct application of the ESD gun to the pins of a connector although it may be a customer requirement. However, if a PCB is not in a conductive enclosure, an air discharge can propagate through the enclosure to the PCB.

FIGURE 5.154 Secondary lightning protection of differential input data circuit where the common-mode input voltage must be limited to close to the supply voltage.

FIGURE 5.155 Protection of secondary power (power provided by one equipment to another equipment).

The human body model (HBM) ESD gun contains a 150 pF capacitor that is charged up and discharges through a 330 pF capacitor in a direct contact test and also through the arc in an air discharge. Figure 5.156 shows a signal input protected by capacitors. Assuming a direct contact to the input of 4 kV and the capacitor is 0.01 μF, then the voltage appearing between input and chassis is 60 V and when the capacitor is 0.1 μF 6 V.

If a series inductor or resistor is added, then the voltage can be much lower. If a high value capacitor will change the operation of the circuit, then other transient protection devices may be used.

If a series resistance of 1 kΩ is added with a 0.01 μF, then the voltage will be 15 V. ESD air discharges have very fast rise times in the order of 100 ps. Parasitic inductance on the signal/power

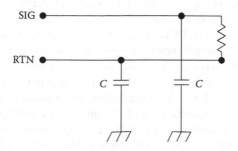

FIGURE 5.156 Electrostatic discharge protection using capacitors.

line tends to slow down this edge, but also when the inductance is in series between the signal and power line, it will increase the peak voltage.

Some of the devices that have been designed to protect against the fast rise time and high energy of EMP will ensure that effective protection against the lower-energy ESD event is achieved. The ESD pulse rise time is typically 15 ns, although the IEC 1000-4-2-specifies 0.8 ns for a contact discharge. IEC 1000-4-2 supplies a method to analyze the radiated fields in an ESD event.

Sometimes, the ESD current is diverted from the susceptible circuit and it is the field radiated as the current flow that results in an upset. For this reason, route the ESD protection circuit and ground as far away from potentially susceptible circuits. Shielding the circuit is another alternative.

The ESD generator must provide a current spike into a low impedance with a rise time of 0.7–1 ns and a pulsewidth of 30–60 ns, and the turn-on time of the majority of EMP devices is 1–2.5 ns. Zener diodes, multilayer varistors, and diode surge arrestors are available with breakdown voltages as low as 6 V and as high as 440 V, with response times from picoseconds to nanoseconds. Connectors are also available with built-in ESD transient suppressors. The European EMC requirements contained in EN50082-1 include an 8 kV air discharge, and in EN50082-2 a 4 kV contact and an 8 kV air discharge, ESD test. IEC 1000-4-2 describes the ESD generator, which has a 150 pF charged up to the test voltage and which is discharged as a single event, through a 330 Ω resistor. In an air discharge, the probe tip is brought close to the EUT as quickly as possible without causing physical damage, and the resistance of the discharge path is the 330 Ω resistor plus the resistance of the resultant plasma. The IEC 1000-4-2 test procedure has the EUT mounted on a conductive tabletop, described as a horizontal coupling plane (HCP) under the EUT. The nonconductive table is located above a ground plane (reference plane). The EUT is insulated from the HCP by a 0.5 mm thick insulator. The EUT is located 0.1 m from a vertical coupling plane (VCP), which is also insulted from the HCP. The HCP is connected to the reference plane by a cable, with a 470 kΩ resistor located at each end, to prevent a buildup of charge. The VCP is also connected separately with an identical cable. The discharge return cable of the ESD generator will be connected to the reference plane (ground plane under the table) at a point close to the EUT. Figure 9.31 shows this test setup. During an air or direct-contact-ESD discharge to the HCP or VCP, the resultant current generates an E field that couples to the EUT. If the EUT enclosure is completely insulating with no exposed metal, for example, fasteners, in either a direct or air discharge to the EUT, the charge is equalized around the nonconductive enclosure, but no noticeable discharge current is detected. In discharges to a metal enclosure, a rapid charging current flow, and in the air discharge a plasma, is seen at the probe tip. The most common problem occurs with discharges close to apertures in a metal enclosure filled with nonconductive material/components, for example, LCD displays, LEDs, touch buttons, keyboards. The field generated by the discharge can couple inside to PCB components through the aperture. Discharges to metal fasteners used to mount nonconductive components may also result in a susceptibility problem. Likewise, discharges to the pins of a connector or even the case of the connector, when it is isolated from the enclosure, are also problem areas. The current flow is predominantly C/M, caused either by the discharge or by the E field resulting from the discharge. Capacitors across the input pins of digital or analog devices can be very effective at increasing immunity, even when a relatively low level. If the capacitor is located between the PCB and the chassis, then it needs to be a relatively high value to be effective at reducing the induced voltage. Although the 330 Û resistor in the generator slows down the charging time for the capacitor, the peak instantaneous charging current is very high and the voltage may be high. For example, in a 4 kV direct discharge, a 10 pF will charge to 3750 V, a 100 pF to 2400 V, a 1000 pF to 520 V, an 0.01 μF to 59 V, and a 0.1 μF to 5.9 V.

If the capacitor is not connected between the PCB ground and the chassis but is used between the input/output pins of a device, the D/M voltage developed due to C/M current flow or the incident E field can be drastically reduced. As with increasing immunity to RF, the capacitor/s should be located as close as possible to the junction/s to be protected. Another use for a capacitor is to slow down the voltage appearing across a slow-acting transient protection device and to give it a chance to clamp the voltage.

Many devices are now available in SM packages for ESD and lightning protection.

The Littelfuse SM712 has two lines each containing a diode and zener in series and is suitable for RS-485. The ESD protection has ±30 kV air, ±30 kV contact lightning 8/20 μs = 4.5 A, and EFT 5/50 ns 50 A. Capacitance is 75 pF and so not suitable for high-speed data.

The Littelfuse SP724 device contains four bipolar SCR/diode clamping circuits with a typical 3 pF capacitance, ESD 8 kV contact, 15 kV air, and lightning 8/20 μs 3 A. The clamp voltage is +Vcc + Vbe and −Vcc − Vbe, peak current 8 A for 1 μs and 2.2 A for 100 μs.

The Littelfuse Surgector thyristor is designed for telecom protection, for example, on tip and ring circuits. The Vmax is 77–400 V and Imax 10 × 1000 μs pulse is 50–100 A and for 2 × 10 μs pulse 320–500 A.

The Microsemi SRLC05 is suitable for !0 Base T Ethernet and USB data rate 900 Mbs. The standoff voltage is 5 V breakdown 5.6 V clamping voltage at 1 A = 8 V, clamping voltage at 5 A = 11 V capacitance 4–6 pF.

One Littelfuse device for high-speed application is the SP0504 S. This contains four diodes to VCC, four diodes to ground, and a zener between Vcc and ground. The specification is ESD = ±12 kV contact, ±12 kV air with a capacitance of 0.85 pF. lightning 8/20 μs = 4.5 A, EFT 5/50 ns 40 A.

Another Littelfuse device with low capacitance suitable for high-speed data and RF is the DP5003. This contains C/M filtering as well as protection diodes. Three C/M chokes and four diodes protect two differential line pairs. The specification is bandwidth >4 GHz, capacitance 0.8–1.3 pF, C/M Z = 32 Ω at 100 MHz. ±15 kV 4-2 Level 4 contact discharge.

The Microsemi TVSF0603 FemtoFarad ESD suppressor has a small 0603 footprint and has an ultralow capacitance of 0.15–0.25 pF and is capable of withstanding a 15 kV 45 A ESD event. These are designed for high-speed data, RF circuits, and mobile devices.

One device designed to protect in input circuits of receivers or the output circuits of transmitters from antenna-coupled ESD events is the AntennaGuard™ manufactured by AVX. These antenna protector chips have capacitance values less than or equal to 12 pF in the 0603 chip size and less than or equal to 3 pF in the 0402 chip size. Antenna protector chips are especially useful in the ESD protection and EMI filtering for high-gain FETs typically used in wireless products. The AntennaGuard may also be used with an inductor to make a low-pass filter. The varistor in the AntennaGuard can turn on in 300–700 ps and provide typical transient suppression of 15 kV air discharge ESD events, reducing the level to survival levels for most FET input preamplifiers.

5.11 ELECTROMAGNETIC PULSE PROTECTION

The primary effect of EMP in shielded cables located outside of buildings, ships, and vehicles is a high-current flow in the shields of cables and a high voltage developed across the inputs of devices connected to these cables. One form of protection is the use of cable shield ground adapters that shunt the high current to ground, for example, as the cable enters the metal wall of the building. The signal or power carried by the shielded cables may then be protected by gas discharge arrestors with fast response times of approximately 1 ns. Secondary protection on LF signal lines and power may be achieved by fast-response semiconductor devices similar to those used for lightning protection. The Polyphaser Corporation provides an EMP or lightning protection device for DC remote/telephone line and tone remote/dedicated line protection. Reliance Comm/Tec manufactures a building entrance terminal that incorporates EMP/lightning protection for 6, 12, or 25 pair tip and ring circuits.

The protection of HF receivers, transmitters, and preamps against EMP is more difficult than for lightning protection, due to the fast rise time of the EMP event. One method is the use of protection tees made up of a shorted quarter-wave section in parallel with the input. The shorted quarter-wave section provides a low impedance to ground at frequencies other than the signal for which the tee is designed. Figure 5.157 shows the schematic of the tee, combined with a decoupling line, and the construction of a 900 MHz tee. The decoupling line is a concentric coaxial line in series with the signal of length equal to a quarter-wave at the signal frequency and may be used without the shorted quarter-wave section for protection. The frequency limits of the tee and decoupling line are from 25 MHz to 18 GHz.

FIGURE 5.157   Tee and decoupling line schematic and construction. (Courtesy of Les Cables de Lyon Alcatel.)

The current induced by an EMP event has been characterized in MIL-STD-461F CS116 as a damped sinusoid with a frequency component extending from 10 kHz to 100 MHz. The maximum C/M current flow is 10 A, which is injected directly into cables (test CS116). Protection against CS116 test levels may be achieved by filters and/or the semiconductor transient suppressors described for secondary lightning protection.

REFERENCES

1. R.K. Keenan. *Digital Design for Interference Specifications.* The Keenan Corporation, Vienna, VA, 1983.
2. E. Kami. Design guide for electromagnetic interference (EMI) reduction in power supplies, MIL-HDBK-241B. Power Electronics Branch, Naval Electronic Systems Command, Department of Defense, Washington, DC, 1983.
3. H.M. Schlicke. *Electromagnetic Compossibility.* Marcel Dekker, New York, 1982.
4. R.B. Cowdell. Don't experiment with ferrite beads. *Electronic Design,* June 7, 1969, 12.
5. Schematics of high immunity interface circuits, PLR2, PLD2, HD-1, HR-1, HAD-1, and HAR-1. CAL Corporation.
6. D.J. Kenneally. RF upset susceptibilities of CMOS and low power Schottky D-type, flip flops. *IEEE International Symposium on Electromagnetic Compatibility,* Denver, CO, May 23–25, 1989.
7. T.F. Fang and J.J. Whalen. Application of the Nonlinear Circuit Analysis Program (NCAP) to predict RFI effects in linear bipolar integrated circuits. *Proceedings of the Third Symposium Technical Exhibition on Electromagnetic Compatibility,* Rotterdam, the Netherlands, May 1–3, 1979.
8. D.S. Britt, D.M. Hockanson, F. Sha, J.L. Drewniak, T.H. Hubing, and T.P. Van Doren. Effects of gapped groundplanes and guard traces on radiated EMI. *IEEE International Symposium on Electromagnetic Compatibility,* Austin, TX, August 18–22, 1997.
9. F.B.J. Leferink and M.J.C.M. van Doom. Inductance of printed circuit board ground plates. *IEEE International Symposium on Electromagnetic Compatibility,* Dallas, TX, August 9–13, 1993.
10. L.O. Johansson. EMC fundamentals: Passive and their EMC characteristics. *Compliance Engineering,* July/August 1998, 15(4), 24–39.
11. H.G. Ghadamabadi and J.J. Whalen. Parasitic capacitance can cause demodulation RFI to differ for inverting and non-inverting operational amplifier circuits.
12. B. Archambeault and J. Diepenbrock. IBM, RTP, NC PCB power decoupling myths debunked. Available on the internet. https://ewh.ieee.org/r3/enc/emcs/archive/2012-10-10b_Decouplingmyths.pdf.
13. Creating ferrite bead models. www.spectrum-oft.com/news/winter2011/ferritebead.shtm.
14. C. Rostamzadeh, F. Grassi, and F Kashefi. Modeling SMT ferrite beads for SPICE Simulation. *IEEE Electromagnetic Compatibility Symposium,* Long Beach, CA, 2011.

FIGURES 18.7

REFERENCES

6 Electromagnetic Shielding

6.1 REFLECTION, ABSORPTION, AND SHIELDING EFFECTIVENESS

6.1.1 REFLECTION IN A PERFECT CONDUCTOR

When an electromagnetic wave impinges on a sheet of metal, the electric field component induces a current flow in the sheet. Some electromagnetic theories of shielding state that the current flow has a current density of $2H_o$ or $2E_o/Z_w$, whereas other electromagnetic theories state that the current is equal to H_o. Figure 6.1 shows the incident electromagnetic wave and the induced current, from which it can be seen that the magnetic intensity produced by the induced current is such as to annihilate the magnetic intensity behind the sheet and to double the intensity in front of the sheet, assuming a perfectly conducting infinitely large sheet. A standing wave is generated in front of the sheet, and this can be seen in Figure 6.1.

No metal is a perfect conductor (i.e., exhibits zero impedance), and thus, complete reflection of an electromagnetic wave is not possible.

6.1.2 TRANSMISSION-LINE THEORY APPLIED TO SHIELDING

Although wave theory may be used to predict the level of shielding obtained by a conductive sheet [1], the concept of wave impedance introduced by Schelkunoff in 1938, and the application of transmission-line theory to shielding, results in a considerably simpler solution. Using transmission-line theory, the reflection is explained by the impedance difference between the incident wave and the metal impedance.

Figure 6.2 illustrates the incident field with impedance Z_w impinging on the near side of the metal shield, which exhibits a metal impedance Z_m or a barrier impedance Z_b, as described in the following sections. When the wave impedance is higher than the metal impedance, some of the incident field is reflected from the near side of the shield, due to the impedance mismatch. As the remaining wave propagates through the shield, it is attenuated. A fraction of the field incident on the far side of the interior of the shield is re-reflected due to the metal- to air-impedance mismatch. The remaining field propagates from the far side of the shield into space. A second re-reflection of the internal field occurs at the near-side metal-to-air interface, and some small fraction of the return trip field leaves the far side of the shield. This small fraction is insignificant when the absorption through the shield is greater than the numeric 5. The re-rereflected field as it leaves the metal is considered to be in phase with, and therefore adds to, the field that has propagated through the metal.

As discussed in Reference [2], some authors state that the wave from the shield on the far side of the incident wave is either 377 Ω or the same impedance as the incident field. Other authors state that there is no reflection coefficient at the far side of the shield: the impedance of the wave is the same as the impedance of the barrier, and the wave impedance approaches 377 Ω only at a distance of $\lambda/2\pi$ in meters from the shield. In the measurements on aluminized Mylar, discussed later, the results indicated that the current in the barrier material was consistent with the value of the H field and that the impedance of the wave emanating from the barrier was the same as the impedance of the barrier, or, in the case of the aluminized Mylar, 2 Ω.

The following equations are used by packaging engineers and designers for small enclosures with seams for which they are totally inapplicable as described in detail in Section 6.2.1.

If the problem is the shielding of a six-sided enclosure with seams, then the analysis shown in Section 6.3.10 is more appropriate. If the problem is an enclosure with apertures, then the analysis described in Section 6.4.1 should be used.

369

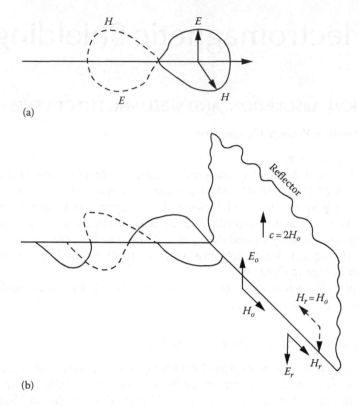

FIGURE 6.1   Plane wave incident on a perfect conductor. (a) Plane wave and (b) resultant wave.

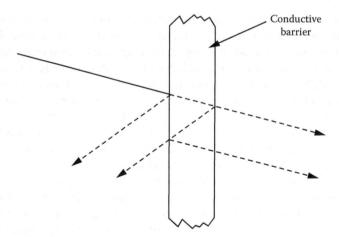

FIGURE 6.2   Transmission losses at a conductive sheet.

6.1.3   METAL IMPEDANCE, SKIN DEPTH, BARRIER IMPEDANCE

The impedance of a metal is made up of resistance and inductance and is a function of its conductivity, permeability, thickness, and the frequency of the current flow in it. The impedance of a metal when its thickness is much greater than the skin depth is given by

$$Z_m = 369 \sqrt{\frac{u_r f}{G_r}} \ \left[\mu\Omega^2 \right] \tag{6.1}$$

where
  μ_r is the permeability relative to copper
  G_r is the conductivity relative to copper
  f is the frequency in megahertz

We use the term Ω^2 to describe surface impedance or resistance. The question is often asked, per square what? Is it inches, centimeters, or meters? The equation for the resistance of a conductor is

$$R = \frac{\rho l}{wt} \tag{6.2}$$

where
  l is the length of the conductor
  ρ is the resistivity
  w is the width of the conductor
  t is the thickness of the conductor

From Equation 6.2, we see that, for a constant thickness and assuming $l = w$, the resistance is constant, regardless of the magnitude of l and of w. Thus, w and l may be 1 mm or 1 m and the resistance is the same. When the Ω^2 term is used to describe surface resistance, the thickness should be specified. When used to describe surface impedance and when the thickness is much greater than the skin depth at all frequencies of interest, the thickness of the conductor may be omitted, for as the frequency of current in a conductor increases, the current density does not remain constant throughout the depth of the metal but is greater at the surface. The skin depth of a metal is defined as the surface depth at a specific frequency at which 63.2% of the current flows. Equation 6.3 shows that the skin depth for any metal decreases with increasing frequency and is lower for metals with high conductivity and permeability:

$$\delta = \frac{1}{\sqrt{\pi f \mu G}} \; [\text{m}] \tag{6.3}$$

where
  μ is the permeability of the metal $= \mu_o \mu_r$
  G is the conductivity in mhos/meter
  f is the frequency in hertz

For copper, where $\mu_r = G_r = 1$,

$$\delta = \frac{0.066}{\sqrt{f}} \; [\text{mm}]$$

where f is the frequency in megahertz. For any metal,

$$\delta = \frac{0.066}{\sqrt{f \mu_r G_r}} \; [\text{mm}]$$

where f is the frequency in megahertz. Table 6.1 shows the skin depth of some of the common metals at several frequencies.

TABLE 6.1
Skin Depth for Some Common Conductors

Conductor	Conductivity [× 10⁷ mhos/m]	Relative Conductivity (G_r)	Relative Permeability (μ^a)	Skin Depth with Frequency (mm)				
				1 kHz	1 MHz	10 MHz	100 MHz [× 10⁻³]ᵇ	1 GHz [× 10⁻³]ᵇ
Aluminum	3.54	0.6	1	2.7	0.085	0.027	8.5	2.7
Copper	5.8	1	1	2.0	0.066	0.02	6.6	2.0
Gold	4.5	0.7	1	2.5	0.079	0.025	7.9	2.5
Tin	0.87	0.15	1	5.4	0.170	0.054	17	5.4
Nickel	1.3	0.23	100	0.43	0.014	0.0045	1.4	0.44
4% silicon iron	0.16	0.029	500	0.55	0.017	0.0055	1.7	0.55
Hot-rolled silicon steel	0.22	0.038	1,500	0.27	0.0087	0.0027	0.87	0.27
Mumetal	0.16	0.029	20,000	0.087	0.0027	0.00087	0.27	0.087

ᵃ At 10 kHz and $B = 0.002$ Wb/m².

ᵇ Expressions in square brackets represent the exponent portion of the number, for example, 3.54×10^7 mhos/m or 3.54×10^{-5} mhos/cm.

Due to the skin depth effect, the barrier impedance of the metal instead of the metal impedance may be required in the evaluation of shielding effectiveness (SE). The barrier impedance of a metal is a function of the ratio of metal thickness to skin depth, in addition to the frequency dependency of the impedance. The barrier impedance of any metal is given by

$$Z_b = 369\sqrt{\frac{\mu_r f}{G_r}}\frac{1}{\left(1-e^{-t/\delta}\right)}\left[\mu\Omega^2\right] \qquad (6.4)$$

where
   f is the frequency in megahertz
   t and δ are in the same units

When $t/\delta \ll 1$,

$$Z_b = \frac{1.414}{0.058G_r t}\left[\mu\Omega^2\right] \qquad (6.5)$$

where
   t is the thickness [mm]
   G_r is the relative conductivity [mhos/m]

Because no metal is a perfect conductor, the current flow in the metal sheet shown in Figure 6.1 is not confined to its surface but extends into the thickness of the metal. The impedance of a pure conductor is generally lower than that of the incident wave, with the possible exception of a current loop in very close proximity to a shield.

6.1.4 Reflection in Practical Conductors

From the explanation of reflection in Section 6.1.1, it may appear that for magnetic fields, where the electric field component is less than that of the magnetic field, the reflection will be less complete than for an electric field, and this is supported by transmission-line theory. The impedance of a field changes with distance and frequency, and the metal impedance of the reflecting sheet is a function of frequency, permeability, and conductivity, as shown in Equation 6.1. The equation for the reflection loss in decibels in an electric field when the metal thickness $t \gg \delta$ is given by

$$R_e = 353.6 + 10\log\left(\frac{G_r}{f^3\mu_r 2.54r_1^2}\right)[dB] \qquad (6.6)$$

where
   f is the frequency [Hz]
   r_1 is the distance from an electric field source [cm]

The equation for the reflection loss in a magnetic field is given by

$$R_m = 20\log\left[\frac{0.181}{r_1}\sqrt{\frac{\mu_r}{G_r f}}+0.053r_1\sqrt{\frac{G_r f}{\mu_r}}+0.354\right][dB] \qquad (6.7)$$

where
   f is the frequency [Hz]
   r_1 is the distance from source [cm]

The equation for the reflection of a plane wave is given by

$$R_p = 168.2 + 10\log\left[\frac{G_r}{\mu_r f}\right] \text{[dB]} \tag{6.8}$$

When the source of emission is neither a low-impedance current loop nor a high-impedance voltage source but a source that exhibits a characteristic impedance, Z_c (e.g., a transmission line), we can use the circuit impedance to calculate the reflection in the near field. The equation for reflection loss using circuit impedance is given by

$$R_c = 20\log\left[\frac{Z_c}{\left(1.48\times10^{-3}\right)\sqrt{\dfrac{\mu_r f}{G_r\times10^6}}}\right] \text{[dB]} \tag{6.9}$$

where f is the frequency in hertz.

6.1.5 ABSORPTION

The variation in current density through a thick metal sheet is shown in Figure 6.3. The radiation from the far surface into space is proportional to the current flow on that surface. Therefore, due to the reduction in current flow, an absorption of the incident wave effectively occurs in the metal. The equation for absorption is independent of wave impedance, proportional to thickness, and inversely proportional to skin depth and is given by

$$A = \left(3.338\times10^{-3}\right)t\sqrt{\mu_r f G_r} \tag{6.10}$$

where
   A is the absorption loss [dB]
   t is the thickness of shield [mils] (1 mil = 1/1000 in. = 0.0254 mm)
   f is the frequency [Hz]

FIGURE 6.3 Variation in current density through a thick metal sheet. (From Broaddus, A. and Kunkel, G., Shielding effectiveness test results of aluminized mylar, *IEEE EMC Symposium Record*, 1992.)

6.1.6 RE-REFLECTION CORRECTION

When the absorption loss in the thickness of the metal is not much greater than 5, the re-reflection within the metal is appreciable and the re-reflection gain should be included in the total SE equation. Often, the term "re-reflection loss or correction factor"* is used to describe the effect of re-reflections, but since this is always a negative number, it subtracts from the total SE and is effectively a gain.

Re-reflection loss, which is always negative, is given as a magnitude in decibels:

$$R_r = 20 \log \left\{ 1 - \left[\left(\frac{Z_m - Z_w}{Z_m + Z_w} \right)^2 (10^{-0.1A} (e^{-j0.227A})) \right] \right\} \tag{6.11}$$

where

Z_m is the metal impedance

A is the absorption loss [dB]

$Z_w = 377 \dfrac{2\pi r}{\lambda}$ when $r < \dfrac{\lambda}{2\pi}$ for a magnetic field

$Z_w = 377 \dfrac{\lambda}{2\pi r}$ when $r < \dfrac{\lambda}{2\pi}$ for an electric field

Z_w is 377 Ω for a plane wave

r is the distance from source [m]

The re-reflection loss is a function of the ratio between the wave impedance Z_w and the metal impedance Z_m and of the absorption loss. For thin metals, where the thickness is less than the skin depth, Z_b should be used instead of Z_m. When Z_w is much greater than Z_m or Z_b, the re-reflection loss becomes an inverse function of the absorption loss only.

Kunkel [31,32] describes measurements and a comparison between the electromagnetic shielding using circuit theory and wave theory.

The conclusion is that the interpretation of the "accepted shielding effectiveness" equation, described in this section, has led the EMC community to the belief that there is a reflection factor for a wave inside the barrier and that the E and H fields are attenuated equally, whereas the measurements and wave theory do not support this interpretation.

6.2 SHIELDING EFFECTIVENESS

SE is defined as the ratio between the field strength at a given distance from the source without the shield interposed and the field strength with the shield interposed. The total SE of a conductive barrier in decibels is the sum of the reflective losses expressed in decibels, the absorption in decibels, and the re-reflection loss in decibels; thus, SE is given by

$$SE = R + A + R_r \, [\text{dB}] \tag{6.12}$$

where R, A, and R_r are in decibels.

The magnetic field reflection loss versus frequency, distance, and ratios of relative conductivity to relative permeability (G_r/μ_r) is shown in the nomogram of Figure 6.4.

The electric field reflection loss versus frequency, distance, and ratios of (G_r/μ_r) is shown in the nomogram of Figure 6.5. The plane-wave reflection loss versus frequency, distance, and ratios of G_r/μ_r is shown in the nomogram of Figure 6.6. The absorption loss versus frequency, thickness, and

---

* As discussed in Section 6.1.2 in measurements on aluminized mylar, 2 Ω, no reflection coefficient was seen.

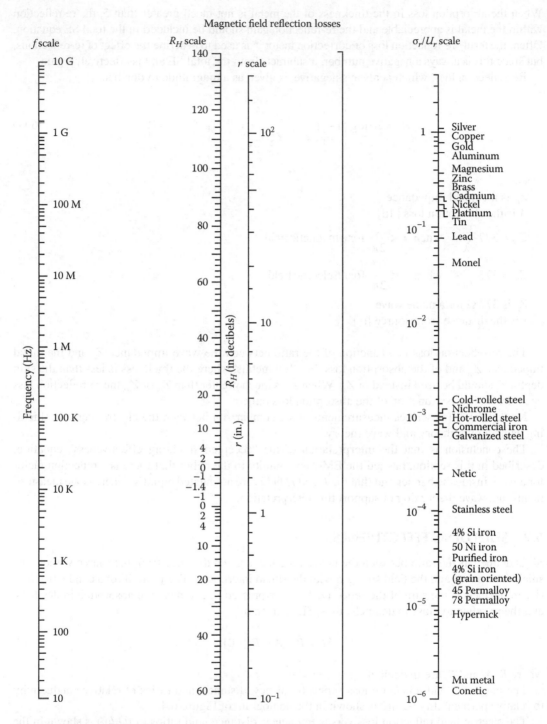

FIGURE 6.4 Magnetic field reflection loss. (From Broaddus, A. and Kunkel, G., Shielding effectiveness test results of aluminized mylar, *IEEE EMC Symposium Record*, 1992. Reprinted from EDN, September 1972. © 1972. Cahners Publishing Company, Division of Reed Publishers, New York.)

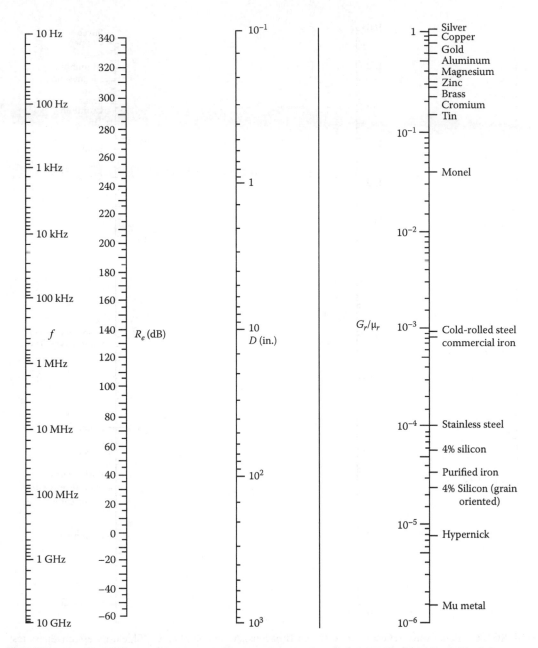

FIGURE 6.5 Electric field reflection loss. (From Broaddus, A. and Kunkel, G., Shielding effectiveness test results of aluminized mylar, *IEEE EMC Symposium Record*, 1992. Reprinted from EDN, September 1972. © 1972. Cahners Publishing Company, Division of Reed Publishers, New York.)

product G_r^*/μ_r is shown in the nomogram of Figure 6.7. All of these nomograms are reproduced from Reference [3]. The nomograms in Figures 6.4 through 6.6 provide for a minimum ratio of G_r/μ_r of 1×10^6, so the SE of very-low-conductivity materials with a low permeability, such as graphite, may be found.

In choosing the location and type of shield, the following rules are useful, though not infallible:

Absorption losses increase with increase in frequency, barrier thickness, barrier permeability, and conductivity, all of which increase the ratio of barrier thickness to skin depth.

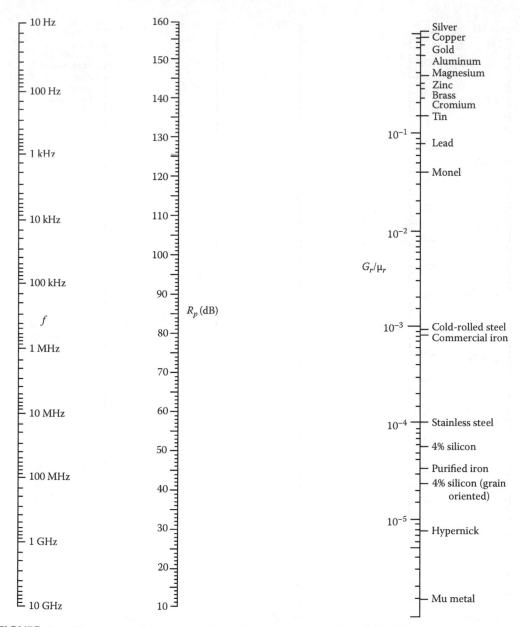

FIGURE 6.6 Plane-wave reflection loss. (From Broaddus, A. and Kunkel, G., Shielding effectiveness test results of aluminized mylar, *IEEE EMC Symposium Record*, 1992. Reprinted from EDN, September 1972. © 1972. Cahners Publishing Company, Division of Reed Publishers, New York.)

Assuming that the shield thickness is greater than the skin depth, the following are valid:

Reflection losses, above 10 kHz, generally increase with an increase in conductivity and a decrease in permeability (i.e., with a decrease in metal impedance).

Reflection, *E* field, increases with a decrease in frequency and a decrease in distance from the source, both of which increase the ratio of wave impedance to metal impedance.

Reflection, *H* field, increases with an increase in frequency and an increase in distance from the source, due to a corresponding increase in wave impedance.

Reflection, plane wave, increases with a decrease in frequency as the metal impedance decreases.

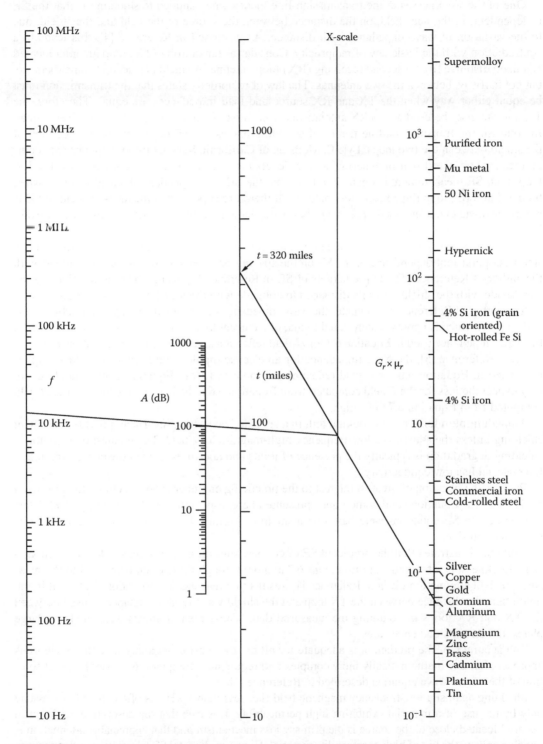

FIGURE 6.7 Absorption loss. (From Broaddus, A. and Kunkel, G., Shielding effectiveness test results of aluminized mylar, *IEEE EMC Symposium Record*, 1992. Reprinted from EDN, September 1972. © 1972. Cahners Publishing Company, Division of Reed Publishers, New York.)

One of the weaknesses of the transmission-line theory when applied to shielding is that the SE is dependent, in the near field, on the distance between the source of the field and the shield, due to the variation in wave impedance with distance. As discussed in Reference [4], this leads to a contradiction with the basic law of reciprocity. Consider a transmitting (TX) loop antenna located at a fixed distance from a second receiving (RX) loop antenna. A shield is placed at some distance, but not halfway, between the two antennas. The law of reciprocity states that the transmission must be equal either way when the TX and RX source and load impedances are equal. Therefore, the TX antenna may be used as the RX antenna and vice versa, with no change in the received signal. In contrast, the transmission-line theory of shielding makes the SE dependent on the location of the shield between the two loops. Lyle E. McBride of California State University, in a private communication, observed that in experiments the Schelkunoff assumptions are correct when the RX loop in an SE measurement is open circuit but that the SE is independent of shield location when both TX and RX loop impedances are matched. It therefore appears that the measurement of SE is not an intrinsic electromagnetic parameter but is dependent on the test setup, as discussed further in Section 6.6.2. Reference [1] compares the prediction of magnetic field shielding for copper, aluminum, and steel shields placed between circular loop antennas and notes in the measurements a small but persistent dependence of the SE on the spacing between the two loops. In common with the author of Reference [5], the predictions of SE in Reference [1], using transmission-line theory, were made with the shield virtually displaced to coincide with the location of the RX loop.

The author of Reference [5] computes the wave impedance of a magnetic field generated by a magnetic dipole, which is approximately equal to that of a current loop if the radius is small, at one-third the wave impedance used in Equation 6.7 for H-field reflection and obtained from Equation 2.29.

From Reference [5], the wave impedance for an electric dipole is three times the wave impedance used in Equation 6.6 for E-field reflection and obtained from Equation 2.30. Therefore, we may expect the loss for the E field computed from Equation 6.6 to be low and the loss for the H field computed from Equation 6.7 to be high.

Thin aluminum foil, such as cooking foil, in a six-sided enclosure, invariably provides sufficient shielding unless the source is a low-frequency high-magnitude H field. The greatest contributor to shielding degradation is typically the presence of joints and apertures and not the enclosure, unless the material has low conductivity.

Perhaps the question of greatest interest to the practicing engineer is how accurate the predicted SE is using the equations and nomographs presented here, especially for the shielding of H fields. The preceding SE equations were based on a shield of infinite size, whereas practical enclosures are limited in size.

Table 6.2 illustrates that the predicted SE of copper, aluminum, and steel shields of $\frac{1}{16}$ in. and $\frac{1}{8}$ in. thickness, as obtained from Equations 6.7 and 6.9 through 6.12, when compared to the measured H-field attenuation given in Reference [1], results in a reasonable correlation. Note that in the prediction, the distance between the TX loop and the shield was the same distance as used between the TX and RX loops in obtaining the measured data. These measurements were made on large plates, not on six-sided enclosures.

The accuracy of the prediction is adequate for all but the most exact analysis. A more rigorous approach, but one mathematically more complex and requiring a computer, is to use the exact solution of the vector wave equation described in Reference [1].

Shielding against a low-frequency magnetic field (i.e., less than 1 kHz) is often quoted as possible only by the use of a material exhibiting high permeability. It is true that the absorption achieved in material located close to the source is the primary loss mechanism and that appreciable attenuation is possible only by the use of high-permeability material. However, the total SE (absorption and reflection) of a 30 mil thick copper sheet ($\mu_r = 1$, $G_r = 1$) is higher at frequencies below 3 kHz than a 30 mil silicon steel sheet [$\mu_r = 1500$, $G_r = 0.038$ ($\mu_r \times G_r = 57$)] at a distance of 38 cm from a magnetic field source.

The reason is that the reflection from the copper sheet, at 38 cm distance, is higher than that from the steel sheet. A thin, 20–50 micron sheet (20 μm = 0.0008 in. = 0.02 mm; 50 μm = 0.002 in. = 0.05 mm)

TABLE 6.2

Predicted and Measured Shielding Effectiveness

F	d (cm)	Shielding Effectiveness (dB)	
		Predicted	Measured
Aluminum $\frac{1}{16}$ in. thick $G_r = 0.35$, $\mu_r = 1$			
100 Hz	10	7.5	5
20 kHz	10	50	42
2 kHz	20	34	25
200 Hz	35	21	12
Aluminum ⅛ in. thick, $G_r = 0.35$, $\mu_r = 1$			
1 kHz	3	18	15
30 kHz	3	66.9	65
100 Hz	30	19	10
1 kHz	30	36.6	28
Copper ⅛ in. thick, $G_r = 1$, $\mu_r = 1$			
100 Hz	3	9.2	8
30 kHz	3	101	98
100 Hz	20	22	17
1 kHz	20	43	38
Steel $\frac{1}{16}$ in. thick, $G_r = 0.17$, $\mu_r = 112$			
100 Hz	3	12	16
10 kHz	3	91	102
1 kHz	20	33	32

with a relative permeability of 200 and a plating of either copper, zinc, nickel, tin, or chrome is available from the Iron-Shield Company, Harrington, IL 60010-0751. This material combines the advantages of high surface conductivity and moderate permeability and may be laminated to various nonconductive materials. The SE is shown in Figure 6.8a.

The reflection loss to a magnetic, electric, field or plane wave is dependent on the surface conductivity. When aluminum is left untreated, aluminum oxide builds up on the surface and the conductivity and reflection loss are reduced. Alodine, Iridite, and Oakite are high-conductivity passivation finishes and do not appreciably reduce the reflection loss, whereas anodizing results in a finish with very low conductivity. Figure 6.8b illustrates the degradation in reflection loss due to these types of finishes. The reference 0 dB line represents clean, unpassivated aluminum, the reflection loss of which varies with frequency but is approximately 80 dB. The reflection loss degradation is specified for an E-field source at low frequency and a plane wave at high frequency.

Coating a metal surface with an insulating finish such as the majority of lacquers and varnishes does not have an adverse effect, whereas certain types of paint containing carbon, graphite, or conductive particles will reduce the reflection loss.

High-permeability sheets and stocks designed to shield against DC to 60 Hz magnetic fields, typically for use around transformers and cathode ray tubes, are available from Eagle Magnetics Co. Ltd., P.O. Box 24283, Indianapolis, IN 46224; Magnetic Shield Corporation, 740 North Thomas Drive, Bensenville, Illinois 60106-1643; and Mμ Shield, P.O. Box 439, Goffstown, New Hampshire 03045. Such materials exhibit relative permeabilities from 20,000 to 300,000 and resistivities of 12–60 $\mu\Omega/cm^3$.

Table 6.3 shows the SE and permeability versus incident magnetic field at DC and 60 Hz of high-permeability material. The equations are useful in calculating the SE, and an example is shown for a 3 in. × 3 in. × 3 in. five-sided box around a source of magnetic field. Table 6.3 and the equations are reproduced by the courtesy of Eagle Magnetic Company. Material with a permeability higher than 1000 tends to be sensitive to frequency and magnetic flux density within the material. Saturation

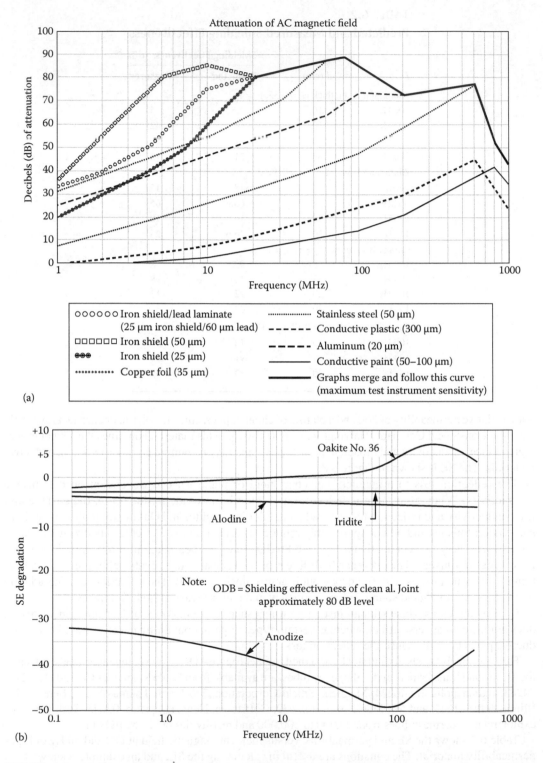

FIGURE 6.8 (a) Examples of electric and magnetic field attenuation for various types of shielding. (Reproduced courtesy of Iron-Shield Company.) (b) Reflection loss degradation due to surface finishes on aluminum.

TABLE 6.3

Shielding Effectiveness and Permeability versus External Field Strength, Frequency, and Thickness

H_1 (Oersted)	B (Gauss)	μ (EFF)	G (AttRat)	SE (D_b)	H_o	μ (EFF)	G (AttRat)	SE (D_b)	H_o
		DC					60 Hz		
	EAGLE AAA.025 in. (thickness)					EAGLE AAA.025 in. (thickness)			
1	150	90,000	375	52	.0028	30,000	125	41	.008
5	750	140,000	565	55	.008	40,000	167	44	.030
10	1500	200,000	835	58	.012	45,000	188	46	.053
25	3750	300,000	1250	63	.020	30,000	125	41	.200
50	7500	Too close to saturation				Too close to saturation			
	EAGLE AAA.031 in. (thickness)					EAGLE AAA.031 in. (thickness)			
1	121	90,000	465	53	.0021	25,000	127	42	.0079
5	605	140,000	710	57	.0071	34,000	173	45	.028
10	1210	200,000	1020	60	.0099	36,000	183	46	.055
25	3025	300,000	1520	62	.0165	26,000	132	43	.190
50	7500	Too close to saturation				Too close to saturation			

Equation #1 $\quad B = \dfrac{2.5DH_1}{2T} = \dfrac{2.5(3)(1)}{2(.025)} = 150$ Gauss

Equation #2 $\quad G = \dfrac{\mu T}{2L} = \dfrac{90,000(.025)}{2(3)} = 375$

Equation #3 SE = 20 log G = 20 log (375) = 52 dB_0

Equation #4 $\quad H_o = \dfrac{H_1}{G} = \dfrac{1}{375} = .0028$ Oersted

H_1 is the internal field (Oersted)
B is the flux that shield sees
μ is the effective permeability
C is the attenuation ratio
SE is the shielding effectiveness (D_b)
H_o is the external field (Oersted)
T is the thickness (inches)

of the material occurs at high flux densities, and the permeability decreases with increasing frequency. When shielding against low-frequency high magnetic fields, the use of a material such as steel followed by a high-permeability material is recommended. The first barrier of steel is designed to absorb the magnetic field, without saturation, to a level at which the second high-permeability barrier does not saturate. To illustrate the importance of choosing the correct shielding material for the job, we offer the following example: A shielded room was constructed, at great expense, of mumetal, with the aim of achieving a high level of shielding. Unfortunately, because the permeability of mumetal decreases rapidly with increasing frequency and the material exhibits a low conductivity, the SE of the room at frequencies in the megahertz region was less than that achieved with a much cheaper room constructed of cold-rolled steel.

6.2.1 Caution in the Use of the Shielding Effectiveness Equation

Many mechanical and packaging engineers apply the preceding equations to six-sided conductive electronic enclosures with a massive over- or underestimation of the true SE.

For a six-sided enclosure made from thin foil with no seams, the theoretical shielding H-field SE with the TX loop antenna 0.2–0.5 m from the enclosure may be much higher than predicted from the H-field reflection equation as shown in Figure 6.89.

TABLE 6.4

Comparison of Transmission Equation for *E*-Field Shielding and an Enclosure with Seams

Frequency [MHz]	SE Using Transmission Line, Equation 6.12	SE Measured and Predicted
200	207	83
500	274	63
700	308	60

For *E*-field shielding, Section 6.3.10 describes a room with seams where the measured SE is much lower than predicted using the transmission-line equations, and this is illustrated in Table 6.4.

6.2.1.1 Isolated Metal Plate with and without Gap

When the source and the receptor are close together, such as on adjacent PCBs, then the measured coupling between source and receptor, described in more detail in Section 11.9, can vary from 13.8 dB at 10 MHz to 0 dB and 5.7 dB at 820 MHz. Although when tried as an EMI fix has proven effective. The term *isolated plate* means it does not contact the inside metal surfaces of the EUT enclosure or any signal or power ground. A resistor with a high value connected between the plate and chassis may be used to avoid an ESD event.

When the source is some distance from the plate, as in a radiated immunity/susceptibility test, then the radio frequency (RF) current is induced across the surface of the plate that reradiates and the level of SE achieved behind the plate is either limited, zero, or negative.

If a plate is divided into two with a gap between the two halves, then very close to the gap, on the side of the plates furthest from the antenna, the *E* field may be higher than without the plates. Further behind the gap, some attenuation of the field may be measured.

For a 60 cm tall plate with a 30 cm width and a vertically polarized incident field, the measured SE at 300 MHz is 0 dB 5 mm behind the plate, 4 dB at 2 cm behind the plate, 6 dB at 4 cm behind the plate, and 10 dB at 7 cm.

At 240 MHz, the SE is 10 dB at 5 mm, 10 dB at 2 cm, 12 dB at 4 cm, and 15 dB at 7 cm.

With a 2 mm wide slot in the center of the 0.6 m × 0.3 m plate, the gain at 300 MHz 5 mm behind the slot is 7 dB, −1.5 dB at 2 cm, −3 dB at 4 cm, and −5 dB at 7 cm. That is, the plate with slot exhibits gain very close to the slot and some limited attenuation further behind. At 240 MHz, the gain at 5 mm behind the slot is 3 dB, at 2 cm 0 dB, at 4 cm −1 dB, and at 7 cm −3 dB.

This means that at 240 MHz the difference between the plate with and without a slot varies between 7 and 13 dB, whereas at 300 MHz the difference is between 7 and 3 dB.

With a 4 mm wide gap, the gain at 300 MHz compared to no plates is 5 dB at 5 mm, 0 dB at 2 cm, 3.5 dB at 4 cm, and 3 dB at 7 cm. At 240 MHz, close to where the plate height is 0.5 lambda, the gain is 10 dB at 5 mm, 6 dB at 2 cm, 3 dB at 4 cm, and 1 dB at 7 cm.

For a 0.5 m × 0.5 m plate without a slot at 600 MHz, the attenuation was 6 dB at 5 mm, 8 dB at 2 cm, 8 dB at 4 cm, and 9 dB at 7 cm behind the plate.

At 300 MHz, where the 0.5 m height corresponded to 0.5 lambda, no attenuation was seen; instead, the *E* field was higher than without the plate, and this increase (gain) was 11 dB at 5 mm, 8 dB at 2 cm, 5 dB at 4 cm, and 2 dB at 7 cm behind the plate.

When the plate measured only 20 cm × 20 cm, then no attenuation nor gain could be measured at either 300 MHz or 240 MHz.

An EUT exposed to a radiated immunity test at 20 V/m failed at 12 V/m, 300 MHz. The most susceptible side was covered by a vertical plate, approximately 0.3 m × 0.3 m in dimension, which was not electrically connected to the baseplate. Both plates and the frame on which they were mounted were anodized and thus isolated, and a gap existed between the baseplate and the vertical plate. With the anodizing removed from the edges of the baseplate and the vertical plate and with aluminum tape with conductive adhesive used to cover the gap and electrically connect the two plates, the EUT passed at 35 V/m with the field incident on the plate side of the EUT, a 9.3 dB improvement.

On an open side of the EUT, not covered by a plate and with the field incident on wiring and a PCB, the EUT passed at 29 V/m.

6.3 NEW SHIELDING MATERIALS: CONDUCTIVE PAINTS AND THERMOPLASTICS, PLASTIC COATINGS, AND GLUE

With an increase in the use of nonconductive materials such as plastics and resins for the enclosures of computers, peripheral equipment, and the like, increased use is also made of conductive coatings and conductive moldable materials for the purpose of shielding. Conductive thermoplastics are available with a base material of nylon, polypropylene, polyethylene, thermoplastic rubber, SMA, ABS, polycarbonate, PES, polystyrene, acetal, etc. The filler may be carbon fibers, carbon black, metallized glass, nickel-coated carbon fibers, or stainless steel fibers. The higher the ratio of filler to base material, the higher the conductivity, up to a limit, but the greater the change in the physical properties of the base material. The most common change is that the material becomes brittle and will break more readily when dropped. The conductivity of conductive thermoplastics is typically lower than that of conductive coatings, with the use of carbon-filled plastic confined to electrostatic protection. The surface resistivity of conductive thermoplastics ranges from 10 Ω^2 to 1 MΩ^2 and the volume resistivity from 1 Ω/cm^3 to 1 MΩ/cm^3. Conductive coatings may be painted on or applied by flame spray, thermospray, or plasma flame spray. The surface resistance of a nickel-based paint versus thickness is shown in Figure 6.9, from which we see that the conductivity increases with increasing thickness in the thickness range of 1.5–3 mm. The manufacturer of the paint, along with other conductive coatings, is Acheson Colloids Company, Post Huron, MI 48060.

The SE of a material is often measured using the MIL-STD-285 test method, in which a TX antenna is located within a small, shielded enclosure that is in turn placed inside a shielded room.

FIGURE 6.9 Resistance versus film thickness. (Reproduced courtesy of Acheson Colloids Company, Port Huron, MI.)

One face of the small, shielded enclosure is removed and covered by a panel of the material under test. An RX antenna is placed inside the shielded room at a distance of 1 m from the panel under test. The *shielding effectiveness* is defined as the ratio of the measured field without the panel to the measured field with the panel closing the aperture. When a loop antenna is used as the TX and RX antennas, the magnetic field SE is measured, whereas when monopole, dipole, or broadband antennas are used, the *E*-field or plane-wave SE is measured, depending on the frequency.

Measurements using MIL-STD 285 can show a good correlation to transfer impedance data as shown in Reference [37]. Another measurement, described in Reference [34] by the same author, compares the DTL-83528 measurement to transfer impedance and proves the DTL-83528 measurements to be function of the antenna locations and therefore the results may be inflated or erroneous.

A shielding material that exhibits a high surface resistance (low conductivity) has a poor SE, often 0 dB, against a low-impedance *H* field and a very much higher SE against high-impedance *E* fields. In Reference [2], tests are described on aluminized Mylar with a 1.4 Ω^2 surface resistance and a thickness of approximately 2×10^{-8} m (5 µm), which is 0.008 skin depths at 10 MHz. The size of the sample was 59 cm × 59 cm in a 2.44 m × 3 m × 2.44 m high enclosure. The SE of both *E* and *H* fields was tested. The *H*-field SE was measured with both an *E*-field source (monopole antenna) and an *H*-field source (loop antenna). With either source, the *H*-field SE was 0 dB from 50 kHz to 10 MHz. The *E*-field SE using the dipole as the source was 90 dB at 100 kHz, 72 dB at 500 kHz, 65 dB at 1 MHz, 51 dB at 5 MHz, and 43 dB at 10 MHz, as shown in Figure 6.11a. The *E*-field SE using the loop antenna as the source was 0 dB from 100 kHz to 400 kHz, 9 dB at 1 MHz, and 28 dB at 10 MHz, as shown in Figure 6.11b.

Manufacturers of conductive coatings, paint, and conductively loaded (composite) materials often publish a predicted or measured SE without mentioning that the data are valid only for an *E* field. Figure 6.10 shows the predicted SE of a thin transparent conductive coating with a surface resistance of 5 and 20 Ω^2, which is typically used as an EMI window in front of a display or cathode ray tubes. The theoretical values of SE shown in Figure 6.10 for the lower-resistance 5 Ω^2 coating are much higher than that seen in the measured *E*-field shielding data for the 1.4 Ω^2 aluminized Mylar material shown in Figure 6.11a and b. The theoretical values in Figure 6.10 ignore the effect

FIGURE 6.10 Theoretical shielding effectiveness of a thin (≪2.5 µm) coating.

FIGURE 6.11 (a) *E*-field shielding effectiveness of 1.4 Ω aluminized Mylar using an electric dipole source. (b) *E*-field shielding effectiveness of 1.4 Ω aluminized Mylar using a loop antenna source. (a, b: From Broaddus, A. and Kunkel, G., Shielding effectiveness test results of aluminized mylar, *IEEE EMC Symposium Record*, 1992. © 1992, IEEE.)

of the termination of the material at the surrounding metal, which is included in the measured values and assumes a high-impedance *E* field.

Reference [6] describes a test on a small enclosure that contained a battery-powered 10 MHz oscillator driving two TTL gates. This oscillator noise source produced harmonics up to 200 MHz and beyond. Twelve different composite and plated plastic enclosures were manufactured and then tested with the noise source inside the enclosure and the *E*-field RX antenna located 3 m from the enclosure. The measurements were comparative with the original polycarbonate nonconductive enclosure used as the control. Table 6.5, from Reference [6], may be used to compare the different materials and coatings.

As so often happens in the application of an EMC improvement or an EMI fix, implementing an apparent improvement sometimes results in a worsening of the situation. Thus, applying a conductive coating to an enclosure may increase the level of radiated emissions. Because the field is to some extent contained by the enclosure, the common-mode current across the PCB increases. The common-mode current on any unshielded cable connected to the PCB will also increase, resulting in increased radiation, especially when the cable length is resonant. Apart from the probability that emissions from unshielded cables connected to the enclosure will increase when the equipment is placed in a conductive enclosure, the radiation from the enclosure itself may increase at certain frequencies. As an example, consider a PCB and switching-power supply mounted on a metal base to which more than one power

TABLE 6.5

Material Evaluation

Sample	Material	Shielding Effectiveness (70–300 MHz) [dB]	Relative Cost
1	Polycarbonate glass filled	0 (control)	1.0
2	Polycarbonate with copper paint	25–30	2.25
3	Polycarbonate with 10% SS/10% glass	20–30	5.0
4	Polycarbonate with 10% Ni graphite	20–30	4.1
5	Polyvinylchloride with 30% NiC fibers	10–20	3.9
6	Polycarbonate with 7% SS fibers	25–35	4.3
7	Polycarbonate with vacuum-deposited aluminum	25–35	2.5
8	Mica	19–25	4.6
9	Polycarbonate with 7% SS fibers	21–31	3.1
10	Polycarbonate with Ni paint	19–27	2.25
11	Polycarbonate with 10% SS fibers	20–32	5.0
12	Polycarbonate with special carbon	12–21	2.9

Source: Bush, D.R., A simple way of evaluating the shielding effectiveness of small enclosures, *IEEE EMC Symposium Record.*

SS, stainless steel.

and signal ground connection is made. The base is surrounded by a plastic enclosure that is coated on the inside with graphite. The graphite is electrically connected to the metal base and, with the doors of the enclosure closed, an electrical connection exists between the graphite on the two halves of the doors. The equipment must meet FCC part 15 requirements placed on type A (commercial) equipment.

Measurements were made of the radiated E-field emissions from the enclosure with the doors open and closed. The graphite acts as an absorber of radiation from sources located on the PCB; thus, some small level of attenuation may be expected with the doors of the enclosure closed. However, the absorption is offset at certain frequencies by radiation from the enclosure as a result of current flow in the graphite coating. The current flow in the graphite is a result of noise current flow in the metal base of the enclosure, due to multiple ground connections, some of which takes the alternative path through the graphite coating. The measured effect of opening or closing the doors was an increase in radiated emissions of 4.5 dB at 48 MHz and 4.7 dB at 107 MHz with the door open and an increase of 6 dB at 88 MHz with the door closed. The level of radiated emissions in this example was brought below the Federal Communications Commission (FCC) limit by a combination of modifications. The radiation from unshielded cables connected to the equipment was reduced by the addition of ferrite baluns (beads) as the cables entered the enclosure. The radiation from both the enclosure and the cables was reduced by a modification in the grounding scheme on the PCB, including a reduction in the number of PCB power ground to metal base connections and the inclusion of an additional decoupling capacitor. The graphite coating was retained as much for ESD protection as for its SE.

Emerson and Cuming manufacture a range of conductive surface coating materials. Some examples follow:

Eccocoat CC 2 is a silver lacquer with a surface resistivity of 0.1 Ω^2 maximum (0.025 mm coating thickness).

Eccocoat CC 40A is an elastomeric silver-filled conductive coating with 0.001 Ω^2 to 0.05 Ω^2 for a 0.025 mm (1 mil) thickness.

Eccocoat CC 35U is an air-dry waterborne, nickel-filled acrylic coating with a surface resistivity of 1 Ω^2 for a 2 mil (0.05 mm) thick coat.

Evershield of West Lafayette, IN, also manufactures coatings. These include the following:

A nickel-filled two-part urethane-based coating at 1.5 Ω^2 (2 mil thickness)
Copper acrylic, 0.3 Ω^2 (1 mil thickness)
Graphite acrylic, 7.5 Ω^2
Silver PVC, 0.01 Ω^2 (1 mil thickness)
Silver acrylic, 0.04 Ω^2 (1 mil thickness)

The conductivity may be found from surface resistance as follows:

$$G = \frac{1}{R/sq \left[\dfrac{2.54t}{1000}\right]} \left[\text{mhos/cm}\right] \tag{6.13}$$

where t is the thickness of the shield [mils].

Manufacturers' SE curves for these materials are invariably for electric field attenuation, which is the easiest type of field to shield against. MIL-STD-285-type measurements on silver- and nickel-loaded paints have shown that the SE is an inverse function of surface resistivity. Because the silver paint tested had a surface resistivity one-tenth that of the nickel, the measured SE was 20 dB greater. Increasing the thickness of the paint from 2 to 4 mils provided an additional 6 dB of shielding, but 6 mils of nickel produced 1–2 dB less shielding. More information on the test results may be obtained from Chomerics, Woburn, MA.

The SE of a flame-sprayed 22-gauge copper mesh panel compared to solid brass and against a magnetic field, measured in a transverse electromagnetic (TEM) cell, is shown in Figure 6.12

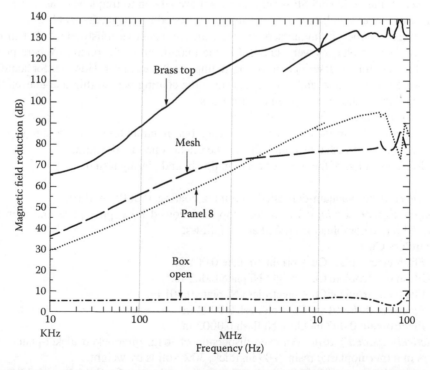

FIGURE 6.12 Magnetic field attenuation of a good flame-sprayed panel, 22-gauge copper mesh and a solid brass cover. (From Hoeft, L.O. et al., Measured magnetic field reduction of copper sprayed panels, *IEEE EMC Symposium Record*, San Diego, CA, September 16–18, 1985. © 1985, IEEE.)

(from Reference [7]). The surface resistance of flame-sprayed copper is 0.0006–0.001 Ω^2 for an unspecified thickness. The panel under test was mounted on a 0.6 m × 0.6 m × 0.2 m deep enclosure mounted on the floor of a TEM cell. The cell when driven by a signal source causes a current flow on the outside of the test enclosure and panel. A magnetic field sensor was located on the outside of the enclosure under test and a second cell inside the enclosure. The magnetic field attenuation of the panel was defined as the ratio of external field to internal field.

Due to the relatively high surface resistance and thinness of most conductive coatings, neither their SE nor their current-carrying capability is as high as that of pure metal. High surface resistance coatings, although useful in meeting FCC, EN55022/11, CS108, or similar commercial radiated emission requirements, may not be suitable for the more stringent military standard requirements and certainly not for use where EMP or indirect lightning strikes must be protected against.

An electroless copper coating that has a surface resistance close to pure metal is available and has been used on military plastic enclosures to meet MIL-STD-461 and DEF STAN EMI requirements.

6.3.1 Magnetic Shielding Effectiveness of Conductive Coatings

Conductive coatings are typically effective at shielding against electrostatic fields and also electromagnetic fields that have a wave impedance close to 377 Ω; however, they are typically less effective at shielding against magnetic fields except for coatings that have a surface resistivity close to pure metal.

The fields very close to the surface of a PCB containing logic are predominantly magnetic and the fields close to switching power supplies are also typically magnetic. Switching power supplies operate at frequencies as low as 100 kHz, and although clocks are typically at a higher frequency, data can contain frequency components at very low frequency.

Thus, for small enclosures, where the source of the field is close to the inside surface of the enclosure, the level of magnetic field SE is very important even down to frequencies as low as 100 kHz.

A number of plaques with conductive coating were tested for magnetic field SE.

Plaques were supplied by manufacturers with their conductive finish added, but in one case, nonconductive plastic plaques were supplied to the manufacturer for plating. Where possible the thickness of the coating is described, as this is an important parameter. However for some samples, these data were not available, and in on one plaque, the coating was so thin and uneven that when held up to the light, it shone through in certain areas.

As expected this material performed poorly.

We had no control over the quality of the plating, but manufacturers were told to supply, where possible, the very best finish they could, that is, the thickest practical coating.

The following is a list of the types of coating, finish, and plating tested:

Aluminum coating. Vacuum-deposited aluminum coating, 0.0005 in. thick.
Electroless copper coatings. Electroless copper followed by copper plate and then nickel plate. The five combinations tested are as follows:
Electroless Cu +
- Flash electroplate Cu + bright Ni plate 0.001 in.
- Flash electroplate Cu + bright Ni plate 0.002 in.
- Flash electroplate Cu + electroless Ni plate 0.001 in.
- Electroplate 0.001 in. Cu + Ni flash 0.0002 in.
- Electroplate 0.002 in. Cu + Ni flash 0.0002 in.
Conductively loaded plastic. An even dispersion of long, interlocked nickel-plated carbon fibers in a thermoplastic resin, 0.09 in. thick, 30% solids by weight.
Conductive silver paint #1. Sprayed silver conductive paint, <0.5 mil thick, with 47% ± 1.5% solids by weight.
Conductive silver paint #2. Sprayed silver paint, 0.4–1.0 mil thick, with 47% solids by weight.

Silver conductive coating. Sprayed silver conductive paint, 0.5–1.5 mil thick, with 50.8% ± 0.5% solids by weight. Low volatile organic chemical content.

Silver-plated, copper conductive EMI coating. Sprayed silver-plated copper conductive paint, 0.6–0.8 mil thick, with 35% solids by weight.

Silver-coated, copper conductive paint. Sprayed silver-coated copper paint, <1 mil thick, with 29% ± 1.5% solids by weight.

Nickel paint. Sprayed nickel flake paint, unknown thickness, 19% nickel, 27% solids by weight.

Tin/zinc alloy conformal coating. High-purity 80% tin/20% zinc alloy applied as a molten metal and cooled to a laminar film with low porosity, 0.001 in. thick.

The tests were of the IEEE-STD-299 type in which the radiated *H* field from an enclosure with an aperture is compared to the radiation with the aperture closed by a conductive plaque. However, because the plaques provided were small, a smaller enclosure (8.5 in. × 4.5 in. × 3.5 in.) than that specified in IEEE-STD-299 was used. Using a small enclosure also made it easier to achieve higher current flow on the inside of the enclosure. The measurements are comparative, and the measurement without plaque is referred to as the reference measurement. The SE in dB is the difference between the reference measurement, measured in dBμV, and the measurement with the plaque, again measured in dBμV.

Two sets of measurements were made, over the lower frequency range from 100 kHz to 10 MHz and from 10 to 156 MHz using different RX antennas for the two ranges. The IEEE-STD-299 test is not of an intrinsic electromagnetic parameter, such as transfer impedance; instead, the test method affects the results. For example, using a small-loop TX antenna inside a small enclosure, the internal current flow on the enclosure affects the magnetic field generated by the loop. Thus, we expect the field generated by the loop to be slightly different from the aperture open and with the aperture closed by a plaque. Also, it is to be expected that the performance of the RX antenna is dependent on the type of antenna used and its proximity to the enclosure with plaque versus enclosure without plaque. Reproducibility in this comparative type of test is very important. It was found that the exact location of the TX and RX loops was critical.

The pressure between the plaque and the enclosure was kept as constant as possible from one test to the next as this was expected to change the test results.

To ensure repeatability, the same plaque was measured three times, and when comparing the results in the first group of plaques tested to the second group, the correlation up to 80 MHz was good. Above 80 MHz, the dips and peaks in SE occurred at different frequencies but the comparison between the SE for the different plaques in the same group was valid. Also group 1 could be compared to group 2 above 80 MHz if a different frequency were chosen. For example, the SE for silver-coated copper paint was 47 dB at 115 MHz in the first group tested and 49.7 dB at 100 MHz in the second group. Table 6.6 shows a comparison of the test result for the silver-coated copper paint between one test and the next.

TABLE 6.6

Comparison between Test Results on the Same Plaque

f (MHz)	SE for Silver-Coated Copper Paint in Group 1 Test	Silver-Coated Copper Paint in Group 2 Test	Δ (dB)
1.0	20	18.5	1.5
10.0	39	36.9	2.1
80.0	45	44.1	0.9
115/100	47	49.6	2.7

Therefore, below 80 MHz, the results can be trusted within a margin of error of maximum 3 dB.

Thus, if the SE between any two plaques is within 3 dB, we must assume that the SE of the plaques is virtually the same.

Group 1 test results showed that a 1 mm aluminum plate provides a significantly better SE than any of the coatings tested. The maximum SE achieved using the aluminum plate was 74 dB at 55 MHz, which is 24 dB higher than the maximum SE attained by any of the coatings. Even a 0.15 mm thick aluminum foil performs slightly better than the best conductive coating, which was one of the conductive silver paints. Not far behind were the silver-coated copper paint and the Sn/Zn pure metal alloy coating. The worst coating was found to be the nickel paint, having a maximum SE of only 9 dB. See Figure 6.13 for a comparative SE plot of all the plaques tested in group 1.

Eight new plaques were tested in the second group, along with three of the plaques tested in group 1—the 1 mm aluminum plate, Sn/Zn pure metal alloy–coated plaque, and silver-coated copper paint plaque. Comparisons of the group 1 and group 2 findings for the pure metal alloy and silver-coated copper paint show reproducibility in the findings. The variation was 6 dB up to 10 MHz and up to 8 dB above. This means that when comparing the plaques of groups 1 and 2, any difference less than 8 dB is not valid.

See Figures 6.14 and 6.15 for these comparisons.

As shown in Figure 6.16, the 1 mm aluminum plate still achieved the highest SE of all plaques tested in group 2. However, there were some conductively coated plaques in group 2 that performed almost as well as the aluminum plate. Four of the coatings made with electroless copper, followed by copper plate and nickel, were as good, if not better than the aluminum plate at low frequency (up to 1 MHz). At high frequency, all five of these electroless copper-coated plaques provided at least 50 dB of SE, and one provided as much as 63.5 dB at 65 MHz. Differences in the SE existed between the electroless copper-coated plaques as well. Refer to Figure 6.17 for a comparison between

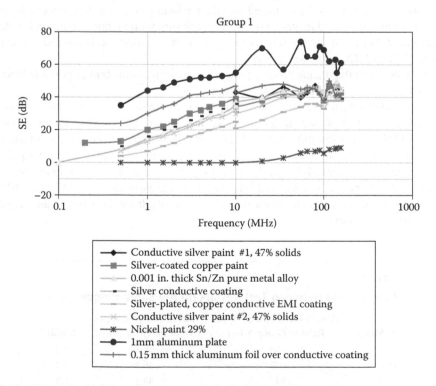

FIGURE 6.13   Magnetic shielding effectiveness of all plaques tested in group 1.

FIGURE 6.14 A comparison of Sn/Zn pure metal alloy–coating shielding effectiveness results.

FIGURE 6.15 A comparison of the results of silver-coated copper paint–coating shielding effectiveness.

the different electroless copper, copper plate, nickel plate plaques. While 0.002 copper electroplate with flash nickel plate is the best at low frequency, flash electroplate copper with 0.002 in. bright nickel plate is the most effective at high frequency. Looking at Figure 6.18, we can also say that bright nickel plating is more effective than electroless nickel plating below 20 MHz.

The worst conductive coating from group 2 was the conductively loaded plastic, with a peak SE of 41.3 dB at 100 MHz. This is what we would expect since this plaque lacks a conductive metal surface for making good electrical contact at the interface with the enclosure.

Figures 6.19 and 6.20 demonstrate an important point about the SE of a plaque—that is, thicker applications provide more attenuation. In Figure 6.19, the difference of 0.001 in. in the thickness of the copper electroplating on the second layer increases the SE of the plaque by as much as 5 dB at some frequencies. Figure 6.20 shows an even more dramatic improvement in SE when a double layer of the tin/zinc alloy coating is applied. At 1.5 MHz, the improvement in SE was as high as 16 dB.

When selecting a coating for EMI shielding, it is also important to keep in mind the susceptibility of the material to chemically degrade. Over time, some metals will oxidize, changing the shielding properties of the material. As shown in Figure 6.21, clearing away the oxidized surface on an aluminum plate using steel wool improves the SE significantly. This underlines the importance of selecting a chemically stable material for plating.

FIGURE 6.16   Magnetic shielding effectiveness of all plaques tested in group 2.

FIGURE 6.17   Magnetic shielding effectiveness differences between the electroless copper, copper plate, and nickel plate plaques.

FIGURE 6.18 Comparison between the shielding effectiveness of bright nickel plating and electroless nickel plating on electroless copper, flash electroplate copper plaques.

FIGURE 6.19 Effect of copper electroplate thickness on shielding effectiveness.

FIGURE 6.20 Effect of tin/zinc alloy coating thickness on shielding effectiveness.

FIGURE 6.21 The effect of buffing the oxidized surface on an aluminum plate.

6.3.2 Correlation between SE and Surface Resistivity

The SE tests described here require great care in the execution, a shielded room, and test equipment. Measurements of the surface resistivity of a material and the SE show a close, but not perfect, correlation. Thus, manufacturers can perform the much simpler surface resistivity test and obtain a good idea of how effective a material will be.

Table 6.7 shows the comparison of the surface resistivity and the SE of different materials at 1 MHz.

The conclusion from this is that conductive coatings vary greatly in their SE and some coatings are nearly as effective as solid aluminum.

The conclusion from the measurements is that a class of electroless copper coatings with copper electroplate and nickel finishes provided by Artcraft show the best SE, almost as effective as a 1 mm aluminum plate.

In a companion investigation, the question was asked: "Can MIL-STD-461 EMI requirements be met with a conductively coated plastic enclosure?" These measurements were on very small enclosures with dimensions of typically 11 cm × 7 cm × 5.5 cm designed to be worn by personnel.

Since that report, EMI qualification testing was performed on equipment containing RF and digital circuits that were housed in a 42 cm wide by 14 cm deep by 50 cm tall plastic enclosure. The enclosure was plated inside and outside with electroplate copper and bright nickel plate by Artcraft. This equipment met the most stringent RE102 limit for Navy Mobile and Army equipment.

TABLE 6.7

Comparison between SE and Surface Resistivity

Material	Surface Resistivity (Ω^2)	SE at 1 MHz (dB)
Sprayed silver-coated copper paint	10 m	20
Conductive silver paint #1	52 m	7
Silver-plated, copper conductive EMI coating	23 m	16
Conductive silver paint #2	26 m	13
Aluminum foil	3.3 m	30
1 mm aluminum plate	33 μm	44
Sprayed nickel paint	11	0

6.3.3 ELECTRIC FIELD SHIELDING EFFECTIVENESS OF CONDUCTIVE COATINGS

An enclosure was used during the testing of all plaques. This enclosure had a large aperture on one side that each plaque was held against during testing, so as to expose the coated side of each plaque to the exterior of the enclosure. Each plaque was held against this aperture using two wooden pieces that were tightly bolted over each side of the plaque on the interior of the enclosure. Inside this enclosure, to detect the transmitted signal, a small bow-tie antenna was used. This antenna was connected to a fiber-optic transmitter that interpreted the signal from the antenna and then transmitted it through a fiber-optic cable that exited the enclosure through a small pipe (waveguide below cutoff) at one end. This cable was connected to a fiber-optic receiver that allowed the signal in the enclosure to be monitored from a distance without influencing the test.

Thirteen different plaque coating types were used during testing. A description of each of these coatings is as follows:

Aluminum coating. Vacuum-deposited aluminum coating, 0.0005 in. thick

Conductive silver paint #1. Sprayed silver conductive paint, <0.5 mil thick, with 47% ± 1.5% solids by weight

Silver-coated, copper conductive paint. Sprayed silver-coated copper paint, <1 mil thick, with 29% ± 1.5% solids by weight

Nickel paint. Sprayed nickel flake paint, unknown thickness, 19% nickel, 27% solids by weight

Silver conductive coating. Sprayed silver conductive paint, 0.5–1.5 mil thick, with 50.8% ± 0.5% solids by weight. Low volatile organic chemical content

Conductive silver paint #2. Sprayed silver paint, 0.4–1.0 mil thick, with 47% solids by weight

Silver-plated, copper conductive EMI coating. Sprayed silver-plated, copper conductive paint, 0.6–0.8 mil thick, with 35% solids by weight

Tin/zinc alloy conformal coating #1. High-purity 80% tin/20% zinc alloy applied as a molten metal and cooled to a laminar film with low porosity, 0.001 in. thick applied on one side of the plaque

Tin/zinc alloy conformal coating #2. High-purity 80% tin/20% zinc alloy applied as a molten metal and cooled to a laminar film with low porosity, 0.002 in. thick applied on one side of the plaque

Tin/zinc alloy conformal coating #3 (second sample of #2). High-purity 80% tin/ 20% zinc alloy applied as a molten metal and cooled to a laminar film with low porosity, 0.002 in. thick applied on both side of the plaque

Conductively loaded plastic. An even dispersion of long, interlocked nickel-plated carbon fibers in a thermoplastic resin, 0.09 in. thick, 30% solids by weight

Electroless copper coating. Electroless copper followed by copper plate and then nickel plate. Flash electroplate Cu + electroless Ni plate 0.001 in.

1 mm Aluminum. 1 mm thick aluminum sheet metal mounted on a plaque

Each plaque was tested over the range of 20–400 MHz. From 20 to 30 MHz, a monopole antenna was used to transmit a signal to the testing enclosure. The enclosure was placed 15 cm away from the monopole during testing. From 30 to 400 MHz, a stripline antenna was used to transmit a signal to the enclosure. The enclosure was placed in the middle of the stripline antenna and was raised 8 cm above the ground plane.

When testing each frequency span, an initial test was done using a plaque that had no conductive coating on either side, as this facilitated a test uninfluenced by any factor other than the absence of a conductive coating on the plaque. During testing, the signal input to the antenna was increased until the signal in the enclosure reached a level just above the minimum detectable limit, which ensures that the logarithmic detector was operating over its high dynamic range. This value was the same during every test completed. The results of the plaque without a coating were then compared with

TABLE 6.8

Recorded Attenuations (in dB) of Each Plaque over Each Frequency Tested

Tested Plaque	Frequency (MHz)										
	20	25	30	225	275	300	320	340	360	380	400
Aluminum coating	—	50	45	41	38	40	41	39	35	44	
Conductive silver paint #1	—	55	60	42	32	36	41	34	29	45	38
Silver-coated copper conductive paint	50	42	40	41	44	44	44	43	41	30	43
Nickel paint	54	47	43	42.5	48	46	47	44.5	38	47	>51
Silver conductive coating	—	53	52	38	36	40	40	40	38	45	42
Conductive silver paint #2	50	45	40	41	44	43	43	40	36	43	34
Silver-plated, copper conductive EMI coating	53	50	50	37	36	37	39	39	38	45	45
Tin/zinc alloy conformal coating #1	48	42	40	41	37	39	39	38	35	50	43
Tin/zinc alloy conformal coating #2	>54	50	52	38	39	38	41	42	40	50	39
Tin/zinc alloy conformal coating #3	53	50	55	38	36.5	48	31	41	43	>58	44
Conductively loaded plastic	37	40	40	36	37	37	36	34	32	37	27
Electroless copper coating	>53	55	60	39	43	42.5	45	43	40	45	37
1 mm aluminum	>53	>58	63	46	44	46	50	52	48	53	44

the results of each tested plaque to find the resultant attenuation of the plaques. This attenuation was found by subtracting the results of the tested plaque with no coating from the results of each plaque tested. These attenuations are shown in Table 6.8.

The attenuations of each plaque were plotted over the tested frequency range after testing. These resulting plots are seen in Figures 6.22 through 6.27.

6.3.4 CONDUCTIVELY COATED PLASTIC ENCLOSURES VERSUS ALUMINUM IN MEETING MIL-STD-461 REQUIREMENTS

Military electronic equipment mounted in vehicles, ships, and planes are commonly housed in metal enclosures for ruggedness and thereby can, when correctly implemented, achieve very high levels of SE. The MIL-STD-461 EMI test requirements such as RE02/RE102, RE01/RE101, RS01/RS101, and RS03/RS103 can be easily met using metal enclosures combined with either properly terminated shielded cables or filtered input/output and power lines.

For equipment carried by a person, such as communications, wireless telephone, or other control and monitoring equipment, the lighter weight of nonmetallic enclosures is a great advantage.

FIGURE 6.22 Resultant attenuation of all conductively coated plaques tested from 20 to 30 MHz.

FIGURE 6.23 Resultant attenuation of all conductively coated plaques tested from 20 to 30 MHz.

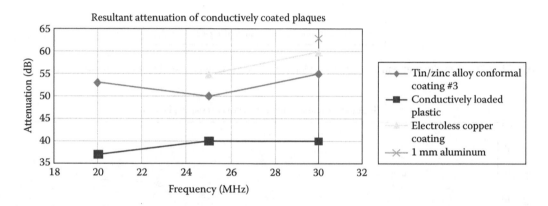

FIGURE 6.24 Resultant attenuation of all conductively coated plaques tested from 20 to 30 MHz.

FIGURE 6.25 Resultant attenuation of all conductively coated plaques tested from 225 to 400 MHz.

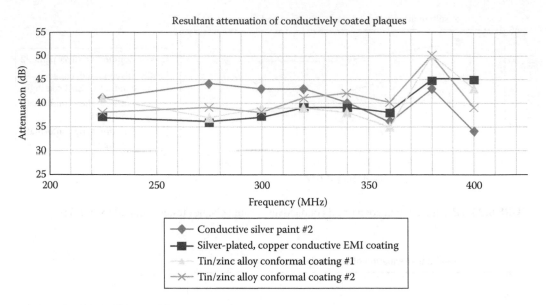

FIGURE 6.26   Resultant attenuation of all conductively coated plaques tested from 225 to 400 MHz.

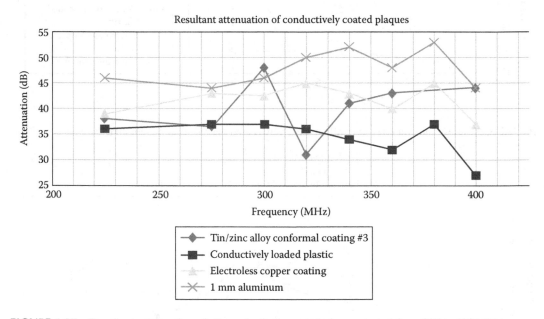

FIGURE 6.27   Resultant attenuation of all conductively coated plaques tested from 225 to 400 MHz.

One other application for nonsolid metallic shielded enclosures is in bomb disposal suits that are typically made of Kevlar, fiberglass, and other fiber materials. The use of ballistic steel in these suits, to provide blast protection, is usually confined to certain areas to keep the weight as low as possible. Some of these suits contain electronics for cooling systems, audio systems, and communications, and these electronics are prime candidates for shielding.

Meeting MIL-STD-461-type EMI requirements, without the use of any shielding in the enclosure, places extreme requirements on the design of electronics. Where highly sensitive analog or high-frequency digital electronic or DC–DC converters are used, it may be simply impossible to meet the emission and susceptibility requirements without shielding.

Traditionally, conductively coated plastic or fiberglass enclosures have been used in commercial products, typically because a molded plastic enclosure is aesthetically more pleasing than a metal enclosure and also to save weight and cost.

With correct PCB layout, common-mode filtering on cables, spread spectrum clocks, or low clock and data frequencies, it is sometimes possible to meet commercial EMI requirements without shielding the enclosure. However, with the increased complexity and speed of logic, using a conductive coating on nonconductive enclosures is becoming a very common practice in order to meet radiated emission limits even in commercial equipment.

To ensure meeting the European EMC directive, medical directive, or automotive directive, equipment must be immune to either 3, 10, or 20 V/m usually from 80 MHz to 1 GHz, but in the case of the medical directive to 2.5 GHz. In order to meet these radiated immunity requirements with analog, RF, or telephone equipment, the use of a shielded enclosure may also be mandatory.

Experience has shown that the conductive coating by itself can be very effective. It is the joint between two halves of an enclosure or between the enclosure and the lid that is the weak link in SE, and this is shown a number of times in this book. This is due to poor contact pressure at the interface, low contact surface area, and gaps due to unevenness in the formed parts or the coating

Only one set of measurements is described within this section, and these are measurements on different types of small enclosures—plastic conductively coated, cast metal, or milled metal.

Although the emphasis is placed on sensitive circuits operating with MIL STD susceptibility test levels, the SE described in this study is just as effective in reducing radiated emissions from circuits that are noise sources.

6.3.5 COATED PLASTIC ENCLOSURE VERSUS METAL ENCLOSURES

A milled plastic enclosure was coated inside and out with a nickel vacuum–deposited film.

The enclosure contained a groove for a gasket and a conductive contact between the two mating halves of the enclosure also existed on either side of this groove for the gasket.

Two enclosures were made one with a film thickness of 0.001 in. (0.0254 mm) and the second with a 0.002 in. (0.0508 mm) coating.

The enclosures were tested for relative SE over the 10 kHz to 250 MHz frequency range. The radiated *H*-field emissions were performed from 10 kHz to 22 MHz using the RE101 test method from 10 kHz to 100 kHz.

From 10 to 250 MHz, the RS03/RS103 test setup was used and an electromagnetic field was generated. Both the *E* field and *H* field inside the enclosure were monitored in this test.

6.3.6 ENCLOSURE TYPES

The dimensions of the different enclosures were either identical or, in the case of bought out enclosures, as close as possible to the manufactured. In all enclosures, either metal screws, nonconductive nylon screws, or nylon washers under metal screws were used to fasten the two halves of the enclosure together. The reason for using nonconductive fasteners was to see the real performance of the joint that otherwise would be masked by the impedance of the fastener in parallel with the transfer impedance of the joint as well as to show the importance of the fasteners. For the cast aluminum and nickel-plated plastic enclosures, measurements were also made with and without EMI gaskets in the joints.

The dimensions of the different enclosures and the locations of fasteners are as follows:

For the milled aluminum enclosure

6.9 cm wide
10.8 cm long
5.4 cm high

3 mm thick

With screws 3.3 cm apart on the long seam and 7.8 cm on the short seam

For the nickel-plated milled plastic enclosure

7.1 cm wide
11.5 cm long
5.4 cm high
Coated on the inside and outside with either 0.001 in. (0.0254 mm) or 0.002 in. (0.0508 mm) thick plating
With screws 3.5 cm apart on the long seam and 6 cm on the short seam

For the aluminum foil–covered plastic enclosure

8.2 cm wide
12 cm long
5.7 cm high
Foil thickness approximately 0.0005 in. (0.0127 mm)
With one screw at each of the four corners

For the cast aluminum enclosure

12 cm wide
12 cm long
6 cm high
3 mm thick
With one screw at each of the four corners

6.3.7 TESTS PERFORMED

A summary of the enclosures tested and the type of field and frequency range is provided in the following sections.

6.3.7.1 RS101: 10–100 kHz

1. Nickel 0.001 in. thick vacuum-deposited plastic enclosure. Nonconductive screws
2. Nickel 0.002 in. thick vacuum-deposited plastic enclosure. Conductive screws
3. Cast aluminum enclosure with four nonconductive mounting screws, due to uneven contact surface gaps that were present in the joint
4. Cast aluminum enclosure with four conductive mounting screws, due to uneven contact surface gaps that were present in the joint
5. Cast aluminum enclosure with conductive mounting screws; wire mesh gasket in the joint
6. Cast aluminum enclosure with conductive mounting screws and gaps; Gore gasket in the joint
7. Plastic enclosure covered in aluminum foil with silver-loaded gasket
8. Milled metal enclosure with same dimensions as 1 and 2. Nonconductive screws
9. Milled metal enclosure with conductive screws

6.3.7.2 H Field: 100 kHz to 10 MHz

1. Nickel 0.002 in. thick vacuum-deposited plastic enclosure; conductive screws; current flows across the seam.
2. Nickel 0.002 in. thick vacuum-deposited plastic enclosure; conductive screws; current flows parallel to the seam.

3. Cast aluminum enclosure with four conductive mounting screws, due to uneven contact surface gaps that were present in the joint.
4. Cast aluminum enclosure with four conductive mounting screws and gaps; Gore gasket.
5. Cast aluminum enclosure with four conductive mounting screws and gaps; wire mesh gasket.
6. Plastic enclosure covered in aluminum foil with silver-loaded elastomeric gasket.
7. Milled metal enclosure with same dimensions as 1 and 2; conductive screws.

6.3.7.3 RS103 Electromagnetic Wave Using *E*-Field and *H*-Field Probes

1. Nickel 0.002 in. thick vacuum-deposited plastic enclosure; conductive screws; no gasket
2. Nickel 0.002 in. thick vacuum-deposited plastic enclosure; conductive screws; silver-loaded elastomeric gasket
3. Cast aluminum enclosure with four conductive mounting screws, due to uneven contact surface gaps that were present in the joint
4. Cast aluminum enclosure with four nonconductive mounting screw and gaps; Gore gasket
5. Milled metal enclosure with same dimensions as 1 and 2; conductive screws

The SE of a material is dependent on the skin depth that is determined by its conductivity, permeability, and frequency as shown in Equation 6.1. Even a thin, highly conductive material, such as aluminum baking foil, should achieve a high level of shielding above approximately 10 MHz. For a thicker pure conductor, this frequency will be much lower. However, the enclosures used to contain electronics must have seams, and as demonstrated in the measurements described in this report, the seams are the weak link in the enclosure.

In many years of experience in the EMI testing of military, commercial, scientific, industrial, space, medical, and automotive equipment, the sources of radiated emissions and susceptibility, in a radiated susceptibility test, are invariably the cables. This is true when the electronics are placed in a conductive enclosure and the size of aperture is relatively small and the ratio of aperture to enclosure size is kept relatively low. To see just how large an aperture is possible without unduly affecting SE, refer to Figure 6.105.

One good example of the shielded cable as the limitation on enclosure SE measurements is the test setup for RS03 described in this report.

Here, the enclosure under test was mounted under a stripline antenna with only the end of the cable connected to the enclosure illuminated by the incident *E* field. To measure the level of cable pickup and therefore the highest SE that could be measured on the enclosures, the milled shielded enclosure, with the highest predicted level of attenuation, was used, and instead of an *E*-field probe or *H*-field probe inside the enclosure, a 50 Ω SMA termination was installed. The coaxial cable was a semiflex type with no apertures, the gold-plated SMA connectors are specified with greater than 90 dB of SE, and the cable outside of the stripline antenna was covered with both high- and low-frequency ferrite baluns. Despite these precautions, the measured SE was as low as 23 dB to a maximum 58 dB. Once a solid copper pipe was used to cover the coaxial cable and beryllium finger stock was used to connect the end of the copper pipe to the enclosure under test, the SE, limited by the test setup coupling and not the enclosure, rose to 50–76 dB.

The reason to mention this is that however effective a conductively coated plastic enclosure is, it is still necessary to terminate the shield of shielded cables to the conductive surface of the plastic with a low-impedance connection. If unshielded cables are used, then it is necessary to terminate common-mode filter capacitors to the conductive surface of the plastic with a low-impedance connection, and it is these connections that often pose the greatest engineering challenges to the mechanical/packaging engineer and not the design of the plastic enclosure. It is also typically these connections that limit the shielding against incident fields and not the enclosure.

6.3.7.4   RS101 Test Setup for Shielding Effectiveness

In this test setup, a 22-turn coil of #18AWG magnetic wire with a diameter of 2.37 cm was contained within each of the enclosures as an RX antenna. An SMA barrel was soldered into a brass plate, and each of the enclosures has an aperture over which the brass plate was placed and held in place by four 2–32 fasteners. A semiflexible cable assembly with SMA connectors on both ends was used to connect from the SMA barrel on the enclosure to a bulkhead-mounted SMA connector on the anechoic chamber wall. The cable assembly has a greater than 90 dB SE up to 18 GHz according to the manufacturer. In addition, low-permeability, high-frequency as well as high-permeability, low-frequency toroidal ferrite baluns were placed on the RX antenna cable to reduce the current flow induced onto the cable by the test field. For this test setup, the cable achieved a sufficiently high attenuation so as not to limit the measurement on the enclosure. This is due to the low SE of the enclosure at these low frequencies so that the coupling between the TX loop and the RX cable becomes insignificant.

Also, the low-frequency magnetic field coupling to a coaxial cable is a function of the eccentricity of the center conductor relative to the shield. When the center conductor is exactly in the center of the coaxial shield, then the magnetic field coupling to the cable will be close to zero and minimum cable coupling exists. As with all tests, a measurement was made of the coupling between the TX loop and the RX antenna cable and its termination to the enclosure alone. With the TX loop 5 cm from the enclosure, the RX loop inside the enclosure was replaced with an SMA 50 Ω load. The coupling between the TX loop and the RX cable was below the noise floor of the measuring system, which included a 20 dB preamplifier.

The TX antenna used was the 20-turn 12 cm diameter TX loop specified in MIL-STD-462 and MIL-STD-462D.

The TX loop has been calibrated using the 51-turn 4 cm diameter RX loop, specified in MIL-STD-461D and MIL-STD-651A-F, at a distance of 5 cm from the TX loop.

During measurements on the enclosure, the TX loop was placed at a distance of 5 cm from the enclosure under test and moved around the enclosure. As expected, maximum coupling was seen when the TX loop was coaxial to the RX loop in the enclosure. Figure 6.28 illustrates the test setup.

FIGURE 6.28   Test setup for RS01/RS101 shielding effectiveness.

Comparison of the *H*-field shielding effectiveness of nickel-plated plastic enclosure to metal enclosures, with and without gaskets

FIGURE 6.29 *H*-field shielding effectiveness of the different enclosures from 10 to 100 kHz in the RS101 test and from 100 kHz to 10 MHz or greater in the *H*-field loop test.

A measurement was made with the TX loop and the RX loop at the appropriate distance apart but without an enclosure. This is the reference unshielded level to that the measurements with all enclosures are compared.

The RX loop is then placed in each of the different enclosures under test, and the measurement is repeated. The difference between the reference level and the level with the loop inside the enclosure is the SE of the enclosure.

The SE in dB is thus 20 log (reference level/level inside enclosure).

A summary of the SE of the different enclosures in the RS101 test is shown in Figure 6.29.

6.3.7.5 *H*-Field Shielding Effectiveness Test Results with and without Gaskets

Due to the thin plating on the conductively coated plastic and the thin aluminum foil on the plastic enclosure, the low-frequency SE is expected to be limited and ranges from 2 to 12 dB over the 10–100 kHz frequency range, as shown in Figure 6.29. The measured SE of the 0.001 in. and 0.002 in. thick nickel plating was virtually identical, and only the results for the 0.002 in. thick plating are provided in Figure 6.29.

The cast metal enclosure has a thickness of 3 mm and should be very effective; however, due to the poor seam, the SE is limited to 15–24 dB over the same frequency range. Adding a wire mesh gasket to the cast enclosure actually reduces the SE from 12 to 15 dB. This effect may seem

FIGURE 6.30 Cast aluminum enclosure with metal screws and nylon washers under nylon screws.

surprising; however, adding the wrong gasket normally results in a degradation compared to no gasket at all. The exception to this rule is when large gaps exist in a seam and this is filled by a conductive gasket. This is further explored in Reference [1].

The cast metal enclosure with a Gore gasket results in a great improvement in SE from 21 to 35 dB.

The milled enclosure without gasket has the best seam, and the result is an SE from 31 to 49 dB over the 10 to 100 kHz frequency range.

6.3.7.6 Role of Fasteners

The type of fastener used to seal the enclosure plays a role in the SE. A conductive fastener will provide an alternative current path to the seam and also result in a higher pressure between the mating halves of the seam than is usually obtainable with nonconductive fasteners.

Figure 6.30 shows the difference in H-field SE for the cast aluminum enclosure with metal screws and the same metal screws insulated from the lid of the enclosure by nylon washers. This test result shows the increased SE when screws bridge the gap. The screws with nylon washers result in the same pressure on the seam as without, and so this effect is eliminated.

Figure 6.31 shows the difference in the SE of the milled enclosure with metal screws and with nylon screws. Both the pressure on the seam will be different, and the effective bridging of the gap by the screws and the difference is anywhere from 15 to 20 dB.

The rule of thumb for a thin 1/16 in. (1.59 mm) thick aluminum enclosure is to have metal fasteners no further than 1 in. (2.54 cm) apart to make an effective EMI seal. For milled enclosures, the distance between fasteners may be as much as 2.5 in. (6.35 cm).

6.3.7.7 Prediction Tool for Magnetic Field Shielding Effectiveness

The low-frequency magnetic field SE of an enclosure without apertures is dependent on the dimensions, thickness of the conductive material, resistivity and permeability of the material, and the presence of gaps in the seams and the DC resistance across the seams as described in Sections 6.5.3–6.6.3.

The cast metal enclosure contains small gaps in the seams due to the use of only four fasteners one at each corner of the enclosure and imperfections in the casting. The measurements of SE mentioned earlier were made with ungasketed seams, with a wire mesh gasket, and a thin elastomeric gasket manufactured by Gore. At the same time, a DC resistance measurement was made across a single

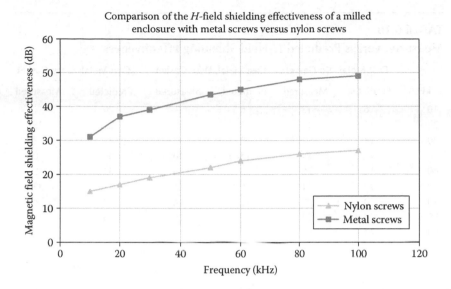

FIGURE 6.31 Milled enclosure with metal screws versus nylon screws.

TABLE 6.9
DC Resistance of Seams in the Cast Aluminum Enclosure

Cast Aluminum Enclosure	Cast Aluminum Enclosure	Cast Aluminum Enclosure
No gasket	Wire mesh gasket	Gore gasket
DC resistance mΩ		
1.0	2.16	0.06

0.12 m long seam. The measurements were made with nylon fasteners, and so the only electrical contact was at the seam. The results of these DC resistance measurements are shown in Table 6.9.

Equations 6.25 through 6.30 are to measure the magnetic field SE of metal enclosures and are available in a computer program online at emcconsultinginc.com.

The data inputs are as follows:

- Dimensions of enclosure
- Length of seams and number of seams
- Contact impedance of the seam
- Inductance of the seam
- DC resistance of the seam
- Relative conductivity of the material
- Relative permeability of the material
- Thickness of the material
- Number of apertures (if any)
- Diameter of apertures

Table 6.10 shows the comparison of the predicted SE with the measured one, and a very good correlation exists.

TABLE 6.10

Measured versus Predicted *H*-Field Shielding Effectiveness

f (kHz)	Cast Metal, No Gasket		Cast Metal, Wire Gasket		Cast Metal, Gore Gasket	
	Predicted	Measured	Predicted	Measured	Predicted	Measured
10	6	13	6	—	20	21
20	11	15	11	—	25	25
30	14	17	13	13	27	28
50	18	19	15	16	31	30
60	19	20	16	15	32	32
70	19	21	16	15	32	33
80	20	22	17	16	33	34
90	20	22	17	16	33	34
100	22	22	18	17	34	35

The measurements were made using metal screws, and so, the DC resistance of the enclosures would be lower than that shown in Table 6.9 The contact resistance (which is multiplied by $\sqrt{f}$), the inductance in Henries, and the DC resistance data were all set to match the measured test results. The values are as follows:

Cast metal enclosure, no gasket

$R_c = 1e-6$
$R = 1\ m\Omega$
$L = 1e-9H$

Cast metal enclosure, wire mesh gasket

$R_c = 5e-9$
$R = 1\ m\Omega$
$L = 3e-9H$

Cast metal enclosure, Gore gasket

$R_c = 1e-9$
$R = 0.2\ m\Omega$
$L = 4e-10H$

As seen in other measurements, adding a wire mesh gasket increases the inductance of the seam, even when apertures exist, and adding a gasket with conductive particles decreases the inductance when small apertures exist in the seam, which are filled by the gasket material.

The measured versus predicted values for a plastic enclosure coated both inside and outside with nickel and with metal screws was provided in Table 6.11. Here, the value of R for the joint resistance was set to 1.0 mΩ, R_c to 1e−6, and L = 1e−9H for the closest match to the measured values as shown in Table 6.11.

The usefulness of these equations is in collecting measured data on different types of enclosure, gasket, and spacing of fasteners that can then be used to predict the magnetic field SE of enclosures with different dimensions.

TABLE 6.11

Measured and Predicted Shielding Effectiveness of a Plastic Enclosure with a .001 in. Thick Nickel Coating

f (kHz)	Predicted Shielding Effectiveness (dB)	Measured Shielding Effectiveness (dB)
30	2	4
50	6	6
60	8	7
100	12	11

6.3.7.8 Magnetic Field Shielding Effectiveness Tests from 100 kHz to 10 MHz or Greater

When an incident H field or the H-field component of an electromagnetic wave impinges on a small enclosure, it sets up a current flow on the outside surface of the enclosure. In the real world, we cannot typically control the direction of the current flow across the seam of the enclosure, and electromagnetic waves are often both vertically and horizontally polarized and may have different angles of incidence due to reflections.

In all of the H-field tests described in this report, the antennas are oriented to ensure maximum current flow across the seam of the enclosure with one exception where the TX antenna was oriented to cause the current flow down the length of the seam.

When a current is induced by an incident field onto the outside surface of a small enclosure, some of the current diffuses through the seams and a current flow occurs on the inside of the enclosure. This internal current flow results in an internal magnetic field in a small enclosure. Regardless of the incident wave impedance, it is the magnetic field SE that is the most important at these frequencies, and typically, the electric field SE is much higher.

In this test setup, a small shielded loop antenna is placed inside the enclosure under test and a 6 cm diameter shielded loop antenna is placed outside the enclosure at a distance of 5 cm from the enclosure.

The TX antenna and RX antenna are both oriented coaxially for maximum coupling and a reference measurement is made between the two antennas without any enclosure. The test is then repeated with the antennas the same distance apart and with the same input levels.

These measurements were made between 100 kHz and 10–18 MHz.

The relative SE of the different enclosures changes with increasing frequency. For example, the nickel-plated plastic enclosure has a higher level of SE above 4 MHz than the cast enclosure without gasket. This is because the cast enclosure has small gaps in the seam and the nickel-coated plastic does not. Also, the cast metal enclosure with wire mesh gasket has a higher level of SE than the cast metal with no gasket in this test setup, and this increases with increasing frequency. The cast metal with Gore gasket is close to the same level as the milled enclosure above 4 MHz. The most surprising result is how good the aluminum foil–covered plastic enclosure, with silver-loaded elastomeric gasket, is with increasing frequency.

Figure 6.29 contains the H-field SE data in the shielded loop test.

As the test results with the 0.001 in. and 0.002 in. thick nickel plating were virtually identical, the results for the 0.002 in. are presented in Figure 6.29.

The conductivity of the enclosure material is normally high enough, and it is the seam that is the weak link even in the relatively thin nickel-plated plastic enclosure. This is shown dramatically in Figure 6.32, which shows the magnetic field SE of this enclosure with the induced current flow across the seam versus the current flow down the length of the seam. The SE at 4 MHz is 40 dB higher when the current does not flow across the seam, proving conclusively that in this enclosure, the seam is the weak link.

FIGURE 6.32 Comparison of the magnetic field shielding effectiveness with the induced current flow across the seam versus down the length of the seam.

6.3.7.9 RS103 Tests on the Enclosures from 10 to 250 MHz

For this test, the enclosures were placed inside the plates of a stripline antenna. The advantage of the stripline for radiated susceptibility tests is that a very high field may be generated for a given input power, the field uniformity is very good, and the field is mostly confined between the plates, meaning that coupling to the cable connected to the enclosure under test is reduced. Despite the use of ferrite baluns, a well-shielded cable, and connector, the maximum SE measurable with this setup was limited. As with all tests, the limit is tested by terminating inside the enclosure, which exhibits the highest predicted SE, using a 50 Ω termination and measuring the coupling to the cable.

Only when the test cable was placed in a solid copper pipe and the pipe terminated to the enclosure under test was the maximum SE sufficiently high.

The wave impedance inside any open transmission line terminated in its characteristic impedance, as long as the medium inside the transmission line has a relative permittivity of 1, is 377 Ω. This is the wave impedance of a plane wave, and the stripline antenna also generates a transverse magnetic wave up to 250 MHz with this impedance.

A small shielded loop antenna and a small E-field probe were used inside the enclosures to measure both the E-field and H-field SE. As seen many times eralier, the E-field SE is much higher in all enclosures than the H-field SE. When the electronic circuits inside the small enclosure are of high impedance, it is the E-field shielding that is important. However, when wire or PCB loops exist and if the termination is a low or even medium impedance, then the H-field shielding is the most important.

Figure 6.33 shows the magnetic field SE from 10 to 250 MHz. The highest attenuation is achieved by the cast aluminum enclosure with Gore gasket. However, without a gasket, the cast enclosure has the lowest attenuation.

The second highest attenuation is achieved by the milled enclosure without gasket, proving that in some cases EMI gaskets are not required. Adding a gasket to the nickel-plated plastic enclosure only increases the SE very slightly, but even without a gasket, the enclosure is as at least as good as and even better above 170 MHz than the cast enclosure with no gaskets. A groove was provided in the rim of the nickel-plated plastic enclosure but the gasket was too thick, and although it compresses enough to ensure a contact, the coating did not make contact to the material on either side of the groove. With a conductive gasket and electrical contact on either side of the gasket, the SE may improve slightly.

FIGURE 6.33 Magnetic field shielding effectiveness measured using a RS103 test setup and a small shielded loop antenna for monitoring.

Figure 6.34 shows the electromagnetic field SE where the source is an electromagnetic wave and the measuring antenna is the small E-field probe. Here, the nickel-coated plastic enclosure has one of the highest levels of attenuation and the milled enclosure with Gore gasket is very slightly lower. The lowest level of attenuation is achieved with the cast aluminum enclosure without gasket, and the milled enclosure, cast enclosure with gasket, and nickel-coated plastic enclosure are all very similar above 150 MHz.

The vacuum deposition of nickel results in a surface conductivity that is very close to pure metal. It can also be expected that electroplated and plasma flame–sprayed plastic will have a very high conductivity. We must consider the measured SE of the nickel-plated plastic enclosure described in this report to be as high as achievable with a conductive coating.

The RS101-radiated susceptibility test starts at a frequency of 30 Hz and even a milled enclosure will have negligible SE at this frequency.

Achieving immunity to the RS01/RS101 test from 10 to 100 kHz is more difficult when using a thin conductive coating compared to the solid metal due to the skin depth effect. Figure 6.33 shows a lower SE over this frequency range. However, even with the milled enclosure, the SE is limited to 32 dB at 10 kHz. Thus, to achieve immunity to the RS01/RS101 test level from 30 Hz to 1 kHz, the same circuit level precautions may have to be made with the conductively coated plastic as for a solid metal enclosure.

From 10 to 100 kHz and even above, some additional care in the design may be required when using the conductively coated enclosure. This involves the reduction of loop areas in PCB traces and wiring and the use of balanced differential inputs and high-impedance loads at low frequency.

6.3.7.10 MIL-STD-461 Comparison

From 10 to 250 MHz, and we can presume beyond, the E field induced into the enclosure during the electromagnetic field test shows the SE of the coated enclosure and the milled enclosure are at the

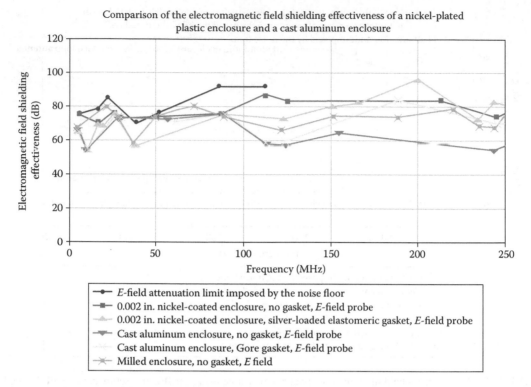

FIGURE 6.34 Electromagnetic wave shielding effectiveness measured using a RS103 test setup and a small electric field probe.

same level or the nickel-coated enclosure is even higher. Thus, achieving immunity to the *E* fields induced during the application of the RS103 test level, which is amplitude modulated with a square wave at 1 kHz, will require the same circuit design for both types of enclosure. For magnetic field shielding (magnetic fields are also induced inside the enclosure during the RS03/RS103 tests), the conductively coated enclosure shows an SE of up to 20 dB lower than the milled aluminum enclosure. This means that some additional "design for immunity" may be required at the circuit level. Any semiconductor junction will rectify an amplitude-modulated RF signal and the level may be heard in an audio signal or result in noise in an analog, RF, or control signal. The most common and effective solution to am demodulation at the circuit level is the use of a low-value capacitor at the input of potentially susceptible circuits and integrated circuits (ICs). The capacitor must be as close to the input pin(s) of the IC as possible and be connected to a solid ground plane, using no or very short and wide PCB traces.

The main problem with achieving sufficient overall shielding with the conductive plastic enclosure is in the technique used to terminate the shields of cables, or the housing of filtered connectors or filter plates to the conductive coating. One technique is to use a small pad of copper or brass mounted inside the enclosure and contacting the conductive coating. The cable shield or connector housing is then terminated to the pad. This pad, which can be thin, has the effect of spreading and increasing the contact area, but it must make good contact, under pressure, to the conductive coating to be effective.

The conclusion is that it is possible, with some minimal additional care in the design, to use a conductively coated plastic enclosure to contain sensitive circuits and be compatible with the military EM environment.

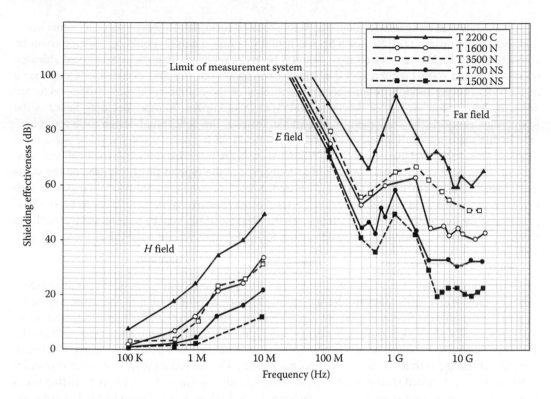

FIGURE 6.35 Shielding effectiveness of conductive fabrics.

6.3.8 Conductive Cloth

Conductive polyester cloth and paper are available. Figure 6.35 (SE for a number of fabrics) is shown by the courtesy of TW Trading International. Up to 26 dB of variation in SE seen in the graph is less a measure of the material characteristics and more a function of the MIL-STD-285 test method. In the test method, the test chamber and shielded room exhibit room resonances, and the reflections in the room cause large variations in the field incident on the test antenna. The material with the lowest SE (T 1500 NS) is a black sheer material with a polyester substrate plated with nickel (14.8% by weight). The material with the highest SE (T 2200 C) is a glossy material with a polyester substrate plated with copper (30% by weight). EMC wallpaper, louvre blind, carpet, and vinyl material are all available, so the construction of a well-furnished shielded room should be feasible!

STM manufactures electrically conductive cloth including flexible lightweight woven rip stop, woven taffeta, knot Tricoat, knot scrim, nonwoven, and knit fabrics typically using a durable nylon base. The material may be used for gaskets, tapes, cable wrap, curtains, doors, tents, and enclosures. The coatings are silver, nickel, or copper.

Section 6.3.15 describes measurements made on a conductive cloth suit.

6.3.9 Hospital Room Shielded Using Aluminum Foil

A local hospital was interested in shielding some sensitive equipment and contemplated installing a shielded room. The room under consideration was the typical laminated wood core panel with galvanized steel on both surfaces, with an RF shielded door, honeycomb air vent panels, and filters for AC power, telephone, and any other signals, meeting the SE requirements of MIL-STD-285 or IEEE 299-1997.

As schedule was important, a room was purchased and installed. However, two interesting questions have been raised: (1) How much SE was actually required and (2) could the existing room be converted to a shielded room and if so would it be less expensive than the conventional shielded room within the room? This report does not set out to answer the first question as factors such as the level of electromagnetic ambient and sensitivity of the equipment to power line conducted noise, E fields, plane waves, or magnetic fields remain unknown. Another deciding factor is the frequencies at which the equipment is susceptible. Instead, the conceptual design for a room covered in thin aluminum foil is described, and the predicted SE for magnetic fields, plane waves, and E fields is measured and predicted. Some estimated costs are provided, although the final cost must depend on the size of the room, the cost of local labor, and the type of finish required on the walls and ceiling. It is worth reiterating that whether the level of predicted shielding achievable with this type of room is adequate depends on the level of the electromagnetic ambient to be shielded against and the susceptibility of the equipment to be shielded. This may be determined by localized shielding of sensitive equipment using aluminum foil or by analysis, although this is more difficult. This susceptibility will almost certainly vary with the type of field and the frequency. One word of warning, medical equipment is often susceptible to the emissions from fluorescent lights, and if either the conventional shielded room or aluminum foil–shielded room is provided with fluorescent lights, it is important to also include alternative lighting such as filament lamps and to use a separate switched circuit for the fluorescent lights. Apart from cost, the aluminum foil has the advantages of being potentially aesthetically more pleasing than a typical shielded room, and in fact if the walls are properly covered and the floor covered in tiles, it will not be obvious that it is a shielded room. Another advantage is that any size room may be shielded without reducing its size, which typically happens when a standard shielded room is placed within an existing room. Also, the existing room may be of any shape and size and still be shielded. If the highest level of shielding is required and cost is not a major concern, then the conventional shielded room would be the obvious choice. Another alternative is a sound absorber drywall, which contains a steel plate and is sold as an RF panel. The available test data for this type of panel show plane-wave SE for a 24 in. × 24 in. panel in the wall of a large shielded room. The plane-wave shielding is almost identical to the 1 m × 0.7 m × 1 m aluminum foil room with stapled joint presented in this report. Unlike the test data provided in this report, no SE for magnetic fields is provided for the drywall material. Other sections in this book show that the field, either plane-wave or magnetic field SE, is a strong function of the joint impedance and not of the conductive material of which the wall is made, and so, the mounting technique for this drywall material would be critical. This material should not be ruled out as a possibility for an inexpensive shielded room, but if magnetic field shielding is a concern, some additional tests on either a room or small box constructed from these panels would be recommended.

6.3.9.1 Conceptual Design of a Room

An existing room could be shielded by adding thin aluminum foil to the walls and ceiling. The foil could be glued in place. At the seams, the foil would be overlapped at the edges. One option is for joints to be stapled at 3.5 cm (1 3/8th of an in.) intervals and sealed with packing tape. This type of joint would be better if the foil is glued to plywood, into which the staples are firmly embedded. The foil could then be covered in either drywall or decorative panels. Alternatively, a thin or thick aluminum or plated steel strip would be used to apply pressure at the seams. If the thin strip is used (0.7 mm thick), then to ensure adequate pressure at the seams, the joint should be over wooden strapping or battens. In the sample tested here, the fasteners were at 15.24 cm (6 in. spacings) over the wooden frame of the enclosure. The implementation of this may involve taking down the drywall and adding additional strapping or plywood at additional cost. The measurements presented in this report do not show a significant difference in SE between the stapled joint and the joint with the metal strip.

The floor cannot be thin foil due to problems of tear and wear. One recommendation is to place thin aluminum panels on the floor, in which case the foil on the wall would be in contact with the panels around the periphery of the room. The aluminum panels would use aluminum or plated steel strip on either side of the joints, which would not need to be overlapped to ensure a good contact as the strips are on both sides of the panels. Around the bottom of the walls, the foil would be placed over the top of the aluminum panel and clamped by the upper strip on the floor. Tiles placed over the floor would then protect this joint.

An option for the door, or doors, is the type of shielded door used in MRI-shielded rooms. This will not have the highest level of SE but it looks like any normal door and so is aesthetically pleasing.

The conceptual design is shown in Figure 6.36.

The aluminum foil described in this report is 0.003 in. (0.076 mm) thick.

The aluminum or plated steel strips may have to be custom made. Once the aluminum foil is on the walls and ceiling, it may be covered with decorative panels or painted. The decorative panels would be aesthetically more pleasing and provide protection against damage to the thin

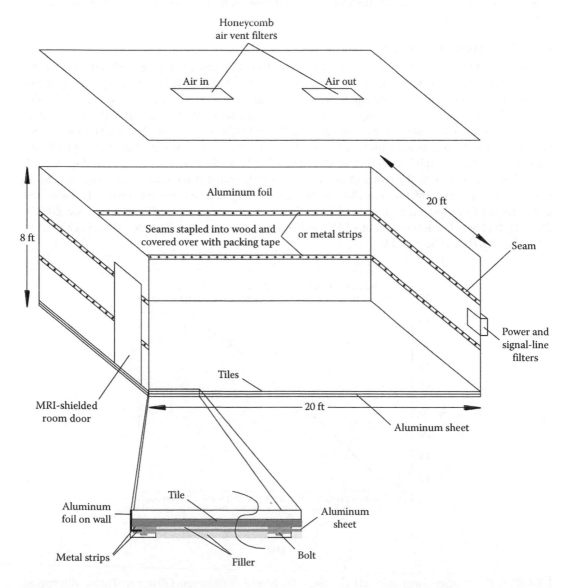

FIGURE 6.36 Conceptual design of room.

aluminum foil. If the area under the aluminum panels on the floor, between the aluminum strips, is covered by a filler material with the same thickness as the strip and glued in place, then the floor tiles would be glued onto the floor. Figure 6.36 shows this thin filler on top of the aluminum between the upper metal strips. The total cost including labor and materials for a 20 ft × 20 ft room with one door and two ventilation panels, but excluding power line and signal filters, is approximately $12,000 to $14,000 CAN. However, this estimate will vary depending on local labor cost, the cost of having aluminum strip manufactured (if used), and the type of floor tile and wall covering. The cost of floor materials and manufacture is not included in this estimate. An architect or mechanical engineer may suggest an alternative design, including the RF drywall material, or may even say that the proposed design is impractical! In contrast, the typical cost for a typical galvanized steel panel 20 ft × 20 ft shielded room meeting the SE specified in MIL-STD-285 or IEEE 299 is $45,000 CAN, and so even with a well-decorated room, the cost of the aluminum room is likely to be much lower than the typical room. To evaluate the performance of such a room, a 1 m × 0.7 m × 1 m version was built and tested. The test results from the small room were used to predict the performance of the 20 ft × 20 ft room.

6.3.9.2 Enclosure under Test

The data from the 1 m × 0.7 m × 1 m box 0.076 mm thick aluminum foil were used to determine the typical SE of the room. The predicted SE of a 20 ft × 20 ft × 8 ft shielded room using aluminum foil on the walls is obtained using the equations and test data in previous sections. If we use a DC resistance of 1.2 μΩ/m and a contact impedance value of 2.2 μΩ/m match the measured SE for the 1 m × 0.7 m × 1 m box with the stapled seam, we arrive at 200 μΩ DC resistance and 0.366 μΩ for the contact impedance for the 6 m wide seam in the room. If we assume 10 such seams in the current path and a room 20 ft wide 8 ft tall and 20 ft in depth, the predicted magnetic field SE is shown in Figure 6.37.

Although it is beyond the scope of this exercise to determine if the level of the predicted SE is high enough, it does show that a significant level of shielding can be achieved. Table 6.12 shows the comparison of the specified SE of a typical shielded room versus the predicted SE of the 20 ft × 20 ft × 8 ft room. The level of SE at 1 MHz was not measured, but based on the 100 dB achieved at 30 MHz, this should be possible.

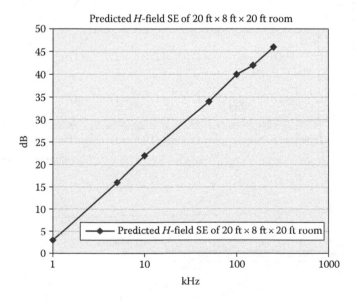

FIGURE 6.37 Predicted magnetic field shielding effectiveness of aluminum foil–covered room with stapled seams.

TABLE 6.12

Comparison between a Standard Room and the Aluminum Foil–Lined Room

Frequency	Type of Field	Standard Room Specification (dB)	Predicted Performance of Aluminum Foil Room (dB)
150 kHz	H	70	42
1 MHz	E	100	100?
30 MHz	E	100	100
400 MHz	Plane wave	100	60
1000 MHz	Plane wave	100	42

6.3.10 PLANE-WAVE SHIELDING OF AN ENCLOSURE WITH SEAMS

The classic approach to the prediction of SE is described in Section 6.1.2 and is based on a transmission line–type analysis of infinitely large sheets of conductive material. In this approach, reflection of the incident wave based on the difference between the wave and metal impedance and absorption through the material are the mechanisms. Apertures and seams are ignored in this type of approach and very high levels of E-field and plane-wave SE are predicted. As an example, we will consider the 0.075 mm (3 mil) thick aluminum foil described in Section 6.5.1, which provides an explanation for the "magic" low-frequency magnetic field SE of thin conductive foil with a relative permeability of 1.

The SE for this thin material, based on the difference between the wave and metal impedance approach, and with a distance to the source of 1.83 m, is shown in Table 6.13.

The levels in Table 6.13 are practically impossible to achieve with the exception of an enclosure with soldered or welded seams. Numerous measurements have shown SE for enclosures with no visible apertures but seams—of anywhere from 100 to 40 dB. In these measurements, the SE decreases with increasing frequency not increasing as shown in Figure 6.37. If an enclosure has apertures, then these may be the weak link, especially at frequencies where the aperture is a resonant length. In addition to the simple equations provided, the effect of resonances on aperture coupling is described in Section 6.4. Books on shielding seem to ignore this aspect and experience has shown that the seam, in the absence of apertures, is always the weak link. This report attempts to provide an explanation for E-field and plane-wave shielding of an enclosure with seams using measured data, provided in Section 6.5.1, and an analysis approach that is valid below enclosure resonance frequencies.

TABLE 6.13

Predicted E-Field and Plane-Wave Shielding of 0.076 mm Thick Aluminum Foil

f (MHz)	SE (dB)
30	131
50	146
100	172
150	191
200	207
300	233
500	274
700	308

6.3.11 MECHANISM FOR *E*-FIELD COUPLING THROUGH AN ENCLOSURE WITH SEAMS

When the magnetic field component of a magnetic field source, *E*-field source, or plane wave is incident on an enclosure, a current is set up on the wall of the enclosure facing the source. In accordance with classical field theory, this current generates a magnetic field that, adding to the incident field, has a magnitude of twice the incident magnetic field. The current density flowing on the enclosure wall is equal to the total magnetic field in A/m divided by the width of the enclosure.

Magnetic field shielding is described in Section 6.5, and the frequencies considered here are predicted to be high for an enclosure without seams. At frequencies where the skin depth is less than the thickness of the material, the majority of the current flows on the surface facing the field source and very little on the inside surface. When the current meets a seam, it flows through the inductance, DC resistance, and contact impedance of the seam. A voltage is developed across this impedance resulting in current flow on the inside surface of the enclosure. The voltage difference across the seam and the current flow generates an *E* field that propagates through the enclosure. Another parameter that has not been characterized is the capacitance across the seam. Even at 900 MHz, the capacitive reactance is likely to be in the order of ohms, which is much higher than the measured impedance due to the inductance and contact impedance of the joint and so we have ignored it here. Figure 6.38 illustrates the coupling.

6.3.12 PREDICTION OF THE INTERNAL *E* FIELD USING A 1 m × 0.7 m × 1 m 0.075 mm THICK ALUMINUM FOIL ENCLOSURE

If we assume an electric field source 0.5 m from the 1 m × 1 m wall of the enclosure, we can predict the incident *H* field for a given current in the source. Using the same source and current, we can then predict the *H* field at a distance of 0.8 m without the enclosure. If we can then compute the *E* field 0.3 m from the inside wall of the enclosure, the SE, given by

$$20\log\left[\frac{E \text{ field without enclosure at } 0.8 \text{ m}}{E \text{ field at } 0.3 \text{ m inside enclosure}}\right], \text{ can be found.}$$

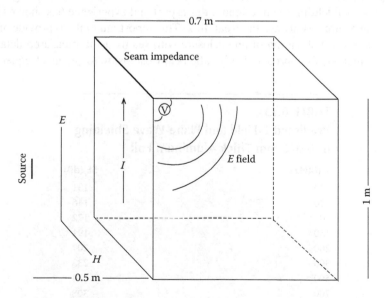

FIGURE 6.38   *E*-field coupling through a seam.

TABLE 6.14
E and *H* Fields Incident on 1 m × 1 m Wall of the Enclosure

Frequency (MHz)	*H* Field Incident on Enclosure (mA/m)	*E* Field Incident on Enclosure (V/m)
30	6.3	6.88
50	6.7	3.8
100	8.7	2.27
150	11	3.1
200	13.9	4.3
300	19.8	6.8
400	25.4	9.2
500	32	11.6
600	36	13.4
700	36	13.4
800	36	13.4
900	36	13.4

Taking arbitrary numbers for the input current of 0.072 A to a 0.262 m long electric current element, representing an electrically short antenna, the *H* field and *E* field at a distance of 0.5 m can be calculated and this is shown in Table 6.14. As SE is a relative number, we can choose any number for the input current as long as it is the same for the shielded and unshielded analyses.

If we take the stapled seam in the 1.0 m × 0.7 m × 1.0 m enclosure described in Section 6.3.9.1 as an example, then seam impedances that most closely match the measured *H*-field SE measurements are $2e^{-12}$ H for the inductance and $3e^{-6}\sqrt{f}$ (Ω) for the contact impedance. We can then calculate the total seam impedance with frequency. From the incident *H* field on the enclosure, we can calculate the current density and the voltage developed across the seam. Table 6.15 shows the voltage across the seam based on the current flow and frequency.

6.3.13 Predicting the *E* Field Inside the Enclosure

We can model the inside of the enclosure using the 4NEC2d numerical electromagnetic code. Ideally, we would use patches to make the front, top, and bottom of the enclosure. However, it is not possible to drive a patch at the extreme edge. An alternative would be to make the walls from wires, but experience has shown that this does not model a conductive surface accurately. Instead, a very simple model of half of the enclosure with no sides or back can be modeled using single wires, which is shown in Figure 6.39. In reality, the enclosure would exhibit resonances resulting in very large changes in *E*-field magnitude as well as spatial changes. The lowest resonant frequency of the 1 m × 0.7 m × 1 m enclosure, from Section 9.5.1, is 212 MHz, and so above this frequency, the *E* field will be very difficult to predict within the actual enclosure using the simple model. Nevertheless, when we compare the measured to predicted fields above 200 MHz, we see a reasonably good correlation and, at all frequencies, certainly infinitely better than the prediction in Figure 6.37.

The Numerical Electromagnetic Code (NEC) program was used to predict the *E* field at a z location of −0.5 and a y location of 0.3, that is, 0.3 m from the front wire and half way up the front wire. This corresponds to the location of the small monopole antenna, used to measure the *E* field from 30 to 200 MHz, and the bow-tie antenna, used to measure the field from 200 to 1000 MHz. The results of these measurements are described in Section 6.3.9.2 shielding a room using aluminum foil. The *E* field for a number of locations corresponding to the length of the antennas was taken, and where a large variation was seen, an average value was computed. The predicted *E* field without enclosure at a location of 0.8 m from the source is divided by the field 0.3 m from the front inside edge of the enclosure to obtain the predicted SE. The predicted SE from

TABLE 6.15

Impedance of Seam and Voltage Developed across Seam

f (MHz)	Inductance of Seam	Inductive Reactance of Seam	R_c	$R_c * ft^2$	Current across Seam	Total Impedance of Seam	Voltage Developed across Seam
3.00E+07	2.00E−12	0.0003768	0.000003	1.64E−02	1.25E−02	1.68E−02	2.10E−04
5.00E+07	2.00E−12	0.000628	0.000003	2.12E−02	1.34E−02	2.18E−02	2.92E−04
1.00E+08	2.00E−12	0.001256	0.000003	3.00E−02	1.71E−02	3.13E−02	5.34E−04
1.50E+08	2.00E−12	0.001884	0.000003	3.67E−02	2.20E−02	3.86E−02	8.50E−04
2.00E+08	2.00E−12	0.002512	0.000003	4.24E−02	2.74E−02	4.49E−02	1.23E−03
2.50E+08	2.00E−12	0.00314	0.000003	4.74E−02	3.32E−02	5.06E−02	1.68E−03
3.00E+08	2.00E−12	0.003768	0.000003	5.20E−02	3.70E−02	5.57E−02	2.06E−03
3.50E+08	2.00E−12	0.004396	0.000003	5.61E−02	4.48E−02	6.05E−02	2.71E−03
4.00E+08	2.00E−12	0.005024	0.000003	6.00E−02	5.08E−02	6.50E−02	3.30E−03
4.50E+08	2.00E−12	0.005652	0.000003	6.36E−02	5.70E−02	6.93E−02	3.95E−03
5.00E+08	2.00E−12	0.00628	0.000003	6.71E−02	6.32E−02	7.34E−02	4.64E−03
5.50E+08	2.00E−12	0.006908	0.000003	7.04E−02	7.18E−02	7.73E−02	5.55E−03
6.00E+08	2.00E−12	0.007536	0.000003	7.35E−02	7.18E−02	8.10E−02	5.82E−03
6.50E+08	2.00E−12	0.008164	0.000003	7.65E−02	7.18E−02	8.46E−02	6.08E−03
7.00E+08	2.00E−12	0.008792	0.000003	7.94E−02	7.18E−02	8.82E−02	6.33E−03
7.50E+08	2.00E−12	0.00942	0.000003	8.22E−02	7.18E−02	9.16E−02	6.58E−03
8.00E+08	2.00E−12	0.010048	0.000003	8.49E−02	7.18E−02	9.49E−02	6.81E−03
8.50E+08	2.00E−12	0.010676	0.000003	8.75E−02	7.18E−02	9.81E−02	7.05E−03
9.00E+08	2.00E−12	0.011304	0.000003	9.00E−02	7.18E−02	1.01E−01	7.27E−03
9.50E+08	2.00E−12	0.011932	0.000003	9.25E−02	7.18E−02	1.04E−01	7.50E−03
1.00E+09	2.00E−12	0.01256	0.000003	9.49E−02	7.18E−02	1.07E−01	7.71E−03

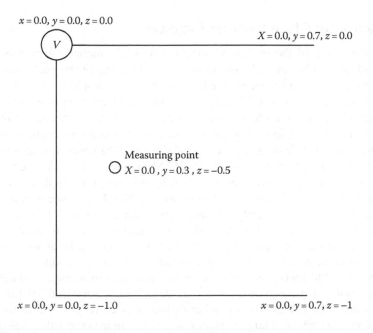

FIGURE 6.39 Geometry of the problem used in the 4NEC2d computer program.

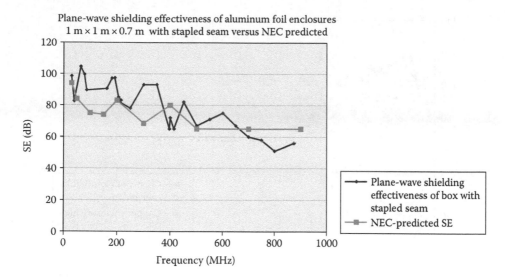

FIGURE 6.40   Comparison between predicted and measured E-field and plane-wave shielding effectiveness.

30 to 900 MHz is compared to the measured one in Figure 6.40, and the correlation is surprisingly good for such a simple model, although that may just mean the errors in measurement and analysis track! Figure 6.40 at least illustrates that the SE of an enclosure with seams is much lower than an enclosure without seams.

6.3.14   E-FIELD AND PLANE-WAVE SHIELDING OF A SMALL ENCLOSURE WITH NO SEAMS

The enclosures constructed from thin aluminum foil with the overlap in the foil at the seams are taped and stapled together, and we have assumed that the weak link is the seams. As we cannot weld thin aluminum foil, a second enclosure was constructed from the same thickness 0.075 mm (3 mil) copper foil and every seam was soldered together. The level of SE was predicted to be very high, limited only by the test setup. In the initial measurements, the SE was very low. It was found that the brass door that was sealed with an EMI gasket leaked badly. Figures 6.105b through 6.105d show how in many cases that an ungasketed seam is better than a seam with an EMI gasket! Obviously, this lesson had not been learnt well enough by the author of the book! The door was removed and a piece of copper foil was clamped behind the brass plate. On testing this configuration, the SE was found to be anywhere from 65 to 118 dB as shown in Figure 6.41. To reduce coupling between the cable connected to the measuring antenna located inside the enclosure and the cable to the TX antenna located inside the anechoic chamber, the power amplifier was moved inside the room and shielded to some extent from the TX antenna by a box of absorber material. The RX antenna cable inside the anechoic chamber was shielded by a copper tube fastened at one end to the anechoic chamber wall and soldered at the other end to the brass plate on the copper enclosure.

The remaining coupling occurred from the door of the anechoic chamber to the RX antenna cables outside of the room. This was despite the fact that the cable from the chamber to the preamplifier was very short and covered in an overbraid. Figure 6.42 is a sketch of the test setup.

The source of coupling was identified by opening the anechoic chamber door, at which point the coupling increased dramatically.

To prove that indeed the seams are the weak link, one seam in the induced current path (the horizontal seam from one side of the enclosure to the other side) was replaced by copper foil overlapped

FIGURE 6.41 Measured shielding effectiveness of soldered up copper enclosure and enclosure with one seam in the copper.

FIGURE 6.42 Sketch of test setup.

by copper foil and held in place by a metal strip. Figure 6.41 shows the decrease in SE due to the seam over the 35 MHz to 1 GHz frequency range and shows categorically the detrimental effect of seam impedance on the SE of the enclosure. The reduction in SE is not as great as with the aluminum foil enclosure. One possible explanation is that the vertical seams in the aluminum foil enclosure are also a possible source of leakage. With the front wall of the copper enclosure replaced by aluminum foil as well as a small section on the top of the enclosure, a seam was made with the aluminum foil overlap.

This seam was again held under pressure either by a metal strip or by stapling. The vertical seams between the aluminum foil and the copper enclosure were stapled together.

The test setup is shown in Figure 6.42. The measured SE of the two types of seams is shown in Figure 6.43. The measured values are almost identical to the measurements made on the aluminum foil enclosure described in Section 6.3.9.2 and is reproduced in Figure 6.44.

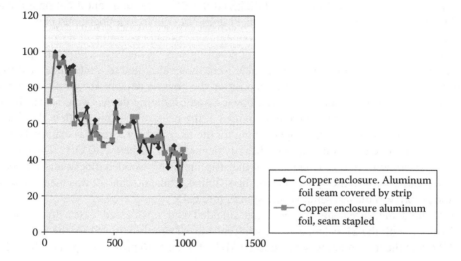

FIGURE 6.43   Measured shielding effectiveness of one seam in the copper box made with aluminum foil.

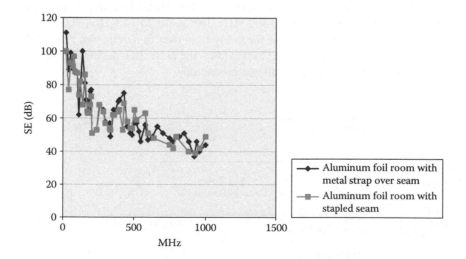

FIGURE 6.44   Measured results for one stapled seam and one strip seam in the aluminum enclosure.

6.3.15 Conductive Cloth Suit

One application of conductive cloth is in the manufacture of clothing for the protection of military personnel from work-related exposure to electromagnetic fields. The purpose of this investigation was to determine the effectiveness of a conductive cloth suit at shielding against such radiation.

The suit contains an aperture for the face, and as the face protrudes beyond the suit, no appreciable level of shielding can be expected for plane waves incident on the face.

In this test, the cloth on either side of the face aperture was folded together and held firmly by plastic clips. Unfortunately, no information was available on the manufacturer of the suit or the conductive fibers used in the suit.

The SE test method used was simply a direct comparison of the fields received from a plane-wave source inside and outside of the suit. A stripline antenna was used as the source of the plane wave. The received field from 40 to 400 MHz was measured using a bow-tie antenna placed inside the stripline and connected to a device that digitized the RF signal and sent it out on a fiber-optic cable. The advantage is that the surrounding field does not couple to a fiber-optic cable, and thus, we eliminate cable pickup from the test measurements as well as penetration of RF currents into the suit carried on a shielded cable. This absence of coupling was verified experimentally prior to this investigation. The data from the fiber-optic cable were then calibrated to a reference measurement in dBμV/m. A polystyrene frame was used to position the bow-tie antenna for consistency between shielded and unshielded measurements. The bow-tie and digitizing device were sealed inside the zippered cloth suit. A very small opening remained at the neck of the suit to allow the fiber-optic cable to pass through; however, the large opening for the face of the wearer was clamped closed for this test. The small opening would not significantly degrade the SE below 10 GHz. During the test, the suit was raised above the ground plane of the stripline using wooden blocks and mounted low enough that it did not touch the upper plate of the stripline. Thus, no contact was made between the suit and the stripline. This test setup is shown in Figure 6.45.

Received field measurements for both the shielded and unshielded cases are presented in Table 6.16 along with the SE calculated at each frequency. The minimum SE of the suit was 27 dB at 400 MHz, and the maximum was 55 dB at 50 MHz. Figure 6.46 provides a plot of the SE versus frequency.

The conductive copper cloth suit provides high SE from 40 to 400 MHz against plane-wave electromagnetic fields over most of the body. Only at the face is the SE virtually zero. One application for the suit has the user exposed to 120 V/m from 40 to 760 MHz. The Canada safety code 6 limit for the general public is 28 V/m from 10 to 300 MHz, increasing to 44 V/m at 760 MHz. Thus, the suit is likely to reduce the 120 V/m to below the safety limit over this frequency range. Wearing the conductive cloth suit is therefore considered to provide some protection during exposure to incident fields. Only the face would need some additional protection.

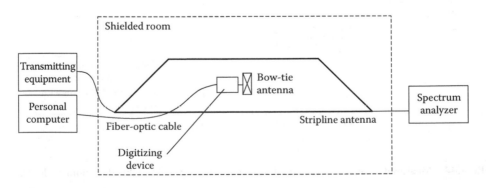

FIGURE 6.45 Basic test setup.

TABLE 6.16
Experimental Results

Frequency (MHz)	Received Field Unshielded (dBµV/m)	Received Field Shielded (dBµV/m)	Shielding Effectiveness (dB)
40	151.6	101.1	50.5
50	151.2	96.2	55
75	146.7	106.2	40.5
100	143.5	111.3	32.2
200	140	91	49
300	128.8	99.3	29.5
400	116.4	89.4	27

FIGURE 6.46 Shielding effectiveness of conductive copper cloth suit with frequency.

6.3.16 CONDUCTIVE ADHESIVES AND ABSORBERS

Conductive adhesives with silicone or epoxy bases filled with silver, silver-plated copper, or copper particles are manufactured by Emerson and Cumings, Tecknit, Chomerics, etc.

The adhesives require that pressure be applied after both surfaces have been mated. One use for adhesives is therefore between the panels of an enclosure that are bolted together. When a sealant must be applied after a joint is made, a conductive caulking is available that may be applied to seal gaps around an enclosure or applied to the outside of a joint. Conductively loaded lubricants are available that may be used between moving surfaces, such as contacts, a rotating shaft, or on a piano hinge, to ensure a low-resistance connection.

In some EMI cases, especially at very high frequencies, shielding by reflection from a very conductive surface is not the best solution, because the incident field is simply reflected somewhere else. The alternative is the use of an absorbing material that converts the incident electromagnetic field into thermal energy inside the material. Microwave-absorber material is available in flexible foam, rigid sheet and stock, castable resin, cloth, plastic sheet, and ferrite tiles.

The only material with appreciable attenuation in the frequency range of 50 MHz to 50 GHz is the ferrite tile. The attenuation achieved by other materials is predominantly in the 3–20 GHz frequency range. In one instance of radiation from slots in a shield on a PCB, the source of emissions was a 2.4 GHz high-current driver. High-level emissions were seen at 2.4, 4.8, 7.2, and 9.6 GHz.

Adding a number of magnetic and dielectric lossy material at the apertures or across the PCB did not measurably reduce emissions. However, adding low-frequency ferrite slabs at the apertures reduced emissions at all frequencies by typically 20 dB. In another instance, where surface currents at 10 GHz flowing on the control cable to a digital attenuator resulted in EMI, the placing of rubber-loaded microwave absorber over the cable and attenuator housing resulted in a significant reduction in the induced EMI level. From measurements, it has been seen that at gigahertz frequencies, waves are launched both vertically and horizontally from a source, which may be one reason that laying absorber on the surface of a PCB may not be effective.

One application of a ferrite tile is to place it behind a wire mesh or honeycomb EMI air vent filter, with a sufficient gap to allow free air flow. At frequencies above that at which the air vent filter becomes an ineffective shield, typically 1 GHz, the ferrite tile will effectively absorb any radiation through the filter. Microwave absorbers are available from the following manufacturers:

Emerson and Cuming
ETS Lindgren
Western Rubber Supply Inc.

6.4   SEAMS, JOINTS, VENTILATION, AND OTHER APERTURES

Inadequate shielding is due rarely to the properties of a conductive material but typically to the presence of seams and apertures. Thus, when measuring radiation from a "quick fix" shield constructed of aluminum foil joined by conductive adhesive copper tape, the predominant source of any radiation is not from the foil but at the copper tape seam. Where an aperture exists in an enclosure manufactured from a conductive material, the SE of the enclosure is adversely affected.

Apertures either are deliberately included in an enclosure for the sake of ventilation, for the ingress of cables, or for viewing displays or are unintentional, such as gaps in seams or joints around enclosure panels, doors, and connectors. The proximity of the source of radiation to the aperture and whether the source is an E field or H field determine the prediction technique used in assessing the attenuation of the enclosure with apertures and the steps required to improve the SE. When the source is physically small inside a large enclosure, such as a cabinet, the coupling through the aperture may be computed from the E-field component of the wave generated by the source. The coupling of an E field through an aperture is illustrated in Figure 6.47. When the enclosure is small, the impedance of the inside of the enclosure is relatively low, and the current flow on the inside of the enclosure may result in H-field coupling through the aperture, as illustrated in Figure 6.48. The current flow on the inside of the enclosure may result from multiple connections of power or signal ground to the enclosure into which noise currents are directly injected, or via magnetic field induction via PCB tracks or unshielded cables in close proximity to the enclosure. When apertures exist, the level of current flow on the enclosure interior surface should be minimized by the use of a single-point enclosure ground, the use of ground planes on PCBs (which are connected to signal or power ground and thus reduce coupling to the enclosure), and the use of shielded cables within the enclosure.

6.4.1   *E*-FIELD COUPLING THROUGH AN APERTURE IN THIN MATERIAL

When the area of the aperture is much less than λ/π, the diameter of the aperture is much greater than the thickness of the material, and the enclosure, if it exists, is very large; the SE of the aperture may be found from the following equations. The dimensions and distance from the source to the shield, D_1, and from the shield to the measuring point d are shown in Figure 6.49. The source of radiation is considered to be a small loop, the E field from which reduces as a function of $1/r^2$ until a distance of $\lambda/2\pi$, after which the reduction is $1/r$ (where r is the distance from the loop to the measuring point).

The *total shielding effectiveness* is defined as the ratio of the E field without shield and aperture to the E field with shield and aperture. When the source-to-aperture distance D_1 is much less than

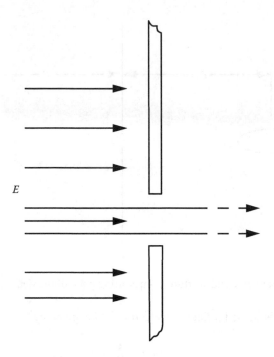

FIGURE 6.47   *E*-field coupling through an aperture.

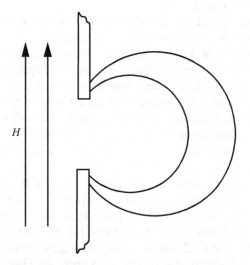

FIGURE 6.48   *H*-field coupling through an aperture.

the aperture-to-measuring-point distance, *D*, point *d* is in the far field, and if the aperture area is much less than λ/π, then the SE is given, from Reference [8], by

$$\text{SE} = \frac{E \text{ at } d \text{ without shield and aperture}}{E \text{ at } d \text{ with shield and aperture}} = \frac{\lambda D_1}{A} \qquad (6.14)$$

where
    A is the aperture area [m²]
    d is the measuring point

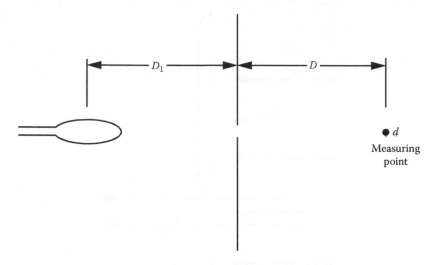

FIGURE 6.49   Source-to-aperture and aperture-to-measuring-point dimensions.

Assuming location d is in the far field, the electric field is given by

$$E = E_o \frac{A}{D\lambda} \tag{6.15}$$

where E_o is the field incident on the aperture.

When E_o is not known, the E field at location d may be computed as the electric field from a loop using Equation 2.26. The distance from the source is given by $r = D_1 + D$. Equation 2.26 is valid when r is in the near or far field. The field E at location d, with shield and aperture, is computed from the field without shield and aperture reduced by the SE given by Equation 6.14.

Equation 6.14 is valid for an electric field source contained within a small enclosure that contains an aperture, when measuring point d is in the far field. E-field-radiated measurements were made in a shielded room from an E-field source with and without a shield that contained an aperture. A solid-wall copper conduit located inside the shielded room was connected to an adapter mounted on the shielded room wall. The conduit was used to shield a coaxial cable connected outside the shielded room to a signal generator. The end of the conduit inside the shielded room was soldered to an 8 cm diameter ground plane, to which the shield of the coaxial cable was also terminated. The center conductor of the coaxial cable was connected to a 5 cm long rod that, with the circular ground plane, formed a monopole antenna. Measurements were made of the far field, inside the shielded room, radiated by the monopole antenna from 100 MHz to 4 GHz. A copper tube 7.5 cm long with a 7.5 cm diameter and a 2.54 cm diameter aperture was constructed. The tube had a lid soldered on at one end and open, with beryllium copper finger stock soldered around the periphery, at the other end. The finger stock was connected under pressure to the monopole ground plane, so the monopole rod was effectively shielded by the can, except for the 2.54 cm aperture. The far-field measurements were repeated with the tube in place, from 100 MHz to 4 GHz. The measured SE showed the variation expected from measurements made in an undamped room; however, the average measured SE was within 5 dB of the predicted at 100 MHz, using Equation 6.14, 15 dB at 400 MHz, and 5 dB at 2 GHz. At 3 GHz and 3.6 GHz, negative SE was seen as a result of enclosure resonance.

When d is in the near field, that is,

$$d < \frac{\lambda}{2\pi} \tag{6.16}$$

$$E = \frac{E_o A}{D^2 \lambda} \tag{6.17}$$

The foregoing equations apply to rectangular apertures but are applicable to circular apertures, with an accuracy of 1 dB, when the apertures are small relative to a wavelength.

The effect of multiple apertures is dependent on the distance between the apertures and the measuring distance. When the distance D to the measuring point d from the apertures is larger than the area of the apertures and the source-to-aperture distance D_1 is much less than D, the radiation is considered to add coherently and the SE is

$$SE = \frac{\lambda D_1}{NA} \tag{6.18}$$

where N is the number of apertures.

In close proximity, in contrast, the radiation from the apertures may be considered to add noncoherently and the SE is approximately

$$SE = \frac{\lambda D_1}{0.316 NA} \tag{6.19}$$

SE measurements were made with an E-field source contained in a 0.39 m × 0.31 m enclosure with a depth of 0.185 m. The face of the enclosure contained 225 × 4 mm diameter apertures. The measured SE was 9 dB higher than predicted, by Equation 6.19, at 100 MHz, 10 dB higher at 300 MHz, and 0 dB at 400 MHz. The SE is inversely proportional to the product of the number of apertures and the aperture area. Thus, separating the aperture with a very fine conductor is not very effective in increasing E-field shielding. For example, the 4 mm diameter apertures were covered in fine wire mesh with 0.7 mm apertures. The measured increase in SE was from 2 to 7 dB, which is close to that predicted using Equation 6.19.

Due to reciprocity, Equation 6.14 is also applicable to the case where a plane wave is incident on a small enclosure with an aperture; in this case, the levels of the plane-wave H-field component shielding and E-field component SE are generally different. One approach, which is described in Reference [9], is to consider the enclosure as a waveguide and to assume a single TE_{10} mode of propagation. The analysis both above and below the cutoff frequency for this mode and both E-field and H-field shielding are calculated as functions of frequency, aperture dimension, multiple apertures, enclosure dimension, wall thickness, and position within the enclosure. In addition, the damping effect and change in resonant frequency as a result of adding electronic assemblies and PCBs inside the enclosure are examined.

A rectangular aperture in an empty rectangular enclosure is represented by an impedance. The radiating source is represented by a voltage V_o with an impedance $Z_o \approx 377 \ \Omega$. The enclosure is represented by a shorted waveguide whose characteristic impedance and propagation constant are Z_g and K_g, respectively. Figure 6.50 shows the dimensions of the enclosure and the equivalent circuit.

To validate the approach, sensors were placed within a range of enclosures, with apertures cut into one wall of the enclosure. A shielded loop antenna was used to measure the H field and a monopole antenna to measure the E field.

The enclosure was located inside a shielded room, and a stripline antenna, which radiates a field with an impedance of ≈377 Ω over the complete operating frequency range, a log periodic antenna, or Bilog antenna, was used to generate the incident field. Figure 6.51 shows the calculated and measured electric field shielding effectiveness, S_E, at the center of a 300 × 120 × 300 mm enclosure with a 100 mm × 5 mm aperture. The correlation between theory and measurement is very good,

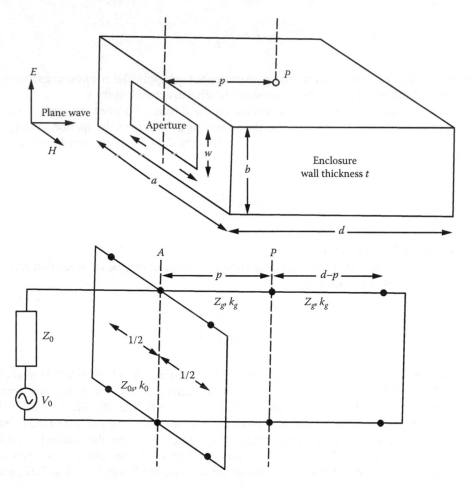

FIGURE 6.50 Rectangular enclosure with aperture and its equivalent circuit.

with variation in the measured S_E with the use of different source antennas and almost certainly because the source is located inside a poorly damped shielded room. Using Equation 6.14 over the 50–500 MHz frequency range, the predicted SE is compared to the data, contained in Figures 6.51 and 6.52; this is shown in Table 6.17. Above 400 MHz, the SE, using Equation 6.14, is higher than that shown in the figures. This is because the enclosure is approaching resonance above 400 MHz.

The SE decreases with increasing frequency, and since the E-field level is higher when closer to the aperture, the SE increases with distance from the aperture. Therefore, locate susceptible circuits on a PCB as far away from the aperture as possible. The 300 mm × 120 tnm × 300 mm enclosure resonates at approximately 700 MHz, and the SE is negative, that is, field enhancement occurs. Figure 6.52 shows the same enclosure but with the aperture increased in size to 200 mm × 30 mm. The resonance and negative shielding can clearly be seen in the measured results. Figure 6.53 shows the S_E for a 222 mm × 55 mm × 146 mm enclosure with 100 × 5 mm aperture. Adding PCBs to an enclosure both changes the resonant frequency and damps the enclosure, as shown in Figure 6.54. However, the PCBs must be mounted close to the inside walls of the enclosure to be effective. If the PCBs are attached to alternative walls of an enclosure, the effective path length increases and the resonant frequency decreases! Try to use the PCBs to divide the enclosure into a number of smaller compartments. The 300 mm × 300 mm × 120 mm enclosure was measured with one, two, and three 160 × 4 mm apertures. For these measurements, the box with a single aperture was used as the calibration standard, which helped iron out the shielded-room resonances in the SE curve.

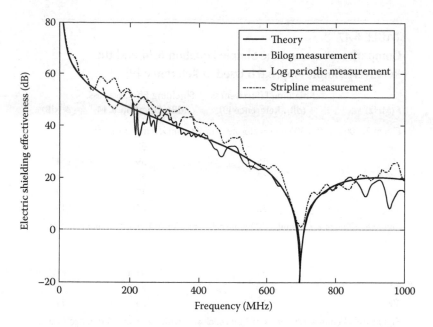

FIGURE 6.51 Calculated and measured S_E at the center of a 300 × 120 × 300 mm enclosure with 100 × 5 mm aperture. (© 1998, IEEE.)

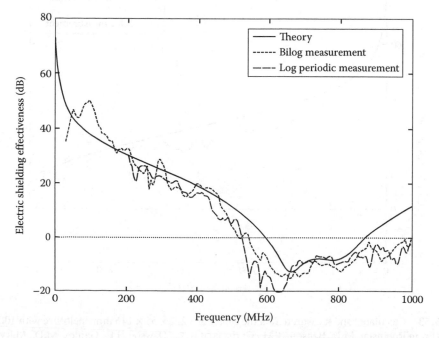

FIGURE 6.52 Calculated and measured S_E at the center of a 300 × 120 × 300 mm enclosure with 200 × 30 mm aperture. (© 1998, IEEE.)

TABLE 6.17

Comparison between the Simple Equation 6.14 and the Relatively Simple Approach Used in Reference [9]

f (MHz)	Shielding Effectiveness (dB), Reference [9]	Shielding Effectiveness (dB), Equation 6.14	Δ (dB)
Figure 7, 300 × 120 × 300 mm enclosure with 100 mm × 5 mm aperture			
50	60	65	5
100	55	59	4
200	49	53	4
400	38	47	9
500	32	45	13
Figure 8, 300 mm × 120 mm × 300 mm enclosure with 200 mm × 30 mm aperture			
50	47	43.5	3.4
100	40	37.5	2.5
200	30	31.4	1.4
400	20	25.5	5.5
500	12	23	11
Figure 10, 483 mm × 120 mm × 483 mm enclosure with 100 mm × 5 mm aperture			
50	60	47.6	12
100	56	63.1	7.2
200	50	57	7
300	42	53.6	11.6

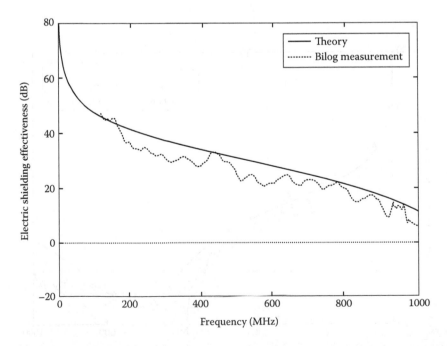

FIGURE 6.53 Calculated and measured S_E at the center of a 222 × 55 × 146 mm enclosure with 100 × 5 mm aperture. (From Robinson, M.P., Benson, T.M., Chrisopoulus, C., Dawson, J.F., Ganley, M.D., Marvin, A.C., Porter, S.J., and Thomas, D.W.P., Analytical formulation for the shielding effectiveness of enclosures with apertures. *IEEE Transactions on Electromagnetic Compatibility*, 40(3), August 1998. © 1998, IEEE.)

FIGURE 6.54 Effect of adding PCBs to an enclosure. (From Robinson, M.P., Benson, T.M., Chrisopoulus, C., Dawson, J.F., Ganley, M.D., Marvin, A.C., Porter, S.J., and Thomas, D.W.P., Analytical formulation for the shielding effectiveness of enclosures with apertures. *IEEE Transactions on Electromagnetic Compatibility*, 40(3), August 1998. © 1998, IEEE.)

TABLE 6.18
Reduction in SE due to the Number of Apertures,
Compared to One, with Each Aperture Size Kept Constant

	SE (dB)	
Number of Apertures	Theory (400 MHz)	Measured (200–600 MHz)
1	0	0
2	5.6	3.7–4.4
3	8.8	6.6–7.7

Increasing the number of apertures was found to reduce the SE. Table 6.18 shows the calculated reduction in S_E at 400 MHz compared to measurements over the 200–600 MHz frequency range.

The analytical solution predicts that both S_E and the magnetic field shielding effectiveness S_H are increased by dividing an aperture up into smaller apertures; thus, the total area is kept the same and the number of apertures increases. Figure 6.55 shows the measured S_E at the center of the 300 × 300 × 120 mm enclosure, with the area divided up into one, two, four, and nine apertures. In each case, the total area was 6000 mm². As predicted, having more, but smaller, apertures improves the shielding. The practical application of the analysis and measurements are discussed in Reference [9] and are, for the *E*-field component:

Long, thin apertures are worse than round or square apertures, because for a typical-size enclosure, the theory predicts that doubling the length of the slot reduces S_E by 12 dB while doubling the height reduces S_E by only about 2 dB.

Doubling the number of apertures reduces S_E by about 6 dB.

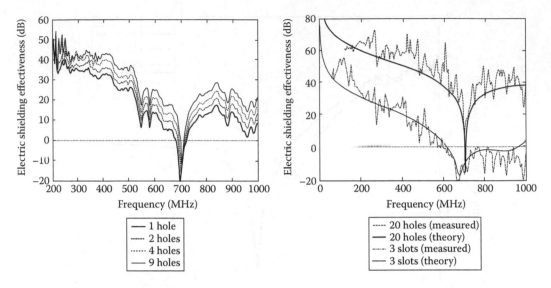

FIGURE 6.55 Measured S_E at the center of a 300 mm × 300 mm × 120 mm enclosure with one, two, four, and nine apertures, with the total area held constant, that is, smaller apertures.

Dividing a slot into two shorter slots increases S_E by about 6 dB.

At subresonant frequencies, doubling the enclosure dimensions while keeping the aperture size and number constant is predicted to increase S_E by about 6 dB, and larger-size enclosures are recommended.

Doubling both the enclosure size and aperture is predicted to reduce S_E by about 6 dB.

Doubling the enclosure size halves the resonant frequency, and to avoid negative SE, a smaller enclosure, or one that is full of electronic components, is preferable.

Reference [9] uses a relatively simple approach to the problem of the SE of rectangular enclosures with rectangular apertures. Reference [23] describes a rigorous full-wave combined electric and magnetic field boundary simulator using a method of moment electrical field integral equation (MOM EFIE). This technique also shows a very good correlation with measurements made in both a full anechoic room with ferrite tiles and a semianechoic chamber in which absorbing cones are used everywhere except for the conducting floor. Despite the good correlation between predicted and measured results, the SEs described in Reference [23] are approximately 5–8 dB higher than obtained by the simple technique described in Reference [9]. The enclosure used for the comparison had the dimensions 50 mm × 500 mm × 500 mm with a 50 mm × 200 mm aperture. Table 6.19 shows the comparison of the two techniques and Equation 6.14 for this relatively large enclosure.

TABLE 6.19

Comparison of Shielding Effectiveness of a 500 mm × 500 mm × 500 mm Enclosure with a 50 mm × 200 mm Aperture

f (MHz)	MOM EFIE (Reference [23]) (dB)	Waveguide (Reference [9]) (dB)	Equation 6.14 (dB)
50	24	17	43
100	24	17	37.5
200	26	18	31.5
300	28	23	28

For *E*-field shielding, the thickness of the material into which the aperture is placed does affect SE, unlike *H*-field shielding. In practice, it is not typically convenient to use the thick material, and so the same effect can be achieved by bending a thin material over to form a flange at apertures or at seams.

In Reference [24], a finite difference time domain analysis of SE is presented using a wire current source placed close to apertures. With a single 25 mm circular aperture and by increasing the thickness of the material from 1 to 12 mm, the analysis predicted an increase of 24 dB in SE. This is approximately 12 dB higher than predicted using the waveguide below cutoff effect shown in Equation 6.21.

When the largest aperture dimension is greater than ¼ wavelength, the aperture may function as a horn or slot antenna and radiate efficiently. For example, a large 6 foot by 6 foot enclosure contained back panels connected to the frame of the enclosure by screws at 6 in. intervals. Both the frame and the panel were painted, and the electrical connection was tenuous, made by some of the screw heads that bit through the paint. The *E* field measured at 1 m distance, with the antenna pointing at the slot between the screws, was higher than when the panels on all four sides of the enclosure were removed. Thus, the enclosure and slots functioned as an antenna with gain.

The SE of an aperture is a function not so much of the width of the gap at high frequencies but of the length of the gap and its depth, depending on the polarization of the wave or the direction of current flow across the gap. Because we are seldom able to determine either the wave polarization or the direction of current flow, we use the largest transverse dimension when estimating worst-case aperture attenuation. A metal enclosure with a removable lid is not an EMI-tight box unless the surfaces are optically flat and no distortion occurs between the lid and the side flanges. In practice, gaps occur between the lid and the side flanges that often are as much as one-half the distance between the fasteners when they are as close as 2 cm apart, and more than half the distance between fasteners when they are much more than 2 cm apart.

The authors of Reference [35] have made additional analysis of the preceding aperture coupling method and noted that "the accuracy of the preceding method is low when the observation point is close to the aperture for two reasons. First, the aperture scattering and diffraction are so huge that Robinson et al.'s method based on the waveguide and equivalent circuit theory does not work near the aperture. The other reason is that accurate results are difficult to obtain due to the presence of many evanescent nodes near the aperture. Robinson's transmission-line model neglects higher-order modes. These higher-order modes will always be excited even if they are evanescent in nature, but their effect is often significant near the aperture. The relative error of SE between Robinson's method and the precise numerical method increases as the distance from the observation point to the aperture p decreases."

Reference [35] uses TLM and FEM analysis to determine the accuracy of the equations.

The minimum distance p from Reference [35] is 34 mm and is almost independent of frequency and the depth of the enclosure below the first resonance of the enclosure.

In the measurements described in Section 6.4.2 aperture coupling from 20 to 30 MHz, the attenuation is much lower than predicted using the equations in Reference [9]. From 575 to 950 MHz, the correlation was better.

In addition, Reference [35] states that in Reference [9], Robertson et al.'s paper, that the aperture must be placed centrally in the front face of the enclosure.

One input to the equations in Reference [9] is the loss factor that is unknown.

A computer program written in C# using the equations in Reference [9] is available from the EMC Consulting website emcconsultinginc.com.

6.4.2 Aperture Coupling Measurements versus the Equations in Reference [9] and the FEKO Program

Measurements were made of the reduction in *E* field inside an enclosure with an aperture.

The enclosure was initially 37 cm wide × 33 cm high and 37 cm deep. Measurements were also made with the enclosure extended in depth to 54 cm.

A monopole antenna and a log periodic antenna were used to generate an electromagnetic wave where the wave is perpendicular with reference to the enclosure wall containing the aperture. Thus, the E field has an angle of 0° with reference to the enclosure wall. This E field is referenced as 0° in the plots and will be generated by antennas removed from the enclosure surface.

The enclosure was placed under a stripline antenna with the enclosure wall containing the aperture facing up to the top plate of the antenna. The E field is now at 90° with reference to the enclosure wall and aperture.

This angle of incidence can be generated by monopole antennas or antennas that generate a vertically polarized wave with reference to the enclosure when mounted on the enclosure.

The monopole and log periodic TX antennas were located at 15 cm from the aperture for all the measurements shown in Figures 6.56 through 6.75.

The enclosure was constructed of copper soldered up with a gasketed door on the side.

The small bow-tie antenna connected to a logarithmic detector, amplifier, digitizer, and fiber-optic driver described in Chapter 2 was used as the RX antenna. The fiber-optic cable was brought through a copper pipe, forming a waveguide below cutoff, which was soldered to the enclosure wall. Measurements were made on the enclosure with no aperture to obtain the maximum level of measured attenuation.

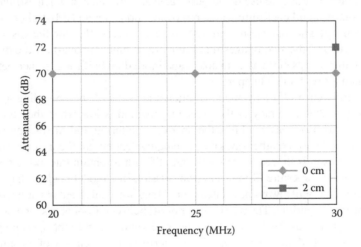

FIGURE 6.56   Attenuation of the enclosure with 1 cm aperture. E field at 90°.

FIGURE 6.57   Attenuation of the enclosure with 1 cm aperture. E field at 90°.

FIGURE 6.58 Attenuation with 1 cm aperture. *E* field at 0°.

FIGURE 6.59 Attenuation of the enclosure with 1.5 cm aperture. *E* field at 0°.

FIGURE 6.60 Attenuation of the enclosure with 1.5 cm aperture. *E* field at 0°.

FIGURE 6.61   Attenuation of enclosure with 1.5 cm apertures. *E* field at 90°.

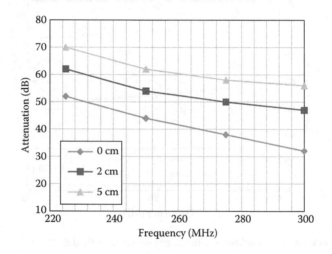

FIGURE 6.62   Attenuation of enclosure with 1.5 cm aperture. *E* field at 90°.

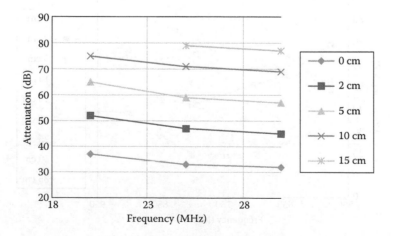

FIGURE 6.63   Attenuation of enclosure with 2 cm aperture. *E* field at 0°.

FIGURE 6.64 Attenuation with 2 cm aperture. *E* field at 0°.

FIGURE 6.65 Attenuation with 2 cm aperture. *E* Field at 90°.

The bow-tie antenna was mounted directly at the aperture at 0 cm and then moved back from the aperture from the initial 0 cm to a maximum of 35 cm.

Measurements were made with no enclosure and with the small RX antenna 15 cm from the TX antenna. Due to the logarithmic nature of the detector, the input level to the TX antenna was reduced so that just above the detectable minimum level was achieved from the RX antenna at all of the frequencies used in the measurements. This achieved the most dynamic range from the detector and was used as the reference level. This minimum reference level was then used for all measurements within the enclosure. The SE is thus the difference between the input level to the TX antenna with no enclosure and that required with the RX antenna within the enclosure.

FIGURE 6.66 Attenuation with 2 cm aperture. E field at 90°.

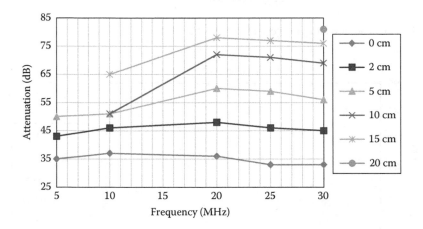

FIGURE 6.67 Attenuation with 2.5 cm aperture. E field at 90°.

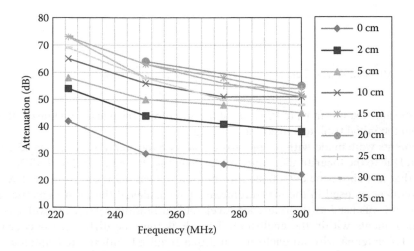

FIGURE 6.68 Attenuation with 2.5 cm aperture. E field at 90°.

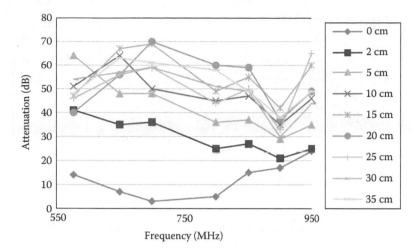

FIGURE 6.69 Attenuation with 2.5 cm aperture. *E* field at 0°.

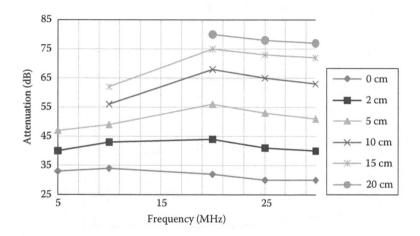

FIGURE 6.70 Attenuation with 3 cm aperture. *E* field at 90°.

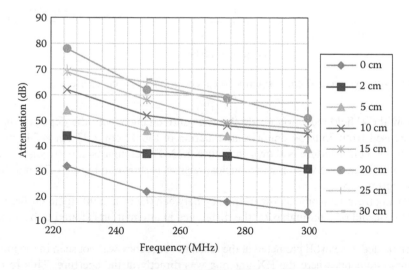

FIGURE 6.71 Attenuation with 3 cm aperture. *E* field at 90°.

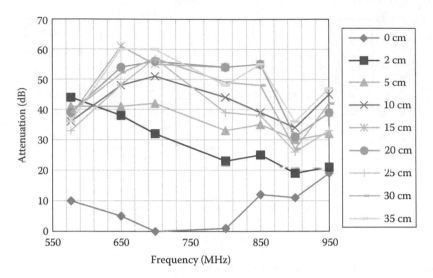

FIGURE 6.72   Attenuation of 3 cm aperture. E field at $0°$.

FIGURE 6.73   Attenuation of 37 cm × 33 cm × 54 cm enclosure with 3 cm aperture.

To see how the E-field reduction inside the enclosure changes with the depth of the enclosure, this was increased from 35 to 54 cm. The attenuation of the 37 cm × 33 cm × 54 cm from 20 to 30 MHz with the 3 cm aperture is shown in Figure 6.73. The comparison between the increases in attenuation with distance between the two enclosure sizes is shown in Figure 6.74 and the reduction in E field in Figure 6.75.

On comparing the attenuation at low frequency between the $0°$ and $90°$ incident fields in Figures 6.59 through 6.61 and 6.70 through 6.74, the difference in attenuation is found to be a maximum of 5 dB.

The dramatic decrease in SE predicted at the resonant frequency was not seen in the measurements except for measurements where the RX antenna was directly at the aperture. This resonance was centered at 750 MHz that corresponds to the second predicted resonance of the enclosure at 760 MHz.

FIGURE 6.74 Attenuation of 35 cm deep versus 54 cm deep enclosure with 3 cm aperture at 30 MHz.

FIGURE 6.75 Reduction with field inside the enclosure—54 cm depth versus 35 cm depth.

Enclosure resonances when the E field inside the enclosure is as higher or higher than the field without the enclosure are not seen with apertures up to 3 cm × 3 cm. Only when the measuring point is at the aperture is the attenuation 0 dB at 750 MHz—the theoretical resonant frequency of the enclosure.

In the equations in Reference [9], the distance between the antenna and the enclosure is not included as input data. In the measurements, this distance was varied at both 20–30 MHz and 575–950 MHz between 0.1 and 1 m. The definition of shielding effectiveness is the difference between the measured level without the enclosure and the enclosure with aperture with the antenna to measuring point distance a constant. The level of attenuation measured with the distance varied from 0.1 to 1 m remained constant, within acceptable measurement error.

Thus, the distance between the antenna and the enclosure in both the near field and far field does not affect the attenuation through the aperture.

With the aperture in the 54 cm deep enclosure being increased from 3 cm × 3 cm to 3 cm × 200 cm, the measurements were repeated.

In addition, the computer program using the equations in Reference [9] as well as a model of the enclosure in FEKO was used to predict the SE.

A shielded copper enclosure with dimensions of 540 × 370 × 330 mm was modeled in the computer program FEKO. A 30 × 200 mm aperture was positioned on one side with a TX antenna in front of it as seen in Figure 6.76. The attenuation of the enclosure was measured at different distances away from the aperture inside the enclosure, including 0, 2, 5 cm, and the center of the enclosure. These FEKO calculations were compared to the same measurements made using a shielded enclosure designed with the same dimensions. Figures 6.77 through 6.80 show this comparison.

An approximate difference of 10 dBm is shown between the calculated FEKO results and the measured results over the range 577–950 MHz. At the frequencies of 695 and 930 MHz, these differences varied greatly when compared to the direct measurements. The measurement was repeated with a 10 kHz step in the swept frequency. The spectrum analyzer sweep was adjusted to measure the 10 kHz step in frequency so that no potential resonances in the enclosure were missed. The FEKO calculation showed a high attenuation at antiresonant frequencies where the measurement showed a resonance. At the center of the enclosure, the measured attenuation was a minimum of −15 dB at 700 MHz, whereas it was −25 dB at 950 MHz as predicted by FEKO. At 5 cm inside the

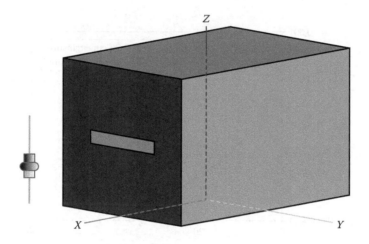

FIGURE 6.76   540 × 370 × 330 mm shielded copper enclosure with a 200 × 30 mm aperture, modeled in FEKO.

FIGURE 6.77   Comparison of the FEKO result and a direct measurement for the attenuation 0 cm inside of a 540 × 370 × 330 mm shielded copper enclosure with a 200 × 30 mm aperture, from 577 to 950 MHz.

FIGURE 6.78 Comparison of the FEKO result and a direct measurement for the attenuation 2 cm inside of a 540 × 370 × 330 mm shielded copper enclosure with a 200 × 30 mm aperture, from 577 to 950 MHz.

FIGURE 6.79 Comparison of the FEKO result and a direct measurement for the attenuation 5 cm inside of a 540 × 370 × 330 mm shielded copper enclosure with a 200 × 30 mm aperture, from 577 to 977 MHz.

FIGURE 6.80 Comparison of the FEKO result and a direct measurement for the attenuation at the center of a 540 × 370 × 330 mm shielded copper enclosure with a 200 × 30 mm aperture, from 577 to 950 MHz.

enclosure, the difference between the FEKO result and the measured one was the greatest, and at 0 cm inside, it was the least.

A comparison was also made between the results for the attenuation of the shielded enclosure from the FEKO calculations, the measurements, and the aperture coupling equations discussed earlier. This comparison for the attenuation in the center of the enclosure is shown in Figure 6.81. The anomalies observed at 695 and 930 MHz of the FEKO result are uncharacteristic of the other two traces of the plot. The results from the aperture coupling equations shared a more similar trend to the measured results than those of FEKO did; however, 10–15 dB differences are seen on either end of the frequency range.

FIGURE 6.81   (See color insert.) Comparison of the FEKO, direct measurement, and aperture coupling equation results for the attenuation at the center of a 540 × 370 × 330 mm shielded copper enclosure with a 200 × 30 mm aperture, from 577 to 977 MHz. Loss factor 0.2 for aperture coupling equation.

The loss factor used in the equations that most closely matched the measurements was at 20–30 MHz, and in the center of the enclosure, FEKO shows 22 dB lower attenuation than measured, and the aperture coupling equations in Reference [9] show a 48 dB higher than the measured. At 5 cm inside the enclosure and from 20 to 30 MHz, the FEKO attenuation was 20 to 22 dB lower than the measured and the aperture coupling equations 20–27 dB higher.

Reference [35] mentions that the aperture equation is not accurately close to the aperture, but we see that even at the center of the aperture, the equations are not accurate at low frequency.

Figure 6.82 shows a close trend between the measured attenuation of a shielded enclosure and the calculated attenuation of a smaller shielded enclosure found using the aperture coupling equations, both with 200 × 30 mm apertures. This close trend was not observed, however, when the measurements were originally compared with the calculated results for an enclosure of equal size, as seen in Figure 6.83. The loss factor included in the aperture-coupling equations changes the attenuation

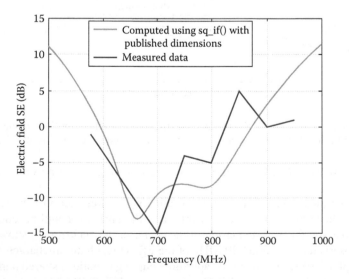

FIGURE 6.82 Measured data of 540 × 370 × 330 mm enclosure plotted against calculated measurements for 300 × 120 × 300 mm enclosure at center. Loss factor 0.

FIGURE 6.83 Measured data of 540 × 370 × 330 mm enclosure plotted against calculated measurements for 540 × 370 × 330 mm enclosure at center. Loss factor 0.

FIGURE 6.84 Measured data of 540 × 370 × 330 mm enclosure plotted against calculated measurements for 540 × 370 × 330 mm enclosure at center with a loss factor of 0.2.

and the frequency at resonance dramatically. Figure 6.83 shows a 27 dB SE at 685 MHz and at resonance of −2 dB with a 0 loss factor. Figure 6.84 shows that with a loss factor of 0.2, the SE at resonance is −9.5 dB and of 2 dB at 685 MHz.

6.4.3 WAVEGUIDE BELOW CUTOFF EFFECT IN THICK MATERIALS

The preceding section examined the SE of an aperture in which the diameter was much greater than the thickness of the material. This section examines the case in which the thickness of the material approaches or is greater than the gap dimensions.

In a waveguide, the dimensions are chosen such that the signal frequency is above the cutoff frequency and the attenuation in the waveguide is negligible. When the frequency of a plane wave incident on a waveguide is below the cutoff frequency of the waveguide, the plane wave is attenuated as it passes down the waveguide. It is just this effect that is used to describe the additional attenuation achieved in an aperture in a thick material.

The additional attenuation of a gap for a plane wave and an E field is given approximately by

$$A = 0.0018\,fl\sqrt{\left(\frac{f_c}{f}\right)^2 - 1}\ \left[\text{dB}\right] \tag{6.20}$$

or

$$A = 0.0018\,fl\left(\frac{f_c}{f}\right)\left[\text{dB}\right] \quad \text{when } \frac{f_c}{f} > 1$$

where
  f is the operating frequency [MHz]
  f_c is the cutoff frequency [MHz]
  l is the gap depth for overlapping members of the thickness of the material for butting members (cm)

For a rectangular gap,

$$f_c = \frac{14{,}980}{g}$$

where g is the largest gap's transverse dimensions, in centimeters. For a round hole,

$$f_c = \frac{17{,}526}{g} \tag{6.21}$$

When $fc \gg f$,

$$A = \begin{cases} 27\dfrac{l}{g}\ \left[\text{dB}\right] & \text{for rectangular gap} \\[2ex] 32\dfrac{l}{g}\ \left[\text{dB}\right] & \text{for a circular gap} \end{cases}$$

When $f \ll g$, the attenuation tends to zero using the waveguide equation, and the attenuation will be that due to an aperture in a thin material, as described in Equation 6.14. Equation 6.20 therefore does not provide the total SE of the metal but the increased attenuation due to the waveguide below the cutoff effect.

For a square array of holes, such as those used for the ventilation of a panel, the equation, when $f_c \gg f$, is

$$A = 20\log\left(\frac{c^2 l}{d^3}\right) + \left(\frac{32t}{d}\right) + 3.8 \left[\text{dB}\right] \qquad (6.22)$$

For a rectangular array $l_1 \times l_2$, use

$$l = \sqrt{l_1 \times l_2}$$

Figure 6.85 shows the dimensions c, d, and l. The attenuation A in this instance is defined as the increased attenuation of the hole pattern compared to the attenuation achieved if the total area $l \times l$ had been removed. The attenuation for multiple holes exhibits, according to Equation 6.22, a higher value than the attenuation of a single aperture to the incident field. This apparent anomaly is explained as follows: The field strength developed across the larger area covered by the multiple holes, expressed in V/m, is higher than the field strength across the smaller area of a single hole, assuming a constant hole diameter, as shown in Equation 6.15.

Equation 6.22, in this or similar form, is commonly found in the EMC literature and indicates that a constant attenuation with frequency exists as long as $f_c \gg f$. However, measurements of the attenuation of multiple holes in a solid 2 cm thick sheet of metal, described in Reference [10], show a measured attenuation that varies as much as 30 dB with frequency and may be 15 dB below that predicted for E fields and plane waves. Equations 6.20 and 6.22 are not valid for the attenuation of H fields.

In addition, the theoretical compensation for multiple apertures contained in Equation 6.22 is not substantiated by test results, which show that the attenuation of multiple apertures is approximated by the $32t/d$ term (i.e., equal to the attenuation of a single aperture).

Again the area of field incident on a number of apertures is greater than for a single aperture, and thus, the field strength appearing on the far side of the apertures will be greater than that of a single aperture.

The waveguide below cutoff effect is used to advantage in the design of honeycomb-shielded air vent panels and in the connection of a fiber-optic communication cable through a tube into a shielded enclosure. Figure 6.86 plots the theoretical attenuation of a ¼ in. wide rectangular honeycomb for several ratios of thickness to width, w.

FIGURE 6.85 Dimensions of array of holes in a panel.

FIGURE 6.86   Theoretical attenuation of a waveguide below cutoff for a number of ratios of t/w.

6.5   ATTENUATION OF ENCLOSURE WITH JOINTS AND APERTURES TO AN H FIELD

6.5.1   ATTENUATION OF A SIX-SIDED THIN FOIL ENCLOSURE WITH SEAMS TO AN H FIELD, WITH VIRTUALLY THE SAME CURRENT INSIDE AND OUTSIDE THE ENCLOSURE

Contrary to expectations and conventional theory, thin aluminum walls can provide significant shielding of magnetic fields. This section examines the measurement data and attempts to explain the mechanism of magnetic field shielding that occurs.

For these tests, an enclosure was built with the dimensions 1 m × 1 m × 0.7 m high. The frame of the box was constructed of wood, and the walls of aluminum foil were 0.076 mm (0.003 in.) thick. Four walls of the enclosure were covered by a continuous sheet of foil clamped together where the ends meet between two strips of brass screwed into wooden supports. The two remaining sides of the enclosure were covered by individual sheets of foil. At the seams, the foil edges were overlapped and stapled every 3 cm to the wooden frame and then covered by packing tape. On one side of the box, the foil is covered in a brass plate. A rectangular aperture is placed in the brass and foil, and this is covered by a door. The door has a Spira gasket around all four sides.

For the first set of measurements, a multiturn RS101 H-field loop was used as the TX antenna and its center was located 0.5 m from the front face of the foil enclosure. The loop was oriented such that the magnetic field lines incident on the front face of the box would be horizontally oriented, resulting in a current that travels vertically on the enclosure wall. A 13 cm H-field loop was used as the RX antenna inside the enclosure and mounted at distances of 0.3, 0.5, and 0.8 m from the front wall of the enclosure.

From 400 Hz to 200 kHz, the field picked up by the RX antenna inside the enclosure was measured at each of the three distances. A current probe was used to monitor the signal going into the TX antenna to ensure that the same field was generated in all of the tests. These measurements were compared to the reference measurements taken at the same distances, but without the shielded enclosure. The difference between the reference measurement and the shielded measurement is the SE.

Similar SE was also found using another technique that compares surface currents. For these measurements, a 560-turn surface current probe was placed at the same relative positions on both the exterior and interior wall of the enclosure illuminated by the TX antenna. The ratio of the

FIGURE 6.87 Measured magnetic field shielding effectiveness of the aluminum foil enclosure versus surface currents.

FIGURE 6.88 Magnetic field shielding effectiveness with distance of receiving antenna inside.

current measured on the outside and inside walls of the enclosure is the SE. Results from both SE tests are shown in Figure 6.87.

In addition to the change in SE with location of the RX antenna inside the enclosure shown in Figure 6.87, a further analysis of the change in SE at the spot frequency of 5.1 kHz with the enclosure far removed from conductive surfaces on a free space range is shown in Figure 6.88 and from 70 to 200 kHz in Figure 6.89. These tests were to eliminate errors due to field perturbation as a result of proximity to conductive surfaces.

Counter to the theory of reciprocity, we observe that the SE we measure from the aluminum foil enclosure is very slightly affected by the location of the RX antenna inside the enclosure. The SE seems to increase the farther we move the RX antenna from TX antenna. Although the field strength from the TX antenna decreases with distance, this falloff should not affect the SE calculation because it is a comparative value. The reduction in H field outside and inside the enclosure is a function of $1/r^3$, where r is the distance from the source. In the reference field measurement, the source is the TX loop, typically 0.5 m from the outside wall of the enclosure. If we postulate that the source for the field inside the enclosure is the current flow on the wall, then we see that the reduction in the resultant field with distance is greater than the field from the TX antenna (the reference field). As an example, the reduction in the reference field from 0.8 m from the source to 1.3 m from the source is 20 log (0.8 m/1.3 m)³ = −12.65 dB. The field inside the enclosure with the TX loop at 0.5 m from the enclosure, and assuming the wall is

FIGURE 6.89 Effect on shielding effectiveness from 70 kHz to 200 kHz of moving receiving antenna. Measurements made outside of building on a free space range.

the source, reduces from 0.3 to 0.8 m as 20 log (0.3 m/0.8 m)^3 = −25.6 dB, a difference of 13 dB. We do not see this magnitude of change in our measurements and we will discuss why this hypothesis is not correct later. We have proved by measurements that the distance of the TX antenna from the wall of the enclosure does not affect the SE as predicted by the theory of reciprocity. Thus, attenuation due to reflection, which depends on wave impedance, does not explain our test results. This independence with source location is very different from the results using the classic approach to magnetic field shielding, which uses a transmission line analogy involving the difference between the incident wave impedance and the metal impedance as well as absorption through the metal to achieve the SE.

With the enclosure and TX loop inside a shielded room, the variation in measured SE with RX antenna location was much greater. With the enclosure just outside of the shielded room and building, far removed from conductive surfaces, this affect was very slight. We, and other investigators, have seen this apparent contradiction to the theory of reciprocity in other measurements, and a number of explanations have been proposed for the dependence in source location, which we do not see. The explanations are based on the real wave impedance of the incident field. Reference [33] explains to some extent this effect.

We also found that the measurements of SE outside of the lab were different above 10 kHz from the measurements inside the lab, and this is shown in Figure 6.90.

To understand the field incident on the enclosure and the eddy currents induced into the enclosure, a thorough investigation of the currents on the exterior of the enclosure was carried out.

FIGURE 6.90 Comparison between shielding effectiveness measured within the lab and outside the lab.

The 560-turn surface current probe was used to measure the pickup on all sides of the enclosure and also to measure the H-field pickup in air at the same relative positions without the enclosure in place. The probe was oriented in both vertical and horizontal orientations to obtain the highest pickup. The highest measurement observed with the probe in air was then given the nominal value of 1, and all other measurements were compared to 1. A sketch of the relative pickup in air at 5 kHz is shown in Figure 6.91, and a sketch of the pickup on the enclosure at 5 kHz is shown in Figure 6.92.

Examining the pickup in air, one might question the validity of the measurements, since some do not seem to reduce as $1/r^3$. However, we must remember that as we move away from the plane of the loop, we must take into account not only H_θ but also H_r. By calculating the composite field at these locations, we find the measurements extremely accurate on all "sides" except for the top.

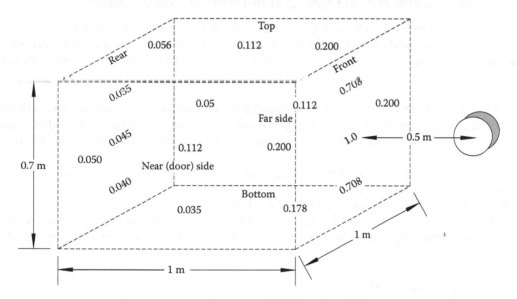

FIGURE 6.91   Relative pickup in air from the surface current probe at 5 kHz.

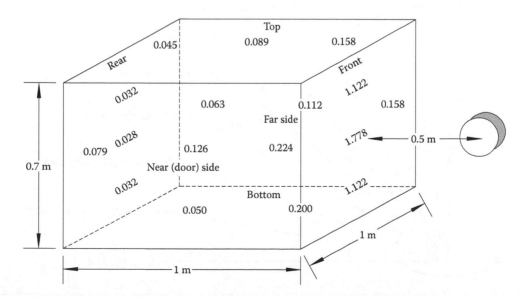

FIGURE 6.92   Relative pickup on the enclosure from the surface current probe at 5 kHz.

In accordance with classic field theory, the total magnetic field on the outside of the enclosure facing the TX antenna is twice that for the incident field. The measurements on the wall facing the TX loop show a field higher than the incident, but not exactly twice the magnitude. The difference between this increase and a doubling in the field is 19 dB, exactly the level of SE at 5 kHz. The same measurement was made on a very large conducting wall versus the field in air with similar results. The relative fields on all other surfaces of the enclosure are incredibly close to the field without the enclosure as shown in Figures 6.91 and 6.92.

6.5.2 Magnetic Fields Generated by the Transmitting Loop and Induced into the Enclosure and the Fields Generated by the Enclosure Currents

The H fields generated by an electrically small-loop antenna are shown in Figure 6.93.

At low frequency and close to the loop, these fields are referred to as "quasi static" and are in the "near field." The magnitude of H_r, when theta is 0°, is twice the magnitude of H_θ, when theta is 90°. Figure 6.93 shows the relative magnitude of these fields with the angle theta varied. Both fields reduce with distance from the loop, in the near field, as a function of $1/r^3$.

As the circumference of the loop is much smaller than a wavelength, the antenna is electrically small, the current around the loop is virtually constant, and the same magnitude of H_θ and H_r generated around the circumference of the loop is constant, as long as the distance and z and y locations are kept constant.

One mechanism for magnetic field shielding is absorption, and this is effective when the metal thickness is greater than the skin depth.

The skin depth for aluminum is shown in Table 6.20.

It can be seen from Table 6.20 that even at 100 kHz, the skin depth is much greater than 0.076 mm—the thickness of the aluminum foil used in the enclosure under test. This means that

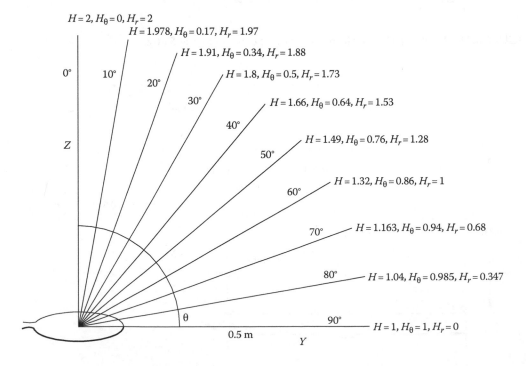

FIGURE 6.93   *H* fields generated by a loop antenna.

TABLE 6.20
Skin Depth of Aluminum

f (Hz)	Skin Depth (mm)
400	4.3
1,000	2.7
5,000	1.2
10,000	0.85
20,000	0.6
30,000	0.49
40,000	0.43
50,000	0.38
60,000	0.35
70,000	0.32
100,000	0.27

virtually the same current flows on the internal wall of the enclosure as on the outside and intuitively it would seem that no magnetic field SE for the foil enclosure is achieved, certainly not by absorption. The mechanism is that the incident field sets up an eddy current in the aluminum that generates in turn a magnetic field.

The field lines cutting the enclosure are shown in Figure 6.94. The field lines inside the enclosure generated by the eddy current are in the opposite phase to the incident field as shown in Figure 6.95 and so "bucks" the incident field.

As the current flow on both sides of the enclosure is in the same phase, the magnetic field from this current on the two sides repel and so the field does not enter the enclosure, as shown in Figure 6.96. This was verified by rotating the RX antenna 90° as shown in Figure 6.97. In this case, the field that is generated inside the enclosure (by the current on the face, top, and on the bottom of the enclosure) does not cut the loop at right angles but is in the plane of the loop and no voltage should be induced. As no field due to the current flow on the sides penetrates the enclosure, the pickup on the RX loop antenna should be negligible and this was confirmed by measurements.

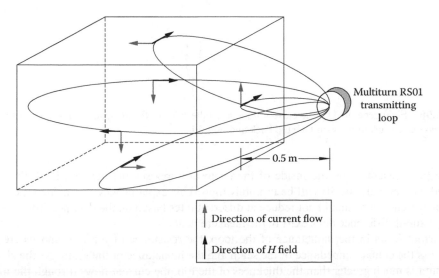

FIGURE 6.94 Magnetic field lines from a loop-transmitting antenna cutting an enclosure.

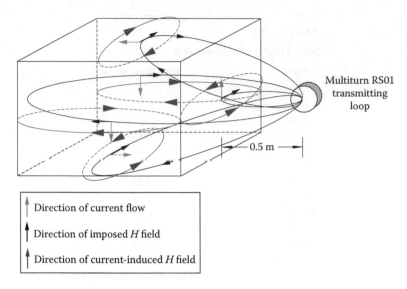

FIGURE 6.95   (See color insert.) Magnetic field lines incident on an enclosure and the magnetic field set up by the current flow on the enclosure.

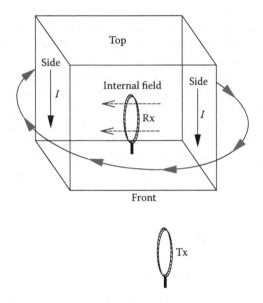

FIGURE 6.96   Field generated by the current flow down the sides of the enclosure does not enter enclosure. Internal field due to current flow on face top back and bottom.

As the field generated on the inside of the enclosure opposes the incident field, the two will cancel and theoretically the SE will be infinitely high. This explains why the incident field and the field inside the enclosure should not reduce at different rates based on the distance from the source. However, a small difference does exist in the measurements.

The current flows in the inductance of the loop, the resistance of the foil, and where a seam exists, across the contact impedance of the seam and the inductance of the seam. As the skin depth of aluminum is much greater than the thickness of the foil, the current flows through the thickness

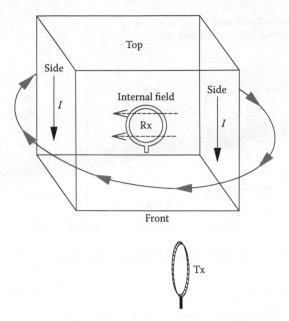

FIGURE 6.97   Internal field cuts the receiving antenna at 0°. Field due to current on the wall does not penetrate the enclosure.

of the foil and no correction for skin depth is needed. The equation for the total DC resistance of the foil enclosure in the current path generating the internal field is

$$R_S = \frac{2L_2 + 2D}{GL_1 t} \tag{6.23}$$

where
  L_2 is the height of the enclosure
  D is the depth of the enclosure
  L_1 is the width of the enclosure
  G is the conductivity in mhos/m
  t is the thickness of the material

Using Equation 6.29 for the SE of the enclosure, the calculated resistance of the aluminum foil enclosure, using the conductivity of aluminum as 3.54e+7 mhos/m, is 1.3 μΩ. At low frequency, the impedance of the seam is lower than the DC resistance of the foil. Only above 10 kHz, do we see that the SE no longer increases monotonically. By using a resistance of 1.3 μΩ, and a contact impedance of 2.2 μΩ, the predicted SE from Equation 6.29 is remarkably close to the measurements as shown in Figure 6.98. By setting the DC resistance and contact resistance to a very low value equal to the material DC resistance, we obtain the theoretical maximum SE achievable with a seam-free enclosure, and this is 26 dB at 5 kHz.

At these frequencies where the skin depth is greater than the thickness of the material, the type of seam is not so important. For example, a seam covered with a strip of metal in a 1 m × 1 m × 1 m enclosure using aluminum foil has a similar magnetic field SE (see Section 6.3.4) when compared to the seam clamped between brass plates described here. However, the type of seam does greatly change the plane-wave SE.

FIGURE 6.98   Measured magnetic field versus theoretical for the 1 m × 1 m × 0.7 m enclosure.

6.5.3 *H*-Field Shielding Effectiveness of Enclosure with Seams and Apertures

When the incident field is magnetic, a different mechanism predominates, especially when the thickness of the material is less than the width of the aperture.

MIL-STD-285 describes a technique for measuring the magnetic field SE of an enclosure, in which small TX and RX loops are oriented such that the magnetic field is tangential to the surface of the enclosure. Consider the case where an aperture exists in a large enclosure and the frequency is high enough that diffusion through the enclosure material is insignificant. The distances between the TX loop and the aperture, D_1, and between the aperture and the RX loop, D, and the orientation of the loops are shown in Figure 6.49.

The magnetic field at the measuring point d, distance D from the shield with aperture, is given, from Reference [20], by

$$H_\theta = \frac{\alpha_m}{2\pi}\left(\frac{1}{D}\right)^3 H_o \cos\phi \sin\theta$$

where
α_m is the magnetic polarizability of an aperture, which equals $4(a^3/3)$ for a circular aperture
a is the radius of the aperture [m]
θ is the angle between the measured magnetic field and the axis of the measurement current loop
ϕ is the angle from the centerline of the aperture (90° from the surface of the enclosure) and the measuring location

Thus, the H_θ measured field is at a maximum when the field cuts the measuring loop at an angle of 90° and the loop is located on the centerline of the aperture (i.e., $\phi = 0°$). Assuming maximum field coupling to the RX loop, the SE of the shield with aperture is given by

$$SE = \left(\frac{3\pi}{4}\right)\left(\frac{D_1}{D_1 + D}\right)^3 \left(\frac{D}{a}\right)^3$$

If $D_1 = D = \dfrac{D_1 + D}{2}$, then

$$SE = 0.295\left(\frac{D}{a}\right)$$

Equation 6.18 for the SE of a shield with aperture to the E field generated by a small loop and the equation for the magnetic field attenuation of a shield with aperture are both functions of distance. This would appear to contradict the law of reciprocity, in which under certain conditions, the TX and RX loops may be interchanged and the received power will remain constant. Hoeft et al. remark [20] that "the shielding effectiveness is a strong function (cubed dependence) of the distance between the shield and measuring location. Thus the shielding effectiveness is not an intrinsic electromagnetic parameter." This means the SE is dependent on the test setup, which includes the distances between the TX loop and the shield and the RX loop and the shield. In contrast, the measurement of the transfer impedance of a shielded cable is an intrinsic function of the cable. For as long as the shield current is measured or known and the transferred voltage is measured, the transfer impedance of the cable may be computed and should be independent of the test method, unlike measurement of the transfer admittance of a cable, which is dependent on the source of electric field and the geometry of the test fixture and is not an intrinsic electromagnetic parameter.

The predicted SE using the equation has been compared, in Reference [20], to measured values. In the measurement, the distance D was fixed at 0.406 m (16 in.), with the aperture radius varied from 0.0076 m (0.3 in.) to 0.05 m (2 in.). The correlation between predicted and measured SE is remarkably good.

The attenuation of enclosures, joints, and apertures to H fields, based on Reference [11], is considered as follows: We assume that the magnetic field (H_{out}) impinging on the outside of an enclosure results in an H field (H_{in}) inside the enclosure:

$$\text{The attenuation of the enclosure is given by } \frac{H_{out}}{H_{in}} \qquad (6.24)$$

The resistance of the enclosure wall or the impedance of the defect (e.g., a joint, aperture, or gasket) is represented by Z_s in Figure 6.99. The inside of the enclosure can be modeled as the inductance of a wide single-turn inductor, given by

$$L = \frac{\mu_o A}{l_1} \qquad (6.25)$$

where
  A is the area of the inside of the box = height × depth
    $= l_2 \times d$
  l_1 = width of the box

The inductive reactance of the inside of the box ($2\pi f L$) is represented in Figure 6.99 as Z_{in}. The H field incident on the box causes a current to flow on the outside of the box. At low frequency where the thickness of the enclosure is less than the skin depth, the principal coupling mode is due to the current diffusion through the enclosure wall. Z_s at low frequency is thus predominantly the AC resistance of the enclosure wall (R_s) given by

$$R_s = \frac{12e^{-t/\delta}}{G_t l_1} \left[\Omega \right] \qquad (6.26)$$

where
  t is the thickness of the enclosure [m]
  G is the conductivity [mhos/m]
  δ is the skin depth [m]

FIGURE 6.99   Enclosure with defects and equivalent circuit.

The conductivity is given by

$$G = G_o G_r$$

where $G_o = 5.8 \times 10^7$ mhos/m. The skin depth is computed from

$$\delta = \frac{1}{\sqrt{\pi f \mu G}} \; [\mathrm{m}]$$

where μ is the permeability [H/m] and is given by

$$\mu = \mu_o \mu_r$$

where $\mu_o = 4\pi \times 10^{-7}$ [H/m].

When the metal contains a number of apertures, the conductivity of the metal is reduced by the apertures. The conductivity is then

$$G \times \frac{\text{Area of panel} \left(l_1 \times l_2 \right) - \text{Total area of apertures}}{\text{Area of panel}} \tag{6.27}$$

The additional components of Z_s when seams and joints are present are as follows:

1. The DC resistance of the joint R, which is typically 15–30 $\mu\Omega$ for a riveted joint but may be as high as 1.5 mΩ for unplated metal.
2. The contact impedance R_c, which has the characteristics of an AC resistance and contains the term $\sqrt{f}$. The contact impedance is dependent on surface smoothness and finish. A typical value of R_c for a riveted joint is $0.06\sqrt{f}$ $\mu\Omega$ and may be as high as $0.25\sqrt{f}$ $\mu\Omega$ for unplated metal.
3. Another component of a joint that contains a gap is the inductance of the gap L, which may have a typical value of 0.5×10^{-12} H.

In addition, screws used to fasten the walls of an enclosure exhibit inductance. Although a complex impedance, the total impedance of a joint may be modeled by a simple series equivalent circuit and is given approximately by

$$R_s + lN(R + R_c\sqrt{f} + 2\pi fL) \tag{2.28}$$

where
$\quad l$ is the length of the joint
$\quad N$ is the number of joints in the current path

When the joint is gasketed, the expression after R_s is replaced by the transfer impedance of the gasket at the frequency of interest.

The attenuation of the enclosure has been expressed as H_{out}/H_{in}. This is equivalent to the ratio I_{out}/I_{in} that is given by

$$\frac{j\dfrac{2\pi f\mu_o A}{l_1}}{R_s + lN\left(R + R_c\sqrt{f} + 2\pi fL\right)} \tag{6.29}$$

where
$\quad A$ is the area of the enclosure $= l_2 \times d$
$\quad l$ is the length of the joint (joint or slot in the current path)
$\quad R$ is the DC resistance of the joint
$\quad R_s$ is the resistance of an enclosure wall as a function of skin depth
$\quad R_c$ is the contact resistance
$\quad L$ is the equivalent series inductance of the slot
$\quad N$ is the number of joints in the current path

The magnitude of the internal H field can be computed from Equation 6.29 when H_{out} is known. The limit in attenuation of an enclosure, with apertures, to an H field is given by

$$\frac{H_{out}}{H_{in}} = 0.445\frac{A}{N\alpha l_1} \tag{6.30}$$

where
$\quad N$ is the number of apertures
$\quad \alpha$ is the magnetic polarizability of the aperture

TABLE 6.21

Worst-Case Values of Magnetic Polarizability for Different-Shape Apertures in an Infinitely Thin Wall

Shape of Aperture	Magnetic Polarizability α
Circle of diameter d or radius r	$\dfrac{d^3}{6}$ or $4\left(\dfrac{r^3}{3}\right)$
Long, narrow ellipse of major axis a and minor axis b	$\dfrac{\pi}{3}\dfrac{a^3}{l_n(4a/b)-l}$
Square of side l	$0.259l^3$
Rectangles of length l and width w	
$l = \dfrac{4}{3}w$	$0.2096l^3$
$l = 2w$	$0.1575l^3$
$l = 5w$	$0.0906l^3$
$l = 10w$	$0.0645l^3$

The worst-case magnetic polarizability of different shapes of aperture is provided in Table 6.21, from Reference [25]. These values assume that the current flows over the longest dimension, the apertures are less than $\frac{1}{6}\lambda$, and the material is infinitely thin.

An approximate thickness correction, in decibels, for round apertures and magnetic fields is $32t/d$ and for square apertures $27.3t/l$, where t is the thickness, d is the diameter, and l is the longest dimension.

As frequency increases, the attenuation of the enclosure, computed from Equation 6.29, increases until the attenuation is limited by the apertures, Equation 6.30, and is then constant with frequency.

Equation 6.30 is applicable to the attenuation of a single-layer wire mesh by assuming that the wire screen behaves like a sheet with square perforations. Measurements of wire have shown that the calculated values are all from 4 to 6 dB lower than measured values.

One type of shield that can be used effectively to shield H-field emissions from a video display is very fine blackened wire. This type of wire mesh has typically 100 openings per inch (OPI) and, although the display is visible, will reduce the brightness of the display and may result in moiré patterns.

The main weak link is at the contact between the enclosure and the wire mesh, due to the contact impedance. The measured surface resistivity of a copper-plated stainless steel mesh with 100 OPI was 1.5 mΩ^2 and that for copper mesh with 100 OPI was 3.3 mΩ^2.

6.5.4 Examples of *H*-Field Attenuation of Enclosures

The application of an insulating compound to a joint may be a requirement for sealing against moisture or to increase the thermal conductivity of the joint. Figure 6.100 illustrates the joint impedance of a test sample made from two pieces of 12 × 24 in. aluminum joined by a 2 in. wide backing strip riveted on 3 in. centers. A sealant was applied to the bond before the strip was riveted, and the configuration is referred to as the "full-bond" sample. The joint resistance of the "joints-only" sample, in which the sealant was applied to the exposed joints after the joint was riveted, is shown in Figure 6.101. Figures 6.101 and 6.102 are reproduced from Reference [22], which describes the experimental and theoretical analysis of the magnetic field attenuation of enclosures. The DC resistance of the full-bond sample was measured to be 35 μΩ, and for the joints-only sample, it was 15 μΩ.

FIGURE 6.100 Measured joint impedance of the "full bond" joint sample. (From Hoeft, L.O., The case for identifying contact impedance as the major electromagnetic hardness degradation factor, *IEEE EMC Symposium Record*, San Diego, CA, September 16–18, 1986. © 1986, IEEE.)

FIGURE 6.101 Measured joint impedance of the "seams-only" joint sample. (From Hoeft, L.O., The case for identifying contact impedance as the major electromagnetic hardness degradation factor, *IEEE EMC Symposium Record*, San Diego, CA, September 16–18, 1986. © 1986, IEEE.)

The magnetic field attenuation of the two samples when mounted in a 0.6 m × 0.6 m × 0.2 m brass box with a thickness of 1.58 mm and a relative conductivity of 0.6 is shown in Figure 6.102.

The equivalent impedances of the two samples that fit the curves when used in Equation 6.29 are as follows:

	R [μΩ]	R_c [μΩ]	L [10^{-12} H]
Joints only	40	$0.03\sqrt{f}$	0.1
Full bond	50	$0.10\sqrt{f}$	0.5

FIGURE 6.102 Measured surface magnetic field reduction of the "full-bond" and "seams-only" joint. (From Hoeft, L.O., The case for identifying contact impedance as the major electromagnetic hardness degradation factor, *IEEE EMC Symposium Record*, San Diego, CA, September 16–18, 1986. © 1986, IEEE.)

FIGURE 6.103 Effect of transverse screw spacing on measured transfer impedance of a large cableway. (From Hoeft, L.O. et al., Predicted shielding effectiveness of apertures in large enclosures as measured by MIL-STD-285 and other methods, *IEEE National Symposium on EMC*, Denver, CO, May 23–25, 1989. © 1986, IEEE.)

The effect of the spacing of the fastener on transfer impedance is illustrated in Figure 6.103. In this example, the number of transverse screws in a large cableway was varied from no screws to one screw per 2 in. spacing. The transfer impedance is measured in a TEM cell in a test configuration similar to that used for magnetic field attenuation measurements. The transfer impedance is a measure of the voltage developed across the seam inside the test enclosure mounted on the floor of the TEM cell, divided by the current flow on the outside of the enclosure and across the joint under test.

The contact impedance and DC resistance of a joint are dependent on the surface finish, as illustrated in Figure 6.104. The unplated aluminum exhibits the lowest level of *H*-field attenuation,

FIGURE 6.104 Magnetic field reduction of a deep drawn aluminum equipment case showing the effects of tin plating and a gasket. (From Hoeft, L.O., The case for identifying contact impedance as the major electromagnetic hardness degradation factor, *IEEE EMC Symposium Record*, San Diego, CA, September 16–18, 1986. © 1986, IEEE.)

whereas tin plating is almost as effective as a conductive gasket. Tin plating is an excellent choice because tin is malleable and tends to flow, filling up the small gaps in the joint.

The components of the equivalent impedance of the seams with different finishes are as follows:

	R [μΩ]	R_c [μΩ]	L [10^{-12} H]
Tin plated	30	$0.015\sqrt{f}$	0.5
Unplated	1500	$0.250\sqrt{f}$	0.5
Gasketed and tin plated	30	$0.001\sqrt{f}$	Negligible

FIGURE 6.105 Magnetic field reduction with apertures (size of enclosure 24 in. × 24 in. × 8 in.). (From Hoeft, L.O., The case for identifying contact impedance as the major electromagnetic hardness degradation factor, *IEEE EMC Symposium Record*, San Diego, CA, September 16–18, 1986. © 1986, IEEE.)

The effect of apertures on magnetic field attenuation is illustrated in Figure 6.105. We see that a hole as large as 4 in. in a 24 in. × 24 in. × 8 in. enclosure still results in a magnetic field attenuation of 50 dB. This explains why, as often seen in practice, a thin metal enclosure with large ventilation apertures, typically 1 in., in close proximity to an *H*-field source, such as a switching power supply/converter, will provide significant shielding. The level of shielding has been determined in practice by field measurements with the enclosure in place and removed.

A large number of small apertures will provide higher attenuation at high frequency compared to a single aperture of the same area. This is illustrated in Figure 6.106. The conductivity of the 0.6 m × 0.6 m × 0.2 m aluminum sheet containing 36,100 $\frac{1}{6}$ in. apertures is lower than the sheet with a single 4 in. aperture; therefore, at frequencies below 30 kHz, the multiple-aperture sample shows a lower attenuation.

Methods for reducing the coupling through ventilation apertures include wire mesh and honeycomb air vent filters. The published attenuation curves for these filters are often theoretical, and the attainable attenuation may be lower. The best type of honeycomb air vent filter is constructed with the metal segments, which make up the honeycomb, welded or soldered together. A more common method has the segments electrically connected via a bead or single spot of conductive adhesive. In addition, the edges of the metal segments are often connected to the frame in which the honeycomb is contained by spring pressure only. In these samples, the contact impedance may be initially high and increases further with time, due to corrosion. The magnetic field attenuation performance of many commercially available honeycomb filters is very poor due to the low conductivity and, therefore, the appreciable skin depth of the honeycomb. It is not uncommon for equipment to fail MIL-STD-461 RE102 or similar radiated emission tests due to magnetic field coupling through poor honeycomb air vent panels, especially when the *H*-field source is in close proximity to the panel.

In one sample of a honeycomb, filter cells were made from aluminum that had been chromated and were bonded by nonconductive adhesive. The electrical contact was made via the relatively high resistance of the chromate finish. The measured SE of the honeycomb was 15 dB at 100 kHz and 3 dB at 14 MHz. With a cadmium plating, the honeycomb achieved an SE of 20 dB at 100 kHz and 55 dB at 14 MHz. A much higher magnetic field SE can be achieved with a steel or brass honeycomb. For optimum SE, steel cells are spot welded together and then tin plated or brass cells are soldered together. An alternative to a honeycomb, when high attenuation of magnetic fields is required, is the wire mesh screen air filter. The mesh must be passivated with a low-resistance finish in order to reduce the skin depth and therefore optimize the SE of the screen.

The impedance of a joint is dependent on the fastening method. The lowest-impedance seam is welded with an impedance close to the base metal when the welding is carefully controlled. The

FIGURE 6.106 Comparison between a large number of small apertures and a single large aperture with the same area.

next preferred method is soldering. Riveted seams provide a low-impedance path at the rivet, for the head of the rivet flows into the surface of the metal. Screws exhibit an inductance of 1–3 nH, depending on the length of the thread from the head to the tapped metal. The impedance of pop rivets is no lower and in some samples higher than that of screws.

The *H*-field SE of an enclosure with aperture changes dramatically at the resonant frequency of the enclosure, even when the aperture is not a resonant length. The analysis described for the *E*-field SE of an enclosure when the source is an external plane wave, described in Reference [9], is also applicable to the *H*-field component shielding. As predicted in Equation 6.30, the SE of an enclosure with aperture to an *H*-field source, above some low frequency, is almost constant with frequency. This is shown in Figures 6.105 and 6.106 and in the theoretical curve in Figure 6.107, up to 400 MHz, and in the theoretical and measured curves in Figure 6.108, up to 500 MHz. However, at the resonant and antiresonant frequencies of the enclosure, the shielding either increases or becomes negative

FIGURE 6.107   Calculated and measured S_H at center in a $300 \times 120 \times 300$ mm enclosure with a 100×5 mm aperture.

FIGURE 6.108   Calculated S_H at three positions in a $300 \times 120 \times 300$ mm enclosure with a 100×5 mm aperture and measured S_H at the center of the enclosure.

(i.e., a field enhancement occurs). These effects are shown in Figures 6.107 and 6.34. Figure 6.108 shows that the magnetic field SE increases (i.e., the magnitude of the *H* field decreases) when the measurement point is further away from the aperture. Therefore, locate susceptible circuits on a PCB as far away from the aperture as possible.

The practical application of the analysis and measurements are discussed in Reference [9] and are, for the *H* field.

6.5.5 Measured Attenuation of a Magnetic Field Source within an Enclosure

To ensure the method of calculating magnetic field attenuation is accurate for the case when the source is inside the enclosure, the following experiment was conducted.

A magnetic field was generated by a rectangular loop placed inside an enclosure with apertures. The size of the enclosure was 36 cm × 25.4 cm × 5 cm deep, and 21 apertures of 1.5 cm × 1.5 cm and one aperture of 8 cm × 2.7 cm were punched out of the enclosure.

In commercial and military EMC requirements, the maximum level of radiated *E* field is often specified. As discussed in Chapter 2, a loop antenna generates both *E*- and *H*-field components and either may be predicted and measured. As a wave couples through the defects in an enclosure, the magnetic and electric field components are not always attenuated to the same extent, and a change in the wave impedance occurs. In this case study, the loop is only 5 mm from the inside of the enclosure, and the wave impedance will be low both inside and outside of the enclosure. Thus, we assume that the wave impedance of the field radiated from the outside of the enclosure is the same as the field from the loop when located outside of the enclosure.

A monopole antenna was located at a distance of 1 m from the loop antenna, and measurements were made of the *E* field with the loop both inside and outside of the enclosure. The antenna factor of the monopole, which is calibrated assuming the far-field impedance of 377 Ω, was corrected for the predicted wave impedance of the field 1 m from the loop antenna. The measured *E* field from the loop located outside of the enclosure was within 6 dB of that predicted, using Equation 2.9.

The current flow induced on the inside of the enclosure by the loop antenna, based on the antenna location, was across the largest dimension, 8 cm, and it was therefore valid to use this dimension in Equation 6.30. The upper limit of the attenuation when computed from Equation 6.30 was dominated by the single 8 cm × 2.7 cm aperture. The 21 1.5 cm apertures placed the upper limit of attenuation at 65 dB, whereas the single large aperture placed it at 42 dB. The measured attenuation was computed from the ratio of *E* field with the loop outside of the enclosure to the measured *E* field with the loop inside the enclosure. A comparison of the predicted and measured attenuations shows an uncharacteristically good correlation:

Frequency [MHz]	Attenuation [dB]	
	Predicted	Measured
12.0	42	45
17.0	42	40
24.0	42	40
27.6	42	48

6.6 GASKETING THEORY, GASKET TRANSFER IMPEDANCE, GASKET TYPES, AND SURFACE FINISH

6.6.1 Gasket Theory

One method of closing seams is to use conductive gasket material. In instances where the seam must be environmentally sealed, some form of gasket material is mandatory.

FIGURE 6.109   Current density through an aperture closed by a gasket.

Figure 6.109 shows the current density through a metal plate with a seam closed by a gasket that has a lower conductivity or permeability than the metal plate. The voltage developed across the gasket may result in a current flow and the generation of a predominantly H field in the enclosure. Alternatively, when the impedance of the enclosure is high, which is more often true of large enclosures at high frequency, a predominantly E field may be generated by the voltage across the gasket.

The impedance of the gasket has components of resistance, contact impedance, and inductance, as does the impedance of the ungasketed joints discussed in the preceding sections. An additional component of gasket material that may become dominant is the capacitance between conductive particles or mesh used in the majority of gaskets.

6.6.2   Gasket Test Methods

To evaluate a gasket's performance under the wide range of wave impedances found in practical applications, radiated testing techniques using E fields, H fields, and plane waves are used. These techniques typically employ one or more metal cavities, with a source of field on one side of the barrier and a measuring antenna/probe on the other side of the barrier. The energy is coupled through the barrier via a seam that contains the gasket under test. The gasket is usually placed under a cover plate that is placed over an aperture in the barrier. The fields coupled through the aperture are measured with and without the cover plate, with and without the gasket under the cover plate, or in some cases with the conductive gasket replaced by a nonconductive gasket and nonconductive fasteners. The presence of conductive fasteners and conductive compression stops can dramatically change the gasket measurement, as discussed later. Also, the position of the antenna in these tests strongly influences the test signal coupled through the antenna. It is highly recommended that the test antenna be moved relative to the aperture and to the surrounding enclosure.

At least 10 test methods exist for measuring the effectiveness of the addition of a gasket to a joint in reducing or preventing electromagnetic leakage through the joint. Of these 10 test methods, only four are considered standardized, although considerable effort has been spent in developing the at-present nonstandard reverberation chamber test methods. These are discussed as alternative techniques. The IEEE Std 1302-1998 "IEEE Guide for the Electromagnetic Characterization of Conductive Gaskets in the Frequency Range of DC to 18 GHz" describes in detail the four standardized measurement techniques as well as five alternative techniques.

MIL-DTL-83528E is a general specification for gasketing material, conductive shielding gasket, electronic, elastomer, and EMI/RFI and specifies IEEE299 for measuring SE. Table 6.22 shows the comparison of the standardized gasket measurement techniques and also shows the repeatability for

TABLE 6.22

Comparison of Standardized Gasket Measurement Techniques

	Measurement Technique			
	ARP 1705-2012	ARP 1173-1988	Def Stan 59-103 (17-Sep-93)	MIL-DTL-83528B (2011)
Measurement principle	Current injection	Aperture attenuation	Aperture attenuation	Aperture attenuation
Parameter measured	Transfer impedance	E/H/plane-wave attenuation	E/H/plane-wave attenuation	E/H/plane-wave attenuation
Unit of measure	dB/Ω m	dB	dB	dB
Test sample configuration	Circular	Rectangular	Circular	Rectangular
Sample size	101.6 mm diameter	300 × 300 mm	409 mm diameter	610 × 610 mm
Frequency range	Test fixture 1705A DC to 2 GHz Test fixture 1705B DC to greater than 2 GHz	400 Hz to 10 GHz	10 kHz to 18 GHz	20 MHz to 10 GHz
Typical dynamic range	60–150 dB	60–100 dB	60–120 dB	60–120 dB
Typical repeatability	± 2 dB	± (6–20) dB	± (6–20) dB	± (6–33) dB

Source: Hoeft, L.O. and Hofstra, J.S., Experimental and theoretical analysis of the magnetic field attenuation of enclosures, *IEEE Transactions*, 30(3), August 1988. © 1988, IEEE.

each of the tests. The ARP 1173-1998, Def Stan 59-103, and MIL-G-83528B show a typical repeatability of between 6 and 20 dB, which indicates that even comparative tests on gasket material using these techniques should be made at the same location and using the same relative antenna locations. The ARP 1173-1998, DEF Stan 59-103, and MIL-G-8352B are measurements of the relative transmission through an aperture, and ARP 1705-1981 is a transfer impedance test. ARP 1173-1988 uses metal spacers between the cover plate and the flange to set up an additional reference measurement. The difference between this reference measurement and the measurement with the gasketed cover in place is referred to as *shielding increase*. Similar modifications to the MIL-G-83528B measurement procedure are in use. In the IEEE Std 299-1997 measurement setup, the reference measurement is made in free space, without the test enclosure, and the comparison is made to the enclosure with the aperture covered by the gasketed plate. This method results in a very high measured SE. In MIL-G-835528B, replaced by MIL-DTL-83528, the reference level is determined by placing the TX and RX antennas in line with the open aperture in the test chamber. This results in a lower reference level and lower SE compared to IEEE Std 299-1997, but the SE is higher than the ARP 1173–1998—"shielding increase." ARP 1173-1998 has been superceded by ARP 1705 "Coaxial Test Procedure to Measure the RF Shielding Characteristics of EMI gasket Materials" 2012. The ARP 1705 test setup is illustrated in Figure 6.110.

One of the alternative test techniques is to use two reverberation chambers or mode-stirred chambers. One of the problems with chamber measurements, seen in all other tests of the transmission through an aperture, is chamber resonances. These dramatically affect the measured SE. In the MSG technique, a paddle wheel is used in each chamber to randomize the field so that the gasket under test is exposed to fields of all angles and polarizations. By selecting the rotation rate of one paddle wheel to be high compared to that of the second, the measurement of maximum coupled fields inside one chamber is a function of the other chamber's tuner position. SE is defined as the ratio of the power that is transferred into the second chamber with only a cover plate over the aperture to the power coupled through the aperture with the gasket under test inserted between the cover plate and the chamber wall. By using two large chambers, testing down to 200 MHz is possible, whereas in the nested stirred-mode method, shown in Figure 6.111, the lower frequency is restricted to 500 MHz because of the size of the smaller test chamber.

FIGURE 6.110 ARP 1705 2012 test setup.

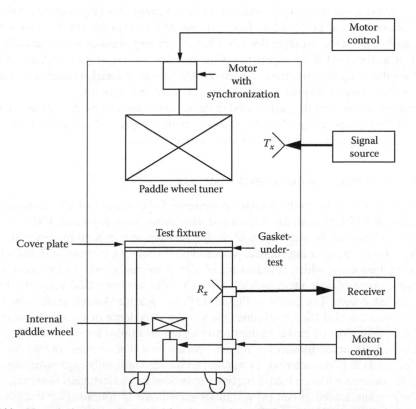

FIGURE 6.111 Nested stirred-mode method for characterizing EMI gaskets. From Hoeft, L.O. and Hofstra, J.S., Experimental and theoretical analysis of the magnetic field attenuation of enclosures, *IEEE Transactions*, 30(3), August 1988. © 1988, IEEE.

One of the major concerns in selection of gasket material is in the interpretation of the manufacturers' test data. Can the results obtained using the various test methods be compared? Reference [26] describes tests made on four different gasket materials using the MIL-G-83528, MSG, transfer impedance, and modified ASTM D 4935-89 test methods. Despite attempts to standardize the tests and to linearize the test data, differences in the shielding performance of the four different gasket materials were up to 50 dB using the different techniques! No two test techniques showed a good correlation, and the shielding performances of the four different gasket materials were also not consistently higher or lower when measured using the four different test techniques. It therefore seems impossible to evaluate the relative performance of gaskets using different test methods. IEEE Std 1302-1998 also states: "Any technique is unlikely to replicate the actual gasket application of interest. Therefore, each of the techniques should be considered as a 'platform' from which minor modifications and adjustments can be made to more closely approximate the application in question."

Kunkel in Reference [34] describes an MIL-DTL-83528 test using a newspaper as the material in the aperture. At 2 GHz, the measured SE varied from 60 to 93 dB depending on the location of the RX antenna inside the chamber!

The SE of a chamber to magnetic fields is also dependent on the size of the chamber.

Transfer impedance does provide an excellent and reproducible indication of the relative quality of the gasket; however, unless the test fixture represents the materials, finish, and dimensions of the equipment of concern, the test results can be misleading, as shown in the comparison of test methods.

The test setup for transfer impedance in accordance with ARP 1705-2012, Figure 6.110, shows a fixture with two cavities. The gap between the two cavities is closed by a metal plate sealed by the gasket under test. A current is caused to flow through an impedance across one face of the metal plate, through the gasketing material, returning on the walls of the driven cavity. A voltage output is taken from the opposite surface of the metal plate, and the ratio of measured voltage to drive current is described as the transfer impedance Z_t. In the test setup described here, the closing pressure was achieved by means of compressed air, so the effect of the impedance of the fasteners was not measured. A number of factors determine the value of Z_t, including the composition of the gasket material, its thickness and width, the impedance of the mating shield joint surface, and the pressure on the gasket.

An alternative to compress the gasket used in the measurements on modern gasket material is to use spacers in the sensing cavity when the gasket is easily compressed and a combination of foam and spacers when the gasket must be tested under a given pressure.

6.6.3 Relative Performance of Gasket Materials

Figure 6.112a through d shows the transfer impedance for a number of commercially available gasket materials at different closing pressures and frequencies, from Reference [14]. The purpose of these graphs is to illustrate the importance of closing pressure on transfer impedance: the higher the pressure, the lower the transfer impedance. Also, sample 2 shows an increase in transfer impedance with increasing frequency, which is indicative of a predominantly inductive transfer impedance. The "porcupine" type of wire mesh gasket, in which wires are imbedded vertically in neoprene foam, exhibits inductance. The gasket in Figure 6.112 is a graphite-loaded gasket and has a much higher DC resistance and at least 1000 times the transfer impedance of the other gaskets from 0.1 to 1 MHz. A graphite-loaded gasket seldom provides any additional level of shielding, especially when conductive fasteners at distances of at least 5 cm are used. The decrease in transfer impedance shown in Figure 6.112d is characteristic of a gasket with a predominantly capacitive transfer impedance, which is common with conductive particles imbedded in an elastomer. However, even above 10 MHz, the graphite-loaded gasket has a transfer impedance 15 times higher than the inductive monel wire mesh shown in Figure 6.112a.

The dimensions of the gasket material are important in determining the impedance presented to current flow; therefore, the gasket transfer impedance is expressed in ohms per meter.

Figure 6.113b shows the current flow around an aperture, and Figure 6.113a shows the aperture closed by a gasketed metal plate.

The current density through the plate is considered to be a uniform 1 A/m. The impedance encountered by the current flow is dependent on the length of material presented to the current flow multiplied by 2, because two lengths of gasketing are in the current path. For example, for gasket material with a transfer impedance of 0.00018 Ω/m, at 100 kHz and a length of 5 cm, the total impedance is

$$\frac{100 \text{ cm}}{5 \text{ cm}} \times 2 \times 0.00018 \ \Omega/\text{m} = 0.0072 \ \Omega$$

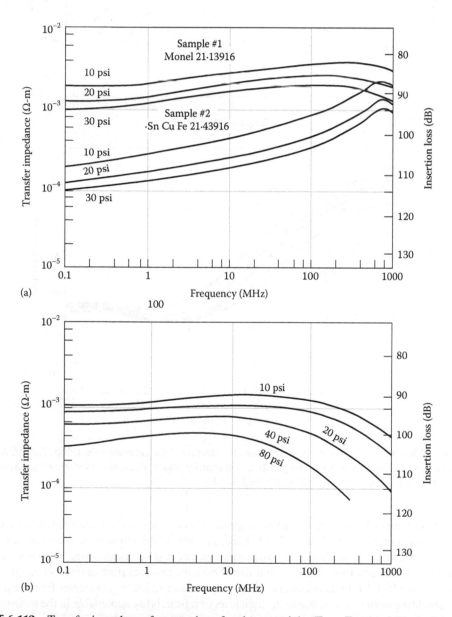

FIGURE 6.112   Transfer impedance for a number of gasket materials. (From Faught, A.N., An introduction to shield joint evaluation using EMI gasket transfer impedances, *IEEE International Symposium on EMC*, New York, September 8–10, 1982. © 1982, IEEE.)   (*Continued*)

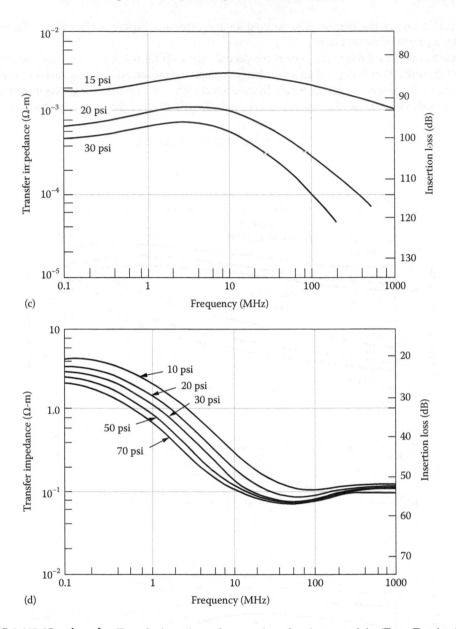

FIGURE 6.112 (*Continued*) Transfer impedance for a number of gasket materials. (From Faught, A.N., An introduction to shield joint evaluation using EMI gasket transfer impedances, *IEEE International Symposium on EMC*, New York, September 8–10, 1982. © 1982, IEEE.)

In practice, many of the manufacturers of gasket material use one of the relative aperture test techniques, based on and including MIL-STD-285. As seen in the attenuation measurement curves on conductive fabric in Figure 6.35, this type of measurement is subject to large variations and, due to dependence on antenna type, location, and size of both the test enclosure and the shielded room, to lack of reproducibility. Transfer impedance, in contrast, is a most useful parameter for the evaluation of gasket shielding performance, due to the high level of repeatability achievable in the measurement.

Experience with the use of gasket material soon teaches that the inclusion of a conductive gasket may unexpectedly reduce the attenuation of an enclosure to external fields and increase the level

(a)   (b)

FIGURE 6.113 Current flow around an aperture and through a gasketed metal plate.

of radiated emissions from the enclosure. When enclosures are constructed from machined metal passivated or plated with a low-resistance finish at least ⅛ in. thick, with approximately ¼ in. overlap at the joint, and fasteners are approximately 1 in. apart, very few gasket materials will improve the performance of the joint and many will degrade it. One gasket material that may improve the performance of the machined enclosure, as measured by the DC resistance of the joint, is thin silver-loaded elastomer when inserted into a groove in the material. The resistance of the joint is at its lowest when the silver conductive gasket is in the form of an O-ring set into a groove in one of the mating surfaces. However, shelf-life measurements on silver-filled silicon elastomer have reported increased resistance with time. In these measurements, in which the material was stored for a year in uncontrolled humidity and temperature, a massive increase of 5000 times the DC resistance and an increase in transfer impedance of 1200 are recorded. Reference [14] presents the measured transfer impedance of a machined enclosure joint at 80 μΩ/m at 50 MHz for the metal-to-metal joint and 7 μΩ/m at 50 MHz for the joint when loaded with a ¼ in. wide, ¹⁄₁₆ in. thick silver-loaded gasket. At 1 MHz, however, the silver-loaded gasket transfer impedance is slightly higher than the metal-to-metal impedance. Reference [27] describes degradation in EMI shielding and mating flange materials after environmental exposure. The test method described in Reference [27] is a relative aperture technique similar to MIL-G-83528B. Two types of gasket material were tested: (1) a silver-loaded copper-filled silicone gasket (Ag–Cu gasket) and (2) a silver-plated aluminum-filled silicone gasket (Ag–Al gasket). Stainless steel bolts and lock washers at a spacing of either 2.4 in. (0.98 cm) or 2.667 in. (1.05 cm) were used to provide compression, with either conductive or nonconductive compression stops to limit the compression on the gasket. One piece of interesting data obtained from the measurements is that the level of SE with the Ag–Cu gasket in place, using conductive compression stops, or with the gasket removed, leaving apertures between the fasteners, is the same up to 10 GHz! This is another illustration that the addition of a gasket material may not be beneficial and may be required only for environmental sealing and not for shielding. However, if the conductive compression stops are replaced by nonconductive stops, the relative level of shielding reduces by up to 60 dB.

Adding two narrow strips of silver-loaded elastomer to a joint in a sheet metal enclosure increased the transfer impedance at 50 MHz from 200 μΩ for the metal-to-metal joint to over 2 mΩ for the gasketed seam. The gasket in the thin sheet metal enclosure was not adequately compressed and provided a small surface area in contact with the metal, which may account for the higher impedance with the gasket than without. However, it is invariably a bad practice to seal a seam between thin metals, especially when the fasteners are more than 1 in. apart, with an incompressible material such as silicone elastomer, because the transfer impedance may be no lower and in fact may be higher than that attained with no gasket.

Gaskets are available that have been shaped to fit between the flange of a bulkhead connector and the enclosure wall. One experience with a connector gasket is related as follows.

The gasket was made from a foam silicone material with vertical monel wires (porcupine gasket material). The wires bite into the metal surface of the enclosure and the connector flange. In the example, the connector was mounted in a large rack. The connector was an MIL-STD type with a conductive cadmium-plated finish. When not in use, the connector was covered by a screw on a cadmium-plated cap. With the cap in place, the radiation from the connector and cap was measured, using a small magnetic field probe. Removing the gasket and relying on the metal-to-metal contact between connector flange and enclosure reduced the level of radiation by 7–13 dB. The most likely cause of the magnetic field was common-mode noise current flowing through the capacitance of the connector pin to case capacitance and diffusing through the impedance presented by the small number of vertical wires in the gasket material. Connecting a cable to the connector and measuring the E-field radiation from the cable over the 30–100 MHz range showed an increase of between 6 and 10 dB with the gasket compared to no gasket material. In this case, a marked reduction in radiation was achieved by adding a shield over the wires inside the enclosure. The shield was connected to the inside of the enclosure and increased the wire-to-enclosure capacitance, which effectively shunted the noise currents to the enclosure and reduced current flow in the connector. In other experiments, the increase in radiation from a cable with gasket material was attributed to the voltage drop across the gasket as a result of current flow on the inside of the shield of the cable and the inside of the enclosure.

Further illustration of the importance of choosing the right gasket for the job, or avoiding the use of a gasket, is shown from the following curves of Figure 6.114a through f, from Reference [16]. The test panel was 20 cm square covering a 10 cm square aperture in the transfer impedance test fixture. Eight #10/32 screws were used to apply pressure to the gasket under test. In addition to the gasket material, a nonconductive neoprene sheet, test numbers 101/201 and 102/202, was tested. In Figure 6.114b, the effect of fastener spacing can be seen in tests with the neoprene spacer. As expected, the higher the number of fasteners, the lower the transfer impedance. With only two screws in the flange, a slot resonance was formed at approximately 380 MHz and the measured transfer impedance increased to 1000 Ω/cm. The brass plate has a lower resistance than the aluminum, so the transfer impedance with the brass plate and no gasket or spacer is lower than with the aluminum. In Figure 6.114c, the monel 195, 203, and 204 gaskets are not much better than the neoprene spacer up to 1 MHz and neither is as good as the bare aluminum-to-aluminum joint up to 100 MHz. In Figure 6.114d, the silver-filled rubber Xecom and porcupine gaskets are worse than the bare aluminum-to-aluminum joint from 0.1 MHz to 1 GHz. The transfer impedance of the stainless steel spira is almost the same as the no-gasket transfer impedance, so none of these gasket materials can be recommended over a bare metal-to-metal contact. From Figure 6.40e, the Xecom silver-filled rubber gasket used with a brass plate is no better than no gasket until above 60 MHz, and at 650 MHz the transfer impedance is 27% lower with than no gasket at all. The only gasket that shows a significant reduction in transfer impedance compared to no gasket is the Sn/Be/Cu spira in either a single- or double-row configuration as shown in Figure 6.114f. With either an aluminum plate or a brass plate, the Sn/Be/Cu spira gasket has a transfer impedance of 1.7% of the no-gasket brass plate at 100 MHz and 0.3% of the aluminum no-gasket plate at 100 MHz. Both the spira and Be/Cu finger stock material are the gaskets of choice when a gap must be filled. Even where environmental sealing is required, both types are available with a rubber or foam strip attached down the length of the gasket.

In life measurements, the increase in transfer impedance for stainless steel spira after storage for 1 year was minimal, and although a tin-plated beryllium spira gasket increased transfer impedance by up to a factor of 8, the transfer impedance after 1 year was still 1/100 of the transfer impedance of the stainless steel spira material. Figure 6.104 illustrates a very strong relationship between the surface resistance of the mating surfaces of an enclosure without a gasket and the magnetic field attenuation. Whereas Reference [28] found that using a shield and cover plate with a

Test number		Gasket material	Width (cm)	Thickness (cm)	Square shape	Circular shape
With brass	With iridited aluminum					
Panel	Panel					
1	-	Solid brass plate	-	0.32	-	-
-	2	Solid (iridited) aluminum place	-	0.25	-	-
100	200	No gasket used	-	-	-	-
101	201	Neoprene (nonconductive) sheet	0.64	0.24	×	
102	202	Neoprene (nonconductive) sheet	0.64	0.48	×	
103	203	Monel mesh - compressed	0.64	0.48	×	
104	204	Monel mesh - compressed	0.64	0.32	×	
105	205	Monel mesh over silicon foam core	0.64	0.32		×
106	206	Stainless steel spiral	0.36	0.36		×
107	207	Tin-placed beryllium copper spiral	0.36	0.36		×
108	-	Two tin-placed beryllium copper spiral	0.72	0.36		×
109	209	Expanded monel sheat	1.0	0.36	×	
110	210	Silver-filled rubber	2.5	0.16	×	
-	300	0.32 cm Aperture aluminum honeycomb	15.0	1.25	×	

(a)

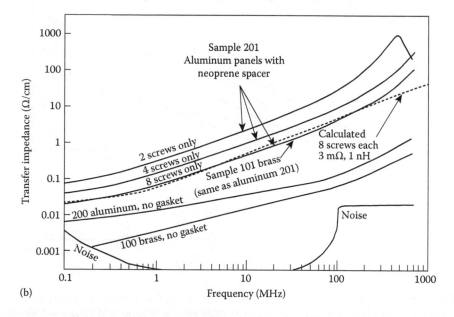

(b)

FIGURE 6.114   (a) Gasket test samples. (From Madle, P.J., Transfer impedance and transfer admittance measurement on gasketed panel assemblies and honeycomb air vent assemblies, *IEEE International Symposium on EMC*, 1976. © 1976, IEEE.) (b) Ungasketed and neoprene spacer transfer impedances. (From Madle, P.J., Transfer impedance and transfer admittance measurement on gasketed panel assemblies and honeycomb air vent assemblies, *IEEE International Symposium on EMC*, 1976. © 1976, IEEE.)   (*Continued*)

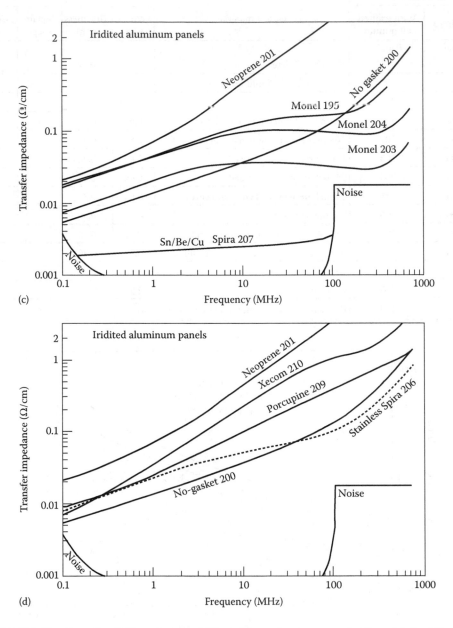

FIGURE 6.114 (*Continued*) (c) Transfer impedances with an aluminum panel. (From Madle, P.J., Transfer impedance and transfer admittance measurement on gasketed panel assemblies and honeycomb air vent assemblies, *IEEE International Symposium on EMC*, 1976. © 1976, IEEE.) (d) Transfer impedances with an aluminum panel. (From Madle, P.J., Transfer impedance and transfer admittance measurement on gasketed panel assemblies and honeycomb air vent assemblies, *IEEE International Symposium on EMC*, 1976. © 1976, IEEE.)

(*Continued*)

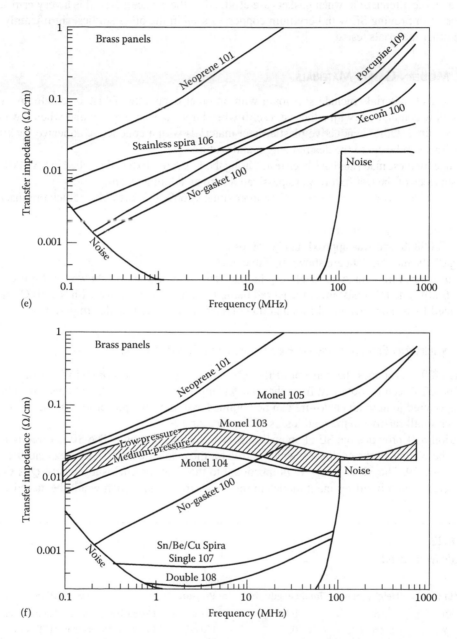

FIGURE 6.114 (*Continued*) (e) Transfer impedances with a brass panel. (From Madle, P.J., Transfer impedance and transfer admittance measurement on gasketed panel assemblies and honeycomb air vent assemblies, *IEEE International Symposium on EMC*, 1976. © 1976, IEEE.) (f) Transfer impedances with a brass panel. (From Madle, P.J., Transfer impedance and transfer admittance measurement on gasketed panel assemblies and honeycomb air vent assemblies, *IEEE International Symposium on EMC*, 1976. © 1976, IEEE.)

surface preparation of either Iridite, tin plate, cadmium plate, or nickel plate and aged for 1 year, a change in the contact resistance of the surface has very little impact on the transfer impedance and therefore on the attenuation, when gaskets are used. Also, the gasket material is a very critical consideration in improving SE with beryllium copper, as seen in the other studies, significantly better than the other materials tested.

6.6.4 MODERN GASKET MATERIALS

Modern gasket materials include soft foam with an outer conductive fabric or conductive rubber with a hollow center, both with adhesive, useful when large gaps are present and when low closure force is required. Other conductive plastic sheet materials with a conductive adhesive backing are available that can be cut to the desired length and width.

A number of these materials have been tested using the 1705A test fixture described in SAE ARP1705. The surfaces of the test fixture were passivated with an Alodine coating.

Compressable gaskets were compressed by approximately 66%, that is, to 34% of their original height.

The pressure on low or noncompressable gaskets was measured using the Fujifilm pressure measurement film kindly supplied by Pressure Metrics LLC, 531 Route 22 East, Suite 323, Whitehouse Station, NJ 08889, and was approximately 150 psi.

The gaskets and test data are shown in Table 6.23.

The transfer impedance of each of the gaskets in ohms/cm is shown in Figures 6.115 through 6.119. The DC resistance, measured using the 1705A test fixture with the 50 Ohm resistor replaced by a short circuit, shows a good correlation with the transfer impedance.

6.6.5 SHIELDING EFFECTIVENESS OF ENCLOSURE WITH GASKETED SEAM

Reference [29] introduces the concept of the *effective transmission width* (ETW) of the gasketed seam. The ETW can be obtained from the transfer impedance of the gasketed aperture. Thus, the SE of a gasketed joint in an enclosure can be obtained assuming the power loss through the gasket aperture is small relative to the wall losses of the enclosure.

Equation 6.31 provides the SE of an enclosure with a gasketed seam. The SE is directly proportional to the volume of the enclosure, which is true for the magnetic field SE of the enclosure shown in Equation 6.29. The SE is inversely proportional to the Q of the enclosure, where Q is the ratio of the energy stored in all resonant modes in the enclosure to that which would be stored with no

TABLE 6.23
Materials Measured

#	Width (mm)	Height (mm)	Total Length (m)	Compression or Pressure	Description
1	9	5	0.294	66.6%	Conductive elastomer, hollow D shaped
2	9	N/A	0.28	150 psi	Carbon-filled cellular PTFE with conductive adhesive (Gore GS21000-010)
3	9	N/A	0.292	150 psi	Nickel-plated conductive polyurethane foam with a conductive adhesive (Gore GS8000-063)
4	9	6	0.284	62%	Conductive fabric over foam With adhesive strip
5	9	N/A	0.242	150 psi	Silver-plated fibers in soft open cell urethane foam No adhesive

FIGURE 6.115 Gasket 1—transfer impedance, ohms/cm.

FIGURE 6.116 Gasket 2—transfer impedance, ohms/cm

FIGURE 6.117 Gasket 3—transfer impedance, ohms/cm.

FIGURE 6.118   Gasket 4—transfer impedance, ohms/cm.

FIGURE 6.119   Gasket 5—transfer impedance, ohms/cm.

losses. The Q of the enclosure is reduced when it is filled with electronics compared to an empty (high-Q) enclosure. The SE is also inversely proportional to the length of the gasket and the square of the magnitude of the transfer impedance of the gasket.

In addition, as we have seen, the SE is also dependent on the location where the field is measured within the enclosure:

$$SE = \frac{3Z_w V}{4l_g} \times Q \,|\, Z_t \,|^2 \tag{6.31}$$

where

Z_w is the wave impedance (m)

V is the volume of the enclosure (m)

l_g is the length of the gasket (m)

Q is the ratio of energy stored in all resonant modes in the enclosure to what would be stored with no losses.

6.6.6 Factors Affecting Gasket Choice

Some of the factors important in the choice of the correct gasket material are as follows.

Level of E, H, or plane-wave attenuation required

> *Maximum current-carrying requirement, resistance of material.* In addition to shielding, one use of gasket material is to provide a low-impedance path for noise current flow. Only certain materials can carry the high currents encountered in a lightning strike or EMP event (EMP typical current density is 3 A/mm).

> *Gap tolerances.* When the gap to be filled does not exhibit a constant depth, a compressible material, with a first choice of finger stock, followed by wire mesh and finally conductive-cloth-over-foam type of gasket.

> *Frequency of opening the aperture.* When a door or removable plate is opened often, then finger stock material, as used on shielded-room doors, is an obvious choice. A silicone foam with wire mesh may also be used. The foam strip is glued to one of the mating surfaces. Obvious though this may appear, experience has shown the importance of ensuring that the adhesive is not used between the wire mesh portion of the gasket and the enclosure!

> *Fastener sizes and spaces.* The distance between fasteners must be small enough to avoid any buckling in the sheet metal material, which will form gaps between the gasket and the enclosure. Fasteners at close distances used with a highly conductive coating on a thick material may eliminate the need for conductive gaskets!

> *Mounting method.* Finger stock gasket material may be soldered or held in place by a metal strip fastened by screws or rivets or clipped over the edge of a frame or door (often small indentations are included in the finger stock clip that mate with small holes around the edge of the frame). When the spring finger of the finger stock touches both of the mating surfaces, the base of the finger stock may be glued in place. Conductive silicone materials may be glued, using conductive adhesive, or held in place until the fasteners are tightened. Conductive silicone may also be formed in place to fill slots in an enclosure. Wire mesh gaskets are available with attached neoprene foam that may be glued to the surface, with the foam acting as an environmental seal.

Compression set

> *Compression deflection.* Nonfoam silicone material is incompressible but will deform under pressure. Wire mesh gaskets are usually compressible, with the exception of sintered metal fiber, mesh in silicone, and wire screen types.

> *Stress relaxation.* Load silicone material to 125% of rated, after which the material will relax to the rated load.

Tensile strength

> *Elongation.* This can increase the resistivity of silicone-loaded material and subsequent reduction in compression. Therefore, limit it to 10%, by the use of compression stops or containing the gasket material in grooves.

> *Temperature range.* Most silicones are good from 55°C to 125°C.

> *Outgassing.* This is usually achieved during the outgassing test on the material. For silicone conductive gaskets, some typical values, when tested for 24 h at 10^6 torr and a temperature of 125°C, are total mass loss 1% or less, collected volatile condensable 0.1%.

> *Gas permeability.* For silicone-loaded material, this is usually less than for the base material.

> *Flammability.* Most gasket material will support a flame but are self-extinguishing.

> *Fungus resistance.* Most gasket materials will support fungus growth, but no internal change will occur.

Cost

Corrosion, oxidization, metal compatibility. Dissimilar metals in contact will result in corrosion. The extent is dependent on the amount of moisture in the atmosphere and the incidence of condensation.

6.7 WAVEGUIDE GASKETS

A gasket material is often included in between the flanges of a waveguide. The reasons to use a conductive elastomer type of gasket are as follows:

1. To pressurize the waveguide to increase its peak power-handling capability
2. To exclude the external environment, dust, or moisture
3. Usually incidentally to seal the joint electromagnetically and reduce electromagnetic radiation

Two basic types of conductive elastomer exist. One is a thin, flat gasket, which is placed as a shim between the waveguide flange surfaces, and the second is either an O- or D-shaped gasket, which is placed in a groove in one of the flange surfaces. With conductive elastomer gaskets, both a conductive metal-to-metal connection between the flange surfaces and a connection between the gasket material and the flange are made. The O- and D-shaped gaskets are suitable for choke-flange and grooved-contact-flange applications. If a thin, flat elastomer type of gasket is placed between the flange surfaces, which are then torqued up, the gasket material will cold-flow out of the joint. This results in the materials entering the waveguide opening and eventually results in a loose joint that will neither pressure seal nor electromagnetically seal. Instead of using a solid elastomer, the cover-flange and flat contact-flange gaskets are die-cut from sheet stock containing an expanded metal reinforcement to eliminate cold flowing of the elastomer material. This type of gasket can be supplied with a slightly raised lip around the iris opening for high-pressure, high-power applications. Standard waveguide gaskets are available that fit the standard UG, CPR, and CMR flanges, and custom gaskets can be made to cover the WR10 to WR23000 range of waveguides. The use of a silver-plated brass mesh gasket can be added to the groove of a flange choke. One reported problem with the fine wire mesh gasket is that the gasket can burn under heavy current, and the power-handling capability of this type of gasket remains unproven. None of the conductive elastomer gasket materials exhibit gasket contact impedances lower than predicted for a metal-to-metal flange impedance; in one case, the gasket contact impedance was 390 times higher. However, none of the available waveguide gaskets use the elastomer material alone to seal the joint; an additional metal-to-metal contact is made. Tech-etch manufactures neoprene or silicone with aluminum and silicone in monel gaskets.

Parker seals manufactures a molded nonconductive gasket imbedded in a metal retainer. The main purpose of these gaskets is to allow pressurization of the waveguide or to environmentally seal the waveguide. Based on the Parker measurements, the RF flange resistance over the 2.6–3.6 GHz frequency range was the highest for pure conductive plastics, just lower for an impregnated wire mesh gasket, with the metal-to-metal flange gasket 1/9 of the impedance of the conductive plastic. Adding a knitted wire mesh gasket to an O-ring groove resulted in a very slightly lower impedance than with the metal-to-metal flange alone. The lowest impedance was a Parker knurled-aluminum gasket with a chemical film CL.3. The RF impedance of this gasket was one-fourth to one-seventh of the value for a bare metal-to-metal flange, so this would result in the lowest level of leakage from the waveguide joint. The most likely reason the serrated type of metal gasket is so successful at reducing the RF resistance is that the ridges are under tremendous pressure, due to the low surface area of the serration and possibly cold flow or weld into the adjacent flange material. The purpose of the nonconductive elastomer part of the Parker seal is for environmental or pressure sealing; if these are not required, then the use of the flange seal without elastomer is likely to be even better. The finish on the ridged gasket also affects the contact resistivity and therefore the RF impedance. The highest contact resistivity is achieved with a transverse ridged brass gasket with a Chem. Film CL3, the next lower resistivity is no finish (bare brass), followed by cadmium plate, gold, and silver, in that order. If a waveguide seam leaks around the edge of the flange, with or without a gasket, the gaps can be filled with conductive caulking, conductive adhesive, or silver-loaded grease. The addition of silver-loaded epoxy to the outside of the joint in a waveguide flange has been reported to reduce emissions by up to 20 dB, despite the epoxy's relatively high-volume resistivity of approximately 2 mΩ/cm. The most probable reason for the silver-loaded epoxy's effectiveness is that the material enters the microscopic gaps in the two surfaces on the outside of the flange faces, and despite the poor conductivity, the skin depth of the epoxy in the 2–12 GHz range is small enough that virtually all of the current flows on the inside surface of the epoxy. One additional technique, applicable only to waveguide sections that contain standing waves, is to position waveguide flanges or components at locations of minimum current flow down the length of the waveguide. The voltage drop across the joint impedance is thereby reduced, as is the emission. The presence of transverse currents in the wall of the waveguide does not result in a voltage across the joint.

6.8 CONDUCTIVE FINISHES, DC RESISTANCE, AND CORROSION EFFECTS ON GASKET MATERIALS

The DC resistance of materials and finishes is often increased by moisture, which increases corrosion, or by oxidization. Table 6.24, from Reference [13], illustrates the initial DC resistance and the effect of 400 and 1000 h at 95% relative humidity.

The DC resistance of a joint without gaps is a good indicator of impedance, for the inductance will be low and the contact impedance, from measurements, appears to track with resistance. Low values of DC resistance cannot be measured with any repeatability by the two-wire measuring technique, available with the majority of multimeters, due to the contact resistance of the probes.

TABLE 6.24
Resistance Measurements on Materials

Material	Finish	Resistance (mΩ)		
		Initial	At 400 h 95% RH	At 1000 h 95% RH
Alum				
2024	Clad/clad	1.3	1.1	2.0
2024	Clean only/clean only	0.11	5.0	30.0
6061	Clean only/clean only	0.02	7.0	13.0
2024	Light chromate conversion/same	0.40	14.0	51.0
6061	Light chromate conversion/same	0.55	11.5	12.0
2024	Heavy chromate conversion/same	1.9	82.0	100.0
6061	Heavy chromate conversion/same	0.42	3.2	5.8
Steel				
1010	Cadmium/cadmium	1.8	2.8	3.0
1010	Cadmium-chromate/same	0.7	1.2	2.5
1010	Silver/silver	0.05	1.2	1.2
1010	Tin/tin	0.01	0.01	0.01
Copper	Clean only/clean only	0.05	1.9	8.1
Copper	Cadmium/cadmium	1.4	3.1	2.7
Copper	Cadmium-chromate/same	0.02	0.4	2.0
Copper	Silver/silver	0.01	0.8	1.3
Copper	Tin/tin	0.01	0.01	0.01

Instead, the four-wire technique described in Section 8.7.4 on bonding must be used. Here, a current is caused to flow across the two metal plates, which are joined together. The injection should be via two wide contacts; here, strips of conductive adhesive copper foil may be used. The voltage-measuring contacts are then placed across the joint or seam, and the DC resistance is the ratio of measured voltage to injected current. Because the voltage-measuring contacts do not supply the current, the effect of contact impedance is greatly reduced.

Measurements of resistance across joints that are passivated by Iridite are in the 10–100 $\mu\Omega$ region, whereas the joint impedance of two zinc dichromate passivated panels held together by pop rivets demonstrated a DC resistance of 1.5–2 mΩ. Anodizing, either hard black or clear, must not be used over contact areas over joints, seams, connector housings, panels, etc., for this finish is virtually an insulator. The surface resistivity of clear anodizing over a conductive surface was measured at a mean value of 5×10^{11} Ω/cm, which is not even adequate for electrostatic discharge. When an enclosure must be anodized, mask those areas where joints are made, panels or doors are located, or connector housings mate. After anodizing, brush Iridite on these areas. The brush Iridite process is electroplating, with one electrode in the form of a brush used to locally apply the Iridite.

Figure 6.104 illustrates the importance of low joint-contact resistance, for we see that the magnetic field attenuation of panels with 1.5 mΩ of DC seam resistance is up to 50 dB less than a panel with a 30 $\mu\Omega$ DC resistance.

The area, surface resistivity, and pressure on a joint affect the contact resistance. Table 6.25 from Reference [30] shows the contact resistance of either bare or coated aluminum alloys. Five types of proprietary conversion coating with three types of plating and two joint compounds were measured. A summary of the findings is as follows: Platings reduce the contact resistance of the joint when compared to bare aluminum, whereas chromate conversion coatings generally increase the contact resistance. Chromate-coating contact resistance varied from test to test and sample to sample over a wide range. Pressure cycling affects the contact resistance from cycle to cycle, with platings remaining reasonably stable and coatings showing random results. Conversion coating in contact with a plating shows a lower contact resistance than a joint with two conversion coatings. The typical reduction with pressure for a nickel bath plating over 6061/6061 with a surface roughness of 32 rms is shown in Figure 6.120.

A lower contact resistance than that shown in Table 6.25 for Iridite has been measured—60 $\mu\Omega$ for a 5.4 cm^2 area. Different formulations are available, so always specify the lowest resistance version. Contact resistance of copper tape and epoxies are shown in Table 6.26. Note that much of these data are presented with different joint areas and the pressure is unknown. However, those materials with a very high contact resistance can be identified.

Magnesium die castings have become popular for structural reasons, and the high-purity magnesium casting alloys AZ91D and AZ91E have demonstrated superior resistance to salt water corrosion, when compared to earlier commercial alloys. Data for different conductive treatments on AZ91D high-pressure die-cast magnesium in $\mu\Omega$ for 1 in.^{2} surfaces after exposure to humidity are shown in Table 6.27. #1, #20, and #21 are all acid chrome based. Neutral chromates, such as dichromate and dilute chrome pickle, have resistivities in the tens of thousands of micro-ohms and should be avoided. Zinc dichromate on aluminum also results in a significantly higher resistivity than electroplated zinc.

The closer the metal is in the galvanic series of metals, the lower the probability of corrosion. Figure 6.121 shows the standard potential of common metals used for shielding. The corrosion potential between silver and aluminum is expected to be high due to the large distance between the metals in the galvanic series. From measurements, the corrosion in a salt spray atmosphere of a silver-loaded silicone elastomer between aluminum plates was not excessive, possibly because the gasket material has a much higher resistance than pure silver, also because the gasket will seal the joint against the ingress of moisture. However, the measured transfer impedance of the

TABLE 6.25

Contact Resistance of 6061-T6 and 7075-T6 at 300 PSI (mΩ) (First Pressure Cycle)

Coating and Surface Roughness	6061-T6/6061-T6 Samples, Joint Area in in.^{2}			6061-T6/7075-T6 Sample, Joint Area in in.^{2}		
	10	5	2.5	10	5	2.5
Bare aluminum—mill	0.443	0.531	0.305	0.288	0.264	1.122
Bare aluminum—32 RMS	—	0.078	—	—	0.072	—
Bare aluminum—18 RMS	—	0.126	—	—	0.072	—
Alodine 1200—mill	4.55	37.1	17.9	6.47	24.7	63.8
Alodine 1200—32 RMS	—	24.3	—	—	34.7	—
Alodine 1200—64 RMS	—	36.8	—	—	8.9	—
Alodine 1500—mill	0.639	5.80	23.4	5.89	5.06	4.82
Alodine 1500—32 RMS	—	2.14	—	—	14.7	—
Alodine 1500—64 RMS	—	13.1	—	—	7.97	—
Alodine 600—mill	5.87	21.6	12.4	8.4	14.9	24.6
Alodine 600—32 RMS	—	10.4	—	—	13.9	—
Alodine 600—64 RMS	—	10.8	—	—	7.9	—
Tin—mill	0.008	0.010	0.013	0.001	0.003	0.001
Tin—32 RMS	—	0.009	—	—	0.001	—
Tin—64 RMS	—	0.012	—	—	0.001	—
Oakite—mill	20.0	14.5	14.8	20.2	32.1	44.8
Iridite—mill	1.12	25.6	74.8	28.5	13.8	35.2
Chromium—mill	0.019	0.016	0.050	—	—	—
Chromium—32 RMS	—	0.017	—	—	—	—
Alodine 1200/tin—mill	0.332	0.249	0.539	0.606	1.01	0.715
Alodine 1200/tin—32 RMS	—	0.336	—	—	—	—
Alodine 1200/tin—64 RMS	—	0.090	—	—	—	—
Nickel bath—mill	0.009	0.023	0.035	0.023	0.006	0.016
Nickel bath—32 RMS	—	0.027	—	—	0.006	—
Nickel bath—64 RMS	—	0.029	—	—	0.008	—
Nickel brush—mill	0.010	0.011	0.008	—	—	—
Nickel brush—32 RMS	—	0.013	—	—	—	—
Nickel brush—64 RMS	—	0.009	—	—	—	—
EJC	—	0.193	—	—	—	—
ALNOX	—	0.018	—	—	—	—

Source: Kountanis, B., Electric contact resistance of conductive coatings on aluminum. *IEEE Electromagnetic Compatibility Symposium Record*, 1970.

silver-loaded silicone material when in contact with Iridited aluminum increased from approximately 10 mΩ/m to 5 Ω/m after a period of 1 year. In equipment designed for space application, negligible change is expected, due to the lack of an atmosphere. However, in ground-based applications, the potentially large increase in transfer impedance should be considered when making the choice of the gasket material.

Tin-plated, stainless steel spiral, and monel mesh gaskets in contact with Iridited, cadmium-plated, and nickel-plated panels exhibit a much lower increase in transfer impedance over a period of 1 year.

In Reference [27], the SE of the Ag–Cu gasket when placed in contact with a stabilized copper-coated aluminum flange did not significantly change when exposed to a sulfur dioxide salt

FIGURE 6.120   Decrease in contact resistance with pressure for nickel bath plating.

TABLE 6.26
Typical Contact Resistance of Coatings, Gaskets, Tapes, and Adhesives

Material	Notes	Contact Resistance (mΩ)	Area (cm^2)
Eccobond solder 57C	Cured at room temperature	201–218	1
Chomerics 584–29 silver-loaded epoxy	Cured at room temperature	106–268	1
Copper tape ribbed, nonconductive adhesive	Moderate finger pressure	13	19
	after 24 h	9.4	19
	Full pressure	37	19
	Low pressure	11.7	19
Spraylat 599-A8219-1 copper conductive coating	Thickness = 2 mils	16.8	8.82
Scotch 3 M tape with conductive adhesive		11.7	19
Conductive fabric over neoprene gasket	Compressed to 2 mm	550	1
Beryllium copper finger stock gasket	Compressed to 2 mm	159	1
Multilayers of very close wire mesh over neoprene	Compressed to 2 mm	105	1
Tin-plated spring steel finger stock	Compressed to 2 mm	8	1

TABLE 6.27

Contact Resistance of Conductive Treatments on AZ91D High Pressure, in μΩ, for 1 in.² Surfaces

| Treatment | Surface Resistance, μΩ/days of Cyclic Humidity Exposure | | |
	Initial	21 days	50 days
Abraded	70–140	140–340	Erratic
#1 Chrome pickle MIL-M-3171, ASTM D 1732	140–200	90–180	90–130
#18 Phosphate	No cond.	No cond.	No cond.
#20 Modified chrome pickle	150–220	130–170	90–150
#23 Stannate immersion	90–160	120–320	70–118
#21 Ferric nitrate bright pickle	50–130	90–120	10–50

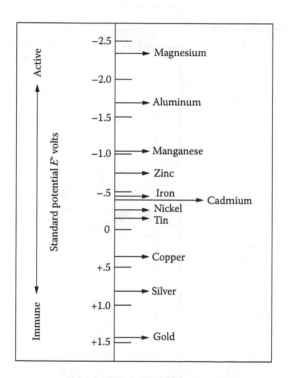

FIGURE 6.121 Standard potential in the electromotive series for common materials.

fog for 192 h. Similarly, the SE of the silver-plated aluminum-filled fluorosilicone (Ag–Al) gasket, mated with the stabilized copper-coated aluminum flange, did not degrade by more than 14 dB when exposed. The volume resistivity of the gasket increased from 0.008 Ω-cm before exposure to 0.014 Ω-cm after. However, the SE of the nickel-coated graphite-filled fluorosilicone (NiGr) gasket material did degrade by up to 36 dB after exposure to SO_2 for 192 h, so at some frequencies the SE of the NiGr gasket was only 24 dB higher than that of the flange without gasket and with nonconductive compression stops, that is, a seam with gaps. The volume resistivity of the NiGr gasket was highly variable, ranging from 0.085 to 0.372 Ω-cm.

The copper coating used in the tests was a two-part urethane coating containing a stabilized-copper filler, copper corrosion inhibitors, or aluminum corrosion inhibitor and an inorganic pigment, which was added over an MIL-C-5541 coating. This type of copper coating showed an initial SE 10–20 dB higher than a sample coated only with a conversion coating per MIL-C-5541, Class 3. Both types of coating with the Ag–Al gasket showed some degradation in SE after 1000 h of salt spray.

Figure 6.122 illustrates methods of excluding moisture from the conductive part of gaskets, thus reducing the probability of corrosion. Some of the many manufacturers of gasket material are as follows: Information on the gasket material is available from Chomerics Ltd., 77 Dragon Court, Woburn, MA 01801. Chomerics produces an EMI/RFI gasket design manual and a gasket catalog. Tech-etch Ltd. of 45 Aldrin Road, Plymouth, MA, produces in addition to elastomer and metal mesh gaskets a range of beryllium copper finger material. Tecknit also manufactures a large range of gasket materials and supplies, a useful EMI shielding design guide, and a catalog, available from Tecknit, 129 Dermody Street, Cranford, NJ 07016. Instrument Specialities is another supplier of finger stock material. Spira, at 12721 Saticoy St. So., Unit B, No. Hollywood, CA 91605, manufactures the stainless steel and tin-plated beryllium copper spiral gasket material.

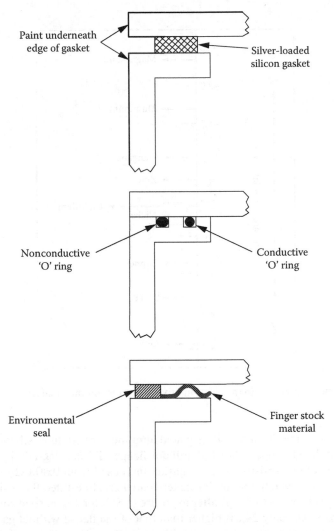

FIGURE 6.122 Methods of excluding moisture from the conductive part of gaskets, thus reducing the possibility of corrosion.

A manufacturer of conductive silver-metallized nylon ripstop fabric bonded to foam is Schlegel, P.O. Box 23197, Rochester, NY 14692. The manufacturer of waveguide gaskets, as well as O-ring and molded gaskets, is Parker Seals, 10567 Jefferson Blvd., Culver City, CA 90230. Two manufacturers of circuit board shields are Leader Tech, 14100 McCormick Drive, Tampa, FL 33626, and Boldt Metronics International, 345 North Erie Drive, Palatine, IL 60067.

6.9 PRACTICAL SHIELDING AND LIMITATION ON EFFECTIVENESS

Our discussion thus far has assumed the use of a totally enclosed box, cabinet, etc. Often, especially between printed circuit boards or to break an enclosure up into compartments, thin metal plates are used that are not totally connected around their peripheries to the inside of the enclosure. Figure 6.123 shows a metal plate inserted between a PCB that is a source of emission and a PCB containing susceptible circuits. The source of emission is likely to be due to switching current flowing around a current loop and thus a source of magnetic field.

Where the receptor circuit or PCB track to be shielded has a small area and is located close to the center of the PCB, the method of shielding described may be adequate, although measurements described in Chapter 11 show 0 dB of attenuation at some frequencies. However, should the metal plate be grounded, either via a ground track on the receptor board or via a length of wire to the enclosure, the EMI situation may be exacerbated by either increased noise currents in the receptor PCB ground or radiation from the ground wire. If the distance between the two boards is kept constant and the metal plate is connected to signal ground on the receptor board, the coupling to the metal plate will be greater due to closer proximity to the source and the potential for EMI may again be increased.

If the metal plate cannot be connected around its periphery to the inside of the enclosure, then there is no reason to ground it at all. Increasing the distance between the source and the receptor board with or without the metal shield may be an adequate solution.

Placing susceptible circuits in small, shielded enclosures on PCBs or in enclosures with shielded input, output, and power cables may be required for low-level high-input-impedance circuits.

Emitter   Shield   Receptor

I_s = Shield current density

FIGURE 6.123 Metal plate shield placed between two PCBs.

For example, a high-input-impedance operational amplifier circuit with a gain of 1000 was used to increase the "pink noise" level from a noise source diode. The output of the amplifier should have been noncoherent noise, whereas the measured signal contained coherent amplitude-modulated signals from local AM transmitters. Placing a small enclosure made of copper foil laminated to cardboard over the circuit and connecting the shield to power ground reduced the interference. However, it was not until a ground plane was added under the enclosure on the track side of the board and electrically connected to the enclosure on the component side, thus completing the shield, that the interference was eliminated.

6.10   COMPARTMENTALIZATION

Placing susceptible circuits or noise sources in a compartment within an enclosure is an option that may be mandatory when dealing with high-level noise sources or low-level analog, video, or RF circuits. Just as with shielded enclosures, the effectiveness of the compartment may be compromised by bringing in power or signal lines that contain noise voltages. Feedthrough filters should be designed into the compartment. The need for the filters may then be assessed by removal, and those filters found unnecessary may be replaced with nonfilter-type feedthroughs.

When apertures exist in the compartment, the levels of E and H fields generated from noise sources should be calculated from the equations of Chapter 2, the coupling through the aperture may be predicted from equations presented in this chapter, and the response of the receptor circuit may be predicted using the techniques described in Chapter 2.

6.11   SHIELDING EFFECTIVENESS OF BUILDINGS

Shielded rooms are designed to achieve specific levels of attenuation and are used principally for EMI measurements and to contain classified electronic information or restricted transmissions. The characteristics of shielded rooms used in EMI measurements are discussed in Chapter 9. In this section, we examine the SE of commercial buildings and residential houses. The reduction in the signal strength of a radio transmission as it propagates inside a building is undesirable when reception is required. Most of us have noticed the reduction in signal strength as a portable radio is moved from a position close to a window to one further inside the building. However, building attenuation is a desired attribute when the source of the field is either high-level broadcast transmissions, lightning, or radar, which have the potential for interference.

In an EMC/EMI investigation that includes a building in the radiated propagation path, data on the average attenuation for different types of buildings are required. When susceptible equipment is contained in the building, the magnitude of the field incident on the building may be predicted from the distance and source characteristics. When the building is used to shield a receptor from a source, the diffraction of the wave around the building must be accounted for, in addition to attenuation through the building. Diffraction and coupling around a structure are discussed in Chapter 10.

As noted, in enclosure shielding, the E-field and H-field components may not be equally attenuated by a material, and the wave impedance may change after transmission through a medium.

In Reference [17], both the E and H fields of ambient broadcast signals were measured inside and outside for seven different types of buildings:

1. Single-family detached residence of the split-level ranch type. *Materials:* wood framed with wood and brick siding, aluminum foil–covered insulation in the ceiling.
2. Single-family detached residence of the wood-framed, raised ranch type. *Materials:* concrete block walls on the lower levels and aluminum siding covering the upper levels; brick veneer front on the main level.

3. Single-story concrete block. *Materials:* concrete blocks with a large storefront window; structure framed with steel columns and a steel bar joist–supported roof; interior divided into cubicles with demountable steel and glass. An extensive array of ducts, pipes, and cables occupied the space above a suspended ceiling.

4. Single-story concrete block. *Materials:* concrete block, steel-framed building constructed on a reinforced-concrete slab. Interior description similar to 3.

5. Four-story office. *Materials:* steel framed with preformed-concrete exterior wall panels; one side covered with metal exterior wall panels. The floors are corrugated steel covered with poured concrete. Interior description similar to 3.

6. Four-story office. *Materials:* steel frame with brick exterior walls; corrugated steel floors covered with poured concrete. Interior description similar to 3.

7. Twenty-story office. *Materials:* steel frame with marble exterior wall panels; interior steel columns spaced approximately 7 m apart; corrugated steel floors covered with poured concrete. The interior is uncluttered, with few interior walls and partitions.

The level of attenuation depends on the interior location at which the measurement was made. Thus, measurements were made at a number of different locations in each building. Table 6.28 provides a summary of the fitted mean, minimum, and maximum measured attenuations of the seven buildings.

TABLE 6.28
Building Attenuation

Building	Frequency	H-Field Mean [dB]	H-Field Limits [dB]	E-Field Mean [dB]	E-Field Limits [dB]
1	20 kHz	0	—	32	—
	1 MHz	0	—	30	—
	500 MHz	0	−5 to 3	0	−5 to 8
2	20 kHz	3	—	22	12 to 33
	1 MHz	0	−5 to 2	9	−8 to 34
	500 MHz	10	2 to 20	12	7 to 20
3	20 kHz	−1	—	—	—
	1 MHz	3	—	18	—
	500 MHz	8	−3 to 19	8	−3 to 15
4	20 kHz	−3			
	1 MHz	12	5 to 25	28	12 to 28
	500 MHz	8	—	10	—
5	20 kHz	2[a]	—	—	—
	1 MHz	3[a]	—	22[a]	—
	500 MHz	6[a]	—	3[a]	—
6	20 kHz	3[a]	—	32[a]	—
	1 MHz	8[a]	—	28[a]	—
	500 MHz	10[a]	—	10[a]	—
7	20 kHz	20[a]	—	40[a]	—
	1 MHz	20[a]	—	35[a]	—
	500 MHz	10[a]	—	10[a]	0–25[a]

[a] Measured 1 m from the outer wall of the buildings. Measurements made 15 m from the outer wall showed attenuation of up to 35 dB for magnetic fields and up to 50 dB for electric fields. Mobile homes provide an average of 28 dB of attenuation over the 20 kHz to 500 MHz frequency range.

TABLE 6.29
Attenuation Correction Factors

		Correction Factor[a] [dB]
Frequency	2569 MHz	1.160
	1550 MHz	0.390
	860 MHz (V)	−0.169
	860 MHz (H)	0.140
Construction	Wood siding	−0.58
	Brick veneer	0.58
Insulation	Blown in ceiling	−0.8
	Ceiling and walls	0.8
Room position	Exposed walls	−0.3
	No exposed walls	0.3

[a] Average attenuation from which the correction factors are added/subtracted is 6.3 dB.

The attenuation of UHF radio signals by buildings has been measured in Reference [18]. Table 6.29 gives a summary of the effects of frequency and building type on the attenuation. The average attenuation from measured data is 6.3 dB. The table shows correction factors that must be added to or subtracted from 6.3 dB to account for frequency, room location, and building materials used in the construction.

When equipment housed in a building is found to be susceptible to a source of high-power radiated emissions, such as a transmitter, and it can be proven that the EMI problem is marginal, then a few decibels of additional attenuation may be all that is required to eliminate EMI. The first step in such an investigation is to prove that the coupling path is truly radiated to the equipment inside the building and not radiated to or conducted on external power or signal lines. The next step is to ensure that a modicum of additional shielding is all that is required.

One method of enhancing the building attenuation is to paint it with conductive paint. In Reference [19], a powdered nickel acrylic base conductive paint was used on concrete or concrete block walls. The measured attenuation after painting increased by 20 dB over the 100–350 MHz frequency range and by 16 dB for the pulsed field from lightning. Additional shielding at windows may be achieved by the use of wire mesh in a frame electrically connected to the conductive paint. A transparent plastic film coated with copper or gold may also provide sufficient shielding at windows but is a more costly solution. Care must be taken to ensure that the conductive side of the film contacts the paint either directly or via a metal window frame.

When a higher level of shielding than achievable with conductive paint is required, the use of sheets of metal foil soldered or taped together on the inside walls and ceiling is an alternative. Earth is a good absorber of electromagnetic radiation; therefore, installing equipment below ground level and/or building up a 2 m thick wall of earth around the room housing the equipment is another potential solution.

Reference 37 describes the SE of reinforced concrete that varies with moisture content, the thickness, the gaps between the rebars, layers of rebars, and frequency but can range from 5 to 140 dB!

6.12   COMPUTER PROGRAM FOR EVALUATING SHIELDING EFFECTIVENESS

```
'Program for evaluating shielding effectiveness
'Function for logarithms to base 10
DEF FNLOG(x)
FN LOG = LOG(x)/LOG(10)
END DEF
```

```
'Initialize constants for PI and speed of light as well as other
'parameters used in the program
C = 3E + 08
PI = 3.14159
f = 100:Gr = .35: Ur = 1: R1 = 8.78: t = 62.5: ANS$ = "H"
GOTO ComputeAttenuation:
AttenuationInputs:
CLS
PRINT "COMPUTATION OF SHIELDING ATTENUATION"
PRINT ""
PRINT "INPUTS:"
PRINT "[A] Compute H or E-field reflection", ANS$
PRINT "[F] Frequency, F", f; "Hz"
PRINT "[D] Distance from conductor, R1", R1; "inches"
PRINT "[T] Thickness of shield, t", t; "mils"
PRINT "[G] Relative conductivity, Gr", Gr
PRINT "[U] Relative permeability, Ur", Ur
PRINT "[X] Exit program"
PRINT "Please select variable you wish to change (A.B.etc)"
PRINT ""
PRINT "OUTPUTS:"
GOSUB Outputs1:
MonopoleLoop:
SEL$ = INKEY$
IF LEN(SEL$)>0 THEN LOCATE 11, 1
IF LEN(SEL$)>0 THEN PRINT SPACE$(70)
LOCATE 11, 3
IF(SEL$ = "A") OR (SEL$ = "a") THEN INPUT "Enter H-field(H)/E-field(E)", ANS$
IF(SEL$ = "F") OR (SEL$ = "f")THEN INPUT "Enter Frequency", f
IF(SEL$ = "G") OR (SEL$ = "g")THEN INPUT "Enter relative conductivity", Gr
IF(SEL$ = "U") OR (SEL$ = "u")THEN INPUT "Enter relative permeability", Ur
IF(SEL$ = "D") OR (SEL$ = "d")THEN INPUT "Enter distance from conductor", R1
IF(SEL$ = "T") OR (SEL$ = "t")THEN INPUT "Enter thickness of shield", t
IF(SEL$ = "X") OR (SEL$ = "x")THEN END
IF LEN(SEL$) >0 THEN GOTO ComputeAttenuation:
GOTO MonopoleLoop:
ComputeAttenuation:
IF(ANS$ = "E")OR(ANS$ = "e")THEN ANS$ = "ELECTRIC(E-field)"
IF(ANS$ = "H")OR(ANS$ - "h")THEN ANS$ - "MAGNETIC(H-field)"
LAMBDA = (C/f) * 39.37'LAMBDA = wavelength in inches
Tcm = t * .00254'Shield thickness in cm for calculation
     'of skin depth
'Reflection of electric field
Re = 353.6 + .(10 * FNLOG(Gr/(f^3 * Ur * R1^2)))
'Reflection of magnetic field
Rm = 20 * FNLOG(((.462/R1) * SQR(Ur/(Gr * f))) + ((.136 * R1)/SQR(Ur/(f * Gr))) + .354)
'Absorption by shield
A = .003334 * t * SQR(Ur * f * Gr)
'Metal impedance
Zm = 369 * SQR((Ur * (f/1000000!))/Gr) * .000001
'Wave impedance
GOSUB ComputeZw: 'Zw is different depending on whether one is looking 'at
magnetic or electric fields
K = ((Zm - Zw)/(Zm + Zw))^2
'K is a function of Zw and is different depending on the type of field
as well
```

```
IF K< 1E-09 THEN K=1E-09'Program sets minimum K at 1.0E-05
'The re-reflection, Rr, will be different depending on type of field since
'it is a function of K
Rr = 20 * FNLOG(1-(K * 10^(-.1 * A) * EXP(-.227 * A)))
SEe = Rr + Re + A
SEm = Rr + Rm + A
GOTO AttenuationInputs:
Outputsl: 1
IF ANS$ = "ELECTRIC(E-field)" THEN GOSUB ElectricFieldOutputs:
IF ANS$ = "MAGNETIC(H-field)" THEN GOSUB MagneticFieldOutputs:
RETURN
ElectricFieldOutputs:
PRINT "Shielding Attenuation"„ SEe;"dB"
PRINT "Wave Impedance"„Zw;"Ohms"
PRINT "Re-reflection attenuation, Rr",Rr;"dB"
PRINT "Reflection of electric field,Re",Re;"dB"
PRINT "Absorption loss,A"„A;"dB"
PRINT
RETURN
MagneticFieldOutputs:
SkinDepth = .0066/SQR(Ur * Gr * (f/1000000!))
Zb = 369 * SQR((Ur * (f/1000000!))/Gr) * (1/(1-EXP(-Tcm/SkinDepth)))
Rrb=20 * FNLOG(1-((((Zb * .000001)-Zw)/((Zb * .000001)+Zw))^2) * 10^
(-.1 * A) * EXP(-.227 * A))
PRINT "Shielding Attenuation"„SEm; "dB"
PRINT "Wave Impedance"„Zw; "Ohms"
PRINT "Re-reflection attenuation,Rr",Rr;"dB"
PRINT "Re-reflection based on Zb,Rrb",Rrb;"dB"
PRINT "Reflection of magnetic field,Rm",Rm;"dB(Based on Zm)"
PRINT "Absorption loss, A"„A;"dB"
IF A > 10 THEN PRINT : PRINT "NOTE: For A > 10: A=10 in calculating Rr"
RETURN
ComputeZw:
IF (ANS$ = "MAGNETIC(H-field)")AND(R1 < LAMBDA/(2 * PI))THEN
    Zw = (377 * 2 * PI * R1)/LAMDDA
END IF
IF (ANS$ = "ELECTRIC(E-field)")AND(R1 < LAMBDA/(2 * PI))THEN
    Zw = (377 * LAMBDA)/(2 * PI * R1)
END IF
IF R1 >= LAMBDA/(2 * PI) THEN Zw=377
RETURN
```

REFERENCES

1. J.R. Moser. Low frequency shielding of a circular loop electromagnetic field source. *IEEE Transactions on Electromagnetic Compatibility*, EMC9(1), March 1967.
2. A. Broaddus and G. Kunkel. Shielding effectiveness test results of aluminized mylar. *IEEE EMC Symposium Record*, 1992.
3. R.B. Cowdell. Nomograms simplify calculations of magnetic shielding effectiveness. *EDN*, September 1972, 44.
4. T. Sjoegren. Shielding effectiveness and wave impedance. *EMC Technology*, July/August 1989.
5. A.C.D. Whitehouse. Screening: New wave impedance for the transmission line analogy. *Proceedings of IEEE*, July 1969, 116(7).
6. D.R. Bush. A simple way of evaluating the shielding effectiveness of small enclosures. *IEEE EMC Symposium Record*.
7. L.O. Hoeft, J.W. Millard, and J.S. Hofstra. Measured magnetic field reduction of copper sprayed panels. *IEEE EMC Symposium Record*, San Diego, CA, September 16–18, 1985.

8. R.K. Keenan. *Digital Design for Interference Specifications*. The Keenan Corporation, Pinellas Park, FL.

9. M.P. Robinson, T.M. Benson, C. Chrisopoulus, J.F. Dawson, M.D. Ganley, A.C. Marvin, S.J. Porter, and D.W.P. Thomas. Analytical formulation for the shielding effectiveness of enclosures with apertures. *IEEE Transactions on Electromagnetic Compatibility*, August 1998, 40(3).

10. H. Bloks. NEMP/EMI shielding. *EMC Technology*, November/December 1986.

11. L.O. Hoeft and J.S. Hofstra. Experimental and theoretical analysis of the magnetic field attenuation of enclosures. *IEEE Transactions*, August 1988, 30(3).

12. L.O. Hoeft. How big a hole is allowable in a shield: Theory and experiment. *1986 IEEE EMC Symposium Record*, San Diego, CA, September 16–18, 1986.

13. E. Groshart. Corrosion control in EMI design. *Second Symposium and Technical Exhibition on Electromagnetic Compatibility*. Montreux, Switzerland, June 28–30, 1977.

14. A.N. Faught. An introduction to shield joint evaluation using EMI gasket transfer impedances. *IEEE International Symposium on EMC*, New York, September 8–10, 1982.

15. R.J. Mohr. Evaluation techniques for EMI seams. *IEEE International Symposium on EMC*, Atlanta, GA, August 25–27.

16. P.J. Madle. Transfer impedance and transfer admittance measurement on gasketed panel assemblies and honeycomb air vent assemblies. *IEEE International Symposium on EMC*, 1976.

17. A.A. Smith. Attenuation of electric and magnetic fields by buildings, *IEEE Transactions on EMC*, August 1978, EMC20(3).

18. P.I. Wells. *IEEE Transactions on Vehicular Technology*, November 1977, VT 26(4).

19. H.E. Coonce and G.E. Marco, *IEEE National Symposium on EMC*, April 24–26, 1984.

20. L.O. Hoeft, T.M. Sales, and J.S. Hofstra. Predicted shielding effectiveness of apertures in large enclosures as measured by MIL-STD-285 and other methods. *IEEE National Symposium on EMC*, Denver, CO, May 23–25, 1989.

21. L.O. Hoeft, J.W. Millard, and J.S. Hofstra. Measured magnetic field reduction of copper sprayed panels. *IEEE EMC Symposium*, Wakefield, MA, August 20–22, 1985.

22. L.O. Hoeft. The case for identifying contact impedance as the major electromagnetic hardness degradation factor. *IEEE EMC Symposium Record*, San Diego, CA, September 16–18, 1986.

23. F. Olyslager, E. Laermans, D. De Zutter, S. Criel, R.D. Smedt, N. Lietaert, and A. De Clercq. *IEEE Transactions on Electromagnetic Compatibility*, August 1999, 41(3).

24. B. Archamneault and C. Brench. Shielded air vent design guidelines from EMI modeling. *IEEE International Symposium Record*, 1993.

25. J.P. Quine. Theoretical formulas for calculating the shielding effectiveness of perforated sheets and wire mesh screens.

26. G.J. Freyer, J. Rowan, and M.O. Hatfield. Gasket shielding performance measurements obtained from four test techniques. *IEEE International Symposium on EMC*, 1994.

27. P. Lessner and D. Inman. Quantities measurement of the degradation of EMI shielding and mating flange materials after environmental exposure. *IEEE International Symposium on EMC*, 1993.

28. G. Kunkel. Corrosion effects on field penetration through apertures. *IEEE Electromagnetic Compatibility Symposium Record*, 1978.

29. IEEE Std 1302-1998. *IEEE Guide for the Electromagnetic Characterization of Conductive Gaskets in the Frequency Range of DC to 18 GHz*. IEEE, Inc., New York, 1998.

30. B. Kountanis. Electric contact resistance of conductive coatings on aluminum. *IEEE Electromagnetic Compatibility Symposium Record*, 1970.

31. G.M. Kunkel. *Schelkunoff's Theory of Shielding Revisited*. Spira Manufacturing Corporation, North Hollywood, CA.

32. G.M. Kunkel. Shielding theory (a critical look). *IEEE Transcations Symposium Record*, 1991.

33. T. Sjoegren. *Shielding Effectiveness and Wave Impedance*. Telub Teknik AB, 35180 Vaxjo, Sweden.

34. G.M. Kunkel. *Transfer Impedance Testing of EMI Gasketed Joints to 10 GHz for Cost Effective Design*. *IEEE International Symposium on Electromagnetic Complatibility*, 2012.

35. E. Liu, P.-A. Du, W. Liu, and D. Ren. Accuracy analysis of shielding effectiveness of enclosures with apertures. A parametric study. *IEEE Transaction on Electromagnetic Compatibility*, December 2014, 56(6).

36. S.-Y. Hyn, J.-K. Du, H.-J. Lee, K.-W. Lee, J.-H. Lee, C. Jung, E.-J. Kim, W. Kim, and J.-G. Yook. Analysis of shielding effectiveness of reinforced concrete against high altitude electromagnetic pulse. *IEEE Transaction on Electromagnetic Compatibility*, December 2014, 56(6).

37. M. Kunkel and G. Kunkel. Comparison between transfer impedance and shielding effectiveness testing, *Electromagnetic Compatibility 1992 Symposium Record IEEE*, 1992.

FIGURE 1.16   Field strength from three cellular phones measured at 20 cm from antenna.

FIGURE 4.58   Attenuation for gap in 2.8 mm wide trace $d = 2$ mm.

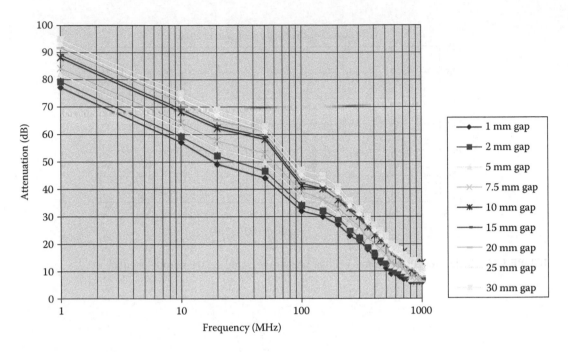

FIGURE 4.59 Attenuation for gap in 5 mm wide trace $d = 4$ mm.

FIGURE 6.81 Comparison of the FEKO, direct measurement, and aperture coupling equation results for the attenuation at the center of a $540 \times 370 \times 330$ mm shielded copper enclosure with a 200×30 mm aperture, from 577 to 977 MHz. Loss factor 0.2 for aperture coupling equation.

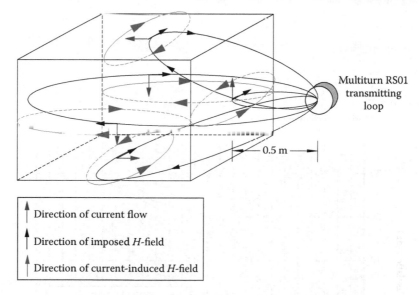

FIGURE 6.95 Magnetic-field lines incident on an enclosure and the magnetic field set up by the current flow on the enclosure.

FIGURE 7.72 Current on the braid wire at the frequencies of interest.

FIGURE 10.20 Horizontal isolation on 26 cm cylinder. (From Weston, D.A., Comparison of techniques for prediction of and measurement of antenna to antenna coupling on an aircraft, *EMC Europe*, York, U.K. © 2011, IEEE.)

FIGURE 11.40 Differential stripline emissions versus the differential microstrip (image plane) emissions from 1 to 8 GHz in the 100 Ω differential impedance tests. Normalized to M2 J3/J4.

FIGURE 11.43 Transmission-line PCB comparison: 1–8 GHz.

FIGURE 11.59 Comparison of the single-sided stripline PCBs relative to SS M1 J3/J4 with the PCB rotated.

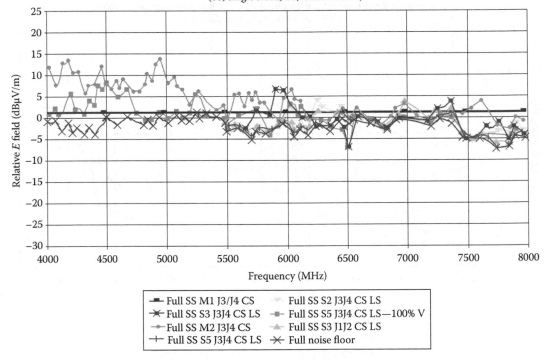

FIGURE 11.62 Comparison of the single-sided Stripline PCBs with loads and connectors poorly shielded relative to SS M1 J3/J4 with the PCB rotated.

7 Cable Shielding, Coupling from *E* and *H* Fields, and Cable Emissions

7.1 INTRODUCTION TO CABLE COUPLING AND EMISSIONS

Cables are a major source of radiated emissions and receptors in electromagnetic (EM) coupling. Commercial equipment that is designed for electromagnetic compatibility (EMC) and that includes a well thought-out grounding scheme and wiring and PCB layout can meet commercial EMC requirements with a number of unshielded cables connected to the equipment. When the equipment enclosure is also unshielded, the sources of emission are likely to be both the enclosure and the cables, although in some cases emissions from cables are higher when the enclosure is shielded.

For equipment that must meet the more stringent MIL-STD-461 RE102 or similar requirements, the use of unshielded cables is the exception. Assuming that equipment does not generate noise voltages, which is typical of analog, video, or low-level radio frequency (RF) circuits with linear power supplies, the level of noise current on unshielded cables may be low enough to ensure meeting the radiated emission requirements. However, the same low-signal-level equipment is likely to be susceptible to noise voltage induced in unshielded cables during exposure to the *E* field generated in the radiated susceptibility RS103 test. In many cases, even equipment with shielded cables fails these tests. This chapter describes some of the reasons why this is possible.

Apart from the EMC requirements, the severity of the EM environment may necessitate the use of shielded cables. Also, some standard interfaces, such as GPIB, MIL-STD 1553, and Ethernet, require the use of shielded cable.

The use of fiber-optic cables may appear to eliminate the need for concern about cable radiation and coupling. However, with the present high cost of fiber-optic links and the need for power interconnections, the use of shielded cables is not quite obsolescent.

7.2 CABLE SHIELDING EFFECTIVENESS/TRANSFER IMPEDANCE

The shielding effectiveness (SE) of a shielded cable is dependent on a number of factors in addition to the cable characteristics: cable length relative to the wavelength of the field incident on the cable or the wavelength of the current flow on the cable core (where this current flow is the source of emission), the wave impedance (i.e., a predominantly *H* field, *E* field, or a plane wave), and the termination of the shield. One of the most common questions is how to terminate the shield of a cable, whether at both ends or at one end only, and if so which end. We shall see indeed that how the shield of a cable is terminated has an effect on its SE. For all types of shielded cable, the SE is not constant with frequency but can be characterized over approximately three frequency ranges:

1. 60 Hz–100 kHz
2. 100 kHz–30 MHz
3. 30 MHz–10 GHz

The physical characteristics of a cable play a large role in its SE. Types of shielded cable are single braid, double braid, triple braid, two types of foil, braid and foil, conduit, semirigid, and flexible corrugated conduit. Many cable manufacturers publish curves of attenuation plotted against frequency. The method of measurement is of importance in the evaluation of these curves. For example, the induction or near-field leakage information is found using probes, current loops, test fixtures, etc., whereas in far-field measurements, large broadband antennas are often used. Some of these test methods are described as follows:

TEM cell: The transverse electromagnetic (TEM) cell is a transmission line used either to generate transverse EM waves or to measure the radiation from a current-carrying cable. A cable may generate radial or axial fields as well as transverse electric (TE) and transverse magnetic (TM) modes, and ideally, all these should be measured. However, it is believed that the TEM mode is the predominant one.

Absorbing clamp: The clamp inductively detects emissions from a cable in the frequency range 30–1000 MHz. The clamp fits over the cable under test (CUT), and the radiation from the shielded cable is compared to the radiation of an unshielded cable carrying the same current. The difference between the two radiated values is the SE of the cable.

Antenna site: The open-field site is most often used to measure the emissions from a current-carrying cable and is of particular use when the cable interconnects two pieces of equipment. Broadband antennas, a spectrum analyzer with a preamplifier, or an electromagnetic interference (EMI) receiver with peak-hold capability are required for this test setup. This test method is often limited, especially at high frequency, by the ambient noise level; it is, however, an effective method of measuring the radiation from a cable, especially in the far field. The height of the cable above the ground plane and the cable length are two important parameters affecting the results obtained using this method. A source of error in this test is the radiation from the signal source and its power lead.

Due to the number of variables, the reproducibility obtained with this test method is relatively poor. When the actual configuration in which the cable will be used, that is, connected to equipment, vehicles, or close to structures and ground, is being simulated, this is the method of choice.

Shielded room test method: This test method minimizes the coupling between the signal source and the receiving antenna by enclosing the receiving antenna inside a shielded room, with the cable entering and exiting the room via feedthrough fixtures connected to the wall of the room. This test method suffers from the same room resonance and reflection errors inherent in the MIL-STD-285 test method.

Reverberation chamber or mode-stirred chamber test method: The reverberating chamber or mode-stirred chamber is well suited to immunity testing and for cables and connectors, because the method is not sensitive to cable layout. Testing on components has been performed up to 40 GHz.

Test fixtures: Many test fixtures have been designed and built to measure the transfer impedance of cables using the methods described in Section 7.2.2. The use of an outer solid tube, with the coaxial CUT placed concentrically inside the tube, the whole forming a triaxial transmission line, is a common test jig. The triaxial assembly is terminated either in its characteristic impedance or in a short circuit, and a current is caused to flow in the CUT. Such a test fixture is capable of transfer impedance measurement up to 3 GHz. However, at these high frequencies, great care is required in the correct termination of the cable, which tends to dominate the transfer impedance, especially when very short lengths of cable are used. Alternatively, any error due to an impedance mismatch must be well understood. In addition, a difference in the propagation delay in the shielded CUT, which will have a core insulation with a relative permittivity greater than 1, and the air-cored test fixture will introduce long-line effects.

One simple transfer impedance test method has been adopted in the European International Electrotechnical Commission Publication IEC 96-1, Radio Frequency Cables. In this method, a current is injected into a shielded cable by an injection cable that forms a transmission line in parallel with the shielded CUT. One possible criticism of test methods confined to current injection on the shield of the test cable is the omission of electric field coupling through the transfer admittance of the cable. Generally, we find the transfer impedance the most useful measure of the SE of a cable in EMI predictions.

The concept of transfer impedance was introduced in Section 6.6, which described gasket transfer impedance. Cable surface transfer impedance is specified in milliohms or ohms per meter length of cable. To obtain the overall shielding picture, we must include the shield termination technique and its transfer impedance. Where the shield connection is via the connector, the transfer impedance of the backshell plus that of the two mating halves of the connector, as well as the receptacle-to-bulkhead transfer impedance, must be included.

The transfer impedance is of use when the current flow on the cable sheath is known. The current flow on the shield may be due to a wave incident on the cable or to the signal current when the shield is the return path for the core current or to common-mode (C/M) noise between the chassis of two units of equipment connected together by the cable shield.

There often is a relationship between the specified SE of a cable and its transfer impedance. One definition of cable SE is the ratio of the current flow on the shield of a cable (I_s) to the current flow on the core/s (I_c), usually expressed in decibels:

$$SE = 20\log\left(\frac{I_s}{I_c}\right) \left[dB\right] \tag{7.1}$$

The core current $I_c = V_{OC}/2R_O$, where R_O is the termination resistance at each end of the shielded cable. When R_O is not given, the assumption that R_O equals the characteristic impedance of the cable will be correct for the majority of the test methods used. V_{OC} is the open-circuit voltage on the core, and for electrically short lengths of cable,

$$V_{OC} = Z_t I_s l \tag{7.2}$$

where *l* is the length of the cable [m].

For samples that are electrically short, one can obtain the SE in terms of the transfer impedance and the termination resistance:

$$SE = 20\log\left(\frac{2R_o}{Z_t} l\right) \tag{7.3}$$

Note: As transfer impedance is expressed in Ohms/m so the shielding effectiveness is for a 1 m length of scale.

Due to the different definitions of SE and the various test methods, the conversion of a manufacturer's SE figure to transfer impedance may not be without error. However, when the transfer impedance is not otherwise available and assuming the cable is terminated in its characteristic impedance:

$$Z_t = \frac{2R_o}{10^{SE/20}} \tag{7.4}$$

If the cable length is not 1 m, then Z_t should be specified for the cable length in question or normalized to the standard Ohms/m.

7.2.1 FREQUENCY DEPENDENCY: 60 Hz–100 kHz

For the majority of flexible coaxial cables, the wall thickness of the shield does not approach one skin depth until approximately 100 kHz. The skin depth is defined as the surface thickness of a metal in which 63% of the current is flowing. For copper, skin depth = 0.0066 f (MHz) [cm].

The transfer impedance below 100 kHz, except for thick-walled shields such as conduit, is approximately equal to the shield resistance R_s.

Other factors important in assessing the SE of a coaxial cable at low frequency are cable inductance and the eccentricity of the inner conductor(s) relative to the outer shield. A coaxial cable exhibits an inductance per unit length of

$$L_c = 0.14 \log \left(\frac{d_s}{d_i} \right) \left[\mu H/ft \right] \tag{7.5}$$

where
d_s is the diameter of the shield
d_i is the diameter of the center core

For a coaxial cable above a ground plane, the inductances of inner and outer conductors are

$$L_i = 0.14 \log \left(\frac{4h}{d_i} \right) \left[\mu H/ft \right] \tag{7.6}$$

$$L_s = 0.14 \log \left(\frac{4h}{d_s} \right) \left[\mu H/ft \right] \tag{7.7}$$

where h is the height above the ground plane. The mutual inductance between the inner and outer conductors is given by

$$L_m = 0.14 \log \left(\frac{4h}{d_s} \right) \left[\mu H/ft \right] \tag{7.8}$$

It can be seen that Equation 7.7 equals Equation 7.8, so the mutual inductance is identically equal to the shield inductance. This can be visualized by considering that all the flux produced by the shield current encircles the center conductor. The mutual inductance is independent of the position of the center conductor relative to the shield. Further, the cable inductance $L_c = L_i - L_s$.

In transmission-line theory, the mutual inductance of a transmission line such as a coaxial cable is normally ignored because the assumption is made that equal and opposite currents flow, and therefore, the mutual inductance between shield and center conductor is effectively zero. Figure 7.1a shows the schematic of a coaxial cable connected at both ends to a ground plane with a signal current flow on the cable. This is a typical shield connection and is effective at shielding against both the ingress and the egress of magnetic radiation, above a given frequency, as well as electric fields. As we shall see later, if either or both ends of the cable shield were removed from the ground plane, then the magnetic field shielding is improved. Figure 7.1b is the equivalent circuit of the cable, and Figure 7.1c is the equivalent open-wire line producing the same external magnetic field as the shielded-wire line produces.

How the attenuation of magnetic fields is achieved is described as follows. The leakage current flow in the ground plane I_L is the current i attenuated by the factor

$$\frac{1}{1 + j\omega \dfrac{L_s}{R_s}} \tag{7.9}$$

From this, it can be seen that at low frequency the cable is ineffective at reducing the magnetic field from the cable because I_L is large. At high frequencies, the magnitude of I_L reduces, the current in the shield approaches the center conductor current due to the mutual inductance between the center conductor and the shield, and the external magnetic field tends to cancel. The position of the center conductor relative to the shield plays a role in the external magnetic field produced by the cable.

FIGURE 7.1   (a) Schematic of a coaxial cable with the shield connected to a ground plane at both ends. (b) Equivalent circuit of the cable. (c) Equivalent open-wire line producing the same external field as a shielded-wire line produces. (From Mohr, R.J., Coupling between open and shielded wire lines over a ground plane, *IEEE Transactions on Electromagnetic Compatibility*, 9(2), 34–45, September 1967. © 1967, IEEE.)

With equal shield and center conductor current and a perfectly concentric cable, the external magnetic field is exactly canceled.

One application of the effect of mutual inductance between pairs, or a number, of wires in close proximity (bundled) is to provide a return wire in the bundle with the supply or signal wires, even when the return is connected to ground at both ends. Then, at higher frequencies, a large proportion of the supply or signal return current will flow in the return wire, even though the DC resistance of the alternative ground path is lower than that of the return wire. The attenuation coefficient *a* is plotted versus (fL_S/R_S) in Figure 7.2. For the low-frequency case, where a current flow is either induced into the shield of a cable due to an incident magnetic field or caused by a C/M voltage between the signal return and the ground plane, the voltage appearing across the shield V_S is equal to

$$V_S = j\omega M I_S + j\omega L_S I_S + I_S R_S \tag{7.10}$$

Since $L_S = M$, then $V_s = R_S I_S$. Therefore, R_S at low frequency equals the transfer impedance. In Figure 7.3a, the current path is shown for C/M noise, and it can be seen that V_S adds to the signal voltage *V* and appears as noise in the signal.

The SE of the coaxial cable against a magnetic field may be seen by comparing the shielded case to an unshielded cable. The unshielded case is shown in Figure 7.3b. Here, the induced current due to the magnetic field flows through the signal and source impedances and generates a noise voltage across the load R_L. With the shield connected, as shown in Figure 7.3c, the current flow in the shield generates a noise voltage across the resistance of the shield in the same way as the C/M current did in Figure 7.3a, and the noise voltage V_S is again equal to $R_S I_S$. If we assume that the magnetic field–induced current flow in the loop is the same for the shielded and unshielded cables and that all of the signal currents return on the inside of the shield, then the SE of the cable is given by the ratio of the load resistance to the shield resistance. Thus, with a load resistance of 1000 Ω and a total shield resistance of 20 mΩ, the SE is 20 log 1000/20 × 10⁻³ = 94 dB. The lower the DC resistance of the shield, the higher the SE of a cable at low frequency. A braided-wire shield will have a lower DC resistance than a foil-type shield; and the thicker the braid or the more braids in the shield, the lower the DC resistance.

As we can see by comparing Figure 7.3b and c, a shielded cable with a tin copper braid shield does exhibit magnetic field SE when both enclosures are connected to ground, even at DC. However, the tin copper braid shield has a relative permeability of 1; if a shield with a higher permeability is used, then

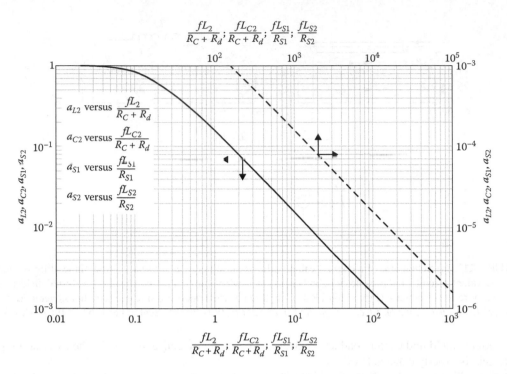

FIGURE 7.2   Attenuation coefficient a_s versus $fLJR_s$. (From Mohr, R.J., Coupling between open and shielded wire lines over a ground plane, *IEEE Transactions on Electromagnetic Compatibility*, 9(2), 34–45, September 1967. © 1967, IEEE.)

an increased magnetic field shielding will be achieved. This improvement is the greatest at DC and at power-line frequencies from 50 to 400 Hz. The Magnetic Shield Corporation, 740 North Thomas Drive, Bensenville, IL 60106-1643, manufactures a shielded four-center conductor cable in which the braid of the shield is manufactured from CO-NETIC AA wire that has a relative permeability of 30,000. Placing cables in a seamless galvanized cold-rolled steel conduit, which has a relative permeability of close to 200, will provide approximately 20 dB of attenuation against 60 Hz magnetic fields.

One practical example of the cable configuration of Figure 7.3a through c is a 19 in. rack with front panels to which shielded cables are connected, or shielded cables external to a vehicle or ship routed above the surface. Another example is in satellites that often consist of a bus on which many electronics units are mounted. The bus often forms a ground plane, and the units are connected to each by shielded cables. At low frequency and for short cables, the cable is much less than a wavelength in length.

In the previous example, the assumption was made that the current flow in the shielded and unshielded cable was the same.

When the incident field is a low-frequency magnetic field, then the calculation of SE based on shield impedance versus load impedance results in a massive overestimation of SE.

This is due to the very large difference in induced current from the incident magnetic field in the shielded versus unshielded cables. In the example of a conduit used here, the predicted SE at low frequency based on the transfer impedance is 70 dB, whereas the actual SE of the configuration is close to 0 dB.

The basic cable configuration is shown in Figure 7.4. The dimensions are those for the 2 cm diameter conduit described later. The SE is often calculated in dB as 20 log the ratio of the two load impedances divided by the combined transfer impedance of the cable and connectors, that is,

$$20\log\frac{2\times RL}{Z_t\text{ total}}$$

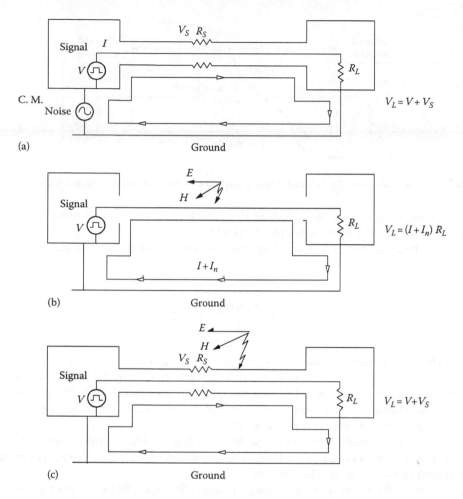

FIGURE 7.3   (a) Current path for common-mode-induced current in a shielded cable. (b) Unshielded case for magnetic waves. (c) Shielded case for magnetic waves.

FIGURE 7.4   Shielded cable configuration.

In this calculation, we assume that the impedance of the ground plane is relatively low, which may not be the case when enclosures are bonded together using ground braid or the front panels are bonded together via the cabinet.

Figure 7.5 shows the coupling to the unshielded center conductor, termed the reference measurement in the conduit SE measurements.

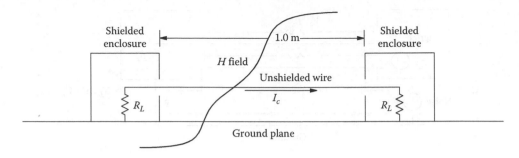

FIGURE 7.5   Unshielded center conductor (used as reference in shielding effectiveness measurements).

The low-frequency transfer impedance of the conduit and its fittings (terminations) has been proven to be equal to the DC resistance that is 40 mΩ.

Assuming the same current is induced into the shielded conduit as induced into the unshielded cable, the magnetic field SE based on the conduit transfer impedance and the two 50 Ω load resistors can be predicted using

$$20\log\frac{2\times50\ \Omega}{0.04\ \Omega}=68\ \text{dB}$$

The cable above a ground can be modeled as a loop-receiving antenna. When the incident magnetic field couples to the loop, a current is induced. This current is limited by the impedance of the loop made up of its inductive reactance, DC resistance, and the termination impedances. At low frequency, the inductance and therefore the impedance of the cable are low, and this is especially true for the low-inductance and low-resistance conduit.

When the magnetic field couples to the unshielded center conductor, the current is limited by the two 50 Ω load resistors and is much lower than the current in the shield, which is terminated to ground at both ends through very low impedances.

Table 7.1 shows the comparison of the current induced into the cable shown in Figure 7.4 compared to the current induced into the center conductor in Figure 7.5, due to a 1 A/m incident magnetic field.

As frequency increases, the transfer impedance of a shielded cable either reduces due to the skin depth, typically from 20 kHz to 2 MHz, or remains constant and begins to increase, typically above 2 MHz, due to porpoising coupling and aperture coupling. The transfer impedance of the conduit

TABLE 7.1
Current in Cable Configuration 1 versus 2

Frequency (MHz)	Current in Shield of Cable (Figure 7.4) I_s (mA)	Current in Center Conductor of Unshielded Cable (Figure 7.5) I_c
0.00015	3.7	1.6 μA
0.001	24.8	11 μA
0.002	49	22 μA
0.005	117	55 μA
0.010	200	110 μA
0.05	349	550 μA
0.1	359	1.1 mA
0.5	363	5.5 mA
1.0	363	11 mA

FIGURE 7.6 Cable low-frequency magnetic field shielding effectiveness based on DC resistance and currents on the shielded versus unshielded cable.

described later is reasonably constant up to 100 MHz and then begins to increase. If we assume that the transfer impedance equals the DC resistance of the cable and fittings up to 1 MHz (a reasonable assumption based on previous measurements), then the SE of the cable to an incident magnetic field can be found from

$$20\log\frac{2\times50\ \Omega\times I_c}{0.04\ \Omega\times I_s}$$

and this is plotted in Figure 7.6.

The measured results on a conduit, presented later, indeed show a very low level of SE from 20 kHz to 10 MHz. The injection method used over this frequency range was a magnetic field injection probe that has some insertion loss (i.e., series equivalent impedance), and therefore, the SE is not as low as shown in Figure 7.6, which assumes coupling from a uniform low-impedance magnetic field. When the measured SE is "corrected" for the difference in the measured current on the shield versus the measured current induced into the center conductor, then the SE is an almost constant 70 dB from 20 kHz to 10 MHz. This is very close to the predicted 68 dB, based on the ratio of load impedance to conduit and fittings transfer impedance.

As the current in the shield of the cable is up to 56 dB higher than the unshielded cable, the voltage induced into the center conductor of the shielded cable is close to the unshielded.

For example, at 50 kHz, the transferred voltage in the 50 Ω load is 550 μA × 50 Ω = 0.0275 V, whereas in the shielded cable, it is 349 mA × 0.04 Ω = 0.0139, and shielding the cable at low frequency is not very effective.

7.2.2 FREQUENCY DEPENDENCY: 100 kHz–22 GHz TRANSFER IMPEDANCE

From the discussion on SE at low frequency, the transfer impedance of a shielded cable was seen to equal the DC resistance of the shield. At some frequency where the current density through the shield is no longer uniform, Z_t no longer equals R_S.

When a current flow is induced by an incident field, the current density is greater on the outside of the shield. When the shield is used as a signal return path or the current is the result of C/M noise on the center conductor and the shield, the current density is greater on the inside of the cable shield. The transfer impedance of several types of cable is shown in Figure 7.7, from which can be

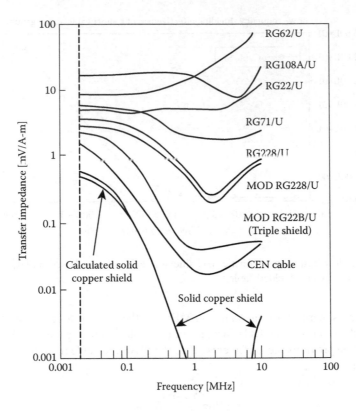

FIGURE 7.7   Transfer impedance of several types of cable.

seen that for a solid copper shield of wall thickness 0.89 mm, the transfer impedance relative to the DC resistance begins to decrease at frequencies as low as 20 kHz. Transfer impedance is defined as

$$Z_t = \frac{V_t}{I_S} l \, [\Omega] \tag{7.11}$$

$$V_t = Z_t I_S l \, [\text{V}] \tag{7.12}$$

where
  V_t is the transferred noise voltage [V]
  I_s is the shield current [A]
  l is the length of the cable

Figure 7.8 shows a shield current flow due, for example, to an incident EM field. Only a percentage of the shield current flows on the inner surface of the shield, and it is this current flow that generates a voltage between the inner surface of the shield and the center conductor. Two basic methods are used to measure the transfer impedance; these are shown in Figures 7.9 and 7.10.

In the test configuration of Figure 7.10, the current flow is provided by a generator connected to both ends of the shield, which simulates the situation where current flows on the outside of the shield, either due to an incident EM wave or due to a C/M current.

As the frequency is increased, the voltage developed across the shield increases due to its inductance; however, this is not the transfer voltage measured at one end of the cable between the center conductor and the inner surface of the shield.

FIGURE 7.8   Shield current flow (transfer impedance).

FIGURE 7.9   Transfer impedance test method 1.

FIGURE 7.10   Transfer impedance test method 2.

The voltage developed from one end of the center conductor to the other end is equal to the shield voltage, with a difference equal to the transferred voltage V_t. It is not necessary to make a physical electrical connection to the inside of the shield when measuring V_i; a connection to the end of the shield, inside or outside, will suffice.

In Figure 7.9, the current flow is between the center conductor and the shield, and V is measured across the ends of the shields.

Some of the test fixtures used to measure surface transfer impedance are the MIL-C-85485 and the IEC 96-1A triaxial fixtures, the quadraxial and the quintaxial. The generic triaxial line looks like a coaxial cable in which the center conductor is the shielded CUT. The drive voltage is connected to the shield of the CUT, and the outer cylinder of the test fixture is used as the current return path. The tri-axial fixture has the internal line, the CUT, terminated in its characteristic impedance at the drive end, and the outer line, the return current line, either terminated in its characteristic impedance or a short circuit at the far end. A detector that has the same characteristic impedance as the cable is used at the far end of the CUT to monitor the transferred voltage. When the outside line is terminated, the upper test frequency is extended, as cable resonances are reduced; however, the long-line effects at high frequency are not eliminated, since the inner-line and outer-line phase velocities are usually different. This difference is due to the higher-core dielectric constant of the CUT compared to the air core outer line. One difficulty with the triaxial fixture is ground loop problems, which interfere with measure-ments at low frequency. These problems can be reduced by the addition of ferrite baluns on each end of the CUT. The quadraxial test fixture will typically have all three lines correctly terminated, and the current in the drive line is relatively independent of frequency. However, due to the extra exterior line, the outer diameter is larger than that of the equivalent triaxial test fixture and non-TEM waves can occur at a lower frequency than with a triaxial fixture. Reference [2] describes two triaxial test meth-ods and one quadraxial test method in more detail. Figure 7.11 shows the measured transfer imped-ance of an RG-58C/U coaxial cable measured using the quadraxial and two triaxial test methods.

Reference [1] describes the predominant high-frequency coupling as "porpoising" coupling in this sample rather than aperture coupling. Thus, the surface electric field makes only a minor con-tribution to the measurement. At low frequency, the triaxial measurements are higher than the quadraxial due to the ground loop problems. At high frequency, the triaxial response is smoother than that of the MIL-C-85485, which has a dielectric constant of 2.6, and the resonance occurs

FIGURE 7.11 Surface transfer impedance of RG-58C/u as measured by quadraxial and triaxial (IEC 96-1A and MIL-C-85485) test fixtures.

at 50 MHz. The IEC 96-1A fixture, which has a dielectric constant of 1, has a resonance above 100 MHz; this is not seen in Figure 7.11.

A test method that in effect makes an easy-to-construct triaxial fixture is the pull-on braid method. This is illustrated in Figure 7.12. The CUT has the shield and center conductor short-circuited at the drive end, and this is connected to a pin in a connector. An additional braid is "milked" on over the insulation of the shielded CUT. The center conductor of the CUT is connected to a pin in the connector at the signal detector end. The milked-on braid shield is then connected to the signal source connector case and the signal detector connector case. The far-end connector pin connects to a detector/measurement device that has the same impedance as the CUT, and the source signal is driven through an impedance equal to the cable impedance. In this triaxial fixture, the CUT is terminated correctly at both ends and the outer cylinder to cable shield is short-circuited. If the CUT is electrically short, then the measurement error is low. At high frequencies, where the cable is electrically long, a correction factor must be made for resonances. The input impedance of the outer coaxial circuit can be measured by the use of a network analyzer and the current computed, but this correction factor can introduce severe errors.

Other test methods include the IEC 96-1 line injection test method, in which an injection wire that is typically a flat copper braid is taped to the CUT. The great advantage in the injection line test method is that it is capable of measuring surface transfer impedance to very high frequencies (above 1 GHz).

The construction of the test fixture is simple and requires only a coaxial cable, injection wire, signal generator, and measuring device. Figures 7.13 and 7.14 show the test setup. The transition from the coaxial injection cable to the injection wire and the transition back can easily be matched to the impedance of the coaxial injection line up to very high frequencies. With a 50 cm coupling length, the phase velocities in the injection circuit and in the CUT are sufficiently well matched to permit measurements up to 3 GHz. With a 10 cm coupling length, measurements up to 20 GHz are

FIGURE 7.12   Simple triaxial-type transfer impedance test fixture using a "milked"-on braid and schematic.

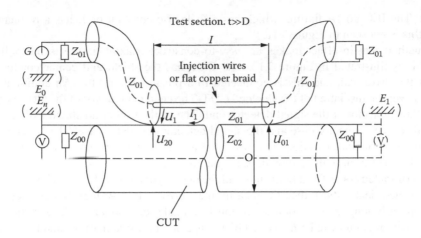

FIGURE 7.13 Schematic diagram of the line injection test setup. (From Fourie, A.P.C., Givati, O., and Clark, A.R., Simple technique for the measurement of the transfer impedance of variable length coaxial interconnecting leads, *IEEE Transactions on EMC*, 40(2), May 1998. © 1998, IEEE.) (Note: New transmission-type-line injection test setup. Subscripts n, f: Near and far end; subscripts 1 and 2: primary (i.e., injection), secondary circuits; Z_0, characteristic impedance of circuits 1 and 2; G, generator; V, voltmeter, receiver; E, usual earthing point for coaxial instruments.)

FIGURE 7.14 Pictorial diagram of the setup used in the line injection test. (From Fourie, A.P.C., Givati, O., and Clark, A.R., Simple technique for the measurement of the transfer impedance of variable length coaxial interconnecting leads, *IEEE Transactions on EMC*, 40(2), May 1998. © 1998, IEEE.) (Note: Complete installation for practical transfer impedance measurements. X, cable under test [CUT]; d, screened room wall; G, generator (synthesizer or tracking generator, etc.); t, test receiver (spectrum analyzer, network analyzer, etc.); g, launcher to injection wire; h, brass tube for additional screening for CUT; i, feeding cables for injection wire (low loss, approximately 0.5 m); k, feeding cable from generator; m, ferrite rings (length approximately 100 mm); n, additional screening for connection between screened room and CUT; p, additional screening for terminating resistance of CUT; q, injection wire.)

feasible; however, at frequencies above even 1 GHz, extreme care must be taken at the transitions and in matching the source and load ends of the cables. Ferrite baluns on the near and far ends of the CUT and on the near and far ends of the injection lines are recommended to reduce C/M current caused by line radiation. And it is recommended that the signal source be placed outside of a shielded room, with the CUT shield terminated at the shielded room wall. A coupling transfer function between the excitation current and the near- and far-end coupling is discussed in the IEC

96-1 standard. One potential source of error is that this method excites only a portion of the circumference of the cable shield. The question of whether this yields the same test results as with other methods is addressed in Reference [2]. Here, the test results obtained with the line injection method were compared with the test results using a quadraxial fixture. It was found that in the frequency range up to 50 MHz, the results were reasonably close and the results up to 1 GHz were more credible using the line injection method.

The IEC 62153-4-4/Ed 2 (2013): Metallic communication cable test methods Part 4-3 Electromagnetic Compatibility (EMC)—Surface transfer impedance—Triaxial method describes two tests one of transfer impedance Z_T and another of the capacitive coupling impedance Z_F referred to late as transfer admittance. It describes in detail the transfer admittance test method, whereas IEC 62153-4-8 describes the capacitive coupling impedance test method.

The author thanks the International Electrotechnical Commission (IEC) for permission to reproduce information from its International Standard IEC 62153-4-3 ed. 2.0 (2013).

All such extracts are copyright of IEC Geneva, Switzerland. All rights reserved. Further information on the IEC is available from www.iec.ch. IEC has no responsibility for the placement and content in which the extracts and contents are reproduced by the author, nor is IEC in any way responsible for the other content or accuracy therein.

The total transfer impedance Z_{TE} equals max $|Z_F \pm Z_T|$.

The transfer impedance is measured by injecting a current on the center conductor that returns on the shield as shown in Figure 7.15. The center conductor(s) is short-circuited to the shield at the far end, and the signal is injected via a resistor R1 that has the same value as the CUT is in series. The transferred voltage is measured across a load resistor at the near end of the shield and at the far end. These resistors equal the characteristic impedance of the outer circuit.

The capacitive coupling impedance is measured by having a current flow on the center conductor(s) but not returning on the shield as shown in Figure 7.16.

The measurement of capacitive coupling is described in IEC 62153-4-8.

The transfer impedance tests can be performed using any of the following:

A vector network analyzer with an S-parameter test set, that is, a full two-port calibration will be established including the connecting cables used to connect the test setup to the test equipment. The reference planes for the calibration are the connector interface of the connecting cables. For networks that respond in a liner manner, they can be completely characterized by parameters at the network terminals without regard to the contents of the network. S-parameters are useful in RF and microwave circuit design in determining current based on input voltage sources and power gain, voltage gain, and loss. The S-parameters are typically measured using a network analyzer.

FIGURE 7.15 Transfer impedance test setup. (Note: $Z_T = U_2/I_1$, where Z_1, Z_2 are the characteristic impedance of the inner and outer circuits; U_1, U_2 are the voltages in the inner and outer circuits (n: near end, f: far end); I_1 is the current inner circuit (n: near end, f: far end); L is the length of cable, respectively, the length of screen under test; λ is the wavelength free space.) (From IEC 62153-4-3 ed 2.0. © 2013, IEC, Geneva, Switzerland. www.iec.ch.)

$$I_{2n} = I_{2f}$$

$$U_{1n} = U_{1f}$$

$$I_{2n} = I_{2f} = (1/2) \times I_2 = I_2/2$$

$$I_2 = I_{2n} + I_{2f}$$

$$Z_F = \frac{U_{2n} + U_{2f}}{I_1} = \frac{2U_{2f}}{I_1} = Z_1 Z_2 \times j\omega C_T$$

FIGURE 7.16 Definition of capacitive coupling impedance. (Note: Z_1, Z_2 are the characteristic impedance of the inner and outer circuits; U_1, U_2 are the voltages in the inner and outer circuits (n: near end, f: far end); I_1 is the current inner circuit (n: near end, f: far end); I_2 is the current outer circuit (n: near end, f: far end); C_T is the coupling capacitance; L is the length of cable, respectively, the length of screen under test; λ is the wavelength free space.) (From IEC 62153-4-3 ed 2.0. © 2013, IEC, Geneva, Switzerland. www.iec.ch.)

A (vector) network analyzer without an S-parameter test set, that is, by using a power splitter, a THRU calibration will be established including the connecting cables used to connect the test setup to the test equipment.

A separate signal generator and receiver, that is, the composite loss of the connecting cables will be measured and the calibration data saved so that the results may be corrected.

The test setup is an apparatus of a triple coaxial form. The cable shield forms both the outer conductor of the inner circuit (the CUT) and the inner conductor of the outer circuit (the test circuit).

In the rigid test setup, the outer conductor of the outer circuit is a well-conductive tube of nonferromagnetic metal (e.g., brass, copper, or aluminum) with a short circuit to the shield on the fed side of the CUTs.

In the flexible setup, the outer conductor of the outer circuit is a tinned copper braid having a coverage >70% and a braid angle <30° that is pulled over the entire length of the CUT.

Figure 7.17 shows the connection to the tube.

The test setup for a vector network analyzer with the S-parameter test setup is shown in Figure 7.18.

The test setup with the network analyzer and a power splitter is shown in Figure 7.19.

The test setup using a signal generator and a receiver is shown in Figure 7.20.

The IEC 62153-4-3 goes on the describe a method with a matched inner circuit with damping resistor in the outer circuit a method with inner circuit with load resistor and outer circuit without damping resistor also a (mismatched)–short-short without damping resistor and a method to determine the impedance of the inner circuit and to determine a impedance matching circuit.

One other simple test method is to place the CUT at a fixed height above a ground plane, which forms a two-wire transmission line with the coaxial cable and its image in the ground. If an RG58 cable is placed with its outer insulation in contact with the ground plane, the characteristic

FIGURE 7.17 Connection to tube. (From IEC 62153-4-3 ed 2.0. © 2013, IEC, Geneva, Switzerland. www.iec.ch.)

FIGURE 7.18 Test setup using a vector network analyzer with the S-parameter test setup. (From IEC 62153-4-3 ed 2.0. © 2013, IEC, Geneva, Switzerland. www.iec.ch.)

impedance of the transmission line is approximately 50 Ω; and at a height of 5 cm above the ground plane, it is approximately 317 Ω. The shield of the transmission line can be excited either by an injection current probe or directly by a signal source with a series impedance equal to the transmission-line impedance. The most common configuration has the shield of the CUT short-circuited at the far end. In this configuration, shown in Figure 7.21a, a current probe is used to monitor current. When the cable is electrically short (less than one-tenth of a wavelength), the current probe measurement can be used directly with the measurement of the transferred voltage to determine transfer impedance. At higher frequencies, a correction factor is required and errors can be introduced. A very short cable length can be used to increase the frequency range; however, the influence of the connector-to-connector and connector-to-bulk head transfer impedances becomes significant. Measurements made on a 50 cm long cable 5 cm above the ground plane have shown a good correlation between a transmission-line method (with direct injection, a short circuit at the far end, and a current probe monitor) to line injection and triaxial methods up to 1 GHz.

A second transmission-line test setup, shown in Figure 7.21b, uses direct injection and terminates the shield of the cable at the far end in the characteristic impedance of the transmission line,

FIGURE 7.19   Test setup using a network analyzer and a power splitter. (From IEC 62153-4-3 ed 2.0. © 2013, IEC, Geneva, Switzerland. www.iec.ch.)

FIGURE 7.20   Test setup using a signal generator and a receiver. (From IEC 62153-4-3 ed 2.0. © 2013, IEC, Geneva, Switzerland, www.iec.ch.)

which reduces the long-line effects. Placing the cable insulation on the ground plane means that the velocity of propagation in the cable-to-ground plane transmission line is approximately equal to that of the shielded cable. This would not be true with a predominantly air space between the cable and the ground. A spacer may be used between the cable and the ground plane that has a similar permittivity to the cable core insulation material to minimize this effect. In this test setup, the addition of the measuring device will present a complex impedance between the shield of the CUT and ground, in parallel with the terminating resistor. To control this impedance, many ferrite baluns should be placed on the shielded cable connecting the measuring device to the termination point. Even with the baluns in place, it is recommended that a network analyzer be used to measure the transmission-line impedance and to detect any major deviation from the required impedance.

In all, five different types of coupling to cables have been identified, as described in References [3,4]. These include *axial transfer impedance* induced by a magnetic field parallel to the cable, such as that generated down the axis of a solenoid or when the magnetic component of an EM field is parallel to the cable, which sets up circumferential currents. It also includes *parallel electric field coupling*, which would occur if the cable were placed between parallel plates at different RF potentials. A third type is *parallel transfer impedance*, caused by a magnetic field loop that penetrates the shield. All of these new couplings disappear for a typical coaxial cable with a single conductor that is concentric. It has been shown that for multiconductor cables, the new coupling modes can induce both C/M and differential-mode (D/M) signals in the cable. The circumferential current flow caused by the axial magnetic will not introduce an EMI voltage in a multiconductor cable in which the conductors are straight, but it will introduce a voltage in twisted conductors, such as those contained in the twisted shielded pair cable.

(a)

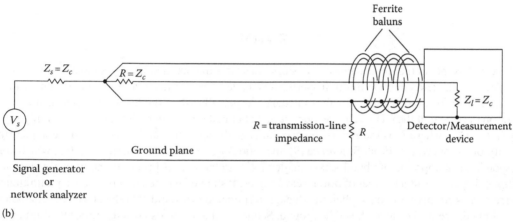

(b)

FIGURE 7.21 (a) Transmission-line transfer impedance test setup with S/C termination and current probe monitoring. (b) Transmission-line transfer impedance test setup with the transmission line terminated in its characteristic impedance.

At low frequency, in practical cables, a magnetic field is generated by the cable, due to eccentricity between the center conductor and the shield. This magnetic field is given by

$$H = \frac{I\Delta r}{2\pi R^2} \qquad (7.13)$$

where
R is the distance between the core and the measurement point
Δr is the eccentricity between the shield and the core

Equation 7.13 assumes equal and opposite currents in the shield and core. For a cable where the transfer impedance decreases with frequency, the current flow on the outer surface, assuming the test method shown in Figure 7.10, decreases, as does the external magnetic field generated by the cable.

From Figure 7.7, we see that the transfer impedance of the solid copper shield continuously reduces with increasing frequency, unlike the braided flexible shield cable, for which it typically begins to increase above 2 MHz. This effect is due primarily to a leakage of the external magnetic field through the apertures in the braid, although electric coupling due to an external electric field can also penetrate the holes in the braid. A second coupling mechanism is referred to as *porpoising coupling*. Porpoising coupling results from current that is pulled into the shield on the strands of wires, or carriers, of which the shield is constructed. This form of coupling has been attributed to the contact impedance between the carriers, which forces some of the current flow to remain on the carrier instead of flowing into the next carrier. If this mechanism is valid, then the porpoising coupling of a cable is expected to increase with an increase in contact impedance with use, due to flexing of the cable that results in loosening of the carriers and due to corrosion. A second cause for porpoising coupling has been proposed: an inductive effect due to the twist in the carriers. Reference [5] says that most cable samples have a surface transfer impedance signature that was indicative of porpoising coupling. Reference [6] notes that the aperture and porpoising coupling should be out of phase and describes a cable with an optimized shield in which wires have been removed and then degraded until the aperture and porpoising coupling components of the transfer impedance tend to cancel.

In open-weave cables, the aperture coupling is often predominant above 1 MHz and becomes more so in the majority of cables above 30 MHz. When aperture coupling is predominant, the transfer impedance of a cable is

$$Z_t = j\omega MA \tag{7.14}$$

where MA is the mutual inductance due to shield apertures that has a nominal value of 3×10^{-10} H/m.

The electric field coupling through apertures is modeled by a transfer admittance, which may not be accounted for in measured transfer impedance, especially when the test fixture injects a current flow on the shield of the CUT. A correction may be obtained for transfer admittance by multiplying Z_t by $(1 + A_e/A_m)$, where A_e is the electric polarizability of the apertures and A_m is the magnetic polarizability of the apertures. From Reference [9], the ratio A_e/A_m for circular apertures is 0.5, which corresponds to an approximate braid weave angle of 40°. Reference [10] provides additional values for elliptical apertures at a number of braid weave angles. It should be noted that the transfer admittance correction is strictly valid only when the shield is terminated to ground at both ends by the characteristic impedance of the cable above the ground. Sections 7.4 and 7.5 also discuss transfer admittance.

With the advent of foil-type cables, one might expect them to achieve the low transfer impedances of a solid copper shield; however, this is not the case. In the construction of early-type foil shields, a spiral wrap was used; later, a longitudinal edge contact was used. At first, the longitudinal joint was insulated, forming a slot. However, most foil braids of recent manufacture are folded at the longitudinal edges. Figure 7.22a shows the transfer impedances of combination braid- and foil-construction shields. Above approximately 50 MHz, the transfer impedance of the combination shield tends to increase. This is due to the AC resistance of the longitudinal joint in the foil. As with the transfer impedance of gasket material, the current is not confined to the outer surface of the foil. But because of an impedance change at the joint, the current diffuses into the center of the foil at the joint.

As the combination cable is flexed, the transfer impedance rises. Figure 7.22b, from Reference [15], shows a large increase from 0.004 Ω to approximately 0.6 Ω for the worst case, unsealed foil-and-braid cable after 49,000 flexures.

7.2.3 MILITARY AND RG CABLE TRANSFER IMPEDANCE

Transfer impedance data are available on both military- and RG-type cables. Figure 7.23a shows the transfer impedance curves for coaxial cables and triaxial cables and Figure 7.23b for twinaxial cables, which is reproduced by the permission of Belden Wire and Cable. It has been found that cables of the same type from different manufacturers may have very different transfer impedances, so the data in Figure 7.23a and b are valid for the cables manufactured by Belden. RG 58 is a very common 50 Ω coaxial cable, and we see a very large variation of up to 20 dBΩ (factor of 10) between cables from different manufacturers, as shown in Figure 7.24, from Reference [11]. Reference [11] describes

FIGURE 7.22 (a) Foil- and braid-type cable transfer impedance. (b) Increase in transfer impedance with flexure. (From Dinallo, M.A. et al., Shielding effectiveness of typical cables from 1 MHz to 1000 MHz, *IEEE International Symposium on Electromagnetic Compatibility*, 1982. © 1979, IEEE.)

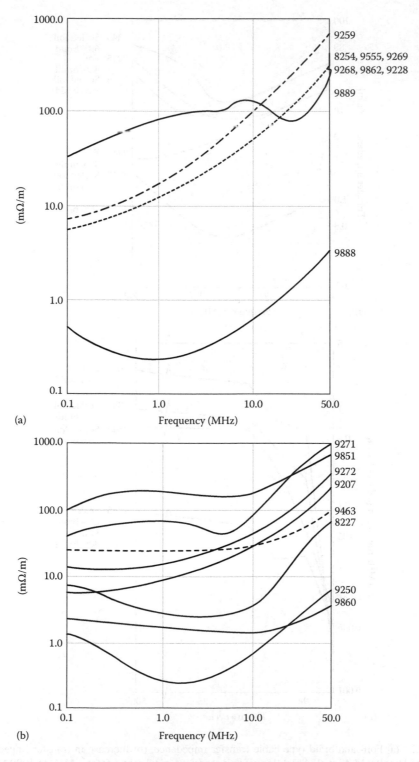

FIGURE 7.23   (a) Transfer impedance of Belden coaxial and triaxial cable. (b) Transfer impedance of Belden twinaxial cable. (Courtesy of Belden Wire and Cable.)

FIGURE 7.24 Transfer impedance of two samples of RG58 from different manufacturers. (From Fourie, A.P.C., *IEEE Trans. EMC*, 40(2), May 1998. © IEEE, 1998.)

one of the transmission-line techniques for the measurement of transfer impedance. For this reason, it is advisable either to obtain the transfer impedance from the cable manufacturer or to use one of the simple test methods, such as the transmission-line or line injection methods, to measure the transfer impedance of the cable of choice.

The shield constructions of cables shown in Figure 7.23a are as follows:

Belden Cable Number	Shield Type
9259, 8254, 9555, 9269, 9268, 9862, 9228	Bare copper braid, 95% coverage
9889	Duofoil™ with #24AWG drain wires
9888	Triax, two bare copper braids, 95% coverage

The shield constructions of cables shown in Figure 7.23b are as follows:

Belden Cable Number	Shield Type
9271	Beldfoil™ with stranded copper drain wire
9272	Tinned copper braid, 93% coverage
9851	Foil with shorting fold and stranded copper drain wire
9207	Tinned copper braid, 95% coverage
9463	Beldfoil with 57% coverage tinned copper braid
8227	Tinned copper braid, 85% coverage
9860	Duofoil with 92% coverage tinned copper braid

The terms *single-, double-, and triple-braid-shielded cable* refer to the number of braids in a shield that are, typically, in electrical contact.

A comparison between the transfer impedances shown in Figures 7.7 and 7.23a and b indicates that the cables with multiple-braid shields exhibit a lower transfer impedance. The upper curve of the RG22B/U in Figure 7.7 is for a double-braid cable, whereas the lower curve for the modified RG22B/U is that for a triple-shield cable.

The term *triaxial* refers to a cable with a center conductor and two shields that are electrically isolated from each other. Triaxial connectors exist that allow the center shield connection to continue through the connector and remain isolated from the enclosure and in which the outer shield is connected to the case of the connector. The triaxial configuration allows the signal return currents to flow on the center shield and the outer shield to be connected to the equipment enclosure. Thus, the transfer voltage does not appear directly in the signal path. The remaining coupling between the inner and outer isolated shields is via electric field coupling through the transfer admittance. This advantage of the triaxial cable over a double-braid cable is considerably reduced if the signal ground is connected to the enclosure at both ends. One example of the correct use of a triaxial cable is shown in the HAR-1 circuit in Figure 5.132, in which the signal is transformer coupled at the receiver end. At high frequencies, it becomes increasingly difficult to isolate signal grounds from the enclosure, due to stray capacitances, so the transfer impedance of triaxial cable in which the shields are connected together at both ends is of interest. The transfer impedances of isolated, single-, double- (triaxial), and triple-braided cables with the braids shorted are shown in Figure 7.25 (from Reference [8]). We see that at frequencies below that at which the cable length is equal to $\lambda/2$, the transfer impedance is lower for the multibraided cables; however, above this frequency, the difference is negligible due to intersheath resonances or different propagation delays down the shields.

Twinax cable is constructed of two-center conductors with an overall shield, and quadrax is a two-center-conductor cable with two isolated shields. A typical use of these cables is in fully balanced or single-ended driver/differential input circuits. Typical shield connections are shown in Figure 7.26a and b. The shielded-cable transfer voltage is C/M; that is, it appears equally on all the center conductors within a shielded cable. Thus, to achieve maximum immunity, the input circuit must exhibit a sufficiently high C/M noise rejection, and any imbalance in the input impedances of the circuit should be minimized. The quadrax cable will exhibit a lower transfer impedance than the twinax by approximately the same magnitude as shown for triax cable. The options for shield connections in Figure 7.26 show the outer shield connected to the enclosure at both ends, for reasons discussed in Section 7.12 on shield termination.

FIGURE 7.25 Comparison of the transfer impedances of isolated, single-, double-, and triple-braided cables, in which the braids are shorted together at both ends. (From Demoulin, B. and Degauque, P., *IEEE Trans. Electromag. Compatibil.*, EMC-22(3), August 1980. © 1967, IEEE.)

FIGURE 7.26 Typical shield terminations for a (a) twinax and (b) quadrax cable. (From Demoulin, B. and Degauque, P., Shielding performance of triply shielded coaxial cables, *IEEE Transactions on Electromagnetic Compatibility*, August 1980. © 1980, IEEE.)

Reference [12] describes a comparison of five cables with shields constructed of copper braid and aluminized plastic foil, with a sixth cable shield constructed of braid alone. These cables were examined for use with 1 GB/s signals. In most of the cables, the braid was on the outside and provided the turn-to-turn contact for the foil shield. In two cables, the braid was inside the foil but still provided the turn-to-turn contact (Figure 7.27b). The core four of the cables was a balanced quad (one pair transmits and one pair receives, with no individual shielding of the pair). In two cables, the pair was individual shielded. In one cable, the metal of the internal shield faced the inside, and

FIGURE 7.27 Cross sections of the cable construction used in the 1 GB/s cables.

turn-to-turn contact was made with a drain wire (Figure 7.27d). In another cable, the metal of the internal foil faced the outside and the braid made the turn-to-turn contact (Figure 7.27c). The highest transfer impedance is that of the shielded twisted pair or inner shield. The measurements showed that the aluminized foil plastic tape did not provide a good EM shield by itself, particularly at high frequencies. This is consistent with experience with inexpensive computer serial and parallel cables, where the shield is an external foil alone with an internal drain wire. Although this type of cable is better than an unshielded type, its SE is not as high as a braid type of cable.

The lowest transfer impedance is with two layers of shielding, one the outside layer and the second the individual shield over the pairs of conductors. Where the plastic foil insulation is between the braid shield and the aluminized layer, resonances occur due to the difference in propagation delay as a result of the plastic between the shielding layers. This effect is shown in Figure 7.25 for insulated shields. Reference [12] concludes with the following:

> The measured transfer impedance of six cable samples that used combinations of braid and foil showed that these cables do not have the classic $R + j\omega M_{12}$ frequency dependence. Above a few MHz, they exhibited a frequency dependence that was approximately proportional to the square root of frequency. This suggests that the coupling mechanism is due to contact resistance somewhere in the shield. The two cables with individually shielded pairs had lower transfer impedances than those that used an overall shield over a balanced quad core. The transfer resistance of the 6 cables ranged from 9 to 21 mΩ This is in the range of single-braid cable shield and is appropriate for 1 GB/s interconnect cables. At 500 MHz, the transfer impedance ranged from a little more than 10 mΩ/m to 135 mΩ/m. The cables with individually shielded pairs were best (12.4 and 21.8 Ω/m).

Figure 7.28 shows the comparison of the transfer impedances at 500 MHz.

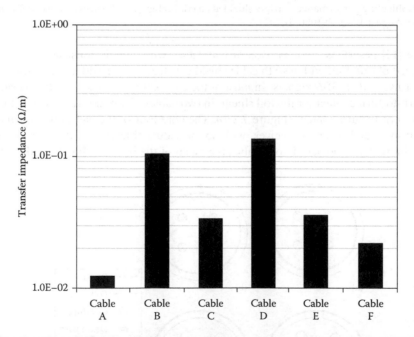

FIGURE 7.28 Transfer impedance (extrapolated) of the six combination braid/foil shields at 500 MHz. (From Hoeft, L.O. and Knighten, J.L., Measured surface transfer impedance of cable shields that use combinations of braid and foil and are used for 1-GB/s Data Transfer, *IEEE EMC Symposium*, 1998. © 1998, IEEE.) A, figure 1d; individually shielded pair with the metal of the foil faced inside (drain wire used for turn-to-turn contact) plus overall shield. B, figure 1a; overall shield, no individual shielding of the pair. C, figure 1b; braid inside the foil, no individual shielding of the pair. D, figure 1b; braid inside the foil, no individual shielding of the pair (cable similar but not identical to C). E, figure 1a; overall shield, no individual shielding of the pair (cable similar but not identical to B). F, figure 1c; individual shielded pair with an overall shield.

Thus far, we have examined the voltage induced in shielded cables when the length of the cable is less than 0.5λ, where λ is the wavelength of the current flow on the shield of the cable. Equation 7.12, which gives the magnitude of the induced voltage, is no longer valid when the cable is longer than 0.5λ, due to so-called long-line effects.

Long-line effects are considered in Section 7.4 (30 MHz–10 GHz), although for long cables, they may occur in the frequency range from 100 kHz to 30 MHz.

Should an unshielded or a shielded cable be the culprit in a susceptibility/immunity problem or result in excessive radiation, the addition of an overbraid may solve the problem. Tin-plated copper braid is sold as shielding and bonding cables with IDs of 3.18 mm to over 25.4 mm. These braids can be attached to a circular connector by the use of a hose clamp, soldered to a brass connector, or clamped under a strain relief on a non-EMI backshell, as described in the section on connectors. The measured transfer impedance of 1 m length of single tubular braid, double tubular braid, or triple tubular braid is shown in Figure 7.29.

Above 30 MHz, the leakage effects are more pronounced, because the holes in the braid of a shielded cable become more effective antennas with increasing frequency. The method of testing cables in the gigahertz region becomes difficult because the geometries must be more rigorously controlled. Forward and backward waves can be generated in the test fixture, and the magnetic field induced in the cable is confined to the TEM mode. Therefore, the test fixture may be far removed from the real-life use of shielded cables where radial, axial, and circumferential fields can be generated by a cable at very high frequencies. Using the line injection test method, the transfer impedances for single- and double-braid cables up to 22 GHz are shown in Figure 7.30 (from Reference [13]). The transfer impedance increases almost constantly up to 20 GHz. Figure 7.31 shows the near- and far-end coupling for a single-braided cable. The far-end transfer impedance increases rapidly above 10 GHz. Reference [13] attributes this to radiation loss through the shield and makes the point that cables used at these frequencies are in some way antennas. Much earlier data on the measured transfer impedance in the gigahertz region are shown in Figure 7.32a for

FIGURE 7.29 Measured surface transfer impedance of 1 m long tin-plated copper tubular braid shields. (From Hoeft, L.O., Measured electromagnetic shielding performance of commonly used cables and connectors, *IEEE Transactions on EMC*, 30(3), August 1988. © 1998, IEEE.)

FIGURE 7.30 Single- and double-braid-shielded cable transfer impedance up to 22 GHz using the line injection test method. (From Hoeft, L.O., *IEEE Trans. EMC*, 30(3), August 1988. © 1992, IEEE.)

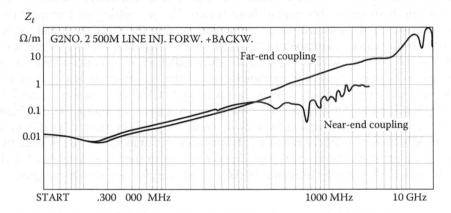

FIGURE 7.31 Single-braided cable showing near- and far-end transfer impedance. (From Hoeft, L.O., *IEEE Trans. EMC*, 30(3), August 1988. © 1992, IEEE.)

single-braid cable and in Figure 7.32b for double-braid cable (from Reference [21]). The transfer impedance of the single-braid RG58C/U is 400 mΩ/cm, or 40 Ω/m at 4 GHz, which is similar to that of the single-braid cable in Figure 7.30, which has a transfer impedance of approximately 30 Ω/m at 6 GHz. The double-braid cable GR cables in Figure 7.32b have an impedance of 40 mΩ/cm, or 4Ω/m, which is much higher than the double-braid-shielded cable in Figure 7.30 with a transfer impedance of 0.2 Ω/m at 6 GHz. However, as we have seen, even samples of the same type of cable show a surprising variation in transfer impedance.

Modern triple-braid stainless steel–armored cables have an SE of approximately 80 dB, which, using Equation 7.3, corresponds to a transfer impedance of 10 mΩ/m.

7.2.4 Shielded Twisted-Pair Transfer Impedance

A common wire type used in avionics, aerospace, and air frames is the M27500 PTFE twisted shielded pair with PTFE insulation.

As no transfer impedance is known for this very common type of cable, a measurement was made on a twisted shielded pair with 24AWG internal conductors and a shield diameter of 3 mm with a jacket insulation approximately 0.3 mm thick.

The C/M voltage induced into both conductors is required when the input connected to the wires is a balanced circuit. When one conductor is connected to ground at one end, which is at some point

FIGURE 7.32 (a) Single-braid coaxial transfer impedance up to 8 GHz. (b) Double-braid coaxial transfer impedance up to 8 GHz.

connected to chassis and likewise at the other end, then the D/M transfer impedance is required. Although it may not reflect the real grounding scheme, the differential impedance measurement was made with one conductor grounded to the shield at both ends.

The measured C/M characteristic impedance of the CUT was 45 Ω, and the D/M was 52 Ω.

The test setup is shown in Figure 7.14. The characteristic impedance of the circuit formed by the insulated 0.5 m long injection wire in contact with the insulation on the shield should have the same characteristic impedance as the injection cables on either side of the 0.5 m length and this is 50 Ω. The shield outside diameter conforms to the shape of the twisted pair, and so to ensure that a good contact is maintained the injection wire was twisted down the length of the shield.

Using a #24AWG injection wire with 0.3 mm PTFE insulation results in an impedance of approximately 52.4 Ω between the wire and the shield. The attenuation in the 1.5 m long injection wire from the signal generator to the 50 Ω load increased monotonically without dips and peaks and was only 5 dB at 1 GHz showing that the injection wire was close to 50 Ω impedance.

To achieve the 45 Ω impedance in the C/M test, a 45 Ω shielded resistor was used at the far end and a 450 Ω resistor was placed across the input impedance of the spectrum analyzer at the near end. As the D/M impedance is close to 50 Ω, the input of the spectrum analyzer was unterminated and the far end resistance was 50 Ω.

Both the far-end and near-end C/M and D/M impedances were measured, and these are shown in Figures 7.33 and 7.34.

7.2.5 Conduit Transfer Impedance

The transfer impedance of a number of conduits was required. The conduits we tested comprised three types.

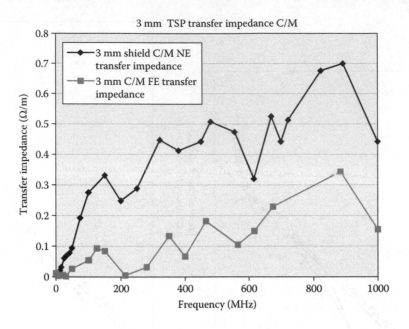

FIGURE 7.33   Common-mode transfer impedance of PTFE-insulated, twisted shielded pair.

FIGURE 7.34   Differential-mode transfer impedance of PTFE-insulated, twisted shielded pair.

The #1 type provides some shielding, #2 much better shielding, and #3 the best shielding. Each conduit we tested was of ½ in. trade size and had an inner diameter of 5/8 in., although other diameters are available.

The #3 is manufactured from a brass spiral wound flexible conduit covered in a tin-plated copper braid. The #2 is a brass spiral wound flexible conduit without braid shield and #1 is aluminum spiral wound with braid. The braid in these cables is fairly open weave and has an optical coverage no greater than 50%. These cables come with threaded connectors "fittings" at both ends and mating threaded fittings that are designed to be mounted on a panel, bulkhead, or enclosure.

The mating fittings can be inserted into ½ in. NPT-threaded holes in the mating surface.

The inner cable or wires are meant to be pulled through the conduit and are not typically part of the conduit construction.

The cables used in the tests were 1 m long. Two types of tests were performed. One is in accordance with IEC 60096-1 and measures the transfer impedance of the conduit alone without including the transfer impedance of the fittings.

The other test is of the SE of the cable and fittings mounted to two soldered-up copper boxes bonded to an underlying ground plane, that is, in a typical cable configuration.

As a cable with a known characteristic impedance is required for the transfer impedance test, a center conductor was manufactured from a ¼ in. diameter solid brass rod. This rod was held in the center of the conduit by small expanded polystyrene plugs about 1 in. in length separated by about 8 in. Figure 7.35 shows the center conductor rod and a wooden plug, which was used at the ends of the cable only, mounted into one of the fittings on a soldered-up copper box, bonded to the ground plane, and Figure 7.36 shows the rod centered in the conduit.

With the rod in the center of the conduit, a coaxial cable has been constructed with a characteristic impedance close to 50 Ω.

We see that data from the SE measurements in measurements described later show a remarkably good correlation to the SE derived from the transfer impedance, over the frequency range where the cable is electrically long and the frequency range where the transfer admittance predominates (100–1000 MHz).

Also, the test setup simulates a realistic cable configuration above a ground plane. An investigation into the "150 Hz to 1 MHz magnetic field coupling to a typical shielded cable above-ground

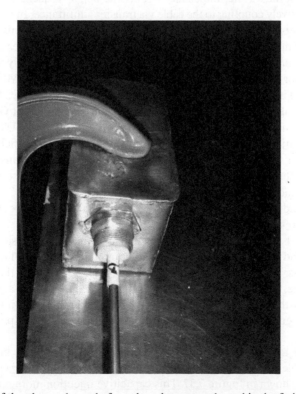

FIGURE 7.35 Photo of the plug at the end of a rod used to center the rod in the fitting.

FIGURE 7.36 Photo of the rod centered in conduit forming a 50 Ω coaxial cable.

plane configuration" illustrates how a shielded cable may achieve a no higher SE than an unshielded cable when exposed to magnetic fields.

These conduit measurements also show that from 20 kHz to 1 MHz, the SE data of the conduit and fittings and the SE derived from combined DC resistance and transfer impedance of the conduit and fittings are virtually identical, when corrected for the different current flows on the shielded versus the unshielded cables in the SE test setup.

The disadvantages of the transfer impedance test are the upper frequency limit and the fact that it does not measure effectively coupling to the cable via transfer admittance. In most setups, the connection between the shield of the cable and the driveline is 0 Ω, and this maximizes the transfer impedance coupling and minimizes the transfer admittance coupling. Many consider the transfer impedance the most important factor and this is true at low frequency, but not necessarily true at high frequency. The upper frequency for most coaxial and triaxial cable test setups is 100 MHz, although 1 GHz is achievable. The method used to obtain the data described in this test is the IEC 96-1 and is useable up to 1 GHz, and theoretically beyond. However, the limit we have achieved is 1 GHz.

The cable transfer impedance test induces a current directly into the CUT, which is part of the test fixture. The predominant form of coupling is porpoising and magnetic field penetration through the apertures, although some coupling of the E field occurs due to the transfer admittance of the cable. The conduit contains an overbraid in which large apertures do exist, although no apparent apertures exist in the underlying conduit.

However, as the conduit is constructed as a helical spiral, a contact impedance almost certainly exists at the joint in the spiral, as discussed later.

Also in the IEC 96-1 test, only the transfer impedance of the cable is tested and not the connectors/fittings at either end of the cable or the bulkhead-mounted connectors/fittings.

The major weakness in using the transfer impedance to derive the SE is the impact of cable configuration. For example, the current flow in a shielded cable connected to a ground plane at either end is typically higher than an unshielded cable terminated with a higher impedance at either end or disconnected from ground at one end. Thus, the SE is lower than the predicted one using the transfer impedance.

The E-field SE uses a capacitive injection probe to create a high electric field between the probe and the cable. The cable was connected to a soldered-up copper box at either end, and the capacitive injection probe is shown in Figure 7.37. This capacitive injection more realistically simulates a

FIGURE 7.37 Photo of the capacitive injection probe on the unshielded rod (reference measurement).

high-level *E* field incident vertically on the cable and the source would typically be from a high-gain antenna and somewhat localized, as the MIL-STD-461F/G RS103 test. The test method maximizes aperture coupling through the cable transfer admittance and at the same time sets up a current down the length of the cable that ensures transfer impedance coupling, both through the magnetic field aperture coupling and porpoising coupling. Although the test setup is not standardized, we have achieved reproducible results using the same type of shielded cable and injection probe in a number of tests.

As the SE test compares the induced levels with and without the shield and is made with the cable electrically long, that is, multiple wavelengths long, the measurement is virtually independent of cable length.

This test method does include the total SE of the cable and of the connectors/fittings at either end of the cable and the mating bulkhead-mounted connectors/fittings.

This test method was used from 200 MHz to 8 GHz.

In addition to the *E*-field test, two magnetic field injection probes were used to inject a current into either the unshielded rod (the reference measurement) or the coaxial cable manufactured from the conduit and rod. One injection probe covered the frequency range from 20 kHz to 1 MHz and the second probe from 1 to 200 MHz. These probes are used in the MIL-STD-461F/G CS114 test and exhibit an insertion loss within specified minimum and maximum values. The investigation into the "150 Hz to 1 MHz magnetic field coupling to a typical shielded cable above a ground plane configuration" covers the magnetic field SE of the configuration where the shield is terminated as a conductive enclosure at both ends of the shield. The current flow in the shielded cable is very much higher than the unshielded cable. This is because the termination impedance of the shield to the enclosures is much lower than the source and load impedances at the ends of the unshielded cable and the magnetic field SE is close to zero at low frequency. Figure 7.4 illustrates the cable configuration. As the current probes exhibit some series impedance, the ratio of current flow in shielded versus unshielded cable is not so high. In the CS114 test, the input power is calibrated and the current on the cable is monitored. The test specifies that the forward power into the probe is maintained at the calibrated level, or the maximum current level for the applicable limit, whichever is less stringent. This means that the current flow in the cable can be lower, and in the case of the unshielded center rod compared to the shielded cable, it is much lower. The test data presented later in this section show the SE both for unequal injection currents (as in the CS114 test) and that for equal currents.

Figure 7.38 shows the test setup with one of the injection probes around the conduit that is connected to a soldered-up copper box at either end with a copper tube soldered to one of the two boxes. The copper tube contains the shielded cable used to measure the voltage induced into the cable.

The transfer impedance test method is included in IEC 96-1 Amendment 2; however, IEC 96-1 contains a major error in the transfer impedance equation on page 41 and does not provide the background information that Section 7.2.2 does. The test setup for the conduit-shielded cable is shown in Figures 7.39 and 7.40.

FIGURE 7.38 Injection probe around the conduit.

FIGURE 7.39 Transfer impedance test setup.

FIGURE 7.40   Cable transfer impedance test setup in anechoic chamber.

In this test, the copper box containing the 50 Ω load is isolated from the underlying ground plane.

The injection circuit is designed to have a characteristic impedance of 50 Ω, and the CUT is also designed to have a characteristic impedance of 50 Ω.

The current is injected via a 4 mm wide braid located above and insulated from the conduit shield and returns to the signal generator via the conduit shield. This injection circuit also has a 50 Ω characteristic impedance. The voltage developed across this segment of the conduit is also measured between the center conductor of the CUT and its shield. This "transferred voltage" is used in calculating the transfer impedance of the cable when measured with a 50 Ω load at the near end and open circuit at the far end of the CUT. When measured with a 50 Ω load at the far end, the measured voltage is multiplied by two to obtain the transferred voltage. As the transfer impedance is specified in ohms/m and the test section is 0.5 m, the transferred voltage is again multiplied by two to obtain the transfer impedance. In the IEC 96-1 test setup, the CUT is extended on either side of the length into which the current is injected, in our case 0.5 m. The voltage developed at the far end of the injection cable is also measured to ensure that the correct value of injection current is used in the test data.

In an earlier discussion, it was recommended that a measurement be made of the insertion loss of the CUT and that a correction be made in the transfer function, and this correction was performed.

The far-end transfer impedance of the 0.5 m length of the #3 conduit was measured as described previously in this section. After correction for the 50 Ω load, the measurement cable attenuation, the injection current, and after normalization to the 1 m length, the measured transfer impedance in ohms/m is shown in Figure 7.41.

At low frequency, the transfer impedance should be the same as the DC resistance of the conduit. The DC resistance of the 1 m length of conduit was 11 mΩ and the transfer impedance from 50 to 100 kHz was measured at 9.5 mΩ, thus validating the transfer impedance measurement at low frequency.

The cable SE without the effect of the transfer impedance of the conduit fittings was required.

The conduit with center rod forms a coaxial cable with a 50 Ω characteristic impedance. At one end of the cable, the measuring equipment provides the termination and this has a characteristic impedance of 50 Ω. At the other end of the conduit, a 50 Ω load is used as the termination. As the cable and load impedances are matched, a value of the SE can be derived from the transfer impedance measurement from the following equation:

$$20\log\frac{2\times RL}{Z_t\ \text{total}}$$

FIGURE 7.41   Transfer impedance of #3 conduit.

FIGURE 7.42   #3 conduit measured shielding effectiveness (SE) from 1 MHz to 8 GHz compared to the limit on SE measurement (test setup capability).

The SE measurements, made with the two current probes and the electric field probe, from 1 MHz to 8 GHz are plotted in Figure 7.42. These measurements include the transfer impedance of the 1 m length of conduit plus the fittings at each end. When making SE measurements, it is important to check that the test setup is not limiting the maximum level of SE that can be measured. To ensure an adequate margin between the measured cable SE and the limit imposed by the test setup, the center conductor of the rod was disconnected from the center conductor of the shielded cable in the copper pipe. This ensured that any voltage induced by C/M current flow on the copper pipe to the shielded room wall would be measured but that the voltage induced into the center conductor of the conduit cable would not be measured. The maximum attenuation limited by the test setup is also plotted in Figure 7.42 and is at least 20 dB higher than the cable-measured SE.

The predicted SE of the conduit alone, calculated from the transfer impedance data, is shown in Figure 7.43, as well as the measured SE using the magnetic field and electric field probes. The SE measurements include the transfer impedance of the fittings.

This additional transfer impedance can explain the approximately 10 dB difference in SE from 50 to 1000 MHz; however, the much lower attenuation from 20 kHz to 50 MHz is due to another mechanism, which is explained later in a description of the "150 Hz to 1 MHz magnetic field coupling to a typical shielded cable above a ground plane configuration."

FIGURE 7.43   Measured shielding effectiveness (SE) of #3 conduit plus fittings (typical for the CS114 test) compared to SE of conduit alone from the transfer impedance measurement.

The SE measurements using the injection probes are made on the shielded and unshielded cables with a constant input power level to the probe, just as in the CS114 test. This means that the current flow on the shielded conduit, terminated to ground with low impedances, is higher than that on the unshielded rod that is terminated at both ends in 50 Ω and the value of SE is lower than that achieved when the current is kept constant. Thus, the measured SE from 20 kHz to 200 MHz can be used to predict the performance of the conduit in a CS114 test.

The input level injected at a constant power level was measured. The current flow on both the conduit and the unshielded rod was measured with a Fischer current probe, useable from 10 kHz to 250 MHz. If the measured SE is corrected for the difference in these two currents from 20 kHz to 50 MHz, then the cable SE and the cable plus fittings SE track more closely and this is shown in Figure 7.44.

FIGURE 7.44   #3 Cable plus fittings shielding effectiveness with corrected for equal currents in the unshielded and shielded measurements.

The DC resistance of the fittings increases the overall transfer impedance of the cable to some extent. The low-frequency-predicted SE can be calculated based on the cable transfer impedance added to the DC resistance of the fittings. If this calculated SE is plotted against the corrected SE, then a better correlation exists, as shown in Figure 7.45, and the SE and the correction for the induced current are validated.

Figures 7.46 and 7.47 show the transfer impedance and the SE derived from the transfer impedance for conduit #2, constructed from a brass spiral without braid. We see that at low frequency, the brass spiral with and without braid is similar. Above about 1 MHz frequency, the transfer impedance of the spiral without braid increases and so the SE, as shown in Figure 7.46, increases.

The high transfer impedance of the #1 conduit caused a great deal of concern initially. It was tested on three separate occasions with the test setup torn down and replaced after each test. The results were within the 4 dB expected due to measurement error. After the final test on the #1 conduit, the #3 conduit was immediately retested with the same setup, and the results were consistent

FIGURE 7.45 Measured and calculated shielding effectiveness from the #3 cable transfer impedance and fittings DC resistance.

FIGURE 7.46 Transfer impedance of #2 conduit.

FIGURE 7.47 Calculated shielding effectiveness from the cable transfer impedance of #2 conduit.

with the previous test on #3. The DC resistance of the #1 conduit was 30 mΩ and that for the #3 was 9 mΩ for the 0.5 m length, and so this easily explains the 10 dB difference in SE at low frequency. However, we have never seen transfer impedance as high as 8 Ω at 700 MHz and SE of only 25 dB with a braid-shielded cable. It was noticed that whereas with #3 and #2 conduit the joints were very stiff, it was seen that the joints in the #1 cable were loose and the cable drooped where the outer insulation was bared back. As the conduit is wound in a spiral, the contact impedance of the joint appears down the length of the cable. Thus, the longitudinal current generated in the transfer impedance test has to jump as many as 125 of these joints over the 0.5 m test length. The optical coverage of the braid is only approximately 50% and the braid may not be well attached to the end fittings, as these are designed to cover the conduit only. Although initially surprising, the test results on the #1 conduit did indeed show a very poor shield compared to #3 and #2, and this is almost certainly the result of the high contact impedance due to loose joints.

Figure 7.46 shows the transfer impedance of the #2 conduit and Figure 7.47 the SE of the #2 conduit. Figure 7.48 shows the transfer impedance of the #1 conduit and Figure 7.49 the SE.

FIGURE 7.48 Transfer impedance of #1 conduit.

FIGURE 7.49   Calculated shielding effectiveness from the cable transfer impedance of #1 conduit.

7.2.6 FLEXIBLE SHIELDS

Instead of a shielded cable, a cable with an absorptive component is often required. The use of absorber over cables (lossy sleeving) is discussed in Section 5.1.10.3.

Judd Wire manufactures cables that include distributed low-pass filters using RF absorptive components. The effectiveness of these at low frequency is unknown.

7.2.6.1 Flexible Conductive Cloth-Shielded Cable Transfer Impedance

Very often, a cable with more flexibility than a braid- or foil-shielded cable is required.

The advantage of replacing the standard tin-plated copper braid or silver-plated copper overbraid shield material with a conductive cloth plated with either nickel plated over silver or nickel plated over copper are weight savings and flexibility.

The test method used is shown in Figure 7.38. The method is also that for the conduit.

The injection circuit is designed to have a characteristic impedance of 50 Ω, and the CUT is also designed to have a characteristic impedance of 50 Ω.

The test method used is described in Reference [1].

The current is injected via a 4 mm wide braid located above and insulated from the cloth shield and returns to the signal generator via the shield of the CUT. This injection circuit also has a 50 Ω characteristic impedance. The voltage developed across this segment of the cloth CUT shield is also measured between the center conductor of the CUT and its shield. This "transferred voltage" is used in calculating the transfer impedance of the cable when measured with a 50 Ω load at the near end and open circuit at the far end of the CUT. The major problems in the use of this conductive cloth material are the very high DC resistance and the low-impedance connection of the cloth to the backshell of the connector.

Although the material was made up into a 50 Ω coaxial cable, it would be an impractical use due to the high DC resistance in the signal path and the high insertion loss. Instead, the shield would be used over signal cables in which the signal return path is contained in the cable bundle.

Both Ni/Ag- and Ni/Cu-plated cloth were tested. The results were very similar and it is the better Ni/Ag material that is reported here. In addition, a second Ni/Ag cable was made up and tested. The DC resistance and transfer impedance of these two cables were very similar up to 200 MHz. Above 200 MHz, the transfer impedance of the first cable was higher. The only difference between the two cables was the length of the conductive cloth shield over which the braid of the injection coaxial cable was bonded. In the second cable, this connection was over a larger area ensuring a lower impedance at this junction.

The second cable was tested more than once over a period of days to ensure reproducibility. The maximum difference in measured transfer impedance over this range of measurements was 2 Ω, which represents a maximum 9.3% (0.77 dB) error, which is well within the measuring equipment accuracy.

In the IEC 96-1 test setup, the CUT is extended on either side of the length into which the current is injected, in our case 0.5 m. The problem here is the very-high-shield DC resistance in series with the 50 Ω load and source impedances, which introduces a high CUT insertion loss. This DC resistance reduces the voltage developed across the 50 Ω measuring equipment and when this voltage is multiplied by two to obtain the open-circuit transferred voltage, V_t, a large error occurs.

At low frequency, up to 2 MHz, this DC resistance is not important in our test setup as the 50 Ω load resistance was removed, and an oscilloscope was used to measure the open-circuit voltage. For the higher-frequency measurements into a 50 Ω load, to reduce the CUT shield resistance, a copper braid was threaded over the CUT on either side of the injection points and this braid was soldered to the SMA connectors. The cloth shield material was then taped tightly to the outside of the copper braid.

This resulted in a lower insertion loss and error in the measurement of voltage into a 50 Ω load versus open circuit. This insertion loss may account for the overlap in the two plots that can be seen at 1 and 2 MHz in Figure 7.53, as the measurements made from 100 Hz to 2 MHz were made open circuit with an oscilloscope and from 1 MHz to 1 GHz with a 50 Ω input spectrum analyzer. However, the oscilloscope measurement is inherently more accurate than the spectrum analyzer (±0.42 dB vs ±2.5 dB), and as the difference between the two plots is only 2.4 Ω (which represents a difference of only 0.76 dB), the difference may be due to measurement error.

Reference [1] recommends that a measurement be made of the insertion loss of the CUT using the technique shown in Figure 7.50 and that a correction be made in the transfer function. The high resistance of the 0.5 m of cloth shield means that a high insertion loss occurs from end to end. However, in the transfer impedance measurement, the transferred voltage is developed across this resistance and the insertion loss is only in the much lower loss of the copper braid–shielded cable attached to the CUT as shown in Figure 7.51.

At low frequency, leakage currents can exist between the signal generator input signal and the EMI measuring equipment, and to avoid this, the signal generator was supplied with AC via an isolation transformer. At higher frequencies, some of the injection current can continue to flow on the outside surface of the injection shielded cable, which then acts as an antenna. To minimize this current, two different types of ferrite baluns—one type effective at low frequency and the other at high frequency— were strung on the injection cable, and these can be seen in the test setup shown in Figure 7.52.

Attenuation measurement of cable

FIGURE 7.50 Recommended measurement setup of insertion loss of cable under test according to Reference [11]. (From Fourie, A.P.C., Givati, O., and Clark, A.R., Simple technique for the measurement of the transfer impedance of variable length coaxial interconnecting leads, *IEEE Transactions on EMC*, 40(2), May 1998. © 1998, IEEE.)

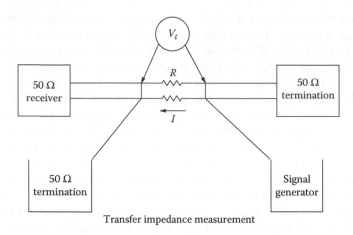

FIGURE 7.51   Transfer impedance measurement setup for the cable under test.

FIGURE 7.52   Photo of the cable under test with copper box on either end and copper tube and injection cable with ferrites.

At low frequency, the transfer impedance of a cable is equal to the DC resistance.

The DC resistance of the 0.5 m of CUT into which the current is injected was measured at 14.8 Ω, and the measured transfer impedance of the cable at 100 Hz is very close at 13.9 Ω.

Two methods exist for the calculation of transfer impedance. One is to measure the input and output power transfer function, using the characteristic impedance of the injection circuit and the CUT circuit. This method takes into account the injection cable insertion loss and the CUT loss. The other one is an alternative, that is, to measure the voltage developed across the 50 Ω termination at the end of the injection current circuit. From this voltage, the injection current can be calculated. The far-end transfer impedance is then given by either the open-circuit voltage developed at the far end of the CUT or the voltage developed across 50 Ω multiplied by two, assuming that both the CUT impedance and load and source impedances are 50 Ω.

The far end is defined as the end furthest away from the injection current point of the CUT.

Both methods were used to calculate the transfer impedance with, as expected, identical results when the injection circuit insertion loss was taken into account.

FIGURE 7.53 Measured transfer impedance of conductive cloth-shielded cable. Plot from 0.0001 to 2 MHz is DC resistance. Plot from 1 to 1000 MHz is the corrected measurements made into 50 Ω load.

The measured transfer impedance from 100 Hz to 1 GHz is provided in Figure 7.53, from which we can see that the value is incredibly high.

The surface resistivity of the cloth is specified as 0.1 Ω^2, and this is exactly the value measured at EMC Consulting. Due to the skin-depth effect, one might expect the SE to increase above approximately 10 MHz. One reason this does not happen may be that although the material has no visible gaps, it is woven and a similar effect to the porpoising coupling seen in braided shielded cables may occur.

Also in the manufacturing of the cloth, it has a longitudinal seam, similar to a foil shielded cable, which is folded over and stitched. And as this is the weak link in the foil cable, so it may be in the cloth cable.

In the construction of the cables, great care was taken to ensure that the injection braid was not directly over the seam.

The equation for transfer impedance in Section 7.4 was used to evaluate the SE is provided by:
Thus, from 10 kHz to 1 MHz, the SE is no more than 10.6 dB.

The hypothesis is that the increase in transfer impedance above 150 MHz is almost certainly due to the increasing impedance of the shield material that is not offset by any attenuation due to skin-depth effect. Thus as the increasing shield impedance reduces the injection current, the voltage drop across the transfer impedance, which is the transferred voltage, does not decrease and the transfer impedance increases.

The transfer impedance of this type of cable is extremely high, but it can be reduced by increasing the diameter of the cable, which reduces the DC resistance. Also at low frequency, a drain wire can be used to further reduce the DC resistance of the shield and the injected current then flows predominantly on the drain wire until its inductive reactance is higher than the DC resistance of the cable shield. The argument can be made that the conductive cloth is redundant and that a drain wire connecting the backshells of two connectors can also provide some shielding when compared to no drain wire.

A practical 1 m long cable was made with a drain wire, and a diameter of 28 mm compared to the 3 mm diameter of the CUT was tested here. The overall SE of this cable when connected to shielded connectors with backshells was 30 dB at 0.2 MHz and 20 dB at 10 MHz. The SE of the CUT is described here at 0.2 MHz (given by 20 log (($2 \times R_o/Z_t) \times l$)) = 20 log ($2 \times 50$ Ω/30 Ω/m) × 1 m = 10.46 dB. The difference in the diameters of the two cables in dB is 20 log (28 mm/3 mm) = 19.4 dB, and the larger diameter cable will have a lower DC resistance. As the cable contains a drain wire, the shield resistance is in parallel with the inductive reactance of the drain wire. If the drain wire reactance is ignored and the CUT-measured SE is corrected only for the increased diameter, the predicted attenuation is 10.46 dB + 19 dB = 30 dB, which is equal to the measured attenuation of the practical 1 m long cable. This shows that the SE at low frequency is strongly dependent on the DC resistance of the cable.

As the cable would be used as an overshield and any signals on the center conductors enclosed by the shield would also contain the signal return, the attenuation is of the C/M current induced into the center conductors.

Although this practical cable shows an improvement over the CUT described here, it is still a long way from the SE achieved by a Belden tin–plated overbraid, which was measured at 71 dB at 0.25 MHz, 58 dB at 10 MHz, and 44d B at 100 MHz.

The SE of the conductive cloth is specified at 83–60 dB over the 200 MHz to 18 GHz frequency range using the MIL-STD-285 test method. This measurement was almost certainly made using a plane-wave source. Intuitively, if a field with a wave impedance of 377 Ω is incident on a material of infinite dimension with a surface resistivity of 0.1 Ω^2, approximately 71 dB of SE can be expected, as the E-field component of the plane wave is reflected. However, if an H field with a wave impedance of 0.1 Ω is incident, the SE may be closer to 0 dB. Using the surface resistivity of the cloth of 0.1 Ω^2, the predicted SE, from measurements at 1 MHz, is only 1.4 dB.

The surface transfer impedance cable test induces a current flow on the shield of the cable and effectively measures predominantly its magnetic field attenuation, and as expected this is low.

7.3 SEMIRIGID CABLE

Semirigid cable is constructed of a solid metal outer sheath, normally copper. The inner insulation is either solid or air with spacers used to support the inner conductor. The cable either is hand-malleable or must be bent by machine. Connectors are soldered directly to the ends of the cable or crimped on.

This type of cable is the best available for the prevention of EMI at high frequency. When properly soldered at the cable-to-connector interface, the only source of ingress or egress of radiation is at the connector mating, due to the transfer impedance of the connector-to-bulkhead interface. Manufacturers offer a quick-connect type of semirigid connector, in which the cable is crimped to the connector, thus adding an additional transfer impedance. In common with all shielded cables, the low-frequency performance is dependent on the thickness and conductivity of the shield and the concentricity of the cable. A new form of solid shield is the "semiflexible" cable, which is a braided shield in which the shield braids have been soldered together. The cable is much more flexible than the semirigid variety, with almost the same SE, but is less flexible than a small-diameter braided cable.

7.4 LONG-LINE EFFECTS

From Equation 7.14, it is seen that when the transfer impedance is dominated by aperture coupling, transfer impedance increases monotonically with increasing frequency. From Equation 7.12, it appears that the transferred voltage is proportional to the transfer impedance, the shield current, and the length of the line and, where Equation 7.14 applies, to frequency; however, this is not true when the line length is greater than 0.5λ. Such a cable is termed *electrically long*. The wavelength of a wave in air is equal to 300/f (MHz), whereas the wavelength of the same wave in a material with a relative permittivity greater than 1 is $300/\left(f(\text{MHz}) \times \sqrt{\varepsilon_r} \right)$. When the shield of a cable is terminated to ground at both ends, the cable is resonant at frequencies for which the line length equals ($k\lambda/2$), where k is an integral multiple, that is, 1, 2, 3,....

The transfer voltage for an electrically long cable in which the shield is terminated to ground at both ends with the characteristic impedance of the cable above ground, derived from Reference [7], is given by

$$V_t = I_s Z_t \frac{\sin \theta}{\theta} \; [\text{V}]$$

(7.15)

where

Z_t is the transfer impedance [Ω/m]
I_s is the shield current [A]

and

$$\theta = \frac{2\pi f}{300} \left(\sqrt{\varepsilon_r} + 1 \right) \frac{l}{2} \; [\text{radians}]$$

where

f is the frequency [MHz]
ε_r is the relative permittivity of the cable sheath
l is the length of cable [m]

Figure 7.54 plots the ratio of $\sqrt{\left(\sin \theta / \theta \right)^2}$, in decibels, for a 20 m long cable with a relative permittivity of 2.2. From Figure 7.54, an envelope reduction of 6 dB per octave or 20 dB per decade is seen, that is, a monotonic decrease with increasing frequency above the first resonance frequency of the cable. Equation 7.14 predicts a monotonic increase in transfer impedance with increasing frequency, so the envelope of the transfer voltage, above the first resonance frequency, should be constant with frequency. One definition of the SE of a cable is the ratio of the shield current I_s to the core (center conductor/s) current I_c with the cable terminated, between core and shield, in its characteristic impedance.

The measured voltage, in decibels referenced to the test fixture voltage of a 1.18 m long RG/58A cable from 1 MHz to 1 GHz, is shown in Figure 7.55. The center reference graticule is −75 dB,

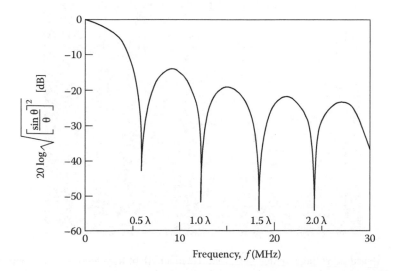

FIGURE 7.54   Plot of 20 log/*R* ((sin θ/θ)²) for a 20 m cable with a sheath relative permittivity of 2.2.

FIGURE 7.55 Measured voltage for a 1.18 m length of RG/58A cable, in decibels, referenced to the test fixture input voltage, as a function of frequency on a log scale. The center reference graticule is at −75 dB, corresponding to a shielding effectiveness of 69 dB. The first marker is at 10 MHz; all other markers are at 100 MHz steps.

which corresponds to an SE of 69 dB. The graticule is graduated 10 dB per division. As the measured voltage increases by 10 dB above the reference graticule, the SE decreases by 10 dB. Thus, at 300 MHz, the SE from Figure 7.55 is 59 dB.

Figure 7.56 plots the measured voltage, in decibels, referenced to the test fixture input for a Raychem-shielded twisted-pair cable 10595-24-2-9 of 1.2 m length. The center graticule is −70 dB, which is equivalent to an SE of 51 dB.

FIGURE 7.56 Measured voltage, in decibels, for a 1.2 m length of Raychem-shielded twisted-pair cable 10595-24-2-9. The center reference graticule is 70 dB, which is equivalent to a shielding effectiveness of 51 dB.

7.5 TRANSFER ADMITTANCE

Sections 7.2.1 and 7.2.2 briefly discussed the transfer admittance of shielded cables. In most cases, the transfer admittance is a secondary phenomenon that can be ignored. When a cable is electrically long, a current is set up on the cable regardless of whether the shield is terminated to ground or not. For an electrically short cable with the shield connected to ground at both ends, an incident field will, depending on the angle of incidence, set up a current flow on the cable. The worst-case figure in these cases is the transfer impedance alone. However, for an electrically short cable on which a high-impedance field is incident, the cable current may be low and the coupling may be via the charge that appears on the cable. This also applies to an electrically short cable connected to an enclosure that is disconnected from the ground. Another situation would be an electrically long cable over which a field in the gigahertz frequency range is incident over a small area of the cable. This would in practice occur when a radar beam is incident on a section of cable. At gigahertz frequencies, the predominant coupling may be of the localized field through the apertures in the shield.

Transfer admittance is not an intrinsic EM parameter, for it is a characteristic of the cable and its surroundings (the test setup in the case of measurements). The through elastance K_T, which is a property of the cable alone, can be used to characterize coaxial cables, and it can be related to the transfer admittance Y_T by

$$K_T = \frac{Y_T}{j\omega C_1 C_2}$$

where
C_1 is the per-unit capacitance of the outer circuit (test fixture to cable shield)
C_2 is the per-unit length capacitance between the two conductors of the coaxial cable

Obviously, Y_T is dependent on the measurement setup, because C_1 will differ from one setup to the other.

Reference [14] describes through-elastance measurements and provides the following conclusions: On coaxial cables of any length and matched at both ends, that is, the cable matched to its characteristic impedance between the center conductor and the shield at both ends, the transfer admittance can be neglected up to 6 MHz and cannot even be measured. However, a cable connected at both ends to a high impedance and submitted to a high-impedance field would be more susceptible to transfer admittance coupling.

In the measurements described in Reference [14], the K_T measurements are very close to the Z_T measurements, and the transfer impedance coupling may be the only one of concern.

7.5.1 SHIELDING EFFECTIVENESS OF LONG CABLES

The transfer impedance data have been applicable to cables 0.5–2 m in length from which the SE can be derived as shown in Equation 7.3, but the SE of longer cables is also of importance. Belden has measured the SE of cables of 30 m length up to 100 MHz and compared this to cables of 1.5 m length using a triaxial test fixture. Reference [28] describes in detail the test fixture and test method. In summary, the input power P1 to the fixture is compared to the output power using a normalization formula.

Figure 7.57 shows the electrical regions of the test chamber and Figure 7.58 the test setup.

As the test chamber is resonant, an envelope is drawn connecting the resonant peaks in the power ratio versus frequency.

Tests were made in a 1.5 m test fixture as well as the 30 m test fixture. The test samples used were either RG-6 or RG-59, and the design is described in Table 7.2 used with Table 7.3.

The test results obtained in the 1.5 m test fixture from 100 to 1000 MHz are shown in Figure 7.59 and from 1 to 100 MHz in Figure 7.60.

A comparison of SE at a 1.5 m length compared to 30 m is shown in Figures 7.61 through 7.64.

FIGURE 7.57 Electrical regions of the test fixture. (Reproduced with kind permission of Belden cable.)

FIGURE 7.58 Test setup. (Reproduced with kind permission of Belden cable.)

TABLE 7.2
Cable Shield Design Data

Cable Number	Foil Type Inner (See Table 7.3)	Braid Coverage (Inner) (%)	Foil Outer (See Table 7.3)	Braid Coverage (Outer)	Shield DCR (mΩ/m)
1	—	95	—	—	9
2	a	60	—	—	31
3	a	80	b	—	15
4	a	60	a	40%	17

Source: Reproduced with kind permission of Belden cable.

TABLE 7.3
Shielding Tape Design

Foil Type	Layer Thickness (mm)			
	Al Foil	Polyester	Al Foil	Width (mm)
A	0.00889	0.02286	0.00889	19.05
B	0.0254	0.02286		25.4

Source: Reproduced with kind permission of Belden cable.

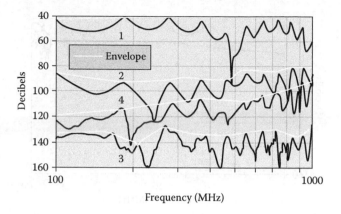

FIGURE 7.59   Shielding effectiveness of cables 1–4 from 100 to 1000 MHz. (Reproduced with kind permission of Belden cable.)

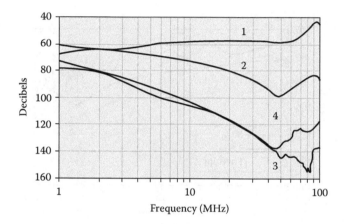

FIGURE 7.60   Shielding effectiveness of cables 1–4 from 1 to 100 MHz. (Reproduced with kind permission of Belden cable.)

FIGURE 7.61   Shielding effectiveness of 1.5 and 30 m lengths of cable #1. (Reproduced with kind permission of Belden cable.)

FIGURE 7.62   Shielding effectiveness of 1.5 and 30 m lengths of cable #2. (Reproduced with kind permission of Belden cable.)

FIGURE 7.63   Shielding effectiveness of 1.5 and 30 m lengths of cable #3. (Reproduced with kind permission of Belden cable.)

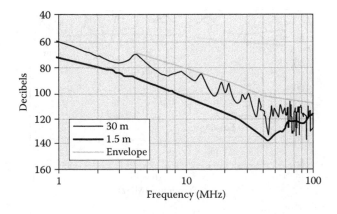

FIGURE 7.64   Shielding effectiveness of 1.5 and 30 m lengths of cable #4. (Reproduced with kind permission of Belden cable.)

From the figures, it can be seen that the difference between the worst-case minimum SE is up to 40 dB, and so if the SE of long cables is required, it is not enough to extrapolate from the data for short cables. Accurate measurements on long cables are difficult due to the very long test fixture required. If an approximate measurement is required, it is possible to lay a length of cable out in a noninductive zigzag at a height above a ground plane that results in a known characteristic impedance, say of 50 Ω. The shield of the cable at the far end is then terminated to the ground plane with a resistor having the same value. The center conductor(s) of the cable are terminated at the far end using a shielded load resistor having the same resistance as the cable impedance. A resistance with the same value is connected between the center conductor and the shield at the near end. The output of the oscillator is applied between the ground plane and the shield at the near end. An oscilloscope with a differential input or CHA – CHB or CHA + CHB inverted capability is connected with one probe on the shield at the near end and the other probe on the center conductor(s) at the near end. A current probe is then placed around the shield at the near end. The oscillator frequency is then varied until the current probe measures a peak level. The current probe should also be moved on the cable to ensure the current is at resonance. The approximate transfer impedance is given twice the transferred voltage divided by the current:

$$Z_t = \frac{2V_t}{I}$$

The oscilloscope will have a high C/M voltage applied due to the oscillator output. Therefore, it is important to connect the CHA and CHB probes together at the injection point of the shield. If the C/M voltage is the same or higher than the transferred voltage, then the test is invalid. This test of the C/M rejection capability of the oscilloscope should be made before constructing the cable test fixture.

7.6   SHIELD TERMINATION EFFECTS ON TRANSFERRED VOLTAGE

One of the most common errors made in the use of shielded cables is to connect the shield to the backshell of the connector or to the enclosure by a "pigtail" (either a length of braid or, more commonly, a length of wire). Alternatively, the pigtail may be connected through a connector pin and then to the inside of the equipment enclosure/chassis. As we will see in Section 7.13 on radiated emissions from cables, the use of a pigtail to connect the shield at both ends of the cable results in emissions close to an unshielded cable. Likewise, the EMI voltage induced in a shielded cable with pigtail connections approaches that of a two-wire unshielded cable. The noise voltage developed across the transfer impedance by the shield current flow I_s is effectively in series with the signal, V_{sig}, in the configuration of Figure 7.65.

If the shield is terminated at the backshell of the connectors and via the connector to the chassis, then the shield current flow I_s is effectively isolated from the signal path, with the exception of current

FIGURE 7.65   Common-mode current inducing a noise voltage (V_t) into the signal.

that diffuses through the cable, couples through the cable apertures, and diffuses through the transfer impedances at the connector. However, if the shield is connected via a length of wire to the chassis and the signal path continues from that point to the load resistor R_L, as shown in Figure 7.66, then the shield current I_s develops a noise voltage across the inductance of the length of wire V_w. Assume a shield current flow of 1 mA (this current may be due to the C/M voltage source as shown or to an incident EM wave) and a 26-gauge wire 2 in. long connecting the shield via a pin in the connector to chassis. Assume also that a moderately effective foil/braid shield is used and that the 26-gauge wire is 1 in. above the chassis and therefore has an inductance of 0.056 µH. Table 7.4 shows the two noise voltages V_t and V_w, the cable transfer impedance, and the inductive reactance of the wire versus frequency.

Table 7.4 illustrates how important it is to terminate the shield to the backshell of the connector. Had a second wire been attached between the shield and the load resistor, then V_w would be effectively isolated from the signal; however, the 2 in. length of wire carrying I_s is still capable of radiating noise to internal circuitry. Also, when the second connection between the load resistor and the shield exists and the load resistor is connected to a signal ground, which is connected at some other location to the enclosure or to chassis ground, then some fraction of the noise current is injected into the signal ground. The optimum connection of a shield is at 360° around the circumference of the braid to a metal backshell. EMI backshells achieve the desired low-impedance connection, typically by clamping the end of the shield around the circumference to the backshell. Backshells are available from Glenair that allow the connection of the shields of a number of cables, using the same connector, to the backshell. Where a connector must be used that does not allow the clamping of the shield to the backshell, the braid may be terminated by one of the following techniques: soldering the braid to the backshell, the use of conductive adhesive (note that the conductive adhesive

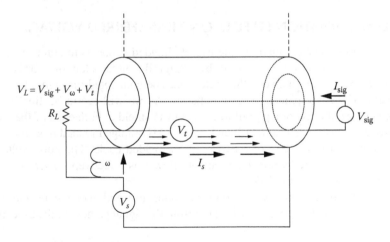

FIGURE 7.66　Shield connection extended into the enclosure.

TABLE 7.4
Noise Voltages V_t and V_w, Cable Transfer Impedance and Inductive Reactance of the Wire versus Frequency

Frequency [MHz]	V_t [µV]	V_w [mV]	Z_t [mΩ]	Z_w [Ω]
0.1	70	0.035	70	0.035
1	50	0.350	50	0.350
10	20	3.5	20	3.5
100	20	35	20	35
300	70	350	70	350

cannot be used as strain relief; use instead a cable clamp or a thin bead of nonconductive epoxy), and the insertion of a canted coil spring between the cable clamp nut and the bared braid, hose clamp. One supplier of a suitable spring is Bal Seal Co., Santa Ana, California. An alternative backshell is a shielded heat-shrink boot, which contacts the connector and the cable shield as the boot is heat-shrunk. The supplier and transfer impedance of a heat shrink boot are provided in Section 7.15.

7.7 COUPLING FROM *E* AND *H* FIELDS

The coupling into a shielded or an unshielded wire disconnected from ground and far removed from a ground plane was examined in Section 2.5.4. Chapter 2 also described the receiving properties of a loop far removed from a ground plane (Section 2.2.5). In this section, we will examine the current flow induced in either an unshielded wire/cable or a shielded wire/cable located above a ground plane. Provided that the height of the wire above the ground plane, h, is much less than the length, l, and $h \ll \lambda$, the single wire can be considered as a two-wire transmission line formed by the wire and its EM image in the ground plane. The terminating impedances of the transmission line are Z_0 and Z_1 and may be close to $0\ \Omega$ when the line is a shielded cable with both ends connected to the ground plane. The characteristic impedance of the line is given by

$$Z_c = \sqrt{\frac{\left(Z_i + j\omega l_e\right)\omega l_e}{jk^2}} \tag{7.16}$$

where

Z_i is the distributed series resistance of the two-wire line [Ω/m]

$k = \omega\sqrt{\mu_o \varepsilon_o} = \dfrac{2\pi}{\lambda}$

$l_e = \dfrac{\mu_o}{\pi} \ln \dfrac{2h}{a}$

$\mu_o = 4\pi \times 10^{-7}$ [H/m]

a is the diameter of wire [m]

When $l \gg h$ and $h \ll \lambda$, the transmission-line theory is expected to be accurate. However, when $hl \ll \lambda$, the circuit theory using the inductance of a rectangular loop of wire will provide a more accurate solution. The impedance of the loop (with image), including termination resistances, is given by

$$Z = \sqrt{2\left(Z_0 + Z_1\right) + \left(j\omega L\right)^2} \tag{7.17}$$

The inductance of a rectangular loop of wire L with image is

$$L = l \ln \frac{4hl}{a(l+d)} + 2h \ln \frac{4hl}{a(2h+d)} + 2d - \frac{7}{4}(l+2h)\,\mu H \tag{7.18}$$

where $d = \sqrt{(2h)2 + l^2}$. When $l \gg h$ and $h \gg a$, then $L = l_e l$.

Equation 7.18 multiplies the physical height h above the ground plane by 2 to account for the EM image and may be used to find the inductance of a wire loop by changing $4h$ to $2h$ and $2h$ to h. The current in the termination impedance $Z_0 = I_0$ and $Z_1 = I_1$. These currents are

$$I_0 = I_1 = \frac{-j\omega 4hB_{in}}{\sqrt{2\left(Z_0 + Z_1\right)^2 + \left(\omega L\right)^2}} \tag{7.19}$$

B_{in} is the incident magnetic flux density, and the illumination is considered as causing worst-case maximum coupling. Maximum coupling occurs for a plane wave when the plane of incidence is coincident with the plane of the loop, with the magnetic field perpendicular to the plane of incidence, as shown in Figure 7.67.

The magnetic flux density B_{in} is related to the magnetic field H by the permeability of free space μ_0 ($4\pi \times 10^{-7}$ [H/m]), that is,

$$B = \mu_0 H$$

When $Z_0 = Z_1 = 0\ \Omega$, which is true for a shielded cable and where the transmission-line equation applies, then

$$I_0 = I_1 = \frac{4E_o h}{Z_c} \tag{7.20}$$

E_o may be found from H by the relationship $E = H \times Z_w$. The angle of incidence of the EM wave does not affect the magnitude of the induced current; however, the magnetic field vector must be perpendicular to the plane of the loop for the assumption of maximum coupling to remain valid. Figure 7.68 shows the comparison of the results of calculations using transmission-line theory and from a computer program (the Numerical Electromagnetic Code, NEC).

Resonance can be seen at 7.5 and 15 MHz, and these correspond to the frequencies where $\lambda/2$ and λ equal the line length of 20 m. Other resonances occur at $l = k\lambda/2$, where k is an integral multiple.

FIGURE 7.67 Magnetic field induction in a transmission line formed by a wire over a ground plane.

FIGURE 7.68 Termination current for a rectangular loop oriented normal to a perfect ground plane and illuminated by a plane wave. (From Degauque, P. and Zeddam, A., Remarks on the transmission line approach to determining the current induced on above-ground cables, *IEEE Transactions on Electromagnetic Compatibility*, 30(1), February 1988. © 1998, IEEE.)

Transmission-line theory does not predict these resonances, though they do exist. The resonances may be calculated based on the termination impedance of the transmission line. In practical shielded cables, the impedance is predominantly the impedance of the vertical section of cable as it terminates to the enclosure. Less typically, the cable does not bend up or down as it connects to the enclosure, in which case the termination impedance is the impedance of the enclosure to ground plane connection, which may be confined to the impedance of the enclosure wall. This impedance may be found from the impedance of a ground plane as described in Section 5.1.2. The impedance of the vertical section of the transmission line is given by

$$Z_v = 60\left(\ln \frac{2h}{a} - 1 \right) \tag{7.21}$$

where
   a is the radius of the vertical section
   h is the height

If the incident plane wave has a vertical *E*-field component, then some coupling to the end sections will occur. However, because we assume that the length of the transmission line is much greater than the height of the line above the ground plane, we ignore any contribution from coupling to the vertical section of the line. Theoretically, when the termination impedance is zero, the termination current is infinitely high. In practice, some terminating impedance will always exist. Figure 7.69 (from Reference [16]) shows the predicted cable current amplitude for a 1 V/m *E* field incident on a 20 m long line of radius 5 mm and at a height of 50 cm above the ground plane. The cable current was predicted by the use of method-of-moments (MOM) antenna theory and by a modification to transmission-line theory described in Reference [16]. We see that as

FIGURE 7.69 Cable current with 0, 1, 10, and 100 Ω termination impedances. (From Degauque, P. and Zeddam, A., *IEEE Trans. Electromag. Compatibil.*, 30(1), February 1988. © 1988, IEEE.)

the termination impedance approaches the characteristic impedance of the cable above a ground plane, the maximum current reduces to that predicted by the transmission-line approach. Often in radiated susceptibility tests, such as the MIL-STD-461 RS03 test where the interconnection cables are located at a height of 5 cm above a ground plane, the equipment is found to be susceptible at frequencies corresponding to the resonant frequency of the interconnection cables. This is especially true for shielded cables with a low-impedance connection of the shield at one or both ends to the ground plane. A simple test, which may be used to verify shielded-cable resonance susceptibility, is to short the shields of the interconnection cables to the ground plane at some distance along the length of the cable and thus shift the resonance frequency. The use of ferrite beads or baluns at the ends of shielded cables will often increase the termination impedance and reduce the cable resonance currents sufficiently to achieve EMC. When this is impractical, the type of shield or number of braids may be changed to increase SE. As described in Section 7.2, there is a limit to the level of shielding achievable in a flexible cable, and in some instances, the only solution is to increase the immunity of the equipment.

Transmission-line theory or the inductance of a loop may be used to find the shield current flow on a shielded cable. The transfer voltage is then found by the use of Equation 7.15. The $(\sin \theta)/\theta$ function of Equation 7.15 is valid for a cable with a matched shield termination only. Therefore, when an unmatched termination exists, the transferred voltage should be multiplied by a factor that accounts for the increase in shield current at resonance. When the cable shield is terminated at both ends to the enclosure, a nominal multiplication factor of 5 may be used. To more accurately account for resonance and termination effects, softwares such as GEM ACS, NEC, MININEC, and a commercially available Radiated Immunity software from EMI Software can be used to model cables above a ground plane, typically as a two-wire transmission line formed by the cable and its EM image in the ground plane. In one example, a cable of length 2.5 m located 5 cm above a ground plane was modeled using GEMACS and the EMI Software Radiated Immunity software. Also, the simple transmission-line equation (Equation 7.20) was used to predict the current flow, but this does not account for cable resonances, which are significant when the termination impedances are low. The termination impedances at both ends of the cable, which represent a typical short circuit, were set at 50 nH in series with 2.5 mΩ. This would be the case for a shielded cable with the shield terminated to the ground plane at each end. The transmission-line and termination impedances were modeled in GEMACS using wire segments and an MOM analysis, with a plane wave incident on the structure. The current flow on the cable with a horizontally polarized field broadside onto the cable was analyzed, as well as a vertically polarized sidefire field, that is, illuminating the vertical end section. In either case, it was the maximum current in any of the wire segments down the length of the cable that was recorded. The same analysis was performed using the EMI Software Radiated Immunity program, which provided a plot of cable current in the cable versus frequency so that the current at resonant and antiresonant frequencies could be seen. It was not clear in the use of this program if the predicted current was an average value or a peak value. The peak current at the first resonant frequencies of 60 and 180 MHz and an incident field of 40 V/m were analyzed.

In addition, cable currents at 1 and 4 GHz and an E field of 3500 V/m, which would be representative of illumination by radar, were analyzed.

The current flow using Equation 7.20 for the horizontal, broadside on, field was also predicted. The height of the cable above the ground plane must be much less than λ for this equation to remain valid, so the prediction at 4 GHz was omitted. Also, Equation 7.20 does not account for current flow at resonance, and this is significant at 60 and 180 MHz, where the termination impedances were lower than the transmission-line impedance. However, at 1 GHz, the termination impedances are high and the transmission-line equation predicts a higher current flow compared to that calculated by GEMACS or the EMI Software Radiated Immunity program. The results of the predictions are shown in Table 7.5, with a good correlation between the GEMACS program and the EMI Software program.

TABLE 7.5

Comparison of Cable Currents Induced by a Plane Wave Incident on a Two-Wire Transmission Line, Using GEMACS, Commercial Radiated Immunity Software, and Equation 7.20

f (MHz)	*E* Field (V/m)	Field Orientation	Current Using TX Line Equation 7.20 (mA)	GEMACS Peak Current (mA)	Radiated Immunity Software (mA)
60	40	Vertical (sidefire)	—	6.1	7
60	40	Horizontal (broadside)	43	320	150.
180	40	Vertical (sidefire)	—	7.4	5.6
180	40	Horizontal (broadside)	43	90	65
1000	3500	Vertical (sidefire)	—	860	500
1000	3500	Horizontal (broadside)	3760	1800	1200
4000	3500	Vertical (sidefire)	—	600	110
4000	3500	Horizontal (broadside)	—	630	300

7.8 USE OF THE NEC PROGRAM TO MODEL THE COUPLING TO CABLES IN, OR CLOSE TO, FREE SPACE CONDITIONS

It would appear that cables in free space would be a rare occurrence. However, a number of examples come to mind.

A helicopter was used to carry a monitoring device at the end of a long cable. The device was used to monitor reflections from the ground, and the cable shield was terminated at the helicopter and at the device. When the cable was vertical, a local AM radio transmitter induced enough current on the shield of the cable to make the device malfunction.

In the design of a proposed Hubble rescue vehicle, a robotic grapple arm made of carbon fiber booms, which carries three cables approximately 24 m (78 ft) long, was part of the design. Close to an isolated cable, situation will exist if the carbon fiber of the boom is nonconductive and has low permittivity, and no additional foil ground plane is used to cover the boom. In reality, the cables in the arm are connected at one end of the equipment that is grounded to the vehicle.

The following measurements also included the attenuation achieved by adding a 20 cm wide foil under the shields of the cables with the shields terminated to the foil at both ends.

Using the NEC provides several advantages over building a physical model to study current coupling. If the predictions made by the software can be believed, then a great deal of effort in building and time can be saved. A computer model would also provide a quick method for tweaking a physical model to achieve better results, without having to start from scratch building a new one. In addition, a valuable application of NEC is in modeling conditions in space, which would be difficult to simulate on earth. Therefore, it is useful to investigate the validity of predictions made by NEC.

With a single cable, only C/M currents are induced. If two or more shielded cables exist and the shields terminate on conductive enclosure at both ends, then the cables form a short-circuited transmission line. The current induced into multiple cables is dependent on the angle of incidence of the wave and can contain C/M and D/M currents. When the difference in phase of the incident field on the cables is zero, then the D/M current is zero as is the termination current. When a difference in phase exists, then a D/M current flows.

For cables over a ground plane with finite width versus no ground plane, some reduction in induced cable current is expected but not as much reduction for a large-dimension ground plane. The reduction based on measurements and the NEC computer code analysis are described later for both sidefire illumination and illumination from above.

A finite ground plane made from aluminum foil has been placed on carbon fiber manipulator arms including the proposed Hubble grapple arm.

The worst case for C/M cable current is when the cables are very far removed from ground and are exactly 0.5l long and the angle of incidence is sidefire, that is, the E field is normal to the cable(s).

That is, as long as the E field is parallel to the cable, maximum C/M current flows. For D/M coupling, the maximum occurs for sidefire or endfire with theta = 0°, as shown in Figure 7.70. When theta = 90°, no D/M current flows; only C/M current flows and it divides between the cables. Close to the isolated cable, situation will exist if the carbon fiber of the arm is nonconductive and has low permittivity.

In the case of the grapple arm, it is possible that the EM environment is characterized by the MIL-STD-461C level of 10 V/m at 6.25 MHz and that it illuminates the arm down its length, but this seems unlikely. It is much more likely that the field is localized and this is discussed later.

The NEC in the 4NEC2D version by Arie was used to predict the worst-case coupling of a plane wave to a resonant 24 m cable for maximum C/M current. Under these conditions and with a 10 V/m incident field, the predicted current flow on a 75 ft (24 m) cable at the resonant frequency of 6.25 MHz (at this frequency the arm is 0.5l long) is 1.83 A.

In addition, measurements were made on a 6.8 m long cable, as well as an NEC program was used to predict the current.

A 6.8 m braid wire was laid out above a set of trestles about 1 m above ground. Approximately, 18 m from the center of the wire, a biconical transmitting antenna was set up in order to produce a close-to-uniform field along the length of the wire, simulating a plane wave. The measurements were made at typical temperatures of −24° in February, and the ground was poorly conducting as it was frozen to a depth of between 1 and 1.25 m. A Fisher current probe was placed at the center of the braid wire to measure the current induced at that location. Many ferrite baluns and tiles were

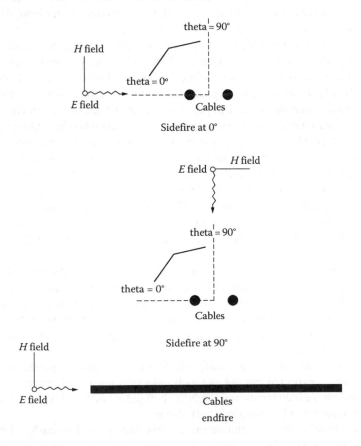

FIGURE 7.70   *E*-field incident angle on the cable.

placed on the cable running from the probe to the measurement equipment to reduce the pickup on the cable from the transmitted field.

During the test, the transmitting antenna swept the 20–260 MHz range using the same input level throughout. The current measured at the current probe was recorded. Later, the field from the transmitting antenna was calibrated and used to calculate the current for a uniform 1 V/m field.

The frequencies at which peak currents are predicted are 0.5λ, 1λ 1.5λ, 2λ…. For the 6.8 m cable over the frequency range of the measurement and NEC analysis, this corresponds to 22 MHz, 44 MHz, 66 MHz, 88 MHz, 110 MHz, 132 MHz, 154 MHz, 176 MHz….

The test setup for this experiment was designed to simulate the NEC model of a 6.8 m wire in free space, illuminated by an incident plane wave along its length. Figure 7.71 shows the NEC geometry for this model. Using the excitation command in NEC, it is assumed that the field applied on the model by the plane wave is 1 V/m and it is sidefire.

The frequencies chosen in the NEC analysis were tweaked until the excitation command generated a peak current at the center of the wire model. A plot of the currents on the wire at the frequencies for peak currents can be found in Figure 7.72.

NEC-predicted peak currents were compared to the measured peak currents at the center of the wire.

FIGURE 7.71   Numerical Electromagnetic Code geometry for the 6.8 m wire in free space.

FIGURE 7.72   **(See color insert.)** Current on the braid wire at the frequencies of interest.

FIGURE 7.73 Numerical Electromagnetic Code predicted current and measured current versus (a) frequency and (b) wavelengths.

Figure 7.73 shows a plot of measured current versus predicted current. The general trend in the peak currents predicted by NEC is similar to that measured in this experiment. If we compare the measured peak at 99.3 MHz to the NEC-predicted peak at 109 MHz, we find an error of only 2.85 dB. Comparing the measured peak at 20 MHz to the NEC-predicted peak at 21.2 MHz, we get an error of 2.04 dB. However, if we compare the measured peak at 175 MHz to the NEC-predicted peak at 197 MHz, we find an error of 11.1 dB, and the measured peak is closest to the theoretical 176 MHz.

We find the NECs' predictions for the magnitude of peaks experimentally good, if we assume an anomaly at 175 MHz during the measurements. The major difference between NEC-predicted current flow on the wire and measured current flow on the wire is the frequency at which we find the peaks. While NEC predicts a peak at 109 MHz close to the 110 MHz for 2.5λ, measurement finds this peak at 99.3 MHz or 2.3λ. NEC also shifts the measured peak at 175 MHz or 4λ to 197 MHz or 4.5λ. After obtaining measurements on the open-area test site (OATS), the NEC analysis was run again at the frequencies where maximum current flow was measured.

The measured currents are shown in Table 7.6 and the NEC-predicted currents in Table 7.7.

An analysis of NEC-predicted current induced into a 12 m cable with and without a ground plane and with a biconical antenna 1 m from one end was performed and compared to measurements. The NEC models are shown in Figure 7.74.

The biconical antenna and the cable were constructed from wires and the finite ground plane by the use of patches.

The current predicted as a result of the NEC analysis was very much lower than the measured one, and the NEC results are not believable.

TABLE 7.6
Measurement Data for the 6.8 m

Frequency (MHz)	Wavelengths (λ)	Current (mA)	Field Applied (mV/m)	Current at 1 V/m (mA)
20	0.5	0.11	2.21	49.05
28.5	0.6	0.05	5.53	8.96
37	0.8	0.04	8.93	4.78
43.6	1.0	0.03	7.98	3.45
60.3	1.4	0.21	26.22	7.97
99.3	2.3	0.61	58.85	10.36
110	2.5	0.27	82.22	3.27
130	2.9	0.28	114.82	2.45
139.7	3.2	0.39	114.82	3.43
175	4.0	0.92	70.83	13.03
204	4.6	0.40	85.01	4.68
229	5.2	0.37	97.72	3.80
249	5.6	0.40	102.09	3.90

TABLE 7.7
Numerical Electromagnetic Code Predictions
for Current at the Center of the 6.8 m Wire

Frequency (MHz)	Wavelengths (λ)	Current (mA)
21.2	0.5	62.07
28.2	0.6	11.59
42.9	1.0	5.31
59.4	1.3	6.86
66.4	1.5	9.43
82	1.9	0.02
97	2.2	0.45
109	2.5	7.46
147	3.3	2.91
153	3.5	4.64
180	4.1	0.12
197	4.5	3.63
240	5.4	2.79
250	5.7	0.27

The measured currents due to the biconical antenna at 1 m distance from the 12 m long cable with no ground plane are shown in Table 7.8.

Adding the finite ground plane does reduce the cable-induced current and the measurements can be used to justify the use of aluminum foil on the carbon fiber manipulator arm.

Table 7.9 compares the reduction in current due to the ground plane in measurement and the NEC-predicted values for theta = 0° and Table 7.10 for theta = 90°.

The frequencies correspond to 0.5λ, 2λ, 4λ, and 8λ.

The conclusion is that NEC is useful in predicting the currents induced into a cable by an incident plane wave down the length of a cable but not for a localized field where a measurement should be used.

FIGURE 7.74 Numerical Electromagnetic Code model of biconical antenna 1 m from a 12 m long cable. Hub3d.out is the cable with no ground plane and theta = 0°, hub3c.out is the cable with a ground plane 5 cm below the cable, and hub3fn.out is the illumination with the ground plane and theta is 90°.

7.9 CABLE SHIELDING EFFECTIVENESS IN GIGAHERTZ

At gigahertz frequencies, it may be more appropriate to conduct an SE test on cables in the configuration in which they are to be used. One test was made of the coupling through a cable when exposed to a local field from 1 to 2 GHz. A capacitive injection clamp was used to inject a high-level E field into the cable, as shown in Figures 7.75 and 7.76. A 20 W TWT amplifier was used to generate the field, and the capacitive injection probe was designed to have a VSWR of

TABLE 7.8

Current Measured into a 12 m Cable with a Biconical Antenna at 1 m Distance and at One End

Frequency (MHz)	Current in Cable (mA)	Incident *E* Field (V/m)	mA/V/m
12.56	17	1.78	9.6
50	7.76	3.8	2.0
100	5.2	7	0.74
200	2.7	13	0.2

TABLE 7.9

Reduction in Induced Current, Sidefire Theta = 0°

Frequency (MHz)	Measured Reduction due to Ground Plane	NEC-Predicted Reduction due to Ground Plane (dB)	Delta (dB)
12.5	16	28	12
50	16	27.6	11.6
100	16	13	3
200	5	0	5

TABLE 7.10

Reduction in Induced Current, Sidefire Theta = 90°

Frequency (MHz)	Measured Reduction due to Ground Plane	NEC-Predicted Reduction due to Ground Plane (dB)	Delta (dB)
12.5	20	15	5
50	15	16	1
100	13	12	1
200	13	4.4	8.6

less than 3:1 to avoid damage to the TWT. Other TWTs require an even lower VSWR. An RG58 coaxial cable of length 2.4 m was set up on 5 cm blocks above a ground plane, with the capacitive injection clamp placed around the cable. A number of locations down the length of the cable were tried for the injection probe. It was found that a large number of locations down the length of the cable resulted in the same cable current and transferred voltage; however, the maximum induced values were highly dependent on moving the injection probe a centimeter or so relative to the cable at any of these locations. The far end of the RG58 was terminated with 50 Ω, while at the opposite end, the center conductor led to a pin in an MIL 38999 connector. The reference to a "38999 connector" implies the combination of a Matrix D38999/24WF35SA female connector, a Bendix JD38999/26WF35PA male connector (mating half), and a Glenair M85049/19-19W06 backshell (mounted on the Bendix connector). The direct termination of the RG58 cable shield to the EMI backshell is not the ideal 360° termination around the periphery of the cable, but it is representative of the type of connection used in the cable harness. An aperture exists between the cable shield and the backshell, as can be seen in Figure 7.78, and coupling from currents flowing on the shield, through this aperture, is almost certainly one reason why the attenuation of the direct connection is so poor. In Reference [17], a conductive elastomer is described that is used to plug the hole in the backshell and to make a connection around the periphery of the shield

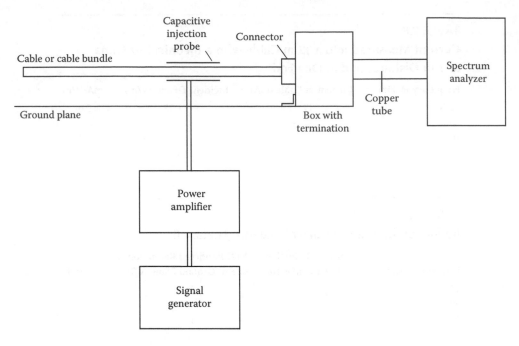

FIGURE 7.75   Capacitive injection clamp setup.

FIGURE 7.76   Injection clamp.

Figure 7.77 shows an improvement at some frequencies with the conductive elastomer compared to an unshielded cable at frequencies from 0.2 to 1.2 GHz.

The 38999 connector was mounted on the end of a metal box that also had an N-type connector mounted on the opposite end. Inside, a short length of RG213 coax was led from the N-type connector to the pin that connected to the RG58 cable. The unshielded length of RG213 cable was kept to a minimum. The box setup can be seen in Figure 7.78. The box lid was fastened with EMI gasket electromagnetically sealing the box, as shown in Figure 7.79. The signal from the N-type connector was taken via a coaxial cable covered in a solid copper conduit to a spectrum analyzer located outside of the shielded room where the measurements were made. The injected signal was swept from 910 MHz to 2 GHz, and the level was recorded on the spectrum analyzer. This test was performed three times in three different configurations. The first was to connect the RG58 shield directly to the

FIGURE 7.77 Voltage induced in (A) shielded cable with a standard electromagnetic interference backshell, (B) unshielded cable, (C) and shielded cable with conductive elastomer backshell. (From Santamaria, J.C. and Haller, L.J., A comparison of unshielded wire to shielded wire with shields terminated using pigtail and shields terminated through a conducting elastomer, *IEEE*, 1995 EMC Symposium Record, 1995. © 1995, IEEE.)

FIGURE 7.78 MIL-38999 connector measurement jig.

pin in the matrix connector to simulate an unshielded wire, and the level from 910 MHz to 2 GHz was recorded. The test was then repeated with the center conductor connected to the pin, and the coax shield terminated directly at the backshell of the connector. Once again, the test was repeated, this time with the shield terminated via a 1 in. pigtail. These two levels were subtracted from the unshielded-wire reference level to obtain the SE of the cable.

The results of this test give the SE achieved with a direct shield termination as well as with a 1 in. pigtail termination using the 38999 connector. The results are shown in Figure 7.80. As shown in the graph, the attenuation achieved with a good shield termination is higher than that achieved with 1 in. pigtails at most frequencies. Above 1400 MHz, the coupling may have been primarily aperture coupling and not transfer impedance coupling due to the long-line limit on transfer

FIGURE 7.79   MIL-38999 connector measurement jig with lid fastened.

impedance. Why are the direct termination and the pigtail better than an unshielded cable from 1700 to 2000 MHz is unclear. However, the test setup is representative of the actual cable, connector, and backshell configuration.

The next test was on a cable with an "ARINC" connector with a provisional copper backshell. This backshell was open, and the copper surround was used to solder the cable shields via the 1 in. pigtail. Because the makeshift backshell was wide open at the back, each conductor in the cable had short unshielded lengths of wire leading from where the shield ended to the pins in the ARINC connector. Radiated coupling to these lengths of wire from the currents set up on the cable shield is very probable. Figure 7.81 shows the attenuation with a 1 in. pigtail and a direct connection of the braid of the shielded cable to the provisional backshell.

7.10   SHIELDING EFFECTIVENESS OF CABLES UP TO 12 GHz

The two main methods of determining the shielding of cables and connectors are SE and transfer impedance.

SE is defined as the ratio of electric field or magnetic field strength at a point before and after the placement of a shield. For a cable, this is typically defined as the voltage or current induced into the center conductors(s) of the cable, by the field, with and without the shield in place. A number of disadvantages exist in the measurement and characterization of SE, and these include the following:

1. SE is not an "intrinsic electromagnetic parameter," which means that its measurement is dependent on the test setup.
2. No unique definition of SE exists.
3. No method exists for independent calibration.

However, data from the SE measurements described here show a remarkably good correlation to the SE derived from the transfer impedance, over the frequency range where the cable is electrically long and the frequency range where the transfer admittance predominates (100–1000 MHz).

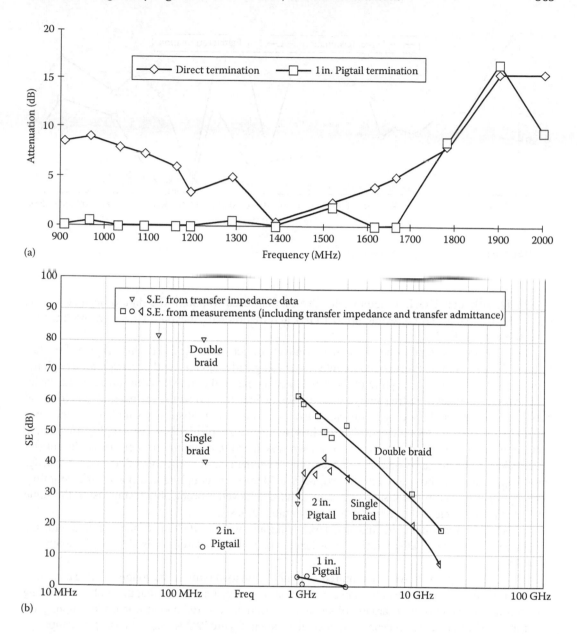

FIGURE 7.80   RG58 shielding effectiveness.

Shelkunoff showed that surface transfer impedance is by contrast an intrinsic EM parameter and its measurement is independent of the test setup. This means that any test method may be used to determine the transfer impedance and the test results should be comparable with the data obtained on the same cable in a different test setup.

The disadvantages of the transfer impedance test are the upper frequency limit and the fact that it does not measure effectively coupling to the cable via transfer admittance. In most setups, the connection between the shield of the cable and the driveline is 0 Ω, and this maximizes the transfer impedance coupling and minimizes the transfer admittance coupling. Many consider the transfer impedance the most important factor and this is true at low frequency, but not necessarily true at high frequency. The upper frequency for most coaxial and triaxial cable test setups is

FIGURE 7.81 Shielding effectiveness of a representative cable harness with ARINC connector and provisional "open" backshell. Direct connection and 1 in. pigtail.

100 MHz, although 1 GHz is achievable. The method used to obtain the data described in this test is the IEC 96-1 and is useable up to 1 GHz and theoretically beyond. However, the limit we have achieved is 1 GHz.

The cable transfer impedance test induces a current directly into the CUT which is part of the test fixture. The predominant form of coupling is porpoising and magnetic field penetration through the apertures, although some coupling of the E field occurs due to the transfer admittance of the cable.

The test method at high frequency uses a capacitive injection probe to create a high electric field between the probe and the cable. This capacitive injection more realistically simulates a high-level E field incident vertically on the cable, and the source would typically be from a high gain antenna and somewhat localized. The test method maximizes aperture coupling through the cable transfer admittance and at the same time sets up a current down the length of the cable that ensures transfer impedance coupling, both through the magnetic field aperture coupling and porpoising coupling. Although the test setup is not standardized, we have achieved reproducible results using the same type of shielded cable and injection probe in a number of tests.

As the SE test compares the induced levels with and without the shield and is made with the cable electrically long, that is, multiple wavelengths long, the measurement is virtually independent of cable length.

At 12 GHz, the probe movement required was only 3 mm to maximize the coupling. Also at numerous locations down the length of the cable, a maxima was seen. This shows that the probe was setting up a resonant cable current regardless of where the injection occurred and also that the coupling via the transfer admittance was independent of cable location. Figure 7.82 shows the basic test setup.

The SE of the Belden 8669 shield braid was measured, as well as an RG58- and a RG214-shielded coaxial cable under an R&D program.

During all the SE tests, the Belden braid when pulled tight had no visible apertures, and so at least 96% optical coverage was achieved. To test the importance of magnetic field (transfer impedance) and electric field (transfer admittance), aperture coupling of the braid on the Belden cable was relaxed somewhat so that small apertures become visible. The SE test was performed again at a highest frequency of 12 GHz and the SE reduced by 10 dB from 25 to 15 dB.

Table 7.11 and Figure 7.83 show the SE from 8 to 12 GHz and how this is derived.

Figure 7.84 shows the SE measurements from 100 MHz to 12 GHz on the Belden 8669 braid, an RG58 single-braid coaxial cable, and an RG214 double-braid-shielded cable. As expected, the RG214 shows the highest SE, although even this double-braid-shielded cable only achieved an SE of 31 dB at 12 GHz. The RG58 had the lowest SE at most frequencies and consistently lower than the RG214, again as expected. Some anomaly was present in the test setup at 6 GHz, and so a plot of

Cable shield braid shielding effectiveness test setup 0.1–12 GHz

FIGURE 7.82 Basic shielding effectiveness setup.

TABLE 7.11

Braid Shielding Effectiveness from 8 to 12 GHz

	Shielding Effectiveness Measurements			
Frequency (GHz)	Shielded Cable (dBuV)	Unshielded Cable (dBuV)	Shielding Effectiveness (dB)	Noise Floor (dBuV)
8	42.5	76	33.5	−7
10	41.7	69	27.3	−6.6
12	34	59	25	−2.7

Using Belden 336 × #34 (.02 mm²) TIN CU Braid 1/2 in. I.D. Tubular.

the trends with the 6 GHz measurement omitted is shown in Figure 7.85. The test results are comparable to the results first obtained using the identical injection probe. On repeating the tests, they were reproducible to at least 6 dB, with the height of the cable above the ground and the tightness of the braid over the center conductor as the most critical variables.

The far-end transfer impedance for the Belden 8669 braid-shielded cable is shown from 0.25 to 100 MHz and 100 to 1000 MHz in Figures 7.86 and 7.87.

The SE test was made with a 50 Ω load at each end of the cable, and so it is possible to predict the SE from the transfer impedance using

$$\text{SE (dB)} = 20 \log (2 \times 50/Z_t)$$

This predicted SE from Z_t and the measured SE above 100 MHz are compared in Table 7.12.

Considering that the two test methods are radically different, it can be seen that above 100 MHz, when the cable becomes electrically long, and especially between 271 and 1000 MHz, the correlation between the two methods is incredibly close.

FIGURE 7.83 (a) Shielding effectiveness measurements from 8 to 12 GHz on the Belden 8669 braid–shielded cable. (b) Shielding effectiveness measurements from 8 to 12 GHz on the Belden 8669 braid–shielded cable.

FIGURE 7.84 Shielding effectiveness data on a 2 and 4.5 m long Belden 8669–shielded cable and RG58 and RG214 coaxial cables.

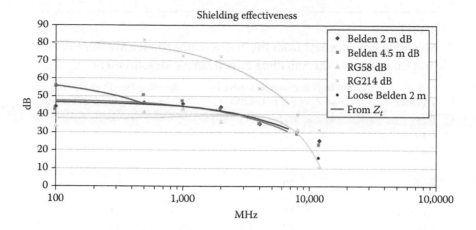

FIGURE 7.85 Shielding effectiveness (SE) trend with the measurement at 6 GHz omitted and compared to Z_t-derived SE.

FIGURE 7.86 Belden 8669 braid–shielded cable far-end transfer impedance of 0.25–100 MHz.

FIGURE 7.87 Belden 8669 braid–shielded cable far-end transfer impedance of 100–1000 MHz.

TABLE 7.12

Shielding Effectiveness from the Transfer Impedance Measurement and Compared to the Shielding Effectiveness Measurement

f (MHz)	Shielding Effectiveness Derived from Z_t (dB) (2 m Length)	Shielding Effectiveness from Measurement (dB) (2 m Length)
0.25	71	—
0.5	71	—
1.0	67	—
10.0	58	—
20.0	56	—
30.0	53	—
100/97.4	56	44
194	47	45
271	48	45
500/478	44	46
900	42	46
1000	43	46

7.11   POLARIZATION AND ANGLE OF INCIDENCE

In the following discussion, we refer to a wave with *horizontal* polarization as one in which the electric field component is parallel to the ground plane and a *vertically* polarized wave as one in which the electric field is perpendicular to the ground plane.

In many cases, we do not know the polarization or the angle of incidence of the wave, and a worst-case coupling of either the magnetic or electric field component must be assumed. One example in which the assumption of worst-case coupling is valid is the RS03 test setup, in which, above 30 MHz, both vertical and horizontal orientations of the E-field-generating antenna must be used. The antenna is located 1 m from the edge of the ground plane. Thus, for frequencies below 50 MHz, the field is near field and exhibits curvature and horizontal, vertical, and radial field vectors. When the RS03 test is conducted in a shielded room, the EM image of the antenna is formed in the ceiling and the angle of incidence of the wave on the interconnection cables is both directly from the antenna and that of the reflected wave from the ceiling. Thus, especially at low frequency, the assumption of worst-case coupling in the RS03 test setup is perfectly valid. In some instances, both the polarization of the wave and the angle of incidence are known. One example is when the orientation and locations of a transmitting antenna that couples to an above-ground signal or power cable are known. The angle of incidence and polarization determine the directivity of the configuration that is used to correct the predicted open-circuit voltage induced in a transmission line above ground. Reference [10] contains directivity patterns for vertical polarization, for which the directivity may be as high as 3.7, and for horizontal polarization, which may have a directivity as high as 2. Reference [10] also provides information on the compensation in the coupling prediction when the transmission line is located above a ground plane with less than perfect conductivity.

7.12   SHIELD TERMINATION TO GROUND

Thus far, we have assumed that the shield of a cable is connected to ground at both ends (symmetrical connection) via the metal walls of an enclosure. We have seen that this configuration provides shielding against an incident magnetic field but that this SE is limited at low frequency. When one or both of the enclosures can be disconnected from ground, greatly increased shielding

FIGURE 7.88 Shield connection for components with low capacitance to ground (chassis).

against a magnetic field may be achieved. However, when it is the shield of the cable that is disconnected from an enclosure at one end and the two enclosures remain connected to the ground plane, the configuration is effectively the same as shown in Figure 7.3b with minimum attenuation of magnetic field–induced voltages. When a shielded cable is used to connect a component such as a potentiometer mounted on a front panel, thermistor, strain gauge, or heating element in which the electrical connections are isolated from chassis ground, then the shield should be connected as shown in Figure 7.88. Here, the shield is isolated from the ground at the component end, and the maximum shielding against both *E* and *H* fields is achieved. The capacitance between the component (and associated connections) and enclosure must be minimized for this to remain valid at high frequencies. As with all guidelines, exceptions exist; one exception in which a shield is deliberately disconnected from the enclosure at low frequencies at one end but effectively connected at RF is described later in this section. In the following discussions, we assume that the shield is connected to the enclosures at both ends of the cable, and isolation of the cable at one end (asymmetrical connection) means that the enclosure is disconnected from ground.

Assuming the conditions for the use of transmission-line theory are met and that the length of the cable is close to or greater than λ/4, Equation 7.20 may still be used to determine the cable current flow for the asymmetrical connection. Figure 7.89a illustrates the current distribution on the shield due to an incident EM wave for the asymmetrical connection and where the cable length is a significant fraction of a wavelength but not greater than λ/4. The effect of removing one end of the ground connection (asymmetrical connection) is merely to move the frequency of resonances; the magnitude of the shield current at nonresonant frequencies is of the same order. The resonances for the asymmetrical connection occur when

$$l = \frac{(2k+1)\lambda}{4} \qquad (7.22)$$

where *k* = 0, 1, 2, 3,.... Thus, resonances for the asymmetrical connection occur for line lengths that equal 0.25λ, 0.75λ, 1.25λ, etc., and for the symmetrical connection when the line length equals 0.5λ, λ, 1.5λ, etc.

Figure 7.89b illustrates that a current does flow on a shield when both ends are disconnected from ground. The current distribution for the symmetrical connection is the same as the distribution shown in Figure 7.89b for the totally disconnected configuration.

Figure 7.90a, from Reference [18], illustrates the voltage induced in a leaky coaxial cable, one in which the transfer impedance increases linearly with frequency, that is,

$$Z_t = R_s + j\omega L_t \qquad (7.23)$$

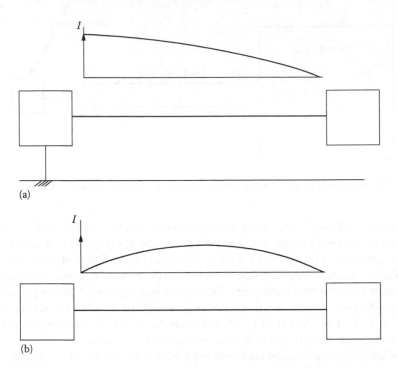

(a)

(b)

FIGURE 7.89 (a) Current flow in a near-resonant length of shielded cable in which one enclosure is connected to ground (asymmetrical connection). (b) Current flow in a cable and enclosures disconnected from ground (or for the symmetrical connection).

and for which $L_t = 16$ nH/m and $R_s = 9$ mΩ/m. The incident EM wave is a $1 = $ V/m plane wave, and the cable has a diameter of 8 mm, a characteristic impedance of 50 Ω, and a length of 15 m and is located 30 cm above a ground plane. The angle of incidence of the plane wave ensures worst-case coupling for all three shield terminations, which are symmetrical, asymmetrical, and no connection. Below 1 MHz, the symmetrical connection exhibits a lower level of shielding than either the asymmetrical or no connection. At 5 MHz, where the cable length is approximately equal to $\lambda/2$, the symmetrical connection achieves the highest level of shielding. In Figure 7.90b, experimental results are shown compared to theoretical, in which the difference in the asymmetrical and symmetrical connections is no greater than 25 dB at 10 kHz in the measured values. This is attributed to additional E-field coupling through the transfer admittance of the shield, which increases the measured EMI voltage, especially for the asymmetrical connection.

In practice, a lower transfer impedance than indicated is achieved above 1 MHz for the majority of cables and, therefore, the SE will typically be higher. When the incident EM wave is not a plane wave but predominantly an electric or magnetic excitation, then at frequencies where the line length is less than the wavelength, the symmetrical or asymmetrical connection of the shields results in different values for the SE of the cable.

For a magnetic field incident on the same cable configuration (i.e., 15 m cable length, etc.), the asymmetric connection achieves a higher level of SE below 1 MHz but a lower level above 5 MHz. For an electric field, the symmetrical connection achieves a higher level of shielding than the asymmetrical at frequencies below 5 MHz, for the 15 m cable length, with little to choose between the two at higher frequencies. Again, measured values of shielding show less difference than the predicted values for the two configurations and for E and H fields, due to the coupling through the transfer admittance of the cable.

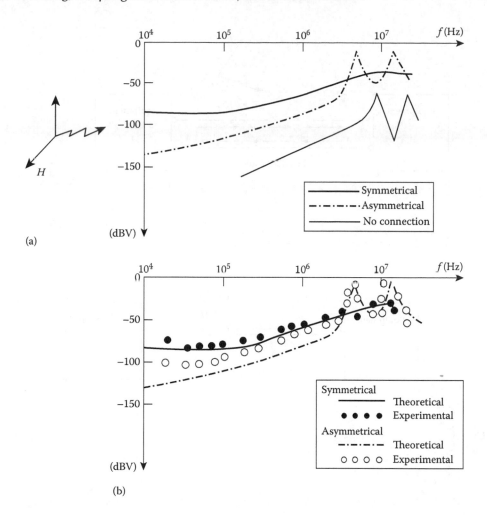

FIGURE 7.90 (a) Electromagnetic interference voltage induced in coaxial cable with asymmetrical, symmetrical, and no connection of the shield and enclosures to ground. (b) Comparison of calculated and measured (EMI) voltages. (From Chandesris, G., Effect of ground connection on the coupling of disturbing signals to a coaxial line, in *Fourth International Zurich Symposium and Technical Exhibition on Electromagnetic Compatibility*, March 10–12, 1981.)

In practice, it is difficult to achieve a true asymmetric connection unless the inner conductors of the shielded cable terminate on a physically small component, such as a transducer, at the end where the shield is floated.

Where considerable circuitry exists, either capacitance-to-ground or a ground connection via a power supply and the line cord safety ground may negate the true asymmetrical connection. The configuration we have considered, of cables running close to a ground plane, is normally achieved only on spacecraft, in vehicles, aircraft, and ships, and in the RS03 test setup.

When the cabinets or cases of two pieces of equipment are connected together via shielded cable and conductive connector backshells, the second connection is often made via additional interface cables or via the two equipment line cords to the safety ground. Where the incident EM field induces current flow on the shields of two or more adjacent interface cables, transmission-line theory may be used to calculate the magnitude of the current flow. In contrast, when a second case-to-case connection is made via safety grounds, a high impedance due to inductance is included in the loop and circuit theory is applicable. These two configurations are shown in Figure 7.91.

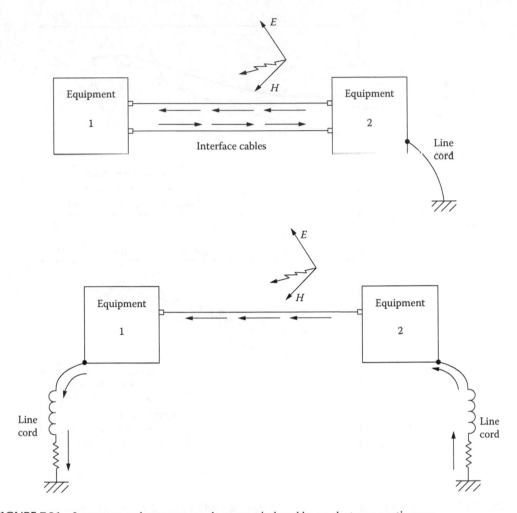

FIGURE 7.91   Loop area and common-mode current induced by an electromagnetic wave.

It should be emphasized that in neither of the situations in Figure 7.91 is the cable shield disconnected from the enclosure; only the enclosure is disconnected from the ground plane. This is also true for Figure 7.89a and b. Where the cable shield is disconnected from the enclosure at one end, the attenuation of the configuration to a magnetic field is negligible. With the shield disconnected, some attenuation to an E field is achieved, especially at low frequency and where circuit impedances are high or where the cable length is much greater than the wavelength, in which case a low level of attenuation to a plane wave is achieved. The level of plane-wave attenuation for a leaky cable will be approximately the same as for the asymmetrical connection, shown in Figure 7.90a, at 5 MHz and above.

One example where the shield of a cable was deliberately disconnected from an enclosure is as follows. The shielded cable was kilometers long and in proximity to a 60 Hz magnetic field. Connecting the shield of the cable to the enclosure at both ends would result in very high levels of 60 Hz current flow on the shield of the cable, which in turn would be coupled via mutual inductance to the center conductors of the cable. Disconnecting the shield of the cable meant that the center conductors of the cable were exposed to the 60 Hz field with virtually no shielding. The signals were differential and terminated in a high impedance at the receiver, so the 60 Hz current was limited by the high input impedance of the receiver. The 60 Hz C/M voltage appearing at the receiver input was rejected by the C/M noise rejection of the receiver. In addition to 60 Hz, the long cable carried currents were induced by FM and AM radio transmissions. The cable was electrically long and achieved the level of attenuation predicted in Equation 7.15, which is independent of the type of shield termination (asymmetrical or symmetrical), as shown in Figure 7.90a. Thus, the type of shield termination appears ideal for the application. The problem is that the shield at the end of the unterminated cable enters the enclosure and, due to a high level of noise voltage between the shield and the enclosure, radiates inside the equipment. One solution is to add an RF capacitor, with a sufficiently low-impedance connection, of approximately 100 pF value, between the end of the shield and the enclosure. The capacitor maintains the isolation of the shield at 60 Hz but effectively connects the shield to ground at RF. The correct location for the capacitor is outside of the enclosure to ensure that the noise currents do not flow on the inside of the enclosure. Not only was the location of the capacitor outside of the enclosure not feasible, but also the connection of the physically small capacitor between the shield of a large cable and chassis was extremely difficult. A potential solution was to use the high permittivity of the cable insulation and, by either bringing the cable through a metal tube or wrapping the cable with metal foil, to form a 100 pF capacitance between the shield of the cable and the enclosure.

7.13 EMISSIONS FROM CABLES AND WIRES

Many of the considerations we have given to coupling from *E* and *H* fields into cables apply to the generation of *E* and *H* fields by cables. One of the more useful axioms to be remembered in EMC work is that where a time-varying current exists, so does an EM field; conversely, where a time-varying field impinges on a conductor, a current flows. The path and, where possible, the magnitude of the current flow must first be identified in predicting the resultant radiation.

Where differential currents flow on a multiconductor cable and the conductors are close together, the amount of radiation from the cable is often less than for a lower-magnitude C/M current that flows in a loop of large area or on an isolated cable. The practicing engineer with little exposure to EMC will often dismiss the potential for radiation from a signal or power interconnection cable on the grounds that the signal- or differential-noise voltage is at a low frequency and, thereby, ignore the C/M contribution. The EMC engineer, on the other hand, may incorrectly ignore D/M currents due to the common experience that the source is so often C/M current flow.

A practical instance of this is a digital data bus connecting two pieces of equipment in which C/M noise voltage exists between the PCBs in the equipment and the equipment case. A current flow may then exist between the case-to-PCB connection and the data bus returning via a ground connection in the far enclosure, as illustrated in Figure 7.92. In the case of shielded cables, the

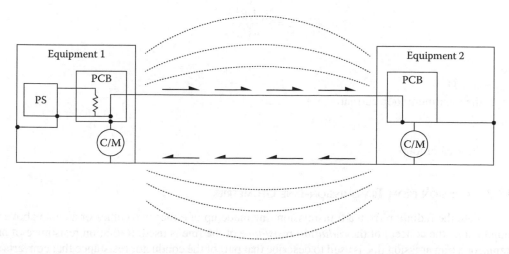

FIGURE 7.92 Radiation as a result of common-mode current generated by noise voltages within equipment.

differential current flow on the center conductors induces only a low level of current on the inside of the shield, whereas any C/M voltage between the center conductors and the shield results in a typically higher current flow on the inside of the shield. Even without a connection of any one of the conductors to the far-end enclosure, a current still flows on the inside of the shield due to the impedance of the transmission line formed by the center conductors and the shield.

When a piece of equipment is connected to ground, for example, via the AC safety ground, it is often C/M currents on the AC power cord that is the major contributor to radiation. As described in Reference [22], experiments with a battery-powered broadband noise source connected via a length of cable to a load in which no ground connection existed still resulted in the detection of C/M current on the interconnection cable. The most likely explanation for this surprising result is displacement current flow via capacitance between the circuit and grounded structures. When a ground connection is made to a similar setup with a radiating loop, a C/M current flows on the ground connection that may be the principal source of radiation, as discussed later and in Sections 11.1 and 11.8.2.

The most accurate method of obtaining the C/M current flow is measurement by the use of a current probe around the shield of a cable or around all of the conductors in an unshielded cable. Alternatively, circuit theory may be used to calculate current flow on cables based on measured or predicted values of C/M voltage and circuit or transmission-line impedance. When the current flow on the inside of the shield is known, the current flow on the outer surface of the shield of a cable may be derived from the methods used to calculate cable-to-cable coupling below 200 kHz and from the transfer impedance of the cable above 200 kHz. The field radiated by the cable may then be calculated by the use of the monopole, electric current element, or current loop equations, whichever are applicable to the cable geometry, discussed in Chapter 2. Alternatively, MOM techniques may be used. In the example shown in Figure 7.92, the geometry of the current path must be known in order to arrive at the loop area.

7.13.1 Emission from Loops

For loops of any shape but that do not fall under the definition of a transmission line, the equations of Section 2.2.3 may be used. These equations are valid when the load impedance at the end of the loop is less than or equal to the transmission-line impedance of the loop, as discussed later. For calculations of the EM wave in the far field, the equations may be simplified to the following with identical results:

$$E = \frac{Z_C I_L \beta^2 A}{4 \check{s} R} \tag{7.24}$$

where
$Z_C = 377 \ \Omega$
I_L is the maximum loop current
$\beta = \dfrac{2\pi}{\lambda}$
A is the area of loop

7.13.2 Emission from Transmission-Line Geometries

To calculate the radiation from a transmission line made up of either two cables or a cable above a ground plane, the concept of the *radiation resistance* of the line is used. Radiation resistance of an antenna or a transmission line is used to describe that part of the conductor resistance that converts a fraction of the power delivered to the load into radiated power. In an efficient antenna, the radiation resistance is designed to be high and the resistance, which converts the input power into heat, is low. In the transmission line, the opposite is true. The radiation resistance for a resonant section of two-wire line of length $\lambda/2$ (or integral multiples of $\lambda/2$), when the line is short-circuited (symmetrical connection), or $\lambda/4$ and other multiples shown in Equation 7.22, when the line is open-circuited or terminated in a load higher than Z_c (asymmetrical connection), is

$$30\beta^2 b^2 \tag{7.25}$$

where b is either the distance between the two-conductor line or twice the height of the cable above a ground plane. The radiated power for the resonant line is given by

$$30\beta^2 b^2 I^2 \tag{7.26}$$

where I is the current flow on the line, either measured or calculated. For the majority of EMI conditions, the radiated noise covers a wide range of frequencies, and resonant line conditions can be expected. For example, when the source of current is converter or digital logic noise, the harmonics may range from kilohertz up to 500 MHz.

Where a single frequency with negligible harmonics is the source of current in the line and the cable length is electrically long, that is, $l > 0.25\lambda$, but either the line is not a resonant length at the frequency of interest or the line is terminated in its characteristic impedance, the radiation resistance equals

$$30\beta b^2 \tag{7.27}$$

and the power is

$$30\beta b^2 I^2 \tag{7.28}$$

The magnetic field at some distance R from the line is

$$H = \sqrt{\frac{Pk}{4\pi R^2 Z_w}} \tag{7.29}$$

where
  P is the radiated power
  Z_w is the wave impedance
  k is the directivity, which is approximately 1.5 for a current loop and resonant line and 1.0 for a
    nonresonant line

and the electric field is

$$E = \sqrt{\frac{Z_w Pk}{4\pi R^2}} \tag{7.30}$$

In the near field, the wave impedance is close to the characteristic impedance Z_c of the transmission line, which may be calculated from Equation 7.16. The wave impedance then changes linearly until the near-field/far-field interface, at which $Z_w = 377\ \Omega$. The E field radiated by the current loop may be obtained from $E = H \times Z_w$. The equation for calculating the wave impedance in the near field is

$$Z_w = \frac{\lambda/2\pi - R}{\lambda/2\pi}(Z_c - 377) + 377 \quad (Z_w < 377\,\Omega)$$

where R is the distance from the radiation source in meters.

Figure 7.93a through c shows the comparison of the calculated and measured magnetic field from a transmission line 2 m long suspended 5 cm above a ground. In the near field, the calculated H field using transmission-line theory produces results closer to the measured values, whereas large errors are obtained for the use of current loop equations. In the far field, the two calculated values tend to converge and the maximum error between the calculated value and the measured one is approximately 6 dB.

FIGURE 7.93   (a) Comparison of calculated and measured fields from a transmission line with distance at 79 MHz. (b) Comparison of calculated and measured fields from a transmission line with distance at 34 MHz.   (*Continued*)

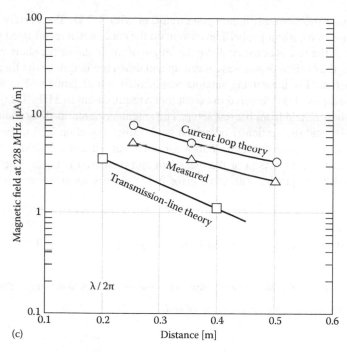

FIGURE 7.93 (*Continued*)   (c) Comparison of calculated and measured fields from a transmission line with distance at 228 MHz.

The criteria for assessing when circuit theory or transmission-line theory is applicable are the same as described in Section 7.7 on coupling into wires. The directivity factor of 1.5 does not take into account the height of the antenna with reference to the transmission line. Even more important when measurements are made in a shielded room, as were those shown in Figure 7.93, are the reflections of the transmission line from the ceiling and wall of the room. Compensation for reflections is discussed in Section 9.3.2.

The transmission-line model is applicable to the RE02 and DO-160 test setups, in which cables are located 5 cm above a ground plane and 10 cm from the edge of the ground plane for a distance of 2 m. When shielded cables are used, the shield is often terminated at the equipment enclosure, which is bonded to the ground plane, at one end and connected to the ground plane, after the 2 m length, at the other end. If we assume that the cable length between the ground connections is 3 m and the relative permittivity of the dielectric is 2, then the cables' first resonant frequency is 35 MHz, at which $\lambda/2 = 3$ m. A typical RE02 narrowband limit at 35 MHz is 22 dBµ V/m measured at a distance of 1 m from the edge of the ground plane. The current flow on the shield of the cable must be below 9 µA at 35 MHz in order to meet the specification limit using Equation 7.28, from which the predicted *H* field at a distance of 1.05 m from the cable is 3.47E-8 A/m. The characteristic impedance of the transmission line is approximately 317 Ω, and thus, the wave impedance 1.05 m from the transmission line is approximately 377 Ω. The predicted *E* field is therefore 3.47 E-8 A/m × 377 = 13 µV/m = 22 dBµ V/m.

In Figure 7.93b, the measured *H* field was approximately 12 dB above the predicted value at a frequency of 34 MHz and a distance of 1 m, based on the measured cable current. Some of the difference may be accounted for by reflections from the ceiling and wall of the shielded room, plus errors in antenna calibration and the damping effect of the current probe on the cable when the cable current is measured. Thus, engineers must set a design goal of less than 1 µA of C/M current on a cable at 20–35 MHz to ensure meeting RE02 limits.

Measurements were also made in a well-damped chamber that, over the frequency range of the measurements, showed a correlation of between 0 and 4 dB to measurements on an OATS. A 2 m long insulated #16AWG wire was raised 5 cm above and 10 cm from the edge of a ground plane and terminated either in a short circuit, in an open circuit, or with 317 Ω. The current on the line was monitored by the use of a current probe. The current on the cable when terminated with the characteristic impedance of the line is constant down the length of the line. For the short-circuit and open-circuit termination, the current probe was moved up and down the length of the line to intercept the highest level of current. The measuring antenna was located at a distance of 1 m from the edge of the ground plane and raised and lowered in height (not a requirement in MIL-STD-462 or DO-160) to measure the highest E field from the test setup. The antenna orientation was changed to intercept both the horizontally and the vertically polarized field. Either Equation 7.26 or Equation 7.27 was used to calculate the power, and Equation 7.30 was used to calculate the E field.

Table 7.13 provides a comparison of the measured and predicted E fields. The accuracy of the prediction is typically acceptable, with some error in the current measurement introduced by the

TABLE 7.13
Predicted and Measured E Field from a Wire above a Ground Plane in an MIL-STD-462/DO-160 Test Setup

Frequency (MHz)	Calculated Level (dBµV/m)	Measured Level (dBµV/m)	Delta (dB)
Short-circuit termination, horizontally polarized field			
83	91.1	85.5	5.6
140	95.6	91.0	4.6
200	96.7	88	8.7
260	96.0	85.0	11
Short-circuit termination, vertically polarized field			
77	90.5	89.0	10.5
146	96.0	89.0	7.0
200	96.7	90.5	6.2
252	96.7	83.5	13.7
Open-circuit termination, horizontally polarized field			
55	85.5	85.5	0.0
118	93.1	86.0	7.1
166	96.1	96.0	0.1
228	94.9	85.0	9.9
293	89.0	87.0	2.0
Open-circuit termination, vertically polarized field			
48	84.4	86.0	1.6
110	92.6	90.0	2.6
172	95.4	93.5	1.9
225	94.8	94.9	0.8
297	89.2	87.0	2.2
Line terminated in its characteristic impedance of 317 Ω, horizontally polarized field			
35	76.2	77.0	0.8
140	78.2	80.0	1.8
233	81.4	77.0	4.4
45	76.3	77.0	0.7
77	77.6	81.0	3.4
112	80.3	75.0	5.3
151	78.5	81.5	3.0
185	80.4	85.5	5.1
216	80.1	84.5	4.4
250	80.7	81.0	0.3

presence of the current probe on the cable. The type and size of antenna used affected the measurement, with a significant difference in the data using a log periodic/biconical versus a reference dipole antenna. A smaller contribution to the error is from the chamber.

7.13.3 EMISSIONS FROM LOOPS WITH AND WITHOUT ATTACHED CABLES

The situation arises where the radiation from a loop that either is disconnected from any ground connection or more commonly is connected to ground is required. One example is wiring or PCB tracks inside an unshielded enclosure to which a ground connection is made.

One practical example of such a loop connected to a square-wave generator is as follows. A 20 cm × 5 cm rectangular loop is connected to a square-wave generator that drives the loop via a 51 Ω resistor with a 2.5 V 1 MHz square wave. The 1 m long cable driving the loop is a double-braid type wrapped four times around a ferrite balun to reduce radiation from the cable. The pulse generator is shielded, and the AC power line is shielded at the generator enclosure. Thus, the prediction is that a low level of current will flow on the cables. The majority of EMI sources are pulses or square waves, and in meeting EMI specifications, we are interested in the frequency domain spectral density. Thus, in an EMC prediction, we must often convert from the time domain to the frequency domain. The maximum current flow in our example loop at the fundamental 1 MHz frequency is 2.5 V/51 Ω = 49 mA. Using the equations of Figure 3.1, the amplitude at the 21st harmonic (21 MHz) is 62 dBμA.

Using this current in the computer program that computes the *E* field from a loop results in a predicted *E* field of 0.28 mV/m, with no compensation for reflections from ceiling and floor. A very approximate correction is to add 6 dB for reflections, resulting in a predicted *E* field of 0.56 mV/m. The *E* field from the loop measured by a monopole antenna located 1 m from the loop is shown in Figure 7.94. The level measured by the monopole is −57 dBm at 21.7 MHz. A 26 dB preamplifier is used, so the output level from the monopole is −57 dBm − 26 dB = −83 dBm. This converts to a voltage of 16 μV. The antenna factor of the monopole is 40, and so, the measured *E* field is

FIGURE 7.94 Output voltage from a monopole antenna with the loop in circuit.

0.64 mV/m. Removing the loop from the end of the cable and replacing it with a near-perfect short is expected to reduce emissions considerably. However, as shown in Figure 7.95, this is not the case. A measurement of the current flow on the AC power-line cable was made using a current probe, with the results as shown in Figure 7.96. The level at 21.7 MHz is −58 dBm, which, accounting for the 26 dB preamplifier, results in a probe output level of −84 dBm = 14 µV. The transfer impedance

FIGURE 7.95 Output voltage from the monopole antenna with a short circuit replacing the loop.

FIGURE 7.96 Output voltage from the current probe placed around the generator AC power cable.

of the current probe at 21 MHz is 0.89 Ω, so the cable current is 15.7 ΩA. Using this current in the computer program for *E* field from an electric current element results in a predicted *E* field of 0.65 mV/m. This example illustrates very nicely the importance of C/M current on cables as the predominant source of radiation, for approximately the same *E* field is generated by the 15.7 µA on the AC cable as by a current of 1.26 mA in the loop. In subsequent measurements, the loop was located in the shielded room close to a tube through which the double-shielded cable was fed, and the generator and power cable were located outside of the room. With this configuration, no emissions were measured with the end of the generator cable shorted.

7.13.3.1 Loop Termination

The loop may be terminated in a low-impedance load, representative of logic gates, or a high-impedance load, representative of analog circuits, or less commonly, an open circuit. The equations for radiation from a current loop assume a short-circuited loop. Therefore, an important question is how applicable the equations are when, as is so often the case, the loop is terminated in an impedance.

The currents flowing in a loop terminated in a load impedance are composed of the uniform current flowing into the load and a nonuniform displacement current that decreases to zero at the end of the loop, as shown in Figure 7.97. The magnitude of the nonuniform current is determined by the input impedance of a short open-circuited transmission line, which, from Reference [10], is given by

$$Z_{in} = \frac{(377 \text{ or } Z_o)\lambda}{2\pi l} \tag{7.31}$$

where *l* is the length of the transmission line, in meters.

The uniform loop current is given by V_{in}/R_t and the nonuniform current by V_{in}/Z_{in}. The *E* field radiated by the nonuniform current may be modeled by two electric current elements separated by a distance *h*, which is the distance between the two conductors of the loop in meters. The direction of current flow in the two elements is out of phase and so are the *E* fields generated by the elements. At some measuring distance in the plane of the loop, the measured *E* field is the difference between the fields generated by the two elements. As an example, consider the 20 cm long by 5 cm wide loop open-circuited at the end. The input impedance of the loop at 30 MHz, from Equation 7.31, is 3 kΩ. The voltage applied to the loop is 0.55 V, so the value of nonuniform current is 0.183 mA. The *E* field at a distance of 1 m from the closest conductor in the loop is required. Using Equation 2.12 for the *E* field at 1 m from an electric current element results in 7.6 mV/m. The *E* field at a distance of 1.05 m from the furthest conductor is 6.555 mV/m. The difference between the *E* fields is 1 mV/m and is the magnitude of the predicted field at 1 m distance.

In the test setup shown in Figure 7.98, the loop was either unterminated or terminated with a short circuit, a 50 Ω, 1 kΩ, or 10 kΩ resistor. The *E* field resulting from either the nonuniform or the uniform currents for each termination was calculated and measured by a monopole antenna at a

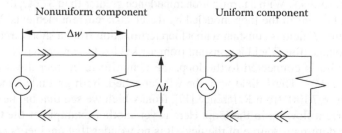

FIGURE 7.97 Uniform (C/M) and nonuniform (D/M) current flow in a loop.

FIGURE 7.98   Loop measurement test setup.

TABLE 7.14

Predicted and Measured E Field from a Loop with Different Terminations

Frequency [MHz]	Termination	Current Loop [mV/m]	Electric Current Elements [mV/m]	Measured [mV/m]
5.8	S/C	0.40	0	0.49
5.8	50 Ω	0.40	0.2	0.49
5.8	1 k Ω	0.02	0.2	0.39
5.8	O/C	0	0.2	0.39
30	S/C	2.4	0	0.90
30	50 Ω	2.4	1	0.90
30	1 k Ω	0.122	1	0.41
30	O/C	0	1	0.41

S/C, short circuit; O/C, open circuit.

---

distance of 1 m from the loop over a frequency range of 1–30 MHz. Table 7.14 provides a summary of the predicted and measured E fields at 5.8 and 30 MHz. With a short circuit or a 50 Ω termination, it is the E field from the uniform loop current that dominates, with minimal contribution from the current element sources. With a termination impedance greater than 377 Ω, the dominant source is the nonuniform current in the loop modeled by the electric current elements. With a fixed loop current, the measured E field is constant with a loop termination from a short circuit to a 50 Ω load; for a fixed input voltage, the E field is constant from a 1 kΩ load to an open circuit.

When a long wire is connected to the loop, the nonuniform current flows on the long wire as shown in Figure 7.99. The E field with the wire attached, both grounded and ungrounded, is reproduced in Figure 7.100, from Reference [19], from which we see that the field is almost independent of the current flowing in the loop. Here is yet another example of the C/M current on a wire acting as the dominant source of the field. It is no wonder that engineers with EMC experience tend to neglect D/M current flow on cables. The danger in doing so is when not all of the

FIGURE 7.99 Nonuniform current flow on a loop connected to a long wire.

	3 m maximum electric field strength [dBμV/m]
50 Ω circuit (dipole prediction):	24.0
50 Ω circuit with ungrounded wire:	35.2
50 Ω circuit with grounded wire:	42.6
800 Ω circuit (dipole prediction):	6.4
800 Ω circuit with ungrounded wire:	35.0
800 Ω circuit with grounded wire:	42.3
10 kΩ circuit (dipole prediction):	6.4
10 kΩ circuit with ungrounded wire:	34.8
10 kΩ circuit with grounded wire:	42.3

FIGURE 7.100 The *E* field from a loop connected to a long wire. (From Hubing, T. and Kaufman, J.F., *IEEE Trans. Electromag. Compatibil.*, 31(1), February 1989. © 1989, IEEE.)

D/M current returns on the cable, for when current returns on the ground, a large loop is formed that may become the major radiation source.

The current flow on the long cable is difficult to predict without the use of the MOM model, as discussed in Reference [19]. The solution when the equipment exists or a breadboard model can be constructed is to measure the current flow on the long cable by the use of the current probe. Once the current is known, the electric current element equation may be used to predict the E field from the cable, as previously described. The electric current element model should be valid only for infinitesimally short lengths of wire. However, when the model is compensated for long lengths of wire, the predicted field is invariably higher than that seen in near-field measurements, and the unmodified model is found to be accurate.

Although a complete shield around the loop will completely eliminate the induction of C/M current from the nonuniform loop current source, an incomplete shield may actually increase the C/M current on the long cable. Reference [19] shows that the impedance of the long wire close to the loop is high, approximately 3 kΩ, 10 cm below the source, and therefore, the inclusion of a ferrite bead on the line reduces the current only slightly.

Although the C/M current flow on interconnection cables is more often the result of C/M voltage between the ground and the conductors or shield of the cable, we see from the foregoing that D/M currents in the cable can induce a C/M current on an attached wire. This is also shown for PCBs in Sections 11.2 and 11.3.

7.14 REDUCTION IN THE EMISSION OF E AND H FIELDS FROM CABLES

The principal EMI reduction technique is to reduce at source. Where the source of current flow is C/M noise voltage, RF capacitors between the signal ground and the chassis ground or the use of filter connectors may be effective. Other techniques, such as the use of beads and baluns, are dealt with in Sections 5.1.10.3 and 5.1.10.4.

The introduction of an impedance (i.e., inductor or resistor in the return current path) is often an effective reduction method. When the current flow on a cable is reduced in both amplitude and frequency content to the extent compatible with normal operation of the interface/power line, the remaining reduction techniques are as follows:

The use of shielded cable exhibiting a low transfer impedance at the frequencies of interest, possibly by the use of double-, triple-, or quad-braided-shielded cable
The use of connectors and backshells with a low transfer impedance
Where used, removal of a "poor" connector bulkhead gasket, such as graphite-loaded, porcupine, wire-screen types, and replacement or installation of an effective gasket, such as thin-sheet or O-ring silver-loaded elastomer
Isolation of the signal return current from the shield by the use of triaxial or twinaxial cable
Reduction in the loop area of the cable (achievable by decreasing the height of a cable above the ground plane or bundling together the cables that carry a D/M current)

When a shielded cable cannot be used, any spare conductors in the cable should be connected to a "clean" ground. If the cable enters a metal enclosure, current flow on the outside of the enclosure is likely to be minimal, whereas current flow on the inside of the enclosure may be considerable. Thus, connection of the spare conductors or the shield of the cable to the outside of the enclosure is likely to be effective, whereas connection of spare conductors or the shield of a cable to the inside of the enclosure or to a noisy signal ground may increase the level of radiation.

7.15 SHIELDED CONNECTORS, BACKSHELLS, AND OTHER SHIELD TERMINATION TECHNIQUES

The effectiveness of a shielded cable is often compromised by the use of a connector that exhibits a high transfer impedance or low SE (i.e., high leakage).

Connectors specifically designed for use in harsh EM environments usually include metal fingers that make electrical contact between the mating halves of the connector and reduce the high transfer impedance inherent in a purely metal-to-metal interface that is not under pressure.

In addition, the bulkhead mounting section of the connector often contains a groove for the inclusion of an O-ring type of EMI gasket. Where the surface is flat, a rectangular gasket may be used of the correct material, as discussed in Section 6.6. The use of the wrong material, as previously noted, may decrease the SE. EMI backshells are designed to connect the shields concentrically around the backshell with a low-impedance connection, with another low impedance at the backshell-to-connector interface. Backshells designed for less-harsh EM environments but that provide some EM protection, for example, to meet FCC levels of emission, are often constructed of nonconductive material coated with a thin layer of deposited conductive material. It should be emphasized that such backshells are not suitable for the conduction of high-level transient current such as that indirectly induced in cables by EMP or a lightning strike. The SE of various types of backshells designed for D-type connectors is plotted in Figures 7.101 and 7.102, from Reference [20]. The test setup is shown in Figure 7.103, in which a 3 ft welded-aluminum cube enclosure contained the test antenna and the CUT. The cable was brought into the enclosure via feedthrough connectors and was terminated in the characteristic impedance of the system.

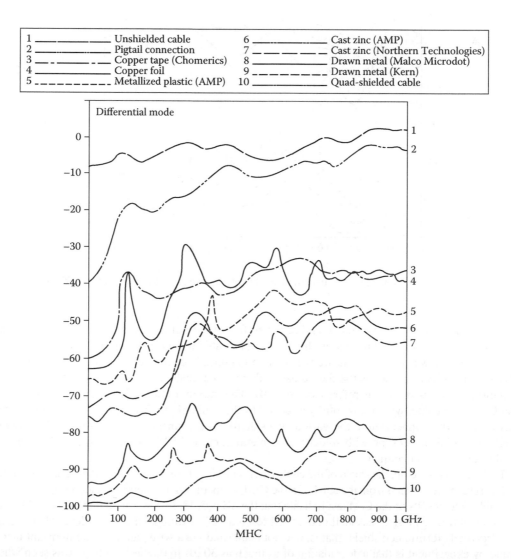

FIGURE 7.101 Differential-mode noise, backshell shielding effectiveness. (From Fernald, D., *ITEM*, 1984.)

1	Unshielded cable	6	Soldered copper backshell
2	Pigtail connection	7	Cast zinc (Nothern Technologies)
3	Metallized plastic (AMP)	8	Drawn metal (Malco Microdot)
4	Copper tape (Chomerics)	9	Drawn metal (Kern)
5	Cast zinc (AMP)	10	Quad-shielded cable

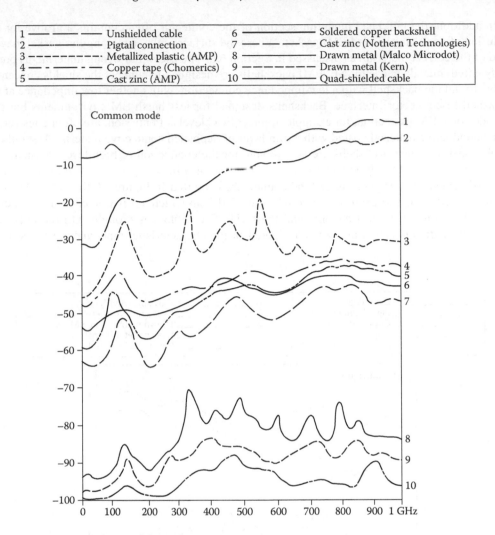

FIGURE 7.102   Common-mode noise, backshell shielding effectiveness. (From Fernald, D., *ITEM*, 1984.)

Antenna reflections were calculated and removed from the final data; however, some of the artifacts of the system are still present in the curves. The cable was used as the radiation source, and the test antenna was used to measure the relative magnitude of the field generated. The zero- or close-to-zero-dB reference is the field measured from the unshielded cable, with shielded cable radiation at minus-dB levels referenced to 0 dB. The maximum attenuation is achieved by the quad-shielded cable, which was solidly attached to the walls of the test chamber. The same cable was used in the connection to the various backshells. The connector used with all of the backshells was a 25-pin D type with solder-plated metal parts and EMI-controlling dimples that are used to ensure good contact.

In the D/M test setup, the shield of the cable carries no signal current, because the signal conductor and its return are isolated from the shield. In the C/M setup, the signal return is made via the shield of the cable. The results of the C/M test indicate that the SE is 5–15 dB lower than that of the D/M test.

The metallized plastic backshell in the C/M test (curve 3) exhibited the lowest level of shielding for a correctly terminated shield, that is, one not connected via a wire pigtail. One important factor found by experiment is that a degradation of as much as 30 dB in the best readings was seen when the backshell was not installed or terminated correctly.

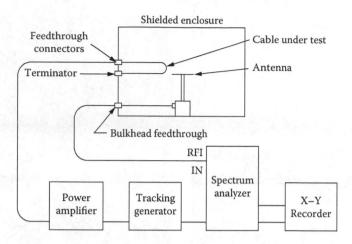

FIGURE 7.103   Backshell shielding effectiveness test setup.

Curve 2 is a clear example of the importance of terminating a shield correctly; here, a (2 in.) pigtail was used to connect the shield to the backshell. As discussed in Section 7.6 on coupling to shielded cables, the termination of the shield with a pigtail results in almost-negligible shielding, especially at high frequencies. Curve 1 is the unshielded reference cable. The curves may best be used by first calculating or measuring the potential radiated emission from an unshielded cable and comparing the predicted EM field at a specified distance—1 m, 3 m, or 30 m (depending on the EMC requirements) to the maximum allowable level contained in the requirements. The reduction due to the use of the quad-shielded cable and different types of backshell may then be assessed and the most cost-effective type of backshell selected for the application. It should be noted that the quad-shielded cable used during the measurements exhibits a high level of shielding; the use of a different type of shielded cable will almost certainly reduce the overall level of shielding.

When MIL-STD-461 or similar radiated susceptibility tests are a requirement or the worst-case EM ambient is known, coupling to the cable with the different types of backshells may be calculated as follows. The combined transfer impedance of the cable and connector may be found very approximately from the SE of the combination and the characteristic impedance of the test setup using Equation 7.4. The current flow on the shield may be calculated using the methods described in this chapter, and the transfer voltage is then found from the transfer impedance.

As a rough guideline, any of the backshells tested, when correctly terminated to the shield, is likely to meet FCC emission limits. For MIL-STD-461 and TEMPEST applications, only the solid metal backshells are likely to suffice. Figure 7.104 shows the cables terminated via very effective "Glenair" backshells and by the use of the ubiquitous pigtail.

Above 1 GHz, connectors and backshells are common sources of leakage. Figure 7.105a shows the leakage characteristics of four types of HF connector (from Reference [21]), and Figure 7.105b and c shows the leakage characteristics translated into transfer impedance. Even the best backshells may be limited by loose-threaded interfaces. In one instance, military equipment failed a radiated emission test, and it was found that the connector backshell was loose. As the backshell was tightened, the emissions could be seen to reduce until they disappeared into the noise level and the equipment passed the emission requirements with a significant safety margin. When the MIL-C-38999 Series III/IV connector-to-backshell interface was loose, the measured transfer impedance was 0.5 Ω at 100 MHz; as the connection was tightened, the transfer impedance was reduced, as shown in Table 7.15.

Figure 7.105d shows the combined transfer impedance of the interfaces of the braid of a shielded cable to the EMI backshell, the backshell to the connector, the two halves of the connector, and the connector to the bulkhead. The connector is a D (non-EMI) type, either gold-plated over brass or

(a)   (b)

FIGURE 7.104   (a) Cable shields terminated in a pigtail. (b) Cable shields terminated in a "Glenair" electromagnetic interference backshell.

zinc-plated (ASTM-B633) over steel. The backshell is manufactured by Glenair. Adding Technit Consil A silver aluminum–filled silicon gasket material between the connector and the bulkhead filled up the gap at the interface and reduced the transfer impedance above 1 GHz, as shown in Figure 7.105d.

An alternative to a solid backshell is a shielded heat-shrink boot. The interior surface of the boot is covered with a conductive coating. As the boot is heat-shrunk, the coating makes electrical contact with the metal of the connector housing and the braid of the shielded cable. A shielded heat-shrink boot for D-type connectors is available from Raychem Canada Ltd., 113 Lindsay Avenue, Dorval, Quebec, H9P 2S6. The transfer impedance of the Raychem boot is 38 mΩ at 10 kHz, increasing almost linearly to approximately 70 mΩ at 100 MHz. As expected, the transfer impedance of the heat-shrink boot is higher than that of the solid-cast Glenair backshell and, therefore, the SE will be lower.

One way of terminating a number of shields to a backshell is via a compressible disk of conducting elastomer material with wire-insertion holes, as described in Reference [17]. This conductive elastomer is pressed against the entry of the backshell and is used to terminate the individual shields. Figure 7.77 shows the attenuation compared to an unshielded cable and a standard backshell.

When a tin-plated copper braid is used over either an unshielded or a shielded cable to improve the SE, one technique for terminating the braid to the connector case is by clamping under a hose clamp. The transfer impedance of this type of termination is shown in Figure 7.106 from Reference [23].

Additional data on the transfer impedance of a 15-pin D-type, N, UHF, BNC, and SMA connectors are shown in Figure 7.107 from Reference [24]. N1 is the N-type connector attached to an RG214 cable and N2 is the N-type attached to a solid tube. This illustrates that it was not the N-type connector transfer impedance that was dominant but rather that of the cable. The SMA connector was attached to a semirigid cable, and the characteristic dip in transfer impedance of a solid-walled cable is seen at 10 MHz. However, above 10 MHz the connector transfer impedance predominates, and it rises to 1.5 mΩ at 1 GHz. Data on a gold-plated SMA connector soldered to a semirigid cable state an attenuation of 100 dB at 1 GHz, which can be converted to a transfer impedance of 0.5 mΩ. It is important to make a solder connection of the SMA connector to a semirigid or a semiflex cable, for a crimped termination exhibits a higher transfer impedance.

The dips around 800 MHz for the N- and D-type connectors are caused by the differences in wave velocities at the inside and outside of the connector under test, which becomes important because of the larger length of these connectors.

Additional data on a 15-position D-type connector show a transfer impedance of 40 mΩ at 100 MHz, 400 mΩ at 1 GHz, and 3 Ω at 3 GHz. These values are much higher than the values in Figure 7.105d but lower than the values in Figure 7.107 at 100 MHz and almost the same at 1 GHz. The problem with transfer impedance measurements on D-type connectors is the typically poor contact area between the connector case and the mounting surface.

FIGURE 7.105 (a) Leakage characteristics of the type BNC, type N, and General Radio Type 874B and Type 874BL connectors. (From Zorzy, J. and Muehlberger, R.F., *The Microwave J.*, November 1961.) (b) Surface transfer impedance of the type N and General Radio Type 874 BL coaxial connectors.   (*Continued*)

(c)

(d)

FIGURE 7.105 (*Continued*) (c) Surface transfer impedance of the Type BNC and General Radio Type 874 B coaxial connectors. (d) Surface transfer impedance of the two halves of a 25-pin D-type connector with "Glenair" backshell.

TABLE 7.15

Effect of Torque on Connector and Backshell Transfer Impedance at 10 and 100 MHz

Torque (in.-lb)	Transfer Impedance at 10 MHz (mΩ)	Transfer Impedance at 100 MHz (mΩ)
25	300	500
50	15	9
100	3	3
150	2	2
200	1	1.5

FIGURE 7.106   Transfer impedance of braid shield terminated to the connector case with a hose clamp. (From Hoeft, L.O. and Hofstra, J.S., Transfer impedance of overlapping braided cable shields and hose clamp shield terminations, *IEEE International Symposium Record*, 1995. © 1995, IEEE.)

In almost all surface transfer impedance measurements of connectors and backshells, a short braid shield is connected to the backshell, and the transfer impedance of this short section of braid is part of the measured surface transfer impedance. In long cable assemblies, it is typically the cable transfer impedance that dominates, and when the braid is very short and when the measured transfer impedance of the connector is very low, the braid may be a significant part of this.

Measurements on the DB-25 from Reference [26] show a large variation in transfer impedance depending on shield termination technique and connector and backshell type.

The surface transfer impedances at 100 MHz of the two best connector and backshells were 9 and 10 mΩ, which compare well with the 5 mΩ shown in Figure 7.105d.

Reference [26] describes the most robust backshell as consisting of a two piece nickel plated die cast back backshell with either a dual cone a compression insert that clamped the cable braid between the two metal cones. The surface transfer impedance of this configuration from Reference [26] is shown in Figure 7.108. The measurements are with and without a plastic-filled metal-mesh gasket under the connector receptacle. The surface transfer impedance of this configuration is higher than that without a gasket and this is to be expected based on the transfer impedance data on monel-wire mesh gaskets shown in Figure 6.40e. Here, the mesh gasket also performed worse than no gasket.

FIGURE 7.107 Transfer impedances of N, SMA, BNC, C, UHF, and 15-pin high-density *D* connector. (From van Horck, F.B.M. et al., *IEEE Trans. Electromag. Compatibil.*, 40(3), August 1998. © 1998, IEEE.)

FIGURE 7.108 Surface transfer impedance of DB-25 Die Cast Connector with a dual-cone braid termination and with and without a wire mesh gasket. (Data from Hoeft, L.O. and Hofstra, J., Electromagnetic shielding provided by selected DB-25 subminiature connectors, Electromagnetic Compatibility 1991 Symposium Record, *IEEE 1991 International Symposium.* © 1991, IEEE.)

FIGURE 7.109 Surface transfer impedance of DB-25 Die Cast Connector with and without a dimpled plug and a compression insert braid termination. (Data from Hoeft, L.O. and Hofstra, J., Electromagnetic shielding provided by selected DB-25 subminiature connectors, Electromagnetic Compatibility 1991 Symposium Record, *IEEE 1991 International Symposium*. © 1991, IEEE.)

In Reference [26], the measured surface transfer impedances of the two-connector backshell combinations that used metal die cast backshells with a compression insert braid termination with and without a dimpled plug are shown in Figure 7.109.

In Reference [26], the measured surface transfer impedance of two connector/backshell tc combinations that used metallized plastic backshells. The metallized plastic backshell braid was terminated either with a set screw or with an internal strain-relief clamp. Its transfer impedance was almost an order of magnitude higher than the set screw termination. This was undoubtedly due to the much poorer contact provided by this braid termination method. The data for this configuration are shown in Figure 7.110.

In Reference [26], the measured surface transfer impedance of a connector/backshell combination that used a metallized plastic backshell with and without a dimpled plug and with a compression insert braid termination is shown in Figure 7.111.

In Reference [26], the final sample used a nonconductive backshell that provided a mechanical termination for the cable. A short length of (2.5 cm) AWG #22 wire provided an electrical termination between the cable braid and the pin 1 of the connector pair. Pin 1 of the receptacle was connected to the connector shell by means of a short length of wire. This sample was referred to as the pigtail braid termination sample.

This connection had a transfer impedance 5000 (74 dB) times the best example.

The data for the sample are shown in Figure 7.112.

Reference [27] presents the surface transfer impedance of braid-to-backshell termination without connectors as well as that for a number of cylindrical connectors.

In Reference [27], the transfer impedance of a typical low-transfer impedance is shown. In addition, the initial and final surface transfer impedances of a sample that used a circular coil spring between cones to apply pressure between the braid and the backshell. The transfer resistance (at low frequency) and transfer impedance are significantly higher than the typical backshell and changed after disassembly/reassembly. This probably results from not using enough torque during assembly. One of the

FIGURE 7.110 Surface transfer impedance of DB-25 Die Cast Connector with plastic backshell and with either a set screw or internal strain relief termination for the braid. (Data from Hoeft, L.O. and Hofstra, J., Electromagnetic shielding provided by selected DB-25 subminiature connectors, Electromagnetic Compatibility 1991 Symposium Record, *IEEE 1991 International Symposium.* © 1991, IEEE.)

FIGURE 7.111 Surface transfer impedance DB-25 Die Cast Connector with plastic backshell with and without a dimpled plug and with a compression braid termination. (Data from Hoeft, L.O. and Hofstra, J., Electromagnetic shielding provided by selected DB-25 subminiature connectors, Electromagnetic Compatibility 1991 Symposium Record, *IEEE 1991 International Symposium.* © 1991, IEEE.)

manufacturers of this type of backshell has recommended that the use of 100 in.-lbs of torque be applied in order to ensure proper electrical performance. This again illustrates the importance of correctly tightening a connector. Figure 7.113 shows the "typical" and two sets of coil spring data.

The MIL-C-26482 series 1 is a circular connector with 3–155 contacts and has a bayonet connection. The transfer impedance for this connector from Reference [27] is shown in Figure 7.114.

FIGURE 7.112 Surface transfer impedance of a nonconducting backshell and a pigtail braid termination. (Data from Hoeft, L.O. and Hofstra, J., Electromagnetic shielding provided by selected DB-25 subminiature connectors, Electromagnetic Compatibility 1991 Symposium Record, *IEEE 1991 International Symposium.* © 1991, IEEE.)

FIGURE 7.113 Typical backshell/braid as well as coil spring surface transfer impedance data. (Data from Hoeft, L.O. and Hofstra, J.S., Electromagnetic shielding of circular and rectangular connectors, backshells, and braid terminations, *IEEE 1992 Regional Symposium on Electromagnetic Compatibility*, © 1992, IEEE.)

The MIL-C-81511 is a circular connector with 3–155 contacts and a bayonet fitting. The surface transfer impedance of this connector from 27 is shown in Figure 7.115.

The MIL-C-28840 is a circular connector with a screw thread fitting and 7–155 contacts. The surface transfer impedance of this connector from Reference [27] is shown in Figure 7.116. The DIN 29729 is a circular connector with a bayonet coupling with 3–128 contacts.

The surface transfer impedance of this connector from Reference [27] is shown in Figure 7.117.

The DPKA/MIL-C-8373 connector comes with 18–131 pins and is available with an EMI backshell. The transfer impedance of a DPKA connector and backshell was measured both with the normal mating force and with the connector mating halves under 70 kg of force. The reduction in transfer impedance and increase in SE due to this force are shown in Figures 7.118 and 7.119.

FIGURE 7.114 Surface transfer impedance of MIL-C-26482 connector/backshell combination. (Data from Hoeft, L.O. and Hofstra, J.S., Electromagnetic shielding of circular and rectangular connectors, backshells, and braid terminations, *IEEE 1992 Regional Symposium on Electromagnetic Compatibility.* © 1992, IEEE.)

FIGURE 7.115 Surface transfer impedance of MIL-C-81511 connector/backshell combination. (Data from Hoeft, L.O. and Hofstra, J.S., Electromagnetic shielding of circular and rectangular connectors, backshells, and braid terminations, *IEEE 1992 Regional Symposium on Electromagnetic Compatibility.* © 1992, IEEE.)

FIGURE 7.116 Surface transfer impedance of MIL-C 28840 connector/backshell combination. (Data from Hoeft, L.O. and Hofstra, J.S., Electromagnetic shielding of circular and rectangular connectors, backshells, and braid terminations, *IEEE 1992 Regional Symposium on Electromagnetic Compatibility.* © 1992, IEEE.)

FIGURE 7.117 Surface transfer impedance of DIN 29729 connector/backshell combination. (Data from Hoeft, L.O. and Hofstra, J.S., Electromagnetic shielding of circular and rectangular connectors, backshells, and braid terminations, *IEEE 1992 Regional Symposium on Electromagnetic Compatibility.* © 1992, IEEE.)

FIGURE 7.118 DPKA connector and backshell transfer impedance with normal mating and under 70 kg of mating force.

When an equipment under test (EUT) fails a radiated emission test with shielded cables connected to metal shell connectors, then the first test should be to test how tightly mated the connectors are.

The ODU AMC Advanced Military Connector features a push–pull locking or breakaway connection is available in six sizes with 4–55 contacts.

ODU-USA Inc, 4010 Adolfo Road, Camarillo, CA 93012-8515, kindly provided a four-contact G38YAR-P04UT00-00L receptacle and an S13YAR-P04XTS0-0000 plug to test the surface transfer impedance, and this is shown in Figure 7.120.

FIGURE 7.119 DPKA connector and backshell under 70 kg mating force shielding effectiveness.

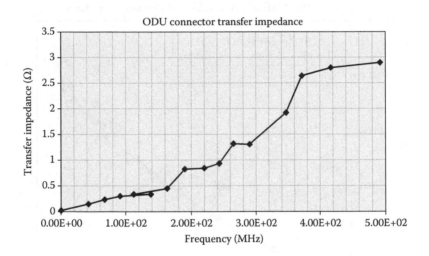

FIGURE 7.120 Surface transfer impedance of the ODU-AMC connector.

7.16   ETHERNET AND USB CONNECTORS

Ethernet and USB ports are now ubiquitous. For an EMC evaluation or in the design of equipment, the EMC characteristics of these ports are of interest; this includes the difference between an unshielded and shielded cable and the location of grounding inside the equipment. Many manufacturers offer the board-mounted female jack connector and these may include magnetics, usually a transformer and a common-mode choke; see Figures 7.121 through 7.123.

Resistors and a capacitor are also included in the connector with the capacitor to be connected to chassis ground. In many cases, this connection is made to a PCB digital ground. The digital ground is very likely to have a C/M voltage on it compared to chassis ground, depending on where the digital ground is connected to chassis and these grounding options and results are shown.

The metal case of the male shielded plug on the cable contacts the inside of the female connector typically by two small spring fingers. All the connectors tested were new and so the pressure on

FIGURE 7.121 Schematic used to determine which pins would be used during the testing of Ethernet port: 08B0-1D1T-06-F.

FIGURE 7.122 Schematic used to determine which pins would be used during the testing of Ethernet ports: J00-0045NL and 7499010122.

FIGURE 7.123 Schematic used to determine which pins would be used during the testing of Ethernet ports: 5-6605758-1 and 6605841-1.

these fingers was at its maximum. It has been observed that with frequent mating and demating, this pressure decreases and equipment may no longer meet a commercial E-field radiation limit. A quick fix is to glue foam rubber under the fingers. However, this is not a viable option when the product is in mass production. It is advisable to inform a manufacturer if this weakness is seen and encourage them to improve the design or to search for a connector with robust fingers.

Based on the measurements, the difference between a shielded and an unshielded cable in terms of reduction in radiation is not high. Much of this may be attributed to the poor connection of the metal case of the plug via the spring fingers.

The radiated emissions, using these different connectors from different manufacturers, were measured to compare relative performance. The purpose was not to compare the level of radiation to military or commercial limits. However, experience has shown that with correct grounding and shield termination, Class B commercial radiation limits can be met, whereas meeting MIL-STD-461 limits may require an Ethernet connector encased in a circular metal connector. The circular plug may have an EMI backshell for the termination of an overall shield.

The metal case of the enclosure containing the oscillator and driver was connected to a table top ground plane, and the cable was located 5 cm above the ground plane and 10 cm from the front of the ground plane. From 30 MHz to 1.2 GHz, a log periodic biconical antenna, connected to a 30 dB gain 0.25 MHz to 1.2 GHz preamplifier, was used 1 m distance from the cable.

A printed circuit board was created for each Ethernet connector pin configuration; four circuit boards were made in total for the Ethernet connectors. Each circuit board was designed to connect the appropriate pins for each Ethernet connector to an oscillator followed by a driver; this sends a signal to the connector and attached cable. The pins chosen for use were decided according to their schematics. For Ethernet connector 08B0-1D1T-06-F, pins 2 and 3 were chosen according to Figure 7.121. For Ethernet connectors J00-0045NL and 7499010122, pins 1 and 2 were chosen according to Figure 7.122. For Ethernet connectors 5-6605758-1 and 6605841-1, pins 5 and 6 were used according to Figure 7.123. For Ethernet connectors 54602-908LF, RJHSE-5380, and RJ45- 8LC2-B, pins 1 and 2 were chosen for convenience, as there were no factors that made any pin choice necessary.

The signal voltage in the cable results in a C/M current flowing on the inside of the shield. This couples through the cable due to the transfer impedance of the cable resulting in an external current that is the predominant source of radiation.

For the unshielded cable, the signal voltage results in current flow between the signal line and the ground. Because the signal is unbalanced, again a C/M current flows.

The rise and fall time and waveshape were identical for each connector. The peak amplitude varied, but only by the worst case having 3.1 dB.

A USB type A connector (690-004-621-013) and USB type B connector (690-004-660-023) were tested along with the Ethernet connectors. The two connectors had different pin configurations, and so, two separate printed circuit boards were constructed for their testing. Both of the connectors did not contain any factors that made any choice of pins necessary, so pins 1 and 2 were chosen for convenience.

Metal plates were made to fit around each of the tested connectors and were secured to the front of the enclosure used so as to reduce the space around the connector. The plate to enclosure had a low-impedance connection. These plates were designed either to fit tightly to ensure contact with the grounding fingers of the connectors if they had them or to allow a small space around the connector so that there was no contact with the plate if the connector had no grounding fingers.

The connectors were tested for radiated emissions over the range of 1 MHz to 1.2 GHz. Multiple grounding configurations were tested for each connector over this range depending on its physical characteristics. If the Ethernet connector was plastic, then two configurations were used: grounded at the source, Figure 7.124, and grounded at the PCB near the connector, Figure 7.125. If the Ethernet connector had a metal case without grounding fingers, it was tested with three grounding configurations: grounded at the source, Figure 7.125, grounded at the connector, Figure 7.126, and grounded at the connector and the PCB near the connector, Figure 7.127. If the Ethernet connector had a metal case and grounding fingers, it was tested with two configurations: grounded at the source with the grounding fingers contacting the enclosure, Figure 7.128, and grounded at the PCB near the connector with the grounding fingers contacting the enclosure, Figure 7.129. A shielded Ethernet cable and an unshielded Ethernet cable was used during testing. The unshielded Ethernet cable were used only for the 54602-908LF connector, while the shielded Ethernet cable was used for every other Ethernet connector tested. Both the shielded and unshielded Ethernet cables were approximately the same length.

The level of attenuation of the different connectors varied significantly with frequency but with the ground connection at the connector and the PCB (best connection) the 08B0-1D1T performed the best overall with the J00-045NL also good.

Both USB connectors had a metal case without grounding fingers, so each was tested with three grounding configurations: grounded at the source, Figure 7.124, grounded at the connector, Figure 7.126, and grounded at the connector and the printed circuit board, Figure 7.127. The USB type A connector was also tested using an unshielded cable, during which it was tested with two

FIGURE 7.124　Connector testing ground configuration 1. Ground connected at the source.

FIGURE 7.125　Connector testing ground configuration 2. Ground connected at PCB near connector.

grounding configurations: the ground at the source, Figure 7.124, and the ground at the printed circuit board, Figure 7.125.

A Corcom filtered connector was tested. This had individual sleeve ferrites on each connection followed by capacitors connected to the case. A version is available with a block ferrite but this also exhibits D/M inductance due to the leakage inductance. Due to the LC filter, the connector exhibits insertion loss and the specification states 1 dB at 1 MHz for the sleeve inductor and 0 dB at 1 MHz for the block inductor.

FIGURE 7.126 Connector testing ground configuration 3. Ground connected at the connector.

FIGURE 7.127 Connector testing ground configuration 4. Ground connected at the printed circuit board and the connector.

The filtered connector RJ45-8LC2-B attenuation performance was poor, as it contains capacitors between the lines and the enclosure. This connector is particularly susceptible to the ground location as C/M noise on the PCB couple out through the capacitors and the ferrites are then ineffective. As expected, with a couple of exceptions grounding the PCB to the enclosure at the source resulted in increased radiation from the Ethernet cable as this increases the C/M voltage impressed upon the lines.

FIGURE 7.128 Connector testing ground configuration 5. Ground connected at the source with grounding fingers contacting the enclosure.

FIGURE 7.129 Connector testing ground configuration 6. Ground connected at the printed circuit board, with grounding fingers contacting the enclosure.

The battery, oscillator, and driver integrated chip were contained in a metal enclosure with a Goretex gasket inserted between the case and the lid.

See Table 7.16 for the physical characteristics of each connector tested.

During testing, all results were compared to a baseline test for each frequency span tested. The baseline used for the Ethernet connectors was the radiated emissions of the 54602-908LF connector, grounded at the source using an unshielded cable, and the baseline chosen for the USB

TABLE 7.16

Physical Characteristics and Manufacturer's Details for Each Connector Tested

S/N	Connector Type	Manufacturer	Connector Material	Grounding Fingers	Magnetic Pin Connector
54602-908LF	Ethernet	FCI	Plastic	No	No
J00-0045NL	Ethernet	Pulse Electronics	Metal	No	Yes
7499010122	Ethernet	Wurth Electronics Inc.	Metal	No	Yes
6605841-1	Ethernet	TRP Connector	Metal	Yes	Yes
5-6605758-1	Ethernet	TRP Connector	Metal	Yes	Yes
08B0-1D1T-06-F	Ethernet	Bel Fuse Inc.	Metal	Yes	Yes
RJHSE-5380	Ethernet	Amphenol Canada Corp.	Metal	Yes	No
RJ45-8LC2-B	Ethernet	TE Connectivity (Corcom)	Metal	No	No, but with connector being filtered
690-004-621-013	USB Type A	EDAC Inc.	Metal	No	No
690-004-660-023	USB Type B	EDAC Inc.	Metal	No	No

connectors was the radiated emissions of the 690-004-621-013 connector, grounded at the source using an unshielded cable. These were chosen to be the baselines because it was hypothesized that the resulting radiated emissions from these connectors and ground configurations should be the highest tested. All of the other connectors and ground configurations were compared to this baseline by subtracting the recorded radiated emissions of the baseline from the recorded radiated emissions from another connector or ground configuration. These differences were then plotted for the comparison of the relative shielding of different connectors and grounding configurations. These plots are seen in Figures 7.130 through 7.158. Table 7.17 shows the corresponding Ethernet connector and ground configuration of each trace seen in Figures 7.130 through 7.150. Table 7.18 shows the corresponding USB connector and ground configuration of each trace seen in Figures 7.151 through 7.158.

At most frequencies, the highest level of emissions when compared to the baseline is with the shielded connector and shielded cable connected to the PCB ground at the source.

The next highest is with the cable and connector grounded at the enclosure. The lowest level of emissions is with the connector connected to the PCB ground and the enclosure at the enclosure.

FIGURE 7.130 Relative radiated emissions of 54602-908LF and J00-0045NL connectors compared to baseline (trace 1) over the span of 1–30 MHz.

FIGURE 7.131 Relative radiated emissions of 54602-908LF and J00-0045NL connectors compared to baseline (trace 1) over the span of 30–140 MHz.

FIGURE 7.132 Relative radiated emissions of 54602-908LF and J00-0045NL connectors compared to baseline (trace 1) over the span of 140–300 MHz.

FIGURE 7.133 Relative radiated emissions of 54602-908LF and J00-0045NL connectors compared to baseline (trace 1) over the span of 230–500 MHz.

FIGURE 7.134 Relative radiated emissions of 54602-908LF and J00-0045NL connectors compared to baseline (trace 1) over the span of 500 MHz–1.2 GHz.

FIGURE 7.135 Relative radiated emissions of 08B0-1D1T-06-F and 6605841-1 connectors compared to baseline (trace 1) over the span of 30–140 MHz.

FIGURE 7.136 Relative radiated emissions of 08B0-1D1T-06-F and 6605841-1 connectors compared to baseline (trace 1) over the span of 140–300 MHz.

FIGURE 7.137   Relative radiated emissions of 08B0-1D1T-06-F and 6605841-1 connectors compared to baseline (trace 1) over the span of 230–500 MHz.

FIGURE 7.138   Relative radiated emissions of 08B0-1D1T-06-F and 6605841-1 connectors compared to baseline (trace 1) over the span of 500 MHz–1.2 GHz.

FIGURE 7.139   Relative radiated emissions of 5-6605758-1 and RJHSE-5380 connectors compared to baseline (trace 1) over the span of 30–140 MHz.

FIGURE 7.140   Relative radiated emissions of 5-6605758-1 and RJHSE-5380 connectors compared to baseline (trace 1) over the span of 140–300 MHz.

FIGURE 7.141   Relative radiated emissions of 5-6605758-1 and RJHSE-5380 connectors compared to baseline (trace 1) over the span of 230–500 MHz.

FIGURE 7.142   Relative radiated emissions of 5-6605758-1 and RJHSE-5380 connectors compared to baseline (trace 1) over the span of 500 MHz–1.2 GHz.

FIGURE 7.143   Relative radiated emissions of 7499010122 connector compared to baseline (trace 1) over the span of 30–140 MHz.

FIGURE 7.144   Relative radiated emissions of 7499010122 connector compared to baseline (trace 1) over the span of 140–300 MHz.

FIGURE 7.145   Relative radiated emissions of 7499010122 connector compared to baseline (trace 1) over the span of 230–500 MHz.

FIGURE 7.146   Relative radiated emissions of 7499010122 connector compared to baseline (trace 1) over the span of 500 MHz–1.2 GHz.

FIGURE 7.147   Relative radiated emissions of RJ45-8LC2-B connector compared to baseline (trace 1) over the span of 30–140 MHz.

FIGURE 7.148   Relative radiated emissions of RJ45-8LC2-B connector compared to baseline (trace 1) over the span of 140–300 MHz.

FIGURE 7.149   Relative radiated emissions of RJ45-8LC2-B connector compared to baseline (trace 1) over the span of 230–500 MHz.

FIGURE 7.150   Relative radiated emissions of RJ45-8LC2-B connector compared to baseline (trace 1) over the span of 500 MHz–1.2 GHz.

FIGURE 7.151   Relative radiated emissions of 690-004-621-013 connector compared to the baseline (trace 20) over the span of 30–140 MHz.

FIGURE 7.152 Relative radiated emissions of 690-004-621-013 connector compared to the baseline (trace 20) over the span of 140–300 MHz.

FIGURE 7.153 Relative radiated emissions of 690-004-621-013 connector compared to the baseline (trace 20) over the span of 230–500 MHz.

FIGURE 7.154 Relative radiated emissions of 690-004-621-013 connector compared to the baseline (trace 20) over the span of 500 MHz–1.2 GHz.

FIGURE 7.155   Relative radiated emissions of 690-004-660-023 connector compared to the baseline (trace 20) over the span of 30–140 MHz.

FIGURE 7.156   Relative radiated emissions of 690-004-660-023 connector compared to the baseline (trace 20) over the span of 140 MHz–300 MHz.

FIGURE 7.157   Relative radiated emissions of 690-004-660-023 connector compared to the baseline (trace 20) over the span of 230 MHz–500 MHz.

FIGURE 7.158 Relative radiated emissions of 690-004-660-023 connector compared to the baseline (trace 20) over the span of 500 MHz–1.2 GHz.

TABLE 7.17
Corresponding Ethernet Connector and Ground Configuration Tested for Each Identified Trace in Figures 7.130 through 7.150

Plot Identity	Connector and Ground Configuration Tested
1	Baseline-54602-908LF unshielded connector with unshielded cable and ground at the source
2	54602-908LF unshielded connector and unshielded cable; radiated emissions when grounded at the source
3	J00-0045NL shielded connector and shielded cable. Radiated emissions when grounded at the source
4	J00-0045NL shielded connector and shielded cable; radiated emissions when grounded at the connector
5	J00-0045NL shielded connector and shielded cable; radiated emissions when grounded at the connector and PCB
6	08B0-1D1T-06-F shielded connector and shielded cable; radiated emissions when grounded at the source and connector and baseline
7	08B0-1D1T-06-F shielded connector and shielded cable; difference between radiated emissions when grounded at the PCB and connector
8	6605841-1 shielded connector and shielded cable; radiated emissions when grounded at the source and connector
9	6605841-1 shielded connector and shielded cable; radiated emissions when grounded at the PCB and connector
10	5-6605758-1 shielded connector and shielded cable; radiated emissions when grounded at the source and connector
11	5-6605758-1 shielded connector and shielded cable; radiated emissions when grounded at the PCB and connector
12	RJHSE-5380 shielded connector and shielded cable; radiated emissions when grounded at the source and connector
13	RJHSE-5380 shielded connector and shielded cable; radiated emissions when grounded at the PCB and connector
14	7499010122 shielded connector and shielded cable; radiated emissions when grounded at the source
15	7499010122 shielded connector and shielded cable; radiated emissions when grounded at the connector
16	7499010122 shielded connector and shielded cable; radiated emissions when grounded at the connector and PCB
17	RJ45-8LC2-B shielded connector and shielded cable; radiated emissions when grounded at the source
18	RJ45-8LC2-B shielded connector and shielded cable; radiated emissions when grounded at the connector
19	RJ45-8LC2-B shielded connector and shielded cable; radiated emissions when grounded at the connector and PCB

TABLE 7.18

Corresponding USB Connector and Ground Configuration Tested for Each Identified Trace in Figures 7.151 through 7.158

Plot Identity #	Connector and Ground Configuration Tested
20	Baseline-690-004-621-013 USB type A connector with ground at source and unshielded cable
21	690-004-621-013 unshielded USB type A connector and unshielded cable; radiated emissions when grounded at the source and grounded at the PCB near connector
22	690-004-621-013 unshielded USB type A connector and shielded cable; radiated emissions when grounded at the source
23	690-004-621-013 unshielded connector USB type A and shielded cable; radiated emissions when grounded at the connector
24	690-004-621-013 unshielded connector USB type A and shielded cable; radiated emissions when grounded at the connector and PCB
25	690-004-660-023 unshielded USB type B connector and shielded cable; radiated emissions when grounded at the source
26	690-004-660-023 unshielded USB type B connector and shielded cable; radiated emissions when grounded at the connector
27	690-004-660-023 unshielded USB type B connector and shielded cable; radiated emissions when grounded at the connector and PCB

The Ethernet connectors show different levels of emissions, but these are not consistent over the full frequency range. For example, plot 7 is of the 08B0-1D1T-06-F with the connector grounded to the PCB and the enclosure. At low frequencies, the emissions are low, but above 500 MHz they are much higher.

Plot 9 is of the 6605841-1 connector grounded at the PCB and the enclosure.

Emissions from 30 to 140 MHz are high, similar to an unshielded cable, but from 700 MHz to 1.2 GHz they are low.

As with the Ethernet connector, the lowest level of emissions from the USB connector at most frequencies is with the metal case connected both to the PCB at the connector and to the enclosure.

Due to the low SE of Ethernet and USB cables and connectors, the assumption is that the transfer impedance of the connectors is high. A metal shell USB receptacle and plug were tested; the contact of the receptacle shell to the enclosure was via two fingers that are compressed to make this contact. The transfer impedance of the USB connector is shown in Figure 7.159 and its value in Ohms, whereas an effective connector has a transfer impedance of milliohms or a fraction of milliohm.

7.17 ALTERNATIVE CABLE SHIELD TERMINATION TECHNIQUES

Although connectors are the most common method for terminating shields to an enclosure, bulkhead, or ground, other methods may be more suitable. For example, when signal and return conductors in shielded cable must be connected to a terminal block, the most common shield termination method is the ubiquitous "pigtail." Other options are available with a lower termination impedance.

One of the simplest and most effective techniques for shield termination is to capture it against the enclosure under a ring. Holes are made in the ring for fasteners. The cable penetrates through a hole in the enclosure. As the ring clamps the shield around 360°, no significant apertures exist and the SE is high. Figure 7.160 illustrates this technique.

A connection that is as good is to use a hose clamp to connect the shield to a nipple mounted on the enclosure as shown in Figure 7.161. Here, the cable and nipple are large, and when a smaller nipple and cable are used, a band may be used instead of the hose clamp.

Another form of shield termination uses a compression fitting, similar to the fitting used in plumbing (in experimental and some commercial applications, we use the plumbing fittings). Here,

FIGURE 7.159 Surface transfer impedance of USB connector.

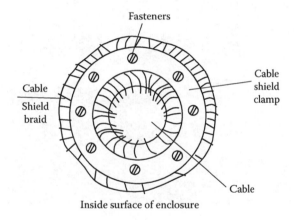

FIGURE 7.160 Clamping the shield under a ring.

FIGURE 7.161 Braid connected with a hose clamp to a nipple mounted on the enclosure.

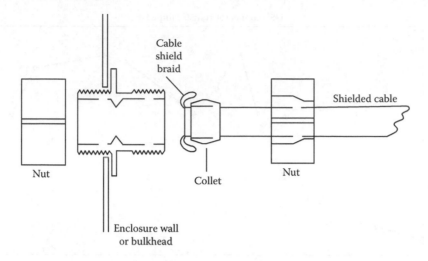

FIGURE 7.162   Collet used to clamp the shield in a fitting.

FIGURE 7.163   Photo of plumbing fittings used for shield termination.

the shield is clamped behind a sleeve or compression collet using a nut and the fitting at the other end may be held against the enclosure wall by a nut, threaded into the wall or soldered. Figure 7.162 shows this technique and Figure 7.163 is a photo of the fittings.

One potential disadvantage is that where the cable is exposed to vibration or flexing, this technique may cut the braid in time.

The next method is less satisfactory as it is easy to leave a gap between the end of the shield and the aperture in the enclosure that will compromise the shielding. However, it is fairly easy to make, and if properly installed, it will typically achieve adequate shielding.

Figure 7.164 illustrates this technique. A metal cable clamp is manufactured with a threaded portion that screws into a threaded enclosure or bulkhead.

A number of cable glands are available, for example, from PFLITSC. These exhibit strain relief, unlike the Figure 7.162 technique, but do not provide a termination for the shield of a cable. However, after the cable clamp section, there is often a section of metal tube through which the cable passes. By opening up the shield and pushing a flexible rubber or neoprene ring between the center conductor(s) and the shield the shield is then larger than the tube inner diameter. By pushing

FIGURE 7.164 Cable clamp screwed to enclosure or bulkhead wall.

FIGURE 7.165 Modifying a cable clamp to achieve a shield termination.

the shield back into the tube, there is now an interference fit between the shield and the inner conductive surface of the tube. This is illustrated in Figure 7.165.

If a cable clamp cannot be used, the braid of the shield can be soldered directly to a solder lug, which is bolted to the chassis keeping the braid length very short.

Another use of the braid to solder lug termination is where the cable enters an enclosure as shown in Figure 7.166.

FIGURE 7.166   Cable enters enclosure.

One design was to have the enclosure manufactured in two halves in the form of a clam shell. The wall material was thin and contained apertures in the form of half circles, and when the two parts of the enclosure were mated, these formed circular apertures through which the cables were routed. The goal was to fill the apertures with circular conductive gasket that contacted the bared shield on the cable. The manufacturing problem was to maintain the location of the gasket between the thin walls and the EMC problem was that the gasket only contacted the enclosure over a very small area and there is no strain relief. One solution is shown in Figure 7.167. However, if a metal enclosure is used, it requires a difficult-to-machine enclosure. An alternative is a plastic enclosure with a very good conductive coating, such as the Artcraft, which was tested with the test results described in Chapter 6. The enclosure can be molded and coated with the conductive coating inside and outside.

When the conductors in a shielded cable connect to a terminal block on a grounded structure, but do not enter an enclosure, it is better to use a cable clamp as shown in Figure 7.168 than a pigtail or solder lug.

The shield termination techniques shown in Figures 7.162 through 7.168 are effective having up to 1 GHz. But increasingly the shield attenuation is degraded due to the radiated coupling from the shield to the unshielded sections of cable. If a cable must terminate at a bulkhead and a connector is not desirable, the bared shield can be soldered to a brass plate connected to a slot cut or milled

FIGURE 7.167   Gasket in tube connected to braid shield.

Terminal
block

Clamp

Exposed
braid

Cable
clamp

Shielded
cable

FIGURE 7.168   Cable shield terminated under a cable clamp.

Brass plate

Shielded cable

Cable braid

soldered to plate
around periphery

Bulkhead

FIGURE 7.169   Cable shield connected to a bulkhead.

into the top of the bulkhead, as shown in Figure 7.169. If the cable must be soldered at either end to a circuit, the cable and brass plate can be manufactured as a cable assembly.

Finger stock may be soldered to the top of the brass plate and will be compressed by the lid to provide a good EMI seal.

Transit systems are used to seal against or exhibit resistance to fire, water, gas, chemicals, explosion, smoke, and vermin and to prove EMC.

They are used in marine, railway, water and chemical plants, offshore platforms, nuclear plant, engine test facilities, tunnel systems, and national defense.

EMI transits are used for the penetration of cables and waveguides and to achieve a conductive bond to bulkheads and to the decks of ships. Multiplug transits (NELSON) are available that can be inserted into pipes with diameters of 2–8 in.; by the use of Nelson insert blocks, these can accommodate a wide range of cables and waveguides.

A poured transit is another technique for bonding the shield of cables or waveguides to a deck or bulkhead. In a poured transit, a conductive compound is used to make electrical contact at the penetration of the cable or waveguide; either sealing compound or a swell strip and impregnation fluid is used to seal the conductive compound from moisture, thereby minimizing corrosion. The major disadvantage of the poured transit is the difficulty of adding or removing a cable.

Another type of transit is the nonpoured rubber block transit, which is constructed of upper and lower module halves of conductive rubber that clamp the cable and are mounted into a frame. By peeling off the inner sheets of the module, it can be modified to fit a number of different cable diameters. The insulation is removed from the cable, exposing the shield, and this exposed area is clamped between the module halves. To make a good electrical contact between the cable shield and the metal frame, a strip of copper tape is placed on the inner surface of each half of the module prior to clamping. The modules are then assembled in rows in the frame. Once the packing space in the frame is full of modules, some of which may be empty and others of which may contain cables, the modules are compressed via a compression plate and the upper space filled with a top packing piece.

Roxtec manufactures a number of EMC inserts with circular and square frames, and one of the inserts is shown in Figure 7.170.

Another manufacturer of pipe and cable modular transits is MCT Brattberg. These are designed for marine and railway applications and provide protection against EMP and EMI.

Hawke EMC transit systems are also designed for EMC, and they have published SE tests using a MIL-STD-285 test method. The published SE is 70 dB at 100 kHz, 95 dB at 1 MHz, 60 dB at 10 MHz, 90 dB at 100 MHz, 60 dB at 1 GHz, and 40 dB at 10 GHz.

The computer programs at the end of Chapter 2 also available in C$^+$ or C^{++} from emc-consultinginc.com, can be used to calculate the field coupling to a loop, transmission line, and a short wire terminated on a ground plane (monopole antenna), as can the emissions from a loop, short wire/cable, or transmission line. Chapter 11 describes case studies of coupling to wires and cable shielding against 2–31,000 V/m fields over the frequency range from 14 kHz to 40 GHz. These techniques can be used to calculate the current flowing through the transfer impedance of the connector and the transferred voltage.

FIGURE 7.170   Photo of Roxtec transit inserts.

7.18 PRACTICAL LEVEL OF CABLE SHIELDING REQUIRED TO MEET MIL-STD/ DO-160C OR COMMERCIAL RADIATED EMISSION REQUIREMENTS

The practicing electronic design engineer who is not a specialist in EMC often requires an idea of the level of cable shielding and the type of cable shield termination required to meet an EMI requirement. To help provide some of this information, a cable SE test was made with a number of different test configurations and cable shield termination techniques. The source of RF current flow on the cable was a battery-supplied clock and driver contained within a well-shielded enclosure that generated the fundamental frequency and harmonics up to 1000 MHz. With the source powered up in the enclosure and without a cable attached, the emissions from the RF source were measured and found to be below the noise floor of the preamplifier and spectrum analyzer over the 40–1000 MHz frequency range. Thus, the source of emissions is confined to the cable alone. A log periodic biconical antenna was used to measure radiated emissions over the 40–1000 MHz frequency range. The clock frequency was 40 MHz, and the driver output was 3.8 V peak, with a rise time of 5 ns and a fall time of 5 ns. An MIL-STD-D38999 connector was mounted on the enclosure and used for the clock signal. A Glenair M85049 non-EMI backshell with a cable strain relief was used on the external connector. A 2.8 m long RG58 50 Ω cable was used to connect the signal to a 50 Ω load at the far end of the cable. The 50 Ω load was contained in a brass enclosure that was soldered on all sides, and the braid of the RG58 cable was soldered to the case around 360°, as the cable entered the brass enclosure. The RG58 shield was thus used as the signal return, and the full signal current flowed on the inside of the shield.

Figure 7.171a shows the connection of the source, shielded cable, and load for the case where the shield braid is connected to the strain-relief clamp on the backshell. Figure 7.171b shows the source and load enclosure connected to the ground plane in the MIL-STD/DO-160 test setup. Figure 7.171c is the typical commercial test setup for radiated emissions contained in ANSI C63.4. The cable shield was terminated at the source end in one of the following four methods:

1. Shield taken through a pin in the D38999 connector and then connected to the inside of the enclosure, using a pigtail in the external connector and inside the enclosure
2. Shield connected using a 1 in. pigtail to a solder lug fastened beneath one of the screws in the backshell strain relief
3. Shield braid connected directly to a solder lug fastened beneath one of the screws in the backshell strain relief
4. Large overbraid clamped to the shield of the RG58 cable using two hose clamps and to the Glenair backshell using a large hose clamp (this termination technique produces results close to an EMI backshell with 360° shield termination)

The first test setup was in accordance with MIL-STD-462 or DO-160, with the two enclosures bonded to a ground plane. The cable was routed 5 cm above and 10 cm from the front edge of a ground plane located 1 m above the conductive floor in a semi-anechoic chamber. The tip of the measurement antenna was located 1 m from the cable. Radiated emission measurements were made at the fundamental and the harmonic frequencies of the 40 MHz clock up to 1000 MHz. The radiated emissions were narrowband in nature, and the measured level was almost independent of the measurement bandwidth.

Figure 7.172 shows the comparison of the emissions with the four shield termination techniques to the MIL-STD-461C Part 2 narrowband limits for aircraft showing curve 1 for the Army and curve 2 for the Air Force and Navy. As the figure shows, only the full overbraid shield termination, #4, comes close to meeting the requirements, and even this "best" method results in emissions that are 5 dB above curve 1 at 120 MHz and 5 dB below curve 2 at 120 MHz. The weak link is not the #4 shield termination technique but the high level of current flowing at the second and third harmonics (80 and 120 MHz) and the limited SE of the RG58 at these frequencies. Although the SE of the

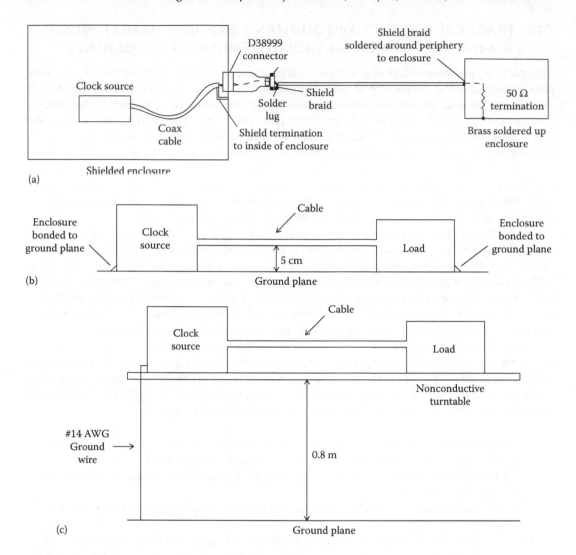

FIGURE 7.171 (a) Enclosure, source, and load connection. (b) MIL-STD-462/DO-160 radiated emission test setup. (c) ANSI C63.4 radiated emission test setup.

cable continues to decrease with increasing frequency, it does reach a limit, due to the long-line effect, whereas the current flow on the cable at these upper harmonics continues to reduce. Another reason for reduced radiation from the cable is that it becomes electrically long and a less effective antenna. The net effect is that the radiated emissions above 120 MHz are well below either the #1 or #2 curve when the full overbraid termination technique is used. For the braid-to-backshell shield termination, #3, the limitation is the cable SE at 120 MHz. Above 120 MHz, the braid termination and the aperture in the backshell at the strain relief are the weak links, and this means that even the #3 termination is above curve 1 at a number of frequencies and even above curve 2 at 280 MHz. The 1 in. pigtail is, as we would expect, less effective than a short braid termination, and bringing the shield through a pin in the connector and attaching this to the inside of the enclosure results in the highest level of emissions, that is, the least effective. One reason is the increased length of the shield connection caused by the external pigtail in the backshell, the pins of the connector and in the internal pigtail.

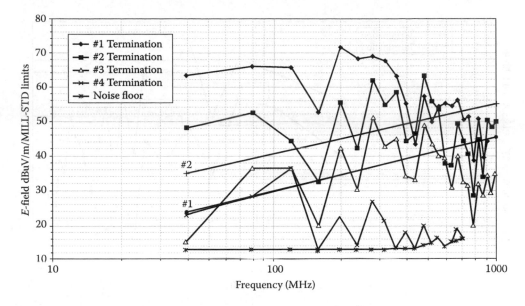

FIGURE 7.172   Radiated emissions compared to MIL-STD-461 C Part 2 with the four types of shield termination.

To reduce emissions, the use of a double-braid-shielded cable will be required to meet the curve 1 requirements. Alternatively, a twisted shielded pair can be used in which the signal return current flows on one of the pair and not on the shield. An even better method is to use an overall braid and to use the RG58 shield as a signal return path. The shield is connected to signal ground inside the enclosure using a pin in the connector. The overall braid shield is connected at both ends of the cable to an EMI backshell. This overall braid is not used to carry signal return current, and thus, the radiation, due to the limited SE in the cable and due to the shield termination, is dramatically reduced. This type of overbraid is ideal for the typically multipin D38999 connector, for the cable bundle inside the shield can contain a mix of unshielded wires, individually shielded wires, and multiconductor shielded cables.

The same RF source cable and enclosure were also measured in the ANSI C63.4 radiated emission test setup on a 3 m OATS. In this test setup, the enclosure containing the source and the enclosure containing the load were mounted on nonconductive turntable located 0.8 m above the OATS ground plane. The 2.8 m long cable was laid out with a horizontal section and also looped down at the back of the table. A 14AWG wire connected the source enclosure to the OATS ground plane. Measurements were made with the antenna oriented both horizontally and vertically and scanned in height to measure the highest emission levels. The turntable was also slowly rotated, again to measure the highest level of emissions. Measurements were made using the quasi-peak detector and a 120 kHz bandwidth.

Figure 7.173 shows the maximum level of emissions with the antenna either vertically or horizontally oriented using the braid shield termination #3 and the full overbraid #4. The radiated emissions are compared to the FCC Part 15 Class A and Class B limits in Figure 7.173. The shield terminations with #3 and with #4 are above the Class B limit, and the best shield termination, #4, is just at the Class A limit at 120 MHz. The problem here is again the signal current return on the shield and the limiting SE of the cable. We know that the #1 and #2 shield termination methods are less effective, and we know by how much, so these were omitted from the OATS tests. It may be surprising that the FCC limits are exceeded, for the measurement distance is 3 m instead of the MIL-STD 1m. Also,

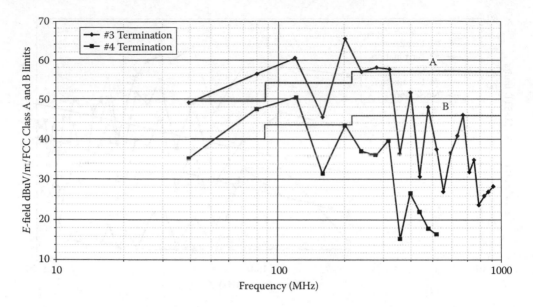

FIGURE 7.173   Radiated emissions measured on a 3 m OATS with two types of shield termination and plotted against the FCC Class A and FCC Class B limits.

MIL-STD-461 curve #1 is 9 dB below the FCC Class B limit at 120 MHz. The reason that radiated emissions are increased in the ANSI C63.4 test setup is that the cable is 0.8 m above a ground plane and the MIL-STD-462 distance is only 5 cm; the closer a cable is routed to ground, the lower the level of radiation.

The measurements in Figures 7.172 and 7.173 are valid for a 40 MHz 3.8 V clock into a 50 Ω load and a 2.8 m long cable. The results may be scaled to your specific signal level and the level of emissions corrected. For example, the emissions may be adjusted based on signal current, which in the test setup is equal to the shield current. Thus, for a 40 MHz clock with a 1 V signal into a 100 Ω load and the same rise and fall times as the 40 MHz clock, the reduction in emissions will be approximately 20 log [(3.8/50)/(1/100)] = 17 dB at any of the harmonics. For different clock frequencies, a Fourier analysis of the current at the different harmonics can be made and compared to a Fourier analysis of the current at the harmonics of the 40 MHz clock pulse used in the test. Even though the harmonics of the two clock pulses do not exactly coincide in frequency, an approximate correction to the emissions shown in Figures 7.172 and 7.173 can be made.

In the previous example, the termination of the shield of a cable was measured with a signal flowing on the center conductor and returning on the shield. One of the most surprising facts for many people is that with no signal flowing on the center conductor(s) and with the shield terminated to the enclosure, the RF currents flowing on the shield may result in emissions that are above the specification limit.

One common method to terminate the shield of a cable is to connect it to a pin in a connector and then terminate it to the inside of a conductive enclosure. Often, the shield termination is connected to a PCB ground plane containing digital circuits and then the PCB ground plane is connected to the inside of the enclosure. In other systems, multiple PCBs connect digital ground to the enclosure. RF currents are set up on the inside of the enclosure due to radiated coupling from PCBs, backplane and internal wiring, as well as multiple ground-plane connections.

This means that a voltage is impressed on the shield even when no signal flows on the center conductor(s) as was the case with the previous example.

7.19 CONNECTION OF SHIELD TO THE OUTSIDE OR INSIDE OF THE ENCLOSURE?

To demonstrate the importance of the shield termination to the enclosure, a brass enclosure was manufactured and soldered together. Inside the enclosure, a battery-powered 25 MHz oscillator was connected to a driver. The output of the driver was connected to a 28 cm loop of wire that was terminated by a 50 Ω resistor back to the driver ground. The output of the driver was a 4.8 V 25 MHz square wave with a rise time of 1.84 ns and a fall time of 2.1 ns.

This loop was not connected to chassis; however, the ground point for the wire and driver was connected to chassis via a 4 cm wire. These connections are shown in Figure 7.174.

In this configuration, only a single-point ground was made and the signal current did not flow in the enclosure only in the loop. The acid test of a soldered or welded enclosure for shielding integrity is to see if it can hold water, which this enclosure could until a 3.5 mm hole was made in the center of the enclosure.

To simulate a shield connection through a pin in a connector, a 4 cm long #22AWG wire loop was soldered 2 cm from the hole to the inside of the enclosure and the far end brought out through the hole. To simulate the connection to the PCB and then to the enclosure, a 10 cm long 22AWG wire was connected to the PCB and brought out through the hole. The same point on the PCB was connected to the inside of the enclosure with a 4 cm long wire. Figure 7.175 shows the wires disconnected from the shield, but in the measurements, one of the wires was connected to the shield.

The enclosure was bonded to a table top ground with a 2 m shielded cable located 10 cm from the front edge and 5 cm above the ground plane in an MIL-STD461 standard test configuration.

A 20–1200 MHz log periodic antenna was located 1 m from the wire. The antenna was connected to a 30 dB, 0.25–1200 MHz low-noise preamplifier that was connected to a 30 Hz to 22 GHz low-noise spectrum analyzer. The antenna, preamplifier, and spectrum analyzer are all calibrated, but prior to the test, a signal generator was used in a sanity check of the spectrum analyzer and preamplifier.

The MIL-STD-461E RE102 limit for fixed-wing internal aircraft with length less than 25 m was chosen to compare emissions too.

The first test was with no wires brought out of the hole and with the shield soldered to the outside of the enclosure. This simulates a low-impedance shield termination such as through an EMI backshell.

The radiated emissions from 25 MHz to 1 GHz were measured and compared to the limit. The plot is shown in Figure 7.176 and the levels in Table 7.19.

FIGURE 7.174   28 cm loop and 4 cm long ground.

FIGURE 7.175   4 cm and 10 cm ground locations.

FIGURE 7.176   Cable shield to outside of enclosure—emission at 436.273 MHz.

TABLE 7.19
Cable Shield to Outside of Enclosure

Test Result Sheet

Sheet	Date	Customer	Equipment	Specification	Test	BW	VBW	Reference	Span
7.1	9-10-2014	R&D	SN	MIL-STD-461E	RE102 Fixed-Wing Internal	9 kHz and 120 kHz			
			LPB, EMC 0.25–1 GHz Preamp, 1 m H&S, 2 m H&S						

Plot	Frequency (MHz)	Measured (dBuV)	A.F. (dB)	Preamp Gain (dB)	Cable Attenuation (dB)	Cable Attenuation (dB)	Corrected (dBuV/m)	RE102 Limit (dBuV/m)	Margin (dB)
	30.000	17.0	18.2	−32.5	0.0	0.1	2.8	34.0	31.2
	100.000	25.7	9.8	−32.8	0.0	0.1	2.8	34.0	31.2
2-1	436.273	29.0	16.7	−32.7	0.0	0.1	13.1	46.8	33.7
	441.000	29.0	17.0	−32.7	0.0	0.1	13.4	46.9	33.5
	818.000	27.0	22.8	−28.2	0.2	0.3	22.1	52.2	30.1
	1000.000	25.7	20.0	−25.0	0.2	0.4	21.3	54.0	32.7

Notes: Horizontal. Cable shield to outside of enclosure = ambient.
Test result sheet 7-1, 25 MHz to 1 GHz.

FIGURE 7.177 Cable shield termination 4 cm inside enclosure—emission at 599.768 MHz.

The levels are at least 30 dB below the limit and are the same as the ambient proving that no RF current flows on the outside of the enclosure and that radiated coupling through the 3.5 mm hole is negligible.

The second test was with the cable shield terminated to the inside of the enclosure via the 4 cm loop. The plot is shown in Figure 7.177, and the levels are shown in Table 7.20.

Here, the emissions are 1.1–3.2 above the limit. These emissions are specific to the test loop and driver. In a practical EUT with many sources of emissions and a poor grounding scheme, emissions may be higher. Likewise with a good layout and grounding, emissions may be lower.

By reducing the length of the insulated internal wire to 2 cm and running it in direct contact with the inside of the enclosure, emission reduces to 9.5 dB below the limit. This is still 20 dB above the emissions with the shield terminated to the outside of the enclosure, and if the wire is brought out through a connector, this configuration may be impractical.

The plot for this configuration is shown in Figure 7.178, and Table 7.21 shows the levels.

In the case of the shield terminated to the enclosure via a 10 cm long wire to the PCB ground plane, the emissions are up to 22 dB above the limit. These are shown in Figure 7.179 and Table 7.22.

The conclusion is that terminating the shield of a cable to the inside of an enclosure versus the outside may easily result in failing EMI emission limits due to radiated coupling to the ground wire and RF currents flowing on the inside of the enclosure.

TABLE 7.20

4 cm Long Connection of Shield-to-Inside Enclosure

Test Result Sheet

Sheet	Date	Customer	Equipment	Specification	Test	BW	VBW	Reference	Span
7.2	10-10-2014	R&D	SN	MIL–STD–461E	RE102 Fixed-Wing Internal	120 kHz			
			LPB, EMC 0.25-1 GHz Preamp, 1 m H&S, 2 m H&S						

Plot	Frequency (MHz)	Measured (dBuV)	A.F. (dB)	Preamp Gain (dB)	Cable Attenuation (dB)	Cable Attenuation (dB)	Corrected (dBuV/m)	RE102 Limit (dBuV/m)	Margin (dB)
	450.000	72.0	17.3	–32.6	0.0	0.1	56.8	47.0	–9.8
	650.000	65.1	19.7	–30.3	0.1	0.2	54.8	50.2	–4.6

Notes: V or H. Cable shield to 4 cm wire inside enclosure.
Test result sheet 7.2, 25 MHz to 1 GHz.

FIGURE 7.178 Cable shield termination 2 cm flat to enclosure—emission at 300.501 MHz.

FIGURE 7.179 Cable shield termination 10 cm—emission at 300.501 MHz.

TABLE 7.21
2 cm Cable Shield Flat-to-Inside Enclosure

Test Result Sheet

Sheet	Date	Customer	Equipment	Specification	Test	BW	VBW	Reference	Span
7.3	10-10-2014	R&D	SN	MIL-STD-461E	RE102 Fixed-Wing Internal	120 kHz			

LPB, EMC 0.25-1 GHz
Preamp, 1 m H&S, 2 m H&S

Plot	Frequency (MHz)	Measured (dBuV)	A.F. (dB)	Preamp Gain (dB)	Cable Attenuation (dB)	Cable Attenuation (dB)	Corrected (dBuV/m)	RE102 Limit (dBuV/m)	Margin (dB)
	100.000	45.1	9.8	-32.8	0.0	0.1	22.2	34.0	11.8
	225.000	47.6	12.4	-33.1	0.0	0.1	27.0	41.0	14.0
2-2	300.501	54.1	12.9	-33.2	0.0	0.1	33.9	43.5	9.6
	450.000	47.3	17.3	-32.6	0.0	0.1	32.1	47.0	14.9
	550.000	48.8	19.0	-31.7	0.1	0.2	36.4	48.8	12.4
	650.000	51.0	19.7	-30.3	0.1	0.2	40.7	50.2	9.5
	700.000	42.6	20.5	-29.7	0.1	0.3	33.3	50.9	17.1
	825.000	44.1	23.1	-28.2	0.2	0.3	39.5	52.3	12.8
	925.000	37.3	23.8	-26.8	0.2	0.4	34.9	53.3	18.4

Notes: Horizontal. Cable shield 2 cm flat to enclosure.
Test result sheet 7.3 25 MHz to 1 GHz.

TABLE 7.22

Cable Shield Connected via 10 cm Internal Wire

Test Result Sheet

Sheet	Date	Customer	Equipment	Specification	Test	BW	VBW	Reference	Span	Margin (dB)
7.4	10-10-2014	R&D	SN	MIL-STD-461E	RE102 Fixed-Wing Internal	120 kHz				
			LPB, EMC 0.25–1 GHz Preamp, 1 m H&S, 2 m H&S							

Plot	Frequency (MHz)	Measured (dBuV)	A.F. (dB)	Preamp Gain (dB)	Cable Attenuation (dB)	Cable Attenuation (dB)	Corrected (dBuV/m)	RE102 Limit (dBuV/m)	Margin (dB)
	75.000	69.5	12.5	−32.8	0.0	0.1	49.3	34.0	−15.3
	100.000	79.0	9.8	−32.8	0.0	0.1	56.1	34.0	−22.1
	300.000	79.8	12.9	−33.2	0.0	0.1	59.6	43.5	−16.1
2-4	300.501	80.1	12.9	−33.2	0.0	0.1	55.9	43.5	−16.4
	350.000	79.1	14.8	−33.0	0.0	0.1	61.0	44.9	−16.1
	450.000	72.0	17.3	−32.6	0.0	0.1	56.8	47.0	−9.8
	650.000	65.1	19.7	−30.3	0.1	0.2	54.8	50.2	−4.6

Notes: V or H. Cable shield to 10 cm wire inside enclosure.
Test result sheet 7-4, 25 MHz to 1 GHz.

REFERENCES

1. J.S. Hofstra and L.O. Hoeft. Measurement of surface transfer impedance of multi-wire cables, connectors and cable assemblies. *Record of the IEEE International Symposium on EMC*, 1992.
2. L.O. Hoeft, T.M. Sala, and W.D. Prather. Experimental and theoretical comparison of the line injection and cylindrical test fixture methods for measuring surface transfer impedance of cables. *IEEE 1998 Symposium Record.*
3. F. Boyde and E. Clavelier. Comparison of coupling mechanisms on multiconductor cables. *IEEE Transactions on Electromagnetic Compatibility*, November 1993, 35(4).
4. F. Broyde and E. Clevalier. Definition, relevance and measurement of the parallel and axial transfer impedance. *IEEE Symposium Record, 1995.*
5. J.S. Hofstra, M.A. Dinallo, and L.O. Hoeft. Measured transfer impedance of braid and convoluted shields. *Record of the IEEE International Symposium on EMC 482-488*, 1982.
6. L.O. Hoeft and J.S. Hofstra. Experimental evidence for porpoising coupling and optimization in braided cable. *Eighth International Zurich Symposium and Technical Exhibition on EMC*, March 7–9, 1989.
7. K.L. Smith. Analysis and measurement of CATV drop cable RF leakage. *IEEE Transactions on Cable Television*, October 1979, CATV-4(4).
8. B. Demoulin and P. Degauque. Shielding performance of triply shielded coaxial cables. *IEEE Transactions on Electromagnetic Compatibility*, August 1980, EMC-22(3).
9. K. Casey. Low frequency electromagnetic penetration of loaded apertures. *IEEE Transactions on Electromagnetic Compatibility*, November 1981, EMC-23(4).
10. E.F. Vance. *Coupling to Shielded Cables.* Robert E. Krieger, Malabar, FL, 1987.
11. A.P.C. Fourie, O. Givati, and A.R. Clark. Simple technique for the measurement of the transfer impedance of variable length coaxial interconnecting leads. *IEEE Transactions on EMC*, May 1998, 40(2).
12. L.O. Hoeft and J.L. Knighten. Measured surface transfer impedance of cable shields that use combinations of braid and foil and are used for 1-Gb/s Data Transfer. *IEEE EMC Symposium*, 1998.
13. L.O. Hoeft. Measured electromagnetic shielding performance of commonly used cables and connectors. *IEEE Transactions on EMC*, August 1988, 30(3).
14. F. Broyde, E. Clavelier, D. Givord, and P. Vallet. Discussion of the relevance of transfer admittance and some through elastance measurement results. *IEEE Transactions on EMC*, November 1993, 35(11).
15. M.A. Dinallo, L.O. Hoeft, J.S. Hofstra, and D. Thomas. Shielding effectiveness of typical cables from 1 MHz to 1000 MHz. *IEEE International Symposium on Electromagnetic Compatibility*, 1982.
16. P. Degauque and A. Zeddam. Remarks on the transmission line approach to determining the current induced on above-ground cables. *IEEE Transactions on Electromagnetic Compatibility*, February 1988, 30(1).
17. J.C. Santamaría and Louis J. Haller. A comparison of unshielded wire to shielded wire with shields terminated using pigtail and shields terminated through a conducting elastomer. *IEEE EMC Symposium Record*, 1995.
18. G. Chandesris. Effect of ground connection on the coupling of disturbing signals to a coaxial line. *Fourth International Zurich Symposium and Technical Exhibition on Electromagnetic Compatibility*, March 10–12, 1981.
19. T. Hubing and J.F. Kaufman. Modeling the electromagnetic radiation from electrically small table top products. *IEEE Transactions on Electromagnetic Compatibility*, February 1989, 31(1).
20. D. Fernald. Comparison of shielding effectiveness of various backshell configurations. *ITEM*, 1984.
21. J. Zorzy and R.F. Muehlberger. RF leakage characteristics of popular coaxial cables and connectors, 500 MHz to 7.5 GHz. *The Microwave Journal*, November 1961.
22. C.R. Paul and D.R. Bush. Radiated emissions from common mode currents. *Proceedings of the IEEE International Symposium on Electromagnetic Compatibility*, 1982.
23. L.O. Hoeft and J.S. Hofstra. Transfer impedance of overlapping braided cable shields and hose clamp shield terminations. *IEEE International Symposium Record*, 1995.
24. F.B.M. van Horck, A.P.J. van Deursen, P.C.T. van der Laan, P.R. Bruins, and B.L.F. Paagmans. A rapid method for measuring the transfer impedance of connectors. *IEEE Transactions on Electromagnetic Compatibility*, August 1998, 40(3).

25. B. Eicher, C. Staeger, and B. Szentkuti. Berne, Heinz Fahrni, Solothun, Technische Mitteilung PTT No3+4, 1988. Published by the Switt PTT Berne. Simple and accurate screening measurements on RF cables up to 3 GHz.

26. L.O. Hoeft and J. Hofstra. Electromagnetic shielding provided by selected DB-25 subminiature connectors. Electromagnetic Compatibility 1991 Symposium Record. *IEEE 1991 International Symposium.*

27. L.O. Hoeft and J.S. Hofstra. Electromagnetic shielding of circular and rectangular connectors, backshells, and braid terminations. *IEEE 1992 Regional Symposium on Electromagnetic Compatibility.* doi: 10.11.09/ISEMC 1992 257571.

28. C.W. Dole and J.W. Kincaid. *Screening Attenuation of Long Cables.* Belden Electronics Division, Richmond, IN.

8 Grounding and Bonding

8.1 INTRODUCTION TO GROUNDING

Perhaps one of the most difficult topics is that of grounding. Even the definition of the term *ground* is not universally accepted, for example, "one of two conductors used to carry a single-ended signal." Or the signal or power return may be termed a *ground*. In the majority of cases, the ground is an electrical connection either to Earth or to a structure that serves a similar function to Earth.

However, we also use the term *floating ground*, which, as its name implies, is either disconnected from the enclosure, Earth, or any other structure or connected via a high impedance. Grounds often serve different functions, and it is common to encounter a number of grounds in equipment, subsystems, or systems. Throughout this book, the function of the ground, for example, safety ground or signal ground, is usually described.

In this chapter, we use the term *signal ground* to describe a reference plane that is as close to an equipotential plane as is feasible. To emphasize the definition, the term *reference ground* is often used. The term *safety ground* refers to a conductor (green wire in America) that has the purpose of conducting fault current and operating a circuit breaker or fuse and in limiting the voltage to ground during a fault. The purpose of the green wire is to protect personnel and, secondarily, the equipment. The definition of *return* used here is "a conductor that has the primary function of returning signal or power back to the source." The problem with this definition is that very often the ground and the return are the same. Thus, an analog signal return is often connected to a ground plane to which the analog supply return is connected.

The term *earth* or *earth ground* is used to describe a low-resistance connection to Earth, often made via rods, plates, or grids. The earth ground may be used to conduct fault and lightning current or to tie a reference ground as close to ground potential as possible. It should not be used for both functions! The term *bond* is used to describe the mechanical connection holding two structures together. Depending on the purpose of the bond, a maximum value of resistance is often specified for the bond.

One of the problems in discussing grounding is the difficulty in describing the geometry of the connection. It is not what you do but how you do it that counts! A schematic may show a ground connection but may not be able to describe the physical realization. As described in Section 5.1.2, above a relatively low frequency (LF) (typically kilohertz), the DC resistance no longer determines the impedance of a connection; it is the total impedance, including AC resistance, and inductive and capacitive reactance. Both parallel and series resonance circuits are formed and are determined by the geometry of the ground connection.

The implementation of the ground connection is extremely important. In Section 7.13, it was seen that the addition of a 2 in. pigtail shield connection to ground reduced the shielding effectiveness of a cable to almost that of an unshielded cable. Likewise, the implementation and location of a signal ground to chassis or enclosure ground connection can make the difference between the equipment's passing or failing EMI requirements or in exhibiting EMI.

It is advisable to consider the LF and high-frequency grounding schemes in equipment and systems separately. It is the LF scheme that is shown in power and grounding diagrams and in grounding schematics.

8.2 SAFETY GROUNDS, EARTH GROUNDS, AND LARGE-SYSTEM GROUNDING

Some of the reasons for grounding are as follows:

Compliance with safety requirements
Lightning protection
Reduction of noise in receiving antennas/equipment
Reduction of common-mode noise between equipment cases/cabinets
Reduction in radiated emissions
Protection against electrostatic buildup
Line-to-ground filter component return
Reduced crosstalk

The question mark after the point "reduction in radiated emissions" serves to remind us that radiated emissions may increase with a second low-impedance ground connection when a primary path to ground exists, as described in Chapter 7. Where possible, the connection of the AC line safety ground should be separated from any conductor that connects equipment grounds together or to a common earth ground.

It is imperative that a grounding system designed to protect against direct lightning strikes be isolated from all other grounds, except for a single connection at the earth ground electrode. The lightning ground must be capable of carrying the lightning strike current without fusing, thereby maintaining the conductive path.

Two examples of large-system grounding schemes are given in Figures 8.1 and 8.2. Figure 8.1 is an ideal large-system grounding scheme in which the steel girders or rebars in the building are bonded together. In some steel-framed buildings, the steel columns are used as part of a lightning protection scheme. Even when the steel frame of a building is not intentionally used as a lightning ground, lightning may punch through a wall of the building and current may flow on the structure. The typical 20,000 A to more than 250,000 A of primary lightning strike current then travels down the steel column, using the easiest path, to earth ground.

If the building is constructed of reinforced concrete, then the current will flow through the rebars to ground. In reinforced concrete, wood, or masonry buildings, it is advisable to provide down conductors on the outside of the building for lightning protection. It should be remembered that for all practical gauge of conductors, the conductor will exhibit an impedance, due to self-inductance, higher than that of a metal plate. Consequently, if the conductor is in close proximity to a large structure or sheet of metal, then the lightning current may well flash over and continue to earth ground in the lower-impedance metal.

In the ideal building ground scheme of Figure 8.1, the steel girders or rebars are welded together, whereas in the typical grounding scheme, no steps are taken to ensure a low-resistance bond between structure members. When a lightning strike occurs to the typical building structure, voltages appear across the ends of the poorly connected metal structure, and an E field is generated. The voltage may be high enough to result in a sideflash to adequately grounded structures, such as cables and equipment enclosures.

A risk protection for lightning protection is described in IEC-62305, which provides for the evaluation of risk based on loss to a structure, due to lightning strikes to earth ground.

Modern installations that need protection from lightning by the appropriate techniques are wind turbines and photovoltaic plants. In References [1,2,5], a number of grounding schemes are described. Field experience suggests that at least two grounding rods should be used per down conductor having a minimum length of 1.5 m with a distance between rods of approximately 2× their length.

The AC safety ground is a common source of radio frequency (RF) voltage induced in the power wiring by AM/FM transmitters or by equipment that is a major source of conducted power-line

FIGURE 8.1   Ideal large-system grounding scheme.

FIGURE 8.2   Typical large-system grounding scheme.

noise, as described in Chapter 1. The technique, shown in Figure 8.1, which reduces safety ground RF current flow, is to connect the safety ground to a low-impedance earth ground at the entry point of the power into the building. The safety code takes precedence over all other grounding rules, but it is possible to tailor the grounding scheme and comply with the code. Most electrical safety codes require connection of the safety ground and neutral to earth ground at the main service panel but do not specify a low-impedance connection.

RF voltages induced on the safety ground by equipment within the building may be reduced by the inclusion of one or more inductors in series. The inductors must be capable of carrying a fault current and thus allow the fuse or circuit breaker to function. The inductors must also exhibit a low DC resistance and low impedance at 50/60 Hz in order to maintain a low touch potential in the event of a short circuit of AC hot or neutral to the enclosure. In the grounding scheme shown in Figure 8.1, it would be correct to assume that the major portion of the fault current would flow into the ground plane and via the #2 or lower gauge wire or plate connecting the ground plane to the earth ground. An obvious question is why have the safety ground connection via the green wire? Why not rely on the grounding scheme to provide the safety ground connection? The reason the green wire must be retained is that the enclosure ground connection, shown in Figure 8.1, may be inadvertently disconnected, whereas the "green wire" is normally disconnected only when the power connector is removed from the outlet. It is for this reason that safety codes often make the green wire mandatory.

The scheme of Figure 8.1 assumes that the equipment is located on separate floors inside the building and, to achieve as low an impedance ground as possible, a #2 (or lower) gauge, wire, or a wide plate of a minimum of 1/4 in. thickness connects the ground planes on the two floors together. A lower impedance may be obtained in a building with steel girders by using the girders as a parallel path to the wire/plate. The problem is that the girders may carry lightning strike currents and will almost certainly carry AC power leakage currents from equipment such as rotating machinery, the housings of which are often electrically connected to the structure. In addition, the frame of the building is an effective antenna at AM broadcast frequencies and thus carries RF currents. For these reasons, it is inadvisable to use the building structure in a grounding scheme.

In Figure 8.2, the typical ground scheme, it is the safety ground and the shields of cables and conductors within the cable that connect the enclosures together. One connection may be made via a signal return in which RF currents and 50–180 Hz leakage currents may flow. The typical large-system ground may be adequate for digital equipment, but for analog/instrumentation equipment, the grounding may result in EMI, as discussed in Section 8.3.2 on multipoint grounding.

The voltages appearing between the earth ground-to-neutral connection within a building and earth ground outside of a building will contain AC power and RFs. For example, in a three-story office building, the measured voltage at a distance of 100 ft between the two earth grounds was 3 V at 60 Hz and approximately 350 mV at 1.1 MHz. When earth grounds are much further apart, a concomitant increase in voltage and current flow in conductors connecting the grounds together may be expected.

Thus far, the assumption has been made that the large system is powered from the same service. When this is not the case, the configuration shown in Figure 8.3a may apply. The AC neutral is connected to safety ground and earth ground in both services, and thus both the safety ground and earth ground carry some of the neutral return current. The safety grounds are connected to the metal enclosures, as often required by the safety code. A problem exists when a signal return is also connected to the enclosure in both pieces of equipment. The connection may not be immediately obvious, especially when the signal return is routed via a PCB to power return, which is in turn connected via a power supply to the enclosure. Whatever the route, some small percentage of the AC neutral return current will flow in the signal return, with the potential for EMI. One solution to the problem is to use two feeders from a service to the two pieces of equipment, in which case the only safety ground-to-neutral connection is at the feeder, as shown in Figure 8.3b. Disconnecting one enclosure from earth ground will reduce earth-borne leakage current.

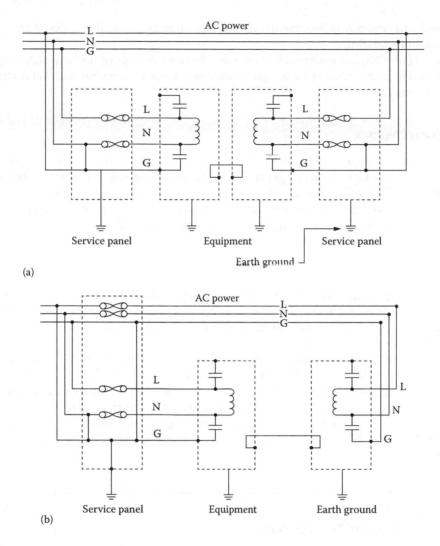

FIGURE 8.3 (a) Equipment connected to separate service panels in which neutral current flows in a signal return. (b) Equipment supplied by separate feeders from a single service panel and with a single structure-to-earth-ground connection.

The paramount concern in any power distribution is to comply with the applicable electrical safety code. The use of isolation transformers on the input side of both of the services shown in Figure 8.3a will interrupt the neutral current flow in the signal return conductor. However, the solution is expensive and may not comply with safety regulations. When the equipment is part of a system and is located some distance from other equipment in the system, it is preferable to use one location as the source of power and to feed the outlying equipment from the single source. Assuming an adequately sized "green" wire is taken to the outlying equipment, no additional earth ground is required and the problem of LF current flow in single-ended signal returns is reduced. Two situations where the isolation of equipment, such as an antenna, from earth grounds is inadvisable are when an earth ground exists for the purpose of lightning protection and when it is located close to the antenna structure. To avoid flashover, the antenna structure should be connected *by a single connection* to the earth ground. The second reason to rethink isolation of equipment from earth is when the equipment is in the proximity of a source of high-power *E* field (e.g., a transmitting antenna). The connection of the equipment structure to a low-impedance earth ground will typically reduce the RF current flow on interconnection cables, thereby reducing the probability of EMI.

The sources of electrical codes seldom consider the requirements of instrumentation systems; however, grounding schemes may be designed that improve signal grounding and that comply with safety regulations.

8.2.1 EARTH GROUND

Grounding to earth is required for safety and to minimize touch potentials due to AC line leakage or equipment breakdown or direct or indirect lightning strikes, or where a receiving or transmitting antenna uses the earth as a counterfoil (ground plane beneath an antenna that forms part of the antenna design). In practice, a low-impedance connection to earth has been found to reduce the incidence of failures in equipment, as reported in Reference [2], pp. 81–107.

8.2.1.1 Earth Resistance

Requirements on the maximum value of earth resistance imposed by insurance, power, and oil companies vary from 1 to 5 Ω. In a survey of U.S. Federal Aviation Administration Air Route Control Centers, Air Traffic Control Towers, and Long-Range Radar Sites, contained in Reference [2], pp. 135–147, the measured earth resistance varied, depending on the site, from 0.1 to 12.8 Ω.

The resistivity of Earth, expressed in ohms per centimeter, is surprisingly high. Table 8.1 shows typical values for different soils. The effect of moisture is shown in Table 8.2, salt content in Table 8.3, and temperature in Table 8.4. The conductivity of the soil is primarily electrolytic, so the higher the concentration of moisture and sodium ions, the lower the resistance. The value of earth resistance is dependent on the area, quantity, and depth of ground rods/plates. A low resistance is achieved because of the large area of the ground shell surrounding the earth electrode, as shown in Figure 8.4.

TABLE 8.1
Resistivities of Different Soils

Soil	Resistivity (Ω-cm)		
	Average	Minimum	Maximum
Fills—ashes, cinders, brine wastes	2,370	590	7,000
Clay, shale, gumbo, loam	4,060	340	16,300
Same—with varying proportions of sand and gravel	15,800	1,020	135,000
Gravel, sand, stones, with little clay or loam	94,000	59,000	458,000
Surface soils, loam, etc.	100	—	5,000
Clay	200	—	10,000
Sand and gravel	5,000	—	100,000
Surface limestone	10,000	—	1,000,000
Limestones	500	—	400,000
Shales	500	—	10,000
Sandstone	2,000	—	200,000
Granites, basalts, etc.		100,000	
Decomposed gneisses	5,000	—	50,000
States, etc.	1,000	—	10,000

Sources: U.S. Bureau of Standards Technical Report 108; Evershed & Vignoles Bulletin 245.

TABLE 8.2

Effect of Moisture Content on Earth Resistivity

	Resistivity (Ω-cm)	
Moisture Content, % by Weight	Top Soil	Sandy Loam
0	$1,000 \times 10^6$	$1,000 \times 10^6$
2.5	250,000	150,000
5	165,000	43,000
10	53,000	22,000
15	21,000	13,000
20	12,000	10,000
30	10,000	8,000

Source: Higgs, P.J., *IEE J.*, 68, 736, February 1930.

TABLE 8.3

Effect of Salt Content on Earth Resistivity

Added Salt, % by Weight of Moisture	Resistivity (Ω-cm)
0	10,700
0.1	1,800
1.0	460
5	190
10	130
20	100

For sandy loam—moisture content, 15% by weight; temperature, 17°C (63°F).

TABLE 8.4

Effect of Temperature on Earth Resistivity[a]

Temperature		Resistivity (Ω-cm)
°C	°F	
20	68	7,200
10	50	9,900
0	32 (water)	13,800
0	32 (ice)	30,000
−5	23	79,000
−15	14	330,000

[a] For sandy loam, 15.2% moisture.

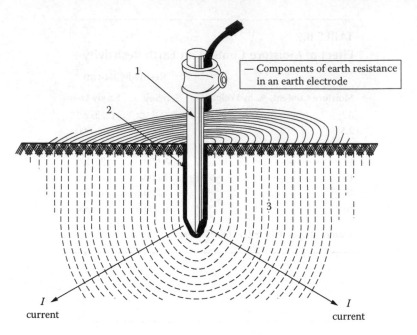

FIGURE 8.4   Ground shell surrounding earth electrode.

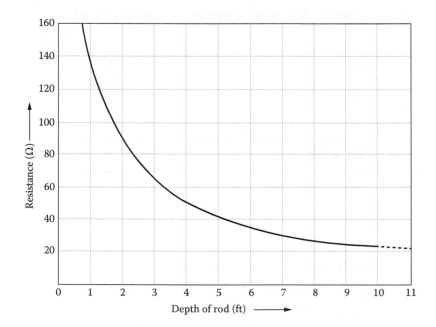

FIGURE 8.5   Effect of depth of electrode on decreasing earth resistance.

The effect of depth of a driven rod on earth resistance is shown in Figure 8.5. As the depth of a ground rod increases, the area of the shell around the rod increases. At a depth of approximately 6 ft, the resistance begins to level out. To decrease the resistance further, use multiple rods, treat the soil with salt, or increase the moisture content. The effect of increasing the number of ground rods is shown in Figure 8.6. A useful nomograph relating the basic factors affecting earth resistance is shown in Figure 8.7.

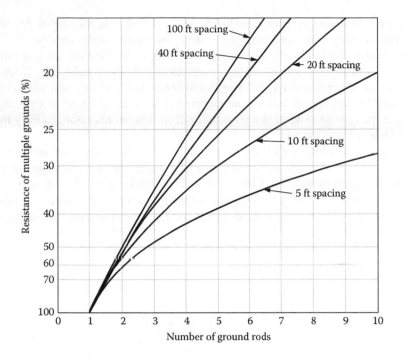

FIGURE 8.6   Effect of increasing the number of ground rods on earth resistance.

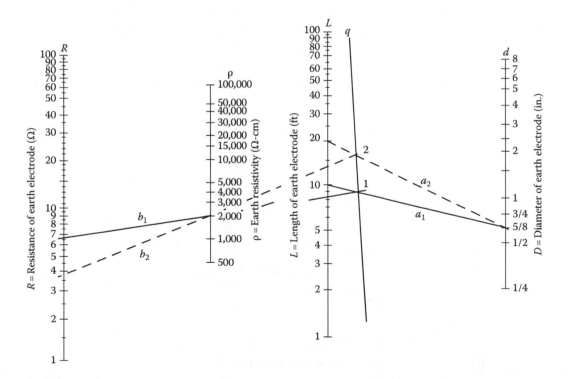

FIGURE 8.7   Nomograph of earth resistance.

The foregoing information is reproduced by the courtesy of Biddle/Megger (Biddlemegger.com) that manufactures a range of ground resistance testers. The earth tester is used in a three-terminal, fall-of-potential earth-resistance test. Two current terminals are provided on the tester and two voltages. A current is caused to flow between the ground electrode connected to the C_1 terminal and a small electrode placed approximately 50 ft away and connected to the C_2 terminal. The resistance between the small electrode and Earth is unimportant, for the requirement is only that sufficient current flows in the ground to develop a potential, and so it is not necessary to bury the electrode more than an inch or 2 into the soil.

The P_1 (potential) terminal of the Megger is connected to the C_1 (current) terminal and the P_2 terminal to a second small electrode located at a 62% of the distance between the two current electrodes (i.e., at approximately 31 ft from the electrode under test). The Megger measures the potential difference between the earth electrode and the P_2 and computes the earth resistance from the current flowing between the two current terminals. The earth resistance test setup is shown in Figure 8.8.

Tables 8.1 through 8.4 and Figures 8.4 through 8.8 are reproduced by the courtesy of Biddle/Megger.

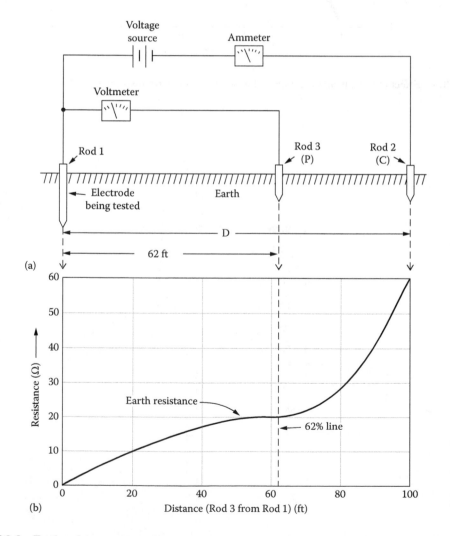

FIGURE 8.8 Earth resistance measurement test setup.

8.3 SIGNAL GROUND AND POWER GROUND

8.3.1 SIGNAL GROUND

An *ideal signal ground* is defined as an equipotential plane that serves as a reference point. In practice, a signal ground is used as a (hopefully) low-impedance path for current to return to the source (see Schemes 8.1 and 8.2). All conductors exhibit impedance, so when current flows in a ground conductor, a voltage drop occurs. A good signal-grounding scheme limits the magnitude of the current flow and the impedance of the path. The major cause of impulsive current flow is signal switching current and changes in IC supply current as the device changes state, and these currents flow back to the source on the power return, which is often the signal return. As we saw in Chapter 4, the closer the proximity of a conductor that carries RF current (signal or power-line noise) to a conductive ground plane, the lower the inductive and capacitive crosstalk. The plane does not need to be grounded or connected to either the signal or power return to be effective. However, ungrounded ground planes will have voltages developed between them and other ground planes and other grounded structures, which in turn can result in radiated coupling or crosstalk.

Section 11.8 describes PCB grounding, with Figure 11.79 illustrating common-mode voltages developed across a ground plane. Section 11.8.2 describes "good" and "bad" PCB ground planes and Section 11.8.3 describes grounding a PCB within an enclosure. In Chapter 5, the isolation or connection of an RF ground plane to digital/analog ground is discussed.

8.3.2 GROUNDING PHILOSOPHY

One of a number of different grounding schemes may be selected; however, very often the designer/systems engineer does not have a choice. The choice may be limited when the grounding scheme is dictated by either a larger system into which equipment or subsystem must fit or by commercially available peripherals to which equipment must connect.

Where a designer may have some control over the grounding scheme is within equipment. However, when standard, RS232, or similar interfaces are used, signal-to-chassis ground connections are likely in the equipment in which the interfaces terminate, and the use of a single-point scheme is not feasible. The basic grounding schemes are as follows:

1. Single point
2. Single-point star
3. Multipoint
4. Single-point DC, multipoint RF
5. Floating ground

In practice, Schemes 1 and 2 become Scheme 4 above some frequency, and Scheme 5 becomes Scheme 3, due to parasitic stray capacitance between circuit board tracks/signal wiring and nearby grounded structure and between signal conductors in cables and shields. Examples of the first four schemes are shown in Figures 8.9 through 8.12.

The aim of a single-point ground at both the board level and the system level is to reduce DC and LF current flow in the structure. The upper frequency beyond which the single-point ground scheme is no longer effective is determined by the magnitude of capacitance between the signal ground and structure or chassis. The single-point DC, multipoint RF ground scheme recognizes the limitation of the single-point scheme and deliberately introduces an RF ground via one or more capacitors in parallel. This scheme has the advantage of reducing common-mode noise voltage on the signal ground at a designated location. One preferred location is at the point of entry of interface signals to equipment. We have seen in Chapter 7, on cable radiation, that the predominant source is common-mode voltage on both shielded and unshielded cables. A second location of choice for the RF (capacitor) ground within equipment is at a common ground point to which a number of

(a)

SCHEME 8.1   Single-point DC, multipoint AC grounding scheme. Power and signal isolated from chassis by at least 10 MΩ. Maximum capacitance or equivalent impedance to chassis is 0.5 μF. Single signal ground connection from unit to unit. Single-point DC, multipoint AC grounding scheme. Power and signal ground isolated from chassis by at least 10 MΩ. Maximum capacitance or equivalent impedance to chassis is 0.5 μF. Single reference signal ground connection from equipment to equipment.   (*Continued*)

(b)

SCHEME 8.1 (*Continued*) Single-point DC, multipoint AC grounding scheme. Power and signal isolated from chassis by at least 10 MΩ. Maximum capacitance or equivalent impedance to chassis is 0.5 μF. Single signal ground connection from unit to unit. Single-point DC, multipoint AC grounding scheme. Power and signal ground isolated from chassis by at least 10 MΩ. Maximum capacitance or equivalent impedance to chassis is 0.5 μF. Single reference signal ground connection from equipment to equipment.

different types of ground are connected. The disadvantage of the multipoint RF grounding scheme is that RF current flow in signal grounds is typically higher than in the single-point ground. Assume that the potential source of common-mode noise current is known, for example, the return of a supply that generates common-mode noise between the return and case/chassis. The introduction of an inductor in the supply return may be effective in reducing the current flow on the signal return or ground plane connected to the return and reduce the voltage developed across the RF ground. Some examples of this technique are provided later. Whenever an inductor and capacitor are used in combination, a resonant circuit is formed and the inclusion of a damping resistor is advisable.

The single-point star-ground scheme is a variation of the single-point scheme that reduces common-mode noise coupling from one piece of equipment to another by the use of isolated interfaces, such as the opto-isolator shown, or signal transformers, or differential-input receivers.

In a similar scheme, the interfaces are isolated and the ground is taken from equipment to equipment as shown in Figure 8.9. This scheme may appear as a great waste of the circuitry required for isolation. However, the signal ground comes as close as possible to the ideal reference ground, which is often termed for neither signal return currents nor power supply return currents flow on the reference ground. Figure 8.13 shows a single-point ground for different PCBs within an enclosure, typically with the use of a bus bar, or a chassis layer on a backplane PCB, to connect the PCB

(a)

SCHEME 8.2 Single-point grounding scheme. Signal ground connected to chassis at one point. Power ground isolated from signal ground and chassis by at least 1 MΩ. Equipment-to-equipment signal grounds isolate single-point grounding scheme. *(Continued)*

(b)

SCHEME 8.2 (*Continued*) Single-point grounding scheme. Signal ground connected to chassis at one point. Power ground isolated from signal ground and chassis by at least 1 MΩ. Equipment-to-equipment signal grounds isolate single-point grounding scheme. (*Continued*)

(c)

SCHEME 8.2 (*Continued*)   Single-point grounding scheme. Signal ground connected to chassis at one point. Power ground isolated from signal ground and chassis by at least 1 MΩ. Equipment-to-equipment signal grounds isolate single-point grounding scheme.

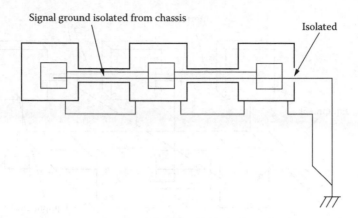

FIGURE 8.9   Single-point ground taken from one piece of equipment to another.

FIGURE 8.10   Single-point DC, multipoint radio-frequency ground.

FIGURE 8.11   Single-point star.

grounds together and to the enclosure. Although at first glance this may not seem to be a single-point ground, the impedance of the ground connection from PCB to PCB and from the bus/backplane to chassis can be kept very low, and so at lower frequencies, this is in effect a single-point ground. In very large systems, such as the space station, a single-point ground scheme becomes unwieldy and a so-called layered single-point scheme is used in which the chassis ground connection is made at a subsystem or logical EMC element (i.e., a group of equipment pieces in fairly close proximity). The use of the term *single point* for this layered scheme is not strictly accurate, however, for the scheme is much different from the multipoint ground.

The multipoint ground philosophy recognizes that the manufacturer of bought-out equipment is obliged to connect safety ground to the chassis and that the signal and secondary power grounds

FIGURE 8.12   Multipoint ground.

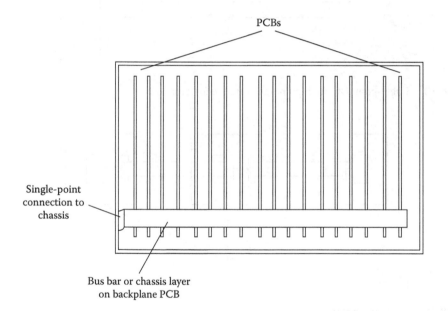

FIGURE 8.13   Single-point ground connecting together PCBs within an enclosure.

are also often connected to the enclosure. In the multipoint scheme, the signal interfaces are usually single ended, with the signal return connected to the enclosure in both pieces of equipment. The scheme by its nature introduces ground loops between the safety grounds and the signal returns. As discussed in Section 8.2, AC safety grounds connected when the power cord is plugged into the receptacle are required by most electrical safety codes.

Many engineers are very concerned about ground loops but these are seldom the problem they are believed to be, especially when equipments are in close proximity.

Many systems made up of computer equipment function correctly with multiple ground loops. Exceptions occur when the neutral return current flows in the signal return, as discussed in

Section 8.2, or large common-mode noise appears between the safety grounds of equipment. Large common-mode voltages are encountered when equipment pieces are located some distance apart and connected to different earth grounds, especially when sources of power-line noise, such as welders, milling machines, elevators, and control and automation equipment, are connected to one of the earth grounds. With the use of single-ended analog/instrumentation interfaces in which the signal return is connected between the two earth grounds, the potential for EMI is high. The use of shielded cables on single-ended circuits or where the shield is the return path for signal currents will not reduce the potential for EMI, because the problem is conducted noise. Multipoint grounds between equipment should ideally be less than 0.1λ in length. If longer, the resonance and antiresonance effects seen with electrically long conductors in Chapter 5 may occur, with resultant high currents or very high impedance.

Decreasing the impedance of the ground connection between equipment pieces located far apart is difficult due to the typically low-impedance source of the noise current and the correspondingly large size of conductor required. The solution may be high common-mode noise immunity differential-interface circuits, as described in Chapter 5 (Figure 5.129a and Figures 5.130 through 5.133).

The multipoint ground scheme illustrated in Figure 8.12 is the same as used in Figure 8.1 for ideal large-system grounding. This scheme reduces the common-mode noise voltage developed between the enclosures of the equipment but does not eliminate ground loops. The interconnections shown between equipment enclosures represent single-ended signal grounds connected to the enclosures at both ends.

Conventional wisdom dictates that multipoint grounding with very short length for the ground conductors be used above 10 MHz, and multipoint grounding is common in equipment containing RF circuits.

At the PCB level, multipoint grounding may not be required, and some advantages from a single-point connection between the PCB ground plane and the enclosure exist. However, single-point grounding is seldom used. More common is the connection of the PCB ground plane of an RF circuit at more than one location around the edge of the PCB to an enclosure provided to shield the circuit. RF engineers have followed this practice for years. One possible reason for multipoint grounding at the PCB level is that the PCB ground plane does not provide a sufficiently low impedance due to crosshatching or interruption of the ground plane by tracks. In this case, the use of a solid ground plane formed by the enclosure may improve the performance of an RF circuit or reduce the radiated emissions from the board.

When other circuits reference the enclosure as ground or when seams and gaps occur in the enclosure, the goal should be to reduce the RF current flow on the inside of the enclosure to the greatest extent. This can be achieved by reducing leakage currents from the circuit and by reducing the radiated coupling to the enclosure. The leakage currents can be reduced by grounding at a single point or several points in close proximity (at the same time reducing the capacitance between the enclosure and the RF ground). Leakage current and radiated emissions can be reduced by good RF design, such as the use of transmission lines formed by the signal and return tracks or by the signal tracks in close proximity to an adequate PCB ground plane. When a signal track is in close proximity to a ground plane, the RF return current tends to concentrate in the ground plane under the track, resulting in a minimum impedance in the return path. Thus, good RF design practice tends to reduce RF leakage current in the enclosure, reduce radiated emissions, and minimize common impedance coupling and conducted EMI problems. Decreasing the capacitance between RF semiconductors and the enclosure to which they are heatsunk can also reduce RF current flow in the enclosure.

Measurements on the shielding effectiveness of small enclosures surrounding RF circuits have shown the importance of the grounding scheme, although the results have not always been as predicted. For example, shielding effectiveness measurements were made on enclosures containing a PCB on which several loads were mounted. An RF signal was connected to the loads via a coaxial connector mounted through the enclosure. Either a single connection from the PCB ground plane to the enclosure was made via the braid of the shielded cable inside the enclosure or the ground plane was connected around the periphery to the enclosure. One measurement was taken on an enclosure made up of an extruded section covered by a lid and capped at the top and bottom by plates. The dimensions of the enclosure were 26.5 cm × 22.5 cm × 4 cm, and the enclosure contained seams but no apertures. The PCB was either mounted in the center of the enclosure for the case when the ground plane was isolated or mounted close to the solid base of the enclosure when the ground plane was connected around the periphery to the enclosure. A comparison of the shielding effectiveness of the single-point connection and the multipoint connection of the ground plane follows with the single-point ground slightly better at most frequencies.

| Frequency [MHz] | Shielding Effectiveness | |
	Single-Point Ground [dB]	Multipoint Ground [dB]
100	73.5	76.0
160	74.0	70.0
195	64.0	57.0

Another enclosure used to shield the same PCB had the dimensions 23.5 cm × 22.5 cm × 1.8 cm. The enclosure was constructed of a solid base plate to which a cover was attached. The defects in the enclosure were the seams plus 38 × 8 mm × 2.2 mm slots in the cover. In this measurement, the ground plane was either connected around the periphery to the enclosure or mounted on 1.8 mm thick nylon washers by nylon screws. The total capacitance between the ground plane and the enclosure with the PCB mounted on the insulating washers was measured at 125 pF. A secondary effect of mounting the PCB on the insulating washers was to move the PCB closer to the cover. Following is the shielding effectiveness of the enclosure with the PCB either single-point or multipoint grounded:

| Frequency [MHz] | Shielding Effectiveness | |
	Single-Point Ground [dB]	Multipoint Ground [dB]
100	66	78
131	39	59
195	64	64
250	86	86
297	84	76

One potential reason for the degradation in shielding effectiveness at some frequencies with the single-point ground scheme is the potential for increased radiated coupling to the cover as the PCB is moved in closer proximity, due to the 1.8 mm thickness of the nylon washers.

One lesson that can be learned from the shielding effectiveness measurements is that the grounding scheme should be chosen on a case-by-case basis. Although multipoint grounding is the method of choice for the interconnection of PCBs, modules, and equipment that contains RF circuits, even at this level exceptions to the rule do exist.

Floating ground schemes are difficult to realize, test, and maintain. In practice, leakage paths exist with megaohms of resistance and the floating ground floats to some undefined potential. The inclusion of a high-value resistor between the ground and chassis will reduce the static potential of the ground and is recommended for static discharge protection. Connections of signal interfaces and peripheral equipment are likely to invalidate the floating ground scheme when signal-to-chassis connections are made in the external equipment. Test devices must have high-impedance differential inputs. The advantage of the floating ground is the very low value of LF currents that flow in the ground. Above some frequency, the ground becomes a multipoint system due to parasitic capacitance. Shielding and grounding RF and digital circuits are discussed in Sections 5.6.2 and 5.6.3, respectively.

A common fallacy is that only one type of grounding scheme is "good," whereas all of the grounding schemes will work very well when properly implemented.

8.3.3 SINGLE-POINT GROUND

The single-point ground scheme is often described as the ideal. How and where the single-point ground is made are vitally important, and in some instances, it is better to make more than one connection, even though this does technically violate the single-point philosophy.

When secondary power (e.g., +5, ±15 V) isolated from the input power return is provided within the equipment, the point of connection of the secondary power return to chassis is considered the single-point ground for the equipment. When DC primary power is used directly by the equipment, the point after the power-line input filter might conveniently be considered the single point for the connection of the different signal grounds within the equipment.

An alternative location for the single-point ground may be chosen, for example, close to an RF component that, by its construction, connects power return and signal return to the conductive enclosure of the component, or at a signal interface where the signal return is connected to chassis, or at an A/D converter where analog and digital grounds are connected together.

It is neither mandatory nor, often, beneficial to make the single-point ground at a power supply within the equipment, although the advantage of this location is that common-mode voltage generated by the supply is shorted by the ground connection. The disadvantage is that common-mode voltages appearing on circuit boards and at the point of entry of interface signals cannot be shorted out by connection of the return to chassis without violating the grounding scheme. One solution when the supply return is isolated from the chassis is to reduce both differential- and common-mode noise from the supply with an L- or T-type filter connected as shown in Figure 8.14. The chassis ground may then be made at the interface circuit. In this configuration, noise currents generated on circuit boards and flowing into the power supply on the boards' ground plane or signal return tracks are reduced by the series inductance of the L- or T-type filter.

Using a nonisolated supply or adding a second chassis ground connection in Figure 8.14 at the supply return will result in the flow of noise currents in the enclosure. These noise currents may be reduced by inclusion of an inductor in the return. When apertures are present in the enclosure and the internal enclosure current is increased, increased aperture coupling must be expected. When the shielding integrity of the enclosure is high, the increased current flow in the enclosure may not significantly increase radiation from the enclosure. The signal grounds within a unit should be divided into the following, where applicable and feasible:

1. Analog/video ground
2. Digital ground
3. RF ground
4. Control signal ground

FIGURE 8.14 Reduction in power supply noise voltage by an *LC* filter, thus allowing a ground chassis connection at the interface.

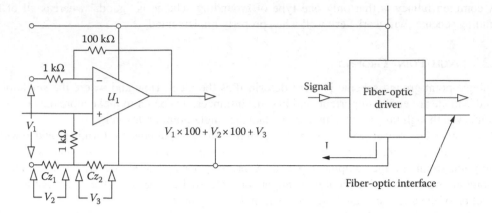

FIGURE 8.15 Common impedance coupling.

These four grounds should be connected to the equipment single-point ground only and shall be isolated from each other to the greatest extent possible, including capacitive coupling between them.

It is easier to design the separation into the system and then interconnect, when necessary, than to attempt to separate after the fact. For example, a PCB layout in which analog and digital grounds are separated may very easily be connected together; in fact, the provision of terminal posts for the purpose is recommended. However, separation of grounds after the PCB is made may be out of the question. Where an A/D converter or similar device requires an analog ground connection and a digital ground, the analog ground single-point connection shall be made at the device in question. Even where device manufacturers supply separate digital and analog ground pins, very often the grounds are either connected within the device or a very low maximum voltage difference between the grounds is recommended.

Common impedance coupling is a frequent source of noise problems in instrumentation, analog, and video circuits. Figure 8.15 illustrates the mechanism of common impedance coupling. The return current from the fiber-optic driver flows into the ground impedance of the U_1 analog circuit resulting in the noise voltages V_2 and V_3. Typically, the sources of ground current are digital or RF circuits.

8.3.3.1 Floating the Analog Ground on the Digital Ground

One technique that has proven very effective in reducing the current flow in analog ground is to float the analog ground on the digital. This scheme is shown in Figure 8.16. The noise currents generated by the logic as it changes state flow back on the digital supply return, with little current flow in the analog ground. The analog circuit supply in Figure 8.16 is an isolating DC-to-DC converter, which is required to achieve maximum isolation between the analog supply return and the chassis and minimum digital supply current in the analog ground. When a supply is used in which the return is connected to the enclosure, the floating analog grounding scheme is still partially effective, due to some level of isolation provided by the impedance of the inductor in the analog supply return, as shown in Figure 8.16.

In the floating analog scheme, large common-mode voltage appears between the enclosure and the analog ground, which often causes concern when measured on an oscilloscope. However, on looking at the output bits of the A/D converter with a nonvarying input voltage, such as zero volts, no change in the status of the bits is seen; thus, a differential noise voltage lower than the weighting of the least significant bit is present. The location of the digital ground to chassis ground is deliberately at the point where a digital interface signal leaves the board. In a practical system, the chassis connection may be made at a connector when the interface is external to the equipment. The reason for this choice of location is to reduce the common-mode voltage on the interface signal and return conductors, thus reducing radiation from the cable. If this is not a consideration and the major concern is to reduce noise at the input of the A/D, then the chassis ground connection should be made at the A/D converter. This is often difficult unless the PCB is directly above chassis, in which case

FIGURE 8.16 Floating the analog ground on the digital ground.

a metal standoff can be used to bring the chassis up to the PCB at that point. Alternatively, if the motherboard/backplane has a chassis connection, this should be brought through as many pins are available on the PCB edge connector, and the A/D should be located at the edge of the board close to the connector to maintain a short connection length to chassis.

8.3.4 Modified Differential Op Amp Circuit

Figure 8.17a illustrates a common type of circuit used as an analog/video/RF amplifier. The output voltage equals the signal voltage times R_2 divided by R_1, plus the contribution of the voltages developed across the ground impedances Z_{g1} and Z_{g2}.

By addition of a single resistor, as shown in Figure 8.17b, the circuit will reduce the magnitude of the ground current and virtually eliminate the noise voltage generated by V_{com}. The signal output voltage is the same as achieved with the circuit in Figure 8.17a for the same differential input voltage, regardless of whether the noninverting input voltage is at the output ground potential or at a positive or negative potential referenced to ground.

When $R_1 = R_2$ and $R_3 = R_4$, the circuit has a gain of unity. The input to R_3 is shown in Figure 8.17b connected to a signal return, in which case the circuit inverts the input signal. Alternatively, the input may be true differential or the input to R_1 may be connected to a signal return and R_3 used as the signal input in which case the circuit is noninverting.

This circuit is invaluable when rejection of common-mode input voltages is required, hence the name *modified differential*. When a higher isolation is required between the two grounds, a true differential circuit, in which the input signals are applied typically to FET input and voltage follower op amps followed by the differential amplifier, shown in Figure 8.17b, may be used.

The limitations on rejection are the common-mode rejection ratio of the op amp and the match in resistor values.

The modified differential circuit may be used as an instrumentation receiver or used to separate analog and digital grounds on a PCB, in which case the input is referenced to analog ground and the output voltage is referenced to digital ground, or to separate system/subsystem grounds.

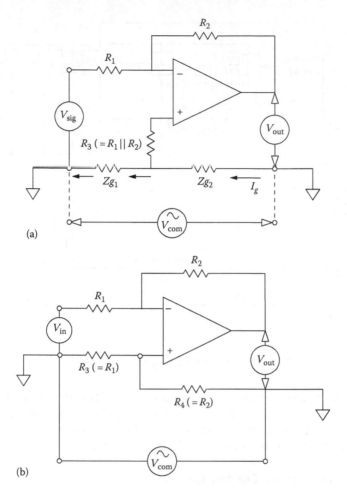

FIGURE 8.17 (a) Typical op amp gain stage or inverter. (b) Modified differential circuit with common-mode noise rejection.

8.4 GUIDELINES FOR SIGNAL GROUNDING

Grounds between equipment shall, where feasible, be DC or LF isolated, normally at the input (receiver) end of the interface. This can be achieved by capacitors, transformers, high-input-impedance amplifiers, and opto-isolators.

High-power stages of a circuit shall be connected the closest to the single-point ground, with the most sensitive circuit the furthest from the single point. This ensures that power return currents do not flow in the ground connection or track used by sensitive circuits and thus minimizes the risk of interference and LF instability, sometimes referred to as *motorboating* in audio circuits.

Always return decoupling capacitors to the appropriate ground. That is, if +10 V is used as the digital circuit supply, do not return a decoupling capacitor connected to +10 V to analog or other ground.

Where possible, separate grounds. It is always possible to connect later.

Always build flexibility into the grounding scheme. Use 0 Ω resistors or jumpers to connect grounds, because these may be removed as desired. Place ground pads on PCBs that can be connected via ground straps to ungrounded piggyback PCBs/memory modules/PCMCIA cards, frames, conductive structures, etc., which are mounted on the PCB.

Always use a solid low-impedance signal ground connection between circuits, backplanes, and PCB, PCB-to-PCB, and different grounds located on the same PCB, when digital or transient signal interfaces exist between these circuits. Keep the signals and ground(s) as close together as possible, and in PCB layouts, keep the ground directly under the signal traces without any interruption in the ground plane, such as slots.

The grounding philosophy should be decided on a system-by-system basis. The manufacturer of bought-out equipment used in a system often connects digital and analog/video grounds to chassis and also, for safety reasons, the AC safety ground to chassis. Thus, a single-point grounding scheme may be impractical. In other units/systems, multipoint grounding is the preferred method.

The physical location and lengths of ground connections shall be included in mechanical/layout diagrams.

When ground connections are much longer than one-tenth of the wavelength of the noise frequency or the ground is in close proximity to chassis, the equivalent circuit of the ground connection must be used in an EMI/EMC evaluation. Section 11.2.1 includes Case Study 11.1, which provides an example of modeling the ground structure.

Also see Section 5.3.3.6 on RF circuit grounding and Section 11.8.

8.5 POWER AND GROUNDING DIAGRAMS

A power and grounding diagram should be constructed showing the connections of primary and secondary returns, as well as the location of filters and the use of shielded or twisted shielded wires within equipment, subsystems, or systems. The power and grounding diagram allows an overview of the realization of the grounding philosophy and, when problems occur, a comparison of the actual implementation to the desired scheme. Very often the two are different! A number of examples of power and grounding for both a single-point ground system comprising a large number of pieces of equipment as well as a single-point DC and multipoint RF system are illustrated in the succeeding pages.

The following power and grounding diagrams are reproduced by the kind permission of Canadian Astronautics Ltd. and the Canadian Space Agency.

8.6 GROUNDING FOR LIGHTNING PROTECTION

8.6.1 LIGHTNING CONDUCTORS

The purpose of a grounding scheme for lightning protection is to carry a direct lightning strike to earth ground with no, or at least minimal, damage to the equipment or structure to be protected. For example, lightning protection may be required for antennas mounted on a mast, buildings, or power/signal lines entering a building.

The IEC standard 62305-1 *Protection against Lightning, Part 1: General principles*, Edition 2, provides good advice.

Buildings may be protected by a mesh of lightning conductors with short vertical rods attached to the top of the building. The mesh is connected to earth ground by vertical conductors. If the building framework is steel, then the steel columns may be used as the vertical conductors. The purpose of lightning conductors is not to discharge a thunder cloud but to divert a lightning strike from the structure to be protected to the lightning protection scheme. The protection scheme typically provides a higher conductivity to ground than the building, and the scheme extends in height above the building. Thus, the probability of a strike at the building location may be higher if the protection were not there. Therefore, install lightning protection only if it is capable of handling a direct strike of at least 25,000 A and preferably one of 300,000 A. Not installing the protection is likely, at least, to reduce the probability of a strike.

The intent of the first pointed lightning protection rods was to dissipate the charge around the structure so that electric fields associated with the charge would not cause electrical breakdown of the air. It was soon found out that the opposite occurred, and lighting strikes to these pointed rods increased. The majority of lighting protection schemes are of this type. As the use of electronics has increased dramatically, the incidence of malfunction or destruction in sensitive equipment due to lightning has increased, and the idea of keeping the very high currents induced in a lightning strike away from the sensitive electronics is appealing. This should be possible by the use of a dissipation array systems (DAS) or a charge transfer system. The theory behind these systems is that they should cause a current to flow from the grounding system due to the ionization of the air around the system. The more the current that is flowing from the system into the air and the longer the duration of this current flow, the higher the space charge. An increased space charge will weaken the electric field and so the probability of an air breakdown is reduced. Early analysis showed that the size of the canopy or umbrella or the quantity of sharpened points would be impractically large. However, these systems have been designed and are in use. Little information on their effectiveness is available, although in Reference [3], the lightning occurrence probability before and after the installation of a DAS remained the same. This was attributed to the fact that the DAS was not installed to the manufacturer's recommendation, caused by the building construction conditions in Japan. Another lightning grounding scheme, made up of a gold-plated ball mounted at the maximum height and connected to a coaxial lightning conductor, was commercially available. This system was computationally modeled and a small physical-scale model of the transmission line was constructed and tested with a lighting generator. In neither the computer model nor the scale model was any improvement seen over a solid ground wire.

Alternatively, a structure may be protected by erecting a mast, rods, or wire above the structure at a sufficient height that the structure falls under an arc of protection, as shown in Figure 8.18. Some lightning protection recommendations show a cone of protection in which a line is drawn at an angle of 45° from the vertical to the ground. Any structure within the cone is considered to be protected from a direct lightning stroke. In practice, lightning strikes have occurred to antennas that are side mounted on a mast within the 45° cone of protection. The National Fire Protection Association recommends instead the arc of protection. Here, the assumption is made that a 150 ft radius circle touches the top of the structure and the ground at 360° points around the structure and that equipment within this volume is protected. Other more conservative models for the arc of protection are used. A lightning strike may be modeled as a constant-current source that must find a path either from a cloud to ground or from ground to a cloud. In practice, a number of strikes occur one right after another. A conductor that carries the strike current must do so without fusing. The majority of lightning strikes generate currents below 75 kA, although strikes as high as 300 kA

FIGURE 8.18 Arc of protection under masts and buildings.

have been recorded (Reference [4]). The minimum conductor size capable of conducting a 75 kA strike with a half-amplitude time of 100 juts without fusing is No. 12 AWG. For a 300 kA, 100 μs half-amplitude strike, the minimum conductor size is No. 6 AWG.

Perhaps more important than the current-carrying capability of the conductor are the DC resistance of the joints and its self-inductance. It is these parameters that determine the voltage appearing across the conductor and across the joints. Sideflashes occur when the lightning conductor is in close proximity to the structure and the voltage differential is too high. When joint resistance is high, arcing may occur across the joint, and local heating therefrom causes the joint either to fuse apart or to weld together. The major cause of high joint resistance is corrosion. To reduce corrosion, the use of dissimilar metals in the lightning conductor scheme should be avoided, and after the joint is made, it should be painted over. Copper and aluminum are the preferred conductor materials, with galvanized steel a third choice. Conductors, other than those buried in the ground or intended to be the first contact point with the air, should be protected, for example, by PVC coating or paint. The inductance of a wide, thin strip should be used where feasible, because its inductance is lower than that of a round conductor when its cross-sectional area is larger, as described in Section 5.1.2. When the vertical conductors are a meter or more apart, the total inductance from the top of the building to the ground is the inductances of the vertical conductors in parallel. It is this total inductance that determines the voltage drop between the air contact point and the ground. If we assume a single 100 ft high grounding conductor of 0.62 in. diameter, then the inductance of the conductor, from Equation 5.4 of Section 5.1.2, is 46 μH. We also assume a lightning current waveform with a 2 μs rise time and a peak current of 25 kA. The voltage developed between the top of the conductor and the ground is approximately 575 kV, from

$$V = L \frac{di}{dt} \tag{8.1}$$

The total inductance of the scheme may be reduced by adding more conductors in parallel. However, when conductors are in close proximity and share the current, the total inductance is not as low as when the conductors are far apart. For, in close proximity, the field from one conductor interacts with the adjacent conductor and results in an increase in inductance. The conductors should be spaced at least 1 m apart to minimize inductance and a maximum of 10 m apart to maximize the protection of the building.

Right-angle bends increase the inductance of the conductor; therefore, add a radius to any bend. When the building has a metal-clad exterior or is of reinforced concrete or steel girder construction, the inductance of the down conductors is reduced by the proximity to the conductive structure, as described in Section 5.1.2.

Further information on the inductance of antenna masts and the application of protection devices is obtainable from Reference [5].

The minimum separation distance between any of the upper components of the protection scheme and the building may be established from the potential difference between them. The impulse breakdown of air is dependent on its ionization and on the geometry of the electrodes. Figure 8.19 plots the minimum clearance for a given impulse voltage for any shape of electrode but assumes nonionization of the air. The source of Figure 8.19 is the International Electrotechnical Commission Document 28 A, "Insulation Coordination of Low-Voltage Systems and Equipment." Bends should be made with as large a radius as possible, and an increased minimum clearance should be used at the bend to reduce the probability of breakdown.

Large mechanical forces are generated on conductors that carry high currents; therefore, all conductors should be mechanically bonded together and anchored firmly to the structure.

Figure 8.20 illustrates a typical grounding scheme for rooftop-mounted cellular or PCS base station antennas. The antenna coaxial cable is brought into a metal cableway B, and at this point, the shield of

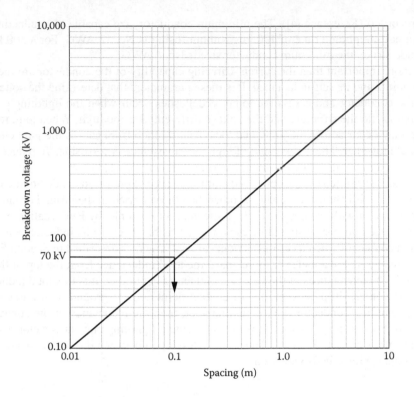

FIGURE 8.19   Curve for the determination of the breakdown voltage in air as a function of spacing.

the cable is bonded to the ground strap/braid/wire C. Also at this point, a combined primary lightning surge arrestor and filter A is inserted in the coaxial cable. The coaxial cable is then taken inside the cableway until the point of entry into the building D. The cableway must be bonded by a bond strap on either side of the seams in the cableway. At least two ground straps/braid or wires are run alongside the cableway and taken down the *outside* of the building to ground rods or, depending on the resistivity of the soil, a ground network, D. If lightning were to strike any of the antennas, the current would flow on the antenna coaxial cable until the junction with the ground wire. The current would then continue flowing on both the ground wire and the outside of the cableway and via the ground straps on the outside of the building into the soil. The voltage developed on the center conductor of the antenna cable is shunted to the shield by the surge protector A, which if it also contains a filter will attenuate the voltage further in amplitude to protect the cellular/PCS equipment. One major mistake that is made is to bond the ground system to a metal vent pipe or the metal of heating/ventilation equipment on the roof and to omit the external ground wires. Metal vent pipes are often connected together inside the building by the use of rubber gaskets, and the lightning strike can then flash over to sensitive equipment within the building and possibly start a fire. If the connection is made to an AC safety ground that penetrates the building, the same flashover or fusing can occur, again with the possibility of fire. Therefore, never bring a lightning protection ground scheme inside a building. The external ground wire/braid can be covered with a plastic cover that has a color similar to that of the building, for example, white or brown, if the aesthetics of the grounding scheme is a problem.

The IEC 62305-1, MIL-HDBK-419A, Chapter 3, and Reference [5] are good sources on grounding for lighting.

When a long conductor is buried in soil and designed for lightning protection, it can be modeled as a transmission line with distributed parameters generating multiple reflections due to impedance mismatches. Reference [6] describes some approximate formulas for the distributed parameters, the velocity, and the terminal voltages developed on such a long buried conductor.

FIGURE 8.20 Typical lightning protection scheme for building-mounted cellular phone/PCS base station antennas. A, primary lightning surge protection and filter; B, metal cableway bonded at seams; C, ground strap, braid, or wire; D, ground rod.

8.6.2 GROUND POTENTIAL

A potential gradient appears across the ground, with equipotential circles spreading out from the point of contact. A lightning strike at an area of the Earth's surface that has a high resistivity tends to charge the area to a potential closer to the cloud's potential, and a high potential gradient exists proceeding away from the immediate area of the strike. The purpose of the lightning protection scheme is to reduce the potential gradient between structures and equipment. A typical ground scheme for buildings and an antenna installation is shown in the following case study.

8.6.3 CASE STUDY 8.1: GROUNDING FOR LIGHTNING PROTECTION AT A COMMUNICATION SITE

The antenna structure shown in Figure 8.21 lies under the arc of protection of the lightning conductor rod, which is constructed of 1.6 cm diameter material with a length of 4.3 m. The minimum distance between the rod and the antenna is 20 cm. Assuming the rod moves in a high wind such that the minimum distance reduces to 15 cm, the impulse breakdown voltage, from Figure 8.19, is 90 kV. Modeling the inductance of the rod and its resistance in parallel with that of the support structure to which it is bonded and using a rise time of 2 μs, the maximum lightning strike current protected by the rod is 60 kA. Thus, the antenna protection scheme is adequate for approximately 80% of lightning strikes.

The lightning conductor is connected to a #2 AWG wire mesh that is also connected to rebars in the concrete base of the antenna and to an adjacent metal fence.

An X-IT rod is used to provide a low-resistance ground connection. The X-IT rod is a metal tube that contains a chemical that leaches out through apertures in the tube and permeates the

FIGURE 8.21 Antenna structure with lightning conductor rod.

surrounding soil. The moisture in the air is used to form an electrode with the chemical, which results in a low resistance. The use of a metal rod immersed in ground that has been soaked in a sodium solution is an alternative way of achieving a low earth resistance. X-IT rods are manufactured by Lyncole (engineering@lyncole.com).

The buildings at the site are clad with cold-rolled steel with a thickness of at least 1 mm. The cladding is constructed of a number of sheets that are bonded together. The DC resistance of the building from the center of the roof to the surrounding ground ring is approximately 5 mΩ. The buildings are thus adequately protected from lightning strikes. The maximum touch potential within the building during a 25 kA strike has been calculated at 125 V.

An approximately 25 m long No. 2 AWG grounding wire connects the buildings to the base of the antenna and is run in a bermed PVC duct along with the antenna interconnection cables.

Several ground conductors within cables and the shields of cables will share the current with the No. 2 AWG grounding wire during a lightning strike. The No. 2 AWG grounding wire ensures that the full lightning current does not flow in the interconnection cables.

During the fast initial transition of the lightning current, the inductance of the cables will limit the current flow. When the earth resistance is 50 Ω or less, the division of the current between the cables and of the ground between the antenna and the buildings becomes significant, especially during the fast initial transition.

The shields of cables are connected to the metal building at the entry point of the cables in order to shunt shield currents to ground. Large ferrite baluns are used on cables after the shield termination point in order to reduce remaining common-mode current flow. The differential-mode voltage induced into the center conductors of cables is reduced by one or more of the methods of transient suppression described in Section 5.8.

A reduction in the induced current in interconnection cables would be achieved by running the cables in a metal conduit, welded at the joints, instead of the PVC pipe. The No. 2 AWG wire should remain outside of the metal conduit, or it may be eliminated, since the metal conduit functions as the ground conductor.

At sites where the electromagnetic ambient was in the volts-per-meter region, the welded metal conduit was required to eliminate EMI, and the improvement in lightning protection was a secondary consideration.

FIGURE 8.22   Communication site grounding scheme.

The grounding system at buildings No. 1 and No. 2, as shown in Figure 8.22, can be described as follows: Three lengths of No. 2 AWG bare copper wire are positioned equidistant beneath the two buildings and cad-welded to the ground ring system (which comprises No. 2 AWG wire around both buildings) and the center leg between the buildings. The wires are buried beneath the Earth to maximize the contact area with the Earth. The two buildings are bonded together via two No. 2 AWG wires of approximately 1.4 m length and spaced 6 m apart. Four X-IT rods are used to connect the ring to Earth via low-resistance paths.

8.7   BONDING

8.7.1   GENERAL

Electrical bonding is a term used for the process of connecting metal structures together in order to achieve a low-resistance contact. Bonding is used in lightning protection schemes to ensure that lightning strike currents may be carried between structures, or structure and conductors, such as lightning arrestors and earth grounding rods. Other reasons for bonding are to provide static protection and in the realization of a ground reference plane. Antenna components also require low-impedance bonds, for currents flow on the junctions between these components. Adequate bonding is vital in minimizing passive intermodulation in antennas. Each of these bonds should be made so that the mechanical and electrical properties of the path are determined by the connected members and not by the interconnection junction. Further, the joint must maintain its properties over an extended period of time, to prevent progressive degradation of the degree of performance initially established by the interconnection. Bonding is concerned with those techniques and procedures necessary to achieve a mechanically strong, low-impedance interconnection between metal objects and to prevent the path thus established from subsequent deterioration through corrosion or mechanical looseness. The resistance of the bond and surface area over which the contact must be made are dependent on the purpose for which the bond is made.

8.7.2   MIL-B-5087, MIL-HDB-419A, AND MIL-STD-464

MIL-B-5087 has been canceled, but it did define a number of different bonds and the requirements for each type, and these requirements have been used on military equipment that is still in the field as well as a number of different space programs. Although MIL-B-5087 has been cancelled and is now contained in MIL-STD-464, MIL-STD-464 still references the bonding classes of MIL-B-5087. MIL-HDBK-419A also provides information on bonding.

These bond types from MIL-B-5087 are as follows:

Class A for antenna installations—No bonding resistance specified, but a negligible impedance over the operating frequency was required.

Class C for current return path—Impedance limited by maximum allowable voltage drop
as shown in Table 8.5.

Class H for shock hazard <0.1 Ω DC.

Class L for lightning protection—Control internal vehicle voltages to 500 V.

Class R for RF potential <2.5 mΩ from electronic units to structure.

Class S for static discharge <1 Ω.

There were also less obvious requirements in MIL-B-5087, such as 2.5 mΩ requirement on connector shells. MIL-STD-464 points out that there is no scientific basis for the 2.5 mΩ requirement for class R bonds. And although the equipment case-to-structure class R requirement is not important in most instances, 2.5 mΩ is still a good number for several electrical bonds, such as terminating shields to connectors and bonding connector to equipment case. It is also a good design value where a good bond is required for other purposes. The other bonding values of MIL-B-5087, for shock protection, current return paths, and static charge, are still valid today.

MIL-STD-464 states that for lightning protection, metallic structural members (aluminum, steel, titanium, and so forth) provide the best opportunity to achieve an electrical bond on the order of 2.5 mΩ. A bond of this level will limit the induced voltage on system cabling to 500 V from lightning strike attachments (200 kA) to system structure.

Bonding resistance is specified and measured, not bonding impedance. The bonding resistance is not necessarily an indicator of the impedance of the bond, which in EMI reduction is often of paramount importance. The impedance of ground planes and rectangular and circular conductors, which may be used as bonding straps or jumpers, is described in Section 5.1.2. As discussed in Chapter 5, stray capacitance, self-inductance, and conductor length result in both series and parallel resonant circuits, and these must be accounted for when considering the bond at RFs.

A poor electrical bond when exposed to two or more high-power fields may result in passive intermodulation and reradiate intermodulation products, as described in Section 10.3.1.

8.7.2.1   MIL-STD-464 Methods of Electrical Bonding

An electrical bond may be made by welding, brazing, soldering, riveting, bolting, conductive adhesive, or conductive grease or by spring finger stock material. The preferred method of bonding

---

TABLE 8.5
Maximum Allowable Voltage Drop for Class C Bonds

Nominal System Voltage	Maximum Allowable Voltage Drop Equipment Operation	
	Continuous	Intermittent
28	1	2
115	4	8
200	7	14

is welding, followed by brazing and soldering. A bolt or rivet is used to connect two conductive surfaces together with sufficient force to provide a low-resistance bond. The thread of a screw or bolt must not be relied upon to provide either a DC current or RF bond.

Conductive grease or spring finger stock material may be used to provide a bond between moving surfaces, for example, conductive grease between spring-loaded contacts, piano hinges, or rotating shafts. When the contact is not under pressure, the bond should, if possible, be backed up by a flexible bonding strap. The bond achieved by a circular spring finger stock around a shaft or conductive grease between the shaft and bearing walls will provide a lower impedance than a bonding strap. When conventional bonding methods, such as bolts, rivets, and screws, cannot be used, then the use of conductive adhesive may be allowed.

8.7.2.2 MIL-HDBK-419A Bonding Practices

MIL-HDB-419A describes in brief the following grounding practices: Equipment emission and susceptibility requirements for proper system operation should be accomplished with the most cost-effective combination of interference reduction techniques. Bonding is an essential element of the interference control effort. MIL-HDBK-419A presents design and construction guidelines to aid in the implementation of effective bonding of equipment circuits, equipment enclosures, and cabling. These guidelines are not intended as step-by-step procedures for meeting EMC specifications. Rather, they are aimed at focusing attention on those principles and techniques that lead to increased compatibility between circuits, assemblies, and equipments.

1. Welded seams should be used wherever possible because they are permanent, offer a low-impedance bond, and achieve the highest degree of RF tightness.
2. Spot welds may be used where RF tightness is not necessary. Spot welding is less desirable than continuous welding because of the tendency for buckling and the possibility of corrosion between welds.
3. Soldering should not be used where high mechanical strength is required; the solder should be supplemented with fasteners, such as screws or bolts.
4. Solder must not be used to form bonds that may be reasonably expected to carry large currents, such as those produced by power-line faults or lightning currents.
5. Fasteners such as bolts, rivets, and screws should not be relied upon to provide the primary current path through a joint.
6. Rivets should be used primarily to provide mechanical strength to soldered bonds.
7. Sheet metal screws should be used only for the fastening of dust covers on equipment or for the attachment of covers to discourage unauthorized access by untrained personnel.
8. Bonds that cannot be made through direct metal-to-metal contact must use auxiliary straps or jumpers. The following precautions should be observed when employing bonding straps or jumpers (see Figure 8.23a and b):
   a. Jumpers should be bonded directly to the basic structure rather than through an adjacent part, as shown in Figure 8.23a.
   b. Jumpers should not be installed two or more in series.
   c. Jumpers should be as short as possible.
   d. Jumpers should not be fastened with self-tapping screws.
   e. Jumpers should be installed so that vibration or motion will not affect the impedance of the bonding path.
   f. Jumpers should be made of tinned copper, cadmium-plated phosphor bronze, aluminum, or cadmium-plated steel.
   g. Mating metals should be selected to offer maximum galvanic compatibility (see Section 8.8.2).
9. Where electrical continuity across the shock mounts is necessary, bonding jumpers should be installed across each shock mount. Jumpers for this application should have a maximum

FIGURE 8.23 (a) Bonding jumper: (A) acceptable and (B) unacceptable uses. (b) Bonding of subassemblies to equipment chassis. (c) Bonding of equipment to mounting surface.   (*Continued*)

Clean to
base metal

Electronic
package

Rack

Clean flange
to base metal

(d)

Clean each mating surface 1/8 in.
around the bushing periphery

Rear of rack

Rear of
electronics
equipment

Dagger pins

(e)

FIGURE 8.23 (*Continued*)   (d) Typical method of bonding equipment flanges to frame or rack. (e) Bonding of rack-mounted equipment employing dagger pins.   (*Continued*)

(f)

(g)

FIGURE 8.23 (*Continued*)   (f) Recommended practices for effective bonding in cabinets. (g) Method of bonding across hinges.   (*Continued*)

Refinish after installation

Connector

Mounting surface

(h)

FIGURE 8.23 (*Continued*)   (h) Bonding of connector to mounting surface.

thickness of 0.06 cm (0.025 in.) so that the damping efficiency of the mount is not impaired. In severe shock and vibration environments, solid straps may be corrugated or flexible, and coarse wire braid may be used.

10. Where RF tightness is required and welded joints cannot be used, the bond surfaces must be machined smooth to establish a high degree of surface contact throughout the joint area. Fasteners must be positioned to maintain uniform pressure throughout the bond area.

11. Chassis-mounted subassemblies should utilize the full mounting area for the bond, as illustrated in Figure 8.23b and c. Separate jumpers should not be used for this purpose.

12. Equipment attached to frames or racks by means of flange-mounted quick-disconnect fasteners must be bonded about the entire flange periphery, as shown in Figure 8.23d. Both the flange surface and the mating rack surface must be cleaned over the entire contact area.

13. Rack-mounted packages employing one or more dagger pins should be bonded as shown in Figure 8.23e.

14. The recommended practices for effective bonding of equipment racks are shown in Figure 8.23f. Bonding between the equipment chassis and the rack is achieved through contact between the equipment front panel and the rack front brackets. These brackets are bonded to the horizontal slide, which in turn is welded to the rack frame. The ground stud at the top of the rack is used to connect the rack structure to the facility ground system.

15. Where hinges are used, establish an alternate electrical path through the use of thin, flexible straps across the hinges, as shown in Figure 8.23g.

16. Standard MS-type connectors and coaxial connectors must be bonded to their respective panels over the entire mating surfaces, as illustrated in Figure 8.23h. Panel surfaces must be cleaned to the base metal for no less than 0.32 cm (1/8 in.) beyond the periphery of the mating connector.

17. In ideal situations, cable shields should be bonded to the connector shell completely around the periphery of the shield with either compression or, preferably, soldered bonds.

18. When an RF-tight joint is required at seams, access covers, removable partitions, or other shield discontinuities, conductive gaskets should be used. They may also be used to improve the bond between irregular or rough bonding surfaces. Gaskets should be sufficiently resilient to allow for frequent opening and closing of the joint and yet be stiff enough to penetrate any nonconductive films on surfaces.

19. Gaskets should be firmly affixed to one of the bond members with screws or conductive cement or by any other means that does not interfere with their operation. The gaskets may be placed in a milled slot to prevent lateral movement.

20. All bonds that are not in readily accessible areas must be protected from corrosion and mechanical deterioration. Corrosion protection should be provided by ensuring galvanic compatibility of metals and by sealing the bonded joint against moisture.

8.7.3   Corrosion, Dissimilar Metals, and Oxidization

When the contact is between dissimilar metals in the presence of moisture, corrosion can occur. The further the metals are apart in the electrochemical series, the greater is the electrolytic reaction (corrosion). One method that may be used to reduce the reaction is to insert a metal that is intermediate in the electrochemical series between the two dissimilar metals to be joined. The intermediate metal may be a surface coating or plating, or it may be a thin piece of metal, such as a washer. In the contact of dissimilar metals, one metal acts as a cathode and the other as an anode. Because a cathode is the source of electron flow, the smaller the cathode relative to the anode, the lower the electron flow and the lower the corrosion. The most effective methods of minimizing electrolytic corrosion are either to use similar metals or to reduce the moisture between dissimilar metals.

Moisture may be excluded by finishing the surfaces after the joint has been made, typically by paint or by plating or caulking. Figure 8.24 shows that it is better to apply the finish over both of the

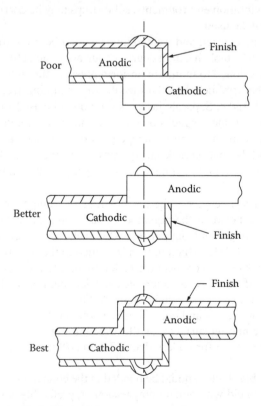

FIGURE 8.24   Finishing over dissimilar metals.

surfaces to be joined. Where this is not possible, finishing the cathodic metal is better than finishing the anodic metal only.

Metals such as aluminum and steel will oxidize and build up a high-resistance coating. The exclusion of air by passivation of the two surfaces to be joined will inhibit the oxide buildup. First, the surfaces must be cleaned and degreased, and in some cases, it is advisable to etch the surface. After cleaning, the surfaces may be coated with a conductive passivation, such as oaktite #36, alodine #1000, iridite #14, or iridite #18P. A chromate finish may be used but does not provide as low a resistance as the preceding finishes. Zinc, tin, or gold plating may also be used to achieve a high level of passivation. Anodizing is usually nonconductive and should be avoided. Cleaning or etching the surfaces and finishing with paint after joining, although not as effective as passivation, will minimize the oxidization at the joint.

Some methods of bonding, reproduced from the *Air Force Systems Command Design Handbook DH 1–4 Electromagnetic Compatibility*, are shown in Figure 8.25a through d.

8.7.4 BONDING TEST METHODS

The use of a two-terminal resistance meter to measure the resistance of a joint will lead to wide fluctuations in the reading, ranging from an open circuit to a fairly low resistance, depending on the pressure used to contact the probes to the joint to be measured. The problem with the

(a)

FIGURE 8.25   (a) Methods of electrical bonding.   *(Continued)*

Sub-note 1(4)   Typical
    method of bonding precipitation
    static discharger assembly
    to exterior of aircraft

Precipitation static
discharger (REF)

Clean aircraft structure
1-1/4 area of mounting base.
Clean base of discharger in
contact with aircraft

Note 2

Refinish as
required

Notes:
1. Install in accordance with manufacturers'
   procedure.
2. Mounting provisions to be specified on
   installation drawing. (See DN 7B2)

Sub-note 1(6)   Typical
    method of bonding with
    dagger pins

Back of rack

Clean to base metal
1-1/4 area of contact

Clean to base metal
both the back of
package and strike plate

Back of
electronic
package

Refinish after
instl. 1-1/2 area
cleaned

Refinish after instl
of strike plate 1-1/2
area cleaned

Note: Clean all dagger pins and strike plate holes after
    installation.

Sub-note 1(5)   Typical
    method of bonding between
    attaching flange of electronic
    package and rack

Clean to
base metal

Electronic
package

Rack

Clean flange
base metal

Sub-note 1(7)   Typical
    method of bonding electronic
    package to rack through
    front attachments

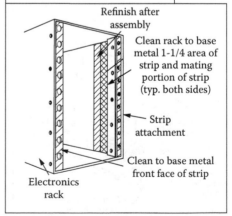

Refinish after
assembly

Clean rack to base
metal 1-1/4 area of
strip and mating
portion of strip
(typ. both sides)

Strip
attachment

Clean to base metal
front face of strip

Electronics
rack

(b)

FIGURE 8.25 (*Continued*)   (b) Methods of electrical bonding.   (*Continued*)

Sub-note 1(14) Connection jumpers

Bolt size:

Bonding - No. 6 and No. 8 screw where edge distance will not permit No. 10

  - 3 16 in. diameter minimum where possible

100 A return-1/4-in. diameter minimum
200 A return-3/8-in. diameter minimum

Note: Electrical bonding to magnesium structure for current return is prohibited

Aluminum alloy screw or bolt
Bonding or current return jumper
Steel lockwasher
Steel washer
Aluminum washer
Aluminum or magnesium structure
Aluminum washer (not used with plate nut)
Steel locknut or plate nut
Refinish after instal 1-1/2 dia of cleaned area
Clean to base metal area 1-1/2 dia of washer

(A)

Bolt sizes:

Bonding - No. 6 and No. 8 screw where edge distance will not permit no. 10 screw

  - 3 16 in. diameter minimum where possible

100 A current return-1/4-in. diameter minimum
200 A current return-3/8-in. diameter minimum

Bonding or current return jumper
Screw or bolt
Steel lockwasher
Steel washer
Plated steel, or CR steel or titanium
Steel locknut or plate
Refinish after instal 1-1/2 dia of cleaned area
Clean to base metal area 1-1/2 dia of term.

(B)

Notes: 1. Not applicable to bonding jumpers used for current return

2. Use bolted connections where jumper is used for current return

3. Ensure that rivet size is equal to the equivalent bolt size

Bonding or current return jumper
Aluminum rivet
Aluminum washer
Aluminum or magnesium alloy structure
Refinish after instal 1-1/2 dia of cleaned area
Clean to base metal area 1-1/2 dia of washer

(C)

(c)

FIGURE 8.25 (*Continued*) (c) Methods of electrical bonding: (A) Bolted to aluminum or magnesium alloy structure, (B) Bolted to alloy steel, plated steel or titanium, (C) Riveted to aluminum or magnesium alloy.

(Continued)

(d)

FIGURE 8.25 (*Continued*)   (d) Methods of electrical bonding.

two-terminal method is that the current is injected into the surface of the material by the same probe as used to measure the voltage, and in reality, it is the probe-to-metal contact impedances of the two probes that are measured. The four-terminal technique injects the current by two terminals and measures the voltage by use of another two terminals, and the measurement is much less sensitive to contact pressure. The resistance is found from Ohm's law by the measurement of the current and the voltage. One method of applying the current and voltage terminals is to solder leads to conductive adhesive copper tape and to apply the tape to the two surfaces. The current terminals should be placed some distance from the joint to be measured, whereas the voltage terminals should be placed close to the joint. The voltage terminals must be placed in the direct path between the two current terminals to minimize measurement errors. The four-terminal measurement technique may be used to make comparative measurements of the surface resistance of metals and finishes. Here, the width of the current and voltage contacts should be constant from measurement

FIGURE 8.26 Four-terminal bonding resistance test method.

to measurement and may be most readily obtained by use of conductive adhesive copper tape of a fixed width for all four terminals and from measurement to measurement. The four-terminal test method is illustrated in Figure 8.26.

8.7.5 GROUNDING SOFTWARE

SES & technologies offer software packages for grounding analysis and design. The CDEGS software package (Current Distribution, Electromagnetic fields, Grounding and Soil Structure) is a set of integrated engineering software tools designed to analyze problems involving grounding/earthing, electromagnetic fields, and electromagnetic interference including AC/DC interference mitigation studies and various aspects of cathodic protection and anode analysis. CDEGS computes conductor currents and electromagnetic fields generated by an arbitrary network of energized conductors anywhere above or below ground for normal, fault, lightning, and transient conditions. CDEGS models simple and multicomponent conductors including bare and coated pipes and pipe-enclosed cable systems buried in complex soil structures. It includes a powerful drawing tool and allows the exchange graphical data with AutoCad and other software supporting DXF, CDEGS.

AutoGroundDesign also by SES is a fully automated software package that can analyze and design grounding systems. It offers powerful and intelligent functions that help electrical engineers design safe horizontal, arbitrarily shaped grounding installations. A multiple-step approach is used for the automated grounding system design. A grid consisting of a buried metal plate, which gives minimum impedance, is used as a starting point to determine if safety limits can be achieved. If yes, a grounding system consisting of a minimum number of conductors is computed. From these results, a preliminary design is then selected based on the SES reference database, other intelligent rules, or as specified manually by the user. Next, the initial design is refined recursively using rule-based techniques and algorithms to improve its performance and meet safety constraints, while reducing the overall cost of the grid. Extensive collections of predefined grids have been analyzed, constructed, and can be easily updated by the user. A strategy has been devised to find an appropriate grid, while at the same time minimizing the size of the database.

REFERENCES

1. C.A. Charalambos, N.D. Kokkinos, and N. Chrisofides. External lightning protection and grounding in large scale photovoltaic applications. *IEEE Transactions on Electromagnetic Compatibility*, April 2014, 56(2).
2. FAA/Georgia Tech. Workshop on the grounding of electronic systems. March 1974. Distributed by National Information Service, U.S. Department of Commerce, Washington, DC.
3. N. Kuwabara, T. Tominaga, M. Kanazawa, and S. Kuramoto. Probability occurrence of estimated lightning surge current at lightning rod before and after installing dissipation array system (DAS). IEEE. NTT Multimedia Networks Laboratories 3-9-11 Midori-cho, Musashino-shi, Tokyo, Japan, 1998.
4. AFSC. *Design Handbook Electromagnetic Compatibility*. DH-1-4. Department of the Air Force, Wright Patterson Air Force Base, Dayton, OH, 1971.
5. *Lightning Protection and Grounding Solutions for Communication Sites*. Polyphaser Corporation, Minden, NY 89423-9000 (775), pp. 782–2511.
6. S. Sekioka and N.D. Hatziargyriou. Approximate formulas for terminal voltages on the grounding conductor. *IEEE Transactions on Electromagnetic Compatibility*, April 2014, 56(2), 444–453.
7. P.J. Higgs. An investigation of Earthing resistance. *IEE Journal*, February 1930, 68, 736.

9 EMI Measurements, Control Requirements, and Test Methods

9.1 INTRODUCTION

Electromagnetic interference (EMI) measurements may be divided into those imposed by EMI requirements, which also impose a test method, and measurements useful in finding the source of EMI problems. These problems include lack of self-compatibility or interference with other systems. Equipment manufacturers may not possess specialized EMI test equipment and therefore wish to make measurements with standard electronic test equipment, such as the oscilloscope. This chapter deals with the use of standard and specialized test equipment and perhaps most importantly with common sources of error in their use. It has been said that one measurement is worth a hundred predictions; however, EMI measurements may significantly affect the parameter measured, resulting in error. The location of antennas may affect the calibration curve, for example, errors greater than 30 dB are common in undamped shielded-room measurements of radiated emissions. A measurement with an oscilloscope may corrupt the measured signal as a result of either the additional ground connection or by radiated pickup on the oscilloscope test leads. Thus, great care has to be exercised in the selection and application of the measurement technique; otherwise, the prediction may be more accurate than the measurement. A design should not be based solely on measurements of performance, such as filter attenuation and shielding effectiveness; instead, the measurement should be used to validate the design. When large discrepancies exist between theory and measurement, both should be examined for error. It is incorrect to assume that measurement data are intrinsically more accurate than predicted values. The measured parameter may not be the intended; for example, small loop antennas may intercept the radial magnetic field and not the intended horizontal when the loop orientation is incorrect. The incorrect test equipment may be used for the measurements, leading to error. For example, a manufacturer of equipment that failed a commercial radiated emission test was using an amplitude modulated (AM) radio with a rod antenna to measure a relative reduction in field after EMI fixes were implemented. The criterion used was the very subjective one of reduction in audio noise. A potential source of error was the radio's automatic gain control (AGC), which may have a 20 dB dynamic range. Thus, a reduction of 20 dB or greater in field strength may be achieved with no perceptible change in measured noise level. Had the AGC voltage been measured by a DVM or small panel meter, a more accurate measurement might have been achieved. However, even this technique has the limitation that the AGC is operative only over a limited range of signal strength.

This chapter describes the use of commonly available, easily manufactured, and relatively inexpensive equipment that may be used for simple diagnostic measurements as well as the use of specialized equipment.

9.1.1 EMI TEST LABORATORIES

It is tempting, when new to EMI measurements, to leave the test setup to the test laboratory, after all they are the experts. However, mistakes can be made and the personnel of even the most reputable and experienced facility can make these.

Most laboratories are conscientious and thorough and carefully follow the EMI test procedure.

683

A few are sloppy, do not follow the test procedure, and even deliberately misinform. Luckily, these are in the minority.

For example, in my experience, two facilities have made the mistake of not including the gain of a preamplifier in a conducted and radiated emission measurement. In one conducted emission measurement, the preamplifier was unknowingly left in circuit from a radiated emission measurement. If an ambient measurement, as required, had been made, it would have been obvious that the ambient was above the limit. Conducted emission and radiated emissions of the ambient with the equipment under test (EUT) powered down must always be made prior to EMI testing. When this laboratory was asked to reproduce a radiated emission measurement, which had been made using a log-periodic biconical antenna with a "Roberts" reference dipole, they used a nonreference dipole that was subsequently found to be uncalibrated and possibly faulty.

Always ensure that equipment used to measure the test level has a current calibration and that a sanity check on the equipment is made.

Most test laboratories follow the cable configuration shown in Comité International Special Des Perturbations Radioélectriques (CISPR) 22, and ANSI-C63.4 (the test method called up in *Federal Communications Commission* (FCC) CFR 47 Part 15) for measuring radiated emissions from table-top and floor-mounted equipment to commercial requirements. In accordance with these requirements, the cables, which leave the turntable, are routed down from the turntable and are placed on the floor sometimes on insulation and other times not depending on the test laboratories interpretation of the requirement. Although not specifically described, the sketches do not show standoffs and so the assumption may be made that the cables are in contact with the floor. This same facility, which did not supply the reference dipole, used a weatherproof open-air test site (OATS) for radiated emissions and routes the cable vertically up from the turntable, and then it is suspended from the roof of the weatherproof building for approximately 3.4 m before being brought down to ground and extending outside of the test area. This may reduce emissions at some frequencies, but in some cases, emissions may be equal or higher in the configuration where the cables are dropped to the floor. With the cables on the ground, the current in the vertical section from the turntable to ground may increase, due to the proximity of the remaining cable to ground, with an increase in emissions at some frequencies.

At the same facility in the conducted emission measurement setup, the cables were not looped down from the table at a distance of 40 cm from a conductive wall and 30 cm from the ground plane as required, and it was obviously standard practice for the laboratory not to comply with the specified layout. In a laboratory in which the use of a measuring tape is not obvious, it is clear that the usual distances specified in EMI measurements, such as 0.1, 0.05, 0.3, 0.4, 1, and 3 m, are not accurately measured. The lack of exact distances may not result in significant errors but does indicate a lack of thoroughness.

This laboratory offers electromagnetic compatibility (EMC) support and EMI problem solving.

In the second of the cases with the preamplifier error, it was programmed in the spectrum analyzer (S/A) and had been ignored since the day the S/A had been purchased. By measuring the internal reference level or connecting a signal generator to the input, the presence of the 20 dB gain preamplifier was immediately obvious.

If witnessing EMI tests, it is imperative that the test setup configuration be examined and compared to the appropriate requirements in the specification. For example, by looking at the conducted emission and radiated emission figures in CISPR 22, ANSI C63.4, DEF STAN, or MIL-STD-461 and comparing these to the test setup used by the test laboratory.

In another instance, a test chamber had met the required field uniformity requirements and had failed an EUT in a radiated immunity/susceptibility measurement. When this measurement was repeated in two other chambers, the EUT passed at the problem frequency. This frequency coincided with one of the field uniformity measurements made in the first chamber that was discarded. Regardless of this error, the chamber was correctly calibrated.

The use of absorber loads in the room may reduce these hot spots.

9.2   TEST EQUIPMENT

All frequency domain limits are expressed in terms of equivalent root mean square (RMS) value of a sine wave as would be indicated by the output of a measurement receiver using peak envelope detection.

9.2.1   OSCILLOSCOPE

The oscilloscope is one of the most common pieces of test equipment found in electronic laboratories. The oscilloscope has both a number of major limitations when used for EMI tests and a number of advantages. A major limitation is that the majority of EMI test requirements place limits specified in the frequency domain. Thus, any oscilloscope measurement of a complex waveform must be converted into the frequency domain in order to compare measurements with limits. For measurements of single frequencies with low harmonic distortion, the oscilloscope may be useful, assuming its sensitivity is adequate. When a number of frequencies are present, measurements of a low amplitude and high frequency superimposed on a high amplitude and low frequency may be impossible. In this case, a frequency-domain measurement taken by using an S/A is the preferred method. When measurements of transient noise in the time domain are required, an oscilloscope is the correct instrument to use. Another limitation is the conversion of common-mode noise into differential across the oscilloscope's high-impedance single-ended input. A simple test for the presence of common-mode noise is to connect the probe ground clip to the signal ground and at the same ground point the probe tip. Although the input to the oscilloscope is thereby shorted out, any common-mode current flow on the shield and center conductor develops a differential voltage across the input impedance of the scope, and the magnitude of the common-mode contribution to the measured noise may be judged. A voltage is developed across the impedance of the shield due to a shield current flow, and it is this voltage that appears across the input of the oscilloscope. In this test, some of the voltage measured may result from radiated pickup by the loop formed between the probe and the ground wire. The S/A also has a single-ended input, but the 50–75 Ω input impedance develops a lower differential voltage from a common-mode current flow than the 1–10 MΩ impedance of the oscilloscope, or the voltage drop across the impedance of the shield is reduced because the input impedance is across the shield impedance. One disadvantage the S/A has is that the low-input impedance may load the signal, especially when the source impedance is high; also, the input is DC coupled, and a decoupling capacitor must be used when measuring signals/noise existing on high DC voltages. Oscilloscope and S/A single-ended inputs connect the signal return to the enclosure of the oscilloscope or S/A and thus to the alternating current (AC) safety ground. This additional ground connection may drastically alter measured noise levels and susceptibility characteristics of the equipment. The single-ended input and chassis ground connection characteristic of the oscilloscope may be useful when these are required to simulate the input of a piece of equipment with the same characteristics. Thus, the single-ended output signal of EUT may be correctly terminated by the use of an oscilloscope. One additional effect is that the inductance of the probe ground wire and the input capacitance of the probe form a resonant circuit that changes the waveshape or amplitude of a measured signal or noise. The resonant circuit may cause ringing on a transient, which may be incorrectly attributed to a circuit under test resonance. The LC circuit forms a low-pass filter and may attenuate the amplitude of high-frequency signals. The inductance of the ground wire should be reduced to a minimum by the use of as short a connection as feasible and by locating it close to the probe. Probes also come with attachments for probing PCB and circuits. These usually have a ground probe in close proximity to the signal probe at the probe tip. One simple device is a metal coil that wraps around the ground ferrule on the tip of the probe with one end of the coil unwrapped and used as a ground probe. It is always advisable to use one of these low-inductance grounds in preference to the probe ground wire. The difference in the measured waveshape can be amazing. Alternatively, a field-effect transistor (FET) input probe with a low-input capacitance may be used. When the oscilloscope single-ended input characteristics are undesirable, the use of a differential input is a solution. Differential plug-ins for oscilloscopes do not necessarily provide the highest level of common-mode noise rejection because of a poor balance between the two inputs caused in part by the use of separate test leads. A higher level of common-mode noise rejection is achieved by the use of a differential scope probe, in which the differential input is located at the probe tip. This type of probe has the advantage of lower pickup on the test lead from the radiated ambient and lower loading and resonance effects. When a differential input is unavailable, the two channels, A and B, of some oscilloscopes may be used in an A − B mode or an inverted A + B mode, which effectively measures differential voltage when the input to one channel is connected to the voltage and the second to the return. When the ground clips are connected together at the probe but not connected to either the ground or the enclosure of the EUT, the ground loop problem inherent in single-ended inputs is eliminated. The common-mode voltage is not totally removed from the measurement using this technique due to the poor balance between the two inputs. The level of common-mode contribution to the measurement may be checked by connecting the tips of both channel A and channel B probes to the signal ground. Some compensation for imbalance may be achieved by adjusting one or both of the compensation trimmer capacitors in the probes for minimum measured level. Radiated pickup on the test probes may be reduced by twisting the probe leads together, which reduces the pickup loop area and tends to cancel the field-induced currents. Another method of differentiating between common-mode and differential-mode noise sources, and of isolating the oscilloscope ground from the signal ground, is to measure the noise current by use of a current probe. In some instances, especially when the circuit impedances are low, the level of noise voltage is low, and it is the noise currents that result in radiation and should be measured (this is often true for current flow in chassis ground). The oscilloscope may be used in relative measurements of electric field by attaching a length of wire to the probe tip. When an oscilloscope has a 50 Ω terminated input or when a 50 Ω external termination is used, the small calibrated loop and other antennas described in Section 2.6 may be used. Antenna or current probe output levels are typically too low for use with an oscilloscope, even on the most sensitive range, and the use of a preamplifier is almost mandatory. Another limitation of the oscilloscope, in addition to the loss of frequency component information, is the limited upper frequency response of the average analog oscilloscope.

One use is to monitor for the presence of the EFT/burst, as simply placing the probe close to the capacitive injection probe will provide a response if the test level is present.

In the CS101 and CS106 series injection test on AC power lines, the oscilloscope may have 120–220 V common-mode voltage applied. Even with the use of the ×10 probe, many oscilloscopes provide erroneous measurements with this level of C/M voltage. The modern digital scope seems to be more susceptible than the older analog. One potential solution is to place the injection in the neutral line, which should have a much lower C/M voltage referenced to chassis/safety ground.

High-impedance probes are available with a 125 MHz, 500 MHz, and 1 GHz bandwidth, 10× and 2× attenuator, 10× probe 10 MΩ input impedance <4 pF capacitance, and 300 V Cat II dynamic range.

A typical active probe may have a bandwidth of up to 2.5 GHz with less than 1 pF input capacitance and ±8 V dynamic range and ±12 V offset.

Oscilloscopes with a frequency bandwidth of up to 2.5 GHz typically have both 1 MΩ and 50 Ω inputs.

For extremely high-frequency oscilloscopes, input impedance is typically much lower than 1 MΩ and the maximum input voltage is much lower.

For example, a 100 GHz 240 GS/s scope has a 120–600 Ω input impedance with a maximum D/M voltage range of ±2 V and C/M ±4 V.

A 20 GHz 10 Gb/s scope, which is ideal for time-domain reflectometry, has a 50 Ω input and a maximum input voltage of ±2 V DC or peak AC.

9.2.2 SPECTRUM ANALYZER

One of the most useful measuring instruments for diagnostic tests, and one that has gained acceptance for use in EMI testing requirements and for certification, is the S/A. An EMI receiver and an S/A are similar in their basic functions. However, a number of important differences exist; these are expanded on further in this chapter. One major difference that makes the S/A superior for diagnostic measurements is the CRT display. Some EMI receivers have the capability of connection to an oscilloscope or a monitor that is used to display amplitude versus frequency, instead of the usual oscilloscope display of amplitude versus time. Most modern S/As have a display at least as good as an oscilloscope. Another feature of the S/A that is different from the EMI receiver is its capability of displaying emissions in a short-duration sweep time (i.e., short-duration changes in amplitude are more easily discernible). To enable a fast sweep rate, the IF filter components must be capable of fast charge and discharge, which results in a filter with a Gaussian shape. The receiver that typically does not allow a fast sweep rate has a rectangular-shaped filter. The advantage of the rectangular filter is greater selectivity due to the very much reduced bandwidth below the 3 dB down point. For example, in comparing the 3–60 dB down bandwidths of the EMI receiver and S/A, the ratio in the EMI receiver is typically 1:2, whereas for the S/A it is 1:14. Thus, assuming a 100 Hz 3 dB bandwidth, the 60 dB bandwidth for the receiver is typically 200 Hz, and for the S/A, it is 1.4 kHz.

A limit on sweep time does exist, and where it is set too short, most S/As display an out-of-calibration condition. The S/A can suffer from compression and overload, just as described in the section on the preamplifier, and the same methods can be used for detection and correction. The S/A and EMI receiver have a built-in variable-input attenuator that may be used to check for compression as follows: By searching over the frequency range of the S/A to find the maximum input level and adjusting the input attenuator so that the displayed amplitude is at maximum, or lower, input compression and overload can be avoided. If on adjusting the input attenuator the displayed magnitude changes, the S/A is compressing. This is because the S/A automatically adjusts the displayed level regardless of the attenuator setting. In compression, the effective gain of the front end changes and so the displayed level does not remain constant. Hewlett–Packard publishes a useful booklet entitled "EMI Measurement Solutions Using the Spectrum Analyzer Receiver," and some of the figures and following information are adapted from that source by kind permission of Hewlett–Packard. A block diagram of the S/A is shown in Figure 9.1.

The S/A is a swept-tuned superheterodyne receiver with an extremely wide input frequency range (typically 10 kHz–1.3 GHz but as wide as 100 Hz–22 GHz). The incoming signal mixes

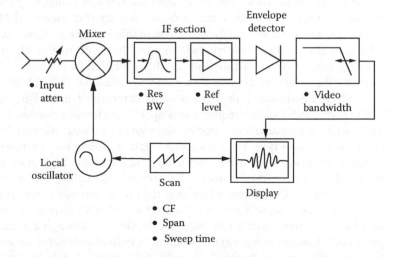

FIGURE 9.1 Spectrum analyzer block diagram. (Courtesy of Hewlett–Packard, Palo Alto, CA.)

with the local oscillator (LO); when the mixed product equals the IF, the signal passes through to a peak detector. Some S/As allow switching in a quasi-peak detector, whereas other S/As allow the connection of an external quasi-peak detector. The front panel of the S/A contains controls for the frequency span, sweep time, resolution, and video bandwidths and reference levels. Additional functions on most S/As include a marker that may be used to read off the frequency and amplitude of a selected component. The accuracy of the displayed frequency is a function of the resolution bandwidth and the frequency span. A typical specification for frequency readout accuracy is ±3% of span and of resolution bandwidth. Therefore, a frequency measured with a 2 GHz span may be displayed with up to ±60 MHz error. For maximum accuracy, the minimum span and resolution bandwidth should be used. On some S/As, a counter is incorporated with the marker function, with which counter accuracy frequency measurements may be made. Maximum or peak hold and storage of one or more sweeps are also useful functions that enable capturing short-duration emissions and comparing before and after emission profiles. The resolution bandwidth control on the S/A has a number of uses, one of which is in reducing the noise floor of the S/A. The narrower the resolution bandwidth, the lower the noise level and the better the displayed signal-to-noise ratio.

Assume that broadband (BB) coherent noise and a single-frequency narrowband (NB) source are both present and are measured with the S/A. As the resolution bandwidth is widened, more of the spectral lines generated by the BB source, which are close together in frequency, will lie within the envelope of the resolution bandwidth, and the displayed signal level will increase. The NB signal level, assuming it is not swamped by the BB signal, will not increase with an increase in bandwidth, for only one frequency is measured in the envelope of the bandwidth.

The resolution bandwidth, sometimes referred to as the IF bandwidth, also determines selectivity and sweep time. Because the S/A traces out its own IF filter shape as it tunes past the signal, the spectral emission lines are not displayed as infinitely narrow lines. Thus, if the resolution bandwidth is changed, the width of the display changes. This is important in differentiating between adjacent signals (selectivity). Hewlett–Packard specifies a 3 dB bandwidth (i.e., the width over which the amplitude is no more than 3 dB down on the maximum amplitude). If two adjacent signals are closer together than the 3 dB bandwidth, they cannot be differentiated in the display. In contrast, if the signals are further apart than the 3 dB bandwidth, they may be differentiated.

One of the definitions of NB or BB impulsive noise refers to the resolution bandwidth of the measuring instrument. In the early MIL-STD-461C requirements, separate NB and BB limits are specified, and in the more recent MIL-STD-461E-G, this is no longer the case. If NB and BB limits are still set (new equipment to be installed on an earlier platform), it is important that typical bandwidths for NB and BB measurements be established and agreed upon by the procuring agency before commencement of tests. In some cases, the agency that imposes EMI measurements will also define the measurement bandwidths. Where this is not true, Tables 9.22 and 9.23 provide typical NB and BB measurement bandwidths. As illustrated in this chapter, whether impulsive noise is displayed as NB or BB depends on the resolution bandwidth. It is important to differentiate between the two types of noise for a number of reasons. The FCC allows a 13 dB relaxation in the conducted emission limit if the noise is determined to be BB, and the early versions of MIL-STD-461, and DO-160 requirements specify different emission levels for BB and NB noise. In diagnostic measurements, tracing the source of emissions may be facilitated by determination of the emission type. For example, a single high-intensity emission may exist in the multiple spectral emission lines from a BB source. If the single high-intensity emission is determined to be NB, then the source is not the BB source, although this may excite the NB signal. The source is instead an oscillation, which may then be traced to a power supply, circuit instability, or a resonant circuit. Signals are characterized as NB with respect to the measuring instrument's bandwidth when there is only one spectral component of the signal contained within the filter band pass. Each spectral component of a signal is individually resolved and displayed in the frequency domain. This means that the frequency of the emission may be read directly off of the X coordinate of the S/A display. Changing the frequency span of the display changes the

Narrowband signals are displayed
as frequency domain spectral lines

Narrowband characteristics:

• Line spacing changes with frequency span

• Line spacing independent of sweep time

• Prf equals spectral line spacing
  (use frequency span control)

FIGURE 9.2   Narrowband display. (Courtesy of Hewlett–Packard, Palo Alto, CA.)

spacing of the spectral emission lines. For example, for a frequency span of 20–200 MHz and 10 divisions, the frequency per division of display is 18 MHz. Thus, two emissions 18 MHz apart will be spaced one division apart. Changing the frequency span to 20–110 MHz increases the spacing to two divisions. Changing the S/A sweep time will not change the spacing when the display is in the frequency domain. The spectral emission lines may be either NB signals (i.e., sine wave sources) or impulsive noise, which is displayed as NB signals and which may be classified as NB when the S/A bandwidth is set to an NB value. For impulsive noise sources displayed in the frequency domain, the pulse repetition frequency (PRF) may be determined by measuring the frequency spacing between the individual spectral lines. In our example, the PRF is 18 MHz. Figure 9.2 illustrates the NB display and the associated NB characteristics.

If we now assume an impulsive noise source at a much lower repetition rate than 18 MHz, for example, 50 kHz, and the resolution bandwidth is increased to include more than one spectral line, then individual spectral lines are no longer resolved and the signals are added under the resolution bandwidth curve. The impulsive noise occurs in the time domain at a specific repetition rate, unlike a continuous wave (CW). Thus, impulsive noise sources, when more than one spectral emission line is covered by the resolution bandwidth, are displayed in the time domain. The amplitude of the display is now equal to the envelope of the $(\sin x)/x$ spectrum.

A scanning analyzer will therefore display a pulse every 1/PRF seconds, with an amplitude proportional to the spectrum envelope amplitude at the frequency to which the analyzer is tuned, and the signal will be displayed as a BB signal. Due to the time-domain display of BB noise, changing the frequency span does not alter the spacing between the pulses as was true of the NB display. Instead, the spacing between the pulses is changed by changing the sweep time. A typical characteristic of a BB display is that the signals appear to "walk" across the CRT because the analyzer sweep time is generally not locked to the PRF of the signal. The PRF is found by taking the reciprocal of the sweep time between the individual pulses. Thus, for a sweep time of 0.2 ms and a one-division spacing between the pulses, the sweep time between pulses is 0.2 ms/10 = 20 μs and the PRF is 1/20 μs = 50 kHz. Note that, as described in Section 3.1.1, the PRF of a pulse with mark/space ratio of unity is often mistakenly calculated at half the actual PRF. This is because the pulses are at zero amplitude at the sidelobes. Figure 9.3 illustrates an impulsive signal, displayed as a BB time-domain signal, and its associated BB characteristics.

(a)

Broadband characteristics:

- Pulse spacing independent of frequency span

- Pulse spacing changes with sweep time

- Period (T) equals spacing of pulse responses (use sweep time control)

(b)   • PRF $= 1/T$

FIGURE 9.3   Broadband display. (Courtesy of Hewlett–Packard, Palo Alto, CA.)

Differentiation between NB and BB signals may also be achieved by changing the video bandwidth, which may produce an averaging of the high-frequency component at the output of the envelope detector. The averaging occurs because the video bandwidth filter is low pass and reduces the amplitude of the high-frequency components. The conditions under which this is achieved are as follows: (1) when the video filter bandwidth is narrower than the resolution filter bandwidth; (2) when the video bandwidth is less than the lowest PRF; (3) the frequency sweep must be slow enough to allow the filters to charge completely; and (4) the S/A must be in the linear amplitude display mode. Average detection reduces the displayed amplitude of BB signals but has no effect on the amplitude values of NB signals. One further test used to differentiate between NB and BB signals is the tuning test. In the tuning test, the S/A is tuned one impulse bandwidth (in the case of VDE requirements) or two impulse bandwidths (in the case of the MIL-STD-462) on either side of the center frequency of the emission under investigation. The impulse bandwidth is defined as the 6 dB filter bandwidth, whereas some S/A manufacturers, such as Hewlett–Packard, quote a 3 dB resolution bandwidth. A change in peak response of 3 dB or less indicates a BB emission, while a change of greater than 3 dB indicates an NB emission. Table 9.1 illustrates the four methods used to differentiate between NB and BB sources. That four methods exist indicates that the differentiation is not easy and that often more than one method must be tried before a conclusion can be made. If the results are inconclusive or contradictory, the tuning test is the preferred method used in arbitration. In some cases, the decision must be made by the EMC engineering authority in charge of tests and equipment compliance or by the agency imposing the EMC requirements.

An illustration of the effect of resolution bandwidth on displayed amplitude is shown in Figure 9.4 for an impulsive signal with a 1 kHz PRF and a 260 kHz main lobe width. At NB settings, the individual spectral lines are captured in the resolution bandwidth, resulting in an NB display. As the receiver bandwidth is increased, more of the spectral lines are captured, the amplitude increases, and the display is BB. Finally, as the bandwidth is further increased, the full signal spectrum is captured by the resolution bandwidth, and any further increase does not result in an increase in amplitude, resulting in a quasi-NB display.

Some of the criticisms of the S/A are its wide input frequency range, which leaves it prone to compression and overload, and its lack of sensitivity, which is to a large extent answered by the use of a preselector and amplifier. The Hewlett–Packard 85685A preselector has a separate 20 Hz–50 MHz

TABLE 9.1
Methods for NB and BB Analysis

Methods	NB	BB	CRT Response
Tuning test "Tune" ΔBW_1	Δ AMPL > 3 dB	Δ AMPL < 3 dB	
PRF test Δ sweep time	No Δ spacing	Δ spacing	
Peak vs. avg det Δ video BW	No Δ AMPL	Δ AMPL	
Bandwidth test Δ resolution BW	No Δ AMPL	Δ AMPL	

Source: Courtesy of Hewlett–Packard, Palo Alto, CA.

FIGURE 9.4 Narrowband and broadband display of impulsive noise with amplitude versus bandwidth. (Courtesy of Hewlett–Packard, Palo Alto, CA.)

input and a 20 MHz–2 GHz input. Although obsolete, it may still be purchased and used. It contains protection against high-voltage transients at the low-frequency input, and the preselector filters include a number of low-pass and tuned band-pass filters. Both high- and low-frequency paths contain preamps with 20 dB gain, which improves the noise figure (NF). The preselector also contains a comb generator that may be used to amplitude-calibrate the system to meet the ±2 dB CISPR specification. An additional advantage of the radio-frequency (RF) preselector is in measuring BB noise that covers a very wide frequency range, for due to the very wide frequency range of the S/A input, a high spectral density may be applied to the mixer, even though the resolution bandwidth filter reduces the spectral density applied to the peak detector. Thus, the dynamic range available, limited by the maximum mixer input level, may not be adequate. By limiting the bandwidth applied to the mixer with the preselector filter, the dynamic range of the instrument may be greatly increased.

For European Commission (EC), European Norm (EN), Industry Canada, and FCC commercial certification tests, the quasi-peak detector is specified; however, if the EUT passes the requirement using a peak-detecting instrument, then the results are acceptable, for, compared to quasi-peak, a peak measurement is worst case. The quasi-peak detector was instituted to attempt to account for the annoyance factor of noise. The annoyance factor of low-repetition-rate noise is considered lower than that of high-repetition noise, and due to the charge and discharge time constant built into the quasi-peak detector, its response is less to low-repetition-rate noise sources. In practice, above a 10 kHz repetition rate, the quasi-peak and peak responses are very close; and this becomes increasingly true as the repetition rate increases above 10 kHz. As the quasi-peak detector incorporates both a charge and discharge time constant, an additional effect is to average out the peak amplitude measured in successive slow scans of a signal that changes amplitude after each scan. This type of variation is very typical of multiple noise sources, which can change from being in phase, and therefore additive, to out of phase, and therefore subtractive, with time. This averaging effect often results in a lower measurement using the quasi-peak detector compared to the peak detector, which has a very much faster charge time.

One important factor in the choice of S/A, preamplifier, and EMI receivers is low production of intermodulation, spurious, and harmonic responses. Additional factors in the choice of the S/A and EMI receiver are flat ±2 dB gain versus frequency; adequate frequency accuracy and stability; wide dynamic range; high sensitivity and low noise floor (EMI receivers have often been quoted as exhibiting lower noise and higher sensitivity than the S/A, but this is no longer true of the more expensive analyzer); outputs for IF, video display, plotter, LO, and audio; acceptance by governing agencies (some do not accept a S/A); computer control capability. The S/A uses a Gaussian-shaped resolution bandwidth filter and the receiver uses an impulse bandwidth. In the measurement of BB noise, a correction factor of approximately 4 dB has to be made to the S/A measurement when compared to the receiver, whereas no correction factor is required for NB noise.

In addition to the scanning or swept S/A described previously, a new development is the real-time S/A. This is ideal for

Capturing changes over both short and long periods such as rare short-duration events
Finding weak signals masked by stronger ones
Finding and analyzing transient and dynamic signals
Capturing spread spectrum and frequency-hopping signals
Testing and diagnosing transient EMI effects

For a S/A to be classified as real time, all information contained within the span of interest must be processed indefinitely without gaps. The analyzer must take all the information in the time-domain waveform and transform it into frequency-domain signals.

FFT S/As have been available for some time but were typically confined to low frequency.

9.2.3 Preamplifiers

Preamplifiers are available with gains of 20–32 dB (numeric 10–40) and NFs from 0.5 to 5 dB. With an NF of 4 dB, assuming the noise level at the 50 Ω input is −152 dBm, with a 10 kHz bandwidth, the noise referenced to the preamplifier input is −148 dBm = 0.009 µV. With a 10 dB NF, the noise referenced to the input is −142 dBm = 0.018 µV.

Thus, with a 10 dB NF preamplifier, it is possible to amplify a 1 µV (−107 dBm) signal level to differentiate between 10 µV (−87 dBm) and 40 µV (−74 dBm), depending on the gain of the preamplifier, and to differentiate between the signal and noise, depending on the resolution bandwidth and the noise floor of the measuring equipment following the preamplifier.

A typical BB preamplifier will have a 10 kHz–1 GHz frequency range with a flatness of 1 dB. The maximum output voltage from a typical preamplifier is 0 dBm (i.e., 225 mV). If the input voltage to the preamplifier is greater than 12 mV, assuming a gain of 26 dB, then compression occurs (i.e., the output of the preamplifier does not increase beyond 225 mV). The problem is that the gain of the preamplifier effectively reduces at all frequencies when compression is present.

As an example, assume a measurement is made with a preamplifier connected to the input of a S/A and a signal is displayed at 10 MHz on a frequency span of 1–500 MHz. Assume the input level to the preamplifier at 10 MHz is 50 µV, at the same time a frequency of 10 kHz and a magnitude of 1 V are input to the preamplifier but not displayed on the S/A. The 1 V level at 10 kHz causes compression in the preamplifier, and the output at 10 MHz is displayed as 200 µV instead of the correct 1000 µV.

One technique that can be used to detect compression, if the input signal of interest is above the noise floor of the S/A, is to bypass the preamplifier (i.e., plug the input signal directly into the S/A), and the displayed signal should reduce by the gain of the preamplifier, which in the example is 26 dB. When the input signal level is too low, a 3–12 dB attenuator may be connected to the input of the preamplifier, after which, in the absence of compression, the signal should be reduced by the appropriate level (3–12 dB). The probability of compression can also be determined by setting the frequency span of the S/A to the frequency range of the preamplifier and noting high-level (i.e., close to 0 dBm) signals at any frequency. The solution to the compression problem is to include a filter, preferably tunable, at the input of the preamplifier, which attenuates the unwanted frequency but not the frequency of interest. Alternatively, the 3–12 dB attenuator may be adequate.

A second result of overload caused by an input signal that is too high is clipping and distortion at the output of the preamplifier. In the frequency domain, the distortion results in an output that exhibits coherently related spectral emissions covering a wide range of frequencies, whereas the input signal may be a single frequency. In some instances, overload can cause the preamplifier to oscillate. Constant attention to signs of, and testing for, compression and overload in preamplifiers, S/As, and receivers must be made to ensure that equipment does not either pass EMI requirements due to compression or fail due to spurious response caused by overload. Preamplifiers are also available that cover the 1–18 GHz frequency range at gains of 9–40 dB. It is important to choose an amplifier with an NF referenced to the input at least 3 dB lower than the lowest signal to be amplified. Generally, the narrower the frequency range of the amplifier, the lower the NF. Often the noise floor is specified in dB instead of (dBµV or µV)/bandwidth or $nV/\sqrt{Hz}$. The noise specified in dB is the NF described, with a typical test method, in Section 5.5.2. Often the NF is referenced to the noise generated by the preamplifier input termination resistor.

Low-noise preamplifiers are especially useful in making measurements at low levels with a S/A, for the preamplifier can drastically increase the signal-to-noise ratio obtainable with the S/A alone.

When measuring DEF STAN DRE01-radiated emissions from 88 to 250 MHz, the limits are usually very low. A typical combination of a 50–500 MHz 17–24 dB gain 0.5–0.6 dB NF preamplifier followed by a 0.25–1000 MHz 30–24 dB gain 4 dB NF is normally required. Fairview Microwave offers a preamplifier with a 0.6 dB NF and a specified 50–500 MHz frequency range. Measurements show that this preamplifier have a gain of 21.8 dB at 500 MHz, 27.5 dB at 30 MHz, 27.5 dB at 40 MHz, and 27.0 dB at 88 MHz.

Space application radiated emission limits often contain low-level notches at microwave frequencies corresponding to receiver frequencies. For measurement in the notches, a 1–11 GHz preamplifier with 17–20 dB gain and NF =1.5–2 dB and/or a 11–18 GHz 17–20 dB gain 1.5–2 dB NF preamplifier followed by a 1–18 GHz 22–30 dB gain preamplifier with a 3.5–5 dB NF is often employed.

These preamplifiers are often static sensitive. Norden Millimeter manufactures a range of reliable microwave preamplifiers.

9.2.4 EMI RECEIVERS

Some of the advantages and disadvantages of the EMI receiver, which is used to be referred to as an EMI meter, compared to the S/A are discussed next.

All the advantages of the preselector are built into the EMI receiver, without the need for an add-on unit with its additional cost and space requirements. In EMI receivers, fundamental mixing is always used, whereas in some S/As, a combination of fundamental mixing and harmonic mixing is used. *Fundamental mixing* means that the LO, which is used in a superheterodyne receiver to downconvert the input signal to the IF, covers the entire input frequency range. Fundamental mixing results in the lowest conversion loss and the lowest NF or highest sensitivity. *Harmonic mixing*, on the other hand, results in lower sensitivity. In measuring impulsive noise, the charge time of the detector is important, and to capture the peak amplitude, the charge time should be less than 1/(10 × IF BW). EMI receiver manufacturers claim that the receiver has a faster charge time than the S/A. This may not be an intrinsic function of the receiver, and in choosing between the receiver and analyzer, a comparison of the charge times should be made. S/As provide an NB CW calibration signal, whereas most receivers contain impulse generators in addition to a CW source. Many receivers contain AM and FM detection, as do some S/As. AM or FM detection is extremely useful in determining the source of a signal, especially in an ambient site survey, and also in determining the susceptibility of the system exposed to the signal. For example, a system may well be more susceptible to an audio frequency AM compared to a CW unmodulated signal. For completeness, phase demodulation should be, but seldom is, provided. In most radiated susceptibility measurements made in a shielded room, the measuring instrument monitoring the susceptibility test level is located outside of the shielded room and is thus not exposed to the test level. For the rare instances where the monitoring equipment is located inside the shielded room, the typically superior shielding of the EMI receiver will be an advantage and will reduce the spurious response of the instrument to the field in which it is immersed. Likewise, where diagnostic measurements are made of radiated or conducted emissions from an EUT with the measuring instrument located in the shielded room, the lower level of radiated emissions typical of an EMI receiver may be an advantage. As with all emission tests, the ambient inside the room, which includes radiation from the measuring instrument, with the EUT powered down should be monitored. One important feature of the S/A or receiver is the capability of programming the instrument, which is of benefit in automating measurements. A typical program will allow entry of the calibration curves of antennas and probes, from which the true noise levels may be calculated and displayed against the specification limits.

The modern receiver combines all the advantages of the S/A and the EMI receiver. For example, the Rohde and Schwarze ESMI receiver has a 20 Hz–26.5 GHz frequency range, extendable to 110 GHz with external mixers.

It has an S/A display and selectable S/A (overview mode) 3 dB resolution bandwidth or an EMI receiver 6 dB resolution bandwidth. Correction factors can be entered into a transducer table to compensate for antenna factors and current probe transfer impedance. The ESMI includes a built-in selectable preamplifier with a gain of 10 dB from 100 Hz to 26.5 GHz. The automatically switched filters include 1 low-pass, 11 fixed band-pass, 4 tunable band-pass, and 1 YIG filter. The 6 dB bandwidths are 10 Hz, 100 Hz, 200 Hz, 1 kHz, 9 kHz, 10 kHz, 100 kHz, 120 kHz, 1 MHz. AM or FM demodulation is provided, and the instrument can be controlled via an RS-232-C, IEEE 488, or parallel (Centronics) interface. Rohde and Schwarze also supply an EMI measurement software package to control the receiver.

In MIL-STD-461F, the requirement is that software needs to be described in the test procedure. For commercial software, identification of the manufacturer, model, and revision must be provided. For locally developed software, the control and methodology must be indicated.

9.2.5 Signal Generator and Power Amplifiers

The best type of signal source in a radiated or conducted susceptibility test is the analog sweep generator or a frequency synthesizer that can sweep continuously over the frequency range of interest. The disadvantage of the analog type of generator is that typically the sweep must be made manually. A frequency synthesizer–based generator, on the other hand, may usually be programmed via push buttons on the front panel or via a computer, using a computer interface. The more modern frequency synthesizer sweep generator is similar to an analog generator and sweeps continuously, whereas the older or simpler types sweep by stepping the frequency. The susceptibility of an EUT often occurs at a single frequency or at a limited number of frequencies. This selectivity is commonly a result of increased current flow in circuits or cables at resonance. Depending on the Q of the resonant circuit, the susceptibility may be seen only at a single frequency, with a sharply decreased response on either side of that frequency. Use of a generator that must be stepped in frequency, despite a small step size, increases the possibility that a resonant frequency at which susceptibility occurs may be missed. A potential disadvantage of the frequency synthesizer generator is the typically high level of harmonics generated compared to an analog generator. Power amplifiers are required to generate the specified level of E field in a radiated susceptibility test and the specified voltage at the input of the EUT in a conducted susceptibility test. Some examples of the frequency range and typical output powers of amplifiers are provided in Table 9.2. An amplifier that is rated at 75, 100 W peak, with a frequency range of 9 kHz–250 MHz, is capable, with the majority of antennas and injection probes, of generating the EN 61000-6-1, IEC 61000-6-2, and EN55024 C/M RF test level and the RF electromagnetic fields.

TABLE 9.2
Typical Power Amplifiers

Frequency Range	Power Output (W)	Manufacturer/s
10 Hz–20 kHz	1,200	Techtron
DC-1000 MHz	50	Amplifier Research
10 kHz–100 MHz	1,250	Amplifier Research
10 kHz–220 MHz	500	Amplifier Research
10–1200 MHz	10	LCF
100–1000 MHz	500/1,000	Amplifier Research
200–1000 MHz	10–400	IFI
10 kHz–1000 MHz	10–100	IFI
1–100 MHz	5,000–15,000	IFI
0.8–2 GHz	50	ITS Electronics Inc.
1–4.2 GHz	10	Amplifier Research
1–2 GHz	200	Amplifier Research
2–4 GHz	200	Amplifier Research
4–8 GHz	200	Amplifier Research
8–18 GHz	200	Amplifier Research
1–2 GHz	20	CPI (formerly Varian)
2–4 GHz	20	CPI
4–8 GHz	20	CPI
8–18 GHz	20	CPI

The power required to generate a specified E field from an antenna is a function of the gain of the antenna and the distance between the antenna and the EUT. MIL-STD 462 and DO-160 specify a distance of 1 m. An equation that provides the minimum power required for a specified E field is

$$P = \frac{E^2 4\pi r^2}{Z_w G}$$

where
r is the distance [m]
Z_w is the wave impedance at the distance r [Ω]
G is the gain of the antenna

Some of the important factors in the choice of a power amplifier are unconditional stability from an open-circuit (O/C) to a short-circuit (S/C) load and any value of inductive or capacitive load. When a power amplifier exhibits a high gain (40 dB or greater), great care must be taken in routing and shielding of the input and output cables. If these cables are too close together or are not adequately shielded, positive feedback may occur, with a potential for the generation of full output voltage across the load. When the load is an antenna, hazardous levels of E field may be generated, even though the input from the signal generator is set to a very low level or even disconnected. Routing the input cable too close to the antenna is a common source of positive feedback. In radiated susceptibility tests, locate the power amplifier outside the shielded room and use a short cable to connect the signal generator to the amplifier. A good power amplifier generates low levels of harmonics, spurious responses, and low levels of BB noise. A front panel meter that indicates power output is a useful function, especially in avoiding input overload and in monitoring instability. Instability and positive feedback can destroy the power amplifier. But even more important, it can destroy the EUT due to the

FIGURE 9.5 Voltage divider used to monitor power amplifier output.

potentially high level of E field. When monitoring the output power of an amplifier by use of a power meter, an EMI receiver, or an S/A, use a power attenuator at the input of the measuring instrument capable of dissipating the full output power of the amplifier or a bidirectional coupler. When using an attenuator, the level of attenuation depends on the input power rating of the measuring instrument. For example, a typical S/A 50 Ω input is rated at 1 W. Consider that the output level of a 400 W power amplifier must be adjusted to a specified level, which is monitored on an S/A. The maximum input level to the S/A is 1 W, and the power amplifier may generate up to 25% more power than the rated value of 400 W (i.e., 500 W). The level of attenuation required is given by 10 log (P_{in}/P_{out}) = 10 log 500/1 = 27 dB. The voltage attenuation required is given using $V = \sqrt{50P}$ as follows:

$$20 \log \frac{\sqrt{50P_{in}}}{\sqrt{50P_{out}}} = 20 \log \frac{\sqrt{50 \times 500}}{\sqrt{50 \times 1}} = 27 \text{ dB}$$

The power and voltage attenuation are the same because the attenuator input impedance and the S/A input impedance are the same, 50 Ω. Thus, in our example, a 26 dB attenuator, which is a common value, rated at 500 W would be used between the output of the power amplifier and the input of the S/A. To be ultrasafe, an additional 3 dB attenuator may be added to ensure achieving at least 27 dB of attenuation.

At low frequency instead of a bidirectional coupler, a voltage divider may be used to monitor the power amplifier output. A suitable circuit is shown in Figure 9.5. Here, a 1000 Ω 5 W noninductive series resistor is used with a 26 dB 2 W minimum attenuator. The frequency response level is from DC to 500 MHz. The disadvantage is that at 100 W, a further 5 W is dissipated in the divider. Above 500 MHz, low-cost bidirectional couplers are available.

9.2.6 CURRENT PROBES

Current probes used in EMC measurements are somewhat different from the current probe used with oscilloscopes. The major difference is that EMC-type probes usually have a larger-diameter opening, thus allowing measurements on larger-diameter cables. They do not measure DC currents but frequencies as low as 20 Hz and as high as 1 GHz. Current probes can accurately measure microamp levels of AC current in the presence of up to 200 A of DC current. When the current probe frequency response lies much above 60 Hz, the probe is capable of measuring low levels of AC current at high frequency in the presence of high levels of 60 Hz AC current. No single current probe can measure over the 20 Hz–1 GHz range of frequencies; instead, a number of probes must be used. Two probes are the minimum required to cover this wide frequency range, with one probe typically covering a 20 Hz–200 MHz frequency range and a second probe covering the range of 20 MHz–1 GHz. The current probe does not contact the conductors over which it is placed; instead, a voltage is generated by the magnetic field surrounding the cable by transformer action. The current probe when placed over a cable that contains a signal line and return or many signal lines and returns measures the common-mode current but not the differential-mode currents. As we have seen, the major contributor to cable-radiated emissions is the common-mode current flowing on cables or cable shields. The calibration of the probe is typically of transfer impedance, which is the voltage developed across a 50 Ω load divided by the current flow measured by the probe.

For example, when the transfer impedance (Z_t) at a specific frequency is specified as 0 dBΩ (i.e., 0 dB above 1 Ω), the transfer impedance is 1 Ω. An example of the use of Z_t follows: If the current measured is 1 μA, then the voltage developed across the 50 Ω load is

$$V = IZ_t = 1 \ \mu A \times 1 \ \Omega = 1 \ \mu V$$

Such a low level of voltage is likely to be in the noise floor of the S/A, and thus a preamplifier must be used. This type of current probe, regardless of whether it is used with an oscilloscope or an S/A, is designed to be loaded with 50 Ω.

When the transfer impedance is provided in dBΩ, this is subtracted from the measurement in dBμV to obtain the current in dBμA. For example, if the transfer impedance is −20 dBΩ and the measured level is 25 dBμV, then the current is 45 dBμA (25 dBμV − 20 dB = 45 dBμA).

If the transfer impedance is 5 dBΩ and the measurement is 25 dBμV, then the current is 20 dBμA.

Although the current probe is typically shielded, it does respond to incident E fields. Thus, any measurement using the current probe should commence with a measurement of the ambient (i.e., with the current probe lying close to the cable to be measured but not placed around it). When the coaxial cable connecting the probe to the measuring equipment is acting as an antenna and responding to radiated fields, the use of baluns on the cable may reduce the level of pickup. Another technique often effective in reducing pickup on the cable is to add an additional overbraid and thus increase the shielding effectiveness of the cable.

Some manufacturers provide a calibration in terms of the transducer factor k, which is equivalent to $1/Z_t$. The transducer factor in dB has to be added to the measured level in dBμV to obtain the current in dBμA.

9.2.7 Magnetic Field Antennas

The construction of these antennas is described in Section 2.6 and their use in Section 9.3.2. The construction of a 13.3 cm shielded loop antenna is described in MIL-STD-462 for RE01 measurements 30 Hz–30 kHz and in MIL-STD-462D:1993 for RE101 measurements from 30 Hz to 100 kHz. Magnetic field antennas are used in a similar manner to the current probe with a preamplifier and S/A. They are shielded or balanced to provide attenuation of the E-field-induced voltage, but the level of attenuation is limited. The magnetic field antenna may be used in close proximity to a shielded enclosure or cables, and the purpose is to "sniff" around apertures and seams for leakage, indicating a faulty joint or increased joint transfer impedance. The H-field antenna is more sensitive to the level of electromagnetic ambient than is the current probe and should be used in as low an ambient as feasible.

9.2.8 BB Antennas

BB antennas used in EMI measurements differ from broadcast antennas in a number of aspects. High directivity in a broadcast antenna is often a desirable parameter, whereas in EMI measurement, antennas with too high directivity are a disadvantage. In radiated emission measurements, the sources of emissions may be 2 m or more apart, which necessitates movement of the antenna and subsequent multiple scans when the antenna directivity is too high. Likewise, in radiated susceptibility tests, the antenna should generate as uniform an E field over the EUT and interconnecting cables as possible with the antenna located as close as 1 m from the EUT. The antennas are designed to cover as wide a frequency range as possible in order to reduce the number of antennas required to cover the typical 14 kHz–18 GHz frequency span used in E-field susceptibility tests. EMI measurement antennas are designed to both radiate and receive, with the exception of active antennas, which are designed to receive only. Thus, calibration curves of both the gain and the antenna factor (AF) of the antenna are provided by the manufacturer, to enable the calculation of the power level required to generate the specified E field and to convert the measured antenna output voltage to an E field. The gain and the antenna factor are either calculated for far-field conditions, measured on an open-field test site, free-space antenna range, and measured at distances of 1, 3, or 10 m from the source. Low-frequency, 10 kHz–30 MHz antennas are often calculated for far-field conditions; thus, when used at a distance of 1 m from a source, the AF may be very much different from the calculated value. The AF is also very sensitive to the incident wave impedance; see further discussion in Section 9.2.8. Using antennas in a shielded room can drastically alter the calibration, due to loading effects caused by proximity to the floor and ceiling of the room as well as the EUT. In addition, room resonances and reflections change the apparent gain and AF of antennas, as discussed in Section 9.5 on shielded rooms. Another factor is the variable-input impedance of the antenna with changing frequency. EMI measurement antennas are specified with a nominal input impedance of 50 Ω, whereas the actual impedance of the antenna may be much different, resulting in a high voltage-standing-wave ratio (VSWR). A high VSWR affects the power amplifier driving the antenna in ways that are unique to the amplifier in use. When the antenna factor calibration is based on measured data, the antenna impedance is accounted for in the calibration, assuming the curve is not idealized (i.e., when the sharp dips and peaks commonly seen in an antenna calibration are not smoothed out). Some figures and curves of antenna factor and gain for typical EMI measurement antennas are shown in Figure 9.6a through d.

(a)

FIGURE 9.6   (a) EMCO Model 3301B active rod E-field antenna (shown without counterpoise). (Courtesy of Electro-Mechanics Co. Ltd., Suwon, South Korea.)   (*Continued*)

(b)

FIGURE 9.6 (*Continued*)   (b) Combined log-periodic and biconical antenna, 20–1300 MHz, with typical antenna factor versus frequency. (Courtesy of EMC Consulting, Inc., Merrickville, Ontario, Canada.)

(*Continued*)

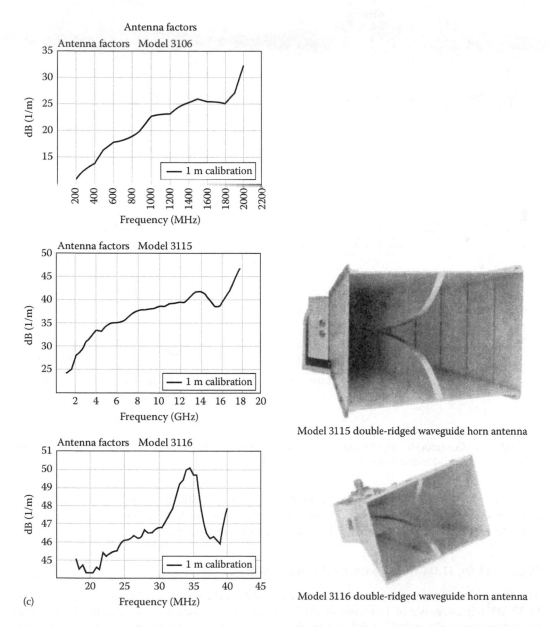

Model 3115 double-ridged waveguide horn antenna

Model 3116 double-ridged waveguide horn antenna

FIGURE 9.6 (*Continued*) (c) Antenna factor versus frequency curve for three models of double-ridged guide antennas. (Courtesy Electro-Mechanics Co. Ltd., Suwon, South Korea.) (*Continued*)

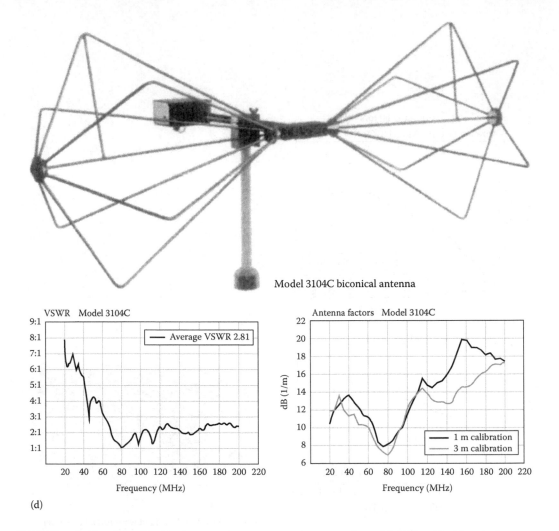

(d)

FIGURE 9.6 (*Continued*) (d) Antenna factor versus frequency for a 20–200 MHz biconical antenna. (Courtesy of Electro-Mechanics Co. Ltd., Suwon, South Korea.)

When using a BB log-periodic biconical type antenna for measuring an emission, which is close to the limit, it is advisable to repeat the measurement using a "Roberts" reference dipole, which is tuned to a specific frequency, and provide a more accurate measurement.

9.2.9 41 in. (1.04 m) MONOPOLE RECEIVING ANTENNA

The 41 in. (1.04 m) monopole antenna is commonly used to measure radiated emissions from 10 kHz to 30 MHz at a distance of 1 m from the EUT and its associated cables. The antenna is only used to measure vertically polarized fields from the EUT. Another use for the monopole is to measure high levels of field, for a safety analysis, antenna calibration, or radiated susceptibility testing. The antenna is invariably used in the near field where the field wave impedance is often higher than the impedance of free space (337 Ω). The near-/far-field transition at 10 kHz is 4777 m, and at 30 MHz, it is 1.59 m. Therefore, for a high impedance source, the wave impedance is only just approaching 377 Ω at 30 MHz at 1 m distance, and it is much higher at lower frequencies. Due to the high wave impedance, the antenna factor (AF) of the antenna, defined as the ratio between the incident field and the voltage developed at the antenna terminal, is very dependent on the load impedance.

Also, the input impedance of the monopole at lower than resonant length frequencies is high and capacitive. For these reasons, to obtain a low antenna factor for measuring very low-level fields, the load impedance should be high, whereas in measuring high levels of field, where the antenna factor can be high, a 50 Ω load impedance can be used.

The antenna factor of the monopole antenna used in MIL-STD-radiated emission measurements, unlike other antennas, is based on the theoretical far-field calibration and not measured values. MIL-STD-461F requires that a signal path check be made on the antenna using a signal generator coupled to the antenna via a capacitor of a specified value.

Thus, the antenna factors of many commercially available 1 m monopoles are based on a theoretical calculation and on the measured gain/loss of the buffer/amplifier. This approach ignores the curvature of the incident field in the near field. It also assumes an infinite ground plane, whereas the typical ground plane is a 24 in. × 24 in. counterpoise, which is either bonded to the underlying ground plane (floor of the chamber) by a wide braid or bonded via an extension of the counterpoise to the metal topped table on which the EUT is bonded or in the case of MIL-STD-461F isolated from the ground. One variable is that different sources of E field can exhibit very different wave impedances, another factor may be the mutual coupling between EUT and antenna.

The typical EMI measurement antenna will have a buffer stage, with or without gain, that converts a high input impedance to a 50 Ω load, which is the normal cable and input impedance of EMI measurement equipment. At low frequency, the load resistance often dominates the sensitivity of the antenna, whereas above typically 0.1 MHz, the ratio of the antenna self-capacitance to the buffer input capacitance, which includes device input capacitance and stray capacitance, determines the antenna factor. Another important parameter for the buffer is the output noise. The MIL-STD 461E, F and the draft G RE102, for a number of applications, have limits with a minimum value of 24 dBμV/m from 2 to 100 MHz. The noise level of the buffer, any subsequent amplifier and the measuring instrument, when measured with a 9–10 kHz bandwidth, and when added to the 1.04 m monopole antenna factor, must be at least 6 dB below this limit.

Although capacitive coupling and mutual inductance between two monopole antennas 1 m apart do not appear to be of importance when calibrating with a low receiving antenna load impedance, it may be significant when the load impedance is high.

The definition of antenna factor (AF) is the ratio of incident E field on the antenna to the voltage developed at the terminal of the antenna $AF = E/V$.

One method of measuring antenna gain is the two identical antenna method described in Section 2.5.3. The measured voltage into a 50 Ω load is made via two paths. One is the radiated path in which a 50 Ω attenuator is used at the input of the transmitting 1 m monopole and a second attenuator at the input of the receiving 1 m monopole. The second is a direct path in which the two cables with the attached attenuators are joined together. The problem with obtaining the antenna factor (AF) from the gain is the unknown value of wave impedance, and also the ratio of wave to antenna impedance may play a role in the conversion from the incident E field to voltage developed in the 50 Ω load. With the two monopoles 1 m apart and using

$$Z_w = 377 \ \Omega \times \lambda/2\pi$$

the calculated AF from the measured gain is close to the theoretical figure at 10 MHz and higher and also the measured, using a calibrated bow tie antenna, is close at 10 MHz and above. However, at lower frequencies, a large difference between theoretical and measured is seen using this calibration technique.

The gain calibration may be made on an OATS with antenna counterpoises either coupled or uncoupled. One calibration setup, which approximates the measurement test setups for the RE02/RE102 in MIL-STD-461E/F, DO-160 section 21, is tested with the counterpoise of one antenna bonded to the conductive tabletop in the anechoic chamber/damped room and with the second

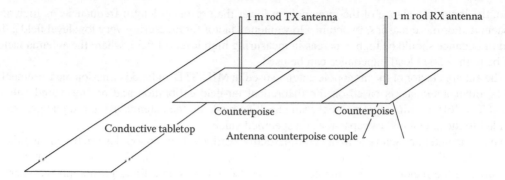

FIGURE 9.7 Test setup with one counterpoise bonded to the conductive tabletop and coupled to the second counterpoise.

antenna coupled to the first as illustrated in Figure 9.7. In this configuration, some differences in AF are seen versus the measurement made on the OATS. Some of this difference may be due to the anechoic chamber/damped room. This despite a very close correlation to the OATS does differ over the frequencies of interest. Another reason for the differences may be due to extending of the counterpoise of one antenna to the conductive tabletop.

Figure 9.8 compares the approximate antenna factors using the two-antenna calibration method with the two antennas either coupled or uncoupled with the receiving antenna loaded with 50 Ω. In the uncoupled case, the counter poise of one antenna is connected to the conductive tabletop and the second antenna counterpoise is bonded to the shielded-room floor using a wide copper plate. These tests were made on the OATS with antennas 1 m apart.

Figure 9.9 illustrates the importance of the extension of the counterpoise and, also if the antenna is to be used either coupled or uncoupled, the importance of a separate calibration for each configuration.

FIGURE 9.8 AF using the measured gain and an approximation for the wave impedance, with the antennas either coupled or uncoupled.

FIGURE 9.9 Comparison between theoretical and measured AF coupled and uncoupled.

The theoretical antenna factor of a receiving monopole above a perfect ground plane, connected to a 50 Ω load, is given by

$$AF = (1 + (Zant/50)) \times 1/Heff$$

where
  Zant is the antenna impedance
  Heff is the antenna effective height

Zant for a monopole is given in Section 2.5.5.
  The effective height describes the current distribution on the antenna. For a physical length of the rod, which equals ¼λ, the current distribution is nearly sinusoidal with an average value of 0.64 for Heff. For an electrically short monopole, Heff is close to 0.5.
  Heff for the monopole may be found from

$$Heff = (\lambda/\pi \tan(H/2))/2$$

where
  $H = \beta \times h = 2\pi h/\lambda$
  h is the physical height of the monopole rod

The reactance and AF in dB for the 1.04 m monopole loaded with 50 Ω are shown in Table 9.3, and the AF plotted in Figure 9.8 is close to the theoretical from 10 to 30 MHz in Figure 9.9.
  These data are presented in Reference [17]. The transmitting monopole was 2.5 m high and the receiving antenna also 2.5 m high and loaded with 50 Ω. The OATS had a ¼ in. wire mesh stretched on a 30 m × 60 m concrete slab. A second measurement was made with a 1 m monopole connected to a matching preamplifier and to a commercial monopole antenna with buffer.

TABLE 9.3
Theoretical Reactance and AF of a 1.04 m
Monopole Loaded with 50 Ω

Frequency [MHz]	Antenna Reactance [Ω]	AF [dB]
0.01	−1,256,259	94
0.02	−628.129	88
0.05	−251,251	80
0.1	−125,625	74
0.5	−25,123	60
1	−12,562	54
5	−2,512	40
10	−1,256	34
15	−837	31
20	−628	28.5
25	−502	26.7
30	−418	25

Using the equations in Section 2.5.3 to obtain the Zant and Heff of the 2.5 m antenna loaded with 50 Ω and comparing the Zant and AF to the data in Reference [17] show a close correlation between the two.

The 10 cm long bow tie is connected to a low-noise differential input with an input resistance of 2 MΩ and an input capacitance of approximately 3 pF. The wiring from bow tie to differential input adds approximately a further 2.2 pF of capacitance. The differential input is converted to a single-sided signal and applied to a detector with a logarithmic response. The detector allows a very high (dB) dynamic range. The output of the detector is connected to an A/D converter, the output of which is translated to a digital data stream, which is the input of a fiber-optic driver. The electronics and battery are contained in a 6 cm × 6 cm × 2.5 cm shielded box, and the only connection to the box is the nonconductive fiber-optic cable. The bow tie is connected to the box using 13 cm long thin wires. Thus, the perturbation in the measured E field by the measuring antenna cable and equipment is kept to a minimum.

Reference [16] describes the bow tie and provides a formula for its calibration, but the exact input capacitance of the differential input and stray capacitances must be known. Instead, the bow tie was placed in the center of a 0.6 m stripline antenna and the antenna factor measured. The E field under the antenna was calculated from the height of the antenna and the input level.

The bow tie antenna is calibrated from the calculated E field under the stripline and the digital output of the bow tie and digitizer. At 10 and 20 kHz, the calibration curve shows a very slight difference compared to the AF above 20 kHz. However, the calibration curves at 0.05, 0.1, 0.5, 1, 5, 10, 15, 20, 25, and 25 MHz are identical.

The theoretical AF measured using the bow tie or the bow tie with 50 Ω load were compared.

One monopole was used as the transmitting antenna and the second as the receiving antenna with the antennas 1 m apart. The voltage across the terminal of the receiving antenna Vt was measured from 10 kHz to 30 MHz. The rod of the receiving antenna was then replaced by the 10 cm calibrated bow tie. As the incident field changes with height above the counterpoise, the measurement of the E field was made with the center of the bow tie at heights of 10, 27.8, 45.6, 63.4, 81.2, and 99 cm. The average value of field E was then used in the AF calculation.

The AF in dB is $20 \log(E/Vt)$.

Figure 9.9 compares the theoretical AF with the measured for both coupled and uncoupled configurations. Again, we see that having the counterpoises coupled or uncoupled affects the AF above 10 MHz.

In a practical MIL-STD or DO-160 test setups, the emission source may be from the EUT bonded to the conductive tabletop. Apart from emissions from apertures, this would also occur if a voltage difference exists between the equipment case and the underlying ground plane. This emission source can, very approximately, be considered a second monopole antenna. One of the reasons for the voltage difference between the equipment case and the ground plane is common-mode current flowing on the attached cables. This common-mode current results in a voltage drop down the length of the cable, which will appear on the equipment enclosure if the cable is shielded and attached to the enclosure.

Even if a ground braid is used to bond the EUT to the tabletop, as required, current flow in the ground braid can result in the voltage drop.

For floor-standing equipment, a typical cable configuration is for the cables to be connected to the top or sides of the enclosure and brought down the sides at a height of 5 cm and then laid out on the ground plane on the floor of the room again at a height of 5 cm. This cable then forms a transmission line with a characteristic impedance determined by the height and diameter of the cable.

The two identical antenna calibration method was used on the OATS with the antennas 1 m apart at 10 MHz and above. Again an approximate wave impedance was used to obtain the AF from the gain. The calibration was first made with the receiving antenna loaded with 50 Ω. A second set of data was obtained with the receiving antenna connected to a buffer. This buffer was different (resulted in a higher AF) than the buffer used for the data in upcoming Figure 9.10.

The difference between the two measurements was subtracted from the AF with 50 Ω load to obtain the high-frequency AF of the monopole with buffer.

The results of these calibrations are shown in Figure 9.11, and again the coupled AF is lower than the uncoupled and shows peaks and troughs.

The three following additional test setups were used to calibrate the monopole. The monopole to monopole setup in the anechoic chamber and the TX line to monopole setup, also in the chamber, approximate the two practical MIL-STD measurement setups discussed, whereas the plate antenna calibration, made outside of the chamber, may more closely approximate a far-field measurement with the exception that the wave impedance will be higher. These measurements were all made with an improved buffer (lower AF than used for the data in Figure 9.12).

FIGURE 9.10 The Reference [17] antennas AF measured at 10 m and the EMC consulting under the plate antenna.

FIGURE 9.11   Comparison of test setups and monopoles with buffers.

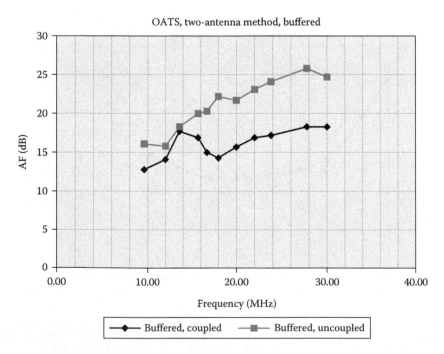

FIGURE 9.12   Two-antenna calibration method on OATS at high frequency, coupled and uncoupled NB buffer with higher AF than used for all other tests.

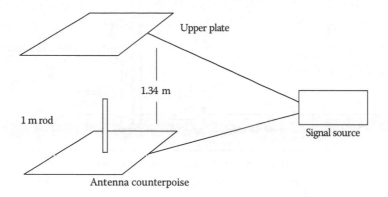

FIGURE 9.13   Plate antenna test setup.

The counterpoise was connected to the return of a signal source and the antenna was placed under a second brass plate with the 24 in. × 24 in. dimensions. This second plate was located at a height of 1.34 m above the antenna counterpoise. The test setup is shown in Figure 9.13.

Due to the small size of the plates of the antenna and the distance between plates, the E field between the plates is not uniform. Therefore, the E field with the rod of the monopole removed was measured with the center of the 10 cm bow tie at heights of 10, 27.8, 45.6, 63.4, 81.2, and 99 cm. The average field strength was then used for the antenna calibration (Figure 9.14).

With the plate antenna measurement, the signal source was changed in frequency over the 10 kHz–5 MHz frequency range. Results above 5 MHz were not believable due to the difficulty of feeding the two plates from the source above 5 MHz. The AF using this test method is shown in Figure 9.11.

As well as the EMC consulting buffer, a commercial (1) 1.04 m rod antenna and buffer, was tested under the plate antenna and the results are also shown in Figure 9.11. The commercial (1) published "typical" AF is also shown. To reduce the noise floor, the input resistance of the EMC consulting buffer is 220 kΩ and this is the reason the AF is higher from 10 to 50 kHz than the commercial. Typical MIL-STD-461 RE102 limit curves limit, which is higher at 10 kHz and in the worst case, reduces log linearly to 2 MHz. Despite the higher AF of the buffer at 10 kHz, the noise floor of the EMC consulting buffer, preamplifier, and S/A, measured with a 1 kHz bandwidth from 10 to 150 kHz, is well below the lowest RE102 limit. Likewise, from 150 kHz to 30 MHz, the noise floor is well below the limit when measured with a 9 kHz bandwidth.

The vertical transmission line shown in Figure 9.10 was constructed and driven by a generator and power amplifier. The line was terminated in the calculated characteristic impedance of the transmission line at the far end.

The receiving monopole was only bonded to the ground plane on the floor of the room in this test. The average E field measured by the bow tie antenna and the voltage from the output of the buffer was used to calculate the AF of the monopole and buffer.

Again, the AF is shown in Figure 9.11.

Reference [17] provides AF data on a 1 m monopole with buffer/preamplifier and a second (2) commercial monopole with buffer measured on an OATS with a 2.5 m transmitting antenna at a distance of 10 m from the 1 m monopoles.

No information on the buffers used in Reference [17] is forthcoming from the authors; however, the buffer/preamp has a 10 dB higher AF than the commercial (2), which means that the commercial antenna may include some gain.

The AF of the commercial monopole with buffer (2), tested in Reference [17], compared to EMC consulting monopole, measured under the plate antenna that is the closest match, is provided in Figure 9.10.

FIGURE 9.14 Test setups. (Reproduced by kind permission from Weston, D.A., *EMC Europe*, Wroclaw, Poland, 2010.)

The preceding investigation shows how coupling the counterpoise of one monopole to the counterpoise of a second monopole or to the conductive table to simulate a MIL-STD-461E RE test setup can significantly change the AF, especially at high frequency.

Also, the AF of the monopole is affected by the proximity and type of *E*-field source as shown in the measurements with a second monopole or a vertical transmission line. These effects may be due to mutual coupling when the receiving antenna is terminated in a high impedance (buffer).

The monopole to monopole and transmission line measured AF may be more representative of the actual MIL-STD test setup for RE, whereas the plate antenna test setup has a closer correlation to measurements made in Reference [17] at a distance of 10 m.

Another important factor is the buffer used with the monopole. The differences in AF between the buffers cannot be explained only by the ratio between input capacitance and antenna capacitance, and so buffer gain or loss must also play a role.

In MIL-STD461F and in the MIL-STD-461G, the requirement is that the counterpoise of the monopole be isolated from the tabletop ground plane and from the underlying ground plane, which as seen previously changes the antenna factor dramatically. In addition, a ferrite with an impedance of 20–30 Ohms at 20 MHz is placed on the vertical section of cable from the antenna to the floor. The cable, after the ferrite, is then bonded to the floor as it touches the floor. The following figures and plots are reproduced from Reference [21] by kind permission of *EMC Europe, EMC*

FIGURE 9.15 Measured AF of monopole with and without counterpoise bonded and with a simulated EUT as the source. (Reproduced by kind permission from ITEM, EMC test & design guide, Plymouth Meeting, PA, 2010.)

Symposium, 2010, Wroclaw, Poland, as well as from Reference [20], "The Interference Technology Magazine (ITEM) 2011 EMC test & design guide 2010."

In Figure 9.15, at many frequencies, the AF of the uncoupled monopole is lower than when coupled. However, with the same input level to the transmitting monopole, the E field measured above the receiving antenna counterpoise was very much lower. This compares the MIL-STD-461E and lower to the latest MIL-STD-461F.

When this difference is added to the uncoupled AF, we see a very large difference in the received level from the same transmitting source and this is shown in Figure 9.15.

We see that coupling the counterpoise of one monopole to the counterpoise of a second monopole to simulate a MIL-STD 461E RE test setup can significantly change the AF, especially at high frequency. In the MIL-STD-461F test setup, where the counterpoise is not connected to ground, the incident field on the antenna and the antenna AF will be very different. Also, when the counterpoise of the receiving monopole is bonded to the floor GP in a shielded room, the AF is different. In addition to the AF difference, the E-field incident on the receiving monopole is significantly higher when the counterpoises are coupled. Thus, for the same E-field source, the measured level may be up to 30 dB higher.

Also, the AF of the monopole is affected by the proximity and type of E-field source as shown in the measurements with a second monopole or a vertical transmission line or from the far-field measurement.

MIL-STD-462 and MIL-STD-461D and E show the counterpoise extended and connected to the table via a bonding strap. This strap is often a metal sheet with the same width as the monopole counterpoise, as specified in MIL-STD-461E. One of the two configurations tested and reported was with a metal sheet bonding strap connecting the antenna counterpoise to a ground plane on the tabletop. MIL-STD-461F categorically states "For rod antenna measurements, electrical bonding of the counterpoise is prohibited," and this was the second configuration tested. The measurements made with the counterpoise bonded versus not bonded result in very different received levels. MIL-STD-461F also modifies the rod antenna height so that the center point of the rod is 1.2 m above the floor ground plane.

In all measurements, the receiving rod antenna cable was loaded with 28 material cores, although this is a requirement only in MIL-STD-461F/G. This provides a high impedance and reduces common-mode currents on the antenna cable.

The E-field incident on the rod antenna, measured with the rod removed, is made using a 10 cm long bow tie antenna driving a fiber-optic cable.

The source is a 0.44 m × 0.38 m metal box, representing the EUT. The box is insulated from the ground plane and connected to the center conductor of an N-type connector mounted in the tabletop ground plane. The front edge of the box is 10 cm from the front edge of the table in a typical MIL STD test setup. The signal is injected between the box and ground plane, which simulates an EUT with an RF potential between the enclosure and ground plane. This potential is often the result of common-mode currents on cables connected to the EUT. Even with the EUT enclosure bonded to the ground plane via a bonding strap, the RF potential may be developed when sufficient current flows to ground in the strap.

The measurements were made in a damped anechoic chamber to simulate a typical MIL-STD-E102 test. This chamber also contains localized ferrite tiles as well as absorber loads and is described in Section 9.5.

Although well damped above 50 MHz, the chamber does exhibit resonances below 50 MHz.

Reference [18] describes the chamber to chamber deviations for five different chambers. These range from one containing the MIL-STD-461F minimum absorber requirements to one containing hybrid absorber in a CISPR 25 compliant chamber.

The results using the plate antenna with a commercially available buffered rod antenna correspond well with the manufacturers' published data. However, when the source of radiation is a vertical transmission line, a second rod antenna or the 0.44 m × 0.39 m enclosure in a MIL-STD-461 RE102 test setup, the measured AF is very different from the far field and dependent on the source.

We measured both the E-field incident on the rod antenna with and without bonding the counterpoise to the table as well as the AF of the rod antenna with and without counterpoise bonded. The most significant data for manufacturers of equipment are the received level with the same input signal level applied to the same source measured with and without bonding the counterpoise to the table.

Here, we assume that the manufacturer uses the same published far-field AF of the rod antenna regardless of the test setup.

Figure 9.15 from Reference [21] shows the AF measured both with and without the counterpoise bonded to the table. As seen previously, the variation in AF below 20 MHz is much greater with the counterpoise not bonded. From 20 to 30 MHz, the AF with the counterpoise not bonded is 6 dB higher, and this is also seen in the previous test results The previous measurements provided the AF calibrated on a free-space range, and the large variations seen in Figure 9.16 below 20 MHz are missing, indicating that these are due to the anechoic chamber.

The E-field incident on the rod antenna is higher with the counterpoise bonded at most frequencies, and this is shown in Figure 9.16.

Using the same buffered rod antenna, the received level with the monopole counterpoise bonded is up to 18 dB higher than with the counterpoise not bonded, and this is shown in Figure 9.17.

Reference [18] describes a traditional (MIL-STD-461E) monopole antenna setup in five different chambers with different types and amounts of absorber. It also describes measurements made with a MIL-STD 461F setup. The source of radiation was either a vertical rod above the tabletop ground plane or a horizontal rod above the tabletop ground plane. From Reference [18], these resonances are attributable to "RF current loops between the counterpoise, ground plane, ground plane to chamber connection point (wall to floor), chamber floor, and the capacitive coupling back to the antenna counterpoise causes resonant conditions." However, in Figure 9.18, the measurements with the counterpoises uncoupled on a free-space range with no underlying ground plane show exactly the same dip followed by peak in AF with approximately the same magnitude as References [18,19].

FIGURE 9.16 Measured *E*-field incident on the antenna counterpoise with the rod removed. (Reproduced by kind permission from ITEM, EMC test & design guide, Plymouth Meeting, PA, 2010.)

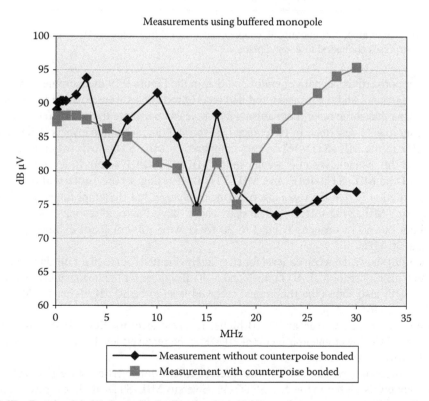

FIGURE 9.17 Received level from the buffered monopole from the same source with and without the antenna counterpoise bonded to the table ground plane. (Reproduced by kind permission from ITEM, EMC test & design guide, Plymouth Meeting, PA, 2010.)

FIGURE 9.18   AF of monopole uncoupled, coupled and with TX line source. MIL-STD-461E configuration with the counterpoises connected to ground plane.

The power cords to the signal generator and S/A in the Figure 9.19 measurements were the only connection to ground, and the transmit rod and receive rod cables were covered in 28 material ferrite. This means that some other mechanisms may at least contribute to the dip and peak.

The measurements described in Reference [18] showed the resonance effects to be greater from 10 to 30 MHz in the MIL-STD-461E setup compared to MIL-STD-461F, whereas our measurements showed the resonances to be greater in the MIL-STD-461F setup. Reference [19] describes an analysis of the MIL-STD-461E and MIL-STD-461F using a bare room without an absorber. Here, the prediction is that the resonant frequencies change, but the amplitudes remain the same when comparing MIL-STD 461E to F. The same paper shows the room resonances with only 10 cm absorber foam but no resonances from 3 to 30 MHz with 100 cm foam exhibiting only 9.5 dB attenuation at 30 MHz.

Reference [18] shows the received level for a vertically oriented source in a well-damped chamber to be from 5 to 8 dB higher for MIL-STD-461F and 5–10 dB higher for the horizontally oriented source.

Reference [18] also shows the theoretical received level to be 10 dB higher for MIL-STD-46E versus MIL-STD-461F setup.

In our measurements, we see an 18 dB higher received level for the MIL-STD-461E setup at 30 MHz with a 5 dB lower antenna factor at 30 MHz, the received level may be 23 dB higher than a MIL-STD-461F measurement.

From measurements, we also see that the received level from a vertical wire 10 cm above a vertical ground plane was higher in the MIL-STD-461E versus MIL-STD-461F test setup. The vertical test setup simulated cables routed down an 18 in. rack.

Above 20 MHz, the output level from a 1.04 m rod antenna is higher with a MIL-STD-461E test setup than MIL-STD 461F. Depending on the source of radiation, we see it may be up to

FIGURE 9.19 AF of monopole uncoupled. No ground plane connection to counterpoise.

23 dB lower in the MIL-STD-461F test. Figure 9.20 shows the combined effect of lower incident *E* field with the counterpoise not bonded to the table as well as the increased antenna factor. This difference in the test data is important information for EMI test personnel and manufacturers of equipment as it means that an EUT that fails RE102 from 10 kHz (2 MHz) to 30 MHz in earlier MIL-STD-461E (and lower) measurements may now pass using the MIL-STD-461F test setup.

As the monopole antenna AF depends on the type of source, it is not an intrinsic factor. A better measuring technique is the use of the small bow tie with detector and fiber-optic cable. As the draft MIL-STD-461G includes the monopole, the probability of a change in measuring technique seems remote.

FIGURE 9.20 Uncoupled AF corrected for *E* field from the same source. MIL-STD-461F configuration counterpoise isolated from ground plane.

9.3 DIAGNOSTIC MEASUREMENTS

Diagnostic measurements are used to locate the source of emissions and the area of susceptibility. In addition, measurements on circuits and equipment may be used in an EMC prediction to provide the magnitude of fields, voltages, or currents on which an analysis is based. Measurements on internal wiring, PCBs, enclosures, and cables may be used in an equipment-level, subsystems-level, or system-level EMC prediction.

9.3.1 RADIATED MEASUREMENT

The most useful antennas for diagnostic measurements are physically small and preferably directional. The small antenna is able to pinpoint the source of radiation to a cable, aperture, or seam in an enclosure and even down to a track or IC on a PCB, when the antenna is sufficiently small. Measurements that are close to the source, and are therefore predominantly near-field, may be made with the small antenna. The advantage of near-field measurements is that the measurements are often less affected by the electromagnetic ambient; however, even near-field measurements may not adequately discriminate against the ambient. For example, assuming equipment must meet the low-level emission requirements of MIL-STD-461, or similar, and the near-field measurements must be made in an unshielded room, it is common for currents induced on cables and equipment by the electromagnetic ambient and conducted on the power cord to the equipment enclosure and cables to result in radiation levels that exceed and mask the equipment emissions. It is therefore imperative, when measurements must be made outside of a shielded room, to compare the levels with the equipment powered down to those with the equipment powered up. It may be necessary either to move the equipment to a quieter location, to lower the ambient by switching off other equipment, or to surround the equipment with absorber foam (useful above 200 MHz), to obtain EUT emissions above ambient measurements.

9.3.2 MAGNETIC FIELD MEASUREMENTS

Some question why make near-field magnetic field measurements when emission specifications typically place limits on the E field? However, because an H field is always present when the fields are time varying, H-field measurements are indicators of the potential E-field magnitude. Also, we see in Section 2.2.5 that the H field reduces more rapidly with increasing distance than the E field, and so this typically predominates at the typical measurement distances of 1, 3, or 10 m. The advantage of H-field measurements, compared to E-field measurements, is lower sensitivity to reflections and room resonances and lower sensitivity of the probe to metallic structure proximity effects.

Figure 9.21 illustrates the typical fields around a cable connected to a piece of equipment. The cable is located above a ground plane on a nonconductive table, which is a typical commercial test configuration. If we assume a common-mode current flow on the cable, then the resultant fields are a longitudinal E field and a circumferential H field. Additional fields are present due to displacement current flow.

Displacement current may flow between the cable and the floor, as well as the ceiling when measurements are made in a shielded room. The additional fields are then an electric field tangential to the cable and a magnetic field in the same plane as the cable. Numerous measurements have shown that it is the longitudinal E field and the circumferential H field that typically predominate. However, this is not always the case, and when making H-field measurements, with the types of loop antenna described in Chapter 2, the loop should be oriented both horizontally and vertically, referenced to the cable under test. Note that the orientation of the plane of the loop must be 90° to the field to be measured. The circumferential H field above a perfectly conducting ground plane results in an image (reflection) in the ground plane that is out of phase with the source, whereas the image due to the horizontal (displacement current) field is in phase with the source. Figure 9.22a illustrates

FIGURE 9.21 Typical fields around a cable connected to a piece of equipment placed on a nonconductive table.

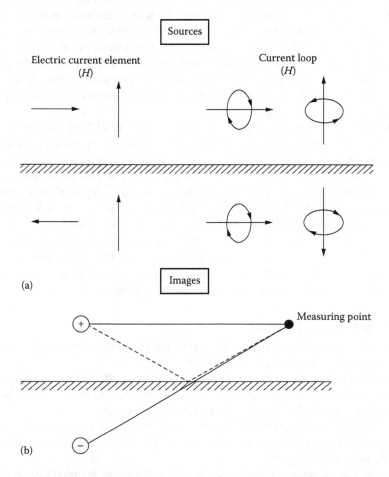

FIGURE 9.22 (a) Orientation and relative phase of the electric current element and current loop magnetic field sources and images in a perfectly conducting ground plane. (b) Direct path, reflected path, and image for a horizontal electric current element.

both the source and the image for the electric current element and the current loop. At distances from the source less than half a wavelength, the source- and image-contributed fields add vectorially, and this addition is illustrated in Figure 9.22b. The total magnetic field is the sum of the source- and image-generated fields. The H field from a cable is approximately that of the electric current element (Equation 2.10), for a loop, it is given by Equation 2.24. When dD is the direct path length and dR is the indirect path, the total field is given by $H(r = dD) + mH(r = dR)$, where r is the distance used in Equation 2.10 or Equation 2.24, $m = \pm 1$, depending on the orientation of the electric current element or loop. For example, a horizontally oriented cable (which may be modeled as an electric current element) generates an image out of phase with the source and $m = -1$. For the displacement current–generated field, the source and image are in phase and $m - +1$. In contrast, when an electric current element is vertical above a ground plane (the monopole antenna is a good example), the magnetic field is circumferential and the E field starts on the rod of the monopole and finishes on the ground plane. Thus, the image is in phase with the source. The relative magnitudes of the vertical (displacement) E field and horizontal (electric current element) E fields may be measured with a small antenna, typically a bow tie from 20 to 250 MHz or a tunable monopole from 250 MHz to 1 GHz. From a measurement of the polarization of the predominant E field, the predominant H field is known. For example, when an antenna is oriented to intercept a horizontally polarized E field and measures a higher level than when vertically polarized, the E field is horizontal, the H field is circumferential, and the image subtracts. When the direct path length dD or the indirect path length dR is greater than half a wavelength, the relative phase of the two fields may be calculated, with the resultant total field the vector sum of these. For example, when the direct path length is different from the reflected path length by exactly half a wavelength, the direct and reflected fields at the measuring point are in phase for the horizontal current element and out of phase for the vertical current element. This attribute is used in an antenna in which a small fixed frequency dipole is located horizontally 1/4 wavelength above a metal reflector. This ensures that the source and image fields in front of the dipole are in phase.

Another measurement technique useful in evaluating the E or H field radiated by a cable is to measure the current flow on the cable by use of a current probe and to use the measured current in Equations 2.12 through 2.14 to obtain the field from a cable far removed from a ground plane or Equations 7.25 through 7.30 for a cable above a ground plane. For greatest accuracy, the "typical" calibration curve provided by most probe manufacturers should be replaced by a calibration of the probe, using a setup representative of the cable under tests. At high frequencies, where the cable length may be greater than half a wavelength, the probe should be moved up and down the cable until a maximum reading is obtained. A source of inaccuracy that must be accepted is the loading effect of the current probe on the cable, which is likely to increase the cable current flow. Figure 9.23 illustrates the use of the current probe in evaluating the radiated emissions from an interface or a power cable. A typical test method for use with the current probe is as follows.

9.3.2.1 Current Probe Test Method

Step 1: The S/A frequency sweep should be set to the frequency range of the probe in use, initially with a 100 kHz resolution bandwidth.

Step 2: A measurement of the ambient with the EUT switched off and the current probe placed on the nonconductive work surface should be made and a record taken using either a plotter or a computer connected to the appropriate S/A digital output.

Step 3: With the EUT switched on and the current probe placed over the cable, monitor the emissions displayed on the S/A and move the current probe on the cable in order to maximize the amplitude displayed.

Step 4: Ensure that no compression has occurred in either the preamplifier or the S/A, as described in Section 9.2.3.

FIGURE 9.23   Test setup for measurement of cable current flow.

Step 5: Where the amplitude displayed on the S/A is in dBm, convert to voltage:

$$V = \sqrt{10^{\text{dBm}/10} * 1 \times 10^{-3} * 50}$$

Step 6: Read the maximum amplitude displayed on the S/A and note the frequency. From the current probe transfer impedance calibration curve, read the transfer impedance at the frequency of the emission under investigation and convert from voltage to current using

$$I = \frac{V}{Z_t}$$

Step 7: Convert the measured current to an E field at the specification distance (SD) using Equation 2.10 or Equations 7.26, 7.28, and 7.30.

The magnitude of the H field from a cable above a ground plane is obtained using Equation 7.28, from which the E-field magnitude may be obtained by multiplying by the wave impedance. Compare the magnitude of the E field to the E field limit. The transfer impedance will be low at some frequencies, which means the current amplitude displayed at those frequencies will also be low. It is therefore important that any peak emissions at frequencies, where Z_t is low, also be converted to current measurements, from which the E field is calculated and compared to the limit. A preamplifier may be needed if the current is low but again differentiation between the EUT generated currents and the ambient must be made.

Repeat steps 1–7 using the remaining current probe(s) required to cover the 14 kHz–1 GHz frequency range for equipment that must meet the MIL-STD requirements or the 30 MHz–1 GHz range for FCC requirements.

Using a combination of current probe measurements and small magnetic field and electric field antennas to conduct measurements in a shielded room, a correlation of, on average, 6 dB and, worst case, 13 dB to the measurements obtained on one specific open-field test site measurements has been achieved. However, a number of different measurements are required to identify room

TABLE 9.4

Reduction in Measured Magnetic Field Compared to Predicted

Freq. [MHz]	$\lambda/2\pi$ [m]	0.1–0.2 m		0.2–0.3 m		0.3–0.5 m	
		Meas. [dB]	Pred. [dB]	Meas. [dB]	Pred. [dB]	Meas. [dB]	Pred. [dB]
20	2.380	18	18	7	10.5		
100	0.470	15	18	4	10.5	5.3	13
220	0.217	9	18	0	3.5		

resonances and reflection, and these are described in detail later in this chapter. Choice of the optimal measuring distance (MD) and correct orientation of the antenna is of paramount importance in achieving maximum accuracy. In the near field, the predicted reduction in the H field from a loop antenna is a function of $1/r^3$, whereas the H-field reduction for an electric current element (cable) is a function of $1/r^2$. From measurements made using a transmitting loop and a receiving loop in a shielded room, it has been found that the near-/far-field interface does not occur abruptly at $\lambda/2\pi$, rather, the $1/r^3$ law changes gradually through $1/r^2$ to $1/r$ in the far field.

Table 9.4 illustrates the reduction in measured magnetic field compared to the predicted, based on a $1/r^3$ reduction in the near field. The calculation used in obtaining the reduction is given in the following example: The initial MD is 0.2 m and the loop is moved to 0.3 m; the reduction in dB is given by

$$20 \log \frac{1}{\left(r_1/r_2\right)^3} = 20 \log \frac{1}{\left(0.2 \text{ m}/0.3 \text{ m}\right)^3} = 10.5 \text{ dB}$$

When near-field measurements are confined to finding a source of emission (e.g., cable or enclosure) or to measuring the effectiveness of EMI fixes, relative measurements are adequate. Often it is required to know how close the emissions from a piece of equipment are to the commercial specification limits, without removing it to an open-field test site or without removing a piece of equipment to a shielded room for MIL-STD-461-type measurements. Using small E- and H-field antennas, we may measure the fields directly, or with the current probe, we can measure current and predict the fields from equipment enclosures and cables. When H-field measurements are made in the near field, we may extrapolate out to 1, 3, 10, or 150 m distances, whichever the specification limit requires. The distance from emission source to measurement is measured and the near-/far-field interface distance (FFID) is calculated. From a cable, the reduction in H field is assumed to be a function of $1/r^2$, and for magnetic field coupling due to current flow on the inside of the enclosure, it is $1/r^3$ out to the FFID distance. For wire sources inside the enclosure, when the circuit impedance is less than 377 Ω, the source is predominantly magnetic and $1/r^3$ should be used. For wires connected to circuits with impedances higher than 377 Ω, the $1/r^2$ law should be used. When the SD is in the near field, the E field can be obtained from the magnitude of the H field calculated at the SD times the wave impedance. When SD is in the far field, to obtain the reduction from the FFID to the SD, use the following equation. When the MD is greater than $\lambda/2\pi$, replace FFID in the equation that follows with MD to obtain the reduction in H field in dB:

$$20 \log \frac{1}{FFID/SD} \tag{9.1}$$

The SD E field in dBμV/m from the FFID magnetic field is

$$E \text{ [dBμV/m]} = H_{FFID} \text{ [dBμA/m]} - SD \text{ reduction [dB]} + 51.5 \text{ dB} \tag{9.2}$$

or numerically

$$E\left[\mu V/m\right] = H_{FFID}\left[\mu A/m\right] \times \frac{FFID}{SD}\left[\text{reduction numeric}\right] \times 377$$

In practice, one or more emissions at the same frequency can emanate from different sources (e.g., more than one cable and aperture and seams in the enclosure). The far field is therefore a composite of these separate sources. The composite H field, H_1, may be calculated from

$$H_t = \sqrt{H_1^2 + H_2^2 + H_3^2} \qquad (9.3)$$

where H_1, H_2, and H_3 are the separate sources.

In practice, if the measurements are conducted over a long time, the sources can add algebraically. In comparisons between measured far E fields and E fields obtained from near H-field measurements, the prediction magnitude is usually lower than the measured. This may be the result of not adding the separate sources of fields, or it may be due to the use of the $1/r^3$ law out to the FFID, whereas the $1/r^3$ law should be modified close to the FFID. Conversely, near E field measurements tend to be high when extrapolated out to a far-field distance and compared to far-field measurements. The major reason for this is thought to be the room resonance and reflections inherent in a shielded-room measurement, described in Section 9.5.1. In addition, the electric current element model is strictly true only for distances of greater than half a wavelength from a ground conductor. Also, these reductions are strictly valid for point sources.

In making H-field measurements on equipment that must meet the low-level MIL-STD-461 specification limits, a preamplifier and an S/A are invariably required. The sensitivity of an oscilloscope is typically not adequate except for measurement 0.5–0.2 m from a source on equipment required to meet the relatively high commercial specification limits.

9.3.3 Conducted Measurements

In this section, we use the term *conducted measurements* to cover the direct measurement of voltage or current or the indirect measurement of current using the current probe. This type of measurement is useful in locating the source of an emission at a specific frequency and in ensuring that a circuit or supply is not oscillating. To infer that a high magnitude of measured voltage or current will necessarily result in a high level of radiation is not necessarily true due to the importance of the circuit geometry in determining radiation levels. Nevertheless, conducted measurements, typically using the oscilloscope, are extremely useful in determining the presence of differential- and common-mode noise. When the equivalent circuit of a coupling path can be determined and the applied voltage or current is measured, the EMI voltage at the receptor may be calculated and the need for modifications to the coupling path can be determined.

9.3.4 Susceptibility/Immunity Tests

Diagnostic susceptibility/immunity testing is useful when a piece of equipment has failed a radiated susceptibility/immunity test requirement and the susceptible cable or enclosure side has to be found. Another use is when searching for an intermittent conducted susceptibility problem. If spikes or a sine wave are injected onto a power or signal interface, which is a likely candidate for examination, and the occurrence of the intermittent fault increases, the chances are that the source of the noise has been found. The injection of an additional noise level is best accomplished by use of an injection transformer wound on a ferrite toroid core. By placing only a few turns on its secondary, the additional low impedance in series with the power or signal line is less likely to affect the circuit under test. The primary should typically have more windings than the secondary and enough to represent a sufficiently high source impedance to the signal generator or power amplifier. Another technique is to directly inject from a signal generator using a probe. The probe has a 50 Ω series resistance followed by a suitable capacitor, depending on frequency. The 50 Ω resistor is to protect the generator from a low-impedance circuit and the capacitor against a DC voltage. Care must be taken when using the probe on a circuit with an AC waveform that this waveform when injected back into the generator via the 50 Ω resistor does not damage the generator. Ideally, the resistor and capacitor are surface mount. The shield of the coaxial cable used to connect from the generator to the probe is connected to a signal ground connection using a very short length. A ground clip as supplied with oscilloscope probes will have the lowest impedance. The generator output should be from +10 dBm to +17 dBm. At +17 dBm, the current injected into a low-impedance circuit will be approximately 30 mA and the rms voltage into a high impedance approximately 1.5 V. Although injecting an additional noise level is indicative of susceptibility, it is no guarantee that the problem has been found. The next step is to measure the noise level and to note the frequencies at which susceptibility occurs. A filter may then be designed to achieve at least 20 dB of attenuation at the susceptibility frequencies, unless these are close to the signal frequencies. If, by the insertion of the filter, the intermittent fault disappears, then the susceptible circuit has been located. If the filter degrades a signal or cannot be permanently incorporated into the equipment, then the source of noise must be found. Clues as to the source are contained in the susceptibility test level and frequencies. In addition, the level and type of noise, differential or common mode, present without the additional injected noise helps in locating the source. If the noise is a common mode, an RF capacitor between the signal or power ground and chassis may suffice. Alternatively, on signal interfaces, an isolation transformer with electrostatic shield between the primary and secondary or an optoisolation circuit may be used to reduce the common mode coupled into the receiver circuit. When the noise is differential, the noise is injected into the signal or power line at some stage, and the location and source must be found.

As described in detail in Chapters 4 and 5, the noise may be injected via radiated coupling, crosstalk, common ground coupling, or a noisy source signal. Radiated susceptibility may be determined by using a loop to generate a magnetic field. The loop is driven by either a source of wide-frequency-range noise, such as a square wave, or a sweep generator that sweeps the frequency over the range at which susceptibility was either demonstrated or suspected. Above approximately 20 MHz, a simple one-turn loop or the balanced loop antenna described in Section 2.6 may be used to generate the field. At low frequency, a multiturn loop should be used to ensure that an adequately high impedance is presented to the signal source. Some power amplifiers are specified as capable of driving a S/C; however, based on the high mortality of power amplifiers, an impedance of at least a few ohms at the lowest frequency is recommended. Certification-radiated susceptibility tests are typically E field, although in the MIL-STD-461 RS04/MIL-STD 461F RS101 test an H field is used. Typically, it does not matter whether the current flow on a cable or enclosure is caused by an H-field or an E-field source during a susceptibility test. If equipment is suspected of susceptibility to specifically a low-frequency E field, then a small wand, monopole, or bow tie antenna may be used; however, it is very difficult to generate adequate E fields with these antennas. Above approximately 250 MHz, small directional antennas, such as the conical log spiral or conical spiral, may be used to radiate a specific location and thus determine the susceptible cable or equipment. The antenna is used in close proximity and pointed at the location under investigation.

Small horn antennas or parallel-plate antennas may be used at microwave frequencies with a TWT amplifier. If the antenna is homemade, it is important to ensure the load is close to 50 Ω over the test frequency to avoid destroying the TWT. Metal oxide, film, or carbon resistors may be used either in series or parallel to modify the antenna impedance to achieve close to 50 Ω.

If a PCB can be tested alone, then one of the miniature antennas shown in Chapter 2 may be applicable. A small toroid inductor will generate a local H field and so can be used as a source of radiation when connected to a signal generator/power amplifier and likewise be used to probe at the PCB level and so act as a receiving antenna.

9.4 COMMERCIAL EMI REQUIREMENTS AND MEASUREMENT

Commercial requirements are typically confined to limits on conducted and radiated emissions. However, with the advent of the EC EMC Directive, mandatory and routine immunity (susceptibility) test requirements were imposed for the first time on commercial equipment. The EMC requirements encompass equipment from RF arc welders to electric drills, household equipment, and digital devices. The definition of *digital devices* is often very broad. For example, the FCC defines a digital device as any equipment containing a clock at a frequency above 9 kHz using digital techniques. Thus, a pinball machine or similar device may qualify. The prime purpose of the FCC requirements contained in subpart B of part 15 of its rules is to reduce emissions to levels that do not interface with nearby radio and TV reception. In a great many countries, people still own a radio or TV or both, and it is to reduce complaints from owners that the majority of commercial EMI requirements are imposed. Even though transmissions are digital, the signal at the antenna is analog and potentially susceptible to transient noise. In this section, only requirements and test methods imposed on digital devices are considered.

9.4.1 FCC RULES ON EMISSIONS FROM DIGITAL DEVICES

Compliance with the FCC requirements is a prerequisite for legal importation and marketing of such apparatus in the United States. The technical requirements for limiting unwanted RF emissions from unintentional radiators, such as computers, videocassette recorders, and microwave ovens, and intentional radiators are contained in Part 15 of the FCC Rules (47 CFR Part 15). The rules for industrial, scientific, and medical (ISM) equipment are contained in Part 18 (47 CFR Part 18).

Specific devices are exempt; these include transportation vehicle, hydropower, digital device used exclusively as industrial, commercial, or medical test equipment or included in an appliance (microwave oven, dishwasher, clothes dryer, air conditioner [central or window], etc.), digital devices that have a power consumption not exceeding 6 nW, joystick/mouse, and devices that use or generate a clock below 1.705 MHz and are not connected to AC power.

Devices that are exempt are subject to the condition that no harmful interference is caused by their operation and that they cease operation upon notification by a Commission representative that the device is causing harmful interference, as described in FCC Section 15.5. In addition, the devices are subject to inspection by the Commission as described in Section 15.29.

Devices that are connected only part of the time via a battery charger or that obtain AC power through another device are *not* exempt from conducted emission measurement. If one device is connected to the public utility AC power and supplies DC power to a second device (or a number of devices), then a conducted emission measurement is made on the AC power-line side, and the composite emissions from all devices, which include those on the DC power lines that are reflected or conducted through the AC–DC power supply, must meet the emission limits. Thus, a wall-socket-mounted AC–DC power supply may have an FCC sticker that says that it meets the FCC requirements and yet that power supply must be retested with a load device for conducted emissions.

As yet the FCC has not set immunity requirements, although section 302 of the Communications Act of 1934, as amended, gives the FCC authority to regulate the immunity of home entertainment products. The FCC also has the authority to regulate any device that in its operation is capable of causing harmful interference to radio communications, that is, a broad power of regulation.

The FCC places different requirements on digital devices, depending on their intended use. Devices marketed for use in a commercial, industrial, or business environment, exclusive of a device marketed for use in the home or by the general public, are classified as Class A. Devices marketed for use in a residential environment, notwithstanding use in a commercial, business, or industrial environment, are classified as Class B. The rationale is that commercial equipment is typically located further away from TV receivers than domestic equipment, which may be used in the next room.

Examples of Class B devices include, but are not limited to, personal computers, calculators, and similar electronic devices that are marketed for use by the general public. The FCC may also classify a device that repeatedly causes harmful interference as a Class B device.

Class A emission limits are higher (less stringent) than Class B limits, which are lower (more stringent).

Tables 9.5 and 9.6 provide the radiated and conducted requirements for Class B equipment and Tables 9.7 and 9.8 for Class A equipment.

For an intentional radiator, the spectrum shall be investigated from the lowest RF generated within the device, without going below 9 kHz,

1. Up to the tenth harmonic or up to 40 GHz, if the intentional radiator operates below 10 GHz
2. Up to the fifth harmonic or up to 100 GHz, whichever is lower, if the intentional radiator operates at or above 10 GHz and below 30 GHz
3. Up to the fifth harmonic of the highest fundamental or up to 200 GHz, whichever is lower, if the intentional radiator operates at or above 30 GHz

TABLE 9.5
FCC Class B (Residential) Radiated Emission Limits

Frequency [MHz]	Distance [m]	Field Strength [μV/m]
30–88	3	100
88–216	3	150
216–960	3	200
Above 960	3	500

Note: Bandwidth of measuring instrument is not less than 100 kHz.

TABLE 9.6
FCC Class B (Residential) Conducted Emission Limits

Frequency [MHz]	Maximum RF Line Voltage [μV]
0.45–30	250

Notes: Bandwidth of measuring instrument is not less than 9 kHz.
CISPR quasi-peak function is used for both conducted and radiated measurements.
FCC allows a 13 dB relaxation when the conducted emission is determined as BB.

TABLE 9.7
FCC Class A (Commercial) Radiated Emission Limits

Frequency [MHz]	Distance [m]	Field Strength [μV/m]
30–88	10	90
88–216	10	150
216–960	10	210
Above 960	10	300

Note: Bandwidth of measuring instrument is not less than 100 kHz.

TABLE 9.8

FCC Class A (Commercial) Conducted Emission Limits

Frequency [MHz]	Maximum RF Line Voltage [μV]
0.45–1.705	1000
1.705–30	3000

Notes: Bandwidth of measuring instrument is not less than 9 kHz.

CISPR quasi-peak function is used for both conducted and radiated measurements.

FCC allows a 13 dB relaxation when the conducted emission is determined as BB.

---

4. Up to the upper frequency limit for the intentional radiator or up to the highest frequency applicable to the digital device, whichever is the higher frequency, if the intentional radiator contains a digital device

For an unintentional radiator, including a digital device, the lowest frequency measured shall not be below the lowest frequency for which a radiated emission limit is specified. The upper frequency limit to be measured, above 1 GHz, is

---

Highest Frequency Generated or Used (MHz)	Upper Frequency of Measurement Range (MHz)
1.705–108	1000
108–500	2000
500–1000	5000
Above 1000	Fifth harmonic of the highest frequency or 40 GHz, whichever is lower

---

On any frequency or frequencies above 1000 MHz, the radiated limits are based on the use of an average detector function (see 47 CFR for further information on peak detector measurements). Unless otherwise specified, measurements above 1000 MHz shall be performed using a minimum resolution bandwidth of 1 MHz. Conducted AC power-line emission measurements are always performed using a CISPR quasi-peak detector.

Any equipment that comes under the definition of digital devices and that is used in the United States must have a representative sample tested. Most digital devices are subject to verification by the manufacturer, which is an equipment authorization process whereby the manufacturer, after testing for compliance, provides information to the user and labels the equipment with the statement required by 47 CFR 15.19.47. CFR 2.901 provides more information on this verification program. The FCC declaration of conformity (DoC) is a procedure where the responsible party makes measurements or takes other necessary steps to ensure that the equipment complies with the appropriate limits. The submittal of units or data is not required by the FCC. The DoC attaches to all items identical to that tested. If a product must be tested and authorized under a DoC, a compliance information statement shall be supplied with the product at the time of marketing or importation.

Equipment verification is a procedure where the manufacturer makes measurements or takes the necessary steps to ensure that the equipment complies with the appropriate technical standards. Submittal of a sample unit or representative data to the Commission demonstrating compliance is not required unless specifically requested by the Commissions.

Examples of devices subject to verification include, but are not limited to, ISM equipment, business Class A computer equipment, and TV and FM receivers. Devices subject only to verification shall be uniquely identified.

DoC is a procedure where the responsible party makes measurements or takes other necessary steps to ensure that the equipment complies with the appropriate technical standards. Submittal of a sample or representative data to the commission demonstrating compliance is not required unless specifically requested by the commission. Examples of devices subject to DoC include, but are not limited to, Class B personal computers and peripherals and CB receivers, super regenerative receivers, and TV interface devices. Devices subject to DoC shall be uniquely identified.

Devices must be tested by a test laboratory accredited by the National Voluntary Accreditation Program (NVLAP), the American Association of Laboratory Accreditation (A2LA), or by an accredited laboratory designated by the Commission under the terms of a negotiated Mutual Recognition Agreement (MRA).

Certification is an equipment authorization issued by the Commission or by a designated telecommunication certification (TCB) based on representations and test data submitted by the applicant.

Examples of devices subject to certification, which must be submitted to the Commission, include, but are not limited to, ultra wideband, software-defined radio, and new technology.

Examples of devices subject to certification and which must be submitted to a TCB are computers and computer peripherals.

Examples of devices subject to certification that may be submitted to either the Commission or a TCB include, but are not limited to cell phones, RF light, microwave oven, RC transmitters, family radio, telemetry transmitters, wireless phones, walkie–talkies.

Once a piece of equipment is authorized any identical piece of equipment is authorized. *Identical* means identical within the variation that can be expected to arise as a result of quantity production techniques.

Obviously, any changes in software or hardware mean the equipment is not identical. However, although it can affect emissions, a change in manufacturer of an IC with the same part number is allowed.

Type acceptance is a route not available for unintentional radiators but is confined to a station authorization. Type acceptance also applies only to units that are identical.

The routes to compliance for manufacturers outside of the United States are verification for Class A digital devices, peripherals, and external switching-power supplies and equipment, and the same is true for almost any Class B switching-power supplies, digital devices, and peripherals. However, if the equipment falls under the category of CB receiver, TV interface device, scanning receiver, and superheat receiver, then the more onerous route of certification applies.

The administrative rules are contained in 47 CFR 2.901.

The FCC label should include the trade name and model number. For an assembly using separate authorized components where the resulting product is not separately tested, the label shall include "Assembled from tested components complete system not tested."

The FCC label should state "This device complies with part 15 of the FCC rules. Operation is subject to the following two conditions: (1) This device may not cause harmful interference and (2) this device must accept any interference received, including interference that may cause undesired operation."

If the Class A limits have been met, then the following, or similar, statement shall be included in the operating manual:

Note: This equipment has been tested and found to comply with the limits for a Class A digital device, pursuant to part 15 of the FCC rules. These limits are designed to provide reasonable protection against harmful interference when the equipment is operated in a commercial environment. This equipment generates, uses, and can radiate RF energy and, if not installed and used in accordance with the instruction manual, may cause harmful interference to radio communications. Operation of this equipment in a residential area is likely to cause harmful interference in which case the user will be required to correct the interference at his or her own expense.

For a Class B digital device or peripheral, the instructions furnished the user shall include the following or similar statement, placed in a prominent location in the text of the manual:

Note: This equipment has been tested and found to comply with the limits for a Class B digital device, pursuant to part 15 of the FCC rules. These limits are designed to provide reasonable protection against harmful interference in a residential installation. This equipment generates, uses, and can radiate RF energy and, if not installed and used in accordance with the instructions, may cause harmful interference to radio communications. However, there is no guarantee that interference will not occur in a particular installation. If this equipment does cause harmful interference to radio or television reception, which can be determined by turning the equipment off and on, the user is encouraged to try to correct the interference by one or more of the following measures:

- *Reorient or relocate the receiving antenna.*
- *Increase the separation between the equipment and receiver.*
- *Connect the equipment into an outlet on a circuit different from that to which the receiver is connected.*
- *Consult the dealer or an experienced radio/TV technician for help.*

The definition of *digital device* is an unintentional radiator that generates and uses timing signals and pulses in excess of 9000 pulses (cycles) per second and uses digital techniques. This includes computers, data processing equipment, and other types of information technology equipment (ITE).

The tests for line-conducted measurements only should show compliance with the U.S. voltages of 110 V/60 cycles. For digital devices with oscillators above 108 MHz, radiated measurements must also be made on frequencies above 1000 MHz using an instrument with an average detector. Both the FCC and CISPR limits below 1000 MHz are based on quasi-peak measurements. FCC measurements above 1000 MHz are based on the use of an average detector with a peak limitation of 20 dB above the average value. Measurements above 1000 MHz are required under 47 CFR 15.33 and must use a minimum of a 1 MHz bandwidth.

CISPR 22 states that if the field strength at 10 m cannot be made because of high ambient noise levels or for other reasons, measurements may be made at a closer distance, for example, 3 m. An inverse proportionality factor of 20 dB per decade should be used to normalize the measured value to the specified distance for determining compliance. However, care should be taken in measurement of large test units at 3 ms and at frequencies near 30 MHz due to the problem of measuring close to the near field. A comparison between the part 15 radiated and conducted emission limits and the CISPR 22/EN55022 limits (as well as other EN limits) is made in upcoming Tables 9.10 through 9.12. The CISPR 22 Class A and Class B limits are identical to the EN 55022 Class A and Class B limits shown in the tables.

Any modification, however minor, to a circuit or software can change the emission profile. For example, a change in software, device function, or grounding shield termination or the addition of a signal interface can significantly change emissions. It is, therefore, recommended that the equipment be retested after any modification. The equipment verification in accordance with FCC CFR 47 applies to identical equipment. However, the FCC, in Public Notice 7, 1982, cautioned that

The manufacturer is cautioned that many changes which on their face seem to be insignificant are, in fact, very significant. A change in the layout of a circuit board or the addition, removal or rerouting of a wire, or even a change in logic, will almost surely change the emission characteristics (both conducted and radiated) of the device. This is particularly true with a device housed in a nonmetallic enclosure. Whether this change in characteristics is enough to throw the product out of compliance can *best* be determined by retesting.

It is clear that retesting is the preferred route, but it is not mandatory, and the notice does not stipulate that this testing needs to be at verification level with plots. Therefore, when the testing of every minor modification to a wide range of products becomes onerous and costly, the modifications may be reviewed by an EMI expert, or "quick look" testing can be performed. This "quick look" test for radiated emissions on an OATS and conducted emissions in a shielded room can be made without plots of the emissions and, depending on equipment complexity and functional operating modes, will typically take no longer than 3 h, with the data presented in tabular form only.

The FCC routinely retests equipment to ensure continuing compliance and is empowered to fine or to seize equipment that is noncompliant. Lack of conformity in equipment is brought to the notice of the FCC through complaints of EMI, referral of competitors, monitoring of trade shows and literature, or the FCC s sampling program. First, the FCC can issue a "Marketing Citation," a letter informing the company that it has apparently violated FCC rules and perhaps requesting that a product sample be sent to the FCC laboratories. If the violations continue, or the equipment is found by the FCC to be in technical violation of the rules, a Notice of Apparent Liability will be issued. It can assess a fine for each occurrence and a higher fine per violator for multiple occurrences. Historically, at this point, Canadian manufacturers have usually opted to recall ail devices that are available from their distributors in the United States and apply EMI fixes.

Many compliance engineers in companies in Canada require their equipment to be 6 dB below the FCC limits, although this is not an FCC requirement, and furthermore require testing of samples in a production run to reduce the risk that any of the exported equipment will fail a retest.

If the "administrative" sanctions are ignored, then the FCC can take "judicial" measures, which are imposed only after the commission becomes convinced that the cited party will not cooperate. The penalty is that

> any person who willfully or knowingly does ... any act, manner, or a thing in the relevant sections (501, 502) of the Communications Act. ... shall, upon conviction thereof, be punished for such offense, by a fine or by imprisonment for a term not exceeding one year.

An even stronger deterrent for manufacturers outside of the United States is an order to stop the importing of the culprit equipment. But how this is covered by the North American Free Trade Agreement is not known.

The test methods for verification are contained in ANSI C63.4-2014, although the FCC has accepted some alternative test methods, described later. The commission *encourages* the use of the ANSI C63.4 procedure for all testing. Any party using other procedures should ensure that such procedures can be relied on to produce measurement results compatible with the FCC measurement procedures. The description of the measurement procedure used in testing the equipment for compliance and a list of the test equipment actually employed shall be made part of an application for certification or included with the data required to be retained by the party responsible for devices authorized by a DoC, notification, or verification. The use of a test site accredited by NVLAP or A2CA is required for the measurements or with a laboratory accredited through a mutual agreement. The FCC has allowed the radiated emission measurement data using a gigahertz transverse electromagnetic (GTEM) cell that has been correlated, usually by use of a computer program, to an OATS. However, the measurement from an OATS shall take precedence, and this is the method the FCC uses. ANSI C63.4 allows the use of a 3 or 10 m site in an anechoic chamber, but this site must meet a more stringent normalized site attenuation (NSA) test than applicable to an OATS. Weatherproof OATS in which either the EUT or the measurement antenna and the EUT are contained in a totally nonconductive enclosure is accepted by the FCC at present. Section 11 of ANSI C63.4 provides specific information on the measurement of ITE and includes information on the operating conditions of hosts, peripherals, and visual display units. Additional information on the placement of tabletop systems, hosts, monitors and keyboards, external peripherals, interface cables, floor-standing equipment, combination tabletop and floor-standing equipment, conducted emission measurements, and radiated emission measurements is provided.

9.4.2 ANTENNA CALIBRATION

ANSI C63.4 specifies the use of a tunable dipole or a dipole tuned only above 80 MHz and set to the 80 MHz tuned length when used between 30 and 80 MHz. Alternatively, BB antennas may be used when correlated to a dipole with an acceptable degree of accuracy. In practice, to achieve reasonable accuracy, the antenna may have to be calibrated against a calibrated dipole using, typically, the three-antenna calibration method. ANSI C63.4 requires that all antennas be individually calibrated to National Institute of Standards and Technology (NIST) or an equivalent standards reference organization. Antennas calibrated to the methods of ANSI C63.5-2006 meet the traceability requirements. ANSI C63.5 describes two fundamental approaches to antenna calibration, both assuming a perfect reference. One approach assumes a perfect antenna whose characteristics are known to a high degree of accuracy by virtue of its standard construction. The "Roberts" tuned dipole antenna is the reference antenna; if the construction drawings are followed carefully, the antenna factor will match the predicted values to within 1 dB. A four-antenna set can be built to cover the 30–1000 MHz frequency range. The antenna factor of any other antenna may be derived by substitution with the reference antenna. In the substitution calibration, the separation distance between the transmitting antenna and the reference antenna or the antenna being calibrated is 10 m. The transmitting antenna can be any antenna, and it should be at least 2 m above the ground. The reference antenna is used as the receiving antenna, which is adjusted in height to be anywhere from 2.5 to 4 m above the ground. The height should be chosen to avoid a null by finding a location where the received signal is either maximum or is varying slowly. After the signal strength is noted with the reference antenna, it is replaced with the antenna being calibrated, located at exactly the same height and position as the reference dipole. (Scanning the receiving antennas in height while recording the maximum received signals will facilitate the measurements, especially at higher frequencies, where many peaks and nulls of the field are present.) The AF of the unknown antenna can be modified by the presence of the ground, and this is minimized by the 2.5–4 m height chosen. But this is neither a free-space calibration nor does it compensate for the presence of the ground. The second approach assumes a perfect, or at least a nearly ideal, "standard site." The standard site will typically be located in a large, open field far from reflecting objects such as trees, buildings, overhead wiring, bumps, or hills and will almost certainly contain an "oversize" ground plane. It would seem that an NSA calibration using identical antennas on a "standard site" is all that is required to obtain their AF. However, ANSI C63.5 restricts the use of NSA measurements on an OATS that is also used for the calibration of the antennas. The argument is that site imperfections may have been erroneously assigned to the antenna factors. C63.5 allows an exception for a large OATS, that is, the near-perfect site described earlier, but the antennas must be calibrated using horizontal polarizations on a propagation path independent of the path used for EMI, and the NSA, measurements. An alternative is to make two NSA measurements on two independent sites or on two independent paths on a large site, for example, two independent 10 m paths on a 30 m test site. The standard site method described in C63.5 requires three site attenuation measurements if three different antennas are used. These measurements are made, with the same distance between antennas, with the height of the fixed antenna the same, and with the scan height of the second antenna the same for all three measurements.

ANSI C63.5 contains tables of the predicted maximum received ground wave field strength for the source antenna heights of 1 and 2 m.

The three different antennas are calibrated in pairs, for example, 1 and 2, 1 and 3, and 2 and 3.

Three equations that contain the measured site attenuations, the frequency and the maximum predicted ground wave field, from the tables, are solved simultaneously to obtain the antenna factors.

If two identical antennas are to be calibrated, then a single measurement is required and a single equation. If any one or two of the antennas are tuned dipoles and the NSA measurement is made at 3 m, then a correction, contained in ANSI C63.5, for one or both dipoles must be made.

When BB antennas, such as the log periodic, are calibrated close to a ground plane, the presence of the ground plane can significantly change the current distribution on the antenna and its input impedance, which results in a very different antenna factor compared to a close-to-free-space calibration. Another way of explaining this effect is that the image of the antenna in the ground plane interferes with the antenna above the ground plane. This effect becomes especially noticeable when the antenna is a wavelength or multiple wavelengths from the ground plane. Thus, with an antenna 1 m above the ground plane, the antenna factor is substantially different from the free-space value at 300 MHz, although the effect is noticeable over the 200–400 MHz frequency range. In the standard site antenna factor, the search antenna height covers the 1–4 m range, so these proximity effects are accounted for in the calibration. A typical "close to free space" calibration, made at a 3 m separation, is achieved when both antennas are at a height of 4 m over a poorly conducting surface, that is, no ground plane, and the surface between the two antennas is covered in large blocks of foam absorber material or ferrite tiles.

The antenna calibration should be performed at the distance to be used; that is, if the measurement distance is 3 m, then a calibration at 3 m is required, and if measurements are also made at 10 m, then a 10 m calibration shall be used. The calibration should be made in accordance with ANSI-C63.5. If this is not the case, difficulty may be experienced in measuring the required open-area NSA. The antennas used for measurements from 1 to 40 GHz should be calibrated and linearly polarized, such as double-ridged guide horns, rectangular waveguide horns, pyramidal horns, optimum-gain horns, and standard-gain horns, although the use of a log-periodic dipole array is also allowed. The beam of the antenna must be large enough to encompass the EUT, or provision shall be made for scanning the EUT. The largest aperture dimension of a horn antenna (D) should be small enough, so the measurement distance in meters is equal to or smaller than $D^2/2\lambda$. Standard-gain horn antennas need not be calibrated beyond whatever is provided by the manufacturer, unless they are damaged or deterioration is suspected. Standard-gain horn antennas have gains fixed by their dimensions and dimensional tolerances. Because they are invariably used in the far field and have a 50 Ω impedance, the simple relationship provided by Equation 2.53 may be used to calculate the antenna factor from the gain.

Moving the measurement antenna over the surfaces of the four sides of the EUT or down the length of an interface cable or another method of scanning of the EUT is required when the EUT is larger than the beam width of the measuring antenna. It is preferred that 1–40 GHz measurements be performed on an OATS or in an absorber-lined room. However, measurements may also be performed where there is adequate clearance, considering the radiation pattern of the EUT, to ensure that reflections from any other objects in the vicinity do not affect the measurements. A conducting ground plane is not required, but one may be used for measurements over 1 GHz.

One of the many reasons for noncompliance in a retest of an EUT may be the use of a BB antenna in the original test, which exhibits high VSWRs at low frequency and resultant large peaks and troughs in the calibration curve. Thus, an appreciable discrepancy may exist between a measurement made with a dipole antenna and with a BB antenna. The commonly achievable average VSWR for a 20–1200 MHz very BB antenna is 2:1, excluding the effect of the proximity of a ground plane to the antenna. However, at low frequency, the VSWR may be as high as 30 at 20 MHz. Radiated emission measurement distances under the FCC rules are 3 m for Class B equipment and 30 m for Class A equipment. ANSI C63.4 allows measurements at 3, 10, or 30 m. Tests shall be made with the antenna positioned in both the horizontal and vertical planes of polarization. The measurement antenna shall be varied in height above the conducting ground plane to obtain the maximum signal strength. For vertical polarizations, the minimum height shall be increased so that the lowest point of the antenna is at least 25 cm above the ground plane.

9.4.3 Test Sites

Tests shall be made in an open, flat area characteristic of cleared, level terrain. Such OATS shall be void of buildings, electric lines, fences, trees, underground cables, pipelines, etc., except as required to perform the tests. A suggested layout for the OATS is in the form of an ellipse with a major diameter of $2F$ and a minor diameter of $\sqrt{3}F$, where F is the measurement distances of either 3, 10, or 30 m.

ANSI C63.4 requires that the OATS used for radiated emission measurements be validated by making horizontal and vertical NSA measurements. Antenna spacings used for making site attenuation measurements shall be the same as the spacings used for the EUT compliance tests at frequencies from 30 to 1000 MHz. The measured NSA data shall be compared to that calculated for an ideal site and the measured NSA shall be within ±4 dB of the theoretical NSA. This ±4 dB tolerance includes instrumentation calibration errors, measurement technique errors, and errors due to site anomalies. C63.4 provides suggestions for actions to be taken if a site fails to meet requirements. The most common error is to make the calibration without at least the minimum specified ground plane. Even damp soil with a high humus content is highly unlikely to meet requirements. Even with a conductive ground plane, such as hardware cloth, the area of the ground plane must be sufficiently large, with maximum dimension apertures of 3 cm and with the sections of the plane properly bonded together.

Reference [23] describes two methods for EMC site qualification, one is the site voltage-standing wave ratio (SVSWR) and the second time-domain reflectometry.

With the VSWR method, the transmission line is the path from the EUT to the receiving antenna, which includes reflections from other objects in the test environment. The test in CISPR 16-1-4 proposes moving the antenna in the chamber or site to six physical locations and recording the maximum and minimum values measured by the receiving antenna.

The errors inherent in this method are the repeatability and reproducibility of the distance and the antenna pattern. As the patterns vary with small increments in angle, so the coupled value varies.

The ratio of reflected to direct signal can be computed and the VSWR will provide a measure of the reflections in the site.

The time-domain method uses vector (voltage and phase) measurements to obtain the SWR with out the need to physically move the antenna. In the time domain, the direct wave will arrive first and the reflected wave later. By applying time gating (a time filter), the direct signal can be separated from the reflected. The actual measurements are done in the frequency domain and are then transformed using inverse Fourier transform. The time-domain data (after parsing) are converted back to frequency domain by which the same reflected wave versus direct wave signals are computed as with the VSWR method.

In the NSA measurement, another source of error is in the calibration of the antenna factor of the transmitting and receiving antennas. It is possible to use the free-space "two identical antenna" calibration method or the "three-antenna method," in which two antennas are identical and the third is the unknown, but the preferred method is the ANSI-C63.5 standard site method. If the free-space AF calibration of a BB antenna is used in the NSA calibration, then the proximity of the fixed-height antenna to the ground plane will almost certainly mean that the required NSA is not achievable at certain frequencies. One step that may help is to change from the 1 m fixed height to 1.5 m, or vice versa.

ANSI C63.4 says that measurements can be made at a location other than an OATS, such as a weather-protected site, an absorber-lined room, a dedicated laboratory, or a factory site, provided the alternative site meets the site attenuation requirements over the volume occupied by the EUT and the minimum conducting ground plane requirements are met.

As of publication of C63.4-1992, weather-protected sites are exempt from the multiple NSA measurements required of alternative test sites. In addition, the use of a GTEM cell with a factor programmed into the calibration of the GTEM to achieve correlation with the OATS attenuation requirements is described in Section 9.5.2.2. For alternative test sites, C63.4 states that a single-point measurement of the NSA measurement is insufficient to pick up possible reflections from the construction and/or RF-absorbing material comprising the walls and ceiling of the facility. For these sites, a "test volume" is defined, and the transmit antenna may have to be placed at various points, up to 20, within the test volume. The points and test method are described in detail in C63.4. The smaller the test volume, the lower the number of calibration points. For example, for a volume no larger than 1 m in depth by 1.5 m in width by 1.5 m in height, a minimum of eight measurement positions may be required.

Weather-protected OATSs have been constructed from fiberglass girders and panels, with nylon or fiberglass nuts and bolts to secure the structure. An alternative is a pressurized rubber tent, plastic radome, or cloth tent with fiberglass poles placed over the turntable on which the EUT is located. With the antenna remaining outside the weatherproof structure, the height of the structure may be limited, typically to the height of the maximum EUT plus the 0.8 m table height. All of these sites can meet the NSA requirements, at least when first constructed and before potential degradation by conductive contaminants.

For a company that owns a totally wooden structure with a tiled roof made of asphalt or wood shingle or ceramic, but not steel or aluminum, a proposal has been made that enables the use of this type of building as a low-cost, covered OATS.

First, a temporary, or permanent, open-field test site area is constructed, using the antenna, antenna mast, and turntable intended for use in the weatherproof building. This site must have a ground plane, typically constructed from hardware cloth, and this may also be subsequently used in the weatherproof building. A typical area for this open-field site is a car park, a field, or a backyard. The open-field test site area must be on even ground, with obstacles such as bushes or trees, and building outside of the "obstruction-free area for site with a turntable" as shown in Figure 5 of ANSI C63.4. An NSA measurement is made on this OATS that must conform to the ANSI C63.4 requirements. A BB swept-frequency noise source (typically battery powered) that covers the full 30–1000 MHz frequency range without any gaps, contained in a conductive shielded enclosure, is constructed. Horizontal and vertical cables are attached to the BB signal output level, and these cables cover the proposed test volume. The vertical cables are draped over the back of the table, with a configuration similar to that in Figure 9.24. The radiated emissions from this dummy EUT may be used to obtain a correlation between the two test sites. Alternatively, the emission test setup used for correlation of the GTEM to an OATS may be used. First, the noise source is measured on the OATS with the antenna moved from 1 to 4 m both horizontally and vertically and the turntable rotated. The same measurement is made on the covered site, with the EUT and cables placed in exactly the same positions (to achieve this, the cables and source can be taped down to a wooden

FIGURE 9.24   Test configuration for tabletop equipment—top view. † LISNs may have to be positioned to the side of the table to meet the criterion that the LISN receptacle must be 80 cm away from the rear of the EUT. (From C63.4-2014, American National Standard for methods of measurement of radio-noise emissions from low voltage electrical and electronic equipment in the range of 9 KHz to 40 GHz, 2014. © 2014, IEEE.)

plane or table). By comparing the two sets of emission measurements, a correlation between the OATS and the less-than-ideal weatherproof site can be obtained, and from this, a correction table can be generated, without the need for the covered site to meet the NSA requirements. However, the covered site should be constructed of nonconductive materials and contain minimum conductive sources of reflections, such as AC power wiring, and contain at least the minimum-dimension ground plane. Although this proposed test site could theoretically achieve as good a correlation as that between a GTEM and an OATS, the FCC will not consider this proposition. Although it is only possible to guess at the reason why this approach is unacceptable, it may be that some modification in the proposed test site, for example, addition of conductive objects, is more likely to occur than any modification in the GTEM cell. Another possible objection is that the volume of the EUT and the orientation of the cables may affect the calibration because reflections from the test area will change. However, a similar error may occur in the way in which the GTEM is excited by the EUT in the standard three orthogonal calibration technique, described in Section 9.5.2.2. Although this type of facility is not acceptable for certification/verification testing, it may be used for comparative measurements when making modifications to a product to reduce radiated emissions.

The minimum ground plane dimensions for AC power conductive measurements are at least 2 m × 2 m in size extending at least 0.5 m beyond the vertical projection of the EUT. If the EUT normally does not make contact with the ground plane, the ground plane shall be covered with insulating materials between 3 and 12 mm thick.

A conducting ground plane is required at radiated emission sites that shall extend at least 1 m beyond the periphery of the EUT and the largest measuring antenna and cover the entire area between the EUT and the antenna. It shall be made of metal, with no holes or gaps, with the largest dimension greater than 3 cm. A larger ground plane may be required for certain test sites and especially if the NSA requirements are not met.

The ambient radio noise and signals, both conducted and radiated, are required to be at least 6 dB below the allowable limit. In a true OATS, this is invariably not possible due to a multiplicity of RF, VHF, and UHF transmitters and BB noise sources. An absorber-lined room and a GTEM cell are two of the measurement techniques that will meet the 6 dB requirement. The alternative recommendations, such as measurements at a closer distance and using a narrower measurement bandwidth for NB and signals masked by BB noise, typically meet with limited success. The recommendation that measurements of critical frequency bands be made during hours when broadcast stations are off the air and at times when the ambient contribution from industrial equipment is reduced to less than the 6 dB level is rarely followed. In one case where this recommendation was followed, a relatively low-level ambient existed at a location at least 55 km from a large town or city and with low TV and radio signal strength. The measured signals from these sources were still well above the limit, even at an MD of 3 m. In addition, a BB ambient noise source was measured during the workday, and it was still present at 1 o'clock, 3 o'clock, and 6 o'clock in the morning. The noise source was a large fan used to dry silage on a farm that was in operation day and night for months. The realistic technique for minimizing the effect of the ambient on the EUT emissions is to go to a low-frequency span and reduced measurement bandwidth to differentiate between the ambient and the EUT emissions. This technique usually requires switching on and off the EUT to determine its contribution to the measured emissions. The frequency of peak emissions detected in the prescan will help in finding the EUT emissions. Because EUT emissions are notoriously unstable in frequency, one alternative technique to identify EUT emissions is to use a small antenna, such as the 20–1000 MHz bow tie, located close (typically 20 cm) to the EUT and cables. The measurements made with this antenna will typically lift EUT emissions well out of the ambient noise and allow identification of the emission in measurements at 3 m.

If an anechoic chamber or semianechoic chamber is available, then a prescan of the radiated emissions can be made, with the antenna typically at a distance of 1 m from the EUT. The prescan is used to identify radiated emissions from the EUT measured on the OATS. If any of the radiated emissions are masked by the ambient, an emission close in frequency but out of the ambient should be measured. This measurement on the OATS at 3 or 10 m can be compared to the prescan measurement of the two frequencies. For example, if an emission at 102 MHz is 6 dB below the limit and a second emission at 102.5 MHz is masked by an FM radio station, then a comparison of these two emissions measured in the anechoic chamber should be made. If the emission at 102.5 MHz is 6 dB below the emission at 102 MHz, then the probability that the 102.5 MHz emission when measured on the OATS will also be at least 6 dB below the limit is high.

Testing at a manufacturer's location or at a user installation is permitted if the equipment cannot be set up on an OATS or alternative test site. In this case, both the equipment and its location are considered the EUT. The radiated emission and conducted emission measurements are considered unique to the installation site. However, if three or more representative locations have been tested, the results may be considered representative of all sites with similar EUTs. The voltage probe, and not the line impedance stabilization network (LISN), shall be used for the conducted emission measurements, and the conducting ground plane shall not be installed for user's installation testing unless one or both are to be a permanent part of the installation.

The table height on which non-floor-standing equipment is placed must be 0.8 m for both AC line-conducted measurements and radiated emissions. Figure 9.25a and b show the test configurations for tabletop ITE equipment. Equipment that can be used as either tabletop or floor-standing equipment shall be tested only in a tabletop configuration. For conducted measurements in a shielded-room, tabletop devices shall be placed on a platform of nominal size, 1 m × 1.5 m, raised 80 cm above the ground plane. The vertical conducting surface of the shielded room shall be located 40 cm to the rear of the EUT. Floor-standing devices shall be placed either directly on the conducting ground plane or on insulating material if the equipment is typically isolated. All other surfaces of tabletop or floor-standing equipment shall be at least 80 cm from any other grounded conducting surface, including the case or cases of one or more LISNs.

Excess interface cable length will be draped over the back edge of the tabletop for tabletop equipment. If any draped cable extends closer than 40 cm to the conducting ground plane, the excess shall be bundled in the center in a serpentine fashion using 30–40 cm lengths to maintain the 40 cm height. If the cables cannot be bundled due to bulk, length, or stiffness, they shall be draped over the back edge of the tabletop unbundled, but in such a way that all portions of the interface cable remain at least 40 cm from the horizontal conducting ground plane, as shown in Figure 9.25a and b.

The system shall be arranged in one typical equipment configuration for the test. In making any tests involving several pieces of tabletop equipment interconnected by cables or wires, it is essential to recognize that the measured levels may be critically dependent on the exact placement of the cables or wires. Thus, preliminary tests shall be carried out while varying cable positions in order to determine the configuration for maximum or near-maximum emission. During manipulation, cables shall not be placed under or on top of the system test components unless such placement is required by the inherent equipment design. This preliminary test, often referred to as a "prescan," is a firm requirement and is often conducted in a shielded room. The cable configuration used during the test is crucial. In the past, the FCC spent approximately half an hour positioning the cable for maximum emissions in the preliminary prescan test, with the antenna positioned close to the equipment. The cables were then taped in position and the equipment taken to the open-field test site for a certification-type test. The prescan is used to determine the emission characteristics of the EUT, and the frequency of peak emissions is noted and is used to help differentiate between the EUT and ambient emissions when measurements are made on the open these are conducted on an open area when these are made on the OATS. Cables are not normally manipulated for floor-standing equipment. Instead, the cables should be laid out as shown in Figures 9.26 and 9.27.

Each EUT current-carrying power lead, except the safety ground, shall be individually connected through a LISN to the input power source. See section 7.2.1 of ANSI C63.4-2014 for more information on the use of the LISN. The LISN uses a combination of inductors and capacitors to present a standard power-line impedance to the EUT. ANSI C63.4-2014 describes two LISNs, both of which will achieve the characteristic impedance of Figure 9.28. One is usable from 10 to 150 kHz, although this may be extended to 30 MHz; the second is usable from 0.15 to 30 MHz.

It is the second LISN that is most commonly used for testing to FCC requirements. A 10 kHz to 30 MHz LISN design can be calibrated and used for FCC/EN tests as well as MIL-STD-461F tests. The voltage measurement instrument (S/A /receiver) with a 50-input termination is connected to a port of one of the two LISNs to measure the noise voltage developed by the EUT across the impedance of the LISN. The "measuring instrument port" of the second LISN must be terminated

Legend:

1. Interconnecting cables that hang closer than 40 cm to the ground plane shall be folded back and forth forming a bundle 30–40 cm long, hanging approximately in the middle between ground plane and table.

2. I/O cables that are connected to a peripheral shall be bundled in center. The end of the cable may be terminated if required using correct terminating impedance. The total length shall not exceed 1 m.

3. EUT connected to one LISN. Unused LISN connectors shall be terminated in 50 Ω. LISN can be placed on top of, or immediately beneath, ground plane.

   3.1 All other equipment powered from second LISN.

   3.2 Multiple outlet strip can be used for multiple power cords of non-EUT equipment.

   3.3 LISN at least 80 cm from nearest part of EUT chassis.

4. Cables of hand-operated devices, such as keyboards, mouses, ect., have to be placed as close as possible to the host.

5. Non-EUT components being tested.

6. Rear of EUT, including peripherals, shall be all aligned and flush with rear of tabletop.

7. Rear of tabletop shall be 40 cm removed from a vertical conducting plane that is bonded to the floor ground plane (see 5.2).

(a)

FIGURE 9.25 (a) Test configuration for tabletop equipment conducted emissions. (From C63.4-2014, American National Standard for methods of measurement of radio-noise emissions from low voltage electrical and electronic equipment in the range of 9 KHz to 40 GHz, 2014. © 2014, IEEE.)

(Continued)

Legend:

1. Interconnecting cables that hang closer than 40 cm to the ground plane shall be folded back and forth forming a bundle 30–40 cm long, hanging approximately in the middle between ground plane and table.

2. I/O cables that are connected to a peripheral shall be bundled in center. The end of the cable may be terminated if required using correct terminating impedance. The total length shall not exceed 1 m.

3. If LISNs are kept in the test setup for radiated emissions, it is preferred that they be installed under the ground plane with the receptacle flush with the ground plane.

4. Cables of hand-operated devices, such as keyboards, mouses, ect., have to be placed as close as possible to the controller.

5. Non-EUT components of EUT system being tested.

6. The rear of all components of the system under test shall be located flush with the rear of the table.

7. No vertical conducting wall used.

8. Power cords drape to the floor and are routed over to receptacle.

(b)

FIGURE 9.25 (Continued) (b) Test configuration for tabletop equipment radiated emissions. (From C63.4-2014, American National Standard for methods of measurement of radio-noise emissions from low voltage electrical and electronic equipment in the range of 9 KHz to 40 GHz, 2014. © 2014, IEEE.)

in 50 Ω. The LISN serves a secondary function of isolating supply-generated noise from the test equipment. Often the LISN is not adequate and an additional power-line filter inserted between the power source and the LISN is required. As with all EMI measurements, a background measurement with the EUT powered down should be conducted prior to conducted emission tests. The use of a calibrated and correctly terminated current probe that fits around each of the current-carrying conductors, separately, may be used when permitted. A LISN should be inserted between the EUT conductors and the mains (power) outlet. The probe is placed between the EUT and the LISN as near the LISN as possible. If an appropriate LISN satisfying the current requirements of the EUT is not commercially available, the LISN may be eliminated and the current probe placed between the

FIGURE 9.26 Test configuration for floor-standing equipment conducted emissions. (From C63.4-2014, American National Standard for methods of measurement of radio-noise emissions from low voltage electrical and electronic equipment in the range of 9 KHz to 40 GHz, 2014. © 2014, IEEE.)

Legend:

1. Excess I/O cables shall be bundled in center. If bundling is not possible, the cables shall be arranged in serpentine fashion. Bundling shall not exceed 40 cm in length.

2. Excess power cords shall be bundled in the center or shortened to appropriate length.

3. I/O cables that are not connected to a peripheral shall be bundled in the center. The end of the cable may be terminated if required using correct terminating impedance. If bundling is not possible, the cable shall be arranged in serpentine fashion.

4. EUT and all cables shall be insulated from ground plane by 3–12 mm of insulating material.

5. EUT connected to one LISN. LISN can be placed on top of, or immediately beneath, ground plane.

    5.1 All other equipment powered from second LISN.

FIGURE 9.27 Test configuration for floor-standing equipment radiated emissions. (From C63.4-2014, American National Standard for methods of measurement of radio-noise emissions from low voltage electrical and electronic equipment in the range of 9 KHz to 40 GHz, 2014. © 2014, IEEE.)

EUT and the mains outlet. Although the current probe measurement without LISN will not achieve good repeatability, the lack of a suitable LISN is one good reason to use the current probe technique. ANSI C63.4-2014, in Appendix F, describes two techniques for calibrating the LISN. One requires a signal generator and receiver/ S/A and the second a network analyzer. Neither method checks the LISN impedance under AC power load conditions.

The measurement of the impedance of the LISN with the maximum AC line current flowing through the LISN may be made at a number of frequencies by use of an impedance meter. The technique involves shorting the two phases of the LISN together (load side) and supplying the input via a

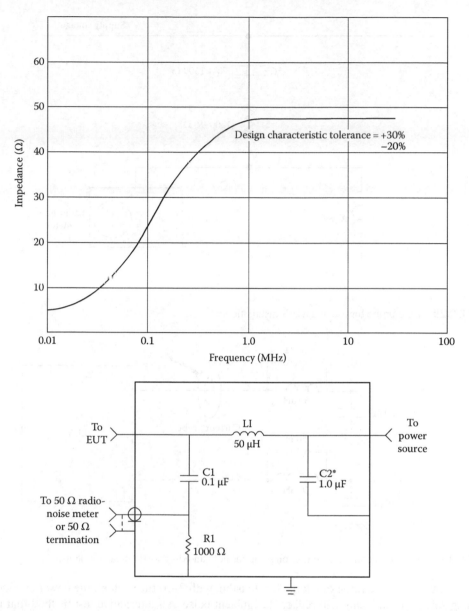

FIGURE 9.28 FCC LISN schematic and impedance versus frequency characteristics. * In some LISNs, a series resistance is included in series with capacitor C2, e.g., CISPR publication 16 (1987) [6]. (From C63.4-2014, American National Standard for methods of measurement of radio-noise emissions from low voltage electrical and electronic equipment in the range of 9 KHz to 40 GHz, 2014. © 2014, IEEE.)

variable low-voltage AC supply (e.g., a variac and isolating step-down transformer). The impedance meter may be connected between the case of the LISN and the shorted outputs of the LISN. It would be wise to test for the presence of a common-mode supply frequency voltage, which may damage the impedance meter before making the connection.

A modified conducted emission test setup that may be utilized at a user's installation using the voltage probe is shown in Figure 9.29. A voltage probe may also be used where the use of a LISN is impossible due to high current requirements of the EUT. In the test setup with the voltage probe, both the EUT and the installation environment are tested. The problem with the test method is that AC power-line-conducted noise from locations outside of the jurisdiction of the user

FIGURE 9.29   Line probe for use at a user's installation.

FIGURE 9.30   Proposed conducted emission setup for field and diagnostic measurements.

is also measured. The addition of a large ferrite balun with 2–10 turns on all the power conductors, including safety ground, can help reduce the ambient noise. An alternative test method that is not approved by the FCC but that has a number of advantages over Figure 9.29 is shown in Figure 9.30. One advantage of the test circuit compared to the LISN is that it may be easily constructed. The inductor may be wound on a metal oxide or ferrite toroid, and the 10 µF capacitor should, where feasible, be an RF feedthrough type. By the choice of a large enough metal oxide toroid, a current capacity of up to 200 A may be achieved. The second advantage of the proposed setup is that the current probe measurement is isolated from power-line-generated noise by the series inductor. The test circuit can provide an impedance within the tolerance limits of Figure 9.28. Although the proposed setup is useful in diagnostic and precertification measurements, only an FCC-type LISN or the line probe, at the user's site, is likely to be accepted by the FCC for certification purposes.

Measurements of radio noise power using the absorbing clamp are allowed only for restricted frequency ranges and if specified in the individual equipment requirements and are not allowed for measurements to FCC requirements.

ANSI C63.4 mentions the use of an artificial hand if the EUT is normally operated in the hand, although this is understood not to be a requirement for FCC testing. The test site location should be in as quiet an electromagnetic environment as feasible. For example, locating a test site close to an airport may be convenient for transportation of equipment, but it is invariably a noisy environment.

No matter how apparently minor changes in an equipment's software program or configuration can make large changes in the emission profile, both radiated and conducted. For example, a piece of equipment appeared to change overnight from within FCC-radiated emission limits to above. The reason was that prior to testing on the second day, the designer had slipped in a new PROM with the updated version of software. The new software initialized interfaces in parallel instead of sequentially and increased emissions during the power-up sequence. Other causes of increased emissions have been traced to a change in the manufacturer of ICs and in a signal interface transformer that was changed from a shielded to an unshielded type.

9.4.3.1 Radiated Emission Test Setup and OATS Case Studies

A small piece of electronic equipment had been tested for radiated emissions on numerous occasions for the manufacturer, throughout the development of the product and on to the production version, and these tests were performed by at least three test personnel. The requirements were the FCC Part 15 and the EN55022 Class B. The software run by the manufacturer was considered "worst case" as it exercised the display, LED, audio, signal interface, and memory. The EUT was powered by a battery and a single cable that contained power, and a serial signal interface was taken down to the battery located on the OATS ground plane and then brought back to the tabletop. The EUT was tested both with the battery terminal connected and disconnected from the ground plane. The cable orientation was as close to that shown in 9(c) of ANSI C63.4 with the power/signal cable brought back up to the nonconductive tabletop and then looped back down again in a serpentine fashion, with the loops at least 40 cm above the ground plane. ANSI C63.4 mentions that

> In order to replicate emission measurements, it is important to carefully arrange not only the system components but also system cables, wires, and AC power cord.

ANSI C63.4 also discusses cable orientation as follows:

> It is essential to recognize that the measured levels may be critically dependent upon the exact placement of the cables and wiring. Thus preliminary tests may be carried out while varying cable positions in order to determine the maximum or near maximum emission level.

In the test site under discussion, the tests were made using a turntable and the radiated emissions have been found to vary dramatically depending on the rotation of the table. After a maximum emission has been found, the cables are then separated or moved within the confines of the recommended ANSI configuration. Again, this manipulation of cables has been performed by at least three different personnel. This reorientation is also only made at one (worst-case frequency), whereas different cable orientations will maximize the emissions at other frequencies. This approach is taken to limit the amount of time required for a test, but by limiting the basic cable orientation to the ANSI C63.4 recommendation, the absolute worst-case emissions may not be found.

The test site, which we shall designate A, was a 3 m OATS, which had been tested in accordance with ANSI C63.4 for NSA. The NSA test had been conducted in 1997 and the measured site attenuation was within ±4 dB of the theoretical NSA provided in ANSI C63.4 with a safety margin, that is, the measured NSA was lower than the ±4 dB tolerance. In all of these tests on site A, the EUT met all of the radiated emission measurements with a safety margin of at least 8.8 dB, below either the FCC or EN55022 limit.

In near-field measurements, close to the EUT and the cable, on all of the different EUTs tested, the measured levels were also consistent with a low-level emitter. The same type of product, but not the identical unit with the same serial number, was then tested in a 3 m semianechoic chamber test site (test site C), which reported emissions up to 3.65 dB above the limit, that is, 12.46 dB higher than the 3 m OATS measurements. At exactly the same frequency, the difference in measurements could have been as high as 25 dB.

At first, different software and different cables were used in the test on site C. However, even when the EUT was retested with the same software and cable, the EUT was above the limit. The EUT was then tested on a 10 m OATS (test site B) and also in a GTEM (test site D). The measurements made on a 3 m OATS and a 10 m OATS often do not correlate to the far-field inverse linear distance extrapolation factor of 20log 10 m/3 m = 10.5 dB. It is common that in moving the EUT location from 3 to 10 m that the reduction at some frequencies is less than 10.5 dB. Thus, an EUT may be just within specification when measured at 3 m and out of specification at 10 m. However, this does not account for the 14 dB difference measured between sites A and B. The correlation between a GTEM and an OATS has a typical mean value of 6 dB with 11–15 dB differences in individual measurements. Unfortunately, OATS to OATS correlation can be similarly poor. In the GTEM measurements, the test engineer made the cable "dormant" by the addition of three ferrites, and so if cables were the main source of emissions, the GTEM measurements should have been the lowest. Also, the test on site C was repeated using a short cable and with the battery placed on the top of the turntable with the same high levels of emissions, and this again led us to believe that cables were not a factor. A comparison of the measurements made on the four test sites and compared to the limit is shown in Table 9.9. Thus, −8.8 dB means that the emission is 8.8 dB below the spec and +5.4 dB is above the spec. A dash at any location in Table 9.9 means that on the specific test site the emission was low and in some case negligible.

These test results did not look good for test site A, and so extensive measurements were made on the site and test equipment. One possibility was that an EMI receiver was used on sites B and C and an S/A was used on site A. A variation in the measurement of BB noise can be expected due to the different shape of the impulse bandwidth used in the receiver and the resolution bandwidth used in the S/A. The analyzer typically requires a 4 dB correction for BB noise. However, the measurements using the S/A did not vary significantly with change in RBW, as the emission was predominantly NB. Also, in the final measurements made on site E (a 3 m OATS almost identical to A), a receiver was used, and a good correlation was seen between sites A and E, not consistent with a 4 dB systemic error.

The test site A measurements were made using two different S/As, which were calibrated and were checked against a calibrated signal source. Quasi-peak and peak measurements were the same,

TABLE 9.9

Worst-Case Radiated Emission Measurements Compared to the Limit for the EUT When Measured on Four Different Test Sites (−dB is below the Limit, +dB Is above the Limit)

f (MHz)	Site A 3 m OATS	Site B 10 m OATS	Site C 3 m Anechoic Chamber	Site D GTEM
197.66	−7.7	—	—	—
204.2	—	+2.6	—	—
206.7	—	—	—	−3.8
208	—	0	—	−5.9
212	—	+4.6	+3.2	—
220	−8.8	+4.4	+3.2	—
228	—	+5.4	+3.65	—
235.88	−20.8	—	−6	—
244.3	—	−1.3	—	—
249.5	−18.8	—	—	—
252	—	−2.1	—	—
279.06	−16.3	—	−12.9	—
368	—	−7.6	—	—

and so the quasi-peak detectors were not a problem. Cable attenuation and preamplifier gain were checked and were okay. Before using the test site, a "sanity check" is made of the antenna, cables, and preamplifier using the ambient from a number of FM radio stations (about the only advantage of the OATS, apart from accuracy), and the levels during all tests on the EUT were within 1 dB of the values measured when the site was first constructed.

The only remaining possibility is the antenna calibration and the NSA characteristics of the site. A log-periodic/biconical antenna and a Roberts dipole were both used during measurements on the EUT, and it seems inconceivable that both would be equally out of specification. Nevertheless, two of the log-periodic biconical antennas were tested on a free-space range, and the coupling between the antennas was exactly as predicted, based on the input signal level, their gain, and antenna factor. Also as expected, the free-space measurement failed the NSA requirements with a difference of 5.7 dB due to the absence of the ground plane.

The equation for A_N is

$$A_N = V_{Direct} - V_{Site} - AF_T - AF_R \text{ (all terms in dB)}$$

where
V_{Direct} is the direct measurement via the cables
V_{Site} is the radiated measurement using the transmit and receive antennas and the same cables
AF_T is the transmit antenna factor
AF_R is the receive antenna factor

As the direct measurement is made with a calibrated output signal generator, this measurement also tests that the S/A is measuring accurately the amplitude of the frequencies of interest, with the expected cable attenuation. NB during radiated measurements the same S/A reference level is used as for direct.

The radiated measurement uses the calibrated antenna factors, which are critical in meeting the NSA requirements, and so if we assume a very good OATS performance, the NSA measurement is also a way of reconfirming the accuracy of the antenna calibration. This was further tested by replacing the biconical/log-periodic receive antennas with two reference dipoles and repeating the NSA measurement.

The difference between measured and theoretical NSA over the frequency range of interest using two log-periodic biconical antennas is shown in Table 9.10 and using two dipoles in Table 9.11. These measurements show that the site is at a worst case within 2.1 dB of the predicted and at the

TABLE 9.10
Delta NSA Using Two Log-Periodic Biconical Antennas

f (MHz)	ΔNSA (dB)
200	0.82
250	1.68
300	2.1

TABLE 9.11
Delta NSA Using Two Dipole Antennas

f (MHz)	ΔNSA (dB)
200	0.31
250	1.69
300	1.9

critical frequency of the EUT (220 MHz) is within approximately 1 dB of the value for NSA, interpolated between 200 and 250 MHz.

Even though site A met all of the requirements, it was decided to test the EUT yet again on a 3 m test site (E). These measurements showed that the EUT passed the requirements, but by only 1.6 dB, unlike the 8.8 dB found on site A. However, one difference in the cable configuration used on test site E was that the cable was taken down from the center of the turntable, and not the edge, and then brought back from the battery and coiled in the center of the turntable in a large loop.

The exact same cable configuration was tested at site A and the difference in emission measurements between site A and E was a worst case 3.8 dB, which is an acceptable difference based on antenna factor calibration errors, NSA errors, measuring instrument errors, and the difference in receiver to S/A bandwidth shape. This final test did show that cable orientation does play a role in the level of emissions, by as much as 2.9 dB in this particular instance.

In the tests conducted by test sites B and C, which showed the highest levels of emissions, the length of cable after the battery was coiled on the ground plane. This cable orientation seems far away from the spirit of the ANSI-recommended layout and may explain the large difference in measurements, although apparently the measurements were the same with the short cable and the battery placed on the top of the turntable.

Table 9.12 shows the emissions in dBμV/m compared to the EN55022 Class B limit.

The problem with a poor correlation between an OATS measurement and a 3 or 10 m semianechoic chamber was addressed in 2000 above and more recently in Reference [22], "Egregious errors in Electromagnetic Radiation Evaluation."

Reference [22] describes a comparison between two 3 m and one 10 m OATS, one GTEM and two semianechoic chambers. One difference between the measurements on all sites was that although the make and model of the EUT were the same, they had different serial numbers.

When emissions were measured at the same frequency at all test facilities, a difference of 12.46 dB between the first 3 m OATS and the 10 m anechoic chamber was seen, and a difference of 14 dB was seen between the first 3 m OATS and the 10 m OATS. In all cases, the first 3 m OATS recorded lower emissions. In all cases, the first 3 m OATS recorded lower emissions. What was really alarming was that at some frequencies, emissions were seen at one site and not at another. This represented a variation of up to 25 dB. The 3 m OATS measurement was then suspect.

The NSA calibration was repeated for the first 3 m OATS at the critical frequencies were large discrepancies were measured using a "Roberts" dipole antenna and the results are shown in Table 9.13. Reproduced by permission of ITEM interference technology.

The calibration of the test equipment, cables, and antennas used in the first 3 m OATS was repeated and all were within specification.

TABLE 9.12

Comparison of Radiated Emissions Measured on Test Site A and Test Site E with the Same EUT, Software, and Cable Orientation

f (MHz)	Test Facility and Antenna	Level (dBμV/m)	Limit (dBμV/m)	Δ (dB)
212	Site E log/bicon	31.4	40	−8.6
212	Site A log/bicon	31.6	40	−8.4
220	Site A dipole	31.2	40	−8.8
220	Site E dipole	35	40	−5
228	Site A log/bicon	35.7	40	−4.3
228	Site E dipole	38.4	40	−1.6
236	Site A log/bicon	36.2	47.5	−11.3
236	Site E log/bicon	37.1	47.5	−10.4

TABLE 9.13

NSA Measurements on 3 m OATS

f (MHz)	NSA (ΔdB)
200	0.31
250	1.68
300	2.1

The cable orientation used at the first 3 m OATS test facility was in accordance with figure 9© of ANSI C63.4, with the power/signal cable brought back up to the nonconductive tabletop and then looped back down in a serpentine fashion with the droops at least 40 cm above the ground plane. Some further manipulations were then made to maximize emissions. ANSI C63.4 mentions that "in order to replicate emission measurements, it is important to carefully arrange not only the system components but also the system cables, wires, and AC power cords."

The same EUT was sent to the second 3 m OATS, and the results were within 7.2 dB. The cable orientation was compared between the two sites, and in the second 3 m site, the cables were draped down from the center of the turntable rather than from the edge in loops. This has the effect of keeping the cables at a fixed distance from the antenna but is not the orientation shown in the figure. The same EUT was then remeasured on the first OATS with the same cable orientation as used on the second OATS, and a better correlation was seen as shown in Table 9.14.

The level of correlation between the two sites is considered good and indicates that, along with the repeat NSA, the first 3 m site can be used for accurate measurements. The measurement between the two sites also reinforces the importance of cable orientation.

It was expected that the article in 2000 and the article in 2010 would result in a lively discussion, comments, recommendations, or disagreement, but no response was elicited. Since 2000, EMC consulting has seen numerous cases, where radiated emission data from different test sites have shown an unacceptable variation. EMC consulting has observed that the correlation between measurements made on two different OATS is often higher than the correlation between OATS and a 3 or 10 m semianechoic chamber.

In one case, a consistent 26 dB was seen between one site and another. Only after the site with the 26 dB higher measurement was asked to use a signal generator to check the test instrument, was the problem found. It was a modern instrument with switchable 26 dB preamplifier, and inclusion of the amplifier was not indicated on the front panel. This situation was in effect since the instrument was first purchased. This illustrates that despite an accredited test facility having a good reputation, the customer should review the test setup against the requirements document and ask

---

TABLE 9.14

Comparison between Emissions from the Same EUT with the Same Cable Orientation on Two Different 3 m OATS

Frequency (MHz)	1st 3 m OATS	2nd OATS	Delta (dB)
212	31.4	31.6	0.2
220	31.2	35	3.8
228	35.7	38.4	2.7
236	36.2	37.1	0.9

Source: Courtesy of ITEM Interference Technology, Plymouth Meeting, PA.

---

the facility to perform a sanity check of antennas, cables, and measuring equipment. Using the FM radio, ambient measurement is a good sanity check on an OATS.

Very recently, we have been working with a manufacturer who has had equipment tested at two 10 m OATS with a good correlation. However, the measurements made in a 3 m anechoic test chamber by a test facility chosen by the customer resulted in some very different emission levels. Not only was there a 38.9 dB variance between the first OATS and the 3 m chamber, but in comparing the emissions, many were higher for vertically polarized fields on one sight and on the other site for horizontally. Table 9.15 compares the emissions against the FCC Class A measured at 10 m on the OATS and 3 m in the chamber. FCC Class A measurements are routinely made at 10 m on the first OATS and two convert these approximately to the 3 m chamber measurements, 20log 10 m/3 m = 10.5 dB was added. This distance correction is strictly true only for a point source. In measurements on large equipment or equipment with cables, we find that the difference between measurements at 10 and 3 m is not always 10.5 dB. However, the error is never as large as the differences shown in Table 9.15.

Despite a close correlation between the two OATS measurements, in which the EUT passed the Class A limits with a 9 dB margin, the customer insisted that the measurements made by the facility with the 3 m semianechoic chamber were correct. The manufacturer had already gone to great lengths in the design and manufacture to achieve the low levels in Table 9.15, and further improvements were required in the design and engineering of the EUT to satisfy the customer. This resulted in more expense for the manufacturer, added to the frustration in knowing that the EUT had almost certainly passed the requirements and a delay due to numerous levels of reengineering.

It is not clear why such a large variance in commercial radiated emission test results exists. If a facilities site has met the NSA calibration, then the OATS and anechoic chamber measurements should be comparable, which is often not the case. Does the anechoic chamber exhibit resonances that are not seen in the NSA calibration? Cable orientation and the time spent in maximizing the cable emissions play a role but surely not to the extent of 25 and 38.9 dB as in the two examples. Also, at high frequency, where the emissions may be sourced by seams in an enclosure, the emissions may be in the form of a narrow beam and table rotation speed does affect the detection of such an emission. The correction of 10.5 dB in going from 10 to 3 m MD is also not always correct,

TABLE 9.15

Comparison between Measurements Made on the Same EUT on a 10 m OATS Compared to Measurements in a 3 m Semianechoic Chamber That Met the NSA for the Chamber

Polarization	Frequency Range (MHz)	OATS 10 m Measurement (dBµV/m)	OATS Corrected to 3 m	3 m Semianaechoic Chamber Measurement (dBµV/m)	FCC Class A Limit Converted to 3 m (dBµV/m)	Delta between 3 m Chamber and OATS Measurement (dB)
H	64	25.7	36.2	39.4	49.5	3.1
V	64	3.7	14.2	53.1	49.5	38.9
H	192	34.5	45	62	54	17.4
V	192	28.5	39	60.8	54	21.8
H	384	27.2	37.7	55.3	57	17.6
V	384	27.2	37.7	55.3	57	17.6
H	704	21.4	31.9	51	57	19.1
V	704	24.4	34.9	49	57	14

Source: Courtesy of ITEM Interference Technology, Plymouth Meeting, PA.

especially for physically large EUTs. In the first OATS described in this report, a sanity check is made of the measurement cable and antenna by using the known levels from FM transmitters in the vicinity. These levels have not changed significantly in the last 12 years, and their use is the only known benefit of the OATs ambient levels.

EMC consulting proposes that some standard form of site validation prior to each test be performed. One suggestion is the mandatory use of an inexpensive battery-powered 30–1000 MHz comb generator connected to a fixed vertical and horizontal orientation and length section of conductor, representing a cable. The source needs to be stable with varying battery voltage and with a shutdown when the voltage falls too low. Expecting a standard EUT cable orientation from all test sites will go some way in correlating emissions between sites. Other options must exist, and the purpose of this section in the book is to stimulate discussion, compare anecdotes, and find the source of errors and a solution to what EMC consulting perceives as a major problem, especially for manufacturers, but also for the reputation of test facilities.

Class B equipment should be measured on a 3 m site and Class A on a 10 m site.

Perhaps, due to the high cost of a properly qualified 10 m chamber, a number of facilities also use the 3 m site for Class A equipment and correct the emissions to a 3 m distance, which is incorrect. To avoid the expense of a 10 chamber, other facilities use a 3 m chamber to characterize emissions and then perform the qualification testing on a 10 m OATS.

This makes the discrimination of the ambient on the 10 m OATS much easier.

9.4.4 CANADIAN REQUIREMENTS

Industry Canada places requirements regulating the sale, offering for sale, and use of ITE in Canada. The requirements place limits on the permissible level of interference from ITE manufactured in Canada or imported into Canada. The latest requirements, applicable after August 31, 2013, are contained in ICES-003, Issue 5. The requirements do *not* apply to

1. ITE used exclusively in any vehicle including motor vehicle and aircraft
2. ITE used exclusively as an electronic control or in a power system, either by a public utility or in an industrial plant/factory
3. ITE used exclusively as industrial, commercial, or medical test equipment
4. ITE used exclusively in an appliance or electrical machinery, for example, dishwasher, clothes dryer, air conditioner (central or window), power tools, electric motors, and generators.
5. Specialized medical ITE, used under the direction of a licensed health-care practitioner, whether in a domestic environment or a health-care facility. Nonspecialized medical ITE retailed to the general public and ITE used by health-care practitioners for purposes other than medical treatment do not qualify for this exemption
6. ITE having a power consumption not exceeding 6 nW.
7. Joystick controllers or similar devices, such as a mouse, used with ITE but which contain only nondigital circuitry or a simple circuit to convert the signal to the format required are viewed as passive add-on devices.
8. ITE in which both the highest frequency generated and the highest frequency used are less than 1.705 MHz and which neither operates from nor contains provision for operation while directly or indirectly connected to the AC power lines.
9. ITE used exclusively in central office telephone equipment operated by a telecommunications common carrier in a central office.
10. ITE used exclusively in broadcasting equipment.

The requirements do not apply to units or models of digital apparatus for which the manufacturer, importer, or owner has been granted a special permission by the minister.

The minister may grant a special permission where

1. The manufacturer, importer, or owner has presented a written application giving
   a. The reasons for the request
   b. An analysis based on sound engineering principles, showing that the unit or model of digital apparatus will not pose a significant risk to radio communications
   c. A guarantee of compliance with all the conditions the minister may set in the special permission
2. The minister is satisfied that the unit or model will not pose a significant risk to radio communication

The special permission is valid only if

1. The unit bears a label stating that it is operating under special permission and setting out the conditions of that special permission and setting out the conditions of that special permission.
2. The unit complies with all conditions set out in the special permission

The minister may revoke or amend the special permission at any time without notice.

ITE is defined as electronic apparatus that generates and uses timing signals or pulses at a rate in excess of 9000 pulses per second and that utilizes *RF* energy for the purpose of performing functions including computations, operations, transformations, recording, filing, sorting, storage, retrieval, and transfer but does not include ISM equipment.

"Class A digital apparatus" means a model of digital apparatus for which, by virtue of its characteristics, it is highly unlikely that any unit will be used in a residential environment, which includes a home business.

"Class B digital apparatus" means any digital apparatus that cannot qualify as Class A digital apparatus.

A representative type of model of each digital apparatus shall be tested in accordance with C108.8-M1983, "Electromagnetic Emissions from Data Processing Equipment and Electronic Office Machines," or alternatively CAN/CSA-CISPR 22-96, "Limits and Methods of Measurement of Radio Disturbance Characteristics of ITE" (the latter is an adoption without modification of CISPR 22:1993, second edition).

The procedural requirements are

A written notice indicating compliance must accompany each unit of digital apparatus to the end user. A suggested text is as follows:
This Class A digital apparatus complies with all the requirements of the Canadian Interference-Causing Equipment regulations.
Cet appereil numérique de le classe A respecte toutes les exigences du Reglement sur le matériel broulleur du Canada.

The notice shall be in the form of a label that is affixed to the device. Where, because of insufficient space or other restrictions, it is not feasible to affix a label to the apparatus, the notice may be in the form of a statement included in the user's manual.

Any suitably equipped laboratory or organization can perform the tests. There are no restrictions as long as the specified test methodology is followed. It is the responsibility of the manufacturer or the importer to ensure the validity of the test results.

The field intensity of radiated emissions may be measured at a distance other than that described in the table of limits but not less than 3 m. The measurement result shall be extrapolated to the prescribed distance as described in C108.8 or CSA-CISPR 22.

The description of the measurement site for both conducted and radiated emission measurements is the same in ANSI C63.4:2003 as published in C108.8-M1983. However, no open-field site calibration test is required by Industry Canada.

The FCC limits are very similar to the Canadian limits. As a result of this, an FCC report that is less than 1 year old can be used in place of a Canadian test report to prove Canadian compliance for radiated or conducted emissions. In the case of radiated emissions, however, the laboratory at which the data were measured must have its OATS approved by Industry Canada.

The conducted emission equipment limits set out in C108.8 are identical to the FCC requirements and so are the radiated emission limits up to 1 GHz. The ICES-003 requirements limit radiated emission measurements to 1 GHz, whereas the FCC requires measurements above 1 GHz as required under 47 CFR 15.33. Some of the minor differences between MP4 and C108.8 have been removed in the adoption by the FCC of ANSI C63.4.1992. Both ANSI C63.4 and the C108.8 test method have the table height set at 0.8 m for conducted and radiated measurements. C108.8 allows the use of a dipole tuned to 80 MHz but used below 80 MHz with appropriate correction factors, which is especially useful for measurements made with vertical polarization. ANSI C63.4 and C108.8 allow measurement closer to the EUT than the specified distance, but not less than 3 m, when the ambient exceeds the limit. A correction factor is then added to the limit to account for the close in measurement. This provision may well be of assistance when the 30 m distance specified for Class A limits apply. In common with the FCC requirements, the preferred measurement instrument is a CISPR type. However, both the FCC and Industry Canada allow the use of an S/A with the quasi-peak detector. Conducted NB emission limits are the same as the FCC limits, and the LISN is the same. Where the FCC allows a 13 dB relaxation for BB noise, C108.8 publishes NB limits (and BB limits that are 13 dB higher). ANSI C63.4-conducted emission test setup complies with that of C108.8 and requires that test site ambient noise level should be at least 6 dB below the FCC limits. C108.8 states that the ambient noise level shall preferably be at least 6 dB below the limit. If the ambient noise level is higher, it must be shown that it does not interfere with the measurements of the source emissions from the EUT. C108.8 also states that the ambient level should not exceed the Class A–radiated emission limits. But if the ambient field is higher than the limit, then a "close in" measurement technique is described.

The author has retested equipment with FCC stickers indicating compliance and found emissions above the limits. The FCC routinely retests equipment and finds certified equipment above limits, and thus it is questionable how effective regulations are unless some policing by the regulating body is undertaken. The unofficial position of Industry Canada is "that computers do not result in EMI," and that may be the reason Industry Canada does not plan to independently retest equipment to ensure continuing compliance on production units.

However, should the operation of equipment result in interference, the radio investigator may elect to investigate whether the equipment meets the appropriate limits. The Department of Industry Radio Communication Act provides for penalties upon summary conviction a fine or 1 year imprisonment or both, in the case of an individual or a corporation.

9.4.5 GERMAN REGULATIONS

EMC standards were described in VDE numbers 0838–0879 in the DIN-VDE series.

Germany in common with many European countries now uses the EC EMC requirements.

9.4.6 CHINESE STANDARDS

China has held EMC requirements since 1999. Certification is a mandatory requirement for all devices and once certified are marked with the China Compulsory Certification (CCC) mark, denoting that the product complies with the standards set by the China Certification and Accreditation (CNCA) organization. All products manufactured in China and imported into China must receive the CCC certification before they can be marketed. This certification can currently only be received in laboratories that have been accredited by the CNCA. There is no MRA held with this organization outside of the country, and as a result, CCC certification can only be received inside China. Due to the length of time it takes to get this certification, and resulting cost, it can be useful to have testing done on a product prior to applying for CCC certification to ensure that it will pass all requirements. CCC certification is based on complying with China's GB standards, and so these limits should be used if testing in advance.

9.4.7   Japanese EMI Requirements on Computing Devices

The control of emission on EDP equipment in Japan is undertaken on a voluntary basis under the supervision of the Voluntary Control Council for Interference (VCCI). Although a voluntary requirement, many Japanese manufacturers comply as a sign of product integrity. Also, equipment manufactured in Japan for export to the United States must meet FCC requirements and is therefore likely to meet VCCI requirements. Equipment imported to Japan is not required to meet VCCI requirements; however, membership in the VCCI is allowed to foreign manufacturers and may well be of value in the marketing of equipment in Japan. The VCCI requirements are modeled on and use the same limits as contained in CISPR 22. Tests on equipment are performed at a VCCI-approved facility, and the test report is submitted to the VCCI, which issues a certificate of compliance. A laboratory can be VCCI certified by VLAC or JAB in Japan, or NVLAP, A2LA, or ACLASS in the United States. Compliant equipment is then labeled with a VCCI mark. The VCCI routinely retests certified equipment and, should it fail, requires the manufacturer to modify the equipment. The members agree to abide by the VCCI decisions, which may include revocation of the certificate on noncompliant equipment. Detailed VCCI regulations are available from Compliance Engineering (508-264-4208) or Dash Straus and Goodhue (978-263-2662).

9.4.8   Taiwanese Standards

The Bureau of Standards, Metrology, and Inspection (BSMI) is in charge of EMC compliance in Taiwan. Regulation is mandatory for manufacturers in Taiwan as well as for products being imported into the country. The standards used are defined by the Chinese National Standards, and these are derived from the IEC standards, with only small variances. An MRA currently exists between the Taiwan and the United States, which states that EMC testing for a product intended for sale in Taiwan may be tested in either country. Testing laboratories used in Taiwan must be certified by the Taiwan Accreditation Foundation, and testing laboratories used in the United States must be certified by the NIST. The BSMI conformity logo must be affixed to all compliant products before they can be marketed.

9.4.9   Australian and New Zealand Standards

The Trans-Tasman Mutual Recognition Arrangement (TTMRA) was agreed upon in 1996 and states that EMC certification in Australia is equivalent to that in New Zealand and vice versa. As a result of this, products can be marketed freely between the two countries without additional testing. Testing for certification can be done within either of the two countries or outside.

EMC standards became mandatory in 1992 in Australia with the passing of the Radio communications act. The standards used in this country are set by the Australian Communication and Media Authority (ACMA). The standards used are a combination of the European unionized standards, the IEC standards, CISPR standards, and Australia and New Zealand standards (AS/NZS). Certification for a product may be granted by the ACMA if it complies with all appropriate standards, at which point it must be marked with the Regulatory Compliance Mark (RCM) to show this compliance. The RCM will be adopted in March 2016, replacing the A-tick and C-tick marks that were applicable to different subsets of equipment so as to create a standardized mark for all compliant equipment.

EMC compliance is governed by Radio Spectrum Management in New Zealand. The process of product certification in this country is very similar to that of Australia, including the use of a combination of EN standards, the IEC standards, CISPR standards, and AS/NZS standards and the use of the RCM to signify compliance. Unlike in Australia, however, a second type of certification is used, which uses its own standards and compliance mark. This second level is not harmonized with Australia like the other, and as a result, products with this certification can only be sold in New Zealand. The mark used for these products is the R-NZ label.

9.4.10 SOUTH KOREA

The EMC compliance standards in South Korea are set and controlled by the Korean Communication Commission. In the aim of speeding the process by which European products can be certified for sale and use in South Korea, the EU-Korea Free Trade Agreement was established in 2011. This agreement changed the EMC standards used in South Korea from country specific standards to those internationally accepted such as IEC standards. As a result of this alignment of standards, manufacturers who already meet these international standards will no longer have to retest with a second set of standards for their products to be available in South Korea. The Korean authorities define two types of devices based on the amount of risk they pose to human health in reference to their EMC. These are low-risk devices, such as industrial equipment, and medium-/high-risk devices, such as computers. A product has to undergo different steps for certification depending on the amount of risk the device poses. A low-risk device can be tested in-house, where as a high-risk device must be tested by a third party. If a product is being tested by a third-party laboratory whether it is tested in South Korea or Europe, it must be a notified laboratory according to the local authorities. Products found compliant can then receive the Korean KC mark to show this compliance.

9.4.11 EUROPEAN ECONOMIC UNION DIRECTIVE 2004/108/EC

The purpose of the directive is to allow free movement of goods between member states of the community and to create an acceptable electromagnetic environment in the Community. This is accomplished by overcoming technical barriers to trade imposed by differing EMC requirements and procedures in different member states.

The proposal from the Commission of the European Community was adopted by the Council of European Communities, which issued the Council Directive of May 3, 1989, on the approximation of the laws of member states relating to EMC. The EMC Directive 2004/108/EC repeals the previous EMC Directive 89/336/EEC and at April 20, 2016, will be replaced by 2014. The original protection requirements are not changed in practice and apply to apparatus and to fixed installations.

The Directive regulates the EMC of equipment. It aims to ensure the functioning of the internal market by requiring equipment to comply with an adequate level of EMC.

Here are the essential requirements.

The following are the general requirements:

Equipment shall be designed and manufactured having regard to the state of the art, as to ensure that

- The electromagnetic disturbance generated does not exceed the level above which radio and telecommunications equipment or other equipment cannot operate as intended
- It has a level of immunity to the electromagnetic disturbance to be expected in its intended use, which allows it to operate without unacceptable degradation of its intended use

The following are the specific requirements for fixed installations:

Installation and intended use of components. A fixed installation shall be installed applying good engineering practices and respecting the information on the intended use of its components with a view to meet the essential requirements as the aforementioned.

Equipment that is in conformity with harmonized standards or parts thereof the references of which have been published in the *Official Journal of the European Union* (OJEU) shall be presumed to be in conformity with the essential requirements covered by those standards or parts thereof.

Compliance of apparatus with the essential requirements shall be demonstrated by means of either of the following conformity assessment procedures:

1. Internal production control as set out in Annex II of the Directive, which states fundamentally the following:
   a. The manufacturer shall perform an EMC assessment of the apparatus, on the basis of the relevant phenomena, with a view to meet the essential requirements.
   b. The manufacturer shall establish the technical documentation. The documentation shall make it possible to assess the apparatus conformity to the relevant requirements and shall include an adequate analysis and assessment of the risk(s). See The EMC Directive for details of the technical documentation requirements.
2. EU type examination that is followed by conformity to type based on internal production control. Fundamentally, this EU type assessment is the part of a conformity assessment procedure in which a notified body examines the technical design of an apparatus and verifies and attests that the technical design of the apparatus meets the essential requirements.

Although the EMC Directive contains a complete list of definitions, the following definition may be of interest:

"Apparatus" means any finished appliance or combination thereof made available on the market as a single functional unit, intended for the end user and liable to generate electromagnetic disturbance, or the performance of which is liable to be affected by such disturbance.

Although the Directive applies to finished apparatus, certain components or subassemblies should, under certain conditions, be considered to be apparatus if they are made available to the end user.

Any "economic operator" (manufacturer, the authorized representative, the importer, and the distributor) that either places apparatus on the market under his own name or trade mark or modifies apparatus in such a way that compliance with this Directive may be affected should be considered to be the "manufacturer" and should assume the obligations of the manufacturer.

Conformity assessment should remain solely the obligation of the manufacturer, ideally the maker of the apparatus as he knows best the details of the equipment design.

When apparatus is available in several configurations, the assessment should cover the configuration, which is most likely to cause maximum disturbance and the configuration most susceptible to disturbance.

CE marking on compliant equipment is mandatory throughout the EC, except for fixed installations. From 1995, the so-called CE mark directive made it optional for most products to include the low-voltage directive on the declaration. From 1997, it became mandatory to do so. Council Directive 2006/95/EC is the low-voltage Council Directive.

The low-voltage directive applies to all electrical equipment designed for use with a voltage rating between 50 and 1000 V for alternating current (AC) and between 75 and 1500 V for direct voltage. The scope of the Directive includes both electrical equipment intended for incorporation into other equipment and equipment intended to be used directly without being incorporated.

These post-1987 directives are termed "New Approach Directives," which means primarily that they do not contain standards and restrict the technical content to the minimum. The OJEU contains the list of applicable EMC requirements. The number of different limits and requirements applicable to specific devices or that are generic is 104 at the moment.

If a specific type of equipment is not covered in one of the requirements, then one of the generic standards may be used.

The directive covers a vast range of apparatus, encompassing as broadly as possible all electrical appliances, systems, and installations whether or not they are connected to the mains (the public utility AC power supply). The directive includes devices connected to the electricity distribution network and the public telecommunications network. No upper or lower limit on the power output or selection of transmission frequencies is imposed. The directive directly covers several sectors of electrical and electronic engineering, in particular household appliances, consumer electronics, industrial manufacturing, information technology, radio communications, and telecommunications apparatus.

Classification as apparatus or fixed installation

The EMC Directive defines equipment as any apparatus or fixed installation. As there are separate provisions for apparatus and fixed installations, it is important that the correct category of equipment be identified.

The devices that are exempt from the EMC Directive are

- Electrical equipment for use in an explosive atmosphere
- Electrical equipment for radiology and medical purposes
- Electrical parts for goods and passenger lifts
- Electricity meters
- Plugs and socket outlets for domestic use
- Electric fence controllers
- Radioelectrical interference
- Specialized electrical equipment, for use on ships, aircraft, or railways, which complies with the safety provisions drawn up by international bodies in which the member states participate

The EMC Directive applies to all devices that contain electric or electronic elements, with the exception of devices whose EMC is regulated through other directives or the following:

Motor vehicles covered by specific Directive 72/245/EEC and 2004/104/EC
Active implantable medical devices by Directive 90/385/EEC
Medical devices by Directive 98/79/EC
Marine equipment if covered by Directive 96/79/EC
Agricultural and forestry tractors covered by Directive 75/322/EEC
Two- or three-wheel motor vehicles within the scope of Directive 97/24/EC

Equipment explicitly excluded from the EMC Directive are

Radio equipment and telecommunications terminal equipment covered by Directive 1995/5 EC (the R&TTE Directive)
Aeronautical products, parts, and appliances referred to in Regulation 1592/2002
Radio equipment used by radio amateurs as defined in International Telecommunication Union (ITU) Radio Regulations

The following are examples of equipment excluded for immunity purposes only from the EMC Directive:

Measuring instruments by Directive 2004/22/EC
Nonautomatic weighing instruments covered by Annex I-8(2) of Directive 90/384/EEC
Equipment, which is inherently benign in terms of EMC, are also excluded as defined in 1.1.4
    of Guide for the EMC Directive 2004/108/EC (February 8, 2010).

Equipment is considered inherently benign in terms of electromagnetic compatibility if the following applies:

Its inherent physical characteristics are such that it is incapable of generating or contributing
    to electromagnetic emissions that exceed a level, allowing radio and telecommunications
    equipment and other equipment to operate as intended, and it will operate without unac-
    ceptable degradation in the presence of the electromagnetic disturbance normally present
    in its intended environment.
Both conditions need to be met in order to classify equipment as inherently benign.

The application of the aforementioned enables the exclusion of the following equipment (exclusion from the application of the EMC Directive, provided that they include no active electronic part(s)):

Cables and cabling cables accessories, considered separately
Equipment containing only resistive loads without any automatic switching device, for example,
    simple domestic heaters with no controls, thermostat, or fan
Batteries and accumulators (without active electronic circuits)
Headphones and loudspeakers without amplification
Pocket lamps without active electronic circuits
Protection equipment that only produces transitory disturbances of short duration during
    the clearing of an S/C fault or an abnormal situation in a circuit and that do not include
    active electronic components, such as fuses and without active electronic parts or active
    components
High-voltage types of equipment in which possible sources of disturbances are only due to
    localized insulation stresses that may be the result of the aging process and are under the
    control of other technical measures included in non-EMC product standards and which do
    not include active electronic components

Here are the illustrative examples:

High-voltage inductors
High-voltage transformers

Other equipment fulfilling the aforementioned criteria are as follows:

Capacitors (e.g., power factor correction capacitors)
Induction motors
Quartz watches (without additional functions, e.g., radio receivers)
Filament lamps (bulbs)
Home and building switches that do not contain any active electronic components
Passive antennas used for TV and radio broadcast reception
Plugs, sockets terminal blocks, etc.

The R&TTE Directive covers most radio equipment and telecommunication terminal equipment and applies identical requirements as the EMC Directive. Any equipment not covered still remains subject to the provisions of the EMC Directive.

Aeronautical products are covered by Regulation (EC) No 1592/2002 and are at least equivalent to the EMC Directive.

Radio equipment used by radio amateurs is excluded unless the equipment is available commercially. Commercially available equipment, which is modified by or kits of components to be assembled by radio amateurs, is not regarded as commercially available.

Amateur radio equipment, which is commercially available, is covered by the R&TTE Directives.

2014/30/EU

A number of amendments are made to the Directive 2004/108/EC in the latest 2014/30/EU Directive:

The main difference between the old and new versions of the directive is that both (apparatus and fixed installations) needs to comply with the requirements of the EMC Directive when it is placed on the market and/or taken into service.

The new legal text makes a clear distinction between the requirements and assessment procedures for apparatus and for fixed installations, respectively (fixed installations can include networks and large machines).

Apparatus on which the CE mark is difficult to fix, such as small units, are exempt only from applying the mark.

Definitions are included for apparatus and fixed installations.

Fixed installations, although they must comply with the protection requirements, require neither an EC DOC nor a CE marking.

Mobile installations are considered apparatus.

For apparatus, there are changes in the documentation and information requirements.

The conformity assessment procedure for apparatus has been simplified to a single procedure. There is no compulsory involvement of a third party, but the manufacturer has the option of presenting his technical documentation to a notified body for assessment.

When deviating from the European harmonized standards or not applying them fully, the manufacturer has to perform an EMC assessment and provide detailed documentary evidence that the apparatus complies with the protection requirements of the EMC Directive.

Apparatus intended for a given fixed installation and not otherwise commercially available may be exempt from the requirements and procedures for apparatus (e.g., EC DOC, and CE marking), provided that certain documentation requirements are met, including precautions to be taken in order not to compromise the EMC characteristics of the fixed installation.

The regulatory role of competent bodies has been removed.

The directive contains requirements for conformity assessment bodies, their subsidiaries, and subcontractors to fulfill the same requirements as a notified body.

In applying the 2014/30/EU Directive, distributors shall be able to supply apparatus that has been placed on the market, namely, stock that is already in the distribution chain, before the date of application of national measures transposing the new Directive.

The standards have been and still are in the process of being developed independent of the directives. A system made up of a number of finished products is considered a final apparatus and must comply with the EMC directive. Components placed on the market for distribution and use in which the direct function is available without further adjustment or connections other than simple ones that can be performed by anyone are considered equivalent to "apparatus" as defined by the EMC directive. Other components that perform a direct function, such as plug-in cards, smart cards, and input/output modules designed for incorporation into computers, are apparatus commonly found in retail outlets and are available to the general public. Once cards of this type are inserted into a PC, they perform a direct function for the user. They must therefore be considered as apparatus and are, consequently, subject to the provisions of the EMC directive. This does not mean that they must necessarily be intrinsically compliant from the EMC point of view in all cases, if this is either impossible or impractical. However, in such cases, they must be designed in such a way that they become fully EMC compliant (emissions and immunity) when they are installed as intended in the apparatus, in any of its possible variants and configurations, without exceptions, and used in the electromagnetic environment determined by the manufacturer. A finished appliance is any device or unit that delivers a function and has its own enclosure. The instructions accompanying the component must include all relevant information and should assume that adjustments or connections can be performed by the end user not aware of the EMC implications. EMI mitigation components should be included when required. These EMI mitigating components may include a shielded cable or the use of a ferrite balun on interconnection cables. It is advisable to include these EMI components with the product if economically viable. If each of the parts of the system bears the CE mark, then the system does not, although the system manufacturer has to provide the system with an EU DOC and therefore should be satisfied that the complete system meets the requirements.

Components and subassemblies intended for incorporation by persons other than the end user into apparatus and/or a subsequent subassembly are not considered to be apparatus and are therefore not covered by the EMC Directive.

Manufacturers may gain a competitive edge by illegally adding a CE mark to equipment for which it is unnecessary. For example, the author has seen a number of cables and directional couplers, which are passive components, mounted to a metal plate to which a CE mark has been added.

A finished appliance is considered an apparatus if it is intended for the end user. When the finished appliance is intended exclusively for an industrial assembly operation for incorporation into other apparatus, it is not considered an apparatus in the sense of the EMC Directive.

The manufacturer needs to perform an EMC assessment of the apparatus in order to ensure that he meets the protection requirements. The manufacturer is fully responsible for applying the appropriate method of assessment, and it is never the responsibility of a third party or an EMC test laboratory. When the assessment determines that the equipment is inherently benign, the apparatus is excluded from the scope of the EMC Directive. The EMC assessment needs to take into account all normal intended operating conditions of the apparatus.

The EMC assessment can be accomplished by the following:

Application of EMC harmonized standards. Here, no third-party laboratory is required to perform testing because the manufacturer may perform these tests.

An EMC assessment where no harmonized standards have been applied and the manufacturer applies his or her own methodology. It is entirely up to the manufacturer to obtain sufficient confidence that the equipment would comply. He or she then makes a legal statement that says in effect "if my product was tested to these harmonized standards it would probably pass." With this route when the EU EMC DoC is signed, it is based on the data/evaluation contained in a technical documentation file.

The EMC Directive contains no legal requirements for performing any EMC laboratory tests.

Mixed assessment using harmonized standards for some requirements and detailed technical assessment for others.

Applying the harmonized standards by testing is equivalent to carrying out the detailed technical EMC assessment.

The technical assessment needs to follow a technical methodology to ensure that the requirements of the EMC Directive are met and the manufacturer will need to provide clear evidence of compliance and to document all steps taken and decisions made to check the conformity of the apparatus for those aspects for which the manufacturer has chosen the technical assessment method.

It used to be thought that if a manufacturer believed the apparatus would meet the EMC Directive that he or she could simply make a DoC, it is clear that a detailed technical assessment is required.

The technical assessment may include testing at some level, EMC design considerations, emission and immunity characteristics, clear specification of relevant sources, and effects of the electromagnetic phenomena covered and compatibility levels applied.

If the manufacturer does not have the services of an EMC competent engineer, then following the harmonized standards route may be much easier, simpler, and safer.

The detailed technical assessment may determine that an apparatus is in conformity. However, test results may be required to prove the validity of the declaration. A manufacturer may be taking a risk in making the declaration without a test, for in Chapter 11, it is demonstrated how difficult it is to lay out a compliant unshielded PCB that contains high-speed devices. And in Section 7.13, it was shown how very low RF current flow on cables can cause equipment to fail radiated emission requirements.

Despite the difficulty of meeting both emission and immunity requirements, and therefore the risk involved in not testing, some, typically smaller, companies who wish to save money have made the technical assessment and the declaration without a test. Some companies have put forward the argument that they have no competitors who make similar equipment in Europe and who can retest the equipment to ensure that it is compliant and therefore of low risk. The member states of the EC could have followed the approach taken by the FCC and implemented policing of the requirements by retesting products. Alternatively, they could have followed the approach taken by Industry Canada, which waits until a complaint of interference by a product occurs. For example, the Swedish enforcement program took three frequency converters off the market in 1997. The German authority verifies conformity declarations regarding their formal correctness and logical plausibility and performs sampling tests of products sold in Germany. During the first 6 months of 1999, 25,384 items were inspected; of these, 3582 products were subjected to EMC testing. Of the tested items, 23.7% (849) did not comply. So, it is obvious that the policing approach has been taken by the EU.

The standards to which the declaration is made must be ENs that have been published in the OJEU and that have been adopted by at least one of the member states. For example, some standards, such as the ENVs, are drafts, and the prENs are prestandards and have not been published in the OJEU as yet.

The directive requires that the DOC be signed by "someone empowered to bind the manufacturer or his authorized representative." Some companies use the signature of the EMC/compliance engineer, although it would appear that unless this person is a director, he may not be empowered to enter into legal commitments.

The question that is most often asked is, since this is a legal document with potentially legal and commercial penalties, who takes the responsibility? Initially, it was thought that the individual signing the document was not responsible but was merely signing on behalf of the company. Later, this was considered incorrect. The present position appears to be that the signatory takes the responsibility.

A good source of additional information is Guide for the EMC Directive 2004/108/EC (February 8, 2010).

The recommended documentation to accompany the equipment is as follows.

For equipment that meet Class A of the EN55022 requirements, the following warning must be included in the instructions for use:

Warning: This is a Class A product. In a domestic environment, this product may cause radio interference, in which case the user may be required to take adequate measures.

An additional certificate does not need to accompany the product, although most manufacturers include it and it is highly recommended, since this is the declaration that customs officials will review. A CE mark alone may result in an investigation. An *example* of the declaration is provided nearby. If the low-voltage directive does not apply, it is better to say so and to state why than to leave it out. The theory is that it is better to add the title of documents with an explanation than to leave out what may appear to be an applicable document. Make sure that the standards to which the declaration is made are the latest version.

Declaration of Conformity
    Application of Council Directive   **EMC Directive** 2004/108/EC
                                            Low-voltage Directive 73/23/EEC
    Standards to which conformity is Declared   EN55022 2010
                                              EN61000-6-1 2007
                                            EN 61000-4-4 2012

Manufacturer's name
Manufacturer's address
Importer's name   **(if available)**
Importer's address   **(if available)**
Type of equipment
Model no.
Serial no. (optional)   **Year of manufacture (optional)**

    I, the undersigned, hereby declare that the aforementioned equipment specified conform to the directives and standards.

    Signed on behalf of the manufacturer and the importer.

Place   **Signature**
Date   **Full name**
                                       Position

Example of DoC

The CE conformity marking that is to be attached to the equipment shall take the following form:

If the CE marking is reduced or enlarged, the proportions shown in the drawing must be respected. The various components of the CE marking must have substantially the same vertical dimension, which may not be less than 5 mm.

In the case of apparatus for which the manufacturer has not applied or has applied in part, the standards referred to in article 7(1) or failing the existence of such standards, the manufacturer or its authorized representative established within the community shall hold at the disposal of the relevant competent authorities, as soon as the apparatus is placed on the market, a technical assessment file. This file shall describe the apparatus, set out the procedures used to ensure conformity of the apparatus with the protection requirements referred to in article 4. It is not required to involve a third party in the file.

The technical assessment is applicable where no relevant standards are available or where the product is too large to be tested in a laboratory. The manufacturer may also want to declare for a product range based on testing one or more products in that range. In this case, a justification as to why these tests represent a coverage of the product range should be made. Another situation in which to use the technical assessment is where products are certified to existing EMC standards not recognized by the EU but that are possibly in compliance with some or all of the requirements of the directive or where harmonized standards have been applied in part only.

If testing has been done to nonharmonized standards, then the technical assessment may make a justification as to why these standards are relevant, how the apparatus is described, and what procedures are used to ensure conformity. The technical assessment will typically have the following contents: description of the apparatus, identification of the apparatus, technical description of the apparatus, technical rationale, detail of significant aspects, and test data.

Using the technical assessment approach without confirming EMI testing is novel and not without difficulty, for as we have seen throughout this book, a design may be adequate in one application but fail in another. An example is a power-line filter that may achieve adequate attenuation when used with one piece of equipment but may exhibit insertion gain when used with a second piece of equipment. Assuming the design approach is adopted, then the need for accurate EMC predictions and competent and experienced EMC engineers to interpret them will increase. Should many manufacturers opt to incorporate EMC into their designs and follow the technical assessment route, the market for a powerful and comprehensive EMC prediction computer program may justify the expense of developing such a tool. An EMC prediction program would aid in standardizing the evaluation of the design if the same program were used by manufacturers in all member states.

For testing to harmonized EN standards, nonaccredited third-party test laboratories can be used. Because only harmonized standards can be applied for the DOC route, it stands to reason that nonaccredited laboratories can be used for the DOC, as they routinely are. The Commission of the EU has indicated to interested parties in the United States that the EU will accept the test data from an independent laboratory outside of an EU country to support the DoC made by the manufacturer of the equipment. The test laboratory must, however, comply with the CISPR 16 testing and measurement techniques, which include the OATS attenuation calibration and compliance with the field uniformity requirements for radiated immunity testing in IEC 61000-4-3. The requirements for apparatus designed for the transmission of radio communications are that it shall be certified by an EC DoC. But in addition, an EC type of examination certificate concerning this apparatus (type approval) shall be provided by a competent body. In Germany, the competent body is the BZT.

Table 9.16 describes the more common normalized standards for EMC, their applicability, and the types of equipment.

Product standards are special limits for emission and immunity for a specific category of devices. They are never less stringent than the generic standards over which they always take precedence when they are available for the specific product family. Special product standards are spoken of when the EMC requirements for a product are embedded in another standard for the product.

Generic standards are applicable to a particular electromagnetic environment. They specify requirements and tests that are used by all products and systems placed in the specified environment, unless superseded by a specific product standard, which takes precedence. The two environments are (1) industry, residential, commercial, and light industry, and (2) industry.

Basic standards define and describe the EMC problem, measurement and test methods, principle measuring instruments, and test setup. They do not provide limits and are typically referenced by the generic and product standards.

CISPR 16-1-1 2014 First part of the new edition: Specification for radio disturbance and immunity measuring apparatus and methods (Part 1 Radio disturbance and immunity measuring apparatus). There is no EN planned to replace CISPR 16.

The EN 61000-x-x standard 1–6 sections are derived from the IEC 801-2-6 standards, which were in turn converted to the IEC 1000-4-x standards. These standards define test methods but do not impose test levels.

The IEC 61000-4-x standards are

IEC 61000-4-1 2006 Overview of IEC 61000-4 series
IEC 61000-4-2 2008 ESD
IEC 61000-4-3 2008 Radio frequency electromagnetic field immunity
IEC 61000-4-4 2012 EFT/Burst immunity
IEC 61000-4-5 2014 Surge immunity
IEC 61000-4-6 2013 Conducted disturbance immunity
IEC 61000-4-7 2009 Power line harmonic emissions

TABLE 9.16

Product and Generic Standards

EN 617 2001	A1 = 2010. Continuous handling equipment and systems. Safety and EMC requirements for the equipment for the storage of bulk materials in silos, bunkers, bins, and hoppers.
EN 618 2002	A1 = 2010. Continuous handling equipment and systems. Safety and EMC requirements for the equipment for the storage of bulk materials except fixed belt conveyors.
EN 619 2002	A1 = 2010. Continuous handling equipment and systems. Safety and EMC requirements for the equipment for the mechanical handling of unit loads.
EN 620 2002	A1 = 2010. Continuous handling equipment and systems. Safety and EMC requirements for fixed belt conveyors for bulk materials.
EN 1155 1997	Building Hardware. Electrically powered hold-open devices for swing doors. Requirements and test methods.
EN 12015 2014	Electromagnetic compatibility. Product family standard for lifts, escalators, and moving walks. Emissions.
EN 12016 2013	Electromagnetic compatibility. Product family standard for lifts, escalators, and moving walks. Immunity.
EN 12895 2000	Industrial trucks. Electromagnetic compatibility.
EN 13241-1 2003	A1 = 2011. Industrial, commercial, and garage doors and gates. Product standard. Part 1, products without fire resistance or smoke control characteristics.
EN 13309 2010	Construction machinery. Electromagnetic compatibility of machines with internal power supply.
EN 14010 2003	A1 = 2009. Safety of machinery. Equipment for power driven parking of motor vehicles. Safety and EMC requirements for design, manufacturing, erection, and commission stages.
EN ISO 14982 2009	Agricultural and forestry machinery. Electromagnetic compatibility. Test methods and acceptance criteria.
EN 16361 2013	Power operated pedestrian doors. Product standard, performance characteristics. Pedestrian door sets other than swing type, initially designed for installation with power operation without resistance to fire and smoke leakage characteristics.
EN 50065-1 2011	Signaling on low-voltage electrical installations in the frequency range 3–148.5 kHz. Part 1, Generic requirements, frequency bands, and electromagnetic disturbances.
EN 50065-2-1 2003	A1 = 2005. AC = 2003. Signaling on low-voltage electrical installations in the frequency range 3–148.5 kHz. Part 2-1, immunity requirements for main communication equipment and systems operating in the range of frequencies 95–148.5 kHz and intended for use in residential, commercial, and light industrial environments.
EN 50065-2-2 2003	A1 = 2005. AC = 2003. Signaling on low-voltage electrical installations in the frequency range 3–148.5 kHz. Part 2-2, immunity requirements for main communication equipment and systems operating in the range of frequencies 95–148.5 kHz and intended for use in industrial environments.
EN 50065-2-3 2003	A1 = 2005. AC = 2003. Signaling on low-voltage electrical installations in the frequency range 3–148.5 kHz. Part 2-3, immunity requirements for main communication equipment and systems operating in the range of frequencies 3–95 kHz and intended for use by electricity suppliers and distributors.
EN 50083-2 2012	Cable networks for television signals, sound signals, and interactive services. Part 2, electromagnetic compatibility for equipment.
EN 50121-1 2006	AC = 2008. Railway applications. Electromagnetic compatibility. Part 1, general.
EN 50121-2 2006	AC = 2008. Railway applications. Electromagnetic compatibility. Part 2, emissions of the whole railway system to the outside world.
EN 50121-3-1 2006	AC = 2008. Railway applications. Electromagnetic compatibility. Part 3-1, rolling stock. Train and complete vehicle.
EN 50121-3-2 2006	AC = 2008. Railway applications. Electromagnetic compatibility. Part 3-2, rolling stock. Apparatus.
EN 50121-4 2006	AC = 2008. Railway applications. Electromagnetic compatibility. Part 4, emission and immunity of the signaling and telecommunications apparatus.
EN 50121-5 2006	AC = 2008. Railway applications. Electromagnetic compatibility. Part 5, emission and immunity of fixed power supply installations and apparatus.

(Continued)

TABLE 9.16 (*Continued*)
Product and Generic Standards

EN 50130-4 2011	Alarm systems. Part 4, electromagnetic compatibility. Product family standard: Immunity requirements for components of fire, intruder, hold up, CCTV, access control, and social alarm systems.
EN 50148 1995	Electronic taximeters.
EN 50270 2006	Electromagnetic compatibility. Electrical apparatus for the detection and measurement of combustible gases, toxic gases, or oxygen.
EN 50293 2012	Road traffic signal systems. Electromagnetic compatibility.
EN 50295 1999	Low-voltage switchgear and control gear. Controller and device interference systems. Actuator sensor interference (AS-i).
EN 50370-1 2005	Electromagnetic compatibility (EMC). Product family standard for machine tools. Part 1, emission.
EN 50370-2 2003	Electromagnetic compatibility (EMC). Product family standard for machine tools. Part 2, immunity.
EN 50412-2-1 2005	AC = 2009. Power-line communication apparatus and systems used in low-voltage installations in the frequency range 1.6–30 MHz. Part 2-1, Residential, commercial, and industrial environment. Immunity requirements.
EN 50428 2005	A1 = 2007. A2 = 2009. Switches for the household and similar fixed electrical installations. Collateral standard. Switches and related accessories for use in Home and Building Electronic Systems (HBES).
EN 50470-1 2006	Electrical metering equipment (ac). Part 1, general requirements, tests and test conditions. Metering equipment (class indexes A, B, and C).
EN 50490 2008	Electrical installations for lighting and beaconing of aerodromes. Technical requirements for aeronautical ground lighting control and monitoring systems. Units for selective switching and monitoring of individual lamps.
EN 50491-5-1 2010	General requirements for Home and Building Electronic Systems (HBES) and Building Automation and Control Systems (BACS). Part 5-1, EMC requirements, conditions and test setup.
EN 50491-5-2 2010	General requirements for Home and Building Electronic Systems (HBES) and Building Automation and Control Systems (BACS). Part 5-2, EMC requirements for HBES/BACS used in industry environment.
EN 50491-5-3 2010	General requirements for Home and Building Electronic Systems (HBES) and Building Automation and Control Systems (BACS). Part 5-3, EMC requirements for HBES/BACS used in residential, commercial, and light industry environment.
EN 50498 2010	Electromagnetic compatibility (EMC). Product family standard for aftermarket electronic equipment in vehicles.
EN 50512 2009	Electrical installations for lighting and beaconing of aerodromes. Advanced Visual Docking Guidance Systems (A-VDGS).
EN 50529-1 2010	EMC Network Standard. Part 1, wire-line telecommunications networks using telephone wires.
EN 50529-2 2010	EMC Network Standard. Part 2, wire-line telecommunications networks using coaxial cables.
EN 50550 2011	A1 = 2014. AC = 2012. Power frequency over voltage protective device for household and similar applications (POP).
EN 50557 2011	Requirements for automatic reclosing devices (ARDs) for circuit breakers-RCBOs-RCCBs for household and similar uses.
EN 50561-1 2013	Power-line communication apparatus used in low-voltage installations. Radio disturbance characteristics. Limits and methods of measurement. Part 1, Apparatus for in-home use.
EN 55011 2009	A1 = 2010. Industrial, scientific, and medical equipment. RF disturbance characteristics. Limits and methods of measurement.
EN 55012 2007	A1 = 2009. Vehicles, boats, and internal combustion engines. Radio disturbance characteristics. Limits and methods of measurement for the protection of off-board receivers. EN 5012 is applicable for giving presumption of conformity under directive 2004/108/EC for those vehicles, boats, and internal combustion engine-driven devices that are not within the scope of directives 95/54/EC, 97/24/EC, 2000/2/EC, or 2004/104/EC.
EN 55013 2013	Sound and television broadcast receivers and associated equipment. Radio disturbance characteristics. Limits and methods of measurement.

(Continued)

TABLE 9.16 (*Continued*)
Product and Generic Standards

EN 55014-1 2006	A1 = 2008. A2 = 2011. Electromagnetic compatibility. Requirements for household appliances, electric tools, and similar apparatus. Part 1, emissions.
EN 55014-2 1997	A1 = 2001. A2 = 2008. AC = 1997. Electromagnetic compatibility. Requirements for household appliances, electric tools, and similar apparatus. Part 2, immunity. Product family standard.
EN 55015 2013	Limits and methods of measurement of radio disturbance characteristics of electrical lighting and similar equipment.
EN 55020 2007	A11 = 2011. Sound and television broadcast receivers and associated equipment. Immunity characteristics. Limits and methods of measurement.
EN 55022 2010	AC = 2011. Information technology equipment. Radio disturbance characteristics. Limits and methods of measurement.
EN 55024 2010	Information technology equipment. Immunity characteristics. Limits and methods of measurement.
EN 55032 2012	AC = 2013. Electromagnetic compatibility of multimedia equipment. Emission requirements.
EN 55103-1 2009	A1 = 2012. Electromagnetic compatibility. Product family standard for audio, video, audiovisual, and entertainment lighting control for professional use. Part 1, emissions.
EN 55103-2 2009	Electromagnetic compatibility. Product family audio, visual, audiovisual, and entertainment lighting control apparatus for professional use. Part 2, immunity.
EN 60034-1 2010	AC = 2010. Rotating electrical machines. Part 1, rating and performance.
EN 60204-31 2013	Safety of machinery. Electrical equipment of machines. Part 31, particular safety and EMC requirements for sewing machines, units, and systems.
EN 60255-26 2013	AC = 2013. Measuring relays and protection equipment. Part 26, electromagnetic compatibility requirements.
EN 60439-1 1999	Low-voltage switchgear and control gear assemblies. Part 1, Type-tested and partially type-tested assemblies.
EN 60669-2-1 2004	A1 = 2008. A12 = 2010. AC = 2007. Switches for household and similar fixed electrical installations. Part 2-1, particular requirements. Electronic switches.
EN 60730-1 2011	Automatic electrical controls for household and similar use. Part 1, general requirements.
EN 60730-2-5 2002	A1 = 2004. A2 = 2008. A11 = 2005. Automatic electrical controls for household and similar use. Part 2-5, particular requirements for automatic electrical burner control systems.
EN 60730-2-6 2008	Automatic electrical controls for household and similar use. Part 2-6, particular requirements for automatic electrical pressure sensing controls including mechanical requirements.
EN 60703-2-7 2010	AC = 2011. Automatic electrical controls for household and similar use. Part 2-7, particular requirements for timers and time switches.
EN 60730-2-8 2002	A1 = 2002. Automatic electrical controls for household and similar use, Part 2-8, particular requirements for electrically operated water valves, including mechanical requirements.
EN 60730-2-9 2010	Automatic electrical controls for household and similar use, Part 2-9, particular requirements for temperature sensing controls.
EN 60730-2-14 1997	A1 = 2001. Automatic electrical controls for household and similar use, Part 2-14, particular requirements for electric actuators.
EN 60730-2-15 2010	Automatic electrical controls for household and similar use, Part 2-15, particular requirements for automatic electrical airflow, water flow, and water-level sensing controls.
EN 60870-2-1 1996	Telecontrol equipment and systems. Part 2, operating conditions. Section 1, power supply and electromagnetic compatibility.
EN 60945 2002	Maritime navigation and radio communication equipment and systems. General requirements. Methods of testing and required test results.
EN 60947-1 2007	A1 = 2010. Low-voltage switchgear and control gear. Part 1, general rules.
EN 60947-2 2006	A1 = 2009. A2 = 2013. Low-voltage switchgear and control gear. Part 2, circuit breakers.
IEC 61000-6-1	Generic Standards Immunity for residential, commercial, and light industrial environments
IEC 61000-6-2	Generic Standards Immunity for industrial environments
IEC 60601-1-2	Immunity for Medical Electrical Equipment
IEC 61000-6-3	Generic standards Emission standard for residential, commercial, and light industrial environments
IEC 61000-6-4	Generic standards Emission standard for industrial environments

IEC 61000-4-8 2008 Power frequency magnetic field immunity

IEC 61000-5-9 2001 Pulse magnetic field immunity

IEC 61000-4-10 2001 Damped oscillatory magnetic field immunity

IEC 61000-4-11 2004 Voltage dips and interruptions immunity

IEC 61000-4-12 2006 Oscillatory waves immunity

IEC 61000-4-13 2009 Harmonics and inter-harmonics emissions

IEC 61000-4-14 2009 Voltage fluctuations immunity <16A per phase

IEC 61000-4-16 2002 C/M disturbance 0 Hz to 150 kHz

IEC 61000-4-17 2009 Ripple on DC power immunity

IEC 61000-4-18 Damped oscillatory wave test techniques

IEC 61000-4-19 Immunity to conducted differential mode disturbance 2 kHz – 150 kHz on AC power port

IEC 61000-4-20 2010 Testing and measurement techniques emission and immunity testing in TEM waveguide

IEC 61000-4-21 2011 Testing and measurement techniques reverberation chamber

IEC 61000-4-22 2010 Radiated emission and immunity testing in fully anechoic chamber

IEC 61000-4-23 2000 Test method for protective devices for HEMP and other radiated disturbance

EN 50147 1997	Anechoic chambers
	Part 1: Shield attenuation measurement
	Part 2: Alternative test site suitability with respect to site attenuation
	Part 3: Shielded anechoic enclosure performance for immunity tests of radiated radio frequency electromagnetic fields
EN 60601-1-2 2	2014 Medical electrical apparatus
	Part 1: General safety requirements, 2nd supplement EMC
EN 61000-2-2 2	2002 Compatibility levels for low-frequency conducted disturbances and signaling in public low-voltage power-supply systems

The European Telecommunications Standards Institute has published a range of product standards and drafts standards that cover radio equipment and systems. Details on these and updates to the preceding standards can be obtained from the web at www.etsi.org/standards.

9.4.11.1 Radiated and Conducted Emissions

The EN radiated and conducted emission limits are very close to the FCC limits. Table 9.17 lists the common EN- and FCC-radiated emissions at the specified distances; Table 9.18 shows the limits converted to 3 m using a far-field inverse linear distance extrapolation factor. NB: The FCC and EN55022 allow these extrapolated 3 m limits to be used for measurements at 3 m, whereas EN55011 requires that the 10 m limits also be applied to measurements at 3 m unless the EUT is considered small in size in which case a measurement at 3 m is allowed. Thus, Table 9.18 is useful for a direct comparison of the different requirements and for comparison of measurements at 3 m to the FCC and EN55022 limits. NB: The FCC also requires measurements above 1000 MHz, under certain conditions in accordance with 47 CFR 15.33, and for these an average detector function is used.

Table 9.19 compares the conducted emission of the EN and FCC requirements above 0.15 MHz, and Figure 9.31 plots the conducted emission limits from 0.15 to 0.5 MHz for the EN 55011 group 1, Class B, the EN 55022 Class B, and the EN 61000-6-3 requirements.

Generic emission requirements apply to electrical and electronic apparatus for which no dedicated product or product family emission standard exists. The meaning of Classes A and B equipment varies from one document to another. For example, in the generic emission requirement EN 61000-6-3, the definition of domestic, commercial, and light industrial locations is locations that are characterized by being supplied directly at low voltage from the public mains.

TABLE 9.17

EN- and FCC-Radiated Emission Limits as Specified

f (MHz)	EN55011 Group 1, Class A EN55022 Class A EN50081-2 Specified Limit at 30 m (dBμV/m)	EN55011 Group 1, Class B EN55022 Class B EN50081-1 Specified Limit at 10 m (dBμV/m)	FCC Class A Specified Limit at 10 m (dBμV/m)	FCC Class B Specified Limit at 3 m (dBμV/m)
30–88	30	30	39	40
88–216	30	30	43.5	43.5
216–230	30	30	46.5	46
230–1000	37	37	46.5–960 MHz 49.5 > 960 MHz	46–960 MHz 54 > 960 MHz

EN	Europaeische Norm (European Standard)
EN55011	Industrial Scientific Medical:
	Group 1, Class A—nondomestic or not connected to low-voltage supply
	Group 1, Class B—domestic or connected to low-voltage supply
	(No increase in the 10 m level allowed for measurements at 3 m)
EN55022	Information technology equipment:
	Class A—commercial
	Class B—domestic
EN50081-1	Generic standard for domestic, commercial, or light industrial
EN50081-2	Generic standard for the industrial environment
FCC	Part 15, Subpart J:
	Class A computing device—commercial
	Class B computing device—residential
	Class A—commercial
	Class B—domestic
EN50081-1	Generic standard for domestic, commercial, or light industrial
FCC	Part 15, Subpart J:
	Class A computing device—commercial
	Class B computing device—residential

Source: Courtesy of EMC Consulting, Inc., Merrickville, Ontario, Canada.

IEC 61000-6-4 covers generic emissions for equipment operating in industrial environments, which are defined as where ISM equipment is present; where heavy inductive or capacitive loads are frequently switched; and where currents and associated magnetic fields are high. EN 55011 covers all ISM equipment and separates them into groups and classes. Group 1 ISM equipment contains all equipment in which there is intentionally generated and/or used conductively coupled *RF* energy that is necessary for the internal functioning of the equipment. Group 1 covers digitally controlled equipment in which the conductively coupled *RF* is a by-product of the digital signals and clocks.

Group 2 ISM equipment includes all equipment in which *RF* energy is intentionally generated and/or used in the form of electromagnet radiation for the treatment of material and spark erosion. The equipment is then divided into Class A equipment that is suitable for use in all establishments *other* than domestic and that directly connected to a low-voltage power supply network that supplies buildings used for domestic purposes. Class B equipment is suitable for use in domestic establishments and in establishments directly connected to a low-voltage power supply network that supplies buildings used for domestic purposes. Unlike the preceding EN requirements, EN 55022, which covers ITE equipment, classifies such equipment as Class B when intended primarily for use in the

TABLE 9.18

EN- and FCC-Radiated Emission Limits Converted to 3 m

f (MHz)	EN55022 Class A — EN50 81-2 at 3 m — Corrected Limit (30–3 m) (dBμV/m)	EN55022 Class B — EN50 81-1 at 3 m — Corrected Limit (10–3 m) (dBμV/m)	FCC Class A at 3 m — Corrected Limit (30–3 m) (dBμV/m)	FCC Class B at 3 m — Specified Limit (No Correction) (dBμV/m)
30–88	50	40	49.5	40
88–216	50	40	54	43.5
216–230	50	40	57	46
230–1000	57	47	57–960 MHz	46–960 MHz
			67.5 > 960 MHz	54 > 960 MHz

EN	Europaeische Norm (European Standard)
EN55011	Industrial Scientific Medical:
	Group 1, Class A—nondomestic or not connected to low-voltage supply
	Group 1, Class B—domestic or connected to low-voltage supply
	(EN 55011 does not allow a reduction in emission levels for measurement at 3 m and so the limits at 30 and 10 m also apply to 3 m)
EN55022	Information technology equipment:
	Class A—commercial
	Class B—domestic
IEC61000-6-1	Generic standard for domestic, commercial, or light industrial
IEC61000-6-2	Generic standard for the industrial environment
FCC	Part 15, Subpart J:
	Class A computing device—commercial
	Class B computing device—residential

Source: Courtesy of EMC Consulting Inc., Merrickville, Ontario, Canada.

domestic environment, which in turn is defined as an environment where the use of broadcast radio and television receivers may be expected within a distance of 10 m of the apparatus concerned. Class A equipment is defined as a category of all other ITE that satisfies the Class A ITE limits but not the Class B limits. Such equipment should not be restricted in its sale, but the Class A warning, described previously, should be included with the equipment.

9.4.11.2 Immunity Tests

Unlike the majority of commercial requirements, the EN requirements include immunity tests. The generic immunity requirement for residential, commercial, and light industry environment, EN 61000-6-2 specifies the following tests:

Radiated immunity: Radiated *RF* energy from 80 to 1000 MHz at 3 V/m unmodulated, with the test performed with a 1 kHz 80% AM using the test setup described in EN 6100-4-3. This is applicable to enclosure port.

Electrostatic discharge: Air discharge 8 kV and contact discharge of 4 kV with test setup IEC EN 61000-4-2, also for enclosure port.

Electrical fast transient (EFT): The requirements for ports for signal lines and control lines and functional earth ports is 0.5 kV, rise time 5 nS, and pulse width 50 nS at a repetition rate of 5 or 100 kHz with the test setup described in IEC 61000-4-4, capacitive clamp. The EFT pulse is an injected common mode and for signal and control lines is only applicable to ports interfacing with cables whose total length according to the manufacturer's

TABLE 9.19

EN and FCC Conducted Emission Limits

f (MHz)	E55011 Group 1, Class A EN55022 Class A IEC/EN6100-6-2 FCC Class A		EN55011 Group 1, Class B EN55022 Class B IEC/EN61000-6-1 FCC Class B	
	Quasi-Peak (dBµV)	Average (dBµV)	Quasi-Peak (dBµV)	Average (dBµV)
0.150–0.450	79	66	See Figure 9.31	See Figure 9.31
0.450–0.500	79	66	See Figure 9.31	See Figure 9.31
0.500–1.600	73	60	56	46
1.600–5	73	60	56	46
5–30	73	60	60	50

EN55011	Industrial Scientific Medical:
	Group 1, Class A—nondomestic or not connected to low-voltage supply Group 1,
	Class B—domestic or connected to low-voltage supply
EN55022	Information Technology Equipment:
	Class A—commercial
	Class B—domestic
EN61000-6-1	Generic standard for domestic, commercial, or light industrial
EN61000-6-2	Generic standard for the industrial environment
FCC	Part 15, Subpart J:
	Class A computing device—commercial
	Class B computing device—residential

Source: Courtesy of EMC Consulting Inc., Merrickville, Ontario, Canada.

functional specification may exceed 3 m. The same EFT pulse is specified for DC power ports with no cable length restrictions; however, the requirement is not applicable to input ports intended for connection to a dedicated battery or a rechargeable battery that must be removed or disconnected from the apparatus for recharging. Apparatus with a DC power input port intended for use with an AC–DC power adapter shall be tested on the AC power input of the AC–DC power adapter specified by the manufacturer or, where none is so specified, using a typical AC–DC power adapter. The test is applicable to DC power input ports intended to be connected permanently to cables longer than 10 m. The EFT pulse for input and output AC power ports is 1 kV, 5 nS rise time, 50 nS pulse width at a repetition rate of 5 or 100 kHz.

Surges: The surge test is applicable to AC and DC power and in some cases to signals. The generator shall have a 1.2 µs rise time and a 50 µs pulse width at 50% amplitude into an open circuit and an 8 µs rise time and 20 µs pulse width into a S/C. The source impedance of the generator is 2 Ω. The charge (O/C) voltage for DC power is 0.5 kV line to ground and line to line. For AC power, the charge voltage is 2 kV line to earth and 1 kV line to line. The line-to-line pulse is injected via an 18 µF capacitor and line-to-ground pulse via a 9 µF capacitor in series with a 10 Ω resistor. The test procedure, including injection methods, is contained in EN 61000-4-5.

RF common mode: Applicable to ports for signal lines and control lines as well as DC and AC power ports and functional earth ports. The RF is an injected common mode at a level of 3 V rms unmodulated, but during the test, this level is modulated at 1 kHz 80% AM. The test methods, including injection methods and calibration method, are included in EN 61000-4-6.

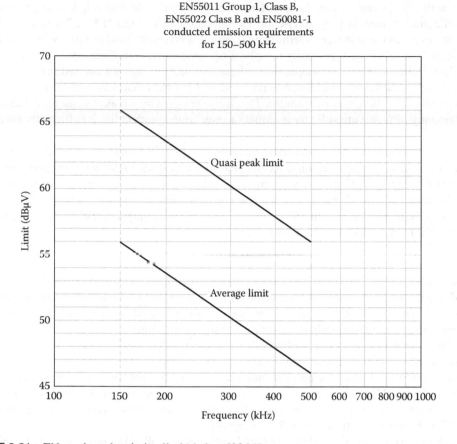

EN55011 Group 1, Class B,
EN55022 Class B and EN50081-1
conducted emission requirements
for 150–500 kHz

FIGURE 9.31 EN conducted emission limits below 500 kHz.

Voltage dips and voltage interruptions: These are applicable to AC power only and include a voltage shift at zero crossing with the test method contained in IEC 61000-4-11. The EUT is tested at 30%, 60%, and >95% below the rated voltage, that is, at 70%, 40%, and <5% of the rated voltage. The duration of the dips is 10, 100, and 5000 mS, respectively. Five dips are performed for each test level at a rate of one dip per minute. The 100% voltage level used is the rated voltage of the equipment. Where a voltage range is specified, the tests shall be performed at the minimum and maximum levels specified. The period is typically from 0.5 to 300 and a 5 second duration.

The voltage dips described in EN 55024 for IT equipment are >95% reduction over a period of 0.5 and a 30% reduction over 25 periods. The voltage interruption is a >95% reduction over 250 periods.

Apparatus shall not become dangerous or unsafe as a result of the application of the tests defined in IEC 61000-6-2. A functional description and a definition of performance criteria, during or as a consequence of the EMC testing, shall be provided by the manufacturer and noted in the test report, based on criteria described in EN 61000-6-1. The criteria are that basically (1) the apparatus shall continue to operate as intended during the test, (2) the apparatus shall continue to operate as intended after the test, and (3) temporary loss of function is allowed, provided the function is self-recoverable or can be restored by the operation of controls. Which criterion applies is typically decided by the criticality of the apparatus. The tests shall be made in the most susceptible operating mode and shall be varied to achieve maximum susceptibility. If the apparatus is part of a system or can be connected to auxiliary apparatus, then the apparatus shall be tested while connected to the minimum configuration necessary to exercise the ports in accordance with EN 55022. EN 61000-6-1 also describes the selection of terminations, inappropriate tests, and immunity test requirements. The rule is that previously compliant equipment that is modified must remain compliant. If the modification changes the configuration and this was not covered in the original declaration, then a new declaration is required, and any tests that have been added since the original declaration must be included. This is one reason why it makes sense to include any immunity tests that are still "for information only" if the product life is sufficiently long.

The generic requirements for industrial environment, IEC/EN 61000-6-2, specify for enclosure ports: the RF field at 10 V/m when unmodulated, but with 80% 1 kHz AM, 80–1000 MHz frequency range, at 900 MHz ± 5 MHz, a 10 V/m unmodulated, but modulated during the test with a 50% duty cycle pulse and 200 Hz repetition rate, the power frequency magnetic field at 30 A/m 50 Hz (applicable only to apparatus containing devices susceptible to magnetic fields), and electrostatic discharge (ESD) with 4 kV contact and 8 kV air discharge.

The EN 55024 requirements for ITE-radiated immunity test level is 3 V/m when unmodulated at 80% AM from 80 to 1000 MHz.

For signal lines and data buses not involved in process control, the IEC 61000-6-2 requirements for conducted disturbance are C/M injected RF, 0.15–80 MHz at 10 V rms, when unmodulated but modulated at 80% AM (1 kHz) with a source impedance of 150 Ω; the EFT requirement is 1 kV peak, and both are applicable only to cables exceeding 3 m in length.

The EN55024 for conducted disturbance is C/M injected 0.15–80 MHz at 3 V when unmodulated but modulated at 80% 1 kHz amplitude.

In IEC 61000-6-2 for ports for process, measurement, and control lines and long bus and control lines, the 10 V rms RF C/M injection applies, as does the EFT, but at a peak voltage of 2 kV.

For DC input and output power ports, as well as AC input and output power ports, the same 10 V rms RF C/M injection and the 2 kV EFT pulse applies.

The informative annex in IEC/EN 61000-6-2 indicates the following tests that may be included in the standard: power frequency C/M injection of 10 V rms at 50 Hz for signal lines and data buses not involved in process control and 20 V rms for process, measurement, and control lines; 2 kV C/M and 1 kV D/M 1.2/50 µs for process, measurement, and control lines, and 0.5 kV D/M and 0.5 kV C/M for DC power and 4 kV C/M and 2 kV D/M for AC power; voltage deviation and voltage variation for DC power and voltage dips, voltage interruptions, voltage fluctuation, and low-frequency harmonics for AC power.

The modulated and unmodulated test levels are defined in ENV 50140, in which if the unmodulated specified test level is 1 V/m rms or 2.8 V p-p, then when AM at 80%, the rms value 1.12 V and the peak-to-peak value is 5.1 V. This increase in level with modulation is important, for many signal generators provide the same output level modulated or unmodulated, and the capability of the power amplifier must include the peak-to-peak levels. It is also important to find out if any E-field-monitoring device is used, whether it measures peak or rms values, and also what the displayed value is.

The emission and immunity tests described in the harmonized standards refer to test methods, test setups, and test location calibration criteria. For radiated emission measurements, the test area must typically be a calibrated OATS, described in ANSI C63.4 2014 and in Appendix A of EN 55022 or a calibrated semianechoic chamber.

9.4.11.3 Emission Tests

In EN 55011, Class A ISM equipment may be measured either on a test site or in situ as determined by the manufacturer. Class B ISM equipment shall be measured on a test site. The test site shall meet an ambient requirement of at least 6 dB below the limit, which is almost impossible for an OATS. The test site for measurements from 9 kHz to 1 GHz shall have a ground plane

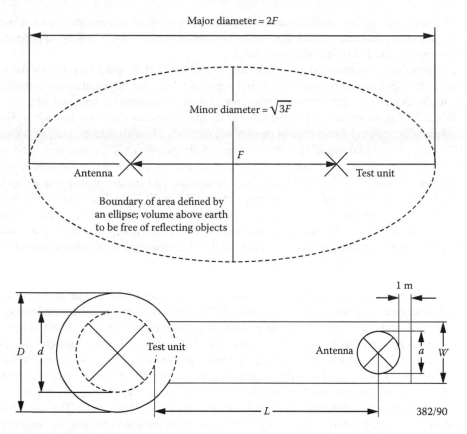

FIGURE 9.32 Characteristics of radiated emissions test site and minimum ground plane for radiated emissions measurement. $D = (d + 2)$ m, where d is the maximum test unit dimension; $W = (a + 1)$ m, where a is the maximum antenna dimension; $L = 10$ m. (Courtesy of CENELEC, Brussels, Belgium.)

for both conducted and radiated measurements with the minimum size and shape as depicted in Figure 9.32. For site validation, CISPR 16-1-4 in 2007 describes a SVSWR test. At the same time, the ANSI C63.4 2005 had developed a draft proposal for a time-domain reflectivity measurement. Equipment shall, if possible, be placed on a turntable and rotated. If a turntable is not possible, the antenna shall be positioned at various points in azimuth for both vertical and horizontal polarizations.

The configuration of the equipment shall be varied to maximize emissions. Interconnecting cables shall be of the type and length specified in the individual equipment requirements. If the length can be varied, the lengths shall be selected to produce maximum radiated emissions. However, it makes no sense to orient cables in a configuration that will never be used in the implementation of the system. For conducted measurements, excess lengths of cable shall be bundled at the approximate center of the cable, with bundles of 30–40 cm in length. Where multiple interface ports are all of the same type, connecting a cable to just one of those ports is sufficient, providing it can be shown that the additional cables would not significantly affect the results. This is difficult to achieve, for the C/M noise on two or more cables can drive against each other, just like the rods of a dipole, and the fields generated by the cables can add or subtract at the location of the antenna. A description of the cable and equipment orientation must be provided so that results can be repeated. If the equipment can perform separately any one of a number of functions, then the equipment shall be tested while performing each of these functions.

Fundamentally, EN 55011 requires that worst-case, maximized emissions be performed. EN 55011 also describes the measuring equipment requirements, antennas, connection to the electricity supply on a test site, "in situ" measurements disposition of medical equipment, types of medical, industrial, scientific, laboratory, and measuring equipment, microwave cooking appliances, and other equipment in the 1–18 GHz frequency band.

For conducted measurements, the 50 Ω/50 μH V-network (LISN) shall be used, or the voltage probe when the LISN cannot be used. EN 55011 supplies additional information on connection of the EUT to the electricity supply network when performing measurements on a test site.

EN 55022:2010 is the emission document for ITE, and it provides the same basic guidelines for EUT configuration, general measurement conditions, methods of measurement, etc., as provided in EN 55011, with some additional EUT configuration details specific to ITE equipment. The test site information is more specific, in as much as a site attenuation measurement requirement is imposed. In general, the site shall be validated by making site attenuation measurements for both horizontal and vertical polarization fields in the frequency range of 30–1000 MHz. The distance between the transmitting antenna and receiving antennas shall be the same as the distance used for radiated emission tests of the EUT. A measurement site shall be considered acceptable if the horizontal and vertical site attenuation measurements are within ±4 dB of the theoretical site attenuation of an ideal site (see also CISPR 16). The problems inherent in achieving the site attenuation requirement are discussed in Section 9.4.3. The test site shall characteristically be flat, free of overhead wires and nearby reflecting structures, and sufficiently large to permit antenna placement at the specified distance and provide adequate separation between antenna, EUT, and reflecting structures. Reflecting structures are defined as those whose construction material is primarily conductive. The test site shown in Figure 9.32 is presented in both EN 55022 and EN 55011, with a minimum alternative test site, shown in Figure 9.33, presented in EN 55022. A conducting ground plane shall extend at least 1 m beyond the periphery of the EUT and the largest measuring antenna and shall cover the entire area between the EUT and the antenna. It should be of metal, with no holes or gaps having dimensions larger than one-tenth of a wavelength at the highest frequency of measurement. The minimum-size ground plane is shown in Figure 9.32. A larger-size conducting ground plane may be required if the site attenuation requirements are not met. As an alternative, tests may be conducted on other test sites that do *not* have the physical characteristics shown in Figure 9.32 or 9.33. Evidence shall be obtained to show that such alternative sites will yield valid results. Annex A describes a method of testing an alternative site whereby a measurement antenna is moved within a volume in both horizontal and vertical polarizations. A second antenna is scanned in height from 1 to 4 m. This alternative site test measurement method could be performed in a semianechoic chamber designed for at least a 3 m test

FIGURE 9.33 Alternative minimum radiated emission test site (EN 55022:1994). *Note:* There shall be no reflecting object inside the volume defined on the ground by the line corresponding to this figure and defined in height by a horizontal plane ≥3 m above the height element of either aerial or equipment under test. (Courtesy of CENELEC, Brussels, Belgium.)

area but would preclude the use of the GTEM cell. However, if the GTEM, or any other test fixture, can be proven to yield valid test results, then its use may be acceptable.

EN 55022 states that tests can be made on one appliance only or on a sample of appliances using a statistical approach, and the requirement then is that 80% of the mass-produced appliances must comply with the limits with at least 80% confidence. EN 55022 states that subsequent tests are necessary from time to time on appliances taken at random from production, especially when only one appliance has been tested. However, although this retest may be highly recommended, it is well to remember that the widely held view is that testing is not mandatory to make a DoC for an appliance.

9.4.11.2.1 Immunity Testing

One of the most common errors when performing immunity tests is the use of an inadequate method of testing any or all the functions of the EUT and an effective method of flagging a failure during or after the test. Digital, analog, video, or RF should flow on all signal interfaces during the test. Digital data should be monitored either by a typical interface device that can flag an error in the data or by a computer that can perform error checking. An alternative to an external device is to loop back the data to the EUT at the end of the signal cable. However, an effective means of flagging the data on the EUT by an audible or visual warning is required. Video data should be displayed on an external monitor, which, if part of the EUT, must remain inside the test setup. If the monitor is used merely to test the video signal, then it may be located outside of the test area. However, adding long cables between a computer and a CRT monitor in order to locate the monitor away from the test area will invariably result in EMI during a radiated immunity or C/M RF conducted immunity test. This typically takes the form of distortion of the image, bars in the image, or corrupted data leading to a corrupted display. In this case, it is probably more realistic to have the monitor located close to the computer, and this is less likely to result in EMI. If a display must be viewed inside a semianechoic chamber during a radiated immunity test, then a door with a wire mesh screen may be used. More convenient is the use of a video camera located inside the room, with an external monitor. In some cases, where a large-diameter "waveguide below cutoff" port is installed in the chamber, a set of mirrors has been used to view the display outside of the chamber.

The EMI induced into a flat screen monitor is typically but not always due to current induced into the signal interface.

If a bar code reader or similar device is connected to the EUT, the bar code should be read and displayed or monitored by the EUT on a repetitive basis. This usually entails the writing of separate software for immunity testing, and this is often required to ensure that all the interface signals are exercised as well as disk drives, alarms, displays, wireless, etc., during the EMI test.

If a writing board or touch screen display requires contact on the surface to read information, this can easily be achieved by a weighted wooden structure containing a suitable pen.

It is not uncommon to be asked to perform immunity tests where no indication of a pass/fail condition is available from the EUT; in this case, there is no sense in performing the test. Another potential problem is to perform an immunity test without a sanity check on the presence of the test level.

As discussed under the EFT test, a data acquisition system with a slow bandwidth or slow acquisition rate can modify waveforms, which appears to be due to the EUT but is in fact either the EFT test level or susceptibility of the data acquisition to the test level.

It is common for radiated and conducted RF immunity tests that the EUT pass/fail performance criterion is A, which means that

> the apparatus shall *continue to operate as intended*. No degradation of performance or loss of function is allowed below a performance level specified by the manufacturer, when the apparatus is used as intended. The performance level may be replaced by a permissible loss of performance. If the minimum performance level or permissible performance loss is not specified by the manufacturer, then either of these may be derived from the product description and documentation and what the user may reasonably expect, from the apparatus if used as intended.

Thus, in testing to the criterion A level, the functional/monitoring test equipment must also continue to function correctly during the EMI test.

For ESD, EFT, and surge tests on either AC/DC power or signal lines, the pass/fail criterion is B. Criterion B is described as

> The apparatus shall continue to operate as intended *after* the test. No degradation of performance or loss of function is allowed below a performance level specified by the manufacturer, when the apparatus is used as intended. The performance level may be replaced by a permissible loss of performance. During the test, degradation of performance, however, is allowed. No change of actual operating state or stored data are allowed. If the minimum performance level or permissible performance loss is not specified by the manufacturer, then either of these may be derived from the product description and documentation and what the user may reasonably expect from the apparatus if used as intended.

Thus, in testing to the criterion B level, the functional/monitoring test equipment must continue to monitor or store the data, which must remain uncorrupted during the test. As described later, this may be very difficult to achieve, and in some cases, a display may momentarily blank or the I/O card in the monitoring equipment may lose the data due to the impressed transient. In this case, as long as data and the correct monitoring of data resumes after the test, the equipment can be said to have passed at criterion B, which is the requirement, even though it is the monitoring device that is susceptible.

Before a radiated immunity test can be performed, the test facility field uniformity test described in IEC 6000-4-3 and ENV 50140 of a Uniform Field Area (UFA) must be performed. This calibration shows the capability of the test facility and the test equipment to generate the required field. At the same time, a database for setting the required field strength for the immunity test is obtained. If a usable test area of 1.5 m × 1.5 m is required, the UFA over this area and at a height of 0.8 m above the floor of the facility must be measured. The transmitting antenna must then have a beam width of at least 1.5 m × 1.5 m. If the EUT and wires are less than 1.5 m × 1.5 m, then the transmitting antenna beam width may also be lower. The UFA is subdivided into a grid spacing of 0.5 m. This results in 16 locations for a 1.5 m × 1.5 m area. The calibration is performed without the EUT present, and 12 of 16 points on a grid over the plane must meet the 0 to +6 dB of the nominal value for not less than 75% of all grid point (Figure 9.34).

For testing larger EUTs, it is required that the intensity of the field at a height of 0.4 m above the ground plane, as well as over the full width and height of the largest EUT, be measured and reported in the test report. Figure 9.35 shows the calibration of the field. In this case, the antenna may have to be moved to a number of positions to achieve a uniform field.

The calibration of the field is performed at 1.8 times the desired field strength, thus for testing at 10 V/m, the calibration is made at 18 V/m. The reason for the 18 V/m is to ensure that the RF power amplifier has the capability of generating the field when AM at 80%.

The harmonics generated by the power amplifier shall be such that any measured field strength shall be at least 6 dB below that of the fundamental.

Before testing, the intensity of the established field strength shall be checked by placing the field sensor at a calibration grid point, and with the field-generating antenna and cables in the same position as used for the calibration, the forward power needed to give the calibrated field strength can be measured. This shall be the same as recorded during the calibration. Spot checks shall be made at a number of calibration grid points over the frequency range 80–1000 MHz. Both polarities shall be checked.

After the calibration has been verified, the test field can be generated using the values obtained from the calibration.

The carrier test level is set so that the specified field strength in V/m is generated by the transmitting antenna. The carrier is then 80% modulated with a 1 kHz sine wave pausing to adjust the RF signal level or to switch oscillators and antennas as necessary. The rate of sweep shall not exceed 1.5×10^{-3} decades/s. Where the frequency range is swept incrementally, the step size shall not exceed 1% of fundamental with linear interpolation between points. Some software that automates

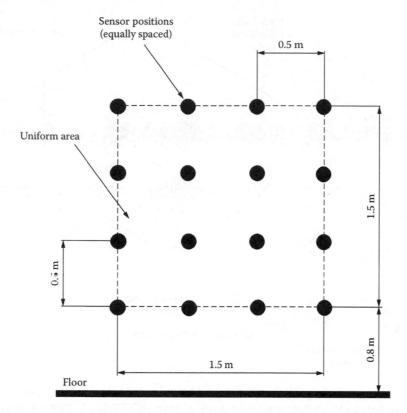

FIGURE 9.34 Calibration of immunity test field. (Courtesy of CENELEC, Brussels, Belgium.)

the sweep will switch from modulated to unmodulated to adjust the forward power to the antenna, and this may result in an extended time for the sweep. It may be faster to determine the modulated forward power level, which will result in the correct unmodulated power level and use this in the adjustment, especially as this eliminates transients in the CW as the switching takes place.

Although the specified sweep rate shall be used for qualification, when performing diagnostic/ fault finding tests, it usually results in excessive test times. Instead, the frequency should be swept over the range at which a failure occurs with the sweep rate increased/dwell time decreased until the failure is no longer detected and then the test performed with using this sweep rate increased by a safety factor of 2–10 or the dwell time decreased by a factor 2–10.

One contentious issue is the generation of transients in the CW carrier during testing. Some test laboratories abruptly switch the carrier on or abruptly change from unmodulated to 80% modulation, for example, when adjusting forward power. This abrupt change creates a transient in the field strength. Based on measurements, this transient, in switching from 0% to 80% modulation, may represent an increase in the E field of approximately 12%. If the EUT displays an error during this transient, should it be construed as a failure? The test procedure only specifies a modulated carrier and does not mention the moment of switching on the carrier, and so this transient appears to be excluded from the test requirements. The counter argument is that transients do occur when transmitters are switched on or off, and so the transients should be part of a valid test. When testing is performed for a specific customer for the EUT, then they would typically make the decision. When not, then the manufacturer does. For test facilities where the modulated carrier is increased gradually from zero without switching off the modulation, the issue does not arise, and it seems unfair that an EUT may fail at one test facility and pass at another.

In the MIL-STD-461F RS103 radiated susceptibility test, the carrier is pulse modulated, and so the transients are part of the test.

FIGURE 9.35 Example of semianechoic chamber suitable for radiated immunity tests. *Note:* Anechoic lining material on walls and ceiling has been omitted for clarity. (Courtesy of CENELEC, Brussels, Belgium.)

The cables shall be attached according to the manufacturers' installation instructions and shall replicate typical installations. If the specified cable length is greater than 3 m, or not specified, then the illuminated length shall be 1 m. This is typically interpreted as meaning the cable length after 1 m may be routed in a conduit or shielded. If a product committee determines excess cable length need to be decoupled (e.g., for cables leaving the test area), then the decoupling shall not impair the operation of the EUT. The EUT shall be in its housing with all covers and access panels in place, unless otherwise stated. If equipment is designed to be mounted in a panel, rack, or cabinet, it shall be tested in this configuration with the EUT cabinet supported on 0.1 m thick insulation.

A metallic ground plane is not required and any supports shall be a nonconducting material. Tabletop equipment is to be placed on a nonconducting 0.8 m high platform.

The dwell time at each frequency shall be not less than the time necessary for the EUT to be exercised and be able to respond but shall in no case be less than 0.5 s.

The horizontally and vertically polarized or isotropic field strength monitor shall be at least 1 m from the field-generating antenna with a 3 m distance preferred.

The manufacturers' specification may define effects on the EUT, which may be considered insignificant and therefore acceptable.

The following classifications may be used as a guide in formulating performance criteria by committees responsible for generic, product, and product family standards, or as a framework for the agreement on performance criteria between the manufacturer and the purchaser, for example, where no suitable generic product or product family exists:

1. Normal performance within the specification limits
2. Temporary loss of function or degradation of performance, which ceases after the disturbance ceases and from which the EUT recovers its normal performance, without operator intervention

3. Temporary degradation or loss of function or performance, which requires operator intervention or system reset
4. Degradation or loss of function, which is not recoverable due to damage of equipment (components) or software, or loss of data

The preferred test facility consists of an absorber-lined shielded enclosure, an example of which is shown in Figure 9.36, in which the anechoic lining material on walls and ceiling has been omitted for clarity. Both the type and construction of the test area and the type of transmitting antenna can lead to errors larger than 6 dB. For example, many anechoic chambers exhibit low-frequency resonances, which can result in a wide variation in E field over the test volume. Addition of absorber loads at strategic locations in the room or ferrite tiles at strategic locations on the walls or any other of the steps discussed in Section 9.51 (9) may be applied. If the transmitting antenna has too high a gain, then the field strength in the main lobe of the antenna can vary by more than 6 dB, and this will be seen in the UFA. This can be solved by moving the transmitting antenna to different locations.

One very important guideline, which should be applied, especially to MIL-STD or any type of radiated susceptibility tests in which there is a ground plane on the table, but which seldom is, is that the test field be calibrated on the empty test area before the test. The test is then performed with the same output power to the field-generating antenna as during the calibration. Control of the field strength during the test is of no value because the field will be distorted and destructive and constructive interference from the EUT, and ground plane when present, and the field may not be correctly measured. Instead, the input power to the antenna should be monitored.

The test sample shall not be placed closer than 1 m to the field-generating antenna, although a test distance of 3 m is preferred.

Annex A of ENV 50140, which is for information, discusses field strength to be expected at the biconical antenna from 20 to 300 MHz and the log-periodic from 80 to 1000 MHz. Circularly polarized antennas may be used only if the output power from the power amplifier is increased by 3 dB. It is recommended that in semianechoic chambers (those with no absorber on the floor), that additional absorber be placed on the floor in the illumination path from antenna to EUT. Striplines, TEM, and (although it is not mentioned) the GTEM cells may be used only if the field homogeneity requirements are met and if the EUT and cable can be arranged as required by the standard. Additionally, wires/cables cannot exceed one-third of the dimension between septum and outer

FIGURE 9.36   C/M injection direct test method for screened (shielded) cables.

conductor. Thus, the TEM, GTEM, and stripline are typically useful for small EUTs. A screened room with limited absorbing material may be used if the field homogeneity requirement and all other requirements are met or an open antenna range providing legal limits on radiating E fields is met and absorbing material is placed on the floor.

9.4.11.2.2   Power-Line Harmonics

The maximum harmonic emissions on AC power lines, single and three phase, drawing up to 16A for all types of equipment are specified in IEC 61000-3-2. There are four classes of equipment A, B, C, and D:

1. Class A is the default classification unless one of the other classes supersedes it.
2. Class B is applicable to portable electrical equipment, including power tools.
3. Class C is for lighting equipment.
4. Class D applies to equipment most likely to cause significant harmonic problems.

Class A equipment must meet limits that are specified in terms of a fixed amount of current for each harmonic, whereas Classes D and C must meet the much stricter requirement of harmonic current per unit of real power. Class D covers high-volume product groups such as TV, PC, and PC monitors (75–600 W). Class D allows the manufacturer to specify test power or current level, providing these are within ±10% of measured values. Applicable to all classes, transitory emissions are measured. All data smoothed via 1.5 s first-order filter. Every window results <150% of limit average during full test <100% of limit. This allows provision for odd harmonics 21-39, permitting the average of some individual harmonics to be >100% (<150%). The test/observation period to be sufficiently long to obtain ±5% repeatability. If too long, select the 2.5 min with maximum harmonics. Harmonic currents less than 0.6% or less than 5 mA shall be ignored.

The individual equipment in a rack or case must be tested individually

No limit for equipment with a rated power of 75 W or less.

The measurement method for both steady state and fluctuating harmonics requires that each individual harmonic from H2 to H40 be calculated over the full duration of the test. If the products power needs are constant, then the profile will not vary over time and the harmonics are easy to measure. If the power needs fluctuate, then the measurement is more complex. IEC 61000-4-7 describes a harmonic analyzer suitable for the test.

In practice, most portable digital equipment and computers easily meet the harmonics limit. However, equipment that takes high-current spikes from the power line may well exceed the limit. The best method to limit these spikes is to use a power factor correction supply. In some cases, an additional series inductor followed by a large value capacitor may average the power-line current sufficiently.

9.4.11.2.3   Flicker

IEC 61000-3-3 specifies limits on power-line flicker, which describes voltage fluctuations on the public power supply. IEC 61000-4-15 specifies the testing and measurement techniques. This variation when applied to a lamp results in light flicker hence the name. EN 61000-3-3 states that tests shall not be made on equipment that is unlikely to produce significant voltage fluctuations of flicker. This exception covers most digital electronics with the possible exception of laser printers and copiers. Otherwise any machine with an electric motor or electromechanical device such as washing machine, refrigerator, dryer is likely to generate flicker and must be tested.

The test must be performed with a standardized value of source impedance. This is specified in IEC 60725. Errors in the measurement can occur when this impedance is higher than specified. This can happen when using a transformer to couple the power or when connecting wires are too long. This can lead to a source impedance higher than the specified 0.4625 Ω.

For flicker test as per IEC 61000-3-3, the maximum permitted voltage drop after cessation of the fluctuation is 3.3%. In addition, the EUT is permitted to exceed this voltage drop for up to 500 mS

The maximum voltage drop at the commencement of the fluctuation is 4%, 6%, or 7% depending on the type of equipment being tested. Handheld equipment, being attended while in use, is allowed up to 7%, while equipment that is cycled a relatively limited times a day is allowed up to 6%. For equipment that cycles power levels on an ongoing basis, the limit is 4%. The test of onset fluctuation must be repeated 24 times. Then, the lowest and highest are discarded, and the average of the remaining 22 values is taken to compare against the permitted limit. The flicker meter measures the rms supply voltage value for each half cycle (10 mS at 50 Hz). The instant the power is turned on, the EUT starts to take current, which in turn causes a voltage drop through the standardized impedance.

9.4.11.2.4   Conducted Immunity C/M RF

The generic conducted immunity (C/M injection), specified for equipment used in industrial environments, is covered by IEC/EN 61000-6-2, Part 2. This references IEC 61000-4-6 for the test method. The generic requirement for conducted immunity on equipment used in residential commercial and light industry is contained in EN 61000-6-1, Part 1. The EN 55024 2010 document describes the immunity requirement on ITE. The test frequency is 0.15–80 MHz AM at 80% at 3 V rms on signal ports and telecommunications ports as long as these cables are longer than 3 m, and DC power ports. The three methods of injection are provided. The direct coupling is for tests on shielded or unshielded cables; the basic test setup for shielded cables is shown in Figure 9.36, and the calibration test setup for unshielded cables is shown in Figure 9.37. The value for R is $N \times 100$ Ω, where N = number of conductors in unshielded cable. An alternative coupling method is via a coupling/decoupling network (CDN). The injection may be made either directly via the CDN, as shown in Figure 9.38, or by an inductive injection EM clamp as shown in Figure 9.39. If a CDN is not available, the method shown in Figure 9.39 can be used with the terminating resistors shown in Figure 9.37 as long as the impedance of the terminating resistors is in accordance with the requirements of ENV 50141 and EN 61000-4-6. EN 61000-4-6 describes a separate setup for level setting.

As with all immunity/susceptibility testing if a problem is detected, it is important to ensure that it is the EUT and not the auxiliary equipment (AE)/test set/monitoring equipment that is susceptible. If the problem is not with the EUT, then additional filtering can be added to the input(s) of the AE.

FIGURE 9.37   Test setup for calibration, direct coupling to nonscreened (unshielded) mains and (power) cables.

FIGURE 9.38   C/M injection directly via the CDN, with CDNs or resistive networks as the loads.

FIGURE 9.39   Test setup for injection via the inductive clamp, with resistive networks, or more typically CDNs, as the loads.

In one test using the EM clamp, it was determined that the AE was susceptible. Additional ferrites were added between the clamp and the AE, which solved the problem. The test facility described in Section 9.1.1, which did not use the correct test method in other tests, objected to the use of the ferrites based on a change in impedance on the cable with the ferrites installed. Whereas, according to IEC 1000-4-6, the "EM clamp (in contrast to the conventional current injection clamp) has directivity ≥10 dB above 10 MHz, so that a dedicated impedance between the common-mode point of the AE and the ground reference plane is no longer required. Above 10 MHz, the behavior of the EM clamp is similar to that of the CDN." CDN is an abbreviation for Coupling–Decoupling Network. As its name implies, the CDN is used to couple the test level into the EUT and decouple the test level from the AE. Thus, as the problem frequency was above 10 MHz, the additional ferrite cores were perfectly allowable.

The injection test setups are described as follows.

Equipment checklist

1. Signal generator 0.15–80 MHz with either an AM capability of 1 kHz at 80% or an external modulation input.
2. Signal generator for 1 kHz. If generator in (1) requires an external modulation source. If generator in (1) does not have an external modulation input, then a modulator is required between the signal generator output and the power amplifier in (3).
3. Power amplifier 0.15–80 MHz. Typically 10–75 W, depending on injection method, with 50 Ω source impedance.
4. Inductive injection clamp, or CDNs, or capacitor/resistor networks (see 6) for direct coupling.
5. Ferrite ring cores (EM clamp) for the direct coupling into screened (shielded) cables test set.
6. 100 and 50 Ω series injection and load resistors and for direct coupling into nonscreened (unshielded) cables <20 nF coupling capacitors.
7. Oscilloscope with at least a 100 MHz bandwidth and a 50 Ω input impedance or a 50 Ω terminating resistor.
8. S/A or EMI receiver.
9. >6 dB power attenuator. Required for CDN injection only.

Setup and operations

For direct coupling into shielded cables and unshielded cables, connect the equipment as shown in Figure 9.36 for shielded cables and Figure 9.37 for unshielded and ensure that the full 3 V or 10 V rms unmodulated can be developed across the 150 Ω load. Because the voltage is measured across only the 50 Ω part of the load and an impedance of 250 Ω is presented in series, the 10 V rms (28 V pk-pk) level is met when the voltage is one-sixth, or 1.67 V rms (4.71 pk-pk measured using the oscilloscope). With the carrier AM at 80%, ensure that the pk-pk voltage increases. EN 61000-4-6 shows a test setup for level setting. Also measure the harmonics using the S/A /receiver and ensure that they are at least 15 dB below the carrier level.

For unshielded cables during the test, the signal is injected in parallel into all the power conductors in the cable via a resistor and capacitor in series. The total value of all of the resistors in parallel (the C/M resistance) is 100 Ω. Thus, if four lines are tested, 4 × 400 Ω resistors are used.

For the coupling directly via the CDN or via the inductive clamp with CDNs or resistor networks:

1. Ensure that the CDN has been calibrated as described in ENV 50141 and meets the Z_{ce} common-mode impedance requirements (with all lines connected together while tested for the impedance) of 150 ± 20 Ω from 0.15 to 26 MHz and 150 + 60/−45 Ω from 26 to 80 MHz when tested in accordance with Figure 7c of ENV 50141:1993.
2. Connect the equipment as shown in Figure 9.38 for direct injection or as shown in Figure 9.39 for injection via the inductive clamp. Ensure that the full 10 V rms unmodulated can be developed across the 300 Ω load presented by the CDNs or CDN and resistor network. Because the voltage is measured across only the 50 Ω output of the CDN and a series impedance of 250 Ω is included in the two CDNs or resistor network, the voltage developed across the 50 Ω termination is one-sixth of the test voltage. Thus, with a 10 V rms (28 V pk-pk) level applied, the test is met when the measured voltage is 1.67 V rms (4.71 V pk-pk measured using the oscilloscope). EN 61000-4-6 shows a setup for level setting. With the carrier AM at 80%, ensure that the pk-pk voltage increases across the 50 Ω load. Also measure the harmonics using the S/A /receiver and ensure that they are at least 15 dB below the carrier level.

When the EUT does not have any conductive surface (such as a terminal contained within a plastic enclosure), a more realistic test setup is with the EUT completely disconnected from the ground plane, and this is shown in Figure 9.40.

FIGURE 9.40 Inductive clamp injection with the EUT disconnected from ground.

FIGURE 9.41 C/M RF injection via the CDN into the power lines.

If a computer or other device is used to monitor the performance of the EUT during immunity testing, then this can be protected from the test level by locating it outside of the shielded room, by the use of ferrite baluns on the cable, or by a filtered connector at the shielded-room entry. Figure 9.41 shows a similar test setup with direct injection of the C/M RF into the power lines.

For coupling via the EM clamp, see Figure A.5.2 of ENV 50141:1993 or Annex A of EN 61000-4-6. The test level is shown in Table 9.20.

9.4.11.2.5 Radiated Immunity H Field

The power frequency H-field immunity requirement is contained in IEC/EN 61000-6-2 and EN 55024 2010 with the test method described in EN 61000-4-8 (IEC 1000-4-8). One calibration test setup, using the MIL-STD 13.3 cm loop antenna available in many test facilities, is shown in Figure 9.42. The transmitting loop is either a 1 m × 1 m single square induction coil connected to a ground plane, a double square loop, 1 m per side and spaced 0.6 m apart, or a single rectangular coil 1 m × 2.6 m, all of which are described in EN 61000-4-8.

The test field in 55024 is 1A 50 or 60 Hz. The calibration method is as follows.

TABLE 9.20
Conducted Immunity Test Levels

Frequency Range 150 kHz to 80 MHz		
Level	V_o dB (μV)	V_o V
1	120	1
2	130	3
3	140	10
X[a]	Special	

[a] X is an open level.

FIGURE 9.42   Calibration test setup for loop antenna.

Equipment checklist

1. AC power source for continuous-mode operation with a current capability of 1–100 A divided by the coil factor (typically 0.833, resulting in a current of 1.2–120 A). The power source is typically the variac and step-down transformer used in the MIL-STD-462 RS02 test.
2. AC power source for short duration, 1–3 s, with a current capability of 300–1000 A divided by the coil factor (typically 350–1200 A).

3. AC power current clamp or the equipment in (7) and (8) or in (8) and (9).
4. AC voltmeter for use with the current clamp.
5. A resistor 0.1 Ω rated at $I^2 \times 0.1$, where I is the test current. At 100 A, the resistor may be constructed of a number of resistors with a total wattage rating of greater than 1 kW.
6. An oscilloscope used for differential measurements (channel A + B or A − B) and an input voltage range of at least 50 V DC or the differential probe with ×10 attenuator, or an AC voltmeter.
7. 1 m × 1 m single square induction coil with ground plane or double 1 m side, 0.6 m spaced square induction coil, or single rectangular coil 1 m × 2.6 m as described in EN 61000-4-8.
8. 13.3 cm shielded loop H-field antenna, as described in MIL-STD-462 RE01, 30 Hz–30 kHz, 1997, and MIL-STD-462D RE101, 30 Hz–100 kHz, 1993.
9. Low-frequency preamplifier, if required, with MIL-STD loop in (8).

Setup and operations
For AC power source continuous current:

1. Connect the variac to the AC power line, with the adjustable winding connected to the high-voltage winding of the step-down transformer.
2. Connect the 0.1 Ω resistor across the low-voltage winding of the step-down transformer with a sufficiently long connection to accommodate the jaws of the current clamp.
3. Connect the current clamp around the low-voltage connection of the step-down transformer and the clamp to the AC voltmeter.
4. Adjust the variac until the voltage measured across the voltmeter is correct for the specified test current.

For AC power frequency test using the OA-ohm resistor and oscilloscope for continuous current:

1. Connect the variac to the AC power line, with the adjustable winding connected to the high-voltage winding of the step-down transformer.
2. Connect the 0.1 Ω resistor across the low-voltage winding of the step-down transformer.
3. Connect the oscilloscope or AC voltmeter across the 0.1 Ω resistor.
4. Adjust the variac until the voltage measured across the voltmeter or measured by the oscilloscope is equal to the specified test current multiplied by 0.1.

For AC power frequency test using the 0.1 Ω resistor and oscilloscope for short-duration current:

1. Connect the short-duration current generator to the 0.1 Ω resistor.
2. Connect the oscilloscope differentially across the 0.1 Ω resistor.
3. Adjust the short-duration current generator until the voltage measured by the oscilloscope is equal to the specified test current multiplied by 0.1.

Calibration of the loop for coil factor:

1. Connect the current source to the induction coil via a 0.1 Ω resistor if the oscilloscope is used to measure current, or directly if a current clamp is used to measure current. The test setup is shown in Figure 9.40.
2. Connect the 13.3 cm loop to the oscilloscope, terminated in a 50 Ω resistor.
3. Adjust the current in the induction coil to a known value, typically 10–50 A.
4. Place the MIL-STD 13.3 cm loop in the center of the induction coil and orient the loop for a maximum voltage reading on the oscilloscope.

5. Use the conversion factor of the MIL-STD 13.3 cm loop or the manufacturer's calibration at the power-line frequency to obtain the H field in amps per meter. For example, assume that the voltage measured from the loop is 11 mV pk-pk, which converts to 3.9 mV rms = 72 dBμV. Assume the conversion factor for the loop is 68.5 dB at the power-line frequency; then, the magnetic field in dBpT is 72 dBμV + 68.5 dB = 140.5 dBpT = 1.06×10^7 pT. The magnetic field in amps per meter is 1.06×10^7 pT $\times 7.936 \times 10^{-7}$ = 8.4 A/m. The coil factor of the transmitting loop is given by the magnetic field strength divided by the input current. With a transmitting induction loop current of 10 A, the coil factor of the transmitting loop is 8.4 A/m/10 A = 0.84.

9.4.11.2.6 Conducted Immunity EFT/B

The extrafast transient/Burst (EFT/B) immunity test is specified in IEC/EN 61000-6-1, IEC/EN 61000-6-2, EN 55024, and other requirements, with the test setup described in IEC /EN61000-4-4. EN 55024 specifies 0.5 kV into signal and telecommunications ports where the cable is longer than 3 m and 0.5 kV in DC power ports. The EFT generator has the following stringent requirements placed upon it: Up to 4 kV O/C, 2 kV when loaded with 50 Ω and 3.8 kV when loaded with 1000 Ω. The waveshape measured across 50 Ω is to be in accordance with Figure 3 of IEC-61000-4-4 (5 ns ± 1.5 nS rise time to 90% amplitude) and (50 ns ± 15 nS at 50% amplitude). The pulse repetition rate is to be (5 or 100 kHz), with a burst duration of (15 ms ± 3 mS at 5 kHz and 0.75 ± 0.15 mS at 100 kHz) and a burst period of (300 ± 60 mSms). A gas discharge tube was shown in earlier versions of the specification, but this is now shown as a switch in IEC 61000-4-4 2012, and an EFT generator with a gas discharge tube is no longer allowed. The reason being that gas discharge tubes have a limited life and are often erratic in triggering. A patented semiconductor device has been manufactured by Behlke Electronic GmbH, 5 Alexander Rd Billerica, MA 01821, United States, Tel.: 1 987 362 3118, USA@behlke.com. This is a very reliable and easy-to-use device for those who want to build their own EFT generator.

The EFT generator can best be calibrated using a 50 and 1000 Ω coaxial calibration network/ high-voltage attenuator. This network will load up to a 2.5 kV pulse with 50 Ω and divide down the voltage so that a fraction of the high-voltage pulse can be developed across the 50 Ω input of a high-speed oscilloscope. The oscilloscope has a −3 dB bandwidth of at least 400 MHz. The test load impedance at 1000 Ω is likely to become a complex network. The characteristics of the test load impedance are as follows:

50 ± 1 Ω
1000 ± 20 Ω (the resistance measurement to be made at DC)

The tolerance of the insertion loss of both test loads shall not exceed as follows:

±1 dB up to 100 MHz
±3 dB from 100 up to 400 MHz

The following parameters shall be measured:

- Peak voltage
- For each of the set voltages of table 2 in IEC 61000-4-4

Measure the output voltage with a 50 Ω load. This measured voltage shall be as specified with a tolerance of ±10%. With the same generator setting, measure the voltage with a 1000 Ω load. This measured voltage shall be as specified with a tolerance of ±20%.

Rise time for all set voltages
Pulse width for all set voltages
Repetition frequency of the pulses within one burst for any one set voltage
Burst duration for any one set voltage
Burst period for any one set voltage

The capacitive coupling clamp is used for tests on signal and control ports. It may also be used on power ports only if the coupling decoupling network cannot be used. Shielded cable should be injected directly. However, if shielded cables are included in a cable bundle along with unshielded cables and the shielded cables cannot be broken out without destroying the cable bundle, then the capacitive injection probe should be used.

The capacitive coupling clamp shall be calibrated with the EFT/B generator connected to one end of the clamp and a 50 Ω termination/attenuator at the far end. An insulated metal transducer plate 120 mm × 1.050 m and 0.5 mm thickness is inserted into the capacitive coupling clamp.

At the EFT/B generator end, this plate is open circuit. At the oscilloscope end, the connection is to a 50 Ω terminator/attenuator. The metal case of the terminator/attenuator to be connected to ground uses a short low-impedance connection.

The distance between the end of the transducer plate and the 50 Ω terminator/attenuator shall not exceed 0.1 m. The characteristics to be met are as follows:

Rise time 5 ± 1.5 nS.
Pulse width 50 ± 15 nS.
Peak voltage 1000 ± 200 V.
The CDN is used for injection into AC/DC power ports.
The CDN shall be calibrated with a calibrated generator.

The output of each of the lines shall be calibrated with a 50 Ω coaxial adapter with all other lines open circuit. The connection between the CDN and the coaxial adapter should be as short as possible but not to exceed 0.1 m. The calibration to be performed at 4 kV is that the rise time shall be 5.5 ± 1.5 nS, the pulse width 45 ± 15 nS, and the peak voltage 2 ± 0.2 kV.

In addition to the calibration of the CDN, the waveform shall be calibrated in common-mode coupling, which means to couple the transients to all lines simultaneously. This also makes sense for the test configuration as the EFT is a result of typically an arc or discharge, which is likely to induce current into all lines. In addition, the capacitive coupling clamp induces common mode into all other ports. However as aforementioned, IEC 61000-4-4 states that the waveform shall be individually calibrated for each coupling line. The interpretation has been that the coupling/decoupling mode requirement is "coupling mode to all lines" and this has changed the test into a common-mode test. This appears to mean that tests on a single power supply line are no longer required. The figure for coupling via the CDN shows all switches open, and so it does not indicate if the test will be common mode or differential mode. It does mention that if a suitable coupler/decoupler network is not available, then "in the case of common and unsymmetric (differential) modes, direct injection using the 33 ± 6.6 nF capacitors is the preferred coupling mode." This seems to leave the door open for both types of injection. Therefore, it is recommended that tests on single lines be performed in addition to the common-mode test in case the requirement may be reinstated or the intent is indeed to perform the single line "differential-mode" test.

Equipment checklist

1. EFT generator of up to 4 kV O/C output and 2 kV when loaded with 50 Ω and 3.8 kV when loaded with 1000 Ω. The waveshape measured across 50 Ω to be in accordance with Figure 3 of IEC 61000-44 (5 ns ± 1.5 nSc rise time to 90% amplitude and 50 ns ± 15 nS at 50% amplitude). The pulse repetition rate to be 5–100 kHz with a burst duration of 15 ms ± 3 mS at 5 kHz or 0.75 ± 0.15 mS at 100 kHz and a burst period of 300 ± 60 mS. If a calibration of the generator cannot be made, ensure that the generator has a calibration certificate and that the calibration was conducted less than a year prior to the proposed test.
2. CDN for AC/DC power supply lines in accordance with IEC 61000-4-4. The typical characteristics of the CDN are decoupling inductor with ferrite >100 μH, coupling capacitor 33 nF.

FIGURE 9.43 ESD test setup.

3. Capacitive coupling clamp for I/O circuits may also be used for AC/DC power when the CDN cannot be used.
4. Oscilloscope and coaxial cable with length of well-insulated wire connected to center conductor of the shielded cable.

The general EFT test setup is shown in Figure 9.43.

A typical capacitive clamp injection between the EUT and an AE is shown in Figure 9.44a and between the EUT and monitoring equipment in Figure 9.44b.

The coupling/decoupling injection setup is shown in Figure 9.44c.

Connect the capacitive clamp or coupling network to a ground plane with a minimum size of 0.8 m × 1 m, as shown in Figure 9.44a and b. The ground reference plane shall project beyond the EUT by at least 0.1 m and so may need to be larger than the minimum specified. Locate the signal/power cable under test through the capacitive clamp and locate the EUT at a distance of less than 1 m between the end of the clamp and the EUT. The clamp has a coaxial connector at one end and a second coaxial connector at the other end. Always connect the output of the EFT generator to the connector closest to the EUT. The minimum distance between the EUT or the capacitive coupling clamp and all other conductive structures other than the ground plane shall be more than 0.5 m. The capacitive coupling clamp is the first choice for the test on I/O or communications ports. If the clamp cannot be used due to mechanical problems (size, cable routing) in the cabling, it may be replaced by a tape or a conductive foil enveloping the lines under test. The capacitance of this coupling arrangement with foil or tape should be equivalent to that of the standard coupling clamp. As an alternative, it might be useful to couple the generator output to the terminals of the lines under test via discrete 100 pF capacitors. One customer required that the capacitive injection probe be used for a shielded cable. The cable shield was an uninsulated over-braid. This was placed in the clamp, and so an S/C was made between the braid and the clamp. Although not the goal of the customer, the setup was close to the specified, where the EFT/B generator is connected directly to the braid.

If only one EUT is to be tested and not the AE, or if a computer or other monitoring equipment is used to monitor the performance of the EUT during the test, it is advisable to locate this equipment outside of the shielded room. The EFT pulse when applied to the capacitive injection probe results in radiation from the probe. This can result in EMI in the monitoring equipment, which is mitigated by locating this equipment outside of the room. Ferrite baluns can be added on the interconnection

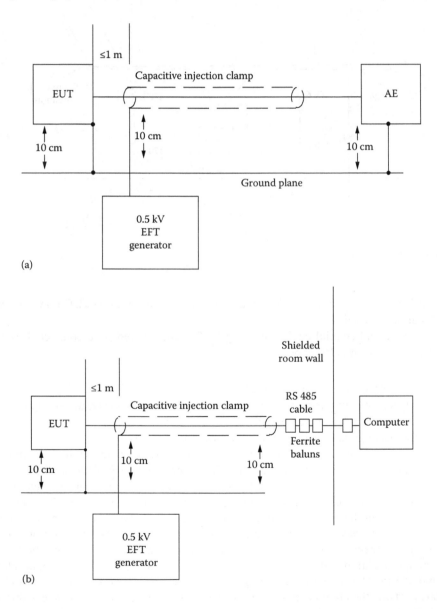

FIGURE 9.44 (a) Typical capacitive clamp injection of the EFT pulse with both EUT and AE tested. (b) Typical capacitive clamp injection of the EFT pulse with the monitoring equipment protected and outside the shielded room.   (*Continued*)

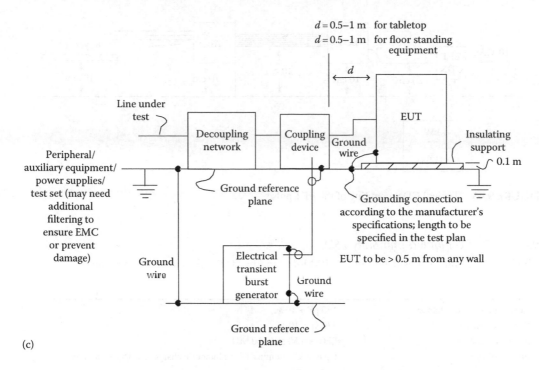

$d = 0.5\text{--}1$ m for tabletop
$d = 0.5\text{--}1$ m for floor standing equipment

(c)

FIGURE 9.44 (Continued)   (c) Capacitive coupling clamp or coupling/decoupling network test setup.

cables or a transient suppressor/filtered connector at the entry of the cable into the shielded room, as shown in Figure 9.44c. Even with these precautions, it is not uncommon for the I/O board on a computer to be destroyed during the EFT test, especially when coupling occurs within the EUT. In this case, use of a fiber-optic coupled I/O board should be considered.

IEC 61000-4-4 shows the test setup for tabletop laboratory type tests with an EUT and AE, for floor-standing equipment where the injection probe is placed over a cable connecting to pieces of EUT, and for equipment with elevated cable entries and an in situ test.

Floor-standing equipment shall be placed on a ground reference plane and shall be insulated by from it by an insulating support with a thickness of 0.1 ± 0.05 m. Floor-standing EUTs and equipment designed to be mounted in other configurations, unless otherwise mentioned, shall be mounted as aforementioned.

All cables to the EUT shall be placed on insulation support 0.1 m above the ground reference plane. Cables not subject to the EFT shall be routed as far as possible from the cable under test to minimize the coupling between the cables.

The latest version of IEC 61000-4-4 contains a numerical model of a Burst pulse.

It is important to ensure that any registered failure is in the EUT and not any peripheral, AE, test set, etc. Although it seems too obvious to state the EFT pulse will appear on all signal and power lines, this will not constitute a failure of the EUT as long as it does not respond to the pulses. In one case, data acquisition equipment was used to monitor a power line and data lines. Due to the slow acquisition speed of the system, the 50 nS EFT spikes were converted into intermittent 2 mS long pulses on the power, indicating that the supply was dropping out. When monitored using a high-frequency oscilloscope, the 2 mS pulses were merely the 50–100 nS EFT spikes at the expected 5 kHz repetition rate.

A typical time period for calibration of the EFT generator is once a year, unless it has sustained some obvious damage, in which case an immediate recalibration is recommended. Figure 9.45 shows the CDN coupling of the EFT pulse.

FIGURE 9.45 Typical CDN coupling of the EFT pulse.

9.4.11.2.7   Conducted Immunity ESD

The ESD generator is described in EN 1000-4-2 and has the following specifications:

Energy storage capacitance	150 pF ± 10%.
Discharge resistance	330 Ω ± 10%.
Charging resistance	Between 50 and 100 MΩ.
Output voltage	Up to 8 kV (nominal) for contact discharge and 15 kV (nominal) for air discharge.
Tolerance of the output voltage indication	±5%.
Polarity of the output voltage	Positive and negative (switchable).
Holding time	At least 5 s.
Discharge, mode of operation	Single discharge (time between successive discharges at least 1 s).
Waveshape of the discharge current	The peak current is reached with a rise time of 0.7–1 ns. At 30 ns, the current is at 53.3% of the peak, and at 60 ns, the current is at 26.6% of the peak.

A test plan shall be made, which should include

Representative operating conditions of the EUT
Whether the EUT should be tested as tabletop or floor standing
The points at which discharges are to be applied
At each point, whether contact or air discharges are to be applied
The test levels to be applied (including intermediate levels)
The number of discharges to be applied to each point

A discharge electrode with a round tip and having specific dimensions is shown for air discharges; a sharp-pointed tip with a specific angle for the tip is shown for contact discharges. A current-sensing transducer, for use with a 1 GHz bandwidth oscilloscope, is described for the calibration of the ESD generator. Dimensions for the transducer are provided in IEC 61000-4-2, and these transducers are commercially available. It is advisable to mount the transducer in the wall of a shielded room, with the ESD generator on the outside and the oscilloscope on the inside. Alternatively, a small faraday cage can be constructed, with the transducer mounted to a face made of solid sheet metal. The oscilloscope can be mounted inside the faraday cage and either viewed through a wire mesh or operated in a storage mode. A typical time period for calibration of the ESD generator is once a year, unless it has sustained some obvious damage, in which case an immediate recalibration is recommended.

The contact discharge should be made to conductive surfaces and to the horizontal coupling plane (HCP) and vertical coupling plane (VCP). Some typical conductive surfaces on an EUT contained in a nonconductive enclosure are screws and other metal fasteners, connector housings, pins in connectors, and metal covers. Some of the most potentially susceptible areas are pins in a connector to which no cable is attached, around a reset switch, and a plasma, LCD, or LED display. Air discharges should be made to insulating surfaces such as video display screens or keyboards. Some of the most potentially susceptible areas are displays and anywhere where a metal fastener, PCB, or other metal is close to the surface in a nonconductive enclosure.

A ground reference plane shall be provided on the floor of the test area. It shall be a metallic sheet (copper or aluminum) of 0.25 mm minimum thickness. Any other metallic materials may be used, but they shall have at least 0.65 mm minimum thickness. The minimum size of the ground reference plane shall be 1 m², and it shall project beyond the EUT or coupling plane by at least 0.5 m.

A distance of 1 m minimum shall be provided between the EUT and the wall of the test facility and any other metallic structure; this requirement may restrict the use of a small shielded room for this test. The EUT shall be connected to the grounding system, in accordance with its installation specifications. No additional grounding connections are allowed. If the EUT is contained in a nonconductive enclosure and DC power is derived via an isolated power supply connected to the AC outlet, no obvious ground connection is made. Often the equipment will interface to a computer or monitor, in which case the ground is often made via these devices, typically via a single-ended digital/analog or video signal interface. Just as with the EFT test, the ESD current can flow via the EUT into the peripheral equipment, resulting in damage. In one instance, the equipment was connected inside a vehicle and the current path to ground was via the power return cable and via the battery terminal to the chassis of the vehicle. Analyzing the path for the ESD event, current flow is very important when solving ESD susceptibility problems. The connection of the earth cable to the ground reference plane and all bondings shall be low impedance. Coupling planes are connected to the reference ground plane with cable that contains 470 kΩ resistors at each end of the cable. The ESD test setup shall consist of a wooden table 0.8 m high standing on the ground reference plane. The HCP is 1.6 m × 0.8 m and shall be placed on the wooden table. The EUT and cables are isolated from the HCP by an insulating support 0.5 mm thick. The ESD generator return is connected to the ground reference plane. The cables used are 2 m in length. The test programs and software, as with all immunity tests, shall be chosen to exercise all normal modes of operation of the EUT. The use of special exercising software is encouraged, but it is permitted only where it can be shown that the EUT is comprehensively exercised. For the ESD test, performance criterion B is required for a pass, so the correct functioning of the EUT after the test has been conducted must be demonstrated.

For direct application of discharges to the EUT, the ESD shall be applied to such points and surfaces of the EUT as are accessible to personnel during normal usage. The test shall be performed with single discharges. On preselected points, at least 10 single discharges (in the most sensitive polarity) shall be applied of each polarity. The interval between discharges is 3 s. Multiple discharges are recommended when attempting to find the source of ESD susceptibility or after modifications designed to harden against ESD. All four test levels to be tested.

The ESD generator is to be held perpendicular to the surface to which the discharge is applied. The discharge return cable is to be kept at a distance of at least 0.2 m from the EUT. For contact discharges, the tip of the discharge electrode shall touch the EUT, before the discharge switch is operated. In the case of air discharges, the round discharge tip of the electrode shall be approached as fast as possible (without causing mechanical damage) to touch the EUT. After each discharge, the discharge electrode shall be removed from the EUT. The generator is then charged for a new, single discharge. The discharge switch, which is used for contact discharge, shall remain closed. At least 10 single discharges of each polarity shall be made to the HCP at points on each side of the EUT. At least 10 single discharges of each polarity shall be made to the center of one vertical edge of the VCP. The VCP is 0.5 m × 0.5 m in size and is placed parallel to and positioned at a distance of 0.1 m

TABLE 9.21
ESD Test Levels

Contact Discharge		Air Discharge	
Level	Test Voltage (kV)	Level	Test Voltage (kV)
1	2	1	2
2	4	2	4
3	6	3	8
4	8	4	15
X[a]	Special	X[a]	Special

[a] X can be any level above, below, or in between the others. The level shall be specified in the dedicated equipment specification. If higher voltages than those shown are specified, special test equipment may be needed.

from the EUT. Figure 9.43 illustrates the ESD test setup. In the case of the air discharge, the ambient conditions shall have the following requirements:

15°–35°.
Relative humidity 30%–60%.
Atmospheric pressure from 86 kPa (860 mbar) to 106 kPa (1060 mbar).
The test levels are shown in Table 9.21.
EN55024 2010 specifies test levels of 4 kV contact and 8 kV air discharge.

9.4.11.2.8 Conducted Immunity Surge Test

The surge generator is referred to as a combination wave (hybrid) generator. The major characteristic of the combination wave generator is that it delivers a 1.2 µs/50 µs voltage surge under O/C conditions and an 8 µs/20 µs current surge into an S/C with an effective output impedance of 2 Ω.

Thus, at a peak O/C voltage of 2 kV, the peak S/C current is 1000 A.

The characteristics of the generator, from IEC 61000-4-5, are as follows:

O/C output voltage	At least as low as 0.5 kV to at least as high as 4.0 kV
Waveform of the surge voltage	Rise time of 1.2 µs ± 30% from 0.1 × Vpk to 0.95 × Vpk; time from 0.1 × Vpk on rising edge to 0.5 × Vpk on falling edge = 50 µs ± 20%
S/C output current	At least as low as 0.25 kV to at least as high as 2.0 kV
Waveform of the surge current	Rise time of 8 µs ± 20% from 0.1 × Ipk to 0.95 × Ipk; time from 0.1 × Ipk on rising edge to 0.5 × Ipk on falling edge = 20 µs ± 20%
Polarity	Positive/negative
Phase shifting	In a range between 0° and 360° versus the AC line phase angle
Repetition rate	At least 1 per min

A generator with floating output shall be used.

Additional resistors of 10 or 40 Ω shall be included to increase the required effective source impedance, as described later, and under these conditions, the O/C voltage waveform and the S/C current waveform in combination with the CDN are no longer as specified.

A 10/700 µS O/C and a 5/320 µS waveform is also specified in ITU-T K series, EN55024 and IEC 60060-1. EN55024 2010 specifies surges of 1 kV peak with a rise time of 10 µS and a pulse width of 700 µS for signal ports and telecommunications ports, where ports are connected directly to outdoor cables. The test level is applied simultaneously to earth (ground).

The EN 55024 surge test on DC power is 0.5 kV 1.2 µS rise time 50 µS pulse width with the test level applied between lines and the earth (ground).

CDNs are used to isolate the power or signal source from the surge, and these should not significantly influence the parameters of the generators, for example, O/C voltage and S/C current.

Capacitive coupling for power supplies allows the surge to be applied line to line or line to earth (ground), while the decoupling network is also connected.

The coupling capacitors are either 9 or 18 µF, and the inductance in the decoupling network is 1.5 mH.

The residual surge voltage on unsurged lines with the EUT disconnected shall not exceed 15% of the applied voltage and shall not exceed 15% on the power supply inputs when the EUT and power supply are disconnected.

Figure 9.46 shows the capacitive coupling into AC/DC power line to line, and Figure 9.47 shows the capacitive injection line to earth.

Capacitive coupling is the preferred method for unbalanced unshielded I/O circuits when there is no influence to the functional communication on that line. The application is via a 0.5 µF capacitor in series with a 40 Ω resistor, in accordance with Figure 10 of IEC 61000-4-5. Coupling via

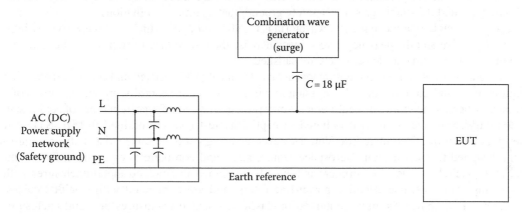

FIGURE 9.46 Capacitive injection of the surge on AC/DC power line to line.

FIGURE 9.47 Capacitive injection of the surge on AC/DC power line to earth.

arrestors is the preferred coupling method for unshielded balanced circuits, as shown in Figure 12 of IEC 61000-4-5.

The injection into shielded cable shall be in accordance with IEC 61000-4-5 Figure 16 for cable shields grounded at each end, Figure 17 for cable shields grounded at one end, and Figure 18 for multiple grounded cables.

Where isolation transformers are used to power one or more units of the EUT, the safety ground connection is made through the shield of the cable. When this shield connection is faulty, the EUT to which the test generator is connected is ungrounded and a dangerous charge can build up on conductive structures connected to the safety ground.

The injection is both differential and common mode with positive and negative polarities.

The number of pulses is five of each polarity at each coupling point. Wait for at least a minute between the application of the pulse to enable any protection devices to recover.

For AC mains, apply the surges line to line. In a three-phase star, apply the pulses L1-L2, L2-L3, L1-L3, L1-N, L2-N, and L3 to neutral, that is, six combinations. For a three-phase delta, apply the pulses L1-L2, L2-L3, and L1-L3, that is, three combinations. For a single phase, one combination.

Also apply the surges line to the safety ground (earth). In a single-phase L-safety ground, N-safety ground, that is, two combinations. In a three-phase delta, L1-safety ground, L2-safety ground, and L3-safety ground, that is, three combinations. In a three-phase star, L1-safety ground, L2-safety ground, L3-safety ground, N-safety ground, that is, four combinations.

The phase angle relative to the AC waveform is 0°, 90°, and 270°, which is at the zero crossings, the peak positive and the peak negative values. Increase the test voltage in steps up to the specified maximum levels so that all lower levels are satisfied.

Manufacturers of commercial equipment are used to meet FCC radiated and conducted emission requirements, and therefore the EN emission requirements do not generally present any more of a problem, whereas the immunity requirements may present a problem for certain types of equipment.

Military/aerospace equipment radiated susceptibility test levels contained in MIL-STD-461 are specified from 1 V/m up to 200 V/m. However, a very great difference exists between equipment designed to meet military/aerospace requirements and commercial equipment. For example, MIL-STD-461 type of equipment is typically contained in shielded metal enclosures with few if any apertures, use effective power-line filters, and are connected with shielded cables. Commercial equipment is often contained in plastic unshielded enclosures or metal enclosures that contain large slots, which often enhance emissions from that location and use unshielded interface cables. Thus, the incident E field in a radiated immunity test results in RF currents that flow on cables and on PCBs inside the enclosure. When equipment uses digital techniques and digital signal interfaces, it is itself a source of high levels of electromagnetic fields. It is self-compatible and therefore typically immune to the specified radiated susceptibility test levels and the resultant RF currents setup in the equipment. The problem occurs with telephone and audio equipment and their associated signal interfaces, in which a high incidence of susceptibility to radiated and conducted test levels.

Sections 5.1.10, 5.5.1, and 5.7 discuss how the immunity of equipment may be increased both in the design and after production.

9.5 SHIELDED ROOMS, ANECHOIC CHAMBERS, TRANSMISSION LINES, AND CELL ANTENNAS

The use of a shielded room for low-level electromagnetic measurements is often based on the high electromagnetic ambient outside of the room; also, the room may be required to contain susceptibility test levels. The use of a shielded room results in errors in the measurement of radiated emissions and in the generation of radiated susceptibility test fields. The cause of errors and some techniques available to reduce them are described in the following section.

9.5.1 SHIELDED-ENCLOSURE INTERNAL FIELDS AND ANTENNA ERRORS

Shielded rooms are used primarily either to contain electromagnetic fields within them, typically for either Tempest requirements or during radiated susceptibility tests, or to reduce the ambient electromagnetic environment to low levels. Two principle types of shielded enclosures are those with low-loss, reflective interior walls and those with lossy interior walls (e.g., the anechoic chamber). When using a shielded enclosure with reflective walls for radiated emission or radiated susceptibility tests, the accuracy of measurements is adversely affected by

1. Enclosure internal reflection
2. Capacitive or inductive coupling of antennas to the conductive walls
3. Distortion, noncoherent, and standing waves due to reflections from the interior surfaces
4. Enclosure excited in the TEM mode
5. Other resonance modes

Some characteristics that affect all of these factors are the shape and size of the enclosure, the position of the test setup, the size and gain of the antenna, and the presence and location of personnel and test equipment and the enclosure.

Errors of as high as ±40 dB are possible in radiated emission measurements made in a shielded room, with a change in field strength of approximately 15 dB due to a 1–2 cm change either in the spacing between the EUT and the antenna or in the location of measuring equipment or personnel. The continued use of nonanechoic chambers for measurements of electromagnetic waves is a puzzle to most antenna engineers. However, shielded rooms have been used for many years for the MIL-STD-462 and DO-160 type of tests.

Both DO-160 and MIL-STD-461 require the use of a ground plane on which nonportable equipment is bonded. Most, but not all, procuring agencies using the foregoing specifications require that the ground plane be connected to a vertical conductive wall and located at a height of 1 m above a second conductive surface (shielded-enclosure floor); however, this does not rule out the use of an anechoic or semianechoic chamber (a chamber in which the floor is not covered with absorber). The resonances of a shielded room with and without equipment and the coupling between antennas located a meter apart compared to the coupling in free space are reproduced in Figure 9.48a and b, respectively, from Reference [3]. Figure 9.49 shows the principal reflection paths in a shielded room that cause a spatial interference pattern due to the phase difference between the primary and reflected wave. It is theoretically possible to measure either an infinitely high wave impedance (i.e., no magnetic field component) or zero wave impedance (i.e., no electric field component) due to the interference. The gain of a current element or short dipole antenna is affected by the proximity of the antenna to the ground plane. The gain of the current element over a ground plane may be found from d, the effective length, and $\beta = 2\pi/\lambda$. When βd is small, the gain of the current element over the ground plane is approximately three times that for an element in isolation. When $\beta d = 2.9$, the gain is a maximum of 6.6 and the length of the antenna d is 0.46λ.

A knowledge of the gain is important when predicting E fields from an antenna located in a shielded room, for susceptibility measurements or for an EMC prediction of radiated emission from a current element close to a ground plane.

After almost 50 years of conducting MIL-STD radiated emission measurements in shielded rooms without damping, the requirement that RF absorber material be used in shielded rooms when performing radiated emission or radiated susceptibility measurements was introduced in MIL-STD-462D, January 11, 1993, although other, unspecified, test sites may be used.

The MIL-STD-462D/MIL-STD-461F absorber requirements, as well as other, often less expensive, methods used to reduce errors inherent in shielded-room measurements, are discussed next.

1. *MIL-STD-462D/MIL-STD-461F absorber requirements.* The RF absorber may be carbon-impregnated foam pyramids or ferrite tiles or a combination of these types of absorber material. The RF absorber shall be placed above, behind, and on both sides of the EUT and behind the

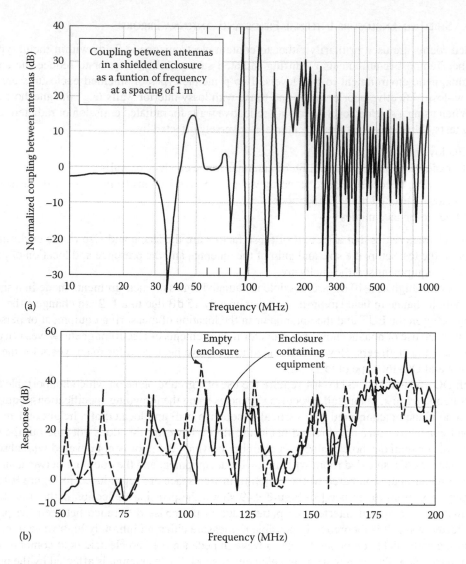

(a)

(b)

FIGURE 9.48 (a) Antenna coupling in a shielded room. (From Marvin, A.C., *IEEE Trans. Electromag. Compat.*, EMC-26(4), 149, November 1984.) (b) Shielded-enclosure resonance effects versus frequency for constant drive with the enclosure empty and with equipment in it. (Marvin, A.C., *IEEE Trans. Electromag. Compat.*, EMC-26(4), 149, November 1984.)

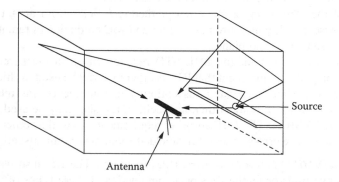

FIGURE 9.49 Multiple-path reflections in a shielded room.

FIGURE 9.50 MIL-STD-462D RF absorber loading diagram.

radiating or receiving antenna, as shown in Figure 9.50. Minimum performance of the material shall be as specified in Table 9.22. The manufacturer's certification of their RF absorber material (basic material only, not installed) is acceptable.

2. *Measurements made in the shielded room at a specific location with antennas calibrated at the same location.* As with all antenna factor measurements, these should be made in the far field and above the frequency where a TEM mode may occur in the room. This technique is most effective when used with high-gain antennas, such as horns, at frequencies above 500 MHz and relies on the fact that the principal resonances in the room are compensated for, to a degree, in the AF calibration, as are reflections. Although less effective, the technique may be applied to BB antennas at frequencies below 500 MHz. The transmitting and receiving antennas should be identical, and the receiving antenna should be located at a specific location in the room. The AF calibration is then valid only for that specific location. The AF is derived using the two-antenna test method described in Section 2.5.3. The accuracy of the AF obtained by use of this method is sensitive to the

TABLE 9.22

Absorption at Normal Incidence of MIL-STD-462/ MIL-STD-461F Specified Material

Frequency (MHz)	Minimum Absorption (dB)
80–250	6
Above 250	10

location of the transmitting antenna, to the location of equipment in the room, and to the exact location of the receiving antenna. Interpretation of the resultant AF is often difficult, due to the sharp dips and peaks in the curve from room effects. Despite these sources of error, the resultant AF is more accurate than that published by the manufacturer, and at the very least room resonances may be identified by examination of the measured AF.

3. *Stirred-mode or reverberation room*. The reverberation room contains a large paddle that moves and changes the resonance frequencies of the room. Figure 9.51a illustrates the setup. The fields inside a reverberation room can be accurately described as isotropic and noncoherent and exhibit a constant, average, uniform field in the large inside volume of the room. The room can be characterized by separating the time and spatial variations, and this can be achieved by filtering the revolution frequency of the moving paddle from the change in frequency caused by the stirred mode. Thus, a calibration curve of the room that compares the measured field strength to the free-space-equivalent field strength can be constructed. The accuracy of the correlation is typically ±6 dB

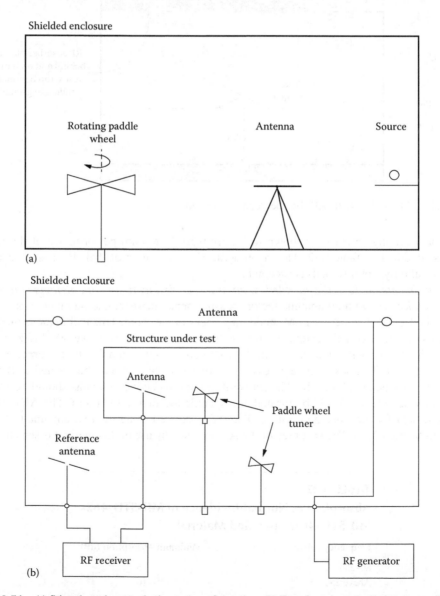

FIGURE 9.51   (a) Stirred-mode perturbation test configuration. (b) Tuned-mode perturbation test configuration.

from 45 MHz to 1 GHz and ±0.5 dB from 1 to 18 GHz. Although a fairly recent development, the use of reverberation chambers has been looked at by CISPR (International Special Commission for Radio Interference) and the U.S. Navy. MIL-STD-1377 describes a tuned-mode test setup that is useful above 200 MHz and is an improvement on MIL-STD-285 for the test of enclosure shielding effectiveness. The tuned-mode test setup is shown in Figure 9.51b. An alternative to the MIL-STD-462 radiated susceptibility test setup (RS03), using a stirred-mode reverberation room useful over the frequency range from 200 MHz to 1 GHz is described as follows: The components used inside the room are shown in Figure 9.51. The room contains a tuner stirrer. Either it is stepped and stopped, with a measurement made during the stopped interval, or a sample-hold technique may be used, with the stirrer in continuous stepped motion. The sampling is then accomplished during the stopped time of the stirrer. The stirrer is rotated through 360° in a minimum of 100 steps. Where fewer than 100 steps are used, a derating correction is required, as described later. The step rate must be slow enough to ensure that the EUT has time to respond to the changing electromagnetic environment. A power meter is used to monitor the output of the signal generator/power amplifier applied to the transmitting antenna. The transmitting antenna may be a matched long-wire antenna, up to the resonant frequency of the long wire. Isolators may be required to protect the signal source and power meter from large reflection coefficients at the test power frequencies and levels. The room is calibrated at a given calibration input power $P(\text{cal})$ that generates a given average calibration E field $E(\text{cal})$. The E field is measured by a receiving antenna and receiver or S/A. To obtain the required input power, $P_{in}(\text{test})$, to develop the specified test field, $E(\text{test})$, the following equation is used:

$$P_{in}\left(\text{test}\right) = P_{in}\left(\text{cal}\right) \times \left(\frac{E\left(\text{test}\right)}{E\left(\text{cal}\right)}\right)^2 + 3\,\text{dB} + \text{step derating}\left(\text{dB}\right)$$

When the number of steps is less than 100, the step derating (given by $10\log(100/N)$, where N is the number of steps), is added. The test frequency is not swept in the stirred-mode test, unlike the standard RS03 test. Instead, the stirrer is swept through 360° at a number of fixed frequencies. Some recommended frequencies are 200, 250, 300, 350, 400, 500, 600, 800, and 1000 MHz.

Other frequencies at which the EUT may be susceptible should be used, if known. The time during which the stirrer is stopped is determined by the time required to complete a functional test on the EUT and to determine if a malfunction has occurred. Thus, test times may be extremely long, especially if the 100 steps per rotation are adhered to. Another disadvantage is that the frequency is not swept, and susceptibility at a specific resonance frequency may be missed.

If the EUT is found to be susceptible, then the level of the E field generated should be closely monitored, using the stirrer position at which susceptibility occurred, if known, and adjusted to the specification level. If the EUT is susceptible at the specification level, the input power is reduced until the EUT functions as normal and the test E field is either measured or calculated. Although a more accurate method of controlling the test E field than the standard RS03 test method, the inaccuracies inherent in measuring the magnitude of the E field with a receiving antenna or probe inside the room remain.

4. *Correction of a shielded room for electric and magnetic dipole excitation.* Below the first resonance frequency of a shielded room, it (the shielded room) can be excited and function as a large TEM cell. The room is typically excited during radiated emission measurements, as per MIL-STD method RE02/RE102, by the EUT, placed on the ground plane, or interconnection cables; this is illustrated in Figure 9.52. A test is described in Reference [10] of the excitation of a shielded room by either an electric field or a magnetic field source in the frequency range of 10 kHz to 30 MHz. It was found that the position of the source of emission on the ground plane (i.e., proximity to the wall) affected the magnetic field source results by approximately 6 dB, whereas the field from an electric source varied by up to typically 25 dB. This relative insensitivity to reflections in the magnetic field

FIGURE 9.52   RE02 radiated emission shielded-room configuration for frequencies below 30 MHz that may excite the room.

has been seen in numerous measurements. One of the characteristics of an excitation mode in a room is that the measurement antenna may be moved from the measurement location to almost any location in the room with very little change in the measured E field. Once the source of emission is discovered (enclosure, aperture, or cable), a correction of the measured results based on the excitation of the TEM mode of the room may be made.

One correction technique that may be applicable is to change the effective dimensions of the room by adding metal structures. If the coupling to the room can be reduced, then the excitation of the room may be reduced. Moving the EUT and cables or even isolating the EUT from the ground plane may effectively reduce the coupling. Although in the radiated emission (RE02/RE102) type of tests, it is a requirement that the EUT be bonded to the ground plane, some measure of the reduction in E field with the room nonresonant may be useful in applying for a waiver. Absorber material placed on the wall behind the EUT and on the ceiling above it, although not very effective at low frequency, can also reduce the coupling to the room.

5. *Long-wire antenna, multiwire transmission lines, and parallel-plate antennas.* The long-wire antenna technique uses the excitation of the room as a TEM cell for radiated susceptibility measurement up to 30 or 50 MHz. The typical configuration is shown in Figure 9.53a. The generation of test fields at low frequency from conventional antennas is difficult. The long-wire antenna, multiwire transmission line, and parallel-plate (stripline) antenna are all capable of generating high E fields, from 14 kH to 30 or 50 MHz for the long-wire/multiwire and up to 200 MHz for the parallel plate, with an E/H ratio of 377 Ω, with the benefit of low cost. A transmission line, such as a coaxial cable or an air dielectric stripline will create an internal 377 Ω field when it is terminated by its characteristic impedance, even when the transmission-line impedance is far from 377 Ω. Because the wave impedance is 377 Ω, only the E or H field need be measured to determine the unknown field. When a transmission line is terminated in other than its characteristic impedance, the field impedance is directly proportional to the load impedance. The author is indebted to a private communication from Reference [5] for a clearer understanding of the relationship between field impedance and circuit impedance ($E/H = V/I$) and that when V/I is replaced by the characteristic impedance of the line, Z_c, the E/H ratio reduces to 377 Ω.

This has a very important implication in generating high E fields using a low-level power amplifier. If a step-up transformer is used between a power amplifier and a long-wire antenna or a stripline antenna, then a higher level of E field can be generated than possible without the transformer, and a closer match to the optimum load, for most power amplifiers, of 40–50 Ω, is achieved.

Suitable transformers are described in an excellent practical book entitled Transmission Line Transformers by Jerry Sevick W2FMI and published by Noble publishing. The book is into its Fourth edition. These transformers provide unbalanced to unbalanced as well as balanced to unbalanced designs with ratios of 1:1.5, 1:2, 1:3, 1:4, 1:9, and 1:16.

FIGURE 9.53 (a) Long-wire antenna in a shielded-room configuration for radiated susceptibility tests in the typical configuration. (b) Long-wire antenna with step-up transformer. (c) Long multiwire unterminated antenna.

The useful frequency range is typically 0.1–10 MHz and 0.5–100 MHz. The power rating of these transformers is as high as 1 kW. One of these designs is available from CWS and is the unbalanced UN-B-800-612-450.

This has impedance rations of 50 Ω:800 Ω, 50 Ω:612 Ω, and 50 Ω:450 Ω. The frequency range is 0.5–15 MHz and is rated at 0.5 kW continuous and 2 kW peak. It is ideal for use with the 2.17 m monopole and the 2.05 m long multiwire antenna.

The multiwire antenna is designed to be placed over equipment cabinets. It is 2.05 m long and 2.1 m high. The frequency range is 10 kHz–30 MHz and can be used with the step-up transformer up to 15 MHz. The antenna is unterminated at the far end. The design is shown in Figure 9.53c. The E field developed depends on the power amplifier rating but can be as high as 50 V/m with a 100 W power amplifier. The multiwire antenna has a much higher field uniformity than the long-wire antenna and can accommodate higher EUTs. The long wire when terminated in the characteristic impedance of the line, it is not nearly as efficient as the long-wire multiwire or stripline, unterminated, with step-up transformer, for it requires a power amplifier of 2 kW to generate 100 V/m.

A small long-wire antenna can be designed with an impedance close to 400 Ω, modeled as a wire over a ground plane, by locating the wire 0.75 m above the tabletop ground plane. The long wire should be approximately 2 m below the ceiling of the room and at least 3 m from the nearest wall, which typically requires moving the table away from a wall. In this configuration using the ×3 step-up ratio of the transformer (impedance change of ×9) and with the long wire terminated in the characteristic impedance of the line, an E field as high as 200 V/m can be generated using a 75 W power amplifier. This configuration is shown in Figure 9.53b. By comparison, a conventional E-field generator requires 800 W to generate a 200 V/m field 1 m from the antenna. Adding a second wire 0.25 m from the first wire will improve the field uniformity of the line, but due to the lower impedance (approximately 280 Ω), the step-up ratio of the transformer is limited to ×2 and approximately 150 W is required to generate 200 V/m. With the long-wire antenna strung across a 2.43 m (8 ft) high room, in the more typical configuration shown in Figure 9.53a, but with the use of a transformer, the E field can be approximately 50 V/m with a distance between wire and ceiling of 0.25 m, and 100 V/m with a distance of 1 m and the 75 W power amplifier. This is approximately 2.8 times the fields generated without the transformer A parallel-plate transmission line, often referred to as a stripline antenna, open TEM cell, or open-cage parallel-plate antenna, achieves high efficiency and a uniform field between the lines. It is terminated in the characteristic impedance of the line and will generate a TEM 377 Ω impedance field between the plates with the E field vertically polarized. If the upper surface of the EUT is equal to or less than one-half the width of the lines, then the upper surface of the EUT can be as close as 3 cm to the upper line when used with an upper frequency of 30 MHz, although this may result in multimoding; that is, more than the TEM is generated. The recommended maximum EUT height is three-fourths of the distance between the lines. The upper useful frequency is given by $\lambda = 2h$, where h is the distance between the plates. In practice, this upper frequency is typically limited to 200 MHz for large lines and may be achieved only with great care in the manufacture of the terminating resistance and with a small EUT.

The length of the line is not limited as long as it is correctly terminated in its characteristic impedance.

Use of the parallel-plate line with a step-up transformer is not as effective as the long-wire antenna due to the lower characteristic impedance of the line. However, a parallel-plate antenna with a 1 m distance between plates and an upper plate width of 0.4 m can be used with a ×2 step-up transformer, which matches the line to the power amplifier reasonably well and can generate an E field of up to 100 V/m with a 75 W power amplifier.

The parallel-plate or stripline antenna has been described in the Society of Automotive Engineers Aerospace Information Report AIR 1209. One of the parallel-plate antennas described has a 1 m wide upper plate with 1/16 in. slots at 1/2 in. centers and an unslotted lower plate. The distance between the upper plate and the lower plate is 1 m, and the characteristic impedance of the line is 120 Ω. The termination impedance is three separated layers of 377 Ω/sq. conductive plastic film.

The film distributes the impedance more evenly at the termination and reduces the coupling to the shielded-room wall from the termination by 36 dB. The recommended grounding of the lower plate is at the termination only over the complete width of the plate. The input to the line is described as a wave launcher rather than a matching section, which has an empirically determined shape.

The recommended frequency range of the line is 0 Hz–30 MHz, and the maximum recommended EUT size is 0.75 times the strip separation.

The second line is a MIL-STD-462 type in which the height-to-width ratio is not unity. Typical characteristic impedance for these lines is anywhere from 80 to 105 Ω, depending on the height-to-width ratio; the normally used frequency range is 14 kHz–30 MHz. Again, the recommended maximum EUT height is 0.75 times the distance between the plates, and the ground plane on the table may be used as the lower plate. It is recommended that the top plate of the line be at least 30 cm from the shielded-room wall. The feed point is tapered to match the 50 Ω power amplifier impedance to the impedance of the line. When used with a matching transformer, this tapered section is not required.

One variation of the stripline is the open TEM cell, in which the input section is tapered up from the drive point, located on the ground plane, to the upper plate and tapered down from the upper plate to the 50 Ω termination, located on the ground plane. An open TEM cell with the dimensions shown in Figure 9.54a is described in Reference [7]. This stripline has the great advantage of very high E fields with a low-power amplifier and a useful upper frequency of 400 MHz. The disadvantage with this antenna is that the maximum size of EUT is 0.5 m × 0.25 m × 0.06 m in height.

Because all of the plate/stripline antennas described in this section are open on the sides, they will radiate and must typically be used in a shielded room. One great disadvantage with using these antennas above 30 MHz is that the radiation from the antenna will result in room resonances that change the E field generated between the lines of the antenna. Reference [7] describes errors as high as 49 dB in the E field generated by the open TEM cell antenna due to room resonances. The techniques described in (9) to damp room resonances with ferrite tiles and absorber loads may be employed. Reference [7] describes surrounding just the open TEM cell antenna with ferrite and small pyramid absorber, with a resultant reduction in the amount of absorber required. With this configuration, with the absorber experimentally optimized for location, the errors were reduced to within a ±3 dB range up to 400 MHz.

Figure 9.54b shows a stripline antenna with a 0.6 m height, which has been calibrated to 220 MHz. Maximum EUT height is 0.4 m and the stripline is used in a damped room.

When irradiating man-worn equipment, or similar, a 2.17 m monopole antenna can be used from 2 to 30 MHz and from 2 to 15 MHz with the step-up transformer.

The design of this antenna is shown in Figure 9.55. The rods are threaded aluminum with the top an expandable rod.

6. *Tuned resonator.* The shielded enclosure becomes a large rectangular resonant cavity at resonant frequencies. The resonant frequencies of an enclosure are given by

$$f[\text{MHz}] = 150\sqrt{\left(\frac{k}{l}\right)^2 + \left(\frac{m}{h}\right)^2 + \left(\frac{n}{w}\right)^2}$$

where
  l is the length of enclosure in meters
  h is the height of enclosure in meters
  w is the width or depth of enclosure in meters
  k, m, n are a positive integer 0, 1, 2, 3, …

Note: Not more than one of k, m, or n can be zero at the same time.

0.03 m

0.2 m — 1.5 m

0.87 m

1.5 m

1.5 m

= 0.02 m

Coaxial-type
connector

0.002 m

0.18 m

Coaxial-type
connector

= 0.005 m

(a)   Maximum EUT dimensions = 0.5 m × 0.25 m × 0.06 m height

(b)

FIGURE 9.54   (a) Open TEM cell antenna usable up to 400 MHz. (b) 0.6 m height stripline useful to 220 MHz.

FIGURE 9.55   2.17 m monopole.

Reference [5] provides an approximation for the maximum number of resonant modes that can exist below a specific frequency:

$$N = 155f^3V$$

where
   f is the frequency [GHz]
   V is the volume of cavity [m³]

A 2.59 m high, 3.2 m wide, and 6.25 m long enclosure theoretically exhibits 68 resonant modes over the 20–200 MHz frequency range. Measurements on the same-size enclosure described in Reference [9] identified 25 resonant frequencies, with the possibility that some resonance peaks and troughs were masked by an overlapping resonance. The technique proposed in Reference [9] to reduce the up to 48 dB variation in measured E field at resonance is to contain the field within a wave trap. A coaxial resonator having a helical inner conductor forms a high "Q" circuit of small physical dimensions. A 7.62 cm section of copper tubing of 5.7 cm inside diameter was constructed with field coupling to the resonator accomplished by a 20.3 cm long antenna soldered to the high-potential end of the inner conductor. The construction enabled the resonator to be tuned manually. The resonator was positioned on a 91.4 cm^2 aluminum plate connected to the enclosure's metal framework. The antenna was oriented for maximum coupling to the vertical E-field component. The exact location of the resonator was found to be unimportant as long as it was located further than 30.48 cm from the walls. Measurements were made using a monopole transmitting antenna and a dipole receiving antenna in a shielded room with and without the tuned resonator and with the same test configuration using an open-field test site. The resonator was tuned to the frequency of the transmitted fields during the test.

Figure 9.56 illustrates a massive reduction of 38 dB at 52 MHz in the resonance field strength due to the tuned resonator, which results in a measurement only 3 dB above that of a specific open-field test site. The maximum deviation between the tuned resonator in the shielded-room measurement and the open-field measurement is 15 dB at 54 MHz.

It would appear that the tuned resonator technique holds the promise of a relatively easy and inexpensive solution to the problem of room resonance. An automatic test setup may be possible in which the scan of the EMI receiver or S/A is synchronized to the tuning of the resonator, which may be driven by a stepper motor or similar device. A computer could be used to control the S/A sweep and a servo system used to drive the resonator and sense its position.

7. *Computing near-field antenna coupling, including reflections, in a shielded room.* Above approximately 50 MHz, near-field coupling between antennas that are physically close (0.25–1 m)

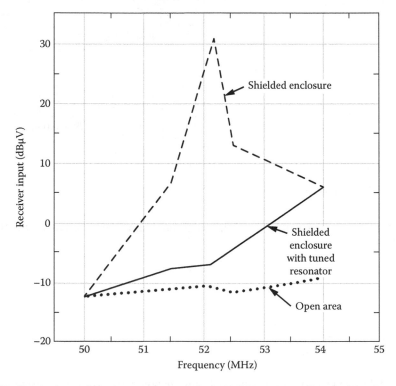

FIGURE 9.56 Comparison of measurements made in a shielded room with and without a tuned resonator and on an open-field test site. (From Dolle, W.C. and Van Stewenberg, G.N., Effects of shielded enclosure resonances on measurement accuracy, *Proceedings of the IEEE International Symposium on Electromagnetic Compatibility*, Anaheim, CA, July 1970.)

can be represented by a mutual impedance. At frequencies up to 150 MHz, the coupling is predominantly capacitive, and above 200 MHz, it is predominantly the radiation mode coupling between the antennas. With the reflections in the shielded-room walls described as images of the antennas, possessing mutual and self-impedance, Reference [10] describes a method of compensating for up to two reflectors (shielded-room surfaces). Alternatively, a computer program such as GEMACS, described in Section 12.4.4, may be used to model the antenna as a system of wires and, where applicable, plates, and the surfaces of the shielded room as ground planes. The composite E field at some distance from the transmitting antenna or the current flow in the load of the receiving antenna may then be computed. It should be pointed out that use of the GEMACS program in this type of application is not without cost in time and effort.

8. *Optoisolated, physically small antennas.* A source of perturbation of the field within a shielded room during radiated emission measurement is the antenna and the metallic connection of the antenna to the measuring equipment. By using small antennas in close proximity to the source of emission, similar to the technique described in Section 9.2.7 but with the use of electrically isolated electric field antennas, the perturbation of the electric field close to the EUT is reduced and the ratio of direct to reflected waves incident on the antenna is increased. Figure 9.57 illustrates the opt-isolated antenna setup and the design and performance is shown in Chapter 2.

9. *Use of ferrite tiles and absorber loads placed at strategically located positions to damp room resonances.* The advantage of this technique is that only a small number of ferrite tiles and small amount of foam absorber material is required, which results in a less expensive modification to the room.

The problem with the standard foam absorber material is that it is placed on the conducting surfaces of the shielded room. At the surface of the room, the E field reduces to zero, and long, typically 36 in., absorber cones are required to obtain any attenuation at frequencies down to 200 MHz. The correct location for the foam absorber is closer into the room and that is exactly the location where blocks of absorber material (absorber loads) are placed. The magnetic field is at a maximum on the surface of the room, and a ferrite tile, which is predominantly an H-field absorber, is ideally suited for mounting on the shielded-room wall. Shielded rooms (semianechoic chambers) have been constructed in which all walls and ceiling are covered in ferrite tiles, often with absorber cones placed on top, and although this approach is effective, it is very costly. When retrofitting a self-supporting room with ferrite tiles, it is advisable to check with a structural engineer or the supplier of the room to provide the loading capability, because ferrite tiles are extremely heavy and can add approximately three times to the weight of a panel. Most rooms are self-supporting with spans of up to 10 ft. If ferrite tiles are added to small rooms, then stiffeners on the walls or ceilings may be all that is required. For rooms with spans larger than 10 ft, an external support made of columns and girders, from which the ceiling panels are suspended, may be required. For the small amount of additional ferrite described in this section, the only precaution may be to ensure that personnel do not climb on top of the self-supporting room.

FIGURE 9.57 Optoisolated small antenna test configuration.

Reference [11] describes the addition of an absorber load to damp resonances in a room, and Reference [12] describes the addition of a very limited number of ferrite tiles. The absorber load alone reduced resonance affected from ±30 dB down to ±15 dB over the frequency range from 50 to 200 MHz. The effect of ferrite tiles was to reduce resonance effects to 10–12 dB over the frequency range from 50 to 230 MHz.

A 12 ft by 12 ft by 8 ft high chamber was loaded with panels 2 ft by 2 ft covered with ferrite tiles. One panel was placed on each side wall, two on the end wall where the table was located, one panel on the rear wall, which contained the door, and three panels on the ceiling. In addition, two movable loads constructed of a wooden container 2 ft by 1.3 ft and 3.25 ft high were filled with absorber foam. Figure 9.58 illustrates one of the wall panels loaded with ferrite and an absorber load. The 12 in. absorber cones are constructed of microwave absorber, which will have no appreciable absorbing properties below 1 GHz and so are not expected to alter the test results below 1 GHz. Measurements were made of the coupling between two antennas, oriented both vertically and horizontally, on the open-field test site and then in the loaded shielded room.

Figure 9.59 illustrates the coupling from 20 to 1000 MHz with the antennas horizontally oriented, and Figure 9.60 illustrates the coupling with the antennas vertically oriented. In Figure 9.59 (horizontal), the only obvious resonances are at 24 MHz, in which a 7 dB difference exists between the open-field and room measurement; at 37 MHz, with a 10 dB difference; and at 50 MHz, where the difference is 7 dB. At all other frequencies, the correlation between the open-field and room measurements is 4–6 dB. In Figure 9.60 (vertical), the worst-case resonance effect is 12 dB at 24 MHz and 10 dB at 49 MHz. At other frequencies, the correlation between open-field and room measurements is remarkably close. The lowest frequency at which the room acts as a resonant enclosure is 58 MHz, and therefore the effects at 24, 37, 49, and 50 MHz are most likely a TEM excitation of the room or possibly a loading effect on the antennas. When compared to antenna coupling in an undamped room, shown in Figure 9.48a, the improvement is remarkable.

FIGURE 9.58 Shielded room with ferrite panel and absorber load for damping.

FIGURE 9.59 Correlation between antenna-to-antenna measurements (horizontally oriented) on an open-field test site and in the loaded shielded room. (Courtesy of EMC Consulting, Inc., Merrickville, Ontario, Canada.)

FIGURE 9.60 Correlation between antenna-to-antenna measurements (vertically oriented) on an open-field test site and in the loaded shielded room. (Courtesy of EMC Consulting, Inc., Merrickville, Ontario, Canada.)

Figure 9.58 illustrates the room with ferrite tile and absorber loads.

10. *Semianechoic and fully anechoic test chambers.* Due to the problems in attempting to isolate EUT emissions from the ambient on an OATS and the problems of antenna calibration on a test site open to the weather, the use of a fully anechoic or semianechoic room may gradually replace OATSs. A room to be used for commercial radiated emission and immunity testing must meet the NSA standards of ANSI C63.4-2014 and the field uniformity requirements of IEC 1000-4-3 or ENV 50140. Using a room for antenna calibration, by comparison, often requires more stringent control of reflections within the room and typically extends above 1 GHz.

Before purchasing a room, the recommendation is that all future possible uses of the room be explored as well as the requirement by the FCC for radiated emission measurements above 1 GHz.

A number of choices exist as to the type of absorber material used on the walls and ceiling of the room. The choices include 1/4 in. to 5/16 in. thick ferrite tiles, grid ferrite tiles, ferrite tile and dielectric combinations, sintered ferrite and ferrite composite combinations, carbon-loaded foam pyramids or wedges, and carbon-loaded foam pyramids and wedges mounted on ferrite tiles.

Ferrite tiles require the least space in a room. Flat tiles can be very effective at low frequency, although the inclusion of a gap between the tiles as small as 0.2 mm can result in a 5 dB degradation in performance. The grid tile has only about 50% of the ferrite material active per polarization and therefore requires twice the volume of a flat tile. The grid has a lower effective permeability, which must be compensated for by increased thickness to obtain the same low-frequency performance. Due to the lower permeability, the degradation in performance due to gaps is less than seen with a flat tile. The grid tile also has a lower permittivity, which improves performance at higher frequency. According to one manufacturer, Toyo, a ferrite grid absorber provides significantly more attenuation above 220 MHz than a flat tile. Toyo describes measurements made on a room with internal dimensions of 8.7 m long × 6 m wide × 5.1 m in height, which is the smallest room size recommended as a 3 m test site, The subject room met the ±4 dB NSA of ANSI C63.4 over the frequency range 30–1000 MHz, with the transmitting antenna located on five different points on a 1.5 m diameter turntable, at 1 m and 2 m heights for horizontal orientation of the antenna, and at 1 m and 1.5 m for vertical orientation. The receiving antenna was scanned in height from 1 to 4 m, with vertical and horizontal orientation, as required in ANSI C63.4.

The room also met the IEC 1000-4-3 field uniformity requirements from 30 to 1000 MHz with the antennas at a distance of 3 m. Sixteen points were measured at 0.5 m intervals, and more than 12 of the points fell within the 0–6 dB window at each measured frequency.

The hybrid combination of ferrite tile and foam absorber can be very effective, however, a word of warning: The correct combination of ferrite tile and dielectric absorber is very difficult to design over the transition frequencies, 30–120 MHz, and a cancelation of performance can be seen. To avoid this potentially severe loss in performance, it is recommended that a properly designed and tested hybrid be purchased.

The use of a multilayered structure made up of layers of different ferrite material has been proposed (Reference [13]), with an improved, lower reflection, compared to flat single-layer tile. The NSA in a 19 m long × 13 m wide × 8.5 m high anechoic chamber fitted with this material was within the ±4 dB requirement, and in a 7 m × 3 m × 3 m room, the field uniformity requirement was met with improved performance compared to a single-layer tile over the 800–1000 MHz frequency range.

The main problem with ferrite tile absorbers is that the performance is not adequate above 1 GHz, although a tile with additional small sections of ferrite mounted "end on" has been tested to 3 GHz.

In the ideal absorber, the impedance of the incident wave and the absorber are the same as free space (377 Ω); the reflection is therefore minimal, and the depth of the absorber is sufficient to absorb all of the incident energy. In practice, this is not the case. Carbon-loaded absorber is usually manufactured in a pyramid or wedge shape, which behaves like a tapered transition with an impedance to a normally incident wave of 377 Ω at the tip increasing gradually over the height. The tapered absorber provides a lower level of reflection than a solid absorber, and a lower amount of absorber material is required.

One problem in the foam absorber is that, despite being flame retardant, it is not completely flameproof, although the use of flameproof material has been proposed. Another problem is that, typically, very long pyramids are required. For example, in a chamber, 9 m long, 6 m wide, and 5.1 m high, a multilayer (Ni–Zn ferrite, 70 mm dielectric Mn–Zn rubber ferrite) absorber met the NSA requirements, whereas a 3 m long pyramidal absorber failed the requirements. Only with a 4 m pyramidal foam absorber were the ±4 dB requirements met.

The recommended maximum reflectivity (level of reflection) of material used in a 3 m test room is less than −18 dB at normal incidence and less than −12 dB at 45° over the full frequency range, although there is no guarantee that lower levels of reflectivity may be required. Rantec manufactures a range of absorber materials, and these include a 1.6 m long polystyrene foam carbon-loaded material mounted on a ferrite tile. This combination provides a higher reflectivity than flat and grid tile from 30 to 300 MHz but a much lower reflectivity from 500 to 1000 MHz. The reflectivity of the Rantec combination, for normal incidence, is −15 dB at 20 MHz, −25 dB at 500 MHz, −30 dB at 1 GHz, and −35 dB at 10 GHz. By contrast, a grid ferrite tile has a reflectivity of −17 dB at 20 MHz, −27 dB at 100 MHz, −25 dB at 500 MHz, but only −15 dB at 1 GHz. According to Rantec, the NSA performance of the combination absorber is within ±3 dB from 30 to 80 MHz. Rantec also manufactures a combination BB ferrite tile and 0.41 m long absorber, which is rated from 30 MHz to 40 GHz and appears ideal for MIL-STD/DO-160 radiated tests and the MIL-STD-462D/E recommended room.

Room resonances are due to the physical parameters of the room; the resonances for a rectangular cavity are described in this section under 6. *Tuned resonators.* By making the walls and ceilings of the room nonparallel (by varying degrees), the amount of pyramidal absorber material required for the room can be reduced, by typically halving the length of the absorber. Reference [14] discusses the analysis of a nonrectangular room.

A problem exists in the use of an anechoic chamber as a 3 or 10 m emission test site at low frequency. In a comparison of emission measurements made on an EUT measured on an OATS and emissions in a semianechoic chamber, significant errors were seen at 50 MHz and are predicted to exist over a 10–60 MHz range. In measurements on a 10 m semianechoic chamber 7.9 m high, 21.3 m long, and 11.6 m wide with 2.4 m long absorber cones, a deviation in the "standard" open-site attenuation performance of 9.3 dB was seen at 45 MHz. The problem is a poor absorber performance at low frequency, which most probably results in a TEM excitation of the room.

Pyramidal absorber cones typically provide adequate attenuation from 1 to 40 GHz but little or no attenuation at low frequencies. Ferrite tile or grid performs well from 30 to 600 MHz but very little attenuation at 1 GHz and above. A hybrid with a base of ferrite tile or grid covered with a pyramidal cone is available. It has been reported that the reflectivity of ferrite tiles and grids can be improved by adding a dielectric layer between the ferrite and the chamber wall. When using a 3 or 10 m chamber for radiated emission measurement in accordance with FCC CFR 47 part 15 or any of the EN requirements, hybrid absorber is often used.

One potential problem with carbon-loaded foam is degradation in both absorptive properties and fire-retardant capability due to water damage. How much if any modification to either the fire-retardant properties or to the absorptive properties of the absorber occurs when the absorber is soaked in water depends on a number of variables, which are described as follows.

Two different methods of adding the carbon and the fire-retardant material to the neoprene latex polyurethane substrate exist. In the two-step process, used, for example, by Rantec, the carbon is soaked into the material in a water-based solution and allowed to dry. The nonorganic salts, which are used as the fire-retardant material, are then added, also in water solution, and allowed to soak the material. By adding the fire-retardant material separately, the carbon is locked in. In tests conducted on this type of absorber, when wall-mounted absorber was wetted but not immersed in water, neither the fire retardant nor RF properties of the absorber were compromised. In one test, when absorber was soaked in cold water for 2 h and allowed to drain naturally, the material was still fire retardant and retained its RF properties.

In the one-step process, used, for example, by Cuming Corporation, the fire-retardant properties and the carbon are both applied in a water-based solvent. Theoretically, when this material is soaked in water, the carbon can migrate down through the material by gravity, resulting in a nonuniform distribution of the absorptive properties. In an extreme case, the carbon can leach out of the foam. The nonuniform distribution of the absorptive properties will, typically, affect only precise antenna measurements in the room and are unlikely to modify radiated emission or radiated susceptibility test results.

Only anecdotal information exists on the performance of this one-step process material: when the material was wetted, it retained its fire-retardant and RF properties, but when soaked, the fire retardancy was compromised.

One other potential source of degradation is if water soaks through concrete onto the material, it can take salts and lime from the concrete into the material and change the properties.

9.5.2 GTEM, TEM, AND OTHER CELLS

TEM-type cells were developed to allow a cost-effective method for radiated susceptibility testing, and their use has been extended to radiated emission measurements and antenna calibration. The cell is a less expensive alternative to a semianechoic chamber, and a number of methods of correlation between the radiated emission results obtained using the GTEM and an OATS have been presented.

9.5.2.1 TEM and Other Cells

The standard dual-port TEM cell, often called a *Crawford cell*, forms a rectangular two-conductor transmission-line structure with a central septum on which the EUT is mounted and with tapered sections at each end. The transmission line is closed and is often called *coaxial*, although it does not have an axis of symmetry. The dual-port cell has an input measuring part and an output measuring port, whereas a single-port TEM has a matched load at the end without a port. The dual-port TEM tapered end sections match the middle section to the 50 Ω coaxial port connectors. The single-port, or wideband, TEM has the shape of a pyramid on its side. The termination is on the wide rectangular side and typically combines a low-frequency lumped load that matches the characteristic impedance of the TEM cell with high-frequency absorber material. The most common form of single-port cell is the GTEM cell, which is discussed in Section 9.5.2.2.

The dual-port TEM cell has a limited bandwidth determined by the generation of higher-order modes in the field. The maximum height of the EUT will limit the field uniformity in a susceptibility/immunity test and change the way in which the EUT couples to the cell. A very approximate maximum useful volume for immunity test in a cell is 0.333 times the distance between the septum and upper surface times 0.333 times the cell width, although this may be pushed to 0.5 × 0.5. A field uniformity of ±1 dB is possible in the classic TEM cell.

The cell is designed to generate a transverse electromagnetic wave with a perpendicular E field and an H field at 90°. The cell is a transmission line terminated in its characteristic impedance, so the field impedance inside the cell is 377 Ω. Because the EUT is mounted, in effect, on the center conductor of the transmission line, a current flows across the EUT. The cell may therefore be used to measure the magnetic field shielding effectiveness of a small enclosure by measuring the internal current, using a small surface current probe inside the enclosure. Typically, the cable connected to the current probe and to the measuring instrument located outside of the cell must be extremely well shielded. This is achieved by containing the cable within a copper tube that is soldered or connected via compression fittings to the enclosure and the wall of the TEM cell. The current probe is then moved to a location outside of the enclosure, and the ratio of measured external to internal current is equal to the magnetic field shielding effectiveness of the enclosure.

The dual-port TEM cell generates higher-order modes than the TEM mode by the nature of the geometry of the cell. When the effective height of the TEM cell is equal to a half a wavelength, the first resonance of the cell occurs; this is the limitation on the useful upper frequency. In addition, when the EUT is placed on the septum, the current around the EUT is perturbed, which also generates higher-order modes. In a standard TEM cell rated at 200 MHz (useful upper frequency), a resonance in the empty cell was seen at 161 MHz with a change in the measured 70 V/m field up to 80 V/m and down to 40 V/m. With an EUT in the cell, the same resonance at 161 MHz was seen and an additional resonance at 190 MHz, at which the 70 V/m field increased up to 120 V/m. Absorber can be placed at locations within the cell to damp these modes, but this interferes with the generation of the required TEM mode. Alternative damping techniques exist, such as cutting slots in the cell wall and backing these slots with absorber and additional shielding as well as a split-up septum. These damping techniques cannot be used to upgrade existing cells but must be built into the new cell. With the use of such damping and a split-up septum, a 0.437 m wide, 0.284 m high, and 1.8 m long TEM cell can be used up to 1 GHz. However, due to the regular shape of the single-port TEM cell, it has no inherent limitation in bandwidth and would appear to be the obvious choice over the dual-port TEM cell.

The TEM cell can be used to measure radiated emissions from an EUT or a PCB. The emitted field from the EUT couples via the transmitting modes of the cell and thereby couples a voltage at the port/s of the cell. Antenna calibration can be accomplished as described in Section 9.5.2.2. The great advantage of the TEM cell for emission measurements is that it is an effective shielded enclosure that reduces the ambient-induced voltages to extremely low levels, typically the noise floor of the measuring equipment. Measurements in a 3 or 10 m anechoic chamber or an OATS are made in the far field, whereas it is the near field that couples predominantly to the TEM cell. If the radiation from the EUT is highly directional, then the coupling to the TEM cell may not be a true measure of the radiation. Also, the radiation characteristics of the EUT are different from those on an OATS, where the only conductive surface in close proximity is the ground plane. The TEM cell measures power transmitted to the output and not the maximum electric field. Acceptable emission measurements can be achieved; for example, the emissions from a PCB located 0.317–0.65 cm from the septum of a TEM cell are recorded in Reference [15]. These emissions are compared to measurements using the EMSCAN™ system and in an absorber-lined shielded room, with a close correlation.

Two very small, DC to 1 GHz and DC to 2.3 GHz, TEM cells with a patented septum design are described by Fischer Custom Communications, Inc., for both emission and IEC 10004-3 immunity testing. Less than 3.7 mW of input power is required to achieve a 10 V/m field and 37 W to achieve 1000 V/m. The dimensions of the area for good field uniformity is 2 in. (5.08 cm) × 2 in. (5.08 cm) × 0.5 in. (1.27 cm), and so the cell is ideal for the measurement of ICs and small PCBs. Fisher also manufactures larger custom TEM cells with a lower useful upper frequency.

Alternative types of test cell, such as the WTEM, TrigaTEM, and Triple TEM, exist but have not proven field uniformity. One alternative, commercially available cell, called the S-LINE, is manufactured by Rhode and Schwarz. Unlike the coaxial construction of the TEM cell, the S-LINE consists of symmetrical, two-wire TEM lines in a shielded enclosure. The lines are fed with the test signal and terminated with their characteristic impedance. A larger S-LINE has 1.5 m × 1 m × 1 m dimensions and fulfills the requirements of IEC 1000-4-3 for field uniformity over a 50 cm × 50 cm area. The cell is useful from 150 kHz to 1 GHz and can generate a 10 V/m field with 20 W of input power at the center of the S-LINE. A smaller model of the S-LINE, with 1 m × 0.7 m × 0.7 m, is also available, and software is available that allows fully automatic testing.

9.5.2.2 GTEM Cells

The limitations as to test volume and operating frequency bandwidth of the TEM cell have been to a large extent solved with the single-port GTEM cell, which was patented in 1991. The asymmetrical septum is designed to match the flared/tapered rectangular-cross-sectional waveguide. The inner conductor is terminated into a hybrid BB -matched load. For frequencies up to about 40–90 MHz, distributed resistors are used; for frequencies above this range, pyramidal foam absorbers are used. The overall field uniformity is better than ±3 dB from 30 to 1000 MHz in a test area behind the large access door. The maximum EUT height is approximately 0.33 times the distance from the ground to the septum, based on the earlier TEM cell research. However, there is no theoretical limit on the size of the GTEM, only a practical one. GTEM cells are available with ground-to-septum heights of 0.5, 1.0, and 1.5 m. A 3.5 m version has been built but has proven to be expensive.

The main purpose in the use of the GTEM in radiated emission measurements is to replace the traditional OATS, due to the problem of electromagnetic ambient, and to provide a less expensive alternative to the semianechoic chamber.

The principle in the use of the GTEM for radiated emission measurements is the development of a mathematical model implemented in software for the direct comparison of data taken in a GTEM to data acquired on an OATS.

Three measurements are made, using the GTEM, with the EUT oriented in three orthogonal positions, centered in the measurement volume of the GTEM. The three positions are assumed in a manner such that the positive x-, y-, and z-axes of the EUT are sequentially interchanged. The frequency, the three voltages measured at each frequency, the height of the EUT under the septum, the height of the septum over the center of the EUT, the measurement distance, and the height scan range used for the test antenna maximization search are inputs to the correlation algorithm, which is described in Reference [16] as follows.

At each frequency, the GTEM correlation algorithm executes the following computations:

Performs a root sum of squares summation of the three orthogonal voltages
Computes the total power emitted by the EUT as determined from the summation of the three
    voltages and the TEM-mode equations for the GTEM
Computes the current excitation of an equivalent Hertzian dipole when excited with that input
    power
Places the hypothetical Hertzian dipole at a specified height over a perfect ground plane
Computes the horizontally and vertically polarized field strength at appropriate height inter-
    vals over the total operator-selected correlation algorithm height, 1–4 m
Selects the maximum field strength (larger) value of the horizontally or vertically polarized
    field strengths over the height range selected
Presents this maximum value for comparison to the chosen EMC specification limit

The main problem seen with measurements in the GTEM and in the correlation to OATS results is that mean value in the variance is 6 dB, with 11–15 dB difference in individual measurements. However, similar differences can be seen in comparing some OATS to other OATS results; see Section 9.4.3.1. This variation from OATS to OATS is particularly disappointing. The main differences between the OATS and GTEM results are that only three faces of the EUT are rotated and measured in the GTEM. A twelve-position measurement consists of a measurement on all six faces of the EUT in both polarizations. A further development is a twelve plus four measurement, which can estimate the directivity of the EUT, from which an estimate of the gain can be obtained and used to replace the value of the gain of the dipole in the correlation algorithm.

A new development is a hyperrotation GTEM, in which the EUT is placed on a gimbaled turntable within the GTEM, which itself can rotate about an axis, thus allowing for two independent degrees of motion between the EUT and the GTEM.

In addition, the effect of the ground plane on the antenna calibration in the OATS measurement should be included in the correlation algorithm, especially for low-gain antennas used at the lower heights in the OATS 1–4 m search. The other big problem is that cable manipulation to maximize emissions is a requirement in the OATS test but is impractical in GTEM tests. The EUT hyperrotation in the GTEM was thought to compensate for cable manipulation, but tests have proven that this is not the case. Measurements of an EUT on an OATS with and without cable manipulation and in a GTEM showed a 1–11 dB change due to cable manipulation.

For MIL-STD 1 m correlation, a single-axis model has been proposed. A standard radiator is installed in the GTEM, and a single-axis voltage measurement is made over a range of frequencies.

The same standard radiator is located in a shielded room in a standard position. An RE02/RE102 test would be performed on the radiator that provides field strength values over the same range of frequencies. The GTEM voltage and E-field levels will provide a calibration factor that can be applied to the GTEM equivalent MIL-STD-461 RE02/RE102 measurement.

The GTEM is gradually gaining acceptance for measurements to FCC and EU requirements. The FCC issued a public notice (September 1993) stating that GTEM measurement data will be accepted under the following limited conditions:

GTEM results will be accepted when the GTEM measurements have demonstrated equivalence to OATS test results.

Acceptable comparison measurements must be filed with the Sampling and Measurement Branch of the FCC with appropriate analysis that demonstrates the GTEM result equivalence with a listed OATS meeting the NSA requirements.

The validity of calculated correlation coefficients must be supported by statistical analysis.

In cases of disagreement, final emission tests will be performed on an OATS.

Measuring antenna factors in a GTEM would appear to result in a good (0.16 dB to maximum 1.74 dB over 300–1000 MHz) correlation to measurements made on ANSI C63.5 range. One GTEM antenna calibration technique has a known voltage applied to the input of the GTEM, with a measurement of the output voltage (V_o) from the antenna. The E field in the GTEM is found from the following simple relationship: E field = input voltage (V_i) divided by the septum-to-floor spacing (h). Since $AF = 20 \log E/V_o$, the antenna factor measured in a GTEM is equal to $20 \log (V_i) - 20 \log (V_o) + 20 \log (1/h)$.

9.6 MILITARY EMI REQUIREMENTS AND MEASUREMENT TECHNIQUES

9.6.1 MIL-STD-461: ELECTROMAGNETIC EMISSION AND SUSCEPTIBILITY REQUIREMENTS FOR THE CONTROL OF EMI

MIL-STD-461 is a document that specifies EMI requirements and test levels on military equipment, although the space industry has also widely used the specifications on nonmilitary equipment. Initially, MIL-STD-461 applied to Navy, Air Force, and Army equipment. The next issue, MIL-STD-461 A, provided notices that applied to individual services. Table 9.23 provides information on which notices of MIL-STD-461 and MIL-STD-462 apply to either all or individual services. With the advent of MIL-STD-461F, the basic test requirements, test methods, and test setups contained in MIL-STD-462 have been updated and included in MIL-STD-461F/G.

The majority of EMI test requirements and test techniques for equipment and subsystems applied by space agencies such as NASA, ESA, and Canadian Space Agency (CSA) are based on MIL-STD-461 and MIL-STD-462. MIL-STD-461 specifies the EMI requirements and limits for military equipment, and MIL-STD-462 specifies the test techniques to be used. System-level requirements are contained in MIL-STD-464, March 18, 1997, and MIL-1541A is the military standard for EMC requirements for space systems applied by the U.S. Air Force. MIL-STD-416F deals with interchangeable modular equipment and states that equipment that is made up of line replaceable modules (LRMs) must be requalified if an LRM is replaced by a new or different model, even if it is a form, fit, and function.

MIL-STD-461F also prohibits the use of shielded power cables unless the platform on which the equipment is to be installed shields the power bus from point of origin to the load. However, MIL-STD-461F specifically states that input (primary) power leads, returns, and wire grounds shall not be shielded.

The basic rule is that the wiring/cable used in testing shall simulate the actual installation and usage.

TABLE 9.23

Services to Which Notices of MIL-STD-461 and 462 Apply

Document	Year	Applicability
MIL-STD-461	1967	All Services
MIL-STD-461A	1968	All Services
Notice 1	1969	All Services
Notice 2	1969	Air Force
Notice 3	1970	Air Force
Notice 4	1971	Army
Notice 5	1973	All Services
Notice 6	1973	All Services
MIL-STD-46IB	1980	All Services
MIL-STD-461C	1986	All Services
MIL-STD-462	1967	All Services
Notice 1	1968	All Services
Notice 2	1970	Air Force
Notice 3	1971	Army
Notice 4	1980	Navy
Notice 5	1986	Navy
MIL-STD-46ID	1993	All Services
MIL-STD-462D	1993	All Services
MIL-STD-46IE	1999	All Services
MIL-STD-461F	2007	All Services
MIL-STD-461G	2015	All Services (draft)

The goal of this section is to provide an overview of the tests but, more important, also of the most common errors that occur in EMI testing. These errors can result in damage to equipment, subsystems, or flight systems, and it is imperative, to avoid damage delay and additional cost, that a thorough understanding of the correct test methods exist and that a detailed test procedure be in place.

Although superseded by MIL-STD-461 D, E, and F MIL-STD-461A to MIL-STD 461C may still be applied, especially for reprocurement of equipment contained on old platforms. If the procuring agency is satisfied that the earlier versions of MIL-STD-461, or a tailored version thereof, adequately covers a specific electromagnetic ambient, then there is no good reason to replace them. Also, if all other equipment on the platform is qualified to MIL-STD-461A to C, then new equipment is also likely to use these earlier versions. Some of the differences between MIL-STD-461A/B/C and D/F will also be discussed. The emission and susceptibility requirements in MIL-STD-461 are designated in accordance with an alphanumeric coding system, where

C = conducted
R = radiated
E = emission
S = susceptibility
UM = unique requirement(s) intended for a miscellaneous, general-purpose piece of equipment or a subsystem

Table 9.24 lists the emissions and susceptibility requirements contained in MIL-STD 46IB Part 1.

MIL-STD 461B contains 10 parts; parts 2–10 apply to specific equipment and subsystem classes and part 1 contains general information. Table 9.25 lists the classes of equipment and which part of MIL-STD-461B is applicable. Different requirements are placed on the different classes.

TABLE 9.24
Emissions and Susceptibility Requirements Contained in MIL-STD 46IB Part 1

Requirement	Description
CE01	Conducted emissions, power, and interconnecting leads, low frequency (up to 15 kHz)
CE03	Conducted emissions, power, and interconnecting leads, 0.015–50 MHz
CE06	Conducted emissions, antenna terminals, 10 kHz–26 GHz
CE07	Conducted emissions, power leads, spikes, time domain
CS01	Conducted susceptibility, power leads, 20 Hz–50 kHz
CS02	Conducted susceptibility, power leads, 0.05–400 MHz
CS03	Intermodulation, 15 kHz–10 GHz
CS04	Rejection of undesired signals, 20 Hz–20 GHz
CS05	Cross modulation, 20 Hz–20 GHz
CS06	Conducted susceptibility, spikes, power leads
CS07	Conducted susceptibility, squelch circuits
CS09	Conducted susceptibility, structure (common-mode) current, 60 Hz–100 kHz
RE01	Radiated emissions, magnetic field, 0.03–50 kHz
RE02	Radiated emissions, electric field, 14 kHz–10 GHz
RE03	Radiated emissions, spurious and harmonics, radiated technique
RS01	Radiated susceptibility, magnetic field, 0.03–50 kHz
RS02	Radiated susceptibility, magnetic induction field, spikes, and power frequencies
RS03	Radiated susceptibility, electric field, 14 kHz–40 GHz
UM03	Radiated emissions, tactical and special-purpose vehicles, and engine-driven equipment
UM04	Conducted emissions and radiated emissions and susceptibility engine generators and associated components UPS and MEP equipment
UM05	Conducted and radiated emissions, commercial electrical and electromechanical equipment

TABLE 9.25
Classes of Equipment and the Applicable Part of MIL-STD 46IB

Class	Description	Applicable Part
A	Equipment and subsystems that must operate compatibly when installed in critical areas	
A1	Aircraft (including associated ground support equipment)	2
A2	Spacecraft and launch vehicles (including associated ground support equipment)	3
A3	Ground facilities (fixed and mobile, including tracked and wheeled vehicles)	4
A4	Surface ships	5
A5	Submarines	6
B	Equipment and subsystems that support the Class A equipment and subsystems but that will not be physically located in critical ground areas. Examples are electronic shop maintenance and test equipment used in noncritical areas; aerospace ground equipment used away from flight lines; theodolites, navaids, and similar equipment used in isolated areas	7
C	Miscellaneous, general-purpose equipment and subsystems not usually associated with a specific platform or installation. Specific items in this class are:	
C1	Tactical and special-purpose vehicles and engine-driven equipment	8
C2	Engine generators and associated components, uninterruptible power sets (UPS) and mobile electric power (MEP) equipment supplying power to or used in critical areas	9
C3	Commercial electrical or electromechanical equipment	10

TABLE 9.26

Categories of Class A1 Equipment and Subsystems

Category	Description
A1a	Air-launched missiles
A1b	Equipment installed on aircraft (internal or external to airframe)
A1c	Aerospace ground equipment required tor the checkout and launch of the aircraft, including electronic test and support equipment
A1d	Trainers and simulators
A1e	Portable medical equipment used for aeromedical airlift
A1f	Aerospace ground equipment used away from the flight line, such as engine test stands and hydraulic test fixtures
A1g	Jet engine accessories

TABLE 9.27

Categories of Class A2 Equipment and Subsystems

Category	Description
A2a	Equipment installed on spacecraft or launch vehicle
A2b	Aerospace ground equipment required for the checkout and launch, including electronic test and support equipment
A2c	Trainers and simulators

The individual classes of equipment are further divided into separate categories, and those of classes A1 and A2 are shown in Tables 9.26 and 9.27, respectively. Which of the requirements shown in Table 9.24 are applicable depends on the category of equipment, and this is shown in Table 9.28 for categories of Class A2 equipment.

CE01 and CE03 requirements are placed on primary power sourced by the vehicle, ship, spacecraft, etc., and on control lines using primary power.

Although MIL-STD-461 includes power to other equipment in the applicability for CE01 and CE03, the requirement is invariably confined to primary power, which is passed on, from one equipment or subsystem to another. When CE01 or CE02 is applied to secondary power, it is typically limited to power supplied to equipment procured under a different contract or from a different manufacturer. Excessive noise on secondary power lines and signals is often detected by interequipment/subsystem susceptibility and failure during radiated emission tests.

For MIL-STD-461F requirements, Table 9.29 describes the emission and susceptibility tests and Table 9.30 summarizes the requirements for equipment and subsystems intended to be installed in or on or launched from various military platforms or installations. When an equipment or subsystem is to be installed in more than one type of platform or installation, it shall comply with the most stringent of the applicable requirements and limits. An *A* entry in the table means the requirement is applicable. An *L* means the applicability of the requirement is limited as specified in the appropriate requirement paragraphs of this standard; the limits are contained herein. An *S* means the procuring activity must specify the applicability and limit requirements in the procurement specification. Absence of an entry means the requirement is not applicable.

If MIL-STD-461 requirements are placed on equipment, it is important to ascertain the class, category, and location of the equipment and the procuring service for these depend on the test levels. The test levels can drive the design of equipment. For example, the radiated

TABLE 9.28

Categories of Class A2 Equipment and Subsystems and Applicable Parts of MIL-STD 461

Requirement	Categories of Class A2 Equipment/Subsystems			Applicable	
	A2a	A2b	A2c	Paragraph	Limit Curve
CE01	T			2	3-1
CE03	Y	Y	Y	3	3-2, 3-3
CE06	Y_L	Y_L		4	
CE07	Y	Y	Y	5	
CS01	Y	T	Y	6	3-4
CS02	Y	Y	Y	7	
CS03	Y_L	Y_L		8	
CS04	Y_L	Y_L		9	3-5
CS05	Y_L	Y_L		10	
CS06	Y	Y		11	3-6
CS07	Y_L	Y_L		12	
RE01	T			13	3-7
RE02	Y	Y	Y	14	3-8, 3-9
RE03	Y_L	Y_L		15	
RS02	Y	Y	Y	16	3-6
RS03	Y	Y	Y	17	

Notes: Y, applicable; Y_L, limited applicability; T, applicable on a case-by-case basis.

TABLE 9.29

Emissions and Susceptibility Requirements Contained in MIL-STD-461F/G

Requirement	Description
CE101	Conducted emissions, power leads, 30 Hz–10 kHz
CE102	Conducted emissions, power leads, 10 kHz–10 MHz
CE106	Conducted emissions, antenna terminal, 10 kHz–40 GHz
CS101	Conducted susceptibility, power leads, 30 Hz–150 kHz
CS103	Conducted susceptibility, antenna port, intermodulation, 15 kHz–10 GHz
CS104	Conducted susceptibility, antenna port, rejection of undesired signals, 30 Hz–20 GHz
CS105	Conducted susceptibility, antenna port, cross modulation, 30 Hz–20 GHz
CS106	Conducted susceptibility, transients, power leads (CS106 removed from MIL-STD-461G)
CS109	Conducted susceptibility, structure current, 60 Hz–100 kHz
CS114	Conducted susceptibility, bulk cable injection, 10 kHz–200 MHz
CS115	Conducted susceptibility, bulk cable injection, impulse excitation
CS116	Conducted susceptibility, damped sinusoidal transients, cables, and power leads, 10 kHz–100 MHz
RE101	Radiated emissions, magnetic field, 30 Hz–100 kHz
RE102	Radiated emissions, electric field, 10 kHz–18 GHz
RE103	Radiated emissions, antenna spurious, and harmonic outputs, 10 kHz–40 GHz
RS101	Radiated susceptibility, magnetic field, 30 Hz–100 kHz
RS103	Radiated susceptibility, electric field, 2 MHz–40 GHz
RS105	Radiated susceptibility, transient electromagnetic field

TABLE 9.30
MIL-STD-461E Requirement Matrix

Equipment and Subsystems Installed in or on or Launched from the Following Platforms or Installations	Requirement Applicability																
	CE101	CE102	CE106	CS101	CS103	CS104	CS105	CS106	CS109	CS114	CS115	CS116	RE101	RE102	RE103	RS101	RS103
Surface ships	A	A	L	A	S	S	S	A	L	A	L	A	A	A	L	A	A
Submarines	A	A	L	A	S	S	S	A	L	A	L	A	A	A	L	A	A
Aircraft, Army, including flight line	A	A	L	A	S	S	S			A	A	A	A	A	L	A	A
Aircraft, Navy	L	A	L	A	S	S	S			A	A	A	L	A	L	L	A
Aircraft, Air Force		A	L	A	S	S	S			A	A	A	L	A	L		A
Space systems, including launch vehicles		A	L	A	S	S	S			A	A	A		A	L		A
Ground, Army		A	L	A	S	S	S			A	A	A		A	L	L	A
Ground, Navy		A	L	A	S	S	S			A	A	A		A	L	A	A
Ground, Air Force		A	L	A	S	S	S	A		A	A	A	A	L		A	A

Notes: A, applicable; L, limited as specified in the individual sections of this standard; S, procuring activity must specify in procurement documentation. No entry means requirement is not applicable.

susceptibility test levels (RS03) for Class A4 surface ships contained in part 5 of MIL-STD 461B are 150 V/m from 14 kHz to above 10 GHz for areas exposed above decks, whereas the limit is only 1 V/m below decks. The EMI emission and susceptibility test levels for category A2a equipment installed on spacecraft that are most frequently applied are provided as follows from MIL-STD-461B part 3:

CE01	30 Hz–2 kHz	130 dBμA = 3.16 A reducing to 86 dBμA = 19.9 mA at 15 kHz
CE03	15 kHz	86 dBμA reducing to 20 dBμA at 2 MHz
	2–50 MHz	20 dBμA = 10 μA
CS01	30 Hz–1.5 kHz	5 V rms or 10% of supply, whichever is less, = 2.8 V rms for 28 V supply
	1.5–50 kHz	Reduces linearly to 1 V rms or 1% of supply voltage, whichever is greater at 50 kHz
CS02	50 kHz–50 MHz (400 MHz)	1 V (rms)
CS06	Spike 1	± 200 V, $t = 10$ μs $\pm 20\%$
	Spike 2	± 100 V, $t = 0.15$ μs $\pm 20\%$
RE01	0.03–50 kHz	Dependent on location and AC or DC current equipment
RE02	NB lowest limit	30 dBμV/m = 31.6 μV/m at 27 MHz
	BB lowest limit	65 dBμV/m/MHz at 200 MHz (see Figure 1.4 for BB RE02 curve)
RS02	Spike 1	200 V, $t = 10$ μs $\pm 20\%$
	Spike 2	100 V, $t = 0.15$ μs $\pm 20\%$
	Power frequency	3A
RS03	Frequency range	E field (V/m)
	14 kHz–30 MHz	10
	30 MHz–10 GHz	5
	10–40 GHz	20

CS06 requirements are relaxed for equipment and subsystems whose power inputs are protected with varistors or similar transient protection devices so that if the equipment is not susceptible to peak voltages equal to the maximum safe level of the device, then the requirement is met. The problem with the relaxation is the transient device may be damaged with the equipment connected to the power line in the installation, assuming voltage transients at the specified levels are present on the line and the source impedance is sufficiently low. If the maximum transient voltage predicted for the line will never be as high as specified, then the requirement should be relaxed to a realistic worst-case level. If the maximum transient voltage does exist, then a realistic CS06 test for equipment with transient protection is to calculate or measure the line impedance between the EUT and the closest source of transient noise and to use this impedance in series with the specified transient voltage minus the breakdown voltage of the device to predict the peak current through the device. The protection device may then be chosen to be capable of withstanding the peak current, and the spike generator output voltage may be applied to the input of the EUT with a series resistor equal to the supply-line impedance. This realistically tests the effectiveness of the protection device and the susceptibility of the equipment.

A safety margin exists between the emission limits and the susceptibility test levels. This safety margin ensures that, with a number of emitters colocated, the radiated ambient, which is a composite of all the emissions, is below the radiated susceptibility test levels. Likewise, the conducted emissions limit ensures that, when a number of units of equipment or subsystems share a power supply, the composite emission is below the susceptibility test levels. In many instances, the safety margin is excessive. For example, if we take the NB RE02 limit at 27 MHz and the susceptibility test level, which is a NB CW field, we could allow 158,489 pieces of equipment 1 m apart, emitting coherently at the RE02 limit, and still achieve a 6 dB safety margin between the RS03 test level and the radiated ambient. It is physically impossible to locate 158,489 pieces of equipment 1 m from each other and highly improbable that each piece of equipment emits at the limit and at the same frequency. Even when equipment and cables are located in close proximity, the level of E field generated by any equipment that has met RE02 requirements is likely to be m V/m and not V/m. For example, a safety margin of approximately 8–70 dB (above 1 MHz), dependent on frequency and source, exists between typical RE02 limits and RS03 test levels, assuming the equipment and cables are colocated 5 cm apart. One justification for imposing high radiated susceptibility test levels is that equipment may be in the proximity of transmitting antennas. The power radiated by antennas at frequencies that are required for the transmission is not included in the RE02 emission requirements. As the location of aircraft, vehicles, and ships changes, so does the electromagnetic ambient, it is reasonable to specify high susceptibility test levels over the 14 kHz–40 GHz frequency range. However, if stationary equipment is located within structures that provide a high level of shielding, such as most aircraft, vehicles, or ships, or when the location of stationary equipment and subsystems relative to transmitters and the frequency and power of the transmissions are known, then the susceptibility test levels should be tailored accordingly. It is the intention of MIL-STD-461 that these levels be tailored for the operational radiated electromagnetic environment, and this includes both friendly and hostile emitters that equipment or a subsystem may encounter during its life cycle. For example, if a spacecraft contains a transmitter at 1.5 GHz, then the susceptibility test levels from typically 1.4–1.6 GHz may be increased, based on the transmitted power and the location of the transmitting antenna relative to equipment on the spacecraft. Test levels should be as realistic as possible, and with this intent, MIL-STD-461 encourages the procuring agency to tailor the levels based on the predicted ambient. This means that any susceptibility test level considered appropriate may be specified. Likewise, the radiated and conducted emission test levels, although dependent on the classification of the equipment, may be tailored to match the environment. It is imperative to establish before the design begins which test levels apply and to design accordingly, just as a good design accounts for other environmental factors, such as vibration and thermal. MIL-STD-461 and -462 apply to units, individual pieces of equipment, and subsystems, but not to systems.

The exact definition of *subsystem* or *system* is open to discussion; however, one definition of a subsystem is one or more units of equipment that provide a function but not the complete function of a system. When a number of units comprise a subsystem, which will be tested as such, it is advisable to budget the specification test levels between the units, especially when the units are procured from different manufacturers. The EMI-budgeted requirements can then form part of the procurement specification. Budgeting must be done realistically; otherwise, the approach falls into disregard. As an example of budgeting, let us assume a subsystem comprising three units. One of the units is a transmitter at 1 GHz, the second is a piece of digital equipment, and the fourth is an analog servo system driving a high-impedance transducer. In budgeting the radiated emission limit, the 1 GHz transmitter is the only device likely to radiate at 1 GHz and harmonics thereof. Thus, this unit will get a major share of the budget at 1 GHz and above, for even though the transmitting antenna is excluded from the test setup, significant radiation at 1 GHz and above must be expected from cables and enclosure. For example, it would be realistic to budget the NB radiated emission limits such that the 1 GHz transmitter level is 2 dB down on the subsystem limit and the other units are 20 dB down. Assuming the analog unit does not contain a switching-power supply, it is unlikely to generate NB or BB conducted noise, whereas the digital equipment is likely to generate both NB and BB conducted noise. Remember, NB noise is not necessarily limited to CW sources but includes harmonically related emission with a PRF above a specific frequency. The 1 GHz transmitter probably contains a switching-power supply for efficiency and thus may generate both BB and NB conducted noise. A reasonable budget for a BB conducted noise is therefore 5 dB down on the subsystem limits for the digital equipment, 8 dB down for the transmitter, and 26 dB for the analog equipment. Should any unit fails the budgeted specification limit but meets the subsystem limit, the manufacturer would apply for a waiver; and if the emissions from the remaining two units were insignificant at the same frequency or frequencies, then the waiver would be granted. Budgeting ensures that a subsystem will meet requirements and place a contractual requirement on unit manufacturers and subcontractors.

It is a common occurrence for equipment to fail at least one or more of the MIL-STD-461 requirements. Equipment that has not been designed with any consideration given to achieving EMC fails most frequently, and some equipment appears to have been designed to fail. Some of the most difficult tests to pass are RE02/RE102, CS01/CS101, and RS03/RS103. RE02/RE102 emission limits are low and are readily exceeded. For example, with no more than 30 μA at 27 MHz flowing on a cable located 5 cm above the ground plane, the typical RE02/RE102 test setup, the RE02/RE102 limit is reached. CS01/CS101 test levels at 30 Hz and above often result in equipment susceptibility. Power-line filters rarely provide significant attenuation at frequencies below 10 kHz, and regulated power supplies attempt to regulate at frequencies from 30 Hz to 1 kHz. Over this frequency range, the phase shift in the supply control loop may cause the supply to oscillate at one or more critical frequencies. Thus, power supplies should be designed, or chosen, with the CS01 test requirements taken into consideration. RS03 levels can result in high current flow on cables, especially at cable resonant frequencies and less frequently may enter enclosures via aperture or seam coupling.

MIL-STD-461C was introduced in 1989 and included test levels that covered the direct and indirect effects of a nuclear electromagnetic pulse. Here are some of the additions to MIL-STD-46 IB contained in MIL-STD-461C:

CS10	Conducted susceptibility, damped sinusoidal transients, pins, and terminals, 10 kHz–100 MHz at a maximum current of 10 A
CS11	Conducted susceptibility, damped sinusoidal transients, cables, 10 kHz–100 MHz at a maximum current of 10A
RS05	Radiated susceptibility, electromagnetic pulse field transient 550 ns wide at the 0.1 × peak field pulse level (peak field is 50 kV/m)

Some of the modifications made to MIL-STD-461B and contained in 461C are

In part 1, the filter capacitor limitation of 1.1 pF maximum for 60 Hz and 0.02 pF maximum for 400 Hz supplies is confined to U.S. Air Force equipment only.

RE01	Relaxed 36 dB in the 30 kHz frequency range
CE07	Requirements limited to 50 μs duration spike.
CE01	Measurement bandwidth limited to power frequency +20% for AC and 75 Hz for DC leads

The major differences between MIL-STD-461B/C and MIL-STD-461F are given in the following sections, describing the different tests.

9.6.2 MIL-STD-462: Measurement of EMI Characteristics

Up until MIL-STD-461E, the MIL-STD-462 document contained the test method to be used in testing to MIL-STD-461 requirements. The intent in this section is not to reproduce the contents of MIL-STD-462, which is now obsolete but may be applicable to new equipment to be installed on older platforms. Instead, a test plan and sample test procedures based on MIL-STD-462 are presented. The test plan and procedures contain widely accepted interpretations of MIL-STD-462 methods and describe common errors encountered in this type of test.

In the typical MIL-STD-462 test configuration, the EUT is mounted on a copper or brass ground plane that is at a height of 1 m above the floor and bonded to one wall of the shielded room in which EMI tests are conducted. This test configuration is acceptable for units installed in a similar configuration. However, when units are mounted in a different configuration, for example, one above the other on a conductive structure or open framework, then the typical MIL-STD-462 test setup, in which units are located side by side, is not representative. Figure 9.61 shows a test setup in which units of equipment are mounted on a metal frame. The equipment and frame comprise the Wind Imaging Interferometer (WINDII), which was manufactured by Canadian Astronautics for the CSA. The instrument is located at some height above the spacecraft and electrically connected to the spacecraft via a bonding strap. The wooden frame on which the instrument is mounted during EMI tests places the instrument at approximately the same height above the shielded-room floor as the instrument is above the spacecraft. One unit of equipment that is physically separated from the instrument and that is mounted directly on the spacecraft structure is placed on the ground plane as shown in Figure 9.62. Thus, the test setup is as representative of the actual spacecraft configuration as possible and ensures that the influence of the metal structure, proximity of units to each other, and interconnection cable lengths are accounted for.

When equipment is rack mounted, the racks are placed on the floor of the shielded room and colocated as close to the actual configuration as possible. Interconnection cables are typically looped down and placed 5 cm above the floor of the room. In radiated emission and susceptibility tests, the antenna is located 1 m from the side of the equipment from which maximum emissions are detected or that is predicted to be most susceptible. Tests may have to be repeated when more than one side of the equipment meets these criteria. In the test setup shown in Figure 9.62, the equipment

FIGURE 9.61 Wind Imaging Interferometer (WINDII) instrument. (Courtesy of The Canadian Space Agency, Saint-Hubert, Quebec, Canada.)

FIGURE 9.62 RE02 and RS03 test setup for the WINDII instrument.

mounted on the ground plane was tested for radiated emissions and susceptibility separate from the instrument on the wooden frame.

The EUT transmitting antennas are excluded from both BB and NB emission tests. The EUT antennas should be replaced by dummy loads located outside of the shielded room. The CE06/CE106 test is intended for measurement of conducted emissions on antenna terminals. Therefore, receiving antennas connected to the EUT are typically excluded from radiated emission measurements and should definitely be excluded from radiated susceptibility measurements. Receiving antennas are excluded from radiated susceptibility tests because the receiver will fail at in-band susceptibility frequencies. Also, damage may occur to the front end of the receiver due to the high incident field. If antennas cannot be removed cover with absorber. The conducted susceptibility tests CS03/CS103, CS04/CS104, and CS05/CS105 are applicable to receive equipment and subsystems and are designed to test for out-of-band response, intermodulation products, and cross modulation. Transmitters (without antenna) are not specifically excluded from RE02 requirements at the frequency or band of frequencies at which the transmitter operates but may be excluded at the discretion of the procuring agency.

9.6.3 Test Plan and Test Procedures

The test plan should outline techniques, procedures, and instrumentation used in verifying that the equipment/subsystem complies with the EMC requirements. The test plan describes the purpose of the test and how the operation of the equipment is chosen to ensure a maximum emission and susceptibility mode of operation. The test procedure details the means of implementation of the tests to be performed in order to demonstrate compliance with the applicable EMC requirements.

9.6.3.1 Test Plan

The performance-monitoring and functional test equipment and method should be described in detail in the qualification test plan. In MIL-STD-461F, the requirement is that software needs to be described in the test procedure. For commercial software, identification of the manufacturer, model, and revision must be provided. For locally developed software, the control and methodology must be indicated.

However, where the qualification test plan describes an invasive form of test (e.g., monitoring of voltage, waveform, or data within the EUT), these tests are not applicable during EMI testing. Instead, a test set or fixture is often used to perform an operational test of the EUT during EMI testing.

The test set should provide stimulus to the EUT where required (e.g., data bus control lines). It should be isolated from the ambient electromagnetic fields during radiated susceptibility tests and should not be a source of radiated emission from the enclosures during radiated emission tests. This isolation and suppression of emissions is typically achieved by locating all functional test equipment outside of the shielded room during EMC tests.

The load represented by the test set should simulate the actual load on the EUT as closely as possible. Where a difference exists between the test set interface circuits and that used within the actual load, the voltage induced during EMC radiated susceptibility tests should be measured at the test set inputs. The level of induced voltage should then be compared to the noise threshold of the actual interface circuits to establish EMC.

Where the circuits are identical, the level of immunity will be determined simply by correct operation of the EUT as determined by the test set. The time required to complete tests on all control lines, data buses, etc., by the test set must be known in order to limit the speed of the frequency scan during radiated emission tests. Where possible, the test set should be designed to replace the need for manual operation of switches, etc.

It is usually not feasible to operate control levers or switches remotely because the presence of an operator close to the EUT introduces large errors into the radiated emission and radiated susceptibility tests, the control levers should remain in one position during radiated emission and radiated susceptibility EMI tests. However, some procuring agencies require the operation of switches during EMC tests, regardless of the errors introduced.

Radiated emissions from interface cables caused by common-mode noise generated within the test set should be reduced to the extent feasible. Installation of filter connector adapters on power and signal connectors on the test set is an effective approach in reducing both common- and differential-mode noise.

A background radiated emission measurement should be made with the EUT powered down and the test set powered up in order to characterize any radiated emissions from the interface cables caused by the test setup.

9.6.3.2 Test Procedures

Test procedures should contain, as a minimum, the following:

1. List of contents
2. Applicable documents
3. Purpose of test
4. Table of frequencies and repetition rate of BB frequencies at which interference may be expected
5. Test sample excitation and performance monitoring
6. Test cables
7. Dummy loads
8. Power supplies
9. Grounding and bonding during tests
10. Procedures for each parameter test stating:
    a. Applicability
    b. Test setup, including detailed drawings
    c. Test points
    d. Limits and frequency range

   e. Operating procedures for the test sample operating and monitoring equipment

   f. Test sample operating modes

   g. Pass/fail criteria for susceptibility tests

   h. Transfer impedance of probes

   i. Antenna factors

11. Test result sheets, including

   a. Procedure step

   b. Specification for limits

   c. Parameter

   d. Test data

   e. Initial of person responsible for test

   f. Quality assurance signature

9.6.4 GENERAL TEST GUIDELINES

1. All test equipment will be certified as calibrated, with verification that the calibration is still current. This information will be entered on the test equipment sheet. The test equipment serial numbers and calibration information shall be entered on this sheet.

2. Where a monopole antenna is used for either radiated emission or radiated susceptibility tests (14 kHz to 20–30 MHz), the antenna counterpoise (ground plane) should be bonded to the ground plane on the table. If the EUT is too large and is mounted on the floor of the shielded room, then the monopole is used uncoupled. Where the monopole has been calibrated coupled and uncoupled (coupled is when two test antenna ground planes are connected together by a counterpoise, and uncoupled is without the counterpoise), care must be taken to use the uncoupled calibration.

3. The test antenna shall be remote from the measuring equipment. The distance shall be determined by two factors.

4. Far enough away to reduce interaction between the antenna and measurement equipment.

5. The distance shall be limited by excessive attenuation in the cable length. Ideally, the measuring equipment shall be located outside of the shielded room.

6. When performing radiated emission measurements, no point of the measuring antenna shall be less than 1 m from the walls of the shielded enclosure or obstruction.

7. For susceptibility measurements, no point of the field-generating and the field-measuring antennas shall be less than 1 m from the walls of the enclosure.

8. MIL-STD-46 IE requires that the enclosures be sufficiently large such that the EUT arrangement requirements and antenna-positioning requirements described in the individual test methods are satisfied.

9. Care should be taken to ensure that all accessory equipment used, such as spectrum monitors, oscilloscopes, earphones, and other equipment used in conjunction with EMI analyzers, do not affect measurement integrity.

10. The side of the equipment emitting the highest level of radiation shall be determined, and that side shall face the receiving antenna.

11. Ambient levels measured during testing with the EUT deenergized shall be at least 6 dB below the applicable limits of MIL-STD-461. Ambient conditions shall be determined prior to the beginning of measurements on the energized EUT and repeated during the progress of the test if there is reason to believe ambient levels have changed. If these conditions cannot be met and the offending sources are outside of the control of the test engineer (i.e., noisy functional test equipment or, despite filtering, noisy power supplies), the test may proceed, provided it can be demonstrated that the EUT provides negligible contributions to the emissions over the frequency range in question.

12. Typical product assurance provisions demand that the assigned quality assurance engineer shall witness the tests and determine, in conjunction with the test conductor, whether or not a failure has occurred. Any failure shall be subject to a test observation record. Prior to the start of testing, a test readiness review shall be held to determine test readiness and compliance. The system shall be visually inspected and approved as a production-representative unit.

13. The measuring equipment should be monitored first for spurious emissions. False data caused by such spurious recordings should be identified on *X-Y* recordings of photos from oscilloscopes.

14. If a high-level signal is present at a frequency outside of the frequency span examined by the EMI receiver or S/A, false measurements may be made as a result of overload and compression of the front end of the receiver, S/A, or preamp (where used). To check for the presence of overload and compression, either bypass the preamp (where used) or insert a 3–20 dB attenuator at the input of the measuring instrument and ensure that the peak emission recorded falls by the appropriate dB value. The S/A/receiver input attenuator may be adjusted to achieve the same attenuation. NB: Where the signal is close to the noise floor, a 3–6 dB attenuator may be more appropriate than the 20 dB.

15. Prior to performing tests, the measurement equipment shall have been switched on for a period of time adequate to allow parameter stabilization. If the operation manual does not specify a period of time, the minimum warm-up time shall be 1 h.

16. BB and NB swept measurements are made with the peak detector function. Spot frequency measurements may be made using an rms detector, provided the signal characteristics are known and errors can be calculated. The following is from MIL-STD-46 IE: A peak detector shall be used for all frequency-domain emission and susceptibility measurements. This device detects the peak value of the modulation envelope in the receiver band pass. Measurement receivers are calibrated in terms of an equivalent root mean square (rms) value of a sine wave that produces the same peak value. When other measurement devices, such as oscilloscopes, nonselective voltmeters, or BB field strength sensors, are used for susceptibility testing, correction factors shall be applied for test signals to adjust the reading to equivalent rms values under the peak of the modulation envelope.

17. All measurement equipment shall be operated as prescribed by the applicable instruction manuals unless specified herein. For test repeatability, all test parameters used to configure the test shall be recorded in the EMI test report. These parameters shall include measurement bandwidths, video bandwidths, and sweep speeds.

The general test guidelines and precautions that are generally applicable or that are contained in MIL-STD-461F are as follows:

1. *Calibration of measuring equipment and antennas.* Test equipment and accessories required for measurement in accordance with this standard shall be calibrated under an approved program in accordance with MIL-STD 45622. In particular, measurement antennas, current probes, field sensors, and other devices used in the measurement loop shall be calibrated initially at least every 2 years unless otherwise specified by the procuring activity, or when damage is apparent. For MIL-STD-461F, once the calibration is performed, then only the system-level check is needed. Antenna factors and current probe transfer impedances shall be determined on an individual basis for each device.

2. *Measurement system test.* At the start of each emission test, the complete test system (including measurement receivers, cables, attenuators, couplers, and so forth) shall be

verified by injecting a known signal, as stated in the individual test method, while monitoring system output for the proper indication.

3. *Antenna factors.* Factors for electric field test antennas shall be determined in accordance with SAE ARP-958

Requirements that are specific to MIL-STD-46 1E are

1. *Excess personnel and equipment.* The test area shall be kept free of unnecessary personnel, equipment, cable racks, and desks. Only the equipment essential to the test being performed shall be in the test area or enclosure. Only personnel actively involved in the test shall be permitted in the enclosure.

2. *RF hazards.* Some tests in this standard will result in electromagnetic fields, which are potentially dangerous to personnel. The permissible exposure levels in ANSI C95.1, in the United States, and safety code 6, in Canada, shall not be exceeded in areas where personnel are present. Safety procedures and devices shall be used to prevent accidental exposure of personnel to RF hazards.

3. *Shock hazard.* Some of the tests require potentially hazardous voltages to be present. Extreme caution must be taken by all personnel to ensure that all safety precautions are observed.

4. *FCC and Industry Canada restrictions.* Some of the tests require high-level signals to be generated that could interfere with normal FCC- or Industry Canada-approved frequency assignments. All such testing should be conducted in a shielded enclosure. Some open-site testing may be feasible if prior FCC or Industry Canada coordination is obtained.

5. *EUT test configurations.* The EUT shall be configured as shown in the general test setups. These setups shall be maintained during all testing unless other direction is given for a particular test method.

6. *Bonding of EUT.* Only the provisions included in the design of the EUT shall be used to bond units such as equipment case and mounting bases together or to the ground plane. When bonding straps are required to complete the test setup, they shall be identical to those specified in the installation drawings.

7. *Shock and vibration isolators.* EUTs shall be secured to mounting bases having shock or vibration isolators if such mounting bases are used in the installation. The bonding straps furnished with the mounting base shall be connected to the ground plane. When mounting bases do not have bonding straps, bonding straps shall not be used in the test setup.

8. *Interconnecting leads and cables.* Individual leads shall be grouped into cables in the same manner as in the actual installation. Total interconnecting cable lengths in the setup shall be the same as in the actual platform installation. If a cable is longer than 10 m, at least 10 m shall be included. When cable lengths are not specified for the installation, cables shall be sufficiently long to satisfy the following conditions specified: At least 2 m (except for cables that are shorter in the actual installation) of each interconnecting cable shall be run parallel to the front boundary of the setup. Remaining cable lengths shall be routed to the back of the setup and shall be placed in a zigzagged arrangement. When the setup includes more than one cable, individual cables shall be separated by 2 cm measured from their outer circumference. For benchtop setups using ground planes, the cable closest to the front boundary shall be placed 10 cm from the front edge of the ground plane. All cables shall be supported 5 cm above the ground plane.

9. *Input power leads.* Two meters of input power leads (including returns) shall be routed parallel to the front edge of the setup in the same manner as the interconnecting leads.

The power leads shall be connected to the LISNs. The LISN replaces the 10 μF capacitors in Figures 9.65 and 9.70 in MIL-STD-461E and above. Power leads that are part of an interconnecting cable shall be separated out at the EUT connector and routed to the LISNs. After the 2 m exposed length, the power leads shall be terminated at the LISNs in as short a distance as possible. The total length of power lead from the EUT electrical connector to the LISNs shall not exceed 2.5 m. All power leads shall be supported 5 cm above the ground plane.

10. *Construction and arrangement of EUT cables.* Electrical cable assemblies shall simulate actual installation and usage. Shielded cables or shielded leads (including power leads and wire grounds) within cables shall be used only if they have been specified in installation drawings. Cables shall be checked against installation requirements to verify proper construction techniques, such as use of twisted pairs, shielding, and shield terminations. Details on the cable construction used for testing shall be included in the EMITP.

11. *Operation of EUT.* During emission measurements, the EUT shall be placed in an operating mode that produces maximum emissions. During susceptibility testing, the EUT shall be placed in its most susceptible operating mode. For EUTs with several available modes (including software-controlled operational modes), a sufficient number shall be tested for emissions and susceptibility such that all circuitry is evaluated. The rationale for modes selected shall be included in the EMITP.

12. *Computer-controlled receivers.* A description of the operations being directed by software for computer-controlled receivers shall be included in the EMITP required by MIL-STD-461. Verification techniques used to demonstrate proper performance of the software shall also be included.

13. *Orientation of EUTs.* EUTs shall be oriented such that surfaces that produce maximum radiated emissions and respond most readily to radiated signals face the measurement antennas. Bench-mounted EUTs shall be located 10 ± 2 cm from the front edge of the ground plane, subject to allowances for providing adequate room for cable arrangement as specified in (10).

14. *Susceptibility monitoring.* The EUT shall be monitored during susceptibility testing for indications of degradation or malfunction. This monitoring is normally accomplished through the use of built-in test, visual displays, aural outputs, and other measurements of signal outputs and interfaces. Monitoring of EUT performance through installation of special circuitry in the EUT is permissible; however, these modifications shall not influence test results.

15. *Use of measurement equipment.* Measurement equipment shall be as specified in the individual test methods of MIL-STD-46 IE. Any frequency selective measurement receiver may be used for performing the testing described in this standard, provided that the receiver characteristics (i.e., sensitivity, selection of bandwidths, detector functions, dynamic range, and frequency of operation) meet the constraints specified in this standard and are sufficient to demonstrate compliance with the applicable limits of MIL-STD-461. Typical instrumentation characteristics may be found in ANSI C63.2.

16. *Safety grounds.* When external terminals, connector pins, or equipment grounding conductors in power cables are available for ground connections and are used in the actual installation, they shall be connected to the ground plane. Arrangements and length to be in accordance with "interconnecting leads and cables."

9.6.4.1 Emission Tests

1. *Operating frequencies for tunable RF equipment.* Measurements shall be performed with the EUT tuned to not less than three frequencies within each tuning band, tuning unit, or range of fixed channels, consisting of one mid-band frequency and a frequency within ±5% from each band or range of channels.

2. *Operating frequencies for spread spectrum equipment.* Operating frequency requirements for two major types of spread spectrum equipment shall be as follows:
   a. *Frequency hopping.* Measurements shall be performed with the EUT utilizing a hop set that contains 30% of the total possible frequencies. The hop set shall be divided equally into three segments at the low end, middle, and high end of the EUT's operational frequency range.
   b. *Direct sequence.* Measurements shall be performed with the EUT processing data at the highest possible data transfer rate.

9.6.4.2 Susceptibility Tests

Frequency scanning. For susceptibility measurements, the entire frequency range for each applicable test shall be scanned. For swept-frequency susceptibility testing, frequency scan rates and frequency step sizes of signal sources shall not exceed the values listed in Table 9.25. The rates and step sizes are specified in terms of a multiplier of the tuned frequency (f_o) of the signal source. *Analog* scans refer to signal sources, which are continuously tuned. Stepped scans refer to signal sources that are sequentially tuned to discrete frequencies. Stepped scans shall dwell at each tuned frequency for a minimum of 1 s. Scan rates and step sizes shall be decreased when necessary to permit observation of a response.

1. *Modulation of susceptibility signals.* Susceptibility test signals above 10 kHz shall be pulse modulated at a 1 kHz rate with a 50% duty cycle unless otherwise specified in an individual test method of this standard.
2. *Thresholds of susceptibility.* When susceptibility indications are noted in EUT operation, a threshold level shall be determined where the susceptible condition is no longer present. Thresholds of susceptibility shall be determined as follows:
   a. When a susceptibility condition is detected, reduce the interference signal until the EUT recovers.
   b. Reduce the interference signal by an additional 6 dB.
   c. Gradually increase the interference signal until the susceptibility condition reoccurs. The resulting level is the threshold of susceptibility.

9.6.5 Typical EMI Receiver or S/A Bandwidths for MIL-STD-461 A-C

The receiver bandwidths used with the typical EMI measurement system for measurement of narrow and BB emissions are shown in Table 9.31. These are based on measurements made of

TABLE 9.31

MIL-STD-461C EMI Receiver NB and BB Bandwidths

Frequency Range	Narrowband	Broadband
Radiated		
20 Hz–5 kHz	3–10 Hz	30–100 Hz
5–50 kHz	30–100 Hz	200 Hz–1 kHz
50 kHz–1 MHz	200 Hz–1 kHz	9–10 kHz
1–200 MHz	9–10 kHz	100 kHz
200 MHz–40 GHz	100 kHz	1 MHz
Conducted		
20 Hz–20 kHz	10–50 Hz	—
20 kHz–2.5 MHz	200 Hz–1 kHz	1–10 kHz
2.5–100 MHz	1–9 kHz	10–50 kHz

the EMI receiver system noise floor and on the worst-case emission requirements. Resolution bandwidths may be changed where necessary, for example, when the measuring equipment bandwidths are different or when the measuring equipment noise floor is close to or above the limit.

The change from recommended bandwidth shall be no greater than a factor of 0.5–2, and the BB bandwidth shall be from 5 to 10 times the NB bandwidth.

9.6.5.1   Determination of Either NB or BB (BB) Emissions

Use the tuning test described in Section 9.2.2 to determine if the emission is NB or BB. If the results of the test are inconclusive, use one or more of the alternative tests. Clarification or final arbitration on the decision should be provided by the EMC engineer or the procuring agency.

Conversion of BB measured values to the normalized BB values of dBµV/m/MHz and dBµV/ MHz. The reference bandwidth of BB measurements is 1 MHz. BB measurements at other bandwidths must be converted to the reference bandwidth. To convert from the measurement bandwidth to the reference bandwidths, add the following to the measured coherent BB value:

$$20 \log \frac{R_{BW}}{M_{BW}}$$

where
R_{BW} is the reference bandwidth
M_{BW} is the measurement bandwidth

As an example, the measurement is a voltage of 25 dBµV at a bandwidth of 10 kHz.

$$20 \log \frac{1\,\text{MHz}}{10\,\text{kHz}} = 40\,\text{dB}$$

The corrected measurement is therefore 25 dBµV + 40 dB = 65 dBµV/MHz.

If the measured increase in level is approximately 10 dB for an increase factor of 10, then the noise is noncoherent (random) BB noise. For non-coherent BB noise, the conversion is

$$10 \log \frac{R_{BW}}{M_{BW}}$$

This conversion for noncoherent BB noise is made by reputable test facilities and is a well-accepted procedure. However, this is not contained in MIL-STD 462 and should be agreed to by the procuring agency or EMC authority.

9.6.5.2   MIL-STD 461F Bandwidths and General Emission Guidelines

1. *Bandwidth correction factors.* No bandwidth correction factors shall be applied to test data due to the use of larger bandwidths.
2. *Emission identification.* All emissions, regardless of characteristics, shall be measured with the measurement receiver bandwidths specified in Table 9.29 and compared against the limits in MIL-STD 641E. Identification of emissions with regard to NB or BB categorization is not applicable.

TABLE 9.32
MIL-STD-461F Bandwidth and Measurement Time

Frequency Range	6 dB Bandwidth	Dwell Time (s)	Minimum Measurement Time, Analog Measurement Receiver
30 Hz–1 kHz	10 Hz	0.15	0.015 s/Hz
1–10 kHz	100 Hz	0.015	0.15 s/kHz
10–250 kHz	1 kHz	0.015	0.015 s/kHz
250 kHz–30 MHz	10 kHz	0.015	1.5 s/MHz
30 MHz–1 GHz	100 kHz	0.015	0.15 s/MHz
Above 1 GHz	1 MHz	0.015	15 s/GHz

3. *Frequency scanning.* For emission measurements, the entire frequency range for each applicable test shall be scanned. Minimum measurement time for analog measurement receivers during emission testing shall be as specified in Table 9.32. Synthesized measurement receivers shall step in one-half bandwidth increments or less, and the measurement dwell time shall be as specified in Table 9.32.

4. *Emission data presentation.* Amplitude versus frequency profiles of emission data shall be automatically and continuously plotted. The applicable limit shall be displayed on the plot. Manually gathered data are not acceptable except for plot verification. The plotted data for emission measurements shall provide a minimum frequency resolution of 1% or twice the measurement receiver bandwidth, whichever is less stringent, and minimum amplitude resolution of 1 dB. The foregoing resolution requirements shall be maintained in the reported results of the EMITR.

The MIL-STD 461F radiated emission measurement bandwidths are shown in Table 9.32.

9.6.6 Measurement to Very Low-Level Radiated Emissions Limits, Typically in Notches for Receivers

When radiated emission specification limits are very low, typically 25 dBμV/m/MHz for BB and 5 dBμV/m for NB, the noise floor, using standard EMI measuring antennas and preamplifiers, may not be sufficiently low. In some cases, the noise floor of the measuring system using standard equipment is well above the specification limit, and even the noise voltage from a 50 Ω resistor into a 50 Ω conjugate load may be higher than the signal developed at the output of the typical receiving antenna. These low-level limits are typically over a narrow band of frequencies corresponding to a receiver bandwidth.

In evaluating the noise floor of the measurement system, it is important to realize that the receiver/S/A measures peak levels, which are displayed as rms values. Most S/A and receiver manufacturers specify an rms noise floor for their equipment, measured using video or digital averaging. The peak value, theoretically, may be infinitely higher; however, in measurements on three different manufactures' S/A s, the peak measured noise, displayed as an rms value, was between 10 and 12 dB above the digital or video bandwidth-averaged noise floor. Hewlett–Packard independently confirmed the approximate 12 dB difference between peak and average.

Another problem in evaluating the noise floor of the measuring system is that the preamplifier NF may be specified, but the noise referenced to the input of the preamp is not given. Very often, this is significantly lower than a 50 Ω resistor, and to use this, 50 Ω value results in an error in the NF calculation.

In measuring low-level NB emissions, the use of averaging will reduce the noise floor without affecting the NB emissions. But this is not allowable for BB emissions measurements, for the

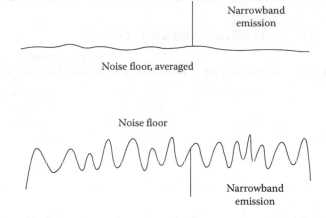

FIGURE 9.63   Use of averaging to bring a narrowband emission out of the measuring system noise floor.

emissions will also be reduced. Figure 9.63 illustrates how averaging can bring an NB emission out of the noise. The noise floor can also be reduced by reducing the resolution bandwidth. But this is not allowable for BB measurements, and often the resolution bandwidth is dictated by the bandwidth of a receiver that is installed on the EUT, which is the reason the notch is imposed.

The BB specification uses a 1 MHz reference impulse bandwidth, whereas many S/As use a 3 dB resolution bandwidth. For example, a Hewlett–Packard S/A requires a correction of 4 dB at a 1 MHz measurement bandwidth to correct from the 3 dB resolution bandwidth to the impulse bandwidth. This correction is subtracted from the measured emission level and also from the measured system noise floor.

Another factor when measuring BB emissions in the notch is the random nature of the measuring system noise floor. If only coherent BB emissions are of concern, then the emissions can be viewed during the troughs and not the peaks of the noise floor display.

The measurement bandwidth used for BB emissions also affects the measurement of coherent noise. Figure 9.64 illustrates how increasing the measurement bandwidth can bring coherent BB emissions out of the noise. But this does not help for quasi-BB (noncoherent) emissions. Figure 9.64 illustrates, for simplicity, that the coherent BB noise does not ride on top of the noncoherent noise but is buried beneath it. Although this is often true, it is not always the case.

Using a typical high-end receiver, the noise floor, using a 1 MHz bandwidth, was measured at 29 dBμV at 2 GHz. Adding a 34 dB preamp at the front end of the receiver increased the measured noise floor by 11 dB–40 dBμV. A typical double-ridged guide horn antenna has an antenna factor of 26.6 dB at 2 GHz, and let us assume that the BB limit is 25 dBμV/m/MHz and the measurement bandwidth is at 1 MHz. The input to the preamp is then 25 dBμV/m/MHz (limit) − 26.6 dB (AF) = −1.6 dBμV. And at the input of the receiver, it is −1.6 dBμV + 34 dB (preamp gain) = 32.4 dBμV, that is, 7.6 dB below the noise floor of the system. A 3 or 10 MHz bandwidth will bring coherent noise further out of the noise floor, but only if the coherent noise covers a 3–10 MHz frequency range.

One antenna that will typically achieve a sufficiently high signal-to-noise ratio for most measurement systems at 2 GHz is a 1.7–2.6 GHz standard-gain horn, which has a gain of at least 20 dB and an AF of 16 dB. When used in conjunction with an HP8566B S/A and a low-noise 50 dB preamp, and applying the 4 dB correction for the 1 MHz resolution bandwidth to 1 MHz impulse bandwidth, the noise floor will be below the BB limit. For example, the measured noise floor of the S/A with 50 dB preamplifier and a 1 MHz bandwidth is 55 dBμV − 4 dB (resolution bandwidth correction) = 51 dBμV. The signal level at the S/A input with 25 dBμV/m/MHz incident on the antenna is 25 dBμV/m/MHz − 16 dB (AF) + 50 dB (preamp gain) = 59 dBμV, that is, 8 dB above the noise floor.

FIGURE 9.64 Effect of the measurement bandwidth on coherent and noncoherent signals, assuming for simplicity that the coherent noise does not ride on top of the noncoherent (measuring system) noise.

Realistically, it may be necessary to relax the requirement that the noise floor be at least 6 dB below the limit level for these extremely low-level RE02 notches.

9.6.7   MIL-STD-461A, B, and C EMI Measurements

The tests most frequently performed are described next.

9.6.7.1   Description of CE01 and CE03 Tests

9.6.7.1.1   Purpose

These tests are designed to measure conducted emissions on all power lines.

This test is applicable to primary power, that is, the input AC or DC power supplied to the equipment or subsystem. The test is not usually applicable to power lines that interconnect equipment or secondary power, although primary power supplied to other equipment may be included in the requirement. Control circuits that use EUT power and grounds and neutrals that are not grounded internally to the subsystem or equipment are included.

CE01 are NB emission limits only, and separate NB and BB limits are imposed for CE03. Signal leads are typically exempt, and when applied, the limits are developed on a case-by-case basis.

9.6.7.1.2   Test Setup

The CE01, CE03 30 Hz–50 MHz test setup is shown in Figure 9.65. All primary power and control leads are tested.

9.6.7.1.3   Test Equipment Required

1. Shielded current probe or probes covering the 30 Hz–50 MHz frequency range
2. EMI receiver or S/A

FIGURE 9.65 CE01, CE03 conducted emission test setup. 1. Low-impedance bond to ground plane. 2. 10 μF RF feedthrough capacitors. 3. 5 cm standoffs. 4. DC bond impedance between the ground plane and the shielded room wall <2.5 m ohm. 5. Current probe. 6. COMM 2 Chassis connection to ground plane <2.5 m ohm.

3. BB preamplifier or preamplifiers with at least 20 dB of gain
4. 10 μF feedthrough capacitors with a characterized impedance over the frequency range of interest
5. Oscilloscope

The following points provide an overview of the CE01, CE03 step-by-step procedure:

1. The equipment test mode, in which maximum emissions are generated, is detected in a "quick scan" test. When more than one test mode results in emissions at different frequencies, the CE01, CE03 test should be repeated in each test mode.
2. Measure the noise pickup on the current probe with the AC and DC power supplies located outside of the shielded room, switched on but with the power supply disconnected from the EUT. This is a measure of the background and the power supply–generated noise. The ambient emissions must be 6 dB below the conducted emission limits.
3. With the current probe around first the supply and then the return, measure the emissions using the NB bandwidth of the receiver/S/A.
4. Repeat Step 3 using the BB measuring bandwidth for the CE03 BB measurement. Change current probes as required to cover the full 30 Hz–50 MHz frequency range.

9.6.7.2 MIL-STD 461F, CE101, and CE103

The general test setup is similar to CE01 and CE03, with the exception of the use of the MIL-STD 461E-G LISN. Note the MIL-STD 461E-G LISN is different than either the low-frequency or high-frequency LISN described in MIL-STD 462 Notice 3.

For CE101 and CE 102, a check on the measurement system is required, and the test setup for this is shown in Figures 9.66 and 9.67, respectively. The CE101 measurement test setup is shown in Figure 9.68 and the CE 102 in Figure 9.69.

MIL-STD-461F has added CE101 to the surface ships, and the Appendix provides notes on the use of a 5 μH LISN, but it must be approved by the procuring agency. A new limit below 150 kHz when using the LISN is provided.

An alternative test procedure for high current loads, especially for 400 Hz loads, is to eliminate the LISN altogether. For AC loads that are drawing harmonic currents below 10 kHz (where the filter may resonate), or for currents higher than the filter rating, it may be necessary to short out the power-line filters on the shielded-room wall.

For very high currents 100–1000 A from a motor generator, a filter is typically required to ensure that noise supplied by the generator does not result in a conducted ambient above the limits.

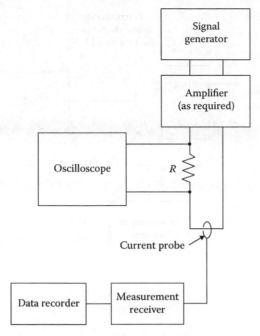

FIGURE 9.66 CE101 measurement system check setup.

FIGURE 9.67 CE 102 measurement system check setup.

This filter will be very large and may require both C/M and D/M components. The best location is close to the generator outside the building with the C/M capacitors in the filter connected to an external ground rod. An alternative is to connect the r C/M capacitors in the filter to the building ground rod at the distribution panel. All installations of this type must comply with the relevant power safety requirements and with adequately sized power connections.

FIGURE 9.68   CE 101 measurement setup.

FIGURE 9.69   CE 102 measurement setup.

The most common errors incurred in conducted emission measurements with a current probe are as follows:

1. Leaving the current probe around the DC or AC power cable when switching on or off the power. The resultant current spike through the probe can generate very high voltages, and there is a danger of damage to the front end of the measuring instrument. *Always disconnect the current probe from the instrument when powering up or down*!

2. The current probe should be calibrated in a test fixture that raises the body of the probe from contact with the return (ground plane) of the test fixture. It is important in measurements to ensure that the body of the current probe is not in contact with the ground plane and if possible is at the same height above it as used in the test fixture.

3. The measuring instrument must be grounded to the ground plane with use of an isolation transformer on the AC power to keep ground currents, typically at harmonics of the AC power supply, from corrupting the measurement level.

4. It is not mandatory to make the measurement in a shielded room, but if this is the chosen location, the power-line filters at the entry point to the room will effectively prevent the external ambient from corrupting the measured level. If measurements must be made outside of the shielded room, the conducted ambient noise, typically from computers, impressed on the AC power line or coupled from the AC input to the output of a DC supply can put the EUT out of specification, despite the presence of the 10 μF feedthrough capacitors or a LISN. The solution is to add a power-line filter at the input of the AC or DC power, before the 10 μF feedthrough capacitors or the LISN.

   The second problem with measurements outside of a shielded room is that the radiated ambient, typically from TV and radio, couples to the power cable. The current probe measures this ambient-induced current, which may put the EUT out of specification. If the level of noise from the EUT is low either side of the ambient emission, when examined with a narrow frequency span and narrow bandwidth, then the emission can be attributed to the ambient, and many procuring agencies or equipment customers will accept the test results. Check first, however. With an unacceptably high ambient, either locate the supply outside of the shielded room and bring the power through the room filters or add a filter between the DC output from the supply and the 10 μF capacitors or LISN.

5. Poor or nonexistent grounding of the EUT, LISN, or 10 μF capacitors to the ground plane. This will reduce common-mode currents and may allow an EUT to pass when the EUT is out of specification.

6. Incorrect characterization of NB and BB emissions. The most common problem is to apply the coherent BB noise correction to noise that can be demonstrated to be noncoherent in converting to the Reference 1 MHz bandwidth.

7. Use of incorrect measurement bandwidth, missing a repeat measurement on the power or return, exceeding calibration frequency range of probe are all the problems that can be avoided by use of a checklist of the tests with a tick box to ensure no test is missed.

8. The application of all emissions, regardless of characterization, to both the NB and BB limits.

9. Software-controlled measurements have a number of potential problems. The majority of these programs do not allow multiple sweeps using peak or maximum hold, but instead take data over a single sweep, a snapshot in effect. Emissions often vary significantly in amplitude from one sweep to the next. With a snapshot of the data, there is no guarantee that the worst-case emissions have been captured. The argument is made that this has become an industry standard, but it seldom results in worst-case emission measurements. A much more serious problem with many computer measurement programs is a very flawed analysis of BB and NB emissions; when confronted, many manufacturers admit to the deficiency in their program. Typically, only one of the four test methods is applied, and even that method can lead to a wrong characterization of the emission.

9.6.7.3 Description of MIL-STD 461B/C RE02 Compared to MIL-STD-461E-G Electric Field Tests

9.6.7.3.1 Purpose

These tests are designed to measure radiated emissions from the equipment or subsystem and all interface cables.

These tests are divided into measurement of NB radiated emissions from 14 kHz to 18 GHz and BB radiated emissions from 14 kHz to 10 GHz (some procuring agencies extend the requirement to 18 GHz).

9.6.7.3.2 RE02 Test Setup

The RE02 NB and BB radiated emission test setup for the EUT is shown in Figure 9.70. This is the typical MIL-STD-462 test setup. The exposed cable is located 5 cm above the ground plane at a distance of 10 ± 2 cm from the front of the ground plane. The cables after the exposed 2 m length may be shielded either by an expandable braid that is taped to the ground plane with electrically conductive adhesive copper tape or simply by copper tape placed over the cables and contacting the ground plane on either side of the cables (NB: Before taping the cables into position, the orientation of the subsystem must be decided to ensure that the area of maximum emission is facing the antenna, as described in the step-by-step test procedure).

Maximum coupling between the subsystem and the monopole (rod) antenna will be ensured by the extension of the counterpoise of the antenna to the ground plane, as shown in Figure 9.70.

9.6.7.3.3 Test Equipment Required

1. Antennas covering the range of 14 kHz–18 GHz. The antennas will typically be of the following types:

Frequency	Antenna Type
14 kHz–20 MHz	Active or passive rod
20–200 MHz	Biconical
200 MHz–1 GHz	Log periodic (double-ridged guide for MIL-STD-461E/F)
1–18 GHz	Double-ridged guide

FIGURE 9.70 RE02 radiated emission test setup. 1. 2 m length of cable 5 cm above the ground plane. 2. 10 μF RF feedthrough capacitors. 3. 5 cm standoffs.

2. EMI receiver and/or S/A with a noise floor no higher than 75 µV over the bandwidth being used.
3. BB preamplifier or preamplifier(s) with at least 20 dB of gain and a noise floor no higher than 10 dB.
4. 10 µF RF feedthrough capacitors.

9.6.7.3.4   Use of Electric or Magnetic Field Probes to Locate Sources of Emission

Electric or magnetic field probes are available, as described in Section 2.6, and used typically with a preamplifier and S/A. These probes may be used to monitor the cables and sides of units to determine the sources of maximum emission. This information is necessary to determine EUT orientation. A second use for the probes is to locate sources of out-of-specification emissions. The probes may be handheld and positioned 7–15 cm from cables, enclosure walls, apertures, etc.

The probes may be calibrated, although the absolute value of H field measured by the probes is not required. The calibration curve does, however, provide an indication of the frequency response of the probe. This may be used as described in the following example. Assume an emission detected at 70 MHz, using a balanced loop antenna, with a second emission 6 dB higher at 80 MHz. The balanced loop sensitivity is 0.4 mV/mA/m at 70 MHz and 1.1 mV/mA/m at 80 MHz (i.e., 8.8 dB higher). Therefore, the 70 MHz emission is 2.8 dB above the 80 MHz emission after correction for probe sensitivity.

9.6.7.3.5   RE02 Test Method and Preparation

The test setup should be in compliance with Figure 9.70. Where RE02 measurements of the applicable levels are routine with the test equipment available, the following calculation of system sensitivity may be omitted.

Where the limits are lower than previously measured or unfamiliar equipment is used, it should be ascertained if the sensitivity of the antennas and measuring equipment ensures that the radiated emission level is 6 dB above the noise floor of the measuring equipment.

Compare the antenna factor of each antenna to be used with the NB radiated emission level over the frequency range of interest. At the frequency or frequencies where the antenna factor is high or the limit is low, make the following calculations:

$$V_{in} \text{ (dB}\mu\text{V)} = \text{dB}\mu\text{V/m} - \text{AF (dB)}$$

where
V_{in} is the input voltage to receiving equipment
dBµV/m is the specification limit at the frequency under consideration
AF (dB) = antenna factor in decibels = $20 \log_{10} \text{AF}$
AF = numeric antenna factor at the frequency of interest

Using V_{in}, ensure that the measuring instrument has sufficient sensitivity, or the noise floor is sufficiently low, at the bandwidths specified in Table 9.27. When this is not the case, either a BB low-noise preamplifier or a more sensitive or lower-noise measuring instrument should be used. This calculation must be repeated for each antenna used and for the BB specification limit, again selecting an appropriate bandwidth from Table 9.27. An example of a typical calculation is as follows.

A NB limit at 40 MHz is 24 dBµV/m. The antenna factor of the antenna used is 3.83, which is 11.67 dB. Therefore,

$$V_{in} = 24 \text{ dB}\mu\text{V/m} - 11.67 \text{ dB} = 12.3 \text{ dB}\mu\text{V} = 4.12 \text{ }\mu\text{V}$$

The NB bandwidth to be used is 10 kHz (refer to Table 9.28 at 40 MHz). At 40 MHz, the noise floor of the measuring instrument is −80 dBm, which is equivalent to 23 µV into 50 Ω (the input impedance of the instrument). V_{in} is thus 15 dB (20 log V_{in}/noise floor) below the noise floor, and the system sensitivity, therefore, is not adequate and a preamplifier with a minimum gain of 21 dB must be used. For conversion to BB unit, add the BB correction given by

$$20 \log \frac{1\,\text{MHz}}{R_{BW}}$$

where R_{BW} is the resolution bandwidth used in the measurement in MHz.

The frequency to which the receiver is tuned will be swept over the frequency range of the antenna in use. The sweep time will be selected to ensure that BB *emissions* occurring at the lowest pulse repetition rate of the system are measured. A maximum EMI receiver scan rate will therefore be imposed. The maximum EMI receiver scan rate is determined by both the EMI receiver bandwidth and the lowest PRF to be intercepted.

NB emissions are either from CW sources or sources at high PRF, and thus the maximum scan rate limitation is not applicable. The scan rate in hertz per second is given by

$$\text{Scan rate} = \frac{\text{EMI receiver bandwidth}\,[\text{Hz}]}{\text{Lowest expected PRF}\,[\text{s}]}$$

An example for a 50 kHz BB bandwidth is

$$\text{Scan rate} = \frac{50\,\text{kHz}}{4.5\,\text{s}} = 11.1\,\text{kHz/s}$$

The slowest sweep time available on a typical S/A is 1500 s; thus, the maximum frequency span that can be swept in 1500 s is given by

$$\text{Scan rate} \times \text{sweep time} = 11.1\,\text{kHz/s} \times 1500\,\text{s} = 16.6\,\text{MHz}$$

The 2.5–30 MHz frequency range must therefore be divided into two sections. The approximate time required for the BB test to cover this frequency range is 41 min. At high frequencies, the limit on scan rate often results in inordinately long test times, and an alternative method must be used.

Two approaches can be used to facilitate the BB RE02 test. One is either to program the S/A via the front panel or to use a computer to automatically change frequency and plot the results. The other approach is to use a fast sweep time and to record the frequency range of interest for a specified number of minutes, using peak (maximum) hold on the S/A. The probability of detecting the harmonically related emissions occurring at a 1 min or faster repetition rate is thus high. The sweep time of the S/A for NB tests may be short, limited only by the S/A. When the sweep time is too short, the S/A displays an out-of-calibration condition. When the EUT has more than one mode of operation, any of which may result in significant emissions, the radiated and conducted emission measurements must be repeated for each mode.

9.6.7.3.6 Expected Frequency of Emissions

The frequencies at which emissions are expected should be tabulated and examined carefully during emission measurements. For example, BB and NB harmonically related emissions are expected at clock, converter, and logic frequencies and harmonics thereof.

9.6.7.3.7 RE02 Step-by-Step Test Procedure

The following three points provide an overview of the RE02 step-by-step procedure.

1. Check for background emissions over the entire frequency band of interest (i.e., 14 kHz $<f>$ 18 GHz). Above 20 MHz, check with both vertical and horizontal polarizations of the antenna.

2. Find the side of the EUT that has the highest emissions (check all test modes).
3. Measure peak emissions for all test modes.

Note:

- For 14 kHz <*f*> 18 GHz, measure peak emissions using the appropriate NB bandwidth of the S/A and compare the measured peak emissions to the limits.
- For 14 kHz <*f*> 10 GHz, measure peak emissions using the appropriate BB bandwidth of the S/A and compare the measured peak emissions to the limits.
- For emissions exceeding the limits, determine whether they are BB or NB by one or more of the methods outlined in Section 9.2.2.
- For *f* > 20 MHz, peak emissions must be measured using vertical and horizontal antenna orientations.
- For 1–18 GHz, measure peak emissions with the antenna located 1 m from the center of the interconnection cables as well as 1 m in front of the EUT.

When the passive or active rod antenna is used, connect the counterpoise of the antenna to the ground plane on which the EUT is mounted using a sheet of copper-clad PCB. The PCB may be electrically bonded to the ground plane and counterpoise by use of conductive adhesive tape. The MIL-STD-462 test setup does not show the counterpoise connected to the ground plane, unlike the DO-160 test setup. Bonding the counterpoise to the ground plane results in a smoother AF, especially at higher frequencies.

MIL-STD 46IE RE 102 does require that the rod antenna counterpoise be used and be electrically bonded to the ground plane. MIL-STD 46IE also requires a test of the measurement path using a signal generator and requires the use of multiple antenna positions above 200 MHz and specific antenna heights of 120 cm above the ground plane for the biconical and double-ridged guide (horn). The measurement bandwidths and dwell times for RE 102 are contained in Table 9.32.

9.6.7.4 Common Errors in RE02/RE102 Measurements

The most common errors incurred in radiated emission measurements follow:

1. Ambient inside the shielded room too high due to RF common-mode current flow on interconnection and power cable. This is typically sourced by functional test equipment or electrical ground support equipment (EGSE) located outside of the shielded room. The solution is always to use D-type connectors on interface connections either at the test equipment/EGSE or at the cable entry into the shielded room. By using D-type connectors, if a problem is seen, a filtered D-type connector adapter (male and female connectors back to back with filter components in between) should be tried. If filter connectors cannot be used on interconnection cables, then the shields of cables should be terminated to the shielded-room wall at the entry through a waveguide and brass wool should be used to connect the shield to the interior of the waveguide and to fill the aperture. It is a fallacy to assume that a waveguide aperture performs as a waveguide below cutoff when cables are passed through the aperture. RF current flow on these cables will couple from the outside ambient into the shielded room on the cables coming through the waveguide. Even nonconductive hose containing tap water will couple RF currents into the room due to the conductivity of the tap water. To keep the RF currents out of the shielded room, they must be shunted to the room wall at the point of entry.
2. Incorrect characterization of NB and BB emissions. The most common problem is to apply the coherent BB noise correction to noise that can be demonstrated to be noncoherent in converting to the Reference 1 MHz bandwidth.
3. Use of incorrect measurement bandwidth, missing a repeat measurement with a different antenna polarization, exceeding calibration frequency range of antenna. All of these

problems can be avoided by use of a checklist of the tests (which includes frequency range, antenna polarization, and use of preamplifier), with tick boxes to ensure no test is missed.

4. Transmitting antennas on the EUT connected to BB power amplifiers are sources of BB noise as well as the intended transmitting frequency. These transmitting antennas are excluded from the RE02 and RE 102 tests. Transmitting antennas must be replaced with dummy loads, and the load connector must have a low transfer impedance to reduce radiation from the load itself. If out-of-band emissions from the transmitter are a problem, these should be measured in a conducted test; they are not part of tests of radiated emission E field to MIL-STD requirements.

5. The application of all emissions, regardless of characterization, to both the NB and BB limits. (Apply only BB emissions to the BB limit and NB emissions to the NB limit.)

6. Nonrepresentative test setup, use of a different type of cable than used in the final version, nonrepresentative EUT, addition of aluminum foil, copper tape, or braid over cables that will not normally be shielded.

7. The argument that cable emissions are not the concern of the manufacturer of the EUT when the cable harness is supplied by the customer. Regardless of the type of cable, shielded, or unshielded, it is the EUT that places noise on the cables, and it is the manufacturer of the EUT who must reduce emissions at source. If a manufacturer believes that emission requirements will not be met with a customer-supplied unshielded cable, he should apply for a deviation prior to testing. The customer may allow the deviation, require the EUT to meet the requirement, or change the cable to a shielded type.

8. Software-controlled measurements have a number of potential problems. The majority of these programs do not allow multiple sweeps using peak or maximum hold, but instead take data over a single sweep, a snapshot in effect. Emissions often vary in amplitude significantly from one sweep to the next. With a snapshot of the data, there is no guarantee that the worst-case emissions have been captured. The argument is made that this has become an industry standard, but it seldom results in worst-case emission measurements. A much more serious problem with many computer measurement programs is a very flawed analysis of BB and NB emissions; when confronted, many manufacturers admit to the deficiency in their program. Typically, only one of the four test methods is applied and any one of the methods can lead to a wrong characterization of the emission.

9.6.7.5 Description of MIL-STD 461B/C RE01, RE04, and MIL-STD 461E RE101 Magnetic Field Tests

9.6.7.5.1 Purpose

These tests are to verify that the magnetic field emissions from the EUT and its associated cabling do not exceed specified requirements.

The RE01 test covers the 30 Hz–30 kHz frequency range, the RE04 test covers the 30 Hz–50 kHz frequency range, and the RE101 covers the 30 Hz–100 kHz frequency range.

For both the RE01 and RE101 tests, the receiving loop is specified in both the original MIL-STD 462 and MIL-STD 461E as follows:

Diameter 13.3 cm
Number of turns 36
Wire 7–41 Litz (7 strand, No. 41 AWG)
Shielding electrostatic

MIL-STD 461F requires a calibration of the measurement equipment using a signal generator. In both RE01 and RE101, the loop sensor is located 7 cm from and moved over the EUT face or cable being probed. Orient the plane of the loop sensor parallel to the EUT faces and parallel to the axis of the cables. The exact distance of 7 cm from the EUT or cable is very important, for small errors in distance can result in large deviations in measurement. To ensure the 7 cm distance is maintained, it is recommended that a 7 cm Styrofoam block or a short length of dowel be strapped to the loop antenna. MIL-STD-461F also requires that if the EUT exceeds the limit at 7 cm that the distance is increased until the emission falls within the specified limit and to record the emissions and the measurement distance.

In the MIL-STD 461 RE04 test, the magnetic field sensor, which must be capable of measuring −40 dBnT at 25 Hz, is located 1 m from the EUT and cables. The axis of the antenna is pointed toward the EUT. A calibrator is required to check the magnetic field sensor. The calibrator is required to generate known magnetic fields at a minimum of 10 frequencies over the 20 Hz–50 kHz range. If a calibrator is not available and the magnetic field sensor has been calibrated within the last 2 years, then this would normally be acceptable.

9.6.7.6   RS03/RS103 Radiated Susceptibility Test

9.6.7.6.1   Purpose

The test is for susceptibility of the equipment or subsystem and interconnection cables to incident E fields and plane waves. The test setup is similar to that for RE02 with the receiving antenna replaced with a transmitting antenna.

9.6.7.6.2   Test Equipment

1. Antennas covering the range of 14 kHz–10 GHz and capable of generating the specified E fields will typically be of the following types:

Frequency	Antenna Type
14 kHz–20 MHz	Passive rod or long wire
or	Stripline
14 kHz–200 MHz	E field generating
20–200 MHz	Biconical
200 MHz–1 GHz	Log-periodic or double ridged guide
1–18 GHz	Double-ridged guide

2. Power amplifiers operating over the frequency range from 14 kHz to 18 GHz. The power rating of the amplifiers must be sufficiently high to generate the specified E fields from the antennas in use.
3. E-field sensors, with remote read capability or one of the following calibrated antennas, may be used to monitor the level of E field:

Frequency	Antenna
14 kHz–40 MHz	1 m rod
40–300 MHz	Tunable dipole
20–1000 MHz	Bow tie
300–800 MHz	Resonant 5–20 cm rod
800 MHz–3 GHz	25 cm conical log spiral
3–8 GHz	15 cm × 2.8 cm conical log spiral
8–18 GHz	6.2 cm × 1 cm conical log spiral

Due to the high levels monitored by the antennas, preamplifiers are typically not required.

4. Signal generators covering the frequency range from 14 kHz to 18 GHz with either internal or external modulation capability.
5. Audio frequency signal generator used to modulate the carrier frequency generator in (4).

9.6.7.6.3 Description of RS03 Test

The instrument is exposed to fields generated by antennas over the frequency range from 14 kHz to 10 GHz. The EUT should demonstrate correct performance when exposed to the specified level of E field. The signal source should be AM at a frequency at which the signal or power interfaces are predicted to be most susceptible.

The power and interface cables are exposed to the incident E field over a length of 2 m. The exposed cable is located 5 cm above the ground plane at a distance of 10 cm from the front of the ground plane.

Interface cables connected to other equipment in the room should be dropped to the metal floor after the 2 m exposed length. The cables on the floor should be as far away from the transmitting antenna as feasible and may be either covered with copper conductive adhesive tape connected on both sides of the cables to the metal section of the floor or contained in expandable braid that is connected to the metal floor section.

The cables, which are arranged on the ground plane, may be shielded, after the 2 m exposed length, either by an expandable braid that is taped to the ground plane with electrically conductive adhesive copper tape or simply by copper tape placed over the cables and contacting the ground plane on either side of the cables.

The antennas are located at a distance of 1 m from the face of the EUT. When small-aperture antennas (1–18 GHz frequency range) are used, the position of the antenna should be moved from directly in front of the EUT to directly in front of the interconnection cables.

MIL-STD 46IE requires RSI03 tests to be conducted from only as low as 2 MHz to 18 GHz and optionally up to 40 GHz.

Circularly polarized fields are not allowed.

There is no implied relationship between RSI03 and the RE 102 limits. RE 102 limits are placed primarily to protect antenna-connected receivers, while RS103 simulates fields resulting from antenna transmissions.

MIL-STD 46IE specifies specific locations for the transmit antennas and for the electric field sensors.

When a receive antenna is used, it is placed in the test setup boundary, and first the signal path is checked out using a signal generator.

The procedure for calibrating the E field using a receive antenna is as follows:

Connect the receive antenna.
Set the signal source to 1 kHz pulse modulation, 50% duty cycle. Using an appropriate transmit antenna and amplifier, establish an electric field at the test start frequency.
Gradually increase the electric field level until it reaches the applicable limit.
Scan the test frequency range and record the required input power levels to the transmit antenna to maintain the required field.
Repeat this procedure whenever the test setup is modified or an antenna is changed.

For testing, remove the receive antenna and reposition the EUT. This calibration technique does lead to errors due to the proximity of the ground plane to the receive antenna, but it does eliminate the error due to the proximity of the EUT to the receiving antenna.

The location of the power amplifier and the input lead from the signal generator are critical. For a monopole antenna, the power amplifier should be located under the counterpoise of the antenna and connected to the antenna with as short a cable as feasible.

When a long-wire antenna is used, and when driven without load, it should also be connected to the power amplifier, via a transformer where required, with as short a cable as feasible. For all antennas, with the possible exception of the monopole, the power amplifier should be located outside the shielded room to avoid feedback from the radiated field onto the input cable of the power amplifier.

MIL-STD-461F requires that the incident E field be measured at the EUT location using physically small, electrically short antennas. Even the use of field sensors located close to the ground plane will result in a measurement very different from the E field generated far away from the ground plane. The recommendation is that the E-field value be monitored 1 m from the transmitting antenna and not at the EUT position, where the presence of the ground plane can result in E fields ranging from close to cancelation to E fields greater than 6 dB above the level without ground plane. When this measurement technique is described in the test plan and test procedure, the procuring agency often accepts the deviation from MIL-STD-461 recommended procedure.

When the power amplifier is located outside of the room, the potential exists for coupling between the antenna feed cable and test equipment cables. Here, ferrite beads on the antenna and/or test cables may alleviate the problem or the use of double-braid shielded cables. Above 30 MHz, tests are made with both horizontally and vertically polarized fields.

The major problem with the RS03 test setup in an undamped room is that the field incident on the EUT will vary due to shielded-room resonances and reflections. As shown in Section 9.5.1, the variation may be as high as 50 dB. The E-field incident on the measuring antenna will also vary, and the AF of the measuring antenna will deviate from the published curve due to proximity effects. If the input signal to the power amplifier is adjusted to maintain the specified E field, as measured by the receiving antenna, the magnitude of the E field at the EUT may be anything from almost zero V/m to potentially destructive levels.

One technique that minimizes the possibility of subjecting the EUT to very high levels of E field is to sweep the test frequency over the range of interest and to move the location of the receiving antenna after each sweep. Because the dips and peaks in the measured E field are caused primarily by reflections, they tend to average out as the antenna location is moved. A common test method is to generate as high an E field as possible, limited only by the power amplifier and antenna. If the equipment is susceptible, then the magnitude of the E field at the susceptibility frequency is adjusted to the specified E field. The potential danger is that the equipment is damaged by the test levels generated. Automated tests often use the output from an E-field probe to control the output level of the power amplifier driving the antenna. The problem with this technique is that at those frequencies at which cancelation of the fields at the probe location occurs, the power amplifier output is at maximum, and again potentially damaging levels of E field are generated. The E field generated by a transmitting antenna at 1 m distance is given by the following equation:

$$E = \sqrt{Z_w \frac{WG}{4\pi r^2}}$$

where
  W is the input power
  G is the gain of the antenna at the frequency of interest
  r is the distance from the antenna [m]

$$Z_w = \frac{377\lambda}{2\pi r} \geq 377$$

  λ is the wavelength [m]

TABLE 9.33

Susceptibility Scanning from MIL-STD-461F

Frequency Range	Analog Scans, Maximum Scan Rates	Stepped Scans, Maximum Step Size
30 Hz–1 MHz	$0.0333f_{o/s}$	$0.05f_o$
1–30 MHz	$0.00667f_{o/s}$	$0.01f_o$
30 MHz–1 GHz	$0.00333f_{o/s}$	$0.005f_o$
1–40 GHz	$0.00167f_{o/s}$	$0.0025f_o$

When the input impedance of the antenna is assumed to be a constant 50 Ω and the distance a constant 1 m, the following simplified equation may be used:

$$E = \frac{V_{in}}{\sqrt{\dfrac{628}{Z_w G}}}$$

In common with other susceptibility tests, the maximum sweep rate should be limited by the ground support equipment (GSE) or functional test equipment test cycle time, that is, the time required to test the EUT and determine if it is susceptible. A guideline for the limit on sweep rate is to sweep the carrier frequency over the frequency range of interest, usually limited by the antenna, stopping at three frequencies per decade for the test cycle duration. Sweep between the held frequencies in a duration equal to the test cycle time. When the test cycle time is much over 1 min, inordinately long test times are required and the dwell and sweep durations should be limited to 3 min. The susceptibility test method should be outlined in the test plan and approved by the procuring agency prior to commencement of tests. If degraded performance is observed during the test, the E field should be decreased to determine the threshold of susceptibility. Receiving antennas are excluded from radiated susceptibility tests and should be replaced by a shielded dummy load.

Log-biconical antennas may develop levels at the 2nd or 3rd harmonic, which are above the fundamental even when the E-field probe is indicating the field from the fundamental and not the harmonics.

The MIL-STD-461F susceptibility scan rates are provided in Table 9.33. Also as a minimum level of absorber is specified in MIL-STD-461E/F although present room resonances are not such a large problem as in an undamped room.

In the CS114 and CS116 tests, the level is applied until the specified current is achieved or the calibration level is reached whichever occurs first. For CS115, the calibrated signal level is applied.

Safety Warning: Where power amplifiers greater than 3 W are used, the test equipment is capable of generating E fields of 200 V/m or higher. E fields above 27.5 V/m are considered hazardous. Therefore, do not remain in the shielded room with the signal generator powered up and connected to the power amplifier and antenna.

9.6.7.7 Most Common Errors Incurred in Radiated Susceptibility Measurements

1. A consensus on the pass/fail criteria has not been reached; that is, a few white spots on a display or some minor distortion may be acceptable, whereas multiple images or major distortion is almost certainly unacceptable. Similarly, some bit error rate in data communication may be acceptable, and this must be agreed on prior to commencement of susceptibility tests.
2. Failing to log the test start and stop times when data are analyzed offline, after the test is complete. The data must also be time tagged to allow comparison with the test times.

3. Inclusion of receiving antennas that are part of the EUT in the test chamber. These antennas and associated receivers should not be part of the susceptibility test. A danger is that the receiver front end may be destroyed by the power induced due to the incident susceptibility test field at in-band or close-to-in-band frequencies. If antennas cannot be disconnected, typically when the EUT is a flight model satellite, then wooden forms covered in absorber loads may be required to cover the antenna. In addition, frequencies at either side of the receiver in-band frequencies should not be radiated in the test.

4. Allowing personnel to view displays, indicator lights, or meters on the EUT in the test chamber during a radiated susceptibility test. Due to the potential danger, a set of mirrors can be set up and use made of the waveguide below cutoff to view the EUT display panel. With no cables penetrating the waveguide, it will function as a waveguide below cutoff, and the field coupled out of the shielded enclosure will be negligible.

5. Coupling between the signal generator input cable and the power amplifier output cable, or between the test field and the input cable to the power amplifier, when this is located in the test chamber. This coupling can result in positive feedback and the generation of very high E fields. Because the frequency of this oscillation may be very different than the test frequency, always monitor the radiated field with a wide frequency span.

6. Nonrepresentative test setup, use of a different type of cable than used in the final version, nonrepresentative EUT, addition of aluminum foil, copper tape, or braid over cables that will not normally be shielded.

7. The argument that cable-induced levels are not the concern of the manufacturer of the EUT when the cable harness is supplied by the customer. Regardless of the type of cable, shielded or unshielded, it is the EUT that must demonstrate immunity to cable-induced levels, and it is the manufacturer of the EUT who must reduce build immunity into the circuit. If a manufacturer believes that susceptibility requirements will not be met with a customer-supplied unshielded cable, it should apply for a deviation prior to testing. The customer may allow the deviation, require the EUT to meet the requirement, or change the cable to a shielded type.

8. Missing a repeat measurement with a different antenna polarization, exceeding calibration frequency range of antenna or power amplifier.

All of these problems can be avoided by use of a checklist of the tests, with tick boxes to ensure no test is missed.

9.6.7.8 Description of MIL-STD 461B/C RS01 and MIL-STD 461F RS101

The RS01 test is specified from 30 Hz to 30 kHz, and the RS101 from 30 Hz to 100 kHz. The test setup is shown in Figure 9.71.

FIGURE 9.71 RS01 measurement test setup.

The radiating loop for RS01 has the following specifications:

12 cm diameter
10 turns of AWG-16 capable of producing 5×10^{-5} T/A at 5 cm from the face (plane) of the loop

The radiating loop for MIL-STD 461D RS101 has the following specifications:

12 cm diameter
20 turns of AWG-12 capable of producing 9.5×10^7 pT/ampere at a distance of 5 cm from the plane of the loop.

In addition, MIL-STD 46IE specifies a loop sensor as follows:

Diameter 4 cm
Number of turns 51
Wire 7–41 Litz
Shielding electrostatic

The MIL-STD 462 method, for the limits in MIL-STD 461B/C, is to position the plane of the radiating loop 5 cm from the plane of the test sample. Flux densities approximately 20–30 dB greater than the applicable limit at the test frequencies shall be applied. At the point where the applied field produces the maximum effect, reduce the field until the performance of the EUT is not affected and record the magnitude of the field.

The MIL-STD 46IE test method is similar, for the plane of the radiating loop is positioned 5 cm from the plane of the EUT's surface, but the initial test requirement is to apply a level at least 10 dB above the applicable limit but not to exceed 183 dBpT. At locations of susceptibility, apply the specified field and move the loop to search for possible locations of susceptibility at the specified magnetic field level.

MIL-STD 46IE requires a calibration using the radiating loop and the sensor loop.

The RS101 test setup is shown in Figure 9.72.

For calibration of the RS01 and RS101 test setups, the following test equipment and procedure are required:

Equipment checklist
For the RS01 test:

1. *H*-field loop as described in MIL-STD-461. The loop is made of 10 turns of # 16AWG on a 4.27 in. (12 cm) diameter. The loop shall be capable of generating a magnetic flux density of 5×10^{-5} T/A at a distance of 5 cm from the face of the loop.
2. 13.3 cm shielded loop *H*-field antenna, as described in MIL-STD-462 RE01, 30 Hz–30 kHz, 1997, and MIL-STD-46 IE, RE101, 30 Hz–100 kHz, 1993.
3. Low-frequency power amplifier at least 100 W from 30 Hz to 100 kHz.
4. 1 Ω, 50 W resistor.
5. An oscilloscope used for differential measurements (channel A + B or A − B) used over the 30 Hz–50 kHz frequency range or the differential probe with × 10 attenuator, or an AC voltmeter.

For the RS101 test:

1. 1 *H*-field loop, made of 20 turns of #12 AWG on a 12 cm diameter. The loop shall be capable of generating 9.5×10^7 pT/A of applied current (9.5×10^5 T/A) at a distance of 5 cm from the plane of the loop.
2. Loop sensor made of 51 turns of 7–41 Litz (7 Strand, No. 41 AWG) electrostatically shielded.
3. Low-frequency current probe (30 Hz–100 kHz).
4. Low-frequency preamplifier (may not be required).

FIGURE 9.72 RS101 measurement test setup.

    5. Low-frequency (20 Hz–100 kHz) S/A or EMI receiver.
    6. Low-frequency signal generator (1 Hz–500 kHz).

Setup and operations
RS101 setup and operation
Antenna checkout:

> For the RS101 transmitting antenna: Connect the equipment as shown in Figure 9.73. Set the signal source to a frequency of 1 kHz and adjust the output to provide a magnetic flux density of 110 dB above one picotesla. This level is achieved when 3.33 mA (10.4 dBm A)

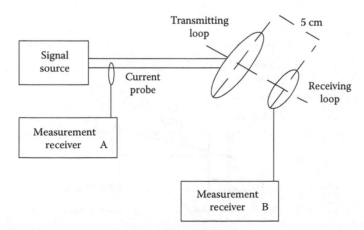

FIGURE 9.73 RS101 antenna test setup for checkout and calibration.

flows through the transmitting loop. This current is set by measuring with the current probe. The voltage measured by the current probe for the 10.4 dBmA is given by

$$V \text{ dBmV} = 10.4 \text{ dBmA} + Z_t(\text{dB}\Omega)$$

For example, with $Z_t = -50 \text{ dB}\Omega$, the value of V dBm with $I = 10.4$ dBmA is -39.6 dBmV $= 20.4$ dBµV $= -86$ dBm. With $Z_t = +18 \text{ dB}\Omega$, the value of V dBm with $I = 10.4$ dBmA is 28.4 dBmV $= -18$ dBm.

If the signal level out of the current probe is not at least 10 dB above the noise floor of the measuring equipment, use the low-frequency preamp between the current probe and S/A.

Measure the voltage output from the loop sensor.

Verify that the output on measurement receiver B is 42 dBpV ± 3 dB.

The setup in Figure 9.73 can be used for calibration of the receiving loop, in which case the frequency is swept over the full frequency range of 30 Hz to 100 kHz, and the receiving loop output is measured and recorded.

RS01 setup and operation
Antenna checkout:

For the RS01 transmitting antenna. Connect the equipment as shown in Figure 9.74. Set the signal source to a frequency of 100 Hz and adjust the output to provide a voltage of 1 V rms (2.829 V pk-pk) across the 1 Ω resistor. Measure the voltage output from the loop sensor using the preamplifier and oscilloscope.

The loop output voltage is given by the voltage measured across the scope (pk-pk converted to rms) divided by the gain of the preamplifier.

FIGURE 9.74 RS01 antenna test setup for checkout and calibration.

Convert this voltage to dBµV. At 100 Hz, the measured level should be approximately 43 dBµV. Adjust the frequency to 1 kHz, measure the voltage across the 1 Ω resistor, and adjust, as required, to 1 V rms. The measured output from the loop, after correcting for the precalibration procedure for the receiving loop. The only difference is that for calibration, the frequency is swept over the 30 Hz–30 kHz frequency range, and the output of the receiving loop is monitored and recorded.

9.6.7.9   Description of MIL-STD 461B/C RS02

9.6.7.9.1   Purpose

This test is to determine the susceptibility to induced AC power frequency and transient magnetic induction fields.

The test is divided into a cable test and a case test, although the case test is seldom a requirement.

In the cable test, a 2 m length of induction cable is wrapped around the cable, and a spike, having the characteristics of the CS06 spike and developed by the CS06 spike generator, is applied to the cable under test. The spike voltage is up to 400 V at 5 µs and up to 200 V at 10 µs. A second test applies a current of 20 A at the power frequency. If AC power is not used on the aircraft, vehicle, or spacecraft, then the AC power test should not be a requirement.

In the spike test at 100 V and 10 µs, the measured current induced in a shielded cable is approximately 30 A. The shield of the cable will attenuate the internally generated common-mode voltage to a certain level.

A shielded cable will not attenuate the voltage induced in the 20 A power test, where the frequency is typically 50–400 Hz, and the signal interface must be immune to the common-mode voltage induced at these frequencies, typically by use of a differential or quasi-differential interface for analog and baseband video.

9.6.7.10   Description of CS01/CS101 and CS02/CS102 Tests

MIL-STD-461E-G does not contain a CS02 test but does contain a CS114 test not contained in earlier versions of MIL-STD.

9.6.7.10.1   Purpose

These tests are designed to measure the susceptibility of AC and DC power leads to audio frequency (AF) and RF ripple voltage and transient voltage.

The AF and RF ripple is applied directly to the AC and DC power leads.

Conducted susceptibility power lines (CS01 and CS02): This test is conducted over the frequency range from 30 Hz to 400 MHz.

If degraded performance is observed during the test, the signal should be decreased to determine the threshold of interference.

The input of the EUT may present a capacitance to the susceptibility test signal. Generation of the test levels may be impossible when the input capacitance is large, even when high-power amplifiers are used. Too high an RF power developed in the EUT input, circuits may result in damage. Consequently, the following alternative test limits are widely accepted and should be incorporated into the test plan and procedure.

The CS01 requirement is also met when the power source, shown in test setup CS01, adjusted to dissipate 40 W into an 0.5 Ω load, cannot develop the required voltage at the EUT input connections, and the test sample is not susceptible to the output of the signal source.

The CS02 requirement is also met when a 3 W source of 50 Ω impedance cannot develop the required voltage at the input power connector, and the test sample is not susceptible to the output of the signal source.

The CS01 injection transformer has a secondary inductance of approximately 1.5 mH. When the transformer is connected into the test circuit with the power amplifier powered down or disconnected, this inductance is in series with the power line and presents a high source impedance to the

EUT. Many DC–DC converters or switching-power supplies are unstable with a high source impedance; therefore, never connect the transformer without the power amplifier powered up. However, with the power amplifier connected prior to powering up the EUT, as the EUT is switched on, the in-rush current can develop a large voltage across the secondary of the transformer. This is stepped up and is applied to the output of the power amplifier, possibly resulting in damage. To avoid both problems, short out the secondary of the transformer until the EUT is powered up and the power amplifier is connected and powered up. REMEMBER TO REMOVE THE SHORT BEFORE AN INPUT SIGNAL IS APPLIED TO THE POWER AMPLIFIER.

Just as it is recommended to perform breadboard level CE101, CE102, and RE102 tests on the power supply filter and power supply, in a suitable shielded enclosure, early in the development phase, so, too, should a CS101 test be performed.

The injection transformer is special and design details are provided in MIL-STD-462. 25–70 V PA transformers are not suitable as the frequency response does not extend down to 30 Hz and the power capability is not high enough.

Solar electronics manufacture a suitable transformer and power amplifier, or they can be rented. A high quality 100 W audio amplifier, such as the Bryston 4B, with a 30–50 kHz frequency response can extend out to 150 kHz at the reduced level. Do not use a Class D switching audio power amplifer as the PWM pulses generated may destroy the EUT!

Any switching spikes generated by the EUT are coupled directly to the output of the power amplifier via the 0.1 pF capacitor in the CS02 test. If these spikes are high, use a low-value series resistor between the output of the power amplifier and the capacitor. Connect bidirectional transorbs, or unidirectional transorbs with diodes, or diodes with voltage ratings above the maximum peak voltage developed by the power amplifier. The 20–50 µH inductor is required to ensure that if the required test level cannot be developed, and it is not due to the low output impedance of the power supply. In some test facilities, the correct test level is never developed; instead, the 3 W calibration level into 50 Ω is routinely applied. However, this approach may result in a serious undertest of the EUT.

If, for example, an EUT powered by a 400 Hz, 440 V AC power supply is tested for CS02, a current of typically 1.1 A flows at 400 Hz into the power amplifier output via the 1 µF coupling capacitor. A simple band-stop filter tuned to 400 Hz can be designed and constructed and can reduce the current to 0.4 mA at 400 Hz.

9.6.7.10.2   Test Setup

The CS01 test setup is shown in Figure 9.75 and the CS02 in Figure 9.76. The coupled noise should be applied to one power bus at a time, with the input voltage level set at the *nominal* power-line voltage unless otherwise specified.

9.6.7.10.3   CS01/CS101 and CS02/CS102 Test Equipment Required

1. Signal generator or generators to cover the frequency range from 30 Hz to 400 MHz
2. Power amplifier minimum 50 W continuous from 30 Hz to 50 kHz (CS01)
3. Power amplifier minimum 3 W continuous from 50 kHz to 400 MHz (CS02)
4. 10 µF RF feedthrough capacitors for MIL-STD-461 up to C and LISNs from MIL-STD-461E-G
5. 1–0.1 µF rated at 1.5 times the peak voltage, or use *x* or *y* capacitors for AC power rated at the AC power-line voltage (CS02)
6. Isolation transformer rated at least 50 W from 30 Hz to 50 kHz with, typically, a primary resistance of 2 Ω and a secondary resistance of 0.5 Ω, turns ratio 2/1 (CS01).

The MIL-STD 461F CS 101 test is similar except for the use of an LISN and a 10 µF capacitor across the line, on the EUT side of the LISN.

A calibration of the test signal across a 0.5 Ω resistor is required prior to testing.

The CS01 transformer must be capable of conducting the DC current required to power the EUT without saturation. The output of the power amplifier may be protected from RF and transients that

FIGURE 9.75   CS01 conducted susceptibility test setup, 30 Hz–50 kHz.

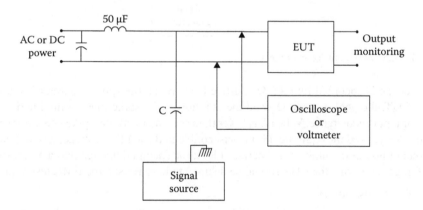

FIGURE 9.76   CS02 conducted susceptibility test setup and connection diagram, 50 kHz–400 MHz. 1. The value of C shall be chosen such that $X_c < 5\,\Omega$ over the test frequency range. 2. Connect the coupling capacitor and the oscilloscope within 5 cm of the power socket mounted on EUT.

may be generated by the EUT, should a failure occur in the instrument during CS01 tests, by the 0.1 pF and 50 pH network shown in setup CS01, Figure 9.75.

MIL-STD 46IE does not specify a differential-mode (CS02-type) test above the CS101 upper frequency limit of 50 kHz.

MIL-STD-461F adds a structure current test, CS109, shown in Figure 9.77, and a bulk cable injection test, CS114, shown in Figure 9.78.

9.6.7.11   Description of CS06 Test

The transient pulse identified in MIL-STD-461 (A, B, or C) applies. The test should be conducted at the nominal power-line voltage. The transient should be applied at a repetition rate of 60 pps for a duration of 5 min and have a pulse width of 10 μs. The test should be applied to the input power leads of the EUT. Under these conditions, the instrument should suffer no degradation, malfunctions, or deviation from specified performance.

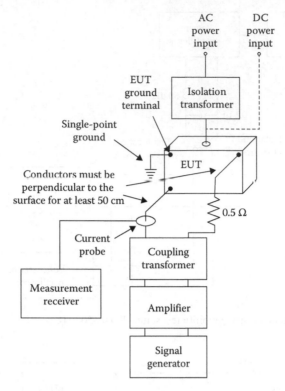

FIGURE 9.77   CS109 structure current test.

The waveshape is specified into a 5 Ω resistive load. When the spike generator is applied to the input of the EUT, the peak voltage (E) and the duration of the spike may be modified by the EUT input filters or input capacitance. When E is lower than specified with the spike generator set to maximum output voltage, and the equipment is not susceptible to the applied voltage, ensure that the spike generator set at maximum output can generate at least the specified voltage into a 5 Ω resistive load. When the foregoing is true, the EUT should be deemed to have passed the EMC test requirement.

9.6.7.11.1   CS06 Test Setup

The proposed test setup is shown in Figure 9.79. An alternative test setup for AC power lines is given in MIL-STD-462 CS06-1, in which the spike generator is connected to an isolation transformer in series with the input power lead. However, many spike generators can function safely when placed across the supply as shown in Figure 9.79.

9.6.7.11.2   Test Equipment

1. 50 µH inductor. The inductor must be capable of handling the AC or DC power current without saturation and with a reduction in inductance to no less than 20 µH.
2. 10 µF RF capacitor.
3. Spike generator with the following specifications:
   a. Pulse width = 10 µs
   b. Pulse repetition rate = 60 pps
   c. Voltage output = not less than the specified voltage into 5 Ω
   d. Output control = adjustable from 0 to the specified voltage
   e. External trigger = 60 pps
   f. Calibrated differential/input oscilloscope with at least 10 MHz bandwidth, adequate sweep rates, and an input voltage rating of the spike voltage $+E$, where E is the power ripple voltage

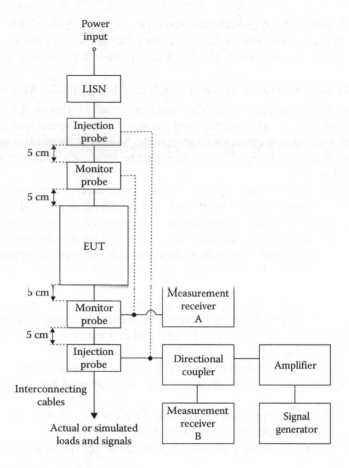

FIGURE 9.78 CS114 bulk cable injection test.

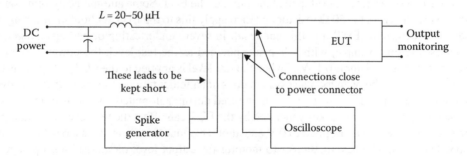

FIGURE 9.79 CS06 conducted susceptibility test setup for spike and power.

After testing with a positive spike, the test should be repeated with a negative spike. Adjust the output voltage from the spike generator to the specified voltage and trigger the oscilloscope from the channel used to view the spike, using AC and normal trigger coupling. At the low repetition rates used, an oscilloscope hood or storage oscilloscope may be required for the spike to be visible. Maintain the spike for the cycle time of the GSE or functional test equipment for 5 min, whichever is longer. When testing on an AC supply, the spike position shall be moved over a full 180° of the AC waveform, either side of the zero crossing point.

MIL-STD 461F requires a bulk cable injection (common-mode) transient test CS115. The injection method is by injection probe, and the setup is similar to CS114.

MIL-STD-461F adds a CS106 transient test that is similar to the above CS06. The only significant difference is that there is no requirement to synchronize spikes to the power line. The spike is 5 μS wide and 400 V in peak amplitude. The 10 μS wide pulse is not specified.

9.6.7.12 Most Common Errors Incurred in Conducted Susceptibility Measurements

1. A consensus on the pass/fail criteria has not been reached; that is, a few white spots on a display or some minor distortion may be acceptable, whereas multiple images or major distortion is almost certainly unacceptable. Similarly, some bit error rate in data communication may be acceptable, and this must be agreed on prior to commencement of susceptibility tests.
2. Failing to log the test start and stop times when data are analyzed offline, after the test is complete. The data must also be time tagged to allow comparison with the test times.
3. Incorrect measurement of the injected signal. The measurement point must be across the power input to the EUT. It must not be across the injection transformer primary or secondary or directly at the output of the power amplifier. Only if the power return is connected to the ground plane *at the EUT* can one of the generator output terminals be connected to the ground plane.
4. In the CS06 test and the CS02 test, use long cables connecting the signal into the EUT. The long cables can reduce the high-frequency signal and degrade the CS06 pulse shape. Inject the signal as close to the EUT as possible and monitor at the injection point, using a differential input to the scope.
5. Nonrepresentative EUT, different power supply, or input power-line filter.
6. Not applying the relaxed CS06 requirements for equipment that contains transient suppressors.
7. Oscillation of the power supply due to inclusion of the CS01 injection transformer with the power amplifier powered down or disconnected (see test method).
8. Injection of the in-rush current spike into the power amplifier via the injection transformer (see test method).
9. Setting up the CS02 level across a 50 Ω resistor. This should be attempted only if the required level cannot be developed at the input of the EUT. Some automated systems set up the test level across the 50 Ω resistor and then apply this to the EUT without monitoring the test level across the EUT, and this can result in severe undertesting or perhaps dangerous overtesting. For example, with 1 V peak specified as the test level, the power developed across the 50 Ω resistor is 0.1 W. If when the test level is applied to the EUT, 1 V peak is not developed, then the open loop test system does not attempt to increase the output test level. On the other hand, if the test level is monitored during automated test measurements in a closed loop system, any spikes generated by the EUT can affect the test level. For example, in one case, the output dropped to 0 V and at other frequencies was at least 6 dB below the specified level. The best method is to monitor the output level on an oscilloscope and to manually control the amplitude.
10. Injection of some fraction of the AC power-line voltage into the power amplifier or, in the case of a closed loop system, into the monitoring port, via the coupling capacitor in the CS02 test (see the test method for a solution).

9.6.7.12.1 MIL-STD-461G

This version retains the isolation of the monopole counterpoise from the table to ground or the floor ground but adds an isolation transformer or optical coupling to the antenna cable under the counterpoise.

A table is now provided for routing of cables, and cable routing for floor-standing equipment is described.

TABLE 9.34

MIL-STD-461G Proposed Bandwidth and Measurement Time

		Dwell Time		Minimum Measurement
Frequency Range	6 dB Bandwidth	Stepped Tuned Receiver[a]	FFT Receiver[b] (Seconds/ Measurement Bandwidth)	Time, Analog Measurement Receiver[a]
30 Hz–1 kHz	10 Hz	0.15	1	0.015 s/Hz
1–10 kHz	100 Hz	0.015	1	0.15 s/kHz
10–150 kHz	1 kHz	0.015	1	0.015 s/kHz
150 kHz–10 MHz	10 kHz	0.015	1	1.5 s/MHz
10–30 MHz	10 kHz	0.015	0.15	1.5 s/MHz
30 MHz–1 GHz	100 kHz	0.015	0.15	0.15 s/MHz
Above 1 GHz	1 MHz	0.015	0.015	15 s/GHz

[a] Alternative scanning technique. Multiple faster sweeps with the use of a maximum hold function may be used if the total scanning time is equal to or greater than the Minimum Measurement Time defined here.

[b] FFT receivers. FFT measurement techniques may be sued provided that FFT operation is in accordance with ANSI C63.2. The user interface of the measurement receiver must allow for the direct input of the parameters in Table 9.34 for both FFT time-domain and frequency stepped modes of measurement in the same manner, without the necessity or opportunity to control FFT functions directly.

CS117 is a new test similar to the Radio Technical Commission for Aeronautics (RTCA) DO-160 of lightning-induced transients onto cables and power leads. No pin injection is required and includes single stroke and multiple burst as described in DO-160G.

CS118 Conducted Susceptibility ESD is a new test based on IEC 61000-4-2. The definition of the ESD gun is as per DO-160 and the calibration target is as per IEC 61000-4-2.

Both contact and air discharge levels are specified from 2 to 15 kV with 15 kV confined to air discharge. The major difference from IEC 61000-4-2 is that in MIL-STD-461G, the EUT shall be electrically bonded in accordance with the normal EUT installation requirements. This maximized the ESD current flow through the EUT to ground. In IEC 61000-4-2, the connection is via 470k resistors. A high current may sill flow in the iEC 61000-4-2 setup due to the capacitance between the EUT and the underlying ground plane.

All RE102 tests now extend to 18 GHz.

CS106 in MIL-STD-461F is deleted in MIL-STD-461G.

No electrical bonding values are specified except the LISN bonded to the table and the conductive table to the wall at ≤2.5 mΩ. The bonding between the units/equipment making up the EUT is not specified but the installation must be typical for the EUT.

In addition to the requirement that the test area must be kept free of unnecessary personnel, equipment, cable racks and desks, etc., the requirement is added to remove all equipment and ancillary gear not required, including antenna, from shielded enclosures not being actively used for a particular subset of radiated tests.

The bandwidth and measuring times are shown in Table 9.34

In the RE102 and RS103 test setups, the antenna height and width are specified.

Although measuring equipment such as EMI receiver, S/A, preamplifiers, oscilloscopes require calibration at least every 2 years, passive devices such as LISN, current probes, receiving, and transmitting antennas must be calibrated initially but then the MIL-STD measurement check prior to a test is sufficient.

When measuring the integrity check on the LISN, the impedance below 2 MHz must also be tested.

The transmitter requirements in CE106 and RE103 for exclusion frequency, transmit power, and harmonics are changed.

9.7 RTCA/DO-160 REQUIREMENTS

The RTCA imposes environmental and test procedures on airborne equipment. These contain power quality and EMI requirements and test methods. The latest version of the requirements document is DO-160G.

The category of the equipment for power quality is A, B, E, or Z, determined by the type of aircraft electrical system. The EMI category is A, B, D, or Z. A is equipment for which interference-free operation is desirable, B is equipment for which interference should be controlled to a tolerable level, and Z is equipment intended for interference-free operation.

The magnetic effect categories are Y, Z, A, B, and C, which apply to installation separation distances of 30–300 cm.

The induced signal susceptibility classes are A, B, C, or Z for the first character, which indicates the test performed and severity level of the test, and the second character C, N, or W indicates the power system operating frequency (constant, Narrow variable, or Wide variable). The RF susceptibility categories are W, Y, V, U, and T, where the W and Y test levels are applicable to locations where equipment and interconnecting wiring is installed in severe electromagnetic environments, such as nonmetallic aircraft or exposed area in metallic aircraft. Category V is defined for moderate environments, such as open areas in an aircraft composed principally of metal. Category U is defined as partially protected environments, such as avionic bays in all-metallic aircraft. Category T is defined as well-protected enclosed areas, such as an enclosed avionics bay in an all-metallic aircraft. The test limits are different for each category of equipment. The test methods and limits are very similar to those contained in the military standards discussed in Section 9.6, with some minor variations. One such variation is that the 10 μs voltage spike is specified at 600 V minimum open-circuit voltage with a 50 Ω source impedance for category A, and for category B, a voltage transient is specified at a much lower amplitude with a 25 μs rise time to peak voltage and 100 μs width from the peak down to the 50% amplitude level, using a low-source-impedance generator. A repetitive spike 10 μs wide at an amplitude twice the rms AC power voltage is also applied from a low-impedance source to category B equipment. If the MIL-STD CS06 spike generator has a 600 V capability, then it can be used for the Class A spike with an additional 50 Ω series resistor. The CS06 spike generator can also be used with series injection for the 10 μs wide pulse. Interconnection cables are subjected to audio frequency magnetic fields from 400 Hz to 15 kHz (category Z) and audio frequency and transient E fields (categories A, B, and Z), induced by wrapping a wire around the cable. For the audio frequency test on category Z equipment, a power source with 30 A capability at 400 Hz reducing down to 0.8 A at 15 kHz is necessary. This would require a large 1.7 kW variable-frequency supply or an impossible 3.6 kW audio power amplifier. Instead, a 500 W, or higher, variable-frequency AC supply used with a 15:1 step-down transformer may be used to generate the high AC currents. In the interconnecting cable spike test for categories A, B, and Z, a 50–1000 μs duration burst of spikes with a repetition rate of 0.2–10 μs at a peak voltage of 600 V is specified. The recommended method of generating the test field is the unsuppressed "chattering" relay circuit. An E-field wrapped-wire-cable test is also specified with a frequency of 380–420 Hz and a maximum voltage of 1800 V/m for category Z equipment and 360 V for category A equipment. Current probes are used to inductively couple RF into cables from 10 kHz to 400 MHz. The injection level is specified in decibels above 1 μA and is determined by the category of the equipment. The recommended test setup requires the use of a monitor probe to monitor the injected current during the test.

Both H fields and E fields are induced into the equipment with the inducing wire swept across the face of the equipment. E-field- and H-field-induced susceptibility tests into equipment are performed at 400 Hz.

These induced signal susceptibility tests are not applicable to power input cables/leads.

An audio frequency–conducted susceptibility test from 750 Hz to 15 kHz, similar to CS01, is specified. If the audio power amplifier cannot develop the specified level, then the test conditions will be adequately met by the use of a power amplifier with a maximum output of 30 W. The impedance of the output transformer used to inject the signal shall be 0.6 $\Omega \pm$ 50%.

Conducted emissions on power lines are measured using the monitor port on the LISN for categories A, B, and Z and a bandwidth described in table of DO-160G. For categories A and Z, a current probe that measures the current flow into the impedance of the LISN may be used. For category A and Z equipment, RF current flow on intersystem interconnecting cable bundles shall be measured using the current probe. The DO 160 5 µH LISN has the same requirements as the MIL-STD 5 µH LISN. The LISN may be manufactured and calibrated or bought out. Both NB and BB conducted emission limits are imposed.

Radiated emission measurements are made from 100 to 6000 MHz for categories B, L, M H, P, and Q. Different limits are applied to the five different categories. Radiation from antennas is excluded in the test setup, just as it is in MIL-STD-461/2 testing. In addition, when testing a transmitter, the selected frequency and ±50% of the band of frequencies between adjacent channels are excluded. An additional control on signals conducted out of antenna terminals is intended but not specified.

Radiated susceptibility E field is generated from 30 to 1215 MHz for categories T, U, and V and from 30 MHz to 18 GHz for categories W and Y. The antenna is located 0.9 m from the edge of the ground plane on which the EUT is mounted, with the EUT and interconnect power cables located 0.1 m in from the front edge of the ground plane. In accordance with DO-160C and D, the antenna shall be located with its lowest point at a height of 0.3 m above the ground plane. Antennas shall be positioned and aimed to establish the field strengths at the EUT and interconnecting wiring. When the beam width of the antenna does not totally cover the EUT and wiring, multiple scans shall be performed. Apertures in the EUT shall be exposed directly to the transmitting antenna. Horizontal and vertical orientations are required for polarized antennas, although this is typically taken as excluding the monopole antenna, long-wire, parallel-plate, or TEM cell, all of which may be used. Parallel-plate or TEM cells matching networks and calibration methods shall be described. Amplitude modulation of the CW source is required from 30 MHz to the upper frequency using a square wave at greater than 90% modulation and 1 kHz frequency. Consideration should be given for other modulations associated with the EUT, such as clock, data, IF, internal processing, or modulation frequencies. The use of field sensors located on the ground plane is proposed, which, as mentioned in Section 9.6.7.6, is prone to error. DO-160 recommends that the field sensors be connected to a control circuit, located outside of the room, using a fiber-optic link. Absorber is required on the surfaces of the shielded enclosure, and the EUT power is connected via the LISNS. When using the anechoic chamber method for radiated susceptibility, all faces of the EUt must be directly exposed to the antenna. In practical terms, this means repeating the test with a new face turned to the antenna. If any face cannot be exposed, then the justification must be included in the test report.

If the power input (DO-160 Section 16) tests are to be performed, then a frequency modulation of the AC power source is required. The surge current requirement necessitates variacs and transformers as well as relays and timers. The AC ripple on the DC power is less stringent than typical CS01 requirements, and the same test equipment may be used.

One of the major differences between MIL-STD-461A-E and DO-160 requirements is that DO-160C–G imposes indirect lightning tests on equipment to test for its capability to withstand effects of lightning-induced electrical transients. The draft version of MIL-STD-461G also adds this type of test taken from the DO-160 requirements. Direct lightning tests are intended to test externally mounted electrical and electronic equipment for the capability to withstand the direct effects of a severe lightning strike. Externally mounted equipment refers to all equipment that is mounted externally to the main skin of the aircraft and includes equipment covered only by a dielectric skin that is an integral part of the equipment. If the dielectric skin is specific to the aircraft and not integral to the equipment, then the equipment is subject to tests specified by the aircraft manufacturer. These lightning tests may be severe, and the test levels are determined by the internal aircraft environments as well as for metallic and nonmetallic aircraft. These environments range from level 1, which is well protected, to level 2, partially protected, to level 3, moderately protected, to levels 4 and 5, which are characterized as severe electromagnetic environments. Each level has a set of waveforms associated with it, with the O/C voltage and S/C current characteristics of the waveform specified. The waveforms are idealized and are described as an exponential/exponential transient in which both the rising edge and the falling edge are in the form of an exponential. The second waveform is a damped sinusoid with a specified frequency, a peak voltage, and current reducing to a specified level after four cycles. Two groups of indirect test may be specified. The PIN test is designed to test for damage tolerance by injecting directly into the pins of the EUT connector, usually between each pin and case ground. For equipment electrically isolated from case and local airframe grounds, a dielectric withstand or hi-pot test is adequate. The second group of tests is the cable bundle tests, which are used to determine if functioning equipment will experience upset or component damage when exposed to a cable-induced transient. These tests are specified as either single stroke or multiple stroke where a first transient is followed by a further 13 strokes each greater than 10 mS and less than 200 mS interval. An alternative injection method is ground injection, where the injection point may be at the EUT case ground or into the case of support equipment. In any of the induced tests, a current-measuring transformer is placed around the cable bundle to measure the bulk (C/M) cable current. When equipment has shielded cables that are not connected to available support equipment, a connector may be mounted on a small metal enclosure, the cable plugged into the enclosure, and a direct injection made between the ground plane, to which the EUT is bonded, and the small connectorized enclosure. The injection method is not so important as ensuring that the current flows on the cable bundle or the correct voltage is applied between the ground plane and the cable. The level of common-mode current induced in the center conductors of shielded cables and some possible transient protection components for different circuits are described in Section 5.7.

The exponential/exponential waveforms are (1) and (4) 6.4 µs rise time falling in 70 µs to 50% amplitude, (2) a 100 ns rise time falling to 0% amplitude after 6.4 µs, and (5) a 50 µs rise time falling in 500 µs to 50% amplitude. The waveforms are specified with an O/C voltage and S/C current. For pin injection testing, waveform (4) ranges from 50 V/10 A to 1600 V/320 A and waveform (5) from 50 V/50 A to 1600 V/1600 A, depending on the specified level (1–5). Pin injection also specifies a damped sinusoid, waveform (3), at the aircraft resonance frequency, or 1 MHz if the resonant frequency is unknown. Waveform (3) varies from 100 V/4 A to 3200 V/128 A, depending on the specified level. For pin injection, waveforms (1) and (2) are specified from 50 V/100 A to 1,600 V/3,200 A, waveform (3), the damped sinusoid, from 100 V/20 A to 3,200 V/640 A, and waveforms (4) and (5) from 50 V/300 A to 1,600 V/10,000 A.

Direct lightning injection tests are conducted with the EUT bonded to a ground plane in accordance with the installation requirement. A high-voltage test is performed on equipment that is covered by a dielectric in which an electrode is at a specified distance from the EUT. Two test methods are described. One has a voltage waveform applied to the test electrode with a rise time of 1.2 µs and a fall time to 50% in 50 µs. The total gap lengths, between the electrode and the EUT, can range from 0.5 to 1.5 mm with voltage ranges of 250–1500 kV. One alternative test method has a linearly increasing voltage applied to the electrode at an average dV/dt of 10,000 kV/µs until the gap breaks down. In this alternative method, the gap can be from 0.5 to 1.5 mm and the voltage from 750 to 2400 kV.

A high-current-arc entry test is specified in which an arc is initiated and enters the EUT. Several components of the current waveform are specified, with only some of the components applicable, depending on the category of the equipment. The peak current in component A is specified at 200 kA for 500 µs, followed by component B with a peak amplitude of 2 kA. Component C is a continuing current of between 200 A and 800 A, followed by a restrike current component D with a 100 kA peak. An alternative conducted entry test has the current injected through the ground plane on which the test object is mounted, representative of the lightning current distribution in an aircraft during a lightning strike. As a minimum, a surface current density of 50 kA/m should be applied. Although the ground plane injected test may appear trivial, in reality, magnetic fields of hundreds of thousands of amps per meter and E fields of hundreds of thousands of volts per meter can be generated, depending on the location of the EUT.

REFERENCES

1. Compliance Engineering.
2. E. Bronaugh and D.R. Kerns. IEEE transactions on electromagnetic compatibility: A new isolated antenna system for electromagnetic emissions measurements in shielded enclosures. *Proceedings of the IEEE International Symposium on Electromagnetic Compatibility*, Atlanta, GA, 1978.
3. A.C. Marvin. The use of screened (shielded) rooms for the identification of radiated mechanisms and the measurement of free-space emissions from electrically small sources. *IEEE Transactions on Electromagnetic Compatibility*, November 1984, EMC-26(4), 149–153.
4. Private communication with Ken Javor, EMC Engineer.
5. B. Audone, L. Bolla, G. Costa, A. Manara, and H. Pues. Design and engineering of a large shielded semi-anechoic chamber meeting the volumetric NSA requirements at 3-m and 10-m transmission length. *Proceedings of the IEEE International Symposium on Electromagnetic Compatibility*, Denver, CO, 1993.
6. W. Bittinger. Properties of open strip lines for EMC measurements. *Proceedings of the IEEE International Symposium on Electromagnetic Compatibility*, Denver, CO, 1993.
7. H.A. Mendez. A new approach to electromagnetic field-strength measurements in shielded enclosures. Wescon Technical Papers Session 19. Los Angeles, CA, August 20–23, 1968.
8. W.C. Dolle and G.N. Van Stewenberg. Effects of shielded enclosure resonances on measurement accuracy. *Proceedings of the IEEE International Symposium on Electromagnetic Compatibility*, Anaheim, CA, July 1970.
9. A.C. Marvin. Near-field antenna coupling theory in a shielded room: The mutual impedance model. *EMC Symposium: Third Symposium and Technical Exhibition on Electromagnetic Compatibility*, Rotterdam, the Netherlands, May 1–3, 1979.
10. A.C. Marvin and A.L. Marvin. Method of damping resonances in a screened room in the frequency range 30 to 200 MHz. EMC Technology, Stuart, FL, July/August 1991.
11. L. Dawson and A.C. Marvin. Damping resonances within a screened enclosure, *IEEE Transactions*, 43(1), 2001.
12. Y. Naito, T. Mizumoto, M. Takahashi, and S. Kunieda. Anechoic chamber having multi-layer electromagnetic wave absorber of sintered ferrite and ferrite composite membranes. *Proceedings of the IEEE International Symposium on Electromagnetic Compatibility*, Chicago, IL, 1994.
13. B. Archambault and K. Chamberlin. Modeling and measurements of an alternative construction technique to reduce shielded room resonance effects. *Proceedings of the IEEE International Symposium on Electromagnetic Compatibility*, Chicago, IL, 1994.
14. J.P. Muccioli, T.M. North, and K.P. Slattery. Investigation of the theoretical basis for using a 1-GHz TEM cell to evaluate the radiated emissions from integrated circuits. *Proceedings of the IEEE International Symposium on Electromagnetic Compatibility*, Santa Clara, CA, 1996.
15. J.D. Osburn and E.L. Bronaugh. Advances in GTEM to OATS correlation models. *Proceedings of the IEEE International Symposium on Electromagnetic Compatibility*, Denver, CO, 1993.
16. H. Trzaska. *Electromagnetic Field Measurement in the Near Field*. Noble Publishing Corporation, Atlanta, GA, 2001.
17. J.H. Kim and J.I. Park. Development of standard monopole antenna factor measurement. *Proceedings of the IEEE International Symposium on Electromagnetic Compatibility*, Austin, TX, May 21–23, 1997.
18. C.W. Fanning. Improving monopole radiated emission measurement accuracy; RF chamber influences, antenna height and counterpoise grounding. *Proceedings of the IEEE International Symposium on Electromagnetic Compatibility*, Kyoto, Japan, 2009.
19. D.D. Swanson. Analysis of MIL-STD-461E and MIL-STD-461F RE102 test setup configurations below 100 MHz. *IEEE International Symposium on Electromagnetic Compatibility*, Detroit, 2008.
20. Interference Technology (ITEM). EMC test & design guide. Plymouth Meeting, PA, 2010.
21. D.A. Weston. High frequency calibration of the 41 inch (1.04m) receiving monopole with and without connecting counterpoises with differenc sources, *EMC Europe*, Wroclaw, Poland, 2010. 1.04 m Rod, Antenna factor and received level in MIL-STD-461E compared to MIL-STD-461F test set up interference technology 2010.
22. D.A. Weston. Egregious errors in electromagnetic radiation evaluation. Interference Technology (ITEM): EMC directory & design guide, Plymouth Meeting, PA, 2010.
23. M.J. Windler, Underwriters Laboratories Inc., and Z. Chen, ETS Lindgren. EMC test site qualifications: Site voltage standing wave ratio versus time domain reflectometry. *In Compliance Magazine*, January 2010.
24. C63.4-2014, American National Standard for methods of measurement of radio-noise emissions from low voltage electrical and electronic equipment in the range of 9 KHz to 40 GHz. IEEE 2014.

10 System EMC and Antenna Coupling

10.1 SYSTEM-LEVEL ELECTROMAGNETIC COMPATIBILITY

A significant amount of the subject matter dealt with in the preceding chapters is equally applicable to equipment, subsystems, and systems. Consider the case of cable-to-cable or cable-to-equipment coupling that may occur between equipment, subsystems, or systems. In this chapter, system-level electromagnetic compatibility (EMC) is emphasized, although many of the topics are equally applicable to subsystems and equipment. For example, antenna-to-antenna or antenna-to-cable coupling can occur within a subsystem and between systems.

The foregoing observations are perhaps obvious, although not always remembered. For example, consider a room containing a system made of a number of pieces of equipment all connected to the same potentially noisy power supply. A decision to incorporate a main power-line filter at the location where the supply enters the room can be defended based on the reduced radiation from the power-line cabling within the room, caused by supply noise, and on the elimination of, or reduction in, filters in the individual equipment. The design or selection of the main filter may be achieved using the methods described in Chapter 5, although these and the case study were applied at the equipment level. For the system-level main filter, the source of noise on the load side of the filter is a composite of the individual equipment noise sources, as is the total load impedance presented by the equipment. When filter components exist in equipment, these must be included in the evaluation of the main power-line filter. The analysis of the performance of the main system-level filter is likely to be more complex than that required for an equipment filter and may be practical only by use of a circuit modeling program such as SPICE. The foregoing is only one example of a great number of EMC design approaches, including shielding, that may be applied equally to circuits, equipment, subsystems, and systems.

10.1.1 MIL-STD SYSTEM-LEVEL REQUIREMENTS

10.1.1.1 General

MIL-STD-461/2 applies to equipment and subsystems only. MIL-STD-464 is a document applicable to all agencies of the Department of Defense in the United States that contains system-level EMC requirements. The associated data item descriptions are DI-EMCS-81540, Electromagnetic Environmental Effects Integration and Analysis Report; DI-EMCS-8151, Electromagnetic Environmental Effects Verification Procedures; and DI-81542, Electromagnetic Effects Verification Report.

Document MIL-STD-1541A is an EMC requirement document for space systems and is applicable only to the U.S. Air Force, although it is used by other agencies, such as NASA.

The purpose of system-level requirements is to ensure that the integrated system design will be electromagnetically compatible with all of the characteristics and modes of operation of the system. The requirements include control of lightning protection, static electricity, and bonding and grounding. Control is achieved by a system electromagnetic compatibility program (EMCP) that is governed by an EMC board. An EMC test is required on the system, as described in this section.

The EMCP includes the necessary approach, planning, technical criteria, and management controls. In MIL-STD-1541, the EMCP requires an EMC control-and-test plan that must be approved by the EMC board.

The system and all associated subsystems/equipment are to be designed for EMC in accordance with the EMC control plan. A system design program shall be implemented that covers the following areas:

Subsystem/equipment criticality categories
Degradation criteria
Interference and susceptibility control
Wiring and cable
Electrical power
Bonding and grounding
Lightning protection
Static electricity
Personnel hazards
EM hazards to explosives and ordnance
External environments
Suppression components

Systems are categorized based on the impact of EMI, or susceptibility malfunction, or degradation of performance.

The system-level EMC control plan is similar to the subsystem and equipment control plan, described in Section 12.1.1, with the following additions:

Methods and requirements for ensuring that sources of interference (subsystems/equipment) do not cause EMI in other subsystems or experience EMI from other sources in the system
Predicted problem areas and proposed methods of approach for solution of problems not resolved by compliance with MIL-STD-461
Design criteria and required tests for lightning protection
Radiation characteristics from system antennas, including fundamental and spurious energy, and antenna-to-antenna coupling

System-level EMI tests are seldom as extensive as those described in MIL-STD-462.

10.1.1.2 MIL-STD-464

This standard establishes electromagnetic environmental effects (E^3) interface requirements and verification criteria for airborne, sea, space, and ground systems, including associated ordnance. MIL-STD-464 supersedes MIL-STD-1818A, MIL-E-6051D, MIL-B-5087B, and MIL-STD-1385B. The general requirements are that the system shall be electromagnetically compatible among all subsystems and equipment within the system and with the environments caused by electromagnetic effects external to the system. Verification shall be accomplished as specified herein on production representative systems. Safety-critical functions shall be verified to be electromagnetically compatible within the system and with external environments prior to use in those environments. Verification shall address all life cycle aspects of the system, including (as applicable) normal in-service operation, checkout, storage, transportation, handling, packaging, loading, unloading, launch, and the normal operating procedures associated with each aspect.

Margins shall be provided based on system operational performance requirements, tolerances in system hardware, and uncertainties involved in verification of system-level design requirements. Safety-critical and mission-critical system functions shall have a margin of at least 6 dB. Ordnance shall have a margin of at least 16.5 dB of maximum no-fire stimulus (MNFS) for safety assurances and 6 dB of MNFS for other applications. Compliance shall be verified by test, analysis, or a combination thereof.

Shipboard internal electromagnetic environment (*EME*): For ship applications, electric fields (peak V/m-rms) below deck from intentional onboard transmitters shall not exceed the following levels:
1. Surface ships
   a. Metallic: 10 V/m from 10 kHz to 18 GHz
   b. Nonmetallic: 10 V/m from 10 kHz to 2 MHz, 50 V/m from 2 MHz to 1 GHz, and 10 V/m from 1 to 18 GHz
2. Submarines: 5 V/m from 10 kHz to 1 GHz

Compliance shall be verified by test of electric fields generated below deck with all antennas (above and below decks) radiating.

Power-line transients: For Navy aircraft and Army aircraft applications, electrical transients of less than 50 μs in duration shall not exceed +50% or −150% of the nominal DC voltage or ±50% of the nominal AC line-to-neutral rms voltage. Compliance shall be verified by test.

Multipaction: For space applications, equipment and subsystems shall be free of multipaction effects. Compliance shall be verified by test and analysis.

Intersystem EMC: The system shall be electromagnetically compatible with its defined external EME such that its system operational performance requirements are met. For systems capable of shipboard operation, Table 10.1 shall be used. For space and launch vehicle systems applications, Table 10.2 shall be used. For ground systems, Table 10.3 shall be used. For all other applications and if the procuring activity has not defined the EME, Table 10.4 shall be used. Intersystem EMC covers compatibility with, but is not limited to, EMEs from like platforms (such as aircraft in formation flying, ship with escort ships, and shelter-to-shelter in ground systems), friendly emitters, and hostile emitters. Compliance shall be verified by system-, subsystem-, and equipment-level tests, analysis, or a combination thereof.

TABLE 10.1

External EME for Systems Capable of Shipboard Operations (Including Topside Equipment and Aircraft Operating for Ships) and Ordnance

Frequency	Environment (V/m-rms)	
	Peak	Average
10 kHz–150 MHz	200	200
150–225 MHz	3,120	270
225–400 MHz	2,830	240
400–700 MHz	4,000	750
700–790 MHz	3,500	240
790–1000 MHz	3,500	610
1–2 GHz	5,670	1000
2–2.7 GHz	21,270	850
2.7–3.6 GHz	27,460	1230
3.6–4 GHz	21,270	850
4–5.4 GHz	15,000	610
5.4–5.9 GHz	15,000	1230
5.9–6 GHz	15,000	610
6–7.9 GHz	12,650	670
7.9–8 GHz	12,650	810
8–14 GHz	21,270	1270
14–18 GHz	21,270	614
18–40 GHz	5,000	750

TABLE 10.2
External EME for Space and Launch Vehicle Systems

Frequency	Environment (V/m-rms)	
	Peak	Average
10 kHz–100 MHz	20	20
100 MHz–1 GHz	100	100
1–10 GHz	200	200
10–40 GHz	20	20

TABLE 10.3
External EME for Ground Systems

Frequency	Environment (V/m-rms)	
	Peak	Average
10 kHz–2 MHz	25	25
2–250 MHz	50	50
250 MHz–1 GHz	1500	50
1–10 GHz	2500	50
10–40 GHz	1500	50

TABLE 10.4
Baseline External EME for All Other Applications

Frequency	Environment (V/m-rms)	
	Peak	Average
10–100 kHz	50	50
100–500 kHz	60	60
500–2 MHz	70	70
2–30 MHz	200	200
30–100 MHz	30	30
100–200 MHz	150	33
200–400 MHz	70	70
400–700 MHz	4020	935
700–1000 MHz	1700	170
1–2 GHz	5000	990
2–4 GHz	6680	840
4–6 GHz	6850	310
6–8 GHz	3600	670
8–12 GHz	3500	1270
12–18 GHz	3500	360
18–40 GHz	2100	750

Lightning: The system shall meet its operational performance requirements for both direct and indirect effects of lightning. Ordnance shall meet its operational performance requirements after experiencing a near strike in an exposed condition and a direct strike in a stored condition. Ordnance shall remain safe during and after experiencing a direct strike in an exposed condition. MIL-STD 464 provides a figure for the direct-effect lightning environment and figures and a table for the indirect-effect lightning environment from a direct strike. In addition, a table is provided for the near-lightning-strike environment. Compliance shall be verified by system-, subsystem-, equipment-, and component- (such as structural coupons and radomes) level tests, analysis, or a combination thereof.

Electromagnetic pulse (EMP): The system shall meet its operational performance requirements after being subjected to the EMP environment. If an EMP environment is not defined by the procuring activity, a figure in MIL-STD-464 shall be used. This requirement is not applicable unless otherwise specified by the procuring activity. Compliance shall be verified by system-, subsystem-, and equipment-level tests, analysis, or a combination thereof.

Nondevelopmental items (NDIs) and commercial items: NDIs and commercial items shall meet EMI interface control requirements suitable for ensuring that system operational performance requirements are met. Compliance shall be verified by test, analysis, or a combination thereof.

Electrostatic charge control: The system shall control and dissipate the buildup of electrostatic charges caused by precipitation static (p-static) effects, fluid flow, air flow, space and launch vehicle charging, and other charge-generating mechanisms to avoid fuel ignition and ordnance hazards, to protect personnel from shock hazards, and to prevent performance degradation or damage to electronics. Compliance shall be verified by test, analysis, inspections, or a combination thereof.

Precipitation static (p-static): The system shall control p-static interference to antenna-connected receivers onboard the system or on the host platform such that system operational performance requirements are met. Compliance shall be verified by test, analysis, inspections, or a combination thereof. For Navy aircraft and Army aircraft applications, p-static protection shall be verified by testing that applies charging levels representative of conditions in the operational environment.

Electromagnetic radiation hazards: The system design shall protect personnel, fuels, and ordnance from hazardous effects of electromagnetic radiation. Compliance shall be verified by test, analysis, inspections, or a combination thereof.

Hazards of Electromagnetic Radiation to Personnel (HERP): The system shall comply with current national criteria for the protection of personnel against the effect of electromagnetic radiation. DOD policy is currently in DoDI 6055.11. Compliance shall be verified by test, analysis, or combination thereof.

Hazards of electromagnetic radiation to fuel (HERF): Fuels shall not be inadvertently ignited by radiated EMEs. The EME includes onboard emitters and the external EME. Compliance shall be verified by test, analysis, inspection, or a combination thereof.

Hazards of electromagnetic radiation to ordnance (HERO): Ordnance with electrically initiated devices shall not be inadvertently ignited during, or experience degraded performance characteristics after, exposure to the external radiated EME of Table 10.1 for either direct-RF-induced actuation or coupling to the associated firing circuits. Compliance shall be verified by system-, subsystem-, and equipment-level tests and analysis. For EMEs in the HF band derived from near-field conditions, verification by test shall use transmitting antennas representative of the types present in the installation.

Electrical bonding: The system, subsystems, and equipment shall include the necessary electrical bonding to meet the E^3 requirements of this standard. Compliance shall be verified by test, analysis, inspections, or a combination thereof for the particular bonding provision.

Power current return path: For systems using structure for power return currents, bonding provisions shall be provided for current return paths for the electrical power sources such that the total voltage drops between the point of regulation for the power system and the electrical loads are within the tolerances of the applicable power quality standard. Compliance shall be verified by analysis of electrical current paths, electrical current levels, and bonding impedance control levels.

Antenna installations: Antennas shall be bonded to obtain required antenna patterns and meet the performance requirements for the antenna. Compliance shall be verified by test, analysis, inspections, or a combination thereof.

Shock and fault protection: Bonding of all exposed electrically conductive items subject to fault condition potentials shall be provided to control shock hazard voltages and allow proper operation of circuit protection devices. Compliance shall be verified by test, analysis, or a combination thereof.

External grounds: The system and associated subsystems shall provide external grounding provisions to control electrical current flow and static charging for the protection of personnel from shock, prevention of inadvertent ignition of ordnance, fuel, and flammable vapors, and protection of hardware from damage. Compliance shall be verified by test, analysis, inspections, or a combination thereof.

Aircraft grounding jacks: Grounding jacks shall be attached to the system to permit connection of grounding cables for fueling, stores management, servicing, and maintenance operations and while parked. ISO 46 contains requirements for interface compatibility. Grounding jacks shall be attached to the system ground reference so that the resistance between the mating plug and the system ground reference does not exceed $1.0\,\Omega$ DC.

Tempest: National security information shall not be compromised by emanations from classified information-processing equipment. Compliance shall be verified by test, analysis, inspections, or a combination thereof. (NSTISSAM TEMPEST/1–92 and NACSEM 5112 provide testing methodology for verifying compliance with TEMPEST requirements.)

Emission control (EMCON): For Army applications, Navy applications, and other systems applications capable of shipboard operation, unintentional electromagnetic radiated emissions shall not exceed -110 dBm/m^2 at 1 nautical mile (-105 dBm/m^2 at 1 km) in any direction from the system over the frequency range from 500 kHz to 40 GHz. Unless otherwise specified by the procuring activity, EMCON shall be activated by a single control function for aircraft. Compliance shall be verified by test and inspection.

Electronic protection (EP): For Army aircraft and Navy aircraft applications, intentional and unintentional electromagnetic radiated emissions in excess of the EMCON limits shall preclude the classification and identification of the system such that system operational performance requirements are met. Unless otherwise specified by the procuring activity, EP shall be activated by a single control function. Compliance shall be verified by test, analysis, inspections, or a combination thereof.

10.1.1.3 MIL-STD-1541A (USAF) EMC Requirements for Space Systems

This standard establishes the EMC requirements for space systems, including frequency management and the related requirements for the electrical and electronic equipment used in space systems. It also includes requirements designed to establish an effective ground reference for the installed equipment and to inhibit adverse electrostatic effects. MIL-1541A defines minimum performance requirements and identifies requirements for system and equipment engineering to achieve EMC and test and analysis to demonstrate compliance with the standard

Lightning: Vehicle and equipment designs shall prevent overstress or damage induced by a lightning strike to the nearest facility-lightning-protective device, by a lightning strike just outside the zone of protection, or by a lightning strike near or above underground cables that are a part of the system.

Triboelectric charging: Disturbances in the performance of launch vehicles, which may arise from electrostatic charging of external surfaces by impact of atmospheric particles and which may lead to corona or streamer discharges and sparking, shall be prevented by bonding all conductive parts and by controlling the charge accumulation on external dielectric surfaces.

Magnetospheric charging: Disturbances in the performance of space vehicles, which may arise from differential charging of external surfaces and internal components by space plasma and which may lead to electrical discharges, shall be minimized by bonding all conductive parts and by controlling the charge accumulation on external dielectric surfaces and in bulk dielectrics.

System and equipment designs shall provide electromagnetic interference safety margins in accordance with the worst-case potential criticality of the effects of interference-induced anomalies. The following categories shall be used:

1. Category I: Serious injury or loss of life, damage to property, or major loss or delay of mission capability
2. Category II: Degradation of mission capability, including any loss of autonomous operational capability
3. Category III: Loss of functions not essential to mission

Electromagnetic interference safety margins shall incorporate allowances for the effects of failures of redundant items, variations in characteristics due to aging, and between like units. Other system requirements are as follows:

1. Category I: 12 dB for qualification, 6 dB for acceptance
2. Category II: 6 dB
3. Category III: 0 dB

Categories I and II safety margins shall be increased by 6 dB if the only practicable method of verification is entirely by analysis based on estimated emission or susceptibility characteristics.

Degradation criteria: Safety margins shall be related to definite degradation criteria, depending on whether the appropriate requirement is freedom from overstress or damage due to continuous or aperiodic interference, autonomous recovery to the state prior to the occurrence of aperiodic interference, or continuous operation within specification limits.

Superposition: The specified safety margins shall be obtained with the combined effects of conducted and radiated broadband (BB) and narrowband emissions.

10.1.1.3.1 System Requirements Evaluation

System analyses shall be performed to validate reductions in the requirements of this standard, to define certain requirements that are commonly peculiar to a system (such as those pertaining to signal and control circuits), and to identify the need for requirements that must be more restrictive or severe than those in the standard. The resulting program-peculiar requirements are subject to the approval of the contracting officer.

Intersystem analysis: Global compatibility with the radiation of all communications–electronics activities, whether earth or space based, shall be evaluated using the techniques described in the NTIA manual and Report R-3046-AF or equivalent. Intersystem compatibility analysis for conductive interfaces shall be done using the methods specified by this MIL-STD-1541A for intrasystem analysis.

Intrasystem analysis time domain: EMC calculations entail finding the peak response of the potentially susceptible circuits in the time domain to several or more extraneous emissions that may be coupled through one or more transfer functions. For steady-state conditions, analyses based on the direct and inverse Fourier transforms, plus the convolution theorem, are appropriate and shall be used. Either amplitude or power spectral density functions may be used.

Frequency and time domain: Steady-state emissions shall be stated in the frequency domain for the same frequency ranges as for the corresponding requirements of MIL-STD-461, as modified by this standard. Oscillatory load-switching transients shall be defined in both the time and frequency domains; nonoscillatory transients may be defined only in the time domain.

Reference requirements: The evaluation shall include all the emission and susceptibility requirements specified in MIL-STD-461, as modified by this standard, for the classes of equipment under consideration. It shall include the applicable technical standards specified in the NTIA manual.

Interference coupling modes: The analyses shall incorporate mathematical models to account for the following:
1. Transverse and common-mode effects in signal and power circuits, including the effects of power source voltage ripple
2. Coupling between circuits in interconnecting cables
3. Bilateral coupling between the system antennas and circuits in interconnecting cables
4. Bilateral coupling between the system antennas, between the equipment enclosures, and between antennas and enclosures

Susceptibility requirements: Susceptibility requirements shall be shown to be sufficient to provide the required integrated safety margins for the applicable frequency range and point safety margins at all frequencies in that range.

Simplifications: Circuits within an interconnecting cable may be represented by only one of each distinct signal classification, as determined by similarity of signal characteristics and the terminations.

Intrasystem analysis program: MIL-STD-1541A has been structured to make best use of the developments that have been realized from the DOD Intrasystem Analysis Program. These benefits include improved analytical methods and programs for computer analysis. The analysis programs described in RADC-TR-74-342 and RADC-TR-82-20 are suitable and well documented.

Analytical methods: The methods presented in the applicable and the referenced documents are representative of the current art pertaining to the topics covered. Accordingly, they may be used as criteria for determining the adequacy and acceptability of any methods used.

Signal and control interface circuits: Emission and susceptibility requirements for signal and control circuits shall be developed and specified as program-peculiar limits. Values suitable for the signal levels and impedances characteristic of the applicable circuits shall be established.

Lightning analysis: Susceptibility to lightning and lightning protection shall be addressed in the analysis.

Magnetospheric charging analysis: Magnetospheric charging susceptibility and protection shall be addressed in the analysis. The program described in NASA CR-135259 shall be used. NASA TP2361 shall be used as a guide in determining applicable design requirements.

Triboelectric charging analysis: For launch vehicles, triboelectric charging shall be addressed.

10.1.1.3.2 Nonstandard Limits

Changes in the detailed emission and susceptibility limits of this standard shall be justified by analysis.

10.1.1.3.3 Consolidated Equipment Requirements

A single set of compatibility requirements shall be established for each piece of equipment. If more than one set is needed (e.g., for items that operate only in orbit, as contrasted with those that also operate in the launch and ascent environments), the differences shall be covered by a special classification method similar to that used in MIL-STD-461. Emission and susceptibility requirements shall be stated directly and explicitly. Susceptibility requirements shall not be specified indirectly in terms of an operating environment.

10.1.1.3.4 Existing Designs and Modifications

To the extent they are complete and suitable, prior analyses and test data shall be used to satisfy the requirements of this standard for new applications or modifications of existing items. The additional system and component analyses and tests shall be centered on defining similarities and differences and evaluating compatibility for the new situation.

10.1.1.3.5 Government-Furnished Equipment

The system emission and susceptibility characteristics shall be such that compatibility, with safety margins, is secured if the government-furnished equipment satisfies the requirements of MIL-STD-461 as supplemented and modified by the equipment.

The detailed intersystem vehicle requirements cover structural materials, electrical ground network, electrical power subsystem referencing, bonding, and lightning protection.

Requirements relating to communications by satellites include surface finish for control of electrostatic charging, interconnecting cables, vehicle electrostatic susceptibility, power bus impedance and transient recovery time, electrical power quality, voltage ripple, spikes, surges, load switching and load faults, and power subsystem faults.

The equipment requirements cover interface safety margins, ground power sources, MIL-STD-461 performance criteria, limits and applicability, electrical power quality and faults, and electrostatic susceptibility.

Test and evaluation includes verification methods, measuring instruments, and system verification (radiated susceptibility, partitioning, lightning, superposition effects).

System qualification tests include selection of test points, category I and II functions, category III functions, test conditions, and acceptance criteria.

MIL-STD-1541 contains a limit on maximum radiated field strength of 0.01 W/cm^2 for human safety.

Some additional limits not contained in MIL-STD-461 are applied to equipment/subsystems in MIL-STD-1541. MIL-STD-1541 recommends that the system be used to test the system and proposes that one or more of the following general test approaches shall be used at the system level:

Inject interference at critical system points at a 6 dB higher level than exists while monitoring other system/subsystem points for improper responses.

Measure the susceptibility of critical system/subsystem circuits for comparison with existing interference levels, to determine if a 6 dB margin exists.

Sensitize the system/subsystem to render it 6 dB more susceptible to interference while monitoring for improper responses.

MIL-STD-1541 also contains system design criteria and requirements.

10.2 ANTENNA-COUPLED EMI

Antenna-to-antenna-coupled EMI is becoming more prevalent with the increased number of transmitting and receiving antennas located on vehicles, ships, and aircraft. With the advent of portable and mobile satellite communication systems, the potential for civilian antenna-coupled EMI is increased. With the satellite Trucker terminal, the potential sources of interference are CB transmitters collocated on the truck and radar or other high-power transmitters the truck may pass by.

When choosing a site for installation of a receiving system, the suitability of the electromagnetic ambient at the location should be assessed. An ambient site survey or prediction should be made for the site, as described in Section 10.3.

A number of potential sources of EMI exist in an environment where a large number of transmitters are in close proximity to the receiving antenna location. Some of the potential sources of EMI are

Harmonic interference, where a harmonic of a transmitter is at the intended receiver frequency

IF interference, where the transmitter is at the receiver IF frequency

Image interference, where the transmitter is at the receiver frequency minus twice the receiver IF frequency

Cross-modulation interference, where a high-level transmission close to the receiver frequency is not sufficiently attenuated by the input filter (here, compression and spurious responses can occur in the receiver)

Adjacent channel interference from a transmitter close enough to lie within the receiver IF bandwidth and the receiver bandwidth

BB noise from a transmitter, which is in band for a receiver

Passive intermodulation interference (PIM) caused by the nonlinearity of metal–insulation–metal or metal–oxide–metal (MOM) interfaces (commonly known as the *rusty bolt effect*)

The analysis of coupling between aircraft-mounted transmitters and receivers is required to apply mitigation techniques or to decide if Simultaneous Operation of Systems (SIMOPS) is possible.

In extreme cases, damage can result to the receiver when the received power is too high. Unless cross modulation or desensitization occurs, the transmitter-tuned frequency, harmonics, and spurious transmitter emissions may possibly be excluded by tuning the receiver to a different frequency. In the case of BB noise covering the complete receiver frequency range, this is not possible. In one case, where 22 antennas are located on a small aircraft, the probability of some level of antenna-to-antenna interference is 100%.

The analysis of antenna-to-antenna coupling may be performed using a number of methods:

1. Measure the fields generated by existing antennas mounted on the aircraft and the coupling both down the fuselage and around the fuselage and wings. Based on the shading/isolation and coupling between existing antennas, an analysis can be performed for any additional antennas, based on frequency range, transmit power, transmit antenna gain, receive antenna gain, and receiver sensitivity.

2. Build a 1/10th-scale model of the aircraft using, for example, wood and copper foil, and locate small antennas on the surface. By increasing the transmit frequencies by a factor of 10 (3 MHz–15.7 GHz), the frequency range of 300 kHz–1.57 GHz for full-scale antennas can be measured. This method has been used for satellites and the EH101 helicopter with good results. The aircraft model has been used with multiple antennas that are either 1/10th scale of the full size or are miniature antennas with similar characteristics to the full scale. These antennas may be either built and calibrated or are commercially available. Some of the 1/10th-scale antennas have been found suitable to represent more than one type of full-scale antenna reducing the number of antennas to be manufactured. The physical

1/10th-scale model of an aircraft has been used for all antenna-to-antenna coupling below 1.8 GHz (18 GHz on the scale model).

3. Use of a numerical model such as the FEKO, General Electromagnetic Model for the Analysis of Complex System (GEMACS), or numeric electromagnetic code (NEC) computer program. FEKO is a suite of tools used for electromagnetic field analysis of 3D structures. The designers claim that the program is ideal for analysis of antenna placement on electrically large platforms such as aircraft. The major potential coupling analysis using FEKO is from an SLAR antenna mounted under the wing on the side of the fuselage and the Maritime Search Radar antenna located under the fuselage and vice versa at 9.375 GHz. This is because it would be difficult to build and test an antenna for 93.75 GHz for use on the 1/10th-scale model. Constructing the full SLAR antenna with fuselage wing and engine nacelle in FEKO is problematic due to the large structure and high frequency, leading to long computation times. Instead, a single E plane sectoral horn was used. To simulate the full 12 in. length of the antenna, 141 such horns would be required and the amount of computing time at 9.735 GHz would be unrealistic. Even for a single horn, fuselage and engine FEKO engineers recommend the use of a 64 bit machine with plenty of RAM for this type of problem. The radiation pattern from a single horn could be analyzed and the data changed to that of the full antenna, but again the amount of work involved and time required would be high and instead the fourth method was used.

4. Full-size mock-up: Full-scale antennas may be used with full-size sections of the aircraft, for example, a section of the aircraft constructed from wood cardboard and aluminum/copper foil. For example, a 9.735 GHz sectoral horn antenna was made and the gain was measured at 9.375 GHz. A full-scale model of the engine and landing gear and a partial fuselage was made and the field reflected from these structures at the location of the Maritime Search Radar was measured. A full-scale calibrated antenna was used to measure the incident power at this location.

The great advantage of the 1/10th model, FEKO, and full-scale model is that the EMI analysis may be performed before work has started on the aircraft installation or even before the aircraft is delivered.

PIM, cross modulation, harmonics, BB, and adjacent channel analysis should be performed for all transmitters and receivers. Finally, after reviewing the results of the analysis or measurements, recommendations can be made for mitigation of any potential EMI and reanalysis if required. The responsible engineer should provide support in the form of recommendations, implementing EMI fixes and consulting for the duration of the program.

With the availability of high-level electromagnetic analysis programs with excellent graphics, it would appear redundant to resort to models or a full-scale mock-up or early measurements on aircraft, yet as we will see in this chapter and in Section 12.4 on computational electromagnetic modeling, errors can and do occur in the use of these programs. The use of measurements or simple antenna-to-antenna transmission equations should always be used as a sanity check on the computer-generated data.

10.2.1 ANTENNA-TO-ANTENNA COUPLING

Although the antennas and coupling examples provided in this section are specifically for aircraft, the same analysis techniques can be used for ship- and vehicle-mounted antennas.

A fundamental potential EMI situation occurs when the frequency of a transmitter is within 20% of the tuned frequency of a receiving system. The minimum frequency separation is approximately 10%, at which greater than 20 dB of cosite isolation is often required. A second EMI situation can arise when one or more out-of-band frequencies result in a spurious response in the receiver.

The distance between the location of the transmitting and receiving antennas is the critical factor in determining if an EMI situation can occur.

In assessing a potential EMI situation for antennas located kilometers apart, the first step is to verify the existence of line-of-sight coupling, based on the Earth's radius.

Where line-of-sight coupling exists, the propagation loss correction for non-line-of-sight (NLS) coupling is 0 dB.

A wave propagates in four principal modes: *direct wave*, which applies to line-of-sight coupling, the *reflected wave* from the intervening surface between the transmit and receive antennas, *surface wave*, and the *skywave*. The skywave is the wave bounced off of the ionized layers in the upper atmosphere. There are daily and seasonal variations in the height of the reflecting layers and also a critical frequency above which vertically directed waves are not reflected by the layers. The surface wave is caused by ground currents set up by the energy as it enters the ground. Below 3 MHz the surface wave predominates; between 3 and 300 MHz the reflected wave, direct wave, and skywave all contribute; and above 300 MHz the direct wave predominates.

When a wave is incident on a curved body, it is diffracted around the body and it is termed a creeping wave, and where the direct wave is no longer seen, it is termed the shadowed surface.

When a computer program is not available, the following approximate calculations can be made:

When the line of sight is blocked by terrain, calculate the distance from the transmitting antenna to the terrain, R_{ter}, the additional factor to be added into the propagation for NLS is $40 \log R_{ter}/R$, where R is the distance between the transmitting and receiving antennas.

The power output and gain of the transmitter at the fundamental and spurious emissions must be known. Where spurious emission levels are unknown, they can be assumed to be approximately 60 dB below the fundamental. The receiver spurious susceptibility level if unknown may, very approximately, be taken as 80 dB above the fundamental susceptibility.

When a frequency misalignment exists between the transmitter fundamental and spurious emissions and the receiver-tuned frequency, then corrections for the receiver and antenna out-of-band responses must be made to the simple in-band antenna-to-antenna coupling prediction.

The first step in the antenna-to-antenna coupling prediction is to calculate the power generated by the transmitting antenna and the coupling path. This coupling factor is typically defined as the transmitted power into an antenna versus the received power at the receiving antenna. When antennas are mounted with clear line of sight, then the greater the distance between the antennas, the lower the coupled power. The antenna gains and cable losses are part of the calculation. When the antennas are shaded by a structure, then the level of shading or power coupling factor can be calculated by comparing the transmitted to the received power with the antenna gains subtracted.

When aircraft engineers want to see the benefit of moving colocated antennas to alternative locations around the fuselage, they may want to calculate the reduction in coupling by relocating the antenna to a path that deliberately or incidentally contains structures that reflect, diffract, or absorb the electromagnetic field and the reduction in dB relative to the original location is required. This is what we term isolation.

In this book, the isolation is defined as the power received by an antenna mounted on one surface of, for example, a cylinder from a transmitting antenna mounted around the cylinder (out of line of sight, shaded, or isolated). We use the distance around the cylinder in the first case to determine the distance between two antennas mounted on the same surface (direct). The ratio between direct and isolated received power provides the level of isolation. As this is the difference between the direct coupled power and the attenuated power as long as the two receiving antennas are the same, then the gains of the transmitting antenna and the receiving antennas are not part of the calculation. As shown later, the direct line-of-sight coupling may be lower than the isolated coupling when the direct and reflected waves in the line of sight tend to cancel at the receiving antenna.

TABLE 10.5

Typical In-Band Gains of Various Types of Antennas

G < 10 dB	10 dB ≥ *G* ≤ 25 dB	*G* > 25 dB
Linear	*Array*	*Array*
Cylindrical	Yagi	Mattress
Biconical	Broadside	
	Curtain	
Dipoles		
Folded dipoles		*Aperture*
Asymmetrical	End-fire curtain	Horns
Sleeve	*Traveling wave*	Reflector
Monopole	Rhombic	Lens
Collinear array	Surface and leaky wave	
Loop		
Aperture, slot	Aperture	
Helix	Horn	
	Corner reflector	
	Log-periodic	
	Conical log-spiral	

Typical in-band gains of antennas are provided in Table 10.5. It must be emphasized that these gains are for the direction of intentional radiation and are readily available from manufacturers. The power radiated from the sidelobes and from the rear of an antenna is more difficult to ascertain if the radiation pattern is unknown. Although worst case, a gain of 0 dB for nonaligned antennas and those with low gain, such as monopole and loops, may be used in the preliminary antenna-to-antenna coupling analysis.

When antennas are colocated on the same surface in relatively close proximity (i.e., not kilometers apart), the basic line-of-sight antenna coupling equations apply, with corrections for reflections. The level of reflected field depends on the angle of incidence of the field and the conductivity or permittivity of the reflecting surface. Case Study 10.1 describes the hazardous zone around a transmitting parabolic reflector antenna and considers the reflection from a high-conductivity and a high-permittivity surface. Case Study 10.1 provides the approach and calculations entailed in predicting field enhancement due to reflections.

When antennas are mounted above a conductive surface, the direct and reflected waves add for a vertically polarized source and subtract for a horizontally polarized source.

Depending on frequency, height, and distance, the coupling to antennas may be reduced by this interference when the direct and reflected waves are out of phase.

Equation 10.1 gives the correction ratio for the composite field due to this constructive and destructive interference for vertical sources and Equation 10.2 gives the composite field for horizontally polarized sources. The predicted field without reflection is multiplied by this value. The equation is valid for phase shifts up to 180° or when the sum of the antenna heights is approximately 2.5λ or less. The assumption is that the surface is perfectly conducting and that the difference in the field strength of the direct and reflected wave is negligible.

$$V = \sqrt{2 + 2\cos\theta} \qquad (10.1)$$

$$H = \sqrt{2 - 2\cos\theta} \qquad (10.2)$$

where

$$\theta \text{ in degrees} = \frac{\sqrt{(h2+h1)^2 + D^2} - \sqrt{(h2-h1)^2 + D^2}}{\lambda} 180$$

D is the distance between the two antennas [m]
$h1$ is the height of the transmitting antenna [m]
$h2$ is height of the receiving antenna [m]
$\lambda = 300/f$ [MHz]
f is the frequency

Typical in-band antenna gains are provided in Table 10.6. It must be emphasized that these gains are for the direction of intentional radiation and are readily available from manufacturers. The power radiated from or received in the sidelobes and from the rear of the antenna is more difficult to determine if the radiation pattern is unknown. In the case of radar antennas, the manufacturer typically provides the gain in the H plane (vertical) and E plane (horizontal). Although worst case, a gain of 0 dBi for nonaligned antennas and those with low gains such as monopole and loops may be used

---

TABLE 10.6
Typical In-Band Antenna Gains

Antenna Type	Application	Frequency Range (MHz)	Gain (dBi)
¼ wave monopole	TCAS	960–1,215	5.28
¼ wave stub	DME	1,025–1,150	–0.95
		1,150–1,220 –1 dB	
Blade	ATC	1,030/1,090	=>0
Equivalent to ¼ wave stub	IFF	1,030/1,090	Approximately –1 at 1030
	RADALT	4,200–4,300	9.5
	Weather radar	9,375	31 dB 5.6°s beam width
Loop	Nav receiver	108–118	0 ± 2 dB
Folded dipole	ILS Glideslope	329–335	0
Wire in stabilizer leading edge	HF radio	3–30	0
Standard gain horns	Measurement	1.7–2.6, 2.6–3.95	10, 15, or 20
		3.7–4.9, 3.95–5.85	
		4.9–7.05, 5.85–8.2	
		7–11, 8.2–12.4	
		10–15, 12.4–18	
		15–22, 18–26.5	
		22–33, 26.5–40	
Long wire with matching unit	HF radio	1.5–30	2.15
Blade	VLF/HF	2–30	–2
		30–50	–4
Blade	V/UHF	30	–23
		88	–14
		108–174	–4
		225–512	2
Monopole with frequency separation (receive only)	AIS	156/163	0

(Continued)

TABLE 10.6 (*Continued*)
Typical In-Band Antenna Gains

Antenna Type	Application	Frequency Range (MHz)	Gain (dBi)
Blade	V/UHF	30	−24
		600	1
	Triband GSM, GPS, and satellite	Downlink 1,525–1,559	2.8 front
	LHCP Omnidirectional	Uplink 1,626.5–1,660.5	−22 rear
	RHCP vertical	GPS 1,575.42	−17.5 side
Coplanar waveguide	Broadband	1,000–10,000	4.5–7
			At 2.7 dB main lobe
			Sidelobe = 0
Microwave patch antenna	Narrowband radio at 900 MHz, 1.2, 1.9, 2.4, 2.7, 3.5, 5.2, 4.9–5.8 GHz	1,000–30,000	7–9
			Sidelobe = −25
Yagi-Uda antenna	TV and communications	300–2,500	7–16.5 dB
Sleeve dipole	Communications and measurement	140–5,000	1.5–3
Folded dipole	Radio communication	1.8–30	Typically 2
		150–1,000	
Ground plane	Radio communication	144–2,500	0
One vertical rod three pointing down			
Micro Radionics blade		800	1.1
		900	1.2
		1,000	3.65
		1,100	4.13
		1,215	5.6
Maritime Search Radar		8,900–9,500	E plane 33 dBi
			10 dBi at ± 4°
			H plane 33 dBi
			10 dBi at 6.5°

in a preliminary antenna-to-antenna coupling analysis. When the coupling is between transmitters and receivers not operating in the same frequencies, then "out-of-band" coupling occurs and the out-of-band antenna gain is required. Most manufacturers do not have this information and few papers exist addressing the out-of-band gain. This is particularly true of high-gain apertures or microstrip antennas. The gain of a 3.75, 15, and 30 cm high monopole coupled to a second monopole on a ground plane has been calculated using GEMACS and verified by the measurement. We see that when the antenna height is a 0.25 wavelength or odd multiples, then the antenna gain is at its highest. When the antenna height is at 0.5 wavelength and odd multiples of a wavelength thereof, the antenna gain is still relatively high. When the antenna is at a wavelength or even multiples of a wavelength, then the gain is at its lowest and this is shown in Figure 10.1. Reference [17] provides the calculated below-band gain of a VHF/FM whip antenna (in-band frequency range 30–88 MHz), a balanced loop antenna (in-band frequency range of 108–122 MHz), and a UHF/AM blade antenna (in-band frequency 225–400 MHz) over the frequency range of 5–25 MHz and this is reproduced in Figure 10.1. Ideally, any antenna used in the out-of-band analysis should be measured for gain in an anechoic chamber. The reason for the chamber is that when low receive levels are measured, concomitant with a low out-of-band gain, coupling to the receiver/spectrum analyzer can occur. One method is to position a calibrated antenna at a measured distance from the transmitting antenna and

FIGURE 10.1 Antenna gains versus length in wavelength for a 3.75, 15, and 30 cm monopole (whip) antenna. (From Joffe, E.B., Out of band response of VHF/UHF airborne antennae, *IEEE International Symposium on Electromagnetic Compatibility*, Denver, CO. © 1989, IEEE.)

record the transmitted level. The antenna under the test then replaces the calibrated antenna and the measurement repeated. By using the same transmit level and transmit and receive cables, these do not form part of the calibration. The gain of the antenna under test is given by

$$\text{Antenna gain} = G_{ref} + (AUT - Ref) \tag{10.3}$$

where

G_{ref} is the gain of the reference antenna (dB)
Ref is the level measured with the reference antenna dBm(W)
AUT is the level measured with the antenna under test dBm(W)

Figure 10.2 provides the measured out-of-band gain of a whip antenna with an in-band operating frequency of 138–174 MHz.

FIGURE 10.2 A 138–174 MHz whip antenna below-band gain.

When the level measured with the antenna is low, concomitant with low antenna gain, the measurement may be of the coupling to the receiving cable. Dual braid cable, such as RG214 with HF and LF ferrites, may be sued at low frequency, and dual braid microwave cable at higher frequencies is covered with lossy sleeving if cable coupling is a problem. The measurements should be made with the antenna under test removed to ensure that the cable coupling level is lower than the AUT level.

Every antenna will exhibit different out-of-band gains, but as an example, the gains of a 1.6 MHz microstrip patch antenna, a dual 2.4 and 5 GHz T section and loop antenna, and a 20 dB 5–11 GHz horn antenna are shown in Figures 10.3 through 10.5.

The gain of the antenna as a function of the effective area A_e and wavelength is given in Equation 2.37. Equation 2.49 provides the effective height of an antenna as a function of

FIGURE 10.3 Low-frequency gain of patch antennas.

FIGURE 10.4 High-frequency gain of 1.6 MHz patch antenna.

FIGURE 10.5 Gain of 20 dB 5–11 GHz horn antenna.

the effective area, antenna impedance, wave impedance, and radiation resistance of the antenna. The effective area of the antenna, defined as the ratio of the received power to the incident power, changes at frequencies on either side of the design frequency of the antenna. The equation for effective area is multiplied by the polarization mismatch factor p and the antenna impedance mismatch factor q to obtain its out-of-band value, and p has a value shown in Table 10.7 for vertical or horizontal and circular polarization mismatches between a transmitting antenna with a gain of less than 10 dB and greater than 10 dB and a receiving antenna with a gain of less than 10 dB and greater than 10 dB.

The effects of q, the antenna impedance mismatch, for a 1 GHz dipole or loop antenna with an average polarization mismatch p of 0.5 are shown in Figure 10.6. At below-band frequencies, A_e is proportional to f^2. At above-band frequencies, A_e is equal to $c^2/8\pi f^2$, where c is the velocity of a wave in free space, in cm/s (300×10^8 cm/s), and f is in Hz. References [4,5] provide further information on antenna out-of-band response.

For a dipole or monopole antenna, the effective area may be found from the effective height; alternatively, the receiving characteristics of the antenna from the effective height may be used. Use of Equation 2.49 for the effective height will include the out-of-band mismatch corrections, such as the antenna and load impedance mismatch and the radiation resistance and wave impedance mismatch.

A PIM, cross modulation, harmonics, BB, and adjacent channel analysis should be performed for all transmitters and receivers. Finally, after reviewing the results of the analysis or measurements,

TABLE 10.7

Polarization Mismatch Factors for Antennas with Gain Greater Than and Less Than 10 dB

Antenna Gain		Polarization	
Receiving (dB)	Transmitting (dB)	Horizontal/Vertical	Circular
<10	<10	0.025	0.5
>10	<10	0.025	0.5
<10	>10	0.025	0.5
>10	>10	0.010	0.5

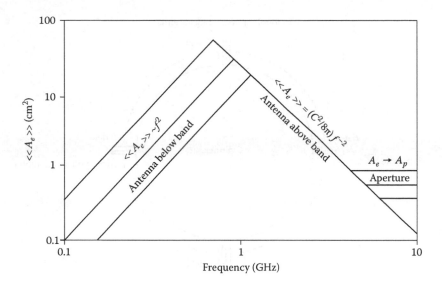

FIGURE 10.6 Effective aperture trends. (From Gonshor, D.V., Attenuation, transmission and reflection of electromagnetic waves by soil, *IEEE International Symposium on Electromagnetic Compatibility*, Atlanta, GA. © 1987, IEEE.)

recommendations can be made for mitigation of any potential EMI and reanalysis if required. The responsible engineer should provide support in the form of recommendations, implementing EMI fixes and consulting for the duration of the program.

Many antennas on an aircraft are either monopole or blade antennas with an approximately isotropic radiation pattern, which means that they radiate equally in the azimuth direction. When one antenna is mounted on one surface of the aircraft and the second is shaded from that antenna, then we need to find either the coupling factor or isolation.

The GEMACS, NEC4, and FEKO computational electromagnetic tools have been used to calculate isolation around cylinders and on a 1/10th-scale as well as full-scale Dash 8 aircraft.

Measurements on a 26 cm diameter 2 m long cylinder, a 1/10th-scale dash 8 aircraft. A mock-up of a section of the full-scale fuselage and the Dash 8 aircraft have been performed and are described in References [18,20].

The field that propagates around a convex body is called a creeping wave. Creeping waves and the entire radiation in the shadow area, where antennas are not line of sight, involve only the refracted field and so waves are bent as they curve around a cylinder. Thus, for surface and creeping waves, a vertical polarized E field, for example, the coupling from one monopole to another, remains vertically polarized with reference to a second monopole. This is shown in Figure 10.7.

The wave propagates around both sides of the cylinder and these waves are in phase. Thus, by constructive interference they are added up.

The General ElectroMagnetic Analysis of Complex Systems is an older hybrid electromagnetic analysis program now available in a Windows version. At present Windows XP or earlier.

It combined the method of moments (MOM), uniform theory of diffraction (UTD), and finite difference formulalisms in the frequency domain. The UTD formalism enables the modeling of large objects and includes the fields scattered by the object. These are found by the use of optic principles, ray tracing, and reflection. UTD also includes diffracted ray paths, including creeping waves and diffraction coefficients and so is ideal for computing the diffraction of a wave and therefore the isolation or power coupling factor around a cylinder.

For the modeling of monopoles on a cylinder, a hybrid uses the MOM method to model the antennas and the UTD method for the coupling.

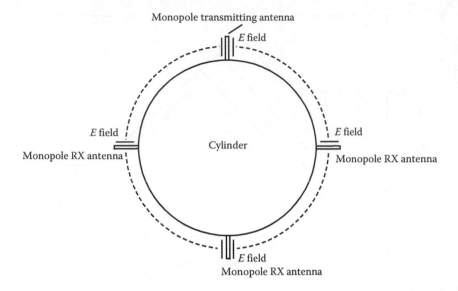

FIGURE 10.7   Perpendicular (vertical) E field coupling around a cylinder.

The coupling around a cylinder with a radius of 10 wavelengths at around 1 GHz has been calculated. This corresponds to a cylinder with a radius of 3 m, diameter 6 m.

The GEMACS program was used to model quarter-wave monopoles with a height of 7.5 cm.

At 90°, the coupling factor (= TX power/RX power + TX antenna gain + RX antenna gain) is equal to 53 dB. At 180°, the coupling factor is 97.8 dB.

The change in coupling factor when the angle around the cylinder is close to 180° is of interest, and Table 10.8 shows the change in coupling factor as a function of angle and wavelength change in circumference of the cylinder.

A 4.2 m diameter cylinder was modeled with 15 cm monopole antennas. The coupling and isolation were calculated with the antennas both 90° and 180° apart.

The results of the GEMACS analysis are shown in Figures 10.8 and 10.9.

A 1/10th-scale model of an aircraft using wood and copper foil was built with small antennas located on the surface. By increasing the transmit frequencies by a factor of 10 (3 MHz to 15.7 GHz), the frequency range of 300 kHz to 1.57 GHz for the existing antennas can be measured. The physical 1/10th-scale model of the aircraft has been used for all antenna-to-antenna coupling below 1.8 GHz (18 GHz on the scale model).

TABLE 10.8

Change in Coupling Factor Close to 180°
at 1 GHz on a 3 m Radius Cylinder

Degrees	Change in Lambda	Coupling Coefficient (dB)
180	0	97.8
177.14	0.5	87
174.28	1	79.4
171.4	1.5	78.7
168.54	2	77.67
165.68	2.5	76.4
162.814	3	75.33

FIGURE 10.8 Monopole isolation analysis on a 4.2 m diameter cylinder.

FIGURE 10.9 Coupling factor analysis on a 4.2 m cylinder.

The model aircraft is shown in Figure 10.10 and three of the 1/10th-scale antennas are shown in Figures 10.11 through 10.13.

The azimuth radiation pattern of a monopole is relatively omnidirectional. When a monopole antenna is mounted on the side of the fuselage of an aircraft, the wave propagates away from the aircraft in the Z direction, a horizontally polarized field, whereas the field is vertically polarized in the Y direction. In this chapter, we are only considering antennas mounted on the aircraft, and so the definition we use for horizontal and vertical polarization is relative to the surface of the aircraft. This convention makes the discussion of polarization simpler. For example, Figure 10.14 shows monopole type antennas (rod or blade) mounted around a cylinder and down the length of the cylinder. The E field diffracted around the cylinder in the Y direction from antenna A to B or from

FIGURE 10.10 1/10th-scale model of the aircraft. (From Weston, D.A., Antenna to antenna coupling on an aircraft using a 1/10th scale model with results compared to the FEKO electromagnetic analysis program, *EMC Europe*, Wroclaw, Poland. © 2010, IEEE.)

FIGURE 10.11 Dual whip antenna 1.25–4 GHz. (From Weston, D.A., Antenna to antenna coupling on an aircraft using a 1/10th scale model with results compared to the FEKO electromagnetic analysis program, *EMC Europe*, Wroclaw, Poland. © 2010, IEEE.)

FIGURE 10.12 Blade antenna 10.8–17.4 GHz. (From Weston, D.A., Antenna to antenna coupling on an aircraft using a 1/10th scale model with results compared to the FEKO electromagnetic analysis program, *EMC Europe*, Wroclaw, Poland. © 2010, IEEE.)

FIGURE 10.13 Microstrip patch antenna 15.75 GHz. (From Weston, D.A., Antenna to antenna coupling on an aircraft using a 1/10th scale model with results compared to the FEKO electromagnetic analysis program, *EMC Europe*, Wroclaw, Poland. © 2010, IEEE.)

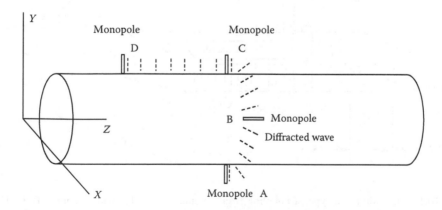

FIGURE 10.14 Monopole antenna coupling around and down a cylinder. All waves are considered vertical with reference to the surface of the cylinder.

A to C remains vertically polarized with reference to the surface and ensures maximum coupling into the rod type antenna, which is also vertical with reference to the cylinder surface. Likewise, the E field and the power incident on antenna D from antenna C in the Z direction, down the length of the fuselage, is vertical with reference to the cylinder surface and again ensures maximum coupling to antenna D.

Often antennas are mounted at an angle with reference to the aircraft surface, and here, compensation in the gain of the antenna must be made to account for a level of cross polarization. The main beam of some antennas, typically microstrip, horn, parabolic, and waveguide slot, points away from the aircraft and the coupling is via the sidelobe of the antenna pattern. In the case of the balanced loop antenna mounted on the stabilizer, D in Figure 10.15, the polarization is horizontal away from the stabilizer and radiated onto the upper surfaces of the aircraft in the X direction.

Figure 10.15 shows a parabolic antenna mounted under the fuselage (antenna G) and a waveguide slot antenna E mounted in the nose cone. The E field generated by antenna G is horizontally polarized with reference to the aircraft under surface and so the coupling is cross polarized with reference to the helical antenna H. Antenna H is mounted vertically with reference to the aircraft surface and slightly round the fuselage (out of line of sight). Although the sidelobes from antenna E in the X direction are horizontally polarized, the E field referenced to the side of the fuselage is vertically polarized and propagates a field at right angles to and down the fuselage. The E field incident on the top and bottom of the fuselage is horizontally polarized with reference to the aircraft surfaces, as shown in Figure 10.15, and so is cross polarized with reference to blade antennas on the top and bottom of the fuselage. Other types of antenna generate circularly polarized waves with the main beam facing away from the aircraft, and so the interaction with a monopole antenna located further down or around the fuselage of the aircraft is more complex.

FIGURE 10.15 Location and type of coupling from antennas. A, J, K, blade antenna; B, C, H, helical antenna; D, balanced loop antenna; E, weather radar; F, SLAR antenna; G, Maritime Search Radar; L, HF wire antenna. (From Weston, D.A., Comparison of techniques for prediction of and measurement of antenna to antenna coupling on an aircraft, *EMC Europe*, York, U.K. © 2011, IEEE.)

10.2.1.1 Measurement Techniques

The measurement techniques on the full-scale aircraft, the 26 cm diameter cylinder, and the 1/10th-scale model aircraft were similar. The test setup is shown in Figure 10.16 from Reference [19].

Figure 10.17 shows the locations of the antennas in the test and the analysis; however, only two antennas, one TX and one RX, were placed at these locations, for example, one TX antenna under the cylinder and one 90° around the cylinder. This RX antenna was then moved to 180° around the cylinder (on top of the fuselage).

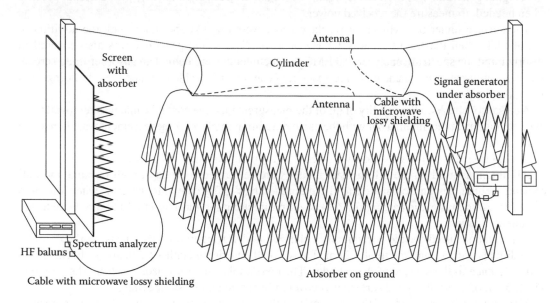

FIGURE 10.16 Basic measurement setup. (From Weston, D.A., Comparison of techniques for prediction of and measurement of antenna to antenna coupling on an aircraft, *EMC Europe*, York, U.K. © 2011, IEEE.)

FIGURE 10.17 The locations of dipole antennas located 2.5 cm above the 26 cm cylinder. Note only one TX and one RX antenna were placed on the cylinder in the measurements and analysis. (From Weston, D.A., Comparison of techniques for prediction of and measurement of antenna to antenna coupling on an aircraft, *EMC Europe*, York, U.K. © 2011, IEEE.)

The cables in the 1/10th model and cylinder could be run inside the fuselage, and this was not always possible with the full scale when the antennas were not mounted to the fuselage.

The cables to the transmit TX and receiver RX antennas were covered in lossy sleeving, which is an effective absorber above 1 GHz. For measurements at 1 GHz and below, the cables were loaded at the instrument and signal generator end with high-frequency material ferrite cores. The purpose of the absorber is not only to reduce cable coupling but also to present a high impedance between the fuselage and ground.

A signal generator was used as the signal source and a spectrum analyzer, with external preamp when needed, to measure the received power.

The surface under the cylinder or fuselage was covered in absorber foam with the addition of ferrite tiles when the frequency was 1 GHz or lower. The signal generator was also covered in absorber and the spectrum analyzer and EMC test technician were behind an absorber lined screen. The measurements were made on a free space range at a considerable distance from buildings and a ground plane.

As a sanity check on the validity of all of the measurements, the coaxial cable was removed from the TX antenna at the antenna but with the cable remaining in place. A measurement was then made of the coupling between the transmitting antenna cable and the receiving antenna. In all cases, the cable coupling was well below the level when the receiving antenna was connected.

All antennas used on the 1/10th-scale model were calibrated for gain at both in-band and out-of-band frequencies and compared to the manufacturers' full-scale antenna gain when this was available. Antennas, which are reported to be omnidirectional in the azimuth direction, were tested for antenna pattern.

The antennas used for the horizontally polarized and vertically polarized tests on the cylinder were dipoles with a total length of 3.2 cm. These could be oriented either vertically or horizontally with reference to the surface of the cylinder. The coax cable shield was isolated from the cylinder. The 2.5 cm of cable above the surface was covered in ferrite and lossy sleeving.

Measurements were made of the field coupled through an aperture in a cylinder that represents a window. The field was measured using a 4 cm long bow tie antenna which was connected to a 10–1000 MHz, 90 dB dynamic range log detector. The DC output of the detector was connected using 0.5 m long leads to the A/D input of a PIC (microcontroller), the output of which was used to drive a fiber optic transmitter. The fiber was the only external connection brought through the totally sealed cylinder via a 6 cm long, 5.5 m inner diameter pipe soldered to the cylinder.

10.2.1.2 Simple Equation for Shading

The equation is published in References [21,22] derived from Hasserjian and Ishimaru. Reference [18] also compares the result of this equation to a basic hybrid moment method and geometrical theory of diffraction (MM/GTD) technique.

The shading function (SF) from References [2,3] is given by

$$SF = \frac{A}{\eta A + \varepsilon} \tag{10.4}$$

where

$$A = pf\theta_s^2 \sqrt{\frac{2\pi}{\lambda D_c}}$$

pf is the radius of the cylinder (m)
θ_s is the angle around the cylinder of propagation path (rad)
λ is the wavelength 300/*f* (MHz)

D_c is the distance of cylindrical segment of propagation path

$\eta = 5.478e{-}3$, for $A < 26$

$\eta = 3.34e{-}3$, for $A \geq 26$

$\varepsilon = 0.5083$, for $A < 26$

$\varepsilon = 0.5621$, for $A \geq 26$

10.2.1.3   Measurements on the 1/10th-Scale Model

Unlike the equation for shading, the measurements on the 1/10th-scale aircraft and the full-scale aircraft include reflection from the wings, tail stabilizer, and, especially for A to B and C, the engine nacelle. The locations of antennas A, B, C, J, and K on the model are shown in Figure 10.15. The reduction is compared to the Friis transmission equation except for antenna K to J, which is from the measurements published in Reference [1]. These measurements are shown in Table 10.9.

10.2.1.3.1   Horizontally and Vertically Polarized Fields Coupling around a 26 cm Cylinder

In measurements on aircraft, we see that horizontally polarized fields, those with the E field parallel to the surface, and the magnetic field normal to the surface also propagate around the fuselage and wings. The GEMACS and FEKO programs and the measurements on the 26 cm cylinder all show coupling of horizontally polarized fields around the cylinder. In Reference [4] the authors talk about two types of creeping wave: the transverse electric (TE) and the transverse magnetic (TM).

The definition is that in TE wave, the electric field is directed mainly along the transverse coordinate, and in TM, it is the magnetic field that is directed along the transverse coordinate. A comparison of the isolation around the 26 cm cylinder is given from the measurements, FEKO, GEMACS, and the SF equation at 1, 2, 5, and 9.375 GHz. The coupling around a 2.6 m cylinder is given by GEMACS and the SF equation. A 26 cm diameter cylinder was chosen because it is easier to manufacture than a 2.6 m cylinder. However, from Reference [1], we see that the measurements and analysis also apply to a 2.6 m cylinder when the frequencies are reduced to 100, 300, 500, and 937.5 MHz. A comparison between the GEMACS Figure 10.18 from 1000 to 5000 MHz and Figure 10.19 from 100 to 500 MHz shows the same level of isolation.

When antennas are mounted above a conductive surface, the direct and reflected waves add for a vertically polarized source and subtract for a horizontally polarized source. For horizontally polarized sources and depending on frequency, height, and distance, the coupling to antennas may be reduced by this interference as shown in Figure 10.20.

We observe the effect of the sum of the direct and reflected wave in the "direct" GEMACS analysis and in the measurement as a negative isolation, as shown in Figure 10.20 for the horizontally polarized wave at 500 MHz.

TABLE 10.9

Reduction in Coupling around the 1/10th-Scale 0.13 m Radius Aircraft Compared to Line-of-Sight Coupling

Scale Frequency (GHz)	Angle around Fuselage (°)	Distance Down Fuselage Direction Z (m)	Reduction in Coupling from Measurements (dB)	Antennas
10.3	110	0.6	7.2	A–B
10.3	75	0.76	9.2	A–C
1.37	180	0.2	18	K–J

Source: Weston, D.A., Electromagnetic ambient inside an aircraft from transmitting antennas mounted on the outside compared to safety levels and radiated susceptibility test levels, *IEEE International Symposium on Electromagnetic Compatibility*, Denver, CO, 2013.

FIGURE 10.18   Vertical isolation on the 26 cm cylinder. (From Weston, D.A., Comparison of techniques for prediction of and measurement of antenna to antenna coupling on an aircraft, *EMC Europe*, York, U.K. © 2011, IEEE.)

FIGURE 10.19   Vertical isolation on the 2.6 m cylinder. (From Weston, D.A., Comparison of techniques for prediction of and measurement of antenna to antenna coupling on an aircraft, *EMC Europe*, York, U.K. © 2011, IEEE.)

Some of the TX and RX antenna frequencies are in band and some of the very high-power transmitters are physically very close to an RX antenna but the frequency is out of band. All model antennas were calibrated for gain at both in-band and out-of-band frequencies and compared to the manufacturers' full-scale antenna gain when this was available. Antennas, which are reported to be omnidirectional in the azimuth direction, were tested for antenna pattern.

The input power induced into the RX antenna on the full-scale aircraft is identical to the 1/10 scale, assuming that the input power to the transmitting antenna and antenna gains are identical, that is, the change in frequency and the change in distance between antennas cancel in calculating induced power. This is because the radiated power from the TX antenna is proportional to the reciprocal of distance squared and the received power level is proportional to the reciprocal of lambda squared, and therefore these then cancel. Thus, as the distance increases by 10, the radiated power density at the receiving antenna is 1/100 less but the power induced into the antenna remains the same.

Horizontal coupling on 26 cm cylinder, FEKO, GEMACS, and measurements

Legend:
- ◆ FEKO H 180°
- ▲ GEMACS H 180°
- ✳ Measurement *H* at 90°
- ■ FEKO H 90°
- ✕ GEMACS H 90°
- ● Measurements at 180°

(Y-axis: Isolation (dB); X-axis: Frequency (MHz))

FIGURE 10.20 **(See color insert.)** Horizontal isolation on 26 cm cylinder. (From Weston, D.A., Comparison of techniques for prediction of and measurement of antenna to antenna coupling on an aircraft, *EMC Europe*, York, U.K. © 2011, IEEE.)

Using the Friis transmission equation, the radiated power density *Pd* at distance is

$$G \times Pin/12.56 \times r^2 \ [\text{W/m}^2] \tag{10.5}$$

And the *Pd* based on received level *Pr* is

$$12.56 \times Pr/\lambda^2 \times G \ [\text{W/m}^2] \tag{10.6}$$

A sanity check on all of the 1/10th-scale measurements was made using these equations. When antennas are close together, the correlation between the measurement of incident power and the predicted coupled power was excellent; see Table 10.12. When one antenna was at the top of the fuselage and another at the bottom, the measured level was approximately 28 dB below that of the level between the two antennas mounted on the top, which is as expected. This is not a true measure of the level of isolation as the two antennas on top were 0.22 m apart (2.2 m on full scale) and from top to bottom 0.5 m (5 m). Also, the frequency in one case was 137 MHz and in the top to bottom case 174 MHz. In other instances, where the type and characteristics of the full-scale receiving antenna are unknown or where a suitable 1/10th-scale model antenna could not be constructed, the level of radiated power at the receiving antenna is provided instead of the received power into the antenna and again the level of isolation can be predicted.

There are numerous possibilities for error using both a physical scale model and mathematical modeling. When consensus is found between the two approaches, the results can be viewed with confidence.

Using FEKO to predict the *E* field from a transmitting antenna mounted on the full-scale aircraft at 174 MHz, an input power of 20 W at a distance of 2 m results in an *E* field of 5.2 V/m = 0.072 W/m². Using this result in Equation 10.5, the antenna gain is –7.44 dBi.

FIGURE 10.21 FEKO model of aircraft. (From Weston, D.A., Antenna to antenna coupling on an aircraft using a 1/10th scale model with results compared to the FEKO electromagnetic analysis program, *EMC Europe*, Wroclaw, Poland, 2010.)

The measured gain of the 1/10th-scale antenna at 1.74 GHz is −8.4 dBi, which is a good correlation.

FEKO is a suite of tools used for electromagnetic field analysis of 3D structures. The geometry of the full-scale FEKO model with antennas mounted on the top of the fuselage is shown in Figure 10.21.

The calculation for coupled power on the full-scale aircraft compensates for the differences between the model and the full-scale aircraft, and these include input power to the transmitting antenna, the gains of the full scale versus 1/10th-scale antennas, cable attenuation in the model measurement, and inclusion of a preamp.

The typical worksheet for an out-of-band calculation is shown in Table 10.10 from Reference [18] and predicts that 36 dBm will be induced into the receiving antenna at 1.03 GHz on the full-scale aircraft when these antennas are only 1 m apart. This high level could result in damage to the front end of the receiver.

A typical worksheet for in-band measurements, where one antenna is on the top of the fuselage and the other on the bottom, is shown in Table 10.11 from Reference [18], and here the coupling from top to bottom is −10 dBm.

TABLE 10.10
Out-of-Band Coupling Analysis Worksheet

Conditions

TX frequency = 1030 MHz IFF (10.3 GHz on model)

RX antenna = Whip on top

TX = Blade antenna on top

RX frequency range 118–137 MHz (1.18–1.37 GHz on model)

TX power into full-scale antenna (W)	1585
TX power into 1/10th-scale antenna (dBm)	10
TX full-scale antenna gain (dBi)	0
TX 1/10th-scale antenna gain (dBi)	0
TX 1/10th-scale antenna cable and attenuator attenuation (dB)	10.5
RX 1/10th-scale antenna cable/attenuator attenuation and preamp gain	−23.9
1/10th-scale antenna received level (dBm)	−13.1
Received level corrected for TX cable and attenuators and preamp gain	−26.5
Predicted input level in full-scale antenna (dBm)	36.0
Correction for RX antenna gain delta, negligible	0

TABLE 10.11

In-Band Coupling Worksheet

Conditions

V/UHF transmitting antenna to acoustic V/UHF receiving antenna

Top to bottom

Top = blade antenna

Bottom = Blade in Figure 10.15

TX power into full-scale antenna (W)	20
TX power into 1/10th-scale antenna (dBm)	−2
TX full-scale antenna gain (dBi)	0
TX 1/10th-scale antenna gain (dBi)	4.00E−01
TX 1/10th-scale antenna cable and attenuator attenuation (dB)	4.3
RX 1/10th-scale antenna cable/attenuator attenuation and preamp gain	−14.7
1/10th-scale antenna received level (dBm)	−47
Received level corrected for cable and TX antennas (dBm)	−5.78E+01
Full-scale RX antenna gain (dBi)	−0.3
1/10th-scale RX antenna gain dB	1.3
Predicted input level in full-scale antenna (dBm)	−10.1
Predicted incident radiated power on full-scale antenna (W/m^2)	4.43E−07
Frequency (MHz)	174

The measurements were made on the top of the fuselage, with the wings and engines attached, between two whip antennas 0.2 m apart. This measurement was corrected to the data on the full-scale aircraft using a similar worksheet to Table 10.11 from Reference [18].

FEKO was used to model the full-scale aircraft and antennas as was the 4NEC2D MOM program. As a sanity check, Equations 10.5 and 10.6 were also used to calculate the induced power, and the results are shown in Table 10.12. Equations 10.5 and 10.6 provide an approximate value as reflections are not included.

The low level predicted by FEKO is to some extent due to an overestimation of the antenna input impedance, which reduces the gain of the transmit and receive antennas. FEKO calculates the input impedance of the whip antenna impedance at 198 Ω, but using the programs in Reference [1], the value is between 44.5 and 47.5 Ω and 4NEC2D comes up with a similar value. This value of impedance is also close to the value given in most antenna books. This illustrates the importance of checking the output data from any electromagnetic analysis program. Chapter 12 describes errors and weaknesses with these programs. Use of a cylinder instead of a wire in FEKO for the antenna rod may provide a more accurate antenna impedance. However, the impedance mismatch does not fully explain the 23 dB difference in measured FEKO received levels. No cable impedance to antenna impedance mismatch is included in the FEKO analysis for the transmit antenna, as the excitation

TABLE 10.12

Aircraft-Mounted Whip Antenna to Whip Antenna Coupling Analysis Comparison

Frequency (MHz)	137
Measured received power (dBm)	18
FEKO predicted (dBm)	−5
Equations 10.7 and 10.8 (dBm)	20
4NEC2D MOM program	24.4

FIGURE 10.22 Measured isolation on the 1/10th-scale model and FEKO analysis on both the 1/10th-scale and full-scale aircraft. (From Weston, D.A., Electromagnetic ambient inside an aircraft from transmitting antennas mounted on the outside compared to safety levels and radiated susceptibility test levels, *IEEE International Symposium on Electromagnetic Compatibility*, Denver, CO. © 2013, IEEE.)

voltage drives the base of the antenna directly, but the input power is reduced compared to a 44 Ω input impedance. For a given input voltage to the transmitting antenna and for a 50 Ω load on the receiving antenna, the transmit and receive antennas' mismatch only accounts for an approximately 15 dB reduction in the received level.

A comparison of the 1/10 scale aircraft measurement as well as the FEKO 1/10 scale and full scale are shown in Figure 10.22.

A similar 26 cm diameter cylinder was modeled using GEMACS with 1.6 cm monopole antennas. The result of this analysis was close to the measured although the GEMACS cylinder did not include the wings.

The comparison is shown in Figure 10.23.

10.2.1.4 Measurements on the Full-Scale Aircraft

When the frequency range of interest is from 4 to 18.75 GHz, then measurements on the 1/10th-scale model are less practical as the model antennas would have to operate from 40 to 187.5 GHz. Instead, measurements on the full-scale aircraft or use of electromagnetic analysis software using the full-scale aircraft dimensions and frequencies are preferable. The limits of FEKO for the full-scale model are discussed in Reference [1].

The most accurate measurements are those made with the flight antennas installed on the aircraft. In some cases, these antennas are not available or are not mounted on the aircraft. In the case examined here, nonflight antennas were used and the coupling results were corrected for the gain and antenna pattern differences between the flight and the measurement antennas. The aperture antennas were mounted

FIGURE 10.23 GEMACS 26 cm cylinder compared to measurements on a scale model aircraft with a 26 cm diameter fuselage.

on wooden structures at approximately the correct location on the fuselage. In the case of blade, whip, and microstrip antennas, the paint on the fuselage was removed around the base of the antenna, which was then taped in place and bonded to the surface. The cables for these antennas were covered in absorber and taped to the aircraft so that the TX and RX antenna cables were as far apart as possible. The measuring instrument was either inside the aircraft or behind a screen covered in absorber material, and the signal generator was on the ground also covered in absorber. To measure the received level, a battery powered spectrum analyzer or an analyzer with the power cables covered in absorber was used.

For the level of coupling, when the aircraft is landing or taking off the undercarriage should be deployed. In flight, it should be retracted. If the landing gear is not recessed in the engine nacelle but mounted on the wings, then covering the landing gear with absorber, making the measurements well away from the hanger, and placing absorber under the aircraft will reduce the measurement errors. If the landing gear is normally retracted into the engine nacelle, then the aircraft engineer may decide that it is possible to support the aircraft by the wing and retract the landing gear.

As with all measurements, a sanity check on cable-to-cable coupling with the receiving antenna removed was made.

Some of the measurements made on the aircraft are as follows:

Weather radar E in Figure 10.15 to receiving antenna B and C at 9.375 GHz
Maritime Search Radar G to helical receiving antenna H at 8.9, 9.375, and 9.5 GHz
SLAR antenna F to Maritime Search Radar G at 9.375 GHz

The angle between the location of the Maritime Search Radar and the location of the H antenna in the H plane is approximately 15°. The gain of the Maritime Search Radar at 15° in the H plane is 28 dB down on the boresight gain, that is, 33 dB − 28 dB is 5 dB.

The G Maritime Search Radar polarization is horizontal. The specified azimuth and elevation gains of the H helical antenna are with the antenna oriented perpendicular to the surface of the fuselage, and so the horizontally polarized E field from the G antenna cuts the H antenna at approximately 76° to the axis of the antenna as it creeps around the fuselage. The gain of the H antenna at a 76° angle is −10.3 dB. From measurements, the difference in received level between horizontal–horizontal coupling and horizontal–vertical coupling (cross polarization) is approximately 6 dB, and this was subtracted from the receiving antenna gain. The actual cross polarization gain for the H antennas is unknown and the 6 dB value can be modified once this is known.

10.2.1.5 Measurements on the Aircraft with Flight Transmitters/Receivers and Antennas Installed

After using measurements prior to the availability of the aircraft, measurements were made using the receivers, transmitters, and antennas delivered with the aircraft. All transmitters were operated at maximum power, and all personnel outside of the aircraft were located at a safe distance.

The fields from transmitters were measured at the window closest to the transmitting antennas using calibrated monopole, bow tie, log-periodic biconical, and double-ridged guide antennas.

The worst-case (highest) fields inside the aircraft are shown in Table 10.13.

Many of the measured levels at the receiving antennas were much lower than predicted. This was typically because the sidelobe from high-gain antennas was overestimated or the transmit or receive out-of-band antenna gain was different than predicted. In many cases, the difference between predicted and measured was greater than 20 dB.

In the case of the weather radar in the nose cone, it was shaded more by the end of the fuselage than expected.

The fields predicted and measured inside the aircraft show levels that are well below the safety limits applied to the general public.

A comparison between some of the predicted and measured fields is shown in Table 10.14.

TABLE 10.13

Highest Level of Fields Measured inside the Aircraft

	E Field V/m
Frequency (MHz)	Pk = Peak
1.5	0.003
2	0.079
30	0.041
100	0.9
250	2.238
512	2.44
1045.5	1.03
1083.35	0.204 Pk
8967.5	7.36 Pk
9375	13.46 Pk

TABLE 10.14

Comparison of Predicted and Measured Received Levels

Transmit Frequency (MHz)	Receive Frequency Range (MHz)	Predicted Receiver Antenna Output Level (dBm)	Measured Receiver Antenna Output Level on the Full-Scale Aircraft (dBm)	Type of Coupling	Analysis Method	Delta (dB)
1030	329–335	−8	−19	Line of sight	1/10th-scale model	11
1030	1,030	−18	−25.8	Line of sight	1/10th-scale model	7.8
9220	0.5–40,000	−22.7	−24	Shaded	Full-scale model	1.3
8900	0.5–40,000	−25.1	−22.7	Shaded	FEKO	2.4
1030	156/163	11	−3.5	Line of sight	1/10th-scale model	14.5
1030	30–600	19.4	11.7	Line of sight	1/10th-scale mode	7.7
1030	110–512	10	−11	Line of sight	1/10th-scale mode	26

10.2.1.6 Aperture Coupling on an Aircraft

The coupling from the distance measuring equipment (DME) transmitter, which uses a quarter-wave monopole antenna, through the closest window in the aircraft was modeled using FEKO for simplicity by a cylinder with aperture. The distance between the antenna and aperture was 2.21 m. Figure 10.24 shows the geometry. As the wings and engine nacelle were some distance away from the window, it was assumed that reflections would not be significant.

The FEKO program is a popular electromagnetic analysis program. Its use for aircraft modeling is described in References [18–20]. It was used to model the problem in Figure 10.24, where the frequency is 1045.5 MHz. As the problem is electrically very large (90 wavelengths), it was solved using the multilevel fast multipole method (MLFMM) simulation. The problem was meshed into 3,270,004 unknowns that required 74.2 GB of memory running for 9.5 h on a workstation with Intel® Xeon® CPU E5-2650 at 2 GHz, 16 parallel processes. Double precision was used to solve the problem to account for a better convergence of the MLFMM algorithm. Instead, if a single precision had been used, the resource requirements will reduce by half (both the runtime and memory).

From FEKO at 1045.5 MHz, the field at the window was 4.56 V/m and the internal field at 30 cm inside the center of the window was 0.983 V/m. The level measured in the aircraft, at the same location, was 1.205 V/m as shown in Table 10.13. Thus, FEKO showed the window attenuation to be 13.3 dB at 30 cm inside the cylinder. Using GEMACS to model the field at line of sight versus 87° around the cylinder showed the isolation was 17.4 dB, which corresponds well with the 30 dB previously predicted by GEMACS for 180° at 1 GHz around a 2.6 m cylinder.

For an HF wire antenna connected to the top of the fuselage and 58° from the center line of the fuselage, as shown in Figure 10.25, the level of E field at the window compared to the line-of-sight

1/4 wave monopole 7.17 cm long

Frequency 1045.5 MHz. Antenna input power = 300 W
Required output data = Internal cylinder field in the center
of the aperture and 30 cm behind aperture

FIGURE 10.24 Geometry of the cylinder, aperture, and the monopole antenna. (From Weston, D.A., Electromagnetic ambient inside an aircraft from transmitting antennas mounted on the outside compared to safety levels and radiated susceptibility test levels, *IEEE International Symposium on Electromagnetic Compatibility*, Denver, CO. © 2013, IEEE.)

TABLE 10.15
Measured Levels inside the Aircraft

Transmitter	Measured Level (V/m)	IEEE C95.1 Averaged Limits for the General Public (V/m)	MIL-STD-461 RS103 Limits for Aircraft Internal Air Force	DO-160 Limits Category S and T (V/m) S	T
TCAS	0.204 peak	52	60	1	5
DME	1.023	51	60	1	5
IFF	0.646 peak	52	60	1	5
Weather radar	0.065	153	60	1	5
Cockpit HF 1 at 2 MHz	0.08	412	20	1	5
Cockpit HF 1 at 30 MHz	0.011	27.5	20	1	5
Cockpit HF 2 at 2 MHz	0.018	412	20	1	5
Cockpit HF2 at 30 MHz	0.003	27.5	20	1	5
Mission HF at 1.5 MHz	0.003	549	20	1	5
Mission HF at 30 MHz	0.041	27.5	20	1	5
Mission V/UHF at 100 MHz	0.9	27.5	20	1	5
Missions V/UHF at 250 MHz	2.24	27.5	20	1	5
Mission UHF at 250 MHz	2.24	27.5	20	1	5
Missions UHF at 512 MHz	2.44		20	1	5
Maritime Search Radar	7.36 peak		60		
SLAR	13.46 peak		60		

field is effectively the same, that is, no appreciable isolation. However, at 30 MHz, the level of attenuation through the window is higher than at 1045.5 MHZ.

FEKO predicts an external field at the window at 30 MHz of 2.29 V/m and 0.115 V/m at 0.3 m inside, that is, an attenuation of 26 dB at 0.3 m. At 0.7 m inside the window, the FEKO predicted field is 0.0134 V/m. The measured level at approximately 0.3 m inside the aircraft was 0.041 V/m at 30 MHz. The measured field in the aircraft at 2 MHz, which was not analyzed or measured, was 0.08 V/m.

The measurements were made on a 1/10 scale cylinder with 77 cm long wire antenna in the 1/10th-scale geometry of Figure 10.19.

A 4 cm long bow tie antenna connected to the digitizer and fiber optic cable and the fiber was the only external connection brought through the totally sealed cylinder via a 6 cm long, 5.5 m inner diameter pipe soldered to the cylinder.

The measurements on the cylinder were made at 300 and 97.8 MHz. The cylinder-measured external field was normalized to the FEKO result of 2.29 V/m and the cylinder measurements, and the measurement on the full-scale aircraft and the data using FEKO are shown in Figure 10.26.

10.2.2 DESENSITIZING OF RECEIVERS BY HIGH-POWER TRANSMITTERS

Receiver desensitization or compression occurs when a strong adjacent signal such as from a transmitter with a frequency offset, but in band, from the signal frequency results in an apparent decrease in the gain of the receiver.

Typically, the result is also an increase in the receiver noise figure.

This means that if the intentional received level is low, then it may disappear into the noise and be completely missed due to desensitization.

The effect is due to the compression in the front-end amplifiers as well as often the mixers and IF amplifier. The desensitization is generally specified as the level of unwanted signal at which an offset (20–100 kHz) signal gives a 3 dB reduction in gain.

FIGURE 10.25 Wire HF antenna geometry. (From Weston, D.A., Electromagnetic ambient inside an aircraft from transmitting antennas mounted on the outside compared to safety levels and radiated susceptibility test levels, *IEEE International Symposium on Electromagnetic Compatibility*, Denver, CO. © 2013, IEEE.)

FIGURE 10.26 Comparison at 30 MHz. (From Weston, D.A., Electromagnetic ambient inside an aircraft from transmitting antennas mounted on the outside compared to safety levels and radiated susceptibility test levels, *IEEE International Symposium on Electromagnetic Compatibility*, Denver, CO. © 2013, IEEE.)

From Reference [23], we see that desensitization of a desired signal at the reference sensitivity level due to an adjacent channel signal is defined as adjacent channel rejection (ACR). The measurement of ACR detailed in ANSI/TIA/EIA-603-A and ANSI/TIA/EIA-102CAAA documents can be used to quantify receiver desensitization at any frequency offset and for higher desired signal levels. Receiver LO sideband noise can heterodyne an undesired signal into the IF passband by mixing with a single high-level signal, typically in excess of −50 dBm, and usually within 500 kHz of the desired signal. This mechanism is often confused with adjacent channel interference.

Receivers will typically have a ≥90 dB rejection of signals that are offset by ≥500 kHz from the desired channel and a ≥80 dB rejection for offsets exceeding approximately 50 kHz.

Excessive desired on-channel signal levels can overload the receiver, usually the result of automatic gain control design limitations. The front end of the receiver can be overloaded by a single high-level unwanted signal, not on the desired channel, typically in excess of −25 dBm, or multiple high-level unwanted signals whose peak instantaneously exceeds −25 dBm. This is known as receiver blocking.

Receiver front-end (RF amplifier) nonlinearity can create intermodulation products on the desired frequency by mixing two or more high-level signals, typically ≥−50 dBm. Figure 10.27 from Reference [23] shows sensitivity-level intermodulation rejection (IMR) for typical mobile receivers, relative to the receivers' reference sensitivity signal level.

The sensitivity performance of a receiver can be measured by the SINAD (signal + noise + distortion), which looks at the degradation of a signal by an interferer. It is the ratio of the total signal power level to the unwanted signal power (noise + distortion).

The SINAD is expressed in dB and is defined as

$$SINAD = 10 \log(SND/ND)$$

where
  SND = combined signal + noise + distortion power level
  ND = combined noise + distortion power level

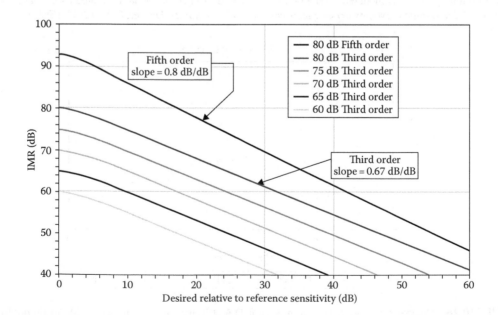

FIGURE 10.27 Receiver IM above reference sensitivity. (From Motorola Interference Technical Appendix, Issue 1.41. February 2002. © 2002, Motorola.)

FIGURE 10.28 IMR performance. (From Motorola Interference Technical Appendix, Issue 1.41. February 2002. © 2002, Motorola.)

In Figure 10.28, the reference sensitivity for 12 dB SINAD is −119 dBm, Cs/N is 4 dB, and the IMR is 80 dB.

Using the TIA IMR test methodology, for example, in Figure 10.28, the −119 dBm produces a 4 dB Cs/N that creates the 12 dB SINAD reference sensitivity. The signal is boosted by 3 dB) −116 dBm) and the equal signal level interferers are increased until 12 dB SINAD is again reached. This indicates that now a 4 dB $Cs/(I + N)$ has been reached but the desired is now −116 dBm. Thus, the composite noise floor is −120 dBm, consisting of −123 dBm from the receiver noise floor and −123 dBm, the equivalent noise from the intermodulating signals. The difference between the desired and the IMR signals (−39 dBm) is the IMR performance of the receiver (80 dB).

Although the desensitization is receiver specific, Figures 10.27 and 10.28 provide examples.

Once the level of narrowband or BB noise coupling from a transmitting antenna is calculated, either the receiver SINAD or receiver desensitization characteristic can be used to calculate the receiver desensitization.

References [26,27] show the desensitizing effects from high-power HF broadcast transmitters on HF airborne receivers.

10.2.3 SIMOPS

The analysis of spurious (spurs), harmonic, and BB emissions from transmitters coupling to receivers can be used to determine if SiMultaneous OPerationS (SIMOPS) can be made.

The spurious and harmonic emissions are discrete, and so in-band interference will only occur when the spur or harmonic is at the receiver-tuned frequency, although desensitization, cross modulation, and generation of spurs can still occur if the transmitted level is high or the transmit and receive frequencies are close.

Typically, the data required to conduct a SIMOPS evaluation are

1. RX and TX antenna gains and, when not omnidirectional in the azimuth, the gain pattern
2. Transmitter output power, spurious emissions, harmonic, and BB noise emissions
3. When antennas are not line of sight, that is, mounted around the fuselage, the power coupling factor minus the gains of the TX and RX antennas (thus providing the power at the input of the receiving antenna)
4. When line of sight, the FRIIS transmission equation with compensation for reflections or an electromagnetic computational program
5. Receiver input sensitivity, preferably in band and out of band or the input level at which the receiver is marginally affected by the EMI level

One of the problems with performing a SIMOPS analysis is to find the out-of-band gain of an antenna. The effect on the receiver of the interfering power level may also be difficult to determine, especially if the receiver contains an input filter.

In cases where transmitter spurs, harmonic, and BB data are not available, or only partially available, then generic data must be used for the BB noise. The conducted emissions contained in CE106 for the spurs and harmonics may be used. CE106 is specified relative to the fundamental, for example, 80 dB down on the fundamental. Thus, for example, 80 dB has been subtracted from the predicted coupling between the transmit and receive antennas at the fundamental.

In one case, a receiver would expect as high as −20 dBm at its input over its 136–174 MHz frequency range without practical impact for the receiver. Another receiver in the 118–400 MHz band, with the desired signal level at nominal sensitivity, shall permit a SINAD of at least 7 dB, with the interferer more than ±10 MHz away from the wanted carrier, at a level of 0 dBm.

Section 10.2.2 provide more information on receiver desensitization.

A typical SIMOPS power coupling chart is shown in Figure 10.29.

10.2.4 Antenna-to-Antenna Coupling Mitigation Techniques

Much of the following information was published in *ITEM* magazine in 2012 and is reproduced by kind permission of *ITEM* [24].

Improving intersystem compatibility begins at the initial design stage. When choosing onboard systems, careful consideration should be given to the transmit and receive frequencies of the desired equipment as well as the output power levels, sensitivity levels, blanking capabilities, and intended usage.

A table of all systems and relevant information should be prepared, and this table can be used to determine potential conflicts and the possibility of SIMOPS. It is important to consider intended usage as some conflicts are quite acceptable. For example, the emergency locator transmitter (ELT) transmits in the same range as regular communication radios; however, the ELT will only be used in the case of a downed aircraft, and under those circumstances, many of the other systems will not be required to operate.

For the coupling path between sources of high electromagnetic fields and antennas where high current or voltages are induced, an EMI mitigation technique can be chosen. The best technique will adequately eliminate, suppress, or minimize EMI with minimum modification or change. One important factor in the decision is the answer to the question, can the best technique be practically implemented on the aircraft. In many cases, the solution is a modification within equipment, which the manufacturer may be unwilling to make.

For systems that are required to operate in the same or similar bandwidths, the primary preventative measure is the antenna placement. Conflicting system antennas can be mounted at opposite ends of the aircraft or in some cases at the top and bottom of the aircraft. This reduction

Maximum coupled power evaluation (dBm) for spurs, harmonics, and broadband noise only					
Transmit Frequency (MHz)	Receive Frequency (MHz)	TX	Pin (dBm)	RX V/UHF 9 Cockpit 30–88, 118–174, 225–400, 400–600 MHz	RX VHF 136–173.5 MHz
108–600	108–600	Mission V/UHF	−139		
30	30–90	Mission HF	−31		
30	30–90	13 Cockpit HF	−68		
291.4	291.4–582	Sonobuoy UHF	−145		
116–137	116–411	Cockpit VHF	−59		
45.33–174	136–173.5	Cockpit V/UHF	−86		
137	137	Cockpit VHF	−60		
136–173.5	136–173.5	Mission V/UHF	−64		
27–30	136–173.5	Cockpit HF	−85		
27–30	136–173.5	Mission HF	−90		
	No SIMOP				
	SIMOP with frequency restriction				
	SIMOP should be without or with minimum restriction				
At discrete frequencies. Interference will occur from spurs and harmonics or BB noise at some frequencies.					

FIGURE 10.29 Typical SIMOP work sheet.

in coupling increases the isolation between the antennas as discussed in earlier sections. Knowledge of the isolation allows an EMC or aircraft engineer to decide if it is worthwhile to reposition two colocated antennas to the top and bottom of a fuselage or wing or increase the distance between them.

Effective reduction in coupling may be achieved for horizontally polarized transmitting antennas that operate over a narrow frequency range and that couple to either a vertically or a horizontally polarized receiving antenna in close proximity. The location of the two antennas may be chosen so that the direct wave and the wave reflected from the surface of the aircraft tend to cancel. This cancellation may be more effective than locating antennas on opposite sides of the fuselage at low frequency.

Many pieces of equipment also include blanking capabilities. For example, a transmitter may provide a DC output signal when transmitting, and this DC output can be used to disconnect receiving signal paths of local sensitive equipment. Similarly, some equipment includes blanking input connections that disconnect the received RF signal when a DC level is present at the blanking input.

10.2.5 Absorber

Absorber can also be placed on surfaces, such as the wings, engine, vertical stabilizer, and tail plane to reduce reflection as well as for damping surface waves. The absorber may be flexible silicone rubber lossy sheet, which is weather proof and can be glued to surfaces. These are typically effective at microwave frequencies, such as used by radar, with some frequency tuned and others effective over a BB of frequencies. Also, a thin ceramic ferrite absorber is available effective from 10 MHz to 1 GHz. One manufacturer is Cuming Corporation. On flush-mounted antennas such as used for GPS and some radar and SATCOM, a circle of absorber may be glued around the antenna.

10.2.6 Filters and an "In-Band" EMI Solution

10.2.6.1 Filters

High-pass, low-pass, bandpass, or band-stop (notch) filters can be used to attenuate out-of-band EMI signals caused by coupling from fields incident on a receiving antenna. To avoid cross modulation or intermodulation in the first stage after the antenna, for example, the front end of a receiver, a low-noise amplifier (LNA), or a low-noise downconverter (LNC), the filter should be connected between the antenna and the first stage. The problem with this location is the additional insertion loss introduced by the filter. Most antenna designers have optimized the gain of the antenna and reduced losses in the RF cable to the antenna, and if the antenna is constructed with a number of PCB elements, then it has combined these PCB elements using a no-loss or low-loss combiner. Even a fraction of a decibel of additional loss as a result of the filter can reduce the G/T of the receiving system, as described in Section 5.5.3. If the system uses an LNA or LNC and the EMI is not caused by an overload in these elements, resulting in cross modulation or intermodulation, then the filter should be located after the LNA or LNC to minimize the effect of the insertion loss of the filter. A number of types of filter are available: Filters with helical and folded helical resonators have been described in the literature, with design guidelines and examples. Helical resonators can be used as bandpass filters between several tens of megahertz and up to 2 GHz. The helical resonator consists of a single-layer solenoid enclosed in a shield, with one lead of the winding short-circuited to the shield and the other end open-circuited. Input and output couplings to the filter are built with taps to the appropriate impedance points on the helix that match the impedance to the source and load. By including an adjustable tuning screw in either side of the helix, it is possible to change the capacitive or inductive load of the helical to tune the center frequency. Helical filters are designed mostly for the Butterworth response. The Chebyshev response can be used, but the insertion loss of the Chebyshev is much higher. A typical loss for a helical filter is approximately 1 dB.

Surface acoustic wave (SAW) filter technology is based on the conversion of electrical signals into SAWs and then back to an electrical output. The frequency range for these filters is 10 MHz to 3 GHz, with a relative passband from 0.01% to 100%. The disadvantage is the minimum 6 dB of insertion loss. In the SAW impedance element filter (IEF), the long SAW transducer (resonator) has an impedance dependence similar to a classic inductance capacitance or quartz resonator, and the SAW itself plays only an auxiliary role. This device can be used as a bandpass filter, and in practice, a number of elements must be connected together to achieve an adequate out-of-band suppression. In the IEF, the insertion loss can be reduced, and in the balanced bridge version, the insertion loss can be limited to 1.5 dB over an 890–915 MHz passband.

Microwave filter Inc. manufactures a wide range of microwave and RF-customized filters from 1 MHz to 26 GHz, including tunable notch filters. These filters include lumped-constant construction, distributed line construction, and cavity designs. Filters from 7 to 11 GHz are available with an insertion loss of 1 dB maximum.

FSY microwave (API Technologies) also manufactures standard low-pass filters from 1 kHz to 18 GHz, high-pass filters from 5 kHz to 0.4 GHz, and bandpass filters from 10 kHz to 10 GHz. Waveguide filters from 3 to 18 GHz are available with a low insertion loss. Allen Avionics manufactures custom filters from 100 Hz to 1 GHz. K&L manufactures a wide range of lumped-component, tubular, and cavity filters. The cavity filters exhibit lower insertion loss and steeper skirt selectivity than tubular filters and are available with two to six sections. The loss is dependent on the number of sections and the 3 dB bandwidth, but it can be as low as 0.5 dB for a two-section filter at 1 GHz with a 20 MHz (2%) bandwidth, or 1.4 dB for a six-section filter at 1 GHz with a 20 MHz bandwidth.

The evanescent TE mode cavity filter (100 MHz to 18 GHz) is available from K&L and is a very low-loss filter. Typical losses are less than a decibel for most designs with bandwidth of 5% or more, with a rule of thumb of 0.1 dB per section. For wider-band filters, losses are even less. For an EMI problem at a specific frequency either one side or the other of the in-band frequencies, the band-width of the filter passband can be greater than 5%, as long as the interferer is well outside of the passband. If the interferer is close to the edge of the receiver passband, then one skirt of a multisec-tion filter can be close to the interferer and still achieve adequate attenuation.

For example, a "real-world" EMI antenna-to-antenna coupling problem was presented to FSY Microwave, Inc. In this case study, the microwave frequency communications link had a center frequency of 1500 MHz and a 3 dB bandwidth of 20 MHz. An interferer at 1475 MHz was at a high enough level to result in compression at the input of the LNA, so the first filter had to be located between the antenna and the LNA. This filter required a minimum of 30 dB attenuation but should also exhibit an insertion loss of maximum 1.5 dB to maintain the G/T of the system. When this problem was presented to FSY Microwave, they kindly customized a standard filter. The affordable five-pole C1500-20-5NN bandpass filter designed by FSY Microwave had the following specifications:

Center frequency = 1500 MHz
Insertion loss = 1.5 dB max at CF (1.0 dB typical)
3 dB bandwidth = 20 MHz min, 24 MHz max
VSWR = 1.5:1 max over 80% of the 3 dB bandwidth
Impedance = 50 Ω in/out
Attenuation = 35 dBc (decibels above carrier) min from DC to 1475 MHz

Typical five-pole 0.035 dB Chebyshev response

Power rating = 20 W CW max
Size = 1.10 in. × 1.75 in. × 5.35 in. excluding connectors
Connectors = type-N female in/out (or as specified)
Temperature range, operating = −30°C to +60°C
Vibration = 10 G's, 5–2000 Hz, sine swept, 3 axes
Shock = 30 G's in 11 ms, 1/2 sine, 3 axes
Relative humidity = Up to 95%, noncondensing

The insertion loss and attenuation characteristics are shown in Figure 10.30.

The five-pole filter is designed to avoid compression in the LNA with a minimum impact on the G/T figure. If additional attenuation of the interferer is required to avoid an EMI problem in the receiver, then a second filter could be included after the LNA. This second filter could have a higher level of attenuation because the increased insertion loss occurs after the gain of the LNA and so will not adversely affect the G/T.

10.2.6.2 In-Band EMI Solution

In some instances where an interferer is very close to being in band, a filter with a very steep skirt can be used; or if the full bandwidth of the system is not required and some loss at the higher or lower frequencies can be accepted, then the skirt of the filter may extend to some small degree into the signal bandwidth. In one example of EMI, whenever an HF radio was operated onboard a ship,

FIGURE 10.30   Attenuation characteristics of the FSY Microwave, Inc., C1500-20-5NN five-pole filter.

the ship's RF log indicated full speed. The RF current on the ship's hull induced by the HF radio fan antenna was coupled to the RF log transducer, used to measure the ship's speed. This problem was considered to be in band; that is, the interferer frequency is exactly at a frequency required by the receiving system for normal communications to exist. In the case of the ship's log, the full speed indicated by the log was well above the speed of any of the ships in which the EMI problem existed. It was thus feasible to incorporate a filter between the RF ship's log transducer and the detector electronics, which would attenuate the HF radio frequency but, as a result, would limit the display of the ship's speed to a more realistic maximum.

It may be possible to reduce the coupling between an interferer antenna and the receiving system by relocation of antennas. For microwave antennas, it may be possible to increase the directivity of the TX, or RX, or both antennas or to add weatherproof microwave absorber between the two antennas. To reduce coupling from the back lobe and sidelobes and reradiation from a dish edge, absorber can be mounted around the periphery of the antennas. The Microwave Filter Company, East Syracuse, NY, manufacture filters and used to offer a dish sidelobe absorber that can be bonded with temporary adhesive to check out its effectiveness and then bonded with permanent glue if it solves the problem.

A solution to an in-band EMI problem exists when the geometry of the problem is fixed, for example, two nonmovable antennas or, in our example, the ship's HF fan antenna and the RF log transducer. In addition, the problem frequency must be fixed, although the proposed solution can deal with a modulated interferer. One additional limitation is that the degree of required suppression of the in-band frequency is not much more than 30 dB. The solution is to pick up the interferer frequency using a highly directional antenna or a transducer that is more sensitive to the magnetic field or electric field generated by the interferer than the intentional signal. Figure 10.31 illustrates the solution to an antenna-to-antenna "in-band" EMI problem. The directional antenna is angled to receive a higher-level signal from the interferer and a much lower level from the intentional receiver signal. A bandpass filter is recommended after the small directional antenna in signal path B to exclude other out-of-band signals. After the filter, a gain stage or attenuator is included to set the appropriate signal level, followed by an adjustable phase shifter that changes the phase, so the interferer in signal path B is 180°

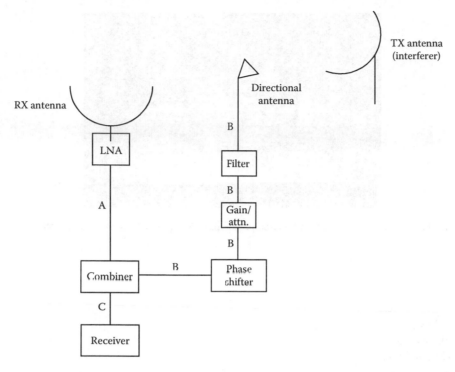

FIGURE 10.31   Antenna-to-antenna coupling with "in-band" EMI solution.

out of phase with the interferer signal coupled into the receiver signal path A. The received signal plus interferer signal, path A, is applied to a summing amplifier or summing resistor network (combiner). The phase-shifted interferer signal, path B, is also applied to the combiner, and the output from the combiner, path C, is the intentional signal, which contains a much attenuated interferer signal. One technique for providing phase shift is the inclusion of a variable delay line in path B.

The phase shifter must be a unique design and built for many frequencies, including microwave, but the Microwave Filter Company, East Syracuse, NY, offers a shifter with an adjustable 5–25 dB typical attenuation and a 180° minimum phase adjustment. The frequency ranges of the different models available are 54–108, 108–216, and 470–890 MHz.

The use of low-pass, bandpass, and band-stop filters at the input of receivers can be effective at reducing compression, generation of spurious response, and desensitization of the receiver. High-power filters are commercially available for use at the output of transmitters, and these are used to reduce harmonics and BB noise from the transmitter.

For transceivers, it is possible to design an input filter that can transmit the output power safely and without excessive attenuation.

As an example, a weather radar transceiver or an SLAR system operating at 9.375 GHz. EMC consulting have designed and built a band-stop filter at 9.375 GHz and another tunable from 700 MHz to 1.1 GHz with 30 dB of attenuation at the center frequency. All of the filters described in this report will fit into the small enclosure shown in Figure 10.32 and have been constructed and tested with either BNC or TNC connectors. This size is ideal for connecting directly at the input of the receiver. A notch filter is also known as a band stop.

Section 5.3.5 describes typical passive signal filters both PCB and connectorized:

9.375 GHz notch (band stop)
8.3 GHz notch
18 GHz PCB band stop

FIGURE 10.32   The enclosure used for band-stop, bandpass, low-pass filters, and the limiter.

Tuneable 700 MHz–1.1 GHz
14 GHz notch
1.03 GHz notch
406 MHz low pass
2 MHz low pass
30 MHz low pass
136–174 bandpass
200 MHz low pass
700 MHz high pass

10.2.6.3   Limiters

Limiters can be used in conjunction with other mitigation techniques or alone. If a high-level signal is being received at the input to a receiver either in band or out of band, a limiter can be used to reduce the input level to an acceptable level without interfering with the desired received signal, unless the two are tuned to exactly the same frequency.

For example, a receiver that is sensitive to −50 dBm but that compresses when an out-of-band signal exceeding 0 dBm is applied may compress when placed in close proximity to a high-power transmitter. The introduction of a 0 dBm limiter would reduce the high-power received levels to a manageable level while allowing the intentional received levels to pass without compression. The insertion loss for a typical limiter is 2 dB from 50 to 500 MHz with a useful frequency range of 5 to 3000 MHz. EMC consulting has manufactured a PCB on which the limiter was mounted and placed in the enclosure shown in Figure 10.32. The input voltage is 28 V, and the limiter is protected against all of the EMC power line conducted susceptibility and power characteristics specified in DO-160 or MIL-STD-461 by additional components mounted on the PCB. The operating temperature range of the limiter and associated components is −54°C to 85°C.

10.2.6.4   RF Switch and Blanking

When transmitters have blanking outputs and receivers have blanking inputs, then these may be connected to blank the receiver when the transmitter is operational. Some additional components may be required to make the control signal levels compatible. Also, when the RF output of the transmitter remains powered up, BB noise is still transmitted. If the transmitter has a blanking output but a receiver does not have a blanking input, an RF switch may be used at the receiver input. One such switch is the Mini-circuits ZFSWA-2-46, which can be used from DC-4.6 GHz. It has maximum 0.8 dB attenuation from DC-200 MHz, 1.3 dB up to 1 GHz, and 2.6 dB from 1 to 4.6 GHz. It requires two signal inputs to switch from "off" to "on," −8 V on one and 0 V on the other or vice versa.

10.2.6.5 Delay and Phase-Shift Cancellation

The topic on in-band EMI solutions for ground-based installation is discussed in Section 10.2.6.2 including cancellation. When the transmit and receive frequency bands overlap, then filters are clearly not applicable. Instead, the interferer is phase shifted 180° and added to the interferer at the input of the receiver.

When the interferer received level is extremely high and may result in damage to the receiver, the limiter may be used. However, even with the limiter in circuit, the level after limiting may be high enough to result in receiver desensitization, cross modulation, or spurious generation.

To cancel or reduce the interferer level a propagation delay, 180° phase shift and summation circuit have been developed. This is shown in the block diagram of Figure 10.33. The output of the transmitter is tapped off with no reduction in power level between the transmitter and antenna. This can be achieved anywhere on the transmit cable from the transmitter end to the transmit

FIGURE 10.33 Delay, phase shift, and summation circuit block diagram. (Reproduced from Weston, D.A., Antenna-to-antenna coupling on an aircraft: Mitigation techniques, *ITEM*, March 21, 2012. With permission.)

FIGURE 10.34 Delay, phase shift, and summation circuit with filter. (Reproduced from Weston, D.A., Antenna-to-antenna coupling on an aircraft: Mitigation techniques, *ITEM*, March 21, 2012. With permission.)

antenna end. The ideal location depends on the total propagation delay of the transmit cable and the coupling path in air between the antennas and the receiver cable, as described later.

The output of the transmitter is attenuated and matched to the impedance of the cable that connects to the circuit at point A. The interferer signal is then passed through a bulk delay. The bulk delay can be in the tens of microseconds, and the only limit is the attenuation inherent in the delay elements and the physical size of the enclosure. In practice, delays up to 500 nS are adequate. For the final selection of the total delay, thin film delay lines can be connected in or out of the circuit and a fine-tuning is performed with a variable capacitor and an inductor. With 0.5 and 3 nS thin film delay circuits, the selectable delay can be from 0.5 to 18 nS in 0.5 nS increments and the tuned delay can be adjusted over 0.5 nS. The total transmitter to circuit cable delay, bulk, and variable delays are chosen to be exactly equal to the transmitter to receiver total path propagation delay. This means that the input signal to the circuit from the receiving antenna and the signal at location B in the circuit are in phase. The signal at point B then undergoes a 180 phase shift, which is virtually frequency independent.

The signal at point C is then adjusted in level and summed with the receiver input signal. This achieves an almost complete cancellation at a specific frequency. At either side of this frequency, the amplitude at the output of the delay circuit is different from the receiving antenna level. This is because the attenuation with frequency of the transmitter to receiver path is different from the phase shifter cancellation circuit. As the intentional signal from the receiving antenna is routed through the circuit, the attenuation of the signal is 1.8 dB. Figure 10.34 shows the path between transmitter and receiver. An input filter is shown in Figure 10.34 but is not mandatory.

The transmit path propagation delay can be determined approximately by transmitting a pulse amplitude–modulated RF from a signal generator, using the modulation pulse to trigger an oscilloscope, and measuring the delay. An external modulator or RF switch may be used if the generator does not allow pulse modulation. The fine-tuning of the delays in the circuit can be accomplished by again using a pulse amplitude modulated signal and comparing the output of the circuit to the input of the receiver. The two channels of the scope used to measure the two signals must have no delay between them (chopped display may be required).

The circuit was adjusted and tested from 1 to 30 MHz, with the maximum attenuation adjusted at 30 MHz. The attenuation achieved is shown in Table 10.16. With the circuit adjusted for 108 MHz, the attenuation from 60 to 108 MHz is shown in Table 10.17. At a single higher frequency such as 152 MHz, the cancellation can result in an attenuation of 28 dB.

The delay circuits within the phase shift and summation circuit have some temperature coefficient as does the permittivity of the transmitter and receiver cables, and this changes the delay. However, the measured change in attenuation of the delay and cancellation circuit was only 1.5 dB from 4°C to 47°C.

TABLE 10.16

Cancellation Attenuation 1–30 MHz

Frequency (MHz)	Attenuation of Transmitter Signal (dB)
1	19
2	19
5	20
10	35
15	26
20	33
25	35
30	>48

Source: Reproduced from Weston, D.A., Antenna-to-antenna coupling on an aircraft: Mitigation techniques, *ITEM*, March 21, 2012. With permission.

TABLE 10.17

Cancellation Attenuation from 60 to 108 MHz

Frequency (MHz)	Attenuation of Transmitter Signal (dB)
60	26
70	26
80	33
90	26
100	35
108	49

Source: Reproduced from Weston, D.A., Antenna-to-antenna coupling on an aircraft: Mitigation techniques, *ITEM*, March 21, 2012. With permission.

The location of the T and the circuit depends on the location of the transmitter, receiver, transmit antenna, and receive antenna. The goal would be to reduce the amount of cable delay and therefore the amount of bulk delay required in the circuit. Two possible configurations are shown in Figure 10.35.

10.2.6.6 Cable Attenuation

Although not a mitigation technique when a cable connecting an antenna that has a low operating frequency carries an out-of-band interfering signal at high frequencies, the cable can exhibit considerable attenuation, thus reducing the level of interference. In some cases, this is so high that an additional filter is not required.

The attenuation of some typical cables is shown in Figures 10.36 through 10.39.

10.2.7 ANTENNA COUPLING FROM LIGHTNING

As lightning strikes the fuselage of an aircraft, very high currents result in high magnetic fields and high electric fields. These can couple into antennas mounted on the fuselage. A direct strike to an antenna is also possible if it is located higher than the fuselage. The major part of a typical blade

FIGURE 10.35 Delay, phase shift, and summation circuit configurations. (Reproduced from Weston, D.A., Antenna-to-antenna coupling on an aircraft: Mitigation techniques, *ITEM*, March 21, 2012. With permission.)

FIGURE 10.36 1 m RG214 attenuation.

antenna is connected to the fuselage with only a short vertical rod that has one end attached to the connector and the other end to the grounded blade. Thus, it is possible for the equipment connected to this type of antenna to survive a direct strike.

Other antennas exhibit a short circuit at the low frequencies inherent in a lightning strike.

The radome on many antennas contains a wire that is intended to carry the lightning current to the fuselage. For antennas mounted in the nose cone of the aircraft, the nose cone commonly also carries conductors for the same reason due to high voltage!.

For example, one-blade antenna specifies a 50 Ω impedance at RF and a DC short but does not specify how much current the short can carry. For other blade antennas, the DC resistance is open circuit. This indicates a capacitive coupling. If this capacitor is low enough in value, it may reduce the lightning-induced transient, assuming that it does not break down due to the high volatage!.

FIGURE 10.37   10 m RG214 cable attenuation.

FIGURE 10.38   1.2 m LM400 cable attenuation.

FIGURE 10.39   12 m LM400 cable attenuation.

Antennas that are most likely to be hit by lightning are long wire antennas and then monopole rod and blade antennas.

We consider an HF wire antenna approximately 10.2 m long and at an average height of 2 m above the fuselage and connected to the vertical stabilizer.

In a 2,000 A lightning strike to the fuselage, the average magnetic field incident on the HF antenna is approximately 10,600 A/m.

The induced voltage is given by

$$V = \mu_o A \frac{dH}{dt} \tag{10.7}$$

where

$\mu_o = 4\pi \times 10^{-7}$ H/m
H is the magnetic field in A/m
A is the loop area $= h \times l$

A 2 μS rise time is assumed.

The open-circuit voltage induced in the antenna is $4\pi \times 10^{-7} \times 10.2$ m $\times 2$ m $\times (10,615$ A/m/2 μS$)$ = 136,000 V.

A second antenna has an average height of 18.4 cm and a length of 41 cm. This antenna is close to the fuselage and is exposed to approximately 190,000 A/m.

Open-circuit voltage $= 10^{-7} \times 0.41$ m $\times 0.18$ m $\times (190,000/$A/m/2 μS$) = 136,000$ V

Lightning protection techniques are described in Section 5.9. Those most suitable for receiver and transmitter protection are gas discharge tubes with and without combination filter and 1/4 wave stub.

The textbook *Lightning Protection of Aircraft* provides a technical guide and a history of the FAA lightning protection regulations [28].

10.2.8 Scattering from Surfaces

When the wavelength of an incident field is much less than the size of the structure, at least five wavelengths in size, then the high-frequency radar cross section (RCS) prediction may be used to calculate the scattered fields from the object. When the object is nonconductive but has permittivity or is not perfectly conducting, some of the incident power is absorbed and the remainder scattered. When the object is perfectly conducting, then all the incident power is scattered.

Figure 10.40 shows the scattering from a cylinder.

As the incident wave excites current on the scatterer, the current results in a reradiation of power. The scatterer acts as an antenna with a capture area $\sigma = $ RCS.

Monostatic or backscattering is the case where the transmitting and receiving antennas are at the same location or are the same antenna, such as a radar antenna. Bistatic scattering is where the transmitting and receiving antennas are at different locations and so is most useful for coupling between antennas on an aircraft.

If the capture area of an object is known, then the ratio of incident power to scattered power can be determined.

The RCS or capture area of an object is determined by the physical size, polarization, and, in some cases, the angle of incidence of the illuminating wave and wavelength.

The σ for a sphere is very simple:

$$\sigma = \pi a^2 \tag{10.8}$$

where a is the area of the sphere and so σ is independent of the wavelength and the angle of the incident field.

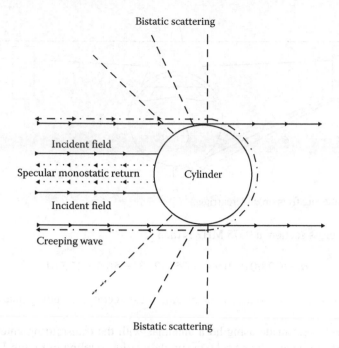

FIGURE 10.40 Cylinder scattering.

The common definition of RCS is

$$\sigma = 4\pi R^2 \left(\frac{E_s^2}{E_i^2} \right) = 4\pi R^2 \left(\frac{P_s}{P_i} \right) \tag{10.9}$$

where

E_s^2 is the scattered E field
E_i^2 is the incident E field
P_s is the scattered power
P_i is the incident power

The monostatic cross section of a cylinder is

$$\sigma = krl^2 \tag{10.10}$$

when

$0°$ is the angle between the cylinder and the incident field
$k = 2\pi/\lambda$
r is the radius
l is the length

In the case of the scattering from the cylindrical structure shown in Figure 10.41, the length of the steel rod in the undercarriage is 2.13 m and the diameter is 0.127 m. The rod is covered by glass fiber panels when the undercarriage is up, and for simplicity, the assumption is that the absorption and reflection from the fiberglass are negligible.

FIGURE 10.41 Scattering from an undercarriage.

The monostatic cross section at 9375 MHz is then

$$\sigma = (6.28/0.032) \times 0.0635 \times 2.13^2 = 56.5 = 17.5 \text{ dB}$$

There are a number of methods to derive the bistatic cross section, including measurements, UTD, and GTD analysis.

The measurements were made using horn antennas with the transmitting antenna at 7 m from the cylinder and the receiving antenna a 3.6 m simulating the coupling in Figure 10.34. The receiving antenna was located so that it faced the cylinder at azimuth angles of 6, 10, 15, and 20 from the specular point where the incident field impinged on the cylinder. In order to rotate the receiving antenna around the cylinder, it was oriented vertically with respect to the ground. A laser was used to align the antennas, and absorber was used on the ground between the antennas to reduce ground reflection. The measurements were also made with the cylinder absent to ensure that the coupling between the two antennas was much lower than the scattered field. The measurements were made with antennas both vertically polarized field at 90° to the length of the tube and 0° to the tube length.

One computer program, which is particularly useful in evaluating bistatic RCs, is the REMCON XFdtd 3D Electromagnetic Simulation Software, and the results using this were compared to the measurement for vertical polarization and are shown in Table 10.18.

The measured results for horizontal polarization are shown in Table 10.19.

Reference [16] describes the use of a 1/48 scale brass model and the NEC (described in Section 12.4.3 to predict the effect of the addition of a high-frequency surface wave radar [HFSWR] onboard a ship). Because of the long wavelength in the HF band (2–30 MHz), the entire ship will radiate as an HF antenna. The HFSWR interacts with the topside superstructure, including existing communication antennas. The NEC yields estimates of the technical parameters of the antennas mounted on ships. These technical parameters of an antenna are completely specified by its impedance, near fields, radiation pattern, and coupling to other antennas. Near fields are used to evaluate

TABLE 10.18
Comparison between Measured and Predicted RCS for Vertical Polarization

Angle Degrees from Specular Point	Measured σ (dB)	REMCON σ
6	3.1	6.5
10	1.1	2.54
15	−0.91	−0.46
20	−3	−3.5

TABLE 10.19

RCS for Horizontal Polarization

Angle Degrees from Specular Point	Measured σ (dB)
6	−3.9
10	−7
15	−8.9

radiation hazards (RADHAZs) to personnel, fuel, and ordnance. Either the NEC program or the brass model measurements may be used to calculate the antenna coupling; in Reference [16], it was the brass model measurements that were used. The two areas of concern are the receiver-adjacent signal (RAS) EMI, in which the receiver signal adjacent to the HFSWR may be interfered with, or the broadband transmitter noise (BTN), which is defined as the portion of the undesired transmitter spectrum (including BTN but not including narrowband spurious emissions) that lies within the nominal passband of the receiver. In this case, RAS was not an issue, but BTN was found to be a major potential problem. The preliminary EMC analysis was based on the assumption that the radar transmitter waveform is a rectangular pulse-train spectrum with a 50% duty cycle, 500 μs period, and 100 kHz bandwidth, as shown in Figure 10.42. The emission spectrum of this radar is therefore a sin *x*/*x* waveform. Assuming this rectangular pulse-train waveform, calculations of the radar transmitting spectrum at 2, 20 MHz (the radar transmitter-tuned frequency), and 30 MHz are −17, 67 dBm (67 dBm = 5 kW transmitted power), and −11 dBm, respectively, as shown in Figure 10.43. Based on the brass model measurements, the antenna couplings at 2, 20, and 30 MHz are assumed to be 25, 30, and 35 dB, respectively. Hence, the received interference power levels at the input of the communication receiver at 2, 20, and 30 MHz are −42, 37, and −46 dBm, respectively. Quasi-minimum noise (QMN) is a reasonable lower-bound estimate of mean levels aboard Navy ships; QMN is defined as some minimum atmospheric noise power that is contaminated with some background local noise. The greater of the QMN and the receiving system noise is used as the EMC designs goal for communication receivers. Comparing these interference power levels with the QMN, that is, the EMC design goal, we found that the excess interference power levels are 44, 152, and 74 dB at 2, 20, and 30 MHz, respectively.

This EMC study shows that if unmitigated, radar transmitter coupling would cause severe problems in communication receivers. There are two potential problems. The first problem is that the radar transmitter will generate enough BB noise to severely interfere with all communication

FIGURE 10.42 Rectangular pulse-train spectral occupancy. (From Li, S.T. et al., EMC study of a shipboard HF surface wave radar, *IEEE Electromagnetic Compatibility Symposium Record*, Atlanta, GA. © 1995, IEEE.)

- If unmitigated, radar transmitter coupling would cause severe problems in communication receivers

		2 7	20 9	30 7
Xmtr power	(dBm)	−17	67	−11
– Antenna coupling	(dB)	−25	−30	−35
Power at receiver input	(dBm)	−42	37	−46
External noise	(dBm)	−86	−115	−120
Excess above external noise	(dBm)	44	152	74

Two problems: ❶ ❷

❶ Off HFSWR frequency, receiver is severely interfered

❷ Could present burnout or saturation problem to some receivers

- Solution to Problem 1 is to filter exciter signal prior to transmission

FIGURE 10.43 Coupling of the transmitted radar signal into HF communication circuits: the problem. (From Li, S.T. et al., EMC study of a shipboard HF surface wave radar, *IEEE Electromagnetic Compatibility Symposium Record*, Atlanta, GA. © 1995, IEEE.)

receivers operated at the entire HF band (2–20 MHz). The second problem is that the radar transmitter could burn out or saturate some EW receivers at radar on-tune frequency (20 MHz).

A solution to the first problem is described in Figure 10.44. We propose that communication receivers will not be interfered with at frequencies more than 5% on either side of the radar transmitter frequency. At 19 MHz, the radar transmitter power is 9 dBm; antenna coupling is −29 dB; BB noise power at receiver input is −19 dBm; QMN is −114 dBm; and excess interference level above

- Proposed requirement: Specify that communication receivers will not be interfered with at frequencies more than 5% on either side of the radar transmitter frequency

  – i.e., HFSWR takes 10% "bite" out of the HF communicator's spectrum

Xmtr power	(dBm)	9
– Antenna coupling	(dB)	−29
Power at receiver input	(dBm)	−19
External noise	(dBm)	−114
Excess above external noise	(dBm)	95

- Hence, need filtering on transmit that will provide 95 dB of attenuation at frequency that is separated by 1000/50 = 20 = 1.3 decades

- Requires filter with

$$\left(\frac{95\ dB}{1.3\ decades}\right)\left(\frac{1\ pole}{20\ dB}\right) \sim 4\ poles$$

Exciter → 2-pole filter → Power amplifier → 2-pole filter → Antenna

FIGURE 10.44 Coupling of transmitted radar signal into HF communication circuits: the solution. (From Li, S.T. et al., EMC study of a shipboard HF surface wave radar, *IEEE Electromagnetic Compatibility Symposium Record*, Atlanta, GA. © 1995, IEEE.)

QMN is 95 dB. Hence, filtering on the radar transmitter was required that will provide 95 dB of attenuation at a frequency that is separated by 1 MHz/50 kHz = 20 = 1.3 decades. A four-pole filter will provide the required 95 dB attenuation.

A second prediction is described in Reference [16] of the interference generated by the shipboard HF communication system coupling into the HFSWR receiver. In this second prediction, in addition to BTN and RAS noise, clutter cross modulation is examined. Clutter is the unwanted echoes, typically from the ground, structures, sea, rain or other precipitation, chaff, birds, insects, and aurora (also known as background return). Clutter cross modulation results when a high-level signal from an onboard communication transmitter transfers its modulation to the clutter through cross modulation occurring in the front end of the receiver.

A 1/48 on-scale brass model was also used in Reference [17] to predict the coupling between a 2–6 MHz transmit fan antenna and a 2–30 MHz receive antenna on a ship. Although the NEC is used to design new antennas and to provide a preliminary determination of the antenna performance and isolation, it is a 1/48th scale that is used to refine the antenna performance and finalize the locations. Typically, the antenna design is not included in the basic design of new construction ships and must be developed after the topside design is fixed. The communication suite, both transmit and receive, is predefined, and even the antenna locations are presented to the antenna designer as a fait accompli. The brass model hull is constructed of laminated wood overlaid with brass. The topside features are machined from brass to 1/10,000 in. tolerances. Antennas are constructed to provide the same electrical characteristics at the increased frequency, due to the scale, but may not have the same physical appearance. Calibrated cables are connected to each cable under test.

The antenna performance covers transfer impedance (voltage–standing-wave ratio [VSWR]), antenna pattern and gain, RADHAZ prevention, and isolation (decoupling) between transmit and receive systems. The VSWR determines the ratio of power input to the antenna versus reflected power from the input of the antenna and is one of the most critical parameters. For a shipboard transmit antenna, VSWR must be below 4:1 across the frequency range of the antenna, including any matching network. However, perturbations introduced by metallic elements in the ship's superstructure often prevent meeting this goal. A shipboard receive antenna does not have the VSWR constraints encountered in transmit antenna design, since the goal is to optimize receive system efficiency. High-frequency receive systems do not necessarily need the most efficient antenna but instead require an antenna that matches receive system performance to the QMN curve. Since it is futile to attempt to receive signals below the QMN, the frequency response of an HF receive antenna should be approximately 10 dB below the QMN to allow for antenna mismatch and cable and system losses.

Overall antenna performance is most easily characterized by the antenna pattern that is the spatial gain of the antenna. The pattern is described as a 3D plot or a number of 2D plots and is usually referenced to a theoretical λ/4 monopole on a perfect ground plane. Antenna patterns are severely affected by surrounding structures, so a perfect antenna is not attainable in a shipboard environment. An acceptable antenna is typically defined as one with an antenna pattern with less than 6 dB nulls at any azimuth over less than 20% of the frequency range of the antenna. A 6 dB null will result in transmission of only 25% of the power in the desired direction.

Reference [16] describes the isolation between the 2–6 MHz fan antenna and the 2–30 MHz receive antenna and shows that the intermodulation interference (IMI) generated in the victim receiver, from inadequate antenna decoupling, was unacceptably high. Relocation of either the transmit or the receive antenna, or both, was the solution offered for the potential EMI problem. After relocation, both the antenna isolation and the VSWR of the antennas were recalculated.

The field strengths were also characterized in close proximity to the transmit antenna, using the 1/48th scale brass mode, during the antenna design phase to allow a RADHAZ assessment. RADHAZs include HERF, HERO, HERP, and RF burn. The maximum field strength levels differ for each type of RADHAZ. The field strength data were used by the antenna engineer to design and locate the antenna, so that it will not exceed allowable levels of HF radiation close to personnel, fuel, and ordnance areas or so it will at least provide warnings of possible dangers.

10.2.9   CASE STUDY 10.1: HAZARDOUS ZONE AROUND A TRANSMITTING PARABOLIC REFLECTOR ANTENNA

10.2.9.1   General

The most accurate method available to ascertain safe distances for personnel from a transmitting antenna is by the measurement using, typically, a calibrated BB probe. A typical test method is described as follows:

1. The BB probe is verified just prior to the measurement by use of a calibrated antenna and signal source at the frequency of interest.
2. The measuring personnel are located outside of the predicted hazardous zone.
3. The antenna is initially operated at below maximum power and the power is increased in steps. Where the measured field strength approaches the safety limit, the personnel then move back from the antenna.

The advantage of the measurement over calculations is that the gain and directivity of an antenna when located close to structures on the ground and the reflection or absorption due to ground and structures are accurately accounted for. In the case where measurements cannot be made, or in order to predict the safety zone before the measurement, a worst-case calculation based on theory and measured data may be made for a specific antenna. To make the calculation less prone to human error and to facilitate changing any of the variables, a computer program was written, and this was used to generate the polar plots of the hazardous zone.

10.2.9.2   Safety Limit

The safety limit used in the calculation is that published by Health and Welfare Canada in "Safety code #6, recommended safety procedures for the installation and use of radio frequency and microwave devices in the frequency range 10 MHz to 300 GHz." The limit for the general public over the 14 GHz frequency range when averaged over a 1 min period is 61.4 V/m, or 0.163 A/m, or 10 W/m^2. Because the limit is for average field strength, the following calculations include a conversion of peak transmitted power into average power.

10.2.9.3   Calculation of the Hazardous Zone

A worst-case scenario is when an antenna is located at a low height above a perfectly conducting ground plane. A more realistic scenario for a parabolic antenna is for it to be mounted at a considerable height above a ground. Two hazard zones are calculated for the Andrew P6-144D antenna, one where the antenna is in close proximity to a perfectly conducting ground plane and the second where the antenna is in close proximity to soil with high permittivity. The case where the antenna is mounted at some height above ground and where structures are located close to the antenna are also discussed.

The worst-case analysis takes the absolute maximum power capability of the transmitter; the minimum attenuation in the antenna cable, feed flange, waveguide, etc.; and the maximum gain for the antenna to predict the E field around the antenna.

Andrew Antenna Company Ltd. publishes a radiation pattern plot for the P6-144D antenna that is given in Figure 10.45. The envelope of measured antenna gains is plotted, resulting in a worst case (i.e., maximum gain). Either of the polarizations that results in maximum gain at any frequency has been chosen, which is consistent with our worst-case approach. The radiation pattern envelope was measured by the Andrew Antenna Company Ltd. with the antenna mounted on a tower with a 60–150 ft height.

From a private communication [10], we learned that the peak power available from the TWT amplifier used to power the P6-144D antenna may be as high as 130 kW (rated peak power = 125 kW). The minimum attenuation in the waveguide and feed flange is unlikely to be less than 1 dB, and this was taken as the worst case. The maximum duty cycle of the transmitted power, from

FIGURE 10.45 P6-144D antenna radiation pattern envelope. (Courtesy of Andrew Antenna Company Limited, Ashburn, VA.)

Reference [10], is 0.1%. The maximum average input power to the antenna (P_{av}) is given by the peak output power of the amplifier (P_{pk}), minus the attenuation, times the duty cycle. Thus, for the case under consideration,

$$P_{av} = P_{pk} - 1 \text{ dB} \times 0.1\% \qquad (10.11)$$

That is,

$$P_{av} = 130 \text{ kW} - 26.74 \text{ kW} = 103.26 \text{ kW} \times \frac{0.1}{100} = 103.26 \text{ W}$$

For a low-gain, low-directivity antenna, such as a dipole, and when the direct radiated path length is close to the reflected path length with the direct and reflected field in phase, the maximum theoretical increase in field strength due to superposition of the two fields is 6 dB in the far field. At close proximity and for vertical polarization that attenuates off-axis radiation due to the vertical directivity of the dipole, an increase of greater than 6 dB may be expected [2].

The 6 dB enhancement ignores any reflections from conductive structures such as buildings and wire fences in close proximity to the antenna, which may further reinforce or reduce the field at any given point. When an antenna is located close to a conductive ground, the characteristics of the antenna may change due to mutual coupling between the antenna and the ground plane, as well as pattern changes, spurious resonances, and other anomalies caused by proximity to the ground. These effects are difficult to predict, so measured results are used in an attempt to derive a worst-case field enhancement due to ground reflections and other factors.

The measurements from which the data were used were made in order to characterize an open-field test site and to arrive at site attenuation. Invariably, such test sites incorporate a conductive ground plane, typically constructed of wire mesh, below at least one of the antennas. The disadvantage of using the results of such measurements is that both a transmitting and a receiving antenna are used, so mutual coupling exists between them, whereas our concern is the *E* field incident on personnel, where the mutual coupling will be much less pronounced. The second disadvantage is that the upper frequency limit for such measurements is 1 GHz, whereas in this case study, the frequency range of interest is from 14 to 15.2 GHz. Because of potential errors, we have used the results of these measurements with a safety factor to provide a worst-case field enhancement. In Reference [3], the distance between the two antennas was chosen to ensure that mutual coupling is not a significant factor.

Assuming that the antenna has been accurately calibrated in an anechoic chamber, the difference between the published antenna factor and that measured on an open-field test site is a measure of the enhancement or reduction in field due to the open field test site characteristics.

From Reference [12], the decrease in antenna factor or increase in field strength at 215 MHz, measured with a log-periodic antenna, due to the open-field characteristics is 6 dB. Additional ambient reflections are included in the measurement. The measurements and calculations given in Reference [4] indicate an enhancement of 5 dB at 950 MHz due to reflections.

Due to the uncertainties involved in the application of such data, a very conservative enhancement of 10 dB was used for the antenna over a conductive surface. This was reduced to 7.5 dB for the situation where a low-conductivity, high-permittivity soil is the ground plane, as discussed in Case 2.

Reference [13] includes a curve (Figure 10.46) of the reflection coefficient versus angle of incidence with vertical and horizontal antenna polarization for average earth, blacktop, and steel. Figure 10.46, reproduced in this report, indicates the importance of including the angle of incidence in the calculation; also, where blacktop is the ground plane, the reflection coefficient is less than for average soil. Figure 10.46 is valid for 30 MHz only, whereas the reflection coefficient at 14.8 GHz is much different, as seen in Case 2.

FIGURE 10.46   Reflection coefficient versus angle of incidence for various materials. (From Gonshor, D.V., Attenuation, transmission and reflection of electromagnetic waves by soil, *IEEE International Symposium on Electromagnetic Compatibility*, Atlanta, GA. © 1987, IEEE.)

Case 1: Antenna located close to a highly conductive ground plane

A wave propagating in air as it impinges on a surface with either high conductivity or high permittivity is reflected as a result of the lower impedance of the surface compared to the wave impedance. We assume that at 14.8 GHz (λ = 2 cm) the reflective surface is in the far field of the antenna, where the wave impedance Z_w = 377 Ω. The barrier impedance of copper is 0.045 Ω/sq at 14.8 GHz, and for steel, it is approximately 1.7 Ω/sq from

$$|Z_m| = 369 \sqrt{\frac{\mu_r f}{\sigma_r}} \, (\mu\Omega/sq) \tag{10.12}$$

where
μ_r is the permeability relative to copper
σ_r is the conductivity relative to copper
f is the frequency (MHz)

Thus, the barrier impedance for both copper and steel is lower than the wave impedance, and the reflection coefficient is at a worst case (Case 1). As discussed earlier, the field enhancement due to reflections and other sources is an estimated worst case of 10 dB; that is,

$$20 \, \log \frac{E_{dir} + E_r}{E_{dir}} = 10 \text{ dB}$$

where
E_{dir} is the predicted direct field
E_r is the reflected, in-phase field incident on the same point

The 10 dB enhancement is also used to account for additional spurious effects, such as antenna loading.
    The equation used in the computer program that enables the generation of the plot of the hazard zone around is derived as follows. The maximum safe E field is 60 V/m, so the maximum plane-wave power density is obtained from

$$W = \frac{E^2}{377} = \frac{60^2}{377}$$
$$= 9.55 \text{ W/m}^2 = 1 \text{ mW/cm}^2 \tag{10.13}$$

The power density, some distance r from the source, is given by

$$W = \frac{P_{av}GF_e}{4\pi r^2} \tag{10.14}$$

where
r is the distance in meters
G is the antenna gain (numeric)
F_e is the numeric power enhancement due to reflections, etc.

From the $W = E^2/377$ equation, it may be seen that a 10 dB increase in E field results in a 10 dB increase in power density. For example, a 10 dB increase in a 10 V/m field (0.26 W-m) results in a 31.6 V/m (2.6 W/m) E field, from

$$20 \, \log \frac{31.6}{10} = 10 \, dB$$

The increase in power density is thus

$$10\log\frac{2.6\text{ W}}{0.26\text{ W/m}} = 10\text{ dB}$$

The safe distance from Equation 10.14 is given by

$$r = \sqrt{\frac{103.3GF_e}{9.55 \times 12.56}} \tag{10.15}$$

The maximum P6-144D antenna gain is 46.3 dBi in the main lobe of the antenna. Figure 10.45 shows the off-axis decrease in gain versus azimuth degrees from main lobe. The antenna gain in dBi was converted into gain numeric and used with Equation 10.15 to calculate the safety zone around the antenna, inside of which is the hazardous zone. Figure 10.47 is a polar plot of the safety distance contour around the antenna. A worked example of the calculation is made as follows to ensure the validity of the computer program.

The on-axis gain of the antenna is 46.3 dB, F_e is 10 dB; thus, the gain numeric times F_e numeric is $42{,}658 \times 10 = 426{,}580$, which results in a safe distance, using Equation 10.15, as follows:

$$r = \sqrt{\frac{103.3 \times 426{,}580}{120}} = 605\text{ m}$$

which agrees with the computer-generated result.

Case 2: Antenna located above a ground with high permittivity

The lowest resistivity of soil is approximately $100\ \Omega$-cm, which is much higher than the $1.72\ \mu\Omega$-cm for copper. The lowest soil impedance at 14.8 GHz, due to the conductivity of the soil, is therefore much higher than $377\ \Omega$, and negligible reflection occurs due to the conductivity of the soil. However, a surface that exhibits a higher permittivity than air also exhibits a lower impedance than $377\ \Omega$. Thus, when a plane wave in air impinges on such a surface, a reflection occurs. The reflection coefficient R_c, from Reference [14], for horizontally polarized waves is given by

$$\left.\frac{E_\tau}{E_i}\right|_H = 1 - \frac{2}{1 + \dfrac{Z_w}{Z_s} \times \dfrac{\cos\Phi}{\cos\theta}} \tag{10.16}$$

and for vertically polarized waves, it is given by

$$\left.\frac{E_\tau}{E_i}\right|_V = 1 + \frac{2}{1 + \dfrac{Z_w}{Z_s} + \dfrac{\cos\Phi}{\cos\theta}} \tag{10.17}$$

where
  Z_w is the wave impedance $= 377\ \Omega$
  Z_s is the soil impedance
  θ is the angle of incidence relative to vertical, but not $90°$
  Φ is the angle of transmitted wave in the soil relative to vertical, but not $90°$

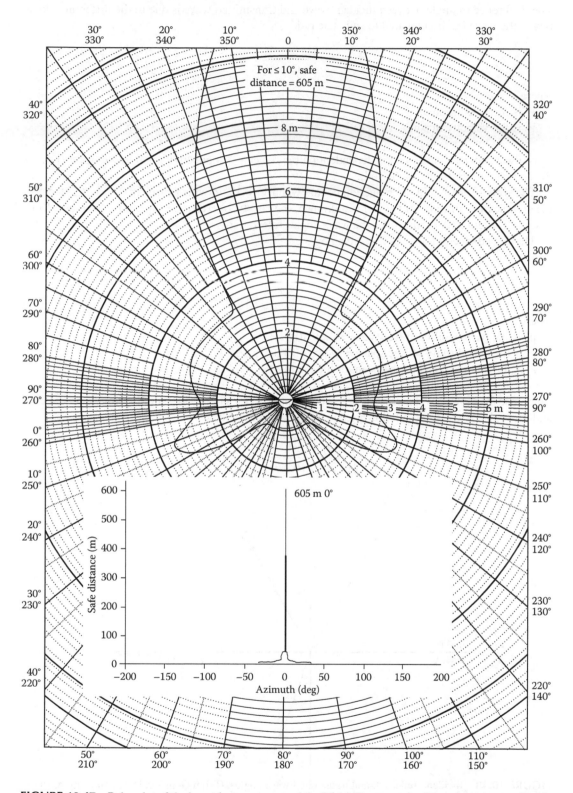

FIGURE 10.47   Polar plot of the hazardous zone around the P6-144D antenna in close proximity to a highly conductive ground.

The difference in angle between incident wave and transmitted wave is due to the different velocities of the wave in air compared to soil. The ratio is

$$\frac{\sin\theta}{\sin\Phi} = \frac{V_1}{V_2}$$

where
  V_1 is the velocity in air
  V_2 is the velocity in soil

V_2 is always lower than V_1 and is dependent on the dielectric conductivity and loss tangent (i.e., on the complex permittivity of the soil).

The worst-case maximum reflection occurs when $V_1 = V_2$, in which case $\sin\theta = \sin\Phi$. Because the complex permittivity of the soil is unknown, the $\cos\Phi/\cos\theta$ term in Equations 10.16 and 10.17 is set to 1. Figure 10.48 illustrates the wave angles referred to.

The impedance of the soil, Z_s, is given by

$$Z_s = \sqrt{\frac{\mu_0}{\varepsilon_0\varepsilon_r}}$$

where
  μ_0 is the permeability of free space = $4\pi \times 10^{-7}$ (H)
  ε_0 is the permittivity of free space = $1/36\pi \times 10^9$ (F)
  ε_r is the relative dielectric constant of the medium

Reference [5] shows a worst-case dielectric constant of 20 for sandy soil (humidity 16.8%) at 10 GHz, and this value is used in the following equation for soil impedance:

$$\sqrt{\frac{4\pi\times10^{-7}}{(1/36\pi\times10^9)20}} = 84.2\ \Omega$$

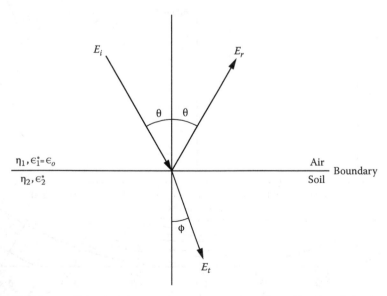

FIGURE 10.48   Incident, reflected, and transmitted wave angles. (From Gonshor, D.V., Attenuation, transmission and reflection of electromagnetic waves by soil, *IEEE International Symposium on Electromagnetic Compatibility*, Atlanta, GA. © 1987, IEEE.)

FIGURE 10.49 E_{tran}/E_{inc} for normal incidence. (From Gonshor, D.V., Attenuation, transmission and reflection of electromagnetic waves by soil, *IEEE International Symposium on Electromagnetic Compatibility*, Atlanta, GA. © 1987, IEEE.)

Using 84.2 Ω in either Equation 10.16 or 10.17 results in a reflection coefficient R_c of 0.635 (i.e., 0.635 of the incident wave is reflected and 0.365 is transmitted or absorbed in the soil). Thus, the ratio of E_{trans}/E_{inc} in decibels is 20 log 0.365 = −8.75 dB, which corresponds well with Figure 10.49. If we assume that the 10 dB (3.162) field enhancement is due to reflection only, ignoring antenna-to-ground proximity effects, using the 10 dB field enhancement, the reflected wave E_r can be found from E_t, the total E field, as follows: $E_t = 3.162E_{dir}$, where E_{dir} is the predicted field strength without enhancement; therefore, E_r, may be found from $E_t = E_{dir} + E_r$ and is $2.162E_{dir}$.

When the reflection coefficient is included in the equation for E_r, we get

$$E_r = E_{dir}\,2.162 \times R_c$$

Therefore,

$$E_r = E_{dir} \times 2.162 \times 0.635 = 1.372E_{dir}$$

$$E_t = E_{dir} + E_r = E_{dir} + 1.372E_{dir} = 2.372E_{dir}$$

Thus, the 10 dB enhancement (F_e) is reduced to 7.5 dB. We have assumed a worst-case normal angle of incidence of the transmitted wave with the soil surface. Be that as it may, from Reference [14], significant reflection also occurs for horizontally polarized waves with low angles of incidence with respect to the surface, and our worst-case approach is therefore valid in close proximity to the antenna. The polar plot of the safety zone with the antenna high-permittivity soil is shown in Figure 10.50.

10.2.9.3.1 Conclusions

In order to err on the side of safety, we have used a worst-case approach in deriving a hazard zone around a specific transmitting antenna. Maximum theoretical field enhancement due to a reflection from highly conductive surfaces is 6 dB. The measurements on open-field test sites where the surface conductivity is lower but where some ambient reflections occur result in measured field

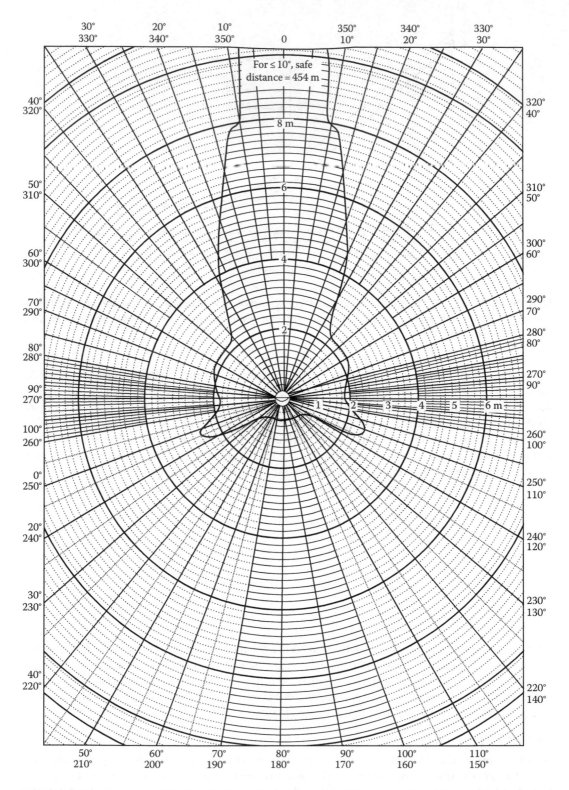

FIGURE 10.50   Polar plot of the hazardous zone around the P6-144D antenna in close proximity to ground with high permittivity.

enhancements of 5 and 6 dB. One major unknown is the effect on the antenna characteristics of proximity to ground. Due to the possibility of these spurious effects, the enhancement was increased to 10 dB for the case where the antenna is located above a highly conductive ground and to 7.5 dB where a high-permittivity ground is present. Where reflective structures exist around the antenna, the enhancement at certain points may be a further 6 dB, and a revised hazard zone may be computed from Equation 10.15.

For a more accurate estimate of the E field at any point around the antenna in the presence of structures, the General Electromagnetic Model for the Analysis of Complex Systems (GEMACS) program may be used. Using GTD formulation, a total field is computed from superposition of fields from direct, reflected, and diffracted sources. The disadvantage of GTD is that only perfectly conductive surfaces may be modeled. The MOM technique available in GEMACS is capable of modeling surface impedance but is not practical for large surfaces at high frequency.

When the antenna is mounted on a mast and the following requirements are met, the area around the mast out to any distance is inherently at a safe level of 60 V/m or less. The requirements are the antenna is at least 4 m above personnel (i.e., approximately 6 m above ground), the antenna center axis is either parallel to the ground or pointing upward, the ground under the main lobe of the antenna does not rise for 605 m, and no vertical conductive surfaces are within close proximity of the antenna or the main lobe of the antenna.

10.3 AMBIENT SITE PREDICTIONS AND SURVEYS

An ambient site prediction or survey is used when a receiving system or potentially susceptible equipment is to be installed at the site. In this section, the ambient due to intentional sources, such as communication transmitters and radar and the coupling to a receiver, is considered.

If the locations of antennas, antenna gains and directivity, transmitter power and frequencies, elevation of the ground, and presence of structures around the site are known, then a prediction of the electromagnetic ambient at a site may be made. A limitation in the prediction approach is that very often the information is limited. For example, the directivity or polar plot of the gain of an antenna is not always available, and some assumptions must be made concerning the sidelobes or backscatter of the antenna. The presence of hills or structures can make a significant difference to the predicted coupling between antennas and to absorption and shading of the receiving antenna; when this information is not available, the accuracy of the prediction suffers. An ambient site survey is the more accurate method of site evaluation. In a survey, a number of antennas, preamplifiers, and, typically, a spectrum analyzer are used to measure the ambient.

Directional measurement antennas are rotated through 360° with both horizontal and vertical polarizations of the antenna. For measurements at frequencies that are in band for the proposed receiver, the measurements should be made, where feasible, with the antenna and system that will be installed at the site. For example, if the system comprises an input filter, an LNC, or a preamplifier and filter, then these should be used in the site evaluation. By using a representative antenna and system, the antenna gain and directivity, filter effectiveness, and out-of-band or spurious responses of the receiving system will be included in the measurement.

This approach is usually feasible only with small antennas and lightweight receiving equipment. When this approach is ruled out, the measurement antenna should have a directivity and gain as close to those of the actual antenna as possible, and the gain and noise floor of the measuring equipment should be at least as good as those of the actual system.

One potential limitation that the ambient site survey has is that not all the transmitters may be operating during the time in which measurements are made. To minimize the risk of missing an emitter, the frequencies of the transmitters should be ascertained and an identification of "off-air" transmitters should be made. Otherwise, measurements should be made over a period of hours and repeated over a period of a day or two to increase the possibility that all transmissions are intercepted.

10.3.1 PASSIVE INTERMODULATION

Intermodulation products are generated when two or more signals mix in devices with nonlinear junctions. When these intermodulation products fall in band for a colocated receiver, an EMI problem may exist.

PIM may occur in any metal structure in proximity to a receiving antenna, such as the antenna structure, railings, towers, or other metallic surfaces.

One common source of a nonlinear junction is either a loose joint or oxidization of metal. A structure that includes ferrous materials, which has a nonlinear magnetic hysteresis, or carbon fiber, which has nonlinear resistivity, may also exhibit PIM and this is often an order of magnitude higher than the joint-generated PIM level.

The composite material on an aircraft is typically glass fiber, aramid fiber (Kevlar), or aluminized glass fiber. The Kevlar and Kevlar should not be a source of PIM, whereas it is unknown if aluminized glass fiber, where the aluminum is woven into the fabric, is a source of PIM.

On an aircraft, the ferrous materials are typically in the landing gear, flap rods/tracks, and door handles with the landing gear and flap rods/track the most likely source.

Figure 10.51 illustrates the two transmitting antenna, the PIM structure and the reradiated power to the receiving antenna.

When a metal surface is painted; passivated with a coating such as Alodine, Iridite, and Oakite; or plated with zinc, the probability of PIM is reduced. When an aircraft is repainted, consideration should be given to passivate critical areas such as the landing gear before repainting.

Composite material may be coated with electroless copper followed by copper plate and finally a nickel flash and then painted. Artcraft of California has coated a number of plastic enclosures that have been qualified to military EMI requirements.

Unfortunately, PIM products have thresholds that often change with temperature and power levels. Typically, the higher the level of radiated power incident on a surface, the higher the level of PIM, although they often behave in an unpredictable way versus power level.

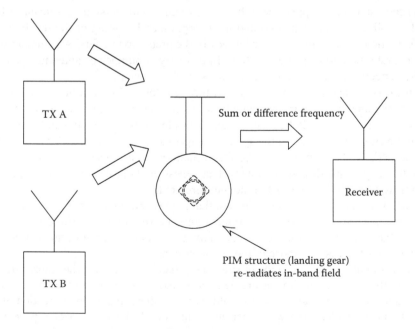

FIGURE 10.51 Illustration of PIM.

Any two sources of emissions of sufficient power may cause intermodulation products, which are then reradiated and detected as a signal. The intermodulation products are generated by $mf_1 + nf_2$, where m and n are positive or negative integers from 1 up. The sum of m and n defines the order, thus $1f_1 \pm 2f_2$ = third order.

Third harmonic (e.g., $3f$) third-order intermodulation products (e.g., $2f_1 \pm f_2$, $2f_2 \pm f_1$) are reported to likely cause problems [25], although second-, third-, and fourth-order intermodulation products (e.g., $f_1 \pm f_2$, $f_1 \pm 2f_2$, $2f_1 \pm f_2$, and $2f_1 \pm 2f_2$) should also be checked in the PIM frequency analysis.

Generally speaking, but not always, the higher the order of intermodulation product, the lower the PIM level. For this reason, we often concentrate on second, third, and fourth order only, especially when the PIM levels are low. Also, as so many potential PIM sources have been identified, the analysis has been confined to only two transmitters operating simultaneously. As the underlying physics of PIM is very complex, time varying, dependent on ambient temperature, and power, among other variables, a heuristic approach has been used by some investigators to determine the potential sources on the aircraft.

Ferromagnetic PIM is an order of magnitude greater than PIM caused by metal oxidization in contacts. The interference may be transmitter related, namely, harmonic, both to the fundamental and master oscillator and/or a BB noise emission. Amplifiers simultaneously amplifying a number of signals may also generate intermodulation products (cross modulation). Receiver related intermodulation products may be the result of signals not sufficiently attenuated by filters (i.e., those frequencies for which cross modulation may be a problem).

The level of PIM versus input power, measured from two samples of ferrous materials as well as stainless steel with a tight new joint and a corroded old joint, is shown in Figure 10.52 from Reference [25]. The predicted level of PIM from a metal to metal junction is shown in Figure 10.52. Both are expressed in dBc. Thus, if the PIM product level is −130 dBc and the input power is 40 dBm, then the PIM received power level is −90 dBm.

Reference 28 investigates the third-order intermodulation level in dBm of graphite fiber composite materials. One of the materials was composed of chopped graphite fiber mat consisting of

FIGURE 10.52 Measured third-order PIM for two samples of ferrous material. (From Joffe, E.B. and Gavan, J., High power broadcast skywave transmitters desensitization effects analysis, *IEEE Electromagnetic Compatibility Symposium Record*, Washington, DC.)

TABLE 10.20

Graphite Material PIM

Material	Product	F_1	F_2	F_{IMI}	Run #1 P_{IMI} (dBm)	Run 21 P_{IMI} (dBm)
Graphite	$2*f_2-f_1$	2.2	4.65	7.1	−101	−108
Composite	$2*f_2-f_1$	2.2	4.65	7.1	−111	−130
Graphite	$2*f_1-f_2$	9.81	16.3	3.32	−110	−117
Composite	$2*f_1-f_2$	9.81	16.3	3.32	−122	−118
Graphite	$2*f_1-f_2$	11.43	29.8	6.94	−116	−96
Composite	$2*f_1-f_2$	11.43	29.8	6.94	−111	−118

TABLE 10.21

Potential Sources of PIM Test Matrix

Case #	Transmitting System A	Aircraft Location	Transmitting System B	Aircraft Location	Frequency Combinations Tested
1	Maritime		SLAR and	B	f_2-f_1
	Search Radar	A	Weather Radar	C	(B − A)
2	Mission V/UHF		Cockpit HF,	E	$f_1 \pm f_2$
		D	Mission HF	F	(A ± B)
4	DME	G	TCAS	H	
	DME	G	ATC	I	$f_1 \pm f_2, 2f_1 \pm 2f_2$
	DME	G	IFF	J	(A ± B, 2A ± 2B)
5	DME	G	TCAS	H	
	DME	G	ATC	I	$2f_1 \pm f_2$
	DME	G	IFF	J	(2A ± B)

a random dispersion of short, nominally 1 in. long fibers impregnated with a brominated vinyles-ter resin. The other was made of a continuous strand, graphite fiber, and plain weave fabric also impregnated with a brominated vinylester resin.

Table 10.20 shows the third-order intermodulation product for $2 \times f_2 - f_1$ and $2 \times f_1 - f_2$. The input power was 30 dBm.

A matrix of potential PIM transmitting systems, which may result in PIM, should be generated and an example is provided in Table 10.21.

A table of test cases should also be generated as shown in Table 10.22.

The distance between transmitters and the PIM structure and from the PIM structure to the receiving antenna is required in the calculation. When the PIM product is not in the line of sight (e.g., from the top of the fuselage to the side), some shading will occur.

The response of the antennas, the separation of the antennas, and the proximity of ferrous material and composite on the actual aircraft will show the potential for PIM and this is provided as follows.

An example of the PIM analysis worksheet for transmitters where the sidelobe or radiation in the elevation is incident on the landing gear, which is stowed beneath fiber glass panels and then reflected to the receiving antenna as shown in Figure 10.41, is provided in Table 10.23.

The transmitter A frequency range is 8900–9500 MHz. The transmitter B frequency is 9325 MHz. The difference frequency when transmitter A is at 9070 MHz is 9325 − 9070 = 255 MHz, which is in band for the receiver.

TABLE 10.22

PIM Test

Maritime Search Radar (A) Frequency and SLAR and Weather Radar (B) (TX at 9375 MHz)	Victim Device Name with Antenna Number (n)	Frequency Range of Potential Conflict (MHz)
8900–9345	Cockpit V\UHF on 30–88, 118–174, 225–400, 400–600 MHz	30–475
8900–9275	Mission V\UHF on 100–512 MHz	100–475
8969–9345	V\UHF Direction finder on 30–406 MHz	30–406
9040–9045	ILS Glideslope on 329–335 MHz	330–335
9219–9212	Automatic Identification System on 156/163 MHz	156–163
9238–9257	Cockpit VHF on 118–137 MHz	118–137
9287–9373	Cockpit HF, Mission HF Mission VLF/HF on 0.1–30 MHz	2–88

The steel in the landing gear is cylindrical and has a length of approximately 2.14 m with a diameter of 0.127 m. Based on the RCS discussed in Section 10.2.8, the scattered field at 3.6 m on the receiving antenna is derived from Equation 10.20.

The incident E field is given by

$$E_{inc} = \sqrt{\text{Power on PIM surface} \times 377} = \sqrt{119 \times 377} = 212 \text{ V/m} \tag{10.18}$$

And the scattered field from Equation 10.9 is given by

$$E_s = \frac{\sigma \times E_{inc}}{4\pi \times R^2} \tag{10.19}$$

Using a worst case σ of 3.1 dB (2.05) at 6°, the E field incident on the receiving antenna is 2.7 V/m and the power density is 0.019 W/m².

The power on the structure as a result of the incident field is given by

$$P = \sigma \times W \tag{10.20}$$

where

σ is the RCS or capture area
W is the incident power density

At 6°, 119 × 2.05 = 244 W = 54 dBm.

From Figure 10.53, the approximate PIM level is −145 dB.

The power at the input of the receiver is −173 dBm, which is so low that no EMI problem exists.

10.3.1.1 Mitigation

Some of the real-world sources of PIM are

- Poor alignments of parts
- Moving structures that are not adequately bonded
- Insufficient or incomplete cleaning of parts before plating

TABLE 10.23
Example of PIM Worksheet

TXA Power Output at Frequency z = 9070	TXA Antenna Gain (dB)	Distance to PIM Surface (m)	Incident Radiated Power on Surface (W/m²) From TXB	TXB Power Output at 9325 MHz	TXB Antenna Gain in Elevation (dB)	Distance from TXB to PIM Surface	Incident Radiated Power on PIM Surface from TXB (W/m²)
8000	Sidelobe = 0	3.6	49	25,000	−10	2.3	119
Scattered Power at RX Antenna	Power in Structure (dBm)	Second-Order PIM Level dB (Difference Frequency)	Reradiated Power (W/m²) at RX Antenna at 255 MHz (W/m²)	Distance between PIM Surface and RX Antenna (m)	Frequency at RX Antenna (m)	Gain of RX Antenna (dB)	Received Power (dBm)
	54	−145	6e-17	7	255	1	−173
Based on Analysis in Section 10.2.8 (W/m²)							
0.019							

FIGURE 10.53 The predicted third-order PIM from a nonlinear junction.

- Contaminated plating tanks
- Poor plating adhesion
- Dissimilar metals in direct contact
- Plating nonuniformly applied and of sufficient thickness (high resistance)
- Material that has not been in the plating bath long enough (high resistance)
- Oxidization

A case study of PIM and other mitigation techniques are described in Section 10.3.2.

10.3.2 CASE STUDY 10.2: ELECTROMAGNETIC AMBIENT SITE PREDICTIONS AND SITE SURVEYS

The emission frequencies of interest in the following site surveys were the fundamental frequency of a satellite communication system at 1544 MHz, the IF at 45 MHz, and the reference at 10 MHz. In addition, field strengths of 1 V/m or greater at any frequency can result in direct coupling to cables and equipment, which can result in EMI. At one site, cross modulation due to a transmitting antenna located 56 m from the proposed site of the satellite communication receiving antenna was predicted as follows.

The transmitter operates at a frequency of 1511.125 MHz at 2.5 W and feeds a yagi antenna with a gain of 17.5 dBi. The transmitting antenna was predicted to be broadside onto the receiving antenna. The sidelobe of the yagi antenna is likely to be, at a minimum, 13 dB down on the main lobe. However, due to the proximity of several towers in the vicinity, the antenna pattern will be affected somewhat and this is an unknown pattern.

Assume then an antenna gain of 17.5 dBi − 13 dB = 4.5 dB. The power density P_d incident on the receiving antenna is given by

$$P_d = \frac{P_{in}G_a}{4\pi R^2} \tag{10.21}$$

where
P_{in} is the antenna input power equal to 2.5 W
G_a is the effective gain of the antenna equal to 4.5 dB
R is the distance equal to 56 m
P_d is the power density in W/m²

Therefore,

$$P_d = 2.5 \text{ W} \times \frac{2.818}{12.56} \times 56^2 = 0.179 \text{ mW/m}^2 = -7.5 \text{ dBm/m}^2$$

The received power (P_r) at the input of a narrowband cavity filter inserted between the receiving antenna and the LNA is given by

$$P_r = \frac{P_d G_b \lambda^2}{4\pi} \tag{10.22}$$

where
G_b is the gain of the receiving antenna equal to 389
λ is the wavelength of the incident power

From Equation 10.22,

$$P_r = \frac{(0.179 \times 389 \times 0.198^2)}{12.56} = 0.21 \text{ mW} = 6.6 \text{ dBm}$$

Assuming the use of the four-section filter, 4C42 − 1544.5/T23.2 − 0.0, then the attenuation provided at 1511.125 MHz is 35 dB from measurements. The input power to the LNA is thus −6.6 dBm + −35 dB = −41.6 dBm. This level is above the −50 dBm at which the LNA compresses, and therefore a reduction of approximately 2 dB in gain is predicted in the 1544.25 MHz received signal. The measurements were made at the same site, the results of which were as follows.

The measurements from 1 to 2 GHz were made with a double-ridged guide (DRG) antenna and a minimum 41 dB gain preamplifier feeding into a spectrum analyzer. The emission at 1.511 GHz from the yagi antenna was measured with the preamplifier both in circuit and out of circuit. The measured level at 1.511 GHz was reduced by 44 dB when the preamplifier was removed, which proves that the preamplifier was not compressing. The reason to include the preamplifier was to ensure that any low level in-band emission from 1540 to 1550 MHz was detected.

The measured field from the yagi antenna was lower than expected, so to confirm the measurement, a second yagi antenna, identical to the transmitting antenna, was used as the measurement antenna. The level with the measurement antenna pointing at the emitter antenna is 14.8 dB above that obtained with the ridged guide antenna, whereas a 9.5 dB increase in level was expected. One possible explanation for the difference in measured level is the different directivities of the two antennas (i.e., the ridged guide antenna is more affected by reflected waves than is the yagi antenna due to its lower directivity). This explanation was confirmed by pointing both antennas away from the emitter, when the ridged guide antenna measurement resulted in a higher level than the yagi.

The antenna that will be used in the satellite receiving system exhibits an 8.5 dB higher gain than the yagi. Therefore, due to its higher gain and directivity, higher peaks and troughs in the received power from emitters, as the antenna sweeps across the sky, compared to the measured levels, are expected. The received power at the measurement antenna for the highest-level emitters at 1511 MHz is calculated as follows: First, the preamp and cables were calibrated using a signal source and spectrum analyzer. The input level to the cable/preamp was adjusted to −44 dBm. The measured output was −4 dBm; thus, the combined gain and insertion loss is 40 dB.

At 1511 MHz, the input level to the spectrum analyzer was measured at −13 dBm; subtracting the gain of the preamp results in an output level from the antenna of −53 dBm.

The gain of the satellite communication system receiving antenna is 17.8 dB higher than that of the measurement antenna; thus, based on the higher receiving antenna gain, the power level at the input of the four-section filter is −35 dBm.

The four-section filter provides an insertion loss of 35 dB at 1.511 GHz; thus, the input power to the LNA is −35 dBm + −35 dB = −70 dBm. The measured input level to the LNA is 28.4 dB below the predicted. The main reason for the discrepancy between predicted and measured levels is that the position of the transmitting antenna was not broadside onto the receiving antenna but pointing away at an angle of approximately 20° from the receiving antenna direction. The maximum level was not measured pointing at the rear of the transmitting antenna but due to a reflection from a nearby hill. This example illustrates the importance of measurements when exact locations of emitters and receptors are not known. The next example of a survey made at a different site also emphasizes the advantage of measurements over predictions.

Passive high-amplitude emission at 1.344 GHz was measured using the DRG antenna with the 40 dB minimum-gain preamplifier. After approximately 1 h of the measurement, the emissions changed. An in-band emission at exactly 1.545 GHz was measured and emissions at either side with spacings of approximately 15 MHz were also measured. This broad combination of emissions increased and then decreased over a 3 s period and then disappeared for approximately 3 s. The peak emissions at 1.344 and 1.305 GHz also increased and decreased in amplitude with an approximately 6 s period.

The source of the 1.305 and 1.3405 GHz emissions was a long-range radar situated east of the antenna measuring location. The power level of the 1.3405 GHz signal was measured at −31.83 dBm without a preamplifier. The loss in the interconnection cable at 1.3405 GHz is approximately 2 dB; thus, the antenna output level was −30 dBm = 1 μW. The gain of the double-ridge guide at 1.36 GHz is approximately 5.6 dB = 3.63 numeric. The incident power density on the DRG antenna is given by

$$P_d = \frac{4\pi P_r}{G\lambda^2} = \frac{12.56 \times 1 \times 10^{-6}}{3.63 \times (300/1340)^2} = 0.069 \text{ mW/m}^2 \tag{10.23}$$

It is possible that the intermodulation seen in the photo of Figure 10.54 was due to the preamplifier oscillating or generating spurious response due to overloading. To measure the susceptibility of the preamplifier to input level, the input signal was attenuated. The only component available that could be used as an attenuator was a 20 ft length of an RG58 coaxial cable.

FIGURE 10.54 Display showing intermodulation product at 1545 MHz.

The reduction in the measured amplitude with the coax cable between the antenna output and the preamplifier at 1.305 GHz was approximately 27 dB. This magnitude of reduction puts the spectral emission lines with 15 MHz spacings close to the noise floor. On close examination, at least four of the spectral emission lines at approximately 1.6, 1.62, 1.72, and 1.84 GHz were seen. Although not conclusive proof, this test indicates that the preamplifier is not the source of the harmonically related emission lines.

The intermodulation was not consistently seen throughout the day, although the high-amplitude 1.34 GHz radar signal was used. The sources of intermodulation are a 15.4 MHz shortwave (S/W) radio, which generates a power density of approximately 2.65 mW/m², and the radar, which generates a power density of 0.069 mW/m². The cause of the in-band frequency is PIM, which occurs only when the S/W radio is used. The intermodulation product is

$$13 \times 15.4\,\text{MHz} + 1344\,\text{MHz} = 1544\,\text{MHz}$$

The location at which intermodulation products are generated is unlikely to be the input of the 41 dB preamplifier because the 15.4 MHz signal level will be drastically reduced by the DRG antenna. The most likely source is a lightning protection scheme that surrounds the top of the building. The measurement antennas were located only 6 ft away from the vertical and horizontal conductors of the protection scheme. Reradiation as a result of passive intermodulation due to nonlinear junctions or the nonlinearity induced by a metal with permeability greater than 1 is typically 40 dB down on the incident power level.

The intermodulation products seen in Figure 10.55 are 30 dB down on the 1.3405 GHz level and approximately 46 dB down on the 15.4 MHz level. Intermodulation will occur only when the 15.4 MHz S/W receiver is operated and the radar is pointed at the building. Therefore, EMI will not always be present. Some of the EMI fixes available are as follows:

Move the S/W transmitter frequency. For example, 16.5, 18, and 28 MHz would be the frequencies to choose for maximum suppression of intermodulation-generated EMI.

Move the S/W antenna at least 500 m from the satellite receiving antenna, which should be moved as far from the lightning protection conductors as feasible.

If the lightning conductor is nonferrous material, paint with an IMI-suppression coating. If ferrous, wrap with 2.5 in. thick all-weather absorber foam.

FIGURE 10.55 Incident field = plane wave. The worst-case angle between the cable axis and the H field is $\beta = 90°$. A mean angle of 45° is taken for the angle between the plane of the loop and the direction of field propagation.

TABLE 10.24

Predicted IM Products Resulting in In-Band Interference Frequencies

N	M	Predicted IM Frequencies (MHz)			Received Frequency f_3
		f_1	f_2	f_3	
9	1	20.8	456	1543.2	−1.299
2	1	12.8	1301.1	1542.7	−1.8
8	1	100.4	742	1545.2	0.7
2	1	100.4	1342	1542.8	−1.7
3	1	151.1	1091.4	1544.7	0.2
4	1	171.8	857.3	1544.5	0
7	1	158.6	434.7	1544.9	0.4
1	1	158.6	1386.2	1544.9	0.3
9	1	171.5	0	1543.5	−1
2	1	448	649.3	1545.3	0.8
8	1	425.3	1859.1	1543.3	−1.2
5	1	438.7	649.3	1544.2	−0.3

If PIM is due to the "rusty bolt effect," one solution is to break all screwed, bolted, or riveted joints in the structure causing PIM and to weld, braze, or cadweld the joints. Alternatively, the joint may be improved by cleaning and passivation of the metal surfaces, which may then be rejoined and sealed against the environment by paint or caulking. In this case study, cleaning the joints would be practically impossible, and the other alternatives listed were considered.

Suppression coatings were obtainable from Omicron Inc., Buffalo, NY, which can no longer be traced. They manufactured two types of IMI-suppression products. Omicron CBA was a polymer-based coating that creates at RF a low-resistance and low-reactance path. This reduces current flow through the nonlinear junction and reduces the level of PIM. The low-reactance path is achieved by metal oxide penetration aids, corrosion inhibitors, chelating agents, and materials of high dielectric constants.

The second suppression agent made by Omicron was the SS-50/SC-60, which was a finely divided conductive material dispersed in specially formulated polymers. The mode of suppression is provided by low-reactance pathways on a molecular level, effectively resulting in joint shunting. According to the manufacturer, laboratory and field tests of Omicron IMI-suppression coatings have shown greater than 40 dB reduction in IMI levels between 0 and 30 MHz.

A search has unfortunately not turned up any other company that manufactures a similar product, although one may well exist.

The 1.344 GHz, 1.305 GHz, and 15.4 MHz are not the only frequencies at which intermodulation may be predicted at the site, as shown in Table 10.24, in which $f_3 = N \times f_1 \pm M \times f_2$.

Frequencies exist at power levels that may be sufficiently high (i.e., 2.2 µW/m² at 171.1 MHz and 1.2 µW/m² at 425.3 MHz) to produce passive intermodulation at a level above the noise floor of the receiving system. Therefore, the potential for PIM at these frequencies also exists.

10.4 CASE STUDY 10.3: COUPLING INTO HV AC LINE FROM HF PHASED-ARRAY RADAR

At one prospective site for a receiver, an ambient site survey was conducted. A source of RF existed that was not recorded in the RF ambient survey because its frequency lies well outside of the 1544 and 468 MHz frequencies of interest. The RF source is backscatter from an HF phased-array radar that operates in the 8–20 MHz frequency band. The average peak-radiated power from the array of 16 log-periodic antennas is 2 kW, and the effective radiated power is nearly 200 kW. From a private communication, it was learned that the level of radiated backscatter 600 m behind the radar antenna is approximately 50 W/m.

In addition to the satellite communication receiver antenna, the received power at the equipment building induces potential EMI voltages in the hydro lines, telephone lines, and an HF whip antenna used for a time receiver (which operates at 7.33, 10, and 14.66 MHz). No calculations are required to predict the high probability that the 8–20 MHz source will interfere with the time receiver, even assuming the source is horizontally polarized and the time code receiver antenna is vertically polarized.

10.4.1 Prediction of Level of EMI

The hydro lines, which carry a voltage of 2400 V, are assumed to be at 12 m above ground with approximately a 1.2-m distance between individual lines. The power lines run parallel to the HF array at a distance of 600 m. The received power at a distance of 600 m from the source can be calculated from

$$P_r = - (32 \text{ dB} + 20 \log_{10} R + 20 \log_{10} f) + P_t \text{ [dBm]}$$

where
  P_r is the received power [dBm]
  P_t is the transmitted power equal to 47 dBm/m
  f is the frequency [MHz]
  R is the distance from the source [km]

Using the worst-case lowest frequency of 8 MHz, the level of received power is

$$-(32 + 20 \log_{10} 0.6 + 20 \log_{10} 8) + 47 = 0 \text{ dBm}$$

Using the far-field value of wave impedance (377 Ω), the E field is calculated at 0.6 V/m and the H field at 1.6 mA/m.

10.4.1.1 Transmission Line

The most accurate method for calculating the EMI picked up on the power line is the use of transmission line theory; although, the theory is valid only for the situation where the height (h) of the line above ground is much less than the wavelength, λ.

The E field increases with height above the ground until the height is approximately 0.5λ, after which the magnitude of the field levels off. One source of error in application of transmission line theory is when this effect is ignored.

The transmission line equation may still be used to calculate the approximate characteristic impedance of the line from Equation 7.16 and the induced current from Equation 7.20 by setting the height to 0.5λ, when this is less than the actual height. At 8 MHz, when 0.5λ is 18.75 m, the 12 m height of the power line may be used in the transmission line equations, whereas if the equations are used at 20 MHz, the height would be set at 7.5 m to provide a limit to the induced voltage. The calculation of differentially induced voltage using these equations is expected to be reasonably accurate because the distance between the wires is less than 0.5λ.

Case Study 12.5 provides worked examples of Equations 7.16 and 7.20, and in the present study, only the results of the calculations are presented.

At 8 MHz and a height of 12 m, the characteristic impedance of the line Z_c is 912 Ω. On using Equation 7.14 to calculate the characteristic impedance between the lines, the distance is set to 0.6 m because the equation multiplies this by 2 in obtaining the impedance of a line above a ground plane. At 8 MHz and a height of 1.2 m, Z_c is 552 Ω. The induced common-mode current for the 0.6 V/m incident E field at 8 MHz is 31 mA. The differential current is 4.5 mA. The transmission line is long and terminated in the unknown impedances of the transformer located at the building and at the substation. If the assumption is made that the load and source impedances are the same as the characteristic impedances of the line, then the common-mode induced voltage is approximately 27 V and the differential-mode voltage is 2.6 V. When the load and source impedances are less than the characteristic impedances of the line, which is likely, the current flow in the lines and terminating impedances at resonant frequencies will be much higher than those calculated. The effects of cable resonance are discussed in Section 7.7.

REFERENCES

1. C.C. Roder. *Link Communication Analysis Algorithm (LINCAL): User's Manual.* ECAC-CR-80-059, Department of Defense, Electromagnetic Compatibility Analysis Center, Annapolis, MD, August 1980.
2. T. Durham. Analysis and measurement of EMI coupling for aircraft mounted antennas at SHF/EHF. *IEEE International Symposium on Electromagnetic Compatibility*, Atlanta, GA, 1987.
3. S.T. Hayes and R. Garver. Out-of-band antenna response. *IEEE International Symposium on Electromagnetic Compatibility*, Atlanta, GA, 1987.
4. D.A. Hill and M.H. Francis. Out-of-band response of antenna arrays. *IEEE International Symposium on Electromagnetic Compatibility*, Atlanta, GA, 1987.
5. S.T. Li, J.W. Rockway, and J.H. Scukantz. Application of design communication algorithm (DECAL) and performance evaluation communication algorithm (PECAL). *IEEE International Symposium on Electromagnetic Compatibility*, San Diego, CA, October 1979.
6. K. Clubb, D. Wheeler, and E. Pappas. *The COSAM II (DECAL/PECAL) Wideband and Narrowband RF Architecture Analysis Program: User's Manual.* ECAC-CR-86-112, February 1987.
7. J.W. Rockway, S.T. Li, D.E. Baran, and W. Kowalyshin. Design communication algorithm (DECAL). *IEEE International Symposium on Electromagnetic Compatibility*, Atlanta, GA, June 1978.
8. L.D. Tromp and M. Rudko. Rusty bolt EMC specification based on nonlinear system identification. *IEEE International Symposium on Electromagnetic Compatibility*, Wakefield, MA, 1985.
9. Private communication with J. Rose, Canadian Astronautics Limited, Ottawa, Ontario, Canada.
10. T. Dvorak. The role of site geometry in metric wave radiation testing. *IEEE International Symposium on Electromagnetic Compatibility*, San Diego, CA, 1986.
11. J. DeMarinas. Antenna calibration as a function of height. *IEEE International Symposium on Electromagnetic Compatibility*, Atlanta, GA, 1987.
12. Studies relating to the design of open field test sites. *IEEE International Symposium on Electromagnetic Compatibility*, Atlanta, GA, 1987.
13. D.V. Gonshor. Attenuation, transmission and reflection of electromagnetic waves by soil. *IEEE International Symposium on Electromagnetic Compatibility*, Atlanta, GA, 1987.
14. S.V. Georgakopoulis, C.A. Balanis, and C.R. Birtcher. Cosite interference between wire antennas on helicopter structures and rotor modulation effects: FDTD versus measurements. *IEEE Transactions on Electromagnetic Compatibility*, August 1999, 41(3), 221–233.
15. S.T. Li, B. Koyama, J.H. Schkantz, Jr., and R.J. Dinger. EMC study of a shipboard HF surface wave radar. *IEEE Electromagnetic Compatibility Symposium Record*, Atlanta, GA, 1995.
16. L.M. Kackson. Small models yield big results. *NARTE News*, 11(2), April–June 1993.
17. E.B. Joffe. Out of band response of VHF/UHF airborne antennae. *IEEE International Symposium on Electromagnetic Compatibility*, Denver, CO, 1989.
18. D.A. Weston. Antenna to antenna coupling on an aircraft using a 1/10th scale model with results compared to the FEKO electromagnetic analysis program. *EMC Europe*, Wroclaw, Poland, 2010.
19. D.A. Weston. Comparison of techniques for prediction of and measurement of antenna to antenna coupling on an aircraft. *EMC Europe*, York, U.K., 2011.
20. D.A. Weston. Electromagnetic ambient inside an aircraft from transmitting antennas mounted on the outside compared to safety levels and radiated susceptibility test levels. *IEEE International Symposium on Electromagnetic Compatibility*, Denver, CO, 2013.
21. M. Donald Siegel. Aircraft antenna-coupled interference analysis. *IEE Transactions on Electromagnetic Compatibility*, June 1969, 11, 85–90.

22. S.A. Davidson and G.A. Thiele. A hybrid method of moments-GTD technique for computing electromagnetic coupling between two monopole antennas on a large cylindrical surface. *IEEE Transactions on Electromagnetic Compatibility*, May 1984, EMC-26, 90–97.
23. Motorola Interference Technical Appendix, Issue 1.41, February 2002.
24. D.A. Weston. Antenna-to-antenna coupling on an aircraft: Mitigation techniques. *ITEM*, March 21, 2012.
25. E.B. Joffe and J. Gavan. High power broadcast skywave transmitters desensitization effects analysis. *IEEE Electromagnetic Compatibility Symposium Record*, Washington, DC, 1990.
26. J. Gavan and E.B. Joffe. An investigation of the desensitization effects by high power HF broadcast transmitters on HF airborne receivers. *IEEE Transactions on Electromagnetic Compatibility*, May 1992, EMC-34, 65–77.
27. P.L. Lui, A.D. Rawlins, and D. Watts. Measurement of intermodulation products generated by structural components. institution of engineering and technology. *Electronics Letters*, 24(16), 1998.
28. J.W. Rockway, J.C. Logan, H.H. Schukantz, and T.A. Danielson. Intermodulation testing of composite materials. Technical document 2705, September 1994 Naval Command. Control and Ocean Surveillance Center.

11 Printed Circuit Boards

11.1 INTRODUCTION

One of the major reasons equipment fails radiated emission requirements is radiation from digital signals, either directly from the printed circuit board (PCB) or from PCB-sourced common-mode (C/M) noise voltage driving attached cables. Even when equipment is contained in a totally shielded enclosure, noise voltage generated across a PCB to which unshielded and unfiltered interface cables are connected can result in a sufficiently high noise current to fail radiated emission requirements. In extreme cases, PCB-generated noise has resulted in equipment failure of commercial radiated emission requirements with shielded cables connected to a PCB metal faceplate, which in turn was connected to a shielded enclosure. If the radiated emission requirements are of the stringent MIL-STD-461 or DO-160, the use of a shielded enclosure is almost mandatory.

When a piece of digital electronic equipment, such as a notebook computer, contains a wireless with an antenna anywhere from 2 to 20 cm from the PCB, then, although a PCB with a low-emission layout is a good starting point, it may not be enough to avoid desensitization or in-band spurious response in the wireless.

The major source of radiation from PCBs and cables is not differential-mode (D/M) current but C/M current, because depending on the geometry, the radiation due to several microamps of C/M current can be as high as several milliamps of D/M current.

Some experts on PCB layout might say that good PCB layout is easy. This may or may not be true, however, with the use of ever-faster PCs and pS fast logic; manufacturers have to go to ever-greater lengths to meet EMI requirements. The need for new techniques in PCB layout has become more important in reducing radiation. That being said, a PCB with the best possible layout can fail due to a single large-scale integration (LSI) chip from which the level of radiation exceeds emission limits or requirements.

Some other factors that affect PCB radiation include the method by which the power supply connection is made to the PCB, PCB-to-PCB or PCB-to-motherboard/backplane grounding, the proximity to adjacent structures (grounded or ungrounded), the electrical connection to ground, and the presence and type of shields. All of these topics are discussed in the following sections.

11.2 PRINCIPLES OF RADIATION FROM PRINTED CIRCUIT BOARDS

It is the current flow in the signal return connection (PCB track or ground plane), due to both displacement current between the signal path and return path and the load current, which develops a voltage in the impedance of the return. This voltage can generate a C/M current, which is the prime source of radiation from the PCB and any attached connection. The displacement current flows as the capacitance of the line is charged and discharged; thus, displacement current is nonuniform, with a maximum at the source end of the line and zero at the load end. Displacement current, and the resultant C/M current on the line, due to the proximity of some conductive structure is more difficult to predict, although Reference [1] does describe a technique. In predicting the radiated emissions from a PCB track, the voltage drop in either signal or power return connections, the typical magnitude of the current spikes, and the impedance of the path must be known as well as the displacement current down the line and to adjacent structures.

That radiation is typically increased with the attachment of a cable on which no differential signal flows and that is connected only to ground is a puzzle to many engineers. Section 7.13.3.1 describes the mechanism in more detail, but briefly: when C/M current exists on a conductive

FIGURE 11.1   PCB tracks with common-mode current flowing on an attached cable.

structure (PCB tracks) and a conductor is connected to the structure, the C/M current flows on the attached conductor. Figure 11.1 illustrates how the C/M current continues to flow on the cable. At frequencies where the length of the attached conductor is less than a wavelength, the radiation increases with frequency until the cable is approximately half a wavelength in length. When the cable becomes electrically long (longer than a wavelength), the radiation from the cable tends to decrease. Attaching a cable to a PCB ground increases the radiation effectiveness of the source structure at low frequency (LF). At high frequency (HF) where the cable, or PCB, is electrically long, the current flow on the structure changes phase down its length, and the cable may be modeled as a series of infinitesimal current sources, some of which are wholly or partially out of phase. The most likely explanation for the constant or reduced radiation at HF is that the composite field from the cable is the sum of the fields from the infinitesimal current sources, which interfere. In Section 7.13.3, it was shown that 15.7 μA flowing on an attached cable at 21 MHz results in the same E field as 1.26 mA of differential current flowing in a small loop. Because C/M current is so often the predominant cause of radiated emissions, EMC engineers tend to ignore the D/M current as a source, just as design engineers ignore the C/M current contribution. However, if a large loop is formed, then the D/M-current-generated emissions can predominate. For example, if not all the D/M current flows on the return conductor, but some leakage current flows on a ground, a large loop may be formed. Adding a complete shield around a small loop, or a PCB, will eliminate the induction of C/M current on the attached cable, assuming the cable is connected to the shield. If the cable is connected through a hole in the shield, then cable C/M currents are often higher than without the shield, due to increased displacement current between the inside of the shield and the loop, or PCB. Likewise, if the shield is only partial around the loop, or PCB, then the C/M current can increase.

Reference [1] describes a method of analyzing a microstrip PCB (signal trace above a ground plane) attached to a shielded enclosure. When the results of the prediction are compared to the test results in Reference [1], scaled to account for the different dimensions of the PCB, the results are extremely close. For example, the radiation from the microstrip is computed at 47.7 dBμV/m and measured at 48.5 dBμV/m, an uncharacteristically high correlation of 0.8 dB. In contrast, one of the computer programs described in Section 11.12 predicts a massive 28 dB higher level of radiation from the same PCB layout as described in Reference [1].

The impedance of a ground trace or ground plane was examined in Section 5.1.2, with calculations based on NBS inductance formulas. Reference [2] introduces the concept of partial inductance and other techniques to describe more accurately the ground return path impedance. The inductance of vias is described in Reference [4]. In a complex PCB layout, many ICs change their state simultaneously, drawing current out of the closest decoupling capacitor and returning it in the ground return path to the capacitor. In addition, digital signal and clock currents flow on traces between ICs and return in the ground return path beneath the traces. Large voltage drops occur in the ground impedance due to these switching currents, which generate C/M currents across the board and out on any attached cables. Upcoming Figure 11.97 illustrates the generation of C/M voltage across a PCB ground plane. Because of the proximity of grounded structures, connection of one PCB to another PCB or to a motherboard/back plane or to a power supply, the C/M current is further enhanced. When the load on a microstrip PCB trace is removed and the configuration is a signal trace above a ground plane, open circuit at the load end, then the radiation reduces at LF but is the same or higher at HF. This effect is discussed in more detail in Section 11.3.3 but is used here to illustrate the importance of the displacement current alone in generating emissions.

The importance of a short- and low-impedance ground path is discussed in Section 11.8.

Certain PCB layouts have been described as low-emission types. To enable a comparison between the different types of layout and different logic families and to test the validity of the "good" layout, the tests described in Section 11.3 were conducted.

11.3 LOW-LEVEL RADIATION PCB LAYOUT: TEST DATA, LAYOUT COMPARISON, AND RECOMMENDATIONS

11.3.1 PCBs Tested

In the first two editions of the book, the measured radiation from PCB materials with a 1.6 mm substrate, with two or three layers, and with 0.7 mm traces was presented. The PCBs had uncontrolled characteristic impedances, as well as complex unmatched loads, with an upper frequency limit of 1 GHz.

Modern PCBs use much thinner traces and many more layers and may have data with frequencies up to tens of GHz.

A number of PCB layouts were manufactured many with controlled impedances, and measurement of the radiation from these layouts was measured.

The data from 14 different PCBs, with and without attached cables, were compared to each other as well as to the FCC and European Union (EU) radiated emission limits.

The dimensions of the PCBs described in this investigation were chosen to be typical for 4–12 layer PCBs and were designed to have either 100 Ω or 50 Ω differential impedance and in some cases to have 50 Ω characteristic impedance between one trace and the ground plane for single-sided measurements.

The 14 PCBs include the following:

- Four types of microstrip/image plane layouts
- Eight types of stripline layouts
- Two types of transmission line layouts

These PCBs were tested with single-sided or differential drivers and, with and without a wire, attached to the return for the transmission line or in the case of the microstrip or striplines to the ground plane at the load end.

The ground in the layouts with differential traces forms an image plane, although it is still referred to as a microstrip.

FIGURE 11.2   Source enclosure, SMA connectors, and a sample PCB.

As described later in the low-level radiation stripline layouts, it was found that, although very short, the vias and traces connecting to the load and the load itself will be a significant source of radiation. Initially, measurements were made with the original four-layer PCBs connector shielded and the load and signal vias on the top of the board and the bottom of the board shielded as well as unshielded. The shield in these four-layer boards was imperfect and the final tests were with three-layer boards having bare copper ground on the top and bottom. With the three-layer board, the copper shield can be soldered around its periphery. Due to the multiplicity of configurations, a minimum of 47 tests were performed over the frequency range of 50 MHz–8 GHz. All of these tests were initially performed using an ECLinPs differential driver as the source. All of the selected devices will have the capability of driving either 100 Ω or 50 Ω loads, and all of the PCB layouts are designed to have either 100 Ω or 50 Ω differential characteristic impedances or 50 Ω single-sided characteristic impedances.

The addition of a wire to the load in the transmission line PCB configuration, or to the underlying ground plane in microstrip or stripline PCBs, simulates the connection of the signal return in a cable or the shield of a shielded pair cable and shows the effect of C/M current flow on cables. The mechanism for this and further modeling and measurements is summarized in Section 11.3.3.

After the initial tests, modifications were made to some of the PCBs, such as drilling out some of the vias down the length of the stripline and increasing the ground on either side of one of the transmission line layouts. The photo in Figure 11.2 shows the enclosure, SMA connectors, and PCB. One of the problems with drilling out vias on a stripline with an overall upper and lower ground plane is that holes are introduced into what was a full plane. This problem will be avoided with the three-layer PCB that will be made with the distances between the vias varied from 3, to 6, to 12 mm from board to board.

11.3.2 Summary of the Best and Worst PCB Configurations in the Differential Configuration

The summary of the worst- to best-case PCB layouts is for a real-world configuration of the stripline configuration where the load resistor and upper signal pads are unshielded, which is provided in Figure 11.3.

This means that the results for the best stripline would be significantly better if the load had a very effective shield over it.

This comparison is designed to help engineers and PCB layout personnel in making a decision on what type of PCB to use. The magnitude of the difference between PCB emissions is described in the following sections.

FIGURE 11.3 Ranking of highest to lowest emitters among the PCB traces.

All dimensions are in thousands of an inch (mils).

Note that all PCBs are of four layers. When the distance between the upper and lower ground planes of the stripline is 21 mil, then the upper ground plane is the second layer down and layer 1 (the top layer) is used to connect to the case of the connector and for the signal pads for the load resistor.

11.3.3 COMMON-MODE CURRENT FLOW ON THE PCBS

Many articles on theoretical PCB radiation discuss only the radiation due to D/M signal current flow, specifically the signal current between the source and the load, returning on either a return trace or a signal ground. This type of analysis on a microstrip PCB often uses a quasi–TEM mode, split into radiation from the horizontal current and the contribution from the vertical currents in the load. Transmission-line equations or full-wave computations based on the method of moments (MOM) are often employed with the capability of analyzing three-dimensional (3D) structures. Another approach is the conjugate gradient–FFT (CG-FFT) method.

This type of analysis results in a 40 dB/decade increase in radiation with frequency up to the frequency where the PCB length is a resonant length. Above resonance, the increase is typically 20 dB/decade.

These results are very different than that presented in these measurements, as well as in measurements described in the first and second editions of the book. The reason is that C/M-current-sourced radiation is ignored in many analyses. In a practical PCB where signal traces are not in the exact center of the board, C/M currents are generated due to lack of symmetry. The C/M current in a microstrip PCB is typically generated by displacement current that flows as the capacitance of the line is charged and discharged. Thus, the displacement current is nonuniform with a maximum at the source end of the line and a zero at the load end. Unlike signal current that is uniform down the length of the board. Although not a typical or practical PCB layout, a symmetrical microstrip with the traces in the center of the board and the source generator buried behind the ground plane powered by a battery and with no attached cables, the C/M currents are generated due to the displacement current. If a single wire is attached to the ground plane of a microstrip PCB, then the C/M current continues to flow on the wire, whereas the D/M current returns in the ground plane. Another source of C/M current when a wire is attached to the microstrip ground plane is the voltage drop in the ground plane due to the returning signal current.

If the microstrip is close to a metal structure such as the enclosure in Figure 11.2, then displacement current flows between the signal trace and the enclosure increasing C/M current flow even further. This is discussed and analyzed in Reference [1]. The CG-FFT program does predict radiation from a microstrip with a high permittivity substrate, like a PCB, but assumes a symmetrical infinitely small source. The radiation as a result of C/M current is ignored and this results in a massive 60 dB underestimation of the fields from the structure!!

It is clear how C/M current is generated from a microstrip, but why does this happen with the stripline PCB? In practical stripline PCBs and in the four-layer boards presented in this initial investigation, the signal vias come up to a load resistor on the top of the board. When the signal is differential, that is, two signal traces in the stripline, then the signal and return are very close to being balanced around ground and the level of radiation is very low, as is the C/M current. When the signal is single sided, one side of the load resistor and its trace are at the ground plane potential and the level of radiation from the load and signal trace is much higher, as is the C/M current.

11.3.4 PCB Test Setup

The frequency of the source was swept using voltage-controlled oscillators controlled by a sawtooth oscillator. The driver was contained in a small metal enclosure that also contained batteries, power supplies, and the swept frequency source. The enclosure lid was sealed using finger stock material and also using dimpled adhesive copper EMI shielding tape. Before every test, the PCB was replaced by two 50 Ω terminations and the radiation from the enclosure and supply leads (ambient) was measured. In all the tests with four-layer boards, this ambient level was always lower than the PCB radiation. At the start of each new day of testing, the copper tape was replaced to ensure shielding integrity. Each PCB was manufactured to contain two circuits (layouts) and have four SMA connectors mounted on it (two per PCB layout). The signal from the source was brought out of the box using semirigid cable and two SMA connectors. The center conductor of each coaxial connector is used for one signal line of a differential pair. Only one of the two connectors is used for testing single-ended drivers and loads. Thus, the signal source in its enclosure can be used with any of the different PCB layouts and is itself not a source of radiation. However, at HF, a significant quantity of emissions was found to be leaking out of the SMA connectors on the board, and thus, the connectors on the boards were shielded part way through the testing. Later it was found that the load resistor and pads on the top of the PCB and the signal pads on the rear of the PCB were also a source of radiation and these were then shielded. The level of shielding achieved with the four-layer boards is limited, and when the three-layer boards are manufactured, the shielding should be much more effective.

The majority of the measurements were made on an open-area test site (OATS) located in an area of relatively low ambient. The receiving antennas were located at a distance of 3 m from the turntable and the site has met all the normalized site attenuation requirements.

At LF, the emissions from some of the stripline PCBs were so low that measurements had to be made in an anechoic chamber with additional damping.

The photos in Figures 11.4 and 11.5 show the enclosure and attached PCB.

For the differential measurements, the PCB was broadside on to the measuring antenna, and the elevation angle of the PCB was varied until a maximum level of emissions was measured. It was expected that varying the azimuth angle, that is, rotating the PCB relative to the antenna should not

FIGURE 11.4 Overall stripline PCB and enclosure with limited shielded connectors and loads.

FIGURE 11.5 Localized stripline with localized upper ground plane PCB and enclosure.

significantly change the emissions level for a PCB without attached wire. However, the C/M and D/M current flow in the PCB varies in phase down the length of the PCB when it is electrically long, and we found that at high frequencies, it was important to rotate the turntable to achieve maximum emissions. As expected, rotating the turntable did not change emissions at LF where the electrical length of the PCB and enclosure is short. For all of the single-sided measurements, both the elevation angle and the azimuth angle of the PCB were adjusted for maximum emissions, whereas with the differential measurements, only the elevation angle was changed. For many of the stripline PCBs, especially with a single-sided signal, maximum radiation occurred with the PCB vertical and with the load facing the antenna. Probing the PCB showed that indeed the load radiated at a higher level than the edge of the stripline, even when ground plane vias were drilled out.

11.3.5   PCB Layouts Tested

11.3.5.1   Basic PCB Description

Seven four-layer PCBs were manufactured, each containing two circuits per PCB.

The basic layouts built on these PCBs are as follows:

1. Two different transmission lines on one PCB
2. Four different microstrips on two PCBs
3. One stripline with localized upper ground plane connected by vias to lower full ground plane
4. Two stripline PCBs with full upper ground plane but with the vias connecting top and bottom ground planes close to the signal traces
5. One stripline with full upper ground plane and with the vias connecting top and bottom located around the periphery of the PCB. One trace 2 cm from the edge and one 5.6 mm from the edge (at edge).

The photo in Figure 11.6 shows the signal and return traces of the transmission line on one side and the transmission line with its traces widened by conductive tape.

The photo in Figure 11.7 shows the M1 J3/J4 microstrip with 100 Ω differential traces and 50 Ω single-sided traces. M1 J1/J2 has different dimensions and a 50 Ω differential impedance.

FIGURE 11.6   Transmission line PCB TX2.

FIGURE 11.7   Microstrip PCB M1.

FIGURE 11.8   Microstrip PCB M2.

The photo in Figure 11.8 shows the M2 J3/J4 microstrip with 100 Ω differential traces and 50 Ω single-sided traces. M1 J1/J2 has different dimensions and a 100 Ω differential impedance.

The photo in Figure 11.9 shows the S2 J3/J4 stripline with 100 Ω differential traces and 50 Ω single-sided traces. S2 J1/J2 has different dimensions and a 50 Ω differential impedance.

The photo in Figure 11.10 shows the S3 J3/J4 and S3 J1/J2 with 100 Ω differential traces and 50 Ω single-sided traces. The only difference is that the traces in J3/J4 are 2 cm from the edge of the PCB and J1/J2 are 5.6 mm from the edge.

The photo in Figure 11.11 shows the S4 J3/J4 with a 50 Ω differential trace and S4 J1/J2 with a 100 Ω differential trace.

The photo in Figure 11.12 shows the S5 J3/J4 with 100 Ω differential and 50 Ω single-sided traces. S5 J1/J2 has a 50 Ω differential trace.

The dimensions for these PCBs are shown in Section 11.3.6.4.

All of the striplines have through holes plated vias tying the upper and lower ground planes together at 3 mm centers.

FIGURE 11.9   Stripline PCB S2 connectors and load with limited shielding.

FIGURE 11.10   Stripline PCB S3 connectors and load with limited shielding.

As it is uncommon in a practical PCB to make a perfectly localized stripline with the vias connecting the bottom and top ground planes at very close proximity (e.g., 3 mm), other tests were performed on the striplines with differential traces with the ground plane vias selectively drilled out to increase the spacing or to create a slot, and these are described for differential and single-sided PCB configurations in Section 11.4.10. Some measurements were made with the load resistor and upper signal pads for the resistor shielded as well as the SMA connector shielded. However, the shielding was very limited and it was found that emissions from under the load shield were higher than the top or edge of the stripline, even with vias drilled out. For this reason, another set of three-layer stripline PCBs were manufactured with bare copper on the upper and lower surfaces. This enables the shields to be terminated as effectively as possible and comparative measurements of the three single-sided striplines are provided later. Instead of the vias drilled out, these three-layer striplines were made with vias at 3, 6, and 12 mm spacing.

FIGURE 11.11 Stripline PCB S4. One-side load and connector shielded the other side unmodified.

FIGURE 11.12 Stripline PCB S5. One-side load and connector with limited shielding.

However, the measurements presented here with the load unshielded are perfectly valid and provide useful data as in a real PCB; traces and ICs will exist outside of the stripline. Thus, the best microstrip may be almost as good as the worst stripline when the load is unshielded.

11.3.6 PCB DIMENSIONS AND TRACE CHARACTERISTIC IMPEDANCE AS BUILT

The goal was to design PCB configurations with differential characteristic impedances of 50 and 100 Ω and where possible also single-sided (trace to ground plane) impedances of 50 Ω. The dimensions were not only dominated by the impedance requirements but were also chosen to be manufacturable using the dimensions of typical multilayer boards. Another factor in the choice of layout was to have as much variation in the different dimensions as possible to see the effect of changes in height, trace spacings, and trace width.

We have assumed a board thickness of 10 mil for two-layer boards and a thickness from 31 mil to a maximum of 125 mil for 12-layer boards. The minimum layer thickness of 5.8 mil is used, although boards with a smaller layer thickness are now possible.

By changing the width of traces and the separation between the traces, the layer thickness may be adjusted to values other than that used in the investigation.

The typical types of PCB configurations are

- Microstrips
- Striplines
- Transmission lines

11.3.6.1 Microstrips

Microstrip PCB 1 has a height between the lower surface of the trace and the upper surface (closest to the trace) of the ground plane of 5.8 mil and so this might be contained in anything from 2-layer to 12-layer boards.

Microstrip PCB 2 has 31 mil between the ground plane and the trace and so could be accommodated in a two-layer board or a four-layer board with 10.3 mil between layers. Using 6.2 mil as the layer thickness, the microstrip 2 could be incorporated into anything from 6- to 12-layer boards.

The earlier investigation described in the first and second editions of this book shows that placing an image plane between the trace and ground plane of a microstrip degrades the performance considerably and is discouraged. This image plane in a typical PCB might be a VCC plane connected to the ground plane with decoupling capacitors. Thus, the layers directly between the signal traces and the ground plane forming the microstrip must be kept clear of copper. In practical PCB layouts, it has been found empirically that the inclusion of an image plane between the upper signal trace and ground plane (which carries the signal return current) in a microstrip invariably degrades the performance. For this reason, the inclusion of an image plane imbedded in a microstrip was not repeated in the data described in this book.

Although for consistency we refer to these PCB layouts as microstrips when measuring with the source connected across the two traces and the load between the two traces at the far end and with no connection of the signal to the ground plane, the configuration is really differential traces above an image plane.

11.3.6.2 Striplines

Three of the four stripline PCBs tested (S2, S3, and S5) were a four-layer design, so that the upper ground plane is the second layer down and the top layer is used for the connector and load resistor connections only. Although the fourth (bottom) layer is ground plane, the upper and lower surfaces of the PCBs are covered in solder mask, and so no bare copper can be used for the shielding.

The striplines were "closed," that is, through hole, plated vias are used down the length of the PCB to stitch together the upper and lower ground planes. In the case of the stripline in which the vias were selectively drilled out, eventually all of the vias are removed and the stripline is what is sometimes referred to as an "open stripline." Stripline 2 is manufactured using local striplines, with the vias forming the stripline close to the traces and with full upper and lower ground planes. This might be typical of boards for which two ground planes have been envisaged but in which local striplines are required. In a practical PCB, other traces and power planes contained on different layers must be kept away from the layers covered by the stripline. Obviously, these layers are available outside of the local stripline. In stripline 2, the traces are located symmetrically between the two ground planes and the height between the ground planes is 21 mil. Using a 10.5 mil thickness, the board could be 4–6 layers, and with a layer thickness of 6.8 mil, the board could be 8–12 layers.

Stripline 3 has the traces placed symmetrically between the ground planes, and the height between the upper and lower ground planes is 21 mils. The upper ground plane is stitched around its periphery forming an overall stripline. Ideally, this would be the upper and lower layers of the PCB, and using a 5 mil thickness, a five-layer board could be manufactured or a four-layer board using 6.2 mil thickness. If the upper and lower layers are just used for integrated circuit (IC) pads and any low-level control or analog signal traces, then a six-layer board is feasible. Adding a third ground plane, two striplines may be formed in an 8- or 10-layer board.

Stripline 4 has local striplines (vias close to the signal traces) but with full upper and lower ground planes. Here the traces are asymmetrical (often referred to as a dual stripline) and 12.3 mil from the upper ground plane and 41.5 mil from the lower ground plane. This means that with 6.2 mil layers the board could be from four layers to as high as twelve using 6.2 mil thickness with the ground planes at layers two and eleven.

Stripline 5 has the same height between the upper and lower ground planes as stripline 2 (21 mil) and the same stack up may be used. However, the upper ground plane is local covering the stripline traces only, and so this layer may be used for other traces outside of the upper ground area.

11.3.6.3 Transmission Lines

Transmission lines are simply traces side by side or one on top of the other and so may be located on a single layer, without ground planes, and so they can be accommodated on a PCB with any number of layers. However, as full ground planes are highly recommended, and this investigation shows why, the transmission lines are typically confined to two-layer PCBs where ground planes are not feasible.

11.3.6.4 Trace Dimensions and Impedances

The POLAR Instruments SI6000 Controlled Impedance Field Solver, which uses advanced field solving methods, was used to calculate PCB trace impedance for both the single-ended and differential designs. Using the SI6000 C program, the trace width and separation were varied until the correct impedance was found. Another approach was to input the height and the required impedance and see what values of trace width and spacing were required to achieve the impedance.

The POLAR Instruments CITS500S Controlled Impedance Test System uses time domain reflectometry to measure the impedance of the finished four-layer PCBs. Test coupons were manufactured with each PCB and the impedance of these coupons was measured. The measured impedance was dependent on the position of the measurement point down the trace and so average values were taken. The test coupons showed variations in the test for the 100 Ω impedance from 97.8 to 108.9 Ω and the 50 Ω impedance from 47.2 to 54.6 Ω. The PCBs chosen for the test had the closest impedances to the specified.

The cross section through each of the PCBs with dimensions and the characteristic impedances are shown in Figures 11.13 through 11.19. Figures 11.6 through 11.12 show the PCBs. All dimensions are in mils.

The photos show the two SMA connectors that are used to connect the differential signals to the PCB. When only one trace is used in a single-sided connection, then only one of the SMA connectors is used, as only one trace is used and the load resistor is between this trace and the ground plane.

FIGURE 11.13 Microstrip 1 dimensions and impedances.

FIGURE 11.14 Microstrip 2 dimensions and impedances.

FIGURE 11.15 Stripline 2 dimensions and impedances.

FIGURE 11.16 Stripline 3 dimensions and impedances.

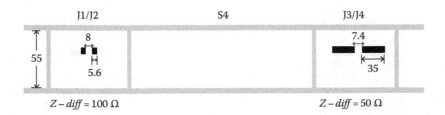

FIGURE 11.17 Stripline 4 dimensions and impedances.

FIGURE 11.18 Stripline 5 dimensions and impedances.

FIGURE 11.19 Transmission line 2 dimensions and impedances.

11.4 SUMMARY OF DIFFERENTIAL CONFIGURATION RESULTS AND CONCLUSIONS AT LOW FREQUENCY

11.4.1 TRACE TYPES

As expected, the stripline traces produced the lowest emissions of the three trace types tested, and transmission-line traces produced the highest emissions. On the OATS, the best stripline under test produced emissions of 60 dB lower than one of the transmission lines at its worst case. In that test, the stripline had very imperfect shield over its load and connectors, and, therefore, these components still contributed to the emissions, and the effectiveness of the stripline layout at reducing emissions in a perfect world (fully shielded stripline) is underestimated. However, in the real world, traces will run on the upper layer outside of the stripline and ICs will be sources of radiation. Microstrip traces generally produced higher emissions than the striplines, but far less than the transmission lines. In some cases a microstrip PCB performed as well as a stripline with an unshielded load. See Figures 11.20 and 11.21 for comparisons of the best trace from each type on the OATS and Figure 11.22 for the worst. Figures 11.23 and 11.24 show the best and worst traces based on 100 Ω differential tests performed in the anechoic chamber. As shown later, the radiation from S2 J3/J4 is initially lower than S4 J1/J2 when measured on the OATS. However, it was found that the most significant source of radiation is from the 50 Ω load on S4 J1/J2. When this was covered in an imperfect copper shield, the emissions dropped below those from S2 J3/J4.

Traces M1 J1/J2, S2 J1/J2, S4 J3/J4, and S5 J1/J2 had 50 Ω differential configuration capability; however, they are yet to be tested. The results from 50 Ω differential testing are presented later.

Based on the LF data collected in this investigation, a stripline PCB has the best likelihood of meeting the most stringent radiated emission requirements. Of the striplines tested, the ones with a localized stripline and a full upper ground plane produced lower emissions than other combinations in the 100 Ω differential tests, especially the localized stripline with localized upper ground plane. However, the predominant source of radiated emissions for the best stripline is almost certainly the load resistor and signal pads and not the stripline. Where radiated emission requirements are less stringent, a microstrip PCB may be sufficient, given that in practical applications, components will not be shielded and thus load radiation will dominate on a stripline layout.

FIGURE 11.20 Comparison of the best transmission line, differential microstrip (image plane), and differential stripline traces over the mid-frequency range for the 100 Ω differential impedance tests on the open-area test site. N.B. The load on S4 J1/J2 is unshielded as in the real world.

FIGURE 11.21 Comparison of the best transmission line, differential microstrip (image plane), and differential stripline above 1 GHz for the 100 Ω differential impedance tests on the open-area test site with emissions normalized to M2 J3/J4. The S4 J1/J2 PCB had very imperfect shields over both its load and connectors.

FIGURE 11.22 Comparison of the worst transmission line, differential microstrip (image plane), and differential stripline traces at low frequency for the 100 Ω differential impedance tests on the open-area test site with emissions normalized to M2 J3/J4. Stripline load was unshielded, as in the real world.

FIGURE 11.23 Comparison of the best transmission line, differential microstrip (image plane), and differential stripline traces at low frequency for the 100 Ω differential impedance tests in the anechoic chamber. The load on S2 J3/J4 was unshielded, as in the real world.

FIGURE 11.24 Comparison of the worst transmission line, differential microstrip (image plane), and differential stripline traces at low frequency for the 100 Ω differential impedance tests in the anechoic chamber. The stripline load was unshielded, as in the real world.

11.4.2 Measured Data and Detailed PCB Comparisons for the Differential PCB Configuration and Transmission Lines

11.4.2.1 Microstrip Comparison

Figure 11.25 shows a comparison of the emissions from the three differential microstrip traces tested on the OATS with 100 Ω differential impedance. We refer to this configuration as a differential microstrip, but in reality it is differential traces above an image plane, as none of the signal current returns in the ground plane. Figure 11.26 gives the same comparison, but with traces M1 J3/J4 and M2 J1/J1 normalized to M2 J3/J4 and then smoothed out. Embedded figures on the plot show the width of the traces, their separation, and their height above the ground plane. From both plots, it is clear that M2 J1/J2 produces higher emissions than M1 J3/J4, by approximately 4–10 dB. The differences between these traces are the width of the traces and their position. M1 J3/J4 has a slightly wider separation between its traces, which are thinner and both traces are closer to the ground plane than those of M2 J1/J2.

Figures 11.25 and 11.26 present the same data in different manners. Smoothing data are done by connecting the peaks and removing the dips from a given dataset. When smoothing is carried out on original data, we end up with an overall trend of the highest emissions for each trace. Trace subtraction data look much different than the original data because dips in trace data become peaks when subtracted from another trace that is relatively flat. If we then smooth the resultant trace, the separation between the original traces becomes emphasized. Refer to Figure 11.27 for a visual explanation.

In comparing M1 J3/J4 to M2 J1/J2, we predict a reduction of 14.5 dB due to height and an increase of 6.5 dB due to an increase in the separation. Therefore, the delta between the two layouts is expected to be 8 dB and the measured one is 4–8 dB.

Unfortunately, we cannot use these relationships to predict the level of reduction for all configurations. For example, the measured reduction between M2 J3/J4 and M1 J3/J4 is 10–15 dB, whereas we would predict a reduction of 30 dB due to height and separation. However, we see that, at least

FIGURE 11.25 Direct comparison of the microstrip PCB trace emissions taken on the open-area test site at low frequency for the 100 Ω differential tests.

FIGURE 11.26 Comparison of the differential microstrip PCB trace emissions, normalized to M2 J3/J4 emissions taken on the OATS at low frequency for the 100 Ω differential tests.

FIGURE 11.27   Smoothing original trace data compared to smoothing trace subtracted data.

from 220 to 310 MHz, the higher the height above the ground plane and the larger the distance between traces, the higher the level of emissions. Also M1 J3/J4, which has the lowest height, is uniformly lower than the other two microstrips at almost every frequency. In the older (as in those from Reference [1]) measurements on microstrips with single-sided signals, we saw a direct relationship between emissions and height, and we see a measured 12–14 dB difference between M1 and M2, as described in Section 11.4.2, and a predicted 14 dB difference due to height and this confirms our earlier measurements on single-sided microstrips.

The resonances in M2 J3/J4 seem to occur at different frequencies than the other microstrips; however, if we were to compare the emissions at the corresponding resonances, M2 J3/J4 would fall between M1 J3/J4 and M2 J1/J2 in terms of emissions. The traces of M2 J3/J4 are as far from the ground plane as those of M2 J1/J2, but they are wider and more widely spaced to ensure the same characteristic impedance.

Measurements taken in the anechoic chamber suggest that M2 J3/J4 produces the highest emissions. However, emissions from M2 J1/J2 were still found to be higher than those of M1 J3/J4, as was also found on the OATS. See Figure 11.31 for a plot comparing the microstrip emissions measured in the anechoic chamber.

Above 1 GHz, only M2 J3/J4 was used for the 100 Ω differential impedance tests. At HF, this trace will represent an average microstrip PCB.

The relative plot on M1 J3/J4 in Figure 11.26 is much higher at 400 MHz than a comparison between the PCBs shown in Figure 11.25 This is because the traces were smoothed at different points in the process, and so in one case the separations are emphasized, whereas in the other the proximities are emphasized. In both plots, M1 J3/J4 is overall lower than the other two microstrips.

From the measurements, we find that attaching a 1 m cable to the ground plane of the differential trace with image plane changes the emissions. Theoretically, for a differential trace, the current in the ground plane is due to the electromagnetic image in the ground plane. This current smears over a larger area than for the single-sided microstrip, where all of the signal return current flows in the ground plane. In the symmetrical differential trace over an image plane configuration, the total voltage drop in the ground plane is theoretically zero and no C/M current flows. However, the two traces are not in the center of the PCB (nonuniform), also differential current results in radiation, especially from the unshielded load resistor, and this couples to an attached cable, and so the composite radiation from the differential traces and the cable is different from the same PCB without cables.

When the cable and PCB are 1 m long, the frequency at which maximum radiation is predicted is by 1 m = 0.5λ. This corresponds to a frequency of 150 MHz. (N.B. When a wire is attached to a large conductive enclosure connected to a ground plane, the maximum emission often occurs when the cable is 0.25λ long.) Figure 11.28 shows the radiation from the M1 J3/J4 PCB with an attached 1 m long cable. The radiation from the PCB with attached wire is lower than that without the wire at some frequencies, which seems counterintuitive. However, as described later, in radiated emissions, predicted and measured from a voltage source in a 0.15 m section of wire (simulating a C/M voltage in a PCB ground) and the effect of an attached wire, this is true as long as the wire is in the same plane as the measuring antenna. When the wire is mounted on a turntable and rotated with reference to the measuring antenna, then the level of emissions with and without the wire is virtually the same as explained in Reference [3].

This leveling off in radiated emissions above the resonant frequency is often seen in measurements of a product with attached cables tested on a turntable. It is contrary to the predictions based only on the D/M current, which predicts an increase of 20 dB/decade with increasing frequency after the resonant frequency.

The emissions from the M1 J3/J4 PCB with and without the wire attached to the ground plane are shown in Figure 11.28 from 200 to 550 MHz with the wire broadside on to the antenna. In these tests, the turntable was not rotated. The length of the wire and PCB in wavelength is provided in Figure 11.29 and for the PCB alone in Figure 11.30.

From 200 to 550 MHz, maximum emissions from the PCB, enclosure, and wire occur when the total length equals 0.5λ, 1.5λ, and 2.5λ and for the PCB alone when its length is equal to 0.5λ.

FIGURE 11.28 Emissions from the M1 J3/J4 microstrip with and without a 1 m wire attached to the ground plane. Turntable was not rotated.

FIGURE 11.29 Emissions from M1 J3/J4 with attached wire versus PCB, enclosure, and wire length in wavelengths.

FIGURE 11.30 Emissions from M1 J3/J4 PCB alone (no attached cable) with PCB length in wavelengths.

A comparison between the three microstrips measured in the anechoic chamber with the antenna located at a distance of 1 m is shown in Figure 11.31.

The emissions from the same microstrip without attached cable and measured on the OATS with the antenna at a distance of 3 m is shown in Figure 11.32.

We would expect the relative level of radiation to be a function of the height of the differential traces above the image in the ground plane as well as the separation between the traces.

FIGURE 11.31 Comparison of differential microstrip PCB trace emissions taken in the anechoic chamber at low frequency for the 100 Ω differential tests.

FIGURE 11.32 Comparison between the microstrip emissions measured on the open-area test site.

As the height increases, the emissions increase, and as the separation between the traces increases, so do the emissions.

The source driver is low-amplitude LVDS, and even with this low-level signal, all of the microstrip PCBs are above the class B limit. Only M2 J1/J2 approaches and exceeds the class A limit. M1 J3/J4 is the best with emissions above the class B limit at only 710 and 850–950 MHz.

11.4.2.2 Differential Trace Stripline Comparison

Figure 11.33 shows a comparison of the emissions from each of the five stripline traces tested with 100 Ω differential impedance at LF. In none of these measurements was the 50 Ω/100 Ω loads shielded. Below 200 MHz, none of the stripline traces radiated above the noise floor on the OATS. For this reason, data below 200 MHz are not presented. Although all stripline traces had very low emissions, S4 J1/J2 was the best of the striplines, especially from 100 to 300 MHz and above 600 MHz. The overall stripline with vias around the outside of the board was expected to be the best performer. In this configuration having the traces 2 cm in from the edge of the PCB reduced emissions compared to 6.4 mm from the edge, which is to be expected. However, the difference was only about 3 dB. In the D/M configuration with both an unshielded load and with a poorly shielded load, the localized striplines with overall upper ground plane were better. The emissions from PCB S2 J1/J2 with a height of 21 thou were lower than the emissions from S4 J1/J2 with a height of 55 thou. For S5 J3/J4, with a localized stripline and localized upper ground plane, from 280 to 400 MHz has emissions were substantially higher than the other traces and this was seen in the early measurements published in editions 1 and 2 of the book.

In the single-sided configuration, the signal return current flows in the upper and lower ground planes and down the length of the vias in the side of the stripline and so the magnetic field shielding

FIGURE 11.33 Comparison of stripline PCB trace emissions for the 100 Ω differential tests at low frequency on the open-area test site. Loads unshielded, as per the real world.

effectiveness of the stripline probably dominates. For the differential stripline, the return current is confined to one of the traces and the magnetic and electric fields incident on the inside surface of the stripline tend to cancel. The shielding effectiveness is likely to be a combination of *E* field and *H* field.

In the original radiated emissions measurements on single-sided striplines with a poor shield, the radiation from the load predominated, despite having a shield of limited effectiveness over the load. As expected, when comparing the emissions on the same PCB for the differential trace configuration versus the single sided, the differential emissions are much lower. In other words, the emissions from the unshielded load connected to a differential signal are much lower than when the load is connected single sided.

On the OATS, S2 J3/J4 produced the next least emissions of up to 400 MHz. Above 400 MHz, S4 J1/J2 was overall lower except from 900 MHz to 1 GHz. The comparisons up to 400 MHz are close to the measurements made in the anechoic chamber up to 400 MHz. As shown in Figure 11.34, when tested in the anechoic chamber, S2 J3/J4 performed slightly better than S4 J1/J2. Both of these traces have local striplines and full upper ground planes.

Anechoic chamber measurements confirm that S5 J3/J4 produces the worst emission profile of all the striplines below 400 MHz, as was also observed on the OATS. Like S2 and S4, S5 J3/J4 also has a local stripline, as shown in Figure 11.33; however, its upper ground plane only extends to the edge of the localized vias. All of the other striplines have full upper ground planes, which is almost certainly the reason that S5 J3/J4 is not as effective at reducing emissions.

In comparing the levels, the measurements on S3 J1/J2 and S3 J3/J4 in Figure 11.35, it can be seen that at HF the layout with the traces 5.8 mm from the edge of the PCB has higher emissions but only by 2–7 dB.

FIGURE 11.34 Comparison of stripline PCB trace emissions for the 100 Ω differential tests at low frequency in the anechoic chamber. Loads unshielded, as per the real world.

FIGURE 11.35 Comparison of S3 PCB trace emissions for the 100 Ω differential tests at low frequency.

It was found that the sides and rear of the SMA connectors mounted on the board contributed to emissions from the stripline and so these were shielded for some of the differential signal measurements (as denoted by CS).

To demonstrate that the load made a significant contribution to PCB trace emissions, a duplicate of PCB S4 J1/J2 was made and a copper shield soldered over its load and over the vias connected to the load on the opposite side of the board. The shield was found to be imperfect; however, the 100 Ω differential impedance LF measurements were taken from this modified board and compared to the data for S4 J1/J2. LS denotes load shielded. Figures 11.36 through 11.38 confirm that indeed the load does contribute to the radiated emissions. Figure 11.38 shows that S4 J1/J2 produced even lower emissions with both its load and connector shielded, no matter how poorly, than the other PCBs.

This result has an impact on PCB board design and has been seen many times in practical, stripline PCB layouts. In many of these boards, the emissions are predominantly from the ICs, connectors' external (to the stripline) traces, and oscillators, with emissions from the stripline at a very low level. This means that radiated emission limits may be exceeded even using the best stripline layout.

As expected, the radiation from traces outside of the stripline (unshielded) increases with increasing frequency as illustrated very well in Figure 11.38 for S4 J1/J2 versus S4 J1/J2 LS.

Taking into account the modified S4 J1/J2 traces, the best stripline turns out to be the S4 J1/J2 PCB with the poor load shield. S4 J1/J2 has a full upper ground plane with the ground planes tacked together on either side of the traces (localized stripline). However, S2 J3/J4 may exhibit lower levels of radiation with the LS. Figure 11.38 is a plot produced by subtracting the S4 J1/J2 LS CS–4 center vias emission data from the other stripline trace emissions measured in the anechoic chamber. The plot is an excellent illustration that the striplines produce similar emission profiles, but with varied field levels.

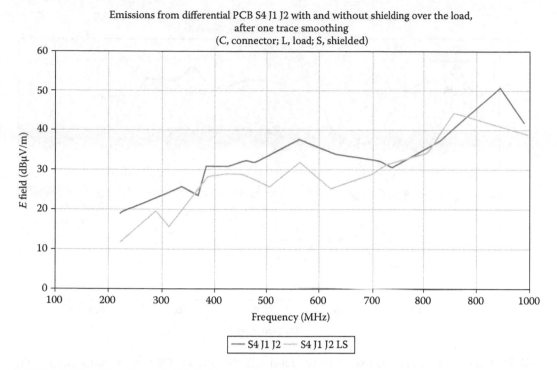

FIGURE 11.36 The effect of shielding (poor shield) the load on differential PCB trace S4 J1/J2 when measuring the radiated emissions on the open-area test site.

FIGURE 11.37 The effect of even a poor shield over the load on differential PCB trace S4 J1/J2 when measuring radiated emissions in the anechoic chamber.

FIGURE 11.38 Differential stripline emissions relative to S4 J1/J2 LS CS measurements taken in the anechoic chamber. *Note:* S4 J1/J2 LS results came from earlier testing before shielding connectors.

In summary, the PCBs with a local stripline, but full upper ground plane, performed the best. The next best was the full upper ground plane with the vias at the edge of the board (the full stripline). The next best was the local stripline with local upper ground plane.

Although the performance of the better striplines may be limited by the unshielded load, S4 may be better with a more effective shield.

Figure 11.38 seems to show that S4 J1/J2 with the LS, but the connector not shielded, has emissions lower than the same configuration with both CS and LS. However, this is not the case as the CS LS S4 J1/J2 had four center vias drilled out and this is why emissions are higher.

Above 1 GHz, three stripline PCBs were tested, S2 J3/J4, S4 J1/J2 LS, with four vias wide gap in the center of the stripline and S5 J3/J4 with 50% of its vias drilled out. All connectors were shielded on these boards before testing began in order to further isolate the emissions coming from the traces. The lowest emission profile of the striplines from 1 to 8 GHz continued to be the S4 J1/J2 trace with load and connector shielded, even with four of its center vias drilled out creating a slot. At 3.55 GHz, emissions from S4 J1/J2 with four center vias drilled out fell below the noise floor. One reason may be that the signal is degraded, that is, rise and fall time increased or amplitude reduced with the center vias drilled out.

Figure 11.39 shows that with the exception of the resonance from 2.3 to 2.8 GHz, the differential stripline PCBs continued to produce lower emissions than the differential microstrip (image plane) PCB from 1 to 4 GHz. Above 4 GHz, the striplines S2 J3/J4 and S5 J3/J4 with 50% of their vias removed have little superiority over the best differential microstrip PCB (which is in reality differential traces over an image plane). S4 J1/J2 LS with four of its center vias drilled continues to produce the least emissions up to 4 GHz, but this PCB had both load and connector shielded. Figure 11.40 gives a comparison of the differential striplines to the microstrip trace above 1 GHz. One conclusion that may be made from these results is that differential signals above a ground radiate at about the same level as differential signals in a stripline probably because the radiation from the load in the striplines predominates at these frequencies.

FIGURE 11.39 Stripline trace emissions from 1 to 8 GHz in the 100 Ω differential impedance tests.

11.4.3 TRANSMISSION-LINE COMPARISON

Two transmission lines were tested with 100 Ω characteristic impedances, TX2 J1/R1 and TX2 J3/R2. Figure 11.41 shows a comparison of the emissions from the transmission-line traces at LF on the OATS, and Figure 11.42 shows the emissions measured in the anechoic chamber. N.B. The abrupt change in emissions at 200 MHz was merely due to a change in antenna. The trace with the highest emissions was TX2 J1/R1, which is made up of three thin, widely separated traces. TX2 J3/R2 only has two thicker traces spaced much closer together.

It would have been possible simply to increase the distance between the two traces to see the importance of keeping the traces close together, but this would have meant that keeping the characteristic impedance of the transmission line at 100 Ω would have been difficult if not impossible. Instead, a single central line was used for the signal with the return current flowing on two traces on either side of the center trace. This meant that the trace separation could be increased from 17.3 thou to 70 thou with the impedance kept at 100 Ω. Using the Numerical Electromagnetic Code (NEC) to model the two structures at 550 and 400 MHz, we saw that the currents were exactly 180° out of phase in the two traces of TX2 J3/R2 and 180° between the center trace and the two outer traces of TX2 J1/R1 exactly as predicted. If we calculate the radiation from the two transmission lines based on the currents and distance of each trace from the antenna, we see TX2 J1/R1 should have a 21 dB higher level of emissions than TX2 J1/R1.

The measurements in Figure 11.41 show a difference of 26.6 dB at 550 MHz and 18.7 dB at 400 MHz. The major difference between our analysis and the measurements is that the two real test configurations have a 7 cm wide, 6 cm high, and 9.5 cm long metal enclosure located 4 cm from the end of the transmission line and 5 mm with the transmission line 5 mm from one edge of the enclosure. Due to this proximity and because the traces are offset from the center of the enclosure, C/M currents will flow due to displacement current between the traces and enclosure.

FIGURE 11.40 **(See color insert.)** Differential stripline emissions versus the differential microstrip (image plane) emissions from 1 to 8 GHz in the 100 Ω differential impedance tests. Normalized to M2 J3/J4.

Both transmission-line PCBs were tested above 1 GHz. Also tested was TX2 J1/R1 with its ground traces widened and moved in closer to the signal traces. From 1 to 4 GHz, TX2 J1/R1 again produced the highest emissions. TX2 J1R1 GMC, the trace with the ground widened and moved in, produced somewhat lower emissions at most frequencies than the unmodified board, but TX2 J3/R2 still performed the best up to 4 GHz. A secondary effect of moving the traces in is that the characteristic impedance of the structure is no longer 100 Ω. The peak emission from TX2 J3/R2 at 2.92 GHz, however, was 4 dB higher than the peak emission from J1/R1 at 2.67 MHz. See Figure 11.43 for the 1–4 GHz transmission line compared with the emissions normalized to trace M2 J3/J4 emissions.

At 4.1 GHz, TX2 J3/R2 took over as the worst transmission-line PCB trace and continued to produce the highest emissions of all the boards up to 8 GHz. At 4.58 GHz, TX2 J3/R2 produced emissions 29 dB higher than TX2 J1/R1 and 21 dB higher than TX 2J1/R1 GMC. This demonstrates that, at least for this layout, above 4 GHz moving the ground in closer to the signal, has no significant effect on the emissions.

11.4.4 Adding a Cable

Adding a cable to the load end of a PCB trace had different effects on the emissions, depending on the trace type. When attached to differential microstrip (image plane) or differential stripline PCBs, the cable had little effect on the emissions. This is expected for differential traces over an image plane or differential traces in a stripline. This is because theoretically the C/M current flow in the

FIGURE 11.41 Transmission-line emissions at low frequency for the 100 Ω differential tests on the open-area test site.

FIGURE 11.42 Transmission-line emissions at low frequency for the 100 Ω differential tests in the anechoic chamber.

FIGURE 11.43 **(See color insert.)** Transmission-line PCB comparison: 1–8 GHz.

ground plane/planes should be zero. However, some C/M current will almost certainly flow in a differential traces over an image plane due to the proximity of the PCB to the metal enclosure and the capacitance between traces and enclosure. This results in a displacement current flow between the traces and enclosure. In a differential stripline without shielded load, some C/M current may flow due to capacitance between the unshielded load traces and the enclosure.

Also, as described in Section 11.4.2.1, when the cable is electrically long, the radiation may be equal to or lower than the PCB without cable when the PCB is not rotated! See Figure 11.44 for "with cable/without cable" comparisons of microstrip emissions and S4 J1/J2 LS CS emission. The microstrips emissions were higher with attached cable from 60 to 170 MHz where the total cable, PCB, and enclosure length were electrically short.

When testing transmission lines, there was a clear increase in the radiation from the PCB when a cable was attached. The addition of a cable to a transmission line sets up a path along which the C/M current in the transmission line can flow, creating an electric field. In the case of microstrips and striplines, adding cable should theoretically have no effect on the emissions when the traces are configured differentially above a ground plane. This is because the current flowing in the traces sets up an image in the ground plane with symmetrical lines of current flow below differential lines. Current flowing in the image should, therefore, cancel and produce a voltage drop of 0 V thereby producing no C/M current flow. In contrast, transmission lines do not have a ground plane, and so the cable is attached to the signal trace at the load end of the PCB. Due to the voltage drops down the length of these traces, C/M currents are set up and flow out on the cable, as the results of this experiment indicate. Figure 11.45 provides a comparison of transmission-line emissions with and without an attached cable and TX2 J3/R2 emissions that are much higher with attached cable.

FIGURE 11.44 Emissions with an attached cable versus without cable for a differential stripline and differential microstrip traces.

FIGURE 11.45 Emissions with an attached cable versus without cable for the transmission-line traces.

FIGURE 11.46 Single-sided microstrip and stripline PCB radiation compared to the EN55022 and FCC Part 15 limits.

It will be seen in Section 11.4.6 that, with a true microstrip, where the signal return current flows under the trace in the ground plane, emissions increase significantly with attached cable before the cable, PCB, and enclosure length become electrically long.

11.4.5 SUMMARY OF SINGLE-SIDED SIGNAL PCB CONFIGURATIONS AT LOW FREQUENCY AND HIGH FREQUENCY

Microstrip and stripline radiated emissions compared to the limits of EN55022 and FCC Part 15 are shown in Figure 11.46 without an attached cable and in Figure 11.47 with a cable. It was found that the stripline PCBs radiate more than a microstrip PCB in the single-sided configuration over the frequency range found in Figure 11.48.

11.4.6 MEASURED DATA AND DETAILED PCB COMPARISONS FOR SINGLE-SIDED CONFIGURATIONS

11.4.6.1 Microstrip with and without Attached Cable

For the D/M connection of the microstrip where the ground plane forms an image plane, the voltage drop in the underlying ground plane is very low and the radiation with and without attached cable is virtually the same (Figure 11.49). However, when the microstrip is driven single sided, the signal current flows on a single signal trace and returns in the ground plane. A voltage drop is generated in the ground plane inductance and the resultant C/M current flows out on the attached cable, resulting in increased radiation. The maximum radiation occurs at the frequency where the combined cable, enclosure, and PCB ground plane length is 0.5 wavelength long. Figure 11.50 shows the relative E fields from 50 to 210 MHz for the M1 J3/J4 PCB and Figure 11.51 for the M2 J3/J4 PCB trace.

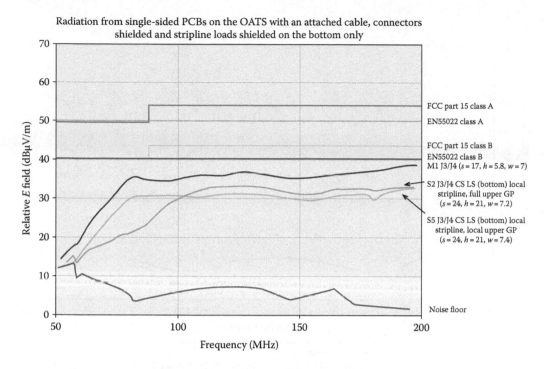

FIGURE 11.47 Single-sided microstrip and stripline PCB radiation with attached cable compared to the EN55022 and FCC Part 15 limits.

FIGURE 11.48 Frequency range where stripline PCBs radiate more than a microstrip PCB in the single-sided configuration.

FIGURE 11.49 Low-frequency radiation from M1 J3/J4 (connector shielded on top) with and without cable attached.

FIGURE 11.50 Low-frequency radiation from M1 J3/J4 (connector unshielded) with and without cable attached.

FIGURE 11.51 Low-frequency radiation from M2 J3/J4 (connector shielded) with and without attached cable.

The cable is 1 m long and the PCB ground plane and the enclosure are 0.165 m long. The measured E field from M2 J3/J4 is maximum at 128 MHz, close to the predicted 127.5 MHz, whereas the maximum from M1 J3/J4 is at 85 MHz.

The relative levels of radiation from 200 to 530 MHz are shown in Figure 11.52 for M1 J3/J4. The measured levels with and without the cable are practically the same. The cable is electrically long above 200 MHz, which means that the current changes phase down the length of the cable. The composite field at some distance from the PCB is due to the fields generated by the incremental currents in the cable.

In the initial measurements, the PCB was broadside on to the antenna and the turntable was not rotated. The results for M2 J1/J2 were also taken with the table rotated and with and without cable. The PCB was also rotated on the table in the "full" radiation measurements. The "full" radiation measurement was the maximum radiation for all possible angles. The E field from a source and a cable were measured and also predicted using the NEC with the wire broadside on to the measuring antenna/field analysis point. In this configuration, the E field reduced with increasing frequency with the wire electrically long. However, when the wire was rotated horizontally around its axis, both the predicted and measured field from the electrically long wire increased slightly with increasing frequency. It is this slight increase, compared to a PCB with no attached wire, which we would expect if the PCB with attached wire had been rotated. In the plots, rotating the PCB with reference to the antenna is designated "full radiation pattern" and achieves the maximum radiation from all possible angles.

The following plots are with connector poorly shielded in some cases, as noted.

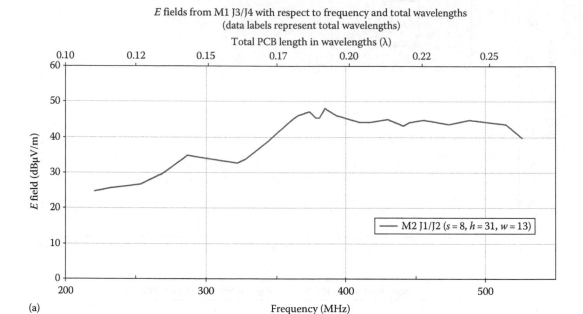

(a)

(b)

FIGURE 11.52 High-frequency radiation from M1 J3/J4 (connector shielded) with and without attached cable.

11.4.6.2 Stripline Comparisons at Low Frequency

The single-sided stripline PCBs generate some C/M current and the radiation increases with the attached cable. This is shown for the different stripline PCBs in Figures 11.53 through 11.56. This increase was not seen or at least was minimal when the signal was differential.

In a PCB containing digital logic and manufactured using good stripline techniques, it is often the ICs, oscillators, connectors, and passive components that are the main sources of emissions compared to the stripline PCB structure, and this is simulated to some extent in the measurements with the load unshielded.

11.4.6.3 Single-Sided Stripline Radiation Compared to Microstrip

Even though the level of shielding achievable with the four-layer boards is limited, it is still somewhat effective. The effect of shielding the 50 Ω load is shown in Figure 11.57 for the S3 J3/J4 board and in Figure 11.58 for the S3 J1/J2 board. This shows that the 50 Ω load on the top of the PCB significantly contributes to the radiation.

The plots from 1 to 2 GHz, 2 to 2.75 GHz, 2.75 to 4 GHz, and 4 to 8 GHz for the microstrips and striplines are shown in Figures 11.59 through 11.62. In these measurements, the connectors on all boards were poorly shielded, denoted as CS, and in some cases the loads were poorly shielded, denoted as LS. In Figure 8.12, the shielded striplines were only just better than M1 J3/J4. In Figures 11.59 and 11.60, the shielded striplines were worse at some frequencies than M1 J3/J4. A comparison between the striplines with loads and connectors fully but imperfectly shielded and no attached cable is shown in Figure 11.63 from 2.7 to 3.7 GHz and in Figure 11.64 from 7.5 to 8 GHz. A comparison between the worst microstrip and the worst stripline is shown in Figure 11.65. That between the best microstrip versus the worst stripline is shown in Figure 11.66, which shows that above

FIGURE 11.53 Low-frequency radiation from S2 J3/J4 (connector only shielded) with and without attached cable.

FIGURE 11.54   Low-frequency radiation from S3 J1/J2 (connector only shielded) with and without attached cable.

FIGURE 11.55   Low-frequency radiation from S3 J3/J4 (connector only shielded) with and without attached cable.

FIGURE 11.56 Low-frequency radiation from S5 J3/J4 (connector only shielded) with and without attached cable.

FIGURE 11.57 Comparison of single-sided S3 J3/J4 with attached cable with and without the 50 Ω load with limited shielding.

FIGURE 11.58 Comparison of the single-sided S3 J1/J2 with attached cable with and without the 50 Ω load with limited shielding.

FIGURE 11.59 **(See color insert.)** Comparison of the single-sided stripline PCBs relative to SS M1 J3/J4 with the PCB rotated.

High-frequency OATS single-sided PCB measurements relative to SS M1 J3/J4 CS baseline data
using full radiation pattern data from PCBs with connectors fully shielded
(SS, single sided; LS, load shielded)

Traces not visible are in the noise

—— Full SS M1 J3J4 CS	—— Full SS S3 J3J4 CS LS
—— Full SS M2 J3J4 CS	—— Full SS S5 J3J4 CS LS
—— Full SS S5 J3J4 CS LS–100% V	—— Full SS S3 J1J2 CS LS
—— Noise floor	—— Full SS S2 J3J4 CS LS

FIGURE 11.60 Comparison of the single-sided stripline PCBs and M2 J3/J4 relative to SS M1 J3/J4 with the PCB rotated.

5 GHz, the stripline is only just better. Probing the striplines showed that the maximum radiation came from the load, thus explaining the similarity between the microstrip and stripline.

The best microstrip has the signal trace close to the ground plane and the worst stripline has a localized upper ground plane with the signal trace close to the edge of this ground plane.

A comparison between the microstrips from 1 to 8 GHz shows that M2 J3/J4 has a signal trace of 31 thou above ground compared to M1 J3/J4 whose trace is 7.8 thou above ground and is shown in Figure 11.67. As expected, keeping the trace close to ground reduces the radiation as the inductance of the ground plane and voltage developed in the ground plane is lower when the trace is close to ground. This was also shown in the measurements described in the first and second editions of this book. The predicted ratio in dB due to the difference in the heights is 15 dB and this difference in radiation is seen at some frequencies.

A comparison of the radiation from the different striplines with load and connector shielded is shown in Figure 11.68 from 1.38 to 2 GHz. Adding a cable without rotating the PCB does not change the emissions at these frequencies.

The importance of rotating the PCB is shown in Figure 11.70 where the emissions increased from the S2 J3/J4 PCB by up to 16 dB with the PCB rotated. Figure 11.70 also shows the emissions from this PCB with the load unshielded. In subsequent measurements made with the vias in the striplines progressively drilled out, all of the PCBs were rotated. It was found that successively drilling out S3 did not change the level of radiation as probing revealed the major source to be the load resistor and traces on top of the board.

The reproducibility of the test setup is shown in Figure 11.70, as well as the effect the of angle. An average variation of about 4 dB is seen in measurements made a year apart, which can

FIGURE 11.61 Comparison of the single-sided stripline PCBs relative to SS M1 J3/J4 with the PCB rotated.

easily be accounted for in measurement accuracy, PCB orientation, receiving antenna calibration, location, height, etc.

Figure 11.69 compares the radiation of single-sided striplines with an attached cable. The stripline S3 has two trace configurations. One has the signal trace 2 cm from the edge of the PCB and the other 5.8 mm. No difference was seen between the two, and the reason again was that the radiation from the unshielded load predominated.

A PCB with a full upper ground plane radiates at a lower level than one with a localized ground plane when the vias are at the same location close to the signal trace over the 1–8 GHz frequency range as well as below 1 GHz. Figure 11.70 shows this from 80 MHz to 1 GHz and Figure 11.71 from 1 to 8 GHz.

In the differential trace stripline, the signal current flows on one trace and returns on the second trace. Any current flow in the stripline ground planes is due to displacement current and the magnetic fields both of which tend to cancel when the signal is balanced around ground. For the single-sided stripline, the signal current returns in the stripline ground planes and the radiation from the stripline increases. This is shown for all the striplines as well as the microstrips in Figures 11.72 through 11.82.

11.4.7 How Much Better Is a Practical Stripline PCB Layout than a Microstrip?

Most of the measurements were made on an OATS with the PCBs oriented at different angles to ensure that the highest level of radiation was measured.

In the previous measurements on the low-level radiation stripline layouts, it was found that, although very short, the vias and traces connecting to the load and the load itself will be a significant source of radiation. Initially, measurements were made on the original four-layer stripline PCB's two configurations. Both had the connector shielded and underlying signal via as well only one had

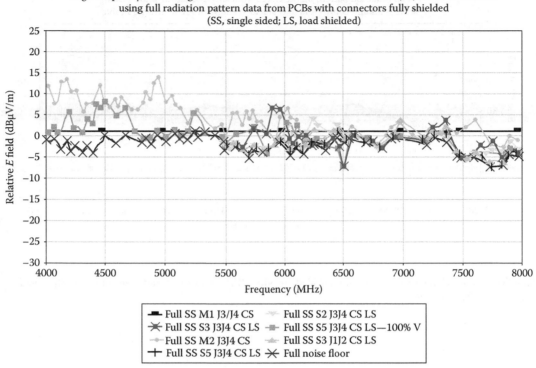

FIGURE 11.62 (See color insert.) Comparison of the single-sided Stripline PCBs with loads and connectors poorly shielded relative to SS M1 J3/J4 with the PCB rotated.

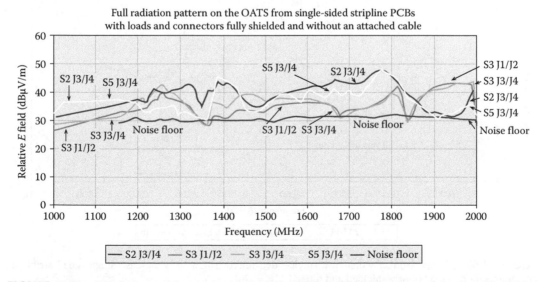

FIGURE 11.63 Comparison of the single-sided stripline PCBs with loads and connectors poorly shielded and with the PCB rotated.

FIGURE 11.64 Comparison of the single-sided stripline PCBs with loads and connectors poorly shielded and with the PCB rotated.

FIGURE 11.65 Comparison of worst-case single-sided microstrip with worst-case single-sided stripline with connector and load poorly shielded and rotated.

FIGURE 11.66 Comparison of best-case single-sided microstrip versus worst-case single-sided stripline with connector and stripline load poorly shielded and rotated.

FIGURE 11.67 Comparison of single-sided microstrips without attached cable.

FIGURE 11.68   Comparison of microstrips with an attached cable.

FIGURE 11.69   Comparison of single-sided striplines with attached cable.

FIGURE 11.70 Comparison of single-sided S2 J3/J4 multiple angle versus single angle and load poorly shielded and unshielded.

FIGURE 11.71 Comparison of single-sided stripline with localized upper plane versus overall upper plane.

FIGURE 11.72 Comparison of single-sided microstrip multiple angles versus single angle and single angle data taken a year apart.

FIGURE 11.73 Comparison of single-sided stripline with trace 5.8 mm from edge versus 2 cm from edge.

the load resistor and underlying signal via shielded. However, the shield in these four-layer boards was imperfect, and the tests on the radiation from three-layer boards having bare copper ground on top and bottom are described further in this chapter.

Four types of striplines were tested and a cross section of these different types is shown in Figures 11.15 through 11.18.

Figure 11.83 is a photo of S3 in the four-layer PCB with the vias tacking the upper and lower ground planes together at 3 mm intervals.

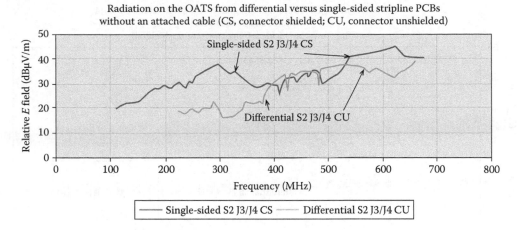

FIGURE 11.74   Comparison of single-sided S2 J3/J4 versus differential.

FIGURE 11.75   Comparison of single-sided S3 J3/J4 versus differential.

11.4.8   EFFECT OF SHIELDING THE CONNECTOR AND LOAD

Even the imperfect shielding achieved with the four-layer board reduced the emissions from the stripline PCBs when driven differentially as shown in Figure 11.84. With a cable attached to the ground plane and the PCB driven single sided, the effect of shielding is even more noticeable as shown in Figures 11.85.

11.4.9   PRACTICAL STRIPLINES VERSUS THE MICROSTRIP

In most practical PCBs, components are mounted on either one or both sides of the board and an edge connector is used for power and signals. Typically, only RF-type boards have a shield over the components and this shield is typically connected down to a lower overall ground plane using vias. Thus, the complete PCB is shielded and this is reasonably effective once the signals and power are brought out on shielded connectors with the metal case of the connector terminated to the wall of the PCB shield. Alternatively, the power can be brought out using feedthrough filters connected to the shield wall.

FIGURE 11.76   Comparison of single-sided S5 J3/J4 versus differential.

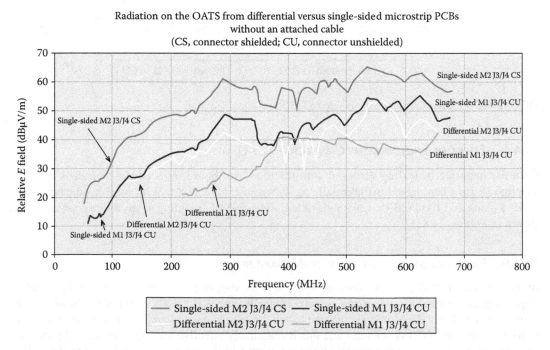

FIGURE 11.77   Comparison of single-sided M2 J3/J4 and M1 J3/J4 versus differential of same PCBs.

FIGURE 11.78 Comparison of single-sided S3 J3/J4 with differential, both with cables.

FIGURE 11.79 Comparison of single-sided S5 J3/J4 stripline with differential, both with attached cables.

FIGURE 11.80   Comparison of single-sided M1 J3/J4 with differential M1 J3/J4.

FIGURE 11.81   Comparison of single-sided S3 J1/J2 (connector shielded) with differential S3 J1/J2 (connector shielded).

FIGURE 11.82   Comparison of single-sided S2 J3/J4 (connector shielded) with differential S2 J3/J4 (connector unshielded).

FIGURE 11.83   S3 in four-layer PCB.

The question posed in this section is as follows: with the stripline load and connector unshielded, which most closely approximates a practical layout, and how effective is the stripline when compared to the microstrip?

Figure 11.86 compares the emissions from the worst case (highest level radiators) for both the transmission line and microstrip and stripline PCBs, when driven differentially. Although we identify the layouts as microstrip, they are in fact two signal traces above an image plane when driven differentially.

Figure 11.87 compares the emissions up to 1 GHz normalized to the microstrip M2 J3J4 that has two 51 mil wide traces separated by 170 mil and these are 31 mil above the ground plane.

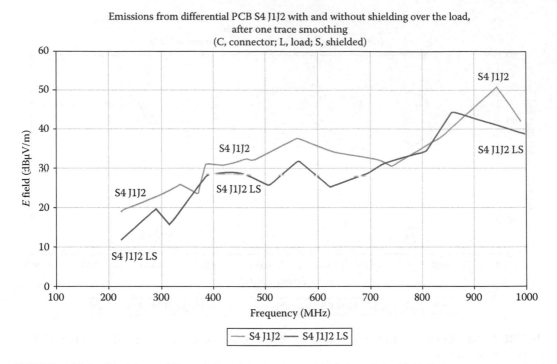

FIGURE 11.84 S4 driven differentially with the load shielded and unshielded (connector remained unshielded).

FIGURE 11.85 S4 driven single sided with attached cable and load shielded and unshielded.

FIGURE 11.86 Comparison of the worst transmission line, differential microstrip (image plane).

FIGURE 11.87 Comparison of the worst transmission line, differential microstrip (image plane) with emissions normalized to the microstrip. Stripline load was unshielded, as in the real world.

FIGURE 11.88 Comparison of the transmission line, microstrip, and stripline.

Figures 11.88 and 11.89 show the best transmission line to the best microstrip and to the best and second-best stripline, respectively, with load unshielded as in the real world.

Figure 11.90 compares the effect of via spacing on a stripline.

From these plots, it can be seen that the level of radiation from the microstrip is not a lot higher than the striplines with the load driven differentially. This means that differential traces where the traces are very close to the image plane is almost as effective as most striplines, and above 900 MHz, the difference is negligible.

Differential stripline trace measurements at LF for the 100 Ω differential impedance tests were performed on an OATS. The stripline load was unshielded as in the real world.

11.4.10 VIA SPACING

A number of spare stripline PCBs were manufactured, and to see the effect of via spacing, some of the vias down the length of the stripline were drilled out. The 18 vias connecting the upper and lower ground planes at the connector and the load remained. These vias are important as they carry the return signal current. The vias down the length of the stripline only make the shielding of the stripline more effective. The original spacing was 3 mm, and by drilling out, every other via PCBs with vias at 6 and 12 mm were made. Unfortunately, not only was the via removed but also a hole was made in the upper and lower ground planes. When driven single sided, the level of radiation from each type of stripline, regardless of whether 3, 6, or 12 mm spacing between the vias, was the same.

In early PCB measurements described in editions 1 and 2 of this book, a difference was seen due to via spacing even with unshielded loads. However, in these boards, the spacing between the upper and lower ground planes was 62.6 mil, whereas most of the present striplines had a 21 mil spacing. Also, the early measurements were made with logic levels at least 10 times higher than ECL using traces with uncontrolled impedances and terminated in typical logic resistive and capacitive loads. The wave shapes across the loads exhibited significant overshoot and ringing.

FIGURE 11.89 Comparison of the best transmission line to the best differential microstrip and the second best differential stripline.

Although a fully shielded stripline would be an impractical board and three layer boards are unpopular, due to warping, it was decided to manufacture three-layer striplines with vias designed into the PCBs at 3, 6, and 12 mm intervals, just to see the effect of via spacing. The three-layer PCBs allowed much better shielding of the connector and load as shown in Figure 11.90. The three-layer boards were driven single sided and had spacing between the upper and lower ground planes of 21 mil (0.5334 mm). The boards tested were the S2, S3, and S5. S4 that has a height of 55 mil (1.4 mm) does not allow a single-sided connection.

The levels from the fully shielded boards were very low and not measurable on an OATS. Instead the anechoic chamber was used. For frequencies of 1 GHz and higher, a double-ridged guide antenna was connected directly to a 30 dB preamplifier. Very low loss cables were then used to connect the preamplifier to the spectrum analyzer outside of the anechoic chamber. In order to make any measurements above the noise floor, the antenna had to be located 0.22 m from the PCB!

The highest level of E field measured at 0.22 m was 11.2 dBμV/m and this would disappear into the noise floor of the measuring equipment and the ambient at a distance of 3 m.

The angle of the PCB relative to the antenna was changed to ensure maximum pick up by the antenna.

No difference was again seen between the 3, 6, and 12 mm spaced via boards. Using a small 3 mm probe, leakage was found at the SMA connectors, around the top of the enclosure and at the power-line filters. The enclosure lid was connected to the body using finger stock and was further sealed with adhesive copper tape. The SMA connector is gold plated and according to the manufacturer has a 100 dB of shielding from 1 to 2 GHz.

Copper tape was soldered over the gap between the lid and the enclosure case, braid was connected over the SMA connector and held in place by a hose clamp, and additional ferrites were connected at the power-line filter. The ferrites were covered in copper adhesive tape connected to the enclosure on all sides.

FIGURE 11.90   Effect of via spacing on the S5 stripline.

With these modifications, the stripline with overall upper and lower ground planes (S3) again showed no difference between 3, 6, and 12 mm spacing. However, S 2 and S5 did show some difference in emissions between the boards.

For S5 the only significant increase was with 12 mm spacing and it may be assumed that the emissions at 3 and 6 mm were at the ambient level. Figure 11.91 shows that the emissions between 1.2 and 1.7 GHz S2 showed some small difference between the 3 and 6 mm boards with a significant increase again from the 12 mm board.

FIGURE 11.91   Effect of via spacing on the S2 stripline.

The measured shielding effectiveness of the striplines is higher than that expected based on calculations. Based on our measurements, we predict that the radiation from components, edge connectors, and short lengths of trace that run outside of the stripline in practical PCBs will predominate over emissions from the stripline PCB structure. Although the distance between vias may appear unimportant in the practical stripline PCB, there exist a number of very important reasons for maximizing the number of vias and these are discussed here.

The differentially driven stripline PCB with the resistive 100 Ω load left unshielded has a lower level of emissions only at certain frequencies when compared to two differential traces close to and over an image plane. At high frequencies, the level of emissions is similar. This configuration with unshielded components more accurately represents the real world.

In the stripline PCB configurations, which have a fully shielded load and connector, the emissions are extremely low, much lower than would be predicted from simple calculations.

As the emissions from the stripline are so low, emissions from connectors, short lengths of unshielded traces outside of the stripline, and ICs will predominate, and the distance between the vias tacking the upper and lower ground planes together is not so important.

However, from observation of numerous PCB layouts, the radiation from a PCB with upper and lower ground planes tied together with a number of vias is invariably lower than a PCB with a single ground plane. Also a PCB with an overall ground plane and an incomplete second ground plane with small islands are only more effective than the PCB with a single ground plane when the islands are connected to the overall ground plane with a number of vias. The reasons for this are many and include the following:

1. When a signal trace is isolated from the ground plane by a power plane, emissions are high.
2. Assume that in five layers of a stripline as shown in Figure 11.92, layer 1 is the upper ground plane, layer 2 is signal, layer 3 is power plane level, layer 4 is signal, and layer 5 is the lower ground plane. If the signal transitions from layer 2 to layer 4, or vice versa, through the power plane, it is very important to add as many ground vias connecting the ground planes, layer 1 and layer 5, around the signal via. This ensures that the signal return current flows close to the signal with reduced radiation.
3. At IC pins, it is most important to via the IC ground pins to both upper and lower ground planes so that signal return currents flow either in both ground planes or in the appropriate ground plane.
4. By adding more ground vias, the goal is to keep the signal return current as close to the signal path as possible.

The fully shielded stripline provides a very high level of shielding—higher than would be predicted from simple calculations. Removing vias down the length of the stripline did not degrade this shielding significantly; however, these vias effect the shielding of the stripline and they are not used to carry signal return current. Based on the leakage from the SMA connector, removing current-carrying vias will almost certainly increase the impedance and the voltage drop across these ground connections at the connector and load.

FIGURE 11.92   Adding ground vias around a signal transition.

11.5 PRACTICAL PCB LAYOUT

Emissions from traces close to ground when compared to a PCB with only signal and power traces have been extensively demonstrated in practical layouts as well as in the measurements shown in Section 11.4.2. As discussed in Section 11.8.2, a ground plane must be solid directly under all data and signal traces. In a complex board, traces run all around the board on multiple layers, and this means that the complete board area must be covered in a solid ground plane that is most conveniently used for digital power and digital signal return. See Section 11.8.2 to find out what constitutes a "good" ground plane. The closer the "hot" traces (traces that carry digital signals and clocks with fast edges over a distance greater than 25 mm) are to the ground plane, the lower the emissions. Therefore, route all the high-speed, long tracks on the layer closest to the ground plane. These hot tracks can be on the layer on either side of the ground plane when this is buried. Tracks carrying less-active signals or that are short can be carried on tracks further away from the ground plane. Keep hot tracks as short as possible and include series resistors at the source end to increase the rise and fall times. Choose buffered input devices to reduce the load on the bus or clock. Where possible, run clocks and data buses in localized striplines in which the edges of the upper ground plane is tacked to the lower ground plane, with vias a maximum of 13 mm apart (closer if possible). If possible extend the stripline over the whole area of the board with the vias located at the periphery of the board.

A common error is to leave large gaps between the vias at locations around the edge of the board. In this case, the emissions are concentrated at these locations, and some enhancement, especially at high frequencies, may be seen. It is possible to use a power plane on one surface of the board with the ground plane on the opposite surface of the board to form a stripline. The two planes are connected together at RF by surface-mount capacitors located at intervals no greater than 25 mm around the periphery of the board. A capacitively coupled stripline is not as effective as a stripline formed by two ground planes connected directly together, and if the capacitors are greater than 25 mm apart, any benefit over a microstrip is greatly reduced. If traces used for low-level signals such as analog, video, or control are present on the board, then these signal tracks may be located on the outside of the stripline ground planes to reduce crosstalk between the digital signals and these low-level signals. However, if these low-level signals reference a noisy ground, namely, the inner surface of the stripline digital ground plane, then the emissions from these tracks may be high, due to C/M noise. If this is a potential problem and the possibility of crosstalk is real, then the solution is to construct an inner stripline dedicated to the low-level-signal tracks and signal return with the associated ICs located directly over this area.

One observation, supported by the Section 11.4.2 measurements and that has been made in measurements on numerous boards, is that the emissions from ICs and their pins can be sufficiently high for the board to fail commercial requirements despite the very best possible PCB layout. In microstrip layouts in which the ground plane is embedded, add a small section of ground plane directly under the IC connected by vias to the underlying ground plane. If the IC has a metal lid, connection of the lid to the ground plane at least at the four corners can reduce emissions, especially at low frequencies. Local shielding of the IC, as discussed in Section 11.9, in the microstrip layouts in which the ground plane is imbedded has the problem of increased radiation from the tracks exiting the shield. If a stripline is used with two outer ground planes, then the connections to the IC pins can be made with vias under the shield to the inner layers. The shield can then be contacted around its periphery to the ground plane, and this configuration is extremely effective. If it saves the use of a metal or conductively coated equipment enclosure, this local shield can be very cost effective.

An image plane may be constructed from aluminum or copper foil enclosed in a nonconductive envelope. When this insulated foil is placed in close proximity to either side of a PCB that contains "hot" traces or ICs that radiate at a high level, the level of radiation can be reduced from that side of the board. When the foil can be wrapped around one edge of the PCB and cover both sides of the PCB, radiation can be reduced from both sides of the PCB. However, the worst-case maximum emissions from the PCB when measured on an OATS, with the PCB rotated through 360° and with the antenna raised and lowered, may not be reduced. This is because the image plane changes the radiation pattern from the PCB and the level of emissions from the gaps at the edge of the PCB may not be reduced, but may even increase at some frequencies.

In one case, PCB-sourced emissions coupled to the antenna of a wireless plugged into the PCMCIA slot of the equipment. The wireless was desensitized due to the broadband noise (typically from data or switching-power supplies) and the harmonics from clocks injected into the antenna. With an image plane covering both sides of a PCB and wrapped around the edge of the PCB facing the antenna, a reduction was seen in the very localized coupling between the PCB and the antenna, and the wireless reception was improved. The overall radiation from the PCB due to the "wrap-around" image plane was not, however, greatly reduced.

The step-by-step design and the rules to follow in laying out a PCB are as follows.

1. Evaluate the EMC requirements placed on a PCB or PCBs. That is, what requirements have to be met? Is crosstalk or common ground impedance coupling to sensitive circuits a potential problem? Is the PCB housed in a conductive enclosure? And is this, or can it be made into, an effective shield? Decide how much emission reduction at source is required. For example, a high-speed PCB with many LSI devices contained in a nonconductive enclosure with attached signal or power cables requires the maximum possible level of source reduction by "good" circuit design and PCB layout. At the extreme, use an LCR circuit to make the clock as close to a sine wave as possible.
2. Locate ICs so that the interconnects are as short as possible. Locate oscillators as close as possible to ICs that use the clock and as far away from signal interface and low-level-signal ICs and tracks as feasible.
3. Minimize path lengths between sources and loads and between loads sharing a common clock or bus, or use shielded stripline over these connections.
4. If microstrip or stripline configurations cannot be used, keep signal and return tracks as close together as feasible.
5. Reduce the drive current on tracks by minimizing loading, typically by the use of buffers at the load end or series resistance.
6. Use the slowest logic possible to achieve the required performance.
7. Increase the decoupling to one capacitor per IC when the ICs are not located close together. Use surface-mount capacitors for decoupling.
8. Choose the location of common ground and chassis ground and isolation of grounds in accordance with the guidelines of Section 11.8.2.
9. Often, placing a low-value capacitor (0.01–0.1 μF) in parallel with a high-value capacitor (1–10 μF) is effective. However, if the capacitors leads are not very short, nothing may be gained.
10. Choose the grounding scheme from Section 11.8.2 that will reduce radiation from the attached cables.
11. Reduce the number of signal tracks that change state simultaneously.
12. If power and return tracks are the source of radiation, add inductors between the power supply and the ICs and increase the number of decoupling capacitors.
13. After the initial layout, use a viewer, such as GerbTool or a Gerber viewer available on the Internet, to trace the routing and component placing, to ensure that none of the design rules has been violated.

If radiation from interface cable predominates and the PCB is located in an enclosure, the multiple connection of the power/signal ground plane to chassis will reduce C/M noise voltage developed across the board. Unfortunately, the RF current flow in the enclosure will increase, radiation from seams and apertures is likely to increase, and the net gain may be zero with a poorly shielded enclosure.

It has been recommended that the width of the power plane be made smaller than the ground plane to reduce fringing fields. Although the mechanism for the reduction is unclear, in measurements on a full stripline with a 40 mil height versus the 1 cm wide stripline, the full-stripline radiation is higher than the 1 cm stripline above 350 MHz. However, in the measurements in Section 11.4.2 where the height is 21 mil, the radiation from overall ground plane is lower than the partial ground plane as shown in Figures 11.33 and 11.34 and the same in Figure 11.39.

Although effective at reducing current loop area, effective decoupling around an IC can increase the magnitude of current spike flowing in the IC as it switches. Inclusion of a resistor or a ferrite bead between the decoupling of the IC and the power pin will reduce the current spike at the expense of increased bounce in the supply voltage (see Figure 5.14). Ferrite beads have also been used on power return pins, but this will inherently result in a higher ground bounce voltage.

Use ICs with the minimum number of internal devices changing state simultaneously.

When using a custom device, ensure a layout with minimum internal path lengths and one that reduces C/M noise generation on the input and output pins. Series resistance may be incorporated within the device, in series with the supply, to limit switching current. The value of the resistor will depend on the tolerance of the device to supply bounce.

11.5.1 High-Frequency (1 GHz and above) Radiation from PCBs

As clock speeds increase, along with the slew rate of the logic, HF PCB radiation also increases. At HF, microstrip or other transmission-line configurations are typically used, with the line terminated in its characteristic impedance Z_c. At these high microwave frequencies, radiation from microstrip discontinuities and lines becomes significant when not masked by C/M current flow-generated radiation. The measured results are often above the predicted results for purely D/M current radiation at 1 GHz, because even at this HF, the C/M current-contributed radiation cannot be ignored! However, as the structure becomes electrically long and the frequencies of interest are above 1 GHz, the D/M current radiation should begin to dominate. Any impedance mismatch at the end of a line will form a discontinuity and increased radiation. A high-impedance termination relative to Z_c will radiate more effectively than a low impedance one, approximately 2.5–19 dB higher. In addition, at higher frequencies, the higher the track above the ground plane, the higher is the radiation. For example, increasing the frequency or the height of the track above the ground plane by a factor of 10 (20 dB) increases the radiated power by a factor of approximately 100 (40 dB) and the E field by approximately 20 dB.

Using the many equations for radiation from discontinuities in a microstrip based on the signal current flow invariably results in predicted levels of E field that are much higher than the measured one from typical printed circuit boards.

It has often been recommended not to daisy-chain the logic connections to a clock but to fan out to clocks, as shown in Figure 11.93. At LF and when the load is not matched to the line, this arrangement has not been demonstrated to be advantageous. However, with matched transmission lines, the fan-out arrangement should be used.

Right-angle bends exhibit increased capacitance, which represents a discontinuity and increased radiation. Use of a 45° bend in a microstrip line, as shown in Figure 11.94, will reduce radiation at very HF. However, for 74F, 74AS, 74S, 74, 74C, 74HC, and ECL100 or similar-speed devices, any reduction in radiation as a result of using the 45° bend is insignificant and, in measurements, was masked by the radiation from the impedance mismatch at the end of the line and the C/M current-contributed radiation from the track.

The theoretical difference in either horizontally or vertically polarized radiation from a straight microstrip compared to one with a 90° bend is ±6 dB at HF, that is, a mean difference of 0 dB. The slight reduction in measured radiation between the 90° bend and the straight microstrip was seen in the horizontal E field and is almost certainly due to the decreased length of the track due to the bend.

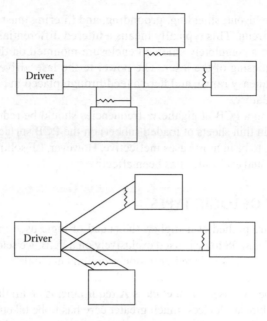

FIGURE 11.93 Daisy-chain versus fan-out of clock or high speed data.

FIGURE 11.94 Forty-five-degree bend in a PCB trace.

The stripline PCB configuration should be used wherever practical. If stripline cannot be used, one technique for reducing radiation from a microstrip is to use differential signals for long computer clock runs on the board and this is demonstrated in Section 11.4.2.2. The two signal traces should be of exactly the same length and be located as close together as allowed. The loads at the far end of the differential lines must be well matched, and this is easier when ECL, ECLlite, or ECLinPs is used. C/M currents can be reduced by running both traces through a PCB-mounted balun. But beware: if the connections to and from the balun are too long, these may radiate more than if the balun is not employed. The FCC requires that radiation from digital devices (unintentional radiators) be measured above 1 GHz to the fifth harmonic of the highest fundamental frequency or to 40 GHz, whichever is lower. Many fiber-optic drivers operate at 2.5 GHz or higher, and measurements up to 25 GHz must be made. When designing a laser driver for fiber optics, the layout, shielding, grounding, and filtering must be at the same level as any medium-power RF circuit. This typically means a filtered differential input to the driver IC, which is contained inside a completely shielded enclosure mounted on the PCB. The enclosure must contact the metal housing of the laser. The power to the laser driver must be well filtered over the 2.5–25 GHz frequency range, and here a feedthrough placed in one wall of the shielded enclosure is almost mandatory.

Although radiation from a PCB at gigahertz frequencies should be reduced by placing microwave absorber (available in thin sheets of loaded rubber) on the PCB surface or over ICs, practical experience has shown this to be in most cases ineffective. However, LF solid ferrite material used to fill apertures in heatsinks and enclosures has been effective.

11.6 COMPARISON OF LOGIC TYPES

As clock and data rates are pushed ever higher, the speed of logic used in digital equipment has increased. When standard CMOS logic is used exclusively on a PCB, we seldom see a problem with meeting class A and even class B radiated emission requirements, unless the designer has done a very poor job of laying out the board. This was not the case with standard 74-type TTL ICs, with which some care had to be exercised to meet class A requirements. With the advent of high-speed CMOS and higher-speed bipolar devices, much greater care has to be taken in the PCB layout, circuit level reduction, and possibly shielding, to reduce emissions to acceptable levels.

The concern has been expressed that use of these new high-speed logic types will result in significant radiation in the 1–10 GHz frequency range, in which the FCC covers in its limits. The highest-speed device that was tested was the ECLinPS device, with a rise and a fall time of between 0.275 and 0.6 ns. The ECLinPS device did not radiate levels at HF above those of the 74F device when the fundamental frequency was low (10 MHz). This is not surprising because the criteria for HF radiation when the fundamental clock frequency is low are the rate of change of voltage and current, that is, the slew rate, and not the rise and fall time alone. As discussed later, the slew rate of the ECLinPS device is very similar to the slew rate of the 74F04 device.

Figure 11.95 illustrates the voltage amplitude in the frequency domain (the spectral emission lines) generated by the 10 MHz clock measured across a 50 Ω load at the output of the ECLinPS device. As the speed of the device increases, so does the maximum frequency at which the device can be used. The maximum frequency of the "ECLinPS Lite" family is 2.2 GHz for flip-flops and 1.4 GHz for buffers with a minimum rise and fall time of 0.1 ns. With a 1.4 GHz fundamental and a 0.1 ns rise and fall time, the first breakpoint in the envelope of spectral emission lines, as shown in Figure 3.5, is at $1/\pi d = 88$ MHz, where d is the pulse width (3.6 ns). After the first breakpoint, the amplitude of the spectral emission lines reduces as a function of 20 dB per decade of increasing frequency. Above the second breakpoint of 3.18 GHz, given by $1/\pi t_r$, the amplitude reduces as a function of 40 dB per decade of frequency.

If the 74F04 and 74AS04 devices had been tested close to their maximum frequencies of 124 MHz, say, at 120 MHz, the levels of emission would have been higher at the same frequency (than the emission from a 10 MHz clock). For example, at a frequency of 360 MHz (the third harmonic of the 120 MHz clock), the 120 MHz clock emission would be 25 dB higher than the emission from the 10 MHz clock and 21 dB higher at 1080 MHz (the ninth harmonic), assuming that the rise and fall times stay constant.

If the ECLinPS device had been operated at a 450 MHz clock frequency instead of 10 MHz, the emission levels at 2.25 GHz would have been approximately 33 dB higher. The increase in emissions corresponds closely to the increase in clock frequency, that is, from 10 to 120 MHz = 21 dB and from 10 to 450 MHz = 33 dB. With the use of the microstrip transmission-line PCB configuration and a 10 MHz clock, none of the logic types resulted in radiation above 1200 MHz, at a level above the noise floor of the measurement system, even with the antenna located 0.55 m from the PCB and with an attached cable on the PCB. Standard CMOS could not be included in

FIGURE 11.95 ECLinPS device frequency-domain voltage measured into a 50 Ω load.

the measurements due to an upper frequency limit below the 10 MHz clock used in the test setup. Table 11.1 illustrates the upper frequency limits of the logic families tested.

The purpose of measuring the radiation from the microstrip PCB configuration with different logic families was twofold. With a comparison available between the 74F04 device, used in all the comparative measurements made on PCB layouts, and other logic types, the radiated emissions from the PCB layouts can be scaled for the other devices. In addition, if a choice was available, the device with the lowest emission level and the required upper frequency limit could be selected. For example, standard TTL is no longer a popular logic family, and in the future it may well be phased out. However, if the typical upper frequency limit of 25 MHz, 13 ns rise, and 6 ns fall times, with

TABLE 11.1
Upper Frequency Limits of Logic Families

Logic Family	Frequency Limit (MHz)
ECLinPS	1200
74F	125
74AS	125
74S	95
74HCT	50
74ALS	35
74LS	33
74	24

propagation delays of 30 ns, are acceptable, then TTL is the best bet, next to standard CMOS, for reduced HF radiation.

The devices tested were all hex inverters, with the exception of the MC100E111 ECLinPS device, which was a 1:9 differential clock driver. The hex inverters were the 74F04, 74AS04, 7404, 74LS04, 74ALS04, 74HCT04, and 74S04 devices. The load for the 74XXX04 devices was a 330 Ω resistor in parallel with a 47 pF capacitor, the same load as used in the PCB radiation measurements. This load is considered typical by many IC manufactures and is often used as the test load. The capacitance of 47 pF corresponds to a fan-out of between 8 and 10. The ECLinPS device was tested with a 47 Ω resistor in parallel with a 4 pF capacitor. According to the manufacturer Motorola, "The input loading capacitance of the device typically measures 1.5 pF and is virtually independent of input fan-out as the device capacitance is less than 5% of the total." The ECLinPS device was also tested with a load of 56 Ω, which matches the characteristic impedance of the microstrip PCB used in the test. The radiation from the ECLinPS with the 47 Ω and 4 pF capacitor load was up to 20 dB higher than with the purely resistive load, and it is the capacitive and resistive load results that are presented here.

The 74AS04 device was also tested over the 200–1000 MHz frequency range with different loads. The radiation from the 74AS04 loaded with a 330 Ω resistor, when compared to the standard 330 Ω resistor in parallel with a 47 pF load, is both higher and lower at different frequencies. In an open circuit, the 74AS04 radiation is up to 10 dB lower, but with a 68 Ω load, it is up to 27 dB lower. The 68 Ω load is an invalid load but illustrates the importance of the amplitude of the output waveform, which reduces considerably with the 68 Ω load, in determining both the LF radiation and the slew rate and, therefore, the HF radiation.

The maximum-voltage amplitude at the fundamental clock frequency determines the LF level of emissions, and the slew rate determines the HF emissions. Table 11.2 plots the maximum voltage excursion during the falling edge of the pulse, V_f, and the voltage measured after the settling time, V.

Table 11.3 shows the rise and fall times over a 3 V excursion during the transition. The more common definition of rise and fall times is the time measured between 20% of the voltage transition and 80% of the voltage transition. Because the voltage transitions varied so greatly from device to device, the traditional measure of the rise and fall time will have little relationship to the voltage and current frequency component generated by the pulse. Since the ECLinPS device maximum excursion is 0.76 V, only the slew rate for this device is shown in Table 11.3. Even the voltage slew rate does not adequately describe the HF components generated by the pulse, because the amplitude of the current pulse plays an important role. For example, the voltage amplitude of the emission at 770 MHz from the ECLinPS device is 1.92 mV, which is lower than that from the 74F04 device at 3.7 mV, whereas the current amplitude from the ECLinPS device at 770 MHz is slightly higher, at 7.9 mA, compared to the 7.3 mA of the 74F04 device.

Tables 11.4 through 11.6 shows the upper frequency limits and rise and fall times of logic.

The measured fall time over the standard 3 V excursion is used to determine the slew rate in Table 11.3.

TABLE 11.2

Maximum Voltage Excursion

Logic Family	V_f	V
74F	6.0	3.8
74AS	6.5	3.58
74HCT	7.0	5.08
74ALS04	5.5	3.88
7404	5.5	3.28
74LS	5.2	3.8
ECLinPS	0.76	0.76

TABLE 11.3
Rise and Fall Times

Logic Family	t_f (ns)	t_r (ns)	Slew Rate (V/ns)
74F04	3.01	4.0	1.0
74AS04	3.28	3.9	0.9
7404	4.36	10.6	0.68
74LS04	4.45	8.67	0.67
74ALS04	3.5	7.27	0.85
74HCT04	2.6	3.29	1.15
74S04	3.05	5.0	0.98
ECLinPS 5	—	—	0.85

TABLE 11.4
Upper Frequency of Logic Families

Logic Family	Frequency Limit (MHz)
HMC (Hittite)	28,000
ECLin PS	1,200
HSTL	1,000
LVPECL	800
LVTTL/LVCMOS	350
74LVT BiCMOS	150
74F	125
74AS	125
74HCT	50
74ALS	35
74LS	33
74	24

TABLE 11.5
Maximum Voltage Excursion

Logic Family	V_f	V_r	Type[a]
HMC (Hittite)	1.45	1.45	D
ECLin PS	0.76	0.76	
HSTL	1.9	1.9	D
LVPECL	0.95	0.95	D
LVTTL/LVCMOS	1.2	1.2	S
74LVT BiCMOS	2.2	2.2	
74F	6	3.8	S
74AS	6.5	3.58	
74HCT	7	5.08	D
74ALS	5.5	3.88	
74LS	5.2	3.8	S
AVC	1.8	1.8	S
GTLP	3.0	3.0	S

[a] D, differential; S, single sided.

TABLE 11.6
Rise and Fall Times

Logic Family	t_f (nS)	t_r (nS)	Slew Rate (V/nS)
HMC (Hittite)	0.014	0.015	103.5
ECLin PS			0.85
HSTL	0.35	0.6	5.43
LVPECL	0.2	0.2	4.75
LVTTL/LVCMOS	1	1	1.2
74F	3.01	4.0	1.0
74AS	3.28	3.9	0.9
74HCT	2.6	3.29	1.15
74ALS	3.5	7.27	0.85
74LS	4.45	8.67	0.67
AVC	10	10	0.18
GTLP	1	1.4	3.0

11.7 CIRCUIT-LEVEL REDUCTION TECHNIQUES

High current spikes flow in the logic signal and return interconnections as the output logic device changes state and charges/discharges the input capacitance of the load device and removes the charge from either a base/emitter junction or a diode. During the transition, the load on the output device is, therefore, predominantly capacitive, and, depending on the geometry of the interconnections, the logic type, and the frequency, the source of radiated emissions is typically a low impedance (i.e., a predominantly magnetic field source). The significance of a low-impedance predominantly H field is discussed in Section 11.9, when examining the shielding effectiveness of a conductive enclosure around a PCB. The current pulse is the predominant source of emissions at frequencies less than approximately 200 MHz. However, emissions created by the voltage transition should not be ignored in an EMC prediction, especially at HF or when the load is of high impedance with low load capacitance. Even with an open circuit at the far end of the PCB, displacement current flows due to the charging and discharging of the capacitance between the signal and return paths. The source of radiation is then predominantly E field, and it is typically the E field, which is specified and measured, in commercial requirements.

Measurements made on a microstrip PCB layout with either an open circuit at the far end or a 330 Ω resistor in parallel with a 47 pF load showed a 10–38 dB increase in radiated level for the microstrip with load below 137 MHz. However, above 137 MHz, the O/C microstrip radiation was from 0 dB to 15 dB higher than the microstrip with load, illustrating the importance of the voltage-generated E field. Some of this increase with an open circuit may be attributable to the decrease in rise and fall times of the unloaded 74F04 driver.

It is the current flow in the signal return connection (PCB track or ground plane) that develops a voltage in the impedance of the return, due to both the displacement current between the signal path and return path and the load current. This voltage can generate a C/M current, which is the prime source of radiation from the PCB and any attached connection.

Many of the measurements on PCB radiation described in this section were made with a cable attached either to the ground plane or to the signal return track at the load end. Thus, in reducing the radiated emissions from a PCB track, the voltage drop in the signal/power return connections and/or the magnitude of the current should be reduced. On an electrically long line, use of the correct value of termination resistance will mask the peaks and troughs in emissions at resonant and antiresonant frequencies. The total magnitude of emissions tends to increase. This may at first appear to be questionable, and it is correct to say that some of the peak emissions due to resonance may decrease with a termination resistor. However, in a typical microstrip transmission line with a 65 Ω characteristic impedance, adding a 65 Ω termination resistor will increase the current flow in the line, especially at LF, where the typical load impedance presented by the device is high. With an increase in current flow, the level of radiation will increase, especially at transmission-line antiresonant frequencies.

If a trace on an existing PCB is found to be a major source of emission at a single or a few frequencies, changing the termination resistance may reduce the problem frequencies, although the emission amplitude at other frequencies often increases. For example, changing the termination resistor from 700 to 600 Ω resulted in a predicted 40 dB reduction in radiated emissions around 300 MHz. This effect occurred with a track length of 51 cm, which equals 1.5λ, at 300 MHz. The source was a 10 MHz clock with a 50% duty cycle and a 10 ns rise time.

In most practical applications, the load on the line at frequencies as high as 300 MHz is a much lower impedance than 600 Ω due to the IC input capacitance. The only way to increase the load impedance at high frequencies is to add series resistance or a series resistor/capacitor combination, if the input characteristics of the logic type allow. One method effective in reducing the magnitude of the current spike is to include series resistance at the output of the driver circuit, as shown in Figure 11.96a. The series resistance is available in a number of different values in a dual in-line package, which saves PCB space. For TTL circuits, the value of the resistor must be limited to 10 Ω or less. A ferrite, available as a surface-mount component, or other form of inductor will limit the magnitude of switching current but tends to resonate and may degrade signal quality. A combination of damping resistance and inductor, as shown in Figure 11.96c, may be the ideal solution.

Do not add a capacitor at the load end of a clock or data signal track because although this will increase the rise and fall times of the voltage step, the magnitude of the current pulse will increase and so will the radiation below approximately 200 MHz. When using very-high-speed devices, $t_r/t_f < 2$ ns, the frequencies above 900 MHz will also be reduced by the addition of a capacitor across the load, as long as the track is electrically short. Nevertheless, the correct location for the capacitor remains at the source end with some series impedance between the source and capacitor. The addition of a capacitor at the driver end *without* a series impedance will increase the radiation from the IC and may be counterproductive. These solutions are illustrated in Figure 11.96a through c.

The lower the fundamental frequency, the slower the edges can be made. In practical terms, regardless of frequency, appreciable reduction in emissions cannot be made lower than the fourth

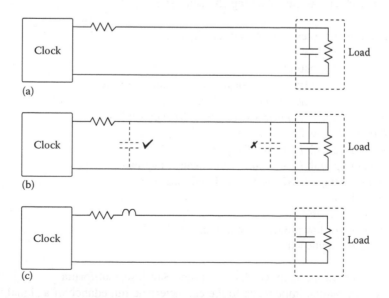

FIGURE 11.96 (a) Added resistance at the driver end. (b) Added capacitance at either the load end (incorrect) or driver end (correct). (c) Added resistance and inductance at the driver end.

harmonic of the fundamental. With present high-speed clocks, this reduction may not be possible at a low-enough frequency. One possibility is to use an LCR network to generate a clock pulse that approximates a sine wave. This solution will result in jitter at the output of the device that uses the clock, and the solution is not practical in circuits susceptible to jitter. It is also very difficult to develop a pure sine wave, because as the device that uses the clock changes state, its input impedance changes and results in a step function in the sine wave. Despite this effect, the sine wave clock can result in a reduction in the emissions at low harmonics of the clock. In one instance, a device selling for $99.00 was contained in a plastic enclosure that had to be sprayed with a conductive coating, at a cost of $20, to meet emission requirements. Changing the clock to a sine wave brought emissions to just below the limit without the need for a shielded enclosure.

Another circuit-level technique for reducing radiated emissions measured by a spectrum analyzer or receiver is to use a spread-spectrum, frequency-hopping, or frequency-wiggle technique. These frequency-shift techniques effectively broadband frequencies to reduce the measured emissions levels from clocks and data, at the fundamental and harmonics. The FCC, and almost certainly other organizations, will accept these techniques in meeting radiated emissions from equipment containing digital components. A second situation in which to implement a spread-spectrum, frequency-hopping, or frequency-wiggle clock is in a computer in which a radio is imbedded or into which a PCMIU card is plugged. The computer clocks and data will generate spurious responses (spurs) in the radio as a result of the power-line and data-line conducted noise and due to the radiated coupling to the radio antenna. Using frequency-sweep techniques will move the spurious response out of the radio-tuned channel. And if this is swept at a low-enough rate, then the reception of packets of error-free data is possible. In addition, no one channel is denied to the user, which may be the case if a fixed spur happens to coincide with the tuned channel. Even if the channel is not totally denied, the area of coverage area for that channel can be severely limited. For more information on radio-induced EMI, see Section 5.6.1. If the clock is generated by a phased lock loop, then the computer can control the loop and move spurs out of band in an intelligent fashion. One commercially available spread-spectrum clock is the IMISM530, manufactured by International Microcircuits, Milpitas, California. This can reduce clock-related EMI by up to 20 dB according to the manufacturer. The IMISM530 applied to existing clock frequencies will modulate that frequency, centering on the input frequency, or up to that frequency and can be used to generate multiples or fractions of that frequency. The clock has the following characteristic:

Replicates and modulates externally applied signals.
3.0–5.5 V operating supply range.
14–120 MHz selectable output frequencies.
Output may be equal to a fraction of, or multiple of, input frequency.
Will accept input frequencies of 14–120 MHz.
TTL- or CMOS-compatible outputs with 6 mA drive capability.
20 pin S S OP or "skinny dip" package.
Frequency spreading with F_m center frequency.
Frequency spreading with F_m maximum frequency of spread.
Compliant with all major CISC, RISC, and DSP processors.
Low short-term jitter.
Synchronous output enable.
Power-down mode for low power consumption.
Locks to externally applied signal.

Another source of spread spectrum clocks is Cypress, San Jose, California.

Matching the termination impedance to the characteristic impedance of a signal track is often said to reduce the magnitude of radiated emissions. In fact, this is seldom the case. Matching the termination is useful in improving signal quality but only serves to increase the current flow in an electrically short signal and return track. An electrically short signal and return track is one in which the propagation delay of the track is low compared to the duration of the logic transition. With increased current flow, the radiation increases.

At LF, the current on an electrically short PCB layout is determined by the load impedance. The load impedance presented by the logic is made up of some series IC package and die lead inductance and the device capacitance and equivalent resistance in parallel. The device components are active and dependent in many devices on the applied logic level. The impedance above $\approx$30 MHz tends to be dominated by the device input capacitance; above $\approx$100 MHz, the impedance is low. At HF, the characteristic impedance of the line plays a role in the magnitude of current flow. With a relatively low load impedance, the higher the impedance of the line, the lower the signal current flow and the lower the radiation. This may be achieved with a constant distance between signal and return tracks by choosing a substrate material with the lowest possible permittivity.

11.8   PCB GROUNDING

11.8.1   COMMON-MODE VOLTAGE DEVELOPED ACROSS A PCB

Figure 11.97 illustrates a typical ground configuration on a PCB containing logic in which power return and signal ground are the same. The transient currents develop voltages in the impedance of the signal return. Due to the correct use of decoupling capacitors, the D/M voltages developed are minimal. The voltages that are developed are common mode, as illustrated by the noise waveforms shown in Figure 11.97 The greater the distance from the chassis ground connection, the higher the C/M noise voltage. In Figure 11.98, the noise voltage developed at IC U6, referenced to point A, the power return, appears equally on the signal return connection (C/M). A C/M voltage will also appear on the return of IC U7, and thus a differential voltage will appear between U2 input and the return of U6.

When the noise pulse duration and amplitude result in noise energy that is higher than the noise energy immunity level of U7, then EMI occurs. In one case of EMI, transients of 1.5 V magnitude

FIGURE 11.97   Common-mode voltage developed across a PCB.

FIGURE 11.98   Poor signal/power return layout.

FIGURE 11.99   Improved signal ground layout.

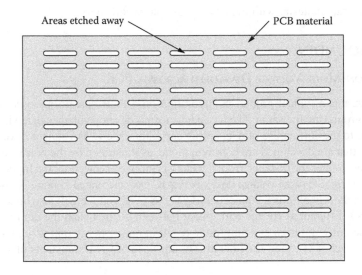

FIGURE 11.100   Typical ground plane with PCB removed for IC pins.

were measured between two ground points on the board and at the input of an IC. One common technique used to reduce the impedance of the signal return is crosshatching, as shown in Figure 11.99. The lowest impedance is achieved by use of a ground plane, the impedance of which is given in Section 5.1.2. As described in Case Study 11.1, the impedance of a practical ground plane made up of connected segments may be much higher than expected.

When a ground plane is used on a board containing ICs, the copper must be removed under the pins of the ICs. Very often the removal takes the form of a slot, as shown in Figure 11.100, and the impedance of the ground plane is dominated by the impedance of the PCB material between the slots, which is considerably higher than that of a solid ground plane. The inductance of the return path is reduced by the proximity of the signal path, and the inductance of a ground plane is much lower than that of a narrow PCB track. It is the lower inductance and, therefore, impedance of the microstrip PCB ground plane that resulted in up to a 36 dB reduction in measured emissions compared to an open transmission line. In the measured configurations, the same distance existed between the signal and return tracks as between the signal track and ground plane in the microstrip, and the only difference was the ground plane.

11.8.2   "GOOD" AND "BAD" PCB GROUND PLANES

Often the ground plane is constructed of cross-hatched tracks. When these comprise short, thick lengths of track, the impedance can be almost as low as for a solid ground plane. However, when the tracks are unduly long and narrow, the impedance will be significantly higher than that of a solid ground plane.

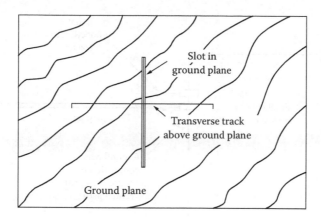

FIGURE 11.101 A slot in the ground plane interrupts the signal return current under the signal track.

The majority of the signal return current flows in the return path directly beneath the signal trace. This means that when the width of the return trace is five times or more, the height of the signal trace above the return trace, an effective ground plane has been achieved for that single signal connection. This concentration of current under the signal trace is again due to mutual inductance. This means that the width of the ground plane can be limited to those areas of the board where signal and clock tracks exist. In most boards, which have tracks over the whole area of the board, the ground plane must also cover the whole area of the board. Figure 11.101 shows a vertical track located on the same layer as the ground plane that creates a slot in the ground plane. Above and at right angles to this ground plane layer track is another signal track. The return signal current in the upper track must now spread out due to the slot in the ground plane, and the ground plane impedance is increased.

In the measurements described in this section, a signal source was made up of a battery, a regulator, a clock oscillator, and a driver IC, all contained in a shielded enclosure, and the signal output was used to drive the PCB. In the construction of the shielded enclosure used to house the source, the enclosure must be connected to the ground planes in the microstrip, stripline, or image plane PCBs. Figure 11.102 illustrates the connection of the enclosure to the signal wire from the source and the ground/image plane. The signal return connection is made from the PCB plane to the feedthrough in the enclosure via a short length of wire, as is the signal connection to the signal trace. To minimize radiation from these short lengths of wire, a small copper doghouse enclosure covers both the feedthrough connections and the lengths of wire. This doghouse extension, which can be

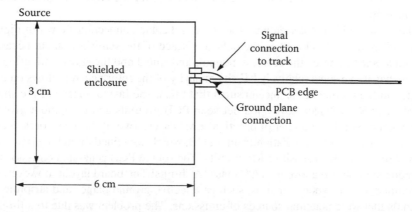

FIGURE 11.102 Separate signal return and the connection between the ground plane and the enclosure.

FIGURE 11.103   Signal return used as the connection between the PCB ground plane and the enclosure.

seen in Figure 11.102, is used to connect the enclosure to the ground/image plane of the PCB. In the connection of Figure 11.102, none of the signal return current flows in the connection between the enclosure and the PCB ground plane.

The grounding experiment involved connecting the enclosure to the PCB via a short length of wire that was connected to the feedthrough and then via the signal return wire to the PCB. This grounding technique is illustrated in Figure 11.103. It can be seen that the signal return current does flow on the connection between the enclosure and the PCB ground plane.

The radiated emissions from the different types of PCB layout were measured with this alteration in grounding. The level of radiation from the low-emission type of PCBs (microstrip/stripline) increased from 56 dB below the high-emission types to 6 dB below these high-emission types. The reason for this dramatic increase is that an RF potential difference is developed between the enclosure and the ground plane(s) or image plane. This potential difference is caused by the signal return current flow on the short (5 mm) length of wire used to connect the PCB to the enclosure. Although the wire is short, it exhibits significant inductance and it is across this inductance that the voltage is developed. In contrast, in the grounding connection shown in Figure 11.102, the enclosure-to-ground-plane connection is low impedance; it does not carry the signal return current and so no RF potential is developed.

Why is the same problem not seen with the semirigid cable connection shown in Figure 11.104. No potential difference is developed on the outside surface of the semirigid shield because of both the mutual inductance between the center conductor and shield and the skin-depth effect. The skin depth ensures that above some relatively LF, the majority of the return current flows on one surface of the shield, and the mutual inductance ensures that this is the inside surface of the shield.

The problem of RF voltages developed between PCB grounds and equipment grounds or the enclosure has been seen in a number of practical EMI cases. Two of these are presented next.

The first case study involved radiation from a PCB, which contained digital and RF circuits, to a nearby PCB, which contained sensitive RF circuits. The source PCB generated emissions that were, at certain frequencies, 40 dB above the FCC class "A" limits! The board layout, however, was exemplary. It contained all the "good" features, such as effective ground planes and stripline configurations for signals that were potential sources of emissions. The problem was due to a thin trace used to connect the RF and digital grounds together at a single point. The thin trace carried the return

FIGURE 11.104   Semirigid cable connection, used with the ECLinPS device, to the microstrip ground plane.

current of a digital signal driving a device on the RF board. The two PCB ground planes and single-point connection are shown in Figure 11.105. The return current generated a voltage drop in the inductance of the return trace, so an RF potential was developed between the ground planes. This is the same problem as seen in the grounding experiment, in which the potential was developed between the ground plane and the enclosure. Many modern PCB antenna designs use a similar layout, driving the antenna at a similar point, and achieve a high gain, that is, efficient radiation.

A significant reduction in radiation was seen simply by widening the thin signal return trace. Widening the trace reduces the inductance and the voltage drop in the return path. The width of the common ground connection between the digital ground and the RF ground is kept narrow to minimize noise current flow from the digital ground into the RF ground, used by sensitive RF circuits. However, the width of this single connection can be increased when sensitive circuits are located far from the connection or when digital signal return current is routed away from the location of the single-point ground. For example, ensure that digital return current does not flow between devices located across (at right angles to) the increased width of the ground plane connection.

An even more effective solution is the use of a semirigid cable connection, as shown in Figure 11.106, or a stripline PCB structure installed across the gap, as shown in Figure 11.107. The stripline is not

FIGURE 11.105   PCB with digital and RF ground planes connected together with a thin track.

FIGURE 11.106   Use of a semirigid cable to connect the ground planes together.

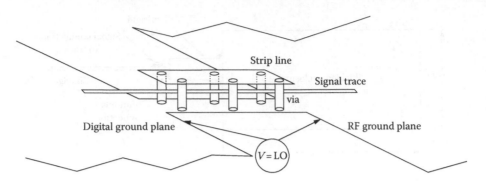

FIGURE 11.107   Stripline connection between two grounds.

as effective as the semirigid cable due to some leakage through the gaps between the vias used to connect the upper and lower ground planes together.

Another potential solution is to disconnect the two ground planes and to use an optoisolator to transfer the digital signal to the RF section, as shown in Figure 11.108. Separating the two ground planes is very effective in reducing current flow between them. However, if an RF potential exists between the planes, as it almost certainly does, then the level of radiation may remain high or may even increase with separated grounds. The solution is the use of the optoisolator, in which the digital return current is forced back to the digital ground plane on a trace, which is disconnected from the RF ground, and the digital and RF grounds are tied together at a single point. Negligible current flows in this common connection and the two ground planes are close together in potential.

Another common example of RF potential developed between PCB ground planes is when signal and signal return connections exist between two or more PCBs. These connections are

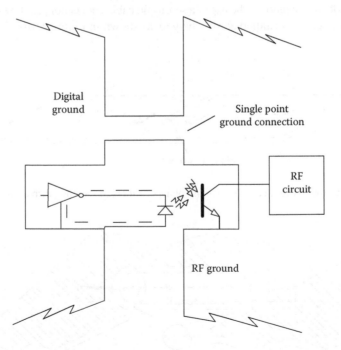

FIGURE 11.108   Use of an optoisolator to keep the signal return current away from the signal point ground connection.

FIGURE 11.109 Piggyback connection of PCBs using pins and sockets.

typically made via pins in edge connectors and a motherboard or back plane. It is important that as many parallel signal return connections as possible be made and that these connections be as short and as wide as possible to reduce the inductance of the signal return and, therefore, the potential difference between grounds. Where possible locate a return pin adjacent to every "hot" signal pin.

Some of the worst violations are seen in boards that are mounted piggyback on a PCB, with signal power and return connected by long, thin pins. This configuration is shown in Figure 11.109. The solution is to use PCB edge connectors for the piggyback PCB and to make the interconnections via a microstrip PCB (single ground plane) or stripline (upper and lower ground planes with tracks sandwiched between the PCB ground planes). If external cables are connected to the two PCBs, then the RF potential between the two ground planes appears on the signal connections as a C/M voltage, regardless of whether the signal is balanced or unbalanced with reference to ground. The two cables result in a very efficient antenna, for the RF potential between them drives one cable against the other, similar to the two rods of a dipole antenna. In this example, the use of finger stock or a similar contact to connect the two ground planes together will reduce the RF potential between them. Ideally, this common connection should be made at the location where the two cables leave the enclosure or at least where they leave the PCB. To reduce radiation from the cables due to a potential difference between the PCB ground plane and the enclosure, finger stock or a direct connection may be made between the ground and enclosure. If a DC connection is not allowed, an RF (capacitive) connection should be made.

Another practical example of EMI was partially due to an RF potential between a ground plane on an approximately 8 cm × 4 cm digital PCB and an RF ground. The digital PCB was located approximately 0.5 cm from a receiver contained in a small, well-shielded case. The small microstrip antenna connected to the receiver was approximately 2 cm from the digital board and was referenced to the receiver case. Due to radiated coupling between the digital board and the antenna, the receiver exhibited desensitization and narrowband and broadband spurious response. Part of the problem was due to the RF potential between the digital ground plane and the RF ground (shielded case). By shorting the case to the digital ground plane at a number of points with finger stock material, the radiated coupling reduced significantly, although not enough to totally cure the problem.

One technique for joining different ground planes on a PCB is shown in Figure 11.110. Here a large surface area is used to make a common ground connection. The common failure with this technique is to take a signal track between the two different ground planes at a location other than the common ground area, as illustrated in Figure 11.110. With this location of the signal track, the return current must flow through the one ground plane down to the common ground connection and back to the source. A large loop is thus formed, and the level of radiation due to this loop is extremely high.

In conclusion, the use of low-emission type of PCB layouts in conjunction with ICs that generate only low-level emissions can reduce radiation to levels where shielding of the PCBs is not necessary

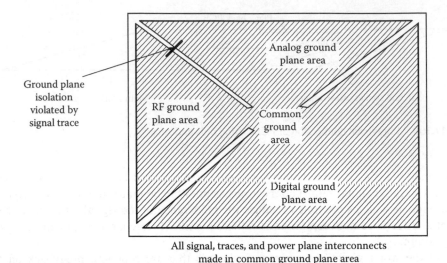

All signal, traces, and power plane interconnects
made in common ground plane area

FIGURE 11.110   Ideal large-area ground connection on a PCB violated by a single trace that is not located over the common ground area.

to meet commercial radiated emission limits. However, when PCB-to-PCB or PCB-to-enclosure/ground connections are of high impedance and carry RF current, then none of these reduction techniques may be sufficient without the use of an effective grounding scheme.

11.8.3   GROUNDING A PCB WITHIN AN ENCLOSURE

Figure 11.111a illustrates a grounding scheme for a board with a signal that is connected to an attached cable. The signal ground of the board-to-enclosure connection is made at the point at which the interface signal exits the board. If this connection represents a sufficiently low impedance, then the noise voltage developed across this ground connection will be low, as will the C/M current flow on the cable.

Many engineers are puzzled by the need for only one connection of the signal ground to enclosure to reduce C/M noise voltage. But by definition, because C/M noise appears equally on the signal and signal ground, it requires only a connection of the signal ground to enclosure to reduce the magnitude of C/M voltage. One secondary effect of this connection is that some of the C/M noise may appear as D/M noise in the signal.

In practical cases, the grounding may be made from the PCB to a connector housing, which is in turn connected to a metal faceplate, which may then be connected via two small screws to the main frame of the enclosure. Due to the relatively high impedance of this connection, a sufficiently high noise voltage can be developed to drive the attached cable with enough current to fail emissions requirements.

If more than one cable is connected to an enclosure and C/M noise voltage appears on one or another of the cables, then one cable can "drive" the other, and fields are generated between the cables in addition to those between cable and enclosure. Try to connect the shields of cables together at a single point with a low impedance connection. If cables are unshielded, connect reference signal grounds within the cables to a single point on a PCB. If shielded, add an additional low-impedance connection between the cable shields. Better still, tie them to a "clean" chassis/enclosure ground, preferably to the outside surface. Digital designers and compliance engineers very often experience out-of-specification emissions over the 40–300 MHz frequency range on boards with attached cables, even when the fundamental frequencies are much lower. If the PCB

FIGURE 11.111 (a) Signal-ground-to-enclosure connection at the location where an unshielded cable exits the enclosure. (b) Shielded balun inside enclosure where an unshielded cable exits the enclosure. (c) RF connection to the enclosure using one or more capacitors. (d) Method of isolating the digital ground from the front panel and at the same time connecting the digital ground to cassis to reduce C/M current on signal interfaces; suitable for shielded cables.

signal ground-to-enclosure connection is made at the wrong location, the C/M voltage developed across the board appears between the cable and the enclosure. The cable then radiates in the same way as a monopole antenna, with the cable current increasing with increasing frequency until the frequency at which half a wavelength equals the length of the cable. Above this resonant frequency, the cable current tends to reduce and the radiated E field exhibits maximum and minimum values.

In other cases where either no connection or a poor connection of signal ground to the enclosure is possible, a C/M inductance wound on a ferrite bead, or toroid, in which both the signal and signal return are passed through the ferrite, will present an impedance to C/M current flow. One location for the balun is where the signal leaves the board, with an alternative location on the cable external to the board. Section 5.1.10.4 discusses the properties of the balun. Avoid threading the signal and return the conductors through separate holes in a ferrite bead, for then signal degradation can occur in the megahertz frequency region.

If the balun is located on a PCB inside an enclosure, radiated coupling from the PCB can induce current in the balun. If this is found to be the case, enclosing the balun in a small, shielded enclosure may help, as shown in Figure 11.111b. Do not, however, connect the enclosure to the noisy digital ground. If a direct connection of the signal ground to chassis is not allowed, then the implementation of an RF ground using one of more capacitors in parallel may be a solution, as shown in Figure 11.111c. Using a C/M inductor in conjunction with the capacitive RF ground can lead to resonance effects, as described in Reference [1], which can be counterproductive. If the location of the C/M inductor or balun is on a high-impedance section of the attached cable, then the level of reduction in C/M current can be negligible, but a direct or RF ground at that point will be very effective. If the balun is located on the cable, moving its location to a low-impedance section can improve the performance.

An alternative approach for digital signals that constantly change state, for example, the 1553 bus, or that go into a high-impedance "off" mode in the absence of a signal is a custom-made transformer that contains an electrostatic shield between primary and secondary. The shield is a turn of conductive foil, aluminum, or copper, wound between the primary and secondary, with the ends of the foil insulated; that is, the shield does not represent a shorted turn. The electrostatic shield must then be connected to a "clean" ground, which is typically the enclosure or chassis. Connecting the shield to the noisy digital ground may actually increase the C/M noise current. With the shielded transformer, C/M noise currents flow via the shield into the ground and not on the cable. If PCBs are contained in unshielded enclosures, then the RF potential developed between the PCBs must be reduced by connecting the ground planes on the PCBs together at the location where interface cables are connected to the board. An RF potential difference between the ground planes of two or more PCBs to which cables are connected results in one cable driving another and a very effective radiating antenna.

The C/M current flow on the power cord may also result in excessively high radiation from PCBs and attached cables. Here the C/M current must be reduced by the use of effective C/M filters or a C/M inductor on the AC safety ground.

In the case where PCBs with a front panel are contained in an enclosure and where the front panels are connected together down the edge with finger stock or spiral gasket or the like, a number of considerations must be made as to the connection of the digital ground to the front panel. Where unshielded cables are connected to the front panel, the decision is relatively easy, for the digital ground should be connected to the front panel with the use of baluns on signal and returns to reduce the C/M current flow. In the case where shielded cables are connected to the front panel, two choices exist and the right decision is not always obvious. If the digital ground is connected to the front panel and also to the chassis at some other location in the equipment, high C/M currents can flow in the front panel, resulting in a voltage drop between the front panel and the rest of the enclosure. In extreme cases, the radiation from the front panel alone, without an attached cable, can be close to class B commercial radiated emission limits. This effect can be demonstrated by using a spectrum analyzer to compare the voltage drop between two locations on the enclosure at a set distance apart and between the enclosure and a front panel. A second method for finding this problem, which provides a very convincing demonstration, is to make a radiated emission measurement, without cables, and then to connect a wire in electrical contact with the front panel. If emissions go up significantly, the cause is either the front panel voltage driving the wire or emissions from an aperture coupling to the wire and causing a current flow on the wire. If, on the other hand, the digital ground is floated from the enclosure, then C/M current on the shielded cable center conductor(s) can result in cable radiation when the cable shield or shield termination or connector backshell is not adequate.

A possible solution to both problems is shown in Figure 11.111d. Here the digital ground is isolated from the front panel. Where a single-ended signal interface must connect to the case of a connector (in this example, a coaxial connector), a balun is used to provide impedance between the digital ground and the front panel. To reduce the C/M voltage applied to the signal interface, the digital ground plane is connected to the chassis before the front panel. In Figure 11.111d, the connection is made as the ground plane is extended to the edge of the PCB and clamped in a metal card guide. To make this connection effective, the card guide must be adequately bonded to the enclosure and the card guide must grip the PCB with spring fingers.

11.9 SHIELDING A PRINTED CIRCUIT BOARD

The radiation in close proximity to a PCB is typically a low-impedance field. In considering the radiation without a shield, our only concern is the field measured at a distance of 3, 10, or 30 m from the PCB, where the field has an impedance of 377 Ω and the relative magnitudes of the E and H fields at the board are unimportant.

When examining the radiation from PCBs contained in a small enclosure (up to a 19 in. rack size), we find that the magnetic field shielding effectiveness of an enclosure is defined as the ratio of field measured at some location from a source without the shield in place to the field with the shield in place. Due to the switching transient currents flowing on a PCB and the close proximity of the PCB to the enclosure, the field impinging on the inside of the enclosure has a high magnetic field component.

The enclosure approximates a loop antenna, which, despite the low impedance of the field close to the source, generates a field with an impedance of 377 Ω in the far field, the location where commercial EMI measurements are made. The electric field shielding effectiveness of a small enclosure may be very high, and the limit is often the magnetic field shielding effectiveness of an enclosure, which results in not meeting EMI requirements.

The magnetic shielding effectiveness of a small enclosure is adversely affected by high joint DC resistance and contact impedance, which depend on the resistance of the surface finish. For example, iridite, oakite, alodyne, electroplated nickel, electroplated tin, and Dow 20 are all low-resistance finishes, whereas stannate and zinc dichromate exhibit impedance 10 times higher, with unplated aluminum 10–30 times higher. A copper-loaded paint on the inside of a nonconductive enclosure will exhibit joint impedance a thousand times higher than the low-impedance finish on a highly conductive material. Surface area also has an impact on the joint impedance. The larger the surface area, the lower the resistance and contact impedance. Small gaps in the joint or seam increase the gap inductance and limit the shielding effectiveness of the enclosure at HF. The presence of apertures also limits shielding effectiveness. For magnetic field shielding, it is better to have many small apertures versus fewer larger apertures, even if the total area of the removed metal is larger. This is discussed in detail in Sections 6.5.3 and 6.6.3. Although we have considered only magnetic field shielding here, at frequencies above 200 MHz and with relatively large apertures the electric field, coupling through the apertures may predominate.

The use of a PCB-mounted shield covering a small section of circuit or IC, which are both high-level emitters, has often been tried, with variable results. If the tracks going through the shield are not C/M filtered, very often the radiation from the tracks connected to the shielded circuit can increase. This effect is very common when adding a shielded enclosure around a PCB to which a cable is attached. If the radiation sourced by the PCB is predominantly from the attached cable, then adding a shield often increases radiation, for the presence of the shield increases C/M current flow on the cable. Connecting the signal return or the shield of the cable to the enclosure where possible or use of the balun or shielded transformer is a possible solution to this problem, as already described in Section 11.8.3.

The use of surface-mount ferrite and/or surface-mount capacitors on all signal and power lines entering the PCB mounted shield will effectively reduce current flow and shunt the current back to the source. The surface-mount capacitors must connect to the shield, because digital ground connections, using tracks, are also sources of C/M noise.

One PCB shield is made in the form of a phosphor bronze fence, which can be bent to any shape. The fence forms the walls of the enclosure, and the top is made of a flat material, which is held in place by spring force. The fence is constructed with pins that go through the board to a ground plane, on the track side of the board, which completes the shield. It is often the poor integrity of the ground plane that compromises the shielding effectiveness. Small gaps are left around the base of the fence to allow components and surface tracks to enter the shield. These apertures limit the shielding effectiveness along with the impedance of the pins of the fence. However, the theoretical magnetic field shielding effectiveness with apertures is still a high 70 dB from 10 to 200 MHz. It is typically the components and tracks entering the shield that result in leakage and a greatly reduced shielding effectiveness. However, in the gigahertz frequency range, the small apertures do result in coupling. Figure 11.112a shows a 97870 custom-made fence without apertures that was effective at solving a problem at 1.5 GHz. The PCB shield shown in Figure 11.112b is compartmentalized, which is a useful technique when sections of PCB that contain sensitive circuits must be shielded from sections of PCB that are potential emitters. The PCB fence material is made by Instrument Specialties Co., Inc., of Delaware Water Gap, PA 18327-0136. Figure 11.112a shows the standard product range from this company.

To demonstrate the importance of low joint and seam impedance, we will consider a number of shields designed to go around PCBs. The first shield is made up of two parts, one, with the dimensions 23 cm × 26.5 cm × 1.5 cm, that covers the component side of the board, and a second, with the dimensions 21 cm × 26.5 cm × 0.6 cm, that covers the track side of the board. The slots in the shield are: one at 0.6 cm × 23 cm at one end of the board and two at 1.5 cm × 9.5 cm at the other end of the board. A ground plane on either side of the board connects the two sections of shield together with vias at intervals of 7 cm. Thus, the board contains both large slots and a high-impedance joint. Figure 11.112c shows the end of the board in which the 23 cm long slot is located.

The technique used to measure the shielding effectiveness entails driving a PCB, laid out with tracks above a poor ground plane, with loads distributed around the PCB. The signal source is located outside of the shielded room, with the enclosure under test located inside the shielded room. One potential source of error is the cable connecting the source to the on-board loads, which can radiate and mask the radiation from the enclosure. To ensure that this did not happen, the cable was contained within a copper tube that was fitted through the shielded-room wall and soldered to a bracket, which was bolted onto the enclosure. To establish a measurement setup noise floor, the best possible enclosure with regard to shielding was constructed. This was a copper enclosure with a lid soldered to the enclosure and with the seams in the sides soldered together so that no seams, apertures, or slots existed in the enclosure. All other enclosures when compared to the soldered copper reference enclosure showed a lower shielding effectiveness from 100 to 300 MHz.

In the case of the shield shown in Figure 11.112c the transfer impedance of the cover-to-cover interface is not confined to an irridited metal-to-metal contact but includes the impedance of the plated-through connection between the ground planes on either side of the board. However, the degradation due to the contact impedance of the seam is swamped by the shielding effectiveness degradation due to the wide slots. The theoretical magnetic field shielding effectiveness is limited

FIGURE 11.112 (a) Instrument Specialties PCB fence material. (b) PCB shield constructed of Instrument Specialties fence material. (c) PCB shield covers, showing long slot down on one end. (d) Shield that totally encloses the PCB. (e) Shield that totally encloses the PCB, with small slots down the edge of the shield.

by the 0.6 cm × 23 cm slot, which is 8 dB at 100 MHz; for the two 1.5 cm × 9.5 cm slots, it is 27 dB. The worst-case measured magnetic field shielding effectiveness of the enclosure at 150 MHz is 3 dB, which is close to the predicted value. One important factor in considering the shielding effectiveness is the directivity of the source, for without a shield, the PCB radiates in all directions, whereas with the shield, the fields are concentrated at the slots in the end of the shield. In a practical installation, the PCB is plugged into a rack that is shielded on five sides. In our example, it is the long slot in the PCB shield, which faces the open front of the enclosure. In this configuration, it is not uncommon to find that the field measured at the front of the rack with a poor PCB shield is higher than with the PCB shield removed!

The shielding effectiveness of the shield shown in Figure 11.112c can be greatly improved by bridging the long slots with a conductor, which contacts the PCB, the ground plane, and the shield. This may be achieved by bending tabs in the enclosure and bolting the ends of the tabs to the ground plane. If the radiation contributed by the long slots is reduced, the example PCB shield is still limited to approximately 30 dB of shielding effectiveness due to the high-impedance seam.

This impedance is dominated by the inductance of the plated-through connections and the screws holding the two covers together. A 7.5 cm slot exists between the screws because the ground planes on either side of the board are connected together only via plated-through holes where the screws go through the board. Although the majority of the current flow on the inside of the shield is caused by the PCB-radiated emissions, if the signal ground planes on either side of the board are less than ideal, some fractions of the noise current can flow on the inside of the shield covers. Therefore, it is inadvisable to connect the shield covers to the signal ground plane around the periphery of the board, instead, leave the shield either isolated or connected at a single point, preferably where connections are made to the board (often, edge connectors). If the shield cannot completely cover the PCB, then the shield cover should contact a track with a width at least equal to the bent edge of the cover. A second, identical track should be fabricated on the other side of the board. Both tracks should follow the contours of the shield edges and be joined together at intervals of approximately 15 mm by vias. The tracks should be either isolated from the signal ground plane or connected at a single point only. The distance of the screws holding the covers together should be no greater than 3 cm. With these modifications made to the PCB shield, shown in Figure 11.112c and the end slots reduced to a centimeter or less, the shielding effectiveness can be increased to at least 40 dB. Figure 11.112d illustrates a PCB shield that totally encloses the PCB, and it is the shield that is plugged into the guide rails of the card cage. This enclosure exhibits a shielding effectiveness of 60–90 dB from 100 to 300 MHz.

Figure 11.112e shows a PCB shield with 0.9 cm long and 0.2 cm wide slots down the edge of the shield. The measured shielding effectiveness of this enclosure is also 60–90 dB over the 100–300 MHz frequency range, although the variation with frequency is different from that of the enclosure shown in Figure 11.112d. The high level of measured shielding effectiveness in the enclosure with small slots is totally in agreement with theory as long as the majority of the noise currents flowing on the inside of the enclosure flow down the 0.6 cm length of the slot and not across the length of the slot. In the two enclosures with a high level of shielding effectiveness, any degradation will be due to unfiltered or unshielded connections on the motherboard or back plane into which the board is plugged.

If a nonconductive enclosure is already manufactured and is designed to enclose one or more PCBs, it is possible to shield by spraying the inside of the enclosure with a conductive coating, such as silver- or copper-loaded paint. With the use of a conductive coating, the lower the surface resistivity, the higher the shielding effectiveness. The weak link is at the joints in the enclosure, which should have as high a contact surface area as feasible to reduce the contact impedance. With the use of conductively coated plastic or plastics loaded with conductive fibers, the measured shielding effectiveness from 50 to 300 MHz is 10–30 dB.

One of the latest forms of PCB shield, called a Conductive Shielding Envelope (patent pending) by the manufacturer, Schlegel, is made of lightweight, pliable, silver-metallized nonwoven fabric. The attenuation published by Schlegel for this type of enclosure ranges from 30 dB at 10 MHz to 60 dB at 100 MHz and stays constant at 60 dB out to 1000 MHz. The attenuation quoted by Schlegel is almost certainly for E field and plane waves, as with the majority of manufacturers of conductive material. The shielding effectiveness for H fields is unknown and is strongly dependent on the effectiveness of the joins in the material.

Use the calculations contained in the PCB radiation report, based on clock frequency, logic type, PCB layout, number and length of PCB tracks, and number of PCBs, to find out if the relatively low shielding effectiveness of the conductive-coated type of enclosure is adequate or if one of the enclosures with higher shielding effectiveness is required.

Placing an isolated shield between PCBs has been found in practice to be somewhat effective. In measurements, the shielding effectiveness was found to be 0 at some frequencies. A 23 cm × 16 cm PCB was shielded from a second 23 cm × 16 cm PCB with a 23 m × 16 cm shield. One PCB contained a clock source connected to traces 3 mm apart and 4 cm long in the center of the PCB. The second PCB contained a similar trace used as a receiving antenna.

The level of shielding effectiveness achieved was as follows;

Frequency (MHz)	Attenuation (dB)
10	13
40	22
60	11
70	18
80	0
90	13
100	8
120	0
140	13
160	14
180	18
230	3
260	7
280	0
330	7
350	12
360	15
430	13
460	13
520	6
590	17
680	24
820	6

11.10 PCB RADIATION, CROSSTALK PREDICTION, AND CAD PROGRAMS

Some of the ideal characteristics of a PCB radiation and crosstalk prediction program is that it be virtually invisible and automatically invoked during CAD PCB layout. The ideal program would calculate the signal integrity and current flow on tracks, based on clock frequency, and predict the near fields as well as maximum horizontal and vertical fields with an accuracy of 6 dB or better at distances of 1 m or greater. Most of the commercially available programs fall short of the ideal. Program capability ranges from a prediction of radiation from a simple structure, such as a two-track transmission line some distance from a ground plane or a microstrip line, based on input clock voltage and rise and fall times, to programs that use schematic or CAD PCB layout as the input data. Accuracies range from worst-case 60 dB to average 20 dB. The programs can be separated into rules-based and simulator programs. Some of the simulator programs do not at present offer an exact prediction of the level of emissions from a PCB but, rather, identify "hot spots" in the layout. The problem with this approach, when used to design a PCB before the prototype has been manufactured, is as follows.

Why bother about the "hot spots" if the unmodified PCB will meet all the radiated emission requirements? This can be determined only by accurate field predictions at a 1, 3, or 10 m distance from the PCB.

11.10.1 Using NEC to Predict Radiation Compared to Measurements

When a wire or cable is connected to a PCB ground, the C/M current flowing in the PCB ground continues to flow on the attached wire. At LF, the wire length may be much less than a wavelength and this is referred to as an electrically short wire.

As the frequency of the source is increased so the level of radiation from the wire increases. For a wire effectively disconnected from ground at both ends, which is the case studied here, the maximum radiation occurs when the wire is a resonant 0.5λ long. This configuration can be considered a half-wave dipole shorted in the middle and this is discussed in Chapter 2. The wire is connected to a PCB contained in a shielded enclosure and the wire exits the enclosure; this can be modeled as a monopole above a finite ground plane (the enclosure) and maximum radiation occurs when the wire is a resonant 0.25λ long.

Thus, for a 1 m long cable attached to a PCB far removed from ground, the maximum frequency of radiation is 150 MHz. For a 1 m long wire attached to a PCB enclosed in a metal enclosure, the maximum frequency of radiation is 75 MHz.

As data and clock speeds increase, the frequencies radiated from attached cables also increase. The FCC in CFR 47, Part 15, Section 15.33, requires that when the highest frequency generated or used in a device is 108–500 MHz, then the upper frequency range of the measurement is 2 GHz. When the highest frequency is 500–1000 MHz, then the upper frequency range of the measurement is 5 GHz, and when the highest frequency is above 1000 MHz, then the upper frequency range of the measurement is the fifth harmonic or 40 GHz, whichever is the lower.

This investigation looks at the level of radiation when the attached wire is electrically long, equal to a wavelength or greater.

In the typical radiated emission setups for commercial requirements contained in ANSI C63.4 and required by the FCC, the use of a turntable is required. This means that the sections down the length of the electrically long cable as it rotates are either closer to or farther away from the measurement antenna. The voltage source in this experiment is a constant frequency and the cable length is changed to obtain lengths from 0.5λ to 5λ.

This investigation considers only the case when the wire is at a distance of 3 m from the antenna and the measurements and predictions are made either with the wire broadside on to the antenna, in the same plane, or the case where the wire and source are rotated, in which case the level of maximum radiation for any angle is measured.

Thus, in one test, the table is rotated and the maximum emission is measured. The length of the wire is then modified and the table again rotated to achieve a maximum level of radiation.

The angle of rotation is noted and the NEC program, 4NEC2 version 5.5.2 by A. V., is used to predict the level of radiation at this angle using the same level of voltage for the source.

One other variable is a potential impedance mismatch between the ground plane impedance and the impedance of the attached wire and this effect is also measured.

11.10.2 PCB Simulation and Test Setup

Assume a logic driver connected in a microstrip PCB configuration connected to a load as shown in Figure 11.113.

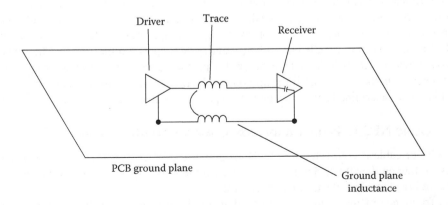

FIGURE 11.113 Logic driver and receiver connected in a microstrip.

As the driver changes state, it drives the input impedance of the receiver that includes an input capacitance. Thus, a current pulse is generated and flows back in the PCB ground. As described in Reference [1], the ground impedance is dominated by the ground plane partial inductance, and although reduced by the mutual inductance between the signal trace and the ground plane, it is significantly higher than was predicted in early EMI textbooks.

As the current flows in this impedance, a voltage is developed in the ground plane and a C/M current flows. When a wire is attached to this ground, the current continues to flow on the attached wire, as shown in Figure 11.114.

The current flows in a very narrow area of the ground plane under the microstrip signal trace. This can be modeled as a voltage source in a short wire as shown in Figure 11.115. If the voltage source is contained in a metal enclosure or has power supplied by a separate power line, then the flowing between the signal trace and this nearby conductive structures will alter the level of radiation. For this reason, a small 1 GHz voltage–controlled oscillator connected to three watch batteries was mounted in the center of a 0.15 m long 22AWG wire.

The measurements were made using a dipole tuned to 1 GHz on a free space range. The wire was raised 1 m above a low conductivity floor that was covered in ferrite tile absorber. The area between the antenna and the wire was also covered in ferrite tile. Thus, any displacement current between the wire and the floor is expected to be minimal and any ground reflection also minimal. The antenna was raised 1.25 m above the floor.

The emissions were predicted using the MOM capability of NEC. The source was fixed in the center of a 0.15 m length made up of 100 segments. The frequency was set to 1 GHz, which made the 0.15 m section equal to 0.5λ. This represents a resonant length PCB ground plane. A wire was attached to one end of the 0.15 m section and this wire was increased in multiples of 0.25λ until the wire and the original 0.15 m section were at a total length of 1.5 m or 5λ long.

Initial measurements were made with the wire in the same plane as the antenna and this orientation was not changed as the length of the wire was varied. The NEC measuring point and the dipole antenna were centered on both the 0.15 m section of wire $y = 0.0$ and at the center of the long wire $y = -0.7543$ m. Figure 11.116 shows the results of modeling and measurements with the antenna at $y = -0.7543$ m. The measuring point was set at a fixed distance of 3 m broadside on to the antenna.

Figure 11.116 shows both a predicted and measured reduction in field as the length of the stationary wire is increased, that is, becomes multiple wavelengths long. This is intuitively what I expected and what I have observed in numerous measurements made on PCBs with and without attached wires (attached to the ground plane on the PCB). The major difference between the true

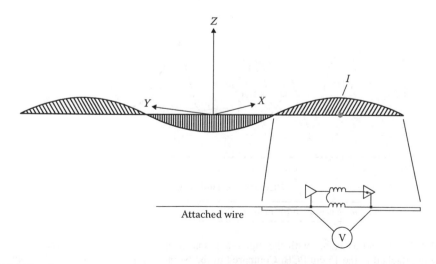

FIGURE 11.114   Current distribution on the wire and source with a total length of 1.5 m.

FIGURE 11.115   Geometry of the setup for stationary wire and source.

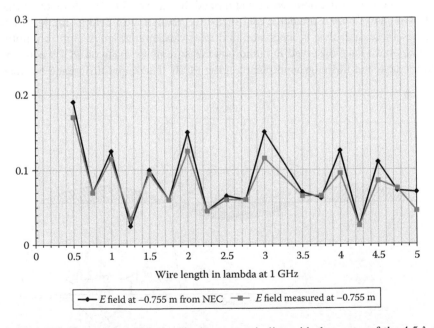

FIGURE 11.116   *E* field measured with the dipole antenna in line with the center of the 4.5 λ (135.9 cm) length of wire attached to the 15 cm PCB. Compared to the Numerical Electromagnetic Code analysis with the measuring point at the same location, that is, at *y* = −0.755 m.

PCB measurements on an OATS and this experiment is that a good ground plane is present on the OATS PCB measurements.

For the second test, the rotation origin centered on the total length of the wire and 0.15 m section. The NEC measuring point and the dipole antenna were set at a fixed distance of 3.365 m to the wire and displaced from center by 0.43 m. For each E-field measurement, the wire and source were rotated with respect to the antenna until the maximum emission was observed. In the NEC geometry, the wire and source were oriented with respect to the antenna at the angles where maximum emissions occurred during measurements. Figure 11.117 depicts the geometry of the test setup.

Figure 11.118 shows the predicted and the measured fields as the length of the wire is increased. The previous investigation with a stationary wire and source indicated that E-field levels measured at the antenna decrease as the length of the wire increases. This trend was not observed when

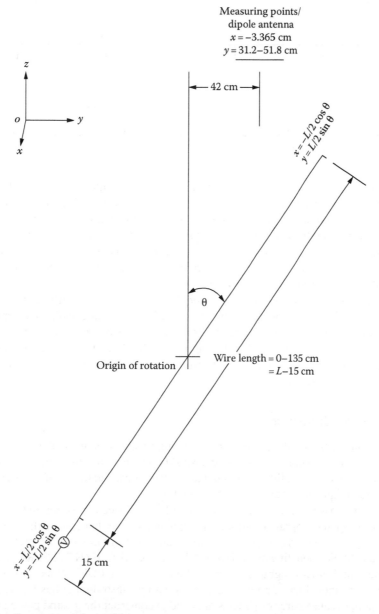

FIGURE 11.117   Geometry of the problem.

FIGURE 11.118 Comparison of the measured E field at the dipole versus the E field predicted by Numerical Electromagnetic Code for the same conditions.

rotating the wire and source. The trend observed in this investigation indicates a slight increase in E-field radiation as the wire length is increased. This can be explained if we take into account the constructive and destructive interferences that occur due to the phase of the currents flowing on the wire. At some orientation, the composite field radiating from the wire will be maximally constructive, and at some angle, they will be maximally destructive, canceling each other out. When we do not rotate the wire, we see the effects of destructive interference and, therefore, do not observe the maximum field strength produced by the wire.

As an adjunct to this investigation, an additional test was performed wherein the attached wire was substituted by 1/2 in. braid wire to simulate the effects of a larger diameter cable. The braid wire produced more radiation. See Figure 11.119 for a comparison of the braid wire to the 22AWG magnet wire. This effect of a mismatch between the smaller diameter wire and the larger diameter braid simulates a mismatch between the PCB ground plane impedance and the attached wire and requires more investigation to quantify it.

11.10.3 Modeling Techniques

Modeling techniques exist for the prediction of radiation from structures, and some can be applied to PCB radiation. These techniques include the MOM and variations thereof, such as the full-wave moment method, with the capability of analyzing 3D structures, and the CG-FFT, FDTD, FEM, and TLM. Another technique is transmission-line modeling.

Due to the complexity of these techniques, they have all been incorporated in computer programs. Use of a computer program eliminates mathematical error but can result in very large errors because of misapplication of the program and some intrinsic limitations in the method. Some of the programs that use MOM modeling are WIRES, NEC, MININEC, EZNEC, FEKO, Efie4ld AB, and GEMACS. These and other programs are described in Section 12.4.3. None of the simple MOM programs allows the modeling of a microstrip PCB with a substrate that has a permittivity higher than 1. In addition, because wires or patches are used to construct the ground plane and current diffuses through these structures, the prediction of the shielding effectiveness of a stripline is in error.

FIGURE 11.119   The effect on the radiated E field of increasing the wire diameter.

One program, described in Section 12.4.3, is the EMAP5 FEM/MOM code, a hybrid finite element/MOM code, which is available from the University of Missouri, Rolla. The EMAP5 program does allow the modeling of the radiation from a microstrip on printed circuit material with a permittivity greater than 1.

The simplest technique, and intrinsically the one least prone to error, is to model a ground plane as a plate to which a wire is connected at one end and then bent to lie above the surface of the ground plane. The far end of the wire cannot be terminated on the ground plane, which forms an image plane, for current cannot flow in a plate. Thus, signal return current in the ground plane is not modeled. Nevertheless this technique of disconnecting the wire at the far end has been used with acceptable results with only the resonant frequencies changing.

The simple application of the MOM method may not be able to account for the presence of 3D structures attached to the PCB layout, although the effect of an attached wire can be analyzed. Such MOM methods do calculate current flow on the structure because of the connection of a single-frequency source.

The MOM programs require lots of computer time. For example, modeling a PCB ground plane with a small number of wires takes a couple of hours of computer time but results in large errors, because the spacing between the wires is large. If patches are used to form the ground plane, then the accuracy is higher but the program can take days to run on a fast Pentium PC.

The CG-FFT program does predict radiation from a microstrip with a high-permittivity substrate but assumes a symmetrical, infinitely small source. Thus, the radiation due to the D/M current flow can be predicted, but not that from C/M current due to an attached structure. This results in a massive 60 dB underestimation of the fields from the structure.

Reference [1] describes an analysis technique compared to measurements on a microstrip PCB attached to a shielded enclosure in a configuration similar to that used to provide the measured data presented in this chapter. The parasitic capacitance, which exists between the microstrip and the box, supplies the return path for the C/M current, just as it does in the test setup described in this report. The reference proposes a moment method of analyzing arbitrarily shaped 3D conductor dielectric composite structures that allow the radiation due to C/M current caused by parasitic capacitance to be calculated. The reference reports a good correlation between the predicted and measured fields at a distance of 3 m from the source. For example, the horizontally polarized field at a distance of 3 m from the source parallel to and in the plane of the microstrip transmission line at a frequency of 150 MHz is 47.7 dBuV/m computed and 48.5 dBuV/m measured. The reference also contains a comparison of predicted and measured fields at different locations around the PCB, with a very good correlation. The excellent correlation indicates that in this specific test setup, the predominant source of C/M current is the presence of the shielded enclosure, with little contribution from imbalance in the source.

When the size of the ground plane beneath a microstrip is very much less than a wavelength, the size affects the radiation only minimally. Therefore, the major parameters of interest are the thickness of the substrate, the length of the track, and the size of the adjacent structure. Based on the length of the track alone, and using a 20 log 0.045/0.18 m correction for length, results in a 12 dB difference, whereas the measured peak difference is 9 dB.

11.10.4 Computer Programs Specifically for PCB and IC Radiation, Crosstalk, Coupling, and SI Prediction

These programs are separated into rule-driven design and analysis types. From anecdotal information on an analysis type of program, the prediction was that the radiation from a microstrip PCB layout is higher than the two-track type of layout, indicating that the C/M current contribution to radiation is ignored. Despite ignoring the C/M current, the prediction was 220 dB higher than the measured results for the microstrip! Perhaps the most surprising result was that the emission at 5 GHz, from a 10 MHz clock source, was only 8 dB lower than the emission at 20 MHz, at the same level as the emission at 200 MHz and 62 dB higher than measured!

Applied Simulation Technology: A range of programs for emulation and modeling are available from Applied Simulation Technology for Windows 32 bit and 64 bit OS.

The emulators modelers are as follows:

ApsimRADIA-WB is a software modeling and simulation program that can predict the radiated emissions from PCB, MCM, cables, and subsystems. The program provides insight into physical placement, electrical specifications, interconnects, and connectors. The engineer can enter physical interconnect locations and simulate their effects from a schematic diagram. The program evaluates the impact of electrical and physical characteristics on EMI using schematic capture, field solvers, and EMI simulators in an integrated system. The inputs to the program are simple voltage or current sources or SPICE transistor-level models, and the interconnect models can be simple impedance or delay lines or more complicated transmission lines extracted from the field solver. The program can simulate nonlinear and linear sources.

Once the design is entered, it is automatically simulated using ApsimRADIA and ApsimSPICE. The results can be viewed graphically or in text form. The results are compared against an EMI standard of choice. The program can predict emission levels measured at some point defined by the user in the far or near field.

ApsimRADIA is an EMC/EMI radiated emission noise simulation.

ApsimSPICE is a circuit simulator with lossy coupled transmission-line capabilities. It can predict noise, reflections, and crosstalk in circuits and systems.

Apsim SKETCH is a schematic capture program that serves as front end to ApsimRADIA-WB. It has special built-in symbols for such entities as multiple coupled transmission lines, connectors, vias, and cables. Traditional symbols for semiconductors and macro models are available.

ApsimOMNI is signal and power/ground modeling and simulation software designed to evaluate nonideal ground planes with holes, slits, and cutouts. As discussed in Section 11.8.2, these imperfections can be major contributors to radiated emissions.

ApsimSPE is a signal integrity, power integrity (PI), and EMI simulator.

Apsim SI PCB is a signal integrity simulator for PCB.

ApsimSI-IC is a signal integrity simulator for ICs.

APsim\PI-LSI is a full-chip-PI simulator considering silicon substrate effects.

ApsimFDTD is a 3D full-wave simulator.

ApsimFDTD-SPICE is a full-wave nonlinear signal integrity and EMI solution simulator.

ApsimFDTD-SPAR is a full-wave electromagnetic solver.

ApsimDelta-I is a ground/power plane noise simulator.

ApsimSPICE is a circuit simulator.

ApsimR-PATH is an EMI path analyzer.

ApsimLCD is an LCD simulator.

The modelers are as follows:

ApsimRLGC is a 2D capacitive field solver.

ApsimIBIS-LCR 2 is for the electrical characterization of IC packages.

ApsimLPG is an LSI model for C/M EMI and SSI.

ApsimIBIS-tool kit is a SPICE to IBIS converter.

ApsimSPAR returns S-parameters.

ApsimPLANE is a 3D inductive field solver.

ApsimCAP3-D is a 3D capacitive field solver.

ApsimMPG is a model package generator.

ApsimIBIS-Translator is an IBIS to SPICE translator.

ApsimTSG is a table SPICE generator.

Cadence SI/PI analysis: The cadence power integrity solutions, based on Sigrity® PowerSI technology, supplies full-wave analysis of IC packages and PCBs.

Sigrity PowerSI is a signal integrity, PI, and design-stage EMI solution. It supports S-parameter model extraction and provides robust frequency domain simulation for the entire IC package and PCB designs.

Sigrity PowerDC is a DC signoff solution for IC package and PCB designs with electrical/thermal cosimulation. Quickly pinpoints IR drop and current hot spots. Automatically finds best remote sense locations.

Sigrity OptimizerPI is an automated board and IC package AC frequency analysis solution. Supports pre- and postlayout decoupling capacitor studies, identifies impedance issues, and suggests placement locations for EMI decoupling capacitors optimized for performance and cost.

Sigrity System Explorer is a general purpose topology exploration tool for the accurate assessment of signal and PI. Topologies can span multiple PCBs and IC packages.

The HyperLinx© program uses a full-wave 3D EM design and verification solution that can be used for PCB and circuit-level simulation and modeling. It uses a MOM implementation with the geometry models of bond wires, solder balls, bumps, vias, traces, and dielectric layers automatically extracted directly from the layout data and meshed to ensure proper handling of the 3D EM engine. It extracts S-parameter models for full system validation and can be used to analyze signal integrity.

EMCoS PCB Vlab features CAD interface, import of PCB Formats, solutions for PCB layouts. It uses a full-wave MOM simulation of PCBs with voluminous elements for the analysis of near electric and magnetic field radiation. Rapid RLC solver included in EMCos PCB Vlab provides parasitic RLC extraction and generation of equivalent circuit for analysis of parasitic effects. The EMCos PCB Vlab supports data exchange with EMC studio for integration into the full simulation process.

Material parameters physical surfaces and dielectric substrate can be applied to PCB traces or regions.

The workflow is as follows: import of PCB data, PCB data preprocessing, assignment of material properties, model construction, parasitic extraction, and export to EMC model.

GREENFIELD 2D from Quantic Products is a parasitic extractor for complex 2D geometries, such as nonplanar photo resists and trace undercuttings caused by etching. Once modeled and solved, users are provided with parasitic results including conductor losses, inductances, capacitances, impedances, and time delays. Other models include cables, connectors, atypical microstrip, and stripline geometries. The software also calculates per unit length self and mutual inductances, mutual capacitances, and capacitances to ground for multiconductor transmission-line systems and calculates resistances for eddy-current losses. Greenfield 2D employs Green's functions and BEM.

ANSYS SIwave-DC is a PI tool as part of the ANSYS 16.0 suite.

AXIEM is a 3D planar electromagnetic analysis tool from National Instruments and is used to characterize and optimize passive components on RF PCBs, modules, LTCCs, MMICs, RFIC, and antennas. AXIEM is schematic driven by means of EM extraction technology; it includes a layout/drawing editor, provides automatic adaptive meshing discrete and fast frequency sweeps, and provides visualization and post-processing results.

OpteM Inspector from Optem is a device, interconnect and substrate extraction tool for submicron digital, analog and mixed-signal IC designs. Both 2D and 3D electromagnetic analyses support conformal geometries and parametric variations of thickness, height, and conductor edges.

The electromagnetic analysis includes crosstalk, interconnect and device parasitics, resistance, and capacitance.

OpteM Inspector features:

Layout extraction includes multi-terminal devices, crossovers, edge and mutual capacitances, and contact resistances.
Library of high-level devices.
Technology independent with user-defined element definitions.
Can analyze the entire design or specific cells of the design.
Operates in a hierarchical flat or mixed mode.
Handles nonorthogonal layouts with 45° angles.
Uses finite element resistance extraction and automated electromagnetic field solvers.
Generates a standard SPICE net list and other formats.
Run time is linear to the size of the layout.
Batch mode processing.

EMX® is an electromagnetic simulator for HF, RF, and mixed-signal ICs from Integrand Software Inc., who claim uncompromised speed and accuracy. The EMX is a planar 3D integral equation solver that uses Maxwell's equations.

It utilizes multipole-accelerated far-field interactions for fast matrix construction, advanced meshing and compression techniques, and a well-conditioned matrix formulation, even at low frequencies.

Its features are as follows:

3D conductors and vias
True volumetric currents to account for skin effect
Accurate sidewall capacitances
Layered and lossy substrates
Substrate and inter-component coupling

EMX is used by foundries and semiconductor design houses for tasks such as the following:

Characterization of passive component libraries
Design of spiral inductors, MOM capacitors, MM capacitors, baluns, and transformers
Design of RF test fixture
Modeling large distributed power transistors
Simulation of high Q passive-package filters
Extraction of complete RF circuits

High-frequency 3D planar EM solvers from Sonnet, Liverpool, NY
The following are 3D Planar Solver products:

Sonnet Suites
Sonnet Pro
Level 3 Gold
Level 2 Silver
Level 2 Basic
Lite plus
Lite
Blink

EN: Sonnet 3D planar electromagnetic analysis engine
This program is particularly useful for modeling PCBs and circuits.
EM® is the electromagnetic analysis engine. It uses a full-wave MOM technique based on Maxwell's equations to perform a true 3D current analysis of predominantly planar 3D structures that includes all parasitic, cross-coupling, enclosure, and package resonance effects. The structures include microstrip, stripline, coplanar waveguide, PCB (single and multiple layers), and combinations with vias, vertical metal sheets, and any number of layers of metal traces imbedded in stratified dielectric material. EM will perform an automatic subsectioning of the circuit, allowing faster analysis to occur, but you can control how that subsectioning takes place. In addition, EM provides the following features:

Complete control of the properties of any number of dielectric layers.
Complete control of the DC and RF conductivity properties of any number of conductors independently.
Output *X*-, *Y*-, or *Z*-parameters in formats compatible with the major circuit theory simulators.
Output equivalent SPICE parameters for use in SPICE circuit simulators (particularly useful for the analysis of crosstalk in digital circuits).
Control circuit port impedance for analysis.
Automatic de-embedding of circuits using TEM-equivalent impedances, if desired.
Include the effects of stray coupling, parasitics, and EM radiation.
Perform the analysis in a true shielded environment, to simulate enclosure effects, or in an open environment.
Size of circuit that can be analyzed is limited only by the size of your computer—software has no circuit size limits.
The FFT-based solver calculates all coupling to full numerical precision.
Dynamic range to over 100 dB.

Number of metal layers depends on level.

Via modeling.

Anisotropic and rough metal modeling available in Sonnet Pro.

Circuit ports depend on level; unlimited for Sonnet Pro and Level3 Gold.

Co-calibrated ports, ideal components, and component VENDOR SMD models depend on level.

Conformal mesh in Sonnet Pro only.

RFIC Micro via array in all levels.

64 bit analysis and parallel processing available in Sonnet Pro only.

DC S parameter in Sonnet Pro only.

A free version of Sonnet® Lite can be downloaded that can be used to analyze planar structures, such as the following:

Microstrip matching networks
Stripline circuits
Via (interlayer or grounded)
Coupled transmission-line analysis
Microwave circuit discontinuities
Broadside-coupled transmission lines
Microstrip or stripline filters
Mounting pad characterization
PCB trace crosstalk analysis
Spiral inductors with bridges
Planar interconnects

A number of example programs are provided along with online manuals and tutorials.

EmcScan Program and EMSCAN tool: The EmcScan tool, from Quantic Laboratories, is a simulator for detecting electromagnetic emissions from PCBs. Designed for use by non-EMC specialists, EmcScan allows PCB designers to identify hot spots and segments of nets likely to cause problems. It simulates current spectra and track current densities as well as the electric and magnetic fields close to a PCB, using Quantic's field-solver techniques.

EMSCAN utilizes 1218 probes in a 29 × 42 array to measure the magnetic field very close to the PCB. It has a resolution of 3.75–0.12 mm resolution. The EMSCAN provides a spatial representation of the RF currents flowing on the board. The hot signals can be traced as well as coupling points between traces. Each loop probe is individually addressable via the control bus.

The RF output of the EMxpert scanner can be connected to a spectrum analyzer to display the frequencies of emissions.

The EMxpert scanner output is connected via an EMxpert adapter to the USB port of a computer that is used to display the PCB and the areas of maximum emissions and the spectrum analyzer. The computer used by EMSCAN is designed and manufactured by ThinkRF.

Reference [3] describes the calibration of the EMSCAN by building a 3 in. loop with a 50 Ω termination resistor. A known RF voltage was applied (therefore the RF current was known in the loop since the loop was kept small relative to the wavelength) and the EMSCAN voltage under the test loop determined. The voltage response to a particular current on the EMSCAN probe array was dependent on the height above the array. Before any predictions of the electric field strength could be made, a complete calibration of the EMSCAN probes was required for each frequency of interest. The results of the calibration showed that the predicted RF currents were consistently higher than the measured data. With a −45 dB correction factor, it was determined that fairly good and repeatable results were obtained.

The frequency of the maximum peaks in emissions was close to those measured on an OATS. There was fairly good agreement overall with the greatest error between the predicted and measured amplitudes on the OATS, which was 7 dB.

The author of Reference [3] concludes

that this amount of deviation between measurement and prediction was considered low enough to allow this technique to be useful. Care should be used whenever using this technique to use the predictions as approximate, and not absolute.

Some amount of deviation should be expected, and if any levels are predicted close to the appropriate limits, extra care should be taken.

As most PCBs have one or more attached cables, radiation from these is likely to predominate at LF and the EMSCAN probes do not measure this.

Also when absolute measurements are made, the EMSCAN requires prior calibration.

It complements Quantic's established Compliance tool, which simulates radiation from a number of PCBs with cables and enclosures. It is not possible to input the dimensions of a simple layout unless this is first entered into a CAD program. The program performs EMC simulation during routing or at the postlayout stage. The EM field simulation and analysis are performed transparently to the user. According to Quantic Laboratories, the program very accurately predicts the relative levels of field and their distribution, although the absolute magnitude of the field at present may not be accurate. However, Quantic is working on this and expects to be able to predict accurately the magnitude of the fields in the near future. For more information, see http://www.quantic-emc.com.

11.10.4.1 Program Comparison and Alternatives

None of the three major manufacturers of the commercially available PCB analysis programs was either able or willing to analyze the microstrip or transmission-line layout examples described in Section 11.5, so the author was not able to compare the performance results to the measured results. With one manufacturer, the predicted results were very different from the results presented in Section 11.5 or Reference [1].

If a PCB prediction or rule-based program is not used, either because such a program is not available or because the results have not been satisfactorily validated, then at the minimum, a viewer program is extremely useful. The most common program is a Gerber viewer, which provides an easy view of the layout, layer by layer, and simplifies the tracing of potentially "hot" tracks throughout the board as well as how effective the ground planes are. All of this can, in a pinch, be accomplished by means of photocopies of the different layers, but the Gerber viewer allows the isolation of individual tracks by highlighting a net and zooming in on a specific area of the PCB. Simple, but generally adequate, Gerber viewers are available on the Internet as freeware. A commercially available viewer and plot program, for use on a PC, with higher capability is the GerbTool View/Plot program from WISE Software Solutions, Beaverton, OR. WISE also manufactures the GerbTool designer, which allows the designer to modify the design, after Gerber databases have been generated, and the GerbTool CAM, which is the full CAM station.

11.10.5 NEAR-FIELD MEASUREMENTS ON PCBs

In measurements of emissions from PCB traces at very close proximity, using extremely small (3 mm wide) probes, high-emission traces can be identified. However, when a larger probe or the small bowtie antenna is used to measure the emissions at a greater distance from the board, typically 10–20 cm, these high-level emissions from a localized area are often seen to reduce rapidly with increasing distance, leaving emissions at other frequencies and from other locations to predominate.

It is often the high emissions measured at a distance of 10 or 20 cm that continue to propagate out to distances of 1, 3, and 10 m. Although the hot spots are a useful indication of problem areas on a board, it is the C/M current generated by a section of board and the area of the antenna formed by the PCB tracks that determine how effectively the board radiates in the far field. When cables are attached to the PCB, these are typically the major source of emissions especially when the cable is at a resonant, or multiple of the resonant, frequency length.

Section 9.3 discusses the use of diagnostic measurements further, and a computer program is available from the website emcconsultinginc.com, which uses the current flow and the cable length to predict the radiation from a distance of 1 m and above.

11.11 PCB DECOUPLING CAPACITORS, IMBEDDED CAPACITANCE, AND THE EBG

The impedance and self-inductance of capacitors with vias and PCB traces is discussed in Section 5.1.10.1 and are applicable when no overall ground or power plane exists on the PCB.

On multilayer boards when these layers are adjacent and then at LF, the impedance is typically lower than a PCB with no overall power and ground planes. This is due to the high-quality capacitance between power and return. Reference [4] describes this capacitance and characterizes it as 2–15 nF for a 6–10 inch square. PCB with adjacent power and return planes. Reference [5] predicts parallel resonances in addition to the series resonances. The series resonance is due to decoupling capacitors and their inductance in series with via and trace inductance. This parallel resonance is due to the board capacitance resonating with the interconnect inductance and the decoupling capacitors.

The power distribution trace inductance no longer serves to isolate the different components as this has been replaced by the planes.

Thus, on PCBs that have adjacent internal power and return planes, all decoupling capacitors are shared in the frequency range in which they are effective. At frequencies above series resonance, the parallel resonances occur.

This parallel resonance results in a high impedance at a frequency between the frequencies for the low impedances due to the decoupling capacitor series resonance. This effect is shown in Figure 11.120, where F_s is the series resonant frequency and F_p the parallel resonant frequency. The schematic and resonances for a 1 µF and 10 nF decoupling capacitor as well as the power bus impedance are shown in Figure 11.121.

The impedance of a board with 65 decoupling capacitors is shown in Figure 11.122.

Reducing the inductance of the interconnects as described in Section 5.1.10.1 is the most effective way of increasing both the series and parallel resonant frequencies. Large vias or multiple vias and using many decoupling capacitors with the same value and same interconnect inductance increases the effective frequency range.

FIGURE 11.120 Power-line impedance of a multiplayer board with n identical decoupling capacitors. (From Hubing, T.H. et al., *IEEE Trans. Electromagn. Compat.*, 37(2), 155, May 1995.)

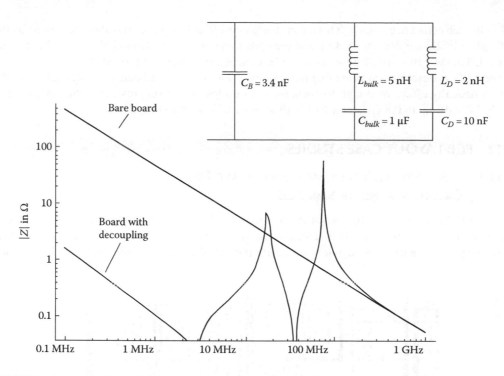

FIGURE 11.121 Power-line impedance of a two decoupling capacitor circuit. (From Hubing, T.H. et al., *IEEE Trans. Electromagn. Compat.*, 37(2), 155, May 1995.)

FIGURE 11.122 Impedance of a board with 65 decoupling capacitors. (From Hubing, T.H. et al., *IEEE Trans. Electromagn. Compat.*, 37(2), 155, May 1995.)

A new approach to reducing power plane noise and radiation is the electromagnetic bandgap (EBG). The EBG creates a 2D periodic structure on the whole or partial power and/or ground planes. There are certain energy bands in the structure that electrons can occupy and others that cannot be occupied (forbidden bands or stopband). The parallel plate–mode electromagnetic waves with a frequency inside the stopband cannot propagate through the EBG material. Various EBG structures with slits, meander lines, and bridges have been used to suppress power plane noise. Some of these use two of the planes in the PCB and others three. A typical size for a single unit cell is 20 mm × 20 mm, and 3–16 such cells are common. This means that a considerable amount of real estate is used by the EBG. Most PCBs are densely packed, and even when the power and ground planes can be utilized for the EBG, the EBG will be penetrated by vias connecting power and ground to ICs.

The EBG is described as providing noise suppression over a wideband, and although true, the EBG is basically effective at high frequencies. For example, noise suppression (reduction in noise) is only 5–24 dB up to 0.8 GHz and 10–90 dB above 1 GHz and up to 5 GHz.

11.12   PCB LAYOUT CASE STUDIES

11.12.1   CASE STUDY 11.1: GROUNDING ANALOG AND DIGITAL CIRCUITS SHARING THE SAME PCB

Figure 11.123a and b shows the two sides of a common PCB layout in which the analog circuits and logic share the same PCB. At first glance, the ground plane under the analog ICs appears adequate, whereas in reality, the layout is a classic example of poor grounding resulting in common

FIGURE 11.123   (a) Example of an analog/digital circuit with ineffective ground plane results in EMI. (b) Reverse side of PCB with ineffective ground plane.

impedance and radiated coupling and crosstalk. Electromagnetic interference was detected on this board by examining the output bits of the 12 bit successive-approximation analog/digital (A/D) converter. The A/D converter was packaged in a 32 pin IC, the pads of which may be seen in Figure 11.123a and b.

When examining the output bits of a successive-approximation converter, the type of noise may often be determined by the distribution of the number of times each bit is set after an A/D conversion. A Gaussian distribution indicates that the source of noise is noncoherent (e.g., thermal or $1/f$), in which case reducing the values of resistors and using low-noise op amps may reduce the noise level. When the noise is coherent (e.g., impulsive or CW noise), the A/D output often exhibits missing codes in which certain bits either are not set or are set infrequently.

In the example presented here, a number of bits, not just the LSB, were toggling. The input of the A/D was connected at the edge connector to analog ground, that is, shorted out, and the differential input to the A/D should have been zero volts. The source of the noise was determined to be the digital logic. On close examination of the analog ground plane in Figure 11.123a, it is seen that the analog ground is separated into isolated islands connected by very thin and often-circuitous tracks. These thin tracks are intended to shield the analog signal from the +5 V power rail that can be seen around the periphery of the PCB in Figure 11.123a. The temporary measure taken to improve the grounding was to fill in the isolated islands, by connection of copper foil, on both sides of the board. A reduction in the number of bits toggling was achieved by the improvement in the ground.

The digital ground connection may be seen in Figure 11.123b between points A and B. The analog ground also connects to both points A and B, so digital noise currents flow in the analog ground. Removing the analog ground connection just before point B effectively floated the analog ground on top of the digital ground, and the number of bits changing state reduced considerably. The thin shields around the analog tracks at location E in Figure 11.123a will provide approximately 6 dB of reduction in capacitive (voltage) crosstalk. Increasing the width of the shield tracks can result in up to 11 dB of reduction. In this crosstalk problem, a low level of noise voltage was measured between the analog ground and the +5 V track, and the major coupling mode is almost certainly inductive. The source is the digital noise currents flowing in the +5 V track. Inductive coupling is unusual in this configuration due to the typically high input impedance of analog circuits. (See Chapter 4 for a description of capacitive and inductive crosstalk.) However, the weighting of the LSB in the 12 bit A/D converter is 1.2 mV, which makes the analog circuits extremely susceptible to noise, especially as a gain stage precedes the A/D converter. The crosstalk problem was solved by cutting the +5 V tracks at locations C and D, shown in Figure 11.123a, and using a wire connection in close proximity to the digital ground track in Figure 11.123b. With the crosstalk problem solved, the noise reduced to a level whereby approximately 1 LSB was toggling.

The remaining problem was found to be in the connection of the metal case of the A/D converter to analog ground. The A/D converter pin at point E is the case connection that was connected by a track to analog ground, although the islands on either side of the track were also analog ground. A voltage was induced in the case of the A/D by radiated coupling from the digital logic circuits and tracks. By making a solid analog ground connection to the case, the remaining EMI disappeared and all output bits of the A/D were at "zero" level.

Although the intra-equipment-level problems were solved, as soon as the analog input was connected to an external piece of equipment EMI reappeared, due to C/M noise voltages between the analog grounds of the two units of equipment. The unit of equipment under investigation was deliverable; thus, only minimum modifications were possible at the PCB level. The immediate solution to the problem was to change the single-ended input to a differential input. This required three video op amp ICs that exhibited a minimum of 30 dB of C/M noise reduction of up to 30 MHz. Consequently, the EMI fix was expensive. It was also difficult to find space for the additional circuitry. The lesson to be learned from this case study is that in designing the printed circuit board layout, the equipment and system with the goal of achieving EMC would have avoided delays in equipment delivery and the costly modifications that had to be implemented in the next version of the equipment.

In summary, the improvements were as follows:

1. The isolated ground islands shown in Figure 11.123a were connected together using copper foil.
2. Connecting the case pin of the 12 bit A/D converter to the improved analog ground eliminated the contribution of radiated coupling EMI.
3. Removing the analog ground connection just before point *B* effectively floated the analog ground on top of the digital ground and eliminated, in conjunction with the improved analog ground, the common ground impedance coupling problem.
4. Disconnecting the +5 V trace that can be seen around the periphery of the board in Figure 11.123a, at points *C* and *D*, and bypassing it by means of a wire running across the board above points *A* and *B* eliminated inductive crosstalk between noise carried on the +5 V line and the sensitive analog traces shown in the right-hand edge of Figure 11.123a.

11.12.2 CASE STUDY 11.2: GOOD GROUNDING TECHNIQUE FOR A VIDEO CIRCUIT ON A PCB

A second video circuit PCB layout is shown in Figure 11.124, the video ground plane of which appears to be as segmented as that shown in Figure 11.123. However, the physical separation in the ground plane is intentional because the modified differential op amp circuit of Figure 8.17b is used to effectively connect the ground plane islands *A* and *B* for differential signals but to isolate the grounds with respect to C/M signals/noise. The modified differential circuit is also

FIGURE 11.124 Example of an analog/digital circuit with effective ground planes connected together by differential amplifier circuits.

used at location C in order to isolate the video ground at the source from the video ground on the video processing board, which is shown in Figure 11.124, point A. The ground plane at B is the digital ground connected to the A/D converter. The advantage of the modified differential circuit in this application is that the resistors connected to one of the two inputs are used to tie the analog and digital grounds together and avoid potentials above the C/M input range of the op amp. Care must be exercised in connecting the video circuit supply return to the video ground. This connection must be made either at the signal source or to plane A on the video processing board, but not both.

The tracks at the side of the board, location D, carry control signals, which are designed not to change state during video processing and A/D conversion. However, these tracks are likely to have C/M noise voltages impressed upon them or to carry C/M currents, with a consequent risk of radiated coupling to the analog circuit. The intent in design verification was to test for the presence of radiated coupling at the prototype board level and then to replace the tracks with six-conductor shielded cables if a problem existed. The shield of the cable would be connected to a "clean" ground, such as structure ground. Tests showed that indeed some radiated coupling existed, but this was at an insignificant level.

The board is used primarily in space applications, with a metal core providing a thermal path to conduct heat away from components. The core is electrically isolated from all circuits but connected to the enclosure, both electrically and thermally, at the card guide.

A noise problem has been encountered in the use of two different boards, both of which contained a metal core connected to the enclosure. When the cores were isolated from the enclosure, the high level of noise, detected again by the toggling of the A/D converter output bits, reduced drastically. The source of the problem is unknown and was initially thought to be RF currents flowing through the PCB core. Assuming that all other PCBs in the same enclosure are sources of noise current flow in the enclosure, isolating the cores of these boards from the enclosure should reduce the noise level on the susceptible board; but it did not. A key to the problem may be the very high capacitance between the core and the analog/video ground plane. The capacitance is dependent on the board and analog ground plane size and has been measured at 3000–5000 pF.

11.12.3 Case Study 11.3: Coupling between Digital Signals and an Analog Signal within the Analog-Shielded Section of a Board

The board shown in Figure 11.125a and b is another example of an analog and digital mix. Figure 11.125a shows the digital section on the left-hand side and the analog ground plane on the right-hand side, with a single trace connecting the two. This board did not suffer from excessive radiation due to this single-point analog-to-digital ground connection, as described in the grounding section, because of the width of the ground connection trace and because it was made on more than one layer and these layers sandwiched the clock and data connections between the digital and analog sections. However, the analog circuit did suffer from pickup due to the high-speed clock entering the analog section. The analog-section ground plane layout was designed to enable the incorporation of a gate type of shield around the analog section. When this shield was incorporated, the level of radiated coupling remained the same. The addition of shielding was ineffective because the source of radiated coupling was the clock signal, which entered the analog section and was connected to the D/A converter. Before the radiated coupling mechanism for EMI was discovered, it was thought that coupling occurred to the ±20 V power supply and analog ground track connections and the problem was caused by C/M current flow on the supply and return connections. The reason that this all-too-common problem was suspected was that instead of making the power supply connection on the right-hand connector adjacent to the analog ground, the circuit designer, against better judgment, was persuaded to allow a connection as shown in Figure 11.125b. This type of connection places the analog supply and return in very close proximity to the digital circuits and is a prime candidate for coupling. However, in this case, coupling to the supply and return was not the problem.

The analog section was required to operate at very high speed, with an analog slew rate of 600 V/μs. The circuit designer had tried a high-Q Chebyshev filter at the output of the analog chain in an attempt to filter out the clock signal; however, this degraded the response of the analog output. Instead, a capacitor was tried at the load end of the analog trace connected to the output of the D/A converter to reduce the load impedance and thereby the EMI voltage induced in the signal. The value of the capacitor was chosen such that the time constant formed by the output impedance of the D/A converter and the capacitor did not increase the slew rate appreciably. The EMI problem was greatly improved and other improvements were possible. However, an executive decision was made to eliminate the voltage-to-frequency converter, which was supplied by the output of the analog section, and to use a digital-to-frequency converter, with the hope that the EMI would disappear. The warning was given that with a high level of C/M noise on the digital data to the digital-to-frequency converter, EMI coupling could occur inside the

Analog ground plane

10.300 ± 0.003 in. between targets

(a) | Layer 1—Primary side | Direct view |

FIGURE 11.125   (a) Analog ground plane and digital section of board.   (*Continued*)

+20, −20 and Agnd planes

10.300 ± 0.003 in. between targets

(b)

| Layer 11—INT8 | Direct view |

FIGURE 11.125 (*Continued*)   (b) Analog power and return tracks inside the digital section.

device; but this was ignored. Sure enough, the EMI was still present, and the designers spent a great deal of time improving the immunity of the digital-to-analog converter, much to the delight of the manufacturer.

11.12.4   CASE STUDY 11.4: OUT-OF-SPECIFICATION RADIATED EMISSIONS FROM TELEPHONE EQUIPMENT

The equipment under test (EUT) was tested for radiated emissions in accordance with EN 55022, class B (Domestic Information Technology Equipment). The original problem was an out-of-specification emission at 185.00 MHz, which was about 6 dB above the limit; however, our own scan of radiated emissions showed the level to be slightly higher, 10 dB above the limit. Therefore, modifications were required to reduce this and several other emissions fairly close to the limit.

11.12.4.1 Circuit-Level EMC Modifications

Several modifications are required to achieve the attenuation of radiated emissions shown in Table 11.7. A shielded power cable is required as well as C/M filtering of the power with 1000 pF capacitors between each line and chassis ground. This is done for every line that is routed down the cable. The shield of the cable must also be terminated properly with a good low-RF impedance to the connector shell as well as to ground at the power supply end. A power-line filter was added between the power supply and the power cable to keep RF out of the power supply and the AC power lines, thereby minimizing radiation from these sources. This was done to prove the effectiveness of the fixes performed on the power cable to the EUT. The connector on the EUT should make contact with the inside of the enclosure to keep coupling to the shield to a minimum. An EMI gasket was also added between the connector and the enclosure to seal the gap on both sub-D connectors on the EUT. The capacitors help by keeping conducted C/M currents off the cable's conductors, especially at lower frequencies. We used surface-mount capacitors; however, capacitors with longer leads would likely do a good job, although the extended lead length will present a greater impedance at higher frequencies.

When the cables were plugged into the serial connectors, a resultant out-of-specification increase in radiation occurred. Surface-mount 470 pF capacitors were connected between the digital ground in each of the connectors and the chassis ground (and thus the conductive coating on the inside of the enclosure), with a resultant decrease in emissions. Because this fix would be extremely difficult to implement, it was decided to try a filtered modular jack that contains capacitors between each line and the metal case of the connector, as well as a ferrite block. One cable was tested alone, because the radiated emissions from this cable were at the highest level of all the four cables tested; thus, if this cable is within spec, the others will be as well. The 470 pF capacitors were removed and the modular jack for the high-emission cable was replaced with a Corcom filtered connector, part number RJ45–8LC2-B. This connector has a metal shield around it, which should be terminated on

TABLE 11.7
Radiated Emissions Measurements before and after Modifications

f (MHz)	Radiated Emissions (dBµV/m) No Modifications	Radiated Emissions (dBµV/m) With Modifications	EN 55022 Class B Limit (dBµV/m)
49.00	35	24	40
51.88	38	20	40
55.74	32	15	40
63.01	28.5	9.5	40
65.90	39	7	40
75.73	29.5	25.5	40
78.83	34	20	40
83.92	34	25	40
101.00	37.5	14.5	40
168.52	34.5	12.5	40
185.00	**50**	18	40
214.55	33	24	40
253.52	44	34	47
386.66	40	22	47

Notes: *Italics* indicates emission level is within 6 dB of the limit (6 dB being our standard safety margin). **Boldface** indicates emission level exceeds the limit.

the inside of the enclosure to the conductive paint. The reduction of the emissions radiated from the cable is 15 dB at 111.00 MHz; however, an out-of-specification emission at 51.88 MHz remained. Adding a 1000 pF capacitor from the digital ground lines of the display connector to the chassis ground drops this emission by about 21 dB. These modifications are required on all of the other modular connectors.

The key to having the modifications to the cables work is that the enclosure is well shielded and the power cable shield and connector shell terminations have low impedance. If not, radiated coupling to the output cables, which will bypass any filtering or high-impedance shield terminations, can easily put emission levels out of spec. Apertures in the enclosure shield can result in this effect. The coating at the two ends of the EUT enclosure exhibited a high impedance, that is, an aperture with respect to electromagnetic fields and especially magnetic; thus, a consistent conductive coating that allows a good contact at the seams is mandatory. The intentional apertures in the EUT enclosure are acceptable, because, due to their small size, only very high frequencies will couple through them. It is the longer gaps in the ends of the enclosure and at improperly connected seams that will allow the undesired coupling to cables at frequencies that can present a problem. In all tests, these open-end sections were covered by aluminum foil connected to the remaining components of the enclosure, which were coated with a relatively low-resistance coating. This aluminum foil also covered most of some bad spots in the conductive paint discovered at the ends of the enclosure.

11.12.4.2 PCB-Level Radiation Reduction

All of the modifications mentioned in Section 11.12.4.1 are costly and difficult to implement. The manufacturer of the equipment was willing to go to a new PCB layout. Weak areas in the existing PCB layout exist, and so going to a new layout was almost certain to reduce radiation. In some instances, a new PCB layout will eliminate the need for a conductive coating on a plastic enclosure.

The problem areas on the PCB were at U40 and U41 in the power supply and around Y1, the 8 MHz oscillator, as well as coupling resulting in C/M currents flowing on cable plugged into J1, J2, J3, and J4. These areas are shown in Figure 11.126a.

The modifications to the PCB layout needed to ensure meeting EN5022 class B are as follows.

U40 and U41 must be moved to a location over the digital ground plane section of the board and not over the analog ground section. The existing digital ground trace, which is taken into the analog section, is not adequate. The supply trace to the digital section of the board should be brought into the power supply area by a plane, and the decoupling capacitors must be connected by vias between the new power plane and the digital ground plane. Likewise, C60 and C61 must be connected directly by vias between the new digital power plane and the digital ground plane. *Do not use traces to connect capacitors* and locate these capacitors close to U40 and U41.

Add 0.1 µF capacitors in parallel with the 47 µF C60 and C61, again using vias between the digital ground and power planes, to improve their performance at HF.

The oscillator Y1 is located over the analog ground plane close to the J1, J2, J3, and J4 connectors. The power and return to Y1 are made with traces, as are the connections to the decoupling capacitor.

Y1 must be moved as far away from J1, J2, J3, and J4 as possible (at least 6 cm) and remain in close proximity to the IC that interfaces to the clock. Y1 must be located over the digital ground plane, with its return pin directly connected, not by a trace, to the digital return plane. The oscillator decoupling capacitor must be connected between the digital power and return traces by vias, not by traces, and in close proximity to Y1.

All clock lines and high-speed data lines must be on inner layers sandwiched between the digital power and return traces and not on the solder sides of the board.

J1, J2, J3, and J4 should be located directly over a section of digital ground plane extending under the connectors. Make provision for a 470 pF capacitor between the digital ground plane and chassis at each of the connectors.

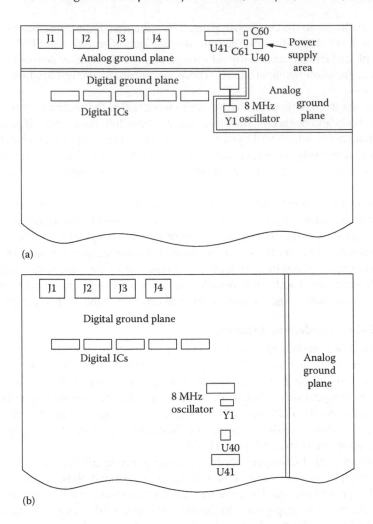

FIGURE 11.126   (a) Problem layout area before modification. (b) Problem layout after modification.

Place only low-level, LF devices and traces close to J1, J2, J3, and J4. Keep high-speed devices and clock and data traces as far away from the connectors and signal traces to the connectors as possible.

The new PCB layout was made, with the improvements made in the critical areas shown in Figure 11.126b.

In formal tests, the equipment met the EN55022 class B requirements without the addition of any of the circuit-level additions, such as filtered connectors and capacitors. However, the customer complained that it only just passed radiated emission requirements, but then admitted that not all the recommended modifications could be made to the board!

11.13   INCREASED PRINTED CIRCUIT BOARD IMMUNITY

With the EU radiated and conducted immunity (susceptibility) requirements present in EN50082-1, EN50082-2, EN55024, EN 6100-4-3, etc., equipment either is subjected to C/M test current injected on signal interfaces and to radiated immunity test fields. In addition to immunity test levels, digital or switching power-supply noise can interfere with low-level signals. The problems of susceptibility of circuits are addressed in Case Studies 11.1 and 11.3 in Section 11.12.

At the PCB layout stage, any layout that reduces emissions will also decrease the level of noise voltage induced in the circuit. For example, a good, solid ground plane will reduce the voltage dropped across it as signal current flows and will reduce emissions. Likewise, if C/M current flows in the ground plane, the voltage drop in a good ground plane is low and the level of common impedance coupling is low.

One special precaution that needs to be taken in laying out a board for maximum immunity is to make signal and signal return track lengths, for differential-input circuits, equal. If they are not equal, any induced C/M RF current will exhibit a different propagation delay, or phase shift, between the signal and the signal return track, resulting in a differential voltage at the input of the receiver. Remember, however, that EMI can also result from a C/M current flow into the input of a single-ended receiver, which is converted to D/M by the differences in signal input impedance and signal ground impedance. The most common and effective solution is to place a low-value capacitor, which has a low impedance at RF and a high impedance at the input signal frequency, between the signal input and the ground. Thus, at RF the signal input has a low impedance to signal ground, and the C/M current, on signal and ground tracks, results in a minimal D/M voltage between the signal and ground.

REFERENCES

1. S. Daijavid and B.J. Rubin. Modeling common mode radiation from 3D structures. *IEEE Transactions on Electromagnetic Compatibility*, February 1992, 34(1).
2. F.B.J. Leferink. Inductance calculations: Methods and equations. *IEEE International Symposium on Electromagnetic Compatibility Record*, Atlanta, GA, 1995.
3. B. Archambeault. Predicting EMI emission levels using EMSCAN. *IEEE International Symposium on Electromagnetic Compatibility Record*, Dallas, TX, 1993.
4. J.R. Miller, J. Novak, and T. Chou. Electrical performance of electronic packaging, *2002 IEEE 11th Topical Meeting*, pp. 123–126, Monterey, CA.
5. T.H. Hubing, L. Drewniak, and T.P. Van Doren. Power bus decoupling on multilayer printed circuit boards. *IEEE Transactions on Electromagnetic Compatibility*, May 1995, 37(2), 155.

12 EMI and EMC Control, Case Studies, EMC Prediction Techniques, and Computational Electromagnetic Modeling

12.1 EMC CONTROL

The emphasis in this book has been on prediction techniques useful in designing EMC into equipment and systems. EMC control ensures that the EMC design is correctly implemented at every stage in the development and manufacture of equipment.

12.1.1 EMC CONTROL PLANS

Control plans are required at both the system and equipment levels. The system-level plan requirements are contained in MIL-STD-464 and equipment/subsystem requirements in the Data Item Description.

The EMC control plan indicates how EMC will be accomplished for a specific equipment, subsystem, or system. Too often the control plan is a general-purpose document reissued with only minor changes for every type of equipment. The control plan should be specific and describe the type of equipment, expected levels and frequencies of emission, level of immunity, electromagnetic interference (EMI) budgeting, and potential waiver requests, or any predicted difficulty in meeting the requirements. Details of the grounding scheme, filtering, shielding, and so on should be provided with design goals and breadboard-level tests.

An important section in the plan, and one increasingly required by procuring agencies, is a detailed EMC prediction that can provide information on the effectiveness of the design and highlight any critical areas. The EMC prediction is typically part of the EMC program plan; however, when this is not a requirement, the prediction should be included in the control plan or as an appendix to it.

The EMC prediction and analysis techniques should be described with validation measurements, where possible, that confirm the applicability of the particular technique to the application. The prediction techniques should be used in potential EM problem areas, for determining degradation criteria and safety margins, tailoring applicable EMC documents, and determining the anticipated EM environment.

The purpose of the plan is twofold: (1) to assure the procuring agency that the equipment will be designed for EMC and (2) to provide guidelines and design goals to designers and systems, mechanical, and packaging engineers.

Once an effective and detailed plan has been created, the next step is to ensure that the plan is used by the responsible engineering people. This is achieved by making one individual responsible for EMC and having that person sit in on design reviews as well as in the process of signing off drawings. The same individual should develop the EMC control program, control and test plans, and the power and grounding diagram. The contents of a typical control plan are as follows:

Scope
    Purpose
    Organization
    Where the EMC activity fits in the contractors' organization
Documentation
    Applicable documents, including the EMC requirements, test methods, grounding and bonding, lightning protection, and safety, as they apply to EMC
    Reference documents, including EMC design guidelines and handbooks
    Documentation requirements
        Test results
        Signal and power interface details
        Power and grounding diagram
        Test matrices
        Test plans
        Test procedures
        Test reports
Management
    EMC responsibility: This should contain the responsibility and authority of the individual who will direct and implement the program and the number of full- or part-time EMC personnel on the program.
    EMC reviews
    Design requirements: Methods and requirements for ensuring that contractor-developed subsystems and equipment will not be adversely affected by interference from sources within the equipment, subsystem, or system or be sources of interference that might adversely affect the operation of other equipment/subsystems. This is typically achieved by meeting MIL-STD-461 requirements. Some of the specific design areas that should be controlled by requirements are as follows:
    Cable design, including wire categorization and criteria for identifying, labeling, and installing
    Interference-generating or susceptible wires, shielding techniques, and wire routing
    Internal wiring and external wiring
    Enclosure structure, shielding, bonding, aperture closure, and gasketing
    Signal interface design goals and tests
    Intraequipment and intrasubsystem design considerations controlled by interface control documents when equipment is procured from different sources
    Intersubsystem design considerations controlled by interface control documents
    Radiation characteristics from system antennas, including fundamental and spurious energy and antenna-to-antenna coupling
    Design criteria and required tests for lightning protection
    Impact of corrosion control requirements on EMC and recommendations for resolution of problem areas
    Methods of implementation of design changes required for EMC

12.1.2 EMC CONTROL PROGRAM PLAN

The program plan provides the data to determine whether all pertinent EMC considerations are being implemented during a program and whether equipment, subsystems, or systems can be expected to operate in the electromagnetic (EM) environment without degradation due to effects from EM energy and without causing degradation to other equipment, subsystems, or systems.

The plan should describe the procedure for managing the EMC program during the project, the EMC organization and responsibilities, lines of authority, responsibilities and role of the Electromagnetic Compatibility Advisory Board (EMCAB), procedures for identifying and resolving potential problems, and a schedule and milestone chart of all EMC activities during the program. The plan should contain the EMC predictions and analysis when this is not a part of the EMC control plan.

The following is an example of one approach that may be taken to ensure EMC in the design of a ship for which a large number of pieces of equipment, not all of which meet MIL-STD-461 or IEC Publication 533 requirements for shipboard equipment, must be chosen. The outline of the program plan is as follows:

1. Characterize the EMC characteristics of the candidate equipment
2. Characterize the predicted EM ambient

Step 1: The following information on the candidate equipment is required:

a. Ship zone in which equipment is located
b. Operational criticality
c. Whether it is an intentional or unintentional transmitter, receiver, or both, for the purpose of EMC
d. EMC requirements met by the equipment, including commercial
e. EM emission profile
f. EM immunity
g. Shielding effectiveness of the equipment enclosure and cable
h. Power requirements and type of supply used
i. Interface types (e.g., RS232, 1553, Ethernet, USB, etc.)
j. Effectiveness of equipment power-line filter, when installed

Step 2: The EM ambient is likely to vary throughout the ship, depending on the proximity to equipment such as transmitters, electrical control rooms, cables, and the level of shielding provided by metal walls and apertures (doors and windows) planned for each zone. In order to derive a typical ambient for each zone, the following factors should be accounted for:

a. The routing of power, control, and transmitter cables in the vicinity of each zone and the nature of the coupling paths between the cables and equipment.
b. Where the radiated emission footprint for a specific equipment is known, it should be used in characterizing the EM ambient. If equipment has met radiated emission limits, then these limits may be used.
c. Where pertinent measured data for ships is available from shipboard EMI surveys, these data may be used.
d. In the absence of either of these types of data, published data similar to that provided in this book for similar environments—such as process control—or computing may be used.
e. The conducted ambient may be derived from shipboard-conducted susceptibility levels contained in IEC publication 533 for commercial ships and MIL-STD-461 for military. These levels should be tailored when information on the actual ambient is known.

With the EM conducted and radiated ambients characterized, a comparison of the EMC characteristic of the equipment to the harshness of the ambient can be made. Potential EMI situations, in which the equipment does not match the ambient, can then be determined. For example, when a computer that meets commercial radiated and conducted emission limits and that exhibits an immunity level adequate for an office environment is intended for operation in a control room, the potential for EMI is high.

A detailed analysis of each potential EMI situation should be performed to ascertain the probability that the problem exists. If the probability of EMI is high, the severity of the degradation due to EMI should be calculated and one of more changes made to ensure EMC. Some of the changes that can be made to achieve EMC are as follows:

a. Relocating equipment within a zone (to separate high emitters from the more susceptible equipment).
b. Adding signal and/or power-line filters to an entire zone or to individual pieces of equipment.
c. Compartmentalizing a zone into "noisy" and "quiet" areas, perhaps by incorporation of conductive doors and walls with feedthrough signal and power-line filtering. A related technique is to shield an individual piece of equipment.
d. Rerouting signal and/or power cables.
e. Increasing the level of shielding on signal and/or power cables and/or their connectors.
f. Where feasible, changing an equipment's interface to a higher-immunity type (e.g., changing from RS232 to RS422 or from an electrical to an optical interface).

The topside antenna-to-antenna and antenna-to-cable coupling must also be examined in the EMC control plan; here the techniques described in Section 10.2.1 are applicable.

12.1.3 QUALITY CONTROL

One of the major reasons why equipment that has been certified as complying with an EMC requirement does not continue to comply in a production version is lack of quality control. The other major factor, already discussed, is modifications to equipment design or software.

Any modifications to a design should be under configuration control from a single authority. Appropriate documentation, such as an engineering change notice (ECN), should be used. The ECN must be reviewed by the responsible EMC authority to evaluate the impact of the change to EMC; once the change is accepted, speedy removal of superseded documents should be accomplished.

Any changes to equipment that has been accepted should be inspected and verified and the equipment retested for compliance with EMC requirements when this is considered necessary by the procuring agency. That apparently minor modifications can result in significant changes in emissions and susceptibility, especially when the changes affect the grounding scheme, must be considered when making the decision to retest or not. Equipment that has had changes incorporated should be marked or reidentified if the magnitude of the change warrants. Inspection records should be maintained that include as a minimum:

Identification of the item
Date of inspection
Number of units inspected
Number of units rejected
Description and cause of discrepancy
Disposition of material
Identification of the inspector

Additional inspection records that include test results and certification of personnel for special processes, such as soldering and plating, should also be maintained. Inspections should be tailored for EMC. For example, if a ferrite core contains a few pinholes, it is likely to be rejected by incoming inspection, although the pinholes will not affect the impedance of the ferrite and will hardly change its physical characteristics. Far more important would be a test of the impedance versus frequency characteristics that affect the performance of the core.

The performance of filters is not necessarily consistent over different production runs and should be individually tested in a setup that approaches as closely as possible the impedances of the actual source and load as well as the type of noise.

Incoming inspection should include a checklist for incoming material, shop travelers, operation sheets, and/or inspection instructions to describe inspection operations during manufacture. Statistical quality control should be implemented on items not 100% inspected. Material subject to deterioration, such as conductive adhesives or caulking, should be periodically inspected and discarded when beyond its useful life.

The periodical training or retraining of inspection personnel, especially when new processes or materials are introduced, should be instigated. Quality control of sources shall be in place, including a list of approved sources and a performance rating system. The latest drawings or specifications should be quoted on purchase orders, with inclusion of quality assurance requirements such as traceability included on the physical optics (PO).

Material not inspected either shall be not released to production or shall be controlled, so that if the item is subsequently found unacceptable, it may be retrieved. All materials should be identified. Materials not released for usage should be segregated in controlled areas with age control and lot identity maintained.

Adequate test procedures should be in existence for qualification tests of engineering models and preproduction and production models of the equipment. Make sure the test procedure is reviewed and accepted by the customer well before the commencement of tests. Records of acceptance tests should be maintained. Control of nonconforming supplies and items shall be such that no out-of-specification component/unit can be incorporated into production equipment. Supplies that cannot be reworked or components that cannot be repaired should either be returned to the manufacturer or be scrapped. Packaging and shipping procedures should be in effect. Sensitive equipment must be adequately identified, specially packed, with the inclusion of an accelerometer to monitor vibration and shock where this is justified.

Measuring and test equipment at a facility should be controlled as follows:

Test procedures for test equipment including calibration limits and description of calibration equipment.
A list of qualified test facilities where equipment is sent should also be maintained.
Standards and calibration equipment used to calibrate test equipment shall be periodically certified.
The test equipment should be kept in an environment with humidity and temperature control and in which no damage or loss of calibration can result.

12.2   EMI INVESTIGATIONS

In an EMI investigation, the approach used is of paramount importance in finding a speedy and cost-effective solution. The approach should be methodical and thoughtful. This is much easier said than done, especially when under pressure to find a quick fix. In some instances, the causes of EMI and its solution are readily apparent and a minimum amount of prediction and diagnostic testing are required. In many cases, however, an immediate answer to the problem is not available, and some analysis and testing are required. One incorrect approach is to try a number of changes in parallel in the hope that at least one will work. I know of one extreme example where equipment was due for delivery but failed an acceptance test due to EMI. The requirement specification was extremely stringent and no waiver was allowed. A systematic approach was not used in finding the source of the problem. Instead, every change imaginable was made to the equipment. Some of these changes may have achieved some small improvement that was masked by other changes, which exacerbated the problem. After 3 years, the problems were solved and the equipment could be shipped. Because no exact records were made of those changes that had effected the improvement, the next version of the equipment exhibited the same problem. Although an extreme case, it serves to illustrate the importance of a systematic and well-documented procedure. More commonly, months may be spent in fixing an EMI problem that a specialist may accomplish in weeks, or the fix involves a costly modification or addition to circuits where a much simpler and cost-effective solution exists. Using the right systematic method requires self-discipline and the cultivation of a step-by-step approach.

One argument for blanket changes is that with additional filtering, shielding, grounding, compartmentalization, and so on, a guaranteed solution exists. However, as we have seen throughout the book, a limit exists to the shielding effectiveness achieved in a practical enclosure, source emissions may be high, and an incorrect filter or grounding scheme may worsen a problem. Unfortunately, the advantage of the systematic over the blanket approach often has to be demonstrated before it is accepted.

It is imperative that the number of variables that are not part of the EMI investigation be reduced to the absolute minimum. Therefore, do not arbitrarily move a cable position or circuit board location without a "before and after" measurement or assessment of the effect on EMI or emission level. Also, do not exchange or modify the software, circuits, equipment power source, installation method, connector or cable type, grounding configuration, or test method and test equipment without a "before and after" assessment of the effect.

The first question to ask is: Is the problem EMI? Or is the equipment malfunctioning due to another cause? Practicing engineers often encounter discussions between software and hardware engineers who believe a problem lies in each other's domain. This is also true of the designer who can find no failure in the hardware or software and assumes the problem is EMI. Therefore if, after investigation, EMI is not found, the cause of the problem may be equipment malfunction.

A typical approach is first to isolate the source of the problem and the coupling path or paths by a process of elimination. As discussed in previous chapters, it is common for both radiated and conducted coupling to coexist. Also, the coupling may appear to be clearly conducted, whereas in reality, the path is radiated or vice versa. In an investigation, every measurement and modification must be documented, with any change in the level of EMI noted, for as seen throughout the book, and especially in the case studies, a modification that results in a worsening of a problem can tell as much about the cause as a change that results in an improvement.

It is often easier to start with the minimum number of circuit boards, cables, and pieces of equipment that exhibit an EMI problem or specification emissions mentioned earlier. Once the source(s) and solution(s) to the problems associated with the minimum configuration have been found, the additional circuits, cables, or equipment can then be connected one after the other and any additional problems solved in a stepwise manner.

Bench tests on individual circuit boards separated from the equipment/system are often helpful in isolating and locating circuit-level emissions and susceptibility. Bench tests are particularly useful when circuit-level problems are masked by equipment-level conducted or radiated emissions. For example, relatively low-noise linear power supplies and low-noise signals from an oscillator/function generator may be used to supply the circuit in a bench test.

Once the source of the problem has been found, one or more solutions can be proposed or temporarily or permanently implemented. One problem is to find a modification acceptable to the engineering manager, the packaging engineer, QA, and production. One or more of these people often want a fix with no modifications at all!

The following case studies are intended to provide examples of effective EMI investigations that have produced timely results with acceptable modifications to equipment and circuits.

12.2.1   CASE STUDY 12.1: EMI INVESTIGATION INTO INTRA-EQUIPMENT SUSCEPTIBILITY

This case study illustrates the importance of understanding the significance of an increase or decrease in EMI when implementing changes to equipment, for once the significance is understood, the source of EMI may be located. The case study also emphasizes the importance of drawing an equivalent circuit in order to find the coupling path.

The equipment under investigation is a spectrometer instrument with an ionization detector contained in a vacuum housing. The output of the ionization detector is fed into the input of a gain of 10 amplifiers, the output of which is applied to a discriminator. The second input to the discriminator is a DC reference level, which is adjusted by a potentiometer to discriminate against noise picked up by the ionization detector and amplified by the gain stage. The problem was that excessive noise resulted in a setting of the reference level, which was so high that low-level signals were also discriminated against. The instrument generated high-power levels at 27 MHz, and this was considered the most likely source of noise.

The required discriminator reference level was 95 mV before discrimination against the noise was achieved. The noise level referenced to the input of the gain stage (assuming the noise source is a 27-MHz sinewave) is therefore:

$$V_{n(\text{pk-pk})} = 2 \times \frac{\text{Reference level}}{\text{Gain of first stage}} = 2 \times \frac{95\,\text{mV}}{10} = 19\,\text{mV pk-pk}$$

When referring to noise level in this case study, the reference level at the input of the discriminator is quoted and not the noise level at the input of the first gain stage.

The gain stage and discriminator equivalent circuit are shown in Figure 12.1a. The differential output of the discriminator is connected to the remaining circuits in the instrument by a ribbon cable with integral shield that was connected to a drain wire in the cable.

Tests and fixes were tried in the following sequence with the following results:

1. Routing the ribbon cable through a ferrite balun had the effect of reducing the noise level from 95 to 75 mV.
2. Adding a shield over the ribbon cable with the shield disconnected from DC return but connected to chassis at both ends increased the noise level from 95 to 103 mV. The main chassis was then isolated from the framework with no decrease in noise level; also, connecting the shield at only one end did not help.
3. Disconnecting one end of the drain wire in the ribbon cable and connecting it to DC return, that is, connecting the shield to the DC return, increased the noise from 95 to 145 mV.
4. Floating the shields of the cables connecting the differential output of the discriminator to the ribbon cable from the ground plane at the discriminator reduced the noise from 95 to 78 mV.

The conclusion to be drawn from the results of the steps 1, 2, 3, and 4 is that a common-mode current is flowing in the ribbon and other interconnection cables, probably in the shields. This conclusion is based on the following:

Adding a chassis-connected overshield that was connected to the ground plane at the discriminator end increases the noise, indicating that the radio frequency (RF) currents are flowing in the chassis and on the shield.

The shield current increases with the application of the overshield because the impedance of the shield and the chassis connection is lower than that of the ribbon cable shield. The balun has the effect of reducing the level of common-mode current by increasing the impedance of the ribbon cable shield.

(a)   Real and parasitic components at the input of the detector pre-amp

(b)

FIGURE 12.1 (a) Equivalent input circuit of the gain stage and discriminator. (b) Simplified equivalent circuit—① 10 μH choke (X_L = 1.8 kΩ) tried at this point. ② The 0.001 μF capacitor combined with its self and lead inductance forms a series resonance circuit at ≈ 27 MHz. Note: All impedances at 30 MHz.

5. In an attempt to reduce RF current flow into the gain stage and discriminator circuits, 0.12-µH chokes were added into the +5 V, −5 V, and return line (because the discriminator output is differential, adding an inductor in the supply return should have no adverse effect); however, the noise increased to 315 mV.

6. Removing the choke in the return but retaining them in the +5-V and −5-V lines, with additional 1-Ω resistors to eliminate the possibility of series resonance, reduced the noise to its original level. This reduction was entirely due to the removal of the choke in the DC return. Clearly, in the overall grounding scheme, the location of the connection of the DC return to the chassis at the detector circuits is important. An attempt was made to measure the chassis current flow in the two ¼-in. ground straps by the use of a current probe. This was found to be impossible due to the approximately 100 mA at 27 MHz measured by the shielded probe when held in the vicinity of the chassis. That is, the probe was measuring the ambient EM field and not conducted currents.

7. Disconnecting the input from the detector to the gain stage reduced the noise level to 13 mV.

8. Replacing the two ¼-in. ground straps connecting the ground plane to the chassis with a 10-µF inductor increased the noise level to 1 V.

9. Reconnecting the ground straps and removing the two high-voltage connections to the detector decreased the noise level by 8 mV.

The results from step 7 indicate that the majority of the noise is not conducted or radiated coupling to the gain stage or discriminator but is picked up by the detector. The detector is well shielded so the coupling is unlikely to be radiated. The capacitance of the vacuum feedthroughs was measured at 8 pF on an LCR meter, and these values as well as lead inductances and the self-inductance of the capacitors were included in the equivalent circuit of Figure 12.1a.

A simplified equivalent circuit showing the impedance of the equivalent circuit to chassis is shown in Figure 12.1b. An important impedance not shown in the equivalent circuit is the impedance of the ground plane. The shield of the ribbon cable is connected to the ground plane and, therefore, RF current flow on the shield flows across the ground plane and to chassis. The majority of the current flows to chassis via the two ¼-in. ground straps. However, alternative paths exist through the approximately 1.8-Ω impedances of the two high-voltage capacitors and the approximately 663-Ω impedance of the feedthroughs. The current path that results in EMI is via the 510-Ω resistor at the input of the gain stage and to chassis through the impedance of the feedthrough.

The preferred approach in combating EMI is to reduce at source. However, in this case study, as in many others, reduction at source is ruled out due to the difficulty and cost of implementing changes in production equipment. Adding a balun on the ribbon cable was found to reduce EMI but not to any great extent.

A better solution would be to incorporate a 510-Ω (or variable) resistor, in series with a 0.001-µF capacitor, in series with an 8-pF capacitor to chassis. The differential input to the gain stage is then developed across the existing resistor, shown in Figure 12.2a. Approximately, equal noise currents are developed in the two resistors, as is noise voltage developed across them. Therefore, approximately equal noise voltages, that is, a common-mode voltage, are applied to both inputs of the gain stage, and the noise appears at a much reduced amplitude at the output of the stage. Another potential solution is to move the physical location of the ground straps to the ribbon cable shield connection and to connect an inductor from the chassis connection to the ground plane. Additional inductors would also be required in the +5-V and −5-V supply connections to the board, because the noise current is common mode. In the location first tried for the connection of the inductor to chassis, a very high voltage was developed across the inductor, and the noise current through the 510-Ω input resistor increased. In the proposed location, a low impedance to chassis for noise currents is achieved by the ground straps, and a high impedance, approximately 1.7 kΩ for a 10-µH inductor, is introduced in the input path of the gain stage. Figure 12.2a and b illustrate the two fixes.

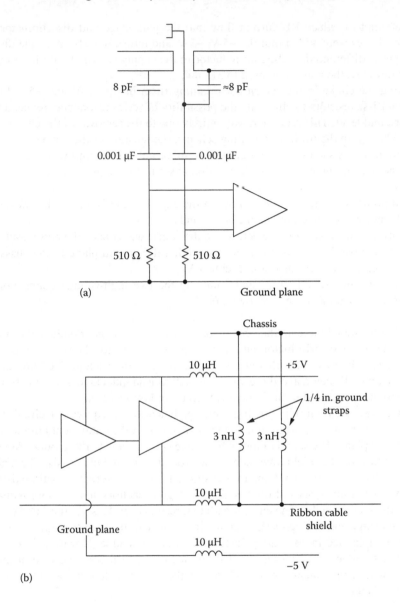

FIGURE 12.2 (a) Developing a common-mode noise voltage instead of differential. (b) Moving the location of the ground straps and adding a 10-μH inductor in the ground plane.

12.2.2   Case Study 12.2: Reducing Radiated Emissions on a Computing Device to FCC Class A Limits

In the initial configuration, the equipment was contained in an enclosure constructed with closed metal sides and a perforated metal top and bottom. The front and rear of the enclosure was open, with a motherboard, located at the rear of the enclosure, to which the PC boards were connected. The boards were plugged in to the front of the enclosure with a metal faceplate mounted on each of the boards. The metal faceplate was not electrically connected to the enclosure, and a gap existed between adjacent faceplates, as illustrated in Figure 12.3.

Two coaxial cables are connected to BNC connectors on the metal faceplate of one of the boards. Both cables are used to carry a 1-MHz signal. A second connector mounted on a second board is used to connect an RS232 cable. The additional DC power and multiconductor cable connections

FIGURE 12.3 Equipment test setup for radiated emissions.

to the equipment are made through the rear of the enclosure. The multiconductor connections terminate on the motherboard inside the enclosure, and none of the cables entering the rear of the enclosure are shielded.

The measurements were made in a shielded room using a tunable dipole and a 6-cm balanced loop antenna, which was used to locate specific sources of emissions, that is, from cables, apertures in the enclosure, and PCBs. A Hewlett-Packard near-field probe was used to locate emissions from individual tracks on PCBs.

The common-mode current on interconnection cables was measured, and the current was then used to predict the radiated emissions from cables.

The equipment was positioned on a nonconductive table at a height of 1 m above the floor of the shielded room. The enclosure ground and digital ground were separately connected to the shielded room floor via two #16 AWG wires, which is the grounding scheme used in the equipment installation.

Measurements of the radiated emissions from the equipment were made with the balanced loop and tunable dipole. The magnetic field measured by the balanced loop was converted to an E-field magnitude at a distance of 30 m (the FCC-specified distance) by a computer program. The dipole measurements were made at the location where the dipole was calibrated in the room, at a distance of 1 m from the equipment. The E field was then extrapolated out to the 30-m distance. The maximum amplitude emissions measured using the dipole and loop antennas were compared. Where a good correlation existed between the two measurement techniques, the maximum amplitude of the two measurements was taken. When a poor correlation existed, the reasons, such as room resonance and reflections, were examined and suitable corrections and an engineering judgment were made based on the results. Table 12.1 shows the source and magnitude of worst-case emissions converted to a distance of 30 m and compared to the FCC limits.

The major sources of emission were the front-panel-terminated coaxial cables and RS232 cables, with predicted emissions from these sources alone as high as 270 µV/m at 30 m. The reasons for the

TABLE 12.1

Measured Fields Converted to a Distance of 30 m and Compared to the Specified Limits

Location	Frequency (MHz)	Measured (Converted to 30 m) (µV/m)	Specified at 30 m (µV/m)
Front of enclosure and cables	49	403	30
Front of enclosure and cables	52	170	30
Front of enclosure and cables	99.8	138	50
Rear of enclosure only	82.7	60	30

high-level emissions are not the differential signals carried by the cables but high common-mode current flow on the coaxial shields and the RS232 unshielded conductors. The current flow is caused by noise voltages appearing between the enclosure, which is connected to the shielded room floor, and the digital ground on the boards, to which the coaxial cable shields and signals are referenced.

The shields could not be directly connected to the enclosure, because DC isolation was a requirement for this type of equipment. Instead, for the purpose of diagnosis, a shield made of aluminum foil was constructed over the front of the equipment, with two insulated islands of aluminum in the shield that were connected to the BNC connectors. The islands were bridged by 1000-pF capacitors with short leads, and thus the connectors and shields of the cables were RF grounded to the enclosure but DC isolated. The resonant frequency of the capacitors is approximately 50 MHz and was chosen to coincide with the worst-case emission at 49 MHz. The maximum emissions from the front of the enclosure and the cables with the addition of the shield and capacitors reduced to 31.7 μV/m at 49 MHz, 100 μV/m at 70.9 MHz, and 113 μV/m at 113 MHz, and it was clear that further improvements were required. Shielding the power cable and adding baluns on the cables exiting from the rear of the enclosure reduced emissions further, with only one frequency remaining too close to the limit.

The out-of-specification frequency was traced to PCB tracks using the HP near-field probe. Ideally, a ground plane connected to digital ground should be incorporated under the tracks in order to reduce board emissions. The addition of a second ground plane connected to the enclosure but isolated from the board is also likely to reduce emissions.

A makeshift ground plane was constructed from aluminum foil, placed on the track side of the board, and isolated from the tracks by cardboard. The ground plane was connected to the enclosure only. A reduction at 59.9 MHz of 8 dB was achieved by the addition of the provisional ground plane.

Further reductions were achieved by adding a ground plane to the motherboard at the rear of the enclosure or shielding the rear of the enclosure with foil. Closing up the apertures at the top of the enclosure reduced emissions slightly at 70.4 MHz but not at 59.9 MHz.

The initial fixes of the 1000-pF capacitors connected to a totally shielded front to the enclosure were considered impractical by the manufacturers and alternative solutions were sought.

In an attempt to reduce the common-mode voltage appearing between digital ground (and therefore all signals) and chassis, 82-pF capacitors were connected between the digital ground connections of all PCBs in the enclosure and chassis. The capacitors were located at the rear of the edge connectors on the motherboard as shown in Figure 12.4.

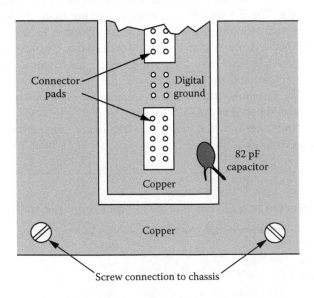

FIGURE 12.4   Method of connecting RF ground capacitors between the digital ground and the enclosure.

The enclosure was improved by adding a metal shield to the rear of the enclosure, after which no backplane was required on the motherboard to reduce emissions from the rear of the enclosure. Additional perforated shields, which allowed air flow, were added to the top and bottom of the card cage inside the enclosure. The unshielded cables at the rear of the enclosure were a source of emissions, and adding a shield around these cables reduced emissions at 83 MHz by 9 dB and at 90 MHz by 8.7 dB. Because these cables were commercially available and were plugged into the equipment, that is, could be removed and replaced by unshielded cables, unshielded cables had to be used in certification tests, in accordance with FCC guidelines.

Some of the conductors in the cables were unused, and these were connected to the enclosure at the equipment. The ideal location for the connection of these conductors is to the outside of the enclosure, which was impractical. Even with the conductors connected to the inside of the enclosure, at the motherboard, radiated emissions were reduced, but not to the same extent as achieved with shielded cables. Ferrite baluns were built into the rear of the enclosure at the location at which the cables entered the enclosure, which resulted in a significant reduction in emissions.

The digital grounds on PCBs were connected to the metal faceplate by 82-pF capacitors, with the faceplate connected to the frame of the card cage via beryllium copper finger stock material. The board, which had benefited from the addition of a ground plane, was relaid out with an additional ground plane layer. These modifications were made to a production piece of equipment, and from measurements made in the shielded room, with the cable arranged for maximum emissions, the prediction was that the equipment would pass FCC limits when measured on an open-field test site. A subsequent certification test, using the identical piece of equipment and identical peripheral equipment on a specific open-field test site, resulted in peak measurements of at least 6 dB below the FCC limits.

An emphasis is laid on the correlation of shielded-room measurements to specific open-field test site measurements because of the up-to-26-dB variation measured in field strength from identical computing systems at six sites and is recorded in Reference [1]. Although identical systems were measured, it is not clear whether exactly the same piece of equipment was measured at all six sites, for if not, a variation must be expected from one piece of equipment to the next.

This case study illustrates that EMI solutions can be found that are acceptable to a manufacturer and that in this case did not require either a totally shielded enclosure or shielded cables.

12.3 EMC PREDICTIONS: GENERAL APPROACH

Too often the management and planning of EMC is not included in the electronic equipment or system design. It is only when equipment is adversely affected by EMI or fails an EMC test that a manufacturer is forced to consider either a redesign or some modification to achieve EMC. It may be argued that the incidence of EMI problems has increased over the past 20 or so years, attributable, among other factors, to an increase in the level of EM ambient, in communications spectrum occupancy, in the number of electronic products, and in the speed of logic and switching-power supplies. Therefore, the traditional approach of relying on good design practice without specifically designing for EMC is proving inadequate.

As a result of the increased incidence of EMI, the statement that you pay for EMC design either now or later (and usually pay more) for EMI solutions has become almost a truism.

An EMC prediction should be made as early in the equipment-, subsystem-, or system-level design as possible and then further refined as the design proceeds. The analysis should take into account the sources of potential problem emissions and the coupling path to a potentially susceptible unit. The geometry of the source, receptor, and coupling elements, such as circuits, cables, or apertures, with values for the impedances in the source to receptor path, are important details to be used in the prediction and, therefore, the information obtained should be as specific as possible. The receptor may be an antenna in a radiated emission prediction or a 10-μF capacitor or LISN in a conducted emission prediction.

The frequency and amplitude of sources and, for an intraequipment, subsystem, or system analysis, the bandwidth and susceptibility of receptor circuits must be known for a prediction of EMC. The sources are both those inherent in the operation of the equipment, subsystem, or system and susceptibility test levels, if these are a requirement of the design. The purpose of the prediction is one or more of the following: to highlight problem areas as early as possible, to aid in cost-effective design, to support a waiver request. Often when specific information is missing, a simple breadboard-level test may be required, for example, of a susceptibility level, attenuation, shielding effectiveness, impedance, or coupling. Where possible, predictions should be validated by measurements.

If the results of a prediction are within 6 dB of a correctly controlled measurement, then the level of accuracy achieved by the prediction is acceptable. More frequently, an EMC prediction will lie within ±20 dB of measurements. When the correlation between the two is worse than 20 dB, either the prediction or the measurement techniques should be examined for sources of error.

In reviewing the first edition of the book, Mr. Jeff Eckert gave the following telling simile: "Two Texans, both blindfolded, are standing in a cow pasture, surrounded by a herd of steer. The first Texan listens intently for a few seconds and drawls, 'I reckon there must be five head of cattle out there.' The second Texan takes a deep whiff of the air and says, 'There's 500 head of cattle on this here range.' They remove their blindfolds and offer mutual congratulations because, after all, both estimates are correct. If you think there is something wrong with that story, consider that the Texans are not cowboys but are EMC engineers from Austin. Welcome to the world of EMC modeling, in which any answer is 'correct' as long as it falls within a multiple or dividend of 10 relative to the true value."

Of course, ±20 dB is far from ideal, but it is the, obviously unpalatable, reality. In 38 years of experience of EMC and in reviewing EMI predictions, this is the typical magnitude of the errors I have seen. In one analysis, the common-mode current flowing on a surface was ignored as a source of radiation and the error was 61 dB. In another instance, in which the current induced by an incident radar pulse on a cable enclosed in a metal conduit on a composite-skin aircraft was calculated, the error was close to 1000 dB! The problem was that the skin depth of the conduit at gigahertz frequencies was calculated and the transfer impedance of the shield-to-connector and connector-to-enclosure mating surfaces was ignored. Because the conduit carried control signals to the aircraft front landing wheel, this error could have resulted in a disaster. When the modeling techniques in Chapter 6, References [9,23], are compared, the correlation between shielding effectivenesses was between 5 and 8 dB, and this was considered by the authors to be acceptable. Also, the errors in using sophisticated PCB electromagnetic radiation computer programs, as discussed in Chapter 11, are unacceptably high. In many cases, errors occur due to a simple lack of reliable data. For instance, the very sophisticated modeling and measurements made on the electric field radiated by the Waves In Space Plasma (WISP) experiment on the space shuttle were based on an input power to the antenna of 250 W into a conjugate 50-Ω load. The reality was that due to the use of an antenna-matching unit, the power amplifier could deliver 1000 V into the dipole antenna when it was not a resonant length, resulting in a potential error of 19 dB. The WISP analysis is described in Case Study 12.5.

The goal throughout this book has been to find simple analysis techniques that achieve an accuracy of 6 dB, and in many instances the measurement errors are of the same order of magnitude, although I imagine even 6 dB would be unacceptable when counting cows!

12.3.1 CASE STUDY 12.3: EMC PREDICTIONS FOR MEETING RTCA-DO 160 REQUIREMENTS ON AN "A" MODEL FIBER-OPTIC PLOTTER

This EMC prediction was based on an examination and simple measurements made at the manufacturer's location. The test equipment available was limited to an oscilloscope with current probe and a digital voltmeter (DVM).

The plotter is constructed of 3/16-in.-thick aluminum and contains a number of cutouts for the purpose of ventilation and for the paper cartridge. The aluminum was anodized, and when the resistance between the front panel and the side panel was measured with the DVM, an open circuit was registered. The CRT and deflection yoke are completely enclosed by a 1/16-in.-thick mumetal case, and this will prove excellent shielding against the deflection-yoke-generated EM field.

High current pulses are generated by the deflection-yoke drive circuits, and a mumetal shield had been installed between the yoke drive PCB and the adjacent stepper motor drive PCB. Due to the high level of currents flowing inside the "A" model plotter, it is mandatory that the plotter case be improved in its function as an EMI shield in order to meet radiated emission levels.

A smaller plotter (the "B" model) that uses approximately 4½-in.-wide paper has been tested to DO-160A EMC levels and passed, except for some narrowband emissions in the radiated emission test. The "B" plotter utilizes EMC gaskets between mating surfaces and a honeycomb ventilation opening for the fan. It also incorporates differential inputs and some input filtering to increase the noise immunity of the control lines. These differential circuits are examined later in this case study.

12.3.1.1 AC-Line Noise Current Measurement

A measurement of noise currents generated by the "A" series plotter was made in order to isolate the source of the major noise generator within the unit, to aid in the selection of a suitable line filter, and to calculate the worst-case radiated emission level.

A PI circuit EMI filter was utilized as shown in Figure 12.5. The filter provides a path for HF current flow via C_2 and C_2' and isolates EMI noise present on the AC line from the current probe.

The AC-line noise current shown in Figures 12.6 and 12.7 was measured with a peak current of 50 mA and the fastest edge at approximately 20 µs. The ramp appearing between the negative and positive transitions has a duration of approximately 3.2 ms, and this corresponds to the horizontal sweep rate.

The stepper motor drive was disconnected to isolate any component of the noise waveform that was caused by the stepper motor current, and the noise waveform was unchanged. It was expected that both the stepper motor drive and the HV supply inverter would cause noise currents; however, when the current probe was connected to the +30 V applied to the HV inverter and to the +24-V supply applied to the stepper motor drive PCB, negligible noise was measured.

With the current probe connected around the +24-V supply wire to the horizontal drive PCB, a large 440-mA current spike was measured, as shown in Figure 12.8. This current spike has the same repetition rate as the noise current measured on the AC line. However, although it was expected that the AC transformer, rectifiers, and 27,000-µF capacitor would modify the transformer primary reflected current pulse, it is unclear why the 3.2-ms ramp and trailing edge currents are not seen on the +24-V line.

FIGURE 12.5 AC-line noise current measurement test setup.

40 mA/div

2 ms/div

FIGURE 12.6   AC-line noise current.

50 mA

50 mA

70 μs

0.5 ms   ≈3.2 ms   4.8 ms

FIGURE 12.7   AC-line noise current with stepper motor and horizontal sweep in operation.

≈8 ms

440 mA

5 μs

10 μs

FIGURE 12.8   Current spike on +24-V line to deflection yoke driver PCB.

12.3.1.2 AC-Line Filter Selection

In choosing a filter, the equivalent load on the AC line presented by the plotter must be known and is shown in Figure 12.9. The load is not constant and changes impedance because the heaters are controlled by triacs.

The purpose of filtering the AC line is to isolate the noise currents from the AC line in order to meet the conducted and radiated emission tests and to isolate the plotter from both the 200-V transient injected directly into the power line and the electromagnetically induced spike. It may be more economical and space saving to achieve the major lower-frequency, that is, 10–200 kHz, filtering at the AC input to the system in which the plotter is used and to filter out at the individual units the high-frequency currents generated by the units.

In case of the "B" model plotter, a PI filter model was employed; however, this type of filter has caused instruments to fail the radiated emission tests due to high current flow in the enclosure.

A T filter is preferred to a PI because this isolates the noise current both from the AC line and from the instrument into the AC line. The only disadvantage of the T filter is that more noise voltage appears across the AC line internal to the unit. The noise voltage can be calculated on the basis that with one T filter installed on the line and a second on the return, approximately 20 µH of inductance is in series with the line and approximately 0.01 µF across the line. The noise current generated by the plotter results in a voltage of $V = L\, di/dt$ across the inductors. Considering the capacitor to be low impedance and using the fastest edge of 20 µs shown in Figure 12.7, the noise voltage is approximately

$$\frac{20\times10^{-6}\,\mathrm{H}\times50\times10^{-3}\,\mathrm{s}}{20\times10^{-6}\,\mathrm{s}} = 50\,\mathrm{mV}$$

FIGURE 12.9 Reflected load on AC line.

Because the AC cabling inside the plotter is shielded, it is not predicted that the noise voltages will cause problems internal to the unit, although care should be taken to continue the shield as far as possible on the cable to minimize the length of unshielded cable.

One area where the AC line is unshielded and where of necessity the plotter case contains an opening for the recording paper is at the platen heater. Possible emissions from this area are considered under radiated emission and radiated susceptibility considerations in the following section.

12.3.1.3 Radiated Emissions Predictions

12.3.1.3.1 Shielding

Based on the area covered and the thickness of the aluminum case, it is feasible to make the unit into a shielded enclosure. The type of finish would require changing from anodizing to Iridite, Oaktite 36, or Alodine 1200, in order to make the surfaces conductive. The fan would require a honeycomb filter, and the ventilation apertures would require wire mesh screens. The corner apertures and other small holes should be covered by copper adhesive tape. The mating surfaces of the front, bottom, back, and sides, which contain apertures between the surfaces, should be made EMI tight by use of EMI gaskets.

The paper cartridge that contains the platen heater, stepper motor drive unit, and paper rolls is inserted into the body of the plotter and is held by a hinge at the bottom and two catches at the top. It is unlikely that a good electrical connection exists between the cartridge and the case, so the cartridge finish should be changed from anodizing to a conductive finish. An EMI gasket should be included at the interface and flexible bonding straps provided between the bottom of the cartridge and the inside of the case. Because emissions levels could increase considerably with the cartridge hinged down, that is, when the paper is changed, to avoid the requirement of testing with the cartridge down, a switch that automatically disconnects the supply when the cartridge is lowered may be a requirement, or an operational requirement that the recorder be switched off before changing the paper may be sufficient.

The worst-case emission emanating from the AC power-line connection to the heaters and radiating via the approximately 1.5×10-in. gap in the shielding at the heater platen has been calculated.

The step function in the noise voltage exhibiting a rise time of approximately 20 μs will produce the highest amplitude in the frequency domain. The spectrum occupancy due to this edge has been calculated, and the broadband component at a frequency of 190 kHz (which is the lowest frequency specified for radiated emission E-field testing) has been obtained using the equation for a step function in Section 3.1.2. The broadband component at 50 kHz is 0.05 μV/Hz and, due to the 20-dB/decade fall in amplitude from the 50-kHz point, the amplitude at 190 kHz is approximately 22 dB down, that is, 0.004 μV/Hz. Using the broadband reference bandwidth of 1 MHz, the voltage is 4 mV/MHz.

The MINCO heater, HR5046A281, which is representative of the type of load presented to the AC line, has a resistance of 103.93 Ω, an impedance measured at exactly the same value, and an inductance of 1.3 μH. The load can thus be considered resistive at approximately 100 Ω, which results in a noise current flow, in the heater cable of 4 mV/MHz/100 Ω = 40 μA/MHz.

The DO-160A limits for broadband-emitted E field measured 3 m from the source at 190 kHz is 98 dBμV/MHz/m, or 80 mV/MHz/m.

A relatively low-noise current flows in the heater wire, which is an imperfect antenna; therefore, the prediction, using the loop antenna or electric current element equations of Sections 2.2.4 and 2.2.5, is that the E field due to this noise source will be at microvolt levels measured 3 m from the source and well below the limit.

Current noise spikes caused by the AC-line triacs have been ignored during this investigation because the triacs are switched at the zero-crossover point. However, if the zero-crossover detector does not supply sufficient gate voltage to switch the triac at the true AC-supply zero point, a large current spike with a rise time in the order of a microsecond may result. This type of current pulse with a high-amplitude high-frequency component would cause problems in meeting the emitted radiation levels, especially because the allowable limits are reduced at higher frequencies. If it is found that the triacs do not switch at exactly zero, one possible solution is to include some additional inductance between the triac and the line filter.

Although the magnetic field produced by the deflection yoke is likely to be well contained by the mumetal shield, some stray field may be emitted via the face of the CRT and via the gap in the shielding at the heater platen.

12.3.1.3.2 Test Setup

Normally individual units comprising a system are tested along with interconnecting cables and with the units spaced out above a ground plane. This configuration is common to both radiated susceptibility and radiated emission levels. The argument can be made that if individual units are normally mounted in a rack that acts as an efficient shielded enclosure, the tests should be conducted with this rack enclosing the units. However, the front of the plotter will not be shielded by the rack, and whichever test setup is decided upon, it is the plotter front panel that is the least well-shielded surface and will be the one facing the antenna during radiated emission and susceptibility tests.

12.3.1.3.3 Radiated Susceptibility Tests: E-Field Susceptibility

The plotter case is recessed behind the paper cartridge and is solid, that is, a good shield; the only wiring inside the cartridge that is exposed to the radiated RF field is the stepper motor drive wires and the heater AC wiring, both of which are predicted to be immune to radiated E fields.

If the units comprising the system are tested with interconnecting cables and power cables exposed to the field, then the effect on the recorder interfaces must be considered. Generally speaking, if the interfaces can function during the interconnection cable–induced susceptibility levels and meet the radiated emission limits, then they are likely to pass the equipment radiated susceptibility tests.

A correctly chosen or designed AC power-line filter can very adequately take care of noise voltage picked up on the AC line during the radiated susceptibility tests.

12.3.1.3.4 Interconnection Cable–Induced Susceptibility Tests

These tests must be carried out on interconnections between units in different racks and may be required between units in the same rack. The tests consist of wrapping spirals of wire around interconnecting bundles of cables and either running a current through the cable or applying a voltage to one end of the cable. The bundle is subjected to current at audio frequency ranging from 30 A/m at 400 Hz to 0.8 A/m at 15 kHz. The bundle is subjected to 360 V/m at 380–400 Hz as well as a voltage spike test.

12.3.1.3.5 AC-Line Noise Immunity

A filter designed for use on a 400-Hz power line is obviously ineffective at filtering AF noise at 400 Hz and not very effective at 15 kHz. The filter will, however, be effective at filtering out the spike that is electromagnetically coupled into the power line.

The plotter utilizes large smoothing capacitors in its DC unregulated supplies and linear regulators in the stabilized supplies and, therefore, no problem is foreseen with power supplies due to the induced AF noise levels. However, if the zero-crossover detection circuit is not sufficiently noise immune, the induced noise may cause misfiring of the triacs, and thus high radiated noise levels within the unit, which may result in interequipment EMI.

12.3.1.3.6 Control-Line Noise Immunity

Cable shielding: A 125-Ω characteristic impedance twisted-shielded-pair cable with a high shielding effectiveness and tight twists in the pair will provide a high degree of immunity to

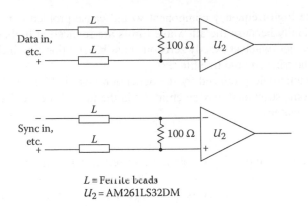

L = Ferrite beads
U_2 = AM261LS32DM

FIGURE 12.10   Differential receiver circuit.

induced noise. An additional overall shield over the entire cable bundle is to be preferred to shielding individual cables, with a concentric connection made to the backshell of the connector and to chassis at both ends.

> *Input and output circuit*: Presuming that the model "B" differential circuits are available for use in the model "A" plotter, it is these circuits that will be considered. The circuits are shown in Figure 12.10. The differential circuit provides a minimum of ± 2.5 V of DC noise immunity and some limited filtering at high frequencies achieved by the ferrite beads, L.

During the spike test, the ferrite beads represent some small inductance and are lossy, but the beads present an insertion loss only over the frequency range 10–100 MHz, which assuming a maximum bead impedance of 50 Ω is limited to 6 dB. The differential circuit in the "B" plotter has been tested with DO-160A radiated susceptibility test levels using twisted-pair unshielded cables of 10–12-ft length. Thus, using shielded cable, the circuits should be sufficiently noise immune, with a good safety margin. However, to be absolutely safe, it would be comparatively easy to build and test a breadboard driver and receiver circuit, using a representative length of shielded cable between the circuits, and to test for noise immunity. The equipment and test circuits necessary to provide the test levels must be available for system-level testing and can be used and tested on the breadboard interface. Radiated susceptibility test requirements may be met with unshielded interface cables in the "A" plotter, and the major reason for the use of shielded cables is to meet radiated emission requirements that the model "B" plotter had failed.

12.3.1.3.7   AC-Line Conducted Noise Susceptibility Tests

These tests comprise an audio frequency level, a spike, and an AC voltage modulation test, all levels to be directly coupled into the AC-line voltage. The AC-line filter will only be effective in reducing the 10-μs transient voltage (spike) test level.

The amplitude and frequency modulation supply levels and the audio frequency conducted susceptibility test levels are not modified by the line filter, and the recorder must function correctly when subjected to the test levels. Of particular importance is the operation of the zero-crossover detect circuit, as previously discussed. In addition to noise, the AC line is subjected to surge voltages up to 180 Vrms for 0.1 s and to undervoltage of a minimum 60 Vrms for 10 s. Correct operation of the unit is determined after the surge or undervoltage is removed. Because the AC-line conducted noise levels and under-/overvoltage levels are easily generated, it is recommended that the plotter be tested at these levels prior to delivery, if possible on a standard plotter with the addition of the chosen line filter.

12.3.1.3.8 Conclusion

To achieve a high degree of confidence that the Model "A" plotter will pass the RTCA/DO-160 EMC requirements, the following measures are recommended:

1. The plotter case should be upgraded to an effective shielded enclosure by changing the finish and using EMI gaskets and honeycomb and mesh air-vent filters.
2. Suitable AC-line filter components should be chosen.
3. The plotter should be subjected to the AC-line conducted noise and over-/undervoltage conditions. Particular consideration should be given during the conducted noise tests to the operation of the zero-crossover detector and triac trigger circuits as well as the high-voltage and stepper motor–drive functions.
4. The manufacturer should breadboard a line driver and receiver and test with RTCA/DO-160 interconnection cable–induced susceptibility test levels.

12.3.2 Case Study 12.4: EMC Predictions on a Power Controller

These predictions were made at the very initial design stage, before the design was finalized and, therefore, concentrates on providing estimated EMI levels and guidelines to be followed in the design. The next iteration of EMC prediction for the equipment will be based on more accurate information on structure and layout.

12.3.2.1 Fields Generated by the Internal 20-kHz Distribution System

The radiated fields generated by the 20-kHz power distribution system internal to the enclosure are predominantly magnetic. The magnetic field strength is to be measured at a distance of 7 cm from the enclosure at points likely to emit high field levels (i.e., seams, slots, and apertures), as required by MIL-STD-462 Method RE01. The magnetic field at low frequency and close to a loop of current-carrying wire is given approximately by

$$H = \frac{I(S/\pi)}{2\left((S/\pi)+r^2\right)^{1.5}} \tag{12.1}$$

where
  H is magnetic field [A/m]
  r is radial distance to point of measurement [m]
  S is area of the loop $= w \times l$ [m²]
  I is magnitude of the current [A]

An average supply current for a wire within an enclosure that carries current to all load power converters is 100 A, and this was used in the calculation. In reality, a number of wires will be used to carry the 100–300 A of current; however, when the wires are bundled together, they may be replaced, as in our analysis, with a single wire. We assume that the supply and return wires are widely separated within the enclosure and that the current-carrying wire is routed around the inside of the enclosure in the form of a loop. The size of the loop is then close to the interior dimensions of the enclosure, which is 0.762 m × 0.432 m.

From Equation 12.1, the magnetic field generated by the current-carrying wire at 10 cm is 135 (164 dBpT) A/m. In order to predict the level of magnetic field measured at a distance of 7 cm from the outside of the enclosure, the shielding effectiveness of a typical enclosure must be estimated at 20 kHz.

The shielding effectiveness of an enclosure to an H field incident on the enclosure may be found from the ratio H_{out}/H_{in}, where H_{out} is the magnetic field on the outside of the enclosure and H_{in} is the

FIGURE 12.11 Geometry of the twisted-pair wiring.

magnetic field on the inside of the enclosure. The equations used to calculate the ratio of H_{out}/H_{in} are presented in Section 6.5.3.

The attenuation is dependent on the enclosure area and will lie in the range 53–60 dB. Assuming a worst case, the magnetic field is attenuated by 53 dB when the enclosure is inserted between the field source and the magnetic field loop. Therefore, the external field is

$$\frac{135}{466.12} = 0.29 \left[\text{A/m} \right] \; (53 \, \text{dB} = 466.12)$$

Converting to dBpT for comparison to the limit: 0.29 A/m = 111 dBpT. Comparing this to the limit at 20 kHz (REO1) of 110 dBpT, this worst-case scenario would be close to the limit.

The preceding data are calculated for a current loop with each arm of the loop running within 3 cm of the enclosure walls. The length of the enclosure was taken as approximately 0.75 m, with a height of 0.38 m and a width of 0.43 m. While a loop may not be a realistic wire layout, it does demonstrate that the generation of high levels of magnetic field is possible and must be controlled in the design.

If the same layout was used but 10 AWG twisted-pair wiring was used to distribute the power, as shown in Figure 12.11, the following formula could be used to predict the magnetic field at 10 cm:

$$H_r = -54.5 \frac{r}{p} - 20 \log \left(\frac{1}{a} \right) - 30 \log p - 10 \log r + 9.8 \; \left[\text{dBgauss/1 A} \right] \tag{12.2}$$

where

a is radii of helixes, inches = 0.08 in.
p is pitch distance, inches = 1.6 in.
r is radial distance from the axis of the pair to the measurement point, inches = 4 in.

This formula gives the magnetic field per ampere of current, which is, in our example, −159 dBgauss/1 A. Therefore, for a 100-A current, the magnetic field is −119 dBgauss = 112 pT = 41 dBpT.

This is a reduction of 123 dB from the magnetic field produced by a large loop of wire. Thus, using twisted-pair wiring would eliminate any problems associated with magnetic field emissions at 20 kHz appearing outside of the enclosure.

12.3.2.2 Fields from Internally Generated Signals above 20 kHz

Radiation from the power distribution above 20 kHz would stem from three fundamental sources:

1. Harmonics of the 20-kHz power supply
2. Digital clocking frequencies and harmonics
3. Intermodulation products of the above frequencies

Magnetic-field emissions are monitored to a maximum frequency limit of 300 kHz. Electric fields are measured from 10 kHz to 10 GHz.

The magnetic field–shielding effectiveness of the enclosure increases with frequency; therefore, it would require a substantial amount of power at any frequency above 20 kHz to exceed the limits. The probability that this would occur is minimal, since the 20-kHz AC power is the only source with sufficient power associated with it by design.

The magnetic field–shielding effectiveness of our example enclosure is as follows:

At 300 kHz = 83.79 dB
At 1 MHz = 95.7 dB

The limit for emissions (magnetic) at 300 kHz (REO1) is 20 dBpT; therefore, an internal source of emissions would have to be 84 dB (84 dB = attenuation of enclosure at 300 kHz) above this limit (104 dBpt) to fail (104 dBpt = 0.13 A/m). Again the assumption is made that this current flows in a large loop of the same dimensions as used in the prediction for the 20-kHz power supply current. Since the total harmonic distortion voltage of the 20-kHz power supply is limited to 2.5% and 300 kHz is the 30th harmonic, there is a low probability of failure to meet REO1 limits above 20 kHz.

Above some frequency, typically 50 MHz, the predominant radiation ceases to be magnetic and becomes either E field or plane wave. As the size of the enclosure becomes a fraction of a wavelength, it may sustain enclosure resonances. Where apertures exist in the enclosure, they can function as slot antennas.

Thus, at high frequencies, meeting REO2 emissions limits becomes the major problem. For our example enclosure without apertures and ignoring the contribution due to cables, a source of considerable energy would have to exist, that is, RF power source, high-power high-speed converter with capacitive or inductive load, within the enclosure to fail REO2 limits.

12.3.2.3   Radiated Emissions from Interconnecting Cables

The radiated emissions from interconnecting cables are caused primarily by common-mode current on the shield of the cable. The length of cable to be tested (see MIL STD 462 METHOD REØ2) is 2 m. From this length, the frequencies of resonance (i.e., $\lambda/2$, $2 \times \lambda/2$, $3 \times \lambda/2$, when the shield is connected to the ground plane at both ends) can be calculated approximately as follows:

Integer of $\lambda/2 = 2$ m	Wavelength (m)	Frequency (MHz) = 300/λ (m)
1	4	75
2	2	! 150
3	1.333	225
4	1	300
etc.		

At a resonant length, the effectiveness of the transmission line as a transmitting antenna is much higher than at nonresonant lengths. This is offset to some extent above the frequency at which $\lambda/2 = 2$ m, that is, 75 MHz, by the change in phase of the field down the transmission line. In addition, the transmission line is positioned not only above the ground but also below the ceiling and in front of the walls of the shielded room, so multiple reflections occur.

The total field at the measuring point is thus the superposition of all field sources. The REØ2 test method requires the location of the test antenna at maximum pickup, that is, the position at which the majority of field sources are in phase. The transmission-line equation used has been validated by tests in an undamped shielded room in which the predicted E field is within +10 dB to −11 dB of the measured E field.

The predicted maximum level of common-mode current allowed on the shield before the limits for narrowband emissions are exceeded for each of the resonant frequencies is given in Table 12.2.

TABLE 12.2

Maximum Common-Mode Current on Interconnecting Cables

Frequency (MHz)	Max. Common-Mode Current (μA)	E Field at 1 m ($\mu V/m$)	Spec E Field at 1 m ($\mu V/m$)
75	22	63.5	63
150	19	109.7	110
225	18	156.0	160
300	17	196.2	200

The maximum common-mode current on the inner conductor(s) of the cable and inside of the shield can be determined using the current on the outside of the shield and the transfer impedance of the specific cable to be used, once this is determined.

12.3.2.4   Conducted Emissions

Conducted emissions are those signals that appear on cables at other than the required operating frequencies. These emissions can either be continuous (CW) or transient in nature. From experience, systems utilizing circuitry that generates emissions rich in harmonic content (i.e., DC–DC converters) or that switch currents are almost certain to fail the requirements. This implies that some type of EMI reduction technique be included in the original design to reduce conducted emissions.

Two fundamental approaches are available, either reduction at source or the inclusion of filters. In extreme cases, both approaches are required. Those circuits that are potential sources of high-level emissions will be designed for emission control that is compatible with performance requirements.

12.3.2.5   Susceptibility of Internal Circuitry to Internally Generated Fields (Crosstalk)

The phenomenon of a system being susceptible to its own radiated emissions inside an enclosure is difficult to predict, because crosstalk and radiated coupling are very dependent on exact geometries. However, a number of estimations of crosstalk using distances of 1 in., 2 in., and 6 in. between power PCB traces and interconnecting cables at a height of 2 in. (5 cm) and 0.06 in. (0.15 cm) above a ground plane were made. The source wire used in the prediction was a 10 AWG circular conductor in order to achieve a low-source resistance. The receptor wire type and gauge are changed, as are the victim source and load impedance, to show the effects of the changes on the level of voltages induced in the victim circuitry.

Ideally, the wires carrying the 20-kHz power and return should be twisted over the complete length. However, due to the large gauge of the wire and the requirement to make terminal or similar connections, some length of the wires will be untwisted. It is the untwisted length that causes maximum H field generation and, therefore, maximum potential crosstalk. The lengths of untwisted 10 gauge wire used in the examples range between 1 in. and 6 in., which are considered practical best- and worst-case lengths. The height above the ground plane in the equations contained in Section 4.3 is also used to describe the distance between the source power wire and its return.

At a very worst case, the average 100-A-load current may be switched as fast as 50 ns, and this was the value used in the initial prediction. Where the switching time is made longer than 50 ns, the peak transient voltage induced in the victim wire is considerably reduced, as shown in Example 12.6.

Example 12.1

Here the coupling is to an unshielded receptor wire over a distance of 1 in., with the receptor circuit exhibiting a source impedance of 250 Ω and a load impedance of 1000 Ω, which is typical for an analog circuit. The peak crosstalk voltage is a massive 24 V when the two wires are only 1 in. apart.

Example 12.2

Here the wires are separated by 2 in., with all other parameters constant. The induced voltage reduces to 13.5 V.

Example 12.3

Increasing the distance between wires to 6 in. decreases the crosstalk voltage to 3.1 V.

Example 12.4

Maintaining the 6 in. distance between wires and increasing the coupled length to 6 in. increases the voltage to 18.6 V.

Example 12.5

Maintaining the 6 in. distance between wires and the 6 in. coupled length but decreasing the distance between the source power and return wires to 0.06 in. (i.e., the insulation of the two wires is touching) and the height of the receptor wire above the chassis to 0.06 in. results in a decrease in the crosstalk voltage to 2 V.

Example 12.6

The same geometry is used as in Example 12.1. But here the rise time is increased to 100 ns, which causes a reduction in crosstalk from 23.8 to 12.2 V. Increasing the rise time further to 1 µs results in a reduction in crosstalk to 1.26 V.

Example 12.7

Here the distance between wires is 1 in., the height above the ground plane is 2 in., the coupled length is a worst-case 6 in., and an RG59A/U shielded cable is used for the victim wire. The load and source impedances are also changed, with the load matched to the coaxial cable characteristic impedance, which is a typical configuration for RF or video circuits. The predicted peak transient is 63 mV.

Example 12.8

This is the same configuration as Example 12.7; only the distance between the source and receptor wires is increased to 6 in. and the maximum peak transient is 8.2 mV.

The transient crosstalk voltage from Example 12.8 (8.2 mV) may be acceptable in high-level RF signals. And even where low-level analog or video signals are concerned, the short-duration spike may not cause an EMI situation because the circuit may not be perturbed by the spike. Where a potential EMI condition does exist, the noise immunity of the receptor circuit may be increased, typically by filtering or/and by increasing the rise time of the source current. Where the coupling is to PCB traces or ICs, the shielding is more difficult.

However, shielding of sensitive circuitry in enclosures and EMI hardening of sensitive circuitry are two approaches that will be considered, even at the PCB level.

12.3.2.6 Susceptibility of Internal Circuitry to Externally Generated Fields

The effectiveness of cables and enclosures are examined separately.

The enclosure: In the section dealing with the enclosure radiation at 20 kHz, a worst-case magnetic field attenuation of 53 dB was arrived at. From the inspection of the formula, it can be seen that the enclosure attenuation will increase with frequency.

The current flow in a wire wrapped around the enclosure during the 20-kHz power frequency test (RS02) is 3 A. The impedance of the enclosure is very low, so approximately 50% of the

3-A wire current flows in the outside of the enclosure. The distance between the loops of wire wrapped around the enclosure is specified as 30 cm, in accordance with MIL-STD-462. The current density in the outside of the enclosure is, therefore, 1.5 A/0.3 m = 5 A/m.

The definition of magnetic field attenuation used in deriving the 53-dB minimum attenuation for the enclosure is the ratio I_{out}/I_{in}, where I_{out} is the current flowing on the outside of the enclosure and I_{in} is the current flowing on the inside of the enclosure. I_{in} is, therefore, 53 dB down on 5 A/m, which is (5 A/m)/446 = 0.011 A/m. The narrowest dimension of the enclosure is approximately 0.43 m, so the current density across this interior dimension is

$$11 \text{ mA/m} \times 0.43 \text{ m} = 4.7 \text{ mA} / 0.43 \text{ m}$$

Assuming a loop of unshielded wire 3 cm from the inside walls of the enclosure and routed around the enclosure, the crosstalk voltage picked up on the wire at 20 kHz will be very approximately 0.6 mV, assuming a load resistance of 1000 Ω and a source resistance of 250 Ω.

Pickup from sources within the enclosure is likely to be higher than the predicted 0.6 mV induced in the RS02 test. Therefore, sensitive signals will be routed via shielded cables. If the wire is used for a control line, 0.6 mV in a control signal is unlikely to cause an EMI problem. The decision on the use of shielded or twisted pair for control lines will be based on the level of the interior ambient. Based on the measurements, the current spike test results in approximately 100 A of current flow in the test wire around the enclosure; however, the duration of the resultant positive and negative spike is short (10 µs). The low-frequency component of the spike has a current amplitude of approximately $2Ad$, where A is the peak amplitude and d is the pulse width. The external current flow is thus $2 \times 100 \times 10 \times 10^{-6}$ = 2 mA/0.43 m, that is, lower than the 20-kHz power frequency test. The high-frequency components of the spike are attenuated to a greater extent than the low-frequency ones, because the shielding effectiveness of the enclosure increases with increasing frequency. We conclude, therefore, that our sample enclosure provides adequate shielding for the RS02 test levels. The levels of common-mode voltage induced in four different types of shielded cables are shown in Table 12.3.

TABLE 12.3

Common-Mode Voltages Induced in Shielded Cables Exposed to RS03 Test Levels (Cable Length = 2 m)

Distance from Antenna (m)	Height above Structure (m)	Cable Type 1[a]	Cable Type 2[b]	Cable Type 3[c]	Cable Type 4[c]	f
1	0.05	3 mV	1 mV	—	—	30 MHz
1	0.05	13 mV	12.6 mV	—	—	7 MHz
1	0.05	1.48 V	0.163 V	0.144 V (4.8 µV)	0.144 V (1.4 µV)	1.05 GHz
1	0.05	7.7 V	0.92 V	0.77 V (1.2 mV)	0.77 V (0.11 mV)	9.975 GHz

Note: Cable type 1 is a standard braided cable exhibiting a transfer impedance of 10 mΩ/m at 100 kHz, 100 mΩ/m at 75 MHz, 600 mΩ/m at 1.05 GHz, and 15 Ω/m at 9.975 GHz. Cable type 2 is a twisted shielded pair, suitable for the 1553 data bus communication link, exhibiting one of the lower transfer impedances (i.e., an average shielded cable). The transfer impedance is 10 mΩ/m at 100 kHz, 100 mΩ/m at 30 MHz, 1 mΩ/m at 75 MHz, 14 mΩ/m at 1.05 GHz, and 133 mΩ/m at 9.975 GHz. Cable type 3 is a double-braided coaxial cable exhibiting a transfer impedance of 7 mΩ/m at 1 GHz and 2 mΩ/m at 9.975 GHz. Cable type 4 is a copper solid-wall tube exhibiting a transfer impedance of 2 mΩ/m at 100 kHz and effectively 0 mΩ/m at 30 MHz and above (i.e., a semirigid cable). Figures in brackets are for an (N (D/U))-type connector.

[a] D-type connector.

[b,c] BNC-type connector.

The impact of these levels of common-mode-transferred voltage (V_t) depends to a large extent on the EMC classification of the circuit class, that is, the sensitivity of the receiver. A potential EMI problem exists when V_t is close to the sensitivity of the receiver. A second potential problem occurs when a wire in the shielded cable is connected, either directly or via a low impedance, to chassis inside the enclosure; in this case, appreciable current can flow due to V_t, and radiation from the wire inside the enclosure may result in EMI.

An important point to note is that it is typical in RS03 tests to modulate the carrier at frequencies of maximum susceptibility, for example, 10 kHz for audio circuits and 1 MHz for the 1553 or similar digital busses. In addition, when high levels of RF are applied to integrated circuits (ICs), which may have a bandwidth much lower than the test frequency, DC offsets may occur due to non-linearities inherent in the IC input stage.

As the design progresses, the implication of the predicted noise voltage should be assessed for each enclosure layout and interface circuit, and appropriate filtering or shielding will be incorporated.

RS02 spike test–induced voltage: When a transient induced voltage is considered the time constant of the source wire controls, the current flow in the source wire. Likewise, when the victim cable time constant is longer than the applied pulsewidth, the induced voltage divides not only between the source and load resistors but also in the cable inductance. Thus, the worst-case-induced voltage is not caused by the 100 V_t > 0.15-µs spike but by the 200-V > 10-µs spike in the shielded cables under consideration due to both the longer time constant of the 10-µs spike and the higher peak voltage. The current flow in the shield of a cable subjected to a 200-V > 10-µs RS02 test has been measured at +30 A/−30 A and, therefore, 30 A was chosen as a typical value for the shield current in the prediction.

The cable crosstalk computer program was used to simulate the source of the spike and to predict the voltage induced in a shielded cable. Initially, the current flowing in an unshielded wire with the same diameter as the shielded cable was modeled, and the current flow in the source wire was adjusted until 30 A flowed in the victim wire. The victim wire was then replaced with the shielded cable, and the voltage induced in the shielded cable with a 30-A peak shield current was predicted by the program. The type of shielded cable used for interface connections on the system is not yet known; therefore, the induced voltage was calculated for a number of different types of shielded cable. Table 12.4 gives the level of induced voltage, which is almost independent of whether the receptor cable is terminated in a high-impedance load or with its characteristic impedance, for the induced voltage is the same for a wide range of receptor load impedances when the emitter source impedance is low.

The voltages shown in Table 12.4 are common mode; thus, where twisted-pair shielded cable with well-balanced differential inputs are used with sufficient common-mode noise rejection to a 10-µs-wide transient, the induced voltages will not cause an EMI problem.

TABLE 12.4
RS02 Test–Induced Voltage

Cable Type	Shield	Common-Mode-Induced Voltage (V)
RG58C/U	Single braid	0.14
RG180/U	Single braid	0.17
RG59A	Single braid	0.12
RG8A/U	Single braid	0.55
RG223/U	Double braid	0.08
RG9B/U	Double braid	0.033

12.3.3   CASE STUDY 12.5: ANTENNA-TO-CABLE COUPLING ON THE SPACE SHUTTLE (ORBITER)

Cables on an experiment that is carried in the cargo bay of the space shuttle are exposed to the E fields generated by an antenna mounted on the experiment. The approximate worst case with the antenna nonresonant, values of E field, generated by the experiment at distances of 1, 2, 4, and 8 m are shown in Table 12.5. (These values are higher than published levels as described in Section 12.3.) The interference voltage induced in the inner conductors of a shielded cable when the cable runs some length in parallel with the antenna and parallel to and above the shuttle cargo bay structure is calculated for a number of cable lengths, heights above the cargo bay, and distances from the antenna.

The cable shield is connected at both ends to the shuttle structure, via the connectors and instrument case. For the sake of these calculations, the shuttle structure is considered a perfectly conducting ground plane. The characteristic impedance of the transmission line Z_c is given by

$$Z_c = \sqrt{\frac{\left(Z^i + j2\pi f l^e\right)\left(2\pi f l^e\right)}{jk^2}} \tag{12.3}$$

where
Z^i is distributed series resistance of the two-wire lines
(Z^i for a 0.5-cm-diameter braided shield is taken as 10 mΩ/m)

$$l^e = \frac{\mu_0}{\pi} \ln\left(\frac{2h}{a}\right) \tag{12.4}$$

$$k = \frac{2\pi}{\lambda}$$

$\mu_o = 4\pi \times 10^{-7}$ [H/m]
f is frequency [Hz]
h is height of cable above the structure

In the frequency range 100 kHz to 30 MHz, the characteristic impedance is constant with frequency and varies with the height of the cable above the structure and the diameter of the cable. For convenience, the illumination of the cable is considered to be plane wave, with the plane of

---

TABLE 12.5

E Field Generated by Experiment at Distances of 1, 2, 4, and 8 m

Frequency	E Field (V/m) at distance r			
	1 m	2 m	4 m	8 m
100 kHz–6 MHz	1000	125	16	2
10 MHz	1000	125	79	39
15 MHz	1000	125	79	39
30 MHz	1000	158	79	39

incidence coincident with the plane of the loop formed by the cable and the structure, using the following values in Equation 12.4 for $h = 10$ cm,

$$l^e = \frac{4\pi \times 10^{-7}}{\pi} \ln\left(\frac{2 \times 0.1}{0.005}\right) = 1.47 \times 10^{-6}$$

The value for k at 30 MHz = 6.28/10 m = 0.628. Therefore, from Equation 12.3, for $h = 1$ cm, $Z_c = 168$ Ω, and for $h = 10$ cm, $Z_c = 441$ Ω.

12.3.3.1 Shield Current

The current flowing in the cable shield due to the incident wave, ignoring re-radiation from the wire structure and resonances in that structure, is given by

$$I = \frac{4E_o h}{Z_c}$$

where E_o is the amplitude of the plane wave.

The shield current is independent of cable length as long as the complete length of the cable is equally illuminated. The shield currents for the two values of height 1 and 10 cm and a cable diameter of 0.5 cm are calculated as follows:

Example 12.9

The cable shield current for a distance of 2 m from the antenna where the E field is = 125 V/m is as follows:

For $h = 1$ cm:

$$I = \frac{4 \times 125\ V \times 0.01}{168\ \Omega} = 30\ mA$$

For $h = 10$ cm:

$$I = \frac{4 \times 125\ V \times 0.1}{441\ \Omega} = 110\ mA$$

N.B. From the foregoing, it is clear that cables should be kept as close to the shuttle structure as feasible when the cables are in the proximity of the antenna.

12.3.3.2 Common-Mode Interference Voltage

The transferred common-mode voltage V_t induced in the center conductors of the shielded cable due to the shield current is a function of the transfer impedance of the shielded cable, which in turn is a function of frequency. Three types of cable are considered:

1. A standard braided cable exhibiting a transfer impedance of 10 mΩ/m at 100 kHz and 100 mΩ/m at 30 MHz
2. A braided multilayer foil type exhibiting a transfer impedance of 8 mΩ/m at 100 kHz and 10 mΩ/m at 30 MHz
3. A copper solid-wall tube exhibiting a transfer impedance of 2 mΩ/m at 100 kHz and effectively 10 mΩ/m at 30 MHz

The common-mode voltage induced is directly proportional to cable length. However, the induced shield current begins to reduce at cable lengths beyond 4 m, for the field strength begins to reduce at these distances from the source.

For a shield current of 30 mA, V_t for cable, 1 per meter, is as follows:

At 100 kHz = $30 \times 10^{-3} \times 10 \times 10^{-3} = 0.3$ mV/m
At 30 MHz = $30 \times 10^{-3} \times 100 \times 10^{-3} = 3$ mV/m

For a shield current of 110 mA, V_t for cable, 1 per meter, is as follows:

At 100 kHz = 1 mV
At 30 MHz = 11 mV

12.3.3.3 Compensation for Cable Resonances

At specific frequencies, resonances in the termination currents due to the short-circuited nature of the transmission line must be expected. The induced voltages shown in Table 12.6 are four times the amount calculated, to compensate for cable resonances.

12.3.3.4 Common-Mode Voltage

The induced voltages appearing across the pair in a twisted shielded pair are common mode; thus, if these voltages are applied to the input of a balanced receiver that has a reasonable common-mode noise immunity, the noise signal is canceled. If the cable is a coaxial type, then the induced noise voltage will appear at the input of the receiver. Triax cable will provide some attenuation of the center core–induced noise voltage.

TABLE 12.6
Interference Voltages with Different Cables and Configurations

Distance from Antenna d (m)	Height above Structure h (cm)	Cable Length L (m)	Cable Type 1, V_t (mV)	Cable Type 2 V_t (mV)	Cable Type 3 V_t (mV)	Frequency
1	1	1	9.6	7.6	1.9	100 kHz
1	1	1	96	0.96	Neg.	30 MHz
1	10	1	36	28	7.2	100 kHz
1	10	1	360	3.6	Neg.	30 MHz
2	1	2	2.4	1.9	0.48	100 kHz
2	1	2	24	0.24	Neg.	30 MHz
2	10	2	8	6.2	1.6	100 kHz
2	10	2	88	0.8	Neg.	30 MHz
4	1	4	2.9	2.3	0.58	100 kHz
4	1	4	29	0.28	Neg.	30 MHz
4	10	4	12	9.2	2.3	100 kHz
4	10	4	120	1.2	Neg.	30 MHz

		Transfer Impedance	
Cable Types	Description	100 kHz (mΩ/m)	30 MHz (mΩ/m)
1	Standard braided	10	100
2	Braided multilayer foil	8	10
3	Copper solid-wall tube	2	–0

V_t, transferred interference voltage; Neg., negligible.

Table 12.6 shows noise voltages induced in the center conductors of shielded cables running parallel to the antenna for distances from the antenna of 1, 2, and 4 m and for cable lengths of 1, 2, and 4 m at frequencies of 100 kHz and 30 MHz and cable heights above the structure of 1 and 10 cm. The transfer impedances of the cables are typical for the three types shown.

12.3.3.5 Conclusions

The values shown in Table 12.6 are approximate values useful for evaluation of any possible EMI problem. For example, with care given to the termination of the shields at the connector, all the proposed options of digital interface circuits available for use in the experiment will provide sufficient immunity to the noise source under consideration.

Should a more exact value for the interference voltage V_t and the frequencies of resonance be required, the physical details of the cable layout and use of a computer program that includes cable resonance effects should be used.

Case Study 12.4 was confined to frequencies below 30 MHz. When high-intensity fields as frequencies above 30 MHz are incident on flexible cable, the interference voltages induced will be higher than those shown in Table 12.2 because the transfer impedance of the cable increases with increasing frequency.

12.3.4 CASE STUDY 12.6: COUPLING FROM RADAR TO THE LANDING CONTROL SIGNAL ON AN AIRCRAFT

In one example, it was predicted that the landing control signal cable used on a composite-skin aircraft would be exposed to an E field of 200 V/m from 14 kHz to 100 MHz, 5100 V/m from 100 MHz to 2 GHz, and 31,000 V/m from 2 to 21 GHz. The reason the signals could be exposed to such high levels is that the composite skin of the aircraft provides little shielding and the aircraft may fly close to high-level radar. The decision was made to run the signal cable in a solid copper conduit with a 0.05-cm wall thickness. Calculations were made based on the skin depth of the copper tube, and a level of shielding in the thousands of decibels was predicted from 2 to 21 GHz. The calculation may have been valid for a tube soldered, welded, or brazed to an enclosure in which the members of the enclosure were similarly joined.

In reality, the copper tube is soldered to an EMI backshell, attached to a connector that is mated to a receptacle mounted on the enclosure. Each of the mechanical interfaces exhibits a transfer impedance as described in Section 7.15. The current at the end of a cable disconnected from ground is zero. However, the cable in question is attached to an enclosure that is several wavelengths long at the higher frequencies of interest. The current flow through the transfer impedances is, therefore, at a maximum at critical frequencies.

Assuming the cable is run at a distance of 1 m or greater from a ground plane or grounded conductor, the approximate voltages induced in the control cable conductors, due to the skin depth of the conduit at low frequency and the transfer impedances of the connections, are given in Table 12.7.

The actual attenuation of the conduit with connector is 54 dB at high frequency and not the thousands of decibels initially predicted! Additional filtering was required at the connector to ensure EMC in this example.

12.3.5 CASE STUDY 12.7: COUPLING FROM AN AM TRANSMITTER TO A SATELLITE COMMUNICATION SYSTEM

In this example, a ground-based system receiving a satellite transmission at a frequency of 1.5 GHz exhibited EMI due to the high EM ambient. The receiving system was located 40 m from a 50-kW transmitter operating at a frequency of either 4.8 or 2.2 MHz. The signal at 1.5 GHz was not directly interfered with due to the frequency separation between the received signal and the interfering

TABLE 12.7

Voltages Induced in Control Cable Conductors due to the Skin Depth of the Conduit at Low Frequency and the Transfer Impedances of the Connections

E Field (V/m)	Frequency	Interference Voltage
200	14 kHz	140 mV
200	7.5 MHz	2.7 mV
5,100	100 MHz to 2 GHz	10 V
31,000	2–21 GHz	64 V

signal, the poor out-of-band response of the antenna, and the inclusion of a bandpass filter between the antenna and down-converter. It was this apparent lack of a potential EMI problem that caused the system to be built so close to a transmitter. However, the receiving system uses a 10–14-kHz time code receiver to obtain the accurate time required to track a satellite and this exhibited EMI. In addition, the antenna control signals and 10-MHz IF signals, carried on shielded cables between the antenna and the IF receiver and computing equipment, were susceptible.

The equipment was housed in a nearby building and the shielded cables carrying the IF and control signals between the antenna pedestal and the building were illuminated by the field generated by the 4.8-MHz or 2.2-MHz transmitter. The incident field induced a current flow on the cable shields, which in turn resulted in a common-mode noise voltage on the center conductor(s) developed by the transfer impedance of the cable and the terminations.

The resultant level of noise voltage developed in the coaxial cable carrying the IF signal, developed at the output of the down-converter, corrupted the IF to the extent that the received signal was lost in the noise. In addition, the noise voltage developed in the shielded cables carrying the signals used to control the antenna drive was sufficiently high that movement of the antenna was inhibited.

The EMI voltage developed on the Omega time code antenna resulted in compression and amplitude modulation in the high-input-impedance preamplifier connected to the antenna, and the integrity of the time code signal was destroyed.

With the receiving system totally inoperable due to EMI, the first reaction of the engineer in charge of the installation was to move the entire system to an alternative site with a lower EM ambient.

Despite the availability of an alternative site with a suitable building, the estimated cost to disassemble, move, and reassemble the equipment was $100,000, with a resultant delay of at least a month in the startup date for the system.

The equipment manufacturer was persuaded to commission an EMI investigation, which resulted in the implementation of modifications and thereby an operational system within 1 week, although the solution to the time code receiver problem took a further 2 weeks.

A local radio technician was able to make a measurement of the field developed by the transmitter, which was approximately 4 V/m. In calculating the V_t from the current flow and the transfer impedance of the types of cables used, the predicted voltage was well below the susceptibility threshold of the interface circuits and signals. This illustrates the importance of conducting a simple EMC prediction. If the results from the prediction indicate that an EMI problem should not exist, then the installation must differ from the assumptions used in the prediction. In this installation, the high transferred voltage in the installation was not developed across the transfer impedance of the shielded cable but across the poor shield termination and the poor ground to which the shield was connected.

The knowledge gained in understanding how to improve the immunity of the system was just as valuable as the EMI cure. The modifications to the equipment went as follows. The time code receiver required a simple filter, the design of which is described in Section 5.3.9, contained in a small, sealed enclosure inserted between the antenna and the preamplifier. The cable-coupled EMI was cured by an improvement in the antenna and building grounding scheme, the shield termination, and the level of cable shielding.

After the proposed modifications were correctly implemented (many EMI fixes are less than effective due to incorrect implementation), the system functioned correctly. It continues to function without EMI 10 years after installation.

A number of lessons were learnt from the EMI investigation, which resulted in the following modifications to the standard version of the system.

The improved grounding scheme and shield termination methods are implemented in all production equipment regardless of the ambient.

Ambient site surveys are now conducted on almost all of the proposed installation sites. The site survey measures the ambient over the frequency range of 10 kHz to 18 GHz and, in addition to the low-frequency type of problem described in this example, examines the potential for in-band EMI at 1.5 GHz, cross-modulation, passive intermodulation (PIM), adjacent channel, harmonic, IF, and image interference. If the ambient site survey results in a prediction of EMI to the time-code receiver, then the antenna filter is installed. For close to in-band EMI problems, the inclusion of a narrower bandwidth filter between the 1.5-GHz antenna and the down-converter may be required. For in-band EMI due to PIM, relocation of the antenna or conductive structures in close proximity may be required, or the use of all-weather absorber foam or anti-PIM coating may be necessary.

When in-band EMI cannot be corrected at the proposed site, an alternative site may be chosen prior to installation.

The site survey is designed to reduce the cost of the equipment and installation because filters, absorber foam, anti-PIM coating, welded conduit for cables, and so on are incorporated only when the EM ambient warrants them. In addition, the estimate of the time required to install the system is more accurate when based on an ambient site survey.

12.3.6 Case Study 12.8: Spurious Response in a Transmitter/Receiver

Switching-power-supply noise resulted in EMI in the equipment that contained a transmitter/receiver. The EMI appeared as spurious emissions during transmission, displaced from the carrier by the power supply switching frequency and harmonics thereof. In the worst case, the spurious emissions were only 6 dB down on the carrier; in this example, a $700,000 late delivery penalty was contained in the contract and the EMI problem was cured 2 weeks prior to delivery.

The choices, common in all EMI problem solving, are to incorporate source reduction, reduction in coupling, an increase in receptor immunity, or a combination of these.

In the example, the EMI solutions included the design and building of a power-supply filter, elimination of a common ground path, and reduction in crosstalk between the output-filtered power lines and unfiltered lines. These solutions were implemented in 6 days, and the equipment passed acceptance testing just prior to the stipulated delivery date.

The transmitter/receiver contained a switching-power supply that provided +5 V for digital logic, ±15 V for RF and analog circuits, and +28 V for RF circuits and the RF power amplifier.

Switching-power supplies are notorious generators of high-level and wide-frequency-range noise. In this case, both radiated and conducted noise sourced by the supply resulted in EMI. The spectrum occupancy of the noise started at the switching frequency of 70 kHz and extended up to 100 MHz.

In the original design, the power supply was not shielded. After the EMI problem had been identified, the manufacturer built a prototype-shielded enclosure that only marginally improved the EMI problem. The prototype enclosure was constructed of an aluminum base plate and a steel cover. A strip type of connector protruded through a 0.105-m aperture cut into the rear of the enclosure, which was 0.11 m wide. The edges of the steel cover at the rear, close to the connector, were

FIGURE 12.12 Original power supply and RF board configuration.

attached to the sides of the cover by pop rivets, as shown in Figure 12.12. The prototype enclosure under test was unpassivated, and the proposed surface finish was the relatively high-resistance zinc dichromate. Both the pop-riveted seams at the rear of the enclosure and the 0.105-m slot were identified as potential leakage paths for radiation. The sources of radiation within the enclosure, such as the transformer and switching semiconductors, emit the highest fields at the switching frequency. Due to the low frequency and the close proximity of the sources to the enclosure, the field impinging on the inside walls of the enclosure is predominantly magnetic.

The inside of the enclosure forms a wide loop, with an impedance determined by the inductance of the loop and the frequency of the currents flowing in it. A current flows as a result of the incident field. Some of the current diffuses through the enclosure material due to its AC resistance. When seams are present, as in our example, the current also diffuses through the seam impedance, consisting of the DC resistance, the contact impedance, and the inductance of the seam. The diffused current then flows in the loop formed by the outside surface of the enclosure and generates a magnetic field. When an aperture is present, the magnetic field couples through the aperture and contributes to the external field. Section 6.5.3 describes the mechanism. We shall compare the shielding effectiveness of three enclosure configurations. One is a totally sealed version with welded seams and dimensions 0.13 m × 0.11 m × 0.24 m. The material used to construct the enclosure has a thickness of 5 mm, a relative permeability of 485, and a relative conductivity of 0.1. The predicted shielding effectiveness of the totally sealed enclosure is 100 dB at 70 kHz. Examining the shielding effectiveness of the same-size enclosure with two pop-riveted seams in the current path and with the use of either unplated metal or the relatively high-resistance zinc dichromate finish, the predicted shielding effectiveness reduces to 31 dB. By adding the 0.105-m aperture for the connector, the shielding effectiveness reduces to approximately 25 dB. The proposed improvements to the enclosure entail closing the aperture for the connector, closing the seams with screw fasteners in close proximity, and passivating the material with a high-conductivity finish such as tin or zinc, which should increase the shielding effectiveness to approximately 55 dB.

The magnetic field measured with the top cover of the enclosure removed, at a distance of 2.4 cm above the transformer, which is the source of maximum emissions, is 6.5 A/m at 70 kHz. The magnetic field measured close to the 0.105-m aperture and the pop-riveted seams was 130–140 mA/m. Thus, the measured effectiveness of 33 dB is close to the predicted effectiveness of 25–31 dB. This close correlation between the measured and predicted attenuations, using the simple analysis method, has been seen in a number of experiments. Therefore, assuming the magnitude of the source had been known, the field at the rear of the enclosure could have been predicted with some accuracy. However, the field at the rear of the power supply was not directly, or the sole, source of EMI. The susceptible circuit was located on the RF board on which a magnetic field was developed as a result of common-mode currents flowing on the interconnections to the board. This current had two sources: one was the leakage field from the power supply enclosure coupling to the interconnections between the power supply and the RF board and the second was the conducted common-mode noise currents generated by the power supply.

The magnitude of the magnetic field at the susceptible location on the RF board was measured to be 13 mA/m at 70 kHz and 0.2 mA/m at 28 MHz. The 28 MHz is a spectral emission with a repetition rate of 70 kHz and is, therefore, also a candidate for the 70-kHz spurious emission EMI. The level of magnetic field at 70 kHz induces a current of approximately 30 µA into a 2 cm × 2 cm loop formed by tracks on the PCB, and the 28-MHz level induces approximately 17 µA in the same-size loop.

In finding the source of EMI, one of the diagnostic methods used in the investigation was to move the interconnection wires away from the rear of the power-supply enclosure. This resulted in reduced coupling to the interconnections, a lower magnetic field on the RF board, and, as a consequence, a lower level of spurious emission.

Numerous attempts were made both to filter the power supplies on the RF board, close to the edge of the board near the connector, and to locally shield susceptible areas of the board, with limited success.

As a result of the pressure to find a quick solution that had a high probability of success, it was decided to improve the shielding effectiveness of the power-supply enclosure and include both common-mode and differential-mode filter components in the power supply on the +5 V, ±15 V, and 28 V supply and return lines. The filter was designed to preclude insertion gain by selecting a resonance frequency much lower than 70 kHz. A second feature of the design was the capability of supplying the high-peak current required during transmission with a minimum reduction in supply voltage and with no generation of transient voltages. The measured attenuation of common-mode and differential-mode noise voltage and common-mode noise current for the filter is shown in Table 12.8.

The filter components were mounted on a PCB contained within a small sub-enclosure that was connected to the inside cover of the power supply as shown in Figure 12.13.

The leakage from the seams of the power supply enclosure was reduced by the use of a high-conductivity finish (both tin and zinc were tried) with fasteners at 2-in. spacings along the bottom of the enclosures. These modifications reduced the AC resistance and contact impedance of the joints, and thus a higher level of shielding effectiveness was achieved. The leakage from the rear of the enclosure, close to the power supply interconnections, was reduced by welding the seams. The modified enclosure, as shown in Figure 12.13, covered the PCB-mounted connector used to make the internal connections to the filter sub-enclosure, and thus the 0.105-m slot for the connector

TABLE 12.8
Filter Performance

Frequency	Attenuation		
	Common-Mode Voltage (dB)	Differential-Mode Voltage (dB)	Common-Mode Current (dB)
70 kHz	50	50	37
3.01 MHz	40	50	22
6.02 MHz	63	—	—
15.05 MHz	—	32	—
17.99 MHz	—	—	22
21.00 MHz	—	50	—
28.00 MHz	—	25	—
29.96 MHz	45	—	—

FIGURE 12.13   Modified power supply and RF board configuration.

was eliminated. Instead, a small aperture was made at the location of the filter sub-enclosure to bring the filtered supply interconnections out of the enclosure.

With the filter and modifications made to the prototype enclosure, the magnetic field measured at the rear of the enclosure close to the interconnections was 12 mA/m at 70 kHz. This represents an improvement of 21 dB over the pop-riveted enclosure with 0.105-m slot and a total shielding effectiveness of 55 dB, which is uncharacteristically close to the predicted value of 55 dB!

The improved enclosure and filter achieved a reduction in the spurious emission to an acceptable 40 dBc, that is, 40 dB down on the carrier, which represents a 34-dB improvement over a nonmodified power supply.

12.4   EMC, COMPUTATIONAL ELECTROMAGNETIC MODELING, AND FIELD SOLVER COMPUTER PROGRAMS

A concise description of computational electromagentics is provided in Reference [1]. Computer programs are useful in the prediction of EMC and in EMI investigations because they reduce the probability of arithmetic error and speed up the computation. The use of a computer program to assist in an EMC analysis is a time-saving tool and may be the only practical method to deal with the multiplicity of potential EMI situations in large systems or complex equipment.

However, the argument may be made that the use of EMC programs requires more engineering judgment, not less. A simple measurement may be possible that will allow validation of the program and increase confidence in its use. Measurements may have already been made, such as those described in Section 10.2.1, that compare the use of FEKO and GEMACS measurements made on antenna-to-antenna coupling.

Another primary limitation on the use of computational electromagnetic modeling (CEM) and EM field solvers is the amount of computing resources or the time required to make a full-wave simulation. This is often determined by the level of geometric complexity, the volume of the objects, or the distance from the objects. Because many codes require that the structure under consideration be divided up into elements that are a fraction of a wavelength in size, the limitation may be on the upper frequency that can be performed using the software. Some guidelines follow.

> If the code uses only transmission-line models, it may calculate only fields generated by the differential-mode currents and ignore the common-mode current as a source.
>
> First perform an analysis/prediction of the potential EMI problem areas and assess the criticality of the problem. Use numerical modeling techniques only when the initial analysis shows a potential EMI problem.

Ensure that a measurement on the equipment or system is not going to provide a faster and more accurate result. Consider the use of a brass-scale model when the system is very large or complex.

The source of radiation must be as simple as possible but must still include the actual return path for grounds.

It may be more efficient to model a chassis using an electromagnet boundary with an equivalent surface impedance, instead of a 3D model.

The dielectric constant of PCB or cable insulation affects resonant frequencies, impedances, and field strength and should be included in the model where possible.

The proximity of conductive surfaces and nonsymmetries can result in common-mode current flow, and here a 3D model may be required for accuracy.

The geometry of the problem should be made as simple as possible, depending on the fidelity required of the analysis.

Many field solvers are available for the analysis of signal integrity in circuits and on PCBs. They are used in timing simulations to compute delays and near-field coupling/crosstalk. These field solvers are often quasi-static, meaning that although the structure guides signals that are fundamentally propagating waves, the interconnect model is usually modeled as a lumped element or transmission line. These can be determined from separately solved values of capacitance from a static electric-field solver and inductance from a static magnetic-field solver, which produce the quasi-static values of inductance and capacitance. For this reason, two separate methods of simulation may be required when both radiation from a circuit and signal quality are required.

Information on computer modeling and a newsletter are available by joining the Applied Computational Electromagnetics Society (ACES) (www.aces-society.org).

The *EMC Analysis Methods and Computational Models* book is published by John Wiley & Sons and contains four computer programs that are supplied on a disk with the book.

These programs are as follows:

1. NULINE: Program to analyze the behavior of an above-ground transmission line excited by either transient or CW-lumped voltage or current source anywhere on the line or by an incident EM field.
2. RISER: Program to compute the voltage and current in a load at one end of a field-excited, above-ground transmission line, modeling the ends (risers) of the line as smaller, vertical transmission lines.
3. LTLINE: Program to compute the load voltages and currents at each end of an above-ground transmission line that is excited by a nearby lightning strike.
4. TOTALFLD: Program to evaluate the total above-ground and below-ground E and H fields produced by an incident EM plane wave. Both CW and transient results for a number of different user-defined waveforms are present.

Sections 11.12 and 11.13 discuss modeling techniques and computer programs specifically for the prediction of PCB radiation, although many of the computer programs described in this section are also capable of modeling PCB radiation.

12.4.1 Simple Computer Programs

Simple programs written in BASIC are appended to several chapters in this book and have been found to be invaluable in predicting cable emission, coupling, and inductive crosstalk problems. These programs written in C+ or C++ or Excel or Basic are available free from EMC Consulting Inc. Many more of the simple equations in the book have been used in computer programs that, as an example, predict attenuation of enclosures with apertures and seams. These and other programs containing the antenna gain and antenna factor equations in Chapter 2 are available online at website www.emconsultinginc.com. As seen in Chapter 11 Section 11.10, it is very important that a benchmark problem be presented to the supplier of a commercially available EM program before purchasing or relying on the output data. The benchmark should have been measured under carefully controlled conditions, using reliable, repeatable test methods, and the measured data compared to the data provided by the EM program. In some instances engineering judgment and experience can be used to test the validity of predicted data.

12.4.2   EMC ANALYSIS PROGRAMS

These are used for radiated emission, conducted emission, radiated susceptibility, and conducted susceptibility predictions and use much simpler tools than the EM analysis programs described in Section 12.4. They typically compare the results to requirements imposed by regulatory authorities such as the military MIL-STD or DEF-STAN, commercial aircraft DO-160, automotive, FCC, and European Union EN requirements.

Some of these programs contain design rules based on engineering experience and possibly simple equations and are the "cook book" type of program and their limitation is the accuracy of their database. Also it is difficult to model complex systems using simple approximations.

One such program is EMI Analyst™ marketed by EMI Software (formerly Spectrasoft) for which an example is given in Section 7.4. All four applications use a combination of circuit analysis and transmission-line theory algorithms to predict the current and voltage at each node and the method of moment (MOM) method to calculate radiated emissions from conductive cables. Eight types of cable may be modeled but not double shielded at present. Braided or solid shields can be used with the shield transfer impedance calculated from dimensional and shield material information.

The radiated susceptibility application uses a transmission-line theory to perform field-to-wire coupling similar to those presented in Section 7.7.

12.4.2.1   The RS Analyst™ is a Radiated Susceptibility Program

This program calculates voltage and current induced in circuits when an EM field strikes the conductors to which the circuits are connected.

It is a design and analysis tool for determining compliance with radiated susceptibility requirements imposed by regulatory agencies.

It is used to assess the ability of circuits to tolerate interference generated by EM fields.

RS Analyst is an integral component of the EMI Analyst suite of EM-interference analysis tools, so analysis models created by RS Analyst can be used when performing other types of analysis in "EMI Analyst."

12.4.2.2   CS Analyst Is a Conducted Susceptibility Program

This program calculates voltage and current amplitude induced in circuits by voltage or current waveforms that are injected on the conductors to which the circuits are connected.

It is a design and analysis tool for determining compliance with conducted susceptibility requirements imposed by regulatory agencies.

It is used to assess the ability of circuits to withstand interference conducted on interconnecting circuit conductors.

CS Analyst is also available from EMI Software.

12.4.2.3   CE Analyst™ Is a Conducted Emission Program

This program calculates the amplitude of RF current and the voltage conducted on electric wiring.

It is a design and analysis tool for determining compliance with conducted emissions requirements imposed by regulatory agencies, or for assessing intercircuit system compatibility.

With CE Analyst from EMI Software, circuits can be analyzed prior to fabrication to determine whether conducted emissions requirements will be satisfied. CE Analyst is ideally suited to predicting conducted emissions for a variety of circuit applications, including the following:

Switch-mode power converters
Power inverters and motor controllers
Digital circuits
High-speed data lines
Interface circuit filtering

The input is in the time domain and so a conversion is required if the input data are from a spectrum analyzer, although the software converts from the time domain to the frequency domain.

Alternatively, the measurement may be repeated using an oscilloscope.

A filter can be modeled using a schematic capture program built into EMI Analyst with the time domain waveform as input.

CE Analyst is an integral component of the EMI Analyst suite of EMI analysis tools, so analysis models created by CE Analyst can be used when performing other types of analysis in "EMI Analyst."

RE Analyst™ is a radiated emissions program that calculates the amplitude of electric fields radiated from conductors carrying RF current.

RE Analyst from EMI Software is designed for the following purposes:

Calculate the amplitude of electric fields radiated from conductors carrying RF currents it is:

Companion program to CE Analyst and is an integral component of the EMI Analyst suite of electromagnetic interference analysis tools.

RE Analyst is a prediction tool for determining compliance with regulatory agency radiated emission requirements and for assessing the level of electric-field radiation produced by circuits.

RE Analyst uses the conducted emission output data produced by CE Analyst as its input. Radiated emissions are calculated using MOM techniques.

RE Analyst incorporates the effects of filtering, shielding, ground plane proximity, and loading.

OPtEM Cable designer is a tool useful in designing and simulating the performance of all types of copper cables including unshielded twisted pair, quad, ribbon, and flex circuits from DC to 10 GHz.

The designer generates a 3D model of the cable and performs an EM-field analysis to extract L, R, C, and G parameters. Further analysis produces S-parameter output that can be presented in the form of insertion loss, return loss, near-end and far-end crosstalk.

The analysis includes skin effect, proximity of conductors eddy currents in shields, and dielectric loss in insulation and generates W-element SPICE models.

12.4.2.4 EMIT (An Antenna-to-Antenna Coupling Program)

The EMIT program from Delcross Technologies predicts cosite interference from antenna-to-antenna coupling. It predicts EMI in complex RF environments containing multiple transmitters and receivers and in-band and out-of-band interference in TX/RX pairs.

The user can customize any number of interference thresholds (and associated colors in the display) to show EMI margin levels. The program provides the following: Markers on result plots immediately show the source of interference including identification of intermodulation products. Plots signal spectra at any point in the signal paths to facilitate analysis and troubleshooting of complex scenarios. Filter results by interference type to display only specific types of problems. It accounts for Broadband TX and RX characteristics, outboard components such as filters (including tracking filters), amplifiers, multiplexers, circulators, isolators, cables, and power dividers. Transmitters and receivers are modeled using EMITs built in parametric radio by importing radio performance data such as transmitter emission masks and receivers selectivity curves.

EMIT includes a library of common RF system models, including GPS, PCS, Blue Force Tracker, SINCGARS, CDL, IFF, and many others.

Streamlined cosite interference workflow provides compatibility with EM simulation tools such as Delcross Savant, ANSYS HFSS, CST studio, FEKO, WIPL-D, and others. RF system, Antenna, and outboard component definitions can be stored for use in future projects and can be shared with other users.

Standard 1-on-1 and N-on-1 analysis modes to predict the effects of multiple site emitters on selected receivers including transmitter and receiver generated intermodulation products. It includes multicore support, where the user can specify the number of cores to use for computations.

Other programs that are specifically designed to model PCB structures are described in Chapter 11.

12.4.3 MOM, MLFMM, FEM, FEM-MOM, GO, PO, UTD, GTD, FEM, BEM, FDTD, PTD, GMT, TLM FIT, CG-FFT, PEEC, ANALYTICAL TECHNIQUES

Often software is described as using full-wave analysis. This solves the complete set of Maxwell's equations without any simplifying assumptions. Typically the fields described are time variant and frequency dependent and include the fields from displacement current, although a simplification can be made for quasi static fields.

Other programs such as physical optics (PO), geometrical optics (GO), and uniform theory of diffraction (UTD) are described as asymptotic using an approximation or expansion. Full-wave techniques for the computation of scattered and diffracted fields are computationally intensive and often impractical for complex designs. As shown in Section 12.6, the full-wave approach alone may be less accurate than a hybrid approach that includes an asymptotic analysis for large structures. Asymptotic methods offer acceptable accuracy to solve scattering problems at short wavelengths but their accuracy for structures that approach a wavelength in size diminishes. The advantage of the asymptotic is reduced computing time.

Three dimensional (3D) modeling takes account all of the changes in a structure whereas two dimensional (2D) takes a slice through a structure. 3D modeling requires longer computational times.

Numerical techniques require much more computing power than simple equations and expert systems.

The most sophisticated program that provides excellent correlation with a benchmark is only as good as the input data, and nine times out of 10, it is an error in input data which is the problem, as described in numerous examples in this book.

To repeat the comment in Section 12.4, it is the common-mode currents that result in the highest level of radiated emissions. This is shown in Section 7.13.3.1. As an example, a common-mode current of 15.7 µA at 21 MHz results in the same radiated field as 1.26 mA of differential current. Predicting the differential mode (signal) current is much easier than the common mode. To predict the common-mode current, every detail of the system including current in ground planes, proximity of circuits to ground, cable and enclosure geometries, and common-mode voltage developed between chassis and circuits and cables must be modeled, which with present software is difficult. It requires the use of a number of techniques including programs with circuit analysis, such as SPICE, signal quality, crosstalk, and conversion of currents to fields.

If a breadboard or engineering model of the circuit in the enclosure, with representative cables and a test facility, is available then engineering-type radiated emission or radiated susceptibility tests will invariably require much less time than EM modeling.

12.4.3.1   FEM Method

The finite element method (FEM) is for solving partial integral or differential equations. It is a variational technique that works by minimizing or maximizing an expression known to be stationary about the true solution. Generally, FEM techniques solve the unknown field quantities by minimizing an energy function.

The FEM technique uses a system of points called *nodes* that make a grid called a *mesh*. The starting point of the method is the subdivision (discretization) of a domain (volume) into small subdomains called *elements*, which are the spaces between the mesh. Although the elements need not be triangles, those subdivisions are called a *triangulation*. An element is described by its vertices and one point on each edge. These points are the nodes, and the FEM mesh is constituted from nodes and elements. The great advantage of FEM is that the domains can be conductive or a dielectric or have the properties of free space. One disadvantage is that the entire volume must be meshed, as opposed to MOM, which only requires the surfaces to be meshed. Often in time-dependent solutions, the FEM mesh is extended to all time levels by a finite difference method.

One weakness with FEM is that it is relatively difficult to model open configurations, that is, where the fields are not known at every point. Absorbing boundary are used to overcome this deficiency.

12.4.3.2   FDTD Method

The finite difference time domain (FDTD) method is a full-wave method using a direct solution of Maxwell's time-dependent (partial differential) equations with boundary conditions directly in the time domain by converting them into finite-difference equations. These equations are then time stepped by alternately calculating the electric and magnetic fields in a 3D grid lattice. One grid contains the points where the magnetic field is calculated and the second grid where the electric field is calculated. Time stepping is continued until a steady state or a desired solution is achieved. Since it is a time-domain method, the solutions can cover a wide frequency range with a single simulation.

As the basic elements are cubes, curved surfaces on a scatter must be staircased, which may require a very small grid size for sharp angles. FDTD techniques with nonrectangular elements have been introduced to solve this problem.

The denseness of the grid is determined by the dimensions of the smallest object to be modeled. Very often, the far field is required. One technique is to make a near field to far field transformation. With a surface S (S may be fictitious) bounding the source of radiation, if the tangential fields on S are known, then these can be expressed as surface currents and these will radiate E and H fields everywhere outside of the boundary S.

Because FDTD is a time-domain technique when a time-domain pulse (such as a Gaussian pulse) is used, a wide frequency range can be solved with only one simulation. This is valuable when resonant frequencies or a broadband result are desired. Thus, a single simulation can determine current, voltage, or fields over a wide frequency range. The frequency-domain results can be obtained by applying a discrete Fourier transform to the time-domain results. Arbitrary signal waveforms can be modeled as they propagate through complex configurations of conductors, dielectric, and lossy nonlinear nonisotropic materials.

The FDTD technique is excellent for transient analysis. It usually does a better job of modeling unbounded problems than FEM and is, therefore, often the method of choice for modeling unbounded complex inhomogeneous geometries. Because FDTD is a time-domain technique that finds the fields everywhere within the space, it can display moving images of the fields as they move throughout the space.

Since FDTD uses a computational domain, this must end somewhere, which needs the establishment of an absorbing boundary to simulate free space. The space must be gridded, and these grids must be small compared to the smallest wavelength and smaller than the smallest feature to be modeled. Very large spaces require very long solution times. Also, models with long, thin features, like wires, are difficult to model because of the very large space taken up by the wire. Because the E or H field is found within the space, the fields at very large distances from a source require very large spaces and excessive computation time. The far-field extensions are available for FDTD but require some postprocessing.

The FDTD technique allows apertures to be determined directly along with shielding effects, since the mesh can be described with different values of conductivity, and so boundaries with and without apertures and seams can be modeled. Also, the fields both inside and outside a structure can be determined.

The only significant disadvantage of FDTD is that, in common with most other techniques, the problem size and the fineness of the mesh will dictate the computation time. The fineness of the grid is determined by the dimensions of the smallest feature to be modeled, and so codes that offer a variation in the mesh size over the structure would have an advantage. Also the entire object, including most of the near field, must be covered.

12.4.3.3   FDFD Method

The finite difference frequency domain (FDFD) method is similar to the FDTD, although closely related to the FEM. Like FDTD, it results from a finite difference approximation of Maxwell's equations, although in this case, it is the time harmonic, differential form, of the equations that are used. Since there is no time stepping, it is necessary to keep the mesh spacing uniform.

Interior problems are best suited to differential equations and so to FDFD. In the GEMACS formulation, modeling a cavity requires that the volume of the cavity be completely contained within a closed surface. The surface may be either perfectly or finitely conducting, but it is always assumed that no EM can pass through the boundary. Within the closed surface, the EM model may be quite complex with barriers and obstacles. All or a portion of the volume may contain dielectric material. Thin wires may be present.

12.4.3.4   MOM or Boundary Element Method

The MOM, also known as the method of weighted residuals, is the most common form of analysis and is available in a great number of programs, described later. The boundary element method (BEM) is also a weighted residual technique. It is essentially a MOM technique whose expansion and weighting functions are defined only for a boundary surface. Most general-purpose MOM electromagnetic modeling codes employ a BEM.

MOM and BEM are full-wave analysis techniques for solving complex integral equations (IEs) by reducing them to a system of simpler linear equations.

The MOM is ideal for coupling to a structure and radiation from a structure that can be modeled by wire segments or metal plates. This is a frequency-domain technique that predicts a single frequency, although much of the available software allows iteration over a number of frequencies by repeating the analysis at a different frequency, with a resultant increase in computing time.

Assume a perfectly electrically conducting (PEC) object on which a field is incident. The incident field sets up a current resulting in scattered fields expressed in terms of integro-differential operators. The total field is then the incident field plus the scattered field. On a PEC body, the incident tangential field must be zero. This leads to the electric field integral equation (EFIE), although the magnetic field integral equation (MFIE) may be used, with both derived from Maxwell's equations. The current on the surface is discretized by dividing them into smaller sections (elements). On each section, the current is approximated by simple functions (expansion/basis) with a continuous current flow between the sections.

With triangular elements upon each adjoining pair of triangles, linear roof-top basis functions are used. Line segments are used to model wires with each segment joined by the linear roof top basis function, with the total current the sum of these functions. The total current is then converted to a system of linear equations to find the vector magnitude of currents from each mesh triangle. The scattered field is generated by the currents on the elements.

Depending on the form of the IE used, the MOM can also be applied to homogenous dielectrics or surfaces that are not perfect conductors.

The MOM is iterative as it uses a residual technique that establishes a set of trial solutions with one or more variable parameters. The residual technique compares the difference between the trial solution and the true solution. The variable parameters are chosen to achieve a best fit of the trial functions based on a minimization of the residuals.

For very large problems, the time required to arrive at the final solution may be very long. The multilevel fast multipole method (MLFMM) is one technique for speeding up the analysis.

An analysis to exactly compute radiation from high-frequency printed circuits may be based on the mixed potential IE with Green's function provided by Sommerfeld integrals.

The physical elements may be wires, patches, infinite ground planes, finite ground plane, Sommer Norton ground, surface meshes, cylinders, spheres, with many NEC programs including graphics.

Visualizing a complex structure is very difficult and has been known to result in wings on an aircraft at odd angles and so the use of a program with graphics is highly desirable.

Very powerful tool errors are possible, and some of these are described in Section 12.6.

The MOM uses a technique known as the method of weighted residuals, which has become synonymous with the MOM and with the "surface integral technique." The MOM technique generally does an excellent job of analyzing unbounded radiation problems and analyzing perfect conductors or structures with a given conductivity. They are not well suited to structures with mixed conductivity and permittivity, although they can be used for homogeneous dielectrics alone or for very specific conductor-dielectric geometries. The MOM technique can be used to analyze a wide variety of 3D EM radiation problems.

The MOM technique requires that a structure be broken down into wires, which are in turn subdivided into small wire segments that are typically no larger than $\lambda/10$. Metal plates are divided into a number of surface patches, which again must be small relative to a wavelength. The current on every wire and every segment is calculated and displayed. The composite E field at any point in space can then be calculated by the contribution from each of the wire segments and patches. The magnitude of the RF current on the structure can be reviewed to ensure that it is not unrealistically large or small. For example, if at some wire location on a closed structure the current changes from a milliamp to an amp value and then immediately returns to a lower value, the structure has almost certainly not been accurately modeled. The MOM allows discrete components to be inserted into a model by simply defining the impedance desired on any given wire segment. Thus, a 50-Ω load can be inserted in the center of a dipole antenna; MOM assumes the current on the wire segment or patch to be the same throughout the conductor's depth. Therefore, using MOM to determine the effects of an aperture with fields both inside and outside is difficult, and often a hybrid approach must be used. Because the MOM analysis is an iterative process, the larger the number of patches and wire segments, the longer the time required for analysis. With modern PCs, running at GHz frequencies, the computation time has reduced significantly.

Modeling large structures with regions of fine detail or complex materials can be impractical using MOM software and a hybrid approach such as FEM/MOM may be required.

12.4.3.5 MLFMM Method

The MLFMM is an alternative formulation to MOM and is applicable to much larger structures making full-wave current-based solutions of electrically large structures possible. The similarity between MOM and MLFMM is that in the FEKO software, the basis functions model the interaction between all triangles. MLFMM groups the basis functions and computes the interactions between these groups. FEKO employs a boxing algorithm that encloses the entire computational space in a single box at the highest level, dividing this box in three dimensions into a maximum of eight child cubes and repeating the process iteratively until the side length of each child cube is approximately a quarter wavelength at the lowest level. Only populated cubes are stores at each level, forming an efficient tree-like data structure. In the MOM framework, the MLFMM is implemented through a process of aggregation, translation, and disaggregation of the different levels. In the source region, the triangles are aggregated and then translated to the observer region where they are disaggregated. Whereas the MOM treats each basis function in isolation, resulting in N^2 scaling of memory requirements whereas the MLFFF requires only $N*\log(N)$ scaling in memory resulting in a more efficient treatment of the same problem. Improvements in the parallel MLFMM formulation achieve high efficiency when distributing a simulation over multiple processors.

This means that MLFMM can achieve orders of magnitude reduction in solution requirements compared to MOM.

12.4.3.6 PEEC Method

Partial element equivalent circuit (PEEC) method is a 3D full-wave technique that combines EM and circuit analyses. It can be used from DC to a maximum frequency determined by the size of the mesh. It extracts the resistance and self and mutual inductance from geometries, which results in a complete circuit solution for 3D geometries. It is ideal for modeling conductors in an enclosure and accounts for skin and proximity effects. Its great advantage is that it is a low-frequency tool.

One program is InCa3D provided by CEDRAT and this is described in Section 12.4.

12.4.3.7 GMT Method

The generalized multipole technique (GMT) is a frequency domain technique that (like the MOM) is based on the method of weighted residuals. The difference is that the expansion functions are analytic solutions of the fields generated by sources located at some distance away from the surface where the boundary condition is enforced.

Instead of charge or current, the expansion functions are spherical-wave solutions corresponding to multipole sources. By locating these sources away from the boundary, the field solutions form a smooth set of expansion functions on the boundary and singularities are avoided.

Like the MOM, a system of linear equations are developed and then solved to determine the coefficients of the expansion functions that yield the best result, which is the field solution.

As no integrals have to be solved, it reduces computing time. The fields may be computed accurately in the near-field region as well as on boundaries. Only the boundaries of the domain have to be discretized.

12.4.3.8 CGM Method

The conjugate gradient method (CGM) is another technique based on the method of weighted residuals and is similar to the MOM but handles the weighting functions differently and in solving the system of linear equations.

The CGM method is best used for large sparse matrices.

12.4.3.9 GTD Method

The geometrical theory of diffraction (GTD) is a high-frequency ray optics approach that has been extended to include diffraction effects from surface discontinuities such as edges and corners and waves creeping around smooth surfaces. The basic geometry objects that can be used in GTD structures are those for which reflection and diffraction coefficients exist and for which it is possible to carry out the ray tracing of the optics approach.

Most software only allows GTD and UTD to apply to flat polygonal plates with minimum edge length in the order of a wavelength or to cylinders and so is not suited to the analysis of curved surfaces.

The GTD predicts a Keller's cone on which infinite numbers of rays can penetrate into the shadow region and accounts for diffraction.

At or near the incident and reflected shadow boundaries, Keller's GTD predict erroneous results.

12.4.3.10   UTD Method

The uniform theory of diffraction (UTD) is an extension of the GTD. Both of these techniques are high-frequency methods. They are accurate only when the dimensions of objects being analyzed are large relative to the wavelength of the field. In general, as the wavelengths of an EM excitation approach zero, the fields can be determined using geometric optics. UTD and GTD are extensions of geometric optics that include the effects of diffraction.

The advantage of UTD over GTD is that by modifying the diffracted field, the erroneous field at the boundary is corrected and the results are correct.

12.4.3.11   PO Method

Physical optics (PO) is useful where large structures must be modeled. PO is an asymptotic high-frequency numerical method of the same nature as UTD.

It is not a precise formulation as it approximates the field on a surface. It integrates the field over the surface and calculates the transmitted or scattered field.

A reasonable correlation is seen between an exact method, used on a standard or canonical cone problem, for an angle of incidence where the complete cone is illuminated but with a larger error in the shadow region.

12.4.3.12   GO Method

The geometrical optics (GO) method like PO and GTD is a ray-tracing method. It is satisfactory for apertures that are large in terms of wavelength.

12.4.3.13   PTD Method

The physical theory of diffraction (PTD) is another ray-tracing method similar to PO, but unlike PO it takes care of the singularity at the edges of a structure.

12.4.3.14   TLM Method

Transmission-line matrix (TLM) is a numerical technique suitable for solving EM problems that may involve nonlinear, inhomogeneous, anisotropic time-dependent material properties, and arbitrary geometries. The method allows the user to compute the time-domain response of 2D-TLM- or 3D-TLM-shaped EM structures to arbitrary excitation in time and space. Instead of interleaving E and H fields, a single grid is established and the nodes of this grid are connected by virtual transmission lines. Excitations at the node sources propagate to adjacent nodes through these transmission lines at each time step. A mesh can be approximated by a lumped element model using inductance and capacitance per unit length of each line in the mesh. Although the time-domain response is computed, the frequency-domain characteristics can be extracted by Fourier transform, and the method is then useful in modeling the response to lightning, EMP, and other events as well as the peak response to a continuous wave incident on a structure.

The excitation source may be an arbitrary function or a continuous waveform. More than one excitation source may exist at points in the mesh. The wave propagation in a shunt-connected TLM network, made up of a mesh of transmission lines, may be shown in which the voltage simulate electric fields and from which the E_y, H_x, and H_z magnitudes can be calculated. In a series-connected TLM mesh in which the node current simulates a magnetic field, the H_y, E_x, and E_z fields may be computed. A total or partial reflection of the wave propagation down the mesh occurs at boundaries that represent a structure. The boundaries may be lossless or lossy homogeneous or lossy inhomogeneous materials. Aperture coupling with an external-source coupling through an aperture in the enclosure or from an internal-source coupling out of the enclosure may be modeled with this technique. Either the E or H field coupling through the aperture may be computed. Complex nonlinear materials are readily modeled, but large problems that demand a fine mesh require excessive amounts of computing time.

An orthogonal combination of the shunt-connected and series-connected network types can be used to compute 3D field problems. The TLM method allows modeling of field magnitudes and wave impedance and the surface wave on a structure.

One of the early problems with the TLM method was in predicting far-field radiation due to the need for a boundary at the mesh that absorbs and simulates free space. However, this problem seems to have been solved in commercial versions of TLM software. But, as with all software, the software company should be asked to run a benchmark prediction, such as far-field radiation from a microstrip configuration, that can be compared to known data before using or purchasing any EM software.

12.4.3.15   CG-FFT Method

The Conjugate Gradient Fast Fourier Transform (CG-FFT) method or EM characterization of highly complex and dense, such as ICs, or electrically large planar structures, a number of fast algorithms have been developed, one of which is the conjugate gradient fast Fourier transform.

12.4.3.16   MOM-FEM

The MOM-FEM hybrid features full coupling between conductors such as wires and metal surfaces in the MOM region and heterogeneous dielectric structures in the FEM region. The MOM solution provides equivalent magnetic and electric currents that form the radiation boundary of the FEM region. FEKO includes MOM-FEM, and according to this the hybrid technique uses the strengths of both MOM and FEM in the following ways: The MOM is used for the efficient modeling of open-boundary radiating structures where no 3D space discretization is required. The FEM is used for the efficient modeling of inhomogeneous dielectric bodies in terms of field distribution.

12.4.3.17   Finite Element Method and Combined FEM/MOM Techniques

The major advantage with FEM is that the electrical and geometric properties of each element can be defined independently. Thus, a large number of small elements can be set up in regions of complex geometry, and fewer, larger elements can be set up in relatively open regions.

In general, FEM excels at modeling complex inhomogeneous structures (components composed of different materials) but is not as good as MOM at modeling unbounded radiation problems. In contrast, the MOM does not model components composed of different materials efficiently. However, the advantage of FEM in describing a dielectric volume can be combined with MOM to solve currents on the surface of (or external to) a volume.

12.4.3.18   GMT

The *generalized multipole technique (GMT)* is a method of solving partial differential (weighted residuals) equations, and like MOM, it is a frequency-domain technique. The solution is approximated by a set of base functions that are analytical solutions of the respective difference equations. However, this method is unique in that the expansion functions are analytic solutions of the fields generated by sources located some distance away from the surface where the boundary condition is being enforced. MOM generally employs expansion functions representing quantities such as charge and current that exist on a boundary surface. The expansion functions of GMT are spherical-wave field solutions, and multiples are in general the most flexible and efficient expansion functions. By locating these sources away from the boundary, the field solutions form a smooth set of expansion functions on the boundary, and singularities on the boundary are avoided. GMT is a frequency-domain method, and solvers for electrostatics or EM scattering can be efficiently built using GMT. It can also be combined with MOM to provide a hybrid approach.

12.4.3.19   MMP

The multiple multipole program (MMP) was proposed in 1980 by Christian Hafner. It is a semi analytical method for numerical field computations that have been applied to EM fields. It is a code for solving EM scattering and guided waves. The amplitudes of the basis fields are computed by a generalized point-matching technique that is efficient, accurate, and robust. MMP "knows" many different sets of basis fields, but multipole fields are considered to be the most useful. Due to its close relations to analytic solutions, MMP is very useful and efficient when accurate and reliable solutions are required.

12.4.3.20 CGM

The *conjugate gradient method (CGM)* is another technique based on the method of weighted residuals. It is very similar conceptually to MOM techniques. Two features that distinguish it from MOM are the way in which the weighting functions are utilized and the method of solving the linear equations. Iterative solution procedures such as CGM are most advantageous when applied to large, sparse matrices.

12.4.4 COMPUTER ELECTROMAGNETIC CODES

The following information on available software and suppliers of software was accurate at the time of going to press. However, the names of software developers/suppliers change, or they are bought out by another company, or even the name of the software may change. Some suppliers of complex software have bought up software packages from other developers and combined them into a hybrid with a different name. Internet sites exist that describe both commercially available software and freeware, and these, or the suppliers' own websites, should be consulted to find out what is currently available.

Also the application and capability of the programs are based on the supplier's information. The programs have not been evaluated with the exception of GEMACS, FEKO, and 4nec2.

12.4.4.1 NEC

The NEC program allows the modeling of EM radiation, antenna performance, radar cross sections with accurate field emission, and field source computation. The NEC program uses MOM techniques for the numerical solution of IEs for the currents induced on a conductive structure by sources or incident fields. The outputs include current distribution on wires, surfaces, coupling between antennas, and near–far-field magnitudes. There are at least four versions available from Nittany Scientific, 1700 Airline Highway, Suite 361, Hollister, CA 95023-5621; PH 408-634-0573; email: sales@nittany-scientific.com. NEC with NEC-2 emerged in 1981 and NEC-4 appeared in 1992. NEC-2 is the highest version of the code under public domain. NEC-2 is a widely used 3D code developed at Lawrence Livermore National Laboratory many years ago. NEC-2 is particularly effective not only for analyzing wire-grid models but it also has some surface patch–modeling capability. NEC-2 is available from Ray Anderson's Unofficial NEC Archives. NEC-4 remains proprietary with Lawrence Livermore National Laboratory and the University of California. It requires a separate license for use and can be purchased from Nittany Scientific who produces a 32-bit Windows version and Roy Lewallen who produces EZNEC Pro that has an option for NEC-4. The program costs approximately $300, and for U.S. citizens an application form must be filled in. For non-U.S. citizens, a customer screening form must be filled in.

Some of the Windows-based versions are NEC-Win Basic, Nec-Win Plus, Nec-Win Pro, of which 4nec2 has excellent graphics, and GNEC. GNEC is available only to licensed users and is available from Nittany Scientific! Nec-Win Basic is designed for beginners, with each of the other versions enhanced by additional commands and geometry visualization, and so on. For more information on the relative capability of all of these versions, contact the supplier online at http://www.nittany-scientific.com. NEC-Win Basic includes the following features:

Simple user interface for easy data entry and configuration
Viewing or printing of graphical representations of antennas
Azimuth and elevation plot generation
The ability to use multitask in the Windows environment

NEC-WIN Basic combines NEC2 with Microsoft Windows and is composed of three parts: NEC-Vu, NEC-Plot, and NEC Geometry. NEC-Geometry is where the antenna is built. The coordinates of the wire are entered along with the specific data concerning frequencies and ground parameters. The following features are included:

Transmission lines
Linear or multiplicative frequency sweeps
Networks
Predefined or define-your-own wire conductivity and gauge
Translate/rotate/scale antennas
Predefined or define-your-own grounds constants
Series/parallel RLC or complex loads
Sommerfeld ground
Voltage/current sources
Linear/radial cliff or radial wire screen grounds
Tabular data for VSWR and input impedance

12.4.4.2 MININEC

MININEC was written because amateur radio enthusiasts who wanted to design antennas did not have access to large-frame computers on which to run NEC. Expert for Windows is ideal for the novice, student, or hobbyist. Expert MININEC Professional for Windows is suitable for experienced students, hobbyists, or professional engineers, whereas Expert MININEC Broadcast Professional for Windows is a tool for advanced students or professional broadcast engineers. The Expert MININEC series is available from EM Scientific Inc. For more information, contact http://www.emsci.com/mininec.htm.

The Expert MININEC Series of computer programs is for the analysis of wire antennas using Microsoft Windows and is available on compatible IBM personal computers (PCs). The accompanying texts provide descriptions of the software and present the theory with relevant examples and validation data. The text for Expert MININEC for Windows is entirely online. Using the MOM, the Expert MININEC Series solves an electric field formulation for the currents on electrically thin wires using a Galerkin procedure with triangular basis functions. This formulation results in an unusually compact and efficient computer algorithm. Radiation patterns, near field, charge distribution, impedance, and other useful parameters are computed from the current solution.

The user interface to Expert MININEC is through Microsoft Windows. Input data screens provide spreadsheet-like entry in individual windows. Output products are displayed in both tabular and graphics forms. The integrated graphics of the Expert MININEC Series include the following:

3D geometry displays with rotation, zoom, and mouse support
3D current and charge displays
Linear, semilog, and log–log plots of currents, coupling, near fields, impedance, and admittance
Linear and polar pattern plots

The actual computational algorithms are implemented in FORTRAN for greater speed and make maximum use of available memory to set array sizes. The formulation has been changed from earlier versions of the MININEC formulation to use triangular basis functions. This results in greater accuracy. The short segment limit is machine accuracy. Square loops and Yagi antennas may be solved with confidence. In addition, a Fresnel reflection coefficient approximation improves the calculation of currents in the vicinity of real ground.

12.4.4.2.1 System Requirements

There are a few minimum requirements to run the Expert MININEC Series. The computer must be an IBM-PC or compatible.

In the display settings, the small font size is recommended.

12.4.4.2.2 Modeling Process

The Expert MININEC Series modeling process has five principal steps:

1. Geometry description definition
2. Electrical description definition
3. Model validation
4. Solution descriptions definition
5. Output display

A wire is subdivided into segments, with currents expanded as triangles centered at adjacent segment junctions. The endpoints of a wire have no triangles. If a second wire is added to the model, the second wire is subdivided into segments, with currents expanded as triangles as in the case of the first wire. In addition, if the second wire is attached to the first wire, a triangle is automatically located at the attachment end. Half of the additional triangle extends onto wire 2 and half onto wire 1. The half of the triangle on wire 1 assumes the dimensions (length and radius of the half segment of wire 1), while the half of the triangle on wire 2 assumes the dimensions of wire 2. Wire 2 overlaps onto wire 1, with a current triangle at the junction end. Additional wires may also overlap onto wire 1. It can be shown that for a junction of N wires, only $(N-1)$ overlaps with associated currents are required to satisfy Kirchhoff's current law. The convention in the MININEC Professional Series is that the overlap occurs onto the earliest wire specified at a junction. A wire junction is established whenever the user-defined coordinates of a wire end are identical to the end coordinates of a wire previously specified.

As mentioned previously, the choice of the number of segments is critical to the validity of the computation. It was suggested that segments around 0.02 wavelengths are a reasonable choice.

Since the conductance and susceptance values both converge as the number of segments increases, the convergence test can be used to determine the accuracy that can be expected for a given segmentation density (i.e., the number of unknowns per wavelength of wire).

Expert MININEC modeling geometry constructs include the following:

Cartesian, cylindrical, and geographic coordinate systems
Meters, centimeters, feet, or inches selection
Straight, helix, arc, and circular wires
Wire meshes
Automated canonical structure meshing
Node coordinate stepping
Symmetry options
Rotational and linear transformations
Numerical Green's function
Automated convergence testing

Electrical description options include the following:
Free space, perfect ground, and imperfect ground environments
Frequency stepping
Loaded wires
Lumped loads
Passive circuits

Transmission lines
Voltage and current sources
Plane-wave source excitation

Solution description options include the following:
Near fields
Radiation pattern
Two-part coupling
Medium-wave array synthesis

Output products are displayed in both tabular and graphics forms. The integrated graphics of Expert MININEC include the following:

3D geometry displays with rotation, zoom, and mouse support
3D currents, charges, and patterns displays
Linear, semilog, and log–log plots of currents, coupling, near fields, impedance, and admittance
Smith chart plots of impedance and admittance
Linear and polar pattern plots

As a summary, Expert MININEC solves the following:
Currents and charges on wires (peak or rms)
Impedance, admittance, S11, and S12
Effective height and current moments
Power and voltage losses
Multiport (antenna-to-antenna) coupling
Near electric and magnetic fields
Radiation patterns (dBi or electric fields, power or directive gain)
Medium-wave array design
Auxiliary calculations of ground wave, stub matching, and tower footing impedance

It has been shown from the selected examples that MININEC gives comparable results to NEC. This is not a complete picture of the comparison of these codes, but it gives the reader a glimpse of the results to be expected. A more thorough analysis of MININEC shows that for a wide variety of problems, MININEC and NEC provide comparable results.

Expert MININEC for Windows: ideal for novices, students, and hobbyists.
Expert MININEC Professional for Windows: suitable for experienced students, hobbyists, and professional engineers.
Expert MININEC Broadcast Professional for Windows: a tool for advanced students and professional broadcast engineers.
Expert MININEC for Windows
File handling
    Open, save, and delete problem files
    Print setup
Geometry constructs
    Geometry points defined in meters, centimeters, feet, inches, or degrees
    Geometry points iteration
    Environmental options, including free space, perfect ground, and real ground straight wires
Electrical constructs
    Frequency stepping
    Ground options
    Lumped loads

Loaded wires
Voltage/current sources
Solution space
    Currents, charges
    Impedance/admittance
    Near electric and magnetic fields
    Radiation pattern
Diagnostics
    List of current nodes
    Geometry guidelines
    Definition evaluation and summary
    Online and context-sensitive help
    Online tutorial
Run options
    Current
    Near fields
    Radiation patterns
    Frequency iteration
Display types
    Text and interfaces to most available spread sheets
    3D geometry, 3D currents/charges, 3D patterns
    Linear plots
    Polar plots for patterns
    Smith chart
Display options
    Admittance, impedance, SI 1, S12
    Effective height and current moments
    Power and voltage losses
    Current/charge on wires (peak or rms)
    Near electric and magnetic fields
    Radiation patterns
Problem limits: 1250 unknowns and 500 wires

12.4.4.3 MMANA-GAL basic

MMANA-GAL basic is an antenna-analyzing tool based on the MOM, which was introduced in MININEC version 3. The BASIC source code of the computation engine is published as a PDS in MININEC. The program uses the MININEC-3 engine modified by Alexandre Schewelew, DL1PBD, and is written in C++. MMNA-GAL is freeware and is co-written by Alexandre Schewelew, Igor Gontcharenko FL2KQ, and Makoto Mori JE3HHT who also owns the copyright.

12.4.4.4 4nec2

The 4nec2 program written by Arie Voors is available on the website free of charge. It has a 3D viewer and was used for many of the EM simulations that provide the data presented in this book.

The geometry includes wires, surface patch, multiple patch surface, cylindrical structures, helix, ground plane, finite ground, perfect ground, and Sommer Norton wire ground. The load may be a series RLC with up to any two blank, parallel RLC with up to any two blank, parallel RLC with input parameters per unit length, impedance with resistance and reactance (when resistance and reactance are used, the impedance does not automatically scale with frequency), and wire conductivity.

It can generate a smith chart and provides an overall geometry check as well as individual checks on segment length, segment radius, segment length/radius, radius at junction, length at junction, and surface patch area.

12.4.4.5 GreenMentor Graphics (Formerly Zeland Software), Wilsonville California

IE3D is a full-wave, MOM, 3D, and planar EM simulation and optimization package for circuit and antenna applications. The original program has been integrated with the Mentor Graphics HyperLynx® product suites and will analyze, in full, EM details.

12.4.4.6 GEMACS Hybrid Solution for Large, Complex Structures

GEMACS allows the modeling of EM radiation and scattering, antenna performance, and radar cross section with accurate field predictions, including sources, reflections, and scattering. Outputs include current distribution on wires, surfaces, coupling between antennas, far/near fields, antenna terminal characteristics, scattering on conductive surfaces, cavity behavior, and aperture coupling.

GEMACS uses the MOM physics model, which is supplemented by GTD and the FEM. Although still referred to as GTD, the formulation is the more accurate UTD. GEMACS models complex structures by a hybrid using combined MOM/GTD/FEM techniques. These three techniques are implemented in GEMACS as follows.

GEMACS inputs are in two categories: the command language, which directs the program execution, and the geometry language, which describes the geometrical properties of the structure being analyzed. The command language controls the analysis used and describes the electric field, voltage or antenna excitation, and the reflected/scattered electric field. The incident, scattered, or total field may be computed for both near- and far-field conditions as a magnitude in a spherical, cylindrical, or Cartesian coordinate system. Two types of structure geometry are defined: thin-wire, wire mesh, or patch geometry used in the MOM analysis and the plate or cylinder geometry used in the GTD analysis. A mix of these is allowed in the MOM/GTD hybrid analysis. In the MOM analysis, the structure under consideration is divided up into a group of thin-wire segments and/or surface patches. The choice of current distribution on these segments and patches is made, in the GEMACS MOM implementation specifically, in order to ease the numerical integration burden. It is now possible to compute the field at any point in space due to the current on any object of the structure, and the contribution from all parts of the structure may be added by superposition to form the total field. The currents on the structure may be due to excitation by plane or spherical waves. The MOM technique is used for electrically small objects. Because the current between the centers of wire segments are approximated by an expansion (basis) function and because near-resonant segments must be avoided, for the sake of accuracy, physical limitations are placed on the segment length and the mesh area. Generally these limitations are as follows:

Mesh circumference <0.5 λ

Mesh length <0.1 λ, although good results have been achieved with long subsection lengths to 0.25 λ and with square mesh to 0.14 λ

Adjacent subsections to differ in length by no more than a factor of 2

Angles between subsections to be no less than 20°

Wire radii <0.001 × subsection length

The wire subsections used in the MOM geometry need not be perfect conductors but may be loaded by either a fixed impedance (as a function of frequency), lumped loads, series or parallel RLC networks, and finite segment conductivity. Quasi-aperture coupling occurs between the apertures in a wire mesh. Patches may be used to reduce aperture coupling and to simulate ground planes, but patches are modeled as conductive surfaces only. The wire segment is specified by coordinates (X_1, Y_1, Z_1) that define one end of the wire segment and (X_2, Y_2, Z_2) that define the other end of the wire segment. The position of the wire segment in the coordinate system allows the construction of complex structures. The geometrical theory of diffraction is a high-frequency ray optics approach that has been extended to include diffraction effects from surface discontinuities and waves creeping around smooth structures. The basic geometry objects that can be used in GTD structures are those for which reflection and diffraction coefficients exist and for which it is possible to carry out the ray tracing of the optics approach. The advantage of GTD is that large objects equal to or greater than a wavelength may be modeled in bulk instead of requiring wire gridding as with MOM. Rather than calculate geometry element currents to obtain field pattern data (the MOM approach), the scattered fields are obtained directly from the sources and geometry by tracing all GO paths from the sources to the field points. The waves are then reflected and diffracted as they follow these paths, from the surfaces, edges, and corners of the elements.

The GTD techniques are used when detailed structure elements are not important and when scattered fields are the only quantities desired. While it is theoretically possible to obtain any level of detail with enough GTD elements, in practice the user is forced to limit either the number of GTD elements or the number of physical interactions that are to be considered. The GTD analysis does not include surface waves, although creeping waves are modeled. A large number of scattering mechanisms are possible, and with GTD, these multiple scattering effects must be explicitly included in the analysis.

With the MOM/GTD hybrid configuration, it is possible to combine the best features of MOM and GTD into a hybrid methodology. The scattering structure is modeled with both MOM and GTD geometry objects. This allows the user to specify detail (where required) by using MOM objects but does not require that the entire geometry be gridded. Hence, the matrix size and number of unknowns are minimized. By use of a hybridized finite differences algorithm with the GTD and MOM formulation, both the interior and exterior problems in the presence of each other are solved. This means that the external field, which may couple through an aperture, excite a MOM or GTD structure, and reflect back out of the aperture, should theoretically be capable of being computed.

If a complex structure such as an aircraft requires a long time to execute, then the alternative is to model a much simpler structure, in which case the fidelity of the analysis could be severely compromised. The latest high-speed PCs will almost certainly reduce the time required, probably to days for a complex structure, but the earlier versions of GEMACS used very little RAM and made extensive use of the hard drive, unlike the latest Windows version. In many cases, for complex structures such as ships and aircraft, a brass scale model complete with transmitting and receiving antennas provides a faster solution when compared to computer analysis.

The latest version of GEMACS7.2 is available from Applied Research Associates and is available for desktop PCs, shared memory computers, and distributed memory computers using MPI (such as Linux clusters and office networks). The latest version of GEMACS incorporates new features, improved support tools, and a higher level of stability.

Due to export control restrictions, GEMACS is currently only available for purchase within the United States and Canada. For Canadian companies working on U.S. military or NASA programs, a Militarily Critical Technical Data Agreement may have to be filled in and submitted to the U.S./Canada Joint Certification Office.

Aperture coupling in GEMACS should be achieved using a FD hybrid; however, in all versions from the earliest to the latest, this function fails when computing the penetration of a field through an aperture into a closed enclosure. However, GEMACS correlated to measurements better than FEKO when predicting a horizontally or vertically polarized wave coupling around a cylinder, as described in Section 12.5.

The task of specifying the geometry and analyzing the output of GEMACS is not easy.

XGAUGE is available and provides a graphic display of input geometry and output fields and currents as color-filled contours. The graphic display capability includes true 3D wire frame and solids, wire-to-patch-to-plate conversions, up to 16 colors (EGA), and pan, zoom, rotate, and translate. The structure may be modeled using GEMACS geometry for input and may be color coded by ID, TAG, Segment. XGAUGE enables model replication and provides GEMACS-readable outputs. The output display includes color mapping of wire/surface currents, color-filled contours for detail, polar and rectangular plots, overlays, and near- and far-field patterns and converts GEMACS outputs to displays.

Commercial customers also receive a copy of SmartView, an enhanced 3D visualizer for geometries, surface/wire currents, and patterns.

The Ultra Corporation, under contract with the Air Force Research Laboratory, has developed a parallel version of GEMACS. The parallel version incorporates recent advances in high-performance computing and enables scalable run-time speedup on distributed processors. The base GEMACS code is highly portable FORTRAN. The parallel version uses the message-passing standard MPI and runs on the widest possible family of parallel machines, from massively parallel supercomputers to networks of workstations or PCs. The code modifications for parallelization are to the structure of the software system, not to the underlying physics. In the low-frequency regime, the computation of the interaction and Green's function matrices are distributed, while a parallel ray-tracing algorithm is used for the high-frequency case. The hybrid situation employs both techniques. The run time is reduced from days to hours while maintaining the accuracy of the original code. The modified GEMACS code can now solve larger, more complex problems and handle higher orders of interactions.

12.4.4.7 FEKO

The numerical techniques available in FEKO are UTD, MLFMM, MOM, FEM PO, FDTD, and GO, and RL-GO (ray launching GO). Hybrids using two of the techniques are possible and these include MOM/PO, MOM/UTD, MOM/GO, MLFMM/PO, MLFMM/GO, MLFMM/FEM, and MOM/FEM. These hybrids are useful for either geometrically complex or electrically large systems. Also the type of analysis can determine the formulation used. An example of a hybrid problem is the use of MOM to model a horn antenna and the PO to model the antenna close to a metal surface.

Surfaces can be PEC or finitely conducting or dielectric.

The finite conductivity surface can be discrete elements R, L, C, Z, or distributed load R, L, C, Z.

There are numerous ways of modeling dielectric objects in FEKO and three of which are the MOM surface equivalence principle, the MOM volume equivalency principle, and the FEM.

There are also special methods to account for thin dielectric, dielectric coating, dielectric half spaces, spheres, and planar multilayer substrate.

One FEKO feature is the capability of modeling a windscreen that allows the inclusion of an arbitrarily shaped windscreens containing multiple dielectric layers. This method solves an imbedded or printed glass antenna without a significant increase in the computational resources.

The coupling generated by an external antenna through an aperture to the inside of a metal enclosure is typically performed using FEM/UTD/PO hybrid methods. In a private communication from Gobinath Gampala from FEKO, USA, and another from Ulrich Jakobus of FEKO, South Africa, they explain that FEKO instead uses the MLFMM to model the antenna and to calculate the currents on the external surface of the enclosure. The internal and external regions are decoupled by using the surface equivalence principle. This generates separate regions (inside and outside of the enclosure) and the fields incident on one would not be present on the other in the formulation, only through the aperture coupling. This method of aperture-coupling analysis is compared to the measurements of an aperture on a cylinder in Chapter 10 with a good correlation.

Some of the applications are as follows:

Wire antennas
Microstrip antennas
Aperture antennas
Reflector antennas
Windscreen antennas
Conformal antennas
Broadband antennas
Array antennas

Lens antennas
Radomes
Antenna placement (radiation pattern)
Antenna placement (coupling)
Biomedical
RADHAZ zones
Periodic structures: FSS, metamaterials
Scattering with plane-wave source
Scattering with localized source
EMC/EMI shielding and coupling
Propagation environment
Cable bundle coupling
Waveguide components
Connectors
Microstrip circuits

The geometry is described by segments (a short length of wire), a cuboid, tetrahedron, polygon, vertex, node (the point where two segments join), a connection point (where a segment is joined to a triangle) an edge, flares, cylinders, and cones. FEKO was the first commercial program to include meshed solid structure versus wires. Conducting surfaces are subdivided into triangles, and wires into segments. For the FEM, the mesh is based on tetrahedral volume elements. Dielectric solid is divided into triangles and thin dielectric sheets are meshed into triangles located along the middle of the sheet. Metal surfaces that employ UTD are not meshed.

GO can be translated using rotate, translate, mirror, align, or scale.

The program flow is as follows:

Models are generally constructed in CADFEKO (used to create and mesh the geometry); EDITFEKO is used to construct advanced models. Next the user runs PREFEKO (processes the model and prepares the input file). The solution kernel is FEKO and the results are viewed from POSTFEKO.

POSTFEKO is also used to visually confirm the correctness of the model before starting a potentially time-consuming FEKO solution. It is used to display the results of the FEKO simulation either as a function of position in a 3D window (with or without the geometry) or in arbitrary 2D graphs. POSTFEKO is updated real time, displaying updated results as they become available.

Multiple models, with their geometry and results can be displayed in a single POSTFEKO session.

12.4.4.8   MefistoTM

Faustus Scientific supplies EM time-domain simulators. The MEFISTo-3D ProTM features text-based import and export, multi-thread architecture, SPICE links, SPICE TLM cosimulation, fully parameterized graphics user interface (GUI), time-domain diakoptics, on-the-fly visualization, time reversal capability, and metamaterial modeling in 32- and 64-bit versions. The MEFISTo-3D NovaTM is a lite version of 3D-Pro. MEFISTO-2D ClassicTM is a basic 2D TLM Solver.

12.4.4.9   APLAC

APLAC, a program for circuit, system, and electromagnetic FDTD simulation and design, is a joint development of the Circuit Theory Laboratory of the Helsinki University of Technology, Aplac Solutions Corporation, Nokia Research Center, and Nokia Mobile Phones. The main analysis modes are of circuits, but it includes an electromagnetic FDTD simulator for solving 3D field problems independently or as part of a circuit design.

12.4.4.10   EZ-EMC

From EMS-PLUS

sales@ems-plus.com

EZ-EMC is a moderately priced full-wave 3D analysis software based on FDTD method with GUI.

Time domain or frequency domain results
Electric and magnetic fields
Applications
Shielding
Gasket performance
Cavity and shielded box resonances
Heatsink emissions analysis
Grounding analysis
Board emissions
Ferrite filter analysis
Antenna applications
Test site analysis (AOTS, GTEM, anechoic chamber, and more)

XFdtd is 3D full-wave finite difference time domain (FDTD method).

XGtd is an accurate model for radar cross section (RCS) analysis based on the method of equivalent currents and physical optics.

Both programs are available from REMCOM.

XGtd has the following characteristics:

Software with a powerful GDU
Scattering amplitudes and excess times of arrival for RCS ray paths
Near–far zone output types: electric field magnitude and phase, complex impulse response, and Poynting vector
Visualization of far-zone ray paths
Visualization of edges considered by the calculation to be part of a curved surface
Full 3D ray tracing, based on UTD with corrections for finite conductivity materials, including thin coatings
Creeping wave diffraction computation for high-fidelity computation of fields in the shadow zones
Compute far-zone radiation patterns including effects of scattering from electrically large objects
Calculate Doppler shift due to motion of transmitter or receiver
Analyze antenna coupling and interference at multiple frequencies
Import complex antenna patterns from Remcon's XFdtd or other sources
Model wideband or narrowband pulses
Generate time or frequency domain output
Analyze propagation in anechoic chamber
Model transmitters at point sources, antenna arrays, distributed sources, and plane waves
3D visualization
Create multiple materials plots, multiple dielectric layers, thin coatings, or radiation absorbent material

12.4.4.11   Penn State FDTD Code

This is a public domain FDTD code developed by *R*. Luebbers and K. Kunz described in their book, *The Finite Difference Time Domain Method for Electromagnetics* published by the CRC Press.

12.4.4.12 EMAP5

This is a hybrid FEM/MOM code designed primarily to simulate EMI sources at the printed circuit board level. EMAP5 is a full-wave EM field solver that combines the method of moments with a vector finite element method. It employs FEM to analyze 3D volumes and uses MOM to analyze the current distribution on the surface of these volumes.

The two methods are coupled through the fields on the dielectric surface. EMAP5 can model three incident plane waves, voltage sources (in the MOM region), and impressed current sources (in the finite element region). The goal of the code is to efficiently model heterogeneous and arbitrarily shaped dielectric bodies attached to one or more conducting bodies. A typical example of such structures is a printed circuit board attached to long wires or cables.

EMAP5 is a 3D numerical EM modeling code developed at the University of Missouri, Rolla. The code can be freely downloaded from the website http://www.emclab.mst.edu/emap.html.

In one case study presented in Reference [8], a structure shown in Figure 12.14 comprised a dielectric volume V_2, which has electrical properties (ε_2, μ_s). The dielectric volume and conductive volume are illuminated by either incident H^i and E^i fields or impressed sources J^{int} and M^{int}. A conductive volume V_3 is enclosed by a conductive surface S_c. The fields within V_3 vanish. V_1, which denotes the volume outside of V_2 and V_3, is assumed to be free space and hence has electrical properties (ε_0, μ_0), (ε_2, μ_2). $(\mathbf{E}_1, \mathbf{H}_1)$ and $(\mathbf{E}_2, \mathbf{H}_2)$ denote the electric and magnetic fields in V_1 and V_2, respectively. The unit normal vectors for S_2 and S_c are defined pointing outward. Tetrahedral elements are used to discretize the dielectric volume V_2. The FEM and the MOM equations are coupled through $\{E_d\}$ and $\{J_d\}$. By solving the two equations, all fields within the FEM region and the surface equivalent currents can be obtained.

The EMAP5 software package includes three major components: SIFT5, EMAP5, and FAR. EMAP5 is developed for research and educational use. It does not have a sophisticated mesh generator or graphic visualization tools. SIFT5 can generate simple meshes for users. EMAP5 is the FEM/MOM field solver. FAR is the far-field calculator. The Standard Input File Translator Version 5 (SIFT5) is designed to generate input files for EMAP5. SIFT5 reads a text file in the SIFT format. Users can describe the structure of interest by using 11 keywords that are discussed in the EMAP5 software.

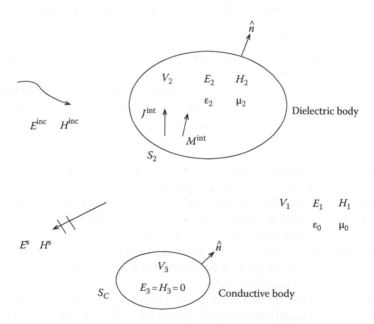

FIGURE 12.14 A dielectric objective and a conductive object illuminated by H^i and E^i fields or impressed sources J^{int} and M^{int}. (From Yun Ji, M.W., Ali, T.H., and Hubing, T., EMC applications of the EMAP5 hybrid FEM/MOM code, *IEEE*, 1998 EMC Symposium Record, 1998. © 1998, IEEE.)

SIFT5 detects inconsistent input parameters in input fields and automatically prompts users to correct them. EMAP5 reads a file generated by SIFT5. EMAP5 will print out fields within areas specified by the *sif* file, one or more output files. All equivalent surface currents *J* and *M* will be printed.

FAR is a program used to calculate the far-field radiation pattern far fields are calculated equivalent surface currents *J* and need two input files. One is the file generated by SIFT5 and the other is the default output file generated by EMAP5. The FAR program will prompt the user to input the following parameters:

The observing distance *R* from the structure, in wavelengths. Usually, the value of *R* should be higher than 20.
The observing θ interval, in degrees, in spherical coordinates.
The observing φ interval, in degrees, in spherical coordinates.
When the MOM part of the code is implemented, EMAP5 must know how triangles are linked to each other. In addition, a rule defining current direction must be specified.
EMAP5 supports the following three kinds of sources:

1. Voltage sources on metal patches
2. Plane-wave sources
3. Current sources within the FEM region

EMAP5 must know whether users need the default output. In addition, EMAP5 must know how many other output files users want.

A nonuniform mesh must be used to discretize some structures, for example, when the width of a trace is very small compared to its length.

Although EMAP5 is a FEM/MOM code, it can model configurations that require only one method to analyze. In this case, only the MOM portion of the code is employed.

A dipole with a source located in a dielectric source is presented as an example. With the dielectric slab set to a relative permittivity of 1.0, the model is a half-wavelength dipole in free space. This problem requires 80 MB of memory and 20 h to run on a Sun Ultra workstation.

12.4.4.13 Students' QuickField

Students' QuickField™ is a 2D finite element simulation package solving electromagnetic, thermal, and stress design simulation with coupled field analysis.

Student's Quickfield is a free but full-featured analysis program with limited number of mesh nodes. If the accuracy or complexity requirements exceed the limitations of the students' version, the capabilities of the professional version should be explored.

12.4.4.14 HFSS

The ANSYS HFSSTM is a complete solution for EM modeling of 3D structures using the FEM tool. It can be used for RF and microwave as well as signal integrity designs.

Based on the geometry, material properties, and the frequency range, HFSS automatically generates the most appropriate, efficient, and accurate mesh for the simulation.

In operation, the HFSS analysis engine subdivides arbitrarily shaped 3D structures into small tetrahedron-shaped elements. Electric fields are represented with higher-order tangential edge-element basis functions within each tetrahedron. The interactions between basis functions are computed using differential forms of Maxwell's equations. These computations yield linear equations that can be represented in a sparse matrix. The solution in turn yields an approximation of the fields as well.

The EM software automatically decreases the size of the finite elements where strength is greatest for increased resolution and accuracy. An improved meshing algorithm uses memory and central processing unit (CPU) resources more practically for reduced computing time and more efficient memory consumption, avoiding the "out-of-memory" messages of previous versions.

To design a structure, an operator enters the desired geometry, adds input and output ports, and assigns material properties (such as the dielectric constant of substrates). The software employs adaptive mesh-refinement techniques to solve to a user-specified level of accuracy. The desired accuracy is a function of computing/solving time, with higher levels of accuracy requiring more CPU time. Display functions provide animated images of EM fields, surface current flow, antenna polar patterns, Smith charts of S-parameter information, and tabular results.

The structure can be a layout design in Cadence, Mentor Graphics, Synopsis, Zken or Altium or mechanical in ProE, STEP, or IGES, which is inputted to ANSYS HFSS via the AnsoftLinks program. The ANSYS HFSS S, Y, Z parameters and fields can be sent to the Ansoft Designer. Another output can be sent to SPICE or RF Tools.

The IE solver is an effective add-on solver for radiation and scattering studies of large, mostly conducting structures. It employs the MOM. Using the HFSS interface, IE shares geometry, material, and certain critical solver technology with HFSS. The E solver also allows the creation of HFSS models that use a hybrid FEM–IE methodology to solve large problems.

HFSS can design and evaluate signal integrity and EM interference in connectors, transmission lines and vias on printed circuit boards, and high-speed components.

HFSS transient is a 3D full-wave transient or time-domain EM-field solver based on the discontinuous Galerkin time-domain method. Applications are time domain reflectometry, ground-penetrating radar electrostatic discharge, and lightning.

ANSYS 16 delivers an integrated and comprehensive multiphysics simulation platform called AIM which is a further realization of ANSYS Workbench.

12.4.4.15 3D MMP

3D MMP is a FORTRAN program by Ch. Hafner of the ETHZ (Swiss Federal Institute of Technology) in Zurich, Switzerland.

The 3D MMP code is described in the book *Electromagnetic Wave Simulator* published by John Wiley & Sons (Christian Hafner and Lars Bomholt, which is out of print.) This version does not contain an eigenvalue solver (for guided waves on cylindrical surfaces). It is restricted to EM scattering (including waveguide discontinuities and similar problems that can be formulated as EM scattering) and antenna design. 3D MMP has a graphic front end for PCs that can be used as a platform for implementing and testing finite difference schemes and other iterative procedures. The 3D MMP includes features for computing gratings and biperiodic structures, multiple excitations, advanced graphics, and so on. An eigenvalue solver for guided waves and resonator computations may still be available. The executable code compiled for Windows NT (Upgrade 3) and text file can be downloaded for free and is available Christian Hafner's server, ETH Zurich.

12.4.4.16 InCa3D

InCa3D from CEDRAT is a conductor impedance and near-field/magnetic-field simulation software. It uses PEEC based on low-frequency formula to extract the electric parameters (resistance, inductance, and self and mutual inductance) from geometries. It assembles R, L, and C values in a matric for solving AC or steady-state Kirchoff's equations. It uses semi-analytical computation by Gauss points accelerated by fast multipole. It takes into account skin and proximity effects and is adapted to model every kind of conductor inside a structure.

It evaluates parasitic phenomena that impact system performance and includes common-mode capacitances and loop inductance.

CAD imports can be CATIA, Pro/Engineer, IGES, STEP or Gerber files and outputs to Portunas, SABER, Modelica, SPICE, and VHDL-AMS. The output data are RLC parameters, coupling with circuit simulators, digital integrity and radiation, current distribution, magnetic flux, Laplace forces, and Joule losses.

It converts time-domain analysis to frequency using FFT.

12.4.4.17   ELECTRO 2D/RS

ELECTRO 2D/RS from Integrated Engineering Software combines the advantages of the BEM and FEM field solvers. It calculates electric field strength, force, torque, transmission-line parameters, and capacitance. Applications are as follows:

Transformers
Insulators, bushings, grounding electrodes
Micro electromechanical systems
High-voltage shields
Power transmission lines
Telecommunications cables
Microstrip and ICs
High-voltage switchgear
Xerographic process modeling
Parts and assemblies subject to electrical fields

12.4.4.18   EMPIRE XPU

EMPIRE XPU from IMST gmbh is based on the FDTD method for 3D modeling. The company claims that due to the adaptive on-the-fly code generation, it exhibits the fastest simulation engine known today.

For structure definition, a GUI is included and several structure import and export formats are supported. Its applications include the following:

Planar, multilayered, and conformal circuits
Components
Antennas
Multi-pin packages
Waveguides
sI/EMC problems

Time signals, scattering parameters, and field animations are generated for a broad frequency range within only one simulation run.

12.4.4.19   AN-SOF

According to the supplier Golden Engineering, AN-SOF is based on an improved version of the MOM. It is designed for the modeling and simulation of antenna systems and general radiating structures. The modeling of the structure can be performed by means of the AN-SOF-specific 3D CAD interface.

AN-XY Chart is for plotting two related quantities, that is Y versus X.

These include frequency-dependent current, voltage, impedance, reflection coefficient, VSWR, radiated power, consumed power, gain, radiation efficiency, and radar cross section. The current distribution on metallic surfaces can also be plotted as a function of position. Besides 2D radiation patterns can be represented for a near or far field as a function of angle or distance.

AN-Smith Chart for the representation of impedances and admittances. Reflection coefficient and VSWR are also shown.

AN-Polar Radiation and scattering patterns versus azimuth or zenith angles can be represented in this polar diagram, from which the width of radiation lobes and front-to-back ratios can be obtained. The represented field quantities include power density, directivity, gain, normalized radiation pattern, total electric field, field polarized components, and RCS.

AN-3D Pattern—A complete view of the radiation and scattering properties of a structure can be achieved with a fully angle-resolved pattern. The task can be accomplished with an AN-3D pattern that implements colored mesh and surface for the visualization of radiation lobes. The represented quantities include the power density, normalized radiation pattern, directivity, gain, total field, linearly and circularly polarized field components, and RCS. Linear and decibel scales are available.

12.4.4.20 EMA3D

EMA3D for EMC is from ElectroMagnetic Applications Inc. EMA is an FDTD solver for calculating lightning, DO-160, EMI from a slot in a box, how fields couple to cables, and how much interference can escape a box.

12.4.4.21 CST Studio Suite

CST Studio Suite from CTS Computer Simulation Technology is variously described as using central finite difference techniques and hence the CST, FDTD, or finite integral technique (FIT). The program is described as capable of simulating individual components and full systems.

12.4.4.22 CST Microwave Studio (CST MWS)

CST Microwave Studio from CTS Computer Simulation Technology is a 3D EM tool for the evaluation of high-frequency components such as antennas, filters, couplers, planar and multilayer structures, and signal integrity and EMC effects.

Beside the flagship module, CST offers the broadly applicable time-domain transient solver and the frequency-domain solver. Filters for the import of specific CAD files and the extraction of SPICE parameters are supplied. In addition, CST MWS can be embedded in various industry standard workflows through the CST user interface.

12.4.4.23 FEST3D

This program from Aurora software is able to analyze different kinds of passive RF structures in waveguide technology. Basically, FET3D is based on the IE technique efficiently solved by the MOM. Additionally, the boundary integral-resonant mode expansion method is employed for extracting the modal chart of complex waveguides with arbitrary cross section. Aurora software claims that FEST3D is able to simulate complex passive devices in extremely short times (in the order of seconds or few minutes), whereas general-purpose software (based on segmentation techniques, such as finite elements or finite differences) can spend several hours for the same calculation.

The further claim is that unlike mode matching techniques, the EM algorithms in FEST3D minimize the problems of relative convergence leading to more confident results.

When many modes are required for an accurate analysis of the component, the technique used in FEST3D results in faster computation time per frequency point.

12.4.4.24 Wave3D

Wave3D from CEMWorks is a MOM solution of the surface-volume-integral equation. Wave3D solves complex problems of electromagnetics for general metal dielectric composite models. This formulation allows the solution of large-scale antenna, scattering and EMC problems including inhomogeneous dielectric structures through acceleration by the broadband fast multipole method.

The numerical solution offered by Wave3D discretizes metal surfaces with triangular mesh while dielectric surfaces are meshed with tetrahedrons.

Applications are as follows:

Antenna analysis and positioning
Absorption in dielectric materials
Absorption in highly inhomogeneous biological tissues
Scattering by complex composite targets
Analysis of dielectric layers such as those found in radome and conformal antenna design

Wave3D uses the fast multipole method (FMM) and the method retains efficiency from DC to X-band and beyond.

The extension of FMM allows efficient EM analysis of large-scale models at low and high frequencies. Excitation sources can be dipoles, plane wave, and internal and external port driving.

Near-field analysis is available in Cartesian and cylindrical and spherical coordinates. Far-field analysis displays include full 3D visualization of scattered fields as well as RCS.

12.4.4.25 WIPL-D Pro

The WIPL-D uses MOM with quadrilateral mesh and high-order basis functions. According to the supplier, these yield highly accurate results of the distribution of currents over surfaces, near-field results, far-field pattern, and circuit parameters.

For electrically large structures, special techniques are applied in order to reduce memory and speed up simulation. Smart reduction option is used for antenna placement setups, and according to the supplier they can decrease memory up to 10 times.

Shadow reduction enables the user to explicitly define parts of the model where the order of approximation will be reduced, such as structures below platforms and bottom parts of vehicles and airplanes. Other programs allow the selection of mesh size to achieve similar results.

Other features include the following:

Calculation of near and/or radiation pattern based on the imported current distribution of selected wires and plates.

Extended limits for low-frequency analysis, and the analysis of small details in complex structures, parts 1/1,000,000 wavelengths in size, is properly handled.

Improved multicore CPU parallelization.

Parallelization of the excitation and near-/far-field calculations by field generators, which enable usage of up to 100,000 sources.

Simulation of transparent multilayered radomes excited by directed arrays of field generators.

Local patch-size settings for selected wires, plates, and objects.

Improved specification of distributed loading: assignment to the objects (in addition to wires and plates) and ability to copy and renumerate defined loadings.

Graphical representation of output results (diagrams with grids, currents in dB, and so on.)

Calculation of specific absorption through integration with WIPL-D time-domain solver.

12.4.4.26 WIPL-D Time-Domain Solver

The time-domain solver is an add-on tool for WIPL-D Pro 3D EM solver. It is intended for transient analysis of 3D structures. It is integrated with WIPL-D Pro and the time domain response is calculated based on the frequency domain simulation in WIPL-D Pro and Fourier transform.

Time-domain excitations are specified from a variety of predefined waveforms such as Gaussian, monocycle, rectangular and exponential pulses, and sinusoidal waves. In addition, time-domain solver can read in any user-specified excitation waveform and calculate the required transient response.

Output results visualization is done within the WIP-D environment, including 2D and 3D graphs.

12.4.4.27 EMPro 3D Simulation Software

EMPro 3D Simulation Software from KEYSIGHT Technologies is a software design platform for EM simulation. It can be used to model components such as high speed and RF IC packages, bondwires, antennas, on-chip and off-chip embedded passives, and PCB interconnects.

It can provide frequency-domain simulation using FEM or time-domain simulation using FDTD.

It features design-flow integration that can create 3D components that can be simulated with 2D circuit layouts and schematics using EM-circuit cosimulation and a GUI that can create 3D structures.

12.4.4.28 Field Precision LLC

Field Precision offers a 2D and 3D package using finite element simulation.
Capabilities include the following:

Electrostatic and high-voltage engineering including calculations of field stress on electrodes and dielectrics, electric field lines, capacitance, and electrostatic forces

Electron guns, injectors, accelerators, and transport systems

RF and microwave technology including resonator design, EM scattering, pulsed power technology, and microwave devices Time-domain and frequency-domain solutions for 3D structures

Electromagnets and permanent magnets; magnetic-field distributions and forces; calculations may include drive coils of any shape, iron with saturation effects, and multiple permanent magnets

X-ray imaging and source design including Monte Carlo simulations of electron, proton and positron interactions in matter from 0.25 kV to 1 GeV, end-to-end design of x-ray sources

Thermal transport in solids and biological media

12.4.4.29 Poynting MW

The Poynting for microwave EM wave analysis software is available from FUJITSU.

It uses the FDTD solver for high-frequency analysis.

It features EM wave analysis and calculates near-field EM fields and EM flows. It calculates far fields including radiated fields and scattered fields. It calculates frequency characteristics such as time-domain waveforms of voltages and currents, S parameters, and impedance.

Applications include the following:

Noise countermeasure evaluation (EMI mitigation): An EMI analysis of a four-layer PCB before and after implementing the countermeasures is analyzed and compared, and the effectiveness of the countermeasures is evaluated.

Parallel computation of near- and far-electric fields is generated by a PCB.

Noise characteristics of a power ground with EM band gap structure.

Metamaterial with a negative refractive index. This is an example of performing an EM simulation of a prism that has a negative refractive index by modeling the metamaterial as a split ring and wire. This uses a new function (numerical wave source) from the new version 3.

Other applications are ESD, mobile phone in a vehicle, combined full wave and IBIS models using the HCS linkage (circuit analysis linkage) option, and Poynting/HCIS IBIS conversion option.

FasterCap: FasterCap is a 2D and 3D multiplatform parallel capacitance fields solver.

FastCap2: FastCap2 is the Windows porting of FastCap.

FastHenry2: Windows porting of the 3D inductance/resistance field solver.

FastModel: Fast model is the 3D viewer and text editor for the above programs.

All these modules are available from fastfieldsolvers.com.

12.4.4.30 PAM-CEM/FD

PAM-CEM/FD from ESI is a 3D Maxwell EM solver based on FDTD technique using a "sugar cube" mesh generator to discretize the model. The user may manage the interaction of external exciting waves with complex structures to deal with radiating wire antennas or to combine such 3D analysis with the CRIPTE software through dedicated susceptibility or radiation coupling procedures. In addition to the sugar cube grids, standard absorbing conditions or perfectly matched layers can be PAM-CEM/HF.

PAM-CEM/HF is another product from ESI that relies on the PTD combining with PO and equivalent edge currents. Applications include high-frequency RADAR scattering.

12.4.4.31   CRIPTE

CRIPTE software from ESI based on EM topology enables the analysis of all conducted characteristics on cable networks. This includes the management of realistic bundles featuring hundreds of wires, splices, sub networks, shielded cables, frequency-dependent lossy dielectric coatings, and buried wires.

12.4.4.32   3D/MTL Coupling

3D/MTL coupling software from ESI provides capabilities to handle realistic models of complex cable networks.

12.4.4.33   EM.TERRANO

EM.TERRANO from EMAG Technologies is a physics-based, site-specific, wave-propagation modeling tool that enables engineers to determine how radio waves propagate in urban, natural, or mixed environments. It uses fully 3D polarimetric and coherent shoot and bounce rays, triangular surface mesher for discretization of arbitrary block parameters, GTD/UTD, intelligent ray tracing, ray reflection, and edge diffraction. It computes transmission through multilayer thin walls, real-time superposition of synchronized transmitters, communication link analysis for super-heterodyne transmitters and receivers, user-defined macromodels for reflection and transmission coefficients of blocks and terrain (imported from other modules or external files), parametric sweeps of scene elements like building properties, radiator heights and rotation angles, or super-heterodyne transmitter and receiver parameters, and statistical analysis of the propagation scene.

12.4.4.34   Efield®

Efield from Efield Solutions provides a number of modules and these include the following:

Efield® FD: Efield FD uses MOM and includes a parallelized frequency domain simulation and is suitable for medium-sized problems on a workstation with parallel processing.

Efield® FD hybrid add-on: Efield® FD hybrid add-on extends the application area to large problems, that is, too large for standard MOM. The module includes both PO and the MLFMM.

Efield® TD: Efield® TD is the basic time-domain module using a FDTD solver.

Efield® TD hybrid: Efield® TD hybrid add-on adds the FDTD–FEM hybrid solver for unconstructed grids. Generating hybrid meshes is an automatic process controlled by the GUI.

The Efield suite includes the following capabilities:

Integrated environment including GUI
CAD import of major formats (CATIA, CADDS, Pro/E, IGES, STEP, Parasolid, ACIS, DXF/
   DWG, STL, VDA-FS, ANSYSY, Inventor, Solidworks, Unigraphics)
Fixing and repair of complex CAD models (gaps, sliver faces, reversed normal, duplicated
   entities)
Surface meshing using triangular elements and volume meshing using Cartesian grids and
   tetrahedral elements
Solver technology in both time and frequency domain including full-wave solutions (MOM,
   MLFMM, FDTD, FEM) approximate (PO) as well as novel hybrid techniques

12.4.4.35   EMA3D

EMA3D for EMC is a simulation software package from EMA.

It has a validation heritage in EMC and EMI and has a CAE interface.

EMA3D is a 3D FDTD solver for Maxwell's equations based on the FDTD method in rectangular coordinates. It has the ability to model wires, thin gaps, and thin surfaces that are smaller than the computational mesh size. Frequency-dependent and anisotropic EM parameters are allowed. Also there are a variety of boundary conditions, sources, and variable mesh sizes available. It has an air conductivity model that considers the plasma fluid nature of air in high electric fields. AMA3D has a parallel solver for workstations and computational clusters.

MHARNESS is a cable harness EM and uses a transmission-line solver that reveals the EM transients on the pin interface to electronics systems. EMA3D and MHARNESS can cosimulate in one self-consistent simulation. MHARNESS can simulate multiple branches, multiple conductors, and multiple shields.

CADfix is a customized version of the CADfix package from ITI Trancendata that is capable of importing geometry from all major CAD formats. CADfix also allows geometry repair and preparation.

12.4.4.36   RF MODULE

RF MODULE is available from COMSOL and is based on the FEM with numerically stable edge elements, also known as vector elements, in combination with state-of-the-art algorithms for preconditioning and iterative solutions of the resulting sparse equation systems. Both the iterative and direct solvers run in parallel on multicore computers. Cluster computing can be utilized by running frequency sweeps, which are distributed per frequency on multiple computers within a cluster for very fast computations or by solving large models with a direct solver using distributed memory (MPI).

The RF module simulates EM fields in 3D, 2D, and 2D axisymmetric, as well as transmission-line equations in 1D and circuit (nondimensional) modeling with SPICE netlists. The 3D formulation is based on the full-wave Maxwell's equations using vector edge elements and includes material property relationships for modeling dielectric, metallic, dispersive, lossy, anisotropic, gyrotropic, and mixed media. The 2D formulations can solve both in-plane and out-of-plane polarizations simultaneously or separately, as well as out-of-plane propagation. The 2D axisymmetric formulations can solve both azimuthal and in-plane fields simultaneously or separately and can solve a known azimuthal mode number.

Both total wave and background wave formulations are available. The full-wave formulation solves for the total fields due to all included sources in the model, while the background-wave formulation assumes a known background field from an external source—a common approach for radar cross section and EM scattering.

Boundary conditions are available for PEC, finite conductivity surfaces, and thin lossy boundaries.

Perfectly matched layers are used to model boundaries to free space. Various excitation boundary conditions exist for modeling ports, rectangular, circular, periodic, coaxial, approximate lumped, user defined and precise numerically computed port excitations are available. You can include boundary conditions representing cable terminations as well as lumped capacitive, inductive, and resistive elements. Line currents and point dipoles are also available for quick prototyping.

Simulations can be set as eigenvalue, frequency domain, and transient problems. Eigenvalue problems can find the resonances and Q-factors of a structure, as well as propagation constants and losses in waveguides.

The equations in all models developed in COMSOL mutiphysics can be coupled such that the EM fields can both affect and be affected by any other physics.

The results are presented using predefined plot of electric and magnetic fields, S parameters, power flow, and losses.

12.4.4.37   FASANT

FASANT from NewFasant is a 3D program for analysis of antennas on board satellites. It uses GO, PO, and UTD. The code has a graphical interface that permits the user to build the geometry and show it from any point of view, situate the antenna, visualize the ray tracing, input all the analysis data, see graphically the result. FASANT can input geometries from many CAD tools such as AutoCAD and Rhino.

Coupling between pairs of antenna, radiation patterns, and near-fields values can be obtained considering direct, reflected, edge diffracted, slope diffracted, corner diffracted, creeping waves, and any combination of these waves.

The results can be represented graphically. The ray tracing, part of it or each component of the ray tracing, can be visualized and computed separately of the rest of the effects.

12.4.5 INTEGRATED ENGINEERING SOFTWARE

This offers in addition to their low frequency/transient solvers described in Section 12.5 the following:

12.4.5.1 SINGULA

SINGULA is a high-frequency program using MOM and FEM. The developer claims that SINGULA's advantage is in solving open regions problems and where modeling boundaries must be precise.

SINGULA calculates near- and far-field results, power and directive gain, radar cross section, axial ratio, input impedance, admittance, and scattering parameters.

Applications include the following:

Wire antennas (monopoles, dipole, yagi, helix, spiral)
Surface antennas (planar strip, spiral, and reflector)
Dielectric antennas (resonant and standing wave) on infinite/finite ground planes
EMC/EMI interactions
Microwave circuits
Microstrip power dividers and filters
MRI RF coils
Power bus structures
Radar cross section
Waveguide filters and transitions
Electromagnetic scattering from dielectric bodies
Conductors
Electromagnetic effects on human bodies

12.4.5.2 CHRONOS

CHRONOS from Integrated Engineering Software is a code using FDTD, MOM, and FEM.

CHRONOS has a smart mesh generator code with the ability to recognize the details of any structure and decide how to distribute the mesh, based on the accuracy requirement of the algorithm. The structure may be meshed either manually or by the smart mesh generator. The results are provided in the time domain and these can be transformed and displayed with different parameters in the frequency domain.

Applications are as follows:

Lightning strikes
Planar and microwave and antenna structures
Wire antennas
Cable junctions
Ultra wide band antennas
Antenna radiation characteristics
Microwave circuits, waveguide, and coaxial structures
EM compatibility and EM interference
Near-field and far-field animations
EM simulation of an airplane and missile
Metamaterial structures

12.5 ELECTROSTATIC, MAGNETOSTATIC, LOW FREQUENCY AND QUASI-STATIC FIELD ANALYSIS

12.5.1 MAXWELL 2D-3D

ANSYS offers Maxwell 2D and Maxwell 3D. ANSYS Maxwell used the finite element method to solve electric or magnetic problems, such as electrostatic, magnetostatic, eddy current, and magnetic transient by solving Maxwell's equations. In Maxwell 3D, the fundamental element is a tetrahedron. For each solver, there is some fundamental defining equation that provides an error evaluation for the solved fields. A value of the percent error energy number is reported by Maxwell after each solution passes and is used as a measure of solution convergence with respect to the adaptively refined mesh. The Ansoft HFSS code is described in Section 12.4.

12.5.2 ANSYS/EMAG

The ANSYS/EMAG program is from the combined development teams of Ansoft and ANSYS. EMAG SO:ID236 and SOLID237 can model magnetostatic, quasi-static time harmonic, and quasi-static time transient magnetic fields. These two elements are formulated using an edge-based magnetic vector potential formulation that allows improved accuracy for low-frequency EM simulation.

Applications include electric motors, solenoids, electromagnets, and generators.

12.5.3 INTEGRATED ENGINEERING SOFTWARE

Integrated software engineering offers the following applications.

12.5.3.1 MAGNETO

This is a 2D/RS magnetic field solver with BEM and FEM solvers. MAGNETO calculates force, torque, flux linkage, and inductance.

Applications are as follows:

Recording heads
Electric (AC/DC) motors
Magnetic shielding
Solenoids and transformers
Magnetizing fixtures
Circuit breakers
Magnetic sensors and instrumentation

12.5.3.2 ELECTRO 2D/RS

ELECTRO 2D/RS electrostatic solver combines the advantage of the BEM and FEM field solvers. It calculates electric field strength, force, torque, transmission-line parameters, and capacitance.

Applications are as follows:

Transformers
Insulators, bushings, grounding electrodes
Micro electromechanical systems
High-voltage shields
Power transmission lines
Telecommunication cables
Microstrip and ICs
High-voltage switchgear
Parts and assemblies subject to electric fields

12.5.3.3 COULOMB

A 3D electric field solver using the BEM method with the FEM solver is also available.

Coulomb calculates electric field strength, force, torque, and capacitance.

Its applications are the same as ELECTRO 2D/RS, but it has a 3D capability.

12.5.3.4 OERSTED

This is a 2D/RS time-harmonic eddy current field solver using the BEM with FEM solver.

Calculations include magnetic field analysis, force, torque, induced voltages, displacement current, flux linkage, induced voltage, power, impedance, and currents.

Applications include the following:

MRI
Nondestructive testing systems
Bus bars, charging fixtures
Induction heating coils
Magnetic recording heads
Magnetic shielding
Coils and transformers
Induction motors

It was not found suitable for a large cryogenic magnet.

12.5.3.5 AMPERES

Amperes is a 3D magnetic field solver using BEM and FEM field solvers.

It can calculate force, torque, flux linkage, and inductance and includes linear, nonlinear, and permanent magnet materials.

Capabilities are as follows:

Magnetizing fixtures
Permanent magnet assemblies
Circuit breakers
Recording heads and enclosures
Electric (AC/DC) motors
Magnetic shielding
Cyclotrons, solenoids, and transformers
Magnetic levitation and bearing systems
Sensors and magnetic instrumentation

12.5.3.6 FARADAY

This is a 3D time-harmonic eddy current analysis based on BEM and FEM solvers.

FARADAY can calculate force, torque, displacement current, flux linkage, induced voltage, power, and impedance.

Applications are as follows:

MRI
Nondestructive testing systems
Bus bars, charging fixtures
Induction heating coils
Magnetic recording heads
Magnetic shielding
Coils and transformers
Induction motors

12.5.3.7 INDUCTO

INDUCTO is a 2D/RS program that provides coupled EM- and thermal-field analyses. The eddy current simulation capabilities of OERSTED are linked to the thermal analysis capabilities of KELVIN to provide complete solution for induction heating analysis.

INDUCT can perform both transient and steady-state simulations. In addition, the OERSTED and KELVIN modules can be used separately when coupled simulations are not required.

12.5.3.8 FLUX

FLUX from CEDRAT. All 2D and 3D low-frequency electric and magnetic analyses are by FEM solver.

Advanced CAD imports, multi-parametric studies, enhanced models for magnetic materials in AC, steady-state simulations, electric-field computations, electric circuit coupling, including some nonlinear models, surface impedance formulation to model skin effects without meshing the conductor's cross section, Portunus-driven cosimulations for inclusion of complex nonlinear models.

Applications are as follows:

Rotating machines
Linear actuators
Transformers
Induction heating devices
Sensors
HV devices
Cables
Nondestructive evaluation

12.5.3.9 Infolytica

A suite of software, including **ElecNet V7**, **MagNet V7**, and **Magnet for SOLIDWORKS**, which are a 2D/3D full-wave EM modeling software employing FEM, are available from Infolytica.

The ElecNet v7 program solves static, AC, and transient electric field and current flow problems. It imports data from SAT, DXF, Pro, E, IGES, STEP, Inventor, and CCATIA and outputs to the OptiNet Automated Design Optimization program.

The MagNet v7 lets you model the performance of any EM or electromechanical device.

The input data sources are not only the same as ElecNet V7 but it also allows input from PSIM, Simulink®, and VHDL-AMS system simulations. Output is again to OptiNet.

MagNet for SOLIDWORKS is a 3D EM-field simulator imbedded in the SOLIDWORKS CAD software. It is a fully integrated add-in to SOLIDWORKS. It enables EM-field simulations and makes geometric modifications to examine their impact on the design without worrying about exporting model data and dealing with compatibility issues [11].

12.6 ERRORS SEEN IN THE USE OF ELECTROMAGNETIC ANALYSIS PROGRAMS

The dangers in overreliance on computer programs are well known and are the belief that the results computed by the program must be right even when the input data are invalid and the application of inappropriate modeling techniques is made because they are embedded in the program. The limitations on a modeling approach are usually frequency and geometry, and when the computer program does not recognize these limitations, the result can be a solution with large errors. The limitation on the computer code used is particularly true for simple programs, whereas the more complex programs require a great deal of decision making by the user, which can also result in large errors. If a simple conductive structure carries a known RF current or is exposed to an incident E field, simple antenna theory may be used to provide a "sanity check" on the radiated field or surface current. If the radiated field or induced current is higher than predicted, then the computer modeling is suspect.

For radiation from antennas, the FRIIS formula may be used, with the approximate gain of the antenna derived from antenna textbooks, as a sanity check.

For example, a simple cookbook-type program asks whether a wall is present between a radiation source and a receptor and, if the answer is yes, applies a fixed attenuation factor to the coupling equation. The type of material of which the wall is made and the presence of apertures, such as windows, as well as distance between source and receptor, will affect the real level of attenuation drastically. This effect is discussed in Section 6.11, in which the attenuation was seen to vary by as much as 40 dB when the fields measured at distances of 1 m from the interior walls of a building were compared to the field 15 m from the interior wall.

The more complex and powerful computer program can calculate the attenuation accurately, based on the conductivity and permittivity of external walls, and account for apertures and internal walls, and so on. The problem with the more complex programs is the amount of time required to accurately input the geometry of the structures and describe the coupling paths required to be modeled. The computing time for large structures at high frequency may also be prohibitive.

Whatever be the type of program used, it is imperative that the magnitude of the results be examined for consistency with either measured results or engineering experience. The modeling method and assumptions made in the program must be available to the user to enable a judgment on the applicability of the method to the problem. The accuracy of the results obtained by the use of a program should be known in order to add a safety factor when a worst-case analysis is required. An overdesign may result when the accuracy of the prediction is low, with a potential increase in cost.

The differential-mode currents in a circuit are typically orders of magnitude higher than the common mode, and it is often difficult to see how these common-mode currents are generated. As discussed in Chapter 11 on PCB radiation and Section 7.13 on emissions from cables and wires, it is often the common-mode currents that are the prime source of radiation, and many commercially available PCB radiation computational programs ignore this important source.

In the case of radiated susceptibility of a transmission line above a finite ground plane in a MIL-STD RE102-type test setup modeled using MOM, the predicted induced current in the line when both the near and far ends of the line were grounded (symmetrical connection) was very different from the measured, whereas with only one end grounded (asymmetrical connection), the correlation with measurements was very good. The asymmetrical connection may be used to derive the currents for the symmetrical connection only the resonant frequency must be corrected.

In Chapter 10, the predicted input impedance of a monopole antenna, using MLFMM, which is a development of the MOM technique, mounted on a cylinder was 198 Ω, very different from that predicted by using antenna equations and a second MOM-based program. The second program showed good correlation to the antenna equations with impedance between 44.5 and 47.5 Ω. The supplier of the MLFMM program obtained similar results with no explanation for the discrepancy.

The input impedance of a rod antenna mounted on a metal enclosure was measured using a network analyzer at 46.9 Ω, whereas a MOM program predicted the impedance at several hundred Ohms.

Two different computer programs were used to model the creeping wave around a cylinder from a monopole antenna attached to the cylinder and a dipole antenna oriented horizontally with respect to the cylinder surface, and these were compared to measurements.

One program modeled the antenna using MOM and the diffraction around the cylinder using UTD, and the second program used MLFMM for both the monopole and to predict the currents on the cylinder and the magnitude of field around the cylinder.

With the wave vertically polarized (the monopole), the MLFMM program differed from the measurement by only 6 dB and the MOM/UTD by only 2 dB. However, with the field horizontally polarized (the dipole), the MLFMM result was 21 dB different from the measurement, whereas the MOM/UTD was within 2 dB.

In Reference [11], three full-wave frequency-domain EM modeling codes were used to model three simple canonical problems. One code employs an FEM solver, another a BEM solver, and the third a hybrid FEM/BEM solver. One of the problems was a pair of perfectly conducting planes with a slightly lossy dielectric material between them. It emulates a printed circuit board power bus (or microstrip patch antenna). The chapter's conclusion is that the codes were capable of generating reasonably accurate solutions of each other (within 3–5 dB); however, at three frequencies the error was as high as 10 dB.

Each of the codes required 29 h to run the PCB power bus structure.

Reference [11] provides interesting comments on all the three codes and the results.

REFERENCES

1. Computational Electromagnetics, Wikipedia.
2. G. Dash and I. Strauss. Digital EMI testing: How bad is it? Studies document variances. *Newswatch EM Compliance Engineering*, Winter 1990, VII(2), 39.
3. S.T. Li, J.W. Rock way, J.C. Logan, and D.W.S. Tarn. *Microcomputer Tools for Communications Engineering*. Artech House Publications, Boston, MA, 1987.
4. R.E. Harrington. *Field Computation by Moment Methods*. Macmillan, New York, 1968.
5. T.E. Baldwin and G.T. Caprara. Intrasystem electromagnetic compatibility program (IEMCAP). *IEEE Transactions on Electromagnetic Compatibility*, 1980, EMC-21.
6. B.E. Keiser. *Principles of Electromagnetic Compatibility*. Artech House Publications, Boston, MA, 1987.
7. W.J.R. Hoefer. *Numerical Techniques for Microwave and Millimeter-Wave Passive Structures*. John Wiley & Sons, New York, 1989, Chapter 8.
8. A. Drozd, A. Pesta, D. Weiner, P. Varshney, and I. Demirkiran. Application and demonstration of a knowledge-based approach to interference rejection for EMC. *IEEE*, 1998 EMC Symposium record.
9. M.W. Yun Ji, T.H. Ali, and T. Hubing. EMC applications of the EMAP5 hybrid FEM/MOM code. *IEEE*, 1998 EMC Symposium Record.
10. T. Hubing. *Survey of Numerical Electromagnetic Modeling Techniques*. University of Missouri-Rolla, Electromagnetic Compatibility Laboratory, Rolla, MO, September 1, 1991.
11. C. Su, X. He, H. Zeng, H. Ke, and T. Hubing. Modeling experiences with full-wave frequency-domain modeling software. *Electromagnetic Compatibility*, 2008 IEEE EMC Symposium record.

REFERENCES

Appendix A: Characteristic Impedance of Conductors, Wires, and Cables

Single-wire line, near ground

For $d \ll h$,

$$Z_o = \left(\frac{138}{\sqrt{\varepsilon_r}} \right) \log_{10} \left(\frac{4h}{d} \right)$$

Two-wire line, differential mode, near ground

For $d \ll D, h$,

$$Z_o = \left(\frac{276}{\sqrt{\varepsilon_r}} \right) \log_{10} \left(\frac{2D}{d} \right) \left(\frac{1}{\sqrt{1 + \left(D/2h \right)^2}} \right)$$

Two-wire line, common mode, near ground

For $d \ll D, h$,

$$Z_o = \left(\frac{69}{\sqrt{\varepsilon_r}} \right) \log_{10} \left(\frac{4h}{d} \right) \sqrt{1 + \left(2h/2D \right)^2}$$

Asymmetric two-wire line

$$Z_o = \left(\frac{277}{\sqrt{\varepsilon_r}} \right) \log_{10} \left(\frac{2D}{\sqrt{d_1 d_2}} \right)$$

Symmetric two-wire line

or twisted pair

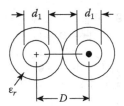

For $D \gg d$,

$$Z_o = \left(\frac{277}{\sqrt{\varepsilon_r}} \right) \log_{10} \left(\frac{2D}{d} \right)$$

Single-wire line between grounded parallel planes, ground return

For $d/h < 0.75$,

$$Z_o = \left(\frac{138}{\sqrt{\varepsilon_r}} \right) \log_{10} \left(\frac{4h}{\pi d} \right)$$

Balanced line between grounded parallel planes

For $d \ll D, h$,

$$Z_o = \left(\frac{277}{\sqrt{\varepsilon_r}} \right) \log_{10} \left(\frac{4h \tanh(\pi D/2h)}{\pi d} \right)$$

Coaxial line

$$Z_o = \left(\frac{138}{\sqrt{\varepsilon_r}} \right) \log_{10} \left(\frac{D}{d} \right)$$

Shielded two-wire line, common mode (sheath return)

For $d \ll D, h$,

$$Z_o = \left(\frac{69}{\sqrt{\varepsilon_r}} \right) \log_{10} \left[\left(\frac{h}{2d \left(h/D \right)^2} \right) \left(1 - \left(\frac{h}{D} \right)^4 \right) \right]$$

Shielded two-wire line, differential mode

For $D \gg d, h \gg d$

$$Z_o = \left(\frac{277}{\sqrt{\varepsilon_r}} \right) \left[\left(\frac{2h}{d} \right) \left(\frac{D^2 - h^2}{D^2 + h^2} \right) \right]$$

Microstrip lines

$$Z_o = \left(\frac{201}{\sqrt{\varepsilon_r + 1.41}} \right) \log_{10} \left(\frac{5.98h}{0.8b + c} \right)$$

Trace-to-trace impedance (no ground plane)

$$Z_o = \left(\frac{277}{\sqrt{\varepsilon_r}} \right) \log_{10} \left(\frac{\pi h}{b + c} \right)$$

Strip lines

$$Z_o = \left(\frac{138}{\sqrt{\varepsilon_r}} \right) \log_{10} \left[\frac{4k}{0.67\pi b(0.8 + c/b)} \right]$$

Flat parallel conductions

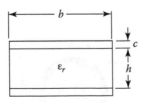

$$Z_o = \left(\frac{867}{\sqrt{\varepsilon_r}} \right) \log_{10} \left(\frac{h}{b} \right)$$

Appendix B: Units and Conversion Factors

Quantity	Units and Relationships between Units
Length	1 inch = 2.54 cm = 25.4 mm = 1000 mil
	1 m = 100 cm = 39.37 inches
	1 Angstrom = 10^{-8} cm
	1 mile = 1.609 km
	1 micron = 1×10^{-6} m = 1×10^4 Angstrom
Mass (kg)	1 kg = 1000 g = 2.2 lb
Area (m^2)	1 m^2 = 10.76 ft^2 = 1550 in.^{2}
Work, energy (W)	1 Joule = 0.738 ft-lb = 0.947827×10^{-3} Btu = 10^7 ergs = 0.239 cal
	1 ft = 1.356×10^7 ergs
	1 Btu = 1.055×10^{10} ergs = 252 cal
Pressure (N/m^2)	1 Newton/m^2 = 1 Pascal = 1 dyne/cm^2
	lb/in. = 6.8947×10^4 dynes/cm^2 = 0.068046 atm
	1 inch mercury pressure = 0.03342 atm
	1 bar = 10^6 dynes/cm^2
Force (N)	1 Newton = 0.225 lb = 1×10^5 dynes
Temperature	$0°C = 32°F = 273.16$ K
	Fahrenheit to Celsius conversion: $T_C = (5/9)(T_F - 32)$
	Celsius to Kelvin conversion: $K = T_C + 273$
Resistivity	1 ohm/cm or 1 ohm/cm^3
Permittivity of free space (ε_0)	$\dfrac{1}{36\pi \times 10^9}$[H/m] = 8.8[pF/m]
Velocity (m/s)	1 m/s = 3.28 ft/s
Magnetic flux ((ϕ))(Wb)	1 Weber = 10^8 Mx = 10^8 lines
	$= 10^8$ gauss-cm^2
Magnetic flux density (B)	1 Tesla (1) = 1 Weber/m^2 = 10^4 Gauss
	1 Gauss = 1 line/cm^2 = 1 Mx/cm^2 = 7.936×10^5 A/m
Magnetic field strength (H)	1 A/m = 0.0125 Oersteds
	1 Oersted = 79.6 A/m
Permeability of free space (μ_o)	$\mu_o = 4\pi \times 10^{-7}$ H/m = 1.256 μH/m, where $\mu = B/H$
Magnetization (M)	1 A/m
Magnetic moment/unit volume	
Magnetic moment	1 A/m^2 = 1 J/(Wb/m^2)
Magnetic motive force (F)	1 amp-turn = 1.257 Gilbert = 1.257 Oersted-cm
Reluctance	1 amp-turn/Weber

dBμV to volts: $V = 10^{((dB\mu V - 120)/20)}$
Volts to dBμV: $dB\mu V = 20\log(V) + 120$
dBV to volts: $V = 10^{(dBV/20)}$
dBV to dBμV: $dB\mu V + 120$
dBμV to dBV: $dBV - 120$

dBm to watts: $W = 10^{((dBm-30)/10)}$

Watts to dBm: $dBm = 10\log(W) + 30$

dBW to watts: $W = 10^{(dBW/10)}$

Watts to dBW: $dBW = 10\log(W)$

dBW to dBm: $dBm = dBW + 30$

dBm to dBμV: $dBμV = dBm - 107$

dBμV to dBm: $dBm = dBμV + 107$

In a 50 Ohm system, dBμA to dBμV: $dBμV = dBμA + 34\ dB$

In a 50 Ohm system, dBμV to dBμA: $dBμA = dBμV - 34\ dB$

Appendix C: Electric Field Strength to Magnetic Field to Power Density Conversions

Far-field conditions

where $Z_w = 377\ \Omega$, $W/m^2 = E^2/Z_w$, $\mu_o = 4\pi \times 10^{-7}$ H/m

1 A/m	= 377	V/m
	= 171	dBμV/m
	= 120	dBμA/m
	= 25.8	dBW/m^2
	= 55.8	dBm/m^2
	= 37.7	mW/m^2
	= 1.26×10^{-6}	T
	= 1.26×10^3	nT
	= 0.0126	Gauss
	= 0.0126	Oersted
1 V/m	= 2.65×10^{-3}	A/m
	= 120	dBμV/m
	= 68.5	dBμA/m
	= 2.65	mA/m
	= −25.7	dBW/m^2
	= 4.3	dBm/m^2
	= −5.73	dBm/cm^2
	= 2.67×10^{-4}	mW/cm^2
	= 2.67×10^{-3}	W/m^2
	= 3.3×10^{-5}	Gauss
	= 3.3×10^{-9}	T
	= 3.33	nT
	= 3.3×10^{-5}	Oersted
1 W/m^2	= 19.4	V/m
	= 5.15×10^{-2}	A/m
	= 0.1	mW/cm^2
1 nT	= 7.936×10^{-4}	A/m
	= 1×10^{-5}	Gauss
1 pT	= 7.936×10^{-7}	A/m
1 T	= 1	Wb/m^2
	= 10^4	Gauss
	= 7.936×10^5	A/m

dBμV/m to V/m: V/m = $10^{(((dB\mu V/m)\ -\ 120)/20)}$

V/m to dBμV/m: dBμV/m = $20\log(V/m) + 120$

dBμV/m to dBμA/m: dBμA/m = dBμV/m − 51.5

dBμA/m to dBμV/m: dBμV/m = dBμA/m + 51.5

dBμA/m to dBpT: dBpT = dBμA/m + 2

dBpT to dBµA/m: dBµA/m = dBpT − 2

1 Gauss	= 1	Oersted
	= 79.6	A/m
	= 0.796	A/cm
	= 1×10^{-4}	T
	= 0.1	mT

Appendix D: Commonly Used Related Formulas

Electric charge, coulombs (C)

$$Q = CV = It$$

Electric energy, joules (J)

$$W = IVt = QC$$

Power density, watts per square meter (W/m^2)

$$P_d = \frac{E^2}{Z_w} = EH$$

Wave velocity in air, meters per second (m/s)

$$v = \frac{1}{\sqrt{\mu_o \varepsilon_o}} = 3 \times 10^8$$

Wave velocity in medium, meters per second (m/s)

$$v = \frac{1}{\sqrt{\mu_r \mu_o \varepsilon_r \varepsilon_o}}$$

Characteristic impedance, ohms (Ω)

$$Z_o = \sqrt{\frac{L}{C}}$$

Velocity, meters per second (m/s)

$$v = \frac{1}{\sqrt{LC}}$$

Inductance, henrys (H)

$$L = \frac{Z_o}{v} = Z_o^2 C$$

Capacitance, farads (F)

$$C = \frac{1}{v^2 L}$$

Wavelength, meters (m)

$$\lambda = \frac{v}{f}$$

Wavelength in air, meters (m)
  where $v = c = 3 \times 10^8$ m/s

$$\lambda = \frac{3 \times 10^8}{f[\text{MHz}]} = \frac{300}{f[\text{MHz}]}$$

Flux density, teslas (T)

$$B = \mu H$$

Resonant frequency of LC circuit, hertz (Hz)

$$f = \frac{1}{2\pi\sqrt{LC}}$$

Reactance of capacitor, ohms (Ω)

$$X_C = \frac{1}{2\pi f C}$$

Reactance of inductor (Ω)

$$X_L = 2\pi f L$$

Impedance of resistance and inductance in series, ohms (Ω)

$$Z = \sqrt{R^2 + \left(2\pi f L\right)^2}$$

Impedance of resistance and capacitance in series, ohms (Ω)

$$Z = \sqrt{R^2 + \left(\frac{1}{2\pi fC}\right)^2}$$

Impedance and resonant frequency of a resistance, inductance, and capacitance in series, ohms, and hertz (Ω, Hz)

$$Z = \sqrt{R^2 + \left(2\pi fL - \frac{1}{2\pi fC}\right)^2}$$

$$f = \frac{1}{2\pi\sqrt{LC}}$$

Impedance of resistance and inductance in parallel, ohms (Ω)

$$Z = \frac{1}{\sqrt{\left(\frac{1}{R}\right)^2 + \left(\frac{1}{2\pi fL}\right)^2}}$$

Resistance and capacitance in parallel, ohms (Ω)

$$Z = \frac{1}{\sqrt{\left(\dfrac{1}{R}\right)^2 + \left(2\pi f C\right)^2}}$$

Impedance and resonant frequency of a resistance in series with inductance and both in parallel with capacitance, ohms, and hertz (Ω, Hz)

$$Z = \frac{R/\left(2\pi f C\right)^2 + j\left[L/\left(2\pi f C\right)^2 - \left(\left(2\pi f\right)L^2\right)/C - R^2/\left(2\pi f C\right)\right]}{R^2 + \left[2\pi f L - 1/\left(2\pi f C\right)^2\right]}$$

$$f_r = \frac{1}{2\pi}\sqrt{\left(\frac{1}{LC}\right) - \left(\frac{R^2}{L^2}\right)}$$

Appendix E: Data on Bare Solid Copper Wire (Dimensions, Weight, and Resistance)

Size AWG	Nominal Diameter [in.][a]	Circular [mil]	Nominal Weight [lb/mil]	Max. Resistance at 20°C (Hard Drawn) [Ω/km][b]
50	0.0010	1.00	0.00303	—
49	0.0011	1.21	0.00366	—
48	0.0012	1.44	0.00436	—
47	0.0014	1.96	0.00593	—
46	0.0016	2.56	0.00775	—
45	0.0018	3.24	0.00981	—
44	0.0020	4.00	0.0121	2700
43	0.0022	4.84	0.0147	2230
42	0.0025	6.25	0.0189	1720
41	0.0028	7.84	0.0237	1380
40	0.0031	9.61	0.0291	1120
39	0.0035	12.2	0.0371	880
38	0.0040	16.0	0.0484	674
37	0.0045	20.2	0.0613	533
36	0.0050	25.0	0.0757	431
35	0.0056	31.4	0.0949	344
34	0.0063	39.7	0.120	272
33	0.0071	50.4	0.153	214
32	0.0080	64.0	0.194	168
31	0.0089	79.2	0.240	136
30	0.0100	100	0.303	108
29	0.0113	128	0.387	84.5
28	0.0126	159	0.481	67.9
27	0.0142	202	0.610	53.5
26	0.0159	253	0.765	42.7
25	0.0179	320	0.970	33.7
24	0.0201	404	0.22	26.7
23	0.0226	611	0.55	21.1
22	0.0253	640	0.94	16.9
21	0.0285	812	0.46	13.3
20	0.0320	1,020	0.10	10.5
19	0.0359	1,290	0.90	8.37
18	0.0403	1,620	4.92	6.64
17	0.0453	2,050	6.21	5.26
16	0.0508	2,580	7.81	4.18
15	0.0571	3,260	9.87	3.31
14	0.0641	4,110	12.4	2.63
13	0.0720	5,180	15.7	2.09

(Continued)

Size AWG	Nominal Diameter [in.][a]	Circular [mil]	Nominal Weight [lb/mil]	Max. Resistance at 20°C (Hard Drawn) [Ω/km][b]
12	0.0808	6,530	19.8	1.65
11	0.0907	8,230	24.9	1.31
10	0.1019	10,380	31.43	1.039
9	0.1144	13,090	39.62	0.8241
8	0.1285	16,510	49.96	0.6532
7	0.1443	20,820	63.03	0.5180
6	0.1620	26,240	79.44	0.4110
5	0.1819	33,090	100.2	0.3260
4	0.2043	41,740	126.3	0.2584
3	0.2294	52,620	159.3	0.2050
2	0.2576	66,360	200.9	0.1625
1	0.2893	83,690	253.3	0,1289
1/0	0.3249	105,600	319.5	0.1022
2/0	0.3648	133,100	402.8	0.08021

[a] To convert from inches to millimeters, multiply by 25.4.

[b] To convert from feet to meters, multiply by 0.3048; to convert from meters to feet, multiply by 3.28.

Appendix F: Permittivity of Materials

F.1 PCB SUBSTRATE PERMITTIVITY

Material	Dielectric Constant, ε_r
Difunctional epoxy (FR4)	4.3–4.8 (depending on manufacturer) approximately 4.6 at 10 MHz and 4.3 at 10 GHz
GX (PTFE/woven glass)	2.4–2.6
BT = bismaleimide–triazine resin	3.7–3.9
Epoxy/PPO	3.95
BT/epoxy	4.0
Rogers RO3003 PTFE/ceramic	3.0
Rogers GETEK	3.8 (50 MHz to 1 GHz)
Rogers Polyflon	2.05 ± 0.05
Rogers No Clad	2.55 ± 0.05
Rogers Polyglide	2.32 ± 0.005
Rogers RO40003C (hydrocarbon/woven glass)	3.38
Rogers RO4000	Selectable 2.55–6.15
Rogers RO3000	Selectable 3–10.2

F.2 COMMON MATERIALS

Material	Dielectric Constant, ε_r
Cellophane	7.6 (50 Hz), 6.7 (1 MHz)
Porcelain (HV electrical)	5.5 (50 Hz to 1 MHz)
Gallium arsenide	12 (1 kHz)
Mica	7 (50 Hz–100 MHz)
Borosilicate glass	5.3 (1 kHz–1 MHz
Fused quartz	3.8 (50 Hz–100 MHz)
Lead glass	6.9 (1 kHz–1 MHz)
Soda glass	7.5 (1 MHz–100 MHz)
Kraft paper	1.8–3.0
Rag (cotton) paper	3.5–4.2
Fiber paper	4.5
Pressboard	3.2
Polyethylene plastic	2.3 (50 Hz–1 GHz)
Polyisobutylene plastic	2.2 (50 Hz–3 GHz)
Poly 4-methylpentene (TPX)	2.1 (100 Hz–10 kHz)
Polyphenylene oxide (PPO)	2.6 (100 Hz–1 MHz)
Polypropylene	2.2 (50 Hz–1 MHz)
Polystyrene	2.6 (50 Hz–1 GHz)
Polytetrafluoroethylene (PTFE), Teflon	2.1 (50 Hz–3 GHz)
Epoxy resin	3.6 (1 kHz), 3.5 (100 MHz)
Melamine resin	4.7 (3 GHz)
Phenolic resin, fabric filled	5.5 (1 MHz)

(Continued)

Material	Dielectric Constant, ε_r
Polyvinyl acetate (plasticized)	4 (1–10 MHz)
Polyvinyl chloride	4 (1–10 MHz)
Natural crepe rubber	2.4 (1–10 MHz)
Bakelite	3.7
Kapton type 100	3.9
Kapton type 150	2.9
Mylar	3.2
Nylon	3.2–5
Lexan	3
Lucite	2.8
Butadiene/styrene rubber, unfilled or compounded	2.5 (50 Hz–100 MHz)
Butyl rubber, unfilled	2.4 (50 Hz–100 MHz)
Chloroprene (neoprene) rubber	6.5 (1 kHz), 5.7 (1 MHz)
Silicone rubber	3.2 (1 kHz), 3.1 (100 MHz)
Balsa wood, 0% relative humidity	1.4 (50 Hz), 1.2 (3 GHz)
Beech wood, 16% relative humidity	9.4 (1 MHz), 8.5 (100 MHz)
Birch wood, 10% relative humidity	3.1 (1–100 MHz)
Douglas fir, 11% relative humidity	3.2 (1–10 MHz)
Scots pine, 15% relative humidity	8.2 (1 MHz), 7.3 (100 MHz)
Walnut, 0% relative humidity	2 (10 MHz)
Walnut, 17% relative humidity	5 (10 MHz)

Index